Introduction to Medicinal Chemistry

Introduction to Medicinal Chemistry

How Drugs Act and Why

Alex Gringauz

New York • Chichester • Weinheim • Brisbane • Singapore • Toronto

Dr. Alex Gringauz
Arnold and Marie Schwartz College of Pharmacy and Health Sciences
Long Island University
75 Dekalb Avenue at University Plaza
Brooklyn, NY 11201

This book is printed on acid-free paper. ∞

Library of Congress Cataloging-in-Publication Data

Gringauz, Alex, 1934-
Introduction to medicinal chemistry : how drugs act and why/by Alex Gringauz.
p. cm.
Includes bibliographical references and index.
ISBN 0-471-18545-0 (alk. paper)
1. Pharmaceutical chemistry. I. Title.
[DNLM: 1. Chemistry, Pharmaceutical. 2. Pharmacology. QV 744 G8671 1996]
RS403.G76 1996
615'.7—dc20
DNLM/DLC
for Library of Congress 95-49331
CIP

Printed in the United States of America

ISBN 0-471-18545-0 Wiley-VCH, Inc.

10 9 8 7 6 5 4 3 2

Preface

This book is intended to be useful, indeed necessary, to students pursuing a career in the health sciences that require a knowledge and understanding of drugs. This includes their rational therapeutic utilization, their shortcomings and hazards, their mechanisms of action, their stabilities in the bottle as well as the in the body, and some grasp of the thinking that goes into drug design and development. The author unabashedly confesses to a highly chemical bias in this presentation, since all aspects of drug comprehension ultimately must be founded on chemistry.

It is the pharmacy student, as the future practitioner of the rapidly changing profession of pharmacy, who will find this book invaluable. It is the pharmacist who is consistently, if not constantly, involved with drugs in all aspects. His or her interest necessarily is not that of the curious bystander. An extensive knowledge of drugs is not only required for professional competency, it is also legally mandated. This requires a background in organic chemistry, biochemistry, and some basic physiology and pharmacology. Since these, and introductory biology courses are somewhat compartmentalized in most undergraduate programs, this book should be viewed as a bridge to the extensive pharmacology and clinical training of the modern pharmacy curriculum.

It is anticipated that this publication will also be used at the early graduate training level, e.g. an M.S. en route to the Ph.D. degree in various pharmaceutical sciences such as medicinal chemistry, pharmacology, and pharmaceutics. The curious bystander who was mentioned previously should not be overlooked. Here one might include the chemist and biologist working in other fields whose interest might be piqued as to what drug chemistry is all about.

This book is not intended for the medicinal chemistry practitioner, one who is practicing its art and science. It also does not purport to make a medicinal chemist out of its reader, although it may be hoped that some will be "turned on" to such a pursuit. Rather, its intent is to explain drugs to the prepared reader as a comprehensive package by combining the necessary biological/physiological concepts with their chemistry so that a more total picture can be seen. After all, drugs must be viewed as complex chemicals used to affect chemical processes in an extremely complex biochemical system: the human mammal. Therefore, any attempt at a complete comprehension of drugs while ignoring their chemistry is an obviously futile endeavor.

Alex Gringauz
Brooklyn, New York

Acknowledgements

No serious effort in life is totally accomplished by oneself. This book is no exception. My wife, Donna, was exceedingly helpful with much of the grunt work, which included adept word processing and very frustrating indexing. The aid was usually cheerfully, and occasionally grudgingly, given. My sons harbored serious doubts regarding the ultimate completion of this effort.

I am also grateful to colleagues who graciously agreed to evaluate portions of the manuscript. Their comments, especially those that were critical, were particularly helpful. I therefore wish to thank Dr. Victoria Roche (Creighton University), Dr. Jack DeRuiter (Auburn University), and Dr. Frank Fillipeli (Drake University). My thanks also to Dr. Ed Immergut for his help and faith in me.

My utmost gratitude, however, must be extended to Steven So, my former student, now colleague and, most of all, friend, without whose creative wizardry at the computer none of the exquisite graphics would have come into existence. I also wish to thank Steve's wife, Mei, for her angelic tolerance of the disruptions this book must have created in her life.

Contents

CHAPTER

1

Basic Considerations of Drug Activity

1.1. Introduction

Voltaire (1694–1778) stated, "Therapeutics is the pouring of a drug of which one knows nothing into a patient of whom one knows less." The medical and pharmaceutical sciences have been working diligently to ameliorate both aspects of the problem. The progress made, especially in the period following World War II, has been impressive, if not astounding. However, there are many important riddles still to be solved and much to be learned.

One of the areas of study has concerned itself with the determination of the factors that affect a drug's activity and the reasons for the effects observed. A relationship between physicochemical properties of a chemical compound and its biological activity has been assumed and sought for more than a century. Our definition of what constitutes physical and chemical properties, however, has been constantly expanding as a result of new ideas, discoveries, and instrumentation. Modern instrumentation in particular has helped to change our outlook on drugs.

1.2. Factors Affecting Bioactivity

The biochemical systems encountered by a drug molecule are extremely complex. Therefore, it should not be surprising that the factors affecting the drug's interactions and contributing to its final effect are also manyfold. The factors may be divided into three categories:

1. Physicochemical properties such as solubility, partition coefficients, and ionization.
2. Chemical structure parameters such as resonance, inductive effect, oxidation potentials, types of bonding, and isosterism.
3. Spatial considerations such as molecular dimensions, interatomic distances, and stereochemistry.

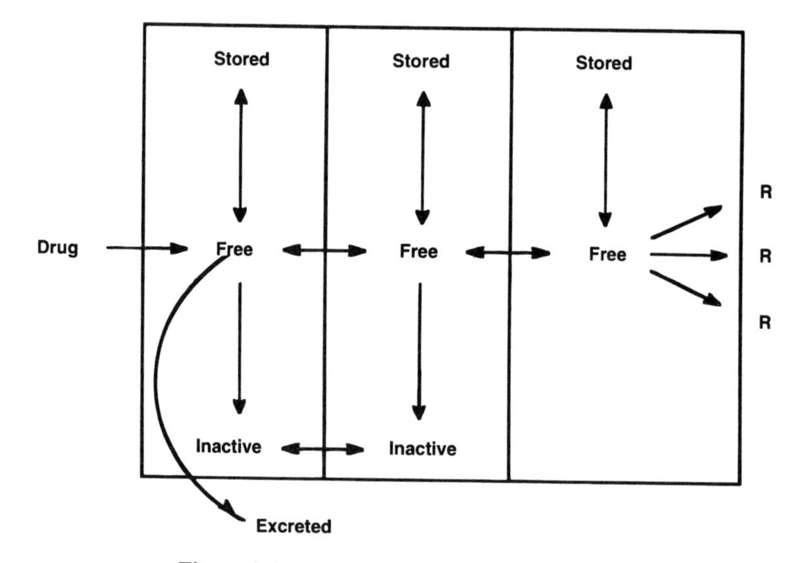

Figure 1-1. The fate of a drug. R is a receptor.

1.2.1. Physicochemical Properties

The physicochemical properties considered in this discussion are important because they all relate to the transport of the drug molecule to its site of action, more than likely a receptor with which the drug will interact in a given tissue or in an invading microorganism. Figure 1-1 represents a simplified distribution chart of a bioactive substance in the body.

A drug given orally or parenterally must traverse several semipermeable membranes before reaching its destination. The efficiency of the passage depends on the solubility characteristics of the drug, that is, its behavior in aqueous solution and toward lipids. Also, note that in each compartment the molecule is subject to various factors tending to decrease the concentration of the active form. Thus the drug may be constantly excreted either directly or following biochemical inactivation. In addition, if the drug is bound in a stored but inactive form, such as to plasma proteins, there tends to be a decrease in its effectiveness. Since it is only the unbound free drug that produces the desired pharmacologic actions, it may be possible to compensate for this phenomenon by increasing the dose.

We are concerned with solubility in polar solvents such as water and in nonpolar solvents such as lipids. More specifically we are interested in a drug's *partition coefficient*, which is the relative solubility between these two phases. Such a coefficient is determined by dissolving the substance in an aqueous solution and equilibrating it by agitation with an organic solvent.[1] The ratio of the concentration of the drug in the two phases is the partition coefficient. Any ratio greater than 0.01 indicates appreciable lipid solubility.

Since most drugs are not structurally similar to normal cellular components, they are not likely to be transported across the membranes by "active transport" mechanisms. Rather, their rate of passage through the lipoprotein membranes is mainly a passive process determined by their degree of lipid solubility, or partition coefficient.

---

[1] Chloroform, olive oil, or 1-octanol to simulate the lipid phase of a biological system.

Solubility is important to bioactivity. Many groups of drugs, especially those with closely related structures, exhibit a direct relationship to solubility (i.e., increased lipid solubility exhibits higher bioactivity). This correlation is true in general anesthetics, local anesthetics, certain antibacterial agents, antiviral agents, and others. Of course, solubility factors are closely related to drug absorption. The degree of absorption is one important determinant of the intensity of the drug's action.

In addition to lipid solubility, another physicochemical property of molecules, which affects solubility directly, is the degree of the drug's electrolytic nature. All chemical compounds can be classified by their electrical conductivity behavior in aqueous solution. When dissolved, inorganic salts will completely dissociate into ions (charged particles). Positively charged ions, which are electron deficient with respect to the neutral atom, are called *cations*, whereas negatively charged ions (carrying excess electrons) are called *anions*. Thus sodium chloride will dissociate, or ionize, yielding sodium ions and chloride ions.

$$NaCl \rightarrow Na^+ + Cl^- \tag{1.1}$$

Substances that ionize completely in solution are considered to be *strong electrolytes*. Compounds that are completely undissociated, but that are still very water soluble, are termed *nonelectrolytes*. They do not ordinarily increase the electrical conductivity of the solution. Examples of nonelectrolytes are such polar organic compounds as sugars, low-molecular-weight alcohols, and urea. A majority of drugs are in a third category, *weak* electrolytes. These substances are only partially ionized in solution. They exist as a mixture of ionized and un-ionized molecular forms. The un-ionized molecular species is the more lipid-soluble form. The ionized portion of such a drug molecule usually has a much lower, often negligible, lipid solubility. Therefore, its passage through membranes frequently approaches insignificant levels. This fact has direct bearing on a drug's capacity for absorption, and therefore activity.

When a drug is a weak acid or a weak base, we find that its lipid solubility is greatly affected by the pH of its environment and by its degree of dissociation, expressed as pKa. The fraction of the total drug concentration that is in the molecular and ionic forms is indicated by the dissociation constant *Ka*. Equations 1.2 and 1.3 illustrate the interaction of weak acids and weak bases with water, which results in dissociation. *A* and *B* represent acids and bases, respectively.

$$\mathrm{HA} + \mathrm{H_2O} \leftrightharpoons \mathrm{H_3O^+} + A^- \tag{1.2}$$

$$\mathrm{BH^+} + \mathrm{H_2O} \leftrightharpoons \mathrm{H_3O^+} + B \tag{1.3}$$

Note that the initial reaction for both substances is shown as a protolytic reaction (protonation) between an acid species and water. The water is present in such large excess that the proton transfer has only a negligible effect on its total concentration. Thus water can be eliminated from the equation without significant error. Our simplified equation for a weak acid now becomes

$$\mathrm{HA} \leftrightharpoons \mathrm{H^+} + A^- \tag{1.4}$$

Applying the law of mass action we obtain the general relationship:

$$\mathrm{Ka} = \frac{[\mathrm{H^+}][A^-]}{[\mathrm{HA}]} \tag{1.5}$$

The equation can be rearranged into the more useful Henderson–Hasselbach equation:

$$\mathrm{pKa} = \mathrm{pH} + \log\frac{C_\mathrm{u}}{C_\mathrm{i}} \tag{1.6}$$

where C_u and C_i represent the concentrations of un-ionized and ionized forms of the drug, respectively. The corresponding relationships for weak bases are:

$$B + \mathrm{H}^+ \leftrightharpoons B\mathrm{H}^+ \tag{1.7}$$

$$\mathrm{Ka} = \frac{[\mathrm{H}^+][B]}{[B\mathrm{H}^+]} \tag{1.8}$$

$$\mathrm{pKa} = \mathrm{pH} + \log\frac{C_\mathrm{i}}{C_\mathrm{u}} \tag{1.9}$$

Weak acids have a higher pKa than stronger ones. Thus, an acid with a pKa of 5 is 100 times weaker than an acid whose pKa is 3; conversely weaker bases have lower pKa values.

It is not surprising that the bioactivity of many weak acids and bases is directly related to their degree of ionization, which in turn is greatly affected by the pH of the medium in which the drug finds itself.

Since many of the drugs we encounter are weak acids or bases, an understanding of their solubility characteristics is important. Because the ionic form is the more water-soluble chemical species and the pH of the solvent environment determines the degree of ionization achieved, it becomes possible, for example, to formulate liquid pharmaceutical products such as injectables, syrups, and elixirs of drugs that would ordinarily be poorly soluble.

Low-molecular-weight carboxylic acids such as acetic acid and propionic are totally water soluble. However, as they go beyond a five-carbon content their solubility decreases rapidly. An interesting example of how advantage can be taken of these factors to form a water-soluble parenteral dosage form of a drug that is highly insoluble is the steroid methylprednisolone (structure I).

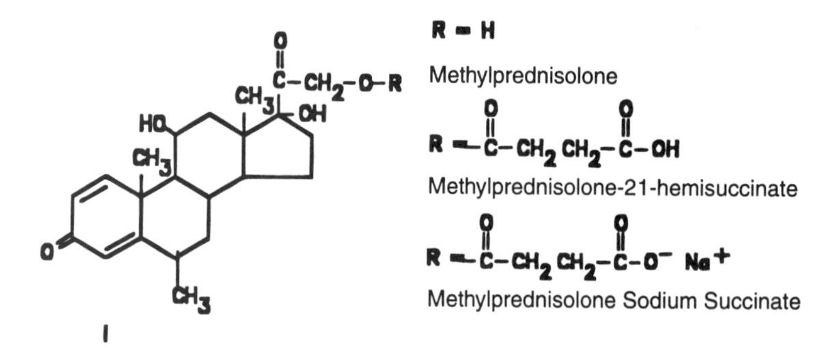

Reacting the drug with succinic anhydride results in the hemisuccinate derivative, obviously now a large 25-carbon carboxylic acid. Its solubility is less than 1 mg/ml. However, by the simple expedient of neutralizing the acidic function and forming the ionic sodium salt the solubility is increased to over 200 mg/ml. This is more than adequate to formulate injectable products of considerable concentrations.

Let us apply these concepts and attempt to make some predictions. The very useful, widely used drug aspirin is a weak acid with a pKa of 3.5. It is usually taken orally. The pH of gastric juice in the stomach is about 1; in the small intestine it is about 6. From which area would the majority of this drug be absorbed into the bloodstream? By applying Equation 1.6 we find that the drug is almost completely un-ionized in the gastric juice.

Since we have already seen that the molecular form of a drug is the lipid-soluble species, we would expect it to be readily absorbed in the stomach, which has lipoprotein membranes in its lining. This is actually the case for many weakly acidic drugs. The converse argument, of course, would apply to weakly basic drugs. We would expect their absorption from the stomach to be poor.

Consider the three barbituric acid derivatives thiopental, secobarbital, and barbital with respective pKa of 7.6, 7.9, and 7.8. These drugs are very weak acids. On the basis of their ionization constants we would expect very little difference in their absorption rates from the stomach, yet the drugs are absorbed at very different rates. The reason becomes apparent when the partition coefficients between chloroform and water are considered. Thiopental's value is over 100, whereas the values of secobarbital and barbital are 23 and 0.7, respectively. Now which would one predict to be the least rapidly absorbed and which the most? By considering only one physicochemical parameter and excluding others, erroneous conclusions can result. Figure 1-2 illustrates a hypothetical relationship of biological activity as a function of pH only.

Studies on the distribution of drugs between the intestine and plasma, between kidney tubules and urine, and between plasma and other body compartments suggests that the important general conclusion that only *lipid-soluble, undissociated forms of a drug passes through membranes* readily. Ionized species usually cannot pass unless a mediated transport system is present for a specific compound (or a close congener) in a given membrane, which is a rare occurrence.

The previous discussion may be an oversimplification since there are some anomalies that are more difficult to explain. For example, almost two thirds of a dose of salicylic acid (pKa 3) is absorbed from the rat stomach in 1 hour at pH 1, as might be expected. However, if the pH is raised to 8, at which point the acid is completely ionized, over one-tenth of the dose is still absorbed. Another possibility that should be kept in mind is that the un-ionized form of some weak electrolyte drugs may have intrinsically poor lipid solubility because of

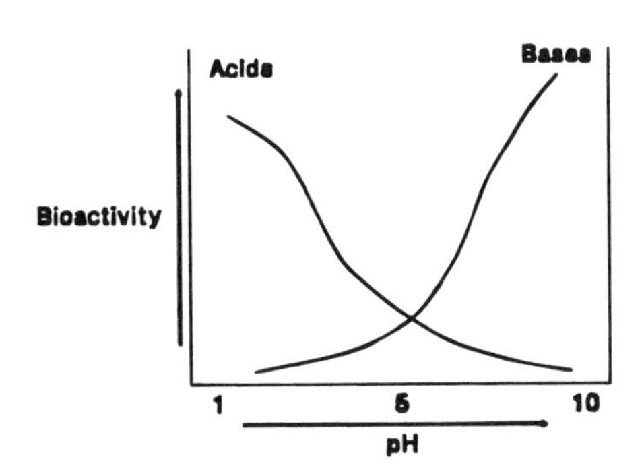

Figure 1-2. Bioactivity as a function of pH.

a high proportion of polar functional groups. The opposite situation, where the ionized form may still have appreciable lipid solubility, where the drug has few polar groups but a relatively large hydrocarbon skeleton, is a possibility. Methyl prednisolone sodium succinate would be an example.

Although lipid solubility at physiological pH enhances a drug's penetrability of a membrane, too much may not necessarily result in increased activity. Many antibacterial sulfonamides exhibit their peak effectiveness at pH values at which they are only approximately half-ionized. These sulfonamides have pKa values in the 6–8 range. The apparent reason is that even though the molecular form can readily penetrate the bacterial membrane, only the anionic form is bacteriostatic once inside. Thus approximately 50% ionization appears to be optimal. Nevertheless, several highly active sulfonamides exist with a pKa considerably outside of this optimal range. Other factors are also presumably involved.

In summary, if bioactivity is caused by ionic forms of drugs, activity will increase as the degree of ionization increases. On the other hand, if undissociated molecules are the active species, then increased ionization will necessarily reduce this activity.

1.2.2. Chemical Structural Aspects

One of the long-term objectives of medicinal chemistry is the establishment of relationships of a drug's structural features to its pharmacological properties [i.e., structure–activity relationships (SARs)]. Although qualitative linkages, often on an intuitive basis, have sometimes been assigned, a quantitative foundation is the goal. Attempts to express pharmacological activity by mathematical means are being made, with some success. Both classical qualitative concepts and the newer more numerical ideas must be taken into consideration to understand drug activities better and, equally important, more rationally to design and then develop new, more effective, and safer drugs. Both aspects will be briefly described here. Some concepts will be developed in somewhat greater detail in subsequent chapters.

1.2.2.1. Resonance and Inductive Effects

Resonance is a concept stating that if we can represent a molecule by two or more structures that differ only in their electron, but not atomic, arrangement then neither (or any) of the representations is satisfactory since the molecule is a *hybrid* of these possible structures. Each structure as depicted contributes to the "real" structure. One advantage of this idea is that it forces us to think of a drug molecule from additional mental angles rather than just those normally printed on a page. Electron density and electron distribution patterns help explain a drug's reactivity.

Unlike the theoretical resonance concept, *inductive effects* are measurable electrostatic phenomena. Inductive effects are caused by actual electron shifts, or displacements, along bonds. These shifts result from attractions exerted by certain groups because of their electronegativity. Thus groups or atoms that attract electrons more strongly than hydrogen have a *negative inductive effect* and tend to displace electron density toward themselves. The halogens are prime examples. Groups with *positive inductive* effect tend to push electrons into the rest of the molecule. These are usually alkyl groups such as methyl and isopropyl. The electronic consequences are a strong influence on physicochemical properties such as acidity. Table 1-1 illustrates this effect. Using formic acid as the prototype, we

Table 1-1. Inductive Effects on Acid Strength

Acid	Formula	pKa
Formic	$HCOOH$	3.76
Acetic	CH_3COOH	4.76
Chloracetic	$ClCH_2COOH$	2.81
Fluoroacetic	FCH_2COOH	2.66

note that the positive inductive effect of the methyl group in acetic acid causes an increase in the pKa by a whole unit, which is a 10-fold *decrease* in acid strength. Conversely, in chloroacetic acid we find the strongly electronegative element chlorine withdrawing electron density from the carboxyl function, which results in a drop of almost two pKa units, or a 100-fold *increase* in acidity. It stands to reason that differences in bioactivity would also occur.

Electronic effects in compounds containing aromatic rings are especially pronounced since inductive effects are readily transmitted through such conjugated systems. An interesting example is found in a pair of carcinolytic nitrogen mustard compounds in which the intensity of activity is dependent on the degree of nucleophilic halogen displacement. In Structure II, the *p*-amino group pushes electrons into the system, resulting in facile chlorine displacement by a nucleophilic species it encounters in a biological environment, therefore exhibiting a cytotoxic effect. The negative induction of the *p*-chloro atom in Structure III has the opposite effect, reducing halogen displacement to less than 10% of that exhibited by Structure II. The result is a decreased bioactivity.

A group of local anesthetics related to procaine (Novocaine) illustrates several of the concepts discussed, particularly the influence of substitutents of the aromatic portion of the molecule on pharmacological effects. The local anesthetic effectiveness of this *p*-aminobenzoic acid ester is believed to be related to the degree of polar character possessed by the ester carbonyl function. Structures IV and V are two resonance forms of procaine, neither of which, as explained, represents the drug. Rather, a hybrid structure, wherein the carbonyl function is partially ionic (not totally as in V), would be a more correct representation. To test this hypothesis it should be possible to vary the anesthetic potency by changing the nature of the para substituent on the benzene ring and relate it to the bond order of the ester carbonyl by measuring its infrared stretching frequency. Such a study was done with interesting results, and is summarized in Table 1-2.

Amino and alkoxy groups are considered electron donors by resonance and thus enhance the dipolar (ionic) character of the C = O group, as can be seen by the infrared frequency values (Table 1-2). Para substituents having a strong electron-withdrawing effect

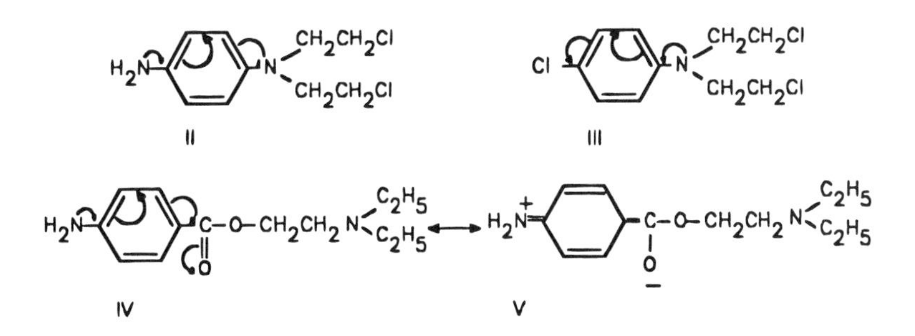

Table 1-2. Relationship of Local Anesthetic Activity and Bond Order

R substituent	IR C = O (cm^{-1})	ED_{50} (mmol[a]/100 mL)
C_2H_5O -	1,708	0.012
CH_3-O -	1,708	0.060
H_2-N -	1,711	0.075
HO -	1,714	0.125
H -	1,727	0.600
O_2N -	1,731	0.740

[a] Effective dose in 50% of guinea pigs tested. (From Galinsky et al. 1963.)

result in the carbonyl function having more double bond character and thus less intense local anesthetic activity.

As was seen with the earlier example of the barbiturates, it is likely that more than a single factor is influencing bioactivity. Since local anesthetics must penetrate a lipoid membrane covering the nerve fibers they affect, then it would be prudent also to consider factors such as lipid solubility. It is likely that the aromatic ring here functions both as a conduit for electron movement and contributes very significantly to the lipid solubility of these compounds. The compound without a para substituent (i.e., only H) is eight times weaker than procaine because the resonance enhancement of the *p*-NH_2 group has been removed. The *p*-NO_2 group, an electron-withdrawing moiety, reduces anesthetic activity an additional 20%. Finally, it is interesting to speculate on the reason for the fivefold greater potency of the *p*-ethoxy over the *p*-methoxy compound. The infrared frequency would indicate no difference in the polarizing ability of the carbonyl by resonance factors. Yet, the additional CH_2 represents a 6% increase in the hydrocarbon nonpolar character of the molecule. It can be safely assumed that this will enhance lipid solubility and therefore penetrability into the nerve fibers to a degree. Is this sufficient, however, to account for a fivefold increase in activity?

1.2.2.2. Oxidation–Reduction Potential

Oxidation–reduction potentials are expressed in volts relative to the standard hydrogen electrode at 0 volts, and represent the tendency of a compound to lose electrons (oxidation) or gain electrons (reduction). Various enzyme systems in respiration reactions (e.g., the cytochromes) utilizing ascorbic acid, hemoglobin, as well as reactions in the Krebs cycle (succinic and fumaric acids) involve electron transfers. Drugs can affect all these systems, and the consequences must be taken into account.

1.2.2.3. Types of Bonding

The types of bonds involved in chemical interactions include the covalent, hydrogen, electrostatic, van der Waals, and hydrophobic bond.

Covalent Bond

The covalent bond is formed by the interaction of electrons from different atoms. If each of the two atoms involved contributes only one electron, then the bond formed and holding them together is known as a *single* or *sigma* (σ) *bond*. If an additional pair of electrons, called *pi* (π) electrons, become involved, then we describe the junction as a *double bond*. Triple bonds also exist. They consist of a σ bond and two π bonds.

Covalent bonds are the bonds commonly encountered in organic molecules. Their strengths—the energies in terms of heat necessary to break them—are considerably greater than the other bonds to be discussed. Energies range from about 33 kilocalories per mole (kcal/mole) to over 200. Table 1-3 lists those of interest in biochemical processes.

At temperatures normally encountered in living matter (30–40°C), bonds stronger than 10 kcal/mole are not likely to be broken by nonenzymatic means. The majority of drugs designed to affect physiological functions do not react with receptors or various biopolymers in our systems by the formation of essentially irreversible covalent bonds. Those few that do are usually very toxic, long-acting, and difficult to control safely from a clinical standpoint. Pharmacodynamic agents that interact at these sites by much weaker, reversible bonding processes (discussed later) are much more desirable. Experimentally, their effects can be easily reversed, usually by simply washing out the loosely bound agents.

In the case of chemotherapeutic agents, drugs intended to rid our body of disease-causing parasites such as bacteria, fungi, protozoa, and helminths (worms), ideal compounds would indeed be those that interact covalently and irreversibly with vital cellular components of the invader and be cytocidal to them. The problem is that these drugs must not be injurious to us, the host (e.g., they must be *selectively toxic*). In many instances we have such "ideal agents." Examples are to be found among many of the antibacterial, antifungal, and other antiparasitic drugs. In the area of cancer chemotherapy, where the cells we seek to destroy are our own, selectivity is, unfortunately, usually not achievable.

Hydrogen Bond

This weak bonding force has a crucial role in stabilizing protein structures as well as in other physiological systems and their interactions with drugs and other exogenous chemical substances. The high intensity of positive character in relation to such a small atomic volume allows it to function as a precarious, yet real, bonding bridge between two electronegative atoms or even sites of electronegativity such as a double bond and π-deficient nitrogen-containing heterocyclic rings. Hydration (by hydrogen bonding) of compounds containing the pyridine ring such as pyridine and nicotinamide leads to water solubility, whereas the analogous benzenoid compounds, benzene and benzamide, exhibit very meager aqueous solubility. As long as the three atoms can assume a linear relationship (e.g., -O-H····N-, -OH····O-), hydrogen bonding becomes feasible. The elements capable of

Table 1-3. Bond Energies of Biochemical Interest

Bond	Energy (kcal/mole)	Bond	Energy (kcal/mole)
H - H	105	C ≡ C	200–230
C - H	77–100[a]	C - N	150
N - N	100	C ≡ N	210
O - H	115	C - F	109
S - H	88	C - Cl	69–95[a]
C - C	77–88[a]	C - Br	54–70[a]
C = C	145–175[a]	O - O	33–40[b]

[a] This value can vary depending on what the other bondings to the carbon atoms are.
[b] Organic peroxides.

bonding that are of primary interest biologically are, of course, O and N. Halogen atoms in drug molecules, particularly F, can also participate.

Since hydrogen bonds are relatively weak (1–5 kcal/mole) they can be easily dissociated, even by warming. The secondary and tertiary structure of enzymes (and other proteins) can be thermally unravelled, leading to destruction of catalytic activity. This protein denaturing process can also be brought about by chemical means such as with urea.

In addition to proteins, biopolymers such as RNA and the double helix of DNA are also stabilized by multiple hydrogen bonds between the base pairs. Hydrogen bonds are effective only over short distances, usually 3 Å or less, and at a very limited range of angles. An approach below 2.4 Å between the atoms bring repulsive forces into play. Thus steric requirements play an important part in hydrogen bonding. The restrictive parameters within which hydrogen bonds must function are considered significant to the specificity of most drug–receptor interactions. Although among the weakest of chemical links, it is hard to overemphasize the importance of the hydrogen bond. It literally holds the molecules of life in shape and plays a vital part in the functioning of enzymes and the action drugs.

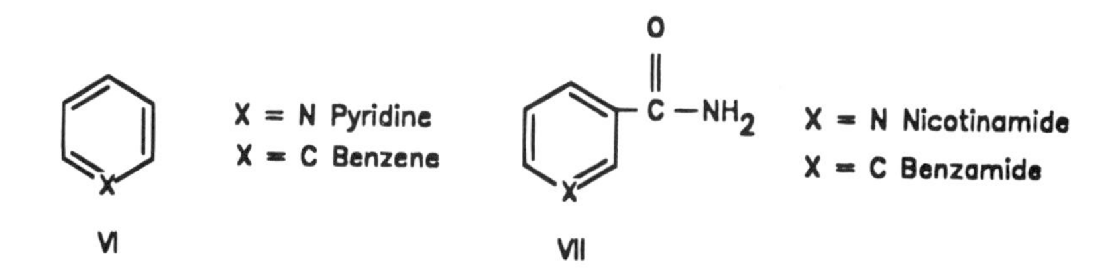

Electrostatic Bonds

Electrostatic bonds result entirely from electrostatic forces and may be between two ions of opposite charge, between an ion and a molecule, or between two molecules. The bond between two ionic species is an *ionic bond* that exerts a binding strength of about 5 kcal/mole. Common salt, Na^+Cl^-, is an example. In a biological solution the most characteristic property of ionic bonds is the ease with which they interchange. Electrostatic bonds may only exist for 10^{-5} sec, on average, because of the large concentrations of ions present in salt solutions. Rapid ion exchange thus results. In spite of their short existence, electrostatic bonds are significant, especially if reinforced by other short-range forces. In the formation of a salt bridge between an amine and a carboxylic acid, a hydrogen bond augments overall bond strength between the ions to as high as 10 kcal/mole (Fig. 1-3).

A second type of electrostatic bond is between an ion and a molecule that is polarized. That is, electrons exist in greater density at one end, or pole, of a neutral (uncharged) molecule, giving that pole a partial negative charge, while the opposite pole acquires a corresponding partial positive charge (these partial charges are noted by the lowercase letter delta, δ^-,δ^+). Such molecules are said to have a *dipole*. A dipole may be caused by an elec-

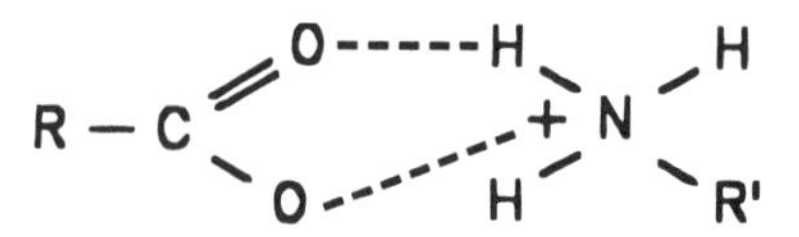

Figure 1-3. An amino-acid salt bridge augmented by a hydrogen bond.

tronegative atom. Its δ^- pole can then attract ions of opposite charge. Dipoles can also be induced by the close approach of other molecules. These *ion–dipole bonds* have strengths ranging from 1 to 7 kcal/mole.

The third type of electrostatic bonding is between two molecules and is caused by *dipole–dipole* interactions.

van der Waals Bond

These bonds, also known as van der Waals or London dispersion forces, are the universal forces of attraction between two molecules. The energy of molecular internal vibrations set up dipoles in the atoms of one molecule that in turn induce dipoles in the atoms of neighboring molecules as long as they are close enough to so affect each other (4–6 Å). Each dipole is only transient (10^{-6} s), but the number occurring is very large so that those that are in phase in adjacent molecules create small (0.5–1.0 kcal/mole) attractive forces between themselves. Unlike electrostatic attractions, whose intensity is inversely proportional to the second power, here we find an inverse proportionality to the sixth power. The significance is that a doubling of the distance between the two atoms decreases the attractive force by a factor of 128. van der Waals forces increase with atomic weight and become significant at 12 or more. Thus the biologically important elements C, N, and O, are included.

As the molecular weight and complexity increase, so do the sites for possible intermolecular contact. It can therefore be expected as a corollary that when intermolecular contact between drug molecules and biopolymers occurs, any interactions feasible would be stabilized to a higher degree than if van der Waals forces were nonexistent. Various complexes of drugs and carcinogenic compounds with DNA are suspected to occur with the aid of these forces.

Hydrophobic Bonding

This term might really be considered a special variant of van der Waals interactions involving either hydrocarbons per se or the hydrocarbon portion of other organic compounds containing oxygen and nitrogen atoms as well. Considering the latter type of molecules, when immersed in water we find the nonpolar hydrocarbon like parts being attracted to each other in a manner that "squeezes out" the water molecules (hence the description, *hydrophobic*). It will be recalled that amino acids have significant portions of the molecule that are of an aliphatic or aromatic nonpolar character. When these amino acids are part of protein chains, these portions can then be involved in hydrophobic interactions with each other, thus supplementing the stabilizing forces supplied to a protein's tertiary and quaternary structure by hydrogen bonding. In addition, these interactions may be between such hydrophobic areas of a protein and that of a drug. Here, too, these forces may be conceived to supplement electrostatic attractions in drug binding to plasma proteins. In cases where proteins are shown to have a significant carrying capacity for low-polarity lipid-soluble compounds such as steroids, hydrophobic interactions, as described, may even be the significant factor. The energy of a hydrophobic interaction contributed by a single CH_2 group has been estimated as less than 1 kcal/mole. However, as the number of such groups increases, the total effect becomes significant. An interesting study that may illustrate this utilized a homologous series of barbiturates to determine protein binding affinity. All six compounds, although having identical acidities (pKa 7.6), showed dramatic differences in fractions of the drugs bound to protein: a low of 5%, where *R* in the following structure was ethyl, to a high of 65%, where *R* increased to *n*-hexyl (VIII).

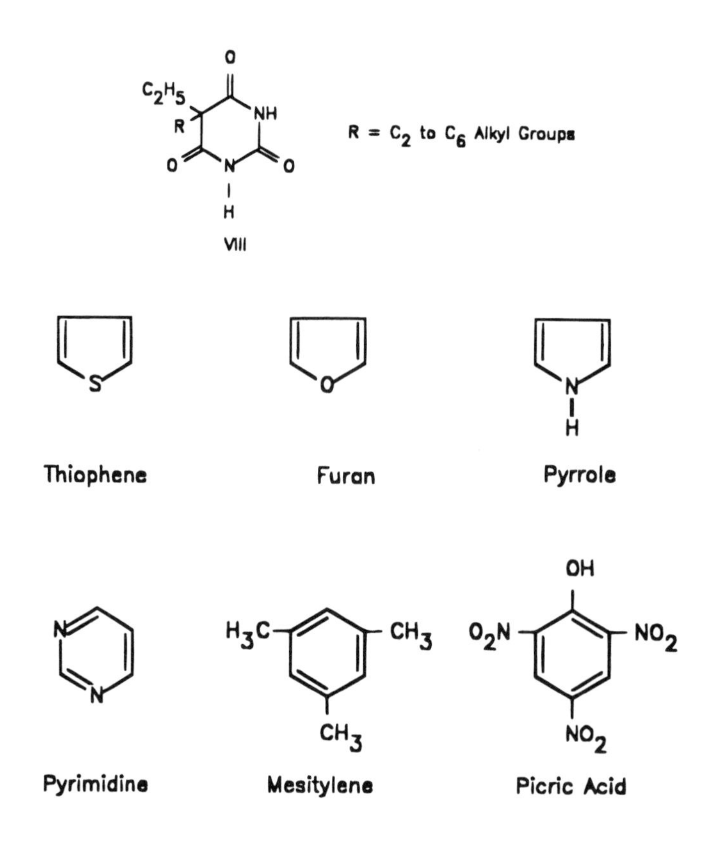

Charge-Transfer Interactions

This is a complex formation process between electron-rich donor molecules and electron-poor acceptor molecules. Energies holding the charge-transfer components rarely exceed 7 kcal/mole. These forces are operative over a 3–3.5 Å distance.

π-Electron-rich heterocyclic ring compounds such as thiophene, furan, and pyrrole, or more complex structures containing them, make excellent donor species for CT complexation; so do aromatic structures having electron donor groups [e.g., 1,3,5-trimethylbenzene (Mesitylene)]. π-Electron-deficient acceptor molecules include pyrimidines and aromatics with strong electron-withdrawing groups such as picric acid.

Drug–receptor interactions undoubtedly involve CT complexation. In some cases they act as the initial step of the drug's mechanism of action. The question arises as to how these various weak forces can be so important to the bioactivity of compounds occurring either naturally or as drugs. The contribution to the stability of interaction of a biopolymer or cell membrane component with a drug molecule by a *single* hydrogen or ionic bond would obviously be inconsequential. One must consider geometric situations where a drug molecule can fit snugly into an active site cleft, produced by the tertiary structure of one enzyme, or the complementary topographical features on a membrane surface that allows relevant functional groups to approach closely. It then becomes possible to conceptualize a summation of many such interactions. What can result, therefore, is adequate available energies to produce stable, although temporary, complexes or adducts of sufficient duration to yield the bioeffects observed. In extreme cases the interactions may approach energetics that rival those of covalent bonds.

1.2.2.4. Isosterism

In the narrowest sense isosteres are compounds, groups, or atoms that have the same number and arrangement of electrons in their outermost electron shell. They can be said to be isoelectronic. The concept was originally proposed by Erlenmeyer about a century ago on observations that some simple molecules could be pair based on certain similarities of chemical and physical properties. For example, C=O and N_2, CO_2 and N_2O, and CH_2N_2 and CH_2=O constitute such pairs. The concept soon expanded to the idea that isosteric replacement of such groups in more complex molecules would also result in similarities of certain properties. The basis of this theory is that the similarities of properties within the vertical groups of elements in the Periodic Table can be applied to whole molecules or molecular fragments. Some evidence for the validity of this notion was the observation that pairs such as benzene and thiophene, thiophene and furan, and even benzene and pyridine exhibited similarities in many physical and chemical properties. The term *ring equivalents* is used to describe the interchangeability of moieties such as -CH=CH- with -S- and -C= with -N= in ring compounds. If one discounts the H atoms, then these isosteric pairs will be seen to be similar in size and in peripheral electron configuration. Table 1-4 can be generated by considering elements in a periodic manner and adding H atoms to give the vertical columns of atoms (and groups) having peripherally isoelectronic arrangements.

Medicinal chemists began to apply the isostere concept to the synthesis of drugs. The utility of isosteric manipulation soon became apparent. Compounds with established drug activity could, by judicious isosteric modifications, be improved either in terms of pharmacologic properties such as efficacy and toxicity, or, less frequently, by more suitable pharmaceutic properties such as stability and bioavailability. Another approach was the synthesis of isosteric analogs of known metabolic intermediates important to cellular life processes. The hope was that they would act as antimetabolites and thus inhibit rapidly growing cancer cells from proliferating or, ideally, kill them. Even though this approach has not resulted in a cure, it has given us some remarkably effective agents. Drugs such as 6-mercaptopurine (SH for OH of adenine) and aminopterin and methotrexate (-NH for OH of folic acid) are two dramatic examples discussed further in Chapter 4.

1.2.2.5. Bioisosterism

Bioisosterism has now become a useful tool in the design of new drugs. In areas other than antimetabolites, improvement in therapeutics can also be achieved. A particularly interesting example is in the area of sulfonylureas used as oral hypoglycemics. The clinical observation that an antibacterial sulfathiazole derivative, IX, lowered blood sugar levels in patients dramatically led medicinal chemists to capitalize on this "discovery" by synthesizing what may be termed the open-ring bioisostere analog, Xa (O for S), resulting in the first such antidiabetic drug, carbutamide. A significantly less toxic isostere, Xb (-CH_3 for

Table 1-4. Isosteric Groupings

- C =	- N =	- O -	- F
	- CH =	- NH -	- OH
		- CH_2-	- NH_2
			- CH_3
			- CF_3

-NH_2), tolbutamide, soon replaced carbutamide in clinical practice. Finally, a third isostere Xc (-Cl for -CH_3), chlorpropamide, was introduced whose attributes were both decreased toxicity, and also extended duration of activity (due to nonoxidizability of the Cl atom compared to the CH_3 group).

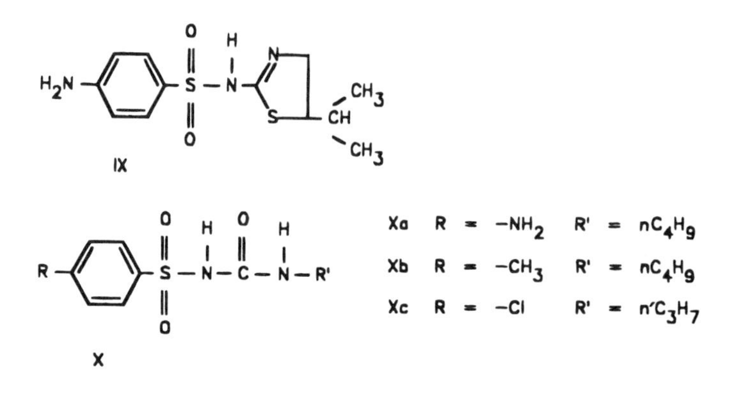

Bioisosteric interchanges, of course, are not limited to univalent functions. In fact, divalent atoms and groups add the additional factor of steric similarity to the equation since the bond angles between the valencies are very similar. The angles for /CH_2\ (111.5°), /NH\ (111°) /S\ (112°), and /O\ (108°) illustrate this point. When a deviation of 3 degrees, depending on the full structural features of the compound, are taken into consideration, it becomes apparent how identical these features can be. A rather convincing example is seen when the bioactivity relationship between the isosteric ester and amide moieties is examined (Fig. 1-4).

In the case of the ester it can be seen that the ether oxygen (C–O–C) has partial double bond character, as shown by the di-ionic resonance form. This, of course, severely restricts any rotation about the carbon–oxygen bond. The result is an actual structure that is rather planar (flat). Similar reasoning shows the amide moiety to be planar due to resonance. It is

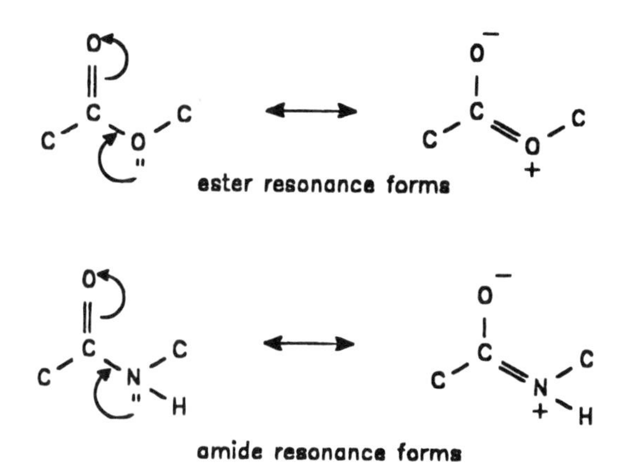

Figure 1-4. Ester-amide isosteric analogy.

therefore not surprising that similar molecules differing primarily in that one is an amide analog of an ester will be isosteric in a bioactive sense. Among the best-known examples are local anesthetics such as procaine (IV) and its amide analog procainamide.[2]

A fascinating example of the application of ester–amide bioisosterism involves the mechanism of protein synthesis in the cell and its blockade (discussed in detail in Chapter 5). Transfer-RNA, a key factor in protein synthesis, has the amino acid tyrosine esterified at the 3′-OH of the terminal adenosine unit (XI). The antibiotic drug, puromycin XII, is an amide analog, being a *p*-methoxy-tyrosinamide of N,N-dimethyladenosine. This transfer-RNA imposter is readily taken up by the system, but it then efficiently blocks protein synthesis. As a result, puromycin is being successfully used as an antibacterial, antiprotozoal, and anticancer agent.

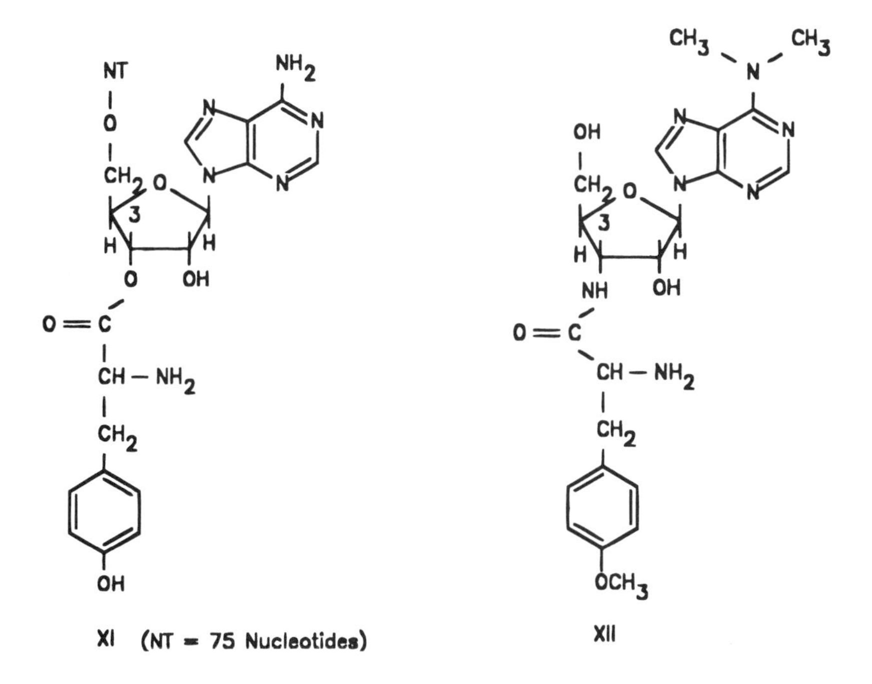

XI (NT = 75 Nucleotides)

XII

Other examples of successful interchanges of O and S and -NH and $-CH_2$ abound in the areas of antihistamines (both H_1 and H_2 types), analgesics, anticholinergic drugs, and psychopharmacological agents, with surprising results in the case of the latter. These will be pointed out in the sections dealing with those drugs.

Equally important isosteric replacements involve trivalent atoms and groups. These include -N= and -C=, especially in ring compounds, and they will be discussed in the appropriate chapters.

The preceding discussion on isosterism dealt mainly with the types of isosteres that may be considered classic (i.e., that is where the molecule bears a more or less readily apparent structural relationship to the parent compound). If one interprets bioisosteric relationships more broadly, less readily apparent similarities may result in useful new drugs. These might aptly be considered nonclassic isosteres.

---

[2] Procainamide, inferior as a local anesthetic, is used primarily as a cardiac antiarrhythmic agent.

One area of minor importance involves group reversal. Thus the narcotic analgesics meperidine (pethidine) XIII and alphaprodine XIV are actually reverse esters, where the first is the ethyl ester of a piperidyl carboxylic acid and the latter is a propanoic acid ester, of a piperidyl alcohol, a more potent drug. In retrospect, the two compounds procainamide and lidocaine XV may be viewed as reverse amides of each other, even though neither drug was originally synthesized on the basis of isosterism. The example of a ring opening procedure leading to improved hypoglycemic agents (IX to X a,b,c) was already mentioned. The opposite, the preparation of ring analogs of open chain active compounds, is also carried out by medicinal chemists as a way of developing better drugs. The skeletal muscle relaxant drug methocarbamol is a carbamate of 3-(*o*-methoxy-phenoxy)-1,2-propanediol, XVI. The oxazolidone ring of metaxolone (Skelaxin) XVII may be viewed as a cyclic carbamate and therefore as a ring isostere of methocarbamol. Another example where the open–closed ring analogy may not be immediately apparent is phenmetrazine, XVIII, and phenylpropanolamine, XIX, which are both anorexiants with differing degrees of central nervous system (CNS) stimulation.

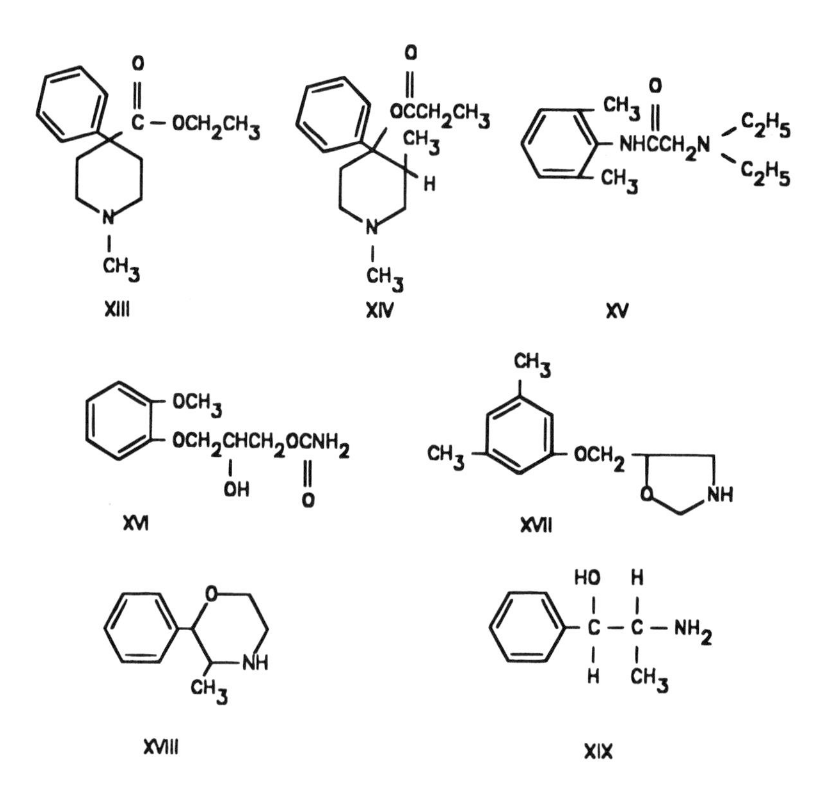

An offshoot of isosterism is the interchange of groups with similar polar effects, particularly acidities. The first successful use involved the carboxyl and sulfonamido groups, -COOH, $-SO_2NHR$. Although serendipitously discovered six decades ago, it resulted in the antibacterial sulfonamides, which several years after their introduction were found to be isosterically antagonistic to the bacterial metabolite *p*-aminobenzoic acid (the reasons for this antagonism will be discussed in a later chapter).

A particularly novel isosteric replacement for the carboxyl group is the tetrazole ring, whose acidity is only somewhat less than most carboxylic acids (pKa 4.9 vs. 4.0–4.4). The ionization and resonance forms of both functional groups are depicted below.

Applications include increased blood cholesterol lowering potency in the tetrazolyl analog of nicotinic acid and improved stability of certain dicarboxylic penicillins in which one carboxyl function has been replace by tetrazol.

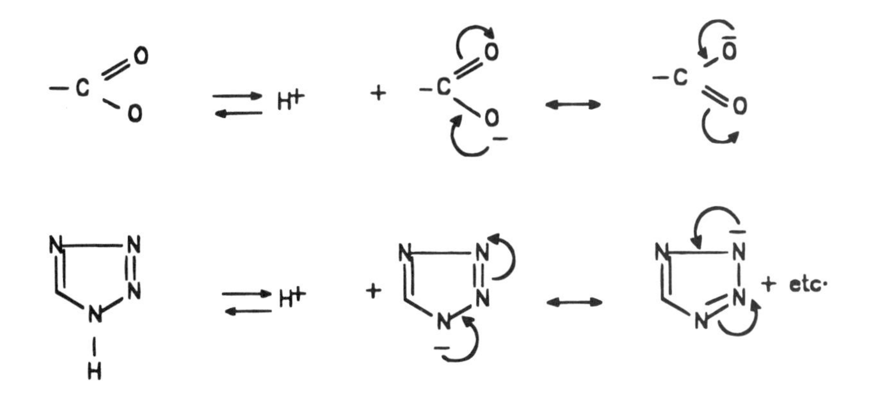

1.2.3. Spatial Considerations

Most drugs must bind or otherwise interact with receptor sites and the reactive sites of enzymes, which have very characteristic and specific topographical (surface structural) features. It is therefore not surprising that the steric requirements for designing effective drug molecules are often quite rigorous. Electronic distribution patterns also determine certain aspects of structure. Thus knowledge of these factors becomes important in understanding drug activity and in rationally designing new agents. In addition, studies of steric requirements have given us insight into the nature of various receptors.

1.2.3.1. Molecular Dimensions and Interatomic Distances

Much of our theoretical knowledge of distances between atoms in molecules, bond angles, diameters of atoms, and even volumes of functional groups has been confirmed and refined by X-ray crystallographic methods, especially diffraction.[3]

Many proteins such as enzymes are coiled into an α-helix. Each amide or peptide is linked to the third amide group down the polypeptide chain by hydrogen bonds. X-ray studies have shown that each complete turn of the coil contains 3.6 amino acid residues. The distance between two such consecutive turns is 5.4–5.5 Å (Fig. 1-5). Furthermore, the distance separating two consecutive peptide bonds in an extended protein chain is 3.6 Å.

It is now widely accepted that the receptors that interact with drugs to produce a drug reaction are largely protein in nature. Thus the two distances between helical turns and peptide bonds assume potential significance in the mechanism of drug action at the molecular level. It is possibly more than coincidence that a considerable number of drugs have functional groups believed to be involved in receptor complexation that are separated by dis-

[3] A brief but excellent discussion can be found in *Remington's Pharmaceutical Sciences* (1985).

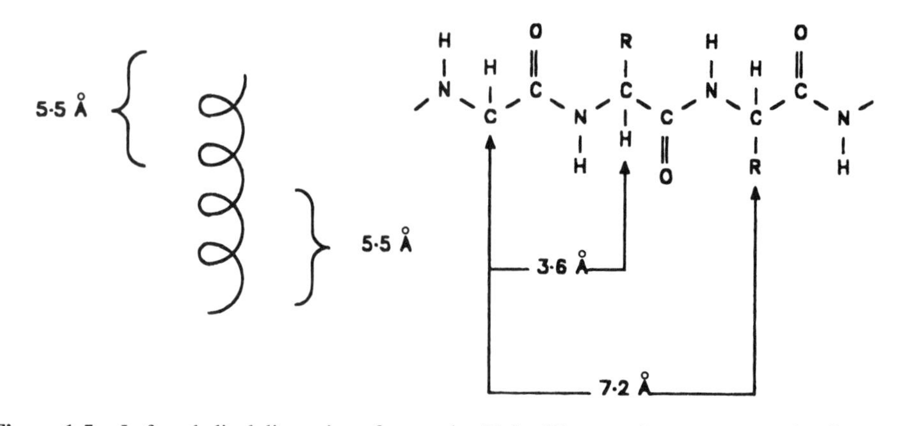

Figure 1-5. Left: α-helical dimension of a protein. Right: Distances between successive *R* groups.

tances that are multiples of the previously stated dimensions. These agents include compounds of such diverse pharmacology as antihistamines (e.g., diphenhydramine), antispasmodics such as adiphenine, cholinergics such as carbachol, and local anesthetics such as procaine. The 3.6 Å distance is also encountered in drugs having a quaternary cationic nitrogen, for example, cholinergics such as carbachol (2 × 3.6 Å = 7.2 Å) and cholinergic blocking agents. In curarelike agents such as decamethonium the distance is a double length of acetylcholine (4 × 3.6 Å = 14.4 Å). There are many other instances.

A striking example of similar molecular dimensions involves the elucidation of the role of *p*-aminobenzoic acid in bacterial biochemistry and the action of sulfanilamide in antagonizing it.

Stereochemistry

Optical isomers are frequently referred to as enantiomorphs or enantiomers. A racemic modification, or a racemate, is a mixture of equal parts of enantiomers (+ and –) and is therefore optically inactive. An optically pure isomer can be racemized, usually by drastic means such as heat, acid, or alkali.

A molecule with two nonidentical asymmetric centers can exist as four stereoisomers (i.e., as two different racemates). Enantiomers from either of the racemates cannot exist in a mirror-image relationship. Such enantiomers are referred to as diastereoisomers, which are illustrated by the drug ephedrine.

Note that D(–) ephedrine and L(+) ephedrine are mirror images of each other, yet they are not superimposable. They represent enantiomers and, if mixed in equimolar proportions, would be a racemate (Racephedrine). The other mirror image pair is D(–) pseudoephedrine and L(+) pseudoephedrine. On the other hand, D(–) ephedrine and D(–) pseudoephedrine are not mirror images and are therefore diastereoisomers. They even have different physical properties, with the former melting at 40°C with a specific rotation $[\alpha]_D = -6°$, while the latter melts at 76°C and rotates $[\alpha]_D = -50°$. Since we perceive interactions of drugs with receptor surfaces with definitive topographical features, it is not surprising the several ephedrine isomers do not possess equal quantitative, and even qualitative, activity. Clinically, D(–) ephedrine is used to a large extent as an antiasthmatic and, formerly, as a pressor amine to restore low blood pressure as a result of trauma and other causes. L(+) pseudoephedrine is used primarily as a nasal decongestant. The ephedrines also have vary-

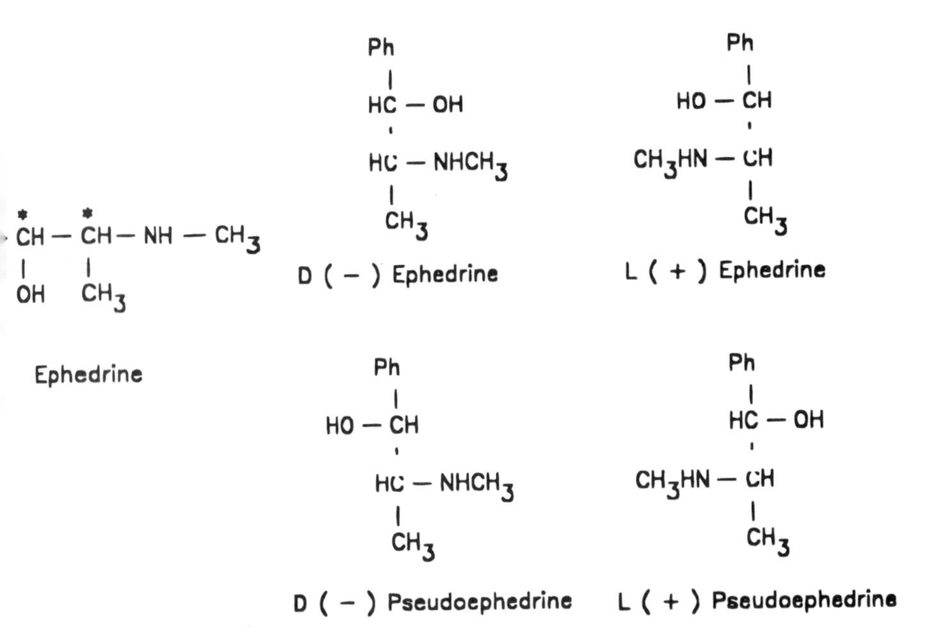

ffer by 180 degrees. Unlike an enantiomeric pair of isomers, *cis* and *trans* isomers in both physical and even chemical properties (e.g., pKa if X is a carboxyl group). ctions with receptors will differ greatly, even to the point of no binding. This should surprising when considering that the distance between X and Y may not be the same two isomers. Examples abound among drug molecules where one isomer only ts the desirable pharmacology.

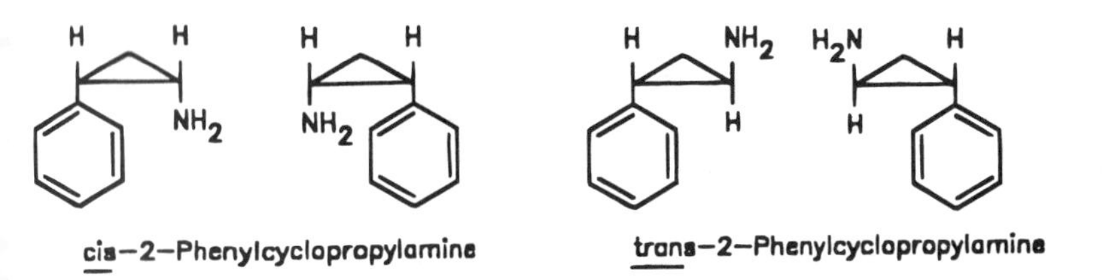

striction of a bond rotation between two carbon atoms resulting in *cis–trans* iso- n can also be imposed by the rigidity of alicyclic rings. Using a simple example such -dimethylcyclopropane, the *cis* isomer finds both methyl groups above (or below) the of the ring; in the *trans* isomer they are situated one above and one below the planar e of the ring. Since it is possible to draw an imaginary plane of symmetry through isomer, there is no chirality,[4] and therefore no optical isomers exist. A more inter- and complex drug example is 2-phenylcyclopropylamine, tranylcypromine ate). The drug, chemically related to amphetamine, is a potent inhibitor of the enzyme

---

ompounds lacking reflection symmetry (i.e., nonsuperimposable mirror images are *chiral* and therefore optically active). Compounds possessing reflection symmetry are identical with their mirror images and be *achiral* and optically inactive.

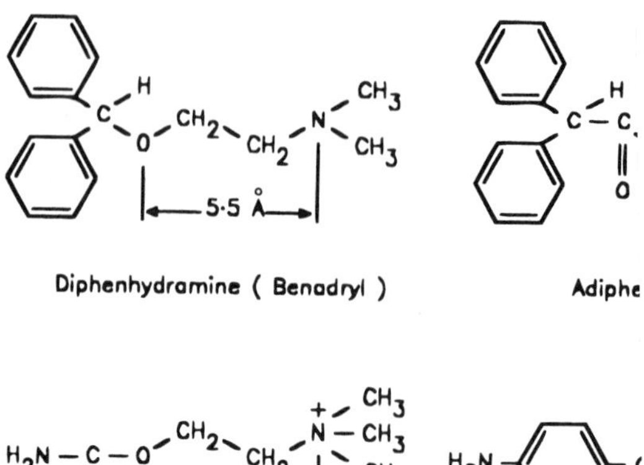

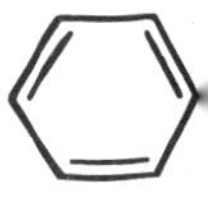

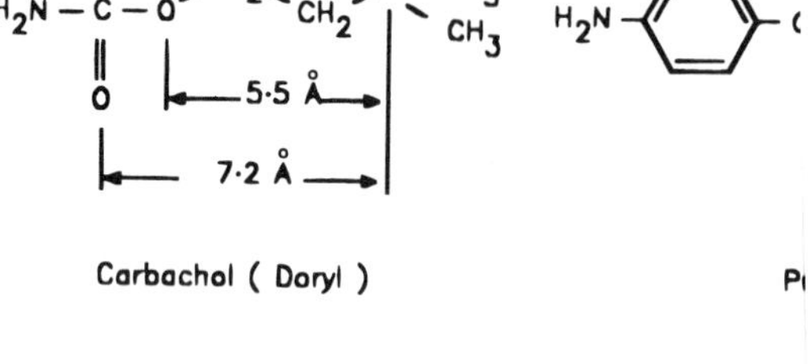

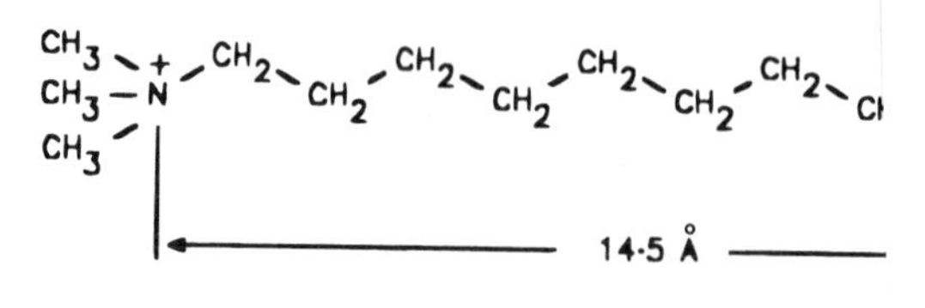

will
diffe
Inter
not b
in th
exhil

ing degrees of CNS stimulant properties. Table 1-5 illustr
pressor activity of the ephedrines and their racemates. As v
cussions, a difference in activity, even to the point of antago
of the same structure is not unique to the ephedrines.

Geometric Isomerism

This represents rotational stability about the ethylenic bo
hybridized carbon atoms. Since free rotation about the carbo
ble, two stable forms of the molecule can now exist. The six
in the same plane. The orientations of substituents *X* and *Y* re

R
meri
as 1,
plan
surfa
eithe
estir
(Par

Table 1-5. Pressor Potencies of the Ephedrines

	Isomer
D(−)	Pseudoephedrine
DL	Pseudoephedrine
L(+)	Pseudoephedrine
L(+)	Ephedrine
DL	Ephedrine
D(−)	Ephedrine

4
can b
said t

monoamine oxidase. Its clinical use is in the treatment of severe depression. The compound exists as two geometric isomers: *cis* and *trans*. Unlike dimethylcyclopropane, however, where reflection symmetry exist for each geometric isomer, here, because of the presence of two different substituents, each isomer is chiral and therefore resolvable into a pair of enantiomers. The *trans* pair is the stronger inhibitor and is marketed as the racemate. The *cis* isomers exhibit much weaker enzyme inhibition. One possibility for this decreased effectiveness is a partial eclipsing of the two functions due to their proximity to each other.

It is important to differentiate the terms *configuration* and *conformation. Configurations* can best be thought of as set in their ways (i.e., stabler relationships of groups to each other in space). *Cis–trans* isomers fit into this category, especially as seen in the case of alicyclic rings such as the 1,2-dimethylcyclopropane and tranylcypromine. Interconversion of these isomers cannot occur without bond breaking. In *conformations* changes of relative spatial positions of substituents are almost totally unhindered because of free rotation around single bonds. Impedance, if any, is very small. Energy barriers are 1–2 kcal/mole. If the groups are bulky, the barriers are somewhat higher, making certain positions [i.e., *conformers* (or rotamers) somewhat more favored]. Such steric effects usually mean that favored conformations are those where such groups are at maximum distances from each other. This is called a *staggered conformation.* However, in cases where such functional groups may be subject to weak interactions such as hydrogen bonds, the stability patterns may shift. Because of the ease of conformer interchangeability, they are impossible to isolate. However, they are useful in conceptualizing drug interactions with receptors and other biopolymeric surfaces. An early idea that evolved was that only one conformer was likely to have the "best fit" to interact maximally with the receptor. This concept was subsequently superseded by the assumption that even in the case of a rigid receptor requirement, the binding drug molecule whose lowest energy (most stable) conformer might not have the best "goodness of fit" could easily change to more favorable conformations. Since energy barriers to such conformational changes are so low, the energies needed to effect the changes are available. They can even be released by the binding process itself. In addition, the probability that receptor and macromolecular surfaces are also flexible further minimized the requirement for a lowest-energy conformer for effective drug–receptor binding (see the later discussion of induced fit theory).

1.3. Theories of Drug Activity

Although there is no single uniform theory that can explain the activity of drugs, several theories have been advanced over the years in attempts to elucidate drug action.

In general we may consider pharmacological action to be of two types. One type is thought to be caused solely by the physicochemical properties of the compound without discernible relationship to chemical structure. These *structurally nonspecific* drugs are usually administered in relatively large doses. It is presumed that such agents form a monomolecular layer over the total area of certain cells in the organism. Examples of such drugs were believed to be general anesthetic gases, depressants such as alcohol and chloral hydrate and antiseptics such as phenol and iodine. *Structurally specific* drugs, on the other hand, have an activity that is believed to emanate primarily from chemical structure. The total cell surface of target tissues or organs is presumably not involved; rather, the interaction of drug molecules entails relatively small, highly specific areas of the cell called *receptors*. These receptors are thought to be localized on, or in, such cell components as enzymes, nucleic acids, and portions of cellular membranes. Since the whole cell surface need not be covered, fewer drug molecules are required to elicit an effect. Thus we would predict the need for much smaller doses. Indeed, in this category we find drugs that are effective in astoundingly small quantities. Even these doses are far in excess of what would be needed for pharmacological effects if the drugs administered, by whatever routes currently used, were not subject to poor absorption, chemical inactivation, dilution by total body distribution, and diversion by excretion, before ever reaching the site of action.

It has been shown in some instances that extremely small fractions of the cellular surfaces need to be affected to elicit a drug response. It can be calculated that the dose of acetylcholine needed to reduce the heartbeat of a toad by 50% would cover only 0.016% of the surface area of the cells in the ventricle of the heart.

High potency (i.e., a low dosage requirement) is not, however, a decisive criterion in categorizing a drug as structurally specific. Many low-potency agents fall into the structurally specific category because of various other factors. A drug may require a high dose due to rapid metabolism to inactive forms, fast excretion, or a high degree of binding to plasma proteins.

The pharmacological effects of structurally specific drugs result from an interaction with a tiny area of certain cells that have functional groups, areas of polarity, and a general topography *complementary* to those of the drug. It is not surprising that we often find that apparently small structural alterations bring about drastic qualitative changes in activity.

1.3.1. Occupancy Theory

The occupancy theory of drug activity was first proposed by Clark (1926) and Gaddum (1926). It was actually an adaptation of the Langmuir isotherm dealing with the adsorption of gases onto metallic surfaces. The basis of their idea was that the law of mass action, operative in gas–solid interactions, could be applied to drug–receptor interactions: that one drug molecule occupies one receptor site, just as gas molecules are adsorbed on active sites of a solid surface. The drug–receptor relationships can be expressed as:

$$R + D \underset{K_2}{\overset{K_1}{\rightleftharpoons}} RD \overset{K_3}{\rightarrow} E \tag{1.10}$$

R represents receptor; D, drug; RD, receptor–drug complex; E, pharmacological effect; K_1, K_2, and K_3 are rate constants. The dissociation constant, K_D, can be calculated at equilibrium as:

$$K_D = \frac{K_2}{K_1}\frac{[R][D]}{[RD]} \tag{1.11}$$

The number of receptors occupied is dependent on the concentration of the drug in a unit area or volume[5] and on the total number of receptors, R_t, in it. As more receptors are

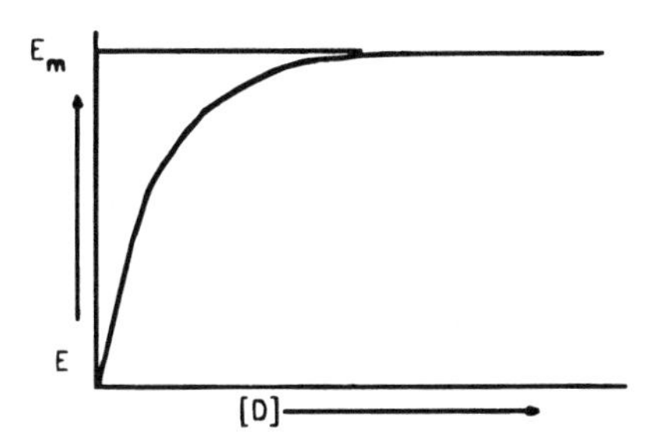

Figure 1-6. Relationship of drug action to drug concentration.

occupied, the pharmacological effect becomes more intense until, when all are filled, a maximal effect, E_m, is achieved.

Finally, the intensity of the effect will be dependent on the drug–receptor complex concentration, $[RD]$,

$$E = K_3[RD] \tag{1.12}$$

and on total receptor concentration

$$[R_t] = [R] + [RD] \tag{1.13}$$

Maximal effect is proportional to total receptor concentration.

$$E_m = K_3[R_t] \tag{1.14}$$

By dividing Equation 1.12 by Equation 1.14

$$\frac{E}{E_m} = \frac{[RD]}{[R_t]} \tag{1.15}$$

it can be shown that

$$E = \frac{E_m[D]}{[D] + K_D} \tag{1.16}$$

Equation 1.16 can be graphically represented in Figure 1-6.

The occupancy theory is most applicable when the drug–receptor interaction is the same for all drugs being considered in a series (i.e., when all drugs produce the same maximal response irrespective of dose).

Soon after this theory was proposed, it was discovered that it had loopholes. With certain series of drugs a maximal response was never achieved even at extremely high doses. The biological effect does not always appear to follow the law of mass action, nor does it seem to be dependent on the affinity of a drug for the receptors.

Ariens (1954) and Stephenson (1956) proposed modifications to the theory in an attempt to explain such anomalous factors. They visualized the drug–receptor interactions as being a two-step phenomenon:

1. Complexation of the drug with its receptor;
2. Production of the effect.

[5] Such a unit volume may be considered a hypothetical compartment.

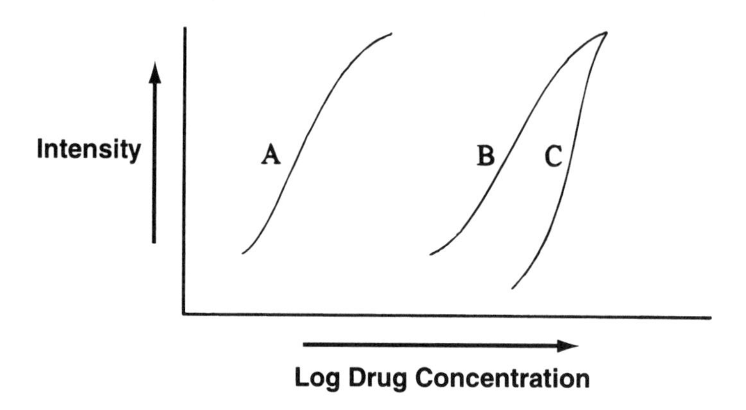

Figure 1-7. Dose–response curves of drugs with equal intrinsic activity.

Affinity of the drug for the receptor alone is not sufficient. The compound must also have what Ariens called *intrinsic activity*.[6] The Ariens–Stephenson idea describes the ability of the drug–receptor complex to produce a biological effect. Their modification brings in the concept of *agonists* and *antagonists*, both of which are thought to have a strong affinity for the receptor and a tendency to form a complex. However, only the agonist gives rise to the stimulus and has intrinsic activity. In the original theory, however, intrinsic activity was assumed to be constant.

Equation 1.17 takes this modification into account by introducing a proportionality constant, α, thus changing Equation 1.12 as follows:

$$E_D = \alpha[RD] = \frac{\alpha[R_t]}{1 + \frac{K_D}{[D]}} \tag{1.17}$$

When $\alpha = 1$, the compound is a full agonist. If the intensity of action of such drugs is plotted against increasing concentrations, there are obtained similarly shaped dose–response curves (Fig. 1-7) having the same maximum values (E_m), but, of course, at different drug concentrations. Thus full agonists, although having different affinities for a given receptor, have the same intrinsic activity.

Partial agonists have α values of less than 1; they do not produce a full maximal effect, although affinities for the receptor remain the same (Fig. 1-8). *Antagonists* have zero α values and show an absence of intrinsic activity. They do have affinity for the receptor sites and will therefore block effects of agonists added subsequently.

In spite of the obvious appeal of the occupancy theory, it fails to explain several important facts of drug action: primarily that drugs vary in their action. It does not solve the puzzle regarding why agonists are active and antagonists only weakly or not at all, even though they occupy the same receptor. Therefore, its main drawback is that it does not suggest a mechanism of action at the molecular level. In the final analysis drug action cannot be totally explained by simple receptor occupation models.

---

[6] Stephenson preferred the term *efficacy*.

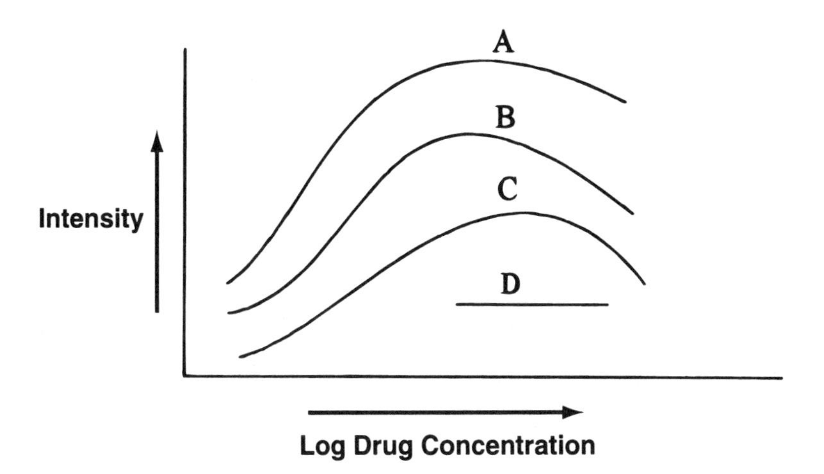

Figure 1-8. Dose–response curves of drugs with different intrinsic activity.

1.3.2. Rate Theory

The rate theory proposed by Paton (1961) as an alternate to the occupancy theory is based on the idea that a drug is effective only at the moment of encounter with its receptor. Thus receptor activation is proportional to the total number of encounters that the drug has with the receptors per unit of time. Pharmacological activity is a function only of the rate of association and disassociation of the drug and receptor. No stable drug–receptor complex is necessary. Agonists are thought to have high rates of association and dissociation; antagonists dissociate slowly, but they may associate quickly.[7]

The rate theory also has serious flaws. Experimentally, we find that agonists often do form relatively stable complexes. The drug phenomena on a molecular level cannot be explained. The question regarding why two similar compounds are antagonistic to each other is also left unanswered.

1.3.3. Induced-Fit Theory

Considerable knowledge exists about the nature of the *active site* of enzymes, their secondary, tertiary, and quaternary structures. In the case of those enzymes that have been obtained in crystalline form and subjected to X-ray analysis, conformational appearance has been deduced. However, the geometry of such an active site, as elucidated on an isolated crystalline enzyme, need not necessarily be complementary to the natural substrate, or drug molecule, to complex and interact with it. Since we visualize such a site as being flexible rather than rigid, the substrate or drug can *induce* such a complementary fit.

Figure 1-9 illustrates in an oversimplified manner how a substrate-induced conformational change in an enzyme might occur. H, +, and – represent the hydrophobic and electrostatic attractions between the substrate or drug molecule and an area on the protein. The resulting complex can dissociate again, and both components can return to their original shapes. The initial geometry of the protein molecule is also critical. The combination with the enzyme seems to induce a change in its conformation, which in turn can result in an

[7] Antagonists frequently cause a brief period of stimulation prior to blockade.

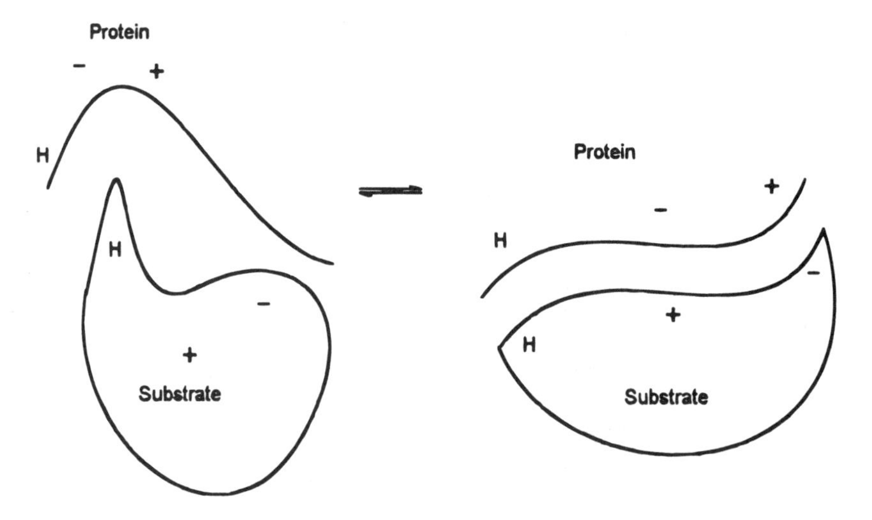

Figure 1-9. Mutual effect of drug and active site on conformation.

active orientation of catalytic groups. The biological effect obtained is not only caused by the relatively strong binding between substrate and enzyme; the effect may actually be caused by the induction of a proper conformational change.

A change in the size, shape, and, therefore, volume of the normal substrate that interacts with the enzyme could likely bring about a change in the proper alignment. This results in molecules that are either more efficient, or even antagonistic. It should also be kept in mind that the induced conformational changes being considered need not be limited to the protein; they may also be brought about in the substrate (drug) molecule resulting in an even more effective interaction. It is essential that an additional point be considered, namely that the interaction of a drug with a receptor protein need not necessarily be produced as a result of complexation with catalytic groups at so-called active sites. Rather, the interaction can also occur at a modifier or allosteric site, which is a regulatory site on enzymes that controls feedback mechanisms.

The induced-fit theory was originally proposed to help explain enzyme–substrate interactions. The subsequent extension of this theory (Koshland, 1961) to explain the mode of drug action seems logical. However, with drugs, biopolymers other than enzymes must also be considered: noncatalytic proteins, for example. Several variations have been added to help explain cooperative effects in which the binding of one type of substrate, not necessarily a drug, may accelerate or otherwise enhance the binding of subsequent types of substrates. The conformationally induced changes in the receptors during the reversible association with a drug can be invoked to help explain, in part at least, the initiation of the pharmacological responses observed.

1.4. Quantitative Aspects of Drug Action: An Overview

The decade of the 1930s was the beginning of attempts to introduce mathematical methods to gain increased understanding of chemical processes such as reaction rates, mechanisms, and certain processes, and to apply these techniques to the art and intuition of prediction. The scientific impetus was the then relatively new discipline of physical organic chemistry pioneered by chemists such as Hammett (1940). These scientists elucidated the SARs of the electronic and steric effects of various substituents on the equilibria of organic reactions, and how they affected the reaction centers on a quantitative basis. Even conceptually separating the inductive and resonance effects was partially successful.

A Hammett relationship might determine what the effects of different substituents on the benzoic acid molecule would have on its pKa value and its *proportionality* to a corresponding set of substituted phenylacetic acids. This can be expressed in logarithmic terms by Equation 1.18, where X is the meta or para substituent and a and b are constants.

$$\log Ka_{\text{benzoic}} - X = a \log Ka_{\text{phenylacetic}} - X + b \tag{1.18}$$

A more complete representation of the relationship would be:

$$\log Ka_{\text{benzoic}} - \log Ka_{\text{benzoic}} - X = a\,(\log Ka_{\text{phenylacetic}} - \log Ka_{\text{phenylacetic}} - X) \tag{1.19}$$

If the left side of Equation 1.19 is plotted against the right side, the proportionality between the free energies of the two equilibria (e.g., the ionizations of the corresponding acids) will be a straight line and is said to be a *linear free energy relationship* (LFER).

Subtracting the log Ka of the substituted acid (in this case a benzoic acid) from the unsubstituted acid gives the value of a constant σ (sigma) representing the electronic contribution of the substituents.

$$\sigma = \log Ka_{\text{benzoic}} - \log Ka_{\text{benzoic}} - X \tag{1.20}$$

and similarly

$$\sigma = a \log Ka_{\text{phenylacetic}} - \log Ka_{\text{phenylacetic}} - X \tag{1.21}$$

and

$$\sigma a = \frac{Ka_{\text{phenylacetic}} - X}{Ka_{\text{phenylacetic}}} \tag{1.22}$$

generalizing this relationship affords

$$\log \frac{K}{K_a} = \rho\sigma \tag{1.23}$$

where K is the rate or equilibrium constant for the substituted compound and Ka for the unsubstituted (reference) compound; ρ (rho), actually the slope of the aforementioned plot, is a proportionality constant representing the sensitivity of the reaction to electron density. Equation 1.23 is the now classic *Hammett equation.* The constants σ, ρ, and E_s which have negative values, are electron-donating groups, while those groups having positive substituent constants are electron withdrawing, or the substituents are not used to determine σ values because of interfering factors such as steric effects. Steric effects, E_s, on LFER can be assessed by a comparison of acid-catalyzed hydrolysis rates of methylacetate, CH_3COOCH_3, with those of substituted carboxymethyl esters, X-$COOCH_3$. E_s is the steric parameter developed by Taft (1956).

$$E_s = \log K_{\text{XCOOCH}_3} - \log K_{\text{CH}_3\text{COOCH}_3} \tag{1.24}$$

The relationships defined by σ, ρ, and E_s are termed *extrathermodynamic* since thermodynamic principles per se do not dictate them. Their utility is in enabling comparison of reaction sensitivities to electronic and steric effects and to study the effects substituents bring to bear on them.

By integrating the previously discussed factors with the partition behavior of organic compounds Hansch (1963, 1964, 1965), Fujita (1964) and others began to apply these principles to quantitative structure–activity relationships (QSAR) of drugs, and this has been improved and expanded with the aid of computers. The hydrophobic parameter π_x within which a given substituent x affects molecular conduct as to drug distribution and transport, as well as aspects of drug–receptor activities, can now be related to a drug's distribution behavior:

$$\pi_x = \log P_x - \log P_{yH} \qquad (1.25)$$

where log P is the logarithm of the 1-octanol-water partition coefficient and y is any parent structure (i.e., the unsubstituted reference compound). Both compounds are assumed to be in the neutral (nonionic) form. One may consider the value of π as indicative, to a degree, of how a substituent contributes to the solubility behavior of a molecule while it is partitioning between lipoidal and aqueous interfaces in the putative compartments it transverses as a drug in order to reach its ultimate site of action. Whether solid surfaces, such as in drug adsorptions on colloidally suspended plasma proteins, are accounted for in the value of π is not clear. What is clear, however, is that taking π and Hammett's σ values into consideration concurrently has allowed useful correlations between biological activities of some compounds with their chemical structure and physical properties. This should not have been unexpected. The fact that a given substituent on a molecule can effect changes in electronic, steric, and hydrophobic properties of a compound has been intuitively understood by organic-medicinal chemists for a long time. Hansch's correlations put these into a more quantifiable, predictive basis—and apply it to biological systems as well. In other words, the conclusion that substituents affect biological properties of a compound because of changes in many (or all) of the physical properties became inescapable. We are probably now on the first rung of the ladder climbing to discover the relationship between biological activity and chemical structure on a quantitative basis.

Much of the earlier published work in this area, especially with aromatic systems, indicates that π and σ values are *constant* and *additive*, especially if no strong steric interactions are involved. Thus, in aromatic compounds with two or more substituents, the total substituent constant is approximately the sum of the individual constants. This applies equally to the π and σ values; π values of substituents not bonded to aromatic systems are numerically different. Table 1-6 is a partial listing of Hansch substituent solubility constants (π) and Hammett's electronic constants (σ). The π constants will differ somewhat from one aromatic system to another. For example, the 4-Cl value of phenoxyacetic acid, $C_6H_5OCH_2COOH$, is +0.70; for phenylacetic acid, C_6H_5-CH_2COOH, it is +0.70; benzoic acid, C_6H_5COOH, is +0.87; for benzyl alcohol, $C_6H_5CH_2OH$, it is +0.86; for phenol, C_6H_5OH, +0.93; and for nitrobenzene it is +0.54. A positive π value for a substituent means that it enhances lipid (or other nonpolar solvent) solubility; a negative value signifies increased solubility in polar solvents. Electron donor groups have negative σ values; positive values indicate electron-attracting properties. The increased lipophilic character imposed by the $-CF_3$ group compared with the corresponding methyl, halogen, or unsubstituted compound has led to its frequent use in drug molecules, where this property is likely to enhance pharmacological effects. The concept was further expanded by the assumption that all three substituents having an effect on drug potency (and efficacy) are independently additive. This gave rise to the following linear Hansch equation:

$$\log \frac{1}{c} = a \log P + bE_s + \rho\sigma + d \qquad (1.26)$$

Table 1-6. Partial List of QSAR Constants

	Aryl π[a,b]				σ[c,d]		E_s[d,j]
3-Cl	0.76[e]	0.68[f]	0.83[g]	*p*-NH_2	–0.66	CH_3	0.00
3-Cl	0.84[h]	1.04[i]		*p*-OC_2H_2	–0.25	OCH_3	0.99
4-Cl	0.70[e]	0.70[f]	0.87[g]	*p*-CH_3	–0.17	OC_2H_5	0.90
4-Cl	0.86[h]	0.93[i]		*m*-CH_3	–0.69	F	0.49
4-CH_3	0.52[e]	0.45[f]	0.42[g]	*p*-F	0.06	Cl	0.19
4-CH_3	0.48[h]	0.48[i]		*p*-Cl	0.22	Br	0.00
4-OH	–0.61[c]	–0.30[g]	–0.85[h]	*m*-F	0.37	I	–0.20
4-OH	–0.87[i]			*m*-Br	0.39	C_6H_5-	–0.90
3-OCH_3	0.12[e]	0.12[i]	0.04[f]	*p*-COOH	0.73	C_2H_5-	–0.07
3-OCH_3	0.14[g]			*m*-OH	0.12		
4-NO_2	0.24[e]	–0.05[g]	0.54[i]	*p*-OHm	–0.37		

[a] For groups conjugated to aromatic systems.
[b] Hansch and Leo (1979).
[c] Hammet (1940).
[d] Taft (1956).
[e] Phenoxyacetic acid.
[f] Phenylacetic acid.
[g] Benzoic acid.
[h] Benzyl alcohol.
[i] Phenol.
[j] o-Substituted benzoates.

C is the concentration of drug producing the biological effect being measured. log A, where A is the biological response relative to a standard or lead compound, can also be used. log P is the substituent constant for solubility (i.e., π), E_s is the Taft constant for steric effects, ρ and σ are as previously defined, a, b, and d are constants of the system.

Free Wilson Analysis is an alternative procedure to the Hansch analysis (Free and Wilson, 1964). The assumption made here is that in a group of related compounds (i.e., a series having substitutions at more than one position), the effect of any given substituent is not dependent on the effects of a substituent at other positions. Measurement of physical properties are not necessary since substituent constants are based on biological activities. Equation 1.27 is the mathematical representation, where A is the log potency of a reference compound,

$$\log\frac{1}{c} = A + \sum_i \sum_j G_{ij} X_i \tag{1.27}$$

G_{ij} is the contribution of activity by the substituent i at the jth position, and X_i represents the presence ($X = 1.0$) or absence ($X = 0$) of substituent at the jth position.

A refinement is given by Equation 1.28, where a_l is the group contribution of the ith substituent, X is as in Equation 1.27, μ is log $1/c$ for the unsubstituted compound. A coefficient with a positive value means that the substituent

$$\log\frac{1}{c} = \sum a_l X_l + \mu \tag{1.28}$$

increases activity (in relation to no substituent, H). The larger the value the more important the position to which the substituent is bound.

Mathematical analysis of the extrathermodynamic relationships exemplified by Hansch and Free–Wilson equations is done by the statistical method of regression analysis, the least-squares fit. With the aid of computers the best fit of a large amount of data to the

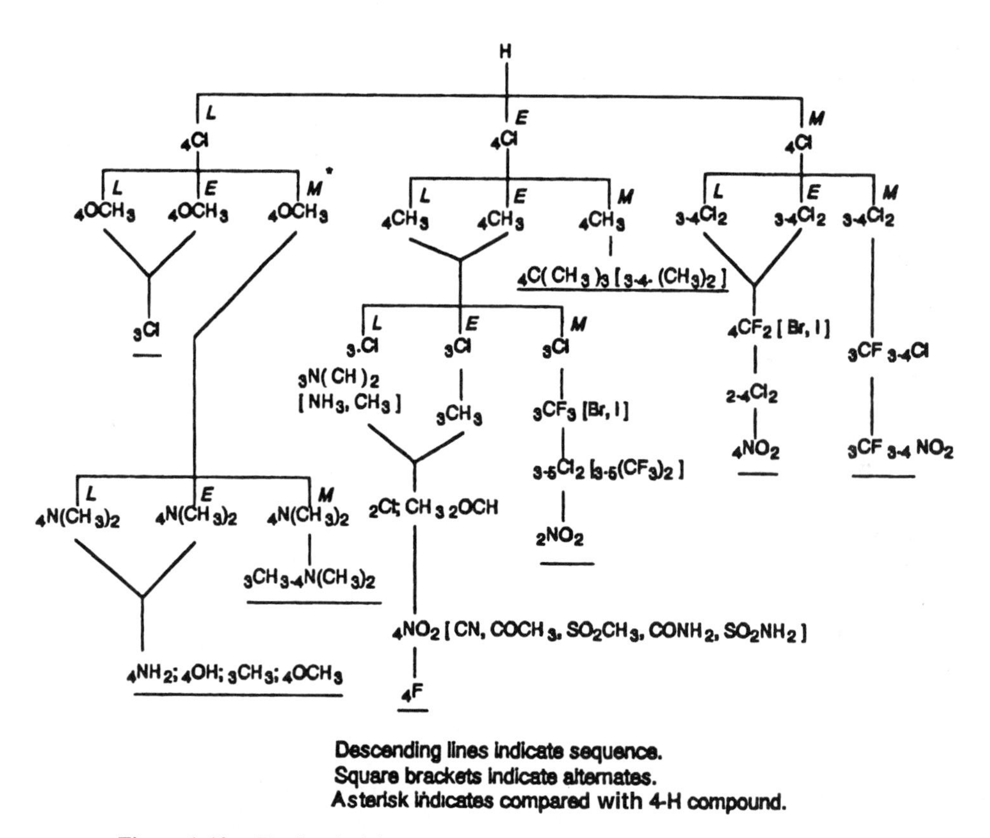

Figure 1-10. Topliss decision tree. *M*, more active; *E*, equiactive; *L*, less active.

equations can be rapidly calculated.[8] This procedure gives useful correlations by both methods in most situations and has been shown to give comparable results.

No doubt other refinements and approaches will be made to quantify structure–activity relationships of drugs to help design new agents and minimize the effort and expense of synthesizing and testing predictably inferior analogs of a *lead compound.* The usefulness of optimizing lead compounds will increase as experience with and understanding of quantitative methods increases, yet it should be understood that QSAR methods do not produce the lead compounds at present; ultimately, lead-generating techniques may evolve. An additional point should be made. The degree of reliability of the methods depend on the accuracy of the biological tests that, of course, have large experimental errors and variability. This may result in correlation errors. The accuracy of biological data, however, may increase in the future as new methods not using animals are developed as they already have been for some toxicological screenings.

An essentially nonmathematical approach to utilizing the basic Hansch concepts to help design drugs was developed by Topliss (1977). It has been called a *decision tree* (Fig. 1-10), by which a lead compound can be efficiently optimized without the use of computers. By preparing several well-chosen analogs of a lead compound, the next several

[8] For an excellent brief exposition of regression analysis, see Martin, 1978, pp. 167–187.

analogs can then be planned based on the biological results of the first batch. An optimum analog can then be frequently arrived at by preparing a total of only a dozen or so compounds. One limitation of this method is that the lead structure have an unfused benzene ring.[9] A scheme for nonaromatic systems also exists. The use of π, σ, and E_s parameters are, of course, considered in the decisions at each step.

As one follows Figure 1-10, the first derivative of the lead compound would be its 4-Cl derivative. Since both π and σ values for *C* are positive (Table 1-6), this is likely to result in the first analog being more active, *M*, than the lead. The next analog, the 3,4-dichloro, should be even more active. Further improvement should then be achieved by the 4-Cl, 3-CF_3 compound, and even by the 3-CF_3, 4-NO_2 derivative. If the 4-Cl derivative were less active than the lead compound (e.g., due to steric effects), then positions other than para may be useful. Substituents with negative π and σ values may be effective at the 4 position (e.g., the –OCH_3). The decision tree also suggests a scheme for equiactive 4-Cl derivatives to follow; an extension of the approach to side-chain alkyl groups as may occur in esters, ketones, amines, and so on. Applications to antiinflammatory agents, diuretics, and amphetamines illustrate the utility of this method.

In summary, in designing new drugs from a discovered lead compound there must be strategy that will allow us to eliminate the synthesis and testing of structures that are likely to be ineffective or otherwise inferior from consideration. It is obvious that one cannot synthesize every conceivable derivative. Consider the hypothetical lead compound XX, where substituents can be put in the five positions indicated. Using only four substituents that are likely to offer good discrimination between steric (E_s), electronic (σ), and partition (π) effects, a total of 1,024 (4^5) compounds would have to be prepared to evaluate the "best." If we decided on expanding the experiment to seven substituents, the number of compounds needed to be synthesized would be an astronomical 16,807. It is clear that a manageable number must be arrived at. QSAR analysis and intuition must now come into play. One approach might be to apply the Free–Wilson method as an initial approach to the synthetic efforts. This will give information on the role of substituents in biological activity of interest and what functional groups to use. The Hansch analysis would then be invoked to separate the three parameters mentioned (π, σ, and E_s). Last, molecular-orbital (MO) calculations might then be used to help determine the electron perturbations likely to produce the desired biological activity. The MO method can be viewed as fine tuning. In a way it constitutes another way of studying the influence of substituents on a compound. MO methods have been described by Kier (1971) and others.

An alternative game plan could be to utilize the Topliss scheme as the initial approach to extract valuable information by the synthesis of a small number of compounds.

Additional approaches to help make more sense, and therefore increased predictability, out of SARs include *pattern recognition*. This method has been applied to the analysis of

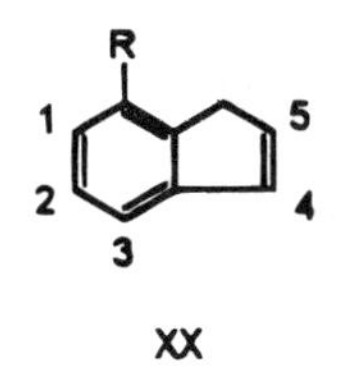

XX

[9] More than one-third of all organic compounds fall into this category.

different spectroscopic data (e.g. NMR or MS) to recognize patterns that appear to relate physicochemical properties to biological effects.

Cluster recognition is a refinement wherein observations of patterns (properties) as data points may cluster at varying distances when graphed and the distances between cluster centers may show relationships or patterns useful in selecting certain substituents for an initial set of analogs of a lead compound to synthesize.

New methods will be proposed; present methods will be refined. However, for a QSAR model to be widely useful, it ultimately must meet one criterion: to predict accurately the biological activity of a molecule before it is synthesized.

1.5. Receptor Concept of Drug Action Mechanisms

1.5.1. Historical Development

The development of the idea that can be called a *drug–receptor theory* required over a century to reach its present modest level of sophistication. It expresses the understanding that a drug exerts its biological activity as a consequence of binding, or interacting in some manner, with a specific receptor in a biological system. Not that people did not wonder how the various powders, potions, and infusions worked their imagined, and occasionally real, magical effects; they undoubtedly came up with "explanations" that at best might elicit a chuckle today. It was not until the nineteenth century that the beginnings of the scientific method and the rudiments of modern chemistry emerged to coalesce into conceptual thought with experimentation. The era of the armchair pseudoscientific theoretician was vanishing.

The credit for the concept of a receptive substance being involved in the physiological action of a drug goes to John N. Langley, a physiologist who in the 1870s developed some experimental support to the speculation of some early pharmacologists that drug action was somehow the result of a chemical reaction of the drug and some cell constituent. His studies dealt with the antagonism between pilocarpine and atropine, which convinced him that these agents compete for some substance in the body. A quotation from his paper on the subject in 1878 states: ". . . We may assume, that there is a substance or substances in the nerve endings or gland cells with which both atropine and pilocarpine are capable of forming compounds . . . according to some law of which their relative mass and chemical affinity for the substance are factors." With hindsight we may consider this a definition of receptor action. Langley suggested that the drugs are in competition to combine with what he believed to be the same protoplasmic material that he called "receptive substance." He expanded his theory more than a quarter of a century later to state that many drugs (and poisons) acted similarly. He also put forward the idea that the receptive substance need not be a separate compound but a "side chain" of, say, a muscle fiber.

Coining the term *receptor*, however, and picturing it in terms of complementarity between a receptor's surface and drug molecule's structure was part of a lengthy presentation by Paul Ehrlich on the contemporary state of chemotherapy in 1909. Ehrlich's earlier work with the selective staining properties of organic dyes on parasitic microorganisms led him to the logical conclusion that drugs (some of them dyes at the time) also have selectivity of effect on certain tissues. He further postulated that mammalian cells contain substances, differing by cell types, which can form insoluble derivatives with drugs as dyes do when fabric fibers are exposed to them, forming insoluble products called lakes (the process of dyeing). Ehrlich visualized receptors as small, chemically definable areas of large molecules. He defined receptors with elegant simplicity: "That combining group of the protoplasmic molecule to which the introduced group is anchored will hereafter be termed *receptor*." Ehrlich's work in the area of chemotherapy involved arsenicals as well as dyes. Because heavy metals such as As and Hg were known to form highly insoluble inorganic precipitates with hydrogen and other soluble sulfides, he argued that mercapto (sulfydryl, -SH) groups were the receptors for arsenic in the trypanosome,[10] and its irreversible blockade (precipitation) by arsenicals causes the organism's death. In spite of his research with cytocidal dyes and heavy metalloorganics whose effects were irreversible, he was aware that other drugs, in areas other than chemotherapy, had only a temporary, reversible effect, and could be easily extracted from tissues. He mentioned the probability that interactions in these situations were probably not "true" chemical combinations. These ideas and pronouncements must today be considered in the context that the understanding of chemical interactions at the time involved only covalent and ionic bonding. The weaker forces reviewed earlier had not yet been developed.

The receptor concept evolved as a result of the need to explain three important characteristics of drugs: (1) Frequently, only extremely small quantities of many drugs are needed to bring about a measurable effect. (2) There is a high chemical specificity among many drugs. For example, optical isomers of the same molecule frequently have different intensities of action or even differing activities. (3) There is considerable biologic specificity among drugs. Thus epinephrine has intense effects on cardiac muscle, yet almost no effect on striated muscle in other areas of the body.

Stereochemical specificity of drugs, when it occurs, was considered some of the early evidence for the existence of receptors. Since enantiomers have identical chemical and physical properties[11] but are nonsuperimposable mirror images of each other, it can be rationally assumed that the three-dimensional shape of the drug is decisive in determining its action. A logical extension of this idea is that at least part of the molecule must be a structure complementary to it.

One of Langley's set of pharmacological experiments may be considered as early evidence supporting the receptor concept. Curare, the South American arrow poison, was known to block impulse transmissions from motor nerves to skeletal muscles. The block was found to be localized at the nerve terminals within the muscle. If the nerve were to be cut, then muscle could still be stimulated chemically by applying nicotine where the nerve had terminated before it was cut. If curare were to be applied first, however, subsequent application of nicotine elicits no response; its effect would be blocked. If such a curare-blocked muscle, whether its motor nerve was still intact or cut, were to be electrically stimulated, it would contract. These observations indicate that both curare and nicotine act upon a substance or area that is neither the nerve nor the muscle. Nicotine, then, probably binds with it, setting off the response resulting in muscle fiber contraction. Curare also combines with it, however, but without triggering a response. Thus curare effectively prevents interaction with nicotine. This specialized substance, or area, became recognized as the *receptor*.

The receptor concept was initially greeted with skepticism. Attempts to isolate receptors as distinct substances failed. However, by the middle 1930s experiments were being performed on a quantitative basis, which began to dispel some of the doubts. Drug–recep-

[10] The protozoan causing trypanosomiasis (sleeping sickness).

[11] This, of course, is only true in a symmetrical chemical environment, which many biological systems are not.

tor interactions could be shown to obey the Law of Mass Action. Quantitative data became more reliable and supported the concept that drug action results from interactions of drugs with specific receptors, either by reversible affiliation or, rarely, by covalent bonding.

Much of the methodology and many of the ideas came by analogy from the biochemical studies of substrate–enzyme interaction research, which preceded much of the drug–receptor work. It is not surprising that some receptors turned out to be enzymes, and these will be considered later (Table 1-7). However, many receptor interactions are known to be more complex than either stimulation of inhibition of an enzyme. Receptor interactions with epinephrine or acetylcholine, for example, are believed to involve integral parts of semipermeable membranes, which may help to explain the partial responses obtained with these agents. Enzymes are frequently involved in the overall mechanisms. The enzyme adenyl cyclase, which produces the "second messenger" cyclic AMP (*c*-AMP) from AMP, is part of a coupling system in certain endogenous ligand–receptor interactions, with epinephrine and acetylcholine being prime examples. It is obvious that drugs that mimic or antagonize these substances are also dependent on the involvement of *c*-AMP.

Receptors need not be enzymatic or even protein in nature. Other biopolymers, or portions of them, can be included. Of course, they must be essential to the integrity of the living cell. Nucleic acids, particularly DNA, are prime targets for several major categories of drugs in the areas of infections and cancer. Planar (flat) polycyclic compounds are capable of "sliding" into the DNA helix between base pairs in a parallel manner, much as a card would be inserted into a deck of playing cards. The drug would be held in by hydrogen bonds, hydrophobic interactions, and charge-transfer forces. The process, called *intercalation*, results in interference with normal DNA replication and therefore imparts cytocidal properties to drugs capable of intercalating. Examples of drugs whose mechanism involves becoming fixed between the purine–pyrimidine "leaves" of DNA are the antimalarial (chloroquine XXI), the tetracyclic anticancer antibiotic (doxorubicin XXII), and antibacterial acridines such as 3,6-diaminoacridine (proflavine XXIII).

The ability of carcinolytic alkylating agents such as the nitrogen mustards (Chapter 4) to covalently cross-link the two DNA chains of the double helix can be viewed in terms of the nucleotide bases: components of DNA acting as receptors for certain drugs.

Table 1-7. Examples of Receptor Substances

Receptor sites	Examples	Representative drugs
Enzymes	Dihydropteroate synthetase	Sulfonamides
	Dihydrofolic acid reductase	Trimethoprin, pyrimethamine, methotrexate
	Hypoxanthine oxidase	Allopurinol
	Acetylcholinesterase	Physostigmine, fluorophate
	Na^+ K^+ ATPase	Cardiac glycosides
	RNA polymerase	Rifamycins
	DNA polymerase	Cytarabine
	Cyclooxygenase	Aspirin
	Transpeptidase	Penicillins, cephalosporins
	Acetylcholine receptors	Carbachol, atropine
Nonenzyme Proteins	Adrenergic receptors	Epinephrine, terbutaline, propanolol
	"Opiate receptors"	Morphine, nalbuphin, nalaxone
Nucleic acids	DNA	Aminoacridines, actinomycin D, doxorubicin

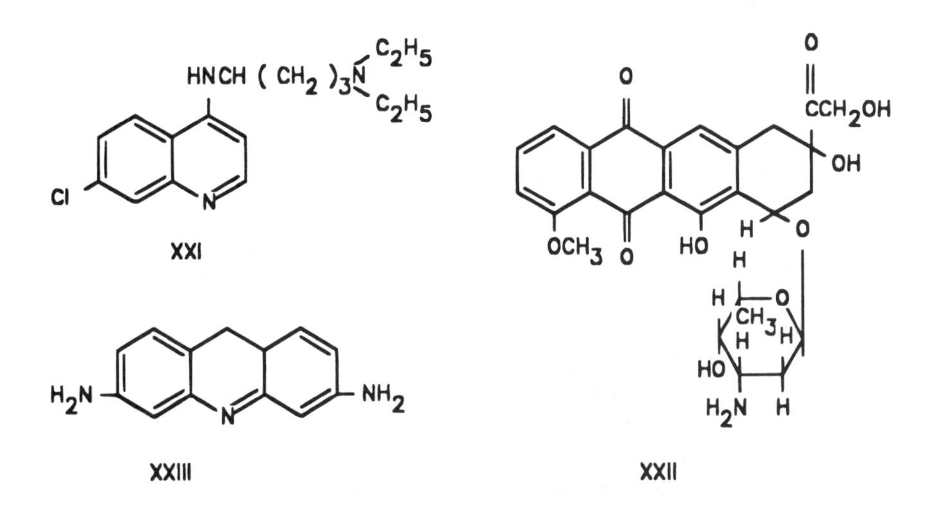

The earlier doubts about the validity of the receptor concept may have been totally dispelled by the isolation from the electric eel, the fairly pure, nonenzymic protein receptor for acetylcholine. The properties, including kinetics, were close to those exhibited by the intact electroplax (the eel's electric organ) from which the receptor was isolated. Other receptors have since been isolated.

1.5.2. Theoretical Aspects of Receptors and Drug Action

The preceding limited discussions regarding receptors and drug activity may be said to be based on classical pharmacology [i.e., interpretation of experimental data derived from the relationships of responses (effectiveness) to doses]. This information was obtained from tissue preparations or whole organs. There really was no way of determining accurately, or at all, the events occurring between the drug's administration and the resulting pharmacological response (i.e., cause and effect). Thus many intuitive assumptions had to be made, sometimes without much verifiable evidence, to arrive at the earlier theories. The information available up to the early 1970s was essentially valid basically up to the cellular level.

The situation changed dramatically over the past 20 years with the development of new techniques in receptor purification and isolation. Receptor binding studies became more accurate, more sophisticated, and more valid, since they are now being carried out almost truly at the molecular level. Using isotopically labeled drugs and endogenous ligands with high specific activity of either radioactive or stable nuclides, combined with the latest high-sensitivity detecting devices, it has become possible to determine molecular concentrations as low as 10^{-12} moles. Thus it has become possible to study actions at receptor sites as molecular events. It is surprising that this has resulted in modifications and refinements of our previous understanding of drug action as well as the promulgation of some new concepts.

The ensuing discussion will deal with that major category of receptors that are essentially components of cellular membranes. For example, the acetylcholine receptor involving skeletal muscles exerts its effect at the end of the motor nerve and its junction with the muscle (neuromuscular junction, see Chapter 7) by a depolarizing action. The fact that receptors are embedded in muscle cell membranes can be surmised by the fact that the contractile effect can be initiated by simply applying acetylcholine to the surface of the muscle preparation; intracellular injection of the agonist produces no effect. A more interesting demonstration involves agents covalently bonded to nonpenetrating polymeric materials, even glass beads, still retaining their biological activity when applied to suspected receptor surfaces.

To help explain why an agonist binding to or interacting with the surface topographical features of a membrane-bound receptor results in an effect at all, it becomes necessary to jettison the idea of a receptor as a single-component functional unit. Rather, we must begin to view it at least as a two-component complex. One component is involved with the recognition of the ligand, either an exogenous drug molecule or an endogenous compound (e.g., a hormone). The specificity of the receptor is thus determined. The second component is considered as mediating the observed response by a catalytic or amplification process. Both components may simply be different sites on one macromolecule so that coupling the response to the recognition might be achieved by molecular perturbation. The other possibility, of course, is that each component is actually a different macromolecule. Coupling may result from mobility of components within the membrane by diffusion processes. A greater likelihood is a coupling mechanism via a third component such as membrane-bound enzymes (e.g., adenyl cyclase or Na^+, K^+-ATPase). Figure 1-11 schematically represents such a three-component system, where the coupling component is within the phospholipid bilayer structure of the membrane and other components outside and inside the cellular membrane.

A modification of this idea suggests that the recognition and amplification components are separate and mobile within the membrane environment, but link together once the ligand binds to the recognition site by a "floating" process initiated by a hormone complexation with that site. If one considers the possibility that such a complex can diffuse along the lateral axis of the membrane and interact with more than one amplification component, then multiple effects caused by one ligand could be explained in some situations.

The classic receptor visualization was of receptor sites binding agonists, and their competitive antagonists as well, to topographically identical sites. It was not difficult to point to groups of agonists with essentially similar qualitative pharmacological properties. In some cases competitive antagonists could be seen to have strong structural resemblance to the agonist. Furthermore, this concept stated that these antagonists could displace the agonist from these sites by virtue of high affinity for these sites, yet, once having done so, elicit no activity of their own. However, it soon became apparent that in many cases competitive antagonists exhibit little or no structural resemblance to their agonist counterparts. Since this fact likely meant that such antagonists were unlikely to bind to the same topographical areas, how then could one explain their frequently much greater affinity for the receptor as a whole? On further examining the agonist–competitive antagonist relationships, we find that in most, but not all, cases close structural similarities exist in the area of chemotherapy. Here antimetabolites of certain important metabolic intermediates

Figure 1-11. Hypothetical membrane-bound multicomponent receptor.

are therapeutically effective. Some drugs with antihormonal and antivitamin activity also fall into the category of structural analogy. However, in the area of pharmacodynamics structural likeness between competitive antagonists and their endogenous or synthetic agonists is not the norm. Thus, anticholinergics such as the antispasmodic drug glycopyrroline when compared with acetylcholine or the synthetic cholinomimetic carbamoylcholine shows a considerable difference in structure over the greater part of the molecule (Fig. 1-12).

Studies going back several decades have established that acetylcholine and synthetic structural analogs designed to mimic its effects (Chapter 7) interact with the cholinergic receptor by essentially electrostatic forces involving the polar, positively charged quaternary nitrogen and carbonyl oxygen, approximately 5.9 Å distant. A competitive antagonist such as glycopyrrolate and others such as propantheline, atropine, and tridihexethyl characteristically contain aromatic and/or alicyclic hydrophobic rings that are unlikely to bind to the sites that polar agonists do. While the positively charged nitrogen atoms of both agonists and antagonists can be seen to form ionic bonds at the same site, another explanation must be devised to account for the remaining interactions. One interesting possibility proposed is the existence of *hydrophobic accessory binding sites* in proximity to those for the agonist (curved line in Fig. 1-12). What the characteristics of such sites might be can only be speculative. Among the possibilities are hydrophobic protein areas of the receptor, areas with a preponderance of amino acids with nonpolar groups folded into appropriately tertiary conformations. A lipid–protein association on or within the membrane is also conceivable.

Another interesting aspect is an important degree of structural similarity among several types of competitive antagonists, or the fact that the same compound can act as a multiple

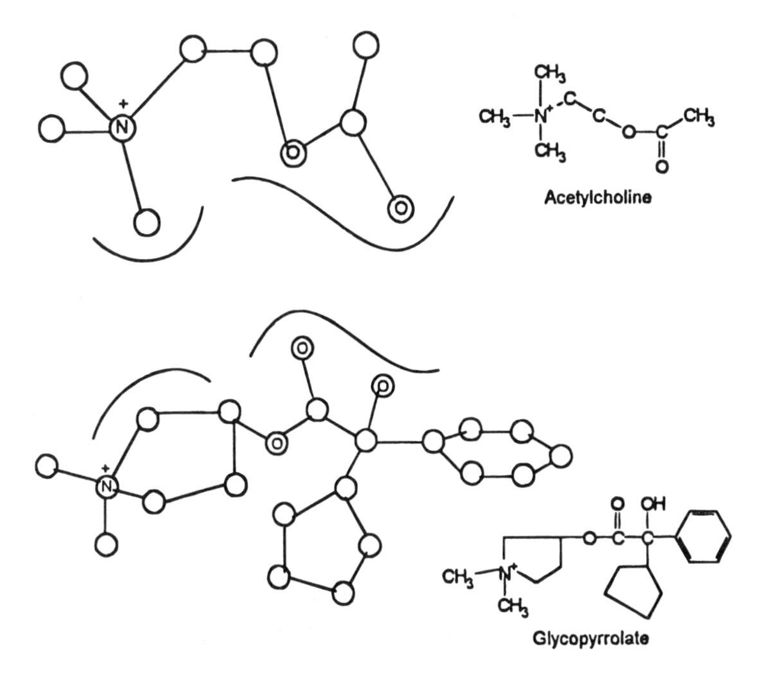

Figure 1-12. Schematic representation of acetylcholine and anticholinergic drug–receptor interactions.

competitive antagonist. For example, diphenhydramine exhibits clinically significant levels of H_1 antihistaminic, anticholinergic, and α-adrenergic blockage, CNS depression, and (at least topically) local anesthetic activity. It would be difficult to accept that diphenhydramine is chemically related to agonists as different as histamine, acetylcholine, norepinephrine, and γ-aminobutyric acid. Acceptance of the accessory-binding concept would increase our understanding of structural incongruity of agonists and competitive antagonists as well as existence of such multifaceted competitive antagonists as diphenhydramine. The breadth of action may be due in part to the fact that diphenhydramine is such a flexible molecule. This may give it the capability to interact hydrophobically with, or accommodate itself to, different accessory sites with facility even though the sites obviously differ. Decreasing this high degree of conformational freedom by the introduction of an ortho methyl group on one of the two aromatic ring brings about a 15-fold decrease in antihistaminic activity while simultaneously increasing anticholinergic activity by an almost equivalent factor of 16. The much bulkier *tert*-butyl group practically eliminates antihistaminic action while maintaining high anticholinergic activity. Complete loss of conformational flexibility is obtained with the cyclic analog nefopam, which exhibits a dramatic increase in skeletal muscle relaxation ability while showing a 90-fold decrease in antihistaminic activity and an 83% decline in anticholinergic character (Fig. 1-13). What we may be seeing here is the fact that increased rigidity increases the possibility of a better (accessory) receptor fit and with it better selectivity of action.

The illustration of diphenhydramine and acetylcholine competition indicates an overlap in that the positively charged nitrogen atoms are assumed to bind the same anionic site on the cholinergic receptor, while the rings interact hydrophobically at a neighboring accessory site. In other situations no such overlap need exist. A competitive antagonist occupying only an accessory site or sites may perturb the agonist receptors sites sufficiently to alter affinities, much as allosteric site substrate binding on enzymes can alter catalytic

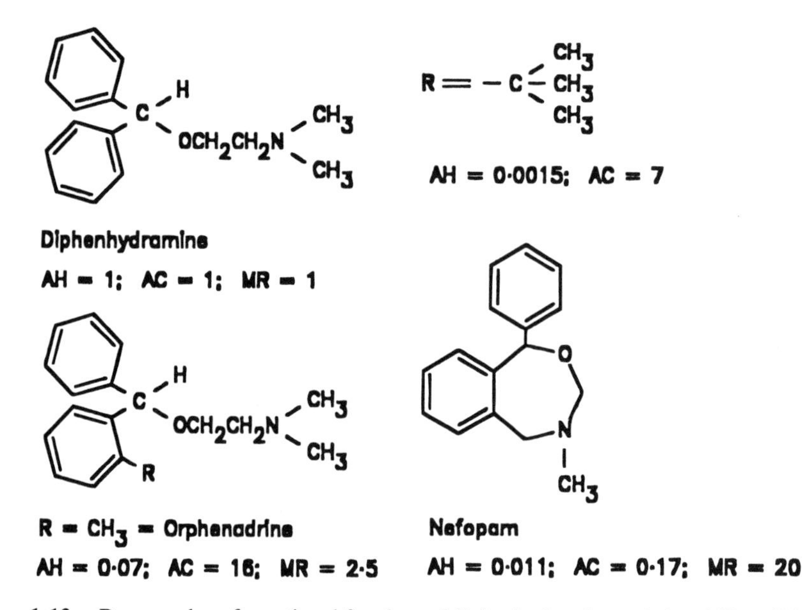

Figure 1-13. Decreased conformational freedom of diphenhydramine activity. AH, antihistaminic; AC, anticholinergic; MR, muscle relaxant activity (see text).

activity. In either case, it can be seen that a competitive antagonist can interfere with normal access by the agonist to its receptor sites.

The *allosteric receptor* is closely related to receptor sites, and it is modeled by analogy to a similar concept in enzymology. Here we visualize the agonist and antagonist binding to separate, nonoverlapping receptor sites that nevertheless affect each other's affinity for their respective ligands in an allosteric manner. Thus a mutual exclusion of binding ability results. Interest is in that situation where the antagonists binding to an allosteric site prevents the agonist from doing so at its receptor. It would, of course, be very difficult to differentiate experimentally between accessory and allosteric competitive interactions. It should be pointed out that whether a particular receptor will be activated or deactivated by these binding interactions is not predictable from these ideas.

From the foregoing discussion, it should not be assumed that competitive antagonists never exhibit structural similarity to agonists involved in affecting physiological functions. A prominent exception to the "rule" is in the field of opiate analgetics, where it will be seen that morphine antagonists show very great structural likeness to morphine and its congeners (Chapter 5), yet at the same time the endogenous ligands for these so-called opiate receptors are peptides, called enkephalins, which are quite dissimilar.

1.5.3. The Two-State Model of Receptor Theory

As the receptor model discussed to this point evolved deficiencies, it also required modifications. While maintaining the idea that drug–receptor interactions result in activation of the receptor from a nonactivated (resting) state R to an activated state R^*, the concept was expanded to state that an equilibrium exists between these two states prior to any interactions with ligands. It is proposed that agonist molecules exhibit affinity only for the activated state and will not bind to the resting state. Similarly, antagonists will not bind to the activated receptor state; rather, they will only interact with the "relaxed" receptor state. Furthermore, the agonist by its presence will bind to the R^* state and will also shift the equilibrium to it, whereas competitive antagonists will displace the equilibrium toward R. Partial agonists are assumed to have affinity for both types of binding sites. Figure 1-14 schematically summarizes the two-state receptor model.

The ratio of affinities for R^* and R needs to be determined to evaluate the intrinsic activity of a partial agonist. This, then, would also indicate the fraction of the receptors in the activated form when sufficient partial agonist is present to saturate all sites. It should be pointed out in considering this two-state model that the two receptor states for agonists and competitive antagonists are assumed to be unrelated (i.e., they each interact only with their putative ligands).[12]

To test the validity of the proposal that in the dual-receptor model the sites for the agonist and corresponding antagonist are actually distinct, experiments to reductively alkylate muscarinic acetylcholine receptors in neural membranes with N-ethylmaleimide, a sulfhydryl-group blocking agent for proteins were carried out. This action distinctly altered the affinity of the receptor preparation toward cholinergic agonists, yet affinity toward antagonists remained unaffected. Other experiments were able to destroy membrane sensitivity of the cholinergic receptor population in the electroplax of the electric eel with

---

[12] For a detailed description and kinetic derivation, see Triggle (1978) and Ariens and Rodrigues De Miranda (1979).

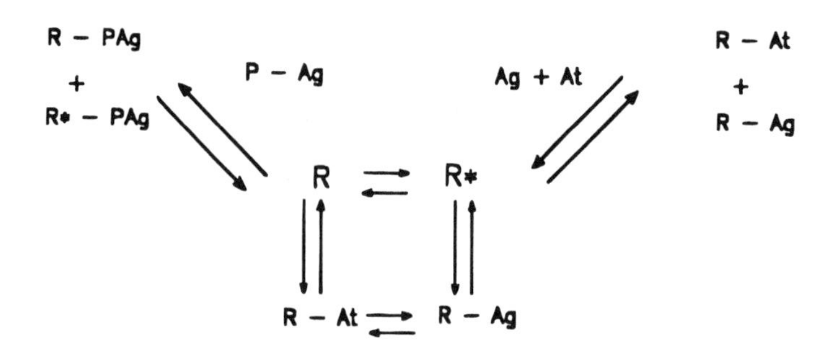

Figure 1-14. Simplified two-state receptor scheme. *R**, activated receptor state; *R*, nonactivated state; Ag, agonist; At, antagonist; P-Ag, partial agonist.

dithiothreitol, which is a reducing agent. This desensitization could be prevented by prior treatment with cholinergic agonists. The potent antagonist *d*-tubocurarine, however, could not protect the receptors against this effect.

An additional factor that should be brought into our thinking about receptors is their mobility within the bilayer membrane structure. Many membrane glycoproteins, presumably including those that are components of receptors that bridge this double layer, could therefore respond to effects on the outside of the cell by effecting, at least transiently, alterations inside the cell. Such transmembrane effects by the nicotinic acetylcholine receptor on the endplate region of skeletal muscle cells were demonstrated. Following acetylcholine release at the motor neuron innervation point and its interaction with the receptor protein, a pore forms across the membrane (for several milliseconds) and allowed a cation flux through it. At some point sufficient ions migrate by this mechanism to depolarize the muscle cell until an action potential results. It can be stated that the effect of acetylcholine on a receptor induces a microfluctuation in the current passing across the membrane. Cooperative receptor interactions involving rapid movement of membrane proteins across a two-dimensional matrix can now be visualized. This type of *receptor mobility* is conceivably involved in various aspects of cell behavior.

1.6. Receptor Characterization

1.6.1. Receptor Isolation

The ultimate in understanding receptors would be to know their structures to the degree that we already know small molecules. To date, our most detailed knowledge comes form X-ray crystallographic studies of those enzymes acting as receptors. Enzymes that have been successfully crystallized while complexed with a molecule of an irreversible inhibitor within its tertiary structure have given us a fascinating "look" at receptor (active-site) topography. It allows identification of the functional groups around an active site with reasonable reliability. To apply similar techniques to the much more complex membrane-bound lipoprotein complexes will be much more difficult because the physical problems of isolating receptor proteins (other than enzymes) are so monumental. Even analytically determining the degrees of purity obtained is very difficult. With enzymes the reaction rates catalyzed can be monitored. In many instances enzymes have even been crystallized. However, monitoring the efficiency of receptor protein isolation has previously not been possible (see following discussion).

The most successful results have been obtained with nicotinic acetylcholine receptors. These have been isolated by complex procedures from the electric organ (electroplax) of several marine animals, particularly *Electrophorus electricus* (electric eel) and *Torpedo marmorata* (the electric ray). A particular set of fortuitous circumstances conspired in this accomplishment. The most important fact that enabled the isolation of the protein carrying the receptor is that the electroplax contains such relatively large quantities of receptor material. The ray has about 1,000 nanomoles per kilogram and the eel up to 100. This represents a 100–1000 times higher concentration than is found in mammalian brain tissue. As much as half the postsynaptic area of the ray's organ is probably receptors. Another fortunate event was the discovery that certain snake venoms, such as α-bungarotoxin of the Indian cobra, are neurotoxic by virtue of their irreversible binding to acetylcholine receptors. This made identification possible. The receptor appears to be an integral part of postsynaptic membrane proteins of the electric organ and must first be solubilized by treatment with detergents such as Triton X-100 or Tween 80. The resulting solution is then treated with radiolabeled toxin to form a toxin–receptor complex. Excess free toxin is eliminated by gel filtration. The toxin–receptor complex can then be removed by elution procedures. How much complex is formed can then be determined by counting the radioactivity. The displacement of the toxin from the complex is achieved by treatment with either cholinergic agonists such as carbachol, or antagonists such as decamethonium or *d*-tubocurarine.

The receptor preparation has been further separated into four or more polypeptide subunits by electrophoresis. Their molecular weights range from 25 to 65 kD.[13] The α-chain appears to be the locus of acetylcholine binding. The function of the other subunits is not clear. It is possible to speculate that as a complete membrane-bound complex they form an ion channel. It is therefore not surprising that the ability to regulate ionic migration disappears once the receptor complex is separated from its membrane anchors. The "receptor site" once isolated as described is, after all, not what existed in the natural state. In this, and in attempts to isolate other receptor types, it is reasonable to assume that the separation from various membrane proteins and associated lipids, conformational changes, even without destructive procedures, may be of sufficient magnitude so the receptor is actually "lost." An extensive discussion on acetylcholine receptor isolations carried out in the middle 1970s is given by Brady (1978).

1.6.2. Receptor–Ligand Binding

It is doubtful that a majority of receptors will ever be isolated and purified to the level of the nicotinic acetylcholine receptor. Even then some important characteristics that the receptor exhibits when embedded in the membrane surrounded by its various ancillary protein and lipid components are lost. Thus studying drug interactions with such receptor isolates will yield limited, possibly erroneous, information to the drug designer. Effects resulting from such interactions on cells, tissues, or organs cannot be evaluated.

The situation is not as hopeless as it might seem. Another methodology circumventing the difficulties of isolation has come to the forefront over the past 20 years or so: the technique of binding ligands specific for one type of receptor even in a preparative environment containing a plethora of binding opportunities. The problems to be overcome to make this approach practical were solved by technical developments. A major hurdle was to

---

[13] Kilodaltons.

determine quantitatively the amount of ligand bound to what must be minuscule receptor concentrations in tissues under study. The only way the binding of vanishingly small amounts of compounds can be measured is by using radioactively labeled ligands with very high specific activities so that significant counts can be obtained. If the ligands utilized are small organic compounds such as drugs or endogenous hormones or neurotransmitters (e.g., norepinephrine, GABA), then radioisotopes such as ^{3}H, ^{14}C, ^{32}P, and ^{35}S can be substituted for the stable atoms in these molecules. In fact, the synthesis of such compounds for investigative purposes has almost become a subspecialty of organic synthesis. Many such radiolabeled ligands are now commercially available for binding studies, drug metabolism investigations, and other radiotracer requirements. Specific activities as high as 30 Ci/nmole are achievable, although tritiated tags are much lower. When the ligand is a large peptide or protein (e.g., α-bungarotoxin or insulin), iodination with ^{125}I or ^{131}I is necessary.

In a typical experiment a small quantity of tissue (<1 mg) is equilibrated (<30 min) with a known quantity of radio-tagged ligand. The ligand-bound material (e.g., a membrane-bound receptor complex) is then separated by rapid filtration or centrifugation, followed by washing with cold buffers to remove all excess, unbound ligand. Scintillation counting (gamma counting) determines the quantity of ligand bound. The limited *specific binding* to the receptor of interest must, of course, now be differentiated from the potentially unlimited *nonspecific binding* that occurs to the various other biopolymers in the tissue sample. This is accomplished by repeating the experiment and using, in addition to the "hot" ligand, a large excess of "cold" unlabeled ligand. This statistically eliminates the likehood of any tagged ligand binding to nonspecific sites. The difference in values obtained ideally represents the amount of specific binding. Repeating these two experiments using a set of labeled ligand concentrations and plotting the results gives the familiar "saturation" curve (Fig. 1-15), showing that the specifically bound ligand reaches a saturation level despite increasing levels of ligand. This, of course, is what would be expected from a limited number of binding sites, in this case specific receptors. This type of data also allows the determination of dissociation constants (K_D) and the number of receptor sites.

The same technique also lends itself to agonist–antagonist studies at the receptor level by using a receptor preparation already containing a specific quantity of labeled ligand and subsequently adding varying amounts of experimental compounds. The change in specific binding (if any) can then be graphed as a variable of the added substance. The concentration at which a 50% inhibition (reduction) of specific binding of the labeled ligand is attained can be determined (IC_{50}). This, then, represents a direct indication of the binding

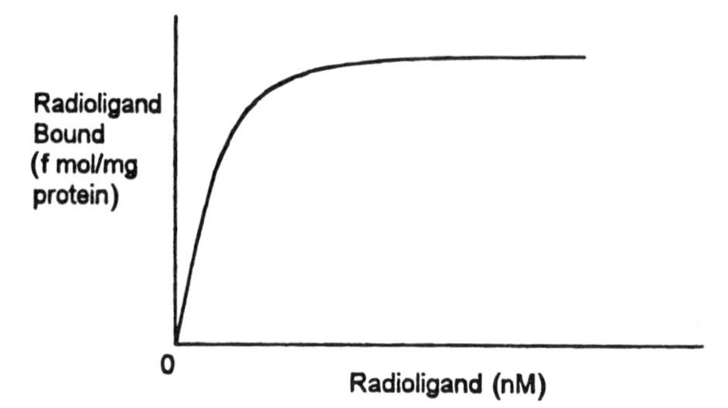

Figure 1-15. Specific binding determination.

effectiveness of a test compound compared with the labeled ligand at its specific receptor. It should be kept in mind that a biological response is not being tested here. Therefore, it cannot be established by these experiments whether the drug being tested is an agonist or antagonist.

One particular asset of IC_{50} determinations is that they represent a direct potency measurement of the drug at the receptor. The unknown quantitative effects of transport, excretion, and metabolic degradation that plague in vivo studies do not apply here. On the other hand, it is precisely these in vivo factors, which also include absorption and membrane penetrability, that determine the ultimate clinical utility of a drug. Thus, even though these in vitro methods allow the determination of the highest potency at the receptor level, this may not always translate into clinical efficacy. Yet, not surprisingly, good correlations have been established. A dramatic example involves the use of neuroleptic agents in the treatment of schizophrenia. One such drug, haloperidol, is known to be an antagonist of the neurotransmitter dopamine at its receptors. An extraordinary correlation was demonstrated in this case between the ability of various neuroleptic drugs to displace ^{3}H-labeled haloperidol from calf-brain tissues with the clinical doses of these drugs needed to control the illness (Fig. 1-16). These types of studies are good evidence to support the contention that dopaminergic defects are at the underlying cause of this mental disease.

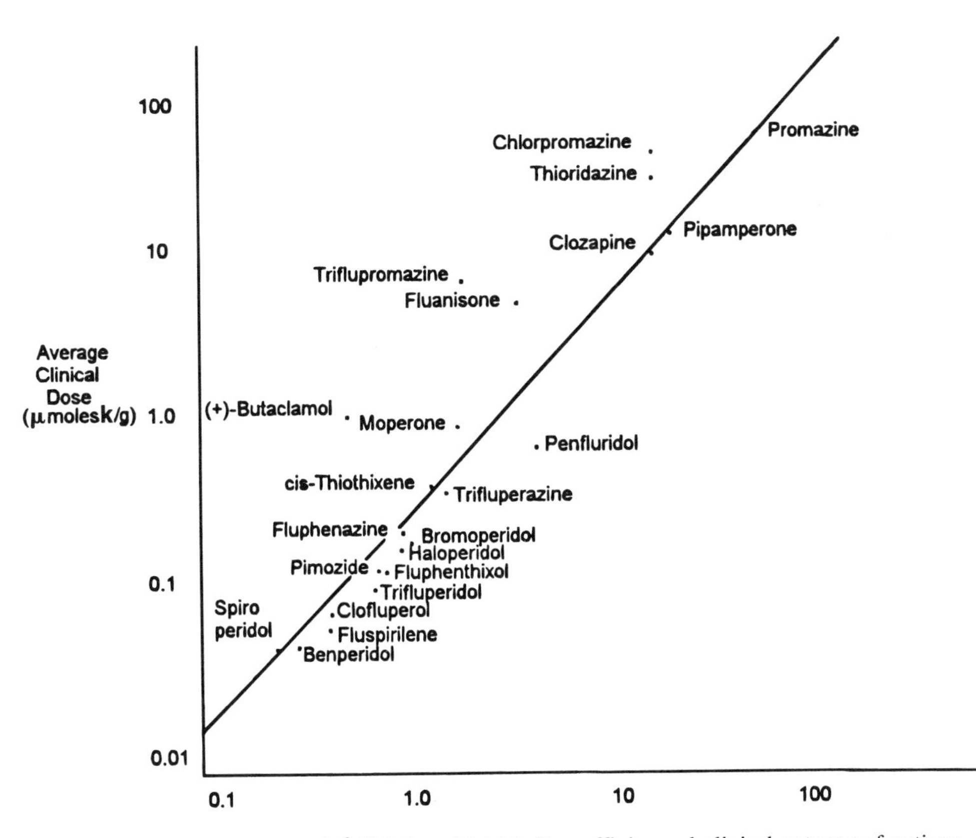

Figure 1-16. Correlation of [^{3}H]-haloperidol binding affinity and clinical potency of antipsychotics. (From Creese et al., 1977, with permission.)

1.6.3. Conformational Methods

To design new and more effective drugs on a more rational and less empirical or intuitive basis, it is necessary to have a clearer picture of the actual topography of the receptor sites involved as well as complete information on the structures of the endogenous compounds and potential drugs that bind to them. The probable impossibility of establishing the drug–receptor structure as it actually is at the time of the event in living tissue by direct methods was alluded to in the previous discussion. At least for the foreseeable future we will have to rely on indirect methods that are becoming increasingly sophisticated. Thus, in addition to the location of atomic nuclei in such an active milieu becoming known, so are the electron densities, which are probably even more important. The detailed study of drugs and other small molecules and their binding to receptors to gain information on the receptor topography has become known as *receptor mapping*. Elucidating what changes in activity result in these small molecules as a function of their chemical and physical properties may, with time, yield a more accurate view of what receptors are and do.

Early attempts at mapping were based primarily on steric fit concepts. For example, the cholinomimetic potencies of various acetylcholine homologs and analogs in animal models were deduced from these data, and certain minimum structural requirements the acetylcholine receptor needs to interact with compounds to elicit certain cholinergic activities were determined. This information was translated into rudimentary "maps." Steric fit, however, is only one aspect of drug–receptor interactions. Electrostatic charges and charge distributions must also play a role in complementarity between ligand and receptor topography. Electrostatic contour maps are now appearing in the literature with greater frequency.

The original idea proposed by Ehrlich at the beginning of this century that a drug fits its specific receptor as a key fits its specific lock has not really been abandoned, but it has been, and still is, undergoing continual evolution towards a more encompassing, complex, and, it is hoped, a more realistic concept. The "key" is now understood to be flexible, possibly to the point of molding itself to the "lock's" requirements. The "lock" may have an undulating "keyhole" or even several, like a stack of locks. Since both the ligand molecules and the receptor's topography varies in response to each other's physicochemical influences as they approach each other, we might even consider a combination lock analogy whose tumblers are not necessarily constant (i.e., perturbation of electrostatic fields). Thus studying bioactive molecules in a state isolated from the real action, such as by spectroscopic analyses of their solutions, will very likely lead to maps based on large approximations. Research thus far indicates that generally bioactive compounds are capable of conformational distortions that permit receptor binding and that changes in conformation and charge distributions are in turn induced in the receptor. It also seems highly probable that the drug's reactivity assumes increased significance as it approaches the receptor.

Some questions are still open. Do drugs producing the same effects interact with the same functional areas in a particular receptor? Not necessarily. Even in cases of structurally similar substrates for certain enzymes, the enzymes have been found to bind to them in different manners. Other questions that have been posed, but as yet not resolved, are whether a drug molecule interacts with its receptor in its preferred (low-energy) conformation, and whether all the molecular atoms believed to interact with complementary areas on the receptor do so at once or in some order of priority based on, possibly, charge magnitudes. It is intellectually more satisfying to propose that there is a limited initial interaction setting off conformational changes followed by additional binding interactions.

It is interesting to follow the status of the question as to whether the lower-energy conformation of a drug molecule is the one more likely to bind to a receptor by considering the opioid analgetics.

The piperidine ring is found as part of the multiring structure of all these compounds, as is an aromatic ring (phenyl) at its 4-position. An early proposal of a hypothetical opiate receptor "predicted" that the aromatic ring in these analgetic compounds will have an axial conformation. The reasoning behind this prediction can best be discerned when considering the most fundamental morphinelike compound as represented by the synthetic drug meperidine. Of the two conformations (XXIV and XXV), only one can be presumed to have the "good fit" to elicit pharmacological activity. From conformational analysis considerations structure XXV would be energetically preferred since in this position the *ortho* hydrogens on the benzene ring would not repulsively interact with the H atom emanating from C-3 of the piperidine ring, as would be the case in conformation XXIV. The less stable form XXV, however, was thought to meet the "fitness" requirements of the receptor's topography. Since these two species are interconvertible at the energy levels of the system, the less favored form was therefore presumably available for receptor binding, possibly even because of receptor interactions. However, synthesis and successful separation of two noninterconvertible conformers (actually in this case configurations) of 1-methyl-4-phenyl-*trans*-decahydro-4-propionoxyquinoline, structures XXVI and XXVII, demonstrated that the aromatic ring conformation was not relevant to analgetic action. Other studies with rigid analogs of meperidine have also shown no significant relations between analgetic effect and the steric position of compounds sharing the 4-phenylpiperidine backbone. These studies did, however, reveal a new wrinkle in this already complicated picture: Two configurations of a compound (and presumably therefore conformations as well) can and do possess some differences in physicochemical properties such as pKa and lipid–water partition values. One implication is that this can mean differences in membrane penetration and tissue distribution rates, resulting in differences of pharmacology such as intensity, speed of onset, and duration of action. The other implication is at the drug–receptor level; namely, that while attempting to relate binding propensities to putative molecular shapes only, we may have been ignoring the role played by the changes in physicochemical properties caused by conformational changes.

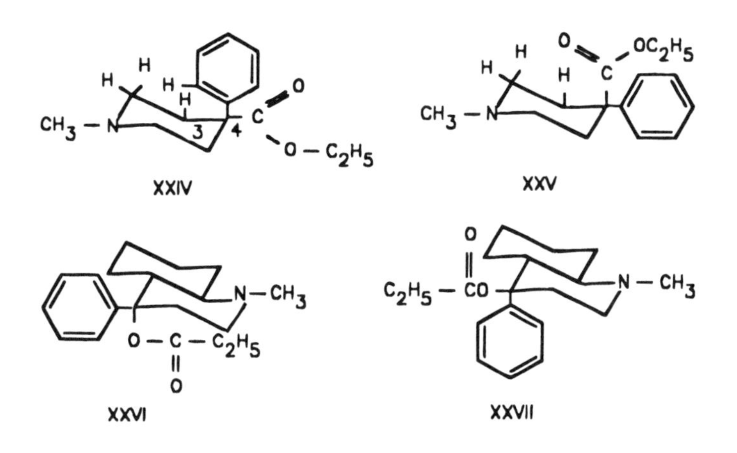

1.6.4. Some Stereochemical Considerations

As a corollary to the previous discussions, some statements regarding stereochemical factors and their relevance to drug action should be made. Organisms function and survive by biochemical and regulatory processes that are stereochemical and, in many instances, stereoselective. Interactions of endogenous ligands with the various biopolymers are the basis of most of these mechanisms. That exogenous bioactive compounds such as drugs, pesticides, and herbicides when entering the biosystems of living organisms are subject to the same stereochemical influences should therefore not be surprising.

Optical isomers frequently exhibit quantitatively different pharmacological activities. A logical presumption is that one optical isomer of an enantiomeric pair shows better receptor interactions. Table 1-8 lists some examples of effect or potency differences of selected drugs.

It is found that by limiting consideration to molecules with only one chiral center and therefore only one pair of enantiomers the usual physical and chemical properties are identical in a symmetrical environment. However, rates of reactions, even reactivity (e.g., metabolism reactions) and binding propensities may differ significantly in an asymmetric bioenvironment. There are cases where no differences are demonstrable. Both + and – cocaine are equipotent local anesthetics. Similarly, both enantiomers of chloroquine are equally effective antimalarial compounds. It is possible that in these instances the centers of asymmetry do not participate in drug–receptor interactions, or, more likely, that the interaction may involve only one or two points of contact.

Table 1-8. Comparison of Stereoisomeric Drugs

Drug	Isomer	Characteristic	Effect/relative potency
Isoproterenol (Isuprel)	*D*-(–)	Bronchodilator	50
	L-(+)	Bronchodilator	1
Norepinephrine	(–)	Bronchodilator	70
	(+)	Bronchodilator	1
Epinephrine	*D*-(–)	Vasoconstrictor	12–15
	L-(+)	Vasoconstrictor	1
Acetyl-α methylcholine	*L*-(+)	Antigout	200
	D-(–)	Antigout	1
Hyoscyamine	(–)	Mydriatic	15–20
	(+)	Mydriatic	1
Muscarine	(+)	Cholinergic (muscarinic)	700
	(–)	Cholinergic (muscarinic)	1
Ascorbic acid	*L*-(–)	Antiscorbutic	Active
	D-(+)	Antiscorbutic	Inactive
Amino acid	(–)	Taste	Tasteless(or bitter)
	(+)	Taste	Sweet
Indomethacin[a]	(+)	Anti-inflammatory	Active
	(–)	Anti-inflammatory	Inactive
Cortisone	(+)	Anti-inflammatory	Active
	(–)	Anti-inflammatory	Inactive
Manserin	(+)	Antidepressant	Active
	(–)	Antidepressant	Inactive
Emetine	(–)	Antiamebic	Very active
	(+)	Antiamebic	Less active
Verapamil	(–)	Smooth muscle effects	7
	(+)	Smooth muscle effects	1

[a] a-Methyl homolog.

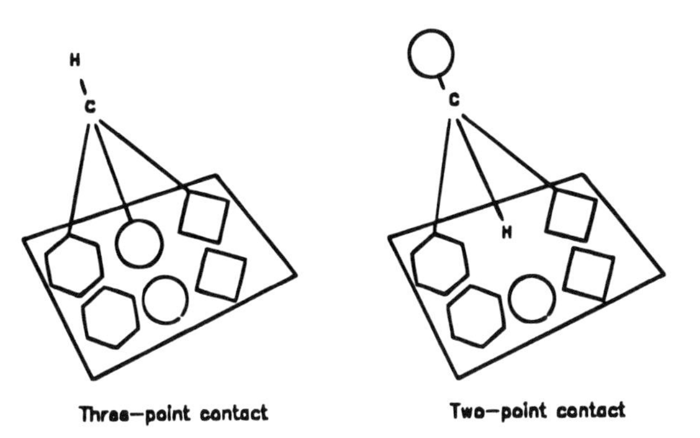

Figure 1-17. Hypothetical receptor interactions of epinephrine isomers. Left: D-(−)-epinephrine. Right: L-(+)-epinephrine.

Epinephrine (XXVIII) illustrates a situation where activity, as well as enantiomer differences, might be explained by a three-point attachment concept. Epinephrine as a hormone and drug has an asymmetric center carrying an OH group on the β-carbon, which is known to be critically involved with receptor site binding in a stereospecific manner. If we assume binding to at least three sites, the aromatic ring, the β-hydroxyl group, and the nitrogen atom, in order to elicit an adrenergic response (Fig. 1-17), then it becomes evident that only one of the two isomers can possibly bind to all three functions. This is a necessary condition for this response. In this case, the more active isomer has the *D*-(−) configuration.

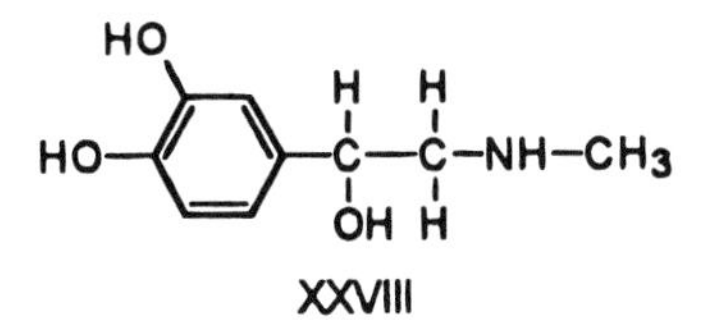

Studies of the differences in biological activities that are displayed by optical isomers, as well as isomers obtained by other steric manipulations of drug molecules, have been an extremely valuable tool in evaluating drug–receptor interactions as well as in attempting to ascertain receptor-surface topography. Additional tools that have been of value are antimetabolic agents, compounds with rigid structures (often complex ring systems), and molecular probes.[14] However, methods that use these tools have been of limited value in helping to deduce the topography of the receptor sites.

It seems improbable that a single, universal theory or concept of drug action is likely to emerge.

---

[14] Molecular probes are small molecules with properties that are measureably changed by interaction with a biopolymer. Nuclear magnetic resonance (NMR) has been an especially useful technique for studying these changes.

References

Ariens, E. J., *Arch. Intern. Pharmacodyn.*, **99**:32, 1954.

Ariens, E. J., Rodriguez De Miranda, J. F., The Receptor Concept: Recent Experimental and Theoretical Developments. In *Recent Advances in Receptor Chemistry*, Gualtieri, F., Giannella, M., Melchiore, C., Eds., Elsevier/North Holland Biomedical Press, New York, 1979, p. 1.

Brady, R. N., Biochemical Isolation of Acetylcholine Receptors. In *Receptors in Pharmacology*, Smithies, J. R., Bradley, R. J., Eds., Marcel Dekker, New York, 1978, pp. 123–64.

Clark, A. J., *Physiol.*, **61**:530, 547, 1926.

Creese, I., Burt, D. R., Snyder, S. H., *Science*, **194**:546, 1976.

Ehrlich, P., *Ber. Deut. Chem. Ges.*, **42**:17, 1909.

Free, S. M., Wilson, J. W., *J. Med. Chem.*, **7**:395, 1964.

Fujita, T., Iwasa, J., Hansch, C., *J. Am. Chem. Soc.*, **86**:5175, 1964.

Gaddum, J. H., *Physiol.*, **61**:141, 1926.

Galinsky, A. M., Gearien, J. E., Perkins, A. J., Susina, S. V., *J. Med. Chem.*, **6**:320, 1963.

Hammett, L. P., *Chem. Rev.*, **17**:125, 1935.

Hammett, L. P., *Physical Organic Chemistry*, McGraw-Hill, New York, 1940.

Hansch, C., et al., *J. Am. Chem. Soc.*, **85**:2817, 1963.

Hansch, C., Fujita, T., *J. Am. Chem. Soc.*, **86**:1616, 1964.

Hansch, C., Steward, A. R., Iwasa, J., *Mol. Pharmacol.*, **1**:87, 1965.

Hansch, C., Leo, A., *Substituent Constants for Correlation Analysis in Chemistry and Biology*, John Wiley & Sons, New York, 1979.

Kier, L. B., *Molecular Orbital Theory in Drug Research*, Academic Press, New York, 1971.

Koshland, D. E., *Biochem. Pharmacol.*, **8**:57, 1961.

Langley, J. N., *J. Physiol.*, **1**:339, 1878.

Martin, Y. C., *Quantitative Drug Design*, Marcel Dekker, New York, 1981, pp. 167–87.

Paton, W. D. M., *Proc. Roy. Soc. (London)*, Ser B, **154**:21, 1961.

Stephenson, R. P., *Br. J. Pharmacol.*, **11**:379, 1956.

Stuper, A. J., Brugger, W. E., Jurs, P. C., *Computer Assisted Studies of Chemical Structure and Biological Activity*, John Wiley & Sons, New York, 1983, p. 179.

Taft, R. W., in *Steric Effects in Organic Chemistry*, Newman, M. S., Ed., John Wiley & Sons, New York, 1956, pp. 556–675.

Topliss, J., *J. Med. Chem.*, **15**:1006, 1972; **20**:463, 1977.

Triggle, D. J., in *Receptors in Pharmacology*, Smithies, J. R., Bradley, R. J., Eds., Marcel Dekker, New York, 1978, pp. 1–65.

Suggested Readings

Pratt, W. B., Taylor, P., Eds., *Principles of Drug Action*; *The Basis of Pharmacology*, 3rd Ed. Churchill Livingstone, New York, 1990.

CHAPTER

2

Mechanisms of Drug Action

2.1. Introduction

Before embarking on a detailed discussion of the major drug categories, certain basic concepts of several principal mechanisms of drug action should be considered. The likelihood of a single all-encompassing mechanism of drug action appears to be extremely remote. The present assumption is that several mechanisms are operative at the molecular level, including:

1. Effects on enzymes: their stimulation, inhibition or alteration.
2. Effects of antimetabolites, derived from the alteration of molecules serving as the normal substrate or template for various biopolymers.
3. Modification of the properties of biological membranes, particularly permeability of cellular membranes.
4. Mechanisms of probably minor importance that may involve the chelating ability of certain compounds.

The action of certain agents whose mechanism of action does not appear related to their chemical structure (structurally nonspecific drugs) may, in some cases, be due to their accumulation in important cellular locations, leading to disorganization either internally, such as in metabolic processes, or externally, by membrane disruption.

One underlying fact should be kept in mind. On a pharmacological and clinical level the physiological events viewed are in actuality the terminal results of a series of successive, as well as simultaneous, biochemical reactions. The introduction of a foreign substance (i.e., a drug) into this complex system, must, if it has any effect at all, result in some form of interference of an essential reaction. Thus the normal physiological event is altered. It must be pointed out, however, that drug activity never induces a previously nonexistent function in a cell. It can only affect an existing one. The realization that the chemistry of the cell is enzymatically controlled led to the predictable conclusion that drugs bring about pharmacological effects by modifying the enzymatically governed reactions. Various early in vitro experiments on isolated enzyme systems, and even on purified enzymes, were carried out successfully. However, it was much more difficult to establish that the drugs that produced the demonstrated effects under such experimental conditions were equally functional in vivo at similar concentrations. Despite considerable effort over many years, the number, or categories, of drugs known to bring about their effects by direct influence on enzymes is relatively small. Nevertheless, several of the enzymes involved are now known to be vital in regulating the synthesis of critical cellular constituents. Some of these enzymes are mammalian, some are unique to pathogenic parasites, and some are found in most life forms. Table 2-1 lists some examples of enzymes whose inhibition by drugs is of therapeutic significance.

2.2. Enzyme Stimulation

Enzymes frequently require coenzymes for optimum activity. The coenzymes are usually vitamins and cofactors, invariably—electrolytes such as mono- and divalent metallic ions (e.g., K^+, Na^+, Ca^{+2}, Mg^{+2}, Zn^{+2}, and Fe^{+2}). These coenzymes activate different enzymes by various means of complexation and stereochemical interactions. A detailed consideration of the mechanisms involved is not within the scope of this discussion. It can be stated, however, that such ions may affect enzymes in one of two ways. Either direct interaction induces changes in the conformation or a charge on the enzyme, or interaction of the cation with an enzyme-inhibiting substance, prevents or minimizes the deactivation.

Enzymes and other polypeptide chains in biosystems have an ample supply of metal-binding groups. These are imidazole rings (in histidine), -SH groups (of cysteine), and

Table 2-1. Enzymes Inhibited by Therapeutic Agents

Enzyme	Drug	Clinical utility
Acetylcholinesterase	Neostigmine,[b] echothiophate[a]	Myasthenia gravis, glaucoma
Alanine racemase[e]	D-cycloserine[a]	Antibacterial
Aldehyde dehydrogenase	Disulfiram[b]	Alcoholism
Aldose reductase	Sorbinol	Diabetes
Angiotensin converting enzyme	Captopril[b]	Hypertension
Carbonic anhydrase	Acetazolamide[b]	Diuretic
Dihydrofolate reductase[c]	Trimethoprin,[b] Pyrimethamine,[b] methotrexate[b]	Antibacterial, antimalarial, anticancer
Dihydropteroate synthetase[c]	Sulfonamides[b]	Antibacterial
DNA polymerase	Cytosine arabinoside[b]	Anticancer and antiviral
Dopa decarboxylase	Carbidopa[a,d]	Parkinson's disease
β-Lactamase[c]	Clavulinic acid[a,e]	Antibacterial
Monoamine oxidase	Tranylcypromine[a]	Depression
Prostaglandin synthetase	Ibuprofen[b]	Analgesic/anti-inflammatory
Ribosylamidotransferase	6-Mercaptopurine[b]	Anticancer
Sodium, potassium ATPase	Digoxin[b]	Heart disease
Thymidine kinase	Idoxuridine[b]	Antiviral
Thymidylate synthetase	5-Fluorouracil[b]	Anticancer
Transpeptidase[c]	Penicillins,[a] Cephalosporins[a]	Antibacterial
Xanthine oxidase	Allopurinol[b]	Gout

[a] Irreversible inhibitor; [b]reversible inhibitor; [c]not present in mammals; [d]only used together with L-Dopa; [e]only in conjunction with penicillins.

-COOH groups of several amino acid residues. Enzymes such as cytochromes and peroxidases, as well as the hemoglobin molecule, have heme iron protein components wherein a Fe^{2+} ion is coordinated in a square planar structure with four pyrrole ring nitrogen atoms. Peptide-hydrolyzing enzymes such as carboxypeptidase need Zn^{2+} for proper functioning. Studies with this enzyme first demonstrated that substrates, aided by metal ion involvement, cause conformational changes.

Other enzymes that catalyze decarboxylation and dehydration reactions require Mg^{2+}. As will be seen, the α-adrenergic receptor, which includes the enzyme adenyl cyclase in association with ATP, utilizes Mg^{2+} to mediate the increased cardiac activity associated with epinephrine. It is therefore apparent that pharmaceutical products that provide these inorganic ions (i.e., mineral supplements), as well as vitamin preparations, should actually be considered drugs whether they are used therapeutically to treat clinical symptoms of deficiency, or as food supplements for the maintenance of good health.

2.3. Enzyme Inhibition

Drugs acting by a mechanism that involves enzyme inhibition may involve definite groups on the protein portion (apoenzyme), which need not necessarily be the active site where normal substrate interactions occur. The reaction may be a weak nonspecific overall adsorption phenomenon. It may involve only a limited modification of the protein conformation (i.e., the tertiary structure), which could be effected by weakening or destroying generally, or locally, hydrogen-bonding capacity. Breakdown of the peptide bonds of the apoenzyme component of the system by actual hydrolysis would be the most drastic form of inhibition. Such total inhibition may be accomplished by therapeutically utilizing proteolytic enzymes such as trypsin and papain. Furthermore, an inhibitor might react with a coenzyme component, thus reducing or eliminating the system's capacity to function. Interaction of such a drug with the total enzyme complex is a possibility. It is also conceivable that an inhibitor may negate the contribution of an activator substance.

It is useful at this time to review briefly some basic aspects of enzyme inhibition and enzyme inhibitors. For a broader review of enzymatic mechanisms and kinetics consult any of several excellent biochemistry textbooks. Enzyme inhibitors are chemical agents capable of modifying an enzyme's capacity to catalyze the reactions of its normal substrates. Effecting such changes of enzyme function by altering the pH, changing the temperature, or subjecting the system to radiation such as UV should properly be considered a denaturation process.

Inhibiting enzymes by chemical means offers many therapeutic possibilities. A drug acting as an inhibitor of a given enzyme can reduce or increase the production of a particular metabolite. Such changes can be effectively achieved if the particular enzymatic step affected in a series of sequential reactions is the rate-controlling step.

Control of parasitic invasions has been successfully achieved with compounds that inhibit key enzymes in pathways vital to the biosynthesis of the building blocks of nucleic acids in these pathogenic microorganisms. This results in their reproductive suppression or death. An alternative mechanism can also exist, particularly if the inhibitor has a close chemical similarity to the enzyme's normal substrate (metabolite). Such an analog compound may still be affected by the enzyme, resulting in a false structural component, which if incorporated into the essential biopolymer at all, will lead to nonfunctioning (or errant) DNA or RNAs. Such drugs are referred to as *antimetabolites*. They have been successfully applied to cancer chemotherapy (Chapter 4) and certain antibacterial drugs. Certain factors must be considered in designing drug molecules to function as antimetabolites. Structural resemblance, in terms of both dimensions and electronic factors, is usually essential. Similarity has sometimes been achieved by functional group substitution, or even a single atom (e.g., 5-fluorouracil vs. uracil). Similarity in electronegativity is sometimes important. At other times comparability of van der Waals radii is more crucial.

2.3.1. Inhibitor Classification

Enzyme inhibition can be effected by two processes. In one the restraint achieved is *reversible* because the forces by which the inhibitor and the enzyme interact are the weak interatomic forces previously considered. This situation may be represented by Eq. 2.1, where E is the enzyme, I is the inhibitor, and EI the enzyme–inhibitor complex presumed to be inactive in that it does not lead to a product or an effect. The other process is an *irreversible* inhibition resulting from *covalent* bond formation, usually with functional groups in the enzyme's active site. This may be simply illustrated by Eq. 2.2.

$$E + I \overset{K_i}{\rightleftarrows} EI \tag{2.1}$$

$$E + I \overset{K}{\rightarrow} EI \tag{2.2}$$

2.3.1.1. *Reversible Inhibition*

Reversible inhibition may be further subclassified as to its *competitive* or *noncompetitive* characteristics. Competitive inhibition occurs when the inhibitor competes with the natural substrate at the enzyme's active site. The reversible enzyme–inhibitor complex formed thus prevents, or decreases, access to the active site by the substrate. Equation 2.3 summarizes these events, where S is the substrate concentration and P is the product, and ultimately an effect. The degree to which the rate of P formation is affected depends on the concentration of inhibitor (I) and the dissociation rate of EI represented by K_i. Smaller numerical values for K_i indicate stronger inhibitor–enzyme binding. In the competitive state inhibition can be overcome by increased levels of a substrate.

$$\begin{array}{l} E + S \overset{K_1}{\leftrightarrows} ES \overset{K_2}{\rightarrow} E + P \\ I \quad\quad K_i \\ \quad EI \end{array} \tag{2.3}$$

An interesting application of competing substrates is in the treatment of accidental poisoning by ethylene glycol (automotive antifreeze). Ethylene glycol is not intrinsically toxic. It is, however, converted to fatally poisonous oxalic acid by a series of oxidative metabolic steps (Eq. 2.4). Giving the victim intoxicating quantities of ethanol, an alcohol and dehydrogenase substrate, competes with the glycol substrate, thus effectively inhibiting the synthesis of the first aldehyde product in the sequence of reactions leading to oxalic acid. The nonmetabolized ethylene glycol will then be harmlessly excreted. Methanol poisoning, leading to the fatally toxic formic acid, can also be successfully antidoted with copious doses of ethanolic beverages (whiskey), using the same rationale.

$$\begin{array}{ccccc} CH_2\text{–}OH & \text{alcohol} & \overset{\displaystyle /\!/}{C}\text{–}H & & COOH \\ | & \xrightarrow{\hspace{2cm}} & / & \longrightarrow\longrightarrow\longrightarrow & | \\ CH_2\text{–}OH & \text{dehydrogenase} & CH_2\text{–}OH & & COOH \\ \text{Ethylene glycol} & & & & \text{Oxalic acid} \end{array} \quad (2.4)$$

Noncompetitive inhibition presents a somewhat modified picture (Eq. 2.5). Here, it is not the formation of the normal enzyme–substrate complex that is inhibited. Rather it is the reversal to the original components (i.e., E and S), which is prevented to a degree indicated by K_i. The inhibitor in such circumstances binds at other than the catalytic sites of the enzyme and therefore does not directly compete with the substrate at all.

$$\begin{array}{l} E + S \rightleftarrows ES \rightarrow E + P \\ I \quad K_i \qquad\quad I \quad K_i \\ EI + S \rightleftarrows ESI \end{array} \quad (2.5)$$

Many of the drugs exerting their pharmacological activity via a reversible enzyme, inhibiting mechanism do so competitively (examples in Table 2-1). Even though compounds synthesized utilizing design rationales based on reversible competitive inhibition have shown potent in vitro activity, they frequently have been disappointing in in vivo animal models and clinically in people. These poor results can vary from no noticeable effects to weak subclinical utility or very short duration of effectiveness. The reasons for such unsatisfactory results, although not always clear, may nevertheless have provable explanations.

Following the discovery that the antibacterial sulfonamides were competitive antimetabolites of an enzymatic substrate (see subsequent discussion), there began what was thought to be a whole new direct approach to rational drug design of chemotherapeutic agents, namely, the structural modification of other essential bacterial and mammalian metabolites, particularly vitamins and amino acids.

After the synthesis and testing of numerous compounds, it became apparent that in the great majority of cases these "antimetabolites" simply did not work under in vivo conditions. With few exceptions, these substances proved ineffective as chemotherapeutic agents.

Many of the compounds evaluated were small, water-soluble molecules and were thus too hydrophilic effectively to penetrate bacterial membranes and reach the target enzymes in meaningful concentrations. In addition, such compounds would be rapidly excreted so that a minimum effective blood level might never be achieved. As the structure I through VI illustrate, many of the compounds synthesized were closely related to the metabolite they were supposed to antagonize. It is not surprising that when antagonism was obtained it was invariably simple competitive inhibition. Effects were transitory since the resultant accumulation of the natural substrate would overcome the inhibition. Other probabilities for the poor results obtained with many of the early antimetabolites may have been that some organisms might not have a nutritional requirement for the corresponding metabolite or, the antimetabolites' affinity for the enzyme in question was insufficient to compete effectively with the metabolite.

Many of these compounds displayed little selectivity in their effects, being toxic to the host as well as the parasite. Pyrithiamine actually produced the symptoms of thiamine deficiency in the test animals. Since many of the enzymatic steps in the biochemical sequence were not known, inhibition of an enzyme not catalyzing the rate-determining step frequently may have minimal effect on the overall pathway.

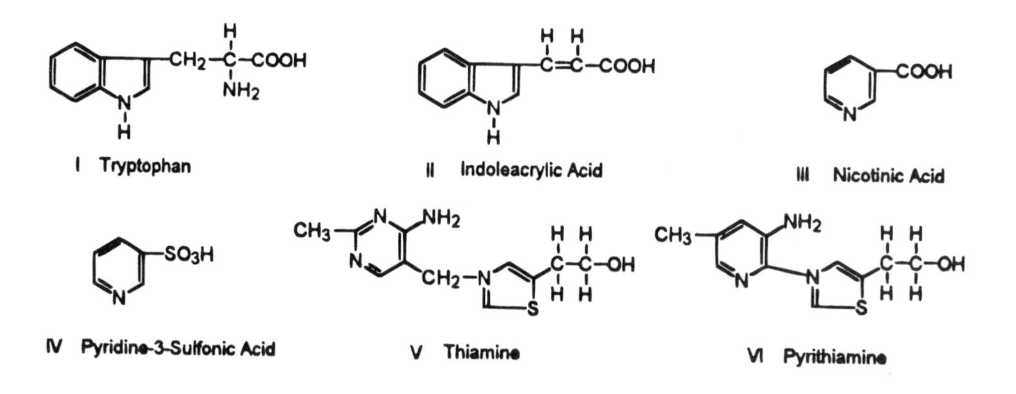

I Tryptophan II Indoleacrylic Acid III Nicotinic Acid

IV Pyridine-3-Sulfonic Acid V Thiamine VI Pyrithiamine

2.3.1.2. Irreversible Inhibition

Irreversible inhibition is effected by covalent bond formation with one or more functional groups, primarily at the active site of the enzyme. Reactions very near the active site, as additional anchoring points, are also known to occur. Such inhibition results in a catalytically inactive enzyme that is unable to interact with its substrate. Irreversible inhibitors can also be subclassified into two categories. Active-site-directed inhibition was a concept first developed by B. R. Baker in the early 1960s. Another concept of irreversible enzyme inhibitors, called K_{cat}, came about in the following decade. Generally, it can be said that irreversible inhibitors have (or can develop) reactive functional groups that alkylate or acylate susceptible groups on the enzyme.

2.3.1.3. Active-Site-Directed Inhibitors

The concept developed by B. R. Baker postulated that it may be possible to inhibit reactive sites of enzymes irreversibly by incorporating alkylating or acylating moieties into molecules otherwise closely resembling that enzyme's substrate. Such covalent bond-forming analogs, still being essentially classic antimetabolites, will assure affinity for the enzyme while permitting facile—and irreversible—reaction with it. Utilizing lactic dehydrogenase as the model enzyme, 4-iodoacetamido salicylic acid (VII) was used to test the hypothesis.

Iodoacetamide, an α-halogenocarbonyl type of alkylating agent, will readily react covalently with -SH sulfhydryl groups. An enzyme's reactive site containing a cysteine residue may therefore be irreversibly inhibited if sufficient concentration can be achieved.

$$\text{Enz–CH}_2\text{–SH} + \text{ICH}_2\overset{\overset{\displaystyle O}{\|}}{\text{C}}\text{NH}_2 \longrightarrow \text{Enz–CH}_2\text{–S–CH}_2\text{–}\overset{\overset{\displaystyle O}{\|}}{\text{C}}\text{–NH}_2 + \text{HI} \tag{2.6}$$

Iodoacetamide itself does not inhibit this enzyme. However, when bonded to the salicylic acid molecule, which has some structural resemblance to lactic acid, irreversible alkylation and irreversible inhibition do take place. Since salicylic acid itself is a reversible inhibitor of lactic acid dehydrogenase, thus apparently achieving the necessary concentrations for this effect, it presumably does so in its derivatized analog as well.

Even though this concept was intensely researched in the 1960s (Baker, B. R., 1967) it has not resulted in widely used drugs. Newer techniques, however, applied to this interesting idea may yet give useful agents.

2.3.1.4. K_{cat} Inhibitors

This type of irreversible inhibitor can be referred to as an irreversible mechanism-based inhibitor (IMBI) and has acquired the not-quite-accurate nickname of *suicide substrate.* Essentially, it is designed to be an analog of a normal substrate, which, once bound to the enzyme's active site because of its structural analogy, is then modified by it to produce a highly active electrophilic group. Covalent bonding to functional groups at the active site by these reactive moieties leads to irreversible inhibition. It is as if the enzyme, by activating this latent function in the imposter substrate, were committing "suicide."

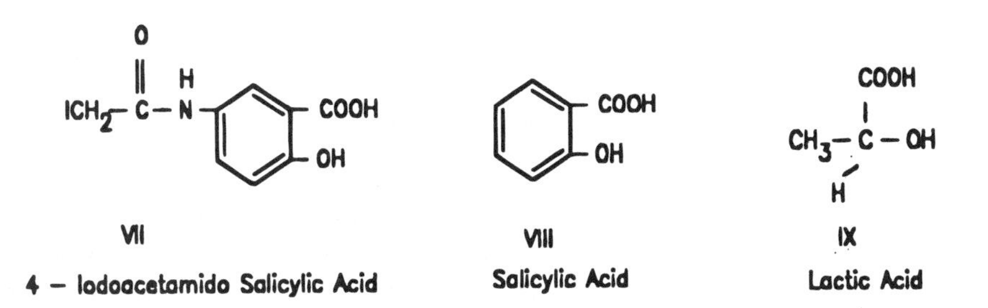

Unlike the active-site-directed inhibitors, K_{cat} inhibitors have intrinsically unreactive functionalities that cannot be activated by other enzymes and nontargeted sites such as membranes and nucleic acids. This limitation of reactivity to only the targeted enzyme affords considerable specificity and, theoretically at least, should result in drugs having very low toxicity. Much of the cellular damage done by random electrophilic attack on other biomolecules is hereby avoided. It is probably more correct, and certainly more descriptive, to refer to such compounds as *enzyme-activated irreversible inhibitors.* Designing such drugs is not a simple task. Many prerequisites are necessary. Among these is an understanding of target enzyme mechanisms, as is a knowledge of the functional groups of the active site of the enzyme. In addition, just any substrate analog capable of binding to the active site is not enough; neither is simply a reactive group capable of covalent interaction. A functionality capable of being "unmasked" to yield such a group enzymatically must be developed.

The most useful, and thus far successful, examples have involved irreversible reactions of nucleophilic functions of an enzyme's reactive site with an enzymatically activated K_{cat} inhibitor of a Michael-type addition reaction. The activation invariably requires participation of the enzyme's prosthetic group (e.g., flavin of monoamine oxidase) or coenzymes such as pyridoxal (vitamin B) as its phosphate, which is associated with several enzymes (e.g., threonine dehydrase, ornithine decarboxylase, α-ketoglutarate transaminase).

A *Michael reaction* is a conjugate addition of a carbanion (a Michael donor) acting as nucleophile, to the β-carbon of an α, β-unsaturated system as represented in Eq. 2.7. The

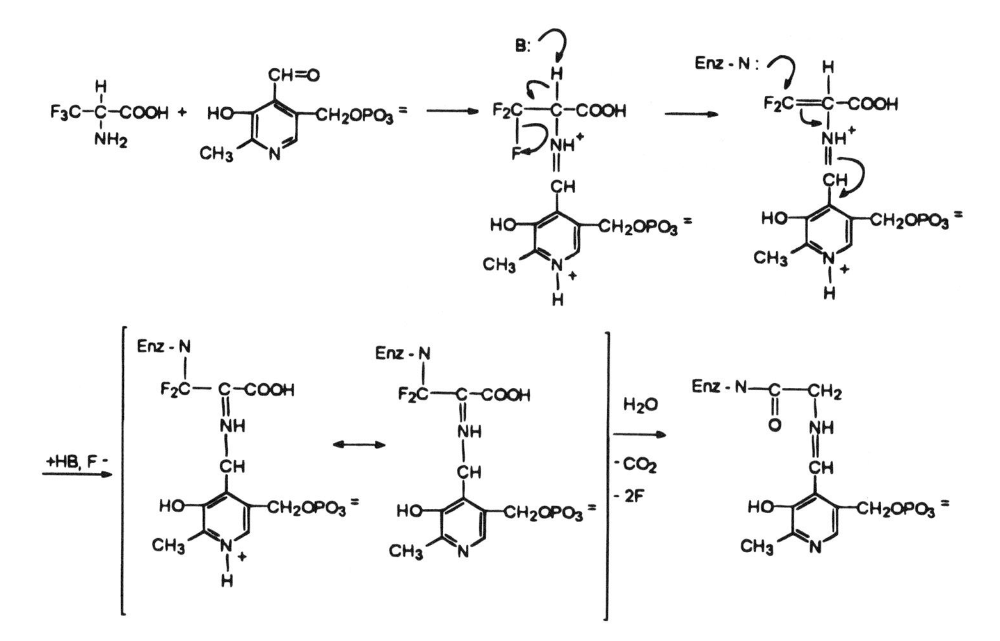

Figure 2-1. Trifluoralanine as K_{cat} inhibitor.

nucleophile (Nu), an electron-rich group in the reactive site of the enzyme, might be the CH_2OH of a serine residue, an SH group of cysteine, or the ε-amino function of lysine. In order to facilitate the "electronic relay" system (illustrated by Eq. 2.7) that will result in the β-carbon becoming electrophilic and then attractive to nucleophilic attack at that point, *X* is usually an oxygen, but can also be a sulfur or even nitrogen atom. The *X* group, by acting as an electron sink, facilitates the electron relay, thus allowing facile alkylation by the nucleophilic carboxyl, nitrile, or halomethyl function.[1]

Nu: H α β C=C–C=X ⟶ Nu H C–C=C–X⁻ ⟶ Nu H C–C–C(=X)– H (2.7)

To function successfully, therefore, a K_{cat} inhibitor must be able to bind to the enzyme's reactive site with high affinity and carry a latent group readily convertible by the enzyme to a function capable in turn of alkylating (or acylating) the site covalently, thus inactivating it permanently. Examples will illustrate this type of potential drug.

Figure 2-1 represents a typical example of an α-amino acid analog, β,β,β-trifluoroalanine, functioning as a K_{cat} inhibitor for alanine in an enzyme requiring pyridoxal (vitamin B_6) as the coenzyme. The α-amino group forms a Schiff base (azomethine) with the aldehyde carbonyl of the vitamin. This would ordinarily be followed by an enzyme-induced proton removal at the α position to the carboxyl, leading to a resonance-stabilized carbanion. Here, however, elimination of a fluorine atom, as F^-, is also observed. The resulting

[1] For a detailed discussion of the Michael reaction, see Bergman, Ginsburg, and Pappa (1959).

product is a highly active *Michael acceptor* that can now readily attack the *Michael donor* (possibly an -SH, -OH, or imidazole) in the reactive site of the enzyme. This forms the resonance-stabilized intermediate (within brackets). This intermediate will in turn rearrange by the loss of two more fluoride ions (by hydrolysis) and by the loss of CO_2. The final product now represents the irreversibly inhibited enzyme.

Figure 2-2 illustrates the irreversible inhibition of the flavine-enzyme monamine oxidase (MAO) by the antihypertensive agent pargyline. This enzyme is important in the catabolism of catechol- and other biogenic amines, including epinephrine and norepinephrine to their corresponding aldehydes. Equation 2.8 illustrates the oxidation of monamines by way of an imino intermediate while the oxidized flavine prosthetic group FAD is simultaneously reduced.

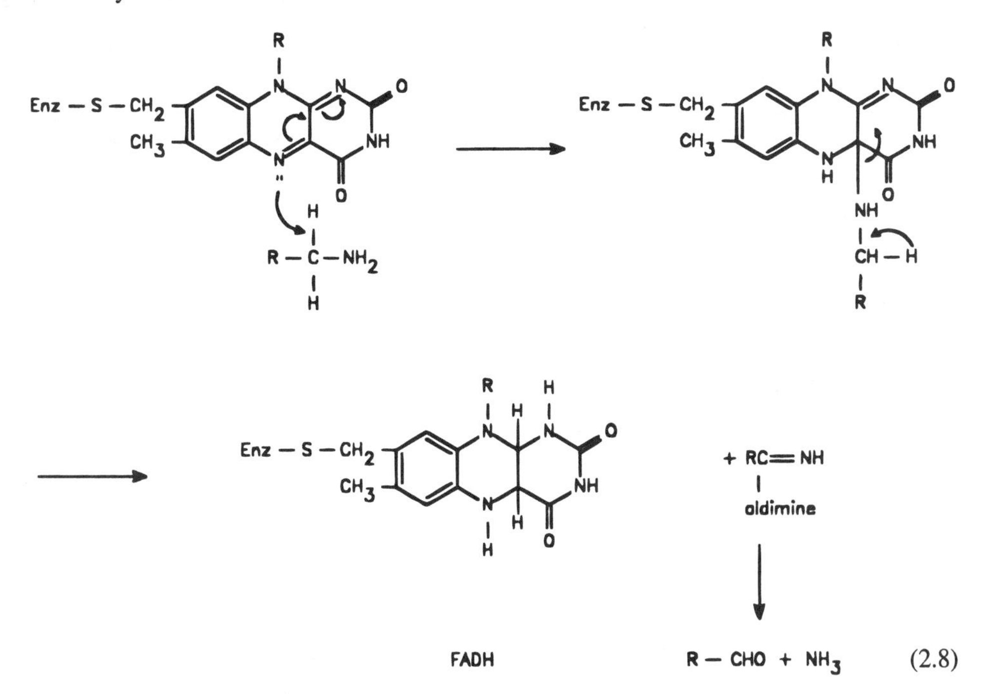

(2.8)

In the case of pargyline, and other β-amino-acetylenic "suicide substrates" of MAO, the mechanism requires initial nucleophilic attack of the flavin component of the enzyme by the drug acting as the Michael acceptor. The antihypertensive action of pargyline, and presumably similar compounds, may be due to the fact that MAO is the first enzyme in the catecholamine metabolic series.

Two diazoketone antibiotics, analogs of the amino acid glutamine (a nitrogen source for the biosynthesis of purines), have potential in cancer chemotherapy by apparently acting as enzyme-activated irreversible inhibitors of a crucial aminotransferase enzyme (Chapter 4). Other potential therapeutic applications are in the area of dopaminergic agents (to treat Parkinson's disease) by developing K_{cat} inhibitors for Dopa decarboxylase. Clavulinic acid, a β-lactam related to the penicillins, can be viewed as a clinically important and potent suicide inhibitor of β-lactamase enzymes elaborated by bacteria that are developing resistance to β-lactam antibiotics (Chapter 6). Other examples involving K_{cat} inhibitors as potential GABAminergic agents, possibly useful as anticonvulsants, will be illustrated later (Chapter 12).

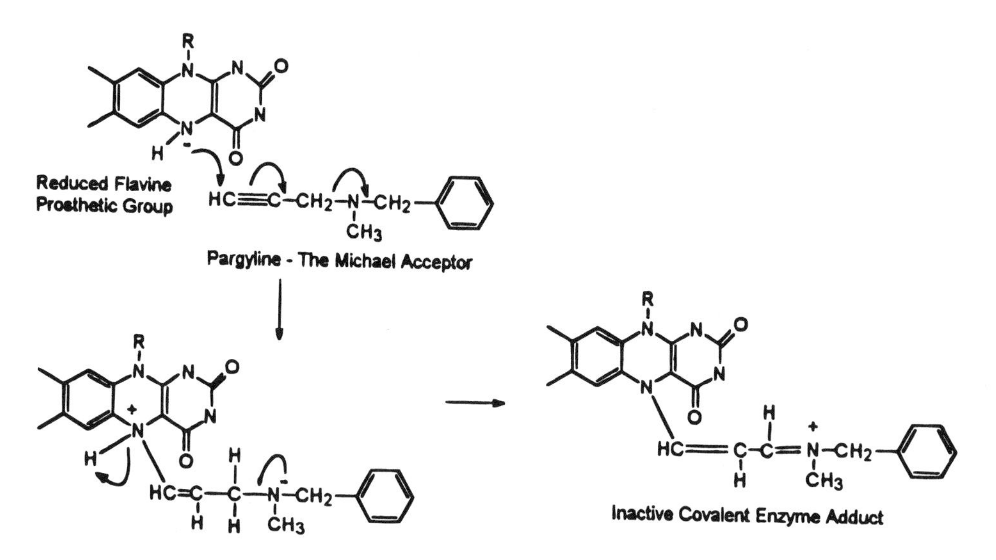

Figure 2-2. Pargyline's inactivation of MAO.

As with other applications of enzymology, suicide enzyme inhibitors as potential drugs are also an outgrowth of this special area of biochemistry.

2.3.1.5. Transition-State Inhibitors

The concept that chemical reactions in most circumstances may involve an unstable (i.e., high-energy) intermediate between the structures of the reactants and the products arose during the fundamental investigations of organic reaction mechanisms (Ingold, 1953). Thus a reaction between two different molecules is perceived to progress through a high-energy activated complex, or *transition state*, arising as a result of molecular collisions of those molecules possessing greater kinetic energy than the majority of molecules in the system. The energy necessary to form this transition state is called the *energy of activation* and represents the energy barrier to the spontaneous occurrence of the reaction.

The transition state, because of its higher energy content and resultant instability, can then break down to the components from which it arose, or to the products of the reaction. It may thus be viewed as being in transition between reactants and products. If we consider the transition-state structure of an enzyme's substrate as the form most tenaciously bound, then it becomes apparent that chemically close, but stable, analogs of such a structure may have greater affinity, and specificity, for the enzyme than ordinary substrate analogs are likely to have.

There are problems with this eloquent concept, however. A transition-state analog, as a matter of practicality, *must* be a stable compound. At best it can only *resemble* the unstable segment of the substrate's actual transition state since that latter's instability is due to partially formed (and unformed) covalent bonds. Our best hope is the development of relatively crude analogs that may nevertheless be superior in affinity to the substrate's ground-state analogs, and yet result in highly effective reversible inhibitors.

It may be useful at this point to differentiate between *transition states* and *unstable intermediates* of a reaction. In the former bonds are in the process of being formed and broken, while in the latter, a chemical species with just-formed bonds exists at lower energy, possibly between the energy peaks of two transition states. An unstable intermediate could conceivably even be isolated, chemically trapped, or at least spectroscopically identified. The transition-state structure *resembles* that of the unstable intermediate. Transition states of enzyme substrates or enzymes are not likely to be as well characterized as those of its reactions occurring utilizing simpler inorganic catalysts in a nonprotein environment. Here, unstable intermediates are much better known. For example the hydrolysis of simpler amides such as acetamide is known to proceed via the addition of H_2O to the carbonyl function, forming a tetrahedral intermediate, which by rapid elimination of ammonia affords the acetic acid product (Eq. 2.9). Similarly, acetaldehyde reacting with methylamine forms a tetrahedral intermediate that then, rapidly losing a water molecule, produces the azomethine product or Schiff base (Eq. 2.10). This understanding can readily be extrapolated to more complex biological systems (e.g., peptide hydrolysis by serine enzymes), as illustrated in (Eq. 2.11).

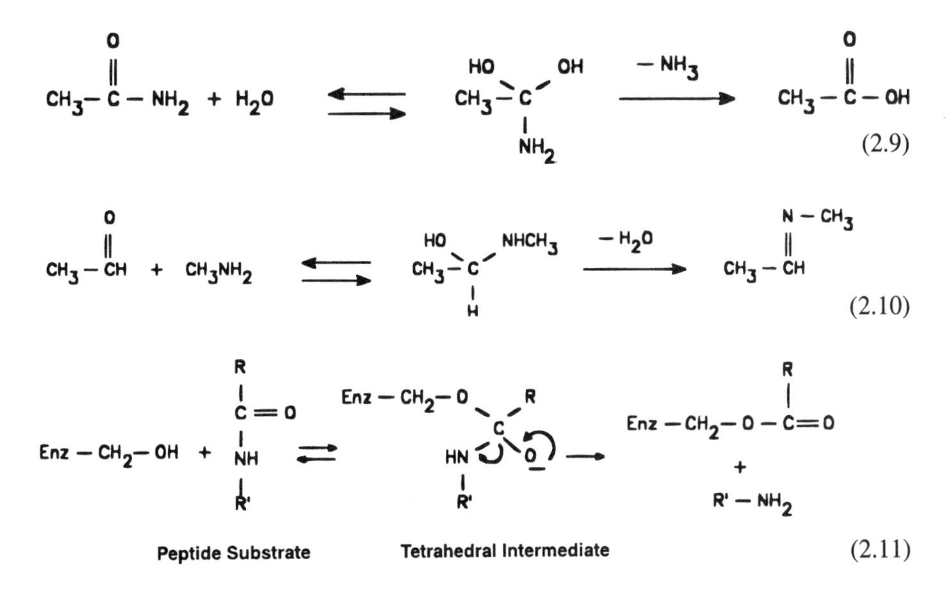

(2.9)

(2.10)

(2.11)

It should therefore be possible to design chemical structures, modeled after known or putative reaction intermediates that resemble postulated transition states. They may exhibit high affinities for the reactive sites of enzymes and therefore function as effective, but reversible, inhibitors. A successful example is the potent and specific cytidine deaminase inhibitor 3,4,5,6-tetrahydrouridine, which effectively blocks the conversion of cytidine to uridine. It was similarly demonstrated that 1,6-dihydro-6-hydroxymethylpurine effectively blocked the deamination of adenine to hypoxanthine by adenine deaminase (Eq. 2.13).

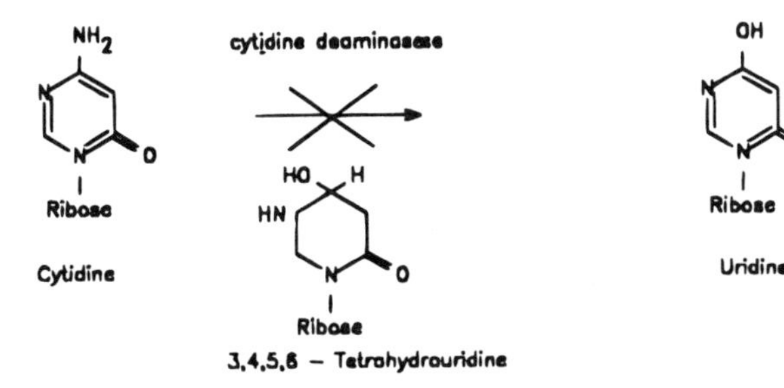

(2.12)

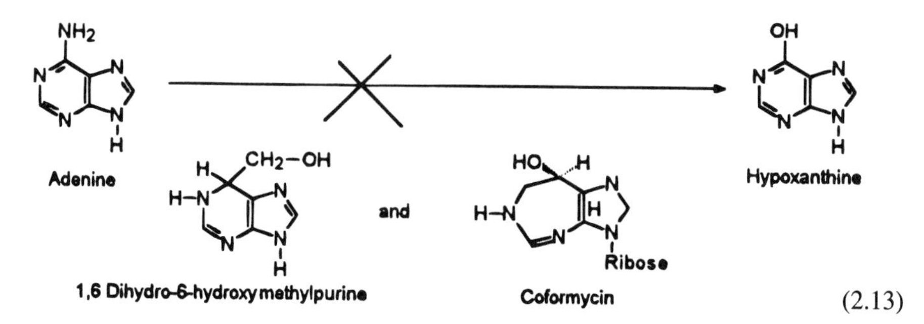

(2.13)

A particularly intuitive application of this concept may be the experimental anticancer drug N-(phosphonoacetyl)-L-aspartate (PALA). The first step in the de novo biosynthesis of the pyrimidine nucleotide formation in the cell involves the condensation of carbamoyl phosphate with L-aspartic acid catalyzed by the enzyme aspartate transcarbamylase (Eq. 2.14).[2] One can postulate a transition state, as shown in Eq. 2.14.

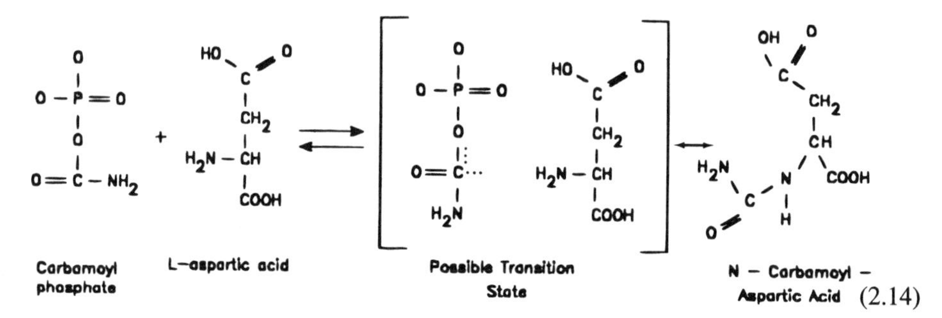

(2.14)

Here the carbon–oxygen bond of carbamoyl phosphate is in the process of breaking, as a nitrogen–carbon bond (from L-aspartic acid's α-amino group to the carbamoyl's carbonyl carbon) is about to be forming. It is further hypothesized that both substrates in this state of transition are more strongly bound (goodness of fit?) than the substrates would be if separate binding affinities were considered. Utilizing such reasoning, PALA was designed and synthesized as a potential transition-state inhibitor for the transcarbamylase-catalyzed reaction. Here a more permanent methylene (CH_2) replaces one of the carbamoyl phosphate's oxygen atoms of the theoretical transition state, while the remainder of the drug's constructional design remains strongly analogous to its hypothetical structure. A compound such as PALA can also be viewed as multisubstrate analog, containing as it does

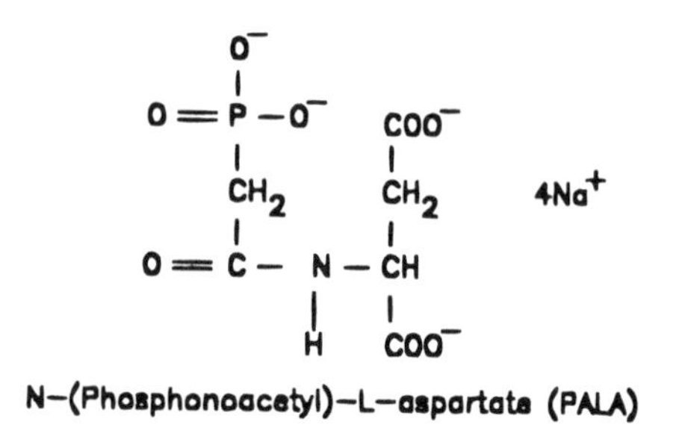

[2] See also Scheme, Chapter 4.

features of both substrates. It should also be noted that inhibition kinetic studies indicate that the drug binds more tightly to the enzyme (ca. 10^3 times) than to the substrates alone. Experimental pharmacology shows that certain lung carcinomas, melanomas, and colonic cancers in rodents respond to this drug.

2.4. Sulfonamides

The concept of antimetabolites arose as a result of the discovery of the antibacterial sulfonamides in the early 1930s. This in turn ushered in the modern era of effective chemotherapy of infectious diseases. A discussion of these early, but still used, antibacterials will illustrate the historical development of the first real success in treating bacterial infections with relatively safe, nontoxic chemical compounds, as well as the concepts of competitive enzyme inhibition and antimetabolites.

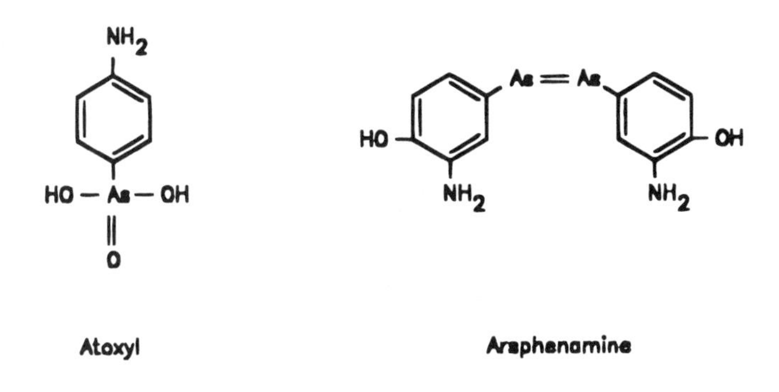

The decades preceding the mid-1930s saw the development of some chemotherapeutic agents that included organic arsenic compounds such as *p*-amino-phenylarsonic acid, wishfully named "Atoxyl" (because it was slightly less toxic than inorganic arsenic compounds), which showed some useful activity in managing the protozoal disease trypanosomiasis. Paul Ehrlichs development of arsphenamine (Salversan, 606) in 1910 as an effective antisyphilitic followed and can now, with over eight decades of hindsight, be viewed as the genesis of a revolution that changed all treatment of infectious diseases. Other early drugs were: the antimony compound tartar emetic (trypanosomiasis and leishmaniasis), the alkaloid emetine from Ipecac root (amebiasis), and the acridines euflavine and proflavine (for bacterial wound infections, topically applied). A less toxic derivative of Atoxyl, tryparsamide was introduced in 1920 to treat sleeping sickness (trypanosomiasis), as was the very effective dye suramin sodium. The next advance, in the late 1920s and very early 1930s, was the development of the synthetic antimalarials, pamaquine (1927), and mepacrine (Atabrine) (1932), the latter of which is still finding utility as an anthelmintic today.

Since most drugs developed up to that time worked primarily against systemic protozoal diseases, it began to look to many workers in the field that effective treatment of systemic bacterial infections with low-molecular-weight synthetic organic compounds introduced into the body may not be achievable; at least there was skepticism. The time for a major breakthrough, however, was at hand. Domagk (1935), testing over 1000 azo dyes synthesized by the Bayer laboratories, demonstrated outstanding antistreptococcal and staphylo-

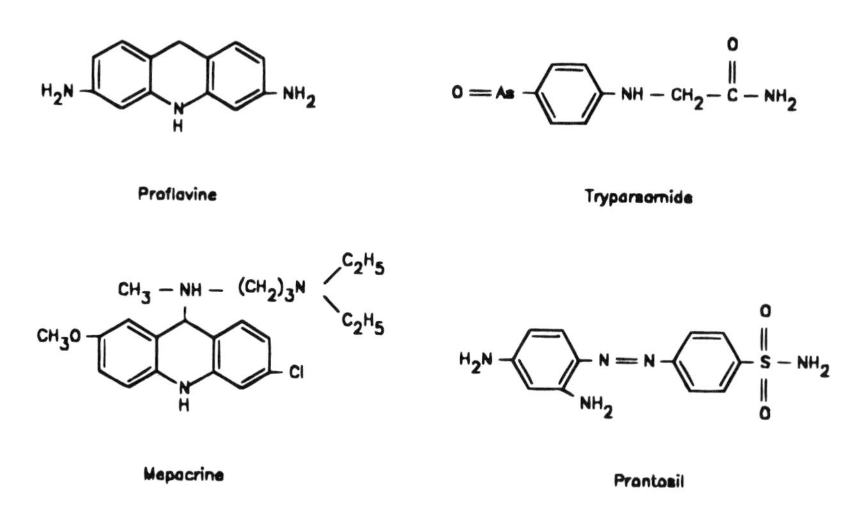

coccal activity for the sulfonamido azo dye prontosil in mice. The rationale for screening sulfonamido (-SO_2NH_2)- containing dyes for possible antibacterial activity was based on the known binding propensity of these dyes to wool fibers that are protein (fast dyes). Analogous strong attachment (covalent bonding?) was assumed likely to bacterial proteins. The drug was introduced into clinical medicine on the basis of successful animal experiments. The curious puzzle that prontosil had no effect on bacterial growth in culture and the basis of its in vivo activity were soon solved. It was found that molecular modifications on the diaminobenzene had no significant effect on antibacterial activity. Furthermore, urine analysis of patients treated with prontosil showed that most, if not all, of the ingested drug was excreted with the azo linkage reductively cleaved, yielding the compound *p*-amino-benzenesulfonamide (sulfanilamide) (detected as its acetyl derivative). Sulfanilamide was thus established as the active antibacterial agent.[3] The mystery of prontosil's in vitro inactivity could now be explained, since the dye had first to be reductively cleaved by hepatic azo-reductase enzymes to the active moiety. Such cleavage, of course, would not occur in the "test tube" environment, unless a reducing agent were added. Thus prontosil had no intrinsic antibacterial action.

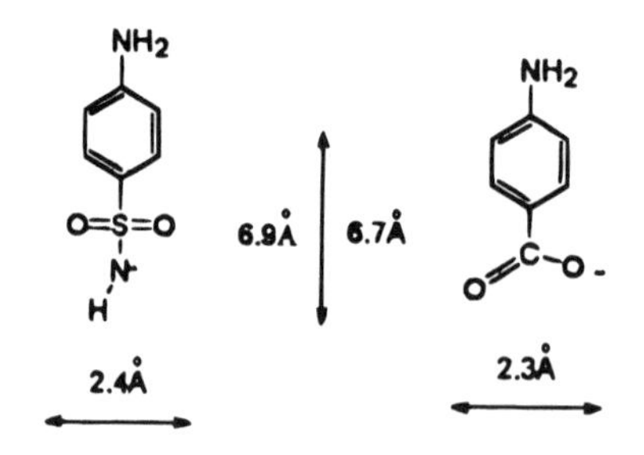

What followed over the next several years was an explosion of sulfonamides, which were essentially derivatives of sulfanilamide. The impetus for this was twofold. One reason, of course, was to capitalize on the lead compound—sulfanilamide—in order to obtain

[3] Prontosil, which today might be viewed as a pro-drug, was soon abandoned.

Table 2-2. Sulfonamides—Physicochemical Properties and Activity

$$\overset{N4}{H_2N}-C_6H_4-\overset{O}{\underset{O}{S}}-\overset{N1}{NH}-R$$

Drug (trade name)	R	pKa	Lipo-solubility (%)	MIC[a] *E. Coli* (μM/L)	$t_{1/2}$ (hours)
Sulfanilamide	H	10.5	10.5	128.0	9
Sulfadiazine (Microsulfon)	N, N	6.4	26.0	0.9	17
Sulfacetamide (Sulamyd)	$-C(=O)-CH_3$	5.4	2.0	2.3	12
Sulfisoxazole (Gantrisin)	H_3C, CH_3, O, N	4.9	4.8	2.2	6
Sulfathiazole	N, S	7.1	15.0	1.6	4
Sulfamethoxazole (Gantanol)	CH_3, O, N	6.0	21.0	0.8	11
Sulfamerazine	CH_3, N, N	7.0	62.0	0.9	24
Sulfamethoxypyridazine (Kynex)[b]	CH_3O, N=N	7.2	70.0	1.0	37
Sulfamethoxine	CH_3O, OCH_3, N, N	6.1	—	0.8	150

[a] Minimum inhibitory concentration in vitro.
[b] No longer used in humans in the United States.

even more effective compounds with lesser side effects. Another factor was that the chemical sulfanilamide was not a patentable drug, since it had been synthesized years earlier. In any case, the commercial competition spawned numerous new and clinically better "sulfas," as they became known. Table 2-2 lists some representative sulfonamides with certain of their physicochemical and biological activities.

It is now known that the ionic form of a sulfonamide is the active antibacterial species (Eq. 2.15). However, its poor lipid solubility would preclude efficient penetration across the essentially lipoidal bacterial membrane. The molecular form, having the much higher lipid solubility, will more readily cross the membrane. However, once within the cell the drug would not be active unless it would ionize to some degree at physiologic pH (7.4).[4]

[4] Recall that a compound with pKa 7.4 would be 50% ionized at pH 7.4.

Thus, for meaningful activity under actual biological conditions, the drug must have a degree of both lipid *and* aqueous solubility. A pKa value that would permit the existence of useful concentrations of both ionic and molecular forms at physiological pH would therefore be desirable.

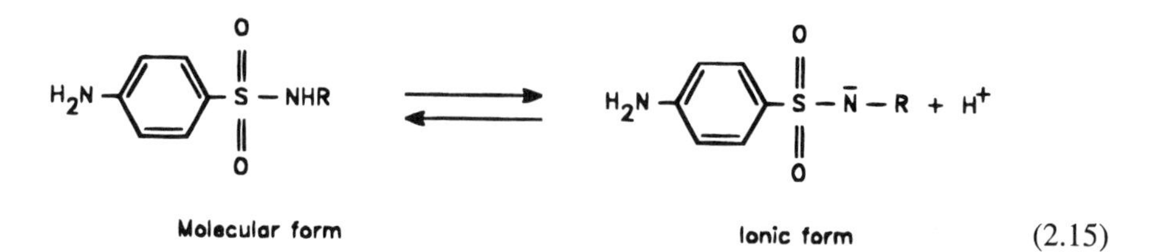

(2.15)

Table 2-2 indicates that sulfanilamide requires 142 times the concentration of sulfadiazine to inhibit the growth of *Escherichia coli.* The lipid solubility of sulfadiazine being only 2.5 times greater than that of sulfanilamide would hardly be enough to explain such a dramatic difference in potency. If the pKa values of the two drugs are examined, however, it can be calculated at pH 7.4 that only 0.03% of sulfanilamide would be ionized, whereas sulfadiazine would be 80% ionic under these conditions. Thus, even though sulfanilamide has an adequate lipid solubility (10.5%) to cross the bacterial membrane, 99.97% of the molecules would remain in the molecular (inactive) form once inside. The difference in lipid solubility between the two drugs would permit the prediction that sulfadiazine would have a longer biological half-life. Higher lipid solubility would result in less rapid excretion via the urinary route, which favors more water-soluble compounds. The data show the prediction to be correct: 17 versus 9 hours.

In the case of sulfamethoxypyridazine we would predict a high degree of activity because of an almost "ideal" pKa value. The very high lipid solubility (70%) would indicate that this drug would be readily stored in fatty tissues of the body and therefore excreted very slowly. Assuming minimal metabolic breakdown, a long duration of action would be expected. Sulfamethoxine's half-life (and duration) is among the longest known.

The low lipid solubility of sulfisoxazole indicates high water solubility and therefore rapid excretion. A short bio half-life would be expected. The pKa value would lead to a good ratio of molecular to ionic forms. Because of its rapid and high concentration in the urinary tract, it is indicated for susceptible urinary tract infections. The closely related sulfamethoxazole, which has a much higher lipid solubility, has almost twice the half-life.

It is now apparent by examining the structures in Table 2-2 that by varying the nature of the *R* substituent on the N^1 nitrogen, pKas and lipid solubility can be considerably varied. Thus the heterocyclic rings found in that position on most clinically used sulfonamides have an electron-withdrawing effect. The resultant decreased electron density of the N^1 atom weakens the N-H bond, thereby increasing the compound's acidity (Eq. 2.15). In addition, the nature of the heteroatoms in the rings and the "organic shrubbery" on them influences overall liposolubility. The ease of synthesizing a large number of chemically very similar compounds and testing their relative effectiveness by a rapid in vitro method presented medicinal chemists with the first meaningful opportunity to evaluate the relationship of structure to a biological activity in a systematic, though empirical, method. The results indicated the necessity for an unsubstituted -NH_2 group para to sulfonamide moiety. N^4 substituents (unless easily removed in vivo) afforded inactive compounds. Isosteric replacement of the benzene by other rings was unfavorable to antibacterial action; additionally, substitutions at other positions in that ring invariably led to inactivity. The -SO_2-C_6H_4-NH_2-*p* (sulfones),

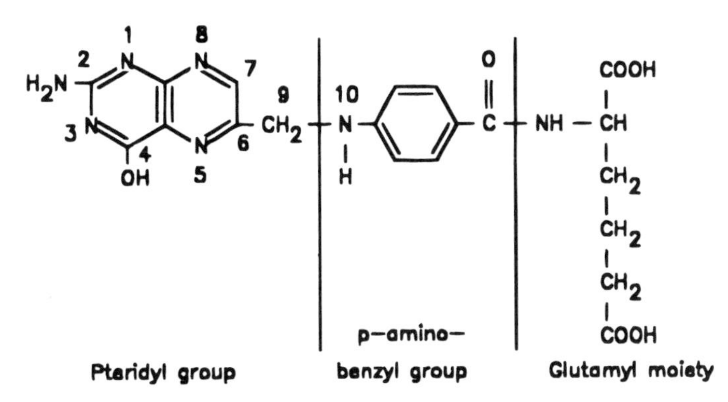

Figure 2-3. Folic acid.

-SO-$C_6H_4NH_2$-*p*, -S-S-C_6H_4-NH_2-*p*, -$CONH_2$, or even -PO_3H_2 groups yielded compounds retaining some antimicrobial properties. Activity approached or exceeded the original sulfonamides in a few instances. Other pioneering studies to relate antibacterial activity to the degree of ionization (pKa), and thus to structural parameters, were carried out. Increased activity was observed as the pKa value reached 6.7 and lessened again with further increases. With time, very active compounds outside this optimal range were discovered.[5]

Unraveling the full mechanism of action of the "sulfas" took over two decades. An early finding was that the action of sulfanilamide against *Streptococcus haemoliticus* can be antagonized by an extract of the bacteria, as well as pus and yeast extracts. It was proposed that the drug was inhibiting an enzyme whose normal substrate was also a low-molecular-weight compound. Subsequent investigations involving extractions and testing of a variety of tissues yielded a weak acid with a diazotable (aromatic) amine, which was identified as *p*-aminobenzoic acid (PABA). It was suggested that PABA was the naturally occurring substrate and that it was an essential metabolite for (sulfonamide-sensitive) bacteria. It was also demonstrated that PABA-containing extracts were able to reverse the sulfanilamide-induced inhibition of bacterial growth in a competitive manner.

A comparison of the structures of PABA (as the carboxylate) and sulfanilamide (as the anion) indicated that this competitive inhibition was due to the amazing congruence of structural and electronic features between the two molecules.

Findings of the wide natural distribution of PABA and the establishment of the structure of folic acid (Fig. 2-3) seemed to confirm Wood's earlier prediction.

A complete understanding of sulfonamide action evolved over a 20-year period. The biosynthesis of the various folates in living cells had to be elucidated; their functions in the "scheme of things" had to be worked out. The following discourse will consider the effects of sulfonamides, as well as that of another group of important enzyme inhibitors—the dihydrofolic acid reductase inhibitors. Figure 2-4 outlines the stratagem as it is presently understood.

Guanine (A) is converted in several steps to the pteridyl alcohol 2-amino-4-hydroxy-6-hydroxymethyl-7,8-dihydropteridine (B). A two-step phosphorylation results in the pyrophosphate (D). The amine function of PABA is now in position to displace nucle-

[5] For a comprehensive review of sulfonamides, see Seydel (1968).

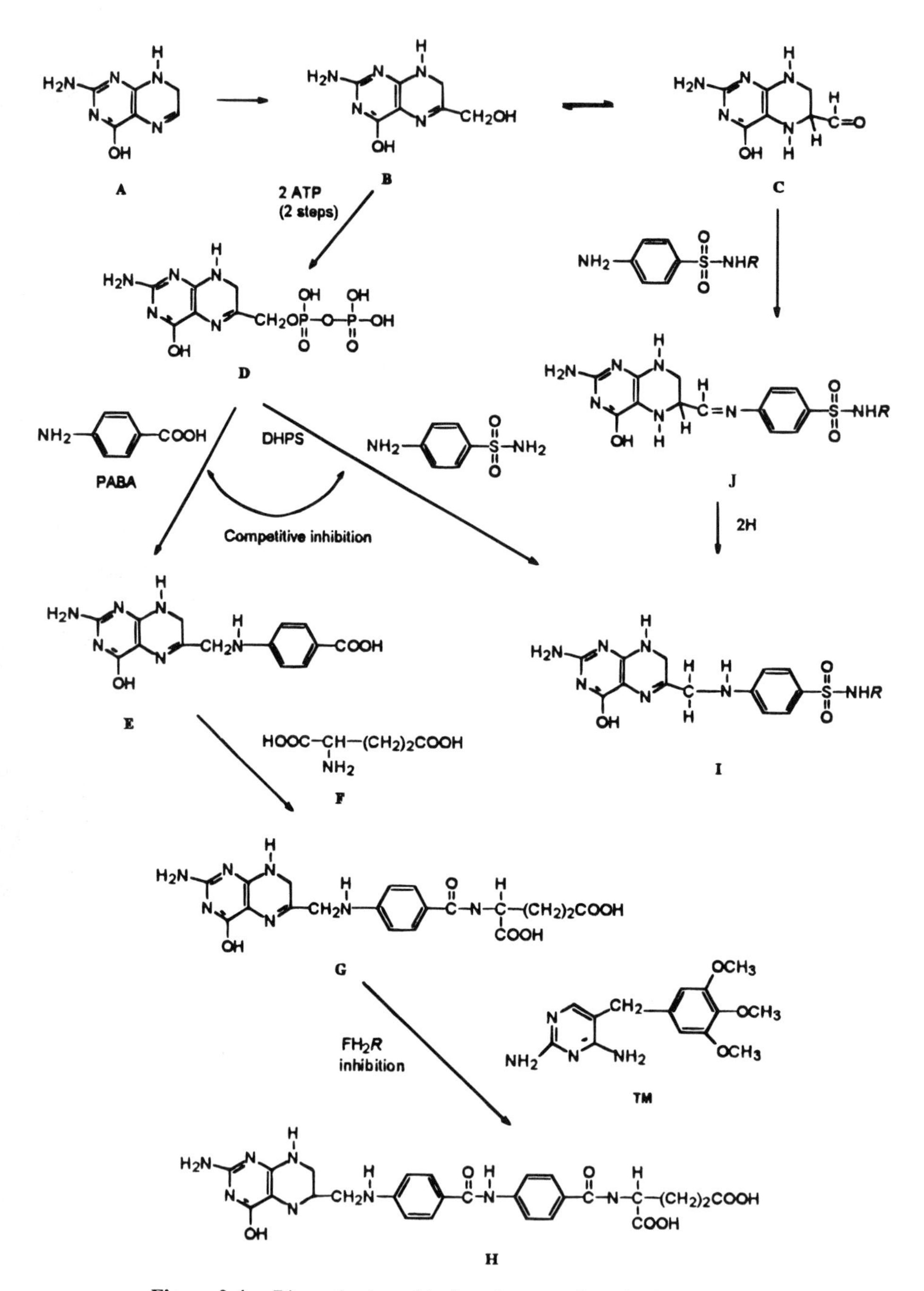

Figure 2-4. Biosynthesis and its interference of tetrahydrofolic acid.

ophilically the pyrophosphate, yielding 7,8-dihydropteroic acid (E). This reaction is catalyzed by the enzyme dihydropteroate synthetase (DHPS). It is at this point that sulfonamides (SA) are believed to compete with PABA by binding to what is probably two sites on the DHPS enzyme that are "reserved" for -NH_2 function of PABA and its carboxylate function, now occupied by the drug's -NH_2 group and sulfonamide anionic moiety, which are also about 6.7 Å apart. The next step involves the incorporation of glutamic acid (F) resulting in 7,8-dihydrofolic acid, FH_2 (G). Finally, the enzyme dihydrofolic acid reductase (DHFR or FH_2R), catalyzes the reduction of FH_2 to 5,6,7,8-tetrahydrofolic acid, FH_4. There exist dihydrofolic acid reductase inhibitors such as trimethoprin (TM) that can prevent this step from occurring, thus preventing FH_4 synthesis as well. As will be seen (Chapter 4), FH_4 is a crucial coenzyme in the biosynthesis of purines and thymidilic acid, and therefore of RNA and DNA. The interference by sulfonamides in this biosynthesis therefore deprives the bacterial cell of the building blocks for new nucleic acids and, ultimately, inhibits bacterial reproduction (bacteriostasis). The selectivity of this cellular toxicity is due to the fact that the mammalian host cells cannot carry out this synthesis but must obtain folic acid from exogenous sources (diet) and reduce it to FH_4 by folate reductase.

The scheme of sulfonamide mechanism as outlined so far fails to answer an important question. If we are really dealing with a classic competitive inhibition by an inhibitor (the sulfa drug) of a substrate (PABA), why is the inhibition not overcome as the level of normal substrate rises under in vivo conditions? Actually, if a sulfonamide and PABA are added to a growing bacterial culture simultaneously, then no growth inhibition is observed. However, delaying the addition of PABA to the sulfa-treated bacterial colony decreased its effectiveness as the drug's antagonist to the point that reversal of growth inhibition ceases after 1 to 2 hours. This means that we are not dealing simply with competitive inhibition; something else must be happening as well.

Several possibilities come to mind. One is that the sulfonamide actually forms a pteridinyl derivative such as I, which in turn may tend to reduce the availability of the pteridyl pyrophosphate ester D, and decrease the production of G. A condensation product of sulfamethoxazole and the pteridyl alcohol B was actually obtained, raising the possibility of a sulfonamide-containing "imposter" of dihydropteroic acid arising. A second viable possibility is to consider a tautometric equilibrium between the pteridine alcohol B and the aldehyde form C, which will very likely form the Schiff base derivative J, with the *p*-amino function of a sulfonamide. This would also lead to a pteridyl alcohol depletion from the path of the FH_2 synthesis (Seydel, 1968).

The fact that the FH_4 synthesis can be blocked at two different enzymic steps (i.e., by sulfonamides at DHPS and by FH_2R inhibitors at that reductive step) resulted in a pharmaceutical product (Co-Trimoxazole) consisting of sulfamethoxazole and trimethoprim. The assumption is that a *sequential double blockade* is at work allowing these two drugs to act synergistically and minimize the emergence of resistance dramatically. Clinical effectiveness appears to bear this out, yet the matter is not readily settled. It was demonstrated that not all FH_2R inhibitors act synergistically with sulfonamides. There is now evidence that sulfonamides act on FH_2R as well as on DHPS, suggesting that as the reason for the actual synergism observed. Thus a simultaneous multiple inhibition, rather than a true sequential blockade, may be operative here.

2.5. Membrane-Active Drugs

2.5.1. Effects on Membranes

In the following discussion the way in which various drugs may act on cellular membranes, and in turn be modified by this action, will be considered. One salient fact must be kept in

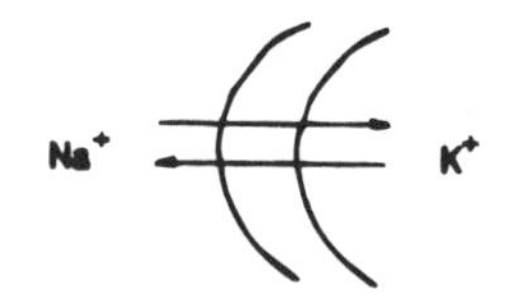

Figure 2-5. Nerve cell membrane.

mind: Membranes pose an osmotic barrier between the internal contents of the cell and its external environment. Thus most studies are concerned with the mechanisms that help to explain the transport of drugs, ions, and biomolecules across membranes. Drug transport mechanisms essentially occur in two stages. The first stage involves the complexation of a drug with some membrane component, possibly an enzyme. The second stage follows with a movement of the complex to the inner surface of the membrane, where the drug may be released into the cell's aqueous interior. Once inside the compound will exert an effect on the cell's metabolism by mechanisms already considered or to be discussed.

Before considering medicinal agents relative to effects on membrane integrity, some basic aspects of membrane physiology will be noted. Figure 2-5 schematically illustrates that the interior of a nerve cell has a higher concentration of potassium ions (K^+) than the extracellular fluid. The opposite is true of sodium ions (Na^+). There is a normal tendency for K^+ to flow out of the cell by diffusion, thus taking the positive charge to the membrane's exterior. The organic anions acting as counterions, to which the membrane is impermeable, cannot follow. Thus a charge separation occurs, resulting in a net negative charge on the membrane's interior. Equilibrium is attained when the positive charge on the outside of the membrane surface is sufficiently distributed to repel further diffusion of K^+ from the cell's interior. External chloride ions (Cl^-) will not enter the cell because of repulsion by the membrane's interior negative charge. The concentration gradient of external Na^+ is usually sufficient to move some of them into the cell. These, then, are non-energy-requiring passive movements. The tendency of Na^+ to be pushed into the cell by sheer weight of numbers is counteracted by the much lower permeability the membrane has for Na^+ than it has for K^+. In addition, there is an energy-requiring active transport system that extrudes three Na^+ ions from the cell for every two K^+s that enter.[6]

This polarized state is particularly characteristic of excitable cells such as nerve, muscle, and myocardial fibers. These cells have a unique feature in that stimulation will reverse the electrical potential across the membrane; that is, the membrane will go through a state of zero potential. During this transitional depolarized state the nature of the permeability is drastically altered with regard to Na^+. They can now suddenly flux into the cell with ease.

When normalcy returns, however, the greatly increased Na^+ concentration results in a positive charge on the inside of the membrane—the reverse of the initial resting state. During this *action potential* K^+ permeability increases and Na^+ begins to leave the cell. Finally, the sodium–potassium pump slowly brings about reversal by pushing Na^+ out until the *resting potential* is restored. Thus both passive diffusion and active transport mechanisms are operative. These changes in electrical potential, although initiated locally, spread to other areas of the cell. This results in impulse transmission in fibrous cells such as myocardial and nerve cells.

---

[6] This is the so-called sodium–potassium pump, catalyzed by adenosine triphosphatase ATP-ase; ATP hydrolysis supplies the energy.

It can now be understood how certain drugs exert their pharmacologic effect—at least in part—by affecting membrane transport processes. Local anesthetics are believed to act on the movement of cations. The anticancer drug methotrexate is known to enter leucocytes by facilitated diffusion. Acetylcholine generally increases membrane potential of myocardial cell membranes by increasing both the time they are polarized and the time they are depolarized, thus helping to restore cardiac arrhythmias to normalcy. Cardiotonic digitalis glycosides inhibit the sodium–potassium pump, thus affecting K^+ influx and Na^+ outflow. The so-called calcium channel blockers are now believed to exert their beneficial cardiac effects by inhibiting calcium ion (Ca^{2+}) influx through "slow channels" into both conductile and contractile myocardial and smooth muscle cells. It is the electromechanical coupling caused by a transmembrane influx of Ca^{2+}, and the activation of Ca^{2+}-dependent ATP-ase, resulting from elevated Ca^{2+} levels in myofibrillar cells, that causes activation of the contractile system in these cells. Inhibiting Ca^{2+} influx through these membranes will decrease hypercontractility such as arrhythmias and affect hemodynamics of blood vessels, thus reducing blood pressure.

It is not surprising that the membranes of prokaryotic cells such as bacteria and fungi have characteristics differing considerably from eukaryotic cells. Compounds that will disrupt the functions of these membranes, but not similarly affect mammalian cells, may be useful as selectively toxic antimicrobials. In fact, there now exist a number of important antifungal and antibacterial drugs that do just that—they disrupt the membranes of these pathogens.

2.5.2. Membrane Disrupters—Antimicrobials

There are antibiotics that disorganize cytoplasmic membranes, resulting in leakage of intracellular constituents, particularly small molecules. The result is a rapid killing action. One such group can be classified as macrocyclic polypeptides. These bacterial products are not, strictly speaking, proteins, because some of the amino acids have the D configuration and several that are never encountered in proteins, for example, α, γ-diaminobutyric acid (DAB).

Tyrothricin, first investigated in 1944, was obtained from *Bacillus brevis*. It is actually a mixture of polypeptides, two of which, *tyrocidin* and *gramicidin*, have been crystallized and sequenced. They are two of a large group of cyclic peptide antibiotics called *tyrocidins*. Their effectiveness is primarily against gram-positive bacteria. Because of toxicity, such as lysis of erythrocytes if given systemically, use is limited to topical application and throat lozenges.

Another group of cyclic peptide antibiotics elaborated by several strains of *Bacillus polymixa* are the *polymixins*. They contain a seven-amino-acid ring portion with a high DAB content, plus a side chain terminating in a methylalkanoic acid chain varying in length from 8 to 10 carbon atoms. Polymixin B, consisting of B_1 and B_2 (Aerosporin), and Colistin, consisting of Polymixin E_1 and E_2, are of clinical interest (Fig. 2-6) since their toxicities are somewhat less than most others in this group. The polymyxins are more effective against gram-negative organisms, even the virulent *Pseudomonas aeruginosa*, which is so resistant to most other antibiotics.

The mechanism of action first involves binding to the plasma membrane. Since it is protected by an elaborate structure in gram-negative bacteria, including an outer membrane, it can be assumed that the drug must disrupt, or somehow penetrate, this barrier first. The result is a disruption of the membrane's integrity, which causes a leakage of phosphates,

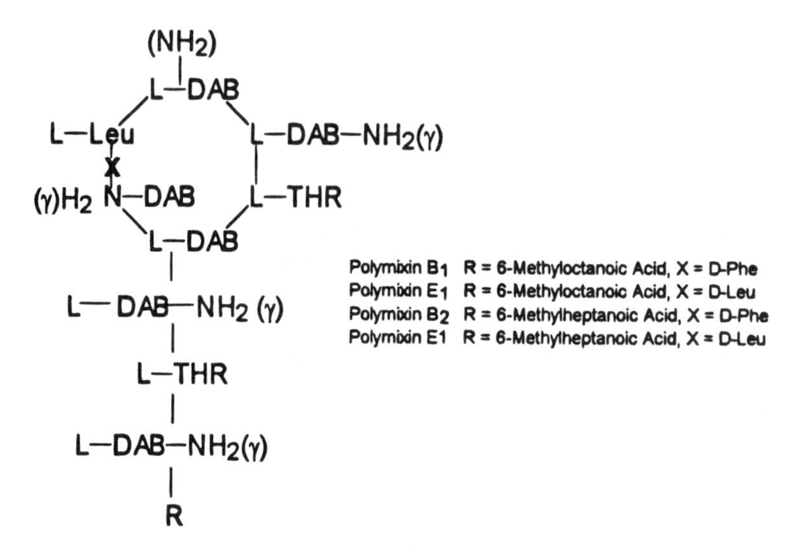

Figure 2-6. Polymixins.

nucleosides, and other small molecules. In addition to having both hydrophilic (peptide) and hydrophobic (alkane) groups, these drugs also have cationic surface-active properties (see later). The antibacterial effects are therefore neutralized by anionic detergents such as soaps.

Polymixins are not orally absorbed. Thus systemic infections by sensitive organisms are treated by intramuscular injections or intrathecally (e.g., spinal meningitis). Toxicities are primarily to the kidneys and transient neurological disturbances, including neuromuscular blockade. Oral use (colistin) is limited to bacterially induced diarrhea. External applications of lotions, ointment, and drops are useful in the treatment of eye and ear infections, as well as burns and other wounds.

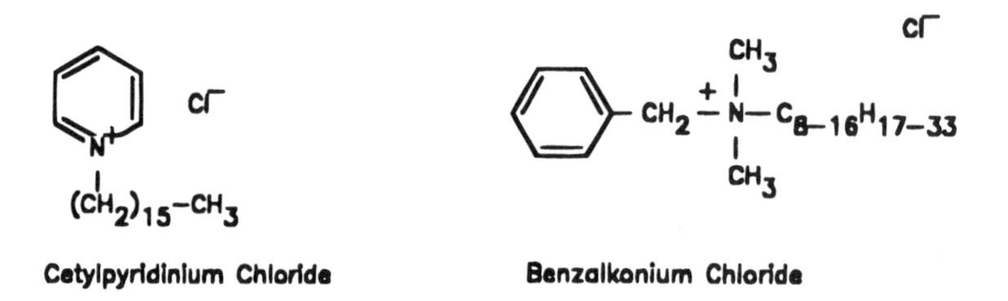

Surfactants, or surface-active agents, exist as cationic, anionic, and nonionic moieties, and share the property of lowering the surface tension of aqueous solutions and, except for the last-mentioned group, also exhibit detergency and antimicrobial properties to varying degrees. Cationic surfactants may be quaternary ammonium, sulfonium, or phosphonium compounds, where the ammonium group represents the most widely used compounds. They usually have bonded, several alkyl and aralkyl functions representing the hydrophobic groups, to the positively charged nitrogen atom—the hydrophilic portion of the molecule. Benzalkonium and cetylpyridinium chlorides are two commonly used agents. These agents are bactericidal at high dilution (sometimes down to less than 1 ppm). They affect most gram-positive and negative bacteria, many fungi and protozoa, but not spore-forming bacteria nor viruses. Their efficiency rises with pH. Following exposure to a quaternary ammonium surfactant, bacterial enzyme and metabolic inhibition can be demonstrated, as well as membrane permeability changes. The latter results in leakage of amino acids, phosphates, and nucleosides, followed by cell death. It seems likely that after initial adsorption on the bacterial cell wall, penetration into it also occurs, allowing for interaction with, and ultimately disorganization of, the plasma membrane.

Ionophoric antibiotics. These are antibiotics that act essentially by producing pores in cellular membranes. Even though they are chemically diverse, they all facilitate the passage of alkali metal cations through hydrophobic complexation. It is particularly the transport mechanism of K^+ in membranes that is of biochemical interest. Because those studied until now show little ability in differentiating between mammalian and microbial membranes, they are unlikely to have clinical antibacterial applications. Therefore, they have been mainly useful as research tools. The antibacterial properties are primarily against gram-positive organisms since the hydrophobic complexes (ionophores) formed do not penetrate the outer membranes of gram-negative bacteria.

The ionophoric antibiotics enhance the uptake of alkali metal ions with varying degrees of specificities. Gramicidins A, B, and C (chemically unrelated to Gramicidin S) show little difference in their ability to induce the uptake of Na^+, K^+, Li^+, or NH_4^+ by cellular membranes. These compounds are linear pentadecapeptides consisting entirely of hydrophobic amino acid residues having no ionizable carboxyl or amino groups. Pore formation by a gramicidin has been postulated as consisting of a helical conformation formed by two antibiotic molecules dimerized at their formyl groups (head to head). The alternating D and L configurations of the amino acid residues would dictate that the interior of the channel (helix) formed would contain the polar groups, permitting more or less parallel hydrogen bonding, thus holding the "structure" together and allowing cations through. The exterior of the pore thus formed would be lipophilic due to hydrophobic alkyl and aryl groups bristling outward perpendicular to the axis of the helix. This arrangement, then, would facilitate the fabricated ionophore gliding into the lipoidal portions of the membrane structure. Even if the channel formed were of a transitory nature (fractional seconds?) its capacity to move ions through it at available voltage gradient could be high.

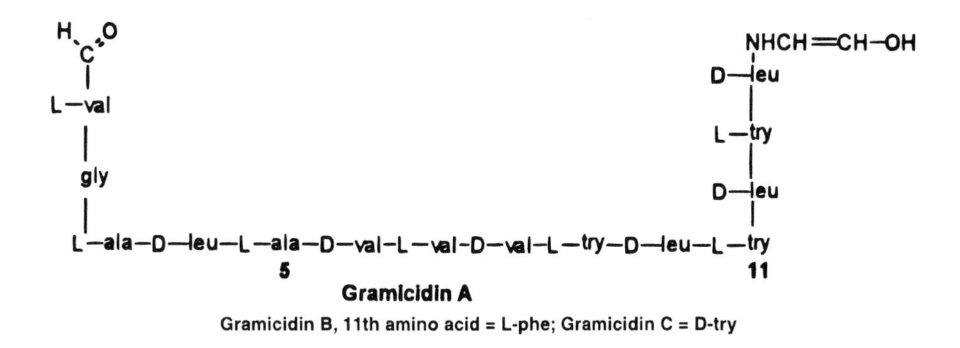

Gramicidin A

Gramicidin B, 11th amino acid = L-phe; Gramicidin C = D-try

Another of the many ionophore antibiotics studied extensively is the cyclic valinomycin containing 12 acid residues of alternating amino acids (peptide linkages) with hydroxy acids (ester linkages) in repeating but sterically alternating D and L, set in units of four. This compound exhibits growth inhibition toward gram-positive bacteria and some fungi. Like

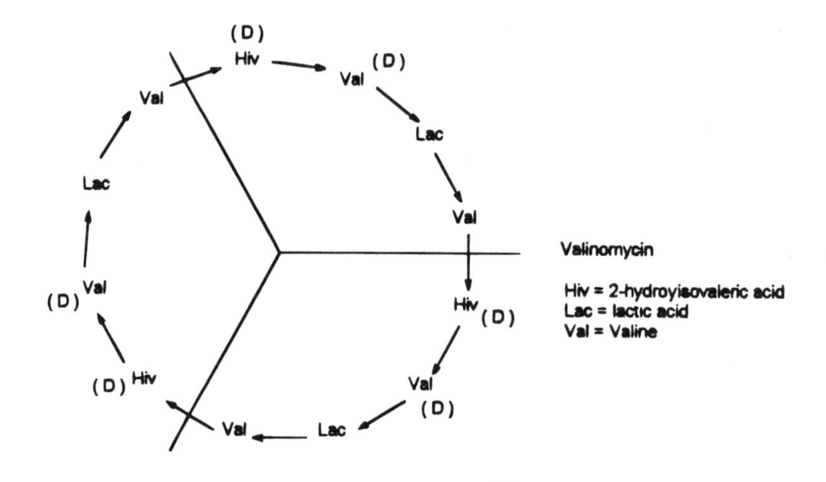

the gramicidins, it is not of clinical utility. Ionophoretically, valinomycin exhibits selectivity for K^+ transport, resulting in the release of H^+ from the cell. It had been proposed that each valinomycin molecule forms a complex with one K^+. Such well-defined crystallized complexes have been isolated and characterized by X-ray crystallography. An octahedral complex of one K^+ coordinated with six oxygen atoms (from the C = O groups of the amide and ester linkages) is formed, where the K^+ has apparently "exchanged" its shell of hydration water molecules for those of the carbonyl oxygens, all of which project to the inside of the ring. Here again, the alternating D and L configurations allow this highly specific conformation of an ionophoric structure to occur.

2.5.2.1. Polyene Antibiotics

Polyene antibiotics are a group of complex, but structurally related, substances that, as their name implies, have many double bonds, usually ranging from four to seven, all in the *trans* configuration and all are conjugated. This portion of the molecule constitutes a relatively rigid and hydrophobic region. Another structural feature shared by these drugs is a large lactone ring ranging from 12 to 38 carbons. Opposite the rigid polyene backbone of the molecule there is a polyhydroxylated region, usually on alternating carbons, that is tightly packed against the polyene area, giving the greater part of the molecule a rod-shaped, rigid appearance. The hydrophilic hydroxylated section, however, does have some flexibility.

Finally, most of these antibiotics have an amino sugar, mycosamine, and a free carboxyl group, thus being zwitterionic. In addition, an aliphatic side chain may be found in some, but not all, of these compounds. As a group the polyenes are highly toxic, unstable, poorly absorbed, and extremely insoluble in water. It is not surprising, therefore, that only a few of the more than 30 known agents have been clinically useful. These, which were obtained from streptomyces species, are: amphotericin B (*S. nodosus*), nystatin (*S. albus* and *noursis*), primaricin (*S. natalensis*), and candicidin (*S. griseus*). Amphotericin B is the only one that is given systemically, usually intravenously.

The mechanism of action of these macrolide antibiotics is now known to affect membrane permeability of sensitive cells. These cells are almost exclusively eukaryotic and therefore do not include bacteria. Fungi, both dermatophytes and yeasts, and, unfortu-

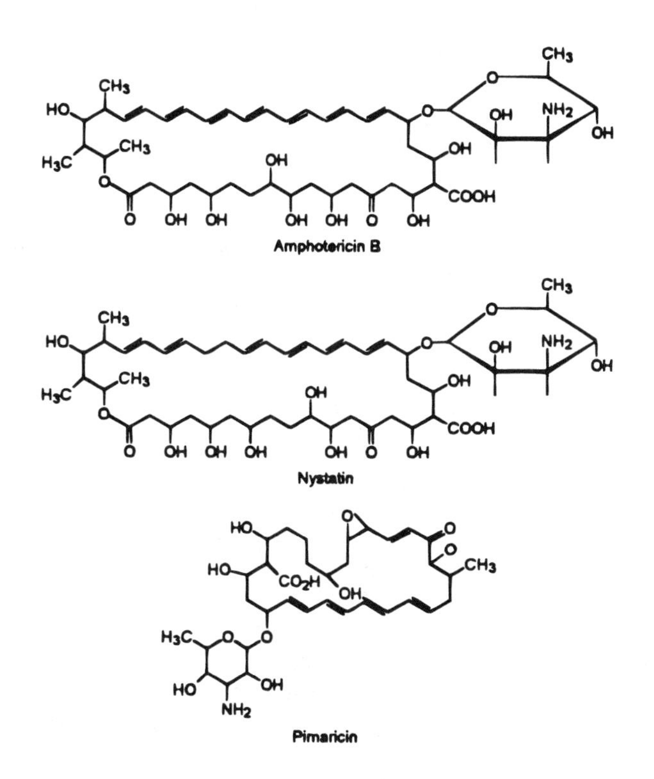

nately, animal cells are sensitive. It is now established that susceptible cells are those containing an appreciable *sterol* content as a membrane component and that the polyenes bind to the membranes (probably the sterols in them) irreversibly. Fungal cell walls are not involved. Following the membrane–polyene complex formation and the resultant alteration in membrane permeability, a leakage of cellular components, beginning with small molecules such as K^+, Na^+, inorganic phosphate, urea, carboxylic and amino acids, is observed. This progresses to nucleotides and ultimately to the larger molecules, particularly enzymes and other proteins. A precipitous drop in pH (from 6.5 to 4.7) also occurs within the cells, causing lysosomal disruption and intracellular lysis. Cell death is the final event.

All cells affected by polyenes contain membrane sterols, and thus, in addition to fungi, protozoa, snails, certain worms, and, of course, mammalian cells, are potentially susceptible. One piece of circumstantial evidence that polyene membrane sterol complexation is responsible for the ultimately observed effects is that their action can be reversed (experimentally) by the addition of exogenous sterols to the system. The tenacious drug–sterol complex formed presumably prevents polyene membrane–sterol interactions. In addition, extracting susceptible membranes with organic solvents such as chloroform eliminated their binding capacities for polyenes completely. Finally, *mycoplasma* species that do not require sterols for growth are resistant to polyene antibiotics. However, if grown in the presence of sterols, which they will incorporate into their membrane structure, *mycoplasma* become fully inhibited by these drugs.

The extent to which the polyene antibiotics with clinical usefulness have tolerable human toxicities is based on a selectivity, which may in turn be due to different sterols in these two types of membranes having different affinities for a particular polyene. The primary sterol in mammalian plasma membranes is cholesterol, while it has been shown to be ergosterol in *Candida* species. In the case of amphotericin B, at least, it has been demonstrated that the basis of the selectivity (of toxicity) is due to a greater binding affinity of the drug for ergosterol than for cholesterol.

Electron microscopic studies using a freeze-etching technique show that amphotericin B causes a large number of pits, depressions, and even craters on the inner surface of membranes. Although this obviously represents membrane damage, at least including lipid disorganization, there appeared no discernible formation of *holes* or enlargement of naturally existing *pores*. Negative evidence of this type does not rule out pore formation, however, particularly if they would be transient in nature. Based on what is known of the chemical features of amphotericin B, nystatin, and other polyenes, and their actions on conductance and permeability of various fungal cell membranes, proposals for such putative pores are very viable. Briefly, the overall dimensions of the polyene drugs are similar to those of the lecithin molecule; the polyhydroxy region is similar to that of a sterol. It is probable that a hydrophobic interaction of the membrane sterol (cholesterol or ergosterol) with the polyunsaturated portion of drug would lead to a sterol–polyene complex, thus decreasing or eliminating the sterol's association with the membrane phospholipids. This newly formed sterol–polyene interaction can be shown by molecular models to be capable of forming a cylindrical channel with a hydrophilic interior. Requirements for such a structure would be satisfied by eight sterol–polyene complexes. The actual formation of such a pore could probably arise by one of several proposed procedures. In either case a pore model, no matter how arrived at, would explain many of the known interactions of membranes with polyenes.

2.5.3. Chelation

Organometallic compounds involve metal-to-carbon bonding. In metal complexes the metal atoms are bonded to electron-donating atoms of the ligand molecule. Chelation involves a process by which ring structures are formed by the combination of an electron-donor compound, the ligand, and a metal ion.

Many chelates occur in biological systems that involve proteins and acids with elements such as iron, magnesium, copper, cobalt, zinc, and manganese. Some of the metals are part of enzyme systems. Several oxidases contain copper; angiotensin-converting enzymes, carboxypeptidase, and the hormone insulin are associated with zinc, and proteolytic enzymes and phosphotases have magnesium. The electron-donor atoms of the ligand molecules are usually nitrogen, oxygen, and sulfur. Thus chelates participate in many absolutely vital cellular functions including oxidation–reduction reactions, oxygen transport (iron-hemoglobin), and carbon dioxide fixation.

It is possible to inhibit or kill pathogenic organisms by introducing a ligand having a greater affinity for an essential metal ion than the natural ligand (e.i., forming a more stable chelate) into a system. The ability of 8-hydroxyquinoline, its derivatives, and isoniazid to chelate iron may, at least in part, explain the mechanism of action of these drugs (Fig. 2-7). The use of chelating agents as antidotes for metal poison is exemplified by dimercaprol, penicillamine, and ethylenediamine tetraacetic acid (EDTA). Dimercaprol has been especially useful in the treatment of poisoning by heavy metals such as gold (Au), mercury (Hg), antimony (Sb), and arsenic (As) (Fig. 2-8).

The tetracycline antibiotics have a high chelating ability for polyvalent cations. This may have some relationship to their mechanism of protein synthesis inhibition, which is

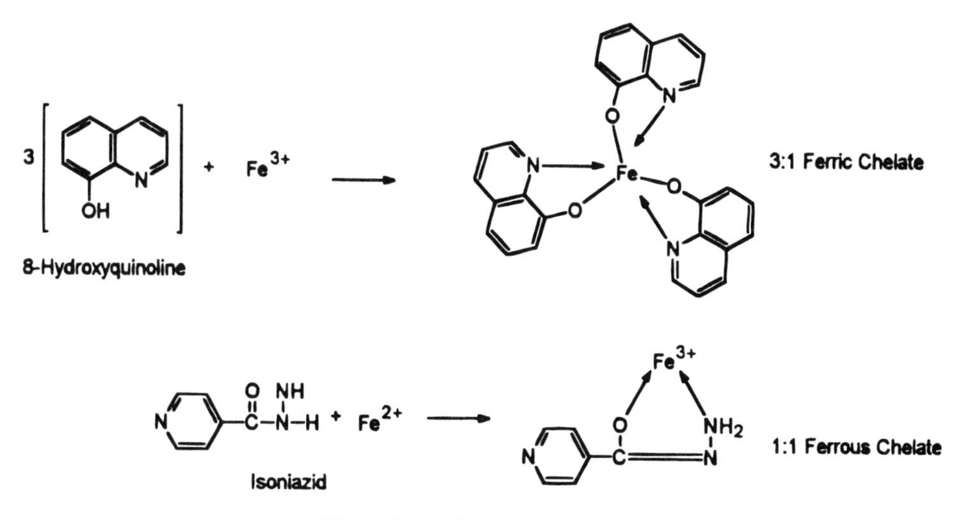

Figure 2-7. Chelations with iron.

believed to significantly involve ribosomal subunit interactions (Chapter 6). A possibility that Mg^{2+} may link the ribosome to the tetracycline via ribosomal RNA phosphate groups and thus negatively affect protein synthesis has been proposed (Franklin and Snow, 1989). Mg^{2+} and Ca^{2+} chelate with the 11,12 β-diketone system as well as the 12 α- and 3-hydroxyl groups and are the likely sites of such potential interactions.

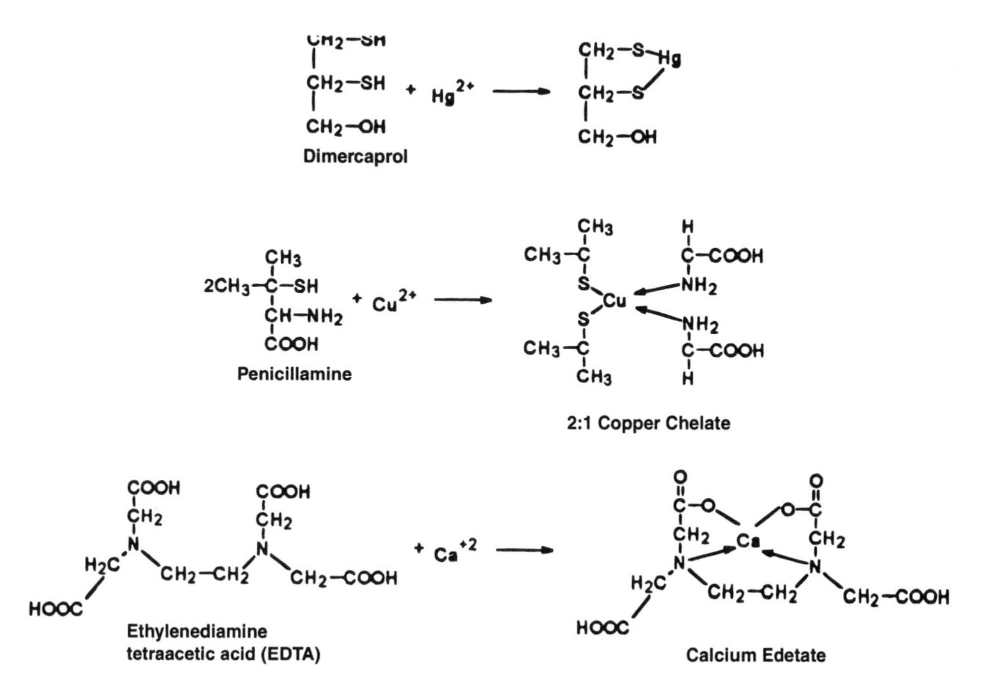

Figure 2-8. Chelation reactions in metal poisoning.

References

Baker R. B., *Design of Active-Site Directed Irreversible Enzyme Inhibitors*. John Wiley & Sons, New York, 1967.

Bergman E., Ginsberg D., Pappo R. *Organic Reactions*, 10, John Wiley & Sons, New York, 1959, p. 179.

Domagk, G. *Deut. Med. Wochenschr.* **62**:250, 1935.

DeKruijff B., Demel R. A. *Biochim. Biophys. Acta* **339**:57, 1974.

Franklin, T. J., Snow G. A. *Biochemistry of Antimicrobial Action*, 4th ed., Chapman and Hall, New York, 1989.

Ingold, C. K. *Structure and Mechanism in Organic Chemistry*, Cornell University Press, Ithaca, 1953.

Prakash N., Schechter, P., Mamont, P., et al. *Life Sci.* **26**:181, 1980.

Seydel, J.K., *J. Pharm. Sci.* **57**:1455, 1968.

Suggested Readings

Lehninger, A. L., Nelson, D. L., Cox, M. *Principles of Biochemistry*, 2nd Ed., Worth, New York, 1993.

Smith, H. J., Williams, H. *Introduction to the Principles of Drug Design*, Wright PSG, Boston, 1983.

CHAPTER 3

Drug Metabolism and Inactivation

3.1. Introduction

Comprehension of a drug is incomplete without consideration of its metabolic fate in the mammalian organism. Knowledge of the drug's biochemical transformations helps us to understand (and to predict) its toxicity and potential interaction with other drugs. The experimental pharmacologist is frequently interested in what the breakdown products are, where they may accumulate, whether or not they are bioactive, and what their routes and rates of excretion are. The medicinal chemist is very interested in the potential that exists for improved drugs, or even new agents, which can be gleaned from a detailed study of the fate of foreign compounds in the body.

From a clinical standpoint drug metabolism should be considered in conjunction with *pharmacokinetics*. This newly developing area of drug kinetics includes the rates and degrees of body distribution and excretion that are a reflection of permeability factors, protein binding, and, of course, metabolism. In fact, some clinical pharmacologists may view the overall disposition of drugs (i.e., metabolism and pharmacokinetics) as significant to a drug's efficacy as its biochemical mechanism of action. If minimally effective levels of drug concentrations are not reachable in a patient, satisfactory therapeutic results may not be achievable. In fact, therapeutic failure can sometimes be attributed to significant variation in drug handling among individual subjects.

Drugs are known that are only active because they are metabolized to bioactive chemical species. An intrinsically active drug is frequently converted to an active metabolite. Well-known examples are the hydrolysis of aspirin to salicylic acid by esterase enzymes, the cleavage of the ethyl ether group of phenacetin to acetaminophen, the demethylation of imipramine to desipramine, and the oxidation of phenylbutazone to oxyphenbutazone. In some cases the precursor drug and its metabolite are both used therapeutically.

Drug metabolism concepts have become useful in the design of new drugs and very important in the rational improvement of existing agents. A knowledge of how the enzyme systems in the body affect various functional groups and at what rates, coupled with an understanding of how pK_a and lipid solubilities affect drug transport across the membranes and thus the quality of the pharmacological effect, allows the medicinal chemist to "mask" easily metabolizable functions. This essentially slows down excretion and extends duration of action. The drugs obtained are sometimes referred to as pro-drugs: compounds converted in vivo to be an active drug. Such pro-drugs can frequently overcome difficulties in absorption, facilitate lipid membrane diffusion of otherwise poorly lipid–soluble compounds, and impede metabolism of the parent drug once inside the cell, organ, or body fluid.

One example of the usefulness of this approach is the derivatization of the phenolic hydroxyl groups of epinephrine with the pivaloyl (trimethylacetyl) group.

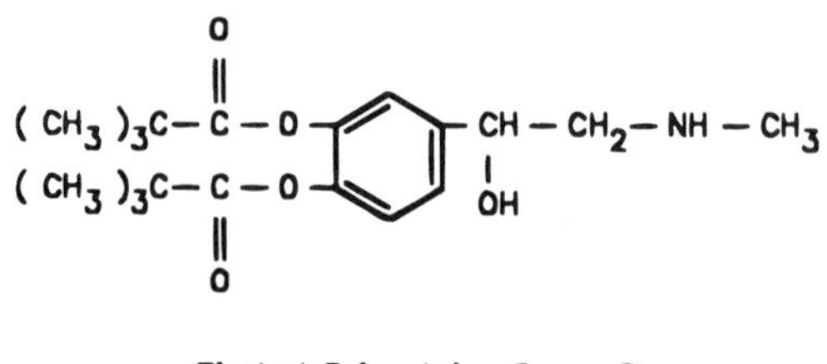

Pivoloyl Epinephrine Pro – Drug

Because of the high aqueous solubility of epinephrine, its usefulness in the treatment of glaucoma is limited. The dipivaloyl derivative, with its higher lipid solubility has a far greater ability to pass through the lipoidal barrier into the cornea, where it is presumably hydrolyzed back to the parent compound. The clinical result of using the pro-drug in this case is a 100-fold increase in therapeutic effectiveness.

The penicillin molecule has been the target of such molecular manipulations to improve certain aspects of its properties. Three such modifications of ampicillin are already in clinical use. One is the pivaloyloxymethyl derivative, pivampicillin, while another is a benzofuranyl derivative, talampicillin, and a third is bacampicillin.

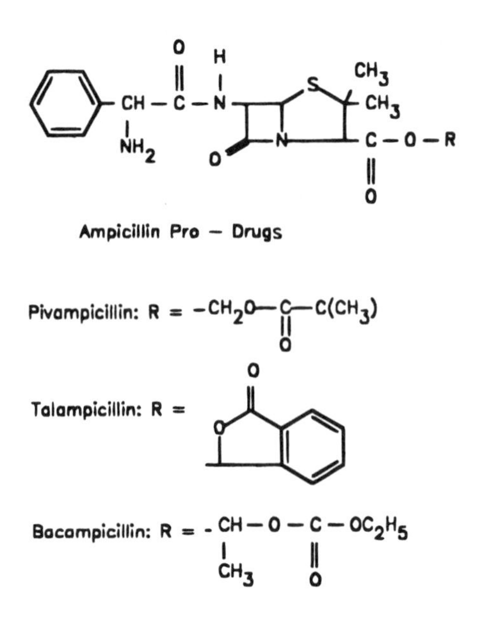

Ampicillin Pro – Drugs

In these cases the amphoteric (zwitterionic) character of ampicillin is eliminated by esterification of the carboxylic acid function. This factor, in addition to the biologically inactive organic moieties, increases the lipid solubility and improves absorption from the gastrointestinal tract considerably. Once in the bloodstream, metabolic hydrolysis of the ester linkage converts the compound back to the active drug—ampicillin. The net result is the achievement of higher drug levels in the blood, and thus more effective therapy. An added bonus is a decrease in gastric side effects, especially diarrhea. Since the pro-drug is intact in the gastrointestinal tract prior to absorption, it does not decrease the local bacterial flora. Such a "sterilization" of the gut ordinarily may result in fungal overgrowth and diarrhea.

The metabolism of drugs and other foreign ingested chemicals, as well as normal chemical constituents of the body, occur in several organs, including the liver, kidneys, intestine, lungs, blood, and skin. The liver is by far the most important. Thus it is the main detoxifying organ.

In the final analysis the intensity of a drug's action and the length of time that this action persists depend to some extent on whether a drug molecule is metabolized, to what extent, and how fast. Some agents are quite resistant to biotransformation and are therefore quite toxic (since they are not readily excreted, or very long acting, or both). The barbiturate barbital (Veronal) is a classic example. It is excreted almost totally unchanged (90%) and very slowly. As a result its sedative effect can persist for days. In fact, measurable quantities of the drug are detectable in the blood for as long as 7 days after the last dose has been administered.

Practically all metabolic reactions are enzyme catalyzed. The result of the biotransformations is invariably the generation of products that are more water soluble than are the starting compounds and that are thus less toxic to the organism because of their rapid excretion. (There are exceptions. Some of the early antibacterial sulfonamides, when acetylated, are actually less soluble than the parent drug and tend to precipitate in the kidney's glomeruli.) Enzyme systems responsible for these reactions might be assigned teleological importance as detoxifying agents. However, it is more logical to assume that their original physiological role in evolution has been (as it still is) to metabolize endogenous substances such as hormones, fatty acids, and steroids.

These enzymes are mainly of low substrate specificity; that is, they affect many compounds that are not necessarily closely related. As chemicals in the environment become more prevalent (including drugs), enzymes attack these also, detoxifying most. Thus this protective function is logically attributed to them and to the organs in which they occur.

3.2. Biotransformations

It may be convenient to subdivide biotransformation into two main categories: metabolic reactions and conjugative reactions. Metabolic reactions involve the introduction of polar functional groups onto the substrate molecule. These compounds, then, are invariably more polar. At this point they may still be bioactive or, more often, pharmacologically inactive. The resultant polar functional groups usually serve as centers for the conjugation reactions that follow. Figure 3-1 summarizes these concepts.

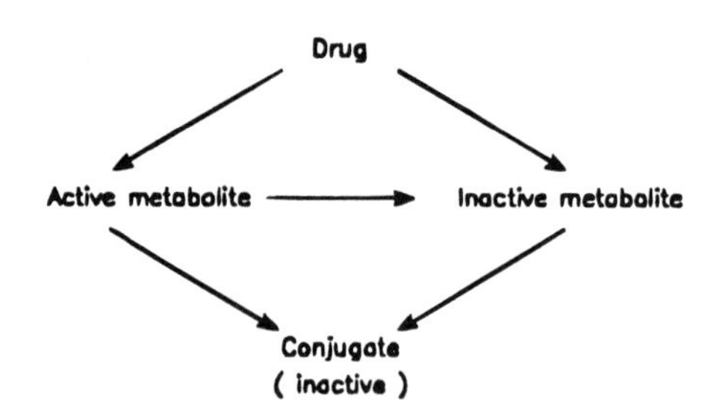

Figure 3-1. General pattern of drug metabolism.

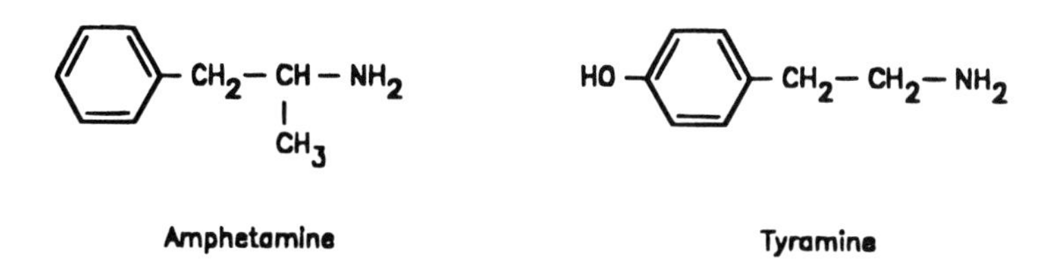

The increased polar nature of the drug metabolites and their conjugates greatly facilitates rates of excretion since tubular reabsorption in the kidneys is considerably decreased. Also, those metabolic products that are acids can now be actively secreted in the proximal renal tubules as well as the biliary system. Even though the primary role of drug metabolism in the liver is one of inactivation followed by clearance, hepatic reactions may occasionally actually activate (pro-) drugs. One interesting example is cyclization of the antimalarial drug chlorguanide (Eq. 3.1); others include the activation of the anticancer drug cyclophosphamide (Chapter 4) and the partial conversion of the analgetic drug codeine to morphine, which is more active (Table 3-1).

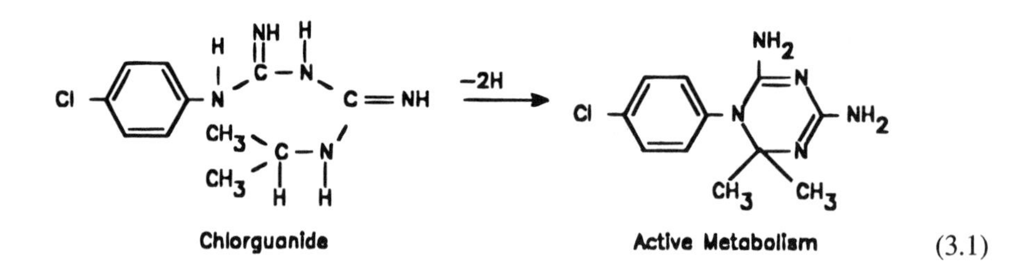

(3.1)

3.3. Metabolic Reactions

Drugs and other types of foreign compounds are metabolized by surprisingly few metabolic reactions, which fall into three general classes:

1. Oxidations
2. Reductions
3. Hydrolyses

Any of these reactions results in a more polar compound by virtue of the functional groups produced, which will be illustrated by actual examples.

Table 3-1. Oxidative Drug Reactions

Type of oxidation	Example

Hydroxylation

Aromatic compounds

Alkyl groups (side chains)

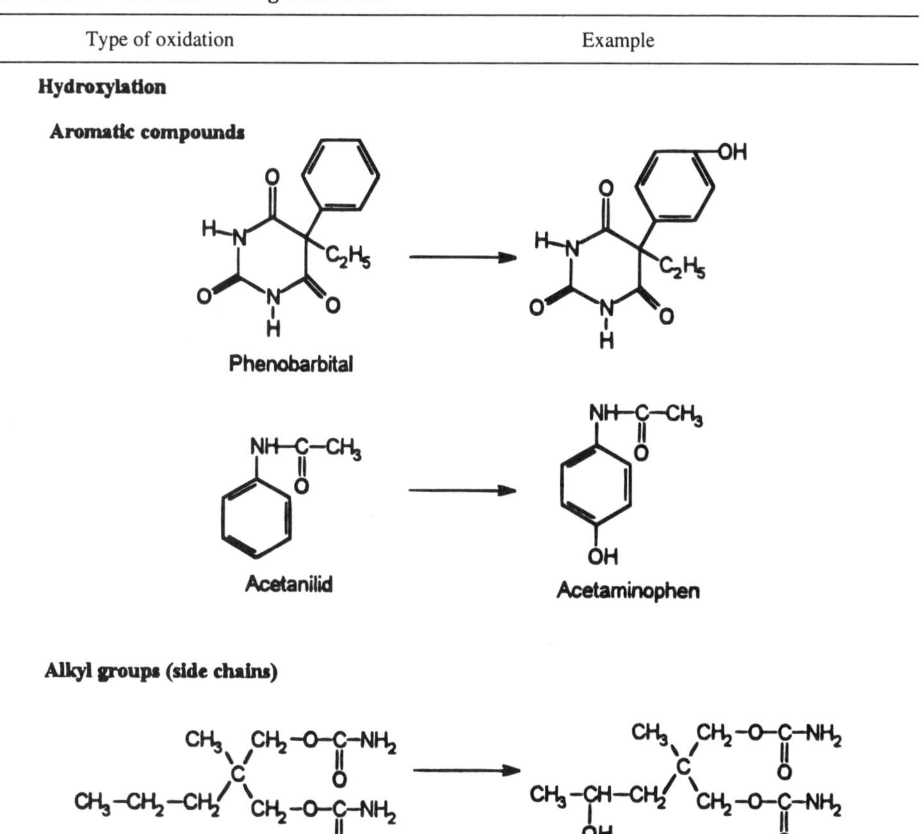

Alkyl to carboxylic groups

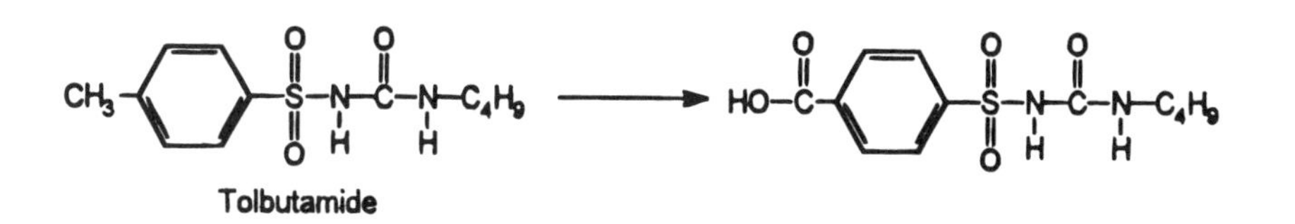

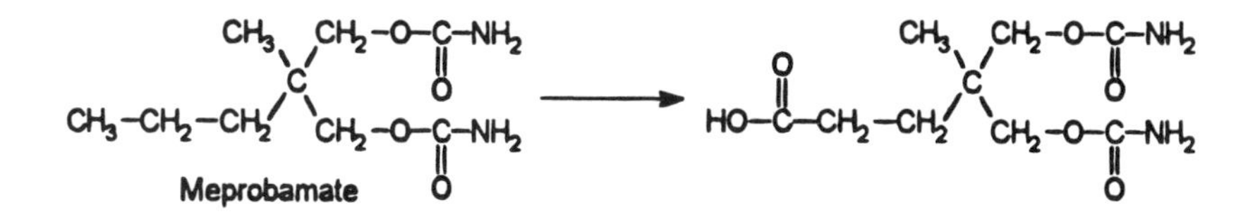

(Continued)

Table 3-1. *(Continued)*

Type of oxidation	Example

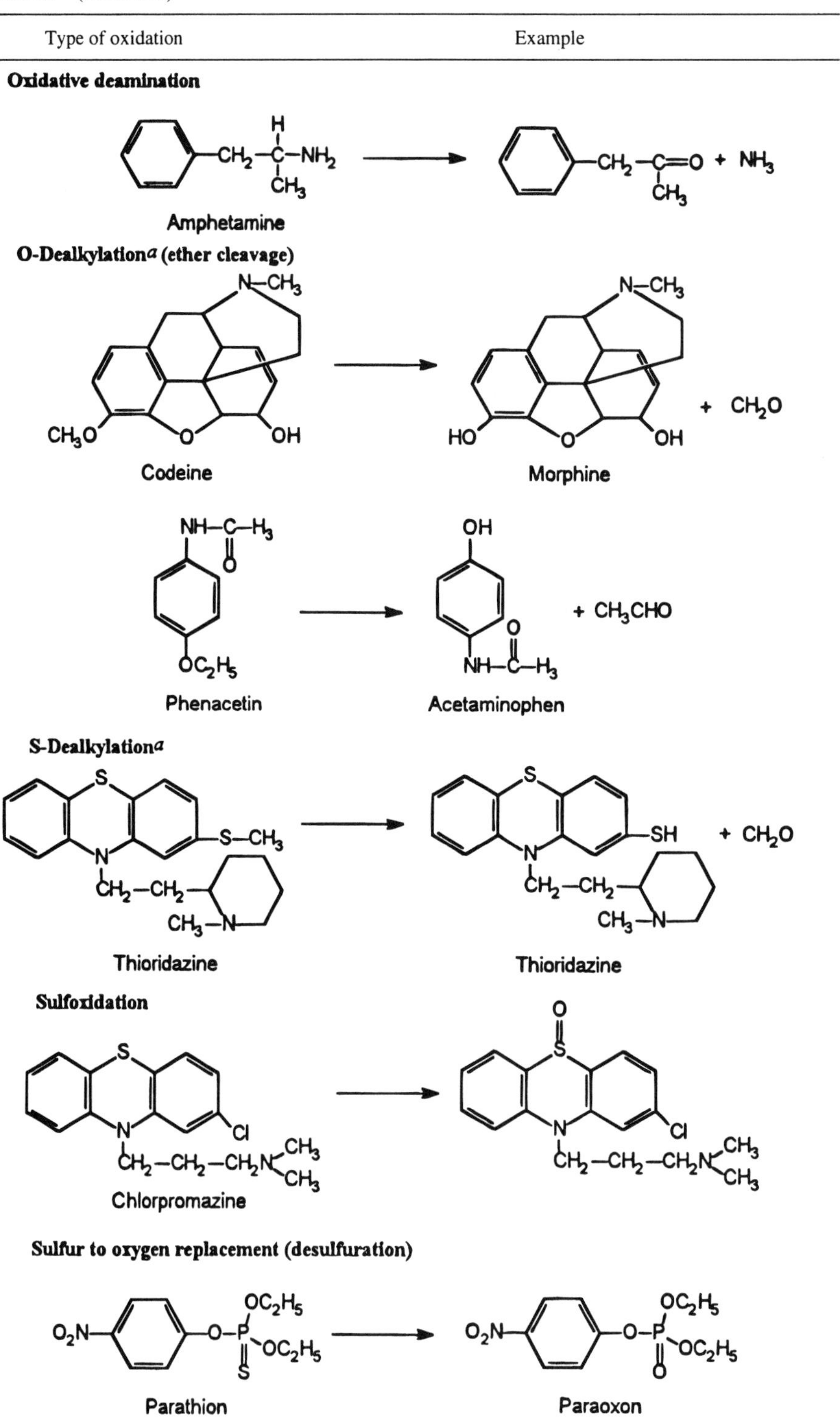

(Continued)

Table 3-1. (*Continued*)

Type of oxidation		Example
Alcohol oxidation		
Cl_3-C-CH_2-OH Trichloroethanol [b]	⟶	$Cl_3-C-C(=O)-OH$ Trichloracetic acid (minor metabolite)

[a] This reaction may be regarded as an ether hydrolysis; however, the cleaved alkyl group is oxidized.
[b] This is actually the active metabolite of the sedative drug chloral hydrate [$Cl_3C-CH(OH_2)$], resulting from its reduction.

3.3.1. Oxidations

Oxidations are probably the most prevalent and general reactions encountered. They are catalyzed by a complex of enzymes that are an integral part of the endoplasmic reticulum of animal liver cells. Homogenizing liver tissue and centrifuging it at speeds in excess of 100,000 times gravity afford a sediment referred to as the *microsomal fraction*. The majority of enzymes capable of metabolizing drugs are in this fraction. Microsomes are actually fragments of the disrupted endoplasmic reticulum of the liver cells. Prior to homogenization this reticulum was a network of microtubules that was in a continuum with both the cellular and nuclear membranes occurring throughout the cell volume. While the protein manufacturing ribosomes give part of the reticulum a granular (microscopic) appearance, the drug-metabolizing enzymes, and other proteins, produced in these ribosomes are found in the smooth endoplasmic reticulum and are in fact tightly bound to it. Only the enzyme glucuronyl transferase is not strongly associated with the microsomal fragments. The attachment of these drug-metabolizing enzymes thus being a lipoprotein system has led to speculative proposals that the foreign substances, such as drugs and many toxins, being generally more lipid in character than most endogenous metabolites, are therefore preferentially acted on by these microsomal enzymes. For example, this might explain why amphetamine is oxidatively deaminated by the microsomal system (Table 3-1) while the structurally similar but more water-soluble endogenous tyramine is not a suitable substrate. It is the extrahepatic enzyme monoamine oxidase (MAO) that catalyzes an analogous reaction of tyramine.

Most of the microsomal reactions can be classified as oxidations by what are referred to as mixed-function oxidases utilizing molecular oxygen and cofactors. The key enzyme is an iron-hemecytochrome P-450, a flavoprotein dependent in its reduction and reoxidation on the NADPH to $NADP^+$ reaction. The 450 notation is based on the 450 nm absorption peak the enzyme exhibits on reaction with carbon monoxide. Thus, drug interactions with this enzyme system can be evaluated by measuring absorption spectra changes.

The microsomal oxidations encountered include aliphatic side chain oxidations, aromatic hydroxylations, N-dealkylations, O- and S-dealkylations (which may also be viewed as ether cleavages), N-oxidations to N-oxides, and N-hydroxylations, and oxidation of divalent sulfur to sulfoxides. Representative examples are given in Table 3-1.

Cytochrome P-450 is also known to function in nonhepatic sites. It has been found in the mucosa of the intestine, in kidneys and in the adrenal cortex. It is believed to hydroxylate endogenous steroids as well as to metabolize some drugs in the cortex.

A given drug can be oxidized at various positions, usually at different rates. Thus the actual metabolic profile is much more complicated than what is indicated by the tables.

For some drugs dozens of metabolites are possible and have frequently been found in various concentrations. In addition, an important oxidative enzyme, MAO, is encountered. Its distribution is more ubiquitous. It is found in intestinal and nerve tissue. Other more specific oxidative enzymes include aromatic hydrocarbon hydroxylases. These oxidize various polycyclic aromatic hydrocarbons and have been related to the carcinogenic nature of some of these compounds. This is a very dramatic example in which oxidative reactions may not result in detoxification but very likely in highly carcinogenic derivatives.

Dehydrogenase enzymes having the ability to oxidize alcohols should be mentioned at this point. Peroxidases and other oxidases are also found outside the microsomal system. These enzymes occur in the soluble fraction of tissue homogenates, as well as in the mitochondria of various types of cells. An interesting iron-containing enzyme, catalase, brings about a very rapid decomposition of hydrogen peroxide. The foaming observed when a 3% solution is poured on a cut as an antiseptic measure is a familiar example of this enzyme's action.

3.3.2. Reductions

Reductions are less commonly encountered than are oxidations, yet they constitute a very important group of biotransformations. An example is the reduction of nitro and azo groups. Nitro- and azoreductase enzymes effect these reactions in the liver. Ketones and aldehydes are reducible to secondary and primary alcohols, respectively. On occasion a carbon–carbon double bond reduction is encountered. Table 3-2 summarizes several of these reactions.

3.3.3. Hydrolyses

Four types of organic compounds are hydrolyzable: esters, amides, nitriles, and ethers. Ethers in biosystems undergo an oxidative cleavage, that is, O dealkylation, which was

Table 3-2. Reductive Drug Reactions

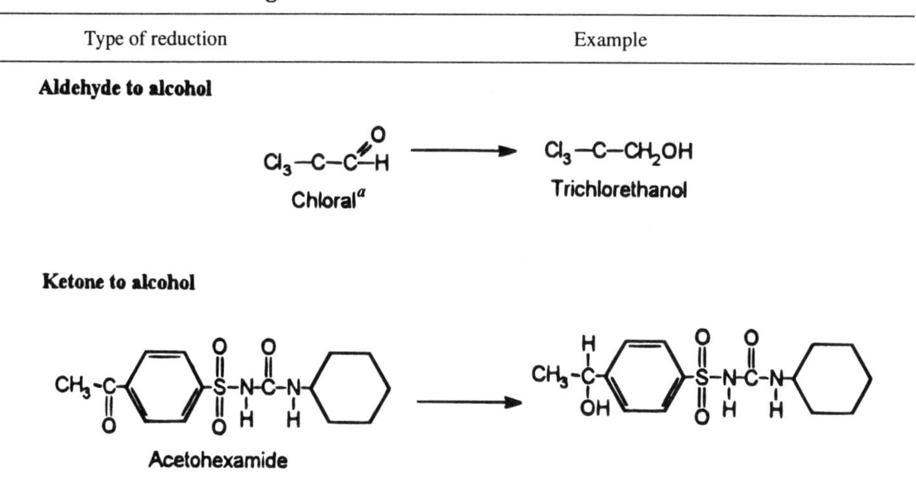

(Continued)

Table 3-2. *(Continued)*

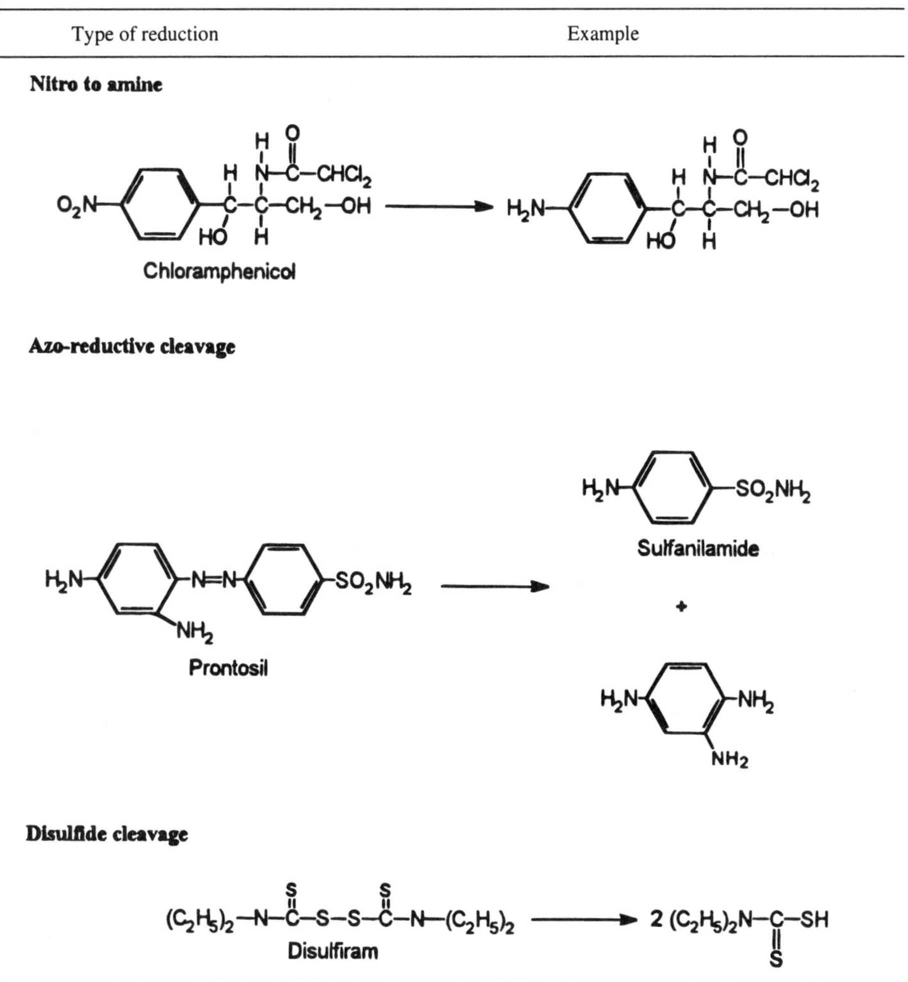

[a] The pharmaceutical form usually used is chloral hydrate.

already considered. Nitriles or cyanides are divided into two categories: simple alkyl and aromatic. Simple alkyl cyanides may result in cleavage to inorganic cyanide ion (CN^-). They exhibit the toxicity characteristic of inorganic cyanide that is metabolized to thiocyanate (CNS^-) by the mitochondrial enzyme rhodanase utilizing thiosulfate ($S_2O_3^{=}$) or β-mercaptopyruvic acid as the sulfur donor. Aromatic nitriles appear to be totally resistant to enzymatic hydrolysis. The aromatic nucleus is instead oxidized in the manner previously described.

Thus the hydrolytic reactions of biological interest, as far as drugs are concerned, are primarily those of esters and amides. In the case of esters, esterase enzymes in the plasma and liver inactivate (detoxify) a relatively nonpolar molecule by splitting it into two polar and therefore water-soluble components, an alcohol and an acid. Analogously amidase enzymes split amides into acids and amines (or ammonia). Table 3-3 gives several examples of drug hydrolyses.

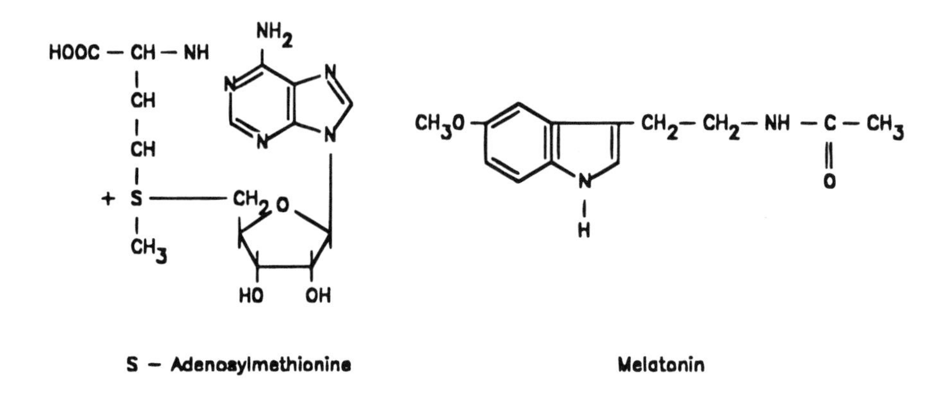

Table 3-3. Drug Hydrolysis

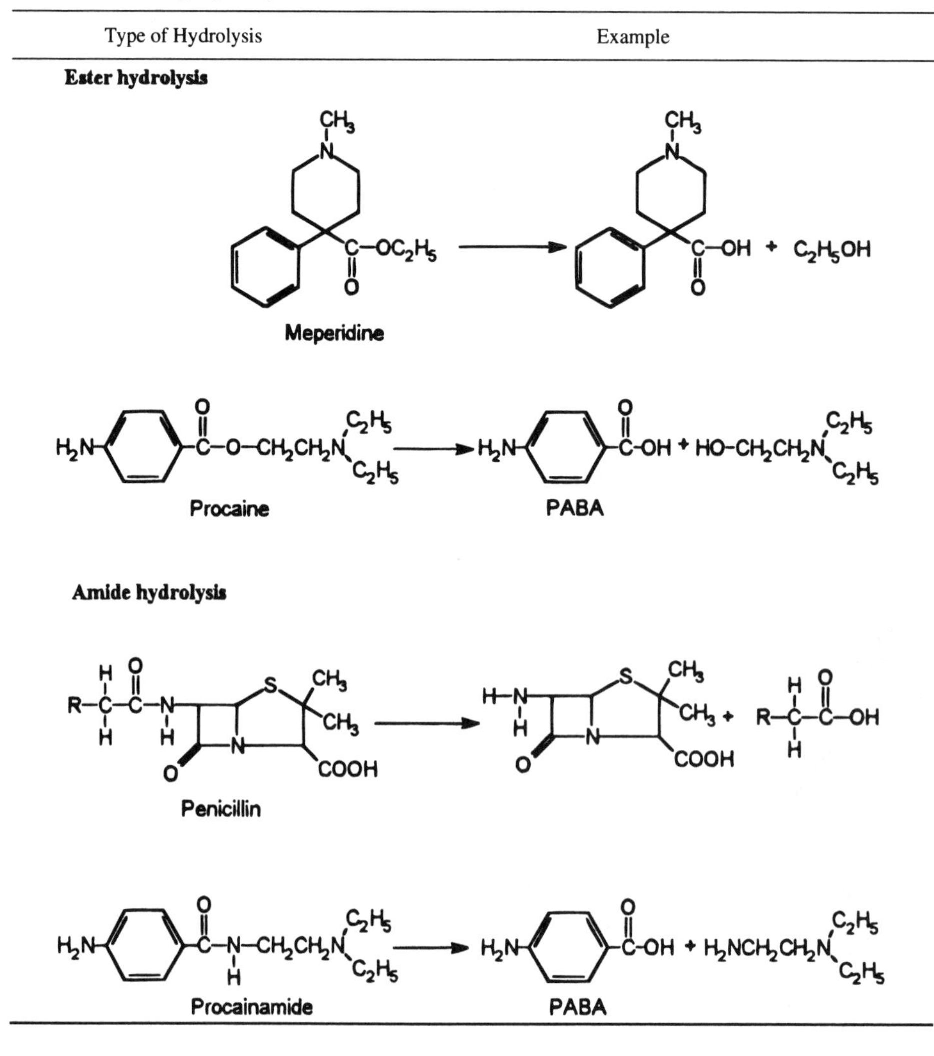

3.4. Conjugation Reactions

Conjugation reactions combine the products of metabolism with certain endogenous constituents, frequently the amino acids glycine and glutamine and glucuronic acid (derived from glucose oxidation). Other conjugates are also encountered. The functional groups that lend themselves to conjugation are carboxyl (COOH), hydroxyl (OH), sulfhydryl (SH), and amino (NH_2). Of course, some drugs already have one or another of these groups and are "ready" for conjugation once absorbed (even though other metabolic reactions on the molecule may produce additional functions for conjugation).

Invariably, the conjugates produced are even more polar and water soluble than are the primary metabolites. For example, glucuronide formation adds three new hydroxyl groups and one carboxyl group. As a result conjugates are of low or no toxicity and almost without exception exhibit no pharmacological activity. The enzymes that catalyze these synthetic reactions are also found mainly in the liver.

The conjugative reactions usually encountered in humans (and other primates) are β-glucuronide formation, sulfate conjugation, glycine conjugation, glutamine conjugation, acetylation, and methylation. Of these glutamine conjugation, primarily encountered with certain aromatic acids, is probably the least general.

The synthesis of glucuronide conjugates is outlined in Figure 3-2. The number 6 carbon of glucose, following formation of uridine-diphospho-α-D-glucose (UDPG), is oxidized by enzymes in the soluble liver fraction to uridine-diphospho α-D-glucuronic acid (UDPGA). In this highly activated form glucuronic acid can now be transferred at the 1 position, with inversion, to the corresponding O-, S-, or N-β-glucuronides, catalyzed by microsomal glucuronyl transferase. A demonstration of C-glucuronidation (in rabbits), involving the acidic terminal ethynyl group of the hypnotic agent ethchlorvynol, has been described.

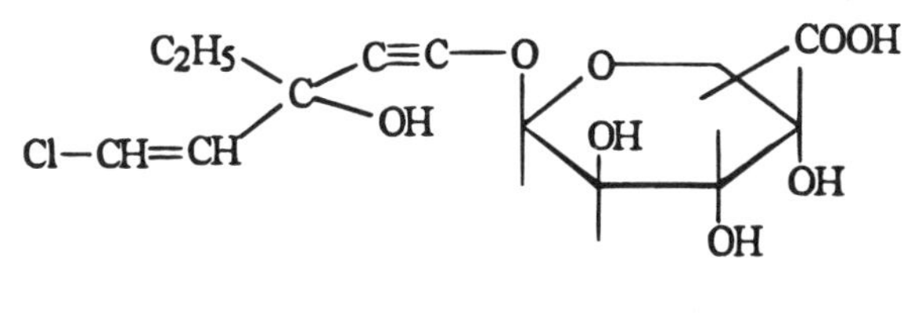

Another conjugative reaction, particularly of phenolic hydroxyl groups, is ethereal sulfate formation. The cytosol portion of liver cells contains enzymes that activate sulfate to a form transferrable to acceptors. Thus 3′-phosphoadenosine-5′-phosphosulfate (PAPS) will sulfate phenols to ethereal sulfates (actually half-esters of sulfuric acid), thus increasing their polarity and excretability greatly (Fig. 3-3).

Conjugation with amino acids (glycine, glutamine) involves the prior formation of coenzyme-A derivatives, which then act as the activated acylating species. The enzymes catalyzing these reactions are found both in the liver and in the gastrointestinal mucosa. The action of acetyl-CoA on amino-function-containing drugs may be viewed as another aspect of acylating conjugation reaction, namely acetylation.

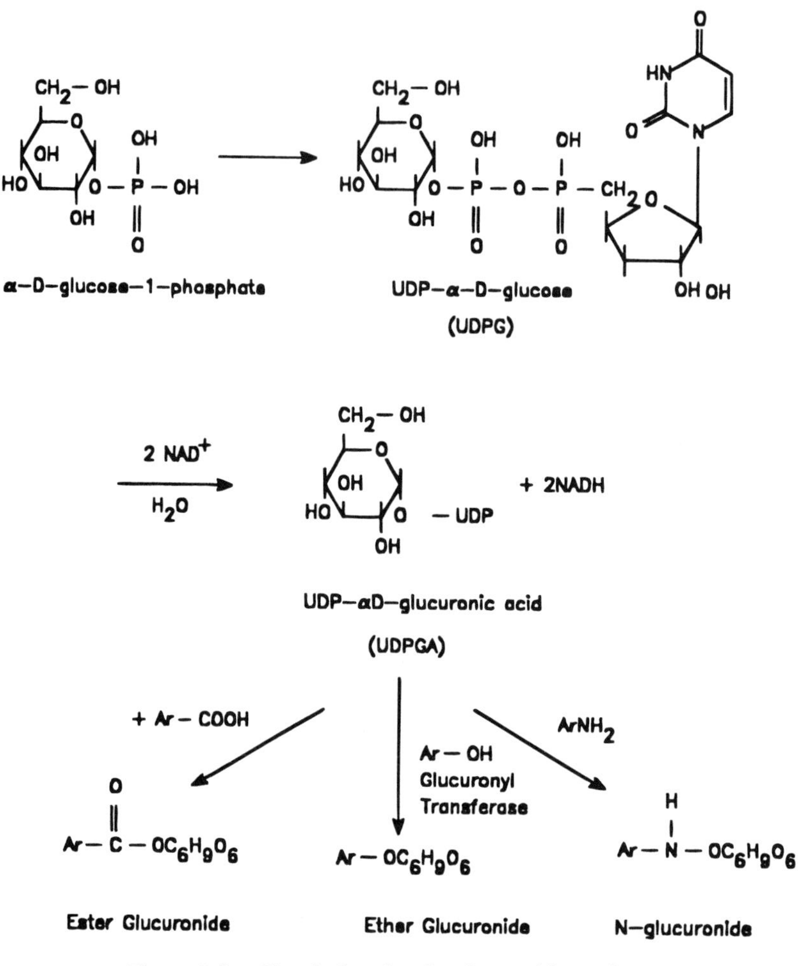

Figure 3-2. Chemical outline for glucoronide syntheses.

Methylation reactions onto oxygen, sulfur, and nitrogen functional groups all utilize S-adenosyl-methionine as the methyl donor, but they are catalyzed by different enzymes in each case. Catechol O-methyl transferase (COMT), a magnesium-dependent enzyme, transfers methyl groups specifically to the *meta* position of catechol derivatives such as epinephrine and isoproterenol (Chapter 9); hydroxyindole O-methyl transferase catalyzes methylation of N-acetylserotonin to melatonin, a pineal gland hormone. Phenyl-ethanol N-methyl transferase is the enzyme responsible for catalyzing the biotransformation of norepinephrine to epinephrine. It will similarly methylate synthetic phenylethanolamine drugs such as the decongestant phenylephrine. Table 3-4 summarizes the more usual conjugation reactions.

The complexity and interrelationships of precursors, bioactive compounds, and their metabolized products can be appreciated by reviewing the biosynthesis and metabolism of catecholamines (Chapter 9). Figures 3-4 and 3-5 illustrate aspects of the metabolic profile of aspirin and chlorpromazine, respectively. The number of metabolites and conjugates possible for chlorpromazine could probably reach 100. Many of them have been identified in humans and experimental animals.

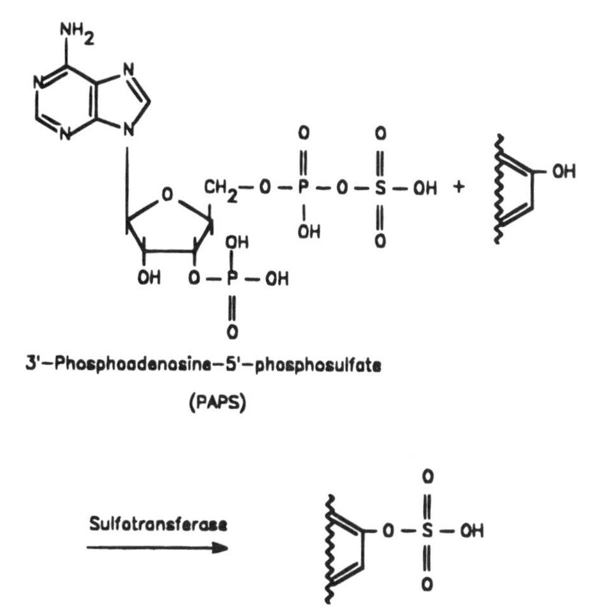

Figure 3-3. Formation of phenolic ethereal sulfate.

Table 3-4. Conjugation Reactions

Type of reaction	Example
Glucuronide synthesis	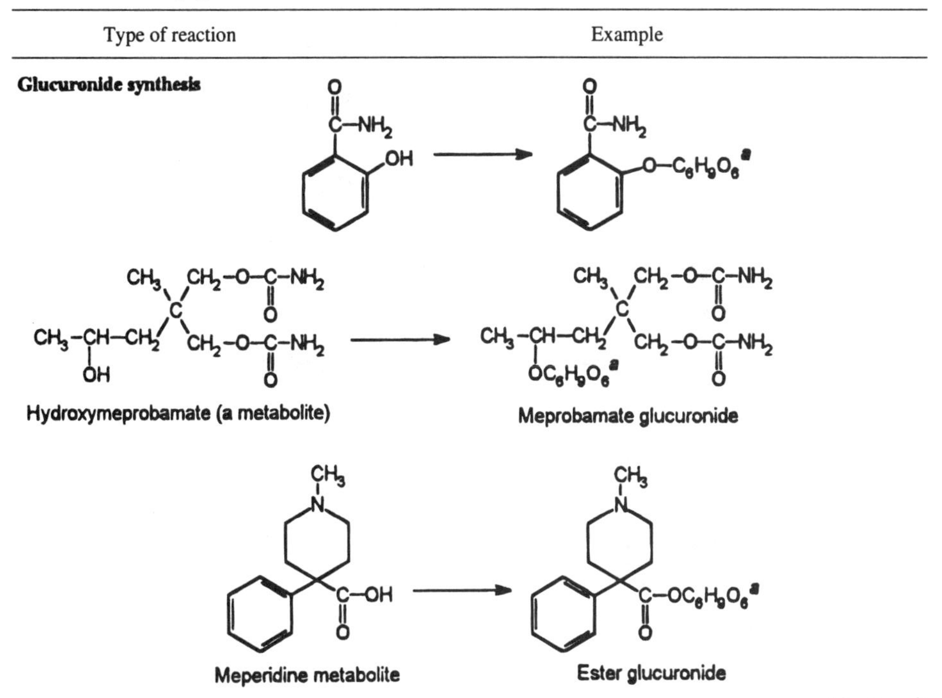

(*Continued*)

Table 3-4. (*Continued*)

Type of reaction	Example

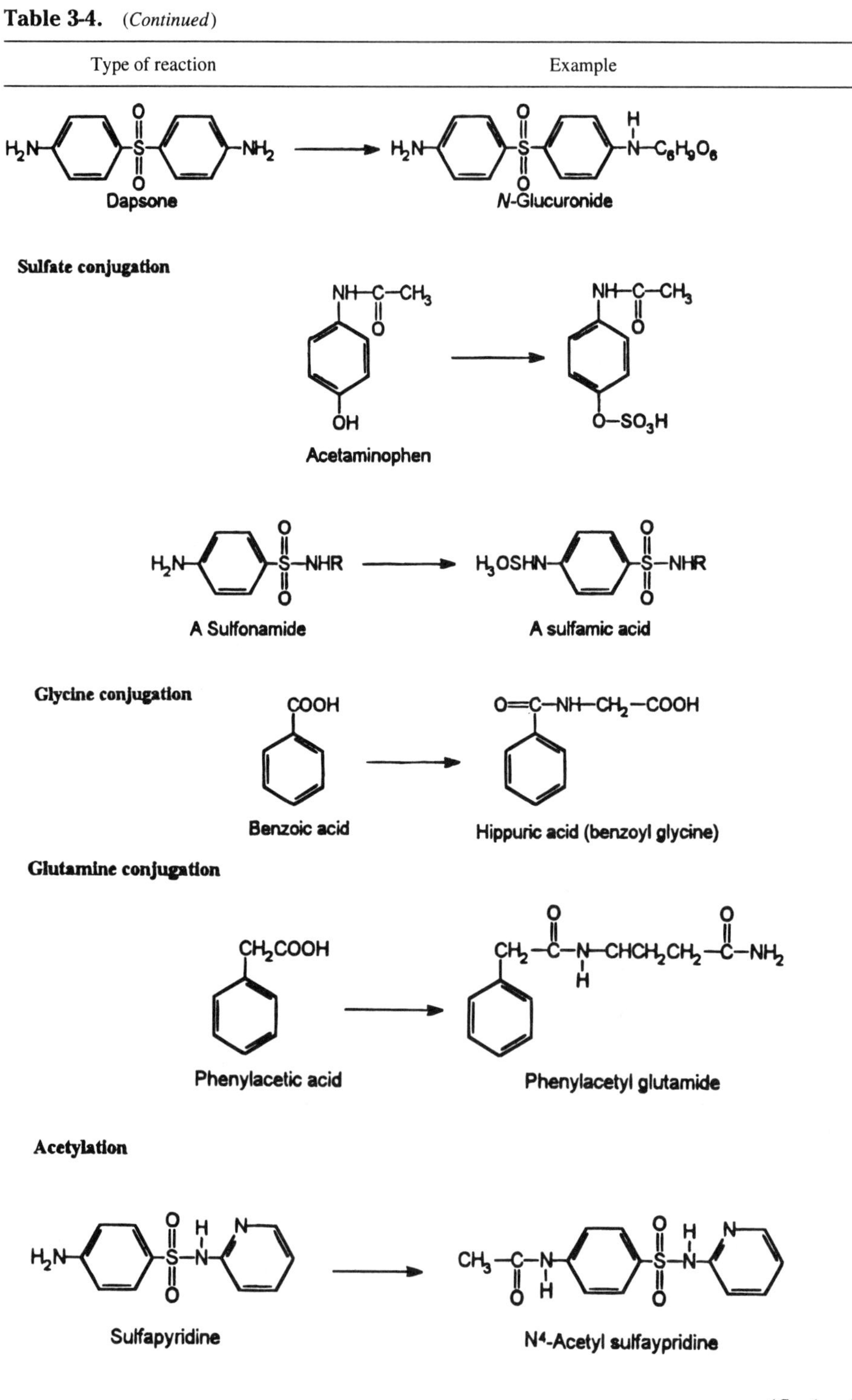

(*Continued*)

Table 3-4. (*Continued*)

Type of reaction	Example

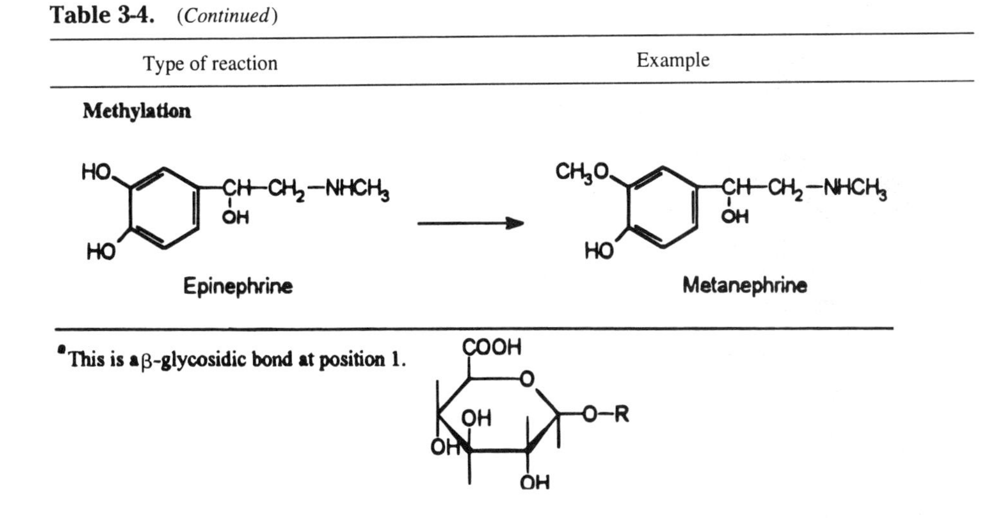

[a]This is a β-glycosidic bond at position 1.

Since most drugs are ingested orally many chemical transformations can and do occur in the intestinal tract before a drug can be absorbed and metabolized further. The great majority of these are catalyzed by enzymes elaborated by intestinal organisms. The gastrointestinal flora of microorganisms is very complex, consisting of at least 60 bacterial species. Thus it is not surprising to find many hydrolyzable functions cleaved in the intestine, including glucuronides, glycine conjugates, ethereal sulfates, amides, esters, and of particular interest, various natural glycosides.

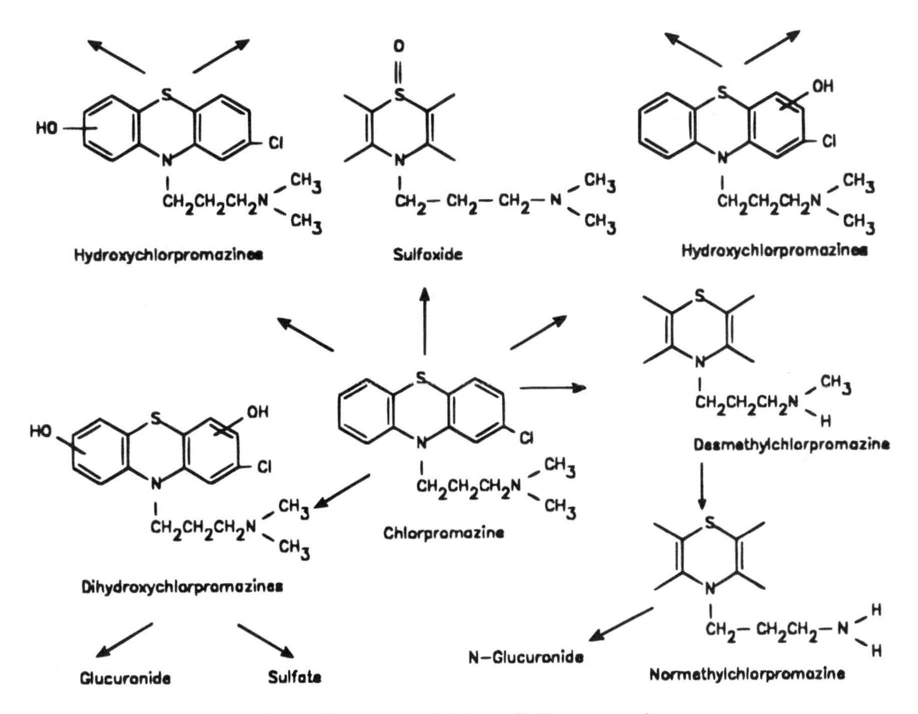

Figure 3-4. Metabolism of chlorpromazine.

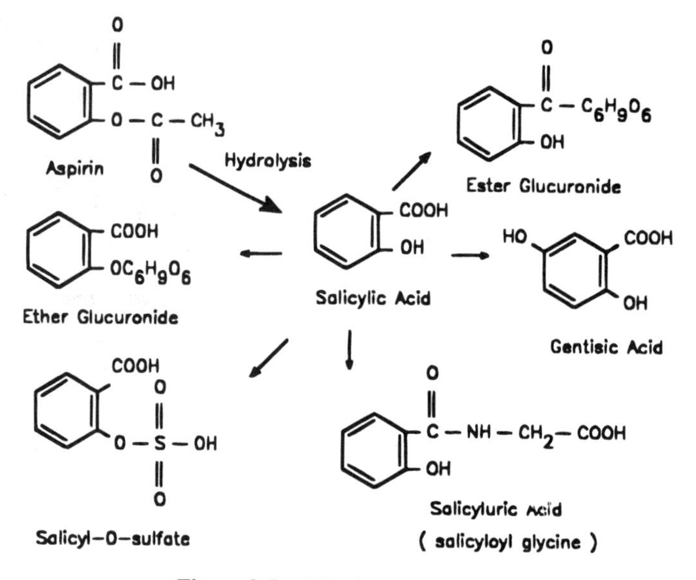

Figure 3-5. Metabolism of aspirin.

For example, it is known that the cathartic glycosides found in senna and cascara sagrada are hydrolyzed in the intestinal tract to the active compound emodin. Therefore, the anthraquinone glycosides are really pro-drugs. In addition, the degree to which the various cardiac digitalis glycosides (Chapter 11) are hydrolyzed in the intestines greatly affects the degree and rates of absorption of these drugs, which may explain the sometimes erratic clinical effects that are observed.[1]

Additional metabolic reactions that have been attributed to the intestinal flora are dehydroxylations (alcoholic), decarboxylations, O-demethylations, nitro reductions, double bond reductions, azo reductions, scission of heterocyclic ring systems, and even aromatization of cyclohexyl carboxylic acids.

In conclusion, those concerned with drugs, be they pharmacists, physicians, pharmacologists, or medicinal chemists, cannot fail to take into account the metabolic fate of a drug molecule once administered to a patient—the ultimate recipient of the benefits of all drug-related research.

[1] Some of these variations in effectiveness, however, have also been shown to be caused by poor drug bioavailability from the tablets as a result of inferior manufacturing procedures.

CHAPTER

4

Anticancer Drugs and Their Mechanism of Action

4.1. Introduction

In this chapter certain aspects of the group of diseases collectively called cancer will be considered. This will aid in an understanding of the problems that exist for the medicinal chemist, the pharmacologist, and ultimately, the clinician. A discussion of the major carcinolytic agents and their mechanisms of actions will then follow.

Cancer is a malignant disease. Among its characteristics is the uncontrolled proliferation of cells, which may be rapid or slow, depending on the particular cancer. Benign tumors may grow rapidly but not be life-threatening. That feature is reserved to cancerous, that is, malignant, tumors. Malignancy is the ability to invade surrounding tissues, resulting in their displacement and ultimate destruction. If that were all that malignancy entailed, it would be highly curable by the intervention of two modalities of treatment: surgical removal of the tumor and radiation therapy. It is the third peculiarity of cancer that accounts for most of the mortality, namely, *metastasis*. Metastasis is a process by which the primary tumor "colonizes" new tumors at sites distant from its point of origin. It now appears that certain malignant cells of the primary tumor (metastatic cells) penetrate the extracellular barrier around the tumor and travel through tissue to either the nearest blood vessel or lymph channel. Then, by secreting basement membrane degrading enzymes, these cells enter the lymphatic system or blood circulation. In the case of blood vessels this is called *intravasation*. Thus, traveling singly or in clumps, these cells now have potential access to almost all parts of the body. Those metastatic cells that survive the rigors of the "trip" then *extravasate* into the new target tissue or organ and begin a new secondary tumor colony—a metastasis.

The clinical dilemma is that by the time a primary tumor is detected and surgically removed, the metastatic process may have already been ongoing for a considerable period of time. In fact, the spread of metastatic colonies is frequently well underway before clinical symptoms of metastasis are evident, in some cases even before symptoms of the primary tumor are manifest. Our ability to detect presymptomatic tumors—primary or metastatic—when they are still extremely small is very limited. For example, a primary lung tumor whose diameter is much less than 1 mm is not detectable by X-rays, yet a tumor of that size already consists of 1 billion malignant cells. Unfortunately, in such a case "early" detection does not significantly alter the prognosis of the disease.

As of now metastases cannot be controlled. Neither can they be detected at a potentially treatable stage in most cases. Even though research in this area is slowly beginning to offer us some understanding of the underlying biochemical processes of metastasis, clinical application of this knowledge is still not feasible.

The view that metastasis is a passive process has now been shown to be false. The requirement of energy, specific surface receptors, and enzymes has been established. Because invasion is an active phenomenon, it presents potentially vulnerable targets for new therapies not yet at hand. Until these are achieved metastasis will remain the main cause of lethality for cancer, since the disease is outside the reach of surgery or local radiation therapy once that stage is attained. We still know precious little about normal cellular growth control mechanisms or their territorial restraint factors. Until this reality is changed our understanding of the behavior patterns of the various types of cancers—and their prognoses—will continue to be empirical.

Various causative factors for cancer have been identified or strongly implicated, comprising genetic disposition, viral infections, radiation, nutritional factors (including deficiencies), and chemical agents.

4.2. Chemical Carcinogenesis

Chemical carcinogenesis (i.e., the production of malignancies by direct application of, or other exposure to, certain chemical substances) has been recognized for many years. However, sophisticated studies of the molecular mechanisms of carcinogenesis have only been undertaken during the past two or three decades. Unfortunately, few definitive answers have emerged.

The picture is a complex one. It is now possible to categorize carcinogens. *Primary* or ultimate carcinogens are those that are biologically active because of their chemical structure. These are usually electrophilic reagents that interact directly and specifically with cellular macromolecules (e.g., DNA). Here, one finds the alkylating agents such as nitrogen mustards, sulfonic esters, ethylene imines (aziridines), unsaturated lactones, chloroalkyl ethers, and peroxides (Fig. 4-1). Paradoxically, this category of alkylating compounds, particularly nitrogen mustards and ethylene imines, contain some of the more effective carcinolytic drugs in clinical use.

Secondary or procarcinogens, which themselves are inert, must first be activated to ultimate carcinogens by host-mediated enzymatic reactions. The host-mediated activation can be by species-specific enzymes. This explained the previously puzzling observations of certain chemicals being carcinogenic in some species and not in others.

A third category of compounds, which apparently are not directly carcinogenic, are *cocarcinogens*. They promote cancer induction by potentiating the effects of ultimate or procarcinogens. Among the best-known cocarcinogens are croton oil, terpenoid principles called phorbols, and their myristate-acetate esters. Application of such promoters to mouse skins previously treated with carcinogenic aromatic polycyclic hydrocarbons speeds up the induction of malignant skin tumors dramatically. It is quite likely that carcinogenic mixtures such as cigarette smoke and coal tars contain relatively greater amounts of cocar-

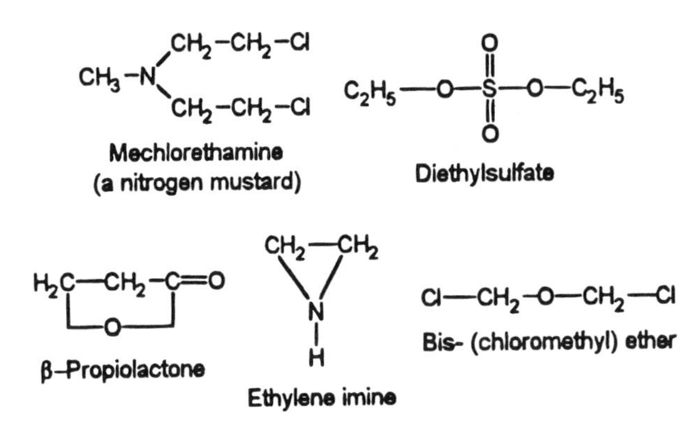

Figure 4-1. Examples of alkylating primary carcinogens.

cinogens than primary carcinogens. Certain endogenous substances such as hormones can help newly formed clones of cancerous cells of organs normally requiring those hormones to grow and develop into tumors. It is well known that testosterone and estrogens, the male and female steroid sex hormones, respectively, are promoters of tumor development in those sex organs whose normal development depends on those hormones. It is now apparent that while some chemicals produce malignancies at the site of contact, the majority of known and suspected carcinogens do so at sites distant from the point of application. Some of the most studied compounds include ethyl carbamate (urethan), 2-naphthylamine, and 2-actylaminofluorene (Fig. 4-2). Of course, natural products are not exempt and also possess cancer-causing properties. Particularly virulent are secondary metabolites of certain microorganisms that are found in the soil or contaminated food, especially if improperly stored. The fungus *Aspergillus flavus* produces a group of toxic polar polycyclic compounds called *aflatoxins*. Aflatoxin B_1 is probably the most hepatotoxic and carcinogenic member of the group (Fig. 4-2). It has been found in peanut products, alcoholic beverages, fruits, cereals, and meat, and has been shown to produce tumors in the liver, stomach, and colon of sheep, pigs, rats, ducks, and even fish when administered in minute quantities. The putative carcinogen produced by AfB_1 is most likely the 2,3-epoxide formed in the liver, since reduction of the 2–3 double bond produces a relatively nontoxic compound.

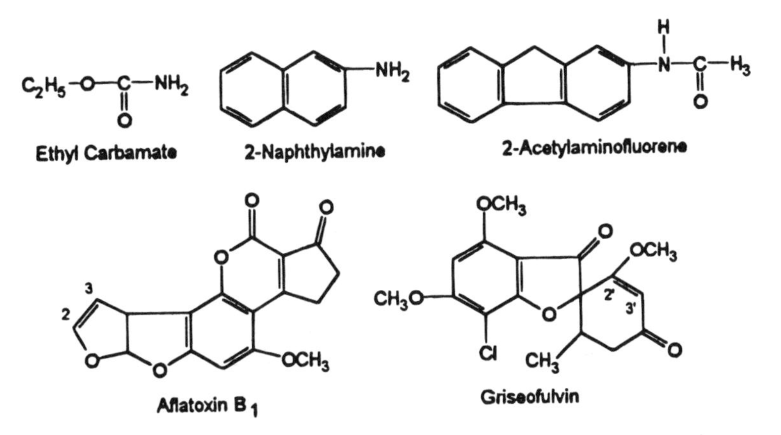

Figure 4-2. Structures of representative carcinogens.

In addition to alkylating agents used to treat cancer that are carcinogens themselves, other types of drugs have been shown to induce cancer in experimental animals. The antifungal antibiotic griseofulvin (a fungal product) has been shown to cause liver tumors in mice. Here, too, the presumed ultimate carcinogen is the 2′,3′-epoxide compound (Fig. 4-2).

The chemically induced carcinogenic process is not fully understood. Cancer is obviously the end result of a series of complex events, each controlled by one or more exogenous and endogenous factors. Whether the carcinogen is primary and an electrophilic species, or must first be converted to one by hepatic metabolism enzymes, it is the electrophiles that are believed to react with specific cellular and molecular receptors, primarily, though not exclusively, with DNA. These interactions are influenced by various stereochemical conditions and competitive inhibitors that have not yet been elucidated. Much of the initial alterations (damage?) to DNA (and RNA) are subject to restoration by enzymatic repair mechanisms. The extent to which this repair fails may represent the formation of abnormal receptors that are now immune to the repair systems. Cells containing them might then represent the early, latent cancer cell. Further growth leads to a tumor. Finally, by an as yet unknown mechanism called *progression* the tumor transforms into an independent undifferentiated malignant neoplasm.

4.3. Cancer and Genetics

Is there a genetic predisposition to cancer? It is now believed that for a normal cell to become malignant, then that cell's DNA must undergo two changes, one of which makes it susceptible to further genetic change. The other alteration, probably a mutation, changes it into a malignant one. It is possible that members of families with a high incidence of a certain cancer may already carry the gene for one of the changes needed to become tumorous. Genetic susceptibility to cancer development varies between individuals. Such a predisposition may be significant in some types of tumors, but not in others. It has been demonstrated that in families with a high incidence of lung cancer genetic susceptibility is important, but probably not more than whether the subject smokes.

4.4. Cancer and Nutrition

Many dietary substances are known to produce cancer. Alcohol is an excellent example of one in common usage. In people who both drink and smoke the risk is considerably magnified. The mouth, esophagus, and pharynx are the most frequent sites. These are the areas of most likely contact for both substances. The liver, of course, is also exposed to alcohol, where it is metabolized. One of seven heavy drinkers develops cirrhosis of the liver. It may be no coincidence that most primary liver cancers in the United States originate in cirrhotic livers; the remainder may be virally induced. Unfortunately, the list of carcinogens occurring in many foods naturally, as additives, and as a result of storage and even cooking processes, is large and growing. Sodium nitrite, a precursor of carcinogens (nitrosamines), occurs in many foods such as vegetables naturally, and in meat products as an additive. Even where foods contain nitrate rather than nitrite, it is secreted into the saliva following absorption, where mouth bacteria readily reduce it to nitrite. Intestinal bacteria additionally produce nitrite from available ammonia.

Anticarcinogens that either directly antagonize carcinogens, or more likely prevent their activation, are also present in many foods naturally or are added to them. Included in this list of "good guys" are vitamins A, E, and C, some Bs, chlorophyll, carotene, butylated hydroxytoluene (BHT), and anisole (BHA). Many of these substances are effective antioxidants that presumably may inhibit oxidative carcinogenic activation. Vitamin C is also a potent inhibitor of nitrosamine formation.

4.5. Radiation

Radiation impinges on us from the environment as background radiation, from radionuclides in the ground and water, and from cosmic radiation. Medical irradiation contributes a small but presumably significant fraction. By far the largest and most potent natural carcinogen is the ultraviolet component of sunlight.

The carcinogenic potential of radiation is due to its action on cellular oxygen, which initially forms an oxygen free radical, $O_2^{\bullet}$. Its reaction with cellular water produces hydroxyl free radicals, $OH^{\bullet}$. It is the OHs that attack DNA, inducing carcinogenesis. An alternate scenario suggests an indirect mechanism whereby radiation, by causing disruption of cellular defenses, permits the expression of the oncogenic potential of certain viruses.

4.6. Viruses and Cancer

The viral theory of cancer induction can probably be traced to the beginning of this century, when Rous showed that cell-free extract of chicken sarcoma could, when injected into healthy chickens, produce the same tumor. It was later discovered that tumor viruses injected into newborn animals having low immunological resistance produce tumors. Since then there have been many reports of viruses isolated from a large variety of animals that cause predictable tumors in different organs when injected into animals of the same or closely related species. Human adenovirus was shown to cause solid tumors when inoculated into newborn hamsters. Experiments based on Koch's famous postulates have repeatedly demonstrated viral causation for certain cancers—in animals. Since such experiments obviously cannot be carried out on human beings, definitive proof that any human cancer is caused by virus exposure is extremely difficult to establish. Indirect "proof" can frequently have an alternate explanation such as viral infection of existing tumors. However, more and more definitive evidence is accumulating for certain cancers. A fascinating example linking a virus and cancer is Burkitt's lymphoma, a tumor affecting African children living in high-rainfall, hot areas—a climate conducive to malaria. It was hypothesized that the lymphoma was produced by a mosquito-transmitted virus. It has since been shown that 99% of African patients with Burkitt's lymphoma have tumor cells marked by Epstein–Barr virus (EB), while in the United States, where the disease is very rare, only 10% of the patients show the EBV antigen. Other major human malignancies where evidence for viral involvement is very strong are nasopharyngeal cancer, Hodgkin's disease, cervical cancer, and certain leukemias. Many oncogenic (tumor-causing) viruses are now recognized and are being intensively studied. It is believed that such viruses integrate their nucleic acid (RNA or DNA) into the host cell's genetic material, bringing about a malignant transformation and continuing the infection by the release of newly synthesized viral particles.

Oncogenic viruses (oncovirus) (i.e., viruses known or believed to cause cancer) are now an accepted fact. Among oncogenic DNA viruses are the papavoviruses, of which the simian virus 40 (SV40) is one the most studied. Also, some adenoviruses and herpes viruses are viable suspects. Following infection of a cell, such a tumor virus integrates into its genome, resulting in a cellular transformation. The existence of oncogenic RNA viruses is also known. The Rous sarcoma virus, producing bird and probably some mammalian malignancies, is the best known. It, and other members of the group of *retroviruses*, contain in their virion, a DNA polymerase that is RNA-directed, usually referred to as *reverse transcriptase*. It should be understood that not all such RNA viruses are carcinogenic, nor are they necessarily otherwise pathogenic—but some are.

A "hot" topic in cancer research is *oncogenes*. These are believed to be genes that once activated can transform a normal cell to a cancerous cell. Such genes are in human cells, and they are probably very close (if not identical) to normal genes. The "oncogene" theory was first proposed by Huebner and Todaro (1969). Two possible mechanisms are presently considered viable and operative. Both require activated oncogenes to appear to begin the process of carcinogenesis. One is that the oncogenes are brought to the normal cell by infecting viruses, most likely RNA-containing retroviruses. The other mechanism supposes that the oncogene is already present in the healthy mammalian cell in an inactive state, possibly even an essential part of that cell's DNA. An activation produced by a mutation or by a chemical carcinogen then initiates the process.[1]

4.7. Cancer Chemotherapy—Special Problems

Certain advantages in the chemotherapy of infectious diseases (bacterial, fungal, or parasitic) do not exist to aid in the treatment of cancer. Most important, the concept of selective toxicity is usually not operative. The often significant biochemical differences between the cells of the host and the invading organism simply do not exist between the cancerous and normal cells of the same tissue. For example, in the case of sulfonamides a biochemical pathway that is nonexistent in mammalian cells is blocked in susceptible bacteria (Chapter 2). Certain antibiotics have a deleterious effect on bacterial cell walls, or a biochemical pathway that mammalian cells do not even have (Chapter 6). Such dramatic variations enable the medicinal chemist to find or synthesize compounds with selective inhibitory capacity. Ordinarily, however, there are no such biochemical or morphological standards from which a selectively toxic carcinolytic agent can be synthesized. Thus the great majority of anticancer drugs bring about a general nonselective interference in cellular processes and cell growth. In fact, many of these cytotoxic agents are almost equally detrimental to normal and neoplastic tissues. Therefore, it is not at all surprising that these drugs produce serious and often debilitating side effects. The usually high doses of these cytotoxic agents that are needed to arrest tumor growth or to obtain a temporary remission of symptoms will also depress bone marrow functions, causing drastic reductions in the number of platelets, leukocytes, and lymphocytes, which in turn lead to greater susceptibility to infections and internal bleeding.

Another advantage manifested in the treatment of infectious disease but not in cancer chemotherapy involves the immunological response of the patient's body to foreign substances. In infectious diseases the drugs utilized are often only bacteriostatic. The body's

---

[1] For an excellent overview of the fundamentals of oncology, see Pitot (1986).

defense mechanisms such as antibody production and phagocytosis then "mop up" the halted invaders. With several exceptions this does not occur with tumors. Many carcinolytic agents are actually immunosuppressants that would eliminate antibody production in those patients capable of producing them. They thus suppress the patient's ability to fight simultaneous infections that may then become the cause of death.

In addition to chemotherapy and surgery, the third method of treating neoplastic disease is radiation therapy. It may be in the form of an outside source of radiation aimed at a tumor, or be administered as a chemical containing radioactive atoms such as iodine (^{125}I, ^{131}I), phosphorous (^{32}P), and others.

In spite of the triple-systems attack—surgery, radiation, and chemotherapy—possible on most types of cancer, with several exceptions, the prognosis for cures is generally not favorable. Relapses from symptom remissions frequently with metastasis and development of resistance to drugs are common. Table 4-1 lists the estimated percentage of 5-year survivors who are apparently free of commonly encountered cancers.

We can synthesize compounds that are designed to interfere with biochemical steps within the cell that are known to be important to its survival. However, currently available drugs have not been able to capitalize successfully on known or still undiscovered biochemical differences between neoplastic and normal cells.

An unusual aspect of cancer is that it is very difficult to consider it as a singular disease entity. Some are rapidly fatal; some only after prolonged periods of time. The rate of incidence of most malignancies increases with age, rising steeply after the fifth decade of life. Yet certain cancers are peculiar to childhood (acute leukemia) and adolescence (Ewing's sarcoma).

Table 4-1. Survival Rates of Treated Cancer

Types of cancer	Survival rates[a]	New cases[b]
Thyroid	93%	10,600
Testis	86	5,000
Uterine lining	85	37,000
Melanoma	80	22,000
Breast	74	119,900[c]
Bladder	73	40,000
Hodgkin's disease	73	6,900
Prostate	70	86,000
Cervix	67	15,000[c]
Colon	52	96,000
Kidney	50	19,700
Rectum	49	42,000
Non-Hodgkin's lymphoma	48	26,500
Ovary	38	18,500
Leukemia	32	49,200
Brain	23	13,700
Stomach	16	24,700
Lung	13	144,000
Esophagus	5	9,400
Pancreas	2	25,200

[a] Five-year survival rates of cases diagnosed during 1976–1981.
[b] Estimated new cases for 1985.
[c] Invasive only.

Sources: National Cancer Institute, American Cancer Society.

The question of what is a cure is not as straightforward as with other diseases. One definition frequently applied to cancer is the nonrecurrence of symptoms for 5 years from the time of initial diagnosis. It is possibly less arbitrary to consider patients cured if death occurs at a rate similar to that of the general population. The Cancer Registry of Norway[2] shows that mortality rates of male lip cancer, after diagnosis, is almost identical to that of the male population in general and is therefore very "curable." However, esophageal cancer in men is rapidly fatal (less than 4% survival after 3 years) and thus essentially not curable.

Many childhood cancers that were rapidly fatal if untreated (e.g., leukemias) today have a high apparent cure rate with suitable therapy, that is, they have remission of symptoms exceeding a decade or more, which may be permanent. Results are less dramatic with older children and adolescents. In the adult population, however, chemotherapy results are not as clear-cut. In fact, in some cases the postsurgical survival rate of patients treated with chemotherapy was the same as that of patients not so treated.

One of the basic tenets of cancer chemotherapy is that all neoplastic cells be eradicated if a cure is to be achieved. Whether the modality used is surgery, radiation, chemotherapy, or, most often, a combination of these, *total cell kill* is the objective of treatment. It has been demonstrated in mice that a single leukemia cell can proliferate to kill an animal. Skipper et al. (1964) showed that after a two-day lag period a single L_{1210} leukemia cell in a mouse proliferated by doubling every 13 hours to 10^9 cells in 19 days, killing the rodent. It was shown that inoculation of larger number of malignant cells into an animal will decrease its survival time in an inverse manner. Skipper et al. (1965) demonstrated that treatment with cytostatic agents increased the host's life span and postulated that this was due to the cell-killing effects of the drugs. With facts such as these it can be seen why total cell kill appears to be necessary for a total cure.

An examination of Figure 4-3 further illustrates the problem. When a tumor of 1 mm^3 is first detectable (30 cell doublings), it already contains 10^9 cells and has completed over 66% of its growth. By this time many tumor cells have shed into the lymph or circulation, some of which undoubtedly already having initiated the metastatic process previously described. Only five more cell doublings (35 total) will produce a 32 cm^3 tumor (1 $foot^3$). An additional five doublings would produce 1 kg mass, a size few hosts are likely to survive.

Can total eradication of malignant cells be achieved? Is it approachable? Would 99.99% be good enough? A given dose of cytotoxic drug kills a fixed percentage of cancerous cells, not a given number. The mathematics is that of first-order kinetics (i.e., total cell kill is approachable, but not achievable). For example, if a patient's tumor containing 1 trillion, 10^{12}, cells with an approximate mass of 1,000 grams were given a drug capable of killing 99.99% of the cells, then it can be calculated that a 100 mg tumor would remain. This would clinically appear as a dramatic remission. The surviving 100 million cells (10^8), however, can, and often do, bring about a relapse and (at 10^{11} cells) ultimate death.

Cytotoxic agents of clinical interest are necessarily those with a high logarithm cell kill ability. Thus a drug that can reduce a cell population of 1 million (10^6) to 10 cells (10^1)—5 log kill—should be capable of decreasing 100,000 cells (10^5) to one cell. This fractional cytocidal effect means that the larger the dose of drug that can be administered, the greater the potential for cell kill. The frequently insurmountable obstacle this presents is that achieving the desired result would require massive doses whose toxicities would them-

---

[2] A country having kept detailed cancer statistics since 1945.

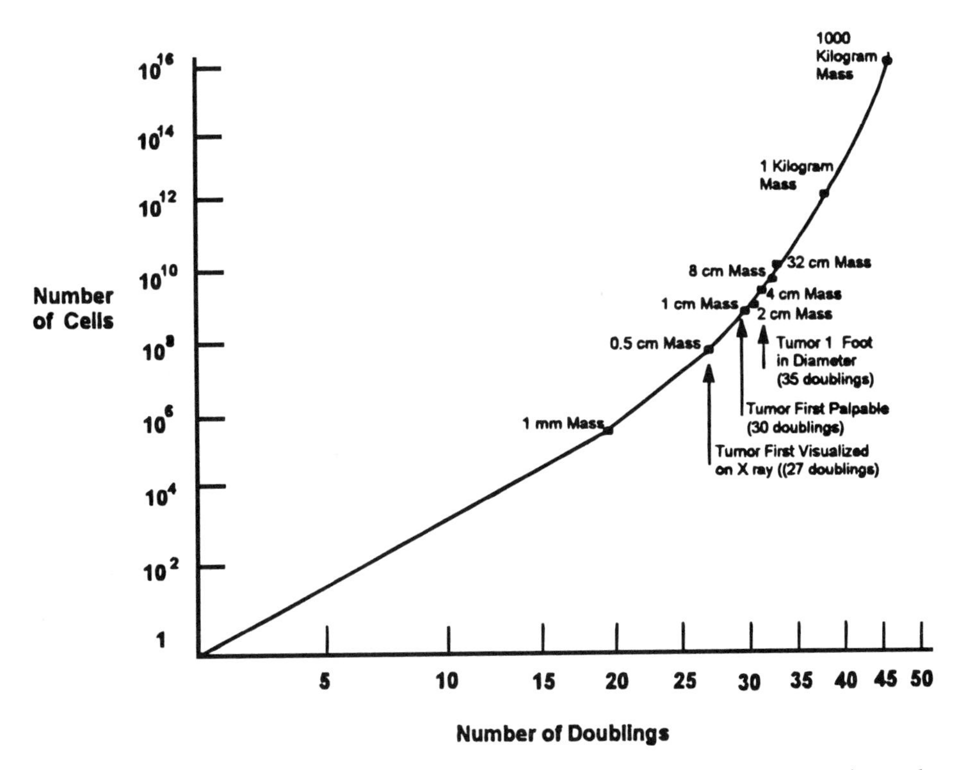

Figure 4-3. Theoretical growth curve relating the number of tumor cells doublings to the number of tumor cells and levels of clinical awareness of tumors (De Vita et al., 1975).

selves contribute to patient mortality. Administering subtoxic doses, even over long periods, would be useless against rapidly proliferating cancer cells.

It is obvious that often when cancer is first diagnosed it is, in terms of cell mass, already at an advanced stage. Even when surgical removal or radiation treatment appears to eliminate a tumor completely there may be millions of cancerous cells circulating in the blood and lymph. In fact, most patients are probably already metastatic at this point.

Are there ways around this dilemma? Ideally, yes. Begin treatment when the tumor is still very small and destruction might be achievable with less toxic doses. However, this would require diagnosis at an unachievable early stage.[3] A more realistic and successful approach is the utilization of drug combinations. If two drugs were given, each with a 50% cell-kill capacity, then, theoretically, a cure should be achievable. The method of multiple drug use (sometimes as many as seven) and their administration in proper doses and timed cycles has greatly advanced the art and science of cancer chemotherapy over the past three decades. In fact, the partly empirically developed therapy protocols, involving effective drug combinations and the sequence and timing of their administration, has been a major reason for the improved picture of cancer therapy. The rationale for combination chemotherapy is to use two or more agents that (1) are individually active against the tumor, (2) exhibit different levels of toxicity, (3) act by different mechanisms, each with its

[3] However, see discussion on monoclonal antibodies following.

own cell-killing ability, and (4) may block two or more points in a given biochemical pathway and thus have an additive, or even synergistic, effect by attacking multiple sites.

4.8. Drug Resistance

Cancer patients who relapse following initial drug-induced remission frequently may not respond to a second round of chemotherapy with drugs that destroyed tumor cells before. The cancer cells that survived the first "battle" have somehow developed resistance to the drugs by various mechanisms. Resistance can be said to arise ultimately by *natural selection*, where the survival of the few naturally resistant cells or organisms (mutants?) in a given population can then multiply in the presence of the drug that killed the bulk of the cells. The probability of a cell or organism being resistant to two drugs, each having a different mechanism of action, has been estimated at $1{:}10^{14}$. Thus concomitant treatment of cancer (or microbial infections) with more than one agent decreases the probability of resistance developing, or at least may postpone its arrival.

Another method by which resistance can emerge is by *gene amplification*. Cells multiply the number of genes that code for the substance the drug inhibits. This is the case in the resistance to methotrexate (MTX) of human leukemia, where the enzyme dihydrofolate reductase (DHFR) is the drug's target. It was found that drug-resistant cancer cells of murine sarcoma and lymphoma have as many as 200 DHFR coding genes per cell, while drug-sensitive cells only have two. The evidence indicates that this gene amplification is actually induced by MTX. Other examples are now known.

The reverse phenomenon, decreased enzyme synthesis, can also be the mechanism of drug resistance. The antimetabolite pro-drug 6-mercaptopurine (6MP) is activated to its nucleotide by inosine-5′-phosphate pyrophosphorylase. The enzyme is deleted in resistant neoplastic cells. Resistance to 5-fluorouracil similarly develops by deletion of the enzyme converting this pro-drug to its active nucleotide. A mechanism of resistance by which a drug is excluded from its site of action can also be operative. This has been established for tetracycline antibiotics. Here the permeability of the cellular membrane to the drug is altered so that it cannot penetrate and accumulate within the target cell. Similarly, it has been demonstrated with such a membrane modification in MTX-resistant leukemia cells in mice.

4.9. Drug Discovery Strategies

A National Cancer Chemotherapy Program was established by the National Cancer Institute (NCI) in 1955. Its purpose was to discover and develop new antitumor agents. The system begins with the acquisition of new compounds, which are then screened in animals, produced and formulated, toxicologically evaluated, and then phased into a series of human clinical trials. Between 1955 and 1982 well over 700,000 substances had been evaluated. Table 4-2 summarizes the types of compounds and extracts involved in this prodigious effort. As large as this number is, efforts of the pharmaceutical industry, research institutes, and universities in the United States, Europe, and other parts of the world must be added to it. Even though the work was largely empirical in the earlier years, it has been put on an increasingly more rational basis as our concepts and methodologies advanced. Developments in enzyme inhibition by suicide inhibitors and transition-state analogs come to mind (Chapter 2). In spite of this extraordinary effort, probably no more than 40 carci-

Table 4-2. National Cancer Chemotherapy Program: 1955–1982[a]

Substances Tested	Numbers Screened
Synthetic compounds	350,000
Fermentation products	200,000
Plant products (extracts)	120,000
Marine animal products	16,000
Biological response modifiers	—

[a] Adapted from Frei, 1982.

nolytic agents with substantial clinical antitumor activity have been discovered, and only several have been marketed.

4.10. The Cell Cycle

Before considering the various carcinolytic drugs, it is important to have some understanding of the cyclical nature of cell growth. It is then possible, at times, to relate drug action to the phases of the cycle so that drug dosing, sequencing, and combinations might be hoped to be put on a less empirical basis. Figure 4-4 represents the fundamental aspects of the cell cycle.

After the cell has divided, there follows a period of relative dormancy. The cell can be considered passive here (i.e., it is not chemically preparing for cell division). This interval corresponds to G_1, with the G_0 subphase probably representing near total inactivity. This is followed by the S phase, in which DNA is actively synthesized. Next is the premitotic phase, G_2, during which synthesis of the cell's proteins and RNA takes place (although RNA synthesis also occurs in G_1). Last in the mitotic phase, M, the cell goes through the stages of mitosis—prophase, metaphase, anaphase, and telophase—culminating in cell division, with the chromosomes having separated and each new daughter cell proceeding to G_1–G_0.

It would be desirable to know the proportion of cells in each phase so as to be able to use the drug(s) most effective at a particular stage. At present this cannot be exactly determined. However, by microscopic examination (mitosis), or by the use of radioactively

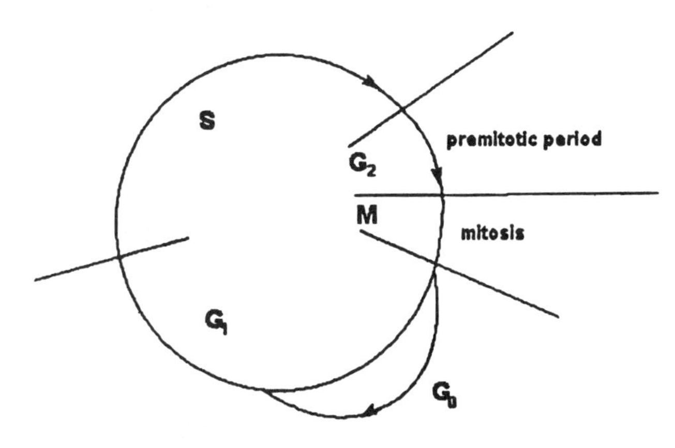

Figure 4-4. Life cycle of cell.

tagged precursors, some crude estimations can be made. It can be logically assumed that the stage of tumor growth is the major determinant of this proportion. Thus it is reasonable that a majority of the cells of a malignancy will be in the S phase during the logarithmic growth stage, while a larger proportion of cells will be in the G_1 or G_0 phase during the later stages of the cancer. The same reasoning also applies to a rapidly proliferating disease such as Burkitt's lymphoma, where a tumor can double in size in one day, as opposed to a slowly growing colonic cancer that may require 80 days to double. It is now known that proper clinical staging of a malignant disease, when combined with a knowledge of cellular phases, cellular kinetics, and the mechanisms of action of the various drugs, leads to vastly improved therapeutic responses. Even though synchronization of cells in a tumor may not be feasible, it may still be possible to administer drugs known to be active at a particular phase, if it can be approximated when cell proliferation peaks within a cycle. This type of *chronotherapy* is now being attempted.

It must be kept in mind that in healthy tissue as well as in tumors there are at any given point in time a definite number of viable cells that can, but do not, proliferate, at least over a period of just hours, if not days. Such a subpopulation of cells will not be significantly affected by those drugs said to be *cell cycle specific* [i.e., those affecting DNA synthesis (S phase) or mitosis (M phase)]. In the case of cancer, these cells could then regenerate a tumor following therapy. The goal of treatment therefore must be to destroy these "resting" tumor cells as well. There exist, then, *cell-cycle-nonspecific* agents that are cytotoxic at all phases of the cell cycle. For this reason cancer chemotherapy protocols usually include drugs in both categories.

At this point it would be helpful briefly to review the steps a cell carries out in the synthesis of cellular components. Beginning from simple precursors such as formic acid, carbon dioxide, and certain amino acids, the purine and pyrimidine building blocks of the nucleic acids are synthesized. These combine with ribose sugars to form nucleosides. Derivatization with phosphoric acid yields nucleotides that polymerize to form the three types of RNA designated as transfer, messenger, and ribosomal RNA. Alternately some nucleotides are reduced to deoxyribonucleotides that polymerize to afford DNA. Through a series of complex steps DNA, which has the genetic messages encoded in it, does transcribe this message to messenger RNA, which in turn translates it to the ribosomes of the cell, where proteins are synthesized based on the code. These proteins are the multitude of enzymes, structural components, and microtubules of the cell. Chemicals are known that can inhibit any of the steps outlined, at least experimentally. Some of these have become clinically useful drugs. Figure 4-5 outlines the steps and gives examples of drugs that affect them.

4.11. Alkylating Agents

Although some ancient Egyptian writings describe the use of crude drugs to treat ulcerating skin tumors, modern cancer chemotherapy is about five decades old. Just before World War II, it was found that mustard gas, and various derivatives, are highly cytotoxic, espe-

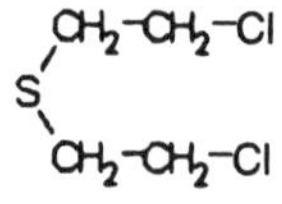

$S(CH_2-CH_2-Cl)_2$

Mustard gas

$R-N(CH_2-CH_2-Cl)_2$

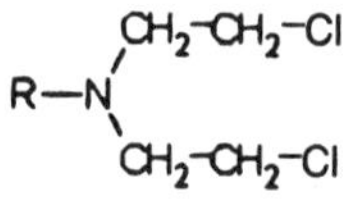

Nitrogen mustard

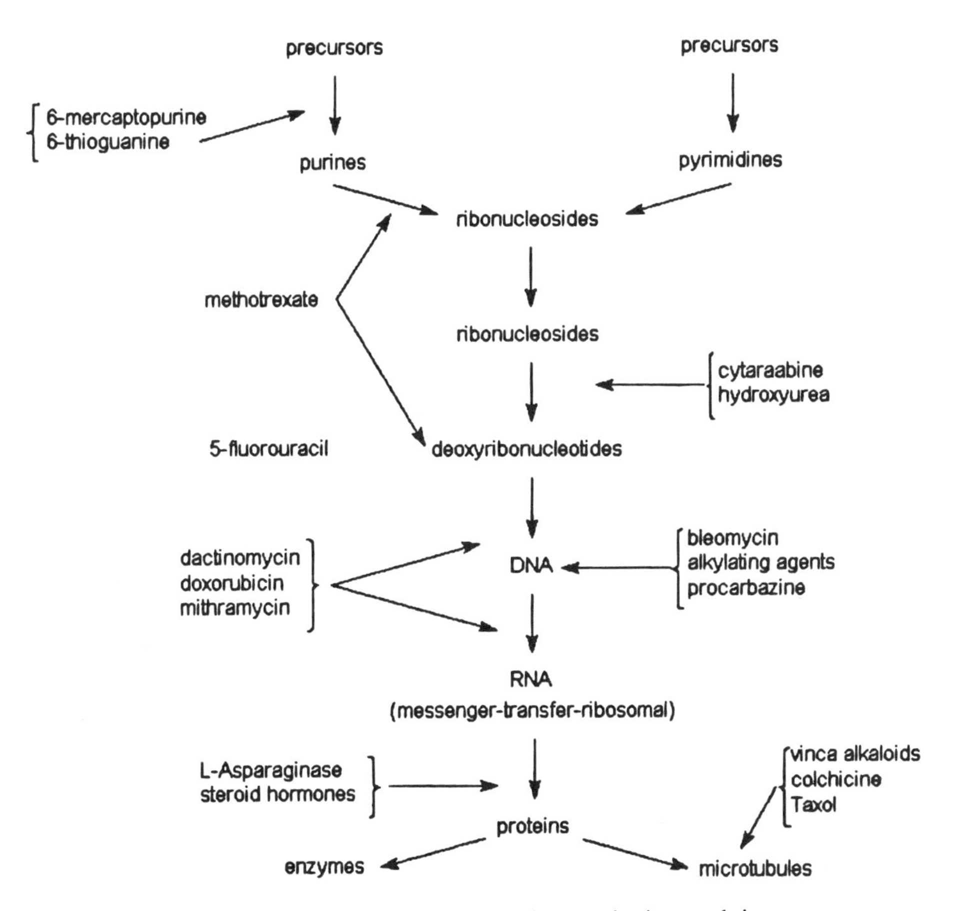

Figure 4-5. Summary of anticancer drug mechanisms and sites.

cially to cells in the blood, lymph, bone marrow, and mucous membranes of the gastrointestinal tract. Nitrogen analogs called nitrogen mustards were initially found to give somewhat favorable results in the treatment of Hodgkin's disease and later in the treatment of acute leukemias in children.

Various derivatives (obtained by structural variations of the *R* group in nitrogen mustard), as well as chemically unrelated organic compounds, were able to bring about temporary remission of clinical symptoms. Once the fact that nitrogen mustards were alkylating agents was appreciated, other types of structures potentially capable of alkylation were developed. There are three major types of alkylating compounds in clinical use—nitrogen mustards, ethylene imines (aziridines), and esters of methane sulfonic acid.

To understand better the mechanism by which alkylating agents in general and nitrogen mustards in particular bring about lethal cellular damage, it is useful to review briefly the

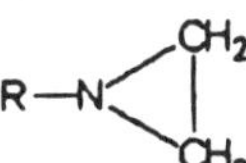

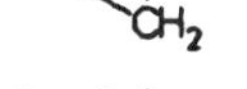

$CH_3-\overset{O}{\underset{O}{\overset{\|}{\underset{\|}{S}}}}-O-R$

Ethylene imines

Esters of methane sulfonic acid

concept of alkylation reactions that fundamentally involve one of two types of nucleophilic substitutions (i.e., an electron-rich center on a molecule, usually a nitrogen atom, displacing an electronegative atom such as a covalently bonded halogen). There is a unimolecular nucleophilic substitution (SN_1) and a bimolecular nucleophilic substitution reaction (SN_2).

The SN_1 mechanism is a two-step sequence of events as follows:

$$R\text{–}X \rightarrow R^+ + X^- \quad (4.1)$$

$$R^+ + Y^- \rightarrow X\text{–}Y \quad (4.2)$$

The first reaction, which is usually the slower reaction and therefore the rate-determining step of the sequence, results in the formation of a carbonium ion, which, because of its high degree of instability, will very rapidly collapse with (i.e., alkylate) a nucleophilic, electron-dense center such as an anion Y^- (Eq. 4.2).[4] The total rate of the reaction does not depend on the nature or concentration of Y.

In an SN_2 mechanism, substitution (alkylation) takes place in a bimolecular one-step process as follows:

$$RX + Y^- \rightarrow RY + X^- \quad (4.3)$$

The rate is dependent on the concentration of both RX and Y^- (hence bimolecular). The more nucleophilic (electron-rich and usually also basic) Y^- is, the more efficient this process is likely to be.

It can be shown that aromatic nitrogen mustards (compounds containing an aromatic ring) usually alkylate via the SN_1 mechanism.

$$\text{AR–}\underset{}{\overset{\text{R}}{\overset{|}{\text{N}}}}\text{–CH}_2\text{–CH}_2\text{–Cl} \longrightarrow \text{AR–}\overset{\text{R}}{\overset{|}{\text{N}}}\text{–CH}_2\text{–CH}_2^+ + \text{Cl}^- \quad (4.4)$$

The carbonium ion thus generated will attack (aklylate) suitable Y^- groups. The mustards can react with free carboxyl (COOH), mercapto (SH), amino (NH_2), phosphate (PO_3H_2), and hydroxyl (OH) groups. The less stable the carbonium is when formed, the less selectivity it will exhibit. Thus highly reactive carbonium ions will also react with H_2O molecules in the environment.

Aliphatic nitrogen mustards form an ethylene immonium ion intermediate:

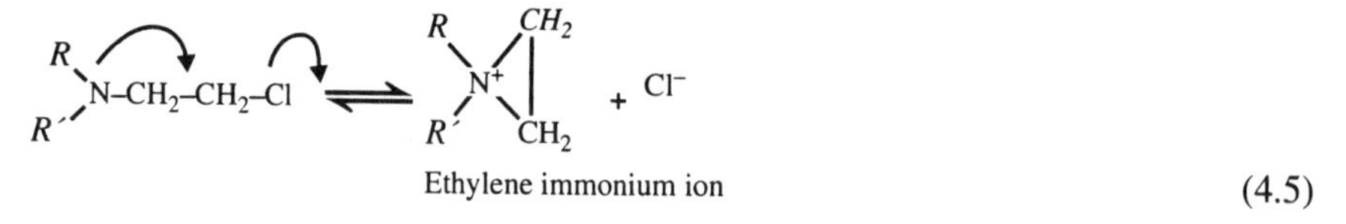

(4.5)

This highly strained three-member ring rearranges by scission to a primary carbonium ion, which in turn reacts with the various nucleophilic groups previously mentioned, including H_2O (Eq. 4.6).

$$\underset{R'}{\overset{R}{>}}\overset{+}{N}\underset{CH_2}{\overset{CH_2}{|}} \longrightarrow \underset{R'}{R\text{–}N\text{–}CH_2} \quad CH_2^+ \quad (4.6)$$

[4] The nucleophilic center need not be a formal anion. It can be an electron-rich atom such as nitrogen in an organic base.

The alkylations that occur in Eqs. 4.2 and 4.3 to form *RY* result in a covalent bond formation that is not reversible under biological conditions. Thus mustard-type compounds alkylating nucleophilic sites on enzymes, other proteins, and RNA and DNA molecules can inactive them and be potentially detrimental to cells. The evidence available suggests, however, that these random reactions with the various cell biopolymers are repairable with one likely exception: DNA. The irreversible alkylations responsible for lethal cell damage are now generally believed to involve DNA, specifically at the nitrogen in position 7 of the guanine nucleotide (Fig. 4-6). In considering a bifunctional mustard as shown, the alkylated nucleotide will have the corresponding structure after repeating the sequences in Eqs. 4.5 and 4.6, forming a new carbonium ion. The second carbonium ion can then either attack a molecule of water (i.e., be hydrolyzed) or aklylate the nitrogen in position 7 or another purine nucleotide, possibly from the adjacent DNA strand of the double helix. This would mean that a bifunctional alkylating agent can effectively cross-link two DNA chains so that they will be unable to separate; each chain then will be inaccessible for the initiation of the synthesis of new DNA. We find that these drugs inhibit DNA synthesis at therapeutic doses, which accounts for their cytotoxic properties. There is currently strong evidence that this direct interaction of the drug with DNA is the major cause for the observed clinical effects.

The drug Busulfan, although a bifunctional alkylating agent, is less reactive because it does not involve a strained ring intermediate. It nevertheless has been shown to cross-link guanine residues, at least in vitro.

$$\underset{\text{Busulfan}}{CH_3\text{-}\overset{O}{\underset{O}{\overset{\|}{\underset{\|}{S}}}}\text{-}O\text{-}(CH_2)_4\text{-}O\text{-}\overset{O}{\underset{O}{\overset{\|}{\underset{\|}{S}}}}\text{-}OCH_3} \longrightarrow {}^+CH_2\text{-}(CH_2)_2\text{-}CH_2^+ + 2\ CH_3\text{-}\overset{O}{\underset{O}{\overset{\|}{\underset{\|}{S}}}}\text{-}O^- \qquad (4.7)$$

The bifunctional alkylating agents also inhibit RNA and protein synthesis, especially at higher doses. Interference with other cellular reactions should not be ruled out. Table 4-3 lists some of the important alkylating agents currently in use.

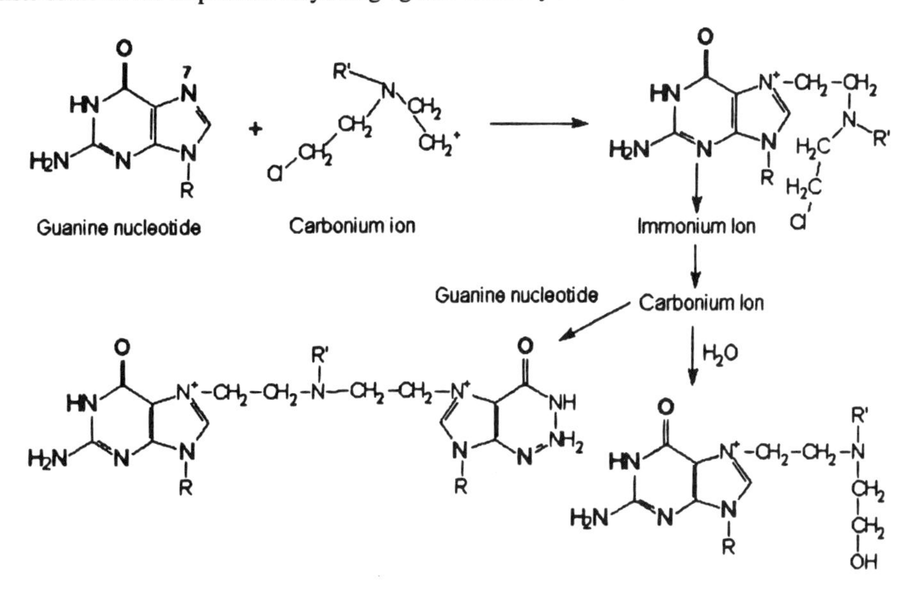

Figure 4-6. Alkylation of nucleotides by nitrogen mustards.

Table 4-3. Currently Used Alkylating Antineoplastic Drugs

Structure	Drug	Major clinical indications
$-CH_2-N(CH_2-CH_2-Cl)_2$	Nitrogen mustard, mechlorethamine (Mustargen), HN_2	Hodgkin's disease and other lymphomas, chronic leukemias, carcinomas
$HOOC-(CH_2)_3-C_6H_4-N(CH_2-CH_2-Cl)_2$	Chlorambucil (Leukeran)	Chronic lymphocytic leukemia, malignant lymphomas
$HOOC-CH(NH_2)-CH_2-C_6H_4-N(CH_2-CH_2-Cl)_2$	Phenylalanine mustard, melphalan (Alkeran)	Multiple myeloma
H, N, O=P—N, O; CH_2-CH_2-Cl, CH_2-CH_2-Cl	Cyclophosphamide (Cytoxan)	Malignant lymphomas, various carcinomas
O, O, P—N-CH_2-CH_2-Cl, N—CH_2-CH_2-Cl	Ifosfamide (Isophosphamide, Ifex)	Lung, breast, ovarian cancer, lymphomas
O, HN, O, N, H; N(CH_2-CH_2-Cl)$_2$	Uracil mustard	Chronic lymphocytic leukemia, malignant lymphomas
S═P—N(CH_2)(CH_2)	Thio-TEPA thiotepa	Carcinomas, chronic leukemias, malignant lymphomas
N, N, N, N, N, N	Triethylenemelamine, TEM	Malignant lymphomas, chronic leukemias
$CH_3-SO_2-O-(CH_2)_4-O-SO_2-CH_3$	Busulfan (Myleran)	Chronic granulocytic leukemia

Nitrogen mustard (mechlorethamine), the earliest mustard alkylating agent introduced, may be viewed as the prototype of the clinical problems encountered with mustards. It has poor selectivity, high toxicity, and produces frequent and intense nausea and vomiting. It is chemically unstable in solution, rapidly reacting with water and reactive compounds in the cell. Administered only intravenously in solutions prepared just before use, it is extremely vesicant, causing severe local reactions to exposed tissues. It is not surprising that many variants have been synthesized to circumvent some of these difficulties. Various theoretical rationales were used in the partially successful attempts to improve mustards. Substituting an aromatic ring for the methyl group can be predicted to increase chemical stability and thereby decrease the rate of alkylation because of the electron-withdrawing effect. This has resulted in compounds such as chlorambucil and melphalan that can even be given orally, and enjoy better absorption and tissue distribution before alkylation is widespread. An additional rationale in the case of melphalan was that the phenylalanine moiety would allow the drug access to the amino acid's membrane transport machinery, thus, one hopes, giving the drug a somewhat better selectivity and therapeutic index. The idea of bonding cytotoxic agents to "natural" carrier molecules is appealing. In the case of uracil mustard there might even have been the hope of a dual mechanism, that of DNA alkylation and a putative antimetabolite activity (see later) due to a 5-substituted uracil. Only alkylating properties are demonstrable. Alkylating functions as part of sugar alcohol molecules is exemplified in two experimental compounds: dibromodulcitol (Mitolactol) and 1,2:5,6-dianhydrogalacticol (DAG).

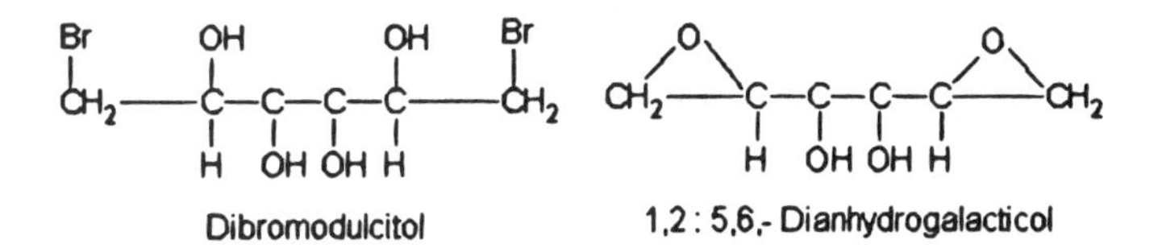

The dibromo compound should most likely be thought of as a pro-drug forming the diepoxide DAG in vivo. It is the very reactive oxirane rings, then, that possess the potent alkylating activity.

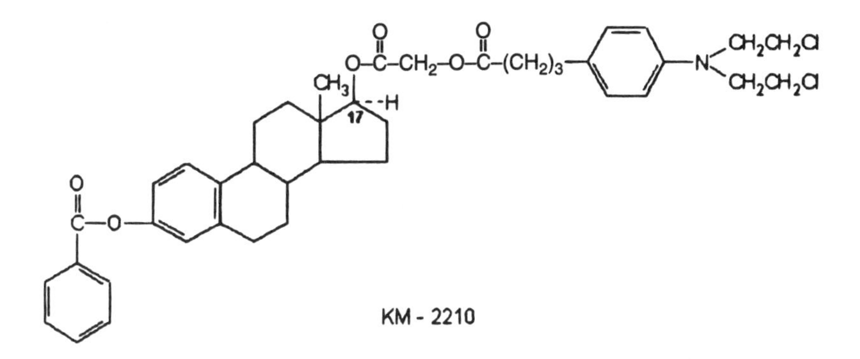

The synthesis of alkylating agent conjugates of steroidal hormones led to compounds with enhanced selectivity due to improved transport properties of the hormone to particular organ tumors presumed to be dependent on these hormones. Thus estramustine (Estracyst), a 3-carbamoyl nitrogen mustard derivative of the potent estrogen estradiol, has shown useful clinical activity against prostate cancer. A 17-estradiol-chlorambucil conjugate (KM-2210) has been reported to have selective activity in animal breast tumor. It seems possible that this drug may exhibit selectivity towards estrogen-receptor-containing tumors.

A special subgroup of alkylating agents are the nitrosoureas (Table 4-4). A feature of these agents is their considerable lipophilicity when compared with the other alkylating compounds. Lomustine and semustine do not even ionize at physiological pH. This permits them to cross more readily the blood–brain barrier and to pass into the cerebrospinal fluid. Furthermore, the nitrosoureas have a dual mechanism of action: alkylation of nucleic acids and carbamoylation of various proteins via the ε-NH_2 group of lysine residues.

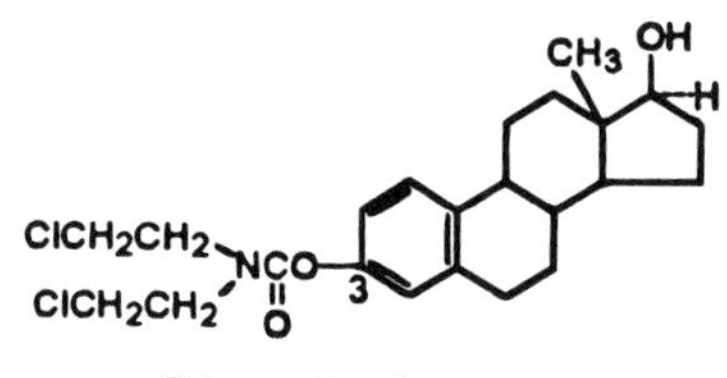

Estramustine (Estracyst)

The nitrosoureas, being unstable in water, and presumably under physiological conditions, decompose into alkylating and carbamoylating chemical species. A scheme was proposed for the decomposition of carmustine (BCNU) (Fig. 4-7). The decomposition of the unstable intermediate oxazolidine leads to a vinyl diazohydroxide, yielding a vinyl

Table 4-4. Carcinolytic Nitrosoureas

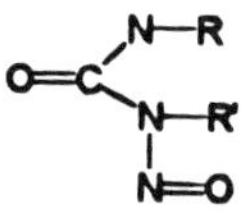

R	*R′*	Name
$-CH_2CH_2Cl$	$-CH_2CH_2Cl$	Carmustine, BCNU, BicNu, chloroethylnitrosourea
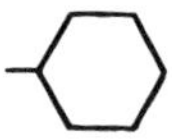	$-CH_2CH_2Cl$	Lomustine, CCNU, CeeNu, chloroethylcyclohexylnitrosourea
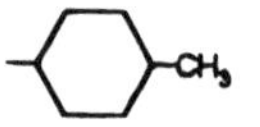	$-CH_2CH_2Cl$	Semustine, MeCCNU, chloroethylmethylcyclohexylnitrosourea
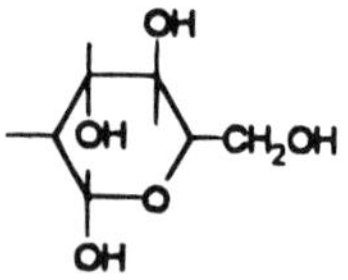	$-CH_3$	Streptozocin, Zanosar
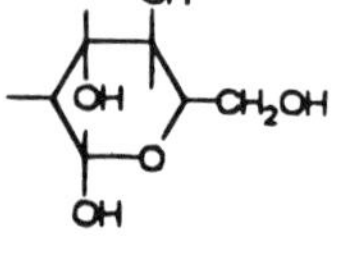	$-CH_2CH_2Cl$	Chlorozotocin, NSC-17824 8 (experimental)

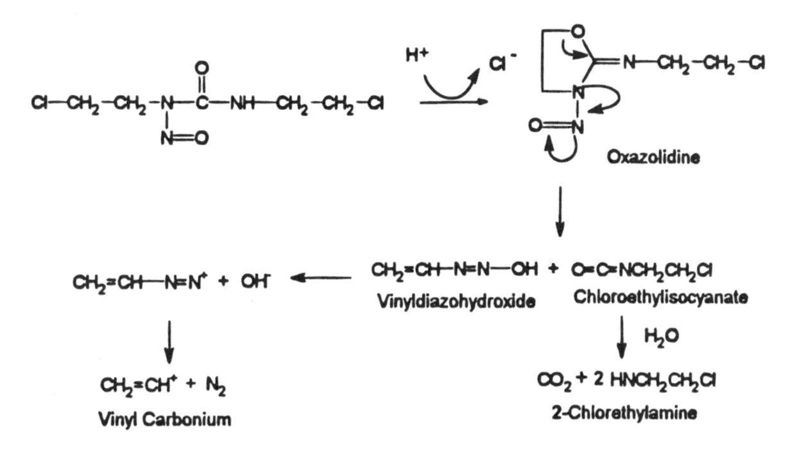

Figure 4-7. Proposed carmustine degradation (Montgomery et al., 1967).

carbonium ion—the alkylating agent. The chloroethyl isocyanate can act as a carbamoylating agent, or, after decomposing to 2-chloroethylamine, become a second alkylating species. A more general scheme for mono-alkyl functional nitrosoureas is outlined in Figure 4-8. Decomposition, possibly initiated by hydroxide ion, affords an isocyanate and hydroxy 2-chlorethyl diazine, which spontaneously decomposes to the 2-chloroethyl carbonium ion, the very active alkylating species that, in a two-step sequence, can actually cross-link DNA.

Because of their highly lipophilic nature, nitrosoureas are particularly useful against malignancies of the central nervous system such as brain tumors (gliomas), where response rates have exceeded 40%. They are also used (in combination therapy) against multiple myeloma, Hodgkin's disease, and non-Hodgkin's lymphoma.

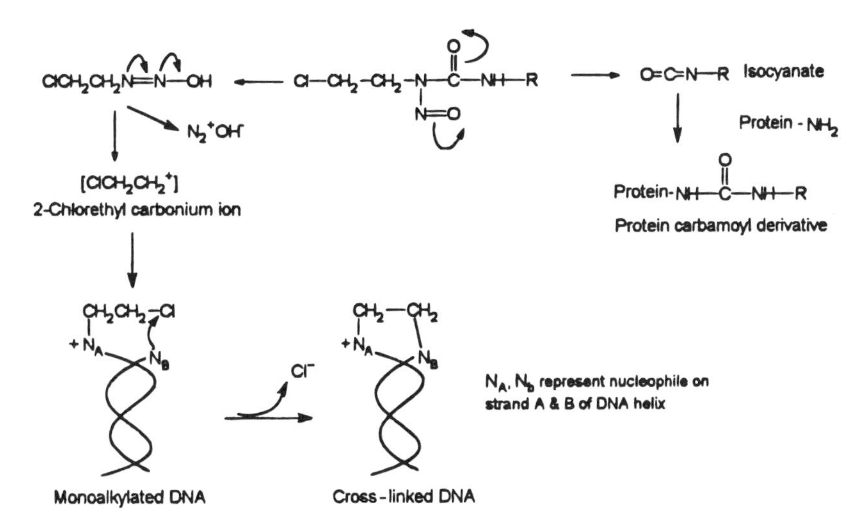

Figure 4-8. Possible nitrosourea degradation scheme with resultant alkylation and carbamoylation (adapted from Reed et al., 1975, and Kohn, 1977).

Streptozotocin is unique both in origin and pharmacology. It is an antibiotic obtained from *Streptomyces achromogenes* and contains a glucosamine component in its structure, which may contribute to its tendency to be taken up into the β-cells of the islets of Langerhans of the pancreas. Where it was initially used to induce diabetes in experimental animals, it is now used to treat pancreatic islet cell carcinoma, resulting in about a 50% response rate, but few survivors past 1 year.

Among alkylating agents cyclophosphamide should be considered as a true pro-drug, requiring extensive metabolic activation to obtain the actual cytotoxic alkylating agent (Fig. 4-9) (see also mitomycin, later). The drug is first oxidized by hepatic mixed-function oxidase (Chapter 3) to yield the 4-hydroxy metabolite. This compound is in tautomeric equilibrium with the acyclic form aldophosphoramide. The 4-hydroxy and acyclic aldehyde metabolites are further enzymatically oxidized to 4-ketocyclophosphamide and carboxyphosphamide, respectively. Neither metabolite is cytotoxic. The ultimate cytotoxic alkylating agent is produced by a beta-elimination reaction of aldophosphoramide, yielding the phosphoramide mustard and the byproduct acrolein, a highly reactive volatile liquid with sharp irritating effects. Since the nonenzymic degradation of aldophosphoramide to the active alkylator most likely occurs extrahepatically (blood, tissues, and, one hopes, in tumor cells), as evidenced by lack of hepatotoxicity, this may in part explain cyclophosphamide's higher specificity of cytotoxicity in the S phase than other alkylators.

In addition to the clinical toxicities encountered with nitrogen mustard in general, cyclophosphamide exhibits considerable urinary tract toxicity in the form of cystitis with hematuria. This has been attributed to an accumulation of the corrosive acrolein in the bladder. The parenteral administration of fluids and acetylcysteine (believed to react with acrolein) prevents cystitis greatly. Another compound, the sodium salt of 2-mercaptoethanesulfonic acid, mesna (Mesnex) -HS-CH_2-CH_2-SO_3H, also minimizes hemorrhagic cystitis due to acrolein by adding across the acrylic double bond of the aldehyde and detoxifying it. A newer analog of cyclophosamide is ifosfamide (Table 4-3). It too is metabolized to an active alkylating species, N,N′-bis(2-chloroethyl)phosphoric acid diamide (ifosphoramide) and acrolein. Another alkylating agent that must be activated to achieve its cytotoxicity is the antibiotic mitomycin C, which is obtained from *Streptomyces caespitosus*. In this case NADPH-dependent enzyme reduction activates the antibiotic to a bifunctional alkylating species. Figure 4-10 outlines the reactions involved in the "bioreductive

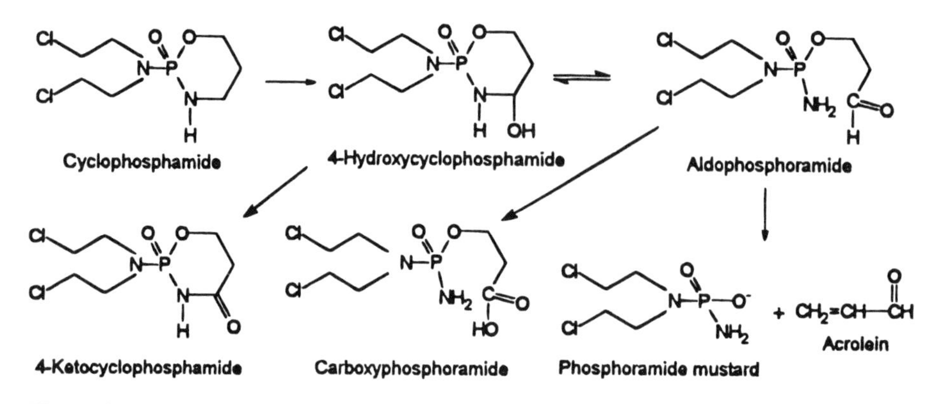

Figure 4-9. Cyclophosphamide metabolism and activation (proposed by Connors et al., 1974).

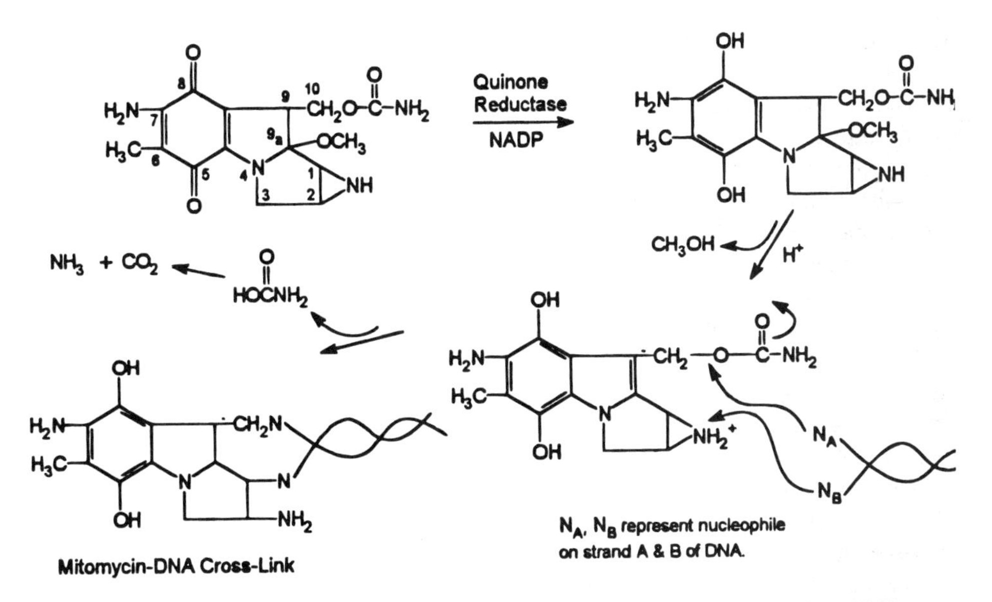

Figure 4-10. Probable mechanism of action of mitomycin C (Iver and Szybalski, 1964).

alkylation" mechanism, by which mitomycin alkyates DNA. Mitomycin C[5] has been shown to require benzoquinone to hydroquinone reduction preceding DNA cross-linkage alkylation. The reduction is effected by a cellular NADPH-dependent quinone reductase system, resulting in spontaneous expulsion of the methoxy group from its tertiary 9a position to form the 9–9a double bond. Protonation of the aziridine ring and the facile leaving of the carbamoyloxy group at position 10 results in two active alkylating carbonium ions at carbons 10 (CH_2^+) and one with which attacking purine and pyrimidine bases of DNA cross-link. Cross-linkage has been shown to be proportional to guanine and cytosine content of the DNA.

Probably the most "unorthodox" cross-linking alkylating agent is an inorganic platinum-coordinating compound *cis*-platinum (II) diamminedichloride, (*cis*-DDP, cisplatin, Platinol). It is a classic square-planar four-coordinate platinum (II) compound.

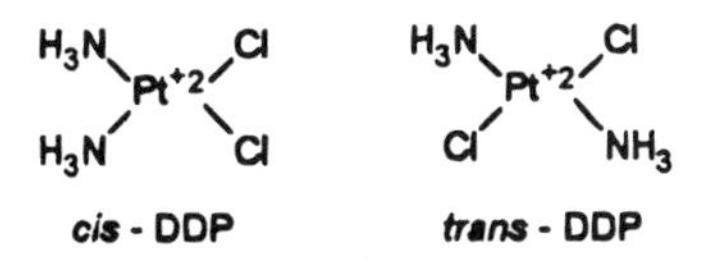

Cisplatin was discovered during electrolysis experiments of bacterial culture media utilizing platinum electrodes. An electrode product exhibited antibacterial activity at very low concentrations. The product, cisplatin, also revealed antitumor activity. Cisplatin cross-links DNA via ring atoms of purines and possibly pyrimidines, with the elimination of Cl^- analogously to the bifunctional alkylating agents discussed. Although *cis*-platinum can be

[5] Mitomycin A and B differ in substituents at positions 7, 9a, and the aziridine nitrogen.

demonstrated to inhibit RNA and protein synthesis in mammalian cells in vitro, it is selective DNA inhibition in vivo that is the primary cause of cytocidal activity. The drug has shown clinical effectiveness against testicular, ovarian, head and neck, and bladder cancer. Combinations with other drugs such as bleomycin, vinblastine, and doxorubicin allow for decreased cisplatin dosages and resultant lessening of toxic effects.

Cisplatin exhibits intense emetic reactions, ototoxicity, and severe cumulative renal toxicity. Some success with "rescue" techniques, using compounds with high heavy metal affinity such as diethyldithiocarbamate, have been reported.

Figure 4-11 depicts a possible sequence whereby cisplatin, forming a reactive aquo intermediate, reacts with the N-7 nitrogen of two guanine bases on two different DNA chains, thus cross-linking them in a manner analogous to the nitrogen mustards. This "mechanism," of course, offers no explanation for the lesser bioactivity of the *trans*-diammin-dichloro-platinum stereoisomer since there seems no apparent reason for a similar reaction sequence not to occur.

The mystery may now be somewhat closer to a solution. Both cisplatin and the *trans*-isomer do share some biological properties, including binding covalently to chromosomal DNA. Both adducts inhibit DNA replication and are therefore cytotoxic. However, the *cis*- isomer seems more potent and appears preferentially more so to malignant cells. Lippard et al. (1985) utilized a circular chromosome from the DNA virus SV40, and discovered that cisplatin was 14 times more effective than the *trans*- isomer in inhibiting DNA synthesis to the same degree, yet the two isomers were bound in the same quantities. Cellular uptake was even at equal rates. However, kinetic measurements showed that after an initial six-hour period *trans* binding was decreased, while *cis* continued to accelerate. After 24 hours, when DNA replication is usually at maximal rates, little *trans* isomer could be found bound. The authors proposed that for geometric reasons the *trans* adduct damage is being more efficiently repaired. Additional X-ray structural determination of cisplatin bound to a dinucleotide of deoxyguanosine (linked via a phosphodiester) clearly shows the cross-linkage to require the two guanine rings to tilt away from the normally stacked relationship of the intact DNA double helical structure. Sherman et al. (1985) therefore proposed that this explains the interference with DNA replication and furthermore that such cross-linkage is, for stereochemical reasons, not possible with *trans*-DDP.

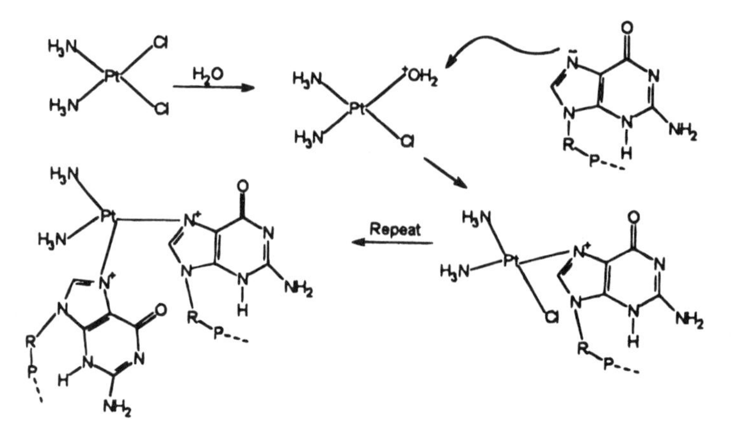

Figure 4-11. Possible cross-link of cisplatin with two guanine bases in DNA.

Since the success of cisplatin has been established, analogs have been synthesized and have reached human clinical trials. Effective compounds with decreased renal toxicity have been reported. Among the more promising leads are derivatives of 1,2-diamminocyclohexano- and cycloheptanodichloroplatinum (II). A malonic derivative showed remission of leukemias and regression of solid tumors. A particularly intriguing new platinum complex is *cis*-1,2-diamminocyclohexylplatinum (II) ascorbate. Unlike cisplatin, here both the *cis and trans* isomer exhibited antitumor activity, and were actually greater than cisplatin in some tumor screens. The activation mechanism by which cisplatin loses its chloride ligands during tissue transport is not possible here, considering the Pt-C bond to ascorbic acid. This raises the possibility of a new differing mechanism—or a rethinking of the currently accepted mechanism for platinum compounds.

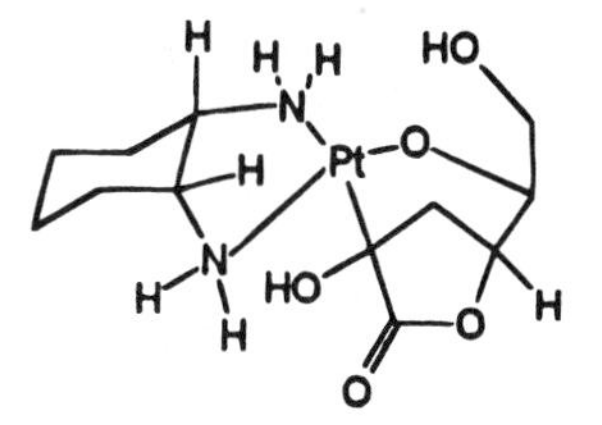

***cis* - 1,2 - Diamminocyclohexylplatinum (II) ascorbate**

Another organometallic platinum drug has been approved in the United States. It is a diammin platinum coordinate with 1,1-cyclobutanedicarboxylic acid named carboplatin (CBDA, Paraplatin). Its advantages over cisplatin are a lesser emetogenic effect as well as a possible decrease in nephro- and ototoxicity.

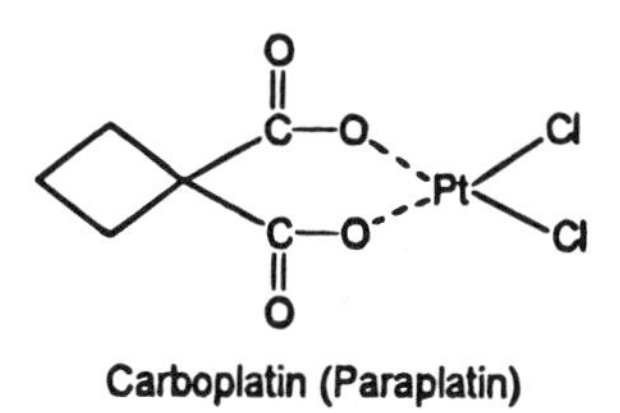

Carboplatin (Paraplatin)

4.12. Antimetabolites

The rationale for utilizing organic compounds structurally similar to normal cellular metabolites as anticancer agents is simple: to interfere with the biosynthesis of these substances within the cell and thus inhibit cell proliferation. Among the precursors that cells require for normal and abnormal growth are the building blocks of nucleic acids, the pyrimidine and purine bases. The biosynthesis of the bases themselves requires, among other steps, the transfer of one-carbon units that are derived from folinic acid (Chapter 2). The synthesis of proteins requires various amino acids. Thus structural analogs of any of these bases and amino acids yield potential folic acid, purine, pyrimidine, and amino acid antagonists of chemical usefulness.[6]

---

[6] Structurally unrelated compounds having the ability to interfere with the various enzymes that are involved in the biosynthesis of nucleosides, nucleotides, and amino acids, and their respective biopolymerization, can be another source of anticancer drugs.

4.12.1. Folic Acid Antagonists

The essential nature of folic acid at the cellular level was already noted in the discussion of the antibacterial sulfonamides (Chapter 2). It was indicated that tetrahydrofolic acid (FH_4) was somehow necessary for the biosynthesis of purines and thymine, which, in turn, are the precursors of the nucleic acids. Consideration of some of these biochemical steps in somewhat greater detail is now necessary.

$$FH_4 \xrightarrow[\text{Hydroxymethylase}]{HOCH_2-CH(NH_2)COOH} N^5,N^{10}\text{–methylene } FH_4 + {}_2HN\text{–}CH_2\ COOH \quad (4.8)$$

$$HS\text{–}CH_2\,CH_2\text{–}CH(NH_2)COOH \xrightarrow[\text{Transferase}]{N^5\text{ - methyl}FH_4} CH_3\,S\text{–}CH_2\,CH_2\text{–}CH(NH_2)COOH + FH_4 \quad (4.9)$$

As shown in Figure 4-12, tetrahydrofolic acid, FH_4 (see Fig. 2-3) picks up a one-carbon unit from L-serine (Eq. 4.8), and is converted to coenzyme N^5,N^{10}-methylene-FH_4, where the CH_2 group bridges N^5 and N^{10} of FH_4. This coenzyme in turn is further oxidized to N^5,N^{10}-methylene-FH_4. Addition of a water molecule across the N^5 nitrogen–carbon bond affords N^{10}-formyl FH_4 or N^5-formyl FH_4. The latter is known as *folinic acid, leucovorin*, or citrovorum factor. Leucovorin is rapidly reducible to N^5-methyl-FH_4, the physiologically circulating folate. An alternate route to N^{10}-formyl-FH_4 is directly from FH_4 by ATP-aided addition of formate. Interconversion thus also affords folinic acid. The various tetrahydrofolate coenzymes are the source for several one-carbon transfers, including the synthesis of L-methionine from L-homocysteine (Eq. 4.9), as will be seen in the biosynthesis of purine (see Fig. 4-14) for C-2 (N^{10}-formyl FH_4) and C-8 (N^5,N^{10}-methenyl FH_4). The conversion of deoxyuridylic acid (dUMP) to deoxythymidylic acid (dTMP) by the transfer of one-carbon fragment to the 5 position of the uracil from coenzyme N^5,N^{10}-methylene-

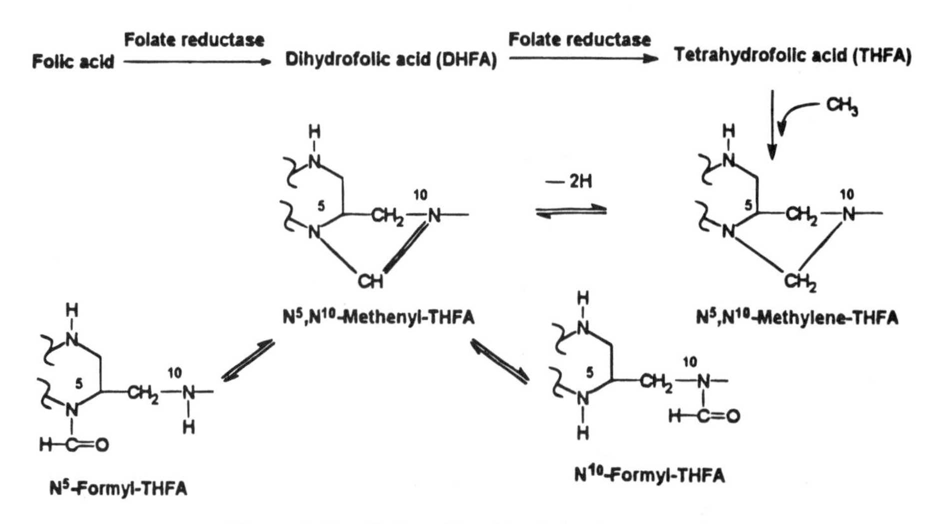

Figure 4-12. Folic acid and its derived coenzymes.

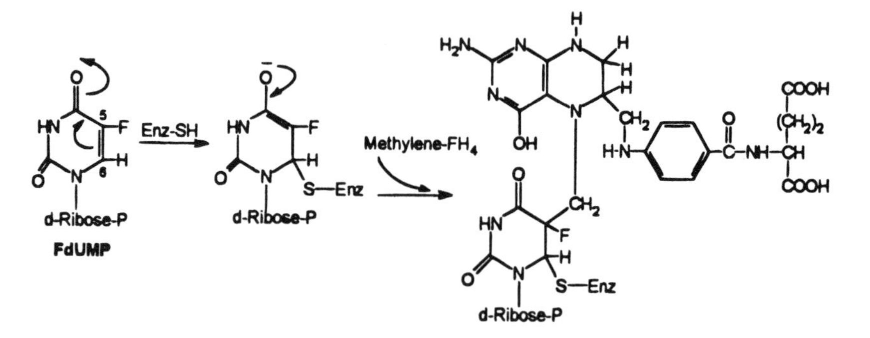

Figure 4-13. Irreversible formation of ternary thymidylate–methylene tetrahydrofolate–deoxyuridine phosphate complex.

FH_4 is carried out by the crucial enzyme thimidylate synthetase (Fig. 4-13). It should be noted that this is a reductive methylation where the tetrahydrofolate coenzyme serves both as a one-carbon donor and reductant. The resultant FH_2 must be reduced back to FH_4 by folate reductase, thus offering further opportunity for DNA synthesis inhibition at this point. Thymidylate synthetase is the sole de novo path to thymidylate; there are no alternate routes. The interest of this point as a target for anticancer agents is obvious.

The isolation in the 1940s of the antianemic factor now known as folic acid and its subsequent synthesis resulted, after many false leads, in the first successful antivitamin antimetabolite aminopterin (Table 4-5). This compound produced sustained remissions in leukemia. The N^{10}-methyl homolog, now designated as MTX (Table 4-5), superseded aminopterin the same year. It is still the only clinically significant antifolate carcinolytic drug.

Methotrexate inhibits the enzyme folate reductase at two steps (Fig. 4-12). Even though the inhibition is technically a competitive one, the enzyme binds MTX much more strongly than the natural substrate, FH_2 (Table 4-5). In practical terms there is no dissociation of the enzyme–drug complex. For thymidylate synthetase to continue to produce thymidylic acid (and therefore DNA), perpetual reduction of FH_2 is essential for cellular survival. The efficiency of MTX inhibition of folate reductase thus indirectly becomes the mechanism by which this drug is so cytotoxic in the S phase. In addition, there is evidence that MTX may, by binding to the tetrahydrofolate coenzyme, also inhibit the thymidylate synthetase

Table 4-5. The Folate Structures and Relative Inhibition Ability of Dihydrofolate Reductase

R_1	R_2	K_m	Name
OH	H	1.1×10^{-7}	Folic acid
OH	H 5,6-Dihydro	1.0×10^{-5}	Dihydrofolic acid
NH_2	H	6×10^{-10}	Aminopterin
NH_2	CH_3	6×10^{-10}	Methotrexate

enzyme. In effect, this would make the MTX molecule a sequential blocker of DNA synthesis, thereby helping to explain its superior carcinolytic properties. In those malignancies where cell generation is rapid, it was shown that normal cells survived 500 times better than lymphoma cells in the presence of MTX doses high enough to achieve maximum cell kill.

Toxicity, as with other antimetabolites, can be severe—even fatal. Oral lesions usually occur first, followed by gastrointestinal lesions, bone marrow depression, baldness, and hyperpigmentation. It is not surprising that these drugs are teratogenic, causing fetal abnormalities and death.

Tumor resistance to MTX has been extensively studied. Several mechanisms are now understood. One is the development of a folate reductase whose drug affinity is considerably reduced. In addition, changes in the plasma membrane decreasing the transport of the drug into cells were found. However, increased enzyme production seems to be of major importance. Drug-induced increases of 200-fold have been reported. It should be possible to overcome, to some degree, resistance due to defective transport, high intracellular reductase levels, and possibly even decreased enzyme binding, by greatly increased tumor exposure through large MTX doses. Cytocidal quantities may even cross the blood–brain barrier and enter the cerebrospinal fluid—if doses are high enough *and* time of drug exposure is kept short enough not to increase mortality. Aborting the invariably fatal high toxicities can, of course, not be accomplished with folic acid or FH_2. These two agents cannot reverse the tenacious MTX–reductase complex (Table 4-5). However, *folinic acid* can rescue the healthy cells by replenishing the MTX-depleted tetrahydrofolate pool. The leucovorin is rapidly reduced to N^5-methyl- FH_4 and upon entering the cell, metabolized to FH_4 (through biosynthesis of methionine, Eq. 4.9). It then becomes available for folate coenzyme conversion and resumption of DNA synthesis.

One of the early applications of this *leucovorin rescue* technique was osteogenic sarcoma. It has since been successfully applied to acute leukemia, non-Hodgkin's lymphoma, and lung cancer.

The doses of MTX used are frequently of the order of 3 g/m^2 (traditional MTX doses are in the 50–200 mg/m^2 range). However, doses as high as 9–30 gm/m^2 have been successfully utilized. Antidoting, or rescuing with leucovorin, usually begins within 24 hours and is given over several days.

4.12.2. Purine Antagonists

Two of the more important purine analogs in use clinically are 6-mercaptopurine and 6-thioguanine. These purine antagonists and glutamine antagonists such as azaserine (Table 4-6) are major antagonists in the biosynthesis of purine bases. Before understanding the mechanism of their action, it is necessary to look at the biosynthesis of inosinic acid, the purine ribonucleotide that is the precursor to both purine bases found in DNA and RNA

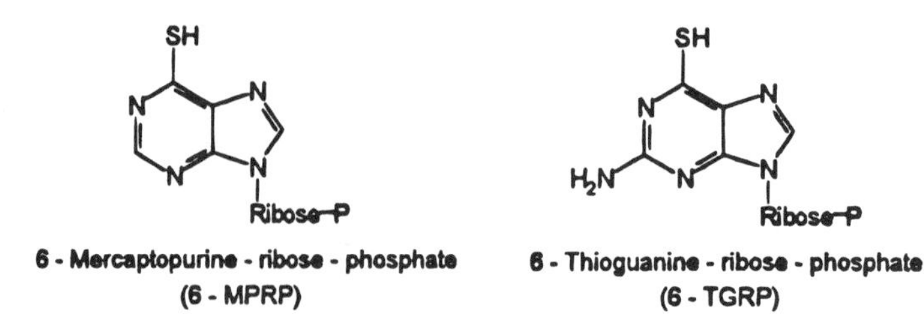

6 - Mercaptopurine - ribose - phosphate (6 - MPRP)

6 - Thioguanine - ribose - phosphate (6 - TGRP)

Table 4-6. Normal Metabolites and Some Clinically Useful Analogs

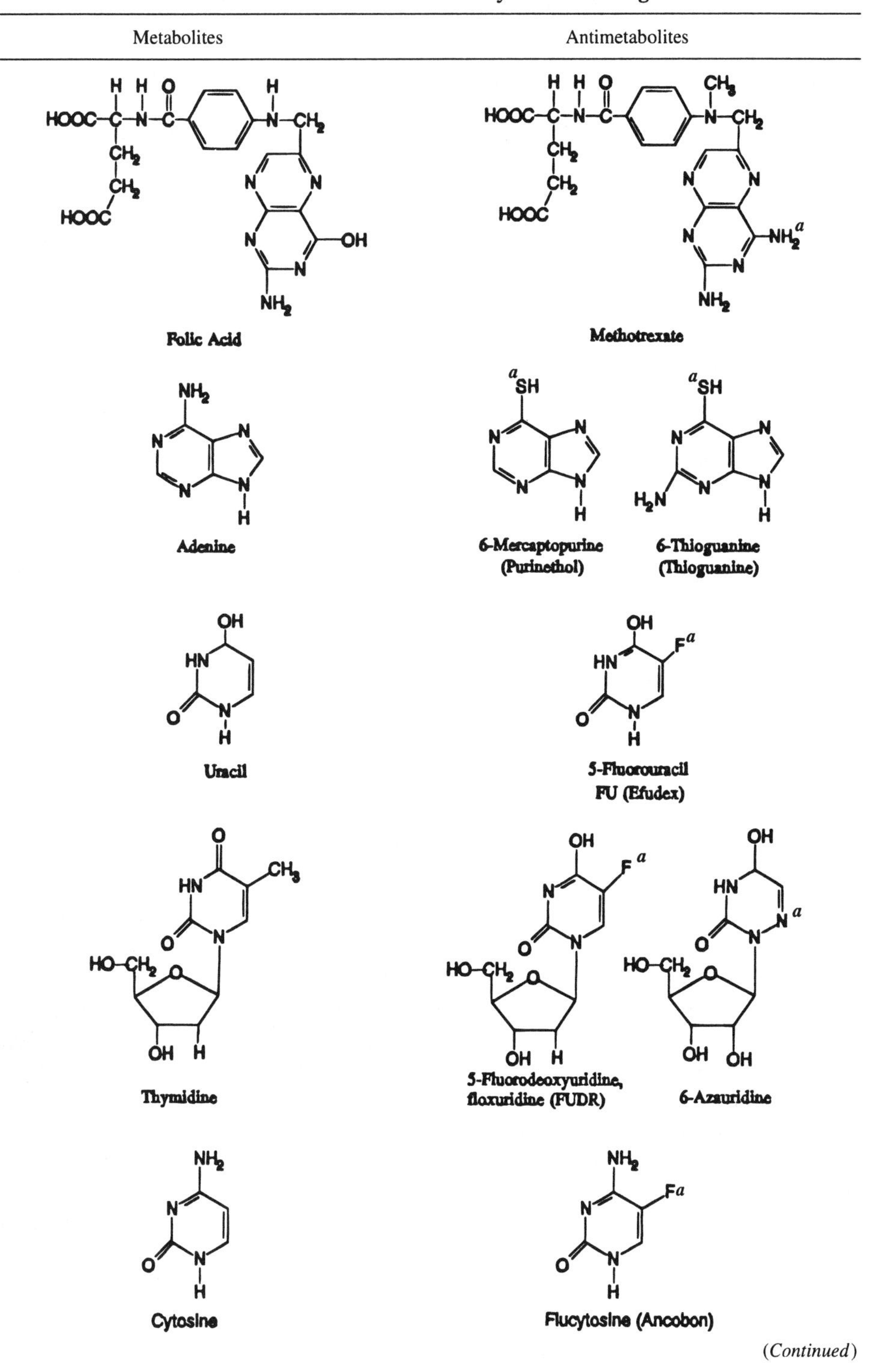

(*Continued*)

Table 4-6. *(Continued)*

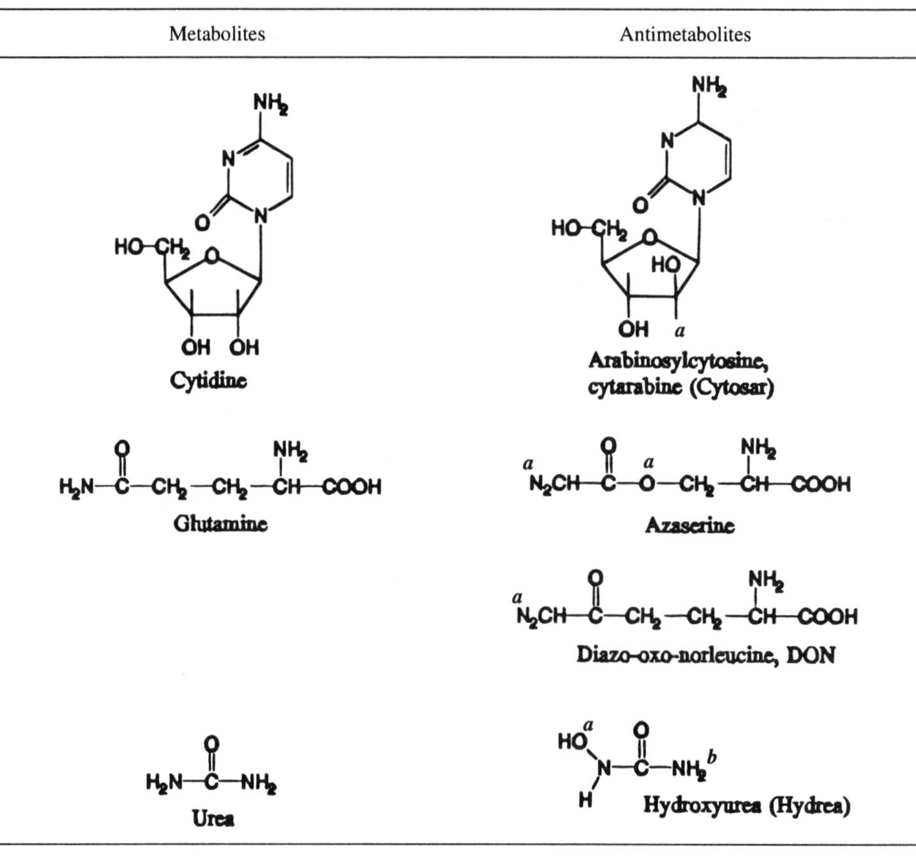

[a] Indicates site where the antimetabolite differs from the metabolite.

[b] Even though a structural analog, hydroxyurea is not an antimetabolite; it actually inhibits reduction of nucleotides.

(Fig. 4-14). Both 6-mercaptopurine and 6-thioguanine are not active purine inhibitors until they are converted to their respective nucleotides by hypoxanthine-guanine phosphoribosyl transferase enzymes in the cell. Since this is an intrinsically destructive process to the cell, it is referred to as a "lethal synthesis" or salvage pathway. Tumor cells or normal cells that have very low levels of this enzyme are resistant to these two agents. Absence of this enzymatic function is not lethal to the cell since the synthesis of purines in the cell can occur by different pathways.

It is frequently found that once a certain level of the final product has been reached in a multistep biochemical synthesis, the product acts as an inhibitor of the enzyme that catalyzes the first step of such a sequence of reactions.[7] Thus such an enzyme is also allosteric. In the case of the purine synthesis, the enzyme that is responsible for the transfer of the NH_2 group of glutamine, which yields the phosphoribosylamine, is inhibited by adenylic and guanylic acids, the end products, or by both sulfur-containing analogs 6-MPRP and 6-TGRP. From Figure 4-15 it can be seen that these drugs also prevent the con-

[7] This is known as *feedback inhibition*.

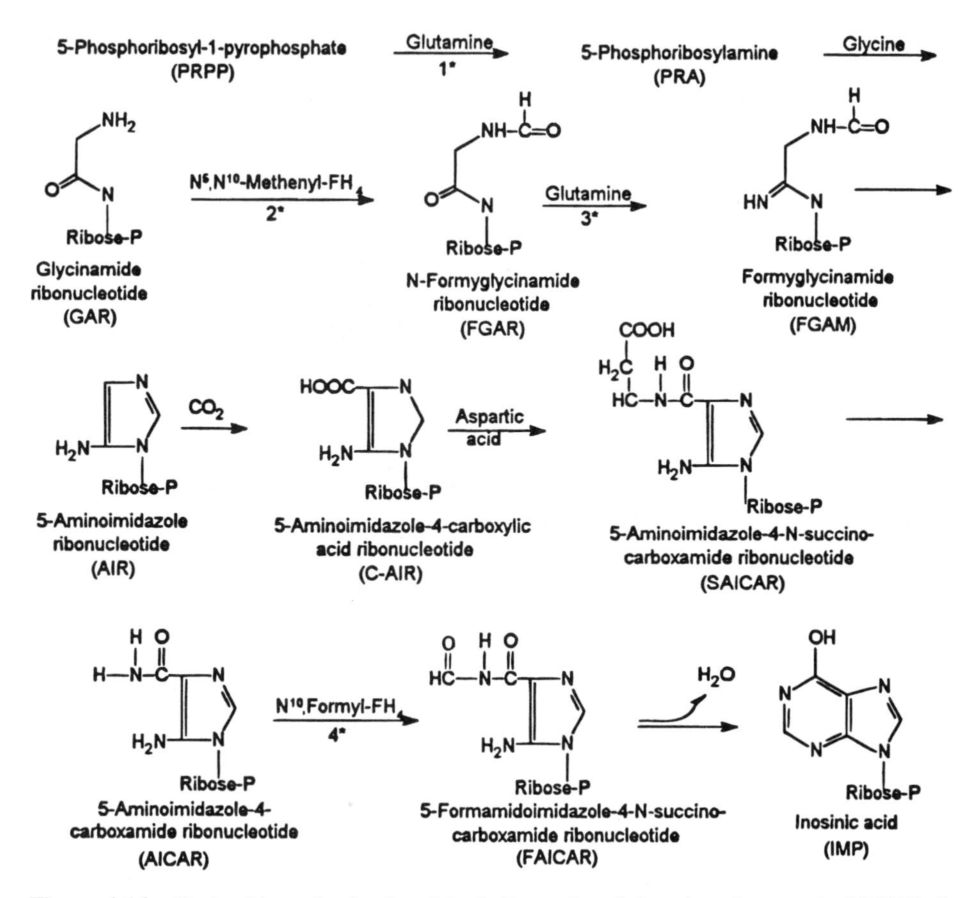

Figure 4-14. Purine biosynthesis. Asterisks indicate site of drug interference; 1, 6-MPRP; 2, methotrexate; 3, azaserine; 4, methotrexate. See text for explanation.

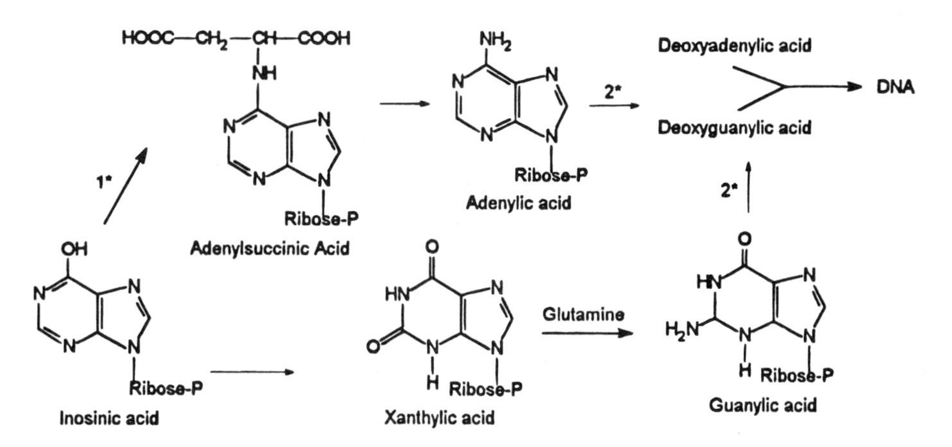

Figure 4-15. Reaction of inosinic acid. Asterisks indicate sites of drug interference: 1, 6-MPRP; 2, hydroxyurea.

version of inosinic acid to both adenylic and guanylic acid. Further examination of the purine biosynthesis will reveal an additional step in which glutamine acts as a donor of a nitrogen atom in position 3 of the purine ring. Since the amidotransferase enzyme at this step is not allosteric, no inhibition by 6-MPRP can occur. The glutamine antagonist azaserine, however, is effective here and also in the first step. In two instances (sites 2 and 4 in Fig. 4-14) a transfer of a one-carbon unit from tetrahydrofolate coenzymes takes place. It can be safely predicted that reduced folate production can be brought about with methotrexate resulting indirectly in purine synthesis inhibition.

Azaserine and related glutamine antagonist antibiotics very strongly inhibit the FGAR-to-FGAM reaction (Fig. 4-14) in bacteria and in transplanted tumors in animals. These drugs have been shown to compete with the natural amino acid for the enzyme catalyzing the transfer of the amino group. The consequence is inhibition of purine synthesis. However, resistance in initially responsive tumors develops. One mechanism implicated involves an increase in the utilization of preformed purines. Unfortunately, the glutamine antagonists have not shown the hoped for clinical effectiveness. Combination with other antimetabolites has offered some hope, however. Azaserine and similar agents have been extremely useful as research tools in helping to clarify the mechanisms involved.

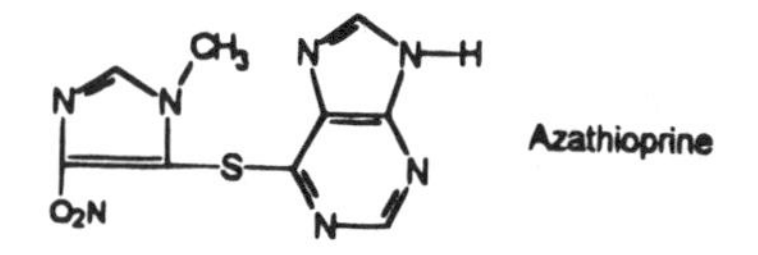

Azathioprine, a 1-methyl-4-nitroimidazolyl derivative of 6MP, deserves mention. Since 6-mercaptopurine exhibits considerable, but short-lived, suppression of the immune reaction, it seemed desirable to slow the rate of excretion in order to achieve a more sustained reaction. This was accomplished with this pro-drug, when the nitro-activated C–S bond is slowly cleaved, releasing the parent drug. Azathioprine has been used in preventing transplant rejections, the graft-versus-host reaction, and as a supplementary treatment for autoimmune diseases such as systemic lupus erythematosus and psoriasis. Table 4-6 lists metabolites and their corresponding antimetabolites.

4.12.3. Pyrimidine Antagonists

By the middle 1950s the idea of utilizing compounds that would inhibit DNA synthesis in cancer chemotherapy had evolved. It was observed that certain rat liver tumors used more uracil in the synthesis of nucleic acids than did normal liver tissue. It therefore became a viable endeavor to look for structural analogs of uracil that would effectively inhibit this uracil utilization and thereby cell proliferation.

Figure 4-16 outlines the biosynthesis of pyrimidines and their conversion to the required deoxyribose triphosphates of uridine and cytidine, the necessary building blocks of RNA. The first step involving the condensation of carbamoyl phosphate with aspartic acid is catalyzed by aspartate transcarbamylase. This enzyme is strongly inhibited by the transition-state inhibitor PALA (Chapter 2). Other steps where drug intervention in the scheme can interfere to inhibit DNA synthesis are indicated.

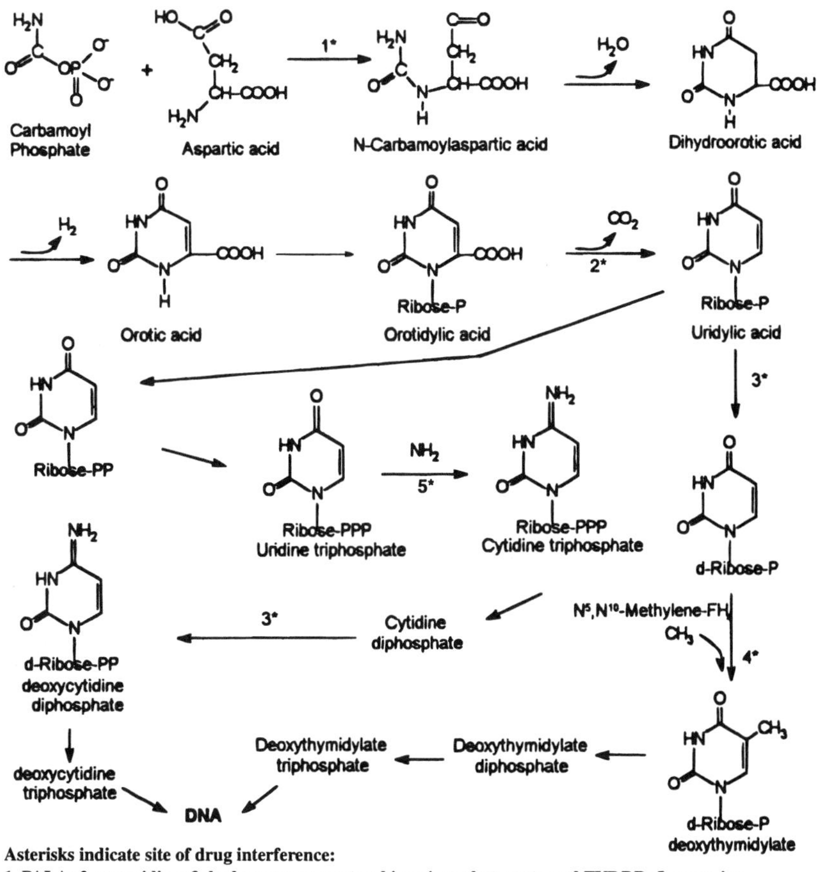

Asterisks indicate site of drug interference:
1, PALA; 2, azauridine; 3, hydroxyurea or cytarabine; 4, methotrexate and FUDRP; 5, azaserine

Figure 4-16. Pyrimidine biosynthesis and interconversions.

The rationale for 5-fluorouracil, 5-FU, was to block DNA synthesis by inhibiting the biosynthesis of thymidylic acid (deoxythymydine monophosphate, dTMP) by virtue of its close structural analogy to uracil. Fluorine, being the smallest atom that could substitute for hydrogen at the 5 position, was assumed to create the smallest possible molecular perturbation and thus be converted to the nucleotide and be accepted by the reactive site of thymidylate synthetase as a substrate imposter. In fact, this was the case. The van der Waals radius of the F atom (1.33 Å) is only slightly larger than that of the H atom (1.20 Å).

As with thiopurines, 5-FU must first be activated by a series of steps to the active cytotoxic agent, 5-fluorodeoxyuridine monophosphate (FdUMP) (Fig. 4-17). Tumor cells deficient in any of the enzymes needed for these conversions (e.g., phosphoribosyltransferase) are likely to be resistant to 5-FU. Since the early discovery that thymidylate synthetase is very sensitive to 5-FU (actually FdUMP), the mechanism of action was believed to be impairment of the conversion of dUMP to dTMP (Fig. 4-13). A more detailed mechanism is now understood to be: in the first step, thymidylate synthetase reacts covalently with C-5 of FdUMP (as if it were the normal substrate, dUMP) via nucleophilic attack of its reactive site's SH group. Methylene-FH_4 cofactor then adds to C-5 of this initial complex.

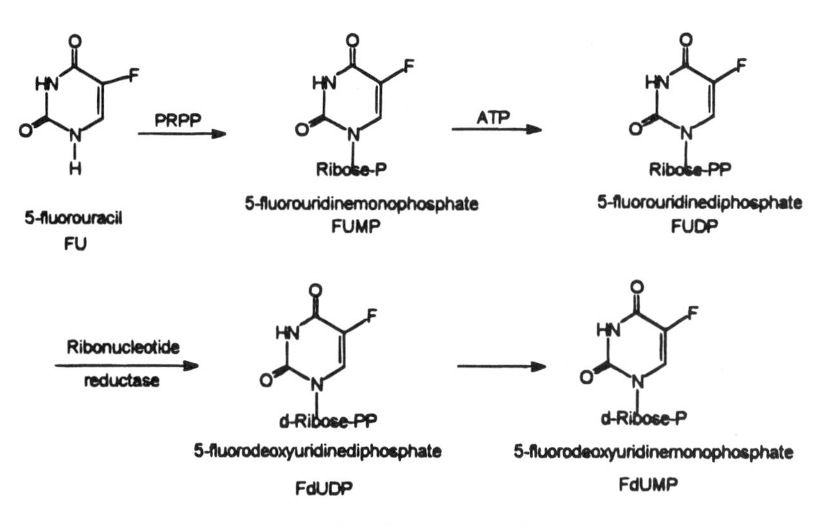

Figure 4-17. Fluorouracil activation.

With the normal substrate the next step would involve a hydride shift to the methylene of the folate and a proton abstraction from C-5, leading to thymidylic acid. Here, however, removal of an F^- from C-5 is impossible. The result is an irreversible blockade of the enzyme by a ternary covalent complex (Fig. 4-13). The synthesis of thymidylic acid (Fig. 4-13), as mentioned earlier, results in FH_2, which must be reduced to FH_4 to keep the scheme going. Since this crucial reaction is so efficiently blocked by MTX, a synergistic effect of this drug, when combined appropriately with 5-FU, has been demonstrated in L1210 mouse leukemia. However, antagonism was also shown, and was explained to be due to impaired enzyme inhibition by the decreased availability of cellular folate. 5-Fluorouridinetriphosphate (FUTP) is incorporated into RNA, which very likely becomes an additional factor in the overall cytotoxicity of 5-FU.

There are other sites in the pyrimidine biosynthesis that are subject to blockade by drugs. Azauridine blocks the decarboxylation of orotodylic acid (Fig. 4-16). It has been useful in some acute leukemias.

A careful examination of the biosynthetic schemes outlined in Figures 4-15, 4-16, and 4-17 will illustrate the key role played by a single enzyme, *ribonucleotide reductase*. It is responsible for the conversion of all ribonucleoside-5′-diphosphates to their 2′-deoxy analogs, which become the precursor building blocks of DNA after conversion to triphosphates. Hydroxyurea and, to a small degree, cytarabine are presently the only important agents capable of inhibiting this enzyme clinically. Hydroxyurea kills in the S phase of the cell cycle. Cytarabine (Cytosine arabinoside, Ara-C, Cytosar) is an antimetabolite where the sugar moiety rather than the base is modified. Arabinose differs from ribose in the stereochemistry of 2′-OH. After activation by conversion to the triphosphate, its major mechanism of cytotoxicity is inhibition of DNA polymerase by competing with cytidine triphosphate, the enzyme's natural substrate. It is an S-phase-specific drug that also has some antiviral activity.

5-Azacytidine, although of somewhat limited clinical utility, is fascinating from a biochemical standpoint. There are reports that this drug is deaminated by cytidine deaminase in bacteria and possibly humans, yet 5-fluorocytosine, which is antifungal because of deamination to 5-fluororuacil, is susceptible fungal cells, is selectively nontoxic in humans

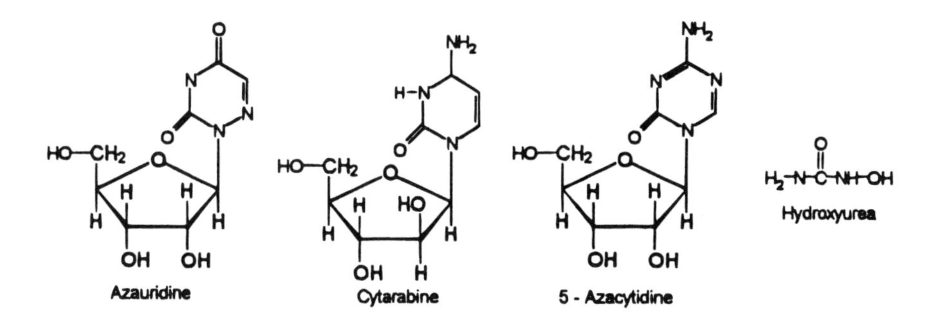

since it is not converted to 5-FU presumably because of an absence (or very low levels) of cytidine deaminase in humans. After activation to the triphosphate, azacytidine is incorporated into RNA and DNA. There is a possibility that the production of such faulty nucleic acids negatively affects the functioning of ribosomal RNA and thus protein synthesis. Inhibition of orotidylic acid decarboxylation may be an additional component of the drug's pyrimidine synthesis blockade.

An interesting situation arises when halogens larger than fluorine are substituted into position 5 of uracil. The van der Waals radius of chlorine is 1.73 Å; of bromine, 1.95 Å. These diameters approach the 2.0 Å radius of the CH_3 group of thymine or thymidine. Because of this similarity in size, 5-chloro- and 5-bromouracil are incorporated into the DNA molecule. 5-Bromouracil incorporation into DNA seems to increase hydrogen bond strength between the two DNA strands, thus impeding strand separation to some degree. This compound also sensitizes DNA to X-radiation.

4.13. Carcinolytic Antibiotics

A clinically important group of anticancer drugs are certain cytotoxic antibiotics that inhibit DNA and/or RNA synthesis by complexing with DNA via an interposing reaction called *intercalation*. The process is now understood to be one by which the planar, polycyclic portions of some drugs can "slide" into the double helical DNA structure in a horizontal manner by interposing between the base layers of DNA, which are stacked at distances of 3.36 Å above each other. Interactions between the purine and pyrimidine bases and the flat, aromatic portions of the intruder molecule are primarily van der Waals. Additional binding forces for particular drugs may include hydrogen bonding between certain groups on the drug molecules and on the DNA bases as well as ionic interactions.

A prominent and intensely studied antibiotic is actinomycin, which is also known as actinomycin D (Cosmogen®). It consists of a phenoxazone ring bonded to two identical cyclical pentapeptides (Fig. 4-18). The phenoxazone ring system intercalates between two successive G = C base pairs of DNA, resulting in a deformed RNA template. DNA transcription along this template is crippled since the polymerase cannot now move along it. Thus RNA chain elongation is now inhibited. In addition, hydrogen bonding exists between the NH_2 group of guanosine and the threonine C = O function.

Actinomycin is particularly effective in the treatment of Wilm's tumor in the kidney (of children) and, as a member of combination protocols, against adult Ewing's and Kaposi's sarcoma. This agent is also useful in MTX-resistant choriocarcinoma. The drug is cell-cycle specific in the S and G_1 phase.

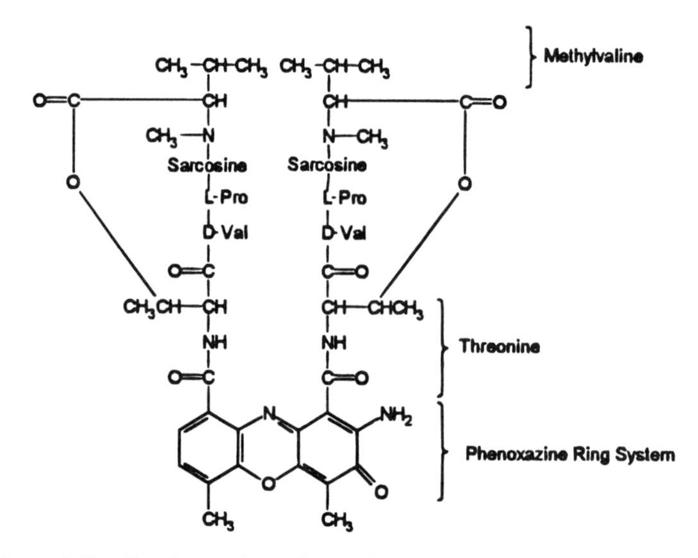

Figure 4-18. Dactinomycin (actinomycin D). (Sarcosine is N-methylglycine.)

4.13.1. The Anthracyclines

Streptomyces coerulerubidos yielded a tetracyclic aminosugar containing an antibiotic named *daunorubicin* (rubidomycin, Cerubidine®), which exhibited actinomycinlike antitumor activity. Severe cardiotoxicity, however, limited its clinical use, particularly as single agent therapy.

Further clinical investigations, however, established its usefulness against acute lymphoblastic, and some nonlymphoblastic leukemias, achieving remissions in patients previously resistant to other agents. Several years later the 14-hydroxy derivative was isolated from *S. peucitus*. It was named *doxorubicin* (Adriamycin®, ADM). This "minor" chemical modification made it into a broad-spectrum carcinolytic agent. A large variety of cancers respond to this drug, including such solid tumors in the breast, head, neck, thyroid, bladder, testes, prostate, and ovary, and Wilm's and Ewing's tumors. Malignant lymphomas and several leukemias have also gone into remission following treatment.[8]

Even though early (Phase I and II) clinical studies reported no cardiomyopathies, serious cardiotoxicity is the main limiting factor in its use. This includes congestive heart failure (CHF), which is delayed until the cumulative dose reaches or exceeds 550 mg/m^2. When CHF occurs it is usually irreversible and fatal. Myelosuppression is another major toxic manifestation.

The exact cytotoxic–carcinolytic mechanism of ADM and daunorubicin is not yet settled. Its strong binding and intercalating ability to DNA from many sources has been repeatedly demonstrated. Because of all of this "evidence," DNA has been widely presumed to be the primary target for ADM's cytotoxicity. However, N-(trifluoroacetyl) adriamycin-14-valerate (Fig. 4-19, AD-32) is a drug undergoing clinical trials, and exhibiting less cardio- and general toxicity than ADM, and is even possibly superior to it, yet this compound is not a pro-drug for ADM. Furthermore, it does not complex with DNA since

[8] The drug also shows activity against gram-positive and -negative bacteria and many fungi.

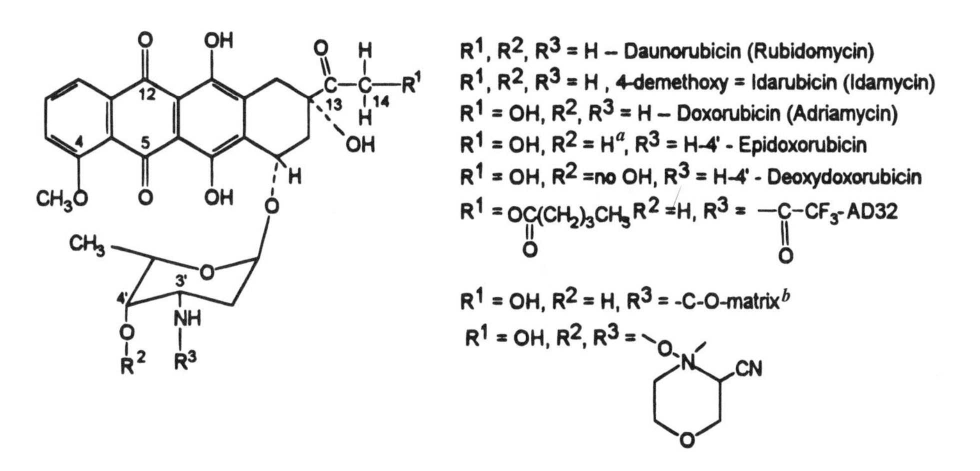

Figure 4-19. Anthracycline carcinolytiscs. [a](OH group is equatorial, i.e., epimeric to ADM); [b]carbonylimidazolyl cross-linked agarose beads.

the 3′-NH_2 group, known to participate in the DNA interaction, is blocked as a trifluoroacetamide. It is interesting that there were SAR reasons proposed to show the AD-32-type anthracycline derivatives should not be effective.

More recently, research indicates that the anthracycline compounds have an inhibitory effect on the intracellular enzyme topoisomerase II (also see Chapter 7). This has been taken as evidence that this enzyme may be the primary target for these drugs. A more balanced view at this time may be that the carcinolytic mechanism of action of the anthracycline antibiotics is related to the binding affinity to DNA (intercalation), inhibition of nucleic acid synthesis, and topoisomerase II interactions.

A fascinating report involves the synthesis of an ADM derivative where the 3′-NH_2 moiety is linked to a polymeric matrix consisting of cross-linked agarose beads linked to the drug via a carbonylimidazole carbamate bridge (Fig. 4-19). The compound, when tested against L1210 leukemia cells, was found to reduce the survival of the cancer cells 100–1000 times more effectively than free ADM, even though no intracellular ADM was detectable with the matrix-linked compound. Endocytosis of the ADM-bound beads is impossible since their diameter—40–120 μm—exceeds that of the leukemia cells considerably, ca. 15 μm. The conclusion, although not ruling out that ADM can adversely affect cell viability by DNA intercalation, is that ADM can also be cytotoxic solely by surface (membrane?) interactions.

In efforts to minimize cardiotoxicity, structural variations on the periphery of the ADM molecule continue. The ADM epimer, 4′-epidoxyrubicin (Fig. 4-19), appears to exhibit a somewhat decreased toxicological spectrum in Phase II human trials. The 4′-deoxy analog (Fig. 4-19) even shows oral activity in mice. A more extensive molecular modification, namely an oxazolidine ring, involving the O from 4′ and N from 3′ representing a cyanomorpholine derivative of ADM, was reported. This and several analogs exhibited intense potency against leukemia cells.

With the clinical introduction of the 4-demethoxy analog of daunorubicin (idarubicin, Idamycin, Fig. 4-19) in 1991 another set of "improvements" were achieved. These include higher antileukemic activity, higher potency, and lower cardiotoxicity than the parent compound daunorubicin. The approved indication is acute myelogenous leukemia.

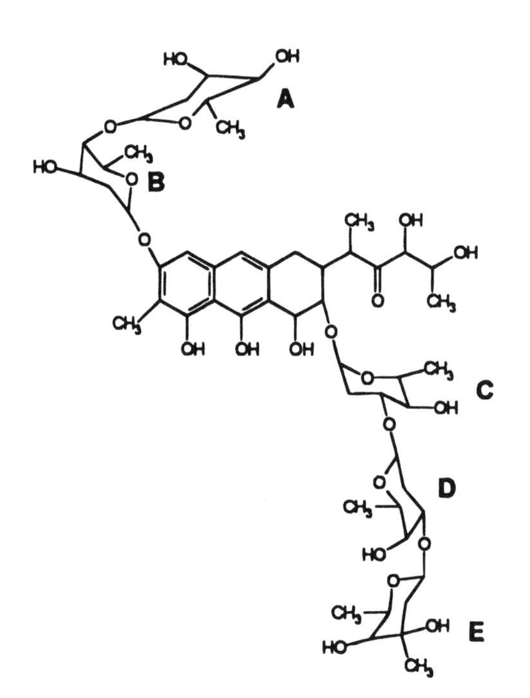

Figure 4-20. Mithramycin (Plicamycin). The aglycone is chromomycine. The sugars are: A, C, and D = olivose; B = oliose, D = mycarose.

4.13.1.1. Mithramycin (Plicamycin)

Mithramycin (Mithracin®) is a weakly acidic complex glycoside of the tetrahydroanthracene chromomycine. It is obtained from *Streptomyces plicatus*. Its ability to inhibit nucleic acid synthesis appears to be primarily on RNA even though binding to DNA has also been demonstrated. Intercalation does not appear to be part of the process. It was proposed that binding of this antibiotic to the DNA template may simply prevent the polymerase enzyme from "reading" it. The result is inhibition of RNA synthesis.

The drug presently has its main carcinolytic indication in the treatment of testicular carcinoma. An unrelated effect, probably on osteoclast bone cells, leads to a significant reduction of plasma calcium at one-tenth the antineoplastic dose. This has been useful in hypercalcemia of malignancy, Paget's disease, and hyperparathyroid conditions.

4.13.1.2. Bleomycin

Bleomycin (Blenoxane®, BLM) is a complex glycopeptide cytotoxic antibiotic obtained from *Streptomyces verticillus*. The BLM product in clinical use is a mixture of BLM A_2, which has a trimethylsulfonium group, and B_2, where the position *R* (Fig. 4-21) is a tetramethylene guanidinium function. Both moieties are strongly basic, which very likely attract that portion of the molecule to the acidic phosphate anionic backbone of DNA. This is in turn believed to permit the bisthiazolyl ring portion sufficient proximity to carry out its ultimate function—cleavage of the phosphodiester linkages of DNA.

The natural antibiotics are obtained as blue copper chelates, but it is the copper-free white forms that are used clinically. The drug is administered metal-free and it chelates

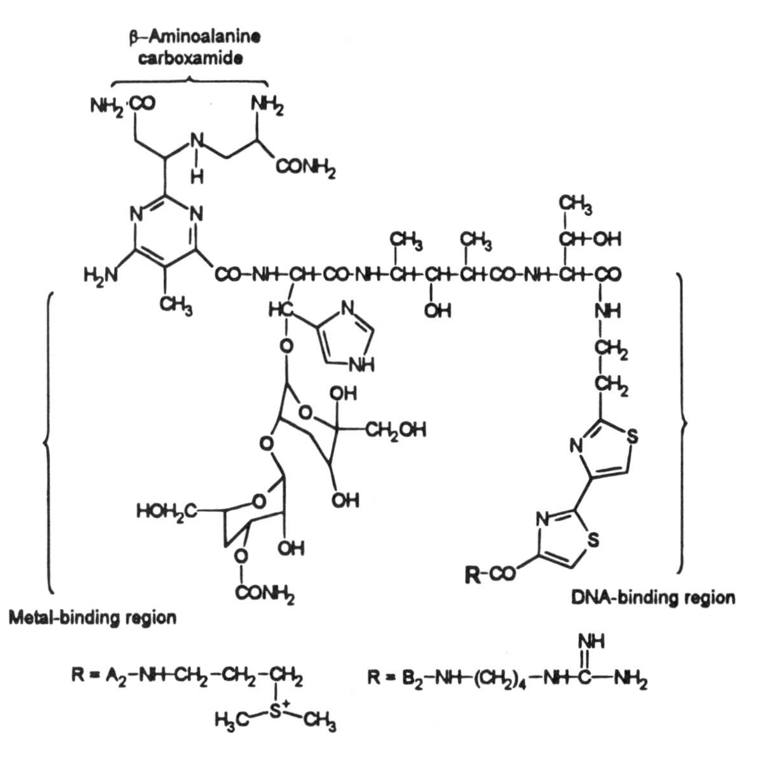

Figure 4-21. The structure of bleomycin A_2 and B_2.

with Fe^{2+} intracellularly in a square pyramidal spatial arrangement involving five of the nitrogen atoms of BLM. It then coordinates its sixth valence with O_2, converting it to a hydroxy-anion free radical: BML Fe^{3-}-OH^- -O^-. It was proposed that this free-radical iron–O_2 complex attacks the 4′-hydrogen atom of the 2′-deoxyribose, leading to an oxidative scission of the 3′–4′ carbon–carbon bond of the sugar ring of the cell's DNA (Eq. 4.10).

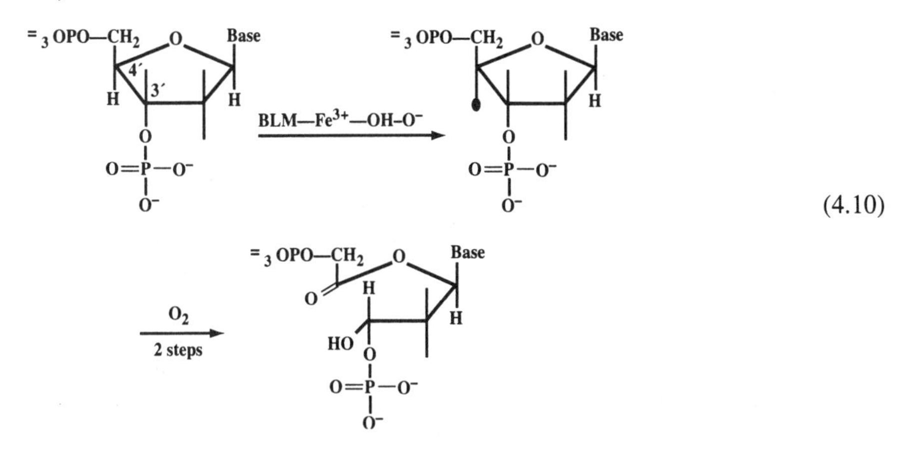

(4.10)

The picture thus emerging is that of a double-blow to the cancer cell. The initial "setup" involves intercalation by the right side of the molecule's bi-thiazole system, probably including electrostatic binding via positively charged atoms (e.g., sulfonium). This step is followed by the left side of the molecule having formed an Fe^{3+}–O_2H free radical complex attacking the 2′-deoxyribose in combination with molecular O_2, resulting in furanose ring cleavage. A series of additional phosphate ester breakdowns follow, yielding a highly cytotoxic thymine-3-propenal from the pyrimidine ring in the original BLM among its degradation products. (Recall acrolein from cyclophosphamide metabolism, Fig. 4-9.)

BLM is also metabolized by bleomycin hydrolase, which is an enzyme catalyzing the hydrolysis of the β-amino-alanine carboxamide (upper left of structure). The carboxylic acid obtained shows decreased DNA affinity. Several hydrolase-resistant BLMs, where the carboxamide nitrogen carries a 3-(α-methyl-benzylamino) propylamine group, exhibited lower pulmonary toxicity.

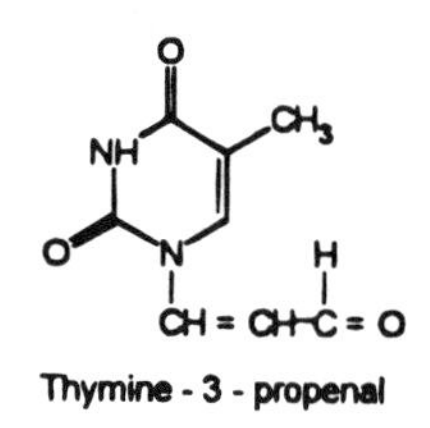

Thymine - 3 - propenal

The clinical usefulness of BLM is against solid tumors such as carcinomas of the testes, head, neck, and uterine cervix. Lymphomas, both Hodgkin's and non-Hodgkin's, also respond.

BLM does not affect the bone marrow. For this reason it is frequently given with cisplatin and/or ADM, which are myelosuppressive. Thus additive carcinolytic activity without additive bone marrow suppression can be achieved.

The most serious side effect of BLM is pulmonary toxicity, which affects about 10% of the patients and accounts for a 1% patient mortality. The pathology includes edema and fibrosis.

BLM also has potential diagnostic utility. Complexes with radioactive T1 or Ga are taken up by certain tumors, such as in the lungs, offering the possibility of enhanced—and earlier—detection.

4.14. Mitotic Inhibitors

Compounds known to inhibit the mitotic phase of the cell cycle offer potential usefulness in the treatment of cancer. Few such drugs, however, have reached clinical application either because of unacceptable toxicity or ineffectiveness. Two antimitotic agents are used for other purposes. The antibiotic griseofulvin is utilized for its antifungal properties. The alkaloid colchicine, although having antitumor activity, is used in the treatment of gout. Colchicine, obtained from the seeds of *Colchicum autumnale*, combines with tubulin reversibly. Tubulin is a protein, normally polymerizing to form microtubules found in animal and plant cells. Of particular interest is their function in chromosomal motion during mitosis. The ketonic area of the tropolone ring portion is the likely binding site. Inhibition of leukocyte microtubules is the probable reason for the alkaloid's strong antiinflammatory effect in gout. Podophyllotoxin, which is obtained from the resin of the May Apple or

Mandrake (*Podophyllum peltatum*) has been in use as a purgative and wart remover for centuries. The active antimitotic (metaphase) inhibition is believed to involve the lactone ring, which is in the *trans* configuration (Fig. 4-22). Epimerization to the *cis* form with alkali results in inactivity. Carcinolytic activity of the natural product is poor. However, some derivatives are more promising. Of particular interest is the epipodophyllotoxin glucopyranoside derivative etoposide (VePesid, VP-16-213), which exhibits metaphase arrest. Microtubule synthesis is not affected. Etoposide is useful against refractory testicular tumors and small-cell lung carcinoma.

The periwinkle plant (*Vinca rosa*), which is indigenous to Florida, has yielded several indole alkaloids. Two with antineoplastic properties are the dimeric indole-dihydroindole derivatives vinblastine (Velban®) and vincristine (Oncovin®). These agents bind to tubulin, arresting mitosis at the metaphase.

Vincristine is effective against acute leukemia in children, frequently used to initiate a treatment that is then switched to more selective agents such as MTX or 6-MP. An alternate approach is to combine it with prednisone to effect a high percentage of complete remissions. Vinblastine, the more active compound, has had much wider clinical application, including solid tumors, especially in combination with such drugs as cisplatin and BLM (e.g., metastatic testicular cancer).

A new semisynthetic analog, 5-nor-anhydrovinblastine (Navelbine), is undergoing clinical trials and exhibits activity against non-small-cell lung cancer, breast and ovarian cancers, and Hodgkin's disease, with remission at 90% in the last. The chemical modification is shortening the distance from the upper (left) indole to the piperidine N on the right by one CH_2 and dehydrating the tertiary alcohol to a double bond. Clinical tolerability is

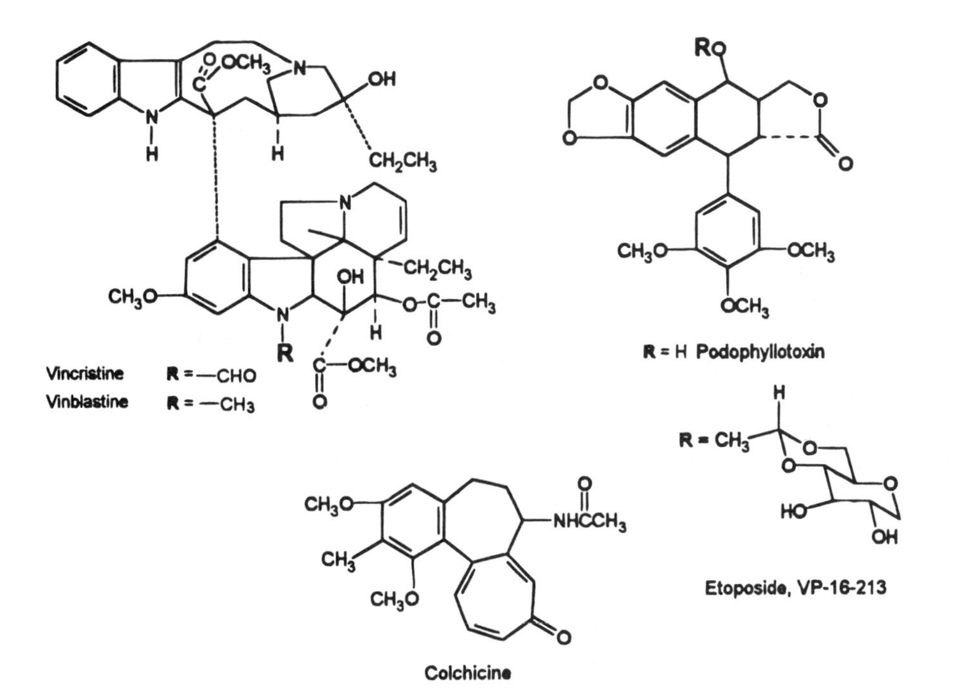

Figure 4-22. Antimitotic drugs.

better, primarily due to decreased hematological toxicity. It is equally important that the drug is orally effective, with a 40% bioavailability.

The newest antimitotic agent to be approved (1993) for clinical use is the diterpenoid taxol (Paclitaxel). It is obtained laboriously from the bark of the Pacific yew tree, *Taxus brevifola*. It requires 11,000 kg of dried bark to yield 1 kg of drug. Although investigated as a crude extract against various leukemias, sarcomas, and lung cancer in the late 1960s, it ultimately evolved into a drug against otherwise drug-refractory cases of ovarian and breast carcinomas.

The medicinal chemistry of taxol indicates that there are three points for esterification on the parent taxane ring system. Both taxol and its active analog taxotere have positions 3 and 10 so occupied. The 7 OH can also be so derivatized to afford active compounds. However, an unesterified C-3 hydroxyl results in an inactive compound. Water-soluble derivatives of C-7 and C-2′ have been synthsized as potential pro-drugs to circumvent the extremely poor solubility of the drug.

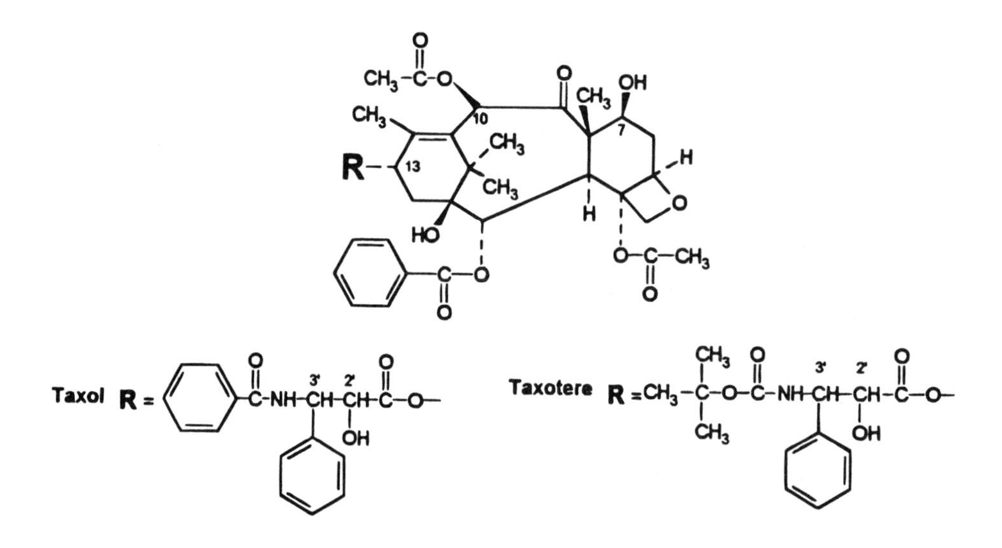

The antimitotic mechanism of taxol differs from the antimicrotubule agents such as colchicine and the vinca alkaloids discussed earlier. Rather than causing disassembly of the microtubules, taxol actually enhances tubulin polymerization. This upsets the normal dynamic equilibrium between soluble tubules, which are dimers, and the microtubule polymers. The stabilization of the latter inhibits mitosis in the latter part of Phase G_2 and M. Although sharing much of the toxicology of many of the anticancer drugs, taxol promises to be an important addition to the cancer armamentarium.

4.15. Hormonal Agents

It has been known for some time that the growth of tumors of primary or secondary sex organs, such as ovaries, breast, and prostate, are dependent on the estrogen and/or androgenic hormones associated with them to a significant degree. This is because carcinomas of these organs continue to exhibit the hormonal requirements of the tissues from which they arose. Depriving the malignant process of the source of these hormones therefore offers a potentially useful method of treatment. The methods available to oncologists are surgical removal of the prostate, testes, and ovaries, inhibition of a particular hormone's biosynthesis (aminoglutethimide, AG), hormonal therapy with estrogenic (ethynyl estradiol, diethylstilbestrol, DES) and androgenic compounds (testosterone propionate), and, more recently, antiestrogens (tamoxifen). All these modalities have resulted in varying percentages of remissions, definite increases in survival times, or, at the very least, considerable palliation of symptoms. Because the mechanisms of these drugs are hormonal, the type of severe drug-induced toxicities (and even mortalities) seen with the alkylating agents, antimetabolites, and cytotoxic antibiotics are not encountered.

Surgical adrenalectomy has been a proved treatment in advanced breast cancer for years. Remissions in about 40% of selected patients are achieved. However, if steroid synthesis could be medically inhibited, it would certainly be safer, particularly in advanced disease patients at high risk from surgery. The drug AG (Cytadren®), which was formerly used as an antiepileptic agent (Elipten®), was withdrawn (1966) because long-term use produced adrenal insufficiency. Subsequent research found that the compound inhibited the adrenal enzymatic conversion of cholesterol to Δ^5-pregnenolone, an early step in the biosynthesis of all adrenal steroids. Other points in steroid biosynthesis are also blocked, including hydroxylations at positions 11, 18, and 21. These oxidative blocks are achieved by inhibition of cytochrome P-450 (Chapter 3). Because the overall blockade reduces available estrogens, the potential utility of this drug in estrogen-dependent tumors is obvious.

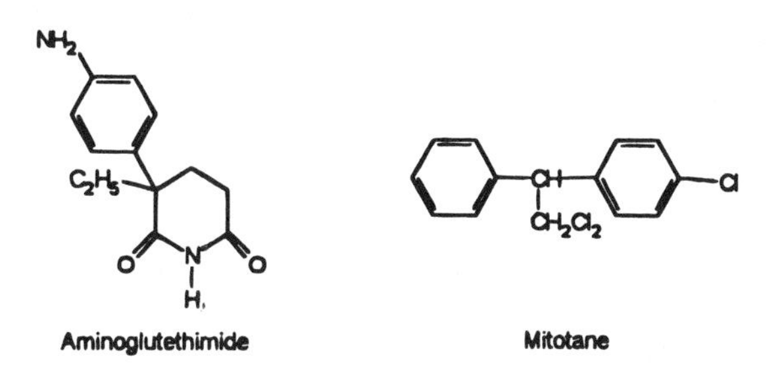

An interesting clinical complication in the use of AG is that the decreased levels of cortisol cause increased secretion of adrenocorticotropic hormone (ACTH) by feedback stimulation of the pituitary gland, which will overcome the steroid synthesis inhibition of AG. This tendency must be overcome by coadministration of hydrocortisone. In spite of considerable clinical literature on the usefulness of AG in the treatment of advanced breast cancer and adrenocortical carcinoma, the only approved indication for AG in the United States (1990) is therapy for Cushing's syndrome.

Mitotane (Lysodren®) is a cytotoxic agent with high selectivity for the adrenal cortex. Mitochondrial damage apparently leads to atrophy of the gland due to cell death. The drug is therefore specific against the rare adrenocortical carcinoma. The palliative response, although high (80%), usually lasts only six months.

4.15.1. Progesterone, Estrogens, and Antiestrogens

It may be useful here to consider briefly the role of estrogens in this complex picture. Estrogen target tissues such as the uterus, vagina, breast, and anterior pituitary bind strongly to the hormone. Studies involving human breast cancer showed a definite relationship between estrogen uptake (i.e., binding) and the response to oophorectomy and adrenalectomy. These intracellular cytoplasmic hormone-binding proteins, now termed *estrogen receptors*, have since been found to exist in up to 65% of breast cancers; postmenopausal women having a much higher incidence and level of receptor-positive tumors than premenopausal women. Negative receptor assays quickly identify patients who will be nonresponsive to hormonal therapy.

The presence of such specific binding proteins in the cells of the various target tissues determines the action of the different steroid hormones, such as androgens, estrogens, progestins, or corticosteroids. The specific steroid–cytoplasmic complex is very strong. Its transformation in the cell leads to translocation into the nucleus, where, following chromatin binding to specific sites, it unmasks particular cistrons.[9] It is probably the steroid–cytoplasmic complex that should properly be considered the actual agonist (or antagonist if an antiestrogen is involved) and the nuclear site as the receptor. Following chromatin binding, RNA-polymerase stimulation leads to the synthesis of specific *m*-RNA, which, on becoming a ribosomal component, codes for the various enzyme regulator and structural proteins. Progesterone-specific binding proteins have also been found in uterine tissue and likely coexist in estrogen-positive tumors. Absence of the latter even in an estrogen-positive tumor may be responsible for resistance to hormonal treatment in those cases.

Androgens are occasionally described as antiestrogens, even though they are not pharmacologically antagonists to estrogens. Since receptors for these agents also exist in estrogen-sensitive tissue, actions on these receptors affect the physiology of these tissues as well. These hormones should therefore be viewed as modifiers of estrogenic tissue responses.

True antiestrogens are exemplified by several triphenylethylene nonsteroidal compounds (Fig. 4-23). They can antagonize estrogen-stimulated uterine growth and estrogen-dependent mammary tumors.[10] These compounds have been shown to bind to estrogen receptors and subsequently migrate to the nucleus, forming what must be considered an antiestrogen–receptor complex. Thus they decrease cytoplasmic estrogen receptor sites without exhibiting estrogenic effects. Dimethylbenzanthracene-induced mammary tumors in rats are dramatically reduced, particularly by tamoxifen and nafoxidine. Tamoxifen (Nolvadex) is used for palliative treatment of advanced breast cancer in postmenopausal women, showing positive estrogen receptor assays. Response rates of 45% are achievable (negative estrogen receptor patients showed no response). Clomiphene (Clomid) achieved a 39% response rate where tumor remission was a mean of 1 year. The drug is also indicated as an ovulatory agent for the treatment of infertility. Nafoxidine, not yet approved in the United States, exhibits somewhat lower response rates than does tamoxifen while having a less desirable adverse effect profile. To effect androgen control in the treatment of nonresectable prostatic cancer involves either orchiectomy (castration), the use of estrogenic drugs, or both. The fact that estrogens can block androgen effects has been known for over four decades; the exact mechanism, however, is still not understood. Of the estrogenic agents available DES is probably the drug of choice. Chlortrianesene (Tace), which is a progesterone, is a long-acting drug

---

[9] A cistron, or gene, is that segment of DNA specifying a complete polypeptide chain.

[10] They also stimulate ovulation in certain women.

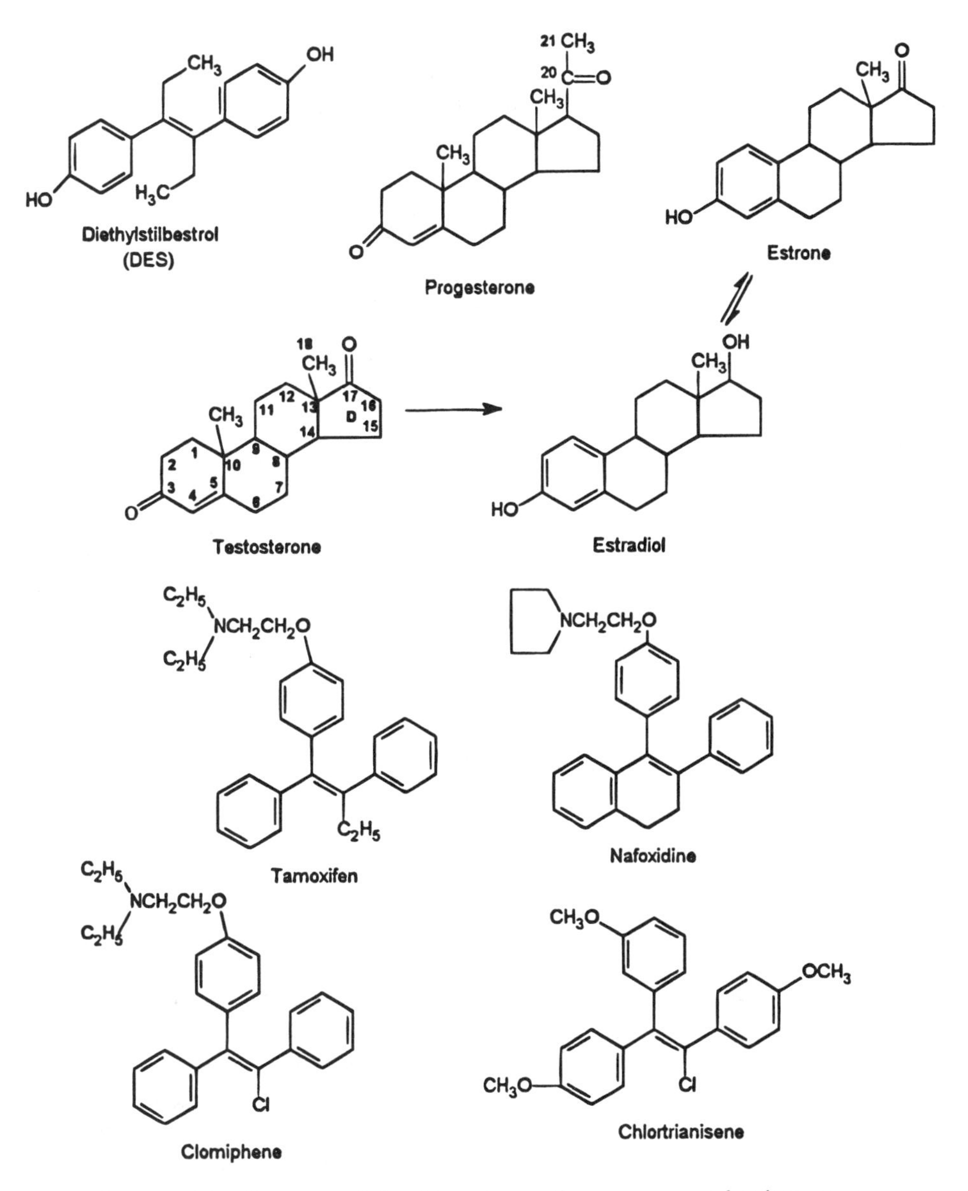

Figure 4-23. Structural relationships of progestins, estrogens, and antiestrogens.

useful in palliative treatment of prostatic carcinoma. It is also used for postpartum breast engorgement.

The use of estrogens against mammary carcinomas at first appears paradoxical in view of the fact that the estrogenic agents have been found to be carcinogenic (e.g., DES, estrone). However, the doses are considerably in excess of physiological needs. It has been suggested that these compounds act directly on tumor cells at high doses, inhibiting the trophic effects of prolactin.

Glucotropic steroids have antianabolic effects on lymphocytes and fibroblasts and are therefore useful in remission induction of chronic and acute lymphocytic leukemia (ALL) as well as lymphomas since the 1940s. Prednisone, particularly in combination drug protocols, now routinely gives complete remissions in 100% of cases. Alkylating-resistant chronic lymphocyte leukemia also responds to these steroids.

4.16. Miscellaneous Carcinolytics

4.16.1. Procarbazine

Procarbazine (Matulane), a hydrazine compound having monoamine oxidase inhibiting activity, is very active against Hodgkin's disease, especially in combination with mustargen, vincristine, and prednisone—the MOPP protocol. The drug has activity against certain lung cancers, myeloma, melanoma, and brain tumors. The drug undergoes complex metabolic activation.

In this form chromosome damage, antimitotic effects, and inhibition of proteins and nucleic acids have all been demonstrated, yet no clear mechanism has emerged. The drug has a short half-life, is rapidly metabolized, develops resistance, and has a high incidence of side effects, including psychic disturbances. Because of its hydrazine structure and MAO inhibition, it is subject to drug interactions with sympathomimetics, tricyclic antidepressants, and tyramine-containing foods. Like the alkylating agents procarbazine is carcinogenic.

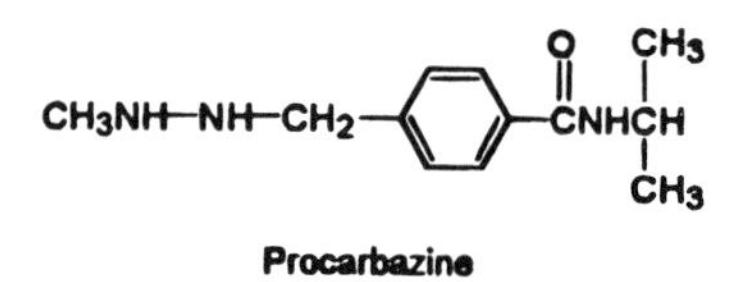

Procarbazine

4.16.2. L-Asparaginase

L-Asparaginase (L-asparagine amino hydrolase, Colaspase®, Elspar®) catalyzes the hydrolysis of asparagine to aspartic acid (and NH_3). Asparagine is an essential amino acid that many tissues synthesize as needed. Some malignancies, particularly ALL in children, obtain it only from exogenous sources. Utilizing the enzyme to degrade the limited amounts of asparagine therefore inhibits protein synthesis, thus in effect making the enzyme indirectly cytotoxic to those malignant cells. The discovery of this (enzyme) factor in guinea pig serum led to more detailed studies and identification of L-asparaginase as the oncolytic factor. The original (wishful) assumption that this qualitative difference between normal and at least certain malignant cells was significant was not borne out since normal tissues are now known to be sensitive to this enzyme. At least some of the toxicities of this drug undoubtedly result from this fact. Not all asparaginases are clinically effective. Only those obtained from *E. coli* and *Erwina casotovara* having K_m values around 10^{-5} (mol. wt. 140,000) are utilized. Purified enzyme preparations also contain varying amounts (1–9%) of glutaminase activity.

The drug's usefulness is essentially limited to ALL and, rarely, myelogenous leukemia. No activity against solid tumors is discernible. Remissions, when achieved, tend to be short.

Toxicity can be extensive. Since the enzyme is a foreign protein, hypersensitivity reactions, including anaphylaxis, are to be expected. Neurotoxicity, hepatotoxicity, hyperglycemia (decreased insulin production?), pancreatitis, and myelosuppression are also encountered.

4.17. Development of New Modalities

There is an increasing awareness of the usefulness of cancer chemotherapy, especially in conjunction with surgery and radiation. Palliation and prolongation of life are more frequently achieved. Except in certain nonsolid tumors, however, cures are still infrequently produced. The best results are achieved primarily in the treatment of neoplasms that have a high proportion of rapidly growing cells in which mitosis and DNA and protein synthesis are actively occurring. Combination therapy has essentially replaced single-entity treatment. It is painfully apparent, however, that simply developing new alkylating agents and DNA or protein synthesis inhibitors will not lead to dramatic improvements. Breakthroughs can only come by filling in the major gaps in our understanding of malignant and metastatic processes.

In the interim less dramatic steps are being taken to improve the overall picture. One interesting idea is to increase the susceptibility of radiation-resistant malignant cells to radiation.

Radiation of localized tumors leads to tumor shrinkage and destruction; however, regrowth usually occurs. It is somewhat paradoxical that those cells within a tumor sufficiently distant from the nearest capillary will eventually die of hypoxia—if the tumor is *not* treated. These are the cells, however, that are resistant to radiation when the neoplasm is so treated. The sensitive cells regress, hypoxic cells reoxygenate, cell growth restarts, and tumor regrowth occurs. Studies have demonstrated that certain compounds are able to increase radiation sensitivity of hypoxic cells without similarly affecting oxygenated cells. These are compounds with high electron reduction potentials, particularly certain nitroheterocycles. The most effective agents developed so far are the nitroimidazoles misonidazole (RO-7-0582) and desmethylisonidazole (RO-05-9963). Clinical studies with misonidazole show good tolerance and tumor penetration.

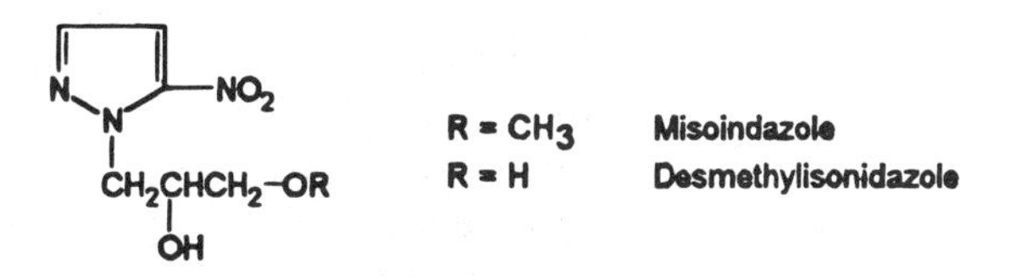

A potentially more promising approach is the application of the new monoclonal antibody (MCA) technology to the problem made possible by their discovery. Research into the production of antibodies against specific tumor cells is being carried out intensely. Even though evidence that MCAs recognize structures that are exclusively tumor-specific is not strong, progress has been made in the areas of gastrointestinal and colorectal malignancies. MCAs are now being used to assess their potential value as therapeutic agents. A preliminary report describes the infusion of monoclonal anti–T-cell antibodies into advanced T-cell lymphoma patients. Clinical results, although gratifying, were temporary. Remissions were not achieved. Favorable, but also temporary, improvement resulted in patients with T-cell leukemias and mycosis fungoides using mouse MCAs directed against lymphoma-associated antigen.

A more imaginative application of MCAs to cancer therapy is to arm them by coupling with radioactive isotopes, thus increasing the localization of radiation damage to cancer cells. For example, it was possible to localize tumors xenografted in immunosuppressed mice with mouse MCA raised against carcinoembryonic antigen (CEA) and teratocarcinomas labeled with^{131} I. Administering radioactively armed MCAs into restricted spaces such as peritoneal, pleural, or pericardial areas enables the achievement of higher local concentrations than by the intravenous route. A tumor-associated MCA armed with^{131} I given to patients with malignant pleural effusion (intrapleurally) and with tumorous pericardium (intraperitoneally) all showed considerable improvement without allergies; only 0.2% radioactivity diffused through the whole body.

Another exciting approach is the use of drug–antibody conjugates. Antibody-targeted "warheads" should be more effective than the shotgun method, where nonselective, highly toxic molecules are poured into a whole system in order to get at a particular target site. Alkylating agents, antimetabolites, and carcinolytic antibiotics are being used. Dactinomycin and Adriamycin coupled to antibodies via the C3 to C4 bond of the aminosugar moiety have retained anticancer effects. Mice with EL4 lymphoma were treated with MCAs raised against this cancer and coupled with MTX (13 moles per mole of antibody). Eight of 11 mice lived 90 days or more, versus 17 days for the controls (Kulkarni et al., 1981). Chlorambucil was coupled to MCAs and showed activity against human colon adenocarcinoma (Bernier et al., 1984). Even toxins such as diphtheria toxin and ricin (from the castor bean) have been successfully coupled to MCAs raised against various tumors.

As tumor specificity of MCAs improves, so will the effectiveness of the drugs coupled to them. It is hoped that the efficiency for the selective defeat of malignant tissues will continue to be improved.

An additional possibility is to reduce the tumor–host imbalance by affecting distribution and uptake to affect the drug selectivity more toward the malignant tissue. This might be accomplished by incorporating cytotoxic compounds into microspheres of various sizes. Particles less than 2 μm are engulfed by macrophages and rapidly cleared from the blood. Depending on their size, however, larger particles will be trapped in particular areas or organs. For example, particles in the 2–12 μm range on intravenous administration will tend to be entrapped in the capillary bed of the lung and liver. Intra-arterial infusions of colloidal particles above 12 μm will block the first capillary bed they encounter, which may be viewed as a temporary "chemoembolism." Such an infarction can produce sustained localized drug action while the biodegradable particles are slowly disintegrating. An example of enhanced dactinomycin activity against subcutaneous sarcoma was described. A particularly intriguing method using microspheres (2 μm) containing Adriamycin and ferromagnetic albumin permitted the achievement of high concentrations of drug (3.7 $\mu g\ g^{-1}$) in tumors by extracorporeal manipulation with a magnetic field. Drug distribution in other tumors did not occur. In addition to albumin, other matrix substances for microspheres have included lipoproteins (with MTX), polyalkylcyanoacrylate (AD), and other substances.

Finally, biological response modifiers should be mentioned. Immunostimulants, a concept predating this century, are making a modern-day comeback with intensive research of interferons, interleukin, and synthetic polynucleotides. Studies involving interleukin-2 described significant responses in patients with melanoma, colorectal, kidney, and lung cancers. This and other stimulant factors are now in clinical use (see Chapter 15).

References

Barich, L. L., Schwartz, J., Barich, D., *Cancer Res.*, **22**:53, 1962.

Bernier, L. G., Page, M., Guadreault, R. C., Joly, L. P., *Brit. J. Cancer,* **49**:245, 1984.

Broome, J. D., *J. Exp. Med.,* **118**:99, 125, 1963.

Connors, T. A., Cox, P. J., Farmer, R. B., et al., *Biochem. Pharmacol.,* **23**:115, 1974.

Deutsch, H. M., Glinski, J. A., Hernandez, M., et al., *J. Med. Chem.,* **32**:788–92, 1989.

DeVita, V. T., Young, R. C., Canellos, G. P., *Cancer,* **35**:98, 1975.

DiMarco, A., Arcamoni, F., Zanino, F., et al., *Antibiotics,* (NY) 3, 1975.

Fox, M., *Nature* (London), **307**:212, 1984.

Frei, E., *Science,* **217**:600, 1982.

Giloni, L., Takeshita, M., Johnson, F., et al., *J. Biol. Chem.,* **256**:8608, 1981.

Huebner, R. J., Todaro, J., *Proc. Natl. Acad. Sci.* (USA), **64**:1087, 1969.

Huggins, C., Grand, L. C., Brillantes, F. P., *Nature,* **189**:204, 1961.

Karnofsky, D. A., Clarkson, B. D., *Ann. Rev. Pharmacol.,* **3**:361, 1963.

Kohn, K. W., *Cancer Res.,* **37**:1450, 1977.

Kulkarni, P. N., Blair, H. H., Ghose, T. I., *Cancer Res.,* **41**:2700, 1981.

Lippard, S. J., Ciccarelli, R. G., Solomon, M. J., Varshavsky, A. J., *Biochemistry,* **24**:7533, 1985b.

McGuire, W. P., *Ann. Intern. Med.,* **111**:273, 1989.

Montgomery, J. A., Ruby, J., McCaleb, G. S., Johnston, T. D., *J. Med. Chem,* **10**:668, 1967.

Pigram, W. J., Fuller, W., Hamilton, L. D., *Nature* (New Biol.), **235**:17, 1972.

Reed, D. J., May, H. E., Boose, R. B., et al., *Cancer Res.,* **35**:568, 1975.

Sherman, S. E., Gibson, A., Wang, H. T., Lippard, S. J., *Science,* **230**: 412, 1985.

Skipper, H. E., Schabel, F. M., Wilcox, W. S., *Cancer Chemother. Rep.,* **35**:1, 1964.

Skipper, H. E., Schabel, F. M., Wilcox, W. S., *Cancer Chemother. Rep.,* **45**:5, 1965.

Tan, C., Tasaka, H., Yu, K. P., et al., *Cancer,* **20**:333, 1967.

Suggested Readings

Farmer, P. B., Walker, J. M., *Molecular Basis of Cancer.* John Wiley & Sons, New York, 1985.

Garner, R. C., Hradec, J., Eds., *Biochemistry of Chemical Carcinogenesis.* Plenum Press, New York, 1989.

Killion, L. J., Fidler, I. J., Mechanisms of Cancer Metastasis, *Arzneim-Forsch.* **39**(8a), 1031–4, 1989 (in English).

La Fond, Re, Ed., *Cancer—The Outlaw Cell,* 2d ed., American Chemical Society, Washington, D.C., 1988.

Pitot, H. C., *Fundamentals of Oncology,* 3d ed. Marcel Dekker, New York, 1986.

Pratt, W. B., *The Anticancer Drugs.* Oxford University Press, New York, 1979.

Reinhardt, D. N., Connors, T. M., Pinedo, H. M., von de Poll, K. W., Eds. *Structure Activity Relationships of Anti-Tumor Agents.* Martinus-Nijhoff, Boston, 1983.

Remers, W. A., *The Chemistry of Antitumor Antibiotics,* Volume Two. Wiley Interscience, New York, 1988.

Remers, W. A., *Antineoplastic Agents.* Wiley Interscience, New York, 1984.

CHAPTER

5

Analgetics and Nonsteroidal Antiinflammatory Agents

5.1. Introduction

The most pervasive, frightening, and important symptom of injury, and much of disease, is pain. Pain should be considered a syndrome of highly unpleasant sensations rather than a symptom.

Everyone knows what pain is, yet it is difficult to define. One concept views pain as a two-component phenomenon: the initial sensation, or actual perception, and the psychological effect, or reaction component. The latter invariably involves mental modifications that account for the variations observed in people. In fact, the same individual may react differently to identical pain stimuli at different times. The intensity of stimulus to which the subject reacts—the so-called pain threshold—varies greatly. The plethora of adjectives used to describe pain illustrates the range of its characteristics. For example, throbbing, gnawing, splitting, stabbing, pinching, crushing, and burning are all expressions associated with the description of pain.

Since pain is both a physiological and psychological phenomenon, many technical problems in its clinical evaluation have not been satisfactorily solved. Because it has such subjective qualities, how can the intensity of pain be quantitatively estimated? How can the degree of relief obtained by various treatment modalities be determined or even compared? With a human subject the investigator is, by necessity, dependent on verbal communication for data. Considering the subject's physiological condition and the placebo effect (where up to one-third of the subjects claim relief having unknowingly received "blank" medication), how reliable are these data?

5.2. Classification of Pain

One classification of pain is to consider it as either *acute* or *chronic*. Since it is now recognized that the etiologies, pathophysiologies, functions, diagnoses, and therefore means of treatment for these two types of pain differ, it is useful to view them on this basis.

Acute pain is a set of unpleasant and emotional experiences often culminating in behavioral responses. Acute pain is, invariably, produced by disease, injury, noxious chemicals, or some physical stimulation (e.g., heat). Much of our knowledge about acute pain has been acquired from studies of experimentally induced pain in laboratory animals or even human volunteers. Clinical situations such as acute dental pain, the pain of parturition, and many postsurgical situations have also been an impetus to research.

Chronic pain, by its persistent and pathological form, appears to have no biological function. It imposes physical, emotional, and social stresses of severe magnitude. The patient's response to chronic pain is very different than to acute pain. Physical deterioration is often seen, and is actually aggravated by resorting to excessive medication. There is some indication that the pain threshold—and therefore tolerance of pain—actually decreases, possibly due to a depletion of endorphins.[1] In some cases the underlying pathology, if any, for chronic pain cannot be found.

Another categorization considers pain from its point of origin. Thus *visceral pain* emanates from nonskeletal parts of the body, such as gastric pain, intestinal cramps, and colic. The so-called nonnarcotic or milder analgetics are usually ineffective in these instances. *Somatic pain* emanates from muscle and bone and includes headaches, sprains, and arthritic pain.

5.3. Classification of Analgetics

The traditional manner of classifying pain-relieving drugs has been on the basis of strong and addictive (or narcotic) compounds and the mild, nonnarcotic agents. Developments have tended to obscure the validity of this separation. The once-believed inseparability of analgetic potency and addiction liability has now been largely overcome. In addition, compounds now exist that chemically would belong in the mild category whose potency exceeds that of some narcotic agents. Furthermore, compounds whose structure is derived from narcotics exhibit strong analgesia yet are not addicting; in fact, they are capable of precipitating withdrawal symptoms in addicts (i.e., they possess narcotic antagonist properties). In this chapter the following modified traditional classification will be used:

1. Mild analgetics
   A. Analgetic–antipyretic drugs
   B. Antiinflammatory analgetics
2. Strong analgetics
   A. Narcotic analgetics
   B. Agonist–antagonist analgetics

Experimental testing for pain relief in humans presents us with an immediate dilemma. Are results obtained from healthy individuals in whom pain has been experimentally inflicted as meaningful as those derived from actual patients experiencing pathological pain? Probably not. Today the most widely recognized method for clinically evaluating analgetic drugs is the double-blind crossover procedure using patients with postoperative pain.

In the employment of the double-blind method, neither the patient nor the investigator knows whether the drug or a placebo is being administered. The crossover concept further ran-

[1] Endorphins are discussed later in this chapter.

Table 5-1. Comparative Analgesia in Mice and Humans[a]

Drug	Mice ED_{50}[b] (mg/kg)	Human dose[b] (total, mg)
Morphine	2.11	10
Codeine	14.2	60–120
Heroin	0.9	3–5
Hydromorphone (Dilaudid)	0.3	2–5
Hydrocodeinone (Dicodid)	3.2	15
Anileridine (Leritine)	3.1	15–30
Methadone (Dolophine)	1.6	10
Meperidine (Demerol)	9.9	50–100
Pentazocine (Talwin)	Inactive	20–30
Aspirin[c]	125	650

[a] Modified from Jacobson 1970.
[b] Subcutaneous in mice, intramuscular in humans.
[c] Oral in mice and humans.

domizes the method so that each patient, by sometimes receiving the drug and at other times the placebo, acts as his or her own control. Investigator bias is ideally also eliminated. Reference compounds (e.g., morphine or aspirin) may sometimes be included in such studies.

It is of interest to note that in carefully controlled studies potent placebo effects are obtained that sometimes exceed 30% of test subjects. This means that one out of three patients experiencing even severe pain who believes to be receiving a potent painkiller while getting an inert substance claims to have partial or even total pain relief. This illustrates how deeply pain is enmeshed with psychological factors. It also points out the difficulties and pitfalls the clinical pharmacologist encounters evaluating analgetic drugs.

In some ways the experimental pharmacologists who test drugs on animals have a simpler task. They do not have to rely on psychologically motivated statements of test subjects. Instead, with a laboratory animal, "pain" thresholds can be reproducibly determined by measuring certain reflex actions caused by noxious stimuli such as heat, pressure, and electric shock. Thus methods such as the mouse hot plate test, the rat tail flick test, and application of electricity to tooth pulp give reproducible end points whose appearance can be delayed in time with analgetic compounds in a more or less direct relationship to drug efficacy or potency. It is interesting that not all analgetic drugs give a positive response to all methods. In some cases false positives are obtained with compounds devoid of analgesia. Nevertheless, much valuable information is derived from such experiments. Table 5-1 gives the potency of representative analgetics in mice expressed as ED_{50} values (effective dose in 50% of test animals) and compares them with the equivalent single effective dose in humans.

5.4. Mild Analgetics

5.4.1. Analgetic–Antipyretics

Drugs in this category have the ability to alleviate mild, and on occasion, severe pain. The agents are also quite effective in reducing fever to normal levels.[2] The drugs most frequently

[2] *Antipyretic* is the term applied to drugs that reduce fever to normal body temperatures. Compounds able to reduce temperature below normal are termed *hypothermic*.

encountered here are acetanilid, acetophenetidin (phenacetin), and acetaminophen (*p*-acetoaminophenol, paracetamol). As Eq. 5.1 illustrates, these three compounds are metabolically interrelated. Both acetanilid and phenacetin are metabolized to the active acetaminophen. Acetanilid–a drug dating back to the late nineteenth century—has not been in use in the United States for several decades. Although still employed elsewhere, it is known to cause cardiac irregularities, jaundice, skin reactions, a high incidence of methemoglobinemia, and, rarely, severe blood dyscrasias, including life-threatening hemolytic anemia.

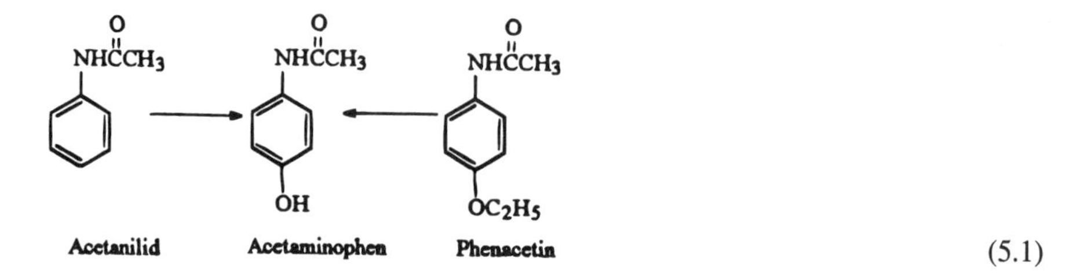

(5.1)

Acetophenetidin, the ethyl ether of acetaminophen, has also been used worldwide for the better part of this century, frequently in combination with aspirin. Because of high-dose long-term abuse in many countries, it developed a reputation of nephrotoxicity. Following evidence that phenacetin caused kidney and liver cancer in the abuser, the drug was banned in the United States in 1982. Thus we are left with acetaminophen as the only aniline-derived analgetic–antipyretic. Even though this drug has no significant antiinflammatory activity, its usefulness in noninflammatory conditions (and fever) approaches that of aspirin. Studies have shown acetaminophen and aspirin to be equianalgetic at the same dose, 650 mg.

5.4.2. Antiinflammatory Analgetics

Antiinflammatory analgetics present an interesting array of chemical compounds. The best-known and most widely used group are the salicylates (i.e., salicylic acid and its various salts and derivatives). Others are derivatives of anthranilic acid (*o*-aminobenzoic acid), which may be considered somewhat isosteric with salicylic acid. Several congeners of the dangerous nineteenth century pyrazolones anti- and aminopyrine are still in use. The introduction of arylacetic acids in the early 1970s has led to a large number of useful antiarthritic–analgetic agents, the structural parameters of which are still being explored. Indole and pyrrole acetic acids, which, of course, are aromatic, should also be considered as a subgroup here. Finally, acidic heterocyclic sulfonamide compounds round out the chemical moieties that offer clinically important antiinflammatory analgetics. Table 5-2 gives examples of agents in each group. It should be pointed out that all these compounds additionally exhibit useful antipyretic properties. Other pharmacological effects will also be discussed.

The analgetic potency of these compounds varies considerably. However, they do share several properties. They are all acidic and exhibit clinically useful antiinflammatory properties. In addition, they share a common mechanism of action, as will be discussed. The nonsalicylate drugs in Table 5-2 are now collectively referred to as *nonsteroidal antiinflammatory drugs* (NSAIDs). The salicylates should actually share this designation.

Aspirin today must be considered the prototype mild analgetic and antiinflammatory agent. NSAIDs are still evaluated by comparison to it. Physicians still consider it the first agent of choice for the treatment of rheumatic conditions.

Table 5-2. Representative Antiinflammatory Analgetics

Structure	Drug[a]	Comments
Salicylates	Salicylic Acid	Used primarily as Na^+, Mg^{2+} and choline salts
	Aspirin Acetylsalicylic acid	Introduced in 1899
	Salicyloyl salicylic acid Salsalate (Disalcid)	A pro-drug of salicylic acid
	Salicylamide	Not metabolized to salicylate in humans
	Diflunisal (Dolobid)	
Anthranilic Acids (Mefenates)		
	Mefenamic acid (Ponstel)	
	Meclofenamic acid (Meclomen)	Marketed as Na salt
Arylacetic Acids		
	Ibuprofen (Motrin)	
	Fenoprofen (Nalfon)	
	Ketoprofen (Orudis)	
	Flubiprofen (Ansaid)	
	Naproxen (Naprosyn)	

(*Continued*)

Table 5-2. *(Continued)*

Structure	Drug[a]	Comments
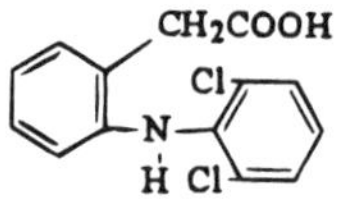	Diclofenac (Voltaren)	
Indolacetic Acids		
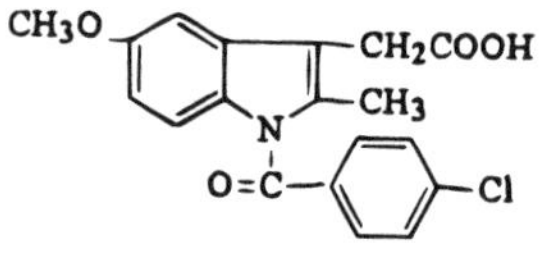	Indomethacin (Indocin)	
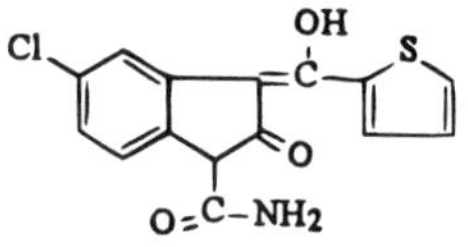	Tenidap (Enablex)	
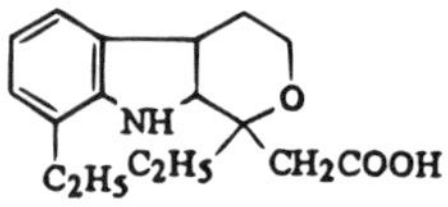	Etodolac (Lodine)	
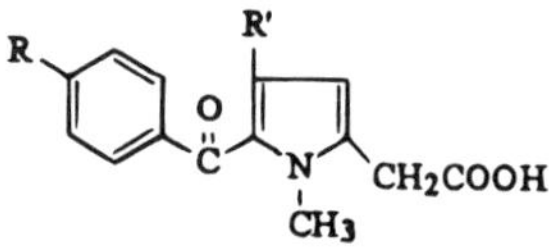	*R* = CH_3 Tolmetin *R'* = H (Tolectin) *R* = Cl Zomepirac[b] *R'* = CH_3 (Zomax)	Oral potency (100 mg) equivalent to meperidine (75 mg)
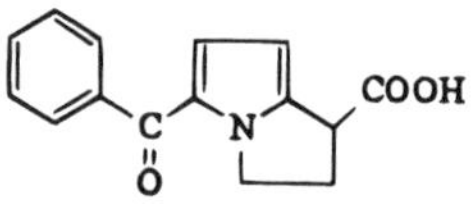	Ketorolac (Toradol) as tromethamine salt	Analgesic efficacy 30–60 mg IM; equivalent to 12 mg morphine or 100 mg merpidine. Also available in 5 and 10 mg oral tablets.
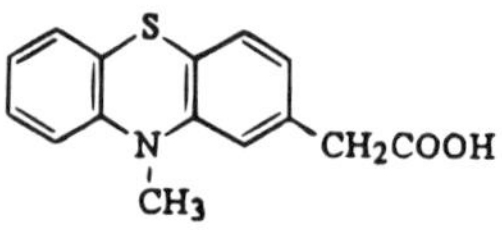	Metiazinic acid	Not marketed in U.S.
Pyrazolidine derivatives		
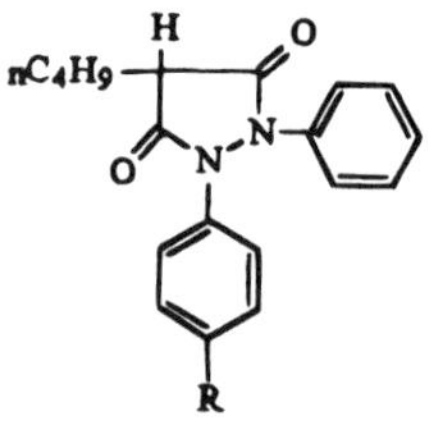	*R* = H, phenylbutazone (Butazolidin) *R* = OH, oxyphenylbutazone[b] (Tanderil)	 An active metabolite

(Continued)

Table 5-2. *(Continued)*

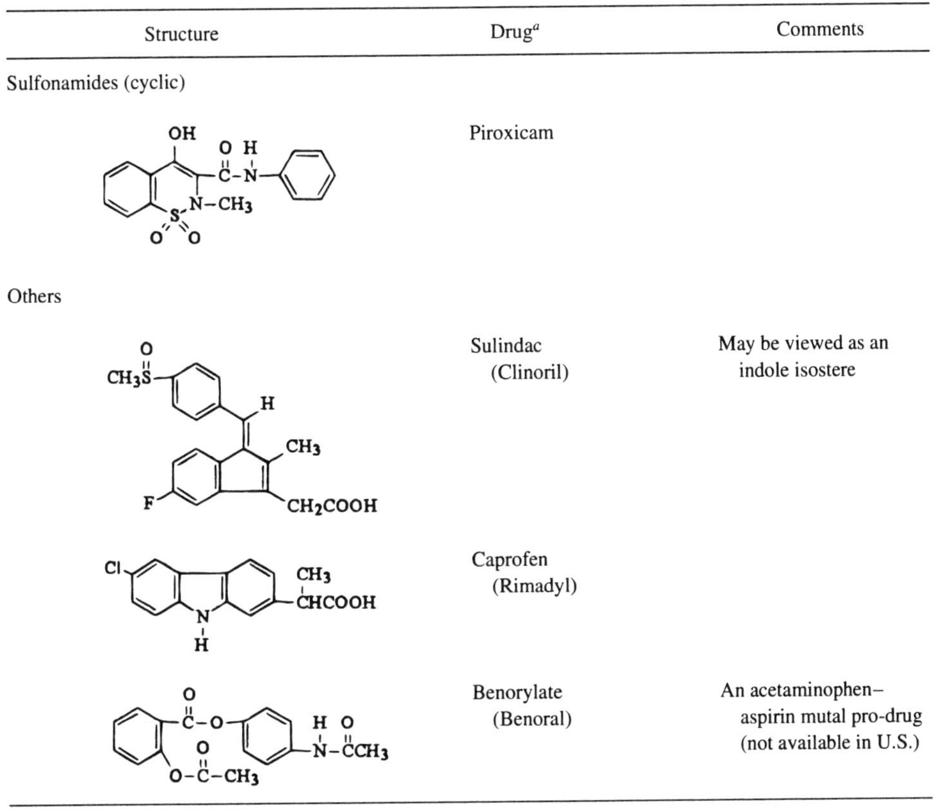

Structure	Drug[a]	Comments
Sulfonamides (cyclic)		
	Piroxicam	
Others		
	Sulindac (Clinoril)	May be viewed as an indole isostere
	Caprofen (Rimadyl)	
	Benorylate (Benoral)	An acetaminophen–aspirin mutal pro-drug (not available in U.S.)

[a] The names in parentheses represent one (of possibly several) trade names. [b] Withdrawn from the U.S. market.

Extracts of willow bark (containing salicin), used to relieve pain and fever, was known to the Greeks in Hippocrates' time. It became used again in England in the mid-eighteenth century when the bitter principle was also recognized as a useful substitute for quinine as an antipyretic. In 1823 Leroux succeeded in isolating the principle in crystalline form. The identification of salicin as a glycoside of salicyl alcohol was accomplished by Piria (1839), who also synthesized salicylic acid (1838) from salicylaldehyde. It had been isolated several years earlier from spirea flowers. Salicylic acid was also synthesized from anthranilic acid (Gerland, 1853). Even though acetylsalicylic acid (later named aspirin) was first prepared by Gilm in 1859, it was introduced by Dreser (1899) into medicine 40 years later in an attempt to circumvent the considerable toxicity and corrosiveness of oral salicylic acid. Figure 5-1 summarizes the historical development of aspirin.

At one time it was believed that aspirin was a pro-drug and owed its activity to the hydrolysis product, salicylic acid. Today salicylate is viewed as an active but inferior metabolite. The pharmacologic properties are now known to be mainly due to the unhydrolyzed parent drug—aspirin.

Attempts to modify aspirin chemically to either circumvent its potential gastrointestinal intolerance, its poor aqueous solubility (1:300), its poor stability toward hydrolysis (k = 0.002 h^{-1}; which makes liquid dosage forms impossible) and, no doubt Bayer's original patent, has led to the synthesis and evaluation of hundreds of compounds over the years. The modifications involved derivatization of the carboxyl function (such as esters remov-

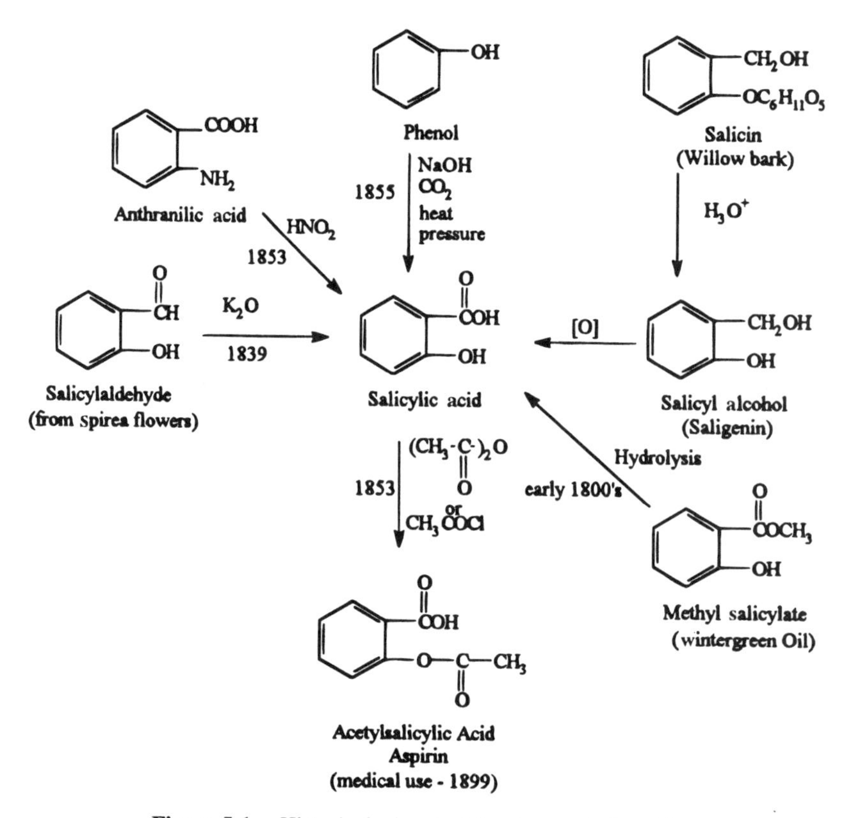

Figure 5-1. Historical–chemical development of aspirin.

able in vivo), variations of the acyl group (to increase stability) or of substitutions into the benzene ring, from a simple alkyl or alkoxy group to fluorinated phenyls such as in diflunisal (Table 5-2). The *ortho* relationship of the acid and phenolic functions, however, seemed inviolate. The acetaminophen ester of aspirin, benorylate (Table 5-2), is a particularly novel example of a pro-drug where both components represent active agents. The drug appears to be devoid of ulcerogenic properties and, following absorption, is split into its two active components by serum esterases. Studies have shown it to be an effective analgetic–antiinflammatory drug.

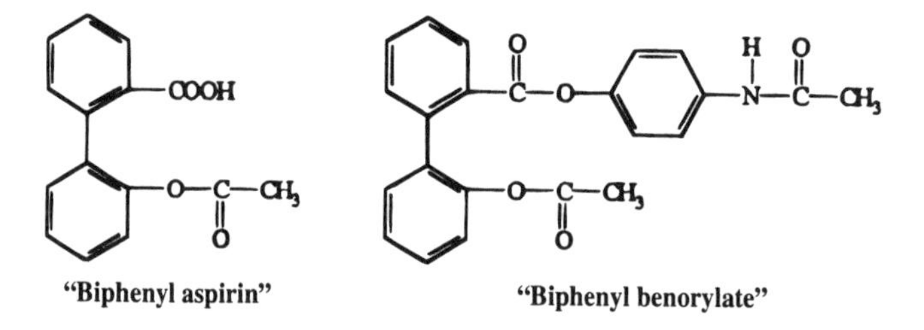

An interesting compound apparently violating the *ortho* relationship was reported (Gringauz, 1976). This compound, which might be considered a "biphenyl aspirin,"

exhibits both antiinflammatory and analgetic properties in rodents. It was also found to be nonulcerogenic in rats. It is not surprising that the acetaminophen ester ("biphenyl benorylate") also exhibited similar pharmacology in rodents. Salicyloylsalicylic acid (salsalate, Table 5-2) is also of interest. This ester of salicylic acid, originally introduced as Diplosal 70 years ago, has recently regained popularity. Its insolubility in gastric fluids (yet facile absorption) gives it nonirritating properties toward the mucosa. Serum esterases then readily metabolize the drug to two salicylate molecules. Finally, a unique replacement for the acetyl group of aspirin is exemplified by 2-phosphonoxybenzoic acid (Fosfosal). The drug is claimed to share all of aspirin's properties, except ulcerogenicity; it is also water soluble.

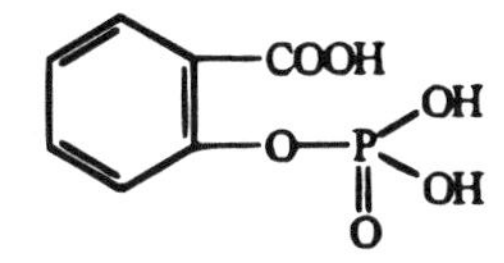

2-phosphonoxybenzoic acid

Antiinflammatory analgetics can have some adverse effects. Because of the commonality of their mechanism of action, aspirin and the NSAIDs as a group also qualitatively share side effects and toxicities. The most common adverse effects are allergic reactions, gastric bleeding, and ulcerogenic potential to the gastrointestinal tract. Since all drugs in this group, the old ones and those newly developed, are frequently compared with aspirin, both as to efficacy and untoward effects, aspirin will here also be considered the prototype. The newer NSAIDS will, of course, wish to justify their existence by attempting to establish clinical superiority to aspirin in either effectiveness or patient tolerance. The "contest" is unlikely to be settled soon.

Allergic manifestation of aspirin is definitely most manifest in asthmatics. An aspirin allergy incidence of over 40% in perennial asthmatics has been demonstrated. It is possible that some asthmatics may be maintaining the disease by using aspirin on a regular basis. Of course, patients so sensitive are just as likely to precipitate allergic symptoms by other compounds listed in Table 5-2, including indomethacin, naproxen, fenoprofen, and ibuprofen. It should be recognized that except in asthmatics, where hypersensitivity can be severe enough to cause an anaphylactic reaction, such a degree of intolerance is rather uncommon in the population as a whole. The 90-year use of aspirin attests to a remarkable safety record. The release of ibuprofen as a nonprescription drug in the United States is also due in large part to the lack of serious problems with this compound in over a decade of clinical experience.

It is curious that it has also been reported that aspirin could relieve asthmatic bronchoconstriction. Earlier, an endogenous substance called Slow Reacting Substance in Anaphylaxis (SRS-A), which is able to contract isolated human bronchial tissue, was reported to be produced during human anaphylactic shock. It was suggested that aspirin antagonism of SRS-A–caused bronchoconstriction might be useful for asthma relief. This suggestion, however, has not been clinically implemented.

Gastric bleeding is another effect of aspirin in particular and NSAIDs in general. This is primarily due to the mechanism of action of these drugs that includes inhibition of platelet aggregation (see explanations following) resulting in bleeding time prolongation. A single dose of 325–650 mg in normal individuals can double bleeding time for several days. In one study of rheumatic patients who had been given only 1 gram of aspirin daily, 85% had been found to have some blood in their stool.

It has been known for some time that aspirin, and the aspirinlike drugs, have a triumvirate of pharmacological properties: antipyresis, analgesia, and antiinflammatory characteristics. It may even be comprehended that the peripheral antiinflammatory effects contributed to an extent to the analgesia in which inflammation was a major factor. A comprehensive picture, however, of the mechanism of action of these compounds and the interrelationship of the pharmacologic effects has only become clear since.

It may be of interest to mention briefly some of the earlier mechanisms proposed. Because histamine always appears at the site of inflammation, inhibition of histamine, or the blockade of tissue response to it, was considered a part of the antiinflammatory mechanism of aspirinlike drugs. It is now known that histamine is of only minor importance in inflammation, occurring during the early phase. Antihistamines are certainly not considered antiinflammatory compounds. Inhibition of serotonin (5-hydroxytryptamine, 5HT) and blockade of tissue response to it was similarly believed to be a viable mechanism for NSAIDs. It is interesting that indomethacin, containing the indole ring system, was synthesized as a potential serotonin antagonist in the erroneous belief that 5HT plays a significant role as an inflammatory mediator. However, few inflammatory reactions are so mediated, even in part. Inhibition of lysosomal enzyme release, antagonism of kinin activity, and the theory that survived the longest, that aspirinlike drugs uncoupled oxidative phosphorylation, have also been proposed. This effect can be shown to occur at salicylate levels used in treating arthritis. The uncoupling affects various ATP-dependent reactions. The fact that potent uncouplers without antiinflammatory action exist negates this concept as a viable explanation of the mechanism of action of the NSAIDs.

5.5. Prostaglandins

Before discussing some of the research on the mechanism of action of aspirin and other acidic NSAIDs, it is necessary to consider the fundamental chemistry of prostaglandins (PGs). Some early work with crude extracts from accessory genital glands (e.g., prostate, hence the name prostaglandins) established such physiological activities as vasodilation and muscle stimulation, as well as that the substance(s) were acidic lipids. However, the difficulties of isolating and identifying the responsible components took two decades and the development of sophisticated instrumentation such as gas chromatography and high-resolution mass spectrometry.

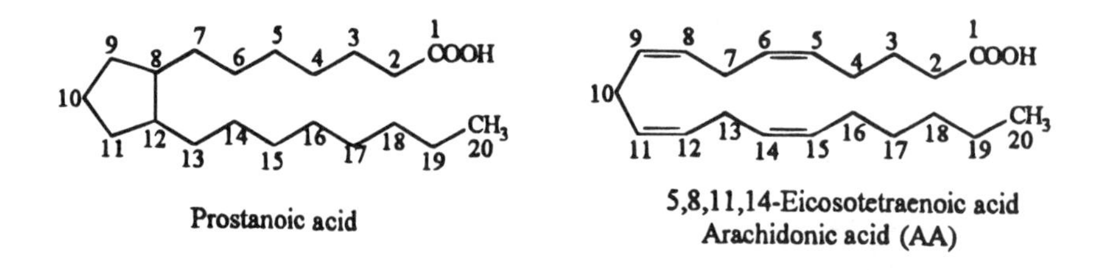

Prostanoic acid

5,8,11,14-Eicosotetraenoic acid
Arachidonic acid (AA)

Prostaglandins are a group of fatty acids derived from the hypothetical 20-carbon *prostanoic acid.* They are known to be ubiquitous, not localized in the prostate gland and seminal fluids. Because PGs are released in many locations and affect biologic processes near their point of liberation, the term *local hormones* has been applied to them. The great potency of PGs would make dissipation throughout the system actually harmful to the whole organism. It is therefore not surprising that mechanisms for their rapid (within minutes) deactivation at or near the locus of release exist.

The factors of very localized production coupled with their brief existence, and therefore low concentrations, has, understandably, made the physiological and chemical studies of PGs difficult. Nevertheless, much of the complexity has already been unravelled.

Before proceeding with a discussion of PGs, a survey of the relevant nomenclature is in order. Figure 5-2 illustrates the differences among the six basic PGs, denoted by letters A through F, in the relationships of the hydroxyl, ketonic functions, and double bonds of the cyclopentane portion of the molecule. Since only the F series of PGs have OH groups at both C-9 and C-11, the stereochemical position of the 9-OH must be indicated as either β, or as found in the natural PGFs, α (i.e., behind the plane of the ring and therefore *trans* to E), as might be obtained synthetically. The bonding of the alkanoic acid chain at C-8 is similarly behind the plane, having a *trans* relationship to the alkane (or alkene) chain bonded at C-12. A numerical subscript will be encountered in those PGs that have additional double bonds on the chains *outside* the cyclopentane ring. Thus 1 indicates one double bond at C-13, 2, two double bonds at C-13 and C-5, and the subscript 3 denotes three double bonds at C-13, 5, and 17. A glance at Figure 5-2 makes these relationships apparent. PGG and PGH represent cyclic endoperoxides. PGIs and the PG-like thromboxanes (TXA and B) are also shown and will be considered.

It is of interest how the presently accepted concept of aspirin and aspirinlike compounds' involvement with PGs came about. The last of the "aspirin theories" to arise proposed that aspirin interfered with leukocyte migration to the site of injury, thus inhibiting the inflammatory process. A "prostaglandin phase" of inflammation, where PGs arose in the exudate of experimentally induced edema after the appearance of histamine and bradykinin, was already known. These two events appeared to coincide in the inflammatory process. Thus, the time was ripe. In studying the mediators responsible for the anaphylactic response in sensitized guinea pig lungs, Piper and Vane (1969) isolated histamine, SRS-A, and a new substance they called "rabbit aorta contracting substance" (RCS), a very unstable material (1 to 2 min) whose release, and presumably production, was selectively inhibited by aspirin-

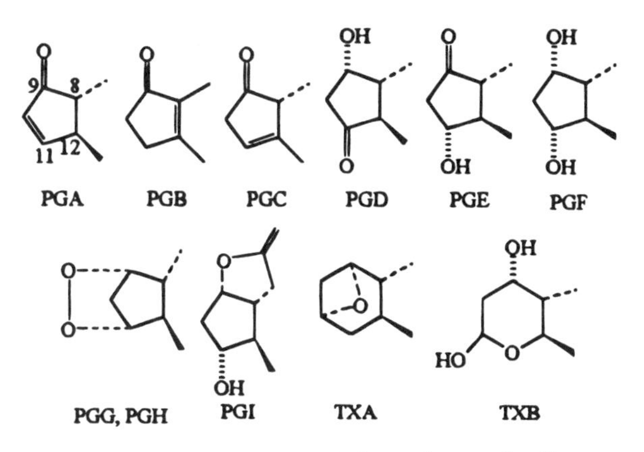

Figure 5-2. Structural variations of prostaglandins.

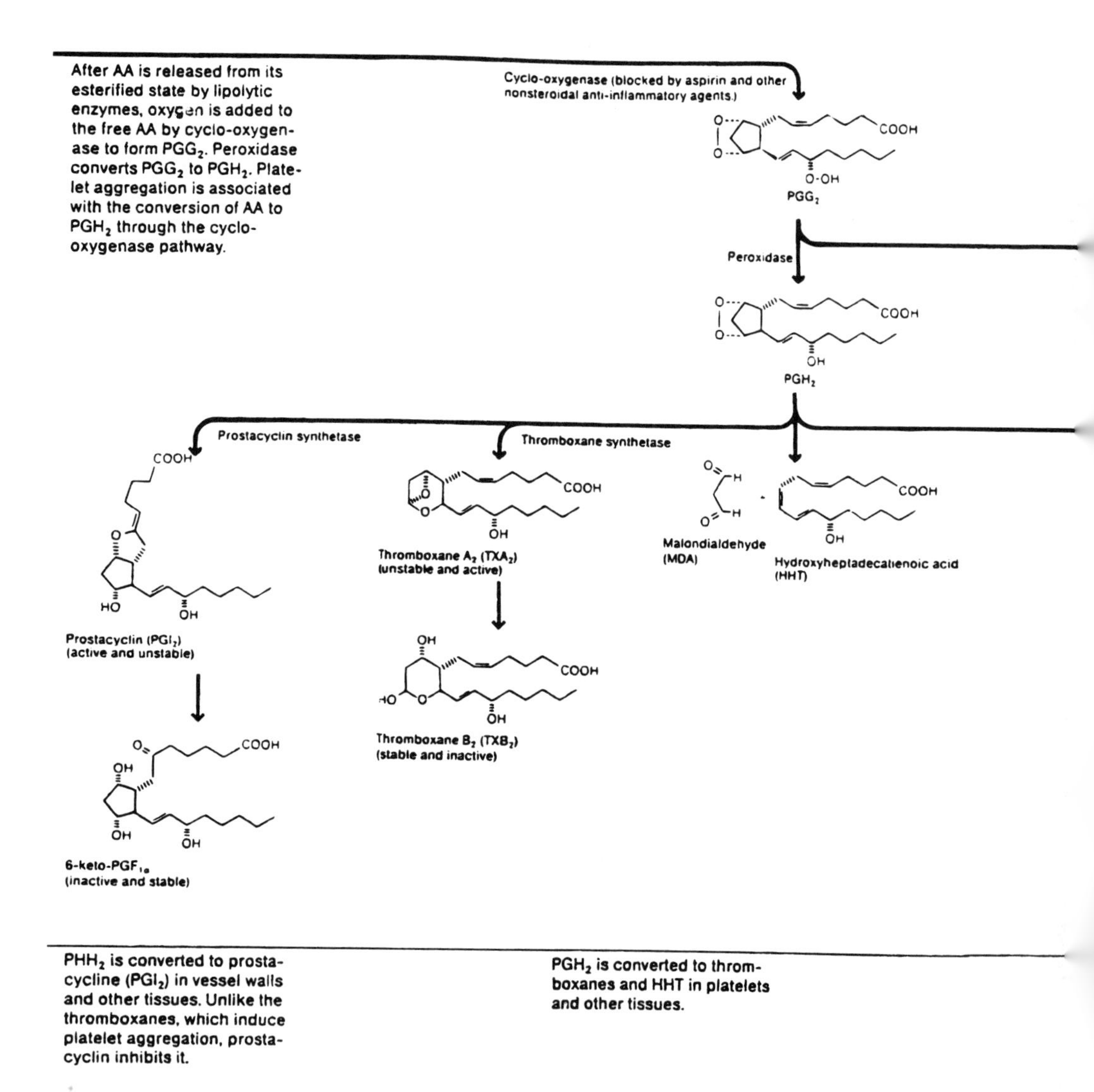

Figure 5.3 Arachidonic acid cascade.

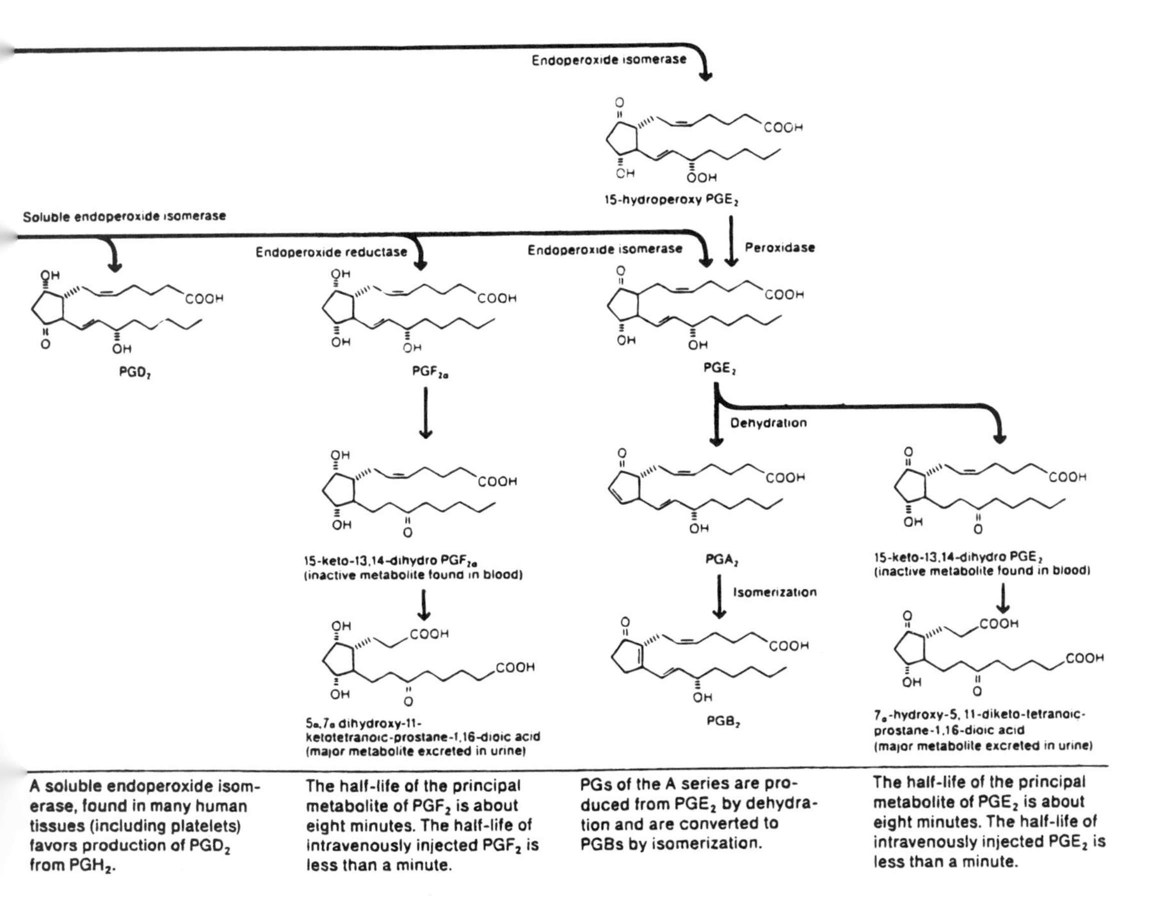

Figure 5.3 *(Continued)*

like drugs. Release of RCS could also be elicited by bradykinin and SRS-A, the cause of bronchoconstriction. This release, as well as the bronchoconstriction, can also be antagonized by aspirinlike drugs. Arachidonic acid (AA), which is the starting material for the PG syntheses (Fig. 5-3), also causes bronchoconstriction and RCS release. This, too, is reversed by aspirin. PG synthesis and release are directly related since a tissue can be stimulated to release more PGs than it contains. The likelihood that aspirinlike drugs inhibit the enzymes catalyzing PG synthesis now appeared very good.

It has already been shown that cell-free homogenates of guinea pig lung can synthesize PGE_2 and $PGF_2\alpha$ from AA. PG activity was determined by testing contraction effectiveness on rat stomach strips (PGE_2) or colon ($PGF_2\alpha$). Incubation of homogenates with AA (30 min) increased $PGF_2\alpha$ activity from 6 μg/ml to 520 μg/ml; PGE_2 levels similarly rose by 500 μg/ml. The stage was now set to determine the actual mechanism of action of aspirin and aspirinlike drugs. Repeating the above experiments with the addition of indomethacin gave a 50% inhibition rate at 0.27 μg/ml (0.75 μmol); aspirin yielded 50% inhibition at 6.3 μg/ml (35 μmol). Thus in regard to $PGF_2\alpha$ indomethacin was 23 times more potent than aspirin (47 times on a molar basis). Similar results were obtained versus PGE_2. Morphine under identical conditions had no effect; hydrocortisone only a slight effect.

There has since been considerable additional evidence that NSAIDs block PG synthesis in many mammalian species. Serum concentrations achieved with clinically used doses of aspirin, indomethacin, and other NSAIDs, even when protein bound to a considerable extent, are sufficient to inhibit PG biosynthesis.

The early work by Vane and co-workers might have left the impression that AA is mainly the precursor source for PGs (at least 15 naturally occurring PGs have been identified and characterized). Subsequent work soon made it apparent that PG synthesis accounts for a small percentage of the total AA metabolism. As can be seen in Figure 5-3, in addition to the PGs, thromboxanes, prostacyclins (PGIs), and 12-hydroxyeicosatetranenoic acids (12HPETE and 12HETE) are also produced. Not shown (but see Fig. 5-5) is that a group of biologically active compounds called leukotrienes (LTs), including SRS-A, also arise from AA. It is now apparent that AA plays a central role as a precursor of a large number of regulators and mediators of various cell functions. In fact the term *arachidonic acid cascade* is now frequently used.[3]

To understand PGs and related compounds better, their physiological functions, potential therapeutic applications, and biosynthesis will now be considered. AA, and other polyunsaturated acids such as 8,11,14-eicosatrienoic acid (dihomo-γ-linoleic acid)—the precursor for the PGA_1-E_1 compounds—do not exist free in most cells. They are primarily esterified to various cellular membrane phospholipids. Generation of free AA from its ester linkage probably occurs as a consequence of cell-surface receptor stimulation and activation of phospholipase enzymes, particularly phospholipase A_2. Next is the key step in the biosynthesis of all the PG, PI, and TX compounds; namely, the addition of molecular O_2 to free AA to form the crucial endoperoxide PGG_2 (Fig. 5-3). Cyclooxygenase, the enzyme catalyzing this addition, is a component of the enzyme complex usually referred to as *prostaglandin G/H synthetase*. It is associated with the microsomal fraction and consists of two membrane-bound enzymes, cyclooxygenase and endoperoxide isomerase, and a solu-

---

[3] Bergstron, Samuelsson, and Vane received the 1982 Nobel Prize for their "discoveries concerning prostaglandins and related biologically active substances." Von Euler's Nobel Prize (1970) was for catecholamine research.

ble endoperoxide peroxidase. These two endoperoxides (G and H) have been isolated and characterized. Their half-life is about 5 min. A single atom of a second O_2 molecule attaches to position 15, forming PGG_2. A peroxidase enzyme then reduces this to PGH_2. Thus, there are two distinct activities catalyzed by PGG/H synthase. Finally, endoperoxide isomerases convert PGH_2 to PGD_2, and PGE_2, while a reductase, by breaking and reducing the 9–11 endoperoxide bridge to two α-OH groups, yields $PGF_2\alpha$. Experiments with isotopic $^{18}O_2$ also established that both ring oxygen atoms of PGs come from a single oxygen molecule and were formed from a common precursor, the cyclically 9–11 bridged endoperoxide PGH_2. (This work also later helped to establish PGI_2 and TXA_2.)

It was established that PG synthetase is inhibited by aspirin and other acidic NSAIDs. It was determined somewhat later that actual inhibition occurs at the active site of the cyclooxygenase component probably by competing with the polyunsaturated acid substrates such as AA. As will be seen, the PG-mediated effects of inflammation, pain, platelet aggregation, and vasoconstriction are all affected by this enzyme inhibition.

It is important to point out that different cyclooxygenase inhibitors may have some variations in their enzyme-restraining mechanism. It is, of course, likely that other polyunsaturated fatty acids, having binding affinities similar to AA (10^{-6}M), will bind to the enzyme, inhibit PG product formation, but will then be displaced by excess AA (i.e., rapid reversible competitive inhibition). The aryl and heteroarylacetic acids (Table 5-2) inhibit cyclooxygenase in essentially that manner. Ibuprofen, whose binding constant of 5×10^{-6}M is similar to that of AA, is a good example. Other workers have shown that binding affinities of this type of compound are affected more by hydrophobic features of the molecules than their ionic characteristics. Ibuprofen's methyl ester binding to the enzyme was significantly different. A large number of NSAIDs, including many of those in Table 5-2, were shown to inhibit the conversion of labeled AA to PGE_2 in in vitro experiments in the 0.1–10 μg/ml range. It is interesting that the activities parallel the potency that these compounds exhibit in the foot-edema assay in the rat.

There is also an irreversible time-dependent inactivation of cyclooxygenase. Here the pharmacodynamic aspects are very different. Effects can persist following a dose even after the drug clears the serum and tissues. Aspirin alone among the clinically used agents has this property, by irreversibly acetylating the OH group of serine at position 530 (Ser530) in the enzyme's active site.

It was known that acetylsalicylic acid (aspirin) can transfer its acetyl group to proteins. To determine if transacetylation also occurs with cyclooxygenase, ^{14}C-tagged aspirin was used (Eq. 5.2). Since at high concentrations aspirin will acetylate most proteins (e.g., serum albumin, immunoglobulins, hemoglobin), acetylation must occur at micromolar levels within minutes for these experiments to be meaningful. It was found that aspirin acetylates a single protein (M.W. 85,000) of human platelets at the same rate at which the drug inhibits platelet function: within 20 min at 50 μM. This also correlates with aspirin's inhibition of cylcooxygenase. If the enzyme is acetylated at a single amino acid residue in the *active site* or at an allosteric point, then the natural substrate, AA, should compete with aspirin and inhibit drug binding. This was also established. Furthermore, the irreversibility or "permanence" of the effect on circulating platelets was also demonstrated.

$$C_6H_4(COOH)-O-\overset{O}{\overset{\|}{C}}-\overset{*}{C}H_3 + Enz-CH_2OH \longrightarrow Enz-CH_2O\overset{O}{\overset{\|}{C}}-\overset{*}{C}H_3 + C_6H_4(COOH)(OH) \quad (5.2)$$

By using site-directed mutagenesis, the Ser530 of PGG/H synthase was replaced with alanine. The K_m for AA and the ID_{50} values for flubriprofen, a fenamate, and aspirin were the same for the mutant and native enzyme. However, only the latter was irreversibly acetylated by aspirin. This indicates that Ser530 is not essential to the reactive site of the enzyme. It was proposed that aspirin's acetylation introduces sufficient bulk at position 530 to interfere with AA binding sterically, thus decreasing PG synthesis (Loll, et al., 1995).

Further experiments later showed that the more potent PG synthetase inhibitor, indomethacin, can inhibit aspirin acetylation at 1 μM levels. However, indomethacin binding is reversible by treatment with aspirin at several 1 hour intervals, which resulted in complete acetylation of the enzyme.

Since it was apparent that PGs accounted for a small fraction of AA's metabolism, efforts were underway to determine what became of the remaining AA. Soon a non-PG metabolite was discovered, generated from the cyclic endoperoxide PGH_2. This product, TXA_2, with a half-life of only 30 sec, is now understood to be a major product of AA metabolism. TXA_2 is metabolized to the more stable, but inactive, TXB_2. Major production sites are the lungs, spleen, and platelets. Its synthesis, like the PGs, is also inhibited by aspirin (irreversibly) and other NSAIDs. TXA_2 turned out to be a very potent platelet aggregant and the most intense arterial vasoconstrictor encountered. It soon became apparent that TXA_2 was the RCS described by Piper and Vane (1969).

Earlier observations, during superfusion assays, of contraction of severed vessels led to further examination of vascular tissues. It is surprising that incubation with endoperoxides resulted in potent vasodilation and platelet aggregation inhibition. The responsible substance was soon identified as PGI_2 or prostacyclin, whose half-life in solution was only 3 min. PGI_2 is the principal AA product of the vascular epithelium, where it dramatically relaxes the smooth musculature of blood vessels and inhibits the TXA_2 induced aggregation of platelets on their surfaces. (The physiological ramifications and pharmacological potentials of these two substances will be considered in detail in Chapter 12.)

The types of PGs produced and released in response to various stimuli have a complex, if not perplexing, array of physiological effects including antagonism to each other. Even though they are dose related, it should be kept in mind that a person's total daily PG production probably does not exceed 2 mg.

Even though much is still to be learned, the following is a brief overview of what is known and has already been applied to therapeutics either by utilizing the natural PGs, synthetic analogs, and derivatives having improved chemical properties (mainly stability), or drugs affecting PGs indirectly such as cyclooxygenase inhibitors (NSAIDs). Stimulators of PGI release or thromboxane synthetase inhibitors are also to be considered.

It may be helpful, although it would certainly be an oversimplification, to consider at least PGEs and PGFs as interacting with two different extracellular receptor modes. In the case of E-type PGs, this interaction stimulates intracellular adenyl cyclase to activate AMP to cyclic AMP (*c*-AMP); F-type PGs activate intracellular GMP to *c*-GMP via guanyl cyclase. The cyclic nucleotides, acting as secondary intracellular messengers in turn, cause two different types of physiological responses, primarily by kinase-catalyzed phosphorylation of enzymes and other proteins. These responses may simply be different (i.e., specific) antagonistic, or sometimes even identical, in instances where the same cylcase is activated or both PGs exhibit affinity for the same receptor. Three complications must be interjected. The cyclic nucleotides are hydrolytically inactivated by the enzyme phosphodiesterase to the corresponding monophosphate nucleotides. This limits the duration and intensity of the PG effects. Drugs that inhibit this enzyme (e.g., theophylline) prolong the effects of the cyclic nucleotides, which may be synergistic or antagonistic. There are very likely PGs that themselves act as intracellular secondary messengers. Figure 5-4 outlines the previously mentioned events. Finally, the metabolic inactivation of the PGs must be considered. These include the dehydration of PGEs to PGAs and the oxidation of the 15-OH to the corresponding inactive 15-keto compounds by 15-hydroxyprostaglandin dehydrogenase (Fig. 5-3). As can be seen (Table 5-3), some of the synthetic analogs used as drugs are designed to protect the 15-OH function against rapid oxidative metabolism, thus resulting in clinically more manageable compounds, primarily with a more prolonged action than the natural PGs. Carboprost and doxaprost, where a methyl group shares C-15 with the labile hydroxyl and gemoprost, where the adjacent C-16 carries two methyl groups, are examples of such protection (Table 5-3). The fragile prostacyclin itself has double protection built into its synthetic analog, 16-methyl 18,18,19,19-tetradehydrocarbocyclin, wherein the 15-OH is protected by a 16-methyl group and the allenic furan oxygen is replaced by a stable saturated carbon atom. Even though the activity of this synthetic compound as a vasodilator and inhibitor of platelet aggregation is reduced compared with PGI_2 itself, this is compensated for to a degree by a more sustained effect.

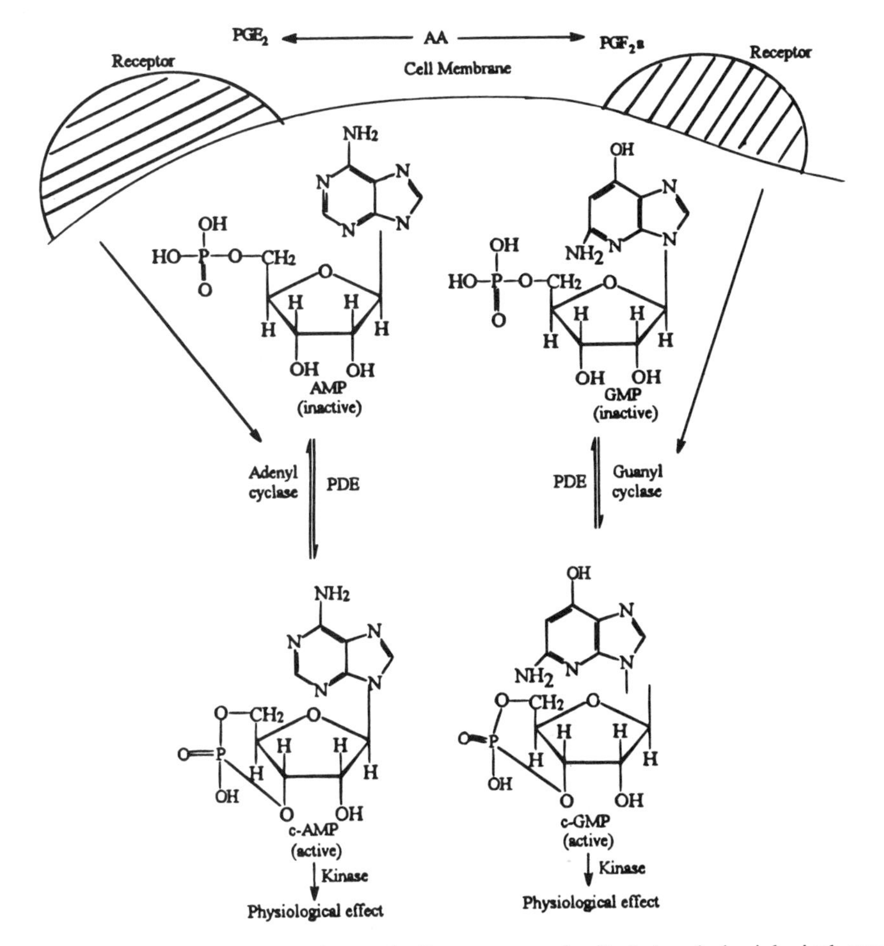

Figure 5-4. Outline of postulated steps leading to prostaglandin-induced physiological events. AMP = adenosine monophosphate; GMP = guanosine monophosphate; PDE = phosphodiesterase.

Table 5-3. Representative Prostaglandin Compounds in Use[a]

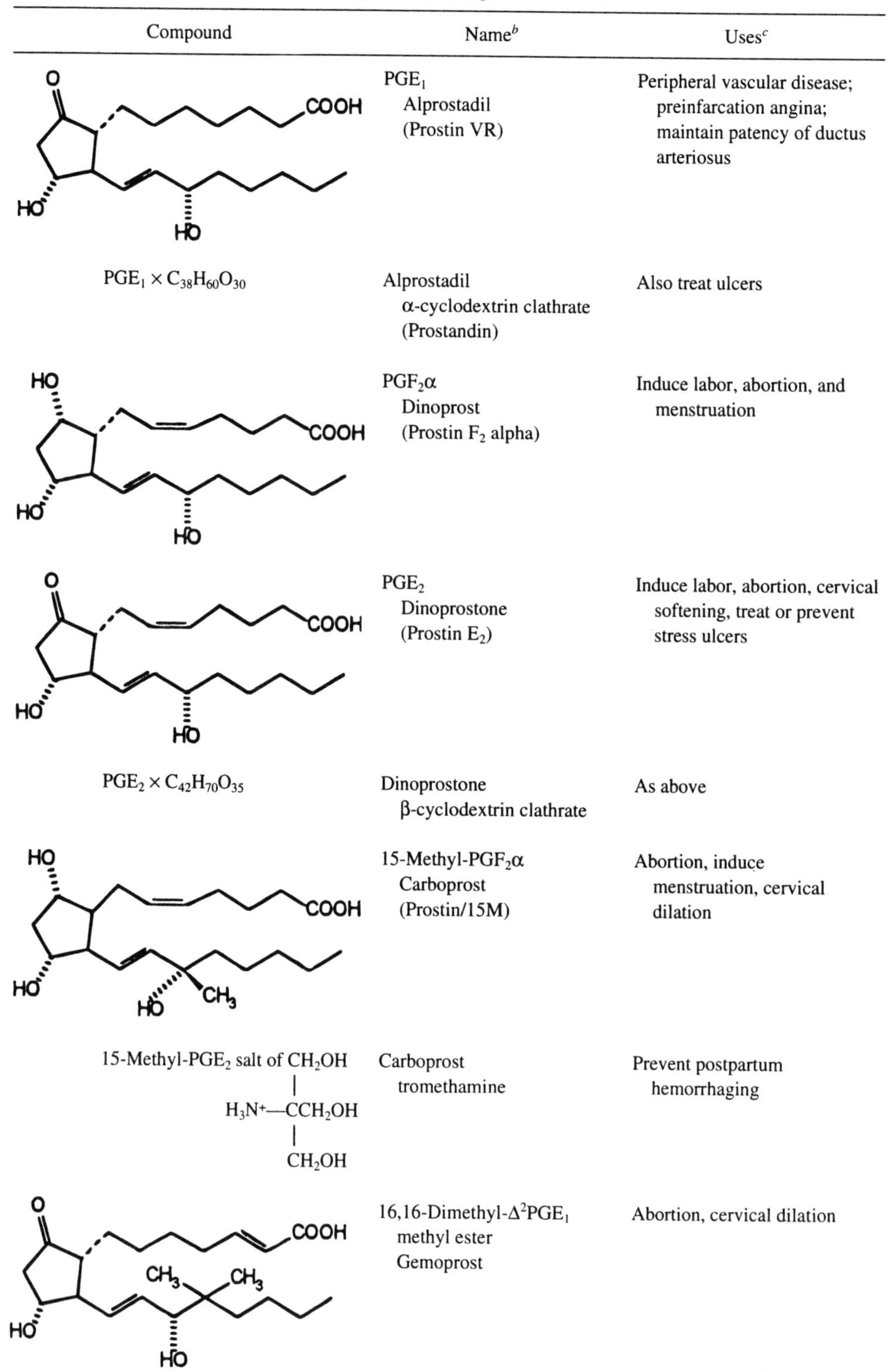

Compound	Name[b]	Uses[c]
	PGE_1 Alprostadil (Prostin VR)	Peripheral vascular disease; preinfarcation angina; maintain patency of ductus arteriosus
$PGE_1 \times C_{38}H_{60}O_{30}$	Alprostadil α-cyclodextrin clathrate (Prostandin)	Also treat ulcers
	$PGF_2\alpha$ Dinoprost (Prostin F_2 alpha)	Induce labor, abortion, and menstruation
	PGE_2 Dinoprostone (Prostin E_2)	Induce labor, abortion, cervical softening, treat or prevent stress ulcers
$PGE_2 \times C_{42}H_{70}O_{35}$	Dinoprostone β-cyclodextrin clathrate	As above
	15-Methyl-$PGF_2\alpha$ Carboprost (Prostin/15M)	Abortion, induce menstruation, cervical dilation
15-Methyl-PGE_2 salt of CH_2OH–$C(CH_2OH)_2$–N^+H_3 (H_3N^+—CCH_2OH with two CH_2OH)	Carboprost tromethamine	Prevent postpartum hemorrhaging
	16,16-Dimethyl-Δ^2PGE_1 methyl ester Gemoprost	Abortion, cervical dilation

(Continued)

Table 5-3. *(Continued)*

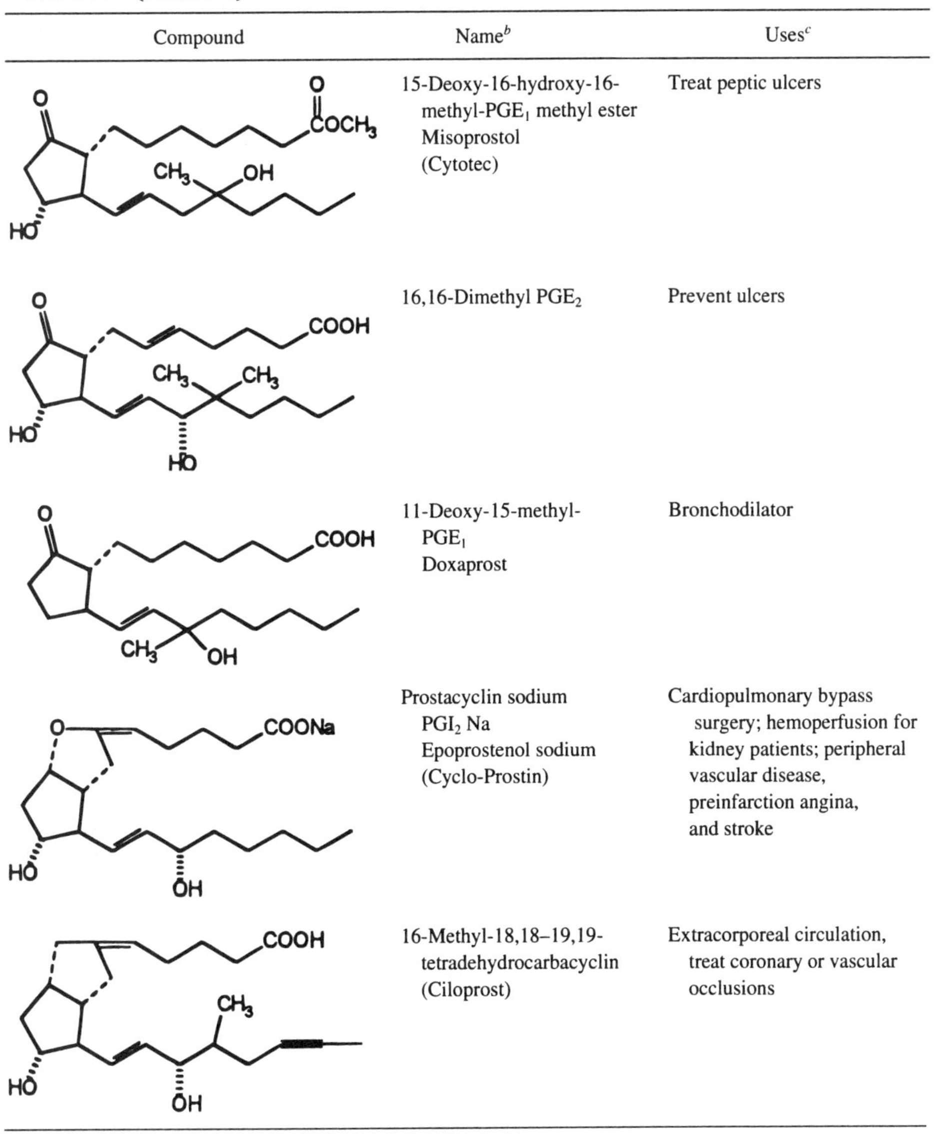

Compound	Name[b]	Uses[c]
	15-Deoxy-16-hydroxy-16-methyl-PGE_1 methyl ester Misoprostol (Cytotec)	Treat peptic ulcers
	16,16-Dimethyl PGE_2	Prevent ulcers
	11-Deoxy-15-methyl-PGE_1 Doxaprost	Bronchodilator
	Prostacyclin sodium PGI_2 Na Epoprostenol sodium (Cyclo-Prostin)	Cardiopulmonary bypass surgery; hemoperfusion for kidney patients; peripheral vascular disease, preinfarction angina, and stroke
	16-Methyl-18,18–19,19-tetradehydrocarbacyclin (Ciloprost)	Extracorporeal circulation, treat coronary or vascular occlusions

[a] Marketed or experimental. [b] Trade names in parentheses. [c] Actual or potential.

Among the effects of PGs to be considered are the following:

1. *Effects on inflammation and pain*: E- and F-series PGs are known to be released in the inflammation and increase capillary permeability, producing edema and painful erythema. PGE_1 can produce this in quantities as low as 1 μg; $PGF_{1\alpha}$ requires about 1 μg. Erythema produced by an intradermal application may persist for 10 h. There is evidence that PGs E_1, E_2, A_2, and $F_{2\alpha}$ stimulate (potentiate?) histamine and bradykinin. Low concentrations of PGs induce hyperalgesia. This appears to be a sensitization of pain receptors. It was also demonstrated that subdermal infusion of PGE_1 only pro-

duced pain if bradykinin or histamine (or both) were added with it. It was shown earlier that aspirin's analgetic effects were peripheral. Now that the drug's ability to prevent PG release during inflammation is known, it is apparent that the analgetic effect is, at least in part, due to decreased pain receptor sensitization by PGs. In fact the Randall–Selitto test for analgetic effectiveness of NSAIDs shows aspirin as ineffectual in uninflamed tissues.

2. *Effects on gastrointestinal tract.* Some types of smooth muscle contract in response to PGEs and Fs. Others, such as circular muscle, are contracted by PGFs but relaxed by PGEs. PGEs and PGAs reduce gastric secretions, including HCl. PGE_1 (as the cyclodextrin clathrate) and synthetic analogs such as the 16,16-dimethyl homolog of PGE_2 (Table 5-3) can afford gastric cytoprotection and heal ulcers.
3. *Cardiovascular and renal effects.* These include increased contractility of heart muscle, dilation of arterioles, and vasoconstriction of other blood vessels. Increased cardiac output and decreased peripheral resistance are observed. However, these vasoactive properties of PGEs and As are greatly surpassed in intensity by PGI_2 and TXA_2. PGE_1 and PGE_2, as well as PGA_1 and PGA_2, can decrease blood pressure, the former likely by a direct effect on smooth muscle, and the latter also involve improved renal functions. An innovative PGE_1–heparin conjugate has demonstrated in vitro ability to inhibit fibrin formation during thrombogenesis and decrease platelet aggregation.
4. *Effects on respiratory tract.* PGEs dilate bronchial muscles. Several synthetic PGE_1 and PGE_2 analogs are currently being evaluated as antiasthmatic bronchodilators (e.g., doxaprost, Table 5-3). PGFs have the opposite effect. $PGF_2\alpha$ sensitivity in asthmatic patients can be high and a clinical dilemma when used to induce abortions in asthmatic women.
5. *Effects on nervous system.* E and F PGs are released from various areas of the central nervous system including the spinal cord, while a $PGF_{2\alpha}$ is present in the cerebrospinal fluid (CSF). When inflammation is present the levels increase. The evidence of PG involvement in thermoregulatory processes is not absolute, yet PGE_1 is a potent pyretic agent when it is injected into cerebral ventricles and the hypothalamus. The temperature increase is instant, dose related, and long lasting (several hours). Periods of elevated body temperatures can be associated with increased CSF levels of PGE_2. Clinical administration of PGE_2 and $PGF_2\alpha$ in abortions can result in fever.

   It was proposed that *c*-AMP acts as the messenger in the activity of PGE_1 on hypothalamic neurons, which in turn leads to fever. Aspirinlike drugs do not abolish the fever produced by PGE_1 injections into a cat's ventricles, but they will abolish fever caused by similar injections of serotonin and pyrogens in general. This, of course, makes sense if one considers the likelihood of E PGs as mediators of fever and that these drugs inhibit the biosynthesis of PGs rather than antagonizes their effects once formed.

   It may now be possible to hypothesize why acetaminophen, like aspirin, is antipyretic, but it exhibits minimal antiinflammatory properties at therapeutic doses. Acetaminophen has been shown to inhibit PG synthetase in brain areas well, but much less so in peripheral areas. This is most likely due to differences in the enzyme sensitivities to these drugs in different parts of the body (isozyme?).
6. *Effects on the reproductive system.* Suffice it to say that PGs cause release of luteinizing hormone, play a part in ovulation, can terminate pregnancy, and initiate parturition. E and F PGs are found in ovaries, fallopian tubes, placenta, menstrual fluid, amniotic fluid (during labor), and in seminal fluid.

7. *Other effects.* There are many, some probably still unknown. However, effects on the kidneys, bladder, atherosclerosis, hemodynamic and hemostatic mechanisms, the skeletal system, association of malignancies with PG-induced hypercalcemia, other malignancy-related metabolic processes, and even glucose metabolism are being investigated.

It is apparent that the effects of natural prostaglandins are multifaceted, complex, and essential. Our understanding is rapidly increasing. This will undoubtedly lead to expanded therapeutic applications.

Research by Vane and others has by now established that inhibition of prostaglandin synthesis is the most plausible mechanism by which aspirin and aspirinlike drugs exert their antiinflammatory, antipyretic, and analgetic action. Even most of the side effects are explicable on this basis.

5.6. The Nonsteroidal Antiinflammatory Drug

The ideal NSAID agent has not yet been developed. Efforts have been continuous since the introduction of aspirin in 1899. For example, the patent literature for 1966 alone reveals disclosures of more than 100 groups (not single compounds) of antiinflammatory structures. The pace of research is not likely to have abated. Unfortunately, only a small number of lead compounds emerge with sufficient potency and adequate tolerance for further development. Chronic toxicity determinations in animals further reduce the number of agents reaching even the preclinical stages of testing.

The saga of the search for a "better aspirin" is particularly illustrative of the rationale involved and the problems encountered. The goal of a highly potent long-acting variant of aspirin with low toxicity has been pursued in both industrial and academic institutions. A long systematic study to determine the optimum structure–activity relationships of salicylates was undertaken to find the best candidate from over 500 salicylates and analogs investigated since 1962 (Hannah et al., 1978). It was found that hydrophobic substituents at the 5 position of salicylic acid improved activity; fluorine atoms on that substituent accentuated the desirable properties even more. The result was the 5-(*p*-fluorophenyl) derivative of aspirin, flufenisal. It exhibited a fourfold increase in antiinflammatory effect, a doubling of analgetic efficacy, a considerable increase in duration, and a 90% decrease in gastric hemorrhagic liability when compared with aspirin in rats.

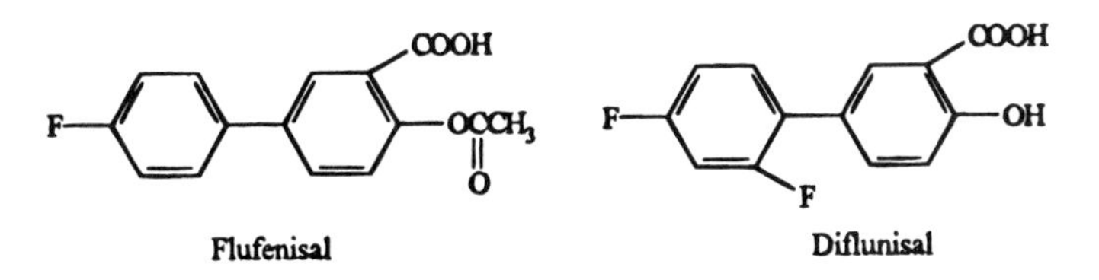

Unfortunately, high-dosage chronic toxicity tests in rats and dogs resulted in gastrointestinal lesions and papillary necrosis. Clinical tests ceased. Reexamination of close analogs of flufenisal led to the selection of the 5-(2,4-difluorophenyl) derivative of salicylic acid—diflunisal. The additional fluorine atom may have compensated for loss of potency attributable to the lack of the O-acetyl group, which in turn probably accounts for the improved safety of this drug. Diflunisal retains certain apparent advantages such as a prolonged (8–12 hrs) analgetic effect. Rat studies using carrageenan-induced foot edema assay show the drug to be nine times more potent as an antiinflammatory agent than aspirin. Its peak analgetic effect in relieving postoperative pain in humans, however, was barely superior. Clinical evaluations in humans sometimes appear to be conflicting. It was found that single doses of 250 mg of diflunisal indistinguishable from a placebo, yet other reports indicated this dose to be more active. Effects on the gastrointestinal mucosa were interesting. In rats, oral aspirin at its ED_{50} dose of 81 mg/kg produces small gastric hemorrhages. Even though diflunisal at 256 mg/kg did not cause intestinal perforations, it did at a dose close to its ED_{50} (520 mg/kg). Studying gastrointestinal blood loss in humans showed diflunisal's effect to be less than aspirin's at therapeutic doses. The theoretically based assumption that diflunisal would be much less likely to interfere with human platelet function than does aspirin was shown to be correct both in terms of aggregation and bleeding time.

In the final analysis the clinician's question as to which drug is best for the patient in pain cannot be answered by experimental pharmacological studies. A clinical study comparing four NSAIDs found individual variations in response so striking that several drugs may need to be tried to find the right one for a particular patient.

Research into the mechanisms of inflammation and the drugs to treat it has given us a much better understanding to apply to patient therapy; yet, many puzzles remain. One in particular is the case with sulfasalazine (SZ), a drug introduced to treat rheumatoid arthritis and ulcerative colitis over five decades ago. Its primary use over this period has been for the latter condition, with the drug being administered both orally and topically (by enema). Even though corticosteroids in high doses are frequently necessary for intense acute attacks, SZ is the safer agent for prolonged therapy.

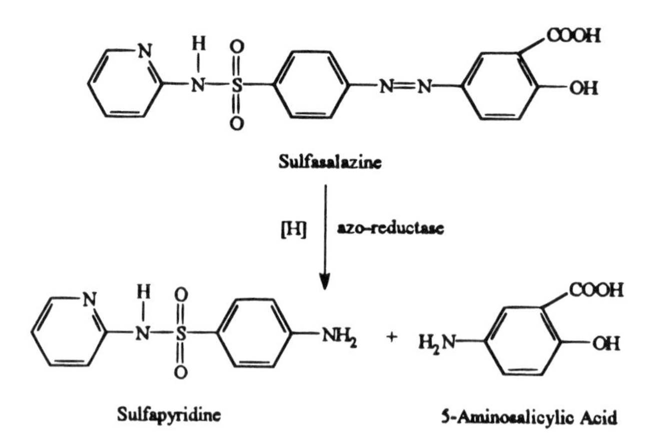

The drug is practically insoluble in water (1 g in 10,000 ml). Thus, being poorly absorbed from the small intestine, SZ reaches the colon intact, where bacterial azo-reductase enzymes cleave the compound to its components sulfapyridine (SP) and 5-aminosalicylic acid (5-ASA). In slow acetylators, where the absorption of SP can lead to plasma levels above 50 μg/ml. It is very likely that it is the sulfonamide that is responsible for the toxic adverse effects associated with SZ. It is ironic that the drug's suppressive effect on ulcerative colitis was at first attributed to the local antibacterial effect of SP. With the discovery of the involvement of PGs in the inflammatory process, inhibition of PG synthesis became the accepted mechanism of SZ. It was not until later, however, that a clinical study actually established that 5-ASA (used alone) is the active component and SP at best should only be considered as a carrier. In fact, studies with enemas and oral dosing have shown that high 5-ASA doses can be successfully used without the usual side effects associated with sulfonamides. Even dramatic remissions were achieved with 5-ASA enemas in distal ulcerative colitis patients previously refractory to SZ and corticosteroids. Another azo precursor of 5-ASA, disodium azodisalicylate, is olsalazine (Dipentum), which has promising characteristics, not the least of which is a prolonged half-life of 4–13 days. Long-term clinical experience will ultimately determine whether it will be the "ideal" oral drug. Rectal biopsies from patients with active disease showed that an increased level of prostaglandin synthesis (PGE_2) exists that was reduced in response to SZ or 5-ASA therapy. Indomethacin and other arylacetic acid NSAIDs, however, did not show good clinical results with this disease. Not even all salicylates appear to be effective.

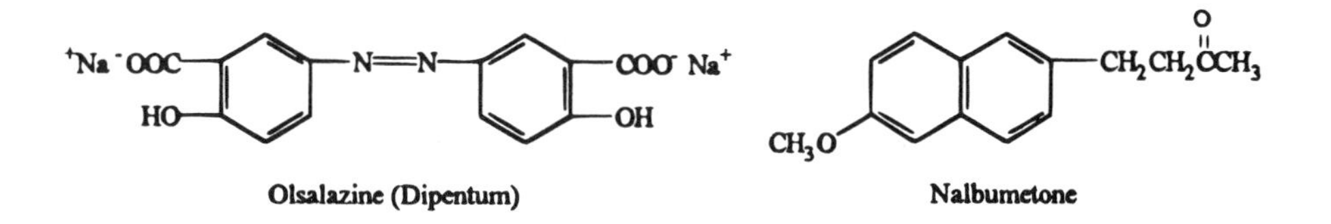

An interesting addition to the NSAID field (1992) is nabumetone (Relafen). This compound, a neutral ketone with extremely weak PG synthesis inhibition, is a pro-drug of 6-methoxy-2-naphthylacetic acid (6-MNA), to which it is rapidly metabolized. Of course, 6-MNA is a homolog of naprosyn. As such it would be expected to share the same mechanism of action and pharmacological profile. It does, including a prolonged half-life, thus allowing infrequent dosing (e.g., 1 g daily as a single dose). The fact that as presented to the gastric mucosa the drug is nonacidic and a poor inhibitor of PG synthetase likely accounts for the low level of gastric irritancy. It is hoped that this feature will become the drug's hallmark.

5.6.1. New Mediators of Inflammation: The Leukotrienes

The mechanism of the antiinflammatory effects of corticosteroids also needs an explanation. A direct inhibitory effect on PG synthesis in cell-free preparations of PG-synthetase could not be demonstrated. Proposals that inhibition of PG release, not synthesis, was occurring were put forward. Steroids were shown to inhibit the release of AA. This limited substrate availability. Inhibition of phospholipase A_2 then became the most likely mechanism for this inhibition. Reduced levels of free AA would also decrease the substrate for the other enzyme in the AA cascade, 5-lipooxygenase (which is not inhibited by aspirinlike drugs, Fig. 5-3). This reduces the effects of the chemotactic substances such as 12-HPETE. If correct, this might explain the superior action of steroids over NSAIDs against inflammatory and allergic conditions. To examine further steroid effects not shared by aspirinlike drugs, on an examination of AA metabolism in polymorphonuclear leucocyte cells (PMNs) known to be crucial to many inflammatory processes was undertaken. It was discovered that the leucocytes metabolized AA primarily by *5-lipooxygenase* to 5-hydroperoxy-6,8,11,14-eicosatetraenoic acid, 5-HPETE (Fig. 5-5), part of which was simply reduced to the 5-hydroxy analog 5-HETE. The major portion was epoxidized to the unstable 5,6-epoxy intermediate LTA, which then becomes the pivotal source for a number of products. These add to a steadily growing family of compounds emanating from the "AA cascade."

The leukotriene nomenclature arises from the fact that these leucocyte-produced molecules all contain a conjugated triene region (even though the compound may have a total of four or more double bonds). Reaction of LTA_4 hydrolase affords 5,12-dihydroxy, 6,14-*cis*-8,10-*trans*-eicosatetraenoic acid, LTB_4 having a set of actions including chemokinetic

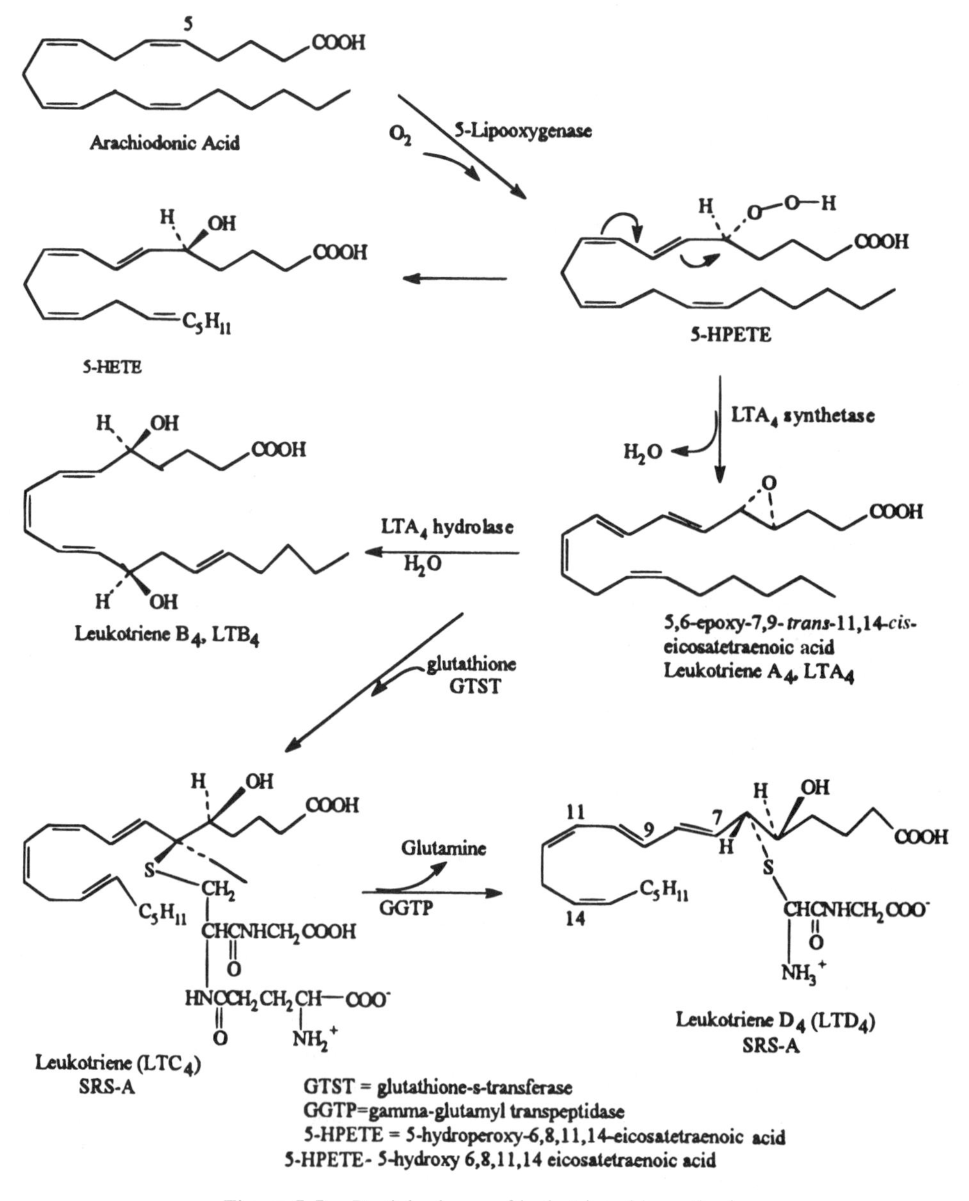

Figure 5-5. Partial scheme of leukotriene biosynthesis.

effects. Receptors for LTB_4 on human PMNs have been shown to exist. The all-*trans* variant LTB_4 and 5-HETE can compete for this receptor. Binding competition correlates with inhibition of the chemotactic property of LTB_4. In addition, LTB_4 exhibits the characteristics of the classic mediator of inflammation. These include attraction of inflammatory cells, stimulation of cellular aggregation, followed by release of their lysosomal enzymes, which in turn precipitate the inflammatory response. Finally, LTB_4 has also been determined to be an agent producing extreme pain sensitivity in inflamed tissue (e.g., rat paws) similar to that seen with bradykinin or PGE_2. Unlike with the latter agents, however, the LTB_4-produced pain is not blocked by cyclooxygenase inhibitors (i.e., NSAIDs). This then opens up a drug development avenue for new types of analgetics; compounds that either inhibit 5-lipooxygenase or antagonize the actions of its metabolic products, particularly LTB_4. In fact, it is already claimed that the mefenamic acid derivative meclofenate sodium (Table 5-2) has a triple action: on both wings of the "AA cascade," namely inhibition of cyclooxygenase *and* lipooxygenase.

Another important product results from the action of the tripeptide glutathione (γ glutamylcysteinylglycine) on LTA_4, whereby the -SH group, catalyzed by glutathione-S-transferase (GTST), nucleophilically attacks the 5,6-epoxide to produce leukotriene C_4 (LTC_4), a thioether derivative. The metabolic product leukotriene D_4 (LTD_4), is then produced by the loss of the glutamic acid residue catalyzed by glutamyl transpeptidase (GTP). The finding of the great similarity of the UV spectra of LTC_4 and LTD_4 to those reported for the potent bronchoconstrictor SRS-A led to the establishment of the identical nature of the two leukotrienes and SRS-A. LTD_4 is the more potent and faster of the two.

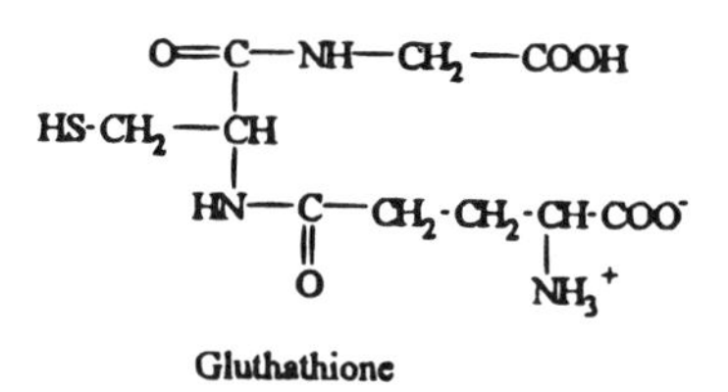

Gluthathione

5.7. Nontraditional Antirheumatoid Drugs

There are a number of drugs currently used to treat rheumatoid arthritis (RA) variously referred to as "disease-modifying" or "slow-acting." Unlike the NSAIDs, they do not produce analgesia or antiinflammatory effects in the traditional sense. Rather, after weeks or even months of therapy, they appear to control symptoms and possibly delay the progress of the malady. Mechanisms of action have not been definitively established for any of them. Adverse effects are frequent and often severe. The carcinolytic drugs, cyclophosphamide, methotrexate, and the 6-mercaptopurine pro-drug azathioprine (Chapter 4), have found increasing use because of the putative benefits of their immunosuppressant properties. In part the rationale for such treatment is the perceived role of immunological factors in rheumatic pathology. The relationship of the development of rheumatoid lesions to the production of immune complexes involving B cells and the necrotic processes with T-cells gives credence to this concept.

The consideration here will concern itself with chrysotherapy (gold), *D*-penicillamine, and the antimalarial drug hydroxychloroquine (Fig. 5-6).

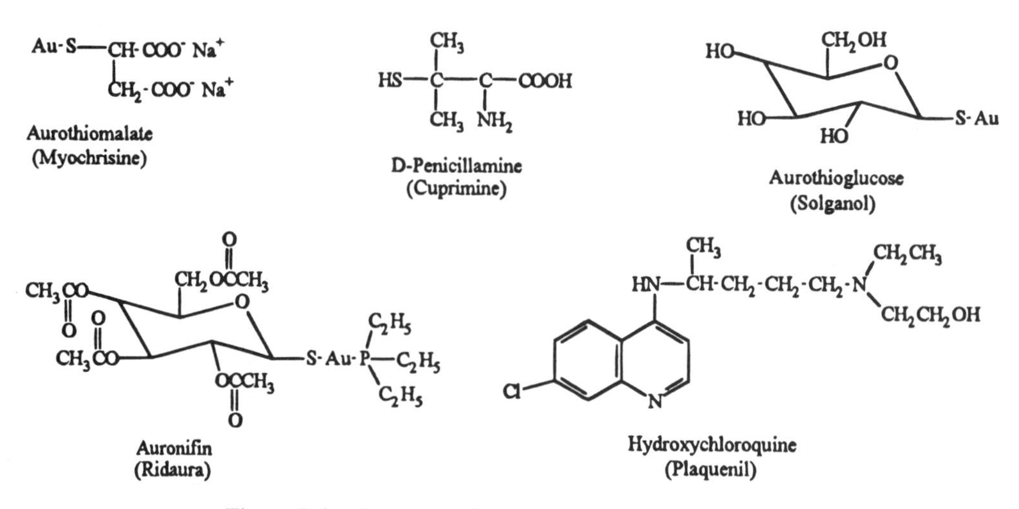

Figure 5-6. Structures of "slow-acting" antirheumatic drugs.

The history of the utility of gold compounds to treat RA should probably be traced to Koch's observation (1890) that gold cyanide inhibited the tubercle bacillus in vitro. At that time this organism was believed to be the causative agent. By using gold thioglucose to treat bacterial endocarditis Lande (1927) observed that joint pain was relieved. Reports of beneficial results in ankylosing spondylitis patients with gold therapy followed. The clinical efficacy of chrysotherapy, however, was not unequivocally established until the results of large-scale multicenter double-blind controlled studies carried out in the late 1950s were reported by the Empire Rheumatism Council (1960) using gold thiomalate. More recently, gold therapy has been shown to slow and even stop the progression of joint degeneration.

The two agents primarily in use until the early 1970s were aurothiomalate (Myochrisine) as an aqueous solution and aurothioglucose (Solganal) as an oil suspension. They are administered by intramuscular injection over a period of weekly doses, usually until a 1.0 g total of gold is reached. Serum gold levels vary considerably and do not seem to have a direct relationship to effectiveness.

A series of orally effective phosphine gold compounds, the most effective of which, auronifin (Ridaura), was then marketed in the United States. This drug, which is prepared from the tetraacetate of aurothioglucose with triphenyl phosphine differs considerably from the gold thiolate compounds, even though it is also a monovalent gold compound. It is monomolecular and crystalline, not amorphous and polymeric. The higher lipophilicity and nonionic character of this drug help explain its oral effectiveness. It is interesting that at effective oral doses serum gold levels were two-thirds less than they were following injected doses of the thiolate drugs; by comparison, kidney gold retention, in animal models, was only one-tenth.

Mechanisms of action of gold compounds are still very much at the speculative stage. An early proposal was antimicrobial activity, based on the demonstrated in vitro inhibition of the tubercle bacillus. This fit in with the belief that RA was a form of tubercular joint disease. Inhibition of lysosomal enzymes, whose release by phagocytic cells is strongly implicated in the pathogenic tissue damage seen in RA, also offers an enticing explanation. Enzymes such as acid phosphatases, β-glucosaminidinase, elastase, and collagenase can be inhibited by aurothiomalate in vitro. Furthermore, auronifin has been shown to inhibit the release of lysosomal enzymes from phagocytic leukocytes in concentrations as low as 1 μM.

Connective tissue diseases can result in protein denaturation whose degradation may be antigenic and causative of RA's inflammation. Gold compounds, interacting with such proteins, may help stabilize them. A protective ability of gold thiomalate toward heat-caused gammaglobulin aggregation can be demonstrated.

Neither the etiology nor the genesis of the pathology of RA have been established. However, immune responsiveness appears to be involved in chronic inflammation. Gold does affect phagocytic activity of PMNs and inhibits cell proliferation that results from the activation of immune processes and connective tissue, including antibody production. Auronifin, in particular, has been shown to decrease elevated serum levels of Igs and rheumatoid factor in arthritic patients.

D-penicillamine (D-3-mercaptovaline, Cuprimine), a breakdown product of penicillin, was, after the discovery of its chelating properties of copper ion (Fig. 2-6), introduced as an antidote to copper poisoning. It was also found useful in the treatment of Wilson's disease, where excess copper accumulation causes liver cell damage. Heavy metal poisoning treatment is not limited to copper. Mercury and lead poisoning are also successfully reversed. Formation of cysteine calculi (cystinuria) can also be reversed with penicillamine by forming a soluble disulfide compound.

The antirheumatic action of D-penicillamine initially had limited application. However, in the early 1970s controlled trials have established the drug's intrinsic activity. There are many similarities to chrysotherapy both in clinical response (lowering of rheumatoid factor serum titer, decreased inflammatory response) and adverse effects (glomerulonephritis). However, some are potentially very serious, particularly aplastic anemia and systemic lupus erythematosus (SLE). The most important hint as to the uncertain mechanism of penicillamine's disease-suppressive properties is probably the considerable decrease in IgM (rheumatoid factor) levels.

Quinoline antimalarials such as hydroxychloroquine (Fig. 5-6) and chloroquine have been found to have antiarthritic properties; however, the onset of clinical improvement, as with penicillamine and gold, takes months. Irreversible retinopathy, including retinal opacity, can be encountered. Lesser toxicities include skin pigmentation and alopecia. Proposals to possible mechanisms of action are speculative at best. It should be emphasized that none of the "slow-action" antiarthritic agents discussed earlier should be considered as initial therapy in RA. The salicylates and other NSAIDs deserve this distinction. If results are unsatisfactory gold may be considered as the subsequent therapeutic step. Penicillamine would be a logical alternate, as would short-term steroids or cytotoxic agents.

Of peripheral interest at this point are osmarins or osmium carbohydrate polymers. These polymers, as well as the more toxic osmium tetroxide, OsO_4, has been in limited and controversial use outside the United States. Their purported long-lasting effect has been attributed to osmium deposits remaining in joints after treatment. Since superoxide ion O_2^- is present in the arthritic joints, which destroys its ability to lubricate joint surfaces, metal complexes catalyzing decomposition of O_2 may account for the reported apparent benefits. The enzyme superoxide dismutase, SDM, a copper-containing enzyme in the liver catalyzing the dismutation of O_2 (Eq. 5.4), has been used to treat arthritis in humans, thus giving some credence to the previous osmium proposal.

$$O_2^- + O_2^- \, 2H \rightarrow H_2O_2 + \tfrac{1}{2}O_2 \qquad (5.4)$$

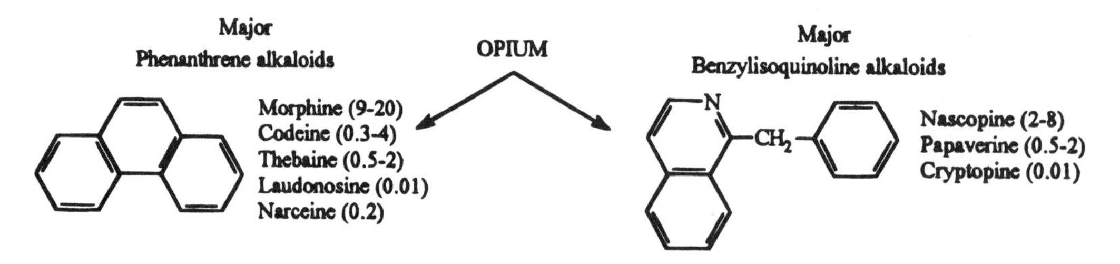

Figure 5-7. Opium and its major alkaloids (percent content in crude opium).

5.8. Opium—and the Strong Analgetics Emanating from It

Opium is the dried exudate of the unripe capsule of *Papaver somniferum*.[4] Varieties of this plant have been cultivated and used for several millennia. Today, it is grown in Turkey, Iran, Afghanistan, Russia, China, Southeast Asia, and Mexico, as well as in other areas of the world, both licitly and illicitly. Crude opium is known to contain over 25 different alkaloids that can be chemically divided into two categories: the phenanthrenes and the benzylisoquinolines (Fig. 5-7). Only morphine, codeine, papaverine, and nascopine are of clinical interest; thebaine is used as a starting compound for some semisynthetic opioids.[5]

This section will deal with some of the general characteristics of narcotic agonist analgetics, several of the opiate antagonists, and the newly emerged, interesting mixed agonist/antagonist group. Aspects of the mechanisms by which these drugs are believed to work, including multiple receptors and the more recently discovered endogenous peptides (incorrectly referred to as "endogenous opiates"), will be considered.

The pharmacologic and social problems of addiction (and tolerance) that so complicate therapy will not be considered to any extent. This is not to minimize the importance of addiction nor the fascinating research in the area, but to keep this chapter to a manageable level. (For an excellent discussion of drug addiction and abuse, see O'Brian, 1996, in Suggested Readings.)

Morphine alkaloid, as the free base and as its hydrochloride and sulfate salts, will be considered the prototype agent in the following discussions. However, it must be understood that the drug's pharmacology is very complex and its multiplicity of actions frequently complicate therapy. Table 5-4 lists the major actions of morphine. Even though the primary interest is in the analgesia the drug offers, it can be seen that at least some of the "side effects" can be useful as adjunct properties (e.g., sedation) or in their own right (e.g., antitussive, decreased peristalsis).

Through the centuries opium was used as an analgetic, sedative, hypnotic, and euphoric in the form of powders, alcoholic extracts (opium tincture, laudanum), syrups, and elixirs. It was even smoked, after the use of tobacco became prevalent.

The alkaloid *morphine* was first reported isolated by the German pharmacist Serturner (1805). Since organic chemistry as we think of it today was not existent, anything con-

---

[4] Opium, from the Greek *opion* or poppy juice; somniferum from the Latin for sleep-producing.

[5] The term *opioid* refers to natural or synthetic compounds with morphinelike or antagonist action. It is frequently used interchangeably with the term *opiate*.

Table 5-4. Pharmacological Profile of Morphine

Action	Comments
a. Analgesia	In doses of 5–15 mg
b. Sedation	Including drowsiness and mental clouding
c. Euphoria	In pain-free persons the opposite, dysphoria, can occur
d. Constriction of pupil	Tolerance to this effect does not occur
e. Antitussive effect	Codeine and heroin are superior
f. Depressed respiration	Frequently cause of death following overdose
g. Nausea and vomiting	By direct stimulation of the chemoreceptor trigger zone; not experienced by all patients
h. Peripheral vasodilation	May result in hypotension; can be dangerous when low blood volume exists
i. Decreased release of stomach hydrochloric acid	
j. Decreased secretion of bile and pancreatic secretions	
k. Decreased peristalsis	With resultant fecal desiccation constitutes a constipating effect
l. Increased biliary tract pressure	In pain due to biliary colic, pain is actually exacerbated
m. Increased tone and contraction of ureter and bladder	Resulting in antidiuretic effect
n. Bronchial constriction	With large dose can aggravate asthma
o. Dilation of cutaneous blood vessels	Flushed warm feeling and sweating may result
p. Tolerance, psychological and physiological dependence	The basis of addiction?

strued as structure determination was then impossible. As the nineteenth century progressed, the molecular formula was ascertained. Toward the end of the century certain structural features became identifiable, such as the existence of two hydroxyl groups (Fig. 5-8), as was the phenolic character of one: the tertiary nature of the nitrogen atom and the existence of a double bond somewhere in the molecule.

The correct structure of morphine was proposed in 1925. Ultimate proof of its correct structure had to await its total synthesis, accomplished three decades later. The lack of the correct structure of morphine did not discourage the synthesis of several morphine congeners and derivatives by chemical reaction with the known peripheral functional groups, specifically the phenolic hydroxyl (C-3), and allylic alcohol (C-6), and the double bond (C-7-8) (Fig. 5-8). Among the derivatives introduced before 1930 and still in common usage today are codeine,[6] ethylmorphine (Dionin), diacetylmorphine (heroin), hydromorphone (Dilaudid), hydrocodone (Dicodid), and methyldihydromorphinone (Metopon).

5.8.1. Chemical Aspects of Morphine

Morphine's poor aqueous solubility as the free alkaloid (about 1:5000) means that the drug is invariably used as the sulfate or hydrochloride salt, with which stable aqueous solutions (1:15 or better) for parenteral or oral use are readily prepared. In this respect, morphine is,

[6] Although obtainable from opium, it is also prepared by methylation of morphine.

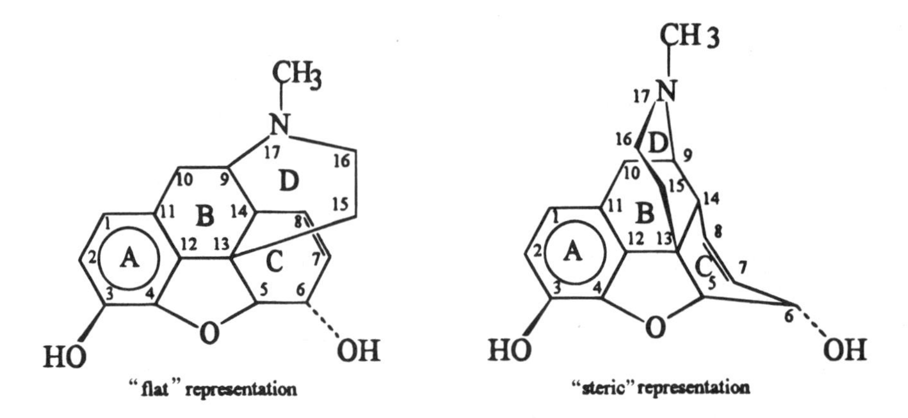

Figure 5-8. Structural representations of morphine. The C ring has the "boat" conformation; the D ring is in the "chair" conformation. Carbons 5, 6, 9, 13, and 14 are chiral.

of course, not unique. Other alkaloids used in medicine, be they opiates or compounds with different pharmacologies, are almost always used as salts in their pharmaceutical preparations.[7]

The early "peripheral" structural modifications previously mentioned were feeble attempts in achieving the almost utopian goal of separating potent analgesia from addiction liability (and other side effects). Heroin was among the first, and in retrospect, the most dramatic failure. Increasing potency proved to be readily achievable. Unfortunately, addictiveness seemed to parallel it, or remain with the new compound. It rarely decreased, much less vanished. It soon became apparent that if a separation of the properties specified were possible, then the sporadic and random synthesis of a few morphinelike compounds would not discover it. A wide-ranging systematic study would be required. This was initiated in 1929 by the National Research Council's Committee on Drug Addiction under the direction of the chemist L. F. Small and pharmacologist N. B. Eddie, whose syntheses and analgetic evaluation of at least 125 morphine derivatives were reported in 1938. In addition to modifications of the intact morphine skeleton, derivatives of compounds that could be viewed as structural "components" of the morphine molecule were also synthesized and evaluated. These "components" were benzofuran, dibenzofuran, the N-isostere carbazole, and phenanthrene. Nothing in this group resulted in useful compounds. Certain empirical structure–activity relationships (SARs), however, did emerge. These included the following regarding the morphinoid nucleus. Methylation of the phenolic OH reduces analgetic activity considerably (thus codeine is about 6 to 10 times less potent). Esterification causes a similar reduction. However, diacetylation (heroin) yields a more potent drug than morphine. This may be due to the fact that the phenolic acetate, which is readily cleaved in vivo yields the high potency intermediate 6-acetylmorphine (Table 5-5). Regarding the 6 position, oxidation of the OH to a ketonic function (particularly if the 7–8 double bond is also reduced) affords a more potent drug (hydromorphone). Epimerizing the 6-OH retains analgetic potency, yet lowers toxicity. 6-Acetylmorphine

[7] Cocaine alkaloid is still occasionally used as an olive oil solution for minor otic surgery. Ephedrine was once available as a nasal decongestant in mineral oil solutions.

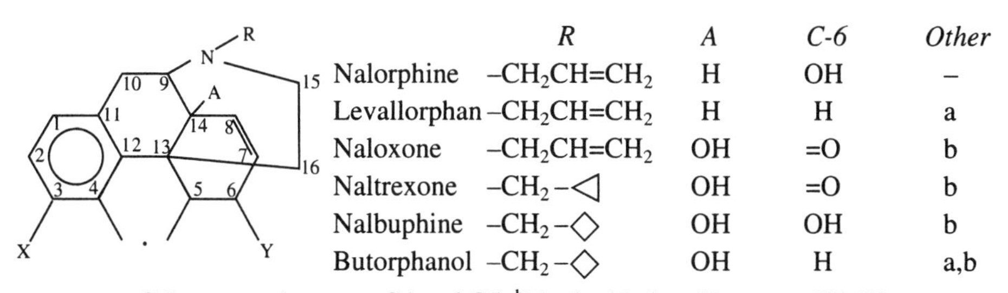

[a]No oxygen between C4 and C5. [b]No double bond between C7-C8.

Figure 5-9. Morphine-derived antagonists and agonists/antagonists.

presents the desirable features of increased potency with decreased toxicity. Table 5-5, however, does not give a clue as to the third parameter in this picture—addictability, which is not eliminated in any of these alterations. Methyldihydromorphinone (Metopon), although being 11 times more potent than morphine (in mice), is 20 times more toxic, and thus does not afford a useful improvement.

An exchange of the N-methyl group with larger alkyl, alkenyl, and arylalkyl function was undertaken. Increasing the size of the alkyl substituent decreased activity with smaller groups (e.g., ethyl) but then increased up to a six-carbon chain. The β-phenethyl group gave the highest activity.

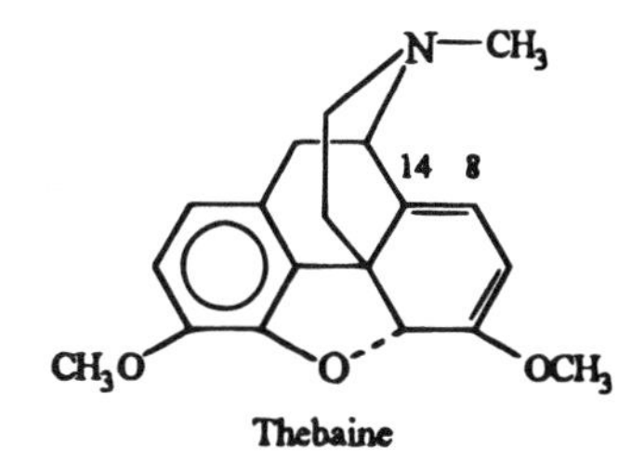

Thebaine

The opium alkaloid thebaine, although therapeutically useless (causes seizures and has little analgetic action), is important as the chemical starting point for the commercial synthesis of methyldihydromorphinone (Metopon), hydrocodone and hydromorphone, oxymorphone, and the very interesting oripavine compounds (see later). Since these drugs are modifications of a natural alkaloid, they may be referred to as *semisynthetic opiates.*

Table 5-5. Toxicity–Analgesia Comparison of Certain Opiates[a]

Drug	LD_{50} (mg/kg)[b]	Analgesia[c]
Morphine	531	0.75
Codeine	241	8.04
α-Isomorphine	890	0.80
6-Acetylmorphine	293	0.18
Hydromorphone	84	0.17
Hydrocodone	86	1.28

[a] Adapted in part from Jacobson et al. (1970). [b] Lethal s.c. dose of 50% of mice. [c] Minimal IM dose in cats.

5.8.2. Totally Synthetic Analgetics

Under consideration here are totally synthetic "stripped down" morphinelike skeletal structures and even simpler fragments of the morphine molecule, initially actually not recognized as such. These are primarily based on the piperidine ring.

Morphinans are basic morphine "skeletals" without the 4–5 ether bridge. The simplest one illustrating analgesia (although weak) is N-methylmorphinan. This compound was synthesized from 2-carbethoxycyclohexanone. This was soon followed by the racemic

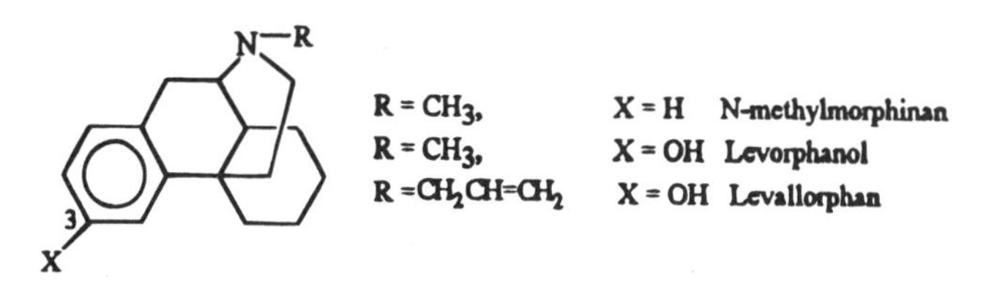

3-hydroxy derivative (racemorphan), whose levo isomer levorphanol (levo-Dromoran) is five times more potent than morphine as an analgetic. The 3-methoxy-derivative, as the (+) isomer (dextromethorphan), although devoid of analgesia, is widely used as a nonhabituating antitussive codeine substitute.

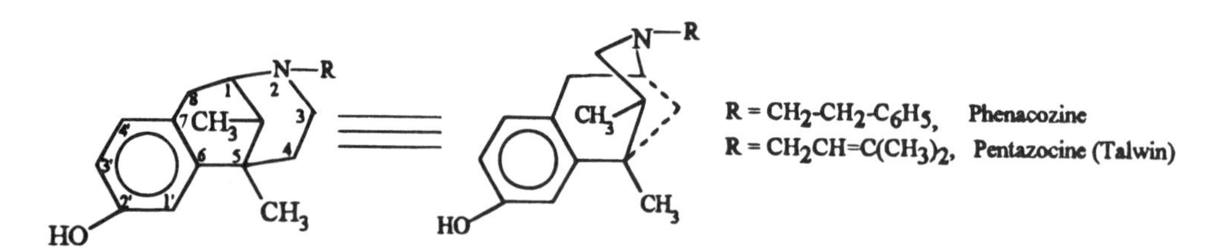

Further simplification of the morphine structure, achieved essentially by deletion of the C ring and substitution of two methyl groups for it, resulted in the 6,7-benzomorphans.[8] Syntheses of these types of compounds resulted in clinically used drugs such as phenazocine, which was marketed briefly in the racemic form, exhibiting at least triple the potency of morphine. Although believed to be less addictive than morphine, it was nevertheless still an addictive drug (see following discussion of antagonist analgetics).

Even though attempts at synthetic analgetics began in the 1920s, the first clinically successful drug was discovered totally serendipitously much later. While investigating a series of aminoesters as potential anticholinergic atropinelike drugs, Eislieb and Schaumann (1938) found several exhibiting analgetic activity as well. The most important, but not only, clinically useful compound to come out of these studies was the ethyl ester of 1-methyl-4-phenylpiperidine-4-carboxylic acid, meperidine (pethidine, Demerol) (Fig. 5-10A). Its structural relationship to morphine was not initially recognized; however, its pharmacology was.

[8] The structure may also be considered as a benzazocine compound.

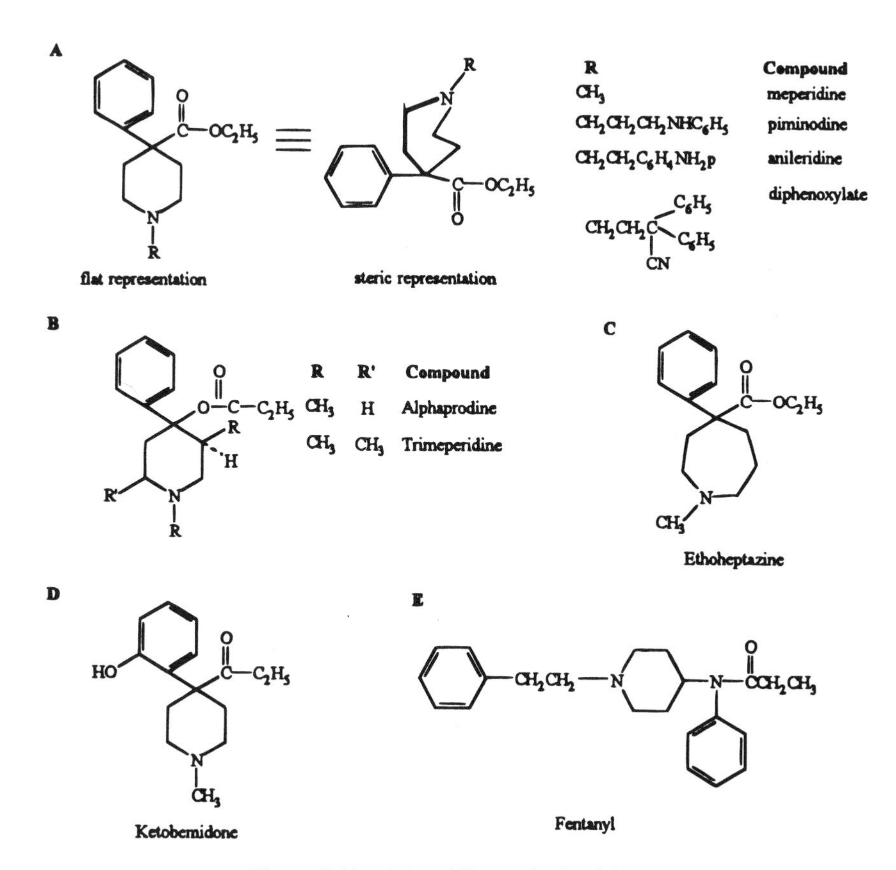

Figure 5-10. Meperidine and related drugs.

With few exceptions the methyl on the nitrogen atom represented the optimal substituent. These exceptions are illustrated by piminodine and anileridine. The latter drug exhibits several times the potency of meperidine with some lesser intensity of side effects.

Diphenoxylate may be viewed as an extension of the meperidine, structural group, or as being derived from the methadone-type molecule (see later). Its analgetic action is negligible, yet it has the opioid properties of suppressing the abstinence symptoms of morphine addicts and definite antitussive properties (similar to codeine). Signs of dependency and suppression of intestinal motility were two other morphinelike properties also reported. It is that last-mentioned property that has led to the marketing of diphenoxylate as an antidiarrheal preparation in combination with atropine to discourage abuse (Lomotil).[9]

It is not surprising that "reversing" the 4-ester function of meperidine afforded effective compounds such as trimeperidine and alphaprodine (Fig. 5-10B), with the latter approaching the potency of morphine. Expansion of the piperidine ring of meperidine to the next higher homolog azepine resulted in ethoheptazine (Fig. 5-10C), which was at first claimed to be equivalent to codeine. However, a double-blind comparative clinical study showed this drug to be significantly inferior to aspirin and actually indistinguishable from a placebo.

---

[9] The atropine is in a subtherapeutic dose.

The considerable number of meperidine analogs prepared and tested showed that other than ester reversal, changes at position 4 usually abolished or decreased activity. An exception, wherein the ester function was altered to ketone and meta hydroxyl added to the phenyl ring, was ketobemidone (Fig. 5-10D), whose potency over meperidine increased five- to sixfold. However, as with molecular "improvements" of other opiatelike compounds, addiction liability increased as well.

A more drastic digression from the by now traditional meperidine model is exemplified by the drug fentanyl (Fig. 5-10E). It is 50–80 times more potent than morphine in humans, but of short duration. It is used frequently in combination with the neuroleptic droperidol, and more recently as a skin patch (Duragesic). Another opiate analgetic synthesized by the Germans during World War II while attempting to develop spasmolytic compounds was methadone, an amino-4,4-diphenyl-3-heptanone, a noncyclic seven-carbon ketone. Its structural similarity to morphine was not appreciated for over 10 years after its introduction as a potent narcotic analgetic. It was probably first recognized by Gero (1954), who, based on inspection of molecular models, pointed out strong similarities to the rigid topology of morphine. Beckett (1956) then postulated a dipole–dipole interaction of the free electron pair of the nitrogen atom with the electron-deficient (i.e., partially positive) carbonyl carbon of methadone, resulting in a favored conformation of the molecule that might be viewed as a "pseudopiperidine" ring (conformation A, Fig. 5-11). Since the pKa of methadone is 8.25, only 11% of the compound would be in the un-ionized state at physiological pH of 7.4. Thus 89% of the compound would be in a protonated form that would also readily conform into a "pseudopiperidine" orientation by an intramolecular ion–dipole

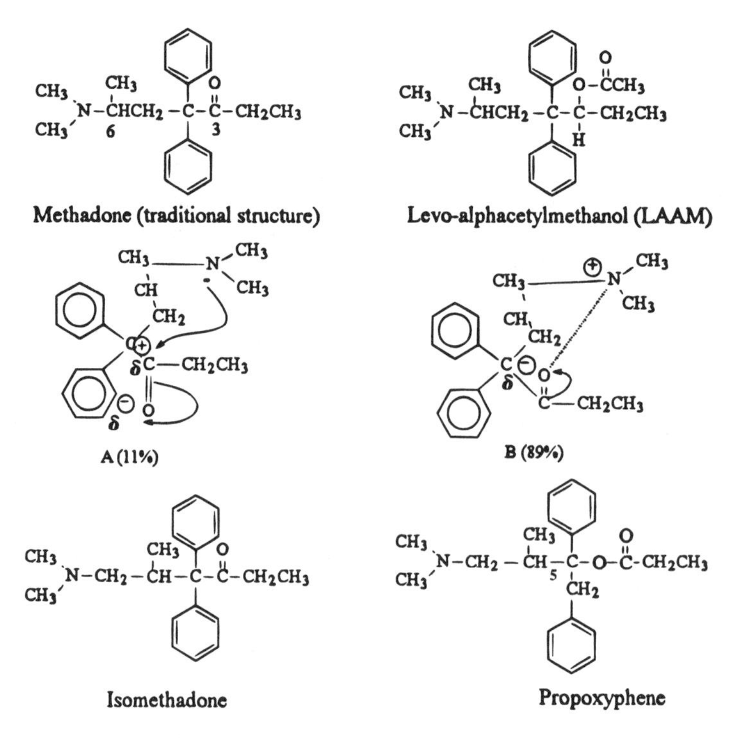

Figure 5-11. Methadone and related drugs.

interaction between the protonated positively charged nitrogen to electronegatively polarized carbonyl oxygen (conformation B, Fig. 5-11).

The original synthesis yielded both methadone and its 5-methyl position isomer isomethadone (Fig. 5-11), which is also an effective pain-relieving compound not used in the United States.

Although it is more toxic than morphine on a weight basis, methadone is more potent. On resolving the racemic compound, the (–) form was found to possess most of the opioid activities. The dextrorotatory isomer retains basically only the antitussive property (the commercial product is the racemic form). The potential advantages of methadone over morphine is its effectiveness via the oral route and considerably longer half-life (24–36 vs. 2–4 hours). This feature, particularly, has led to the drug's extensive use in the "treatment" of heroin addicts to suppress the latters' abstinence syndrome by maintenance of patients with methadone only.

Reducing the keto function of methadone and acetylating the resultant hydroxyl function afforded another long-acting congener, levo-alpha-acetylmethadol (LAAM) (Fig. 5-11). This drug is being used in physically dependent opiate abusers, suppressing withdrawal symptoms by dosing at intervals as long as 72 hours.

A congener of isomethadone, propoxyphene (Darvon), was synthesized and introduced as a hoped-for nonaddicting, safer drug for medium severity pain. In view of present knowledge with this compound, the earlier claims that tolerance did not develop even on prolonged administration, and that doses in up to 50 times therapeutic range have not been fatal, stretches credulity to the utmost. The "absence" of any structural resemblance to morphine was even emphasized.

The earlier euphoria about the drug's safety has long since given way to an understanding that suicidal and accidental deaths due to this drug are reaching epidemic proportions in the United States and exceeded 115 in just one Canadian province (Anonymous, 1979). Furthermore, clinical studies have shown propoxyphene to be inferior to aspirin as an analgetic and possibly indistinguishable from placebos (Gilbert and Moertel, 1972). It appears to have no therapeutic advantage over less toxic and more effective analgetics.

5.9. Narcotic Antagonists

Prior to the early 1940s no satisfactory pharmacologic antidote for opiate overdoses and poisoning existed. Treatment was essentially the same as for other CNS and respiratory depressants (i.e., with CNS stimulants and supportive physical methods). Some antagonism to the respiratory depression of morphine had been achieved by replacing the N-CH_3 group of codeine with an allylic function (-CH_2-CH=CH_2) during the 1920s. This fact was subsequently utilized as a rationale for the introduction of N-allylnormorphine (Nalorphine, Nalline, Fig. 5-9) as the first opiate antagonist into clinical practice. Nalorphine antagonizes most of the pharmacological properties of morphine, including analgesia, euphoria, and the potentially deadly respiratory depression. Respiratory depression brought on by overdoses of alcohol, barbiturates, tranquilizers, or general anesthetics is not reversed.

Since nalorphine failed to demonstrate analgetic properties by the usual animal screening tests, its potential as an analgetic in humans was at first not pursued. In a simplistic clinical experiment to determine if coadministration of morphine and nalorphine in varying proportions would retain analgesia while preventing addiction, it was discovered that nalorphine had potent analgetic properties in humans. It was also shown that nalorphine does not support opiate dependence nor does it produce habituation itself. Even though nalorphine's high incidence of dysphoria and bizarre hallucinations precluded its clinical use as an analgetic, it led to a new approach in analgetic research: the search for narcotic antagonists with quality analgetic properties. The type of N-substituents conferring antagonist properties to morphines, morphinans, and benzmorphans are outlined in the following:

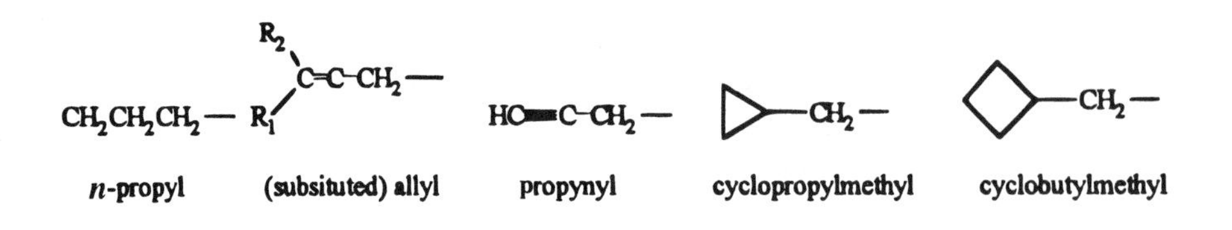

The corresponding morphinan, levallorphan (Lofran, Fig. 5-9), shows all of nalorphine's properties (including the ability to cause respiratory depression in excess doses). It is, however, five times more potent than nalorphine.

Corresponding analogs of the oxymorphone series yielded superior, "pure" narcotic antagonists. Naloxone (N-allyl-14-hydroxynordihydromorphinone, Narcan) (Fig. 5-9) is now considered the new prototype of the almost ideal opiate antagonist. Naloxone does not appear to possess agonist properties. Its potency as an antagonist is greater than nalorphine or levallorphan, and unlike these agents does not produce respiratory depression on its own. Because of this fact respiratory depression produced by opiate overdoses can be readily differentiated from that produced by sedative-hypnotics and other types of CNS depressants. Naloxone even reverses the effects of propoxyphene and the mixed agonist/antagonist agents (see the following).

The onset of action of Naloxone following IV administration is rapid (<2 min) but of short duration. Thus victims overdosing with longer-acting narcotics (particularly methadone) require repeated dosing.

The subsequent introduction of naloxone's cyclopropylmethyl analog, naltrexone (Trexan, Fig. 5-9), with its ready oral absorption and considerably longer duration of action has made possible the use of a pure antagonist in the treatment of narcotic addicts. Naltrexone in once-a-day oral doses of 50 mg will produce satisfactory clinical blockade of intravenously administered opiates.[10] In fact, a single dose of 100 or 150 mg can be used on a once, every other, or third-day basis. This drug may represent a small advance in the battle against opiate dependency. Naltrexone has been approved (1994) to obtund the cravings of alcohol abusers.

5.10. Agonist/Antagonist Analgetics

The disclosure that the antagonist nalorphine was a potent analgetic led to the synthesis of a number of 6,7-benzmorphans that were clinically evaluated as to analgesia, psychotomimetic effects, and the existence of other opioid effects. Phenazocine, which demonstrated high potency but no antagonism, was already mentioned. Cyclazocine, with

[10] This dose will antagonize the effect of 25 mg of heroin.

a 40-fold potency over morphine and high antagonist efficacy, looked promising. However, psychotomimetic effects dampened its luster. Although a weaker analgetic, pentazocine (Talwin) appeared to offer a more acceptable separation of unwanted effects. This, and its very low addiction potential, led to its clinical introduction in 1967 as the first drug in this category.

Pentazocine, although two to three times weaker as an analgetic than morphine, became useful because of its extremely low dependence potential. It does, however, share some of morphine's undesirable side effects, particularly respiratory embarrassment in the upper therapeutic dosage ranges. At single doses of 60 mg or more the dysphoric and hallucinogenic effects first noticed with nalorphine also become apparent with pentazocine. These are reversible with naloxone (but not nalorphine).

Two more compounds were introduced in the late 1970s. Nalbuphine (Nubain) (Fig. 5-9), the cyclobutylmethyl analog of oxymorphone, is equipotent with morphine in humans, with a rapid onset and somewhat longer duration of action. The cardiovascular effects such as increased blood pressure and cardiac work seen with pentazocine are not readily apparent here. Even though respiratory depression at therapeutic doses is about the same as seen with morphine, it does not, as with morphine, increase with ascending doses; rather it reaches a plateau. Its antagonist effectiveness is considerably less than nalorphine's. Side effects at the lower effective doses are few. At seven times therapeutic doses (70 mg) dysphoria and thought aberrations do occur. The to-date almost nonexistent abuse potential finds this drug not scheduled under the Controlled Substances Act (USA). It can be said that nalbuphine represents the achievement of the decades-long search for an essentially nonaddicting opiate with the potency of morphine.

The structural similarity of nalbuphine to naltrexone could not be greater, yet the latter exhibits no analgesia. The reason for this still awaits an explanation.

Butorphanol (Stadol) (Fig. 5-9) is a cyclobutylmethylmorphinan derivative. It may be viewed as the next step, at least with regard to potency (five times morphine). However, its cardiovascular effects resemble those of pentazocine. It may have a slightly higher abuse potential than nalbuphine. In morphine addicts butorphanol does not precipitate a withdrawal syndrome. Thus its antagonist properties are apparently very weak.

The next level of achievement is represented by the introduction of buprenorphine into clinical practice.[11] The development of this highly potent analgetic with mixed agonist/antagonist properties is indicative of the extensive years of research required to reach the present level of sophistication.

It was realized early that the morphine-related alkaloid thebaine, because of the conjugated diene in ring C, was the most chemically reactive of the opium alkaloids. A large variety of dienophiles readily underwent a Diels–Alder reaction with thebaine, affording several adduct compounds with opioid properties, including analgesia.

The syntheses involved facile addition of methyl vinyl ketone to thebaine resulted in an adduct containing an ethylene bridge across C-14 to C-6 and a methyl ketone at position 7 (Fig. 5-12). This "thebaine adduct" was equianalgetic with morphine in spite of the 3-methoxy group as in codeine. Treatment of this compound with Grignard reagents yielded a homologous series of tertiary alcohols. Most of these had analgetic potency superior to morphine, the highest (Fig. 5-12, $R' = CH_3$, $R = CH_2CH_2C_6H_5$) being 300 times more analgetic. Demethylation of the 3-methoxy group (with NaOH in boiling ethylene glycol so as

[11] In 1978 in England; 1985 in the United States.

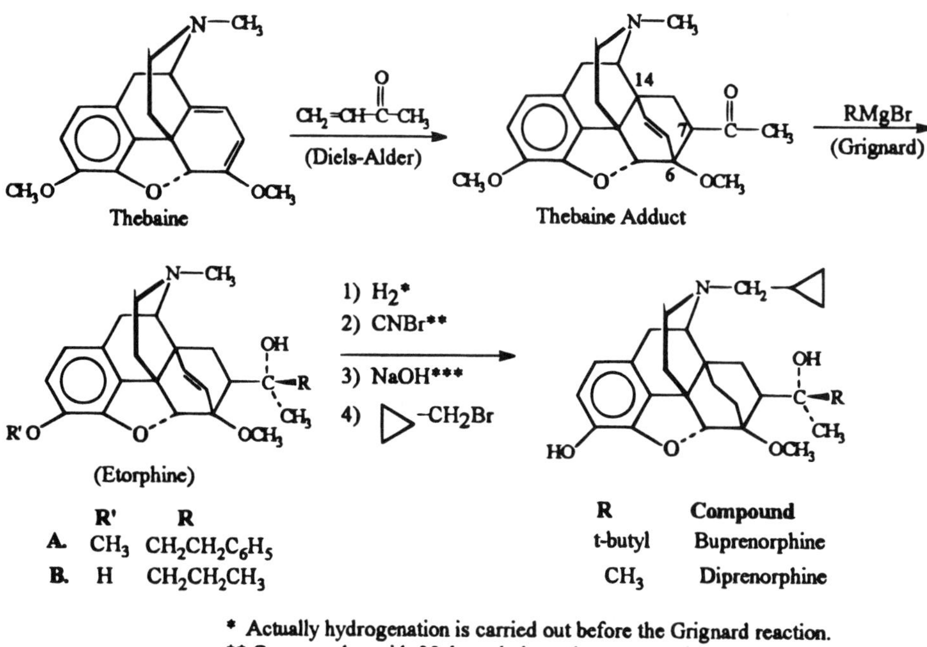

Figure 5-12. The oripavine compounds.

not to affect the tertiary OH) resulted in corresponding phenolic compounds with extremely high potency. The most astounding compound, etorphine (Fig. 5-12), revealed potencies up to 10,000 times that of morphine in animal models. It is interesting that in doses higher than required to produce analgesia (but still small) etorphine is capable of immobilizing even large animals. It soon became widely used in animal conservation efforts requiring the safe capture of animals as large as elephants in the wild.

As extraordinary as these potencies were, these N-methyl compounds exhibited full opiate agonist properties including, unfortunately, addiction liability. The next logical step would be to replace the methyl on the nitrogen atom with the substituents that have afforded either antagonists and/or mixed agonist/antagonists with other morphine-type structures. Although various groups were utilized, the cyclopropylmethyl function yielded the most interesting drugs. Thus the "thebaine adduct" (Fig. 5-12) is hydrogenated to produce a saturated ethano bridge function. Cyanogen bromide (von Braun reaction) N-demethylates the product; the NaOH/ethylene glycol procedure cleaves the phenolic methoxy function. Finally, cyclopropylmethyl bromide realkylates the nitrogen. Diprenorphine, where $R = CH_3$, is not surprisingly, a "pure" antagonist devoid of analgetic action. Although also a potent antagonist, the R = ethyl homolog exhibits strong analgesia (at least in one screening test). The *n*-propyl and *n*-butyl members of this series were primarily potent agonists. Branching of the R group, however, led to increased antagonism; the *tert*-butyl compound, buprenorphine (BP, Buprenex) (Fig. 5-12), finally being 50 times more analgetic than morphine and, in the rat, at least, appearing not to have antagonist properties. In the mouse, however, BP showed strong antagonist properties. The high potency of these compounds and the contradictory pharmacology as to agonism, antagonism, or a mixture of these illustrates the difficulties of evaluating these drugs and, much less satisfactorily, explaining these apparent anomalies at this time. Even more interesting, as well as clinically significant, is BP's effect on gastrointestinal motility and respiration in humans. It is particularly important that the drug's effect on respiratory depression actually decreases as the dosage increases. This may represent an increased safety level. It is also of interest that BP, because of its high lipophilicity, has a slow onset—and long duration of action. If, as suspected, this represents a slowness of opiate receptor binding (not weak affinity) and a resultant slow dissociation value, it may help explain the apparent lack of dependence potential encountered to date. One supportive piece of evidence for the slow drug–receptor dissociation is that while respiratory depression that BP may cause can be prevented by preadministering naloxone, it is difficult to reverse it even with high doses of naloxone, if the depression is produced first.

The discovery that it is an opiate antagonist—nalorphine also had agonist, and therefore analgetic, properties—failed to prove or disprove the original question of that early study. Was it possible to determine the pharmacological effects of mixing an agonist and antagonist in a single dose and find an optimum ratio for these two drugs that will retain potent analgesia while preventing the development of dependence, tolerance, and even abuse? The demonstration of opiate receptors and their subtypes (see later), the development of methodology to determine their regional brain distribution, and the ability to study differentially the binding capacities of agonist and antagonists to them, have combined to enable us to answer such a question with some sophistication.

The drug that may answer this question is currently undergoing clinical evaluation as an analgetic for postoperative pain.

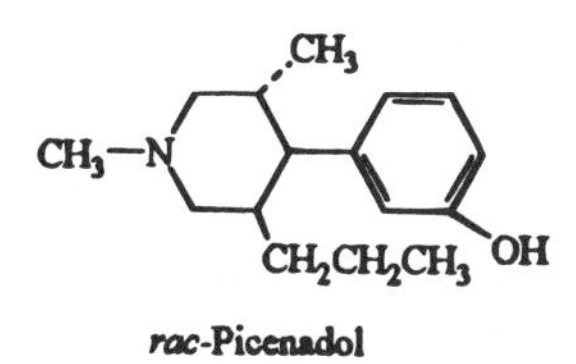

rac-Picenadol

The racemate of picenadol exhibits agonist/antagonist properties. However, it was shown that the dextrorotary isomer possesses agonist, and the *levo* enantiomer the antagonist activity, with some weak agonism. The analgetic action could be shown to result from strong binding of the (+)-isomer to the μ receptor, probably with some contribution from the weak agonist activity of the (–)-isomer. The isomers of the racemate were shown to possess a μ/δ ratio of 1.0 for the *dextro*-isomer, and 0.5 for the *levo*-isomer. κ-Receptor activity could not be demonstrated for either optical form.

Both morphine and picenadol exhibited the expected profile of opiate-type side effects in human subjects. It is surprising, however, that naloxone reversed these only in the case of morphine. It is also of interest that picenadol possesses a higher affinity for the μ-opiate receptor than does morphine. The antagonist activity of the racemate (presumably from the *levo*-isomer) gives the drug a lesser propensity to produce significant respiratory depression than does morphine. Preclinical animal studies demonstrated that picenadol

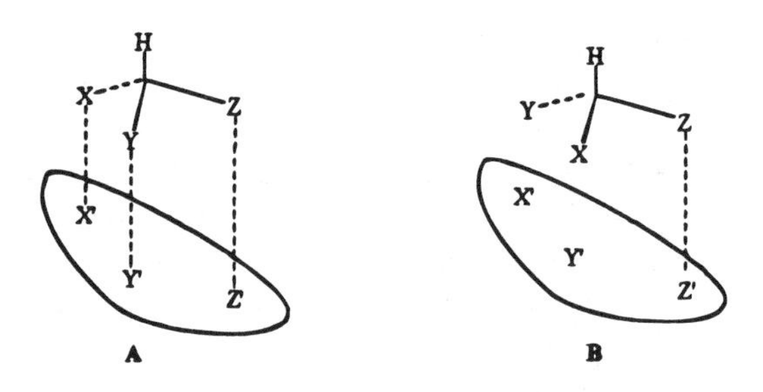

Figure 5-13. Optical isomerism and biological activity.

did not suppress morphine withdrawal symptoms; neither did it initiate them. These and other differences between morphine and picenadol have been explained on the basis of binding-site differences as well as the anticholinergic activity of the latter drug. More extensive human trials will establish whether picenadol will offer any clinical advantages over morphine.

5.11. The Opiate Receptor

The existence of binding sites specific to opiates had been assumed for many years. Some of the early "indirect" evidence leading to this receptor hypothesis included the relatively high potency of the narcotic drugs and the structural and stereospecificity exhibited by them. Even with the synthetic analgetics biological effects are usually found in those optical isomers stereochemically related to the natural (–) morphine molecule.[12]

The concept involved here may be understood by considering Figure 5-13. It will be recalled that enantiomers are nonsuperimposable mirror image structures in which all interatomic distances are the same. If a three-point interaction between a drug and its receptor is assumed, it can be seen that only in "drug" A is the configuration correct to bind all three groups (*X*, *Y*, *Z*) to complementary areas on the receptor surface. It can be said that the receptor can "distinguish" between the two isomers.

The fact that apparently small changes in structures (specifically around the nitrogen atom) resulted in very specific and effective opiate antagonists, of course, added to the strength of the receptor concept.

Beckett and Casy (1954) made the first proposal for the analgetic receptor (Fig. 5-14) based on a detailed study of the geometric features of most morphine-type drugs known then, including morphinans, meperidine, and methadone-type compounds. With later updating (Casy, 1975) the proposal included an anionic site interacting with the cationic (protonated) nitrogen atom; a flat surface accommodating the planar aromatic ring, and a cavity site accommodating the C-15–C-16 carbon atoms of the piperidine ring. The tremendous increases in analgetic potency encountered with the oripavines (e.g., etorphine) would necessitate the addition of a stereospecific lipophilic site to the Beckett–Casy model. The C-19 3° alcohol function may provide additional hydrogen bonding to the receptor's proteinaceous topography, or, alternatively, to the C-6 methoxy group, locking

---

[12] When (+) morphine was synthesized, it was found to be devoid of analgesia.

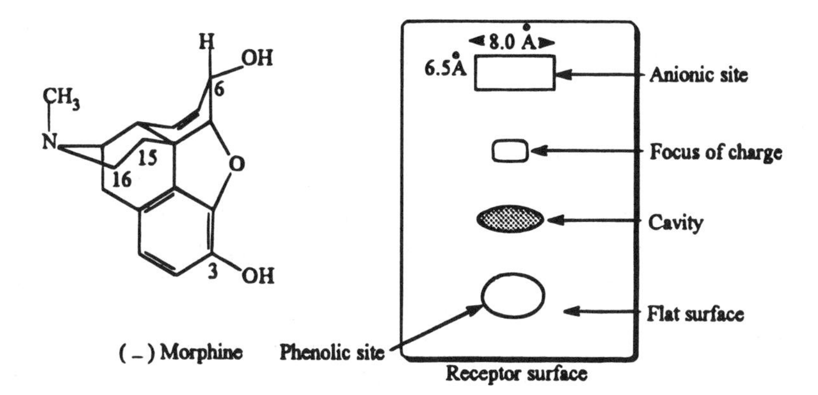

Figure 5-14. Relationship of morphine to Beckett–Casy receptor.

in a very favorable conformation. The aliphatic (or aralkyl) functions on C-19 (*R* group, Fig. 5-12) would provide the extra receptor-binding capacity to the lipophilic site, thus possibly accounting for the quantum jump in potency. The significance of this lipophilic site would shortly be confirmed with the discovery of the natural ligands for these then-still-putative receptors—the enkephalins (see later).

With the discovery of stereospecific opiate binding sites in rat brain homogenate by several groups (Pert and Snyder, 1973b; Simon et al., 1973), the existence of pharmacologic opiate receptors was put on a sound scientific footing. Additional supportive evidence was produced by determining the relative effectiveness of known agonist analgetics, antagonists, and mixed agonists/antagonists in reducing the stereospecific binding of tritium-labeled naloxone at rat brain sites (actually rat brain homogenates) (Table 5-6). The pharmacologi-

Table 5-6. Relative Potencies of Drugs in Reducing Stereospecific ^{3}H-Naloxone Binding to Rat Brain Homogenate

Drug	ED_{50} (nM)[a]	No effect at 0.1 mM
(–)-Etorphine	0.3	Phenobarbital
(–)-Etonitazene	0.5	Norepinephrine
Levallorphan	1	Atropine
Levorphanol	2	Pilocarpine
(–)-Nalorphine	3	Arecholine
(–)-Morphine	7	Colchicine
(–)-Cyclazocine	10	γ-Aminobutyric acid
(–)-Naloxone	10	Bicuculline
(–)-Hydromorphone	20	Serotonin
(–)-Methadone	30	Carbamylcholine
(+/–)-Pentazocine	50	Neostigmine
(+)-Methadone	300	Hemicholinium
Meperidine	1,000	Histamine
(+/–)-Propoxyphene	1,000	Glycine
(+)-3-Hydroxy-N-allyl-morphinan	7,000	Glutamic acid
Dextromorphan	8,000	89-Tetrahydrocannabinol
(–)-Codeine	20,000	Acetylsalicylic acid
(–)-Oxycodone	30,000	Caffeine

[a] ED_{50} = concentration of drug required to inhibit stereospecific ^{3}H-naloxone binding by 50%. (From Pert and Snyder, 1973b.)

cal potencies of these opioid compounds in either producing or antagonizing analgesia generally parallel closely with their determined affinities for naloxone binding sites. One large discrepancy appears to be that etorphine is only 23 times as "potent" as morphine, not the 6000 + times previously mentioned. The explanation may reside in the fact that etorphine has been shown to enter the CNS 300 times more efficiently than does morphine from the circulatory system, thus accounting for the actual in vivo difference of 6,000 (300 × 20).

Although codeine in clinical use is generally thought of as about one-fifth the potency of morphine, the data here show a very weak comparative binding affinity. If one considers the possibility that codeine may actually be a pro-drug for morphine, being slowly metabolized by O-demethylation to it, the data may not be the anomaly it appears to be. In fact, it had demonstrated that codeine's intraperitoneally produced analgesia is not achievable by direct administration into the brains of animals. Table 5-6 also indicates the specificity of the opiate receptor by demonstrating that a large array of nonopiate compounds, even including the analgetic aspirin, do not in fact bind to the same regions even at 1 million times higher concentrations.

Figure 5-15 shows an almost ideal relationship (correlation coefficient 0.98) between the kinetics of receptor binding assays (dissociation constants K_D) and at least one physiological effect of opioids: inhibition of electrically produced contraction of guinea pig intestine by agonists (or reversal by antagonists). The data can be interpreted to show that opiate binding affinities to the receptor tissue can predict relative potencies of their analgesia and other physiological effects and additionally that the binding areas *are* the physiologically active receptors.

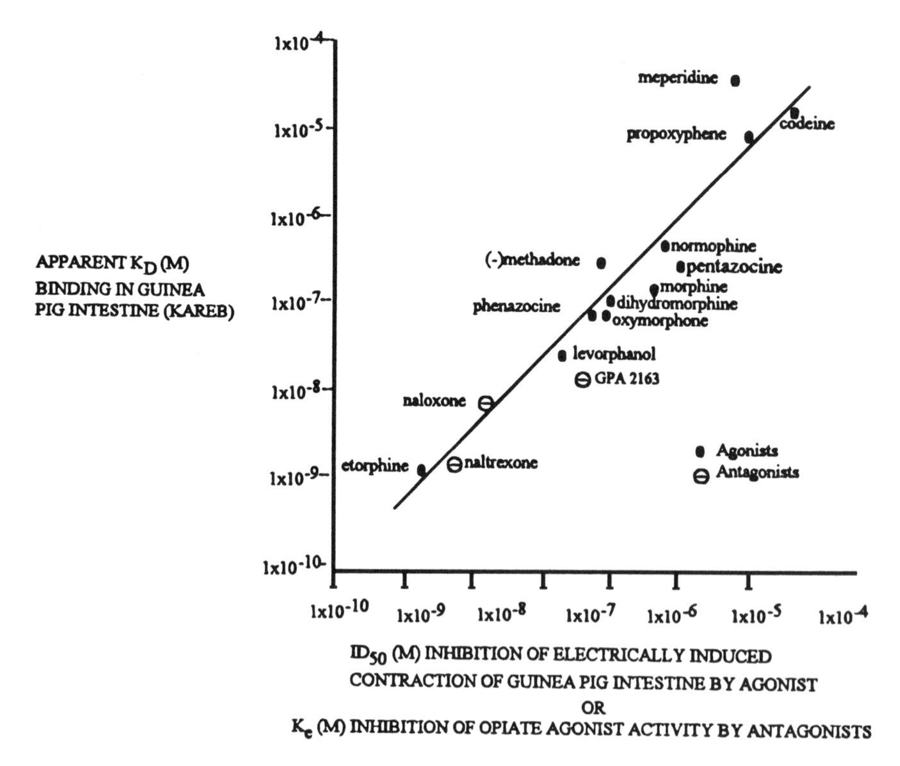

Figure 5-15. Correlation of receptor binding and opiate activity (Creese and Snyder, 1975).

Studies on the regional distribution of the opioid receptors in the brain of monkeys and humans show, as might be expected, the limbic system, particularly the amygdala, with the highest concentrations. The medial and anterior hypothalamus are also high. (The presence of receptors in the intestinal area has been already mentioned.)

Studies with high-specific-activity[13] ^{3}H-naloxone, when compared with those with lower specific activity, revealed the existence of two types of binding opiate receptors: low affinity (K_D = 60 μM) and high affinity (K_D = 0.6–0.9 μM), a differential up to 100. Addition of 100 nM of Na^+ to the test system apparently increases the high-affinity binding sites, but has no effect on the low-affinity sites. With the agonist ^{3}H-dihydromorphine high- and low-affinity functions also exist. However, here Na^+ abolishes high-affinity binding almost totally and decreases low-affinity binding by half.

5.11.1. Effects of Sodium

The two-component binding of naloxone and dihydromorphine (i.e., high and low affinity) shows that Na^+ has a selective effect on the binding functions. These binding differences appear to be significant, especially in the case of the high-affinity sites. The distinct nature of these was shown by the fact that protein-modifying reagents selectively eliminated the high-affinity sites.

Simon et al. (1973) reported a marked inhibition of H^3-etorphine binding by increasing sodium ion concentrations; Pert and Snyder (1973a) the same year found that no such inhibition was demonstrable with naloxone. This Na^+-mediated differential binding effect on the binding of agonists and antagonists to opiate receptor preparations was then established to be a general difference. In fact, in the presence of salt, binding of all agonists is *decreased* while antagonists are usually actually increased. It was further established with etorphine and naltrexone that the binding variations are due to changes in affinities and not to alterations in the number of binding sites being unmasked. The differentiating effects of Na^+ described are unique to Na^+. Other alkali and alkaline earth ions as well as cations such as choline reduce binding of agonists and antagonists indiscriminately.[14]

Simon proposed that opiate receptors exist in two (or more) conformations, depending on the absence or presence of Na^+, and that the one in the presence of Na^+ has a lower affinity for agonists. The various binding studies cited involving competition of agonists for bound labeled antagonists that require higher concentrations in the presence of Na^+ (ED_{50}) actually represent changes in affinity.

It now becomes possible to begin to grapple with the complex opiate pharmacology on a molecular basis and to explain perplexing facts. Unlike in other systems, opiate agonists and antagonists have strikingly similar structures. In addition, "minor" chemical modifications can give dramatic shifts in relative agonist/antagonist properties. Finally, dramatic differences in absolute potencies exist, such as one molecule of nalorphine being able to antagonize 200 molecules of morphine, yet both compounds are known to bind to (i.e., compete for) the same receptor. The critical variable was the Na^+. Thus as little as 1 nM of Na^+ can increase ^{3}H-naloxone binding by 60%, or decrease the ^{3}H-dihydromorphine by 30%.

---

[13] High degrees of radioactivity per mg of compound.

[14] Actually Li^+ shows some sodiumlike effects, possibly due to the similar diameter of the two ions in the hydrated state.

A method was developed to determine the extent to which addition of sodium to incubation media decreases the ability of cold (nonradioactive) test compounds to inhibit ^{3}H-naloxone binding. When this was expressed as a "sodium response ratio," by dividing the relative affinity of the opiate receptor binding in the presence of 100 nM of sodium and correlating it with actual in vivo properties (Fig. 5-16), agonists exhibited large decreases in inhibitory potency. The highly potent agonist etorphine showed a loss of only 12; however, the weak propoxyphene showed a loss of 60. The "pure" antagonists naloxone, naltrexone, and diprenophine predictably exhibited a response ratio of unity. Antagonists "adulterated" with agonist activity such as levallorphan and nalorphine had their ability to displace naloxone reduced two- to threefold; actual mixed agonist/antagonists were in the 3–7 range (e.g., pentazocine). It is interesting that 5-phenylbenzmorphan (GPA 2163) has a sodium response ratio of less than 1. What is the significance of such a superantagonist?

The sodium effect has the practical effect of being an excellent and rapid predictor of the relative agonist-to-antagonist properties of newly synthesized compounds. Some theoretical conclusions that might be drawn from these studies is that the opiate receptor is an equilibrium of two interconvertible conformations. In the presence of sodium, one has a much higher affinity for antagonists (having more binding sites in that conformation?). The absence of sodium would produce a state with higher agonist activity. This would go a long way toward explaining why much smaller doses of antagonists (having presumably much greater affinity for the receptor in vivo) can counteract the pharmacological effects of much larger doses of agonist drugs. This "concept" can also help one understand how apparently small changes in an opiate's structure could, by even slight affinity shifts, produce considerable changes in the agonist/antagonist ratio.

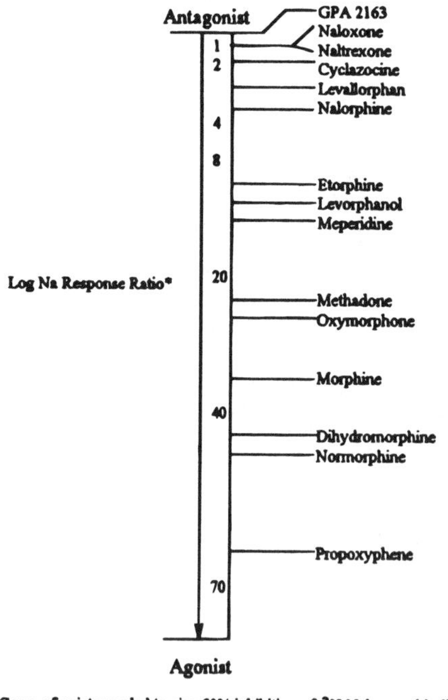

Figure 5-16. Sodium sensitivity of nonradioactive opiates.

5.12 Endogenous Opiate Receptor Ligands

The receptor work described previously now sets the stage to answer an interesting question: What are the receptors for plant alkaloids doing in the mammalian CNS? A reasonable hypothesis is that opiate receptors are part of a neurotransmitter or neuromodulator system that functions through the mediation of some naturally occurring morphinelike substance (i.e., an endogenous ligand). To prove this hypothesis, it would be necessary to find and identify this ligand or ligands in the mammalian CNS and to establish agonist activity at opiate receptor sites. Chemical detective work soon solved part of the riddle. A substance termed *enkephalin* (Greek for "from the head") was isolated first from rat and then from pig brain, and characterized as two similar but distinct pentapeptides identified as H-Tyr-Gly-Gly-Phe-Met-OH (methionine enkephalin) and H-Tyr-Gly-Gly-Phe-Leu-OH (leucine enkephalin).

Working with camel pituitary extracts, a 31-amino-acid polypeptide with potent analgetic properties was isolated and sequenced; it was β-endorphin. (*Endorphin*, a contraction of endogenous morphine, although not totally satisfactory, has become the widely used generic term for all opioid peptides.) The substance showed opiate like activity in vivo. It could also produce catatonia and hypothermia in animals. The pituitary hormone β-lipotropin, which plays a part in fat mobilization, is a 91-amino-acid polypeptide whose amino acid sequence 60–91 is the same as β-endorphin. It was also noted that amino acids 61–65 of the hormone were identical to methionine enkephalin (thus also the same as the first five positions of β-endorphin). Independent studies with both endogenous and synthetic β-endorphin established its analgetic properties. Also isolated were two additional peptides from pig pituitary. One, α-endorphin, corresponds to sequence 61–76 of β-lipotropin and shows analgetic as well as tranquilizing effects in animals; γ-endorphin corresponds to amino acids 61–77 of β-lipotropin. It evokes violent behavior when injected in rats.

It is known that electrical stimulation of the brain can relieve pain in animals and humans. Studies have shown marked similarities between such stimulation-induced analgesia and opiate-produced analgesia, including naloxone reversibility. All these factors led to speculation that endogenous ligands for these receptors must exist. The opiate peptides became the prime candidates. Questions in regard to the relative importance of endorphins and enkephalins, and differences in their distribution, arose. Work in this "mapping-out" process involves histochemical fluorescence procedures, immunoassays, bioassays, and, of course, receptor-binding studies.

It appears that the enkephalins are not metabolic products. There is likely a single β-endorphin system associated with lipotropin within cell bodies in long axons innervating the midbrain and limbic system. Enkephalin-containing cells and nerves were found throughout the brain and spinal cord. Enkephalin-rich areas appear to coexist with dopaminergic, noradrenergic, serotonergic, and substance-P neuronal systems. Opiate receptors and enkaphalin-containing fibers show a parallel distribution in many areas. There are considerable enkephalinergic fibers in the gastrointestinal tract, where a 17-amino-acid peptide has been discovered in the pig duodenum. These findings further explain the effectiveness opiates have in the treatment of diarrhea and dysentery. There may be real and potential clinical relevance as well (e.g., the area of addiction and possible development of new analgetics, including synthetic analogs of the enkephalins themselves) (see Table 5-7).

Direct injection of the two endogenously discovered enkephalins (methionine and leucine) into the periaqueductal gray (PAG) matter of rats demonstrated that they pos-

Table 5-7. Chemical Modification of Enkephalins

	Opiate receptor affinity	Analgesia potency[a]
1. Tyr-Gly-Gly-Phe-Met (methionine enkephalin)	1	1
2. Tyr-Gly-Gly-Phe-Leu (leucine enkephalin)	0.5	0.5
3. Tyr-D-Ala-Gly-Gly-Phe-Met	1	100
4. Tyr-D-Ala-Gly-Phe-Pro-Amide	na	1,500
5. Tyr-D-Ala-Gly-Phe-Met-ol[b]	5	1,600
6. Tyr-D-Ala-Gly-Phe-Met(O)ol[c]	2	9,600
7. Tyr-D-Ala-Gly-Mephe-Met(O)ol[d]	11	28,800
8. β-Endorphin	4	1,200
9. Morphine	2	33

na = not available.

[a] Potency relative to methionine enkephalin; receptor affinities obtained by binding studies with ^{3}H-naloxone analgesic carried out in rats following peptide injection into brain periaqueductal gray matter.

[b] The terminal carboxyl group is reduced to an alcohol function.

[c] The sulfur atom of methionine is oxidized to sulfoxide;

[d] phenylalanine here is N-methylated. (Adapted from Snyder, S. H., The Brain's Own Opiates, *Chem. Eng. News*, Nov. 28, 1977, p. 31).

sessed analgetic properties that were readily antagonized by naloxone. The dose required was relatively high and duration of analgesia short, apparently due to rapid enzymatic degradation. Intravenously and orally, these two compounds are ineffective since they cannot cross the blood–brain barrier (BBB), nor are they significantly absorbed. The highly charged nature of these peptides and their hydrolytic sensitivity, respectively, accounting for their poor penetrating and short-lasting low-potency effect, were soon improved by structural modifications (Table 5-7). For example, the synthetic enkephalin analog 4 exhibits a 1,500-fold increase in analgetic potency over methionine enkephalin following direct brain injection by virtue of increased stability to hydrolysis. This is achieved by the first glycine unit being replaced by a D-alanine component, an amino acid with an "unnatural" configuration that is resistant to attack by the enkephalinase enzymes. Furthermore, the terminal methionine residue being replaced by the amide form of proline eliminates the polar charge there, permitting this compound limited BBB penetration. Thus this peptide exhibits analgesia even following IV administration. In fact, it is five times more potent than IV morphine. Peptide 7, FK33-824, by the intracranial route, exhibits analgesia an astounding 28,000 times that of met-enkephalin (900 times morphine). Intravenously, it is also several times more potent than morphine and, most surprisingly, even has an oral effectiveness about one-fifth that of morphine, indicating some gastric stability. It is a pure agonist and has produced dependence in rhesus monkeys. Withdrawal symptoms are precipitated, and analgesia is reversed by naloxone.

β-Endorphin is also analgetic. It is by far more stable, both in the brain, where analgesia may persist for hours, and in the blood ($t_{1/2}$ = 10 min). Since β-endorphin is larger (31 amino acids), its more complex tertiary structure likely protects the N-terminal tyrosine residue more effectively from cleavage by aminopeptidases. With enkephalins, on the other hand, swift degradation to the inactive tetrapeptide effectively precludes transport to a distant site of action such as the CNS. As a corollary, it can therefore be assumed that enkephalin-containing neurons are also the sites of enkephalin biosynthesis.

A study reported rapid and prolonged analgesia in all of 14 obstetric patients who received a dose of only 1 mg synthetic human β-endorphin intrathecally at time of delivery. Since this peptide does not cross the BBB (or placenta), no fetal respiratory depression accompanied the procedure, nor were CNS or cardiovascular effects observed.

Biochemical and neurophysiological functions and effects of these polypeptides have also been intensively studied. The initial excitement and interest in these endogenous peptides was due to their morphinomimetic properties (i.e., their endogenous control of pain). It soon became apparent that their physiologic role was far more diversified. It is known that these compounds pervade numerous neurological tissues, and not only the nerve cell. Among the possibilities being investigated are that the endorphin/enkephalin system is involved in: (1) analgesia and pain modulation, (2) development of tolerance and addiction, (3) sleep, (4) respiration, (5) sexual activity, and (6) endocrine regulation. All are functions known to be affected by opioid alkaloids such as morphine.

Since β-endorphin is located within the hypothalamus and the pituitary, and has a relatively longer duration of action, it tends to be viewed as a neurohormone. Enkephalins, on the other hand, are more extensively distributed, are very rapidly degraded, and are primarily located in synaptosomal areas. The additional observation that enkephalin release following depolarization of brain (and intestinal) tissues is calcium dependent makes it more realistic to categorize them as neurotransmitters or modulators of synaptic function. Binding sites (receptors) for opioids are found, particularly in synaptosomal brain fractions. The enkephalins are located in neurons whose distribution correlates well with that of the receptors. In fact, regional distribution of peptides and their receptors are closely parallel, as would be predicted for a neurotransmitter system.

Enkephalins exist in numerous areas of the CNS and are highly concentrated in the limbic system and pain pathways. These endogenous opioidlike peptides are present in areas of the CNS presumed to be related to the perception of pain in particular. Laminae I and II of the spinal cord, the spinal trigeminal nucleus, the periaqueductal gray areas of the brain, the preventricular gray areas, and the medullar raphe nuclei are related to movement, mood, and behavior (globus pallidus, stria terminalis, locus ceruleus), and to the regulation of neuroendocrinological functions (median eminence).

Electrical stimulation of the periaqueductal gray matter has produced naloxone-reversible analgesia that displays cross-tolerance with morphine analgesia and releases endogenous opioid peptides into the CSF. Enkephalins also exist in cells whose functions relate to bowel motility (nerve plexuses and exocrine glands of the stomach and intestine).

Furthermore, enkephalins are also located in nonneuronal tissues such as the paracrine cells of the GI tract, the adrenal medulla, and the salivary glands. Met-enkephalin detectable in plasma probably arises from the adrenal medulla, and is secreted into the circulation.

5.13. Multiple Opiate Receptors

Early attempts to explain opiate analgesia by involvement of the receptor concept was followed by later demonstration of the existence of opiate receptors in the early 1970s in their characterization by sophisticated binding studies. This led to the realization that opiate receptors show considerable heterogeneity, and that they are not homogenous. They exist as subtypes with different properties and, in some cases, differ in anatomical loci. Initial work by Martin and co-workers (1976) with the spinal dog, utilizing morphine and certain

Table 5-8. Characteristics of Some Opioid Receptors

Receptor	Prototype agonist[a] ligand	Syndrome in spinal dog
μ	Morphine	Miosis, bradycardia, hypothermia, depressed nociceptive response, indifference to environmental stimuli
γ	Allylnoremetazocine (SKF 10,047)	Mydriasis, tachycardia, tachypnea, mania, abstinence in morphine-dependent animals
κ	Ketocyclazocine	Miosis, flexor reflex depression, sedation, no change in pulse rate

[a] Effects by all three drugs are antagonized by naltrexone inidicating their agonist characteristics. (Adapted from Marion et al., 1976).

benzmorphan analogs (e.g., ketocyclazocine), led to the postulation of at least three distinct types of opiate receptors; μ, σ, and k. (Receptor pluralism with distinct functional manifestation is not new and will be encountered with cholinergic, adrenergic, dopamine, and histamine receptors in subsequent chapters.).

Martin considers morphine as the typical mu (μ) agonist, ketocyclazocine as the typical agonist for the kappa (κ) receptor, and allylnormetazocine (SKF, 10,047), the prototypical agonist for the putative sigma (σ) receptor (Table 5-8). It is of interest that ketocyclazocine produces withdrawal symptoms following chronic dosing that are different from those produced by morphine. It does produce analgesia, but it requires 20 times more naloxone to antagonize its action than does morphine.[15] (Martin has even suggested high-affinity σ_1 and low-affinity σ_2 subtypes.) A fourth receptor, delta (δ), was proposed. Here, enkephalin was demonstrably more potent than morphine in inhibiting electrically induced muscle contraction in mouse vas deferens, an effect opposite that noted with guinea pig ileum, which is assumed to have μ receptors. A later study showing that the mouse (but not the rat) vas deferens was very sensitive to β-endorphin, but only slightly so to morphine and enkephalins, led to the proposal of a fifth receptor, epsilon (ε).

Evidence of multiple receptors for various hormones and neurotransmitters has previously been based primarily on pharmacological experiments. However, much of the accumulating evidence is also biochemical, particularly with opiate receptors. It includes binding studies using labeled μ and δ agonists and antagonists such as ^{3}H-D-Ala^2-D-leu^5-enkephalin.[16]

Studies using cyclic enkephalin analogs indicate that differences between μ and δ receptors may be in the distance between the binding regions on the receptor surface. It has also been demonstrated by the use of fluorescent tags on enkephalin analogs that opiate receptors occur in clusters on certain cell surfaces. The findings that μ and δ are on the same neuron may be even more significant.

A rather surprising discovery was that certain benzodiazepines that do not bind to benzodiazepine receptors instead exhibit moderate to high affinity for opiate receptors, with

[15] Naloxone is considered a competitive antagonist at all three receptor types, though not equally effective.

[16] Superscripts 2 and 5 indicate the positions in the enkephalin where amino acid substitutions have been made.

some selectivity for receptor subpopulations. These fascinating compounds exhibit the classic (opiate) analgesia by the hot-plate test rather than the muscle-relaxant pseudoanalgesia shown by diazepam.

References

Anonymous, *Ontario Medical Review* (February), 1979.

Beckett, A. H., *J. Pharm. Pharmacol.* **8**:848, 1956.

Beckett, A. H., Casy, A. F., *J. Pharm. Pharmacol.* **6**:986, 1954.

Casy, A. F., in *A Guide to Molecular Pharmacology-Toxicology*, Part I, Creese, I., Snyder, S. H., *J. Pharmacol. Exp. Therap.* **194**:205, 1975.

Creese, I. and Snyder, S. H., *J. Pharm. Exp. Ther.* **186**: 317, 1975.

Dreser, H., *Arch. Ges. Physiol.* **76**:306, 1899.

Eislib, O., Schauman, O. *Dtsch. med. Woch.* **65**: 967, 1938.

Gerland, L., *Ann.* **87**:149, 1853.

Gero, A., *Science* **119**:112, 1954.

Gilbert, P. E., Moertel, C. G., Ahman, D. S., Taylor, W. F., et al., *N. Engl. J. Med.* **286**:813, 1972.

Gilm, V., *Ann.* **150**:9, 1859.

Gringauz, A., *J. Pharm. Sci.* **65**: 291, 1976.

Hannah, J., Ruyle, W. V., Jones, H., et al., *J. Med. Chem.* **21**:1093, 1978.

Jacobson, A. E., May, E. L., Sargent, L. J., in *Medicinal Chemistry*, Burger, A., Ed., 3d ed., John Wiley & Sons, New York, 1970, pp. 1329, 1334.

Lande, K., *Munch. Med. Wochenschr.* **74**:1132, 1927.

Leroux, C., *Ann. Chem. Phys.* **23**:440, 1823.

Loll, P. J., Picot, D., Gravito, M., *Nature Struct. Biol.*, **2**:637, 1995.

Martin, W. K., et al., *J. Pharmacol. Exp. Ther.* **197**:512, 1976.

Pert, C. B., Snyder, S. H., *Science* **179**:1011, 1973a.

Pert, C. B., Snyder, S. H., *Proc. Natl. Acad. Sci.* (USA) **70**:2243, 1973b.

Piper, P. J., Vane, J. R., *Nature* (London) **223**:29, 1969.

Piria, A., *Ann. Chem. Phys.* **69**:298, 1838.

Piria, A., *Ann.* **30**:151, 189, 1839.

Serturner, F. W., *Trommsorff's J. Pharmazie* **13**:234, 1805.

Simon, E. J., Hiller, J. M., and Edelman, I., *Proc. Nat. Acad. Sci.* **70**: 1947, 1973.

Von Euler, U. S., *Arch. Exp. Pathol. Pharmak.* **175**:78, 1934; *J. Physiol.* **88**:213, 1936.

Suggested Readings

Bond, M. R., *Pain, Its Nature, Analysis and Treatment*, 2nd ed. Churchill Livingston, New York, 1984.

Imlah, N., *Addiction-Substance Abuse and Dependency*, Sigman Press, Wilmslow, 1989. O'Brien, C. P., Drug Addiction and Drug Abuse. In *Goodman and Gilman's The Pharmacological Basis of Therapeutics*, 9th ed.

Hardman, J. G., Limbird, L. E., Molinoff, P. B., Ruddon, R. W., Gilman, A. G., Eds. Pergomon Press, New York, 1996, pp. 557–77.

Johnson, R. S., Milne, G. M., Analgetics, in *Burger's Medicinal Chemistry*, Wolff, M., Ed., 4th ed., Part III. John Wiley & Sons, New York, 1981, pp. 699–758.

Lednicer, D., Ed. *Central Analgetics*, John Wiley & Sons, New York, 1982.

Lombardino, J. G., Ed. *Nonsteroidal Antiinflammatory Drugs*, John Wiley & Sons, New York, 1985.

Melzack, R., The Tragedy of Needless Pain. *Sci. Amer.* **262**:27, 1990.

Robinson, H. J., Vane, J. R., Eds. *Prostaglandin Synthetase Inhibitors*, Raven Press, New York, 1974.

Snyder, S. H., Drug and Neurotransmitter Receptors in the Brain. *Science* **224**:22–31, 1984.

Snyder, S. H., Mathyse, S., *Opiate Receptor Mechanisms*. The MIT Press, Cambridge, MA, 1975.

Wall, P. D., Melzak, R., Eds. *Textbook of Pain*, 2nd ed. Churchill Livingston, New York, 1989.

CHAPTER

6

Antimicrobial Drugs I

6.1. The Antibiotics

It is difficult to title correctly a discussion that attempts to deal with the broad area of drugs utilized to treat all the types of infections that parasitize humanity. Antiparasitic might be technically more correct because it would include all organisms from the microscopic in size to the 75-foot tapeworm, but the term *parasite* has acquired a much narrower meaning as commonly used. The title *antimicrobial* was selected even though drugs controlling helminthic (worm) infestations will be considered. Similarly, antiviral compounds will also be discussed, even if viruses are not "microbes" as commonly viewed.

The discussions will be primarily based on mechanisms of action such as inhibition of bacterial cell wall or protein synthesis. These are extremely complex biochemical processes. Drugs exist in each category that interfere at different points of the multistep procedure. Some drugs, such as the antifungals, will be considered as a group even though their mechanisms of action vary.

Several drugs have been presented in earlier discussions. The antimetabolite sulfonamides and sulfones, the membrane-affecting polypeptide antibiotics including gramicidin S, and tyrocidin, as well as the ionophoric gramicidin A and B, and valinomycin were included in Chapter 2. These will not be reconsidered.

6.2. Cell Wall Synthesis Inhibitors

Inhibitors of bacterial cell wall synthesis include some of the most widely used antibiotics. Also found in this category are several interesting but lesser known compounds with narrower utility.

The function of the microbial cell wall is as a protective external support to the cell. Its rigid polymeric material functions as a continuous envelope to preserve the protoplast's integrity, something the delicate protoplasmic membrane cannot adequately do in the

Table 6-1. Some Frequently Encountered Pathogenic Bacteria

Gram Negative	Gram Positive	Anaerobes
Escherichia coli	Staphylococcus aureus	Bacillus fragilis
Klebsiella pneumonia	Staphylococcus epidermidis	Clostridium sp
Klebsiella sp	Streptococcus pyogenee	Bacteroides sp
Hemophilus influenzae	Streptococcus agalactiae	Peptococcus sp
Neisseric meningitidis	Streptococcus pneumoniae	Peptostreptococcus sp
Neisseria gonorrhoeae	Streptococcus faecalis	
Proteus vulgaris and mirabilis		
M morgani		
Providencia rettgeri		
Enterobacter sp		
Citrobacter sp		
Serratia sp		
Salmonella sp		
Shigella sp		

hypotonic environment in which the cell functions. Gram-positive bacteria[1] attain internal osmotic pressures approaching 20 atmospheres, whereas Gm-negative organisms exhibit pressures of 5–6 atmospheres, which, although less, is still quite high. Such pressures in the absence of a cell wall lead to rupture (lysis) of the membrane and hence cell death. The cell wall effectively prevents this. As the cell grows, more cell wall must be synthesized to accommodate the additional protoplasmic material. Drugs that can prevent this synthesis in *actively growing* cells invariably cause rupture,[2] since the membrane alone cannot contain the cellular contents.

The composition and characteristics of gram-positive and gram-negative bacterial cell wall structures differ yet share the fundamental "backbone" of the *peptidoglycan* or *murein*. It basically consists of parallel polysaccharide chains cross-linked by short peptide chains. The recurring unit of this polysaccharide consists of a disaccharide of N-acetylglucosamine in a $\beta 1 \rightarrow 4$ linkage with N-acetylmuramic acid (N-acetylglucosamine linked to lactic acid via its C-3 hydroxyl group).

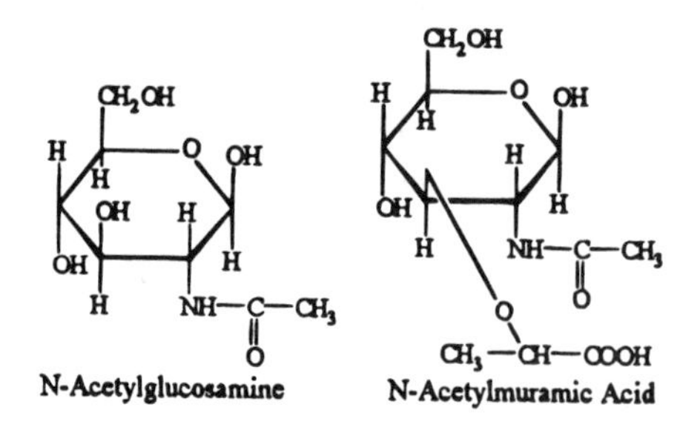

The carboxyl group of the lactic acid residue is in turn linked through a peptide bond to a short polypeptide. Thus a peptidoglycan chain might be represented as shown in Figure 6-1. The long parallel polysaccharide chains are cross-linked by their peptide chains. Thus the

---

[1] Those accepting the bacterial stain developed by Gram in the nineteenth century.

[2] Or drastic cellular shape changes where internal pressures are less severe, such as gram-negative bacteria.

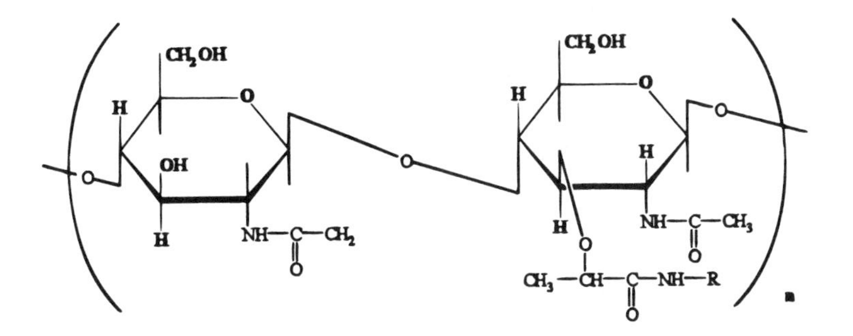

Figure 6-1. A peptidoglycan structure. (*R* is an L-alanyl-D-glutamyl-L-lysyl-D-alanine; D-amino acid residues occur here.)

terminal *D*-alanine of the peptide side chain of one muropeptide is attached to an adjacent polysaccharide either directly or through another short polypeptide chain's ε-NH_2 group of lysine (as in *Staphylococcus areus*, Fig. 6-2). The result is a three-dimensional lattice-type structure that envelopes the cell as a net. This "net" is closed on all sides, making a completely covalent structure. No ionic or hydrogen bonds are involved. This meshwork can be tightly "woven" (e.g., *S. aureus*) or be loose (e.g., *E. coli*), depending on the nature of the cross-linkages. Cell wall structures for gram-positive organisms such as *S. aureus* are usually less complex than those of gram-negative bacteria. The elucidation of the *S. aureus* cell wall by Strominger et al. (1967) represented a milestone achievement.

At least four chemotypes of peptidoglycan structures are now known to exist. Figure 6-2 shows that two tetrapeptides from parallel polysaccharide chains are cross-bridged. The third amino acid is lysine, as illustrated. It can also be a diamino diacid such as meso-diaminopimelic (A_2pm). The fourth amino acid is always *D*-alanine. If one considers the amino acid residue(s) acting as the cross-bridge as a unit *X*, then the complete peptidoglycan of chemotype I is where *X* is absent. That is, there exists a direct linkage between the carboxyl of the *D*-ala of one tetrapeptide and the terminal (ω) amino function of the third residue of the next tetrapeptide (either lysine or A_2pm). This is the case with gram-negative bacteria and bacilli. In chemotype II, *X* is only one amino acid residue such as glycine or *L*-asparagine, or a short pentapeptide such as the five glycines in *S. aureus* (Fig. 6-2). Peptidoglycans, where the cross-bridge is a tetrapeptide identical to that of the tetrapeptide

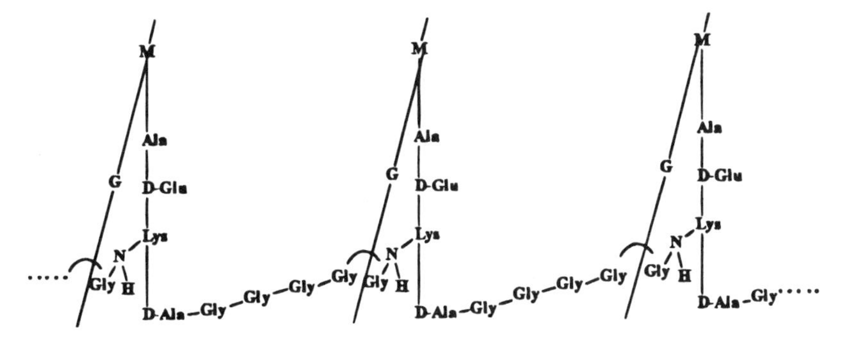

Figure 6-2. A section of the cell wall of *S. aureus*. (*M* = N-acetylmuramic acid, *G* = N-acetylglucosamine.)

bonded to the N-acetylmuramic acid (*M*), represents chemotype III. Chemotype IV is the case where the *X* bridge is a diamino acid or a diamino acid-containing peptide that is between the α-COOH group of glutamic acid and the α-COOH of D-ala.

Cell walls contain a variety of additional components that differ from species to species, some characteristic of gram-positive, some of gram-negative organisms. In the former are teichoic and teichuronic acids. *Teichoic acids* are chains of alternating glycerol or ributol residues with phosphoric acid.[3] *Teichuronic acids* appear in situations where insufficient phosphate levels prevent teichoic acid synthesis. Polysaccharides (from monomers other than glucose) are sometimes encountered in gram-positive bacteria only.

Substances found almost exclusively in gram-negative bacterial cell walls include lipopolysaccharides, amphoteric substances containing lipids, amino sugars, and ethanolamine phosphate. They are a significant component of the outer membrane of Gm-negative bacteria (see later). Their function (and that of teichoic acid in gram-positive cell walls) is probably the concentration, by an ion exchange process, of Mg^{2+} so critical to many cell wall membrane-bound enzymes. Other major chemical components found in gram-negative cell walls are lipids, proteins, and lipoproteins. The proteins—in the outer membrane—provide pores permitting relatively small water-soluble substances, including some antibiotics, to diffuse through to the cell wall itself. These porins are essentially transmembrane aqueous channels.

A relatively recent bacteriological recognition is that gram-negative bacteria have *two* membranes. The inner cytoplasmic membrane that all cells have[4] and another outer membrane about 10 nm apart: The interior or *periplasmic* space contains the peptidoglycan cross-linked cell wall, which, by the way, is thinner than its gram-positive counterpart. As was previously pointed out, it is this outer membrane that incorporates the characteristically gram-negative components such as lipopolysaccharides, phospholipids, and proteins. The outer membrane's interior is hydrophobic.

After the structure of the bacterial cell wall became known, it was possible to determine the chemistry of its biosynthesis. For convenience the steps involved will be divided into three stages: synthesis of starting materials (precursors), transfer of precursors to membrane-bound anchor, and polymerization of the netlike cell wall by cross-linking outside the cytoplasmic membrane.

6.3. Cell Wall Biosynthesis

6.3.1. Stage I—Formation of Starting Materials

Formation of starting materials for cell wall synthesis begins with two metabolic substances normally found in all life forms: N-acetylglucosamine 1-phosphate and the pyrimidine nucleotide uridine triphosphate (UTP) (see Fig. 6-3). Condensation of these two compounds by elimination of pyrophosphate affords uridine-diphospho-N-acetylglucosamine (UDPNAG). Reaction with phosphoenolpyruvic acid (PEP, the activated form of the enol tautomer of pyruvic acid),[5] catalyzed by a specific transferase, yields the 3-O-enolic ether.

---

[3] When associated with membrane, they are referred to as *lipoteichoic acids.*

[4] For detailed review, see Singer and Nicholson, 1972; Singer, 1974; and Nikaido and Nakae, 1979.

[5] Available from the Emden–Meyerhof pathway of glucose metabolism.

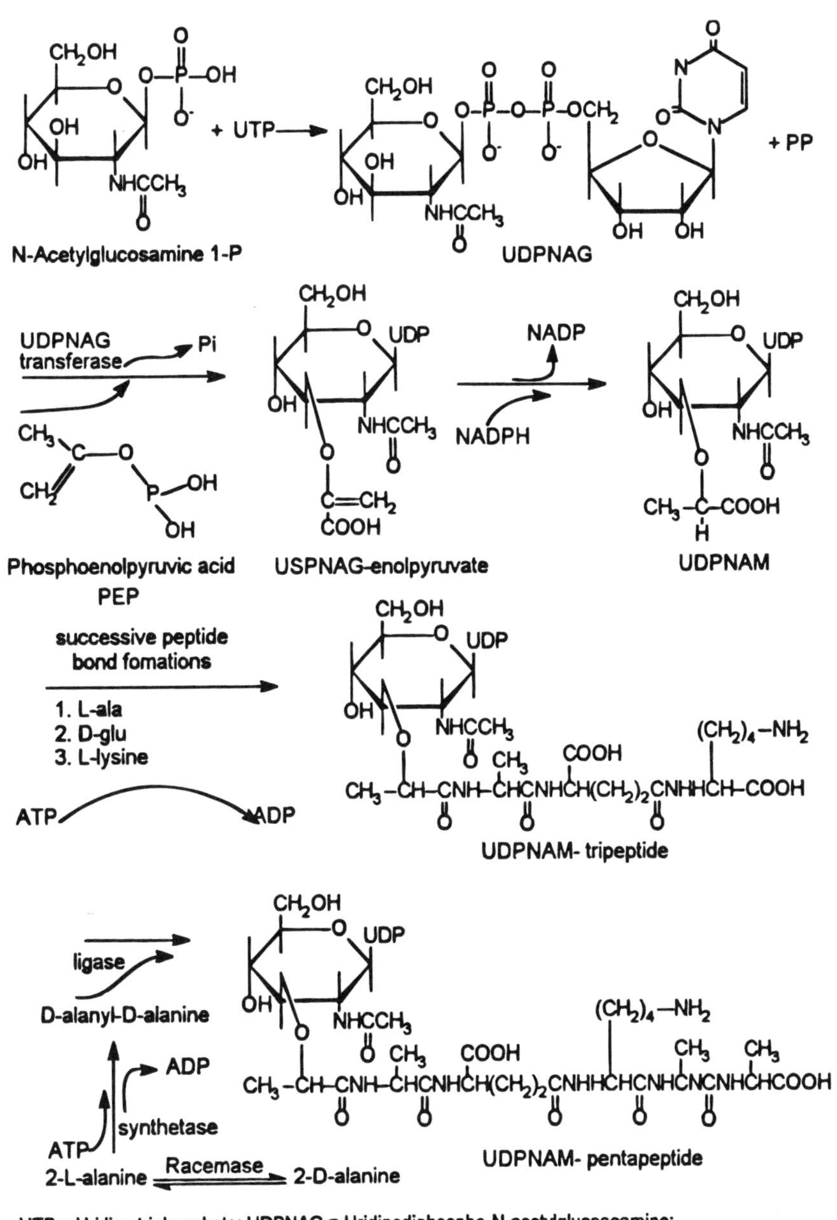

Figure 6-3. Stage 1—cell wall synthesis (*S. aureus*).

Reduction of the double bond by an NADPH-utilizing reductase enzyme gives the desired N-acetylmuramic acid (still as the uridine derivative). The muramic acid nucleotide then forms a tripeptide at the carboxyl group in sequence beginning with L-alanine followed by D-glutamic acid and L-lysine, utilizing ATP and enzymes specific for each amino acid and substrate. The penultimate and ultimate amino acids, both D-alanine, are not added in sequence to form the pentapeptide; rather, they are added in a single step, utilizing the dipeptide *D*-alanyl-*D*-alanine, which in turn was synthesized from two D-alanine molecules by the specific enzyme D-alanyl-D-alanine synthetase. The source of D-alanine is racemization of L-alanine by a racemase enzyme.

There are three antibiotics that interfere with cell wall biosynthesis at this first stage.

Fosfomycin (Phosphonomycin, Fosfocin), produced by *Streptomyces fridiae* and other *S.* species, is 2R(*cis*)-(3-methyl-2-oxyranyl)phosphonic acid. Its mechanism involves peptidoglycan synthesis inhibition. It was introduced clinically in the early 1970s in Europe. The drug inhibits the enol-pyruvyl transferase that catalyzes the transfer of PEP in the formation of UDPNAG-enolpyruvate (Fig. 6-3). The inactivation of the enzyme by the antibiotic (Eq. 6.1) may be viewed in classic chemical terms, where the oxirane oxygen becomes protonated and the C-2 carbon is then nucleophilically attacked by the cysteine residue of the enzyme's active site, forming a covalent, catalytically inactive product. The structural analogy of the drug to PEP is not very good, which probably accounts for the fact that the several mammalian enzymatically catalyzed reactions utilizing PEP (e.g., pyruvate kinase) are not inhibited. In fosfomycin, it is the highly strained 2C–O bond that is particularly susceptible when protonated on the oxygen, at which the reaction takes place.

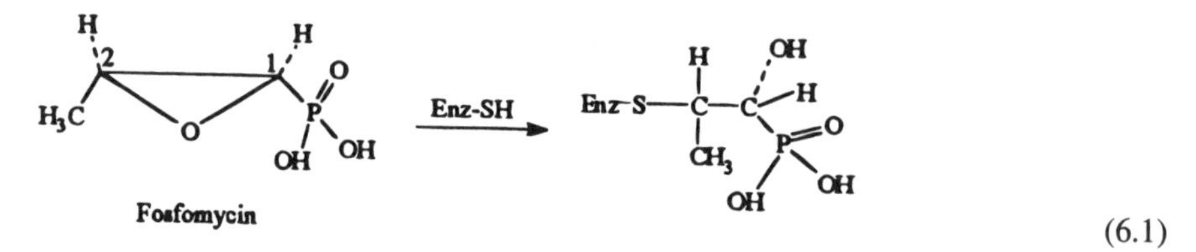

(6.1)

Although active against gram-positive (Gm+) and gram-negative (Gm–) organisms, fosfomycin has so far been utilized primarily against Gm+ infections. Among its potential attributes are low inherent toxicity and no cross-allergenicity or cross-resistance with other antibiotics. In fact, the drug appears to be synergistic with tetracyclines and chloramphenicol.

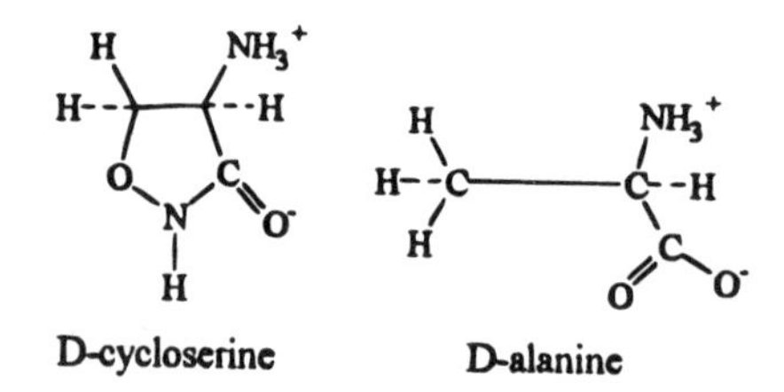

Another interesting but little utilized drug inhibiting cell wall synthesis during Stage I is cycloserine (oxamycin, Seromycin). It is a broad-spectrum antibiotic produced by *Streptomyces orchidaceus* (but now made synthetically). Its use over the past three decades has been limited as a second-line antitubercular agent. Chemically, D-cycloserine (D-4-amino-3-isoxazolidone) might be viewed as a cyclic "analog" of D-alanine.[6] The drug competitively inhibits both alanine racemase and D-alanyl-D-alanine synthetase, thus effectively doubly blocking the extension of UDPNAM-tripeptide to the pentapeptide (Fig.

[6] In spite of the drug's name indicating a relationship to serine.

6-3). There are some unusual factors involved here. As was previously discussed (Chapter 2), compounds that are simply competitive inhibitors of a natural substrate are not usually successful as drugs. This accounts for the fact that "rational" design of antimetabolites as chemotherapeutic agents following the earlier explosion of sulfonamides resulted mostly in failures. We have (with synthetase) a situation where the affinity of the enzyme for the drug is actually 100 times that for the normal substrate, D-alanine (with *S. aureus*), i.e., a high K_m/K_r ratio. It is as if the drug "looks" more like D-alanine to the enzyme's binding site than D-alanine does. In a way that is what was proposed. The rigid isoxazole ring does not permit the drug to assume different conformations while the nonrigid D-alanine, of course, exists in the obviously very favorable conformation (which is the drug's only possible configuration) only fleetingly. It should be mentioned that D-cycloserine enters susceptible bacterial cells by active transport rather than by simple diffusion so that intracellular levels exceed those of the surrounding medium. This fact further helps to explain the effectiveness of the compound.

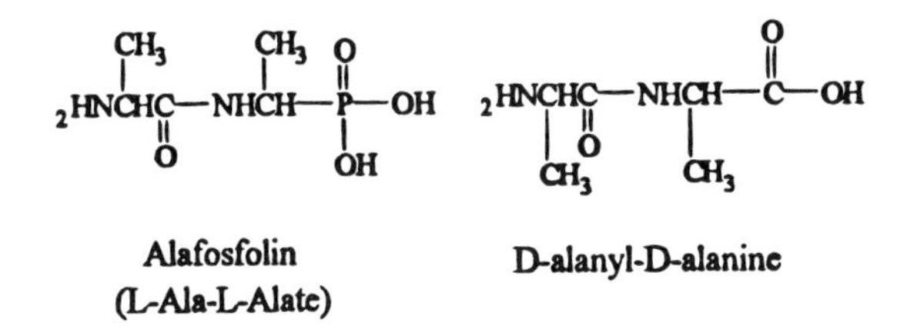

A third compound that inhibits peptidoglycan synthesis that is of particular theoretical interest is alafosfolin (alaphosphin). It is L-alanyl-L-1-aminoethyl phosphonic acid, rationally designed to be a "peptide mimetic" of alanine dipeptide (Allen et al., 1978). This was among the more promising of a group of phosphonopeptides. The C-terminal alanine moiety as been substituted by L-1-aminoethyl phosphonic acid, Ala-P (an alanine isostere?). Incubation of the drug with Gm+ bacteria led to an accumulation of UDPNAM-tripeptide, thus indicating interference with the addition *or* synthesis of the D-alanine dipeptide. The evidence indicates that is the L-Ala-P that is the actual inhibitor of the racemizing enzyme. It was later established that other L-amino acids could successfully replace L-alanine as the N-terminus (e.g., Met, Phe, Val, Lys, and Pro). Furthermore, alafosfolin, using the dipeptide transport system, penetrates the membranes, where intracellular aminopeptidases hydrolyze the drug to ALA-P (the active moiety) and L-alanine. Since Ala-P does not diffuse across bacterial membranes in either direction, high internal concentrations are achieved, accounting for high antibacterial activity particularly with Gm– bacteria (*E. coli* MIC = 0.03 µg/ml).[7] Thus, in effect alafosfin should be viewed as a pro-drug.

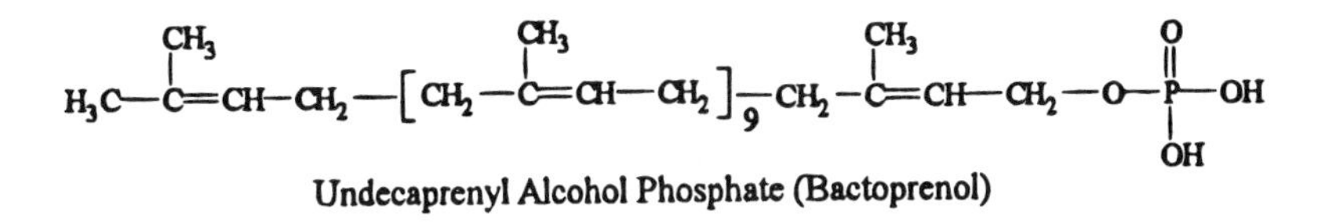

[7] MIC is minimum inhibitory concentration in µg (or µmole)/ml of growth medium. Lower concentration denotes higher potency in vitro.

6.3.2. Stage II—Peptidoglycan Synthesis

The reactions in this phase are not as completely understood, although the overall sequence of events has been defined and is outlined in Figure 6-4. All steps are catalyzed by membrane-bound enzymes. The initial event is a linkage, via pyrophosphate bonding, between the final product of Stage I, the uridyl-diphosphate-N-acetylmuramic acid pentapeptide, and a phospholipid *membrane-bound* carrier, bactoprenol (Strominger et al., 1967). It is a C_{55} polyisoprenoid alcohol (containing 11 isoprene units) esterified with phosphoric acid. It is a lipoidal anchoring group, which assures that the subsequent sequence of events occur in the interior of the cell membrane. It is also necessary, in an as yet not clearly defined manner, for the transfer of the completed precursor through the membrane for final assembly *outside* the cell. The second sugar moiety, uridinediphospho-N-acetylglucosamine, is now incorporated by glycosidation to form a β-1 $\rightarrow$ 4 glycosidic bond between the two glucose-derived units (see Figs. 6-1 and 6-4). The uridine nucleotides released in these reactions can, of course, be rephosphorylated to UTP and be recycled into Stage I. In the case of *S. aureus* five glycines are now added in sequence to the ε-amino function of the lysine residue, each carried by the specific glycyl-*t*RNA. Ribosomes, as used in protein synthesis, are not involved (see subsequent discussion). Unlike in protein synthesis, glycyl peptide bond formations are here made in reverse (i.e., from the N-terminal).

As previously mentioned, bridging groups other than glycine occur. These include alanine, serine, or threonine. In bacteria where the lysine is replaced by *meso*-2, 6-diaminopimelic acid, cross-linking is directly between the D-alanine of the tetrapeptide side chain and the 6-amino group of the pimelic acid residue.[8] Other bridgings are also known.

The complete disaccharide–decapeptide monomer unit, which, on movement through the membrane, is transferred following pyrophosphatase cleavage to a not yet fully identified acceptor (most likely the growing peptidoglycan chain on the outside of the membrane), exists at this point. Continuous β-1 $\rightarrow$ 4 glycosidic bond formation occurs here. Separation from the membrane-bound anchor leaves undecaprenyl phosphate, which regenerates the original phosphate–alcohol ester on hydrolysis by phosphotase. It can then repeat the cycle.

There are several antibiotics that affect Stage II reactions. Bacitracin, which is isolated from *Bacillus subtilis*, is active against many important Gm+ bacteria. However, its nephrotoxicity precludes its systemic administration. It has therefore been relegated almost exclusively to the treatment of superficial skin infections and occasionally the eye. The drug is generally used in ointment form, sometimes in combination with other antibiotics such as polymyxin and neomycin.

Bacitracin contains a cyclic peptide as part of a complex structure of 12 amino acid residues (if one considers the thiazole ring as cysteine derived). Bacitracin-treated bacterial cultures accumulate precursor peptidoglycan chains, as do the penicillins. However, the mechanism is very different. Its prevention of lysine inclusion into the murein structure is probably of lesser significance. Later demonstration show that drug binding to the membrane-bound bactoprene phosphate and the metal ions (Zn^{2+} are particularly involved) is more likely to be important to the antibacterial mechanism, namely, inhibition of dephosphorylation of the phospholipid carrier.

---

[8] This is commonly seen in rod-shaped Gm– organisms.

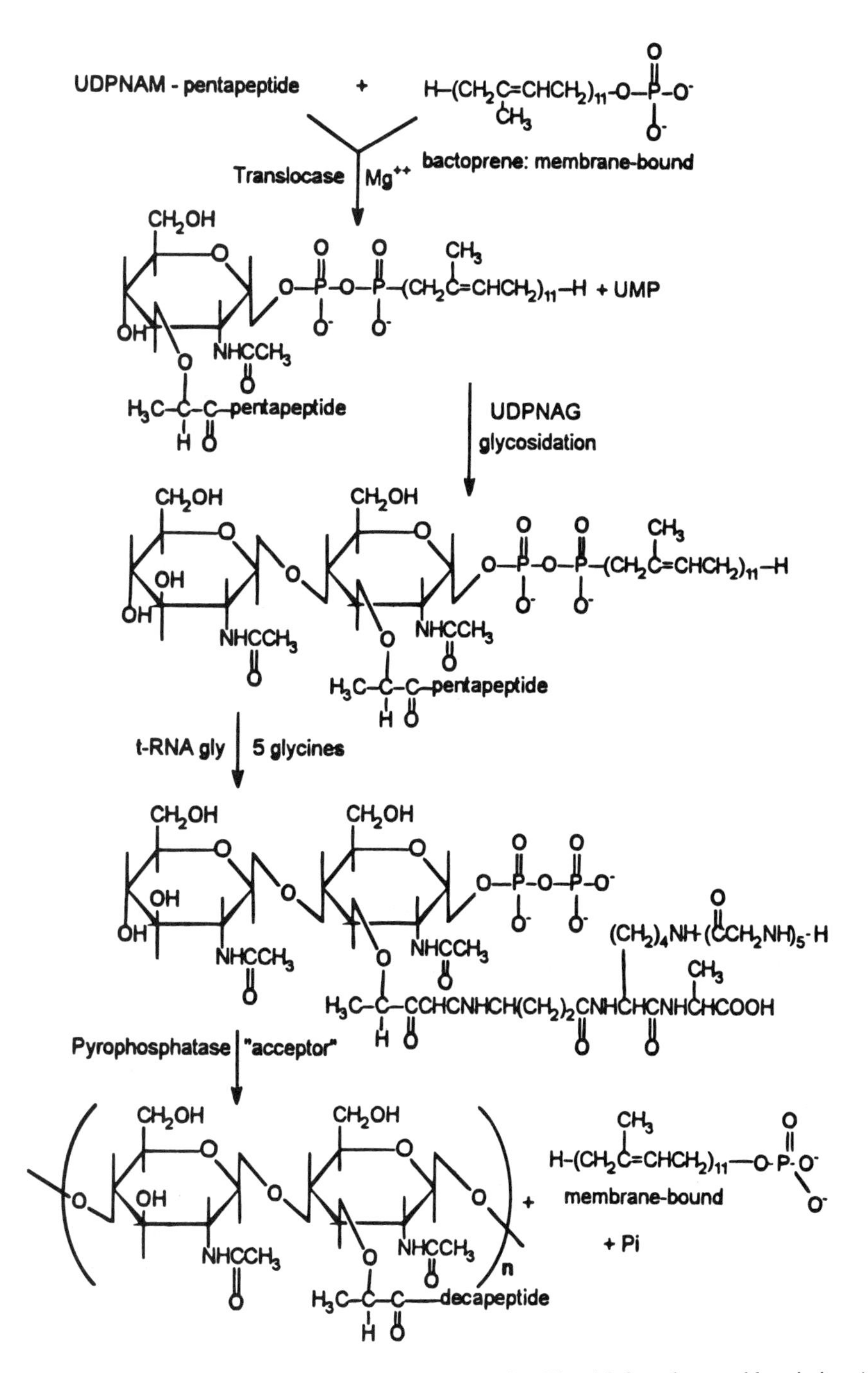

Figure 6-4. Stage II—cell wall synthesis (*S. aureus*). (See Fig. 6.3 for relevant abbreviations.)

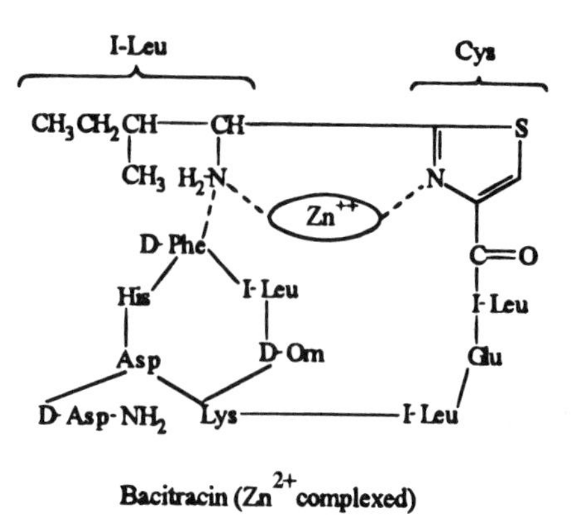

Bacitracin (Zn^{2+} complexed)

By far the clinically more important drug interfering with Stage II synthesis is the antibiotic vancomycin (Vanocin), which was first introduced in the mid-1950s. It is obtained from *S. orientalis*. Vancomycin is a polycyclic glycopeptide molecule whose complex structure was not fully determined until 1978 (Sheldrick et al., 1978). The hexose components are glucose and an amino sugar named vancosamine. They do not seem to be relevant to the mechanism of action. Among the amino acid residues are two unusual β-hydroxy-*m*-chlorotyrosines, aspartic acid, N-methylleucine, and two phenyl glycines with phenolic hydroxyl groups. It is an interaction of the *D*-alanyl-*D*-alanine portion of the forming mucopeptide—most likely involving strong, but not covalent, bonding with the hydroxylated phenyl glycine residues of the antibiotic that constitutes its mechanism. The separation of the murein component to the outside of the membrane is thus impaired and cell wall synthesis is inhibited. Vancomycin is predictably bactericidal.

Vancomycin was initially used primarily in staphylococcal infections that were resistant to the early penicillins. More widespread use of the drug was inappropriate because of its nephro- and ototoxicity, phlebitis, and fever. The advent of penicillinase-resistant penicillins in the early 1960s (see subsequent discussion) seemed to have made vancomycin obsolete. The clinical comeback of this antibiotic can be attributed to several factors: the practical elimination of the previously mentioned side effects by new purification processes, the accelerating resistance problems with many of the newer penicillins (and other antibiotics), the apparent absence of emerging resistance to vancomycin of even multiple resistant Gm+ and anaerobic organisms, and the synergism of this drug with the aminoglycosides. All have restored this drug to a primary position in therapeutics.

6.3.3. Stage III—Peptidoglycan Cross-Link

The final steps in cell wall synthesis involve cross-linkage of the peptidoglycan chains. This converts water-soluble polymeric substances (due to many polar functions) with mobility into a tough, insoluble, and inflexible material.[9] In simplest terms this is a *transpeptidation* reaction where the end amine function of the pentaglycine side chain forms a new peptide bond at the expense of the terminal D-alanyl-D-alanine linkage of a

---

[9] Effects similar to those seen in cross-linkage of linear polymers in the development of industrial plastics.

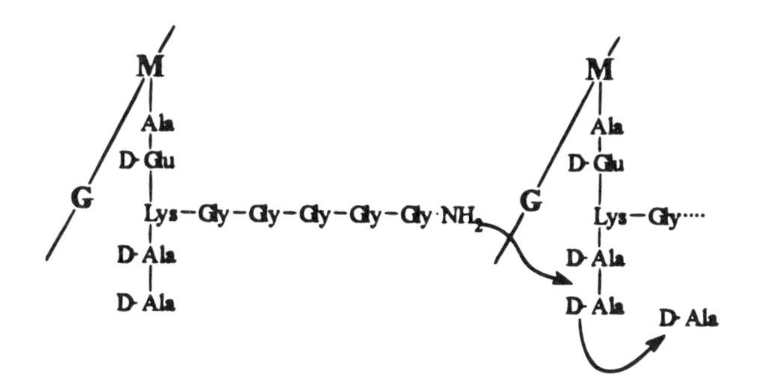

Figure 6-5. Transpeptidation leading to cross-linked peptidoglycan (in *S. aureus*).

neighboring peptidoglycan chain. The enzyme catalyzing these unique extracellular events is a *transpeptidase*. Figure 6-5 illustrates this reaction (Tipper and Strominger, 1965). More recent evidence indicates a multistep process involving first an enzyme–substrate complex formation leading to cleavage of the terminal D-ala (by D-alanyl-D-alanine carboxypeptidase). The resulting complex of the enzyme with the penultimate D-alanyl residue then reacts with the glycyl NH_2 group forming the new peptide bond, (i.e., transpeptidase activity). This last reaction in the lengthy sequence of cell wall synthesis in bacteria is irreversibly inhibited by a large, now diverse, group of bacteriocidal antibiotics that share one fundamental structural feature—a four-membered β-lactam ring.

Before the manner in which these drugs inhibit this climactic step in the development of the bacterium can be further discussed, the chemistry of the β-lactam antibiotics will be considered.

6.4. The β-Lactam Ring—The Enchanting Structure

Just as a lactone is an internal ester (i.e., both the carboxylic and hydroxylic functions emanate from the same molecule), so is a lactam an internal cyclic amide. If the amine function in the amino acid is on the number 3 (or β) carbon, the resulting product can be termed a β-lactam (Eq. 6.2).[10] Although the strain on this molecule (due to the *ca.* 90° bond angles) is not as severe as in the three-membered aziridine ring (see alkylating agents, Chapter 4), one would readily expect this ring structure to possess a certain degree of instability, resulting in facile ring cleavage, in spite of the predicted resistance of amides toward hydrolysis due to resonance stabilization (see Fig. 2-4).

$$\underset{NH_2}{\overset{\beta}{R-CH}}-\overset{\alpha}{CH_2}-COOH \longrightarrow \text{β-lactam ring } (\alpha\text{-C, } \beta\text{-C–R, NH, C=O}) + H_2O \qquad (6.2)$$

Much of the work and theoretical discussions as to the intricacies of the molecular mechanism (and chemical stabilities) of β-lactam antibiotics have previously revolved

[10] It can also be named as an azetidine derivative (e.g., 2-azetidinone, or 2-keto-azetidine).

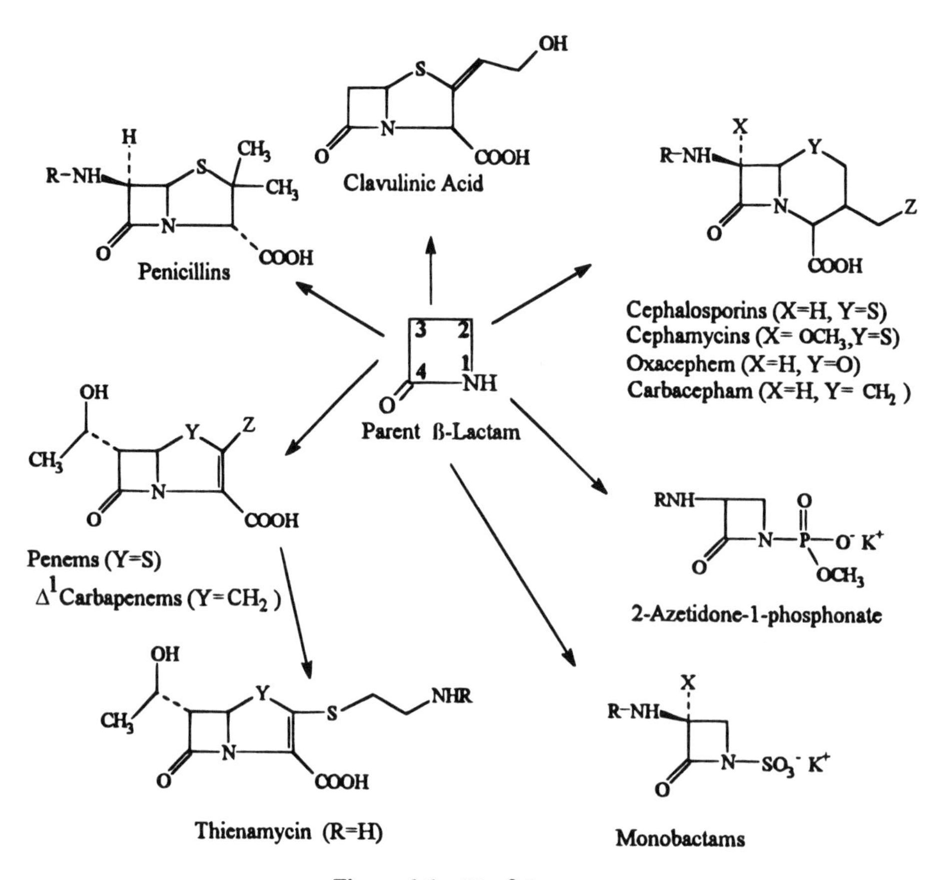

Figure 6-6. The β-lactams.

around the penicillins and cephalosporins. The reasons for this are primarily historical. Penicillins were the first to be discovered, characterized, and clinically utilized (middle 1940s), followed by the closely related cephalosporins during the next decade. The chemically simple monobactams and even carbapenems, did not arrive on the scene until the 1970s (Fig. 6-6).

The penicillin molecule consists of the azetidinone (or β-lactam) ring fused to the 5-membered thiazolidine ring system. The ring system, a penam,[11] was unknown at the time of the initial research on penicillins. In fact, at one point there were several proposed structures for the drug; one even devoid of a sulfur atom. The correct structure was finally elucidated chemically and supported by X-ray diffraction in the mid-1940s involving collaborative work during World War II in Great Britain and the United States.

The credit for the actual discovery of penicillin is usually given to Alexander Fleming. Although he described some of its characteristics in solution,[12] such as its poor stability toward heat and alkali, its low toxicity (in animals), and the fact that it was bacteriocidal to some bacteria (e.g., pyogenic cocci) but not others (e.g., enterococci, influenza bacterium),

[11] 4-Thia-1-azabicyclo[3.2.0] heptane.

[12] Actually, filtrates of broth cultures of the P. notatum mold.

he did not recognize its potential for the treatment of systemic infections. Rather he suggested it as an "... efficient antiseptic for application to, or injection into, areas infected with penicillin-sensitive microbes" (Fleming, 1929). Subsequently, although not able to isolate or even purify the unstable substance, he was able to show its effectiveness against many Gm+ bacteria, and its usefulness in the *topical* treatment of infected wounds. Significant progress was not made until Chain, Florey, et al. (1940) reported that a partially purified penicillin protected mice from lethal quantities of various Gm+ organisms when injected. Soon thereafter, therapeutic effectiveness and low toxicity in humans were demonstrated. The chemotherapy revolution that started with the sulfonamides less than a decade earlier shifted into a higher gear, from which it is still accelerating today.

It is an irony that one reason for the time lag in capitalizing on Flemings work for almost a decade was the appearance of the sulfonamides in 1935. These undoubtedly galvanized many of the medicinal chemists and pharmacologists into investigations. The work of Chain, Florey, and others finally resulted in the recognition of penicillins almost magical systemic effectiveness. By then World War II was already under way. As a result, much of the successful work of that period was not published until the war was over.

During this period the correct structures for the natural penicillins had been established. The original *Penicillium notatum* mold cultured on a yeast extract medium was found to produce a mixture of up to five penicillins that were given letter designations—G, X, F, K (see penicillins 1–4, Table 6-2). The yields were poor when large-scale fermentations were undertaken. Changing to a more productive species, *P. chrysogenum*, modifying the culture medium by addition of metal salts and side chain precursors, particularly phenylacetic acid to emphasize penicillin G, and developing new submerged fermentation techniques with aeration and high—power stirring resulted in higher yields and easier purification.[13]

Benzyl penicillin (Penicillin G) was chosen as the penicillin to emphasize primarily because of high effectiveness and the ability to obtain high yields (by the addition of $C_6H_5CH_2COOH$). Serum protein binding was also low in comparison to the other penicillins. Penicillin G is orally absorbed, yet its ready destruction by acidity (gastric juice) made its oral use a wasteful extravagance in the early days when production costs were high. In later years when costs became minimal, giving high oral doses, to compensate for gastric destruction of two-thirds of the dose, became common practice.

The fact that several "natural" penicillins in the original mix, once separated, were found to vary in potency and in what are today termed *biopharmaceutical properties*, led to attempts to produce new penicillins by manipulating the fermentation medium through the addition of other types of precursor acids to vary the *R*-acyl groups. Even though the method worked to a limited degree, the limitation for precursors was that they be essentially monosubstituted arylacetic acids. The resultant products differed little from the natural penicillins in antibacterial spectrum or potency. No dramatic improvements were apparent. Nevertheless, at least three new penicillins were added to the list from which the physician could choose: butylthiomethyl ($R = CH_3CH_2CH_2CH_2SCH_2$-, penicillin S), allylthiomethyl ($R = CH_2 = CHCH_2$-S-CH_2-, penicillin O), and phenoxymethyl ($R = C_6H_5OCH_2$-, penicillin V). The only one still in use today is penicillin V (Table 6-2). Its main attribute is acid stability (but not toward penicillinase), making oral use more predictable. Its antistaphyloccocal activity, however, is the same as that of penicillin G.

[13] By several-stories-high and several-feet-wide chromatography columns.

Table 6-2. The Penicillins

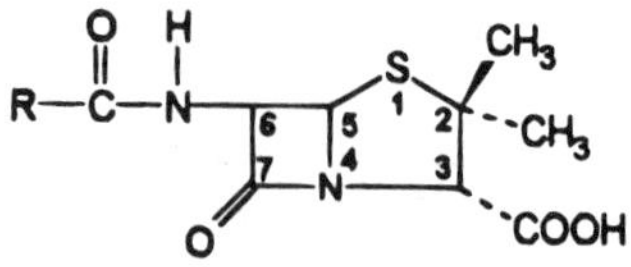

Penicillin	*R*	Acid stable	Penicillinase resistant	Broad spectrum
1. Benzyl (G)	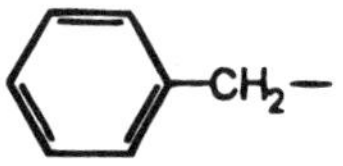 $C_6H_5-CH_2-$	No	No	No
2. p-Hydroxybenzyl	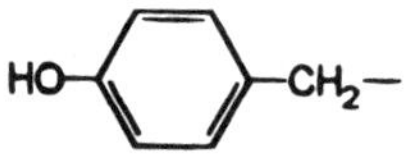 $HO-C_6H_4-CH_2-$	No	No	No
3. 2-Pentenyl (F)	$CH_3(CH_2)_2CH{=}CH-CH_2-$	No	No	No
4. *n*-Heptyl (K)	$CH_3(CH_2)_5-CH_2-$	No	No	No
5. Phenoxymethyl (V)	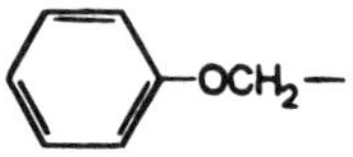 $C_6H_5-OCH_2-$	Yes	No	No
6. Phenethicillin	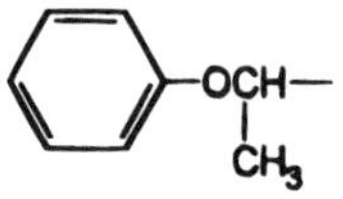 $C_6H_5-OCH(CH_3)-$	Yes	No	No
7. 2,6-Dimethoxyphenyl (Methicillin)	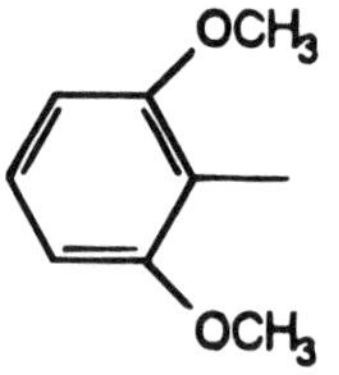 OCH_3, OCH_3	No	Yes	No
8. 5-Methyl-2-phenyl-4-isoxazolyl (Oxacillin)	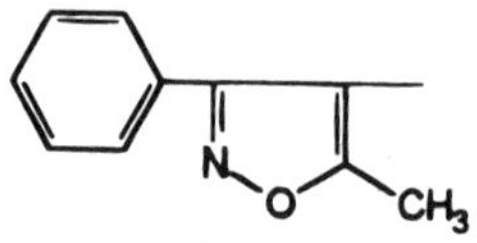 N, O, CH_3	Yes	Yes	No
9. 3-(2-chlorophenyl)-5-methyl-4-isoxazolyl (Cloxacillin)	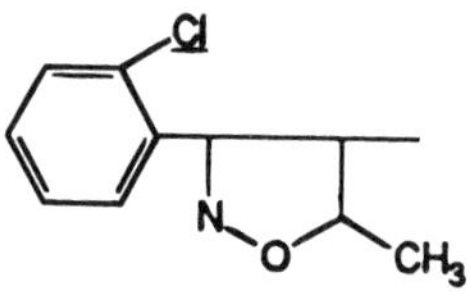 Cl, N, O, CH_3	Yes	Yes	No

(*Continued*)

Table 6-2. *(Continued)*

Penicillin	*R*	Acid stable	Penicillinase resistant	Broad spectrum
10. 3-(2,6-dichlorophenyl)-5-methyl-4-isoxazolyl (Dicloxacillin)	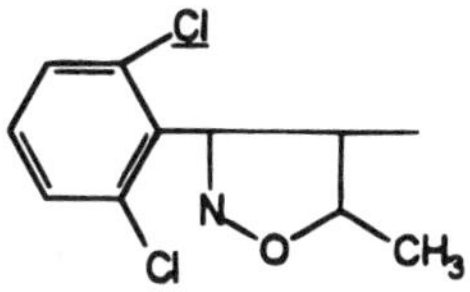 Cl, Cl, N, O, CH_3	Yes	Yes	No
11. 3-(2-chloro-6-fluorophenyl)-5-methylisoxazolyl (Flucloxacillino)	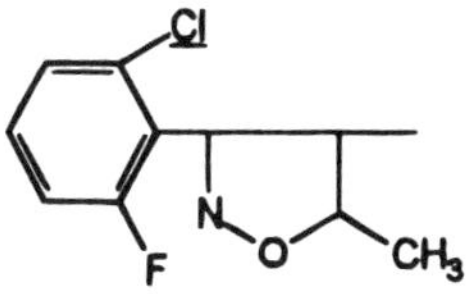 Cl, F, N, O, CH_3	Yes	Yes	No
12. 2-Ethoxynaphthyl (Nafcillin)	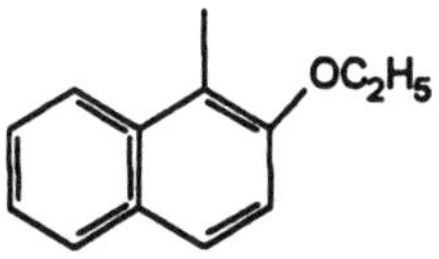 OC_2H_5	Yes	Yes	No
13. D-α-Aminobenzyl (Ampicillin)[a]	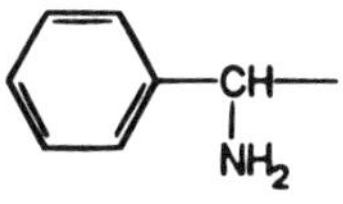 CH—, NH_2	Yes	No	Yes
14. D-α-Amino-*p*-hydroxybenzyl (Amoxicillin)	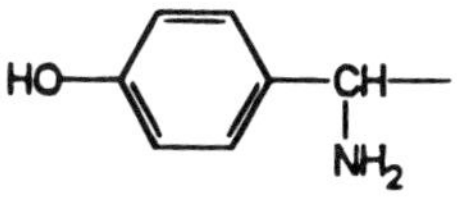 HO, CH—, NH_2	Yes	No	Yes
15. 1-Aminocyclohexyl (Cyclacillin)	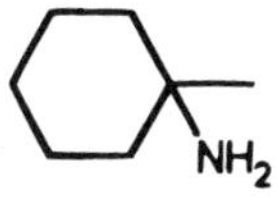 NH_2	Yes	No	Yes
16. D-α-Carboxybenzyl (Carbenicillin)[a]	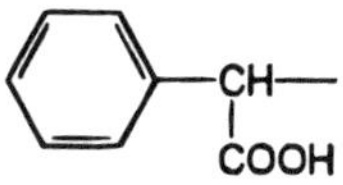 CH—, COOH	No	No	Yes
17. α-Carboxythienyl (Ticarcillin)	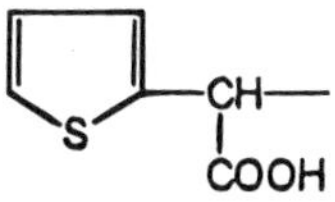 S, CH—, COOH	No	No	Yes

(Continued)

Table 6-2. *(Continued)*

Penicillin	*R*	Acid stable	Penicillinase resistant	Broad spectrum
18. N-(4-ethyl-2,3-dioxo-1-piperazinylcarbonyl)-α-aminobenzyl (Piperacillin)	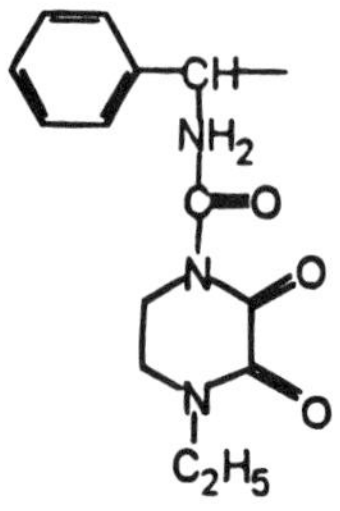	No	No	Very
19. N-(2-keto-1-imidazolidinyl-carbonyl-α-aminobenzyl (Azlocillin)	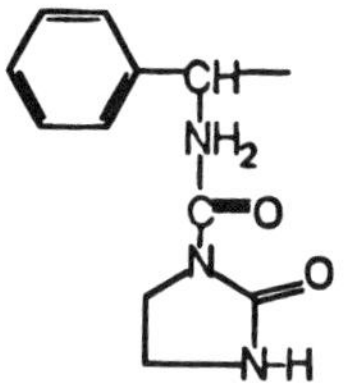	No	No	Very
20. N-([2-keto-3-methyl-sulfonyl]-1-imidazolidinyl-carbonyl-α-aminobenzyl (Meczlocillin)	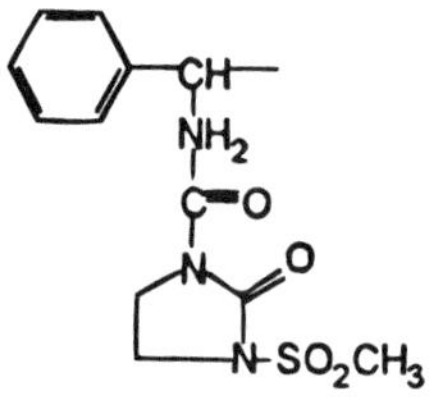	No	No	Very
21. 6-β-(hexahydro-1H-azepin-1)-yl-methylene amino-penicillanic acid (Mecillinam/Amdinocillin)	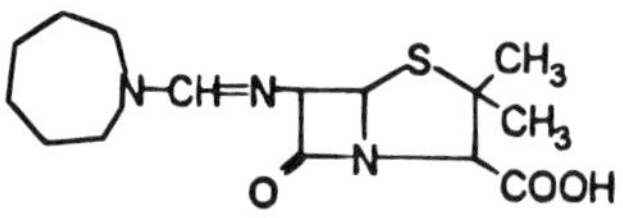	No	Some	See Text

[a] See Figure 6-9 for pro-drugs thereof.

Utilizing various phenyl- or phenoxyacetic acids with substituents in the aromatic ring did not offer significant improvements.

Attempts at modifying parts of the penicillin molecule other than the acyl side chain at position 6 seemed even more futile. Oxidation of the sulfur atom to the sulfone or sulfoxide, although giving the molecule increased stability in an acid environment, retains only a fraction of the parent molecule's antibacterial effectiveness. Any substituent at position 5 of the penam ring system results in a totally inactive compound. Simple esterification of the 3-carboxyl function such as the methyl ester resulted in no antibacterial activity. Neither does benzylpenicillin alcohol, where the carboxyl is reduced to the methylol group (CH_2-OH). Benzylpenicillin carboxamide retains only 25% of the parent compound's activity. The only manipulations of the sacrosanct carboxyl group that retains 90% of the original activity is the thio carboxyl isostere -COSH and certain labile esters (see pro-drug discussion following). Any rearrangement of the penam ring or cleavage of the β-lactam bond yields the totally inactive penicilloic acid derivative, which, of course, is the basis of both in vitro and in vivo inactivation.

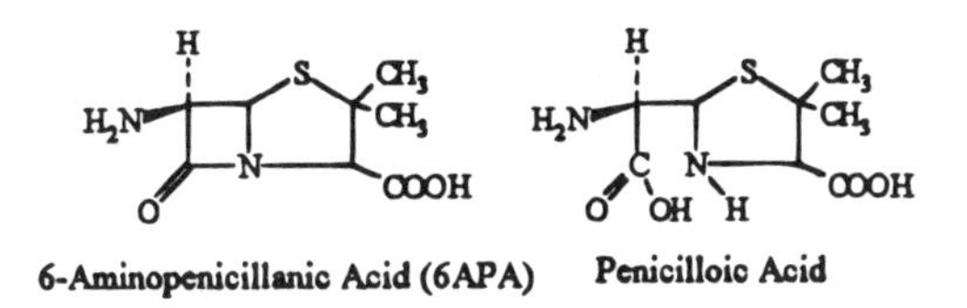

6-Aminopenicillanic Acid (6APA) Penicilloic Acid

The recognition that bacterial enzymes named *penicillinases* could hydrolyze penicillins to the inactive penicilloic acid occurred early. The potential seriousness of this event became evident from an early British report that incidence of penicillinase-producing (and therefore penicillin-resistant) staphylococci had reached above 50% (1948). The problem in United States hospitals soon approached 75%. Hospital-acquired or *nosocomial* (from the Greek nosokomeion—hospital) infections resistant to one or more antibiotics now occur with alarming frequency worldwide.

One other useful, but relatively minor, development in penicillin chemistry should be mentioned: the development of slightly soluble penicillin G salts with organic bases. These tended to be more stable, acted as repository parenteral dosage forms, and gave more uniform prolonged blood levels of the drug. One was the procaine salt, prepared form the sodium salt of benzyl penicillin. An aqueous solubility of 1:250 gives stable suspensions offering prolonged blood levels following intramuscular injection. The other, benzathine penicillin G, is the salt of N,N′-dibenzylethylene-diamine with two penicillin G molecules. Its extremely low solubility (1:5,000) results in very slow absorption from intramuscular injection depots, but offers prolonged therapeutic plasma levels of penicillin in excess of 1 week.

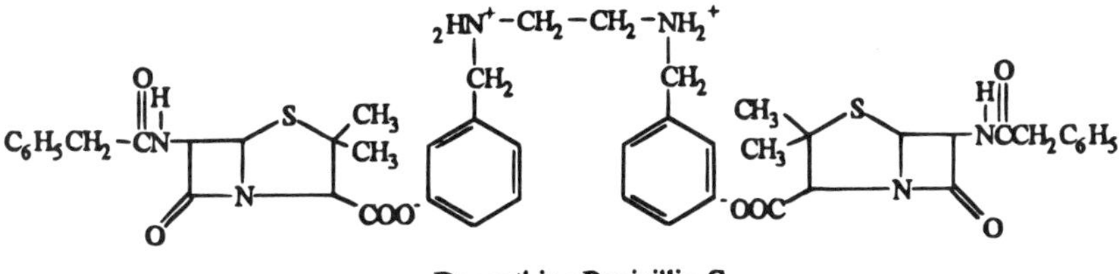

Benzathine Penicillin G

Although useful, increased acid stability (Penicillin V) and prolonged duration (amine salts) did not address the increasing resistance problem, nor did it produce the desirable goal of broadening the antibacterial spectrum of penicillins, especially into the Gm– realm.

The time for a breakthrough was ripe. It occurred in the isolation of 6-aminopenicillanic acid, 6-APA, by Batchelor et al. (1959). The compound was isolated from *P. chrysogenum* fermentations deprived of the normally supplied precursors such as phenylacetic acid. The availability of this "penicillin nucleus" in meaningful quantities now made possible the synthesis of an unlimited number of penicillins by acylating the β-6-amino group with an infinite number of acids, a process that has continued unabated.

It is parenthetically of historical interest that the long-sought total synthesis of a penicillin (V) had been achieved 2 years earlier by Sheehan and Henery-Logan (1957) in an elegant sequence of reactions, with the key procedure being the mild neutral cyclization of the corresponding penicilloic acid to the β-lactam structure with N,N′-dicyclohexylcarbodiimide (DCCI). They also synthesized 6-APA. However, the complex synthesis and overall low yield was not applicable to large-scale production.

The ability to deacylate the benzyl or phenoxymethyl group of penicillins G or V selectively by microbial amidase enzymes, yielding large quantities of 6-APA, were reported by four groups of workers in the United States, England, and Germany the next year. Today all the so-called semisynthetic penicillins are produced by acylating 6-APA, which in turn is produced in large quantities by amidase hydrolysis of the inexpensive fermentation-produced penicillins G and V. In fact the bulk of penicillin G produced is utilized as a raw material for other penicillins (via 6-APA), only a small fraction being diverted for therapy.

Figure 6-7 illustrates the three methods utilized to acylate 6-APA to active penicillins. The first is a reaction with the desired acid anhydride or acid chloride in the presence of a suitable hydrogen chloride scavenger. A second method is the direct coupling of the acid with the $-NH_2$ function by the use of DCCI in a solvent such as CCl_4. Here, as in the Sheehan and Henery-Logan (1957) application in closing the β-lactam ring, the facility of this mild dehydration reaction is based on the formation of the very stable, and highly insoluble (in CCl_4) dicyclohexylurea by the addition of H_2O across the carbodiimide structure. The third process is enzymatic and utilizes amidase obtained from *E. coli*. It is interesting that at pH 5 the reaction produces coupling to form the amide linkage, while the same enzyme at pH 8 would catalyze the reverse reaction, namely, hydrolysis of the 6-acyl side chains.

This development almost immediately began to help solve problems of the early penicillins: susceptibility to penicillinase inactivation (i.e., resistance) and narrow antibacterial spectrum (primarily Gm+).

One of the first attempts to stabilize penicillins toward penicillinase destruction was the drug methicillin (Table 6-2, No. 7). Success was achieved here by the obvious expedient of utilizing *steric hindrance* (i.e., placing the relatively bulky methoxy groups in the 2 and 6 positions of benzoic acid). By omitting the CH_2 group of penicillin G the "blockers" are thus in greater proximity to the labile β-lactam moiety of the drug. Methicillin is hydrolyzed by

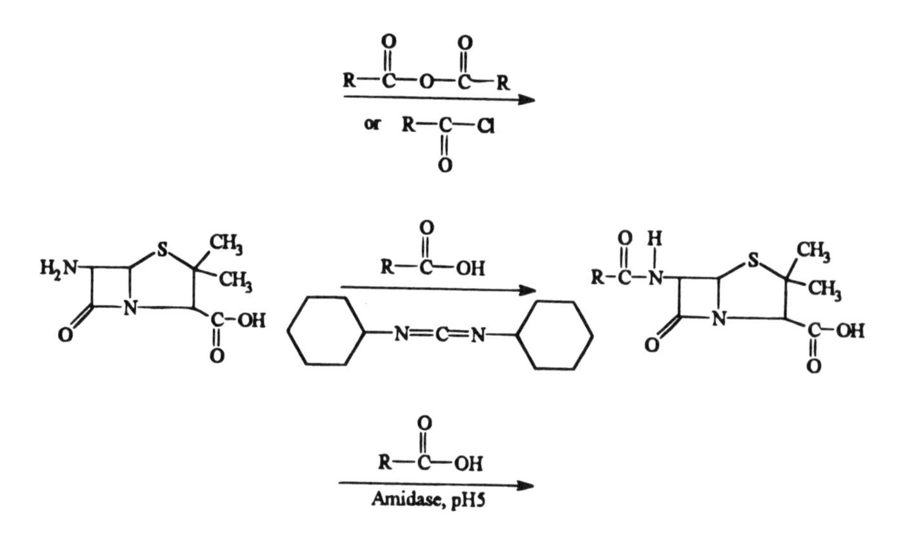

Figure 6-7. Acylations of 6-aminopenicillanic acid.

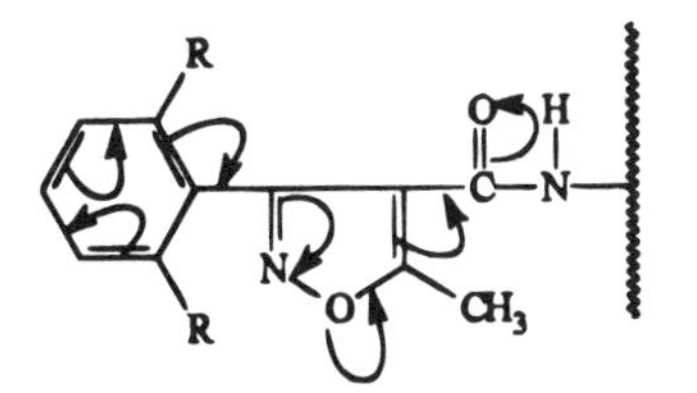

Figure 6-8. Resonance stabilization of oxazole penicillins.

staphylococcal β-lactamase at a rate less than 1% that of penicillin G. The drug thus became the first penicillinase-resistant penicillin in clinical use. It is still the prototype today.

Methicillin is not acid stable, possibly because the very small protonic catalytic species attacking the carbonyl oxygen of azetidinone (the initial step in the acid-catalyzed hydrolytic mechanism) is not effectively blocked as is the presumably large nucleophilic catalyst—the enzyme's reactive site. It is interesting that reinstating the methylene group, as in 2, 6-dimethoxybenzyl-penicillin, reverts the molecule to about the same β-lactamase susceptibility as penicillin G, thus demonstrating the nearness requirement here for effective steric blockage. The lack of absorption following oral administration limits methicillin to parenteral use. Other sterically hindered analogs, particularly those containing an isoxazole ring (penicillins 8–11, Table 6-2) were found to be acid stable as well as penicillinase resistant.

In the case of the isoxazole penicillins there is (in addition to the steric blockade as exemplified in methicillin), the conjugative factor involving the amido, isoxazolyl, and phenyl moieties (Fig. 6–8). These four oxacillins have become the mainstay for oral treatment of penicillin-(G and V)-resistant staphylococcal infections and should be reserved for that purpose. Routine use against nonpenicillinase producing Gm+ organisms is not advisable since their effectiveness is inferior to penicillins G and V (partially due to very high serum protein binding, 95–98%). These compounds have been shown to have low affinity for the enzyme as well as low reaction rates, indicating competitive inhibition of the enzyme. Nafcillin is another early penicillinase-resistant compound that is, however, not acid stable. It appears to have undependable oral absorption.

More recently the whole picture of antibiotic resistance has become better understood as a very complex phenomenon. Even with β-lactam antibiotics, resistance factors other than the mere production of hydrolytic β-lactamase enzymes are operative. In the late 1960s methicillin-resistant staphylococci began to appear as nosocomial infections. In fact these resistant strains show multiple resistance both to β-lactams as well as to antibiotics with different structures and different mechanisms of action.[14]

The acylation of 6-APA with α,α-disubstituted aryl acetic acids yielded penicillins carrying a substituent on the α or benzylic carbon of the *R* side chain. Those of particular interest were penicillins carrying the polar $-NH_2$ group (i.e., α-aminobenzyl penicillins) (Table 6-2, No. 13). The fact that this compound was acid stable and orally absorbed was not particularly novel (see penicillin V). Its antistaphylococcal activity was 25–60% less than that of penicillin G (the D-isomer was more active than the racemic D, L). However, its antibacterial spectrum was dramatically different. Ampicillin's spectrum, unlike previous penicillins, encompassed many important Gm– bacteria. Species of *Hemophilus, Proteus, Salmonella, Shigella*, and *Escherichia* succumbed to ampicillin. Even though resistant

[14] The methicillin-resistant staphylococcal bacteria do respond to Vancomycin, however.

penicillinase-producing strains have arisen, ampicillin retains wide clinical application, primarily because it is the nosocomial infections that are immune to the drug. Many, if not most, *E. coli* infections not acquired in an institutional setting are still readily treatable with this antibiotic.

Ampicillin, which has both an amino and carboxyl function, is zwitterionic and therefore incompletely absorbed from the intestinal and gastric areas. The broader antibacterial spectrum coupled with the unabsorbed function of the drug in the intestinal tract is believed responsible for the observed 8–10% incidence of diarrhea. To circumvent both problems research has led to the development of several ampicillin pro-drugs (Fig. 6-9). Three of these are esters of the carboxyl group, which, unlike earlier, poorly soluble "depot" esters of penicillin G (e.g., the β-diethylaminoethyl ester) that hydrolyzed slowly to produce the parent drug, are quite soluble. Since the zwitterionic characteristic is eliminated by esterification and the added lipophilic nature of the compound increased as well by the addition of "organic shrubbery," these pro-drugs tend to be more completely absorbed and, once in the serum, rapidly hydrolyzed to ampicillin by serum esterases. This would lead to increased serum levels. Talampicillin, the phthalidyl ester, for example, hydrolyzes in less than 15 minutes in human blood at body temperature, and reaches peak serum levels following oral administration that are three times higher than an equivalent dose of ampicillin itself. As mentioned, penicillins without the free 3-carboxyl function are inactive. Therefore, a decreased incidence of diarrhea might be predicted on oral use of these ester pro-drugs since the intestinal bacterial flora should not be significantly altered. In fact, studies did show a 50% reduction in this gastric side effect.

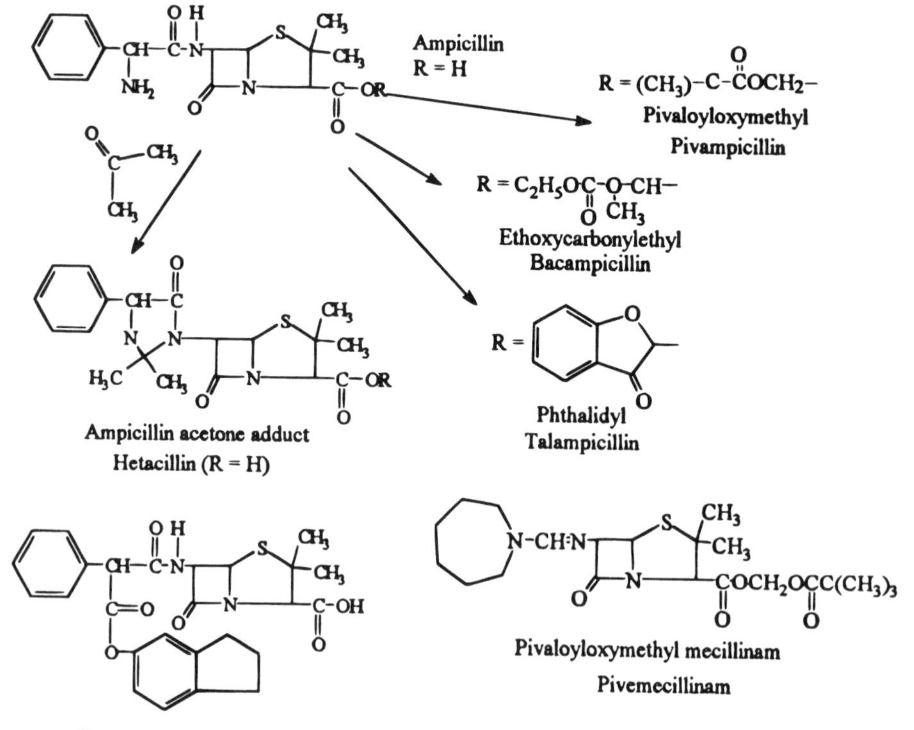

Figure 6-9. Pro-drugs of some penicillins.

Hetacillin, an acetone adduct of ampicillin, formed by the elimination of a water molecule (the α-amino and amido functions supplying the hydrogen atoms and the carbonyl of acetone the oxygen) is a chemically novel way to mask the α-amine function. Its instability, even in aqueous solution (reverting to ampicillin and acetone), makes it difficult to reconcile with reports of double ampicillin peak serum levels compared with oral ampicillin. It appears that no clinical advantage should accrue from the oral administration of this pro-drug. A report that hetacillin in solution is much more stable than ampicillin sodium (whose solution is alkaline) may make this pro-drug the preferable form for intravenous administration.

Cyclacillin (Table 6-2, No. 15) may be viewed as a cycloalkyl analog of ampicillin. Its main advantage might be higher peak serum levels. However, its effectiveness against Gm– organisms is lower. Amoxicillin (Table 6-2, No. 14), the *p*-hydroxy analog of ampicillin, with an identical antibacterial spectrum,[15] also has superior oral absorption, achieving blood levels more than double those of ampicillin.

6.5. Antipseudomonal Penicillins

Several Gm– bacterial infections, particularly when acquired in an institutional setting, are highly resistant to treatment and are therefore frequently life threatening. The causative organisms are *Pseudomonas aeruginosa* and certain species of *Proteus*. Even though the infections usually respond to the aminoglycoside antibiotics, the toxicity potential of these drugs makes them less than ideal. Several penicillins now exist that can be effective.

As was seen with the α-aminopenicillins (ampicillin, amoxicillin), introducing an ionizable group such as -NH_2 at the α-benzyl position widened the spectrum of penicillins considerably into the Gm– domain. It seemed chemically prudent therefore to evaluate another function, namely, the carboxyl group. First reported in 1967, carbenicillin (Table 6-2, No. 16) represented a significant expansion of the penicillin bridgehead into the more troublesome Gm– territory. Explanations for this spectral expansion might include the possibility that the COOH function endows the molecule with improved cell wall or membrane penetrability. More likely is the fact of greater resistance to the β-lactamases produced by those Gm– bacteria that are susceptible to ampicillin. Thus the enzyme produced by *P. aeruginosa* hydrolyzes carbenicillin at one-twenty-fifth the rate of ampicillin. Carbenicillin thus became the first β-lactam antibiotic to exhibit clinical antipseudomonal activity at achievable blood levels. Its acid instability precludes oral use (except as the indanyl pro-drug, Fig. 6-10). The acid instability is due to decarboxylation to penicillin G.[16]

Clinically, there are several shortcomings to carbenicillin. Its potency against *P. aeruginosa* is relatively poor (MIC 60–120 μg/ml), necessitating doses as high as 30 g per day, thus engendering potential toxicity. Since it is available as the disodium salt, daily intake of Na can be 3.2 g (141 mEq), a large sodium load for cardiac patients on a low-sodium regimen. The bioisosteric thienyl analog, ticarcillin, which has a similar spectrum, is somewhat more potent, thus permitting equivalent efficacy at lower doses (15–18 g/day). Sodium loads, however, are still in excess of 2 g.

Penicillins 18, 19, and 20 (Table 6-2) represent the next chemicologic step toward the "ideal" penicillin, wherein the free amino function of ampicillin is variably substituted to

---

[15] With the possible exception of *Shigella* in vivo.

[16] Carbenicillin can be viewed as a phenyl malonic acid derivative, thus is prone to CO_2 loss.

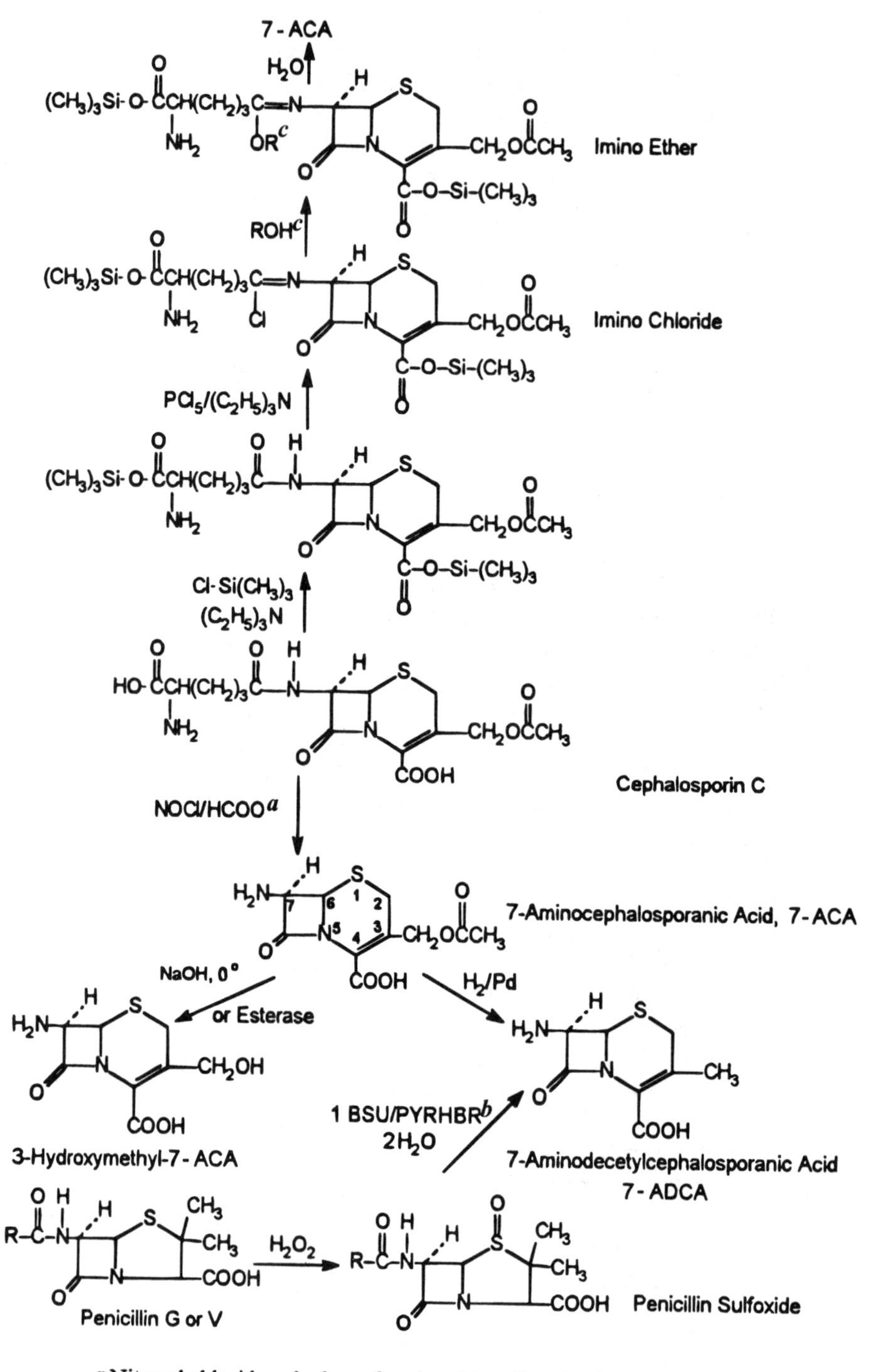

a Nitrosyl chloride-anhydrous formic acid, a diazotization mixture.
b Bis-trimethylsilyl-urea-pyridine HBr.
c R = CH_3 or C_2H_5.

Figure 6-10. Synthetic aspects of cephalosporin precursors.

produce the so-called ureidopenicillins (because of the urea structure). These have the most extended antibacterial spectrum thus far, particularly in the Gm– area, including *P. aeruginosa*. Acid stability of all three compounds is poor, making parenteral administration the only viable route. Piperacillin is likely the most active antipseudomonal penicillin available. Like all penicillins, the ureidopenicillins do not penetrate the intact, noninflamed meninges well.

6.6. Penicillin-Binding Proteins

The unique biochemical aspects of amdinocillin (compound 21, Table 6-2) can be better comprehended following a consideration of the current understanding of bacterial membrane-associated penicillin-binding proteins (PBPs) and their relationship to the transpeptidation reaction.

Membrane proteins are participants in most membrane-mediated dynamic processes. Lipids carry out the nondynamic function of compartmentalizing the various protein groupings by acting as selective barriers to the permeability of substrates.

Some membrane proteins are readily extractable by mild salt solutions. Others, more tenaciously enmeshed, must be solubilized prior to extraction and separation. This requires treatment with detergents such as the anionic sodium dodecyl sulfate (SDS) or the quaternary cationic dodecyltrimethylammonium bromide (DTAB).

Membrane proteins can thus be separated (and visualized) following disruption of protein interactions with other protein or lipid membrane components, by gel electrophoresis. One useful method is the layering of an SDS extract solution onto a polyacrylamide gel followed by application of an electrical field. The resulting differences in electrophoretic mobility then separates the individual proteins based on their mass (not charge). This is called *SDS-polyacrylamide-gel electrophoresis* (SDS-PAGE).

Experiments demonstrating the binding of penicillin (G) to bacteria began in the late 1940s. Later experiments using ^{35}S-labeled penicillin showed by binding studies that concentrations of drug-saturating high-affinity sites in susceptible bacteria correlated with MICs. Furthermore, that covalent bonding of the penicillin group was by ester linkage established by the facts that neither excess antibiotic, β-lactamase enzyme, nor boiling phenol or SDS regenerated the penicillin molecule.

Characterization of the multiple penicillin binding sites and the details of peptidoglycan biosynthesis reactions in the intact organism was not successful. However, separation of the protein components—the PBPs—as described earlier would allow the study of β-lactam antibiotics and the various enzymes at the molecular level.

The working pothesis for the mechanism of action of *trans*-peptidation was that of an acyl–enzyme intermediate. Penicylloyl derivatives of the easily solubilized D,D-carboxypeptidase enzyme of several bacteria including *E. coli* were shown to be transferable to simple acceptors such as hydroxylamine, illustrating the enzyme's continued active conformation. This fact led to the use of hydroxylamine to elute active PBPs from affinity chromatograph columns, where they were first bound to a penicillin (e.g., ampicillin) (Blumberg and Strominger, 1974).[17] Such experiments showed the existence of numerous

[17] This reference, an extensive review of penicillin interactions with PBPs and other penicillin-sensitive enzymes, should be consulted by the interested reader.

PBPs in bacteria in two groupings. One consists of soluble, low-molecular-weight PBPs having D,D-carboxypeptidase (D-alanine carboxypeptidase, DD-Cbase) activity, which is frequently the major penicillin-binding component. In most organisms no physiological functions have been assigned to them. The other larger and less soluble group that has higher molecular weights whose functions are also still unknown are the likely *lethal targets* of β-lactam antibiotics. The presumption is that they are the transpeptidases.

The methods for visualizing PBPs, membrane isolation, incubation with radiotagged penicillins, and SDS denaturation, followed by SDS-PAGE separation, have been developed, particularly with *E. coli.*

Extensive investigations of the properties of PBPs of *E. coli*, strain K12, were carried out. The results are indicative of our present understanding of PBPs and will be outlined. It should be stated that it is now understood that penicillin-sensitive enzymes such as DD-Cbase, peptidoglycan transpeptidase, and endopeptidases identified earlier are almost certainly identical with the PBPs under discussion here. Multiple PBPs have been discovered in all bacterial membranes studied. It is also now apparent that the interactions of β-lactam antibiotics with bacteria can result in one or more effects on the physiology and structure of the cell. Thus inhibition of cell division can be observed; so can lysis, bulge formation, or even the development of ovoid cell forms stable to osmosis.

Table 6-3 illustrates (for *E. coli*) the roles, if any, played by the seven PBPs in cell wall synthesis and cell morphology. Each is a distinct penicillin-sensitive enzyme whose inhibition has a particular consequence to the cell. For example, inactivation of PBP3 inhibits cell division and growth into filamentous cells. Death follows. PBP2 inactivation (the only specific target of mecillinam) in *E. coli* leads to spherical cell formation followed by lysis. Even though PBP2 is only 0.7% of total PBPs, amdinocillin's inhibition of it produces

Table 6-3. Characteristics and Functions of PBPs of *E. Coli* K-12[a]

PBP	Apparent mol. weight	% Binding of penicillin	Molecules per cell	Probable functions	Results of inactivation	Inhibitor[c] β-lactams
1A	92,000	8.1	200	Transpeptidation, transglycosylation for cell wall elongation	Rapid cell lysis[b]	Cephaloridine
1B[d]	90,000		250			
2	66,000	0.7	20	Transpeptidation	Non-growing spheroplasts	Mecillinam, clavulinic acid, thienamycin
3	60,000	1.9	50	Transpeptidation, septum cross-l wall synthesis	Nonseptate filamentous cells	Cephalexin, cefuroxine, aztronam
4	49,000	4.0	110	Secondary transpeptidation D,D-carboxypeptidase	Delayed transpeptidation absent	—
5	42,000	64.7	1800	A D-alanine carboxypeptidase	None	—
6	40,000	57.0	20.6	A D-alanine carboxypeptidase	None	—

[a] Compiled from Spratt, 1977 and Spratt, 1983.
[b] Both 1A and 1B inactivated.
[c] Representative examples.
[d] A third protein identified in one study.

potent cell killing of many Gm-organisms. Combining this drug with other broad-spectrum β-lactam compounds (e.g., ampicillin, cefamandol) produces a synergistic effect. Thienamycins and clavulinic acid also have been shown to bind primarily to PBP2, resulting in the same morphological cellular changes as produced by amdinocillin (i.e., osmotically stable spheres that lyse after a given time).

PBP 1A and 1B, the proteins of highest molecular weight, were originally thought to be a single component, PBP1, but were, with higher-resolution procedures, shown instead to be at least two compounds of similar molecular weights, with PBP 1B being the major component.[18]

The evidence thus far indicates the PBP1 set as functioning in cylindrical cell wall synthesis. PBP 1A has the higher affinity for β-lactam compounds. Inhibition of PBP 1A function alone by these antibiotics is not necessarily lethal to the cell, however. It is imperative that both PBPs be inhibited to effect cell lysis. The two enzymes catalyze transglycosylation and transpeptidation utilizing the bactoprene-P-P-disaccharide-pentapeptide (see Fig. 6-4).

PBP 4, 5, and 6 are D-alanine carboxypeptidases. It has been demonstrated with mutant *E. coli* lacking one, or even two, of these enzymes that viable cells still exist. The still somewhat incomplete picture evolved so far is that the bactericidal effect of β-lactam antibiotics is accomplished by disrupting cell wall synthesis by inhibiting a group of enzymes whose functions include the cross-linking of D-alanyl peptides to peptidoglycan chains of a continuously enlarging cell wall. The higher-molecular-weight proteins (PBPs 1A, 1B, 2, and 3) are absolutely necessary to cell wall production. Inactivating any one leads to death. This is very rapid. In the case of 1A and 1B the activities of these may possibly be coupled to *autolysins* (peptidoglycan hydrolases) that normally degrade cell walls in a delicate balancing act with new wall synthesis.

The soluble DD-Cbases catalyze transfer reactions between donor and acceptor peptides. These enzymes also hydrolyze donor peptides at the same time as transpeptidation is occurring.

The cross-linking process may be better understood if viewed as a two-step process: carboxypeptidation (Eq. 6.3) and transpeptidation (Eq. 6.4). The β-lactam drugs inhibit these reactions because of their structural analogy to the D-alanyl-D-alanine section of the peptidoglycan (Blumberg and Strominger, 1974), where the O=C-N corresponds to the peptide bond. It follows that the drug then can acylate the enzyme in a manner similar to Eq. 6.3. Thus the possibility of reversibility exists. Kinetic studies have lent corroboration to this idea. Using a β-lactam–sensitive bifunctional carboxypeptidase-transpeptidase from *Streptomyces* R61 and R39 enzymes, it was shown that the enzyme *E* binds reversibly to the β-lactam to form an initial complex *EI* (Eq. 6.5),

$$\text{Enzyme} + R\text{-D-ala-D-ala-COOH} \rightleftharpoons \text{Enz-}\underset{\|}{\overset{}{\text{C}}}\text{-Ala}R + \text{D-ala-COOH} \quad (\text{C=O}) \tag{6.3}$$

$$\text{Enz-C(=O)-Ala}R' + R\text{-NH}_2 \longrightarrow R'\text{NHC(=O)-Ala-R} + \text{Enzyme} \tag{6.4}$$

$$E + I \underset{K_2}{\overset{K_1}{\rightleftharpoons}} EI \xrightarrow{K_3} EI^* \xrightarrow{K_4} E + X + Y \tag{6.5}$$

---

[18] PBP 1 has, under more vigorous conditions, been further resolved into three bands.

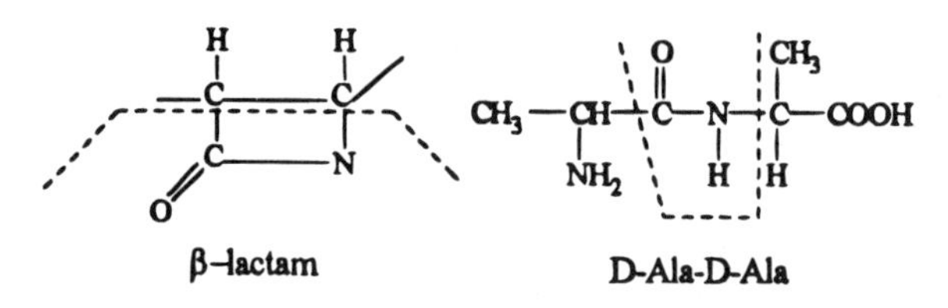

which subsequently transforms to the irreversible complex EI^*. On breakdown the antibiotic fragments ($X + Y$) regenerate the enzyme in active form. At concentrations of inhibitor I, the formation rate EI depends on the K_3/K_1 ratio. Free enzyme levels, of course, depend on antibiotic levels and the values of K_4 and K_3/K_1. It is apparent that an effective inhibitor (i.e., a good drug) should have a low K_4 and high K_3/K_1 value. If this model is really valid, then it may help explain why inhibition can at times seem to be irreversible (low K_4 value).

The covalent catalysis progresses by formation of an impermanent acyl–enzyme intermediate. It and the β-lactam binding involve the same serine residue of the enzyme's active site. The three-dimensional features of a penicillin-sensitive D-alanyl-carboxypeptidase-transpeptidase and exact binding site of a penicillin and cephalosporin have been established by X-ray crystallographic studies. One of the results of these experiments is to provide physical support for Tipper and Strominger's (1965) original hypothesis that β-lactam antibiotics and the normal substrate bind on the same locus of the transpeptidase.

6.6.1. The Cephalosporins—Chemistry

The cephalosporins are the second major group of β-lactam antibiotics. They differ from penicillins by having the azetidone ring fused to a dihydrothiazine ring rather than a thiazolidine. The other difference, which is more significant from a medicinal chemistry standpoint, is the existence of a functional group at position 3 of the fused ring system. This now allows for molecular variations to be introduced at the 7-NH_2 group, as in the penicillins, as well as to effect changes in properties by diversifying the moieties at position 3. As 6-APA became the precursor for the synthesis of all the semisynthetic penicillins, so does the corresponding 7-aminocephalosporanic acid (7-ACA) serve the analogous function for the cephalosporins.

The first member of the newer series of β-lactams was isolated in 1956 (Abraham and Newton) from extracts of *Cephalosporium acremonium*, a sewage fungus initially identified by Brotzu. The structure, and its differences from penicillins, were described later. It was named *Cephalosporin C*. The novel features were an aminoadipic acyl group at position 7 and an acetoxymethyl at position 3. Cephalosporin C exhibited disappointing antibacterial activity. Even the modest molecular alterations that were achieved by the addition of a limited number of exogenous phenylacetic acids to produce useful penicillins from penicillium molds could not be duplicated in this case. Thus it was imperative to develop usable methods to deacylate this clinically useless cephalosporin to 7-ACA. There are now at least two methods available to produce 7-ACA in commercial quantities (see later). It soon became apparent that varying the 7-acyl substituent gave useful new drugs with broader antibacterial spectra than did the earlier penicillins. In addition, many of these new drugs showed high resistance to β-lactamase enzymes. Variations of the substituent at position 3 affected such factors as absorption rates, tissue distribution, and metabolic features.

The production of 7-ACA and related precursors from Cephalosporin C is outlined in Figure 6-10. The antibiotic itself is now produced from mutant strains of *C. acremonium*.

Mild acid hydrolysis yielded only small quantities of the desired product plus 3-hydroxymethyl-7ACA and lactone formed by the 4-COOH with the 3′-OH group. A more efficient synthesis requires protecting the two carboxyl groups of Cephalosporin C by silylation. The compound is then chlorinated with PCl_5, which activates the amide linkage into an imino chloride. Alcoholosis with anhydrous methanol or ethanol affords the iminoether (imidate), which readily hydrolyzes with water, yielding the free amine and the 4-carboxyl group.

Another procedure, later improved by the use of anhydrous formic acid rather than H_2O as the solvent, involved the use of nitrosyl chloride as a diazotizing agent. Followed by hydrolysis of the intermediate, this procedure results in 7-ACA. More recently, the discovery of acylase enzymes and their utilization in immobilized form produced usable enzyme technology for cleavage of the C-7 side chain.

Deacetylation of 7-ACA to 3-hydroxymethyl-7ACA can be achieved with cold aqueous NaOH in less than 1 hour or, on a larger scale, enzymatically as in the synthesis of cefuroxime.

7-ACA can be converted to 7-aminodesacetylcephalosporanic acid (7-ADCA) by hydrogenolysis of 7-ACA over Pd catalyst. 7-ADCA is then the starting compound for the orally effective 3-methyl cephalosporins: cefadroxyl, cephalexin, and cephadrine (Table 6-4), which all have the 3-methyl group as a common feature.

A different approach to 7-ADCA is by ring expansion of the penicillin structure. This is achieved by first oxidizing penicillin G or V to its sulfoxide, followed by a ring expansion rearrangement whereby one of the two methyl groups inserts into the thiazolidine ring, resulting in the thiazine ring. The process is catalyzed by pyridine hydrobromide on the silyl-protected penicillin sulfoxide. Hydrolysis of the silyl ester yields 7ADCA.

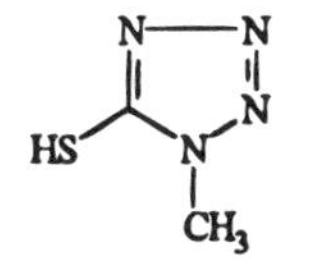

5-Mercapto-1-methyltetrazole

Figure 6-11 offers examples of semisynthetic cephalosporin syntheses. Thus treatment of 7-ACA with 2-thenylacetyl chloride in the presence of triethylamine (as an HCl scavenger) affords cephalothin. Even though cefazolin can be prepared in two steps from 7-ACA with 1H-tetrazolyl-1-acetyl chloride, then nucleophilically displacing the 3′-acetoxy group with 5-methyl-1,3,4-thiadiazole-2 thiol-, a claimed 90% yield one-step reaction that involves treating 7-ACA with a 1-mole equivalent of 5-methyl-1,3,4-thiadiazolyl (1H-tetrazolyl-1) thioacetate in aqueous acetone, is also available. The preparation of cefamandol (No. 9) includes the nucleophilic displacement of the 3′-acetoxy function by refluxing with an acetone solution containing excess 5-mercapto-1-methyltetrazole. This step can be carried out by a heterocyclothiomethyl compound on either cephalosporin C, 7-ACA, or acylated 7-ACA. Cephapirin can be synthesized by acylating the 7-NH_2 group of 7-ACA with bromoacetyl bromide, then displacing the active Br atom with 4-pyridinethiol. Finally, cephalexin can be prepared from 7-ACDA whose -NH_2 and COOH are first

Table 6-4. The Cephalosporins

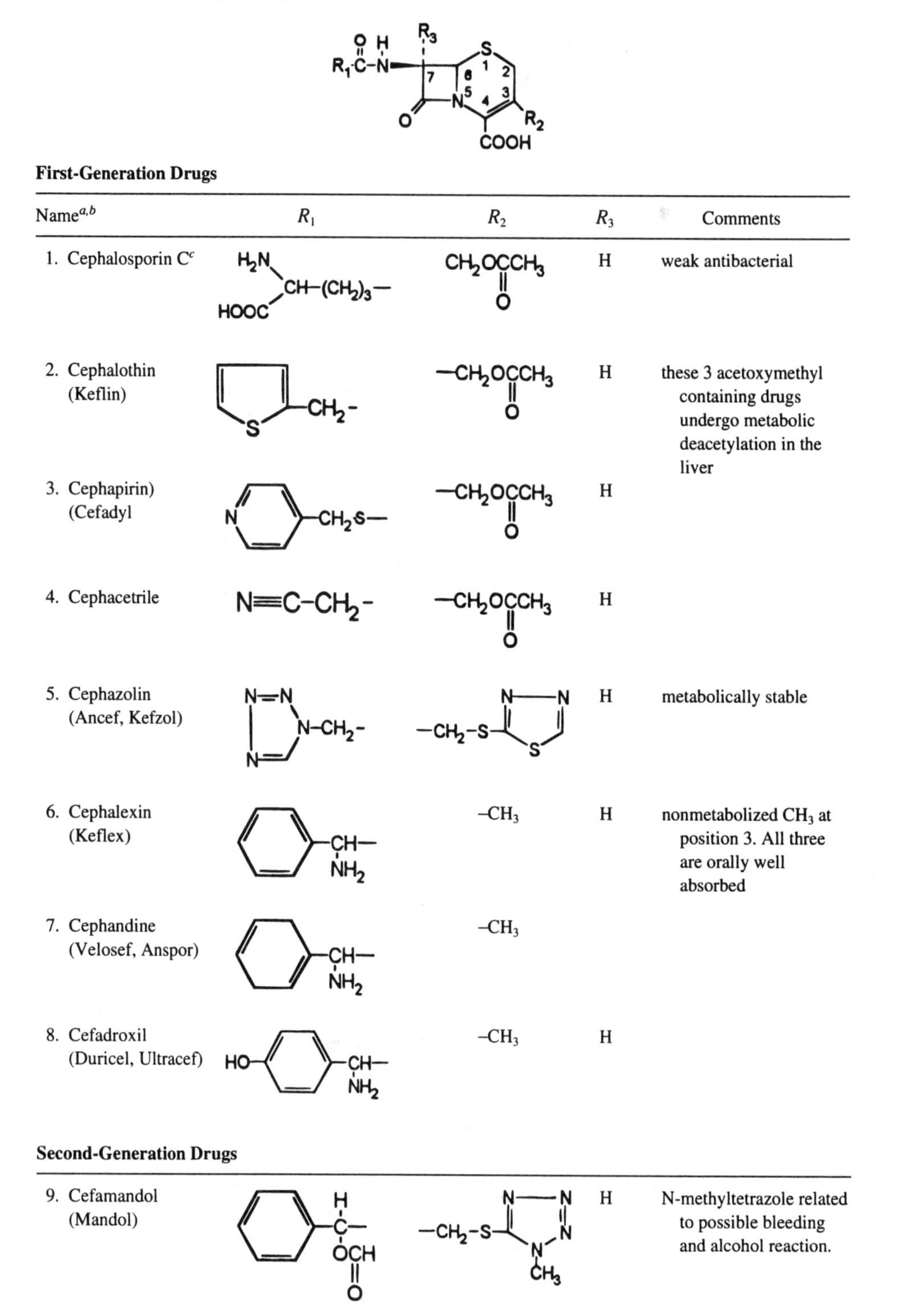

First-Generation Drugs

Name[a,b]	R_1	R_2	R_3	Comments
1. Cephalosporin C[c]	H_2N(HOOC)CH–$(CH_2)_3$–	CH_2OCCH_3 (C=O)	H	weak antibacterial
2. Cephalothin (Keflin)	2-thienyl–CH_2–	–CH_2OCCH_3 (C=O)	H	these 3 acetoxymethyl containing drugs undergo metabolic deacetylation in the liver
3. Cephapirin) (Cefadyl	4-pyridyl–CH_2S–	–CH_2OCCH_3 (C=O)	H	
4. Cephacetrile	N≡C–CH_2–	–CH_2OCCH_3 (C=O)	H	
5. Cephazolin (Ancef, Kefzol)	tetrazolyl N–CH_2–	–CH_2–S–thiadiazolyl	H	metabolically stable
6. Cephalexin (Keflex)	phenyl–CH(NH_2)–	–CH_3	H	nonmetabolized CH_3 at position 3. All three are orally well absorbed
7. Cephandine (Velosef, Anspor)	cyclohexadienyl–CH(NH_2)–	–CH_3		
8. Cefadroxil (Duricel, Ultracef)	HO–phenyl–CH(NH_2)–	–CH_3	H	

Second-Generation Drugs

Name[a,b]	R_1	R_2	R_3	Comments
9. Cefamandol (Mandol)	phenyl–C(H)(OCH=O)–	–CH_2–S–(N-methyltetrazolyl)	H	N-methyltetrazole related to possible bleeding and alcohol reaction.

(Continued)

Table 6-4. *(Continued)*

Name[a,b]	R_1	R_2	R_3	Comments
10. Cefuroxime (Zinacef)	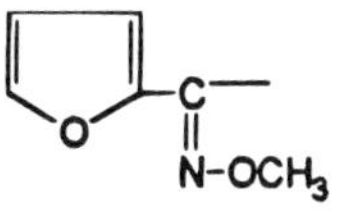 furan-2-yl–C(=N-OCH$_3$)–	$-CH_2OC(=O)NH_2$	H	R_2 carbamate resistant to hydrolysis
11. Ceforanide (Precef)	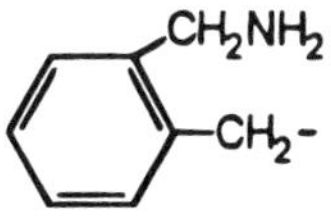 2-(CH_2NH_2)phenyl–CH_2–	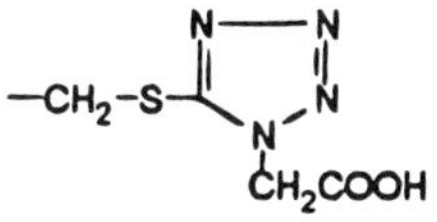 $-CH_2$-S-tetrazolyl (N–CH_2COOH)	H	
12. Cefoxitin (Mefoxin)	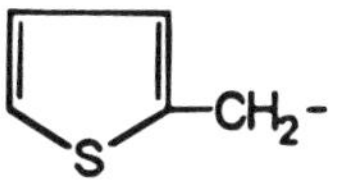 thiophen-2-yl–CH_2–	$-CH_2OC(=O)NH_2$	OCH_3	7-methoxy cephalosporins sometimes referred to as cephamycins
13. Cefotetan (Cefotan)	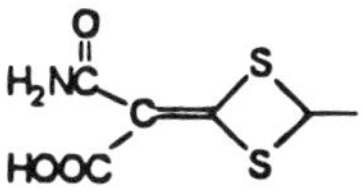 $H_2NC(=O)$(HOOC)C=C(1,3-dithietane)–	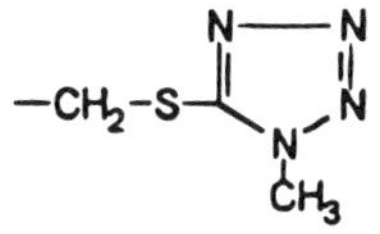 $-CH_2$-S-tetrazolyl (N–CH_3)	OCH_3	
14. Cefaclor) (Ceclor	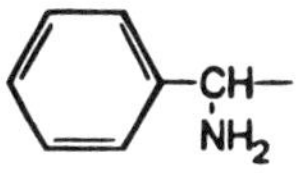 phenyl–CH(NH_2)–	–Cl	H	orally active

Third-Generation Drugs

Name[a,b]	R_1	R_2	R_3	Comments
15. Cefotaxime (Claforan)	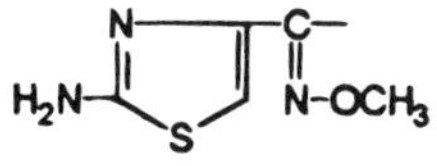 2-H_2N-thiazol-4-yl–C(=N-OCH$_3$)–	$-CH_2OC(=O)CH_3$	H	acetoxy group metabolized up to 25%; metabolite 90% less active
16. Ceftizoxime (Cefizox)	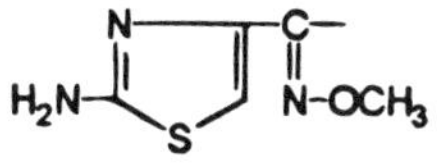 2-H_2N-thiazol-4-yl–C(=N-OCH$_3$)–	H	H	metabolically resistant cefotaxime analog
17. Ceftrioxone (Rocephin)	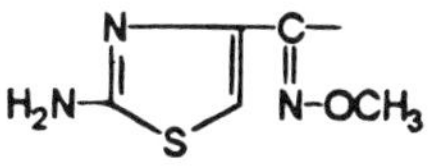 2-H_2N-thiazol-4-yl–C(=N-OCH$_3$)–	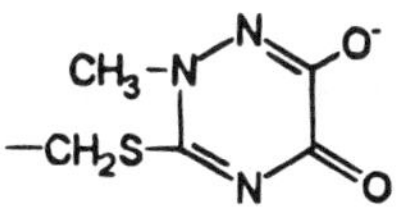 $-CH_2S$-triazinyl (CH_3-N, O^-, =O)	H	
18. Ceftazidime (Fortaz)	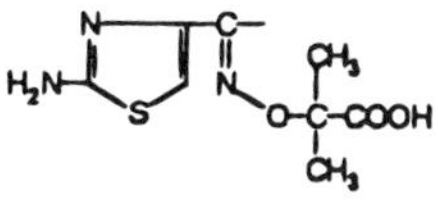 2-H_2N-thiazol-4-yl–C(=N–O–C(CH_3)$_2$–COOH)–	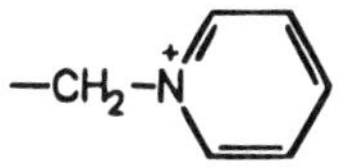 $-CH_2-N^+$(pyridinium)		not metabolized
19. Moxalactam (Moxam)	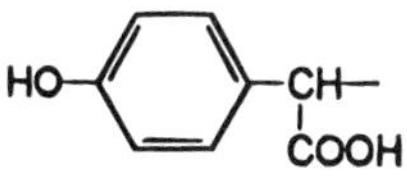 HO-phenyl–CH(COOH)–	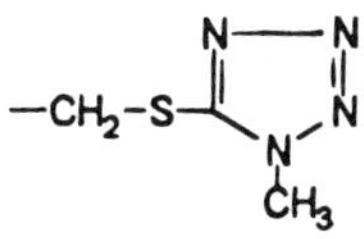 $-CH_2$-S-tetrazolyl (N–CH_3)	OCH_3	an oxacephem, O replaces S atom in thiazine ring

(Continued)

Table 6-4. *(Continued)*

Name[a,b]	R_1	R_2	R_3	Comments
20. Cefoperazone (Cefobid)	HO–C_6H_4–CH– (NH–C=O–N-(4-C_2H_5-2,3-dioxopiperazinyl))	$-CH_2-S-$(1-CH_3-tetrazol-5-yl)	H	not metabolized
21. Cefixime (Suprax)	H_2N-thiazolyl–C(=N-OCH_2COOH)–	$-CH=CH_2$	H	orally active
22. Ceftibuten	H_2N-thiazolyl–C(=CH–CH_2COOH)–	H	H	orally active
23. Cefprozil (Cefzil)	HO–C_6H_4–CH(NH_2)–	$-CH=CH-CH_3$	H	orally active

[a] Names in parentheses U.S. trade names.
[b] A 1975 ruling changed cephalosporin generic nomenclature for *new* drugs from "ph" to "f".
[c] See text regarding utility of this compound.

protected with trimethylsilyl chloride. The secondary amine function is then acylated with phenylglycyl chloride hydrochloride, which affords the desired product after deblocking with aqueous ammonia.

Table 6-4 lists cephalosporins available in the United States in a more or less chronological order of their introduction, usually termed as *generations*. This has become a useful classification since newer β-lactams were introduced primarily to overcome the shortcomings of the earlier compounds. Thus later agents tended to have broader antibacterial spectra, increased resistance to β-lactamase inactivation, less toxicity, or some combination of these features.

The first-generation cephalosporins, although straddling the Gm+–Gm– fence to some degree, are generally most active against Gm+ bacteria, particularly staphylococci and streptococci. Exceptions, such as methicillin-resistant staphylococci, many enterococci, and penicillin-resistant staphylococci, are now frequently encountered. In the Gm– domain one finds most *E. coli, Klebsiella pneumoniae*, and *Proteus mirabilis* susceptible unless they are hospital acquired. Anaerobes are generally sensitive. *Bacteroides fragilis*, however, is resistant.

It should be noted that cephadrine, cephalexin, and cefadroxyl in this group (all with the nonhydrolyzable methyl group in the 3 position) achieve useful therapeutic blood levels on oral administration.

The second-generation drugs have a broader spectrum of action against Gm– bacteria, but interestingly tend to be less useful against Gm+ organisms than the first-generation agents. The primary reasons for this spectral improvement is increased PBP affinity and

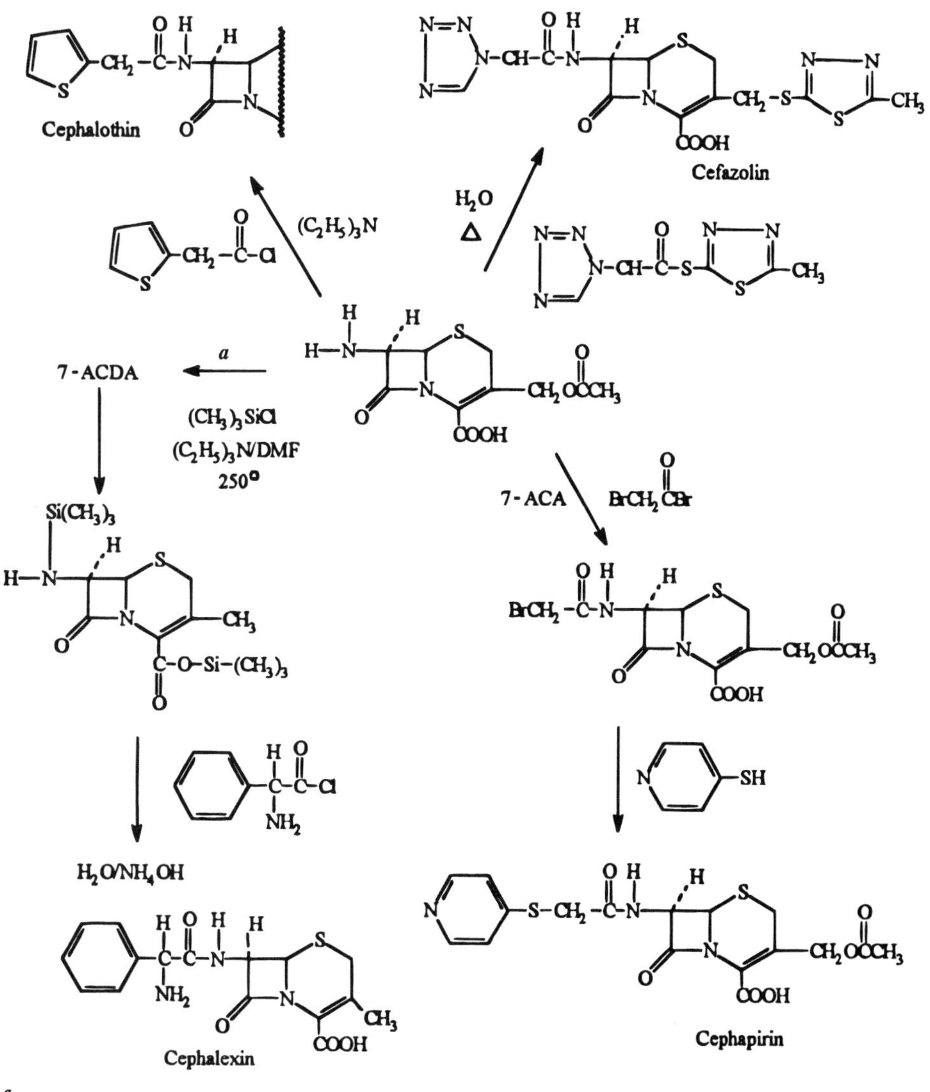

[a] See Figure 6-10 for synthesis.

Figure 6-11. Representative cephalosporin syntheses.

better resistance to the ravages of β-lactamases. Better outer-membrane penetration in Gm– bacteria is also an important factor.

Third-generation cephalosporins represent additional advantages in clinical situations not satisfactorily addressed by the earlier drugs. Excellent activity against Gm– bacilli, in some cases even against *P. aeruginosa*, are seen (e.g., ceftazidime, No. 18). Additional cephalosporins against this obstinate and dangerous organisms will follow. Moxalactam (No. 19) and ceftriaxone (No. 17) have the additional attribute of good penetration into the cerebrospinal fluid, CSF. Both drugs also possess long half-lives, particularly the latter (ca. 8 hours). It is interesting that ceftrioxone distributes well into extravascular tissue, a desirable property not readily predictable considering its high (95%) serum protein binding. Cefixime (No. 21) is the first orally active third-generation cephalosporin with high stabil-

Table 6-5. Relationship of Hydrolysis Rates of 2-Thienylacetaminocephalosporins Intrinsic Activity[a]

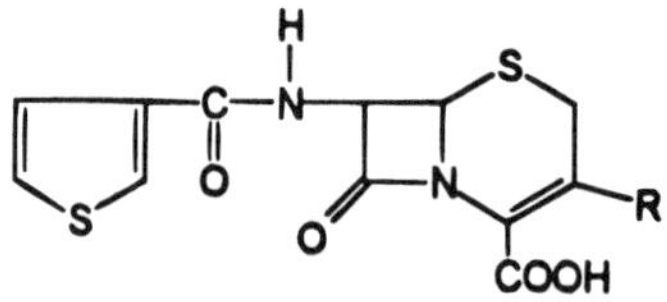

R	$K \times 10^{-5,b}$	*K. pneumonia*[c]	*E. aerogenes*[c]	*E. coli*[c]
–Cl (Cefaclor)	13.1	0.8	0.9	21.2
$CH_2OC(=O)CH_3$ (Cephalothin)	9.7	1.0	3.0	15
$–CH_3$	1.07	10	30	116

[a] Adapted from Indelicato et al., 1977.
[b] β-Lactam opening, pH 10 (NaOH).
[c] MIC (μg/ml).

ity against β-lactamase. Even though absorption is only about 50%, the relatively low protein binding (65%) helps give this drug potent effectiveness at low doses administered as infrequently as once a day.

It is important not to consider the chemical reaction of hydrolysis as monolithic in an in vivo environment. Cephalosporins particularly illustrate this point. Much has been said of the hydrolytic instability of β-lactam antibiotics, yet penicillins and cephalosphorins are generally excreted with the β-lactam nucleus preserved. Thus it is curious that much of an oral dose of penicillin G is hydrolyzed by acid catalysis in the stomach to the corresponding penicilloic acid. However, that portion of the dose that is absorbed from gastric areas intact remains so in the blood, tissue, and urinary fluids, yet, in the presence of β-lactamase-producing bacteria, the chemically "unprotected" β-lactam ring is very susceptible to hydrolysis by nucleophilic catalysis. As was seen with the penicillins, by utilizing steric hindrance and inductive effects, it is possible to preserve, to a degree, the integrity of the β-lactam structure and obtain the so-called β-lactamase-resistant compounds. Therefore, there are two stabilities to consider. The first is the stability of the β-lactam ring. One study has demonstrated that the intrinsic activity of cephalosporins is directly proportional to the rate at which the ring is opened by nucleophilic catalysis, at least against Gm– bacteria. Table 6-5 relates to OH-induced hydrolysis rates of three 2-thienylacetaminocephalosporins with the antibacterial activity against several organisms. There was proposed a general scheme for such ring scission. In the case of cefaclor ($Y = Cl$), a subsequent elimination of chloride occurs (Eq. 6.6). Where $Y = CH_3$ there, of course, exists a nonleaving group. The nucleophile may be neighboring α-aminobenzyl function such as in cephalexin, cephadrine, or cefadroxil (Nos. 6, 7, 8—Table 6-4). Here, catalysis would be *intramolecular.*[19]

[19] An analog ($Y = CH_3$) was isolated from cephalexin (Dinner, 1977).

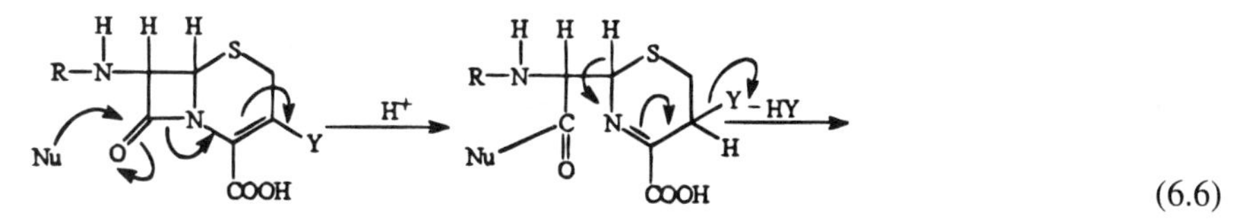

(6.6)

Drug design studies were undertaken to increase the stability of the β-lactam ring against β-lactamases whose poor effectiveness held down the breadth of the Gm– spectrum in many of the first-generation compounds. The discovery of the cephamycin type of cephalosporin such as cefoxitin (No. 12) led to the recognition of the utility of the 7-α-methoxy function dramatically to increase resistance to β-lactamase degradation (see later). Just as was seen in the penicillins, enhancement of Gm– activity can be achieved by increased polarity of the 7-acyl side chain through the introduction of amino or carboxyl functions on the α-carbon.

Since such an α-NH_2 group can participate intramolecularly in the β-lactam ring-opening reaction, it is converted to a methoxime (methoximine) function, resulting in considerable ring stability.[20] Even though both the *syn* and *anti* isomers usually show β-lactamase stability, the *syn* configuration is generally superior. Some cephalosporins carry only the 7-α-methoxy (No. 12, 13, 20) or oximine functions (No. 15, 16, 18, Table 6-6).

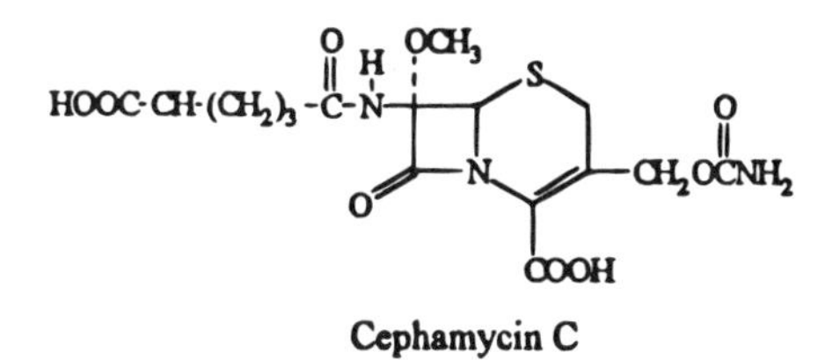

Cephamycin C

The cephamycins, or 7-α-methoxycephalosporins, were first reported by Nagarajan et al. (1971) and then soon by others. Unlike previous β-lactams, they were obtained from *Streptomyces* species, which are bacteria rather than molds. Of the eight or more cephamycins studied cephamycin C is of most interest. The aminoadipoyl group at C-7 is the same as encountered with Cephalosporin C (Table 6-4, No. 1). In addition to the characteristic 7-α-methoxy group of all cephamycins, the existence of a 3-carbamoylmethyl function is unique to naturally occurring products (even other cephamycin do not have it).[21] Cephamycin C was both more stable and potent than the A and B analogs (carrying cinnamoylmethyl groups at C-3). Its antibacterial spectrum was primarily Gm–. A comparison of β-lactamase resistance between cephamycin C and cephalosporin C was dramatic: The cephamycin (A, B, and C) were either totally resistant or several orders more resistant to β-lactamases than the cephalosporin, depending on the virulence of the enzyme studied. Thus cephamycin C was selected for molecular modifications to optimize the desirable qualities and develop new ones. Exchanging the 7-α-adipoyl group with a thienyl moiety produces cefoxitin (Fig. 6-11, No. 12). Although analogous to cephalothin (No. 2),

---

[20] Oximes are hydroxylamine derivatives of ketonic functions that can exist in *syn* and *anti* configurations (roughly equivalent to *cis* and *trans*, or Z and E).

[21] See also physostigmine, Chapter 7.

Table 6-6. Substituent Functions of Cephalosporins and Their Effects on Chemical Stabilities

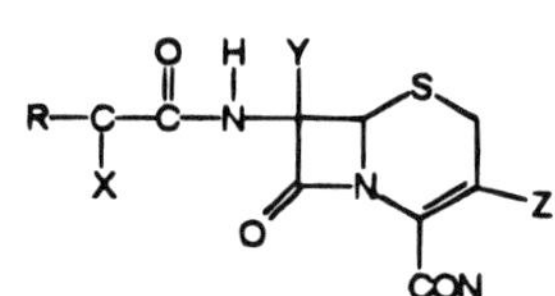

Increased β-Lactamase Stability   Increased Metabolic Stability

X	*Y*	*Z*	Example[a]
$=N{-}OCH_3$	—	$-CH_2OC(=O)NH_2$	Cefuroxime (10)
$=N{-}OCH_3$	-OCH_3	$-CH_2-S-$(1-CH_3-tetrazol-5-yl)	Cefotetan (13)
$=N{-}OCH_3$	—	**H**	Ceftizoxime (16)
$=N{-}OCH_3$	—	$-CH_2OC(=O)-CH_3$[b] (metabolized)	Cefotaxime (15)
COOH	-OCH_3	$-CH_2-S-$(1-CH_3-tetrazol-5-yl)	Moxalactam (19)
—	-OCH_3	$-CH_2OC(=O)NH_2$	Cefoxitin (12)
$-NH-C(=O)-$N(2,3-dioxo-4-C_2H_5-piperazin-1-yl)	—	$-CH_2-S-$(1-CH_3-tetrazol-5-yl)	Cefoperazone (20)
$=N{-}OCH_3$	—	CH_2S-(2-CH_3-4-C_2H_5-5-oxo-6-O^--triazinyl)	Ceftriaoxone (17)
$=N-O-C(CH_3)_2-COOH$	—	$-CH_2-N$(pyridinium)	Ceftazidime (18)

[a] The numbers in parentheses refer to Table 6-4.

[b] Deacetylated to an active metabolite.

the carbamoyl group at C-3 replaces the acetoxy function of the latter. The result is a new antibiotic with high potency against Gm+ bacteria that also retained and improved the Gm– spectrum of cephamycin C. The effect of the thienyl ring, which is already known, is again demonstrated. It is surprising, however, that the thiophene ring also enhanced the already considerable β-lactamase resistance of the cephamycin C 175 times. One can now view the cefoxitin molecule in terms of its contributing parts: The cephem ring system, which is responsible for intrinsic bactericidal activity, is augmented to a broad spectrum—Gm+, Gm–, including Gm– anaerobes—by its thienyl function. The 7-α-methoxy group protects the ring against β-lactamase and the 3′-carbamoyl moiety, by virtue of its resistance to hydrolysis, prevents the drug's rapid metabolism to inactive or less active compounds.

Cefotetan (Table 6-4, No. 13) was the second cephamycin introduced into clinical practice. The two drugs are bacteriologically quite similar against anaerobes. It is somewhat more potent against Gm– bacteria (neither is useful against *Pseudomonas*) and less useful against Gm+ organisms. Cefotetan's main advantage is probably against *Enterobacteriaceae* species, being 8–16 times more potent. Its three times longer half-life (0.8 vs. 3.0 hours) may permit less frequent dosing.

Metabolic stability in cephalosporins is achieved primarily by varying the chemistry on the C-3 position from the hydrolyzable acetoxymethyl function seen in several of the first-generation drugs. Hydrolysis to the hydroxymethyl product invariably results in an inactive or much weaker antibacterial. Table 6-6 illustrates the variations utilized. Ceftizoxime solves the problem by simply having no substituent at C-3. As a result no metabolism occurs.

Altering the chemistry of the 3-position on the cephem nucleus affects both the serum half-life and thus the duration of action, as well as other aspects of the drug's pharmacokinetics. The 1-methyl-tetrazolyl (MTZ) group can also vary protein-binding affinity. In addition, introduction into broad-spectrum cephalosporins appears to give these drugs high serum levels. Converting the 1-methyl group into a polar acidic derivative (e.g., -COOH) gives particularly high levels rapidly and keeps them at above MIC levels for many hours. Ceforanide (No. 11), for example, reaches peak levels within 30 minutes and remains at useful blood concentrations for up to 12 hours. Unfortunately, the MTZ group has also been associated with some untoward reactions: the so-called disulfiram reaction (an intolerance of alcohol) and a more serious hypothrombinemia causing bleeding.

Moxalactam, a totally synthetic drug, is technically not a cephalosporin since the thiazine sulfur atom has been isosterically replaced by an oxygen atom. It is thus the only clinically represented oxacephem, but is generally viewed as a cephalosporin-type drug. Moxalactam represents cumulatively some of the useful structural attributes of various related antibiotics.

First, the oxygen for sulfur trade alone increased intrinsic activity several times. The β-lactamase stability of the 7-α-methoxy group seen in the cephamycins was incorporated here. The effectiveness against Gm– bacteria so notable in carbenicillin and ticarcillin that is attributable to the α-COOH on the acyl side chain at the 6-β position is utilized here in the corresponding 7-β position. As in amoxicillin the *p*-OH group on the benzene ring is applied in this drug to improve blood levels and to increase the half-life. Finally, the use of the MTZ group at position 3, found so useful in several of the third-generation cephalosporins, undoubtedly here, too, expands the Gm– spectrum—especially in conjunction with the aforementioned carboxyl group. The net result is a highly effective agent, keeping in mind the several potential adverse effects mentioned earlier.

6.7. Other Bicyclic β-Lactams

There are some other interesting bicyclic β-lactams under investigation that have clinical potential. One such new family of carbapenems, where the sulfur atom has been replaced by carbon (see Fig. 6-6), are the *olivanic acids*. As will be noted the sulfur atom is still part of the molecule, but it is outside the bicyclic β-lactam structure.

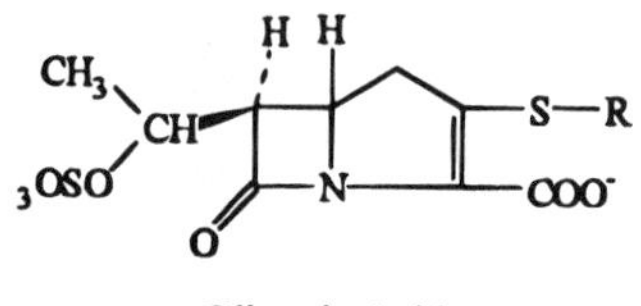

Olivanic Acids

These antibiotics have been isolated from cultures of *S. olivaceus* and have been found to have β-lactamase-inhibiting as well as antibacterial properties (Brown et al., 1977; Butterworth et al., 1979).

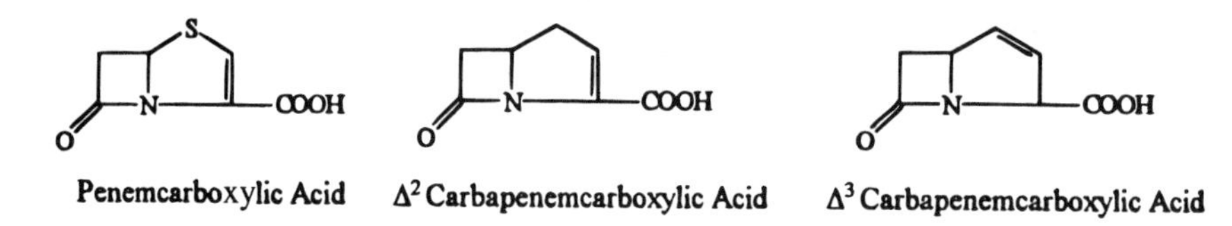

Penemcarboxylic Acid Δ^2 Carbapenemcarboxylic Acid Δ^3 Carbapenemcarboxylic Acid

One of the more exciting carbapenems introduced clinically in the United States (1986) probably is the formimidoyl derivative of thienamycin, imipenem (Fig. 6-12). Thienamycin was isolated from culture broths of *S. cattleya*. An examination of the structure shows obvious similarities to the penicillin cephalosporin β-lactams: the bicyclic geometry and the COOH group. The differences, however, are significant. There is a lack of an acyl function at the β-6 position. Instead, one finds an apparently simpler nonacyl 1-hydroxyethyl group. As the 7-α-OCH_3 in cephamycins, this group also protects the β-lactam linkage admirably against the β-lactamases. The position of the sulfur atom (as an alkylthio side chain) outside the ring must be noted. It undoubtedly helps to increase antipseudomonal activity well into the clinically useful range (*P. aeruginosa*, MIC = 3.0 μg/ml). Finally, the difference in the 2–3 double bond (DB) of olivanic acids and thienamycin and its absence from the penicillin nucleus needs to be considered. It can be deduced (by examining molecular models) that the DB in the fused carbapenem ring system causes increased ring strain and consequently much higher lactam instability, yet merely shifting the DB to the isomeric 3–4 position of thienamycin leads to an inactive compound. The β-lactam amide bond was found to be less reactive. By simplifying the β-lactam further to penem carboxylic acid and its two isomeric carbapenem analogs, it was found that the Δ^2carbapenemcarboxylic acid only was inactive, yet all three are highly strained molecules. The amide bond in all three compounds was hydrolyzable with about equal ease as in the two active compounds. The inactive thienamycin DB isomer and Δ^2carbapenemcarboxylic acid share the same nucleus and DB position. It seems that not all the

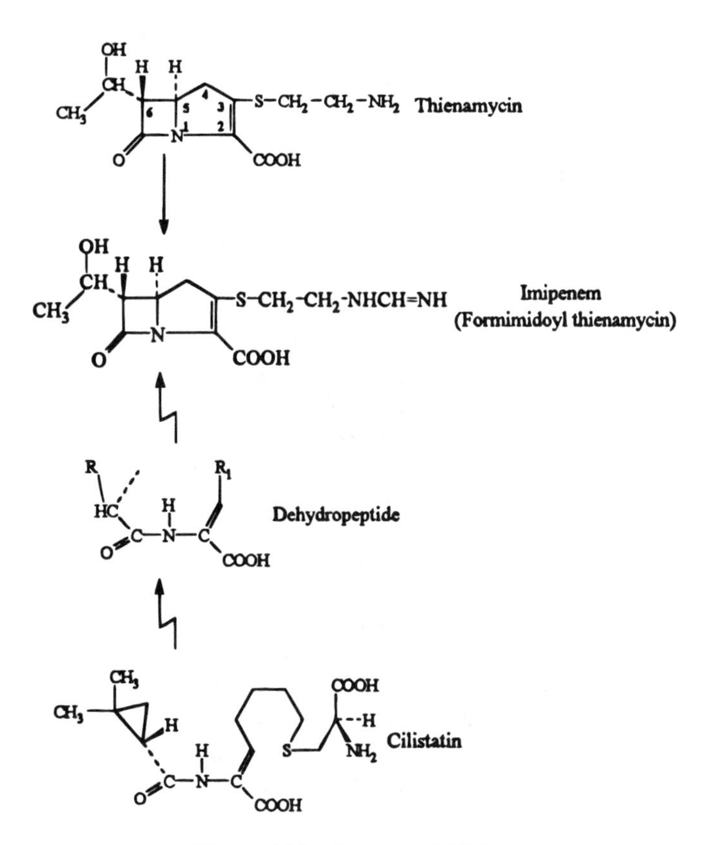

Figure 6-12. Imipenem/Cilistatin

geometric factors involved in the antibiotic action (or enzyme interactions) of β-lactam drugs have been fully explained. It has been suggested that the "goodness of fit" between such drugs and the enzyme is active sites may be the primary factor.

Experience has now shown thienamycin to have the broadest antibacterial spectrum of any drug currently in use, including most aerobic and anaerobic Gm+ and Gm– bacteria (notable exceptions are methicillin-resistant staphylococci and *P. cepacia*). Thienamycin was found effective in 98% of 31,000 bacterial isolates against which it was tested. The β-lactam was found to be easily hydrolyzed at alkaline pH (above 8). It reacts readily with nucleophiles such as hydroxylamine, cysteine, and even with its own primary amine group, which leads to accelerating inactivation as the drug concentration goes up. An inactive dimer is presumably formed. Even in neutral solutions (intramolecular) degradation is 10 times more rapid than penicillin G. That problem was finally solved by derivatizing the self-flagellating amine function as a formimidoyl function, now called *imipenem*. This compound is both stable and even more active than the parent compound.

The very broad antibacterial spectrum of imipenem is explicable when its stability is viewed in the presence of β-lactamases mediated by drug-resistance plasmids (mainly penicillinases) and chromosome (primarily cephalosporinases). Imipenem was shown to be stable to enzymes of both types (except *P. maltophilia*). A comparison of MICs for imipenem, cefazolin, and ampicillin dramatically illustrates the clinical potential for this newer drug. Thus while the latter two agents are totally resistant (MIC > 800 μg/ml), imipenem shows high activity (MIC 1.56 μg/ml), especially when compared with marginally effective β-lactam antibiotics. The clinician, however, should be cautious of the opportunistic appearance of the two resistant species of pseudomonas: *cepacia* and *maltophilia.*

In spite of the high stability of imipenem toward almost all β-lactamases, urinary levels of the intact drug in several animal species and humans were low due to apparent metabolic breakdown of the β-lactam ring. Incomplete urinary recovery of certain cephalosporins has been shown to be due to hepatic enzyme hydrolysis of the 3-acetyl side chains (e.g., cephalothin and cefotaxime). The integrity of the β-lactam ring, however, was not affected; in fact, the metabolites retained partial antibacterial activity. The reason for β-lactam ring hydrolysis of thienamycins and other carbapenems (e.g., olivanic acids) was found to be the membrane-bound renal enzyme dehydropeptidase I (DHP I), which is one of several known renal dehydropeptidases. The enzyme is located in the proximal tubular epithelium, where it catalyzes the hydrolysis of 60–95% of the drug following glomerular filtration of imipenem.

Examination of the structures of this new class of carbapenem β-lactams such as thienamycins indicates an intriguing similarity to dehydropeptides such as glycyldehydrophenylalanine (Fig. 6-12, $R = CH_2$, $R^1 = C_6H_5$). The DB is in the same relative position to the peptide bond as are the carbapenems. β-Lactam antibiotics not exhibiting such a structural homology to dehydropeptidases such as penicillin G or aztreonam (see later) showed no measurable hydrolysis by DHP-I. The development of a specific reversible inhibitor of this enzyme, *cilistatin* (Fig. 6-2), when combined with imipenem in equal dosage (in milligrams)[22] solved the clinical problem. This combination product now permits the use of imipenem against both systemic and severe urinary tract infection (UTI).

6.8. Monobactams

With the discovery of monocyclic compounds in 1981 the trend toward the simplification of β-lactam antibiotic structures has reached the ultimate. The term *monobactam* describes both their source and chemistry (i.e., *mono*cyclic *bact*erially produced β-lact*am* antibiotics). This new class of drugs was discovered independently in Japan and the United States. Unlike the original serendipitous discovery of penicillins, monobactams were found on the basis of rational screens developed to seek out β-lactam–containing compounds from microbial sources. The Japanese groups utilized as their screen mutated strains of *E. coli* (PG8) lacking both chromosomal β-lactamase and PBP-1B and a *P. aeruginosa* PSC-sensitive mutant strain IFO 3080 (Imada et al., 1981). These mutant organisms, which are supersensitive to β-lactam antibiotics, were thus used to screen for novel, natural β-lactams. Soil samples containing several bacterial strains yielded β-lactam antibiotics, the most interesting being sulfazecin and isosulfazecin, its epimer; the bacteria were two new species of *Pseudomonas*.

The American group (Sykes et al., 1981) utilized a strain of *Bacillus licheniformis*, which is exquisitely sensitive to β-lactam–containing compounds, as the screening tool. Seven related monobactams were characterized from 1 million bacterial isolates. These

---

[22] Trademarked as Primaxin®.

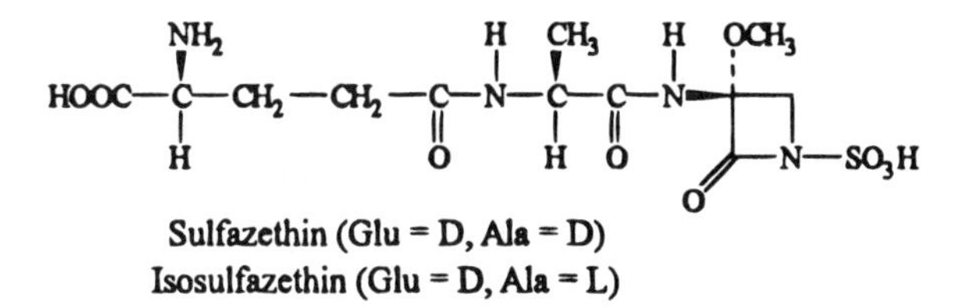

were obtained from various Gm– bacteria, including species from the genera *Pseudomonas, Gluconobacter*, and *Acetobacter*. All these monobactams contain the 2-oxoazetidine-1-sulfonic acid moiety (Fig. 6-13). To maintain the analogy to the previously discussed β-lactam antibiotics, these new compounds may be viewed as derivatives of 3-aminomonobactamic acid.

This natural monobactam had several shortcomings as a potential drug. It showed poor antibacterial properties and chemical instability. Removal of the methoxy function actually improved physiological stability; SQ 26396 (Fig. 6-13). Reasoning from the successful SARs developed for cephalosporins (and penicillins) suggested analogous acyl side chains and substitutions at position 3. "Borrowing" the aminothiazole oxime of ceftazidime (No. 18, Table 6-4) to replace the methyl group of SQ 26180 afforded SQ 81402, which is a compound with high activity against a broad range of Gm– organisms, including *P. aeruginosa*. Because of the previous elimination of the 3-α-methoxy group (to improve chemical stability), this compound now lacked β-lactamase resistance. This problem proved minor. Placing a 4-α-methyl group onto the β-lactam ring resulted in a compound with good intrinsic activity, high β-lactamase, and superior chemical stability—SQ 26726 (Fig. 6-13). This drug, aztreonam, became the first monobactam introduced into clinical use (Italy, 1984).

Aztreonam is clinically active against most Gm– organisms, including *P. aeruginosa* (but not *P. maltophila*). It shows no meaningful activity against Gm+ and anaerobic bacteria. *Neisseria* and *Hemophilus* are exceptions, however. It has high stability in the presence of most chromosomally and plasmid-mediated β-lactamases of Gm– organisms, actually

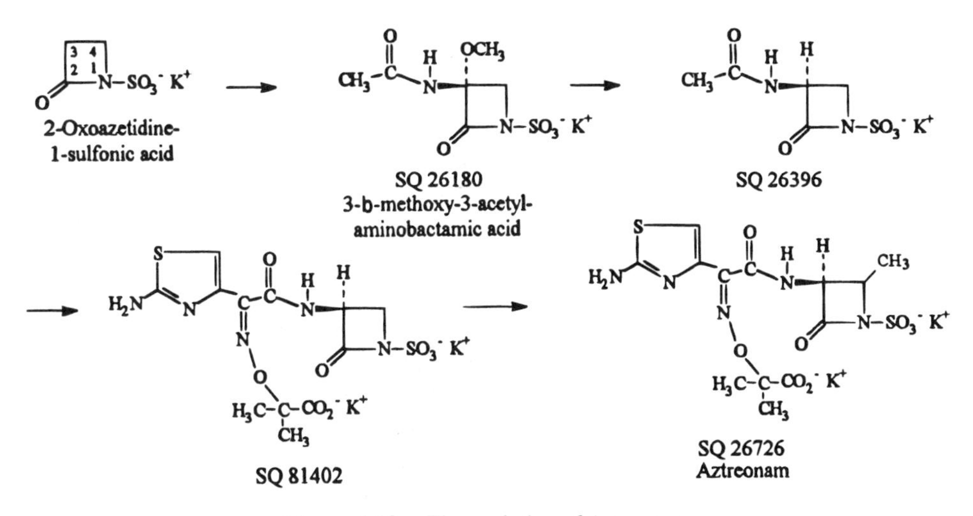

Figure 6-13. The evolution of Aztreonam.

having a demonstrated low affinity (high K_m) for these enzymes. With respect to *P. aeruginosa* it was shown that aztreonam inhibited 250 strains of this organism resistant to piperacillin, cefoperazone, and most aminoglycosides, at MICs of less than 6 μg/ml.

It is interesting that 3-acyl side chains analogous to those of the cephalosporins show closely similar antibacterial activity against Gm– organisms. It is curious that the analogy even extends to Gm+ organisms; therefore, the penicillin G (i.e., benzyl) analogous monobactam exhibits such activity. The phenylglycyl (ampicillin, cefaclor) monobactam analog has been synthesized to determine if it too would be orally bioavailable. The compound was inactive against both Gm+ and Gm– bacteria. The reason is undoubtedly the chemical instability due to intramolecular nucleophilic attack by the amino group of the side chain, on the carbonyl group of the β-lactam. The resultant 2,5-piperazine-dione has actually been isolated and identified (Eq. 6.7). Arylglycyl cephalosporins behave analogously.

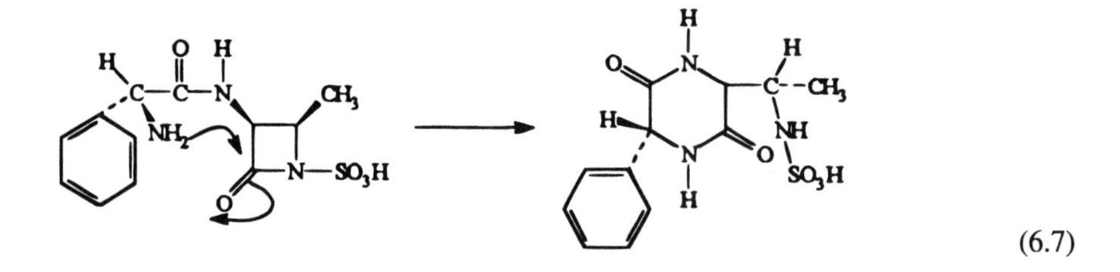

(6.7)

Prior to the discovery of monobactams and the earlier norcardicin (monocyclic β-lactams with meager antibacterial action that was not improved by structural manipulations), explanations for the reactivity of the β-lactam ring of the various β-lactam antibiotics revolved around its fusion—at its 1 and 4 positions—to an additional ring. The bicyclic structure, common to all these compounds, was then thought to activate (i.e., destabilize) the β-lactam ring by disallowing it *coplanarity* with its 1 and 4 substituents. This inhibits the resonance of the amide function responsible for the hydrolytic stability of less stressed amide-containing rings or acyclic compounds. Since the 1 and 4 substituents of the monobactams are not part of a second ring restraining system, that reasoning obviously cannot apply here. In fact, molecular models indicate that substituents on all 4 corners of azetidine are *coplanar*. Thus β-lactam activation of monobactams must be due to other factors.

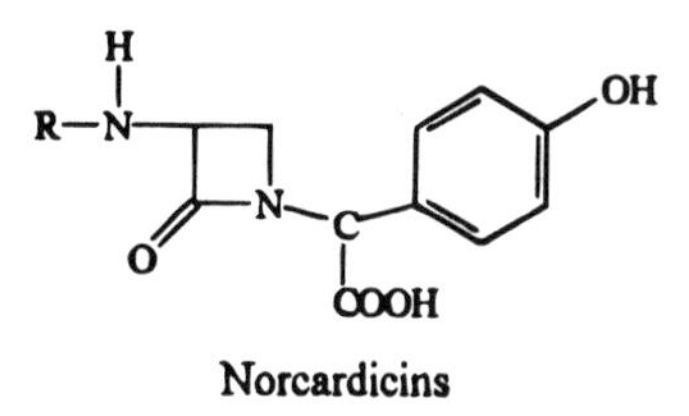

Norcardicins

Suspicion necessarily turns to the electronegative and charged sulfonic acid function. In fact, monobactams without an N-substituent is both bacteriologically and chemically unreactive. Thus it appears that the negatively charged acid function on the nitrogen atom of these monobactams has the dual function of aiding substrate recognition *and* activating the β-lactam ring. It was suggested that resonance suppression through non-planarity is not the main factor; electron-withdrawing groups, such as the sulfur atom in penicillins, by reducing the amide nitrogens basicity, makes it easier for it to leave the oxygen atom (i.e., hydrolyze). Therefore, it is the electronegativity of the substituent that contributes to penicillin's ease of cleavage. It is reasonable, then, to state that electron-withdrawing groups on the nitrogen, the 1-position, are equally capable of ring activation and that the stronger the electron-withdrawing ability (pKas) is the better. A comparison of 3-aminomonobactams with 2-azetidone-1-phosphonate (see Fig. 6-6) (methyl phosphonic acid being much weaker) shows the latter compound to be less active.

It should be understood that if the transpeptidase enzyme and β-lactam compound do not have an optimum "goodness of fit," β-lactam ring activation in and of itself will not result in an antibacterial compound. It was demonstrated that when superimposed penicillin and monobactam molecular models show good congruence of side chains, lactam bondings, and electronegative functions such as SO_3H. The high β-lactamase stability of monobactams tends to confirm this. It is also interesting that the methoxylated monobactams (and cephalosporins) are more resistant to β-lactamase-catalyzed hydrolysis than are those without, yet they are much more susceptible to base-catalyzed hydrolysis. The explanation that reactivity is less important than "goodness of fit" is very appealing.

6.9. β-Lactamase Enzymes

The existence of resistance to β-lactam antibiotics by bacterial production of enzymes that rapidly catalyze the hydrolysis of the β-lactam ring to yield inactive products was recognized soon after the introduction of the early penicillins.[23] As clinical resistance to penicillins became a problem of serious proportions, efforts were devoted to the development of compounds with improved stabilities to these enzymes (now generically referred to as β-*lactamases* or, more specifically, as *penicillinases* or *cephalosporinases*). Penicillins such as methicillin and the isoxazoyl penicillins (Table 6-2) alleviated the problem to a certain extent. However, the difficulties continued to increase, particularly with Gm– organism. The picture improved with the introduction of the second- and third-generation cephalosporins and, most recently, imipenem and aztreonam, drugs with extremely high resistance toward enzyme-catalyzed hydrolysis. Another approach over the past several years has been the expatiation of enzyme inhibitors (see later).

6.9.1. β-Lactamases

Over 80 different β-lactamases are now known. One classification is a system that divides the enzymes into three classes: A, B, and C. Classes A and C are active-site serine enzymes. The serine residue in class A enzymes is at position 70. This class contains four major β-lactamases: 749/C (from *B. licheniformis*), PC1 (from *S. aureus*), 569/H β-lactamase I (from *B. cereus*), and PBR322 and RTEM (from *E. coli*). As with other serine-type hydrolytic enzymes (acetylcholinesterase, trypsin), the mechanism of action requires initial formation of an acylated enzyme, in this case acylation of ser-70 followed by hydrolysis of the derivative to regenerate the enzyme:

---

[23] Although other types of resistance may occur (e.g., PBP mutations, autolytic enzyme deficiency, and decreased drug penetration), β-lactam C–N bond hydrolysis is most significant to clinical resistance.

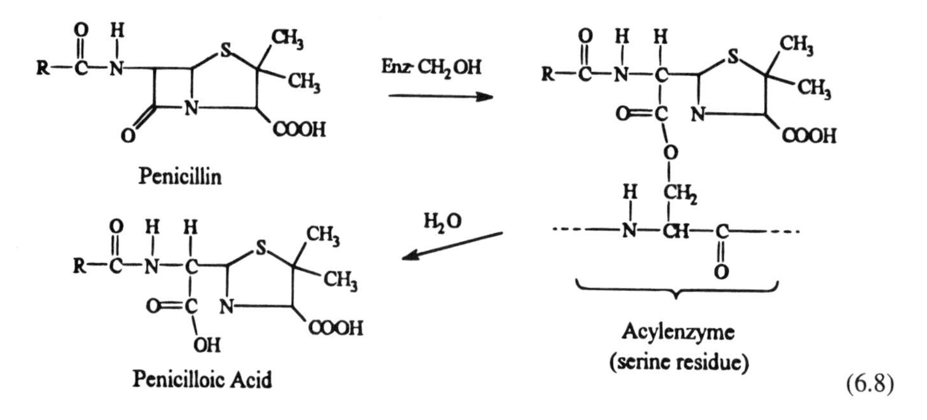

(6.8)

Class C β-lactamases have a serine-80 active site. These enzymes are chromosomally encoded. One type is specified by *amp*-C gene of *E. coli* K-12; another is elaborated by *P. aeruginosa* (Jaurin and Grundstrom, 1981). Class B β-lactamases are metalloenzymes containing thio (SH) groups as the ligand binding to Zn^{2+}, the metal ion necessary for β-lactamase II to function. This enzyme, 569/H/9 from *B. cereus*, has a particularly broad substrate spectrum, even hydrolyzing at significant rates three β-lactam antibiotics resistant to most other β-lactamases: imipenem, cefoxitin, and moxalactam. Its unique Zn^{2+} requirement sets it apart from other β-lactamases. Useful inhibitors for it have not yet been found. An organism elaborating this enzyme is likely to be resistant to all β-lactam antibiotics.

Since the structure of relatively few β-lactamase enzymes have been totally characterized, another method of enzyme classification has been developed by Sykes (1982). Here, Class I contains enzymes that are active primarily against cephalosporins. The gene for their production is in chromosomes; many can be induced by cephalosporins. Class II are mainly penicillinases (i.e., more active against penicillins). They are also chromosomally mediated. Class III β-lactamases are comparable against both types of β-lactam antibiotics. Genes for this class are found in the *R* plasmids, which are the extrachromosomal "pieces" of genetic material transferable between bacteria (of different species) by conjugation. Class IV β-lactamases are chromosomally mediated, but they have substrate characteristics similar to III, and both may therefore be considered "broad-spectrum" β-lactamases. Class V enzymes primarily attack penicillins including methicillin and isoxazolyl penicillins. Finally, there is the group of penicillinases produced by Gm+ bacteria. The genes for this class are in those plasmids transferred by transduction. The preceding is just a brief outline. Additional subdivisions exist for each class. Gm– plasmid (*R* factors)-determined β-lactamases (Class III and V) are particularly complex. Types designated as TEM, particularly TEM-1, occur most frequently and are widely distributed. Their complex amino acid and nucleotide sequence of the gene encoding this enzyme are known. β-Lactamases OXA-1, OXA-2, and OXA-3 are capable of hydrolyzing methicillin and isoxazolyl penicillins at rates faster than penicillin G. Additional Gm– plasmid-mediated β-lactamases have since been recognized, each with distinct species characteristics. Thus SHV and HMS β-lactamases are determined by plasmids of enterobacteria; at least four PSE types are known to be pseudomonas-specific.

Fascinating reports have appeared that seem to support a theory that β-lactamases may have evolved from PBPs, specifically from D-alanyl-D-alanine cleaving peptidase (DD-peptidase). The active-site DD-peptidase of *S.**R61 has been crystallized and the three-dimensional structure and penicillin-binding site determined by very high (2.8 Å) resolution X-ray crystallography. Later the Class A β-lactamase of *B. licheniformis* was similarly characterized. Computer-graphic protein mapping of three-dimensional and primary structures of the DD-peptidase and the β-lactamase allowed comparisons, previously not possible, that indicated good matching of both strands of β-sheeting and helical structures of the two molecules. Overlaying showed seven of eight helices in close congruity. Helices of each enzyme had similar angles and distances. In both cases the active site serine was near the NH_2-terminus.

6.10. β-Lactamase Inhibitors

The concept of synergistically overcoming β-lactamase resistance by concomitant use of a β-lactam antibiotic resistant to β-lactamase hydrolysis with one sensitive to it is theoretically sound. The expectation is that the sensitive compound would be spared destruction and would add its bacterial effect to that of the resistant compound. Even though in vitro results seemed promising, clinically the synergism achieved was not enough.

Examination of the diversity of β-lactamases indicates that β-lactam compounds may exhibit the properties of substrates or inhibitors. Compounds such as cefoxitin, moxalactam, cefuroxime, and ceftizoxime (Table 6-4) have low V_{max} and K_m values for β-lactamases elaborated by Gm– organisms. In such situations these drugs may be considered to be inhibitory substrates that offer significant clinical advantages. Several penicillins (e.g., the isoxazolyl derivatives) also seem to be inhibitory substrates. Thienamycin (Figs. 6-6 and 6-12) and olivanic acids, which are resistant to most β-lactamases, also have inhibitory properties. As a group then such compounds can be regarded as substrates that bind well yet turn over at a low rate. The agents considered earlier, however, are not likely to solve the problem. Experience shows that clinical strains of bacteria resistant to the "best" drugs of yesterday arise. For example, a β-lactamase from *B. fragilis* capable of hydrolyzing cefotaxime has been reported; isoxazolyl penicillins do not inhibit plasmid-mediated TEM β-lactamases.

The answer may be in discovering more specific mechanism-based β-lactamase inhibitors. It is curious that the best drugs developed so far are also β-lactam compounds (Fig. 6-14). One very interesting inhibitor is penicillanic acid sulfone (sulbactam, Fig. 6-14). This drug acts as a substrate (becomes hydrolyzed) and as an irreversible inactivator of TEM-type β-lactamase (covalently binding to it via molecular cleavage).

Many in vitro studies have firmly established a synergism between sulbactam and various β-lactam antibiotics including piperacillin, amdinocillin, ampicillin, and penicillin G. These studies included bacterial strains resistant to ampicillin or penicillin G alone, which became sensitive upon addition of sulbactam. Carbenicillin-resistant *Pseudomonas*, however, was not potentiated.

A fixed-ratio product of cefoperazone and sulbactam has been introduced into clinical practice. The 3-pivaloyloxy-methylester pro-drug of sulbactam has been evaluated in humans and found to be orally absorbed and then cleaved in vivo to the parent compound. A "mutual" pro-drug, sultamicilin (Fig. 6-14) representing an oral drug that is a mixed acetal of ampicillin and sulbactam (a formaldehyde hydrate ester) was evaluated. After oral absorption the compound is hydrolyzed by serum esterases, yielding a 1:1 mixture of ampicillin and the β-lactamase inhibitor. A similar mutual pro-drug of amdinocillin (VD 1825) has also been tested (Fig. 6-14). In addition to the obvious advantage of synergism that both these drugs offer in the treatment of ampicillin- and amdinocillin-resistant bacteria, these double esters, by virtue of their good oral absorption, offer an additional clinical

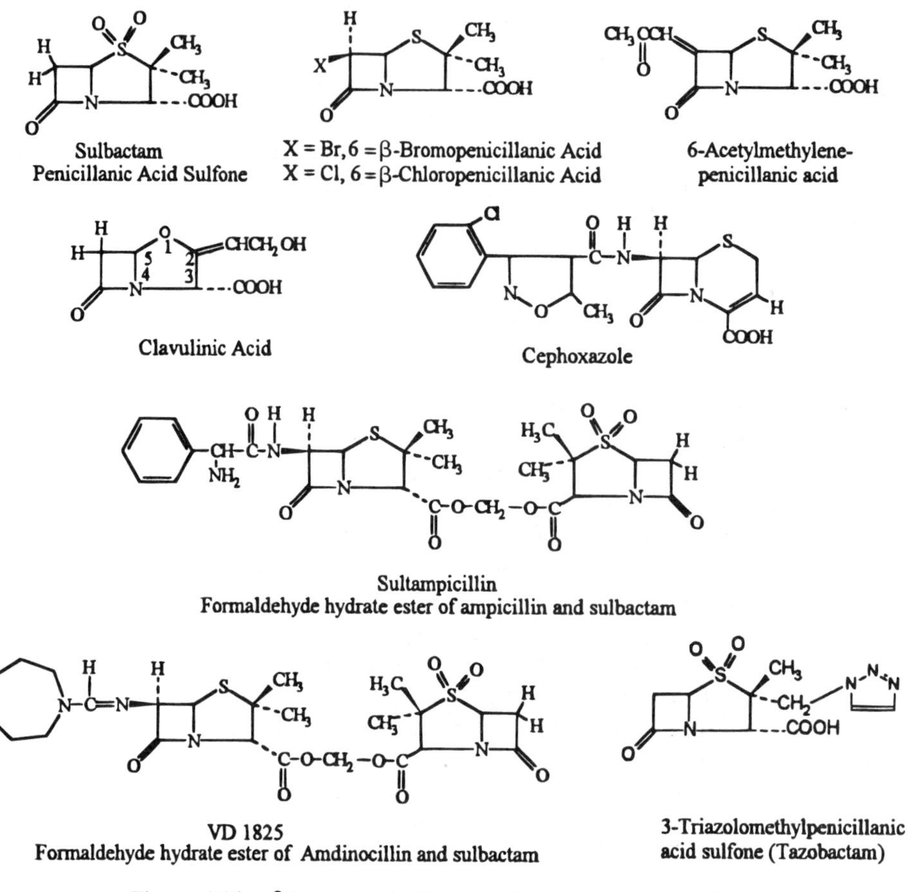

Figure 6-14. β-lactamase inhibitors and several mutual pro-drugs.

advantage. They circumvent the difficulty of poor oral bioavailability of the two antibiotics and the inhibitor if administered singly. Gastric stability (at pH 2) was relatively good, whereas serum and tissue hydrolysis rates (at pH 7.4) were considerably higher. The survival of the product in the stomach acid thus obviously contributes to efficacy.

6-Halogenated penicillanic acids such as 6-β-bromo, chloro, and iodo penicillanic acids (Fig. 6-14) are mechanistically particularly interesting β-lactamases inhibitors (α-isomers are inactive). Figure 6-15 outlines a possible mechanism by which 6-β-bromo-penicillanic acid inactivates a β-lactamase by essentially forming a relatively stable acylenzyme. The nucleophilic attack by the SER-70 OH group (aided by participation of an ancillary second functional group in the active site) forms the classic oxyanionic intermediate followed by irreversible cleavage of the β-lactam ring. Scission of the C-5–S bond allows the anionic sulfide ion to attack the C-5 position, displacing the Br atom and producing a 1,4-thiazine ring. Double-bond shifting leads to the 2,3-dehydro-1,4-thiazine acylated enzyme, which is inactive.

The stability of the acylated enzyme may be rationalized to be due to vinylogous structure of the urethane-type product formed. Analogous mechanisms have been proposed for sulbactam and clavulinic acid. Clavulinic acid can also form an acylated equilibrium between an enamine and imino intermediate. The nonvinylogous, and presumably, some-

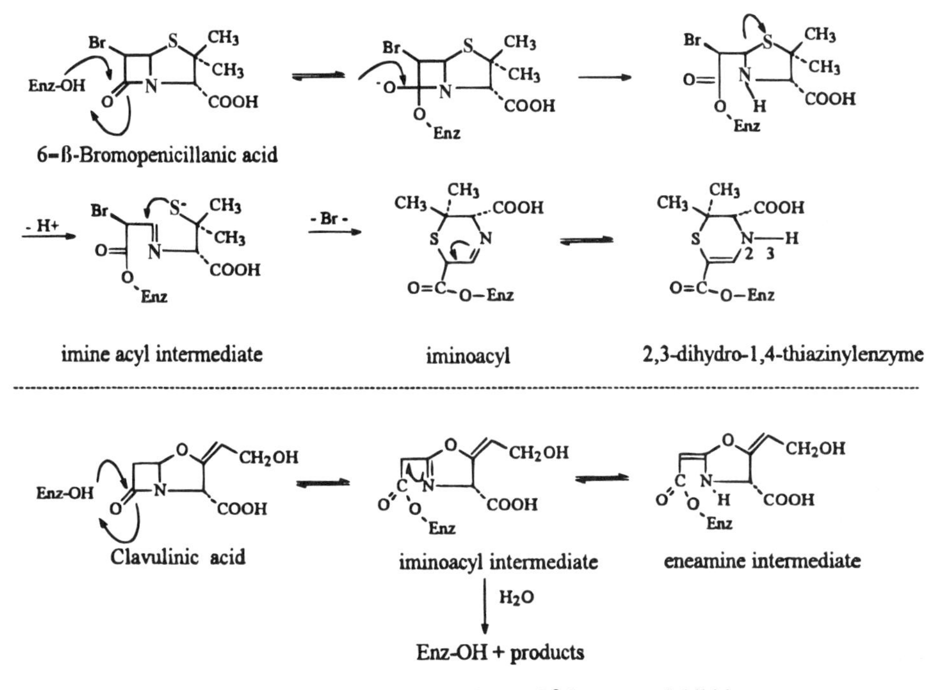

Figure 6-15. Possible mechanisms of β-lactamase inhibition.

what less stable product is slowly but ultimately hydrolyzed, restoring the enzyme (plus degradation products). With a 160-minute half-life (pH 7.3, 37°C) this reaction is sufficiently sluggish to produce enzyme inhibition with dramatic clinical results.

Clavulinic acid inhibits a wide range of β-lactamases, including the chromosomally (e.g., Class II and IV) and plasmid-mediated enzymes responsible for much of the Gm– bacterial resistance encountered. β-Lactamases of Gm+ organisms are also effectively inhibited. For example, a β-lactamase producing *S. aureus* that is totally resistant to ampicillin (MIC > 500 μg/ml) can produce an MIC of only 0.8 μg/ml when co-administered with 1 μg of clavulinic acid. This, and other impressive results, have now resulted in commercial products such as the mutual pro-drugs (Fig. 6-14) as well as physical mixtures of clavulinic acids with amoxicillin (Augmentin®) and ticarcillin (Timentin®). Another newly introduced inhibitor, 3-triazolomethylpenicillanic acid sulfone (Tazobactam) is available as a combination with piperacillin (Tazocillin, Zosyn).

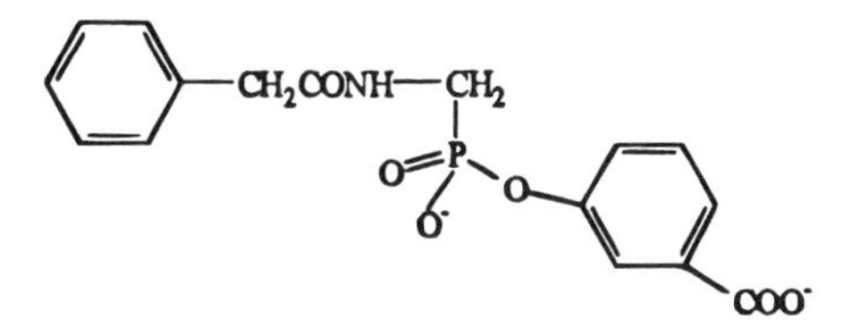

m-Carboxyphenyl phenylacetamidomethylphosphonate

A report describes a member of a new class of very effective non-β-lactam β-lactamase inhibitors. This inhibitor is a monoester of *m*-hydroxybenzoic acid and phenylacetamidomethylphosphonic acid. It appears to be specific against Class C enzymes (e.g., from *E. colaceae*). The compound showed little inhibition of β-lactamases of Class A and B categories. This phosphonate ester appears to act as a transition-state analog inhibitor (see PALA, Chapter 2). Incubation with penicillin G in the presence of this potential drug reduced enzyme activity to zero within 10 minutes.

6.11. Antibiotics Inhibiting Protein Synthesis

6.11.1. Protein Synthesis—An Overview

The antibiotics to be considered here inhibit bacterial protein synthesis as their primary mechanism.[24] As with cell wall synthesis we are dealing with a complex series of biochemical steps that are now fairly well understood. Again, there are drugs that interfere with protein construction at different sites in the sequence of events. In considering the molecular drama, it will be helpful to reacquaint ourselves with the main characters.

Deoxyribonucleic Acid (DNA). The structure of DNA is well known. Its function is to transmit genetic information. The characteristics of all cells are determined by their proteins, particularly the enzymes. The characteristics of a protein are in turn determined by its unique sequence of amino acid residues. DNA contains the master plan for protein formation in its base sequence. The particular sequence of the bases adenine (A), thymine (T), guanine (G), and cytosine (C) thus represent a code. Since all proteins consist of 20 different L-amino acids, the *genetic code* then directs the varying sequences specific for a given protein.

How is this accomplished? The DNA molecule is double stranded. One strand (chain) is complementary to the other; that is, the sequence of bases in one strand determines the sequence in the other. As a result, adenine is always opposite thymine—held in juxtaposition by hydrogen bonding—and cytosine is opposite guanine. The bases pair *only* in this manner.

The question then arises regarding the way in which four different bases (or nucleotides) determine the sequence (code) for 20 amino acids. A code based on two bases is not sufficient since such pairs yield only 16 possible combinations (4 × 4). A minimum of three bases is necessary. The *triplet* code give 64 possible combinations (4 × 4 × 4). As we shall see, each triplet represents a particular amino acid. Some amino acids are actually coded by only one triplet *codon*, whereas others are coded by several. The code is said to be degenerate (Table 6-7).

Messenger Ribonucleic Acid (mRNA). Since DNA is found almost exclusively in the nucleus while protein synthesis occurs in the cytoplasm of the cell, a method must exist by which coding information contained in DNA is transmitted to the site of active protein formation. This messenger, a ribonucleic acid (*m*RNA), is single stranded.

The synthesis of *m*RNA begins with the unraveling of the double-stranded DNA. One of the strands then serves as a template on which single-stranded *m*RNA is synthesized,

---

[24] We are concerned with agents that affect protein synthesis per se. Obviously the antimetabolites previously considered, by interfering with the synthesis of nucleic acids, indirectly and ultimately prevent protein formation.

Table 6-7. Nucleotide Code for Amino Acids

Amino acid	Coding triplets[a]
Alanine	GCU GCC GCA GCG
Arginine	CGU CGC CGA CGG AGA AGG
Asparagine	AAU AAC
Aspartic acid	GAU GAC
Cysteine	UGU UGC
Glutamic acid	GAA GAG
Glutamine	CAA CAG
Glycine	CGU CGC GGA GGG
Histidine	CAU CAC
Isoleucine	AAU AUC AUA
Leucine	UUA UUG CUU CUC CUA CUG
Lysine	AAA AAG
Methionine	AUG
Phenylalanine	UUU UUC
Proline	CCU CCC CCA CCG
Serine	UCU UCC UCA UCG AGU AGC
Threonine	ACU ACC ACA ACG
Tryptophan	UGG
Tyrosine	UAU UAC
Valine	GUU GUC GUA GUG

[a] There are also codons for protein synthesis initiation (AUG) and polypeptide chain termination (UAG, UGA, and UAA).

becoming complementary to it. The synthesis is catalyzed by RNA polymerase (transcriptase). Thus *m*RNA has the code *transcribed* onto it. The new molecule leaves the nucleus and travels to the site of protein formation after transcription, where in turn it acts as a template itself. The function of *m*RNA, then, is to convey the information of the amino acid sequence of the protein to be synthesized from the nuclear DNA to the sites of protein synthesis—the ribosome. There, the communication will be decoded.

Ribosomes. Ribosomes are the protein factories. They are particles consisting of proteins and another type of RNA called *ribosomal* RNA (*r*RNA). The fact that ribosomes are the loci of protein synthesis was only discovered in the early 1950s as a result of incorporation studies with radioactively tagged amino acids. Now, ribosomes are known to be large ribonucleoprotein particles on which the message carried by *m*RNA is *translated* into the protein's amino acid sequence.

A bacterial cell may contain about 10,000 such particles. In animal cells as many as 1–10 million ribosomes are found that are usually associated with the internal membranes of the endoplasmic reticulum. Ribosomes can be isolated from other cell components by subjecting tissue extracts in the ultracentrifuge to speeds equivalent to more than 100,000 times gravity. This treatment shows that ribosomes consist of subunits with different sedimentation values (S). Table 6-8 shows some typical values for ribosomes, their component subunits, and associated types of *r*RNA. The various subunits are associated with varying numbers of different proteins that are believed to be involved in certain aspects of protein synthesis. Bacterial ribosomes may contain up to 37% protein, while plant and animal ribosomes consist of approximately equal amounts of RNA and protein material.

The two subunits of a ribosome have different binding properties. For example, the 30S subunit of *E. coli* binds *m*RNA even in the absence of the 50S subunit. The 50S subunit will not bind with *m*RNA in the absence of the 30S subunit, but will bind to *t*RNA in

Table 6-8. Ribosomal Sedimentation Characteristics

Ribosomes	Subunits	*r*RNA	Mol wt.
Bacterial cells[a] 70S[b]	50S	5S	40,000
	30S	16S	550,000
		23S	1,100,000
Plant cells 80S	60S	5S	40,000
	40S	16S–18S	700,000
		25S	1,300,000
Animal cells 80S	60S	5S	40,000
	40S	18S	700,000
		28S–29S	1,800,000

[a] Ribosomes in mitochondria and chloroplasts are also 70S size.
[b] S = Svedberg unit, a measure of size.

a nonspecific manner. The resulting complex will bind specifically to a particular *t*RNA (see later).

Transfer RNA (tRNA). The function of *t*RNA is to *transfer* activated amino acids to the ribosome for assembly and incorporation into growing polypeptide chains. *t*RNA is the smallest of the three types of RNA (about 75 nucleotides). A given *t*RNA is a specific carrier for *only* 1 of the 20 amino acids. Before the amino acid can be bound to a *t*RNA, it must be first activated, or energized, by ATP, its own specific enzyme aminoacyl, *t*RNA synthetase, and Mn^{2+}:

$$\text{ATP + Amino Acid} \xrightarrow[\text{Mn}^{2+}]{t\text{RNA synthetase}} \text{Amino Acid-AMP-enzyme complex + Pyrophosphate} \qquad (6.9)$$

$$\text{Amino Acid-AMP-enzyme complex} \xrightleftharpoons[]{t\text{RNA} \;\rightarrow\; \text{AMP + Enzyme}} \text{Aminoacyl-}t\text{RNA} \qquad (6.10)$$

Before the activated amino acid is ready to be incorporated into the growing protein chain at a definite point, it must first find its specific position on the *m*RNA template. This requires prior bonding of the activated amino acid onto *t*RNA.

As shown in Figure 6-16, the activated amino acid is attached to the end of its specific *t*RNA molecule. The function of this aminoacyl *t*RNA molecule is to place the amino acid that it is carrying in the proper sequence position on the template. Triplets of nucleotides—called anticodons—in the *t*RNA molecule (Fig. 6-16) are attracted to the complementary codon of *m*RNA). The amino acid carried by its specific *t*RNA is thus brought to the correct position on the *m*RNA codon. Since the anticodon on the *t*RNA is identical to the codon on DNA (except thymine replaces uracil), the DNA "directs" the amino acids in the protein-forming chain. The process of protein biosynthesis can now be considered.

6.11.2. Initiation

The process of protein biosynthesis begins at the amino terminal and proceeds in sequence to the carboxy end of the finished protein. Initiation begins with the formation of a complex that consists of the 30S ribosomal subunit, *m*RNA (carrying the complete code for the particular protein as previously described), and three proteins designated as F1, F2, and F3. The *m*RNA–30S subunit complex, in the presence of initiation factors F1, F2, guano-

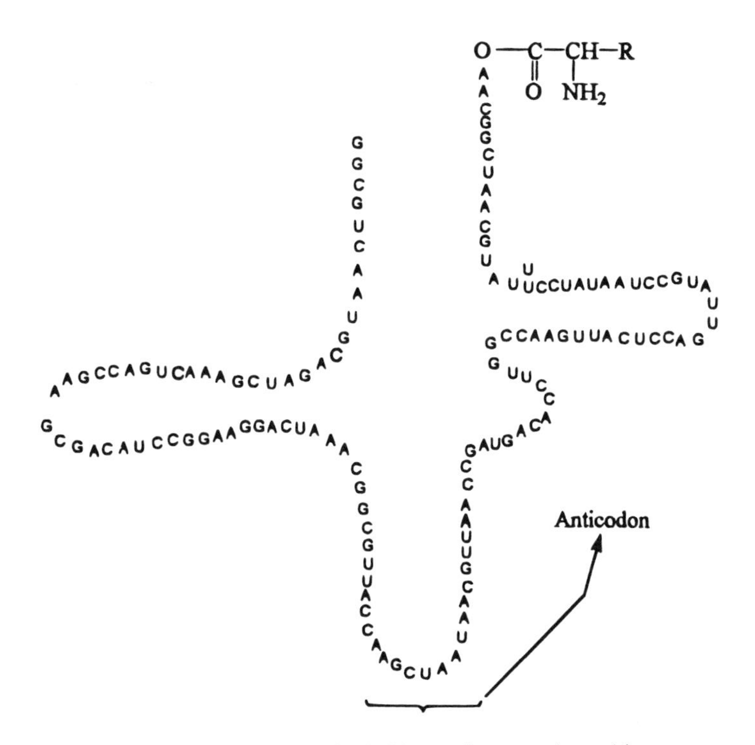

Figure 6-16. Transfer RNA carrying an amino acid.

sine triphosphate (GTP), and Mg^{2-}, binds to N-formylmethionyl *t*RNA[25] (a special *t*RNA for initiation) at the initiation codon, AUG, on *m*RNA. The anticodon triplet (Table 6-8) is put into the proper position relative to the starting signal (AUG) with the aid of the third protein initiation factor. The 50S subunit subsequently associates, forming the complete 70S ribosomal complex.[26] The initiation process is summarized in Figure 6-17. The donor site is occupied by formylmethionyl *t*RNA. The acceptor site at this point is ready to accept the first aminoacyl *t*RNA as directed by the codon adjacent to the initiator, AUG (Fig. 6-17).

6.11.3. Elongation

The elongation process of the peptide chain occurs in three stages:

1. Codon-directed binding of aminoacyl *t*RNA to acceptor site.
2. Peptidyl transfer from the peptidyl *t*RNA to newly bound aminoacyl *t*RNA on the acceptor site.

---

[25] In bacteria the first N-terminal amino acid for all proteins is formylmethionine. This special amino acid is later removed from the amino terminal of the completed protein.

[26] It is of interest to point out that, even though ordinary methionyl *t*RNA and formylmethionyl *t*RNA have the same anticodon, only the latter is effective in bacteria. However, in higher plant and animal cells the unformylated *t*RNA acts as the specific initiator.

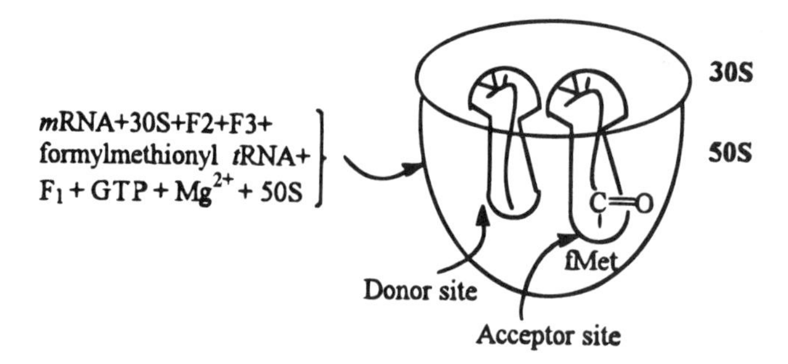

Figure 6-17. Initiation step of bacterial protein synthesis. (30S and 50S are ribosomal subunits; fMet is the formylmethionyl group.)

3. Translocation of synthesized peptidyl *t*RNA and *m*RNA from acceptor to donor site so that the acceptor site can again accept the next incoming aminoacyl *t*RNA that contains the appropriate triplet codon. The translocation of the peptidyl *t*RNA on the 50S subunit involves at least two other protein factors, Tu and Ts, in addition to the peptidyl transferase enzyme.

After the attachment of the initiation unit, formylmethionyl *t*RNA, the next aminoacyl *t*RNA is inserted into the ribosomal acceptor site with the aid of a transferase enzyme. The α-amino group of this aminoacyl *t*RNA rapidly reacts with the activated carboxyl terminal residue of the initiator, or peptidyl *t*RNA, component. The new peptide bond is thus formed. The *t*RNAs must have their anticodons properly aligned opposite their respective triplets (i.e., codon) on the *m*RNA. The amino acids attached to them are then adjacent to each other on the 50S ribosomal subunit. At this point the peptide synthetase enzyme can catalyze the amide bond formation.

6.11.4. Termination

The synthesis of a protein, of course, does not continue ad infinitum, but it must somehow "know" when completion of the polypeptide chain has been achieved. Furthermore, it must be able to leave the manufacturing site and travel to the rest of the cell to perform its proper functions in the organism. The first step of the termination process involves cognizance of a terminal signal in the *m*RNA. This occurs when a *t*RNA with one of the corresponding anticodons binds to the acceptor site. At this point three protein factors are required. The factors R2 and S are involved in the recognition process for terminator codons. The release of the finished product involves a hydrolysis reaction. First, a peptidyl transfer reaction occurs in which the last peptidyl *t*RNA is transferred onto the hydroxyl group of water instead of onto a new incoming aminoacyl *t*RNA (as during elongation). A release factor, TR, then cleaves the residual *t*RNA from the donor site. Finally, the full 70S ribosome separates from the *m*RNA and dissociates into its component 30S and 50S subunits. These components are free to enter the cycle again. At this point the formylmethionyl residue (the initiator), which frees the NH_2 of the finished product, must be removed. The removal is probably effected enzymatically in two steps: removal of the formyl group followed by cleavage of the amino acid methionine. Figure 6-18 schematically summarizes the process of protein synthesis.

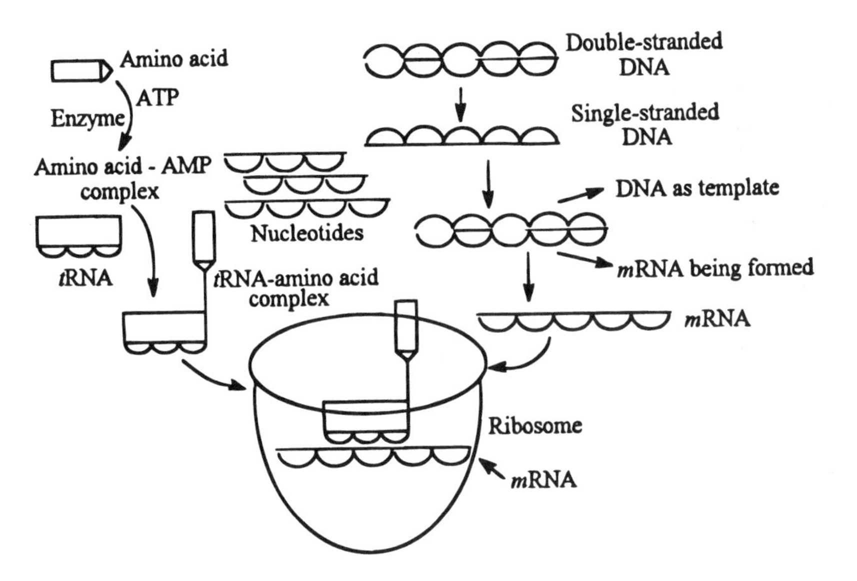

Figure 6-18. Protein synthesis.

A protein will have two alternatives after its synthesis is completed, depending on its function. It may remain within the cell in which it was formed, which is the case for most enzymes since they are needed for the cell's own biochemical function. If the protein is a hormone (e.g., insulin) or an enzyme with a specific purpose outside the cell of origin (e.g., digestive enzymes such as pepsin), it will leave the cell through the membrane. Proteins are continually degraded or "worn out," as well as synthesized, and must therefore be continually replaced by new protein production. The way in which the cell knows when to begin protein synthesis and how many proteins to make is still one of the major biological puzzles. The mechanism for the control of protein synthesis is only beginning to be understood. Such control may be exercised at the level of transcription (DNA → *m*RNA) or translation (*m*RNA → protein). Genes that function to repress the synthesis of messenger molecules are currently being studied.

Just as we have seen that cell wall synthesis inhibition as an antibacterial mechanism can be achieved by various drugs at different points in the complex sequence of biochemical events known as cell wall synthesis, so do different drugs exist that inhibit protein synthesis at various stages. In general, one of three areas appear to be involved. The first area involves aminoacyl-*t*RNA formation. Inhibition of this step has been demonstrated by various amino acid analogs that esterify to the *t*RNA terminus "meant for" the normal amino acids. Examples of such analogs are norleucine, N-ethyl-glycine, and an unsaturated version of proline. These imposter amino acids then continue into the synthesis and incorporate into counterfeit proteins. Several antibiotics that compete for incorporation into *t*RNAs are known. None of the compounds in this category, however, has sufficient species specificity (i.e., selectively toxic) to be of clinical interest.

Inhibitors of initiation–complex formation and *t*RNA-ribosome interactions represent important groups of therapeutic antibiotics, specifically the tetracyclines and aminoglycosides. The third area of protein synthesis that is significantly affected by drugs is peptide bond formation and translocation. Here, chemically diverse antibiotics such as chloramphenicol, lincomycin, and erythromycin are found.

6.11.5. The Tetracyclines

The tetracycline antibiotics, a system of four fused rings, may be considered to be the derivatives of octahydronaphthacene. They were obtained from several *Streptomyces* species in the late 1940s. Duggar (1948) first discovered chlortetracycline (CTC) in 1947 from *S. aureofaciens*. Oxytetracycline (OTC, 5-hydroxytetracycline) soon followed from *S. rimosus*. Tetracycline (TC), the chemical parent of the series, has been obtained by hydrogenolysis of CTC or by fermentation in a low-chloride environment. Table 6-9 gives the structures of these and additional chemically modified (semisynthetic) tetracyclines. These compounds were the first low-toxicity broad-spectrum antibiotics introduced into clinical practice. The spectrum includes Gm+ and Gm– bacteria, mycoplasma, chlamydia, rickettsia, and even some protozoa causing malaria. With the advent of highly specific β-lactam antibiotics and increasing resistance, tetracyclines are not the first-line drugs they once were. They remain important, however, for particular indications. These include respiratory, rickettsial, some Gm–, and certain mixed infections. Acne Vulgaris also responds well.

Early explanations of the mechanism of action of tetracyclines involved their demonstrated ability to inhibit various bacterial enzymes that catalyzed such biochemically essential reactions as glucose oxidation and oxidative phosphorylations. Their ability effectively to chelate di- and trivalent metallic ions was also invoked in the theorization. However, at clinically achievable serum levels the bacteriostatic effects of tetracyclines are now accepted to involve primarily direct inhibition of protein synthesis.

In a cell-free system inhibition can be shown to occur on both 70S and 80S ribosomes. However, in a more realistic in vitro setting intact prokaryotic (i.e., bacterial) cells are much more sensitive. The reason for this selectivity is that tetracyclines are actively transported into bacterial but not mammalian cells. In Gm– bacteria, at least, the more water-soluble compounds seem to cross through membrane channels (pores). The more lipid-soluble drugs (particularly MNC, Table 6-9) diffuse more readily through the lipoidal phases of the membranes. This energy-coupled process then leads to intracellular antibiotic accumulations.

Table 6-9. Tetracyclines

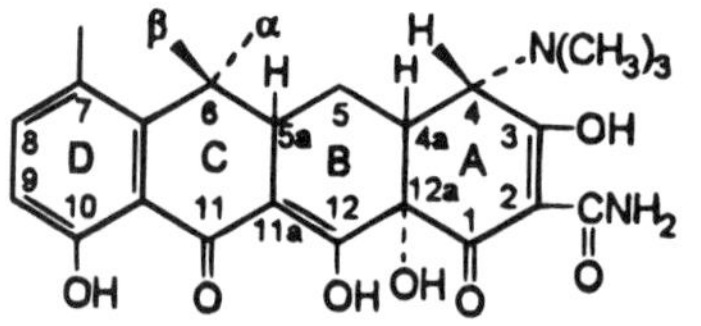

β α H H H $N(CH_3)_2$ 7 6 5 4 3 8 D C 5a B 4a A OH 9 10 11 11a 12 12a 2 1 CNH_2 OH O OH OH O O

Drug	Trade name[a]	7	6-α	6-β	5
Chlortetracycline (CTC)	Aueromycin	Cl	CH_3	OH	H
Oxytetracycline (OTC)	Terramycin	H	CH_3	OH	OH
Tetracycline (TC)	Achromcyin	H	CH_3	OH	H
Demeclocycline (DMCTC)	Declomycin	Cl	H	OH	H
Doxycycline (DC)	Vibramycin	H	CH_3	H	OH
Methacycline (MC)	Rondomycin	H	CH_2	CH_2	OH
Minocycline (MNC)	Minocin	$N(CH_3)_2$	H	H	H

[a] Only one representative (U.S.) trade name is given.

Protein synthesis inhibition results from reversible binding of the tetracycline drugs to the 30S ribosomal subunit. This in turn prevents the attachment of aminoacyl-*t*RNAs to the acceptor site of the ribosomal structure (Fig. 6-17). The ribosomal *m*RNA complex is thus effectively precluded from initiating protein synthesis. Tetracyclines do not interfere with actual peptide bond formation, nor with the translocation process. In vitro studies with 70S ribosomes using photoaffinity techniques demonstrate binding to the 4S and 18S protein components of the 30S subunits. It is likely that these are the actual binding sites.

Resistance to tetracyclines appears to be mainly as a result of reduced bacterial membrane permeability. In these organisms, both Gm+ and Gm– drug transport is blocked. This blockade is determined by extrachromosomal plasmids, which can be induced. In fact, exposing bacteria to subinhibitory tetracycline levels will increase the resistance. In addition, it was shown with *E. coli* that resistance may also, in part, be a result of increased efflux of tetracycline antibiotics out of the cell by an energy, using system within the cell. Finally, bacteria possessing drug-resistant ribosomes in which ribosomal proteins of the 30S subunits have mutated also exist. *B. subtilis* is such an organism.

6.11.5.1. Structure–Activity Relationships

Structural modification of the tetracyclines obtained by fermentation has resulted in very few additional clinically useful semisynthetic drugs. However, those developed (DC, MC, MNC) have some advantageous features over the original compounds. Also, from data of both in vitro and in vivo studies on many bacteria using a large number of tetracyclines obtained by fermentation and synthetic alterations, it became apparent what the structural requirements for good activity are. Table 6-10 summarizes the salient features that

Table 6-10. Summary of Empirical Tetracycline SARs[a]

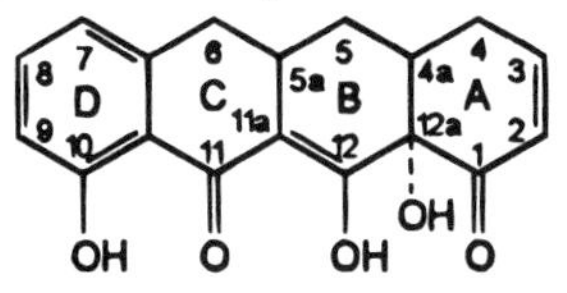

C#	Structural modifications[b]	Effects
1.	Any modification	No antibacterial activity
2.	Acetyl group only	Slight activity retained
3.	Any modification	No antibacterial activity
4.	α-Dimethylamino essential	$NHCH_3$ retains most activity
5.	OH, $CH_3C(=O)$-, RC(=O)-O-, HC(=O)-O-	All retain activity
5a.	Loss of H (together with C^6-OH)	Inactive degradation product
6.	Remove OH, CH_3 or both	Active, more stable drugs
7.	Cl, Br, NO_2, $(CH_3)_2N$-	Activity retained
8.	Little information available	—
9.	Cl and CH_3	Decreased activity (compared to 7 substitution)
10, 11, 11a, 12.	"Inviolate zone" including C-1	Any change lowers or eliminates all activity
12a.	Epimerizing a-OH or deoxy	Decreased activity

[a] Also refer to structure in Fig. 6-23.
[b] Relative to tetracycline.

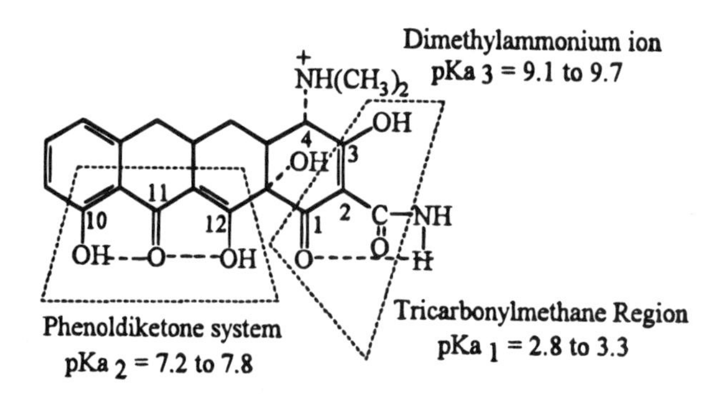

Figure 6-19. Tetracycline "activity" regions.

enhance, reduce, or eliminate activity.[27] The chemistry around C-4 is particularly interesting. The most "naked" tetracycline exhibiting antimicrobial activity is 6-demethyl-6-deoxy-4-dedimethylamino-tetracycline. This compound, which has no nitrogen moiety at C-4, has some *in vitro* activity against several Gm+ organisms. In vivo activity is absent. Addition of a dimethylamino group at C-4 (6-demethyl-6-deoxytetracycline) is the simplest tetracycline having in vivo activity. Also of particular interest from an SAR standpoint is the structural integrity of the "inviolate zone" encompassing C10, C11, C12, and C1. The oxygen functions in this region generate two distinct ultraviolet absorbing chromophoric groups. One group of the π-electron region is the tricarbonylmethane moiety comprising C1, C2, and C3 (Fig. 6-19); the other is the phenoldiketone system consisting of C10, C11, and C12. The interposition of the 12a-hydroxyl function ensures the electron separation of these two areas. Modification of these chromophores by aromatization of ring C, cleavage of a ring, extension, or blocking of the chromophore all lead to essentially inactive compounds.

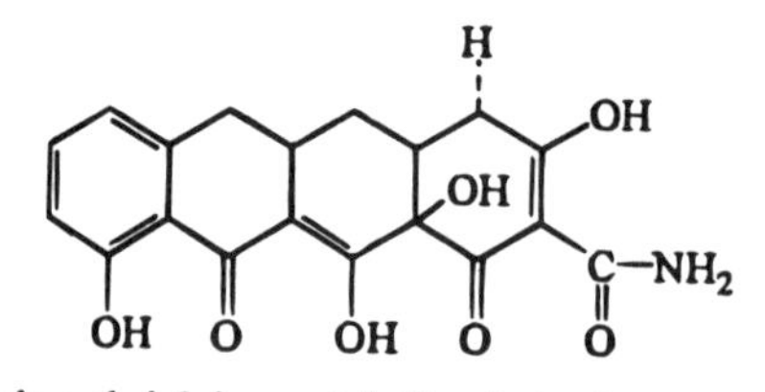

6-demethyl-6-deoxy-4-dedimethylaminotetracycline

Aromatization of ring C is actually a chemical instability occurring as a dehydration reaction under acidic conditions and, of course, accelerated by elevated temperatures (Fig. 6-20). The elements of H_2O derive from the 6-OH and 5a-H atom affording a 5a–6 DB. The 11a–12 DB spontaneously isomerizes to the 11–11a position yielding anhydrotetracycline (ATC). This inactive degradation product is therefore formed in aging TC-containing products, and its rate of formation is increased if improperly stored (e.g.,

[27] An extensive discussion will be found in Blackwood and English (1977).

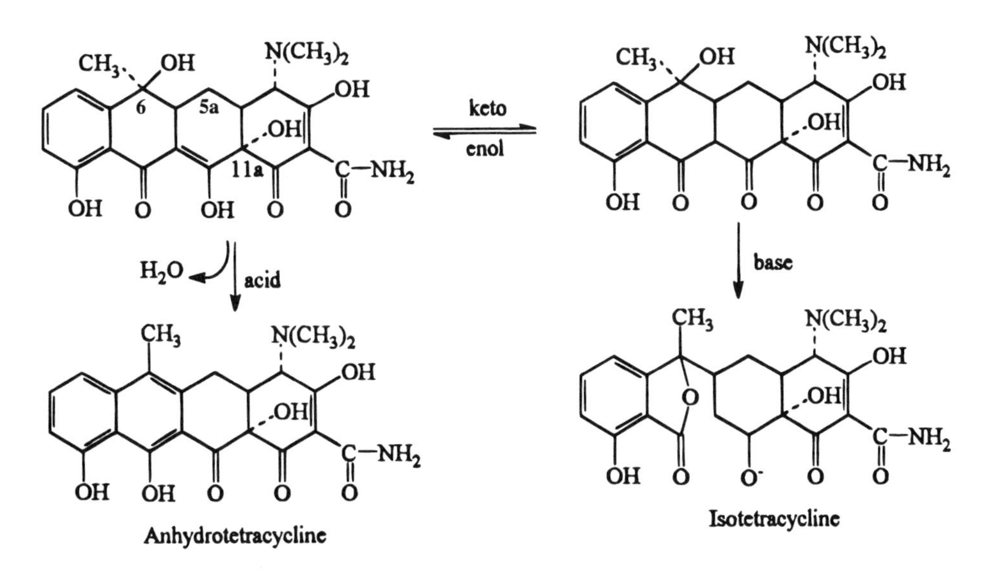

Figure 6-20. Chemical instabilities of tetracycline.

elevated temperatures, increased acidity). ATC[28] has been shown to be nephrotoxic. Another degradation reaction is ring cleavage resulting in isotetracycline (ITC). Following tautomerization of TC to the 12-keto form, base catalyzes β-ketone cleavage of the 11–11a bond followed by lactonization of the 6-OH with the 11-keto function (Fig. 6-20). It will be noted that both these degradation reactions involve the 6-OH group. The semisynthetic tetracyclines, MNC, MC, and DC (Table 6-9), are therefore not subject to these instabilities.

Elimination of the 12a-OH function would extend the chromophore (by producing an additional DB at 12a-1) and eliminate activity. Substituting a halogen into position 11a would constitute a chromophoric block. Both of the preceding actions would predictably result in considerably decreased activity.

Another instability leading to a dramatic decrease of antibacterial action, to which all clinically used tetracyclines are subject, is epimerization of the "natural" 4-α-dimethylamino group A to the β-epimer B (Eq. 6.11). Under acidic conditions a 1:2 equilibrium is established in solution within a day. This occurs in a variety of solvents, especially acetic acid. Anions also tend to support this process. Divalent ions that chelate tetracyclines, particularly Ca^{2+}, facilitate the reversal of the epimerization from the epi to the natural isomer.

Even though any modification of C-2 leads to inferior activity, or none at all, the pyrrolidinomethyl derivative of the 2-carboxamide (rolitetracycline) appears to be an exception. However, on closer scrutiny this Mannich reaction product (Eq. 6.12) will be seen as a prodrug of TC. The compound is 2,500 times more water soluble than the parent TC compound (comparing free bases), to which it hydrolyzes in vivo. Thus its activity is essentially that of the parent compound.

---

[28] Actually anhydro-4-epitetracycline.

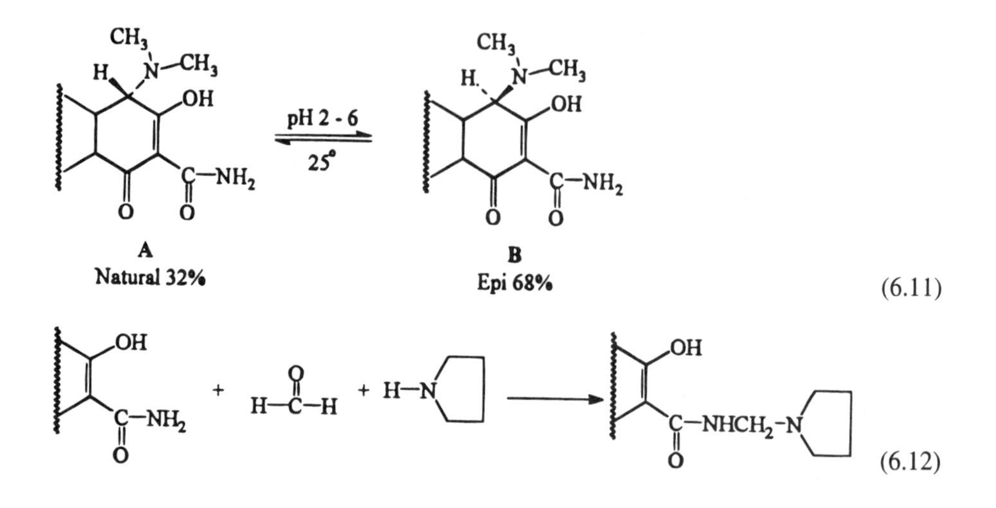

6.11.5.2. Relationships of Chemistry to Pharmaceutics

Tetracyclines are amphoteric compounds that can exist as acid or base salts as well as in the zwitterionic state, depending on the pH of the environment. They may be viewed as tribasic acids, with the 4-dimethylammonium ion (a conjugate acid of the dimethylamino group) being the weakest (pKa_3 9.1–9.7). The phenoldiketone system accounts for the pKa_2 range of 7.2–7.8 and the tricarbonylmethane moiety generates the strongest acidity (pKa_1 2.8–3.3). Figure 6-19 summarizes this information. Minocycline, which has an additional diethylamino group at C-7, generates a fourth pKa value, namely 5.0.[29]

The pH-partition hypothesis relating to membrane penetrability was outlined in Chapter 1. This hypothesis would allow one to make reasonable assumptions of a drug's behavior whose pKa and aqueous–lipid partition were determined. The amphoteric nature of tetracyclines with their multiple pKas, however, presents a more complex situation than simply improving lipid solubility by the deletion of polar hydroxyl groups from positions such as 5 and 6 (e.g., DC, MC, MNC) or the addition of a halogen on C-7 (e.g., DMCTC, CTC).

The early tetracyclines, particularly TC, exhibited erratic oral absorption. In some instances this was later related to dosage formulations where purportedly "inert" fillers were found to contain divalent cations such as Mg^{2+} and Ca^{2+}, which are now known to chelate the drug and actually decrease solubility; so would administration of calcium-containing products (milk, antacids) or iron preparations. Foods can generally interfere with the efficient absorption of the earlier tetracyclines. This effect appears to be much less significant with the newer agents such as DC and MNC. Even when taken on an empty stomach, TC and OTC have a much slower absorption *rate*, with blood levels not reaching maxima for 4 hours, whereas DC and MNC attain it within 2 hours.

Studies relating gastric and duodenal absorption to pH-partition behavior and the various ionizations or pKas have given some insight to the puzzle. Like all amphoteric substances, tetracyclines will be ionized over the whole pH range: cationic at very acidic pH values, anionic in the more alkaline range, and zwitterionic in the neutral range, which

---

[29] This is similar to N,N-dimethyl-aniline (pKa 4.6), as might be expected.

would necessarily include the isoelectric point. Tetracycline was predictably too polar to pass through a totally lipid phase such as cyclohexane in a partition study. However, biological membranes, which are lipoprotein in nature, are more polar and are therefore more closely simulated in partition studies by 1-octanol. Here TC partitioned well (i.e., transferred) in a 5.2–7.4 pH range, but not at the low pH encountered in the stomach, where little absorption occurs. Utilizing *n*-octanol and phosphate buffers, it was demonstrated that the highest concentration of zwitterion[30] exists in the 4–7 pH range, where it was shown that the most lipid-soluble species existed. Since this partially corresponds to duodenal pH values from which the bulk of the TC dose is absorbed, it is possible to suggest that passive in vivo absorption occurs. The fact that structural modifications that would be expected to increase lipoidal solubilities of TC (e.g., removal of 6-OH, 5-OH, addition of 7-Cl) are also much better and more rapidly absorbed (MNC, MC) lends credence to this concept. It was further postulated that the possibility of *intramolecular ion-pair* formation between the adjacent ammonium cation and the tricarbonylmethane ion thus resulted in a relatively lipid-soluble species. Unfortunately, a later X-ray crystallographic study showed that such an interaction between the two oppositely charged ionic regions is not likely. An alternate suggestion was that a neutral totally uncharged species is the one transversing the membrane, even though it is postulated to exist only as a tiny fraction of all the possible tetracycline forms (Purich et al., 1973). However, although verifying through partition studies with simpler ampholytes such as ampicillin and L-phenylalanine that neutral forms at the isoelectric point are a tiny fraction of the molecular species,[31] they have nevertheless theoretically calculated from a "tautomeric equilibrium constant" that such an un-ionized form may be 41% of the total in the case of tetracycline.

Whatever the actual situation is, the relative concentrations of the ionic (and possibly nonionic) forms that exist at physiological pH have a definite bearing on the clinical effectiveness of the tetracyclines. The contribution to lipid solubility that structural changes effect may be gleaned by considering the absence of the 6-OH group in doxycycline and minocycline, whose oral absorptions approach 100%. MNC also achieves the best tissue and organ penetration (including the brain) of all the tetracyclines. This fact, when considered with the stability toward acids and alkali degradation, may well account for the lower doses, longer duration of action, and overall better therapeutic response of these semisynthetic antibiotics. A case may also be made for the improved antimicrobial spectrum seen.

6.11.6. The Aminoglycosides

This antibiotic group of drugs should more correctly be termed *aminoglycosidic-aminocyclitols*, since all these compounds consist of an *aminocyclitol* ring that is glycosidically linked to several cyclic amino sugars. However, the name *aminoglycosides* has been so firmly established by usage that it predominates over the more accurate aminocyclitol and will be used here as well.

As a class these drugs are very basic, forming highly water-soluble stable salts; they are usually marketed as sulfates. Aminoglycosides, which are polycationic in the pH of the

---

[30] Representing an ionized tricarbonylmethane region, an un-ionized phenolic diketone and a cationic dimethylammonium function.

[31] Which is in agreement with some standard textbooks.

small intestines, are not significantly absorbed from the gut and must therefore be administered parenterally. They penetrate cells and the blood–brain barrier poorly. They do, however, penetrate bacterial membranes by a complex series of biochemical events that only now are beginning to be understood. Selective toxicity may be partially due to these factors.

The aminoglycosides first penetrate into the periplasmic area by diffusing through the outer membrane's porin-lined aqueous channels. To enter the bacterial cytoplasm the drug transfers across the inner membrane of Gm– organisms by utilizing the energy-dependent electron transport associated with oxidative phosphorylation. As injury to the plasma membrane develops, the entry rate actually increases. Once in the cytoplasm the drug can interact with the ribosomes to inhibit protein synthesis (see later).

The aminoglycosides are among the oldest antibiotics; the first one, *streptomycin* (SM), was discovered by Waksman (1944). Hundreds are known; about a dozen are in clinical use worldwide. The majority have been obtained as fermentation products from the actinomycetes genus *Streptomyces* (the suffix of the generic names are spelled *-mycins*). Included in this group are SM, neomycin (NM), paromomycin (PM), kanamycin (KM), and tobramycin (TM). A smaller but equally important group are obtained from *Micromonospora* species (spelled *-micins*). The gentamicins (GM) and sisomicins (SSM) are in this category.

As a group aminoglycosides may be viewed as broad-spectrum drugs against aerobic Gm– and Gm+ bacteria. Those effective against *Pseudomonas* are particularly important clinically. SM's usefulness has endured primarily because of activity against the tuberculosis-causing bacillus. Aminoglycosides are ineffective against anaerobes, rickettsiae, fungi, or viruses. With one exception they are bacteriocidal. Paromomycin also affects *Entameba histolytica*, a protozoan.

6.11.6.1. Chemical Aspects

Until the advent of high-resolution mass spectrometry, NMR spectroscopy (C-13), and high-pressure liquid chromatography, characterization of amino glycosides was very difficult. Most could only be obtained in amorphous form. Their melting points were not useful. Ultraviolet and infrared spectra were not particularly characteristic. Optical rotations of structurally related compounds were similar. Paper and thin-layer chromatography were about the only useful tools. Structure elucidation was primarily by acid hydrolysis to establish the identity of the amino sugars and aminocyclitol. Careful oxidation degradations helped establish their bonding points. More recently, however, the structures of several new gentamicins have been totally determined by spectroscopy.

A chemical method of categorizing the aminoglycosides is on the basis of the type of aminocyclitol substitution patterns. There are two aminocyclitols encountered: streptidine, a diguanidine derivative of a 1,3-diaminohexose (streptamine), and 2-deoxystreptamine (2-DOS). The streptidine-derived group contains streptomycin (obtained from *S. griseus*) and the now-obsolete dihydrostreptomycin (*S. humidus*). The 2-deoxystreptamine–derived drugs can be divided into two groups: one where substitution exists on the two geminal hydroxyls (4,5), the other on the two nonadjacent OH groups (4,6).[32] Figures 6-21 and 6-22 outline the structural relationships of the clinically used (U.S.) drugs.

[32] Monosubstituted deoxystreptamine compounds also exist.

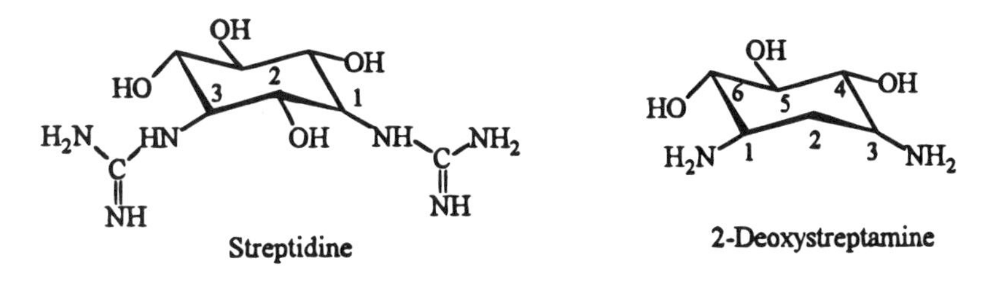

The only drug of interest in the streptidine-derived group is SM. Curiously, *S. griseus* is sensitive to the drug. It must be presumed that resistant cultures are producing the agent. Dihydrostreptomycin, which is produced by *S. humidus* or hydrogenation of the C-3′ aldehyde on the streptose component of SM (Fig. 6-21), has comparable activity and better stability. However, its significantly higher ototoxicity (even irreversible deafness) than the other aminoglycosides makes its use unjustifiable. Ototoxicity, a feature shared by *all* aminoglycosides, as well as potentially serious nephrotoxicity (tubular cell damage and glomerular dysfunction), have tended to inhibit the use of these drugs for uncomplicated infections where less toxic antibiotics are effective. A partial explanation for these adverse effects may lie in some unusual pharmacokinetics. For most aminoglycosides, the average $t\frac{1}{2}$ is 2–3 hours. However, a slower phase, due to gradual release of drug from various tissue-binding sites, also exists. Thus, half-lives in the inner ear can exceed plasma fourfold; in renal sites they may reach hundreds of hours. In addition, the toxicities encountered therefore should not be unexpected when considering the prolonged half-life in cases of renal failure. One other unique side effect should be mentioned: neuromuscular blockade. Although now rarely encountered during therapy when doses may be high, it can occur when patients may also receive nondepolarizing neuromuscular blocking agents (Chapter 8) during surgery. A paralytic condition may result. Otherwise, patients with myasthenia gravis may be particularly at risk.

Although SM is active against many Gm– and Gm+ organisms, it has limited use today. Its main attribute (in conjunction with other agents) is against tuberculosis and diseases

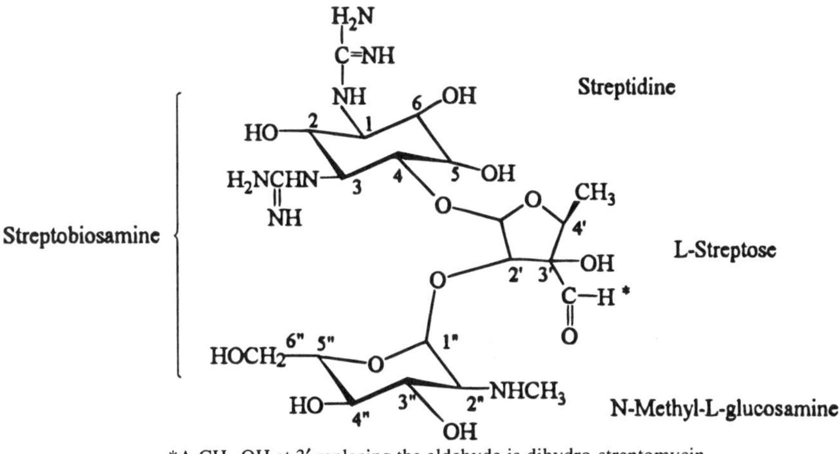

*A CH_2-OH at 3′ replacing the aldehyde is dihydro-streptomycin.

Figure 6-21. Aminoglycosides derived from streptidine-streptomycin (SM).

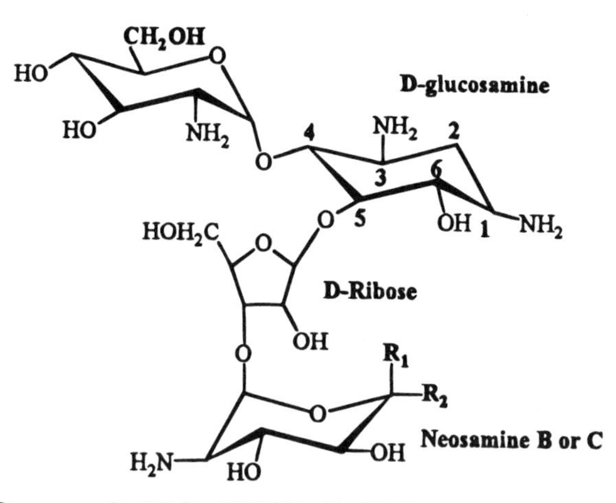

Paromomycin: (**I**, R_1=CH_2NH_2 ,R_2=H; **II**, R_1=H, R_2= CH_2NH_2)

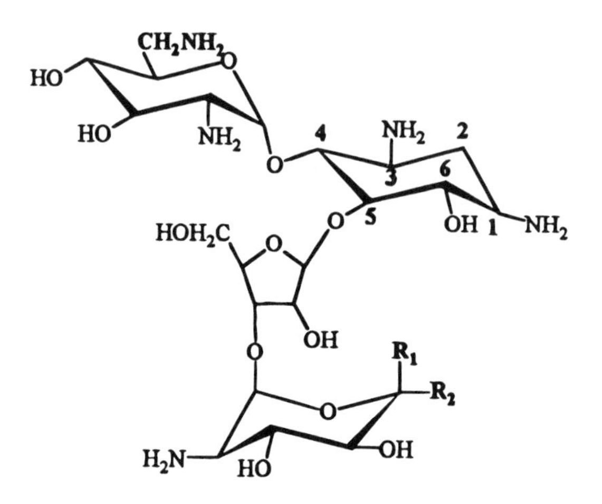

Neomycin: (**B**, R_1= CH_2NH_2, R_2=H; **C**, R_1=H, R_2 = CH_2NH_2)

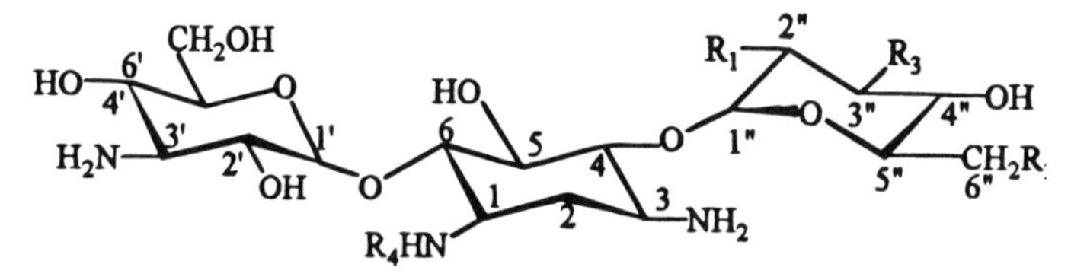

Kanamycin A: R_1= OH, R_2= NH_2, R_3= OH, R_4= H
Kanamycin B: R_1=NH_2, R_2=NH_2, R_3=OH, R_4=H
Kanamycin C: R_1= NH_2, R_2= OH R_3= OH, R_4= H
Amikacin: R_1= OH, R_2= NH_2, R_3= OH, R_4= $C(=O)-CH(OH)CH_2CH_2NH_2$
Tobramycin: R_1= NH_2, R_2= NH, R_3= H, R_4=H

Figure 6-22 (A). Aminoglycosides derived from 2-deoxystreptamine.

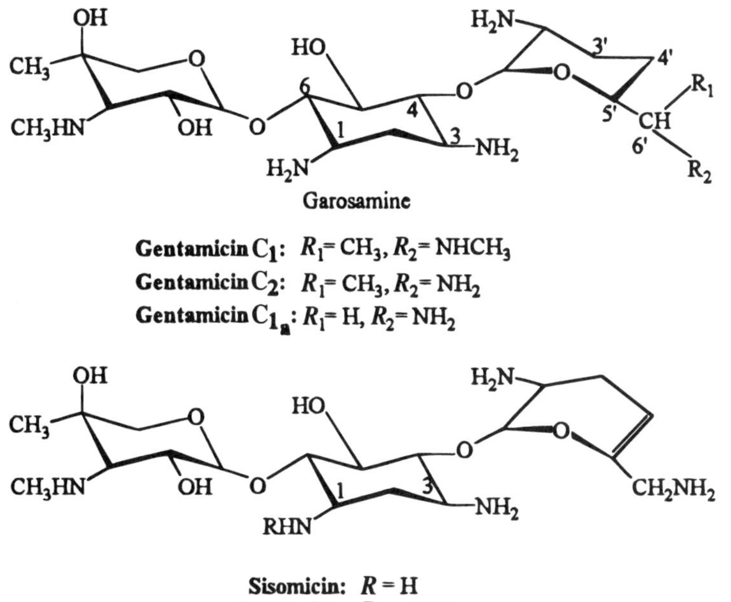

Figure 6-22 (B). *(Continued)*

such as bubonic plague, tularemia, and brucellosis, the latter in combination with tetracycline. Combinations with penicillins have been useful synergistically against certain Gm+ and entero streptococci.

PM (a mixture of PM I and PM II, Fig. 6-22) is obtained from *S. rimosus var. paromomycinus.*[33] The drug is active against clinically important *Shigella, Salmonella*, and *E. coli*, and has been useful against dysentery and gastroenteritis caused by these organisms since no intestinal absorption occurs. However, rapid resistance and relapse has led to the use of other drugs. Its principal use in the United States is treatment of intestinal (but not hepatic) amebiasis and tapeworm infestation in a secondary capacity (Chapter 7).

The closely related neomycin, obtained from *S. fradiae*, as a mixture of epimers NM B and C (Fig. 6-22), also exhibited a broad antibacterial spectrum. However, high nephrotoxicity precludes systemic use. Oral administration for "gut sterilization" preceding bowel surgery is actually a local application, as is, of course, the use of NM in a number of topical creams and lotions, usually in combination with bacitracin and polymyxin.

The kanamycins are 4,6-disubstituted 2-DOS derivatives obtained from *S. kanamyceticus*. The United States product is primarily KM-A; KM-B is also available in Japan (aminodeoxykanamycin). KM was first isolated in 1957. Kanamycin, although effective against both Gm– and Gm+ bacteria as well as against mycobacteria, is now commonly resistant, especially to important Gm– organisms such as *Pseudomonas* and *Serratia*. Its use has

---

[33] Presumably a variety different from the source of oxytetracycline.

been superseded by TM and GM. TM, which is produced by *S. tenebrarius* is the C-3″-deoxy derivative of KM-B, active against many serious Gm– bacteria, particularly *P. aeruginosa, Serratia,* and *Eneterobacteria.*

Two *Micromonospora* species—*purpurea* and *echinospora*—produce the gentamicin C complex, a roughly equal mixture of three closely related antibiotics designated as GM C_1, C_{1a}, and C_2. Its broad spectrum is similar to the other aminoglycosides; of particular interest is its high activity against bacteria such as *P. aeruginosa, Enterobacter, Klebsiella, Serratia, E. coli,* and *Proteus*. The drug is also synergistic with penicillins.

Sisomicin, which is a GM analog (Fig. 6-22), differs in having an unusual unsaturated amino sugar, is obtained from *M. invoensis* and is similar to GM in pharmacology and spectrum. It may not offer any overall advantage, being inactivated by the same enzymes. The semisynthetic C-1-N-ethyl-sisomicin homolog *netilimycin,* however, is resistant to these enzymes, thus being a drug effective against some GM– resistant *E. coli, Klebsiella,* and others.

6.11.6.2. Mechanism of Action

The mechanism of action of SM has been investigated longer, and in greater detail, than the other aminoglycosides. The drug's effects are complex and include alterations of membrane permeability and therefore drug transport and inhibition of protein synthesis. Even misreading of the genetic code by drug-affected ribosomes has been demonstrated in vitro and in vivo. An interesting observed fact is that certain bacteria are actually dependent on SM for growth. One study (1967) described this as an in vivo restriction of the genetic code translation process in a mutant strain that is corrected by the drug.

The single binding site for SM on the 30S ribosomal subunit is the 23S core particle of that subunit. A variety of reactions can follow from this event. One is the induction of conformational change in the complete 70S ribosome. Polyribosomes may break down, and the resulting monosomes could reassociate with *m*RNA, resulting in abnormal initiation complexes, thus inhibiting normal synthesis.

Other work indicated a dual effect by SM on protein synthesis: a rapid initial effect on the elongation of the polypeptide chain and, subsequently, chain initiation blockade, as evidenced by the fact that initiation sites on *m*RNA associate with ribosomes in the presence of SM.

The mechanism of protein synthesis inhibition exhibited by aminoglycosides other than SM differ in particular details. It was shown that KM, NM, and GM have multiple ribosomal binding sites rather than the single site to which SM binds. Translocation on bacterial ribosomes is also inhibited by these agents.

The aminoglycosides are bacteriocidal. Other antibiotics whose mechanism of action involves inhibition of protein synthesis (tetracycline, the macrolides, lincomycin, etc.) are invariably bacteriostatic. The reason for this difference is not known. In fact, the reason that protein inhibition by aminoglycosides should be a cell-killing process has not been satisfactorily addressed. The accumulation of "nonsense proteins" due to misreading of *m*RNA has been shown not to be the reason. If ribosomal binding were an irreversible process, lethality might be comprehensible; SM does not bind irreversibly.

Another point that has defied explanation is this: SM is bacteriocidal only during periods of high rates of protein synthesis. Thus other protein inhibitors such as tetracyclines or chloramphenicol should interfere with its action, yet this theoretical drug interaction is not necessarily clinically contraindicated. In fact, a combination of streptomycin and tetracycline is the treatment of choice for glanders and brucellosis.

6.11.6.3. Resistance

The development of bacterial resistance to the action of antimicrobial agents seems to be universal. The biochemical basis of this phenomenon with the aminoglycosides has been related to particular structural features of these drugs. Discovering active compounds in which one or another of the features is absent, or synthetically modifying them, has given us a better understanding of aminoglycoside SARs and several semisynthetic agents with superior qualities.

There are potentially three resistance mechanisms in bacteria operative against the aminoglycosides. One is mutation of the organism resulting in altered ribosomes no longer capable of binding to the antibiotic. Such a spontaneous mutation against SM was demonstrated in a specific ribosomal protein—1 of 21 in the 30S subunit. Such mutations are rarely encountered in clinical situations and tend to leave the organism sensitive to other aminoglycosides.

A second mechanism of resistance, accounting for about one-fifth of clinical occurrences, is decreased permeability.[34] The energy-dependent process of active intracellular drug accumulation is apparently not operative. The resultant suppression is then seen in all the aminoglycosides, necessitating the use of a different class of drug. Occurrences are mainly with *P. aeruginosa* and *E. coli.*

In one study with 200 strains of *P. aeruginosa* the majority of resistant strains were low level (MIC < 500 µg/ml). These were shown to have a decreased uptake ability for SM, yet their ribosomes were sensitive to the drug. One strain, however, was highly resistant (MIC $> 20{,}000$ µg/ml). Its ribosomes had no binding affinity for the drug.

The most frequently encountered mechanism of resistance to aminoglycosides is undoubtedly due to chemical derivatization of particular functional groups by *R*-factor–mediated enzymes, that is, coded by self-replicating extrachromosomal loops, or plasmids. This mechanism is primarily responsible for the spread of resistance and is therefore the one of most clinical significance.

There are three types of enzymes known to produce drug-inactivating derivatives: One type *phosphorylates* hydroxyl groups, a second *nucleotidates* hydroxyl functions (mostly adenylation), and the third *acetylates* amino groups (Table 6-11). The phosphorylating and adenylating enzymes both use ATP as the source of phosphate and adenylic acid, respectively. Acetyl-CoA is the basis of the acetylation enzyme (which is different from the enzyme acting on chloramphenicol). It should be understood that the various enzymes have different specificities [e.g., AAD (3")] and only work on streptidine antibiotics, while ANT (2") affects several 2-DOS compounds, but not their semisynthetic analogs (Table 6-11).

Since kanamycin B is subject to all three modes of inactivation, it is useful to illustrate them on this compound (Fig. 6-23). The drug, however, is not as useless as it might seem given that some of the reactions are not commonly encountered in clinical isolates, yet may be induced in in vitro experiments (e.g., 6^{T}N-acetyltransferase). In fact, acetylations are generally seen only in species of *Providencia* and *Proteus*. Others, such as 2′-O-phosphorylations, occur primarily in *S. aureus*, for which these drugs are not indicated. The most troublesome inactivating reaction is phosphorylation of the 3′-OH position (APH 3′) I and II (Table 6-11) of both KMs and NMs, particularly by *P. aeruginosa*. The gentamicins, which lack 3′(and 4′) OHs, are obviously not involved in this resistance mechanism.

---

[34] See also tetracycline discussion.

Table 6-11. Aminoglycoside Inactivating Enzymes[a]

Enzyme name, trivial	Abbreviation	Activity
		Nucleotidylation
aminoglycoside 3″-adenyl transferase[b]	AAD(3″)	3″-OH of SM[c]
ag6-adenyltransferase	AAD(6)	Adenylation of 6-OH of SM
ag2″-nucleotidyltransferase	ANT(2″)	Nucleotidylation of 2″-OH of KM, TM, SSM (but not AK and NTN)
ag4′-nucleotidyltransferase	ANT(4′)	Nucleotidylation of 4′-OH of TM, NM, and AK
		Phosphorylation
ag3″-phosphotransferase	APH(3″)	3″-OH of SM
ag6-phosphotransferase	APH(6)	6-OH of dihydro-SM
ag3′-phosphotransferase I	APH(3′)I	3′-OH of KM, NM
ag3′-phosphotransferase II	APH(3′)II	3′-OH of KM, NM
ag2′-phosphotransferase	APH(2′)	2′-OH of KM, GM
		Acetylation
ag6′-acetyltransferase	AAC(6′)	6′-NH_2 of SSM, AK, NM, KM , GMC_{In}
ag3′-acetyltransferase I	AAC(3)I	3′NH_2 of SSM, GM (but not TM, AK, NTM)
ag3-acetyltransferase III	AAC(3)II	3-NH of GM, TM, NM, NTM (but not AK)
ag2′-acetyltransferase	AAC(2′)	2′-NH_2 of GM, SSM, TM

[a] Partially adapted from Mitsuhashi (1975).
[b] Aminoglycoside term hereafter abbreviated ag.
[c] AK = amikacin, GM = gentamicins, KM = kanamycins, NM = neomycins, NTM = netilimicin, SM = streptomycin, SSM = sisomicin, TM = tobramycin.

Tobramycin is also missing the 3′-OH. Therefore, GMs and TM should have good activity against kanamycin-resistant organisms with APT (3′) and ANT (4′) enzymes. Substituents sterically close to the C6′ position can also afford protection against acetylation there by AAC (6′). Acetylation of the 3-NH_2 of the 2-DOS ring is a clinically important site to resistance [(AAC(3) I and II Table 6-11)]. The drugs susceptible here are GM, TM, and NM, but *not* amikacin. AK is a semisynthetic kanamycin-A derivative where the C-1-NH_2 is acylated with *L*-α-hydroxy-α-aminobutyric acid (HABA). This confers resistance to inactivation at this position, particularly against *P. aeruginosa* strains not susceptible to the other aminoglycosides. AK turned out to be something of a breakthrough. In addition to protecting the C-3-NH_2 position, the compounds C-3′-OH and C-2″OH also were inhibited from attack by phosphorylating and adenylating enzymes, respectively, resulting in a superior drug for many infections resistant to the traditional aminoglycosides. It has been suggested that AK not be used as a first-line drug, but that it be held in reserve to postpone the appearance of resistance to it.

A semisynthetic aminoglycoside whose design was also based on knowledge of "resistance" chemistry is the C-1-N-ethylsisomicin, netilimicin. This alkylation protects the drug against ANT (2″) and AAC(3)I.

Although an aminocyclitol, spectinomycin is not strictly speaking an aminoglycoside (Fig. 6-24). The antibiotic, which is isolated from *S. spectabilis*, has an unusual fused

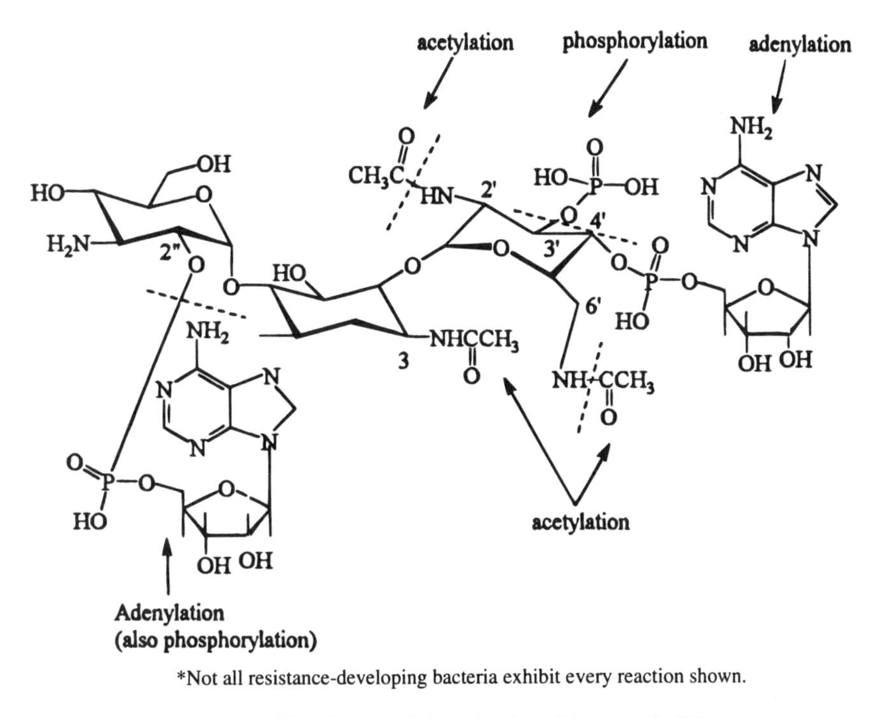

*Not all resistance-developing bacteria exhibit every reaction shown.

Figure 6-23. Enzymatic inactivation of kanamycin B*.

ring structure. Its mechanism of protein synthesis inhibition also involves 30S subunit binding; however, the drug is only bacteriostatic, possibly because the initial binding is more easily reversible. It was postulated that inhibition of polypeptide elongation occurred during translocation. Even though its spectrum is broad, because of rapid development of resistance to most Gm– bacteria, both by ribosomal mutation and adenylation at the 6-OH position, the agent is used primarily against uncomplicated gonorrheal infections, particularly when due to penicillinase-producing strains or in penicillin-sensitive patients.

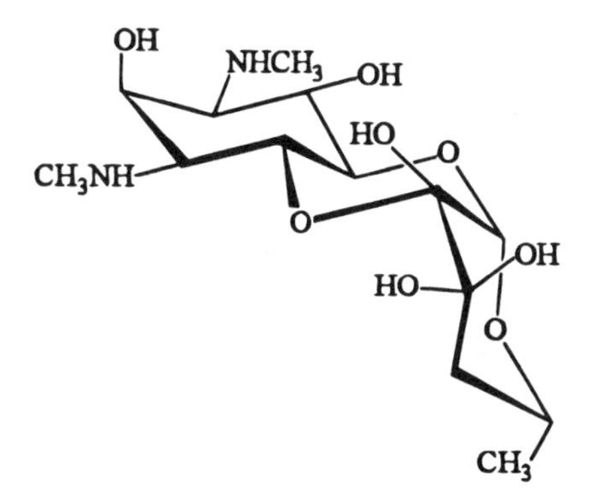

Figure 6-24. Spectinomycin.

6.11.7. The Macrolides

The following discussion will briefly outline two categories of macrolide (large ring system) antibiotics: the lactone macrolides that include clinically useful erythromycins and several related compounds and the ansamacrolides, the most prominent members being the rifamcyins.

The macrolide lactone antibiotics are isolated primarily from the genus *Streptomyces*. Their antibacterial activity is mostly in the Gm+ spectrum. They are classified on the basis of the size of the macrocyclic lactone rings, the *aglycone* component of the compound. These are 12-, 14-, or 16-membered rings, glycosidically linked to one or more amino sugars, thus making these polyfunctional compounds basic (Fig. 6-25).

The 12-membered ring macrolides are not represented by clinical agents in the United States. The most prominent members of the 14-unit rings are the erythromycins (EM) and oleandomycins. Oleandomycin (which forms a spiro-oxirane with the C8 of EM-A) and its triacetate, has been discontinued in the United States.

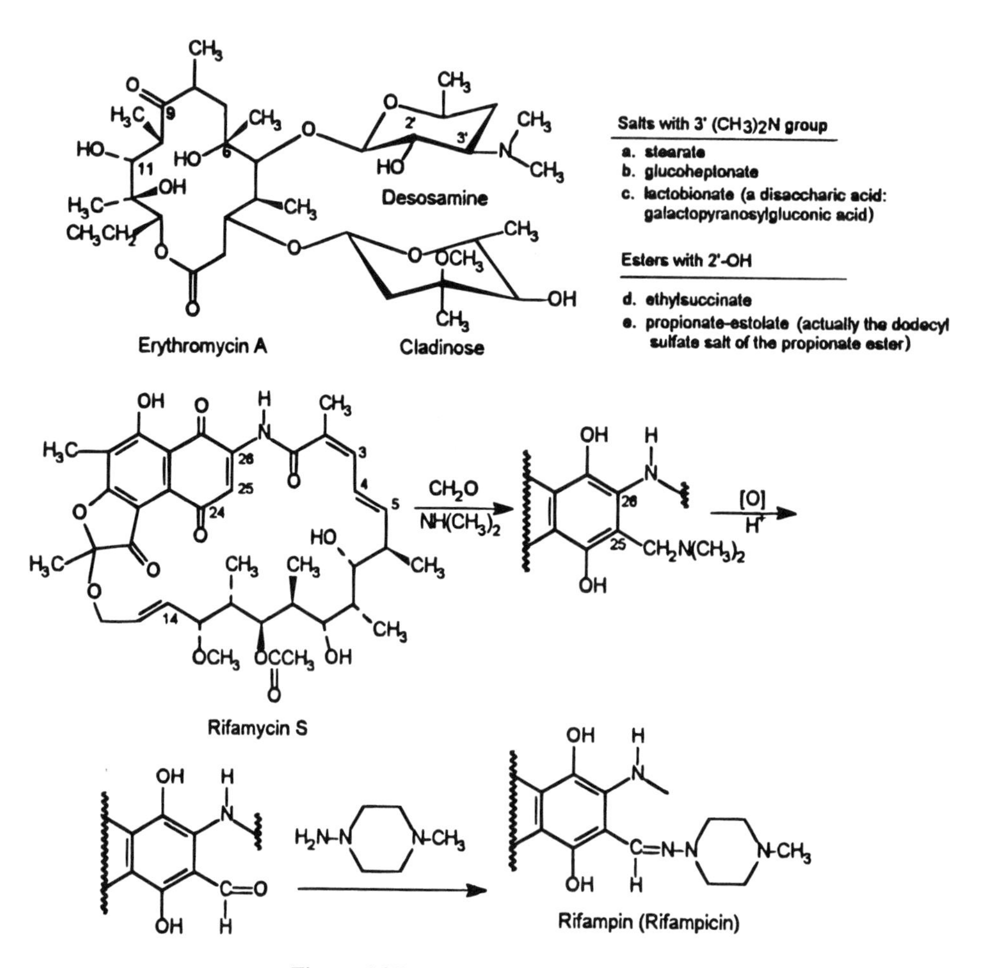

Figure 6-25. The macrolide antibiotics.

Sixteen-membered macrolides are represented clinically by *spiramycin*. It is listed in the French Pharmacopoeia (but not in the United States). It has a similar antibacterial spectrum, but claims higher MIC.

Erythromycin-A (EM) obtained from *S. erythreus* has utility in a wide variety of Gm+ infections, including sexually transmitted diseases such as syphilis, chlamydia, and gonorrhea. The drug can be considered for penicillin-allergic patients. In an outbreak of *Legionnaire's disease*, EM was found to be the only antibiotic to reverse the morbidity of this new disease.

EM is used in six forms. The free base itself is orally absorbed, but it must be enteric coated since it has poor acid stability due to intramolecular ketal formation between the 9 keto group and the 6- and/or 12-OH function (Fig. 6-25). Three salts formed by stearic, glucoheptonic, and lactobionic acid with the dimethylamino function at C-3′ of the desosamine sugar are available. The latter two, freely soluble in water, allow for parenteral administration. The stearate, an insoluble salt, is believed to have better acid stability than the base. In addition, there are two esters. By esterifying the 2′-OH of the desosamine sugar with 3-chloroformylpropionate ($ClOCH_2CH_2CO_2C_2H_5$), erythromycin ethylsuccinate is obtained. Because it is tasteless it is formulated in flavored pediatric suspensions. Finally, erythromycin estolate, which is an ester-salt, is prepared by derivatizing the 2′-OH of desosamine to the propionate ester, forming the hydrochloride and then transposing the Cl (by metathesis) with sodium dodecyl sulfate to give the dodecyl sulfate salt of erythromycin propionate, a relatively insoluble compound. Its attributes are claimed to be resistance to acid degradation and its better absorption, even in the presence of food.

There has been research in developing semisynthetic EM derivatives to improve efficacy by a variety of chemical modifications. Examples are the addition of an F atom at C-8 and esterifications at C-11. Attempts to eliminate the acid sensitivity by manipulating the C-9 keto function have shown promise. One example is formation of a cyclic ketal with the C-9 keto and the C-11-OH functions. Another involves the formation of the 9-oxime (with hydroxylamine) and its derivatization with (2-methoxyethoxy)methyl chloride to form the corresponding oxime-ether (Eq. 6.13). The resultant drug, roxithromycin, exhibits pharmacology and antibacterial spectrum similar to EM.

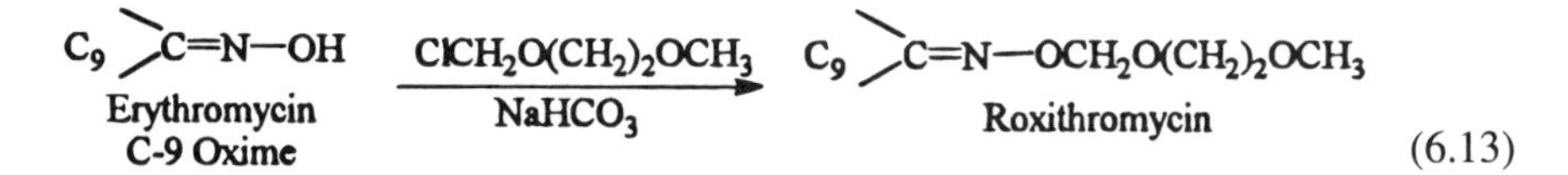

(6.13)

Animal experiments indicate a sixfold increase in potency over EM on oral administration that is attributed to the absence of inactivation by internal ketalization. It is interesting that simpler oximie derivatives, although also stable, showed weaker antibacterial action.

A report (1991) may lead to the ultimate solution of the acid instability problem: the production of 6-deoxyerythromycin A by the use of a mutant strain of *Saccharopolyspora erythrea*. This was accomplished by a targeted disruption of the gene *ery* F, believed to code for the P450 hydroxylase enzyme, which normally produces the C-6 OH function. This new "mutant" antibiotic was found to be acid stable and as effective as EM-A in mice (orally) against several pathogenic bacteria.

The 6-O-methyl derivative has been added to this antibiotic group (Clarithromycin, Biaxin, 1991). The 14-hydroxy metabolite is also active. Its spectrum is not significantly superior to the parent drug (e.g., methicilin-resistant staph are also resistant). Dosing frequency is more convenient, however, and it appears to have fewer gastric side effects.

A more significant improvement (1992) is represented by azithromycin (Zithromax). This nitrogen-containing macrolide ring, named azalide, contains a methylated nitrogen at position 9. It has unique pharmacokinetics, allowing cures after only five daily doses. It may have better gastric stability as well.

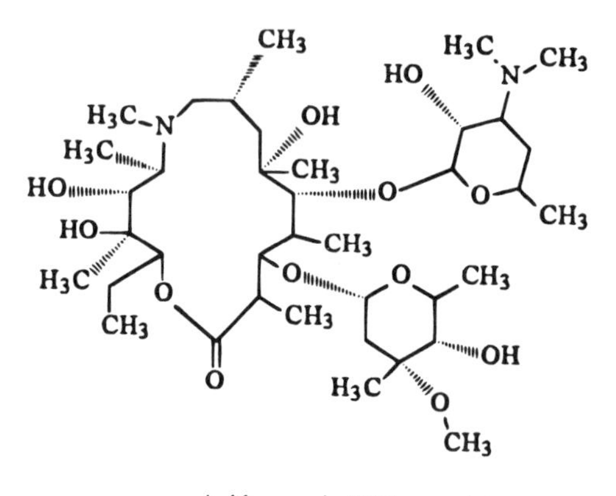

Azithromycin (Zithromycin)

6.11.7.1. Mechanism of Action

Erythromycins bind reversibly with a single high-affinity site on the 50S subunit of susceptible bacterial ribosomes. The site appears to be proteins L-15 and L-16, two of the 34 proteins constituting the ribosomal protein mass of the 50S unit. Removal of several L-16 proteins (by LiCl extraction) from a 50S subunit eliminates its affinity for EM; peptidyl transfer ability is also eliminated. Restoring the L-16 protein alone reestablishes both functions. By itself L-16 has no EM binding capacity; L-15 possesses both the capacity to bind EM and to effect peptidyl transfer, participated in some way by L-16. Both events occur on the P-site. Whether the bacteriostatic antimicrobial action of EM is due to the drugs inhibition of peptide bond formation or by the prevention of its translocation following peptide formation has not been established. To clarify the picture somewhat, perhaps it should be pointed out which aspects of protein synthesis are not affected by EM. They are amino acid activation, synthesis of the amino acid *t*RNA derivative, ribosomal association with *m*RNA, and reassociation of the 30S and 50S subunits to the complete ribosome.

Resistance to EM is encountered intrinsically in such Gm– organisms as *E. coli*, where a single amino acid mutation of the L-4 protein of the 50S subunit is responsible. Induction of resistance to *S. aureus* has also been demonstrated in subinhibitory concentrations. Once induced the resistance is to all macrolides. Induction of resistance has not been demonstrated in 16-membered macrolides such as spiramycin.

6.11.8. Ansamycins

These large macrolide antibiotics are characterized by having a so-called *ansa* bridge, an aliphatic chain connecting two nonadjacent positions on a fused aromatic nucleus. The two known groups differ in that aromatic nucleus. One group has a benezoquinone function, whereas the other contains a naphthaquinoid structure. This latter category has the largest number of ansamacrolides, including the clinically important rifamycins (RF), with some isolated from *S. mediterranei* (RFO), and the others from *Nocardia mediterranei* strains and mutants (S, SV, Y, L, and W). A chemically related group of about 10 streptovaricins is obtained from *S. spectabilis* strains. Few clinical evaluations can be found for these. Several derivatives, however, have been found to inhibit the growth of tumor cells. *N. mediterranei* yields five active rifamycins (A through E), from which only B can be obtained as a crystalline and stable product. A mutant strain was then found that produced only rifamycin B on the addition of diethylbarbituric acid (Chapter 11), probably by suppressing certain metabolic pathways in the organism. Rifamycin B can then be easily converted to several of the others. For example, oxidation with H_2O_2 gives O; hydrolysis gives the very active S, which affords SV on reduction with ascorbic acid.

The rifamycins are generally active against Gm+ bacteria and mycobacteria, especially *M. tuberculosis*. In high concentrations Gm– organisms can be affected. Some antiviral properties have also been established. In spite of the fact that RF-B, O, and S have yielded many derivatives to improve absorption and the spectrum, only three are currently in widespread use. Rifamide [a diethylamido derivative of RF-B (at C-24)] and Rifamycin SV (where the quinone portion of RF-S is reduced to the hydroquinone) are available overseas; the third drug, rifampicin,[35] is available worldwide and in the United States. It is synthesized from RF-S by preparing the 25-dimethylaminomethyl Mannich derivative, which on acidic oxidation yields the 25-formyl analog, rifaldehyde. Reaction with 1-amino-4-methylpiperazine affords the hydrazone rifampicin (rifampin, Fig. 6-25). A large number of rifaldehyde-derived hydrazones, oximes, and Schiff bases showed good activity; however, rifampicin had the best therapeutic profile. In the United States the drug is used primarily to treat tuberculosis (MIC < 1 μg/ml) in conjunction with other antitubercular drugs (Chapter 7) to prevent emergent resistance.

Although stable as a solid, rifampin has certain predictable instabilities in solution. For example, in alkaline solution the hydroquinone hydroxyls (C-24 and 27) would be expected to readily oxidize.[36] The acetates at C-11 and C-22 are likely to hydrolyze slowly with time. Most important, acidic solutions would hydrolyze the C-25 hydrazone, regenerating the inactive, and also unstable, rifaldehyde.

The mechanism of action of rifamycins involves primarily a strong, but noncovalent, interaction with DNA-dependent RNA polymerase enzyme in sensitive bacterial cells. The mammalian enzyme is not affected, which explains the selective toxicity; neither is it mutated to resistant organisms. RNA polymerase has two components. The core enzyme contains polypeptide subunits α, β, β_1, and i and a σ factor, which are needed for recognition of RNA synthesis initiation sites. The drug binds to the β subunit of the complete enzyme only. The result is effective inhibition of RNA synthesis. It is of interest that many rifampinlike hydrazine derivatives were also found to be potent inhibitors of reverse transcriptase and shown to have antiviral properties.

6.11.9. The Lincosaminides

This small group of antibiotics shares an alkyl 6-amino-6-8-dideoxy-1-thio-α-D-galacto-octopyranoside, bonded to a proline function at C-6 (Fig. 6-26). Only one compound, lincomycin (LM), and a semisynthetic derivative are in clinical use. LM was first isolated and

---

[35] U.S. generic name is rifampin.

[36] Just as the hydroquinone in photographic developers does.

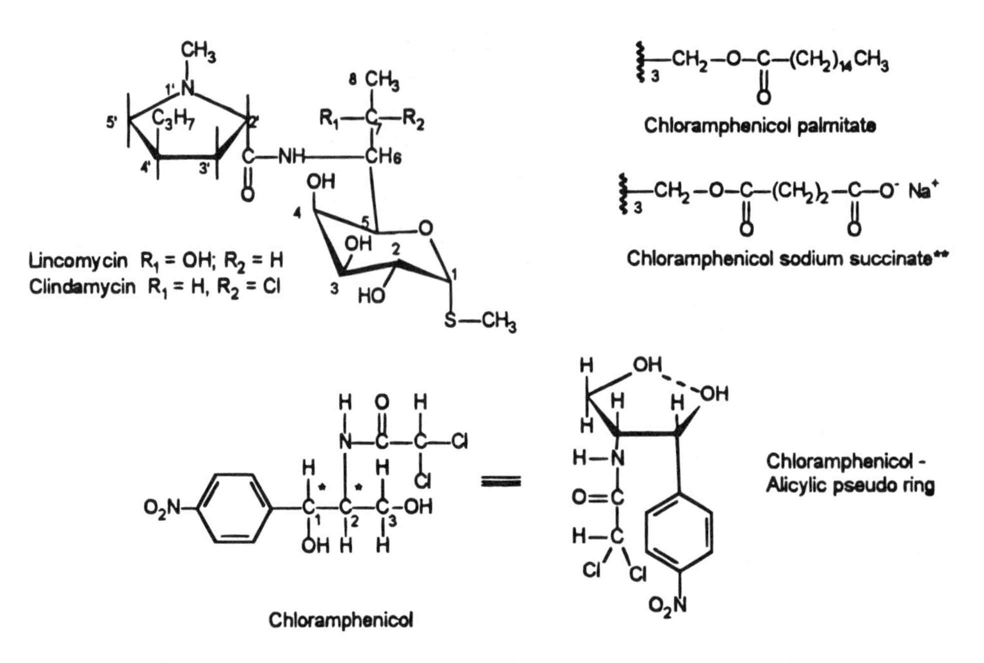

Figure 6-26. The lincosaminides and chloramphenicol. *Chiral centers; **or hemisuccinate.

described in 1962 and was originally obtained from *S. lincolnensis*, although it now can be produced from other species as well. LM is active primarily against Gm+ organisms such as streptococci, staphylococci, and pneumococci. Gm– aerobic bacteria are generally intrinsically resistant. Because of a likelihood of an intestinal condition called *pseudomembraneous colitis*, the major use for LM and its 7-chlorinated derivative clindamycin (CLM) are relegated to the treatment of dangerous anaerobic infections such as *B. fragilis* and *Clostridia* species.

Alkyl substituents on the proline-derived *n*-propylhygric acid nitrogen and the thiolincose sugar sulfur atom of the drug are essential for biological activity, although they may be larger than methyl. Removal of either or both alkyl groups leads to clinically inactive compounds. Extensive structural variations at positions 1, 4′, 1′, and 7 resulted in the development of the 7-chloro-7-deoxy derivative. The conversion can be achieved with $SOCl_2$ followed by alkaline methanolysis of the 3–4 cyclic sulfite of CLM obtained first. Triphenylphosphine dichloride [$(C_6H_5)_3PCl_2$] also effected the conversion. The halogen replacement of the C-7 OH of LM had a profound effect on antibacterial activity, exhibiting increased potency on most organisms tested. The 16-fold increase against the Gm– *E. coli* (MIC 400 vs. 25 μg/ml) was particularly dramatic. The 7-iodo analog was 32 times better. Since CLM was found to be much more lipid soluble than LM, it is tempting to invoke better membrane penetrability as the reason. Human oral absorption was also faster, with blood levels peaking within 30 minutes. The drug was also found to penetrate several plasmodial species, including strains resistant to standard antimalarial drugs such as quinine and chloroquine (Chapter 7).

Clindamycin illustrates how a simple modification, halogenation, of a drug can have such profound effects on the biological activity of an already useful compound: broadened antibacterial spectrum, increased absorption efficiency, and even a new antiprotozoal activity.

Esterifying the 2-hydroxy function with palmitoyl chloride gave the insoluble—and tasteless—plamitate as a pro-drug used in pediatric suspension. Ready in vivo hydrolysis of the ester released the active drug following rapid absorption. Injectable dosage forms contain the soluble clindamycin-2-phosphate; so does a topical solution and gel.

The mechanism of action of LM and CLM is protein synthesis inhibition. The site of action, as with erythromycins, is binding to the 50S ribosomal subunit, which in turn prevents phenylalanine-*t*RNA from binding to the ribosome–messenger complex. Erythromycin interferes with this binding; LM inhibits chloramphenicol binding (see later), leading to the possibility of an overlap, if not identity, of binding sites for the three chemically different antibiotics. Inhibition of peptide chain initiation is probably the main effect of the drug.

6.11.10. Chloramphenicol

This broad-spectrum antibiotic was originally obtained from *S. venezuelae*. Its spectrum includes Gm+ and Gm– bacteria, anaerobes in both categories, and rickettsiae. The drug is active against cholera, lymphogranuloma venereum, and psitacosis. It is particularly useful against typhoid fever and other systemic *Salmonella* infections. The main reason for the caution in using the drug today is the possibility of developing fatal aplastic anemia and other blood dyscrasias such as agranulocytosis. The drug is reserved for situations where the benefits exceed the risk. Included are typhoid fever and bacterial meningitis, which are unresponsive to other agents either because of resistance or inability to cross a noninflamed meninges. Chemically, the drug has functional groups rarely, if ever, encountered in natural products, such as the aromatic NO_2 group and the dichloroacetamide group on C-2 (Fig. 6-26).

The two asymmetric centers (C-1, C-2) allow for four possible stereoisomers: the *threo* and *erythro* geometric isomers each generate a + and – enantiomer; however, only the naturally occurring D (-) *threo* configuration is active.

An electron-withdrawing group in the *para* position of the benzene ring is essential. Other electron-withdrawing substituents in the *para* position may be active, but less so. The methylsulfonyl (CH_3SO_2-), found in thiamphenicol, offers no particular advantage, although it shares most of chloramphenicol's properties. The only isosteric replacement for benzene that exhibited reasonable activity was the thienyl.

Replacing the two Cl atoms with Br on the acetamide group on C-2 showed about 80% of the activity. The monochloroacetamide retained 40%; even the acetamido group without halogens exhibited slight antibacterial properties. In the final analysis, after considerable SAR research, it appears that the structure supplied by *S. venezuelae* was the optimum structure. This is in contrast to what is seen with other antibiotics where molecular modifications resulted in dramatic improvements.

To mask the bitter taste for the pediatric patient, esterification of the primary alcohol, C-3, with palmitoyl chloride affords the water-insoluble tasteless palmitate. Its hydrolysis in the duodenum releases the parent chloramphenicol. It is interesting that serum levels vary considerably among the several crystalline forms of this compound. For example, a suspension of polymorph B was reported to give blood levels seven times higher than a preparation containing only polymorph A. It was also found that polymorph B was hydrolyzed 50 times faster by pancreatic lipase (the palmitate ester, being a fatty acid ester is a good substrate for the enzyme). More recently, an attempt to explain these puzzling phenomena proposed differences in the surface electrostatic charge (zeta potential) between the polymorphs as a reason. An amorphous form also exists. Since the poor solubility of chloramphenicol (2.5 mg/ml) does not allow for reasonable volume parenteral solutions, a freely soluble pro-drug is prepared by reaction with succinic anhydride to yield the hemisuccinate, which on neutralization with NaOH solution, results in sodium salt of the half-ester. This is formulated into an intravenous/intramuscular product.

The bacteriostatic mechanism of action of chloramphenicol involves reversible binding to the 50S ribosomal subunit—at sites near to or overlapping with those for lincomycin and erythromycin. It appears to compete with those drugs at these sites. Aminoacyl *t*RNA is thus prevented from binding its amino acid to the ribosomal binding site. It has not been definitely established whether the drug bonds to the acceptor site or the peptidyl site, which is necessary for chain elongation. Evidence may favor the acceptor site, inhibiting peptidyl transferase activity.

The D (-) threo configuration is theoretically favorably positioned to allow the formation of a pseudo-alicyclic ring by hydrogen bonding between the two hydroxyls (Fig. 6-26). The resemblance to the furanose ribose ring of nucleosides is apparent, leading to suggestions the drug might mimic them. However, spectroscopic (NMR and IR) data failed to show such postulated hydrogen bonding. Furthermore, keto analogs of chloramphenicol are still active.

References

Abraham, E. P., Newton, G. G. F., *Biochem. J.* **63**:628, 1956.

Allen, J. G., Atherton, F. R., Hall, M. J., et al., *Nature* (London) **272**:56, 1978.

Batchelor, F. R., Doyle, F. P., Naylet, J. H. C., Rolinson, G. N., *Nature* **183**:257, 1959.

Blackwood, R. K., English, A. R., in *Structure-Activity Relationships among the Semisynthetic Antibiotics*, Perlman, D., Ed., Academic Press, New York, 1977, pp. 397–426.

Blumberg, P. M., Strominger, J. L., *Bacteriol. Rev.* **38**:291, 1974.

Brown, A. G., Corbett, D. F., Ellington, A. J., Howarth, T. T., *J. Chem. Soc. Chem. Comm.* 1977:523.

Butterworth, D., Cole, M., Hanscomb, G., Rolinson, G. N., *J. Antibiot.* **32**:287, 1979.

Chain, E., Florey, H. W., Gardner, Ad., et al., *Lancet* ii:226, 1940.

Dinner, A., *J. Med. Chem.*, **20**:963, 1977.

Duggar, B. M., *Ann. NY Acad. Sci.* **51**:177, 1948.

Fleming, A., *Brit. J. Exp. Pathol.* **10**:226, 1929.

Imada, A., Kitano, K., Kintaka, M., et al., *Nature* **289**:590, 1981.

Indelicato, J. M., Dinner, A., Peters, L. R., Wilham, W. L., *J. Med. Chem.* **20**:961, 1977.

Jaurin, B., Grundstrom, T., *Proc. Natl. Acad. Sci.* (USA) **78**:4897, 1981.

Lehninger, A. L., Nelson, D. L., Cox, M. M., *Principles of Biochemistry*, 2d ed., Worth Publishing, New York, 1993, p. 113.

Mitsuhashi, S., in *Drug-Inactivating Enzymes and Antibiotic Resistance*, Mitsuhasi, S., Rosival, L., Kromery, V., Eds., Springer Verlag, Berlin, 1975, p. 1175.

Nagarajan, R., Boeck, L. D., Gorman, M., et al., *J. Am. Chem. Soc.* **93**:2308, 1971.

Nikaido, H., Nakae, T., *Adv. Microb. Physiol.* **20**:164, 1979.

Purich, E. D., Colaizzi, J. L., Poust, R. I., *J. Pharm. Sci.* **62**:545, 1973.

Sheehan, J. C., Henery-Logan, K. R., *J. Am. Chem. Soc.* **79**:1262, 1957.

Sheldrick, G. M., Jones, P. G., Kennard, O., et al., *Nature* **271**:223, 1978.

Singer, S. J., Nicholson, G. L., *Science* **175**:720, 1972.

Singer, S. J., *Ann. Rev. Biochem.* **43**:805, 1974.

Spratt, B. G., *Eur. J. Biochem.* **72**:341, 1977.

Spratt, B. G., *J. Gen. Microbiol.* **129**:1247, 1983.

Strominger, J. L., et al., *Biochemistry* **6**:906, 921, 930, 1967.

Sykes, B. B., *J. Infectious. Dis.* **145**:762, 1982.

Sykes, B. B., Cimarusti, C. M., Bonner, D. P., et al., *Nature* **291**:489, 1981.

Tipper, D. J., Strominger, J. L., *Proc. Natl. Acad. Sci.* USA **54**:1133, 1965.

Waksman, Schatz, A., Bugie, E. and Waksman, S., *Proc. Soc. Exper. Biol. Med.* **55**:66, 1944.

Suggested Readings

Franklin, T. J., Show, G. A., *Biochemistry of Antimicrobial Action*, 4th ed., Chapman-Hall, New York, 1989.

Lehninger, A. L., Nelson, D. I., Cox, M. M., *Principles of Biochemistry*, 2d ed., Worth Publishers, New York, 1993 (Chapter 26).

Queener, S. F., Webber, J. A., Queener, S. W., Eds., *Beta-Lactam Antibiotics for Clinical Use*, Marcel Dekker, New York, 1986.

CHAPTER 7

Antimicrobial Drugs II

This chapter will concern itself with antimicrobial agents (as broadly defined earlier) of synthetic origin.[1] Unlike the semisynthetic antibiotics, which were obtained from biological sources and then modified by one or more synthetic steps, these compounds are synthetically "created." In many cases the prototype compound was discovered by a more or less random screening process (e.g., sulfanilamide, Chapter 2) and then, by first applying empirical and now more scientifically predictive SAR methods, optimized, or at least improved. The ultimate goal, of course, is to design a potential new drug from theoretical biochemical concepts, synthesize it, and find it to have the desired antimicrobial and pharmaceutical properties. When our abilities reach that point, we will have extended Paul Ehrlich's goals of selective chemotherapy to the level he must have envisioned over eight decades ago.

Synthetic antibacterial compounds of clinical value have been relatively uncommon. The sulfonamides were the first highly successful group. The others will now be considered.

7.1. The 4-Quinolones

Not all the members of this group are, technically, quinolones, because naphthyridine, pyridopyrimidine, and cinnoline derivatives (Fig. 7-1), which each contain additional nitrogen atoms in one or the other of the two fused rings, are also represented. It may be useful to consider them all quinoline-type compounds. Thus one may view naphthyridine drugs as 8-aza-4-quinolones, the pyridopyrimidine ring compounds as 6,8-diaza-4-quinolones and the cinnoline system as a 2-aza-4-quinolone. The following compromise "generic" structure may depict the common denominator to all the active compounds.

[1] With a few exceptions, such as the alkaloid emetine and the antiparasitic ivermectin.

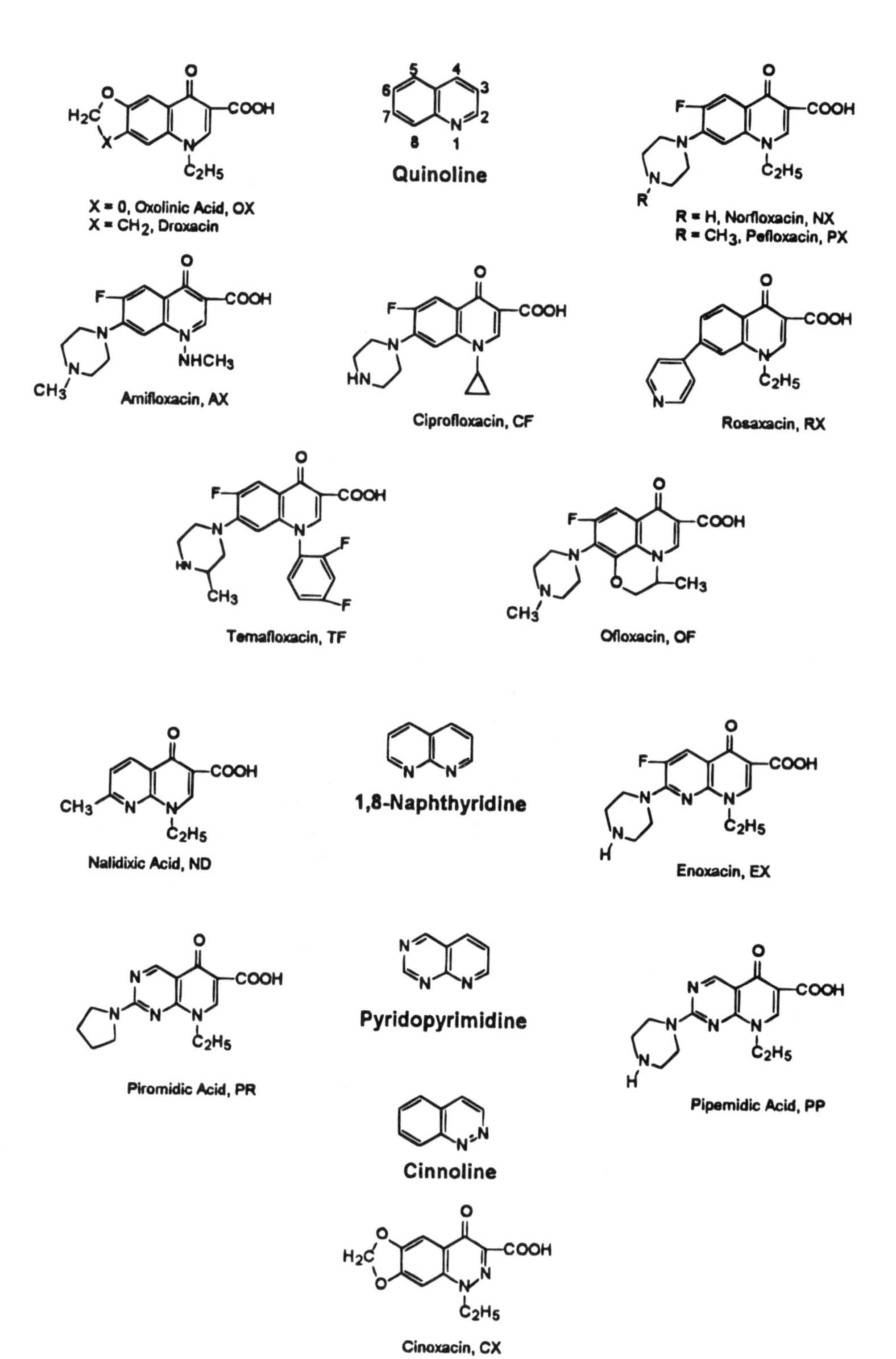

Figure 7-1. The chemistry of the 4-quinolones.

Nalidixic acid (ND) (NeGram) (Fig. 7-1), the prototype of this family of drugs, was synthesized as the result of the discovery that an impurity, isolated during the preparation of the antimalarial chloroquine,[2] had significant antibacterial activity. It was introduced into therapy in 1964. The antibacterial spectrum, however, was relatively narrow and did not include

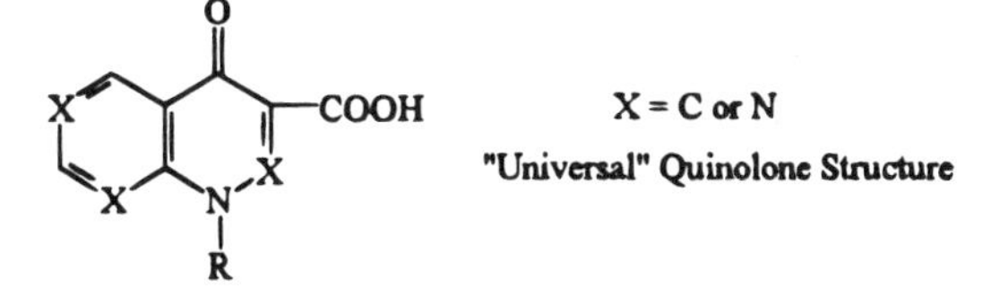

Pseudomonas, anaerobes, and Gm+ bacteria. Nevertheless, the drug, and two congeners introduced a decade or more later—oxolinic acid, OX and cinoxacin, CX (Cinobac) (Fig. 7-1), became useful primarily in the treatment of recurrent and uncomplicated urinary tract infections caused by susceptible organisms. Even though it is five times more potent, OX has the same spectrum as the other two drugs.[3] ND, OX, and CX constitute the first generation of 4-quinolones, and have been studied extensively as to their chemical properties, mechanisms of action, and resistance. The latter two characteristics are qualitatively the same for the newer generation of compounds. Their pharmacological and therapeutic properties, however, are superior.

Though a polar compound, nalidixic acid is only sparingly soluble in water (<1 mg/ml) and alcohol. Yet its polar nature prevents the achievement of antibacterially effective plasma levels following oral administration, in spite of large doses (4 g) and 96% absorption efficiency from the gastrointestinal (GI) tract. Obviously high urinary tract levels are achieved. Its 7-hydroxymethyl metabolite does not alter the situation, even though it is more active than ND itself and may actually be responsible for most of the activity in the urinary tract where both compounds accumulate. The pharmacology of CX and OX is essentially parallel in that urinary tract levels achieved also exceed the MICs needed for a therapeutic response. With the increased appearance of resistance and adverse reactions over the years and an inability to affect systemic infections even when caused by susceptible bacteria these first generation compounds have been relegated to lesser importance.

Congeners carrying an alicyclic amine function at position 7 of the 4-quinolone structure (such as rosaxacin, RX, piromidic acid, PR, and pipemidic acid, PP, (Fig. 7-1) have each shown improvement over the parent drug in some pharmacologic aspect. Neither of these three alicylic agents has yet been introduced in the United States. Yet, if considered as second-generation compounds, these may serve as a "bridge" to the third-generation drugs (see later). PR was found to have good in vitro and in vivo activity against staphylococci and many Gm– bacteria except *P. aeruginosa*, thus having an improved spectrum compared with ND, which does not affect Gm+ organisms. Several years later the synthesis of PP was reported (one of 74 compounds in the series) to possess significant antipseudomonal in vitro activity (MIC 3 μg/ml). It is not readily apparent why an exchange of the pyrrolidino group at position 7 for piperazine should bring about this sought-after improvement. It is even more puzzling that almost all attempted substituents on the unsubstituted nitrogen of the piperazinyl group of PP produced a profound drop in antipseudomonal activity and, in most cases, activity against *E. coli* and *S. auerus* as well. Yet in the third-generation quinolones (see later) an N-methyl substituent afforded clinically useful drugs with high activity. It is very likely that such anomalies would not have been predictable by present QSAR concepts.

[2] Actually the 7-chloroquinoline analog of the then yet unsynthesized nalidixic acid.

[3] Oxolinic acid is no longer available in the United States.

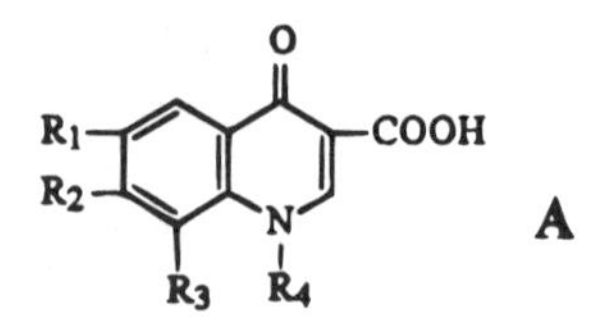

Further attempts to amplify the antibacterial activity were undertaken by additional structural modifications. An extensive SAR study in which 78 compounds of structure A were synthesized and evaluated by QSAR methods helped to determine what substituents would give optimum results.

The SARs were initially analyzed in terms of the effects of substituents at R_1, R_2, R_3, and R_4 singly, and the resultant data were then applied to analogs substituted at two or three positions. Equations for obtaining MIC values were derived for substituents at each of the preceding positions and then calculated utilizing physicochemical parameters such as π, σ, and steric factors such as Es (positions R_1) and R_4 (for R_3). Physicochemical constants for R_2 did not correlate well with activity. However, the piperazinyl group at this position had already been established as the most promising with PP. Keeping R_4 constant as C_2H_5, R_1 was varied by substituting H, F, Cl, NO_2, Br, CH_3, OCH_3, or I. The results indicated that activity changes with these substituent variations were primarily influenced by the size of the group; that is, steric effects were operative. The optimum dimension was intermediate to the F and Cl atoms. The R_3 substituent also exhibited steric effects with optimum size being that of the Cl atom or CH_3 group.

The relationship between activity and the nature of the R_2 sustituents is not clear-cut. Because of a 10- to 30-fold increase of antibacterial activity following the introduction of piperazino, pyrrolidino, and several acyclic substituents of smaller size [(e.g., $N(CH_3)_2$, OCH_3)], the activity was thought to relate to electron-donation capability. In evaluating all the physicochemical factors, the reason for the biological results obtained were not predictable. However, when variable factors obtained for R_2 substituents were combined in one equation with the results of the derivatives substituted at R_1 and R_3, the prediction emerged that particular multiple substituents sets (i.e., R_1, R_2 and R_3 combined) would yield optimum activity. Preparing such polysubstituted compounds proved this prediction. The compound culled from this extensive study was norfloxacin (NX) (Noroxin) (Fig. 7-1), a drug whose antibacterial potency was 16 to 500 times higher than ND (depending on the species).

In regard to the R_4 substituent on the nitrogen atom, it should be pointed out that it was constant as C_2H_5 in this study, even though it was shown earlier that other sterically similar groups such as $CH = CH_2$ and CH_2CH_2F, were also effective. It was not until later that sterically bulkier substituents such as cyclopropyl (e.g., CF) and even *p*-fluorophenyl (structure B) afforded significant increases in activity. Further research also showed that R_3

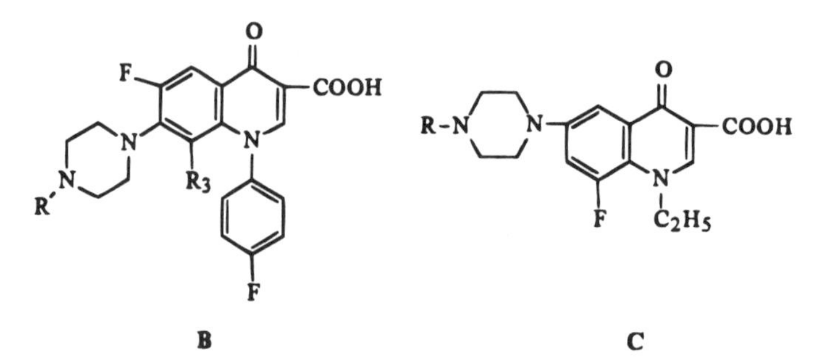

substituents such as Cl or CH_3 could further augment activity. In fact, OF might be viewed as having both a bulkier substituent on the nitrogen atom (equivalent to an isopropyl group) and an -OCH_2- moiety as the R_3 substituent.

It is of interest that NX was not the most active compound in the study; it was an OF analog where the oxygen at C-8 is bonded to a CH_2. This compound, which is based on QSAR correlations and in vitro data, possesses the optimum structure that should produce the highest overall antibacterial activity, yet factors other than in vitro activity must also enter the decision-making process when developing a drug for clinical use. These factors would be low oral toxicity, high metabolic and pharmaceutical in vitro stability, the pharmacokinetic profile, and even the cost of large-scale synthesis.

The uniqueness and importance of the F substituent at the 6 position of the 4-quinolone structure was exemplified showing that a series of quinolones carrying the piperazinyl function at that position, along with the F at C-8 (structure C), were devoid of activity against *S. aureus, E. coli*, and *P. aeroginosa*. The reason for this surprising effect from "simply" switching substituents is not apparent.

Ciprofloxacin (CF) (Cipro) was the next drug introduced to the clinic, followed by ofloxacin (OX) (Floxin) in 1991. The others, perfloxacin (PX), amifloxacin (AX), and enoxacin (EX) are currently at various stages of clinical evaluation, or are available in other countries. As a group these drugs are helping to solve various clinical problems, particularly nosocomial infections resistant to ND and most β-lactam antibiotics, including *P. aeruginosa*. They may ultimately avoid the difficulties associated with the currently necessary use of aminoglycosides in patients with resistant *E. coli* strains, methicillin-resistant staphylococci, and penicillinase-producing *Neisseria gonorrhea* (PPNG). The oral effectiveness of the new quinolones avoids parenteral dosing and even extended hospitalization.

To illustrate the effectiveness of CF, for example, the MIC against even PPNG was 0.5–3 μg/ml. In one clinical study of 100 cases of acute male gonorrhea, 99 were cured with a single 100 mg dose of CF. In the case of *P. aeruginosa*, NX was more effective than carbenicillin, tetracycline, trimethoprin, and cotrimozaxole. The same organisms resistant to 128 μg/ml of sulfamethoxazole responded to 4 μg/ml of NX.

Resistance to the new 4-quinolones is not a significant problem so far. Spontaneous resistance occurs in less than 10^{11} colony-forming units. Bacterial resistance to ND in bacterial chromosomes arise by two mechanisms. One is the result of mutation of the A subunit of DNA-gyrase (see later), which causes a high level of resistance; the other involves membrane alteration at the B subunit, which causes low-level resistance. The feared plasmid-mediated resistance (transfer of resistance) often seen with other antibiotics has not been encountered so far. Thus the only route for development of resistance in these drugs is by mutation.

7.1.1. Mechanism of Action

Nalidixic acid and all its analogs act by selectively inhibiting bacterial DNA synthesis. The actual biochemical target of the quinolones is the enzyme DNA-gyrase, a topoisomerase enzyme found only in procaryotic cells. This accounts for the selective toxicity of these drugs.

At this point a brief review of certain DNA characteristics may be helpful to an understanding of the mechanisms of action of the 4-quinolones. Although the base pairing in DNA from all natural sources studied is constant, several different conformational variations of the double helix have been identified. These result from the rotational freedom inherent in the single bonds present in the ribose and phosphate backbone. The three conformational forms are called A, B, and Z. The geometry of these forms differs in the conformations of the glycosidic bonding to deoxycytidine and deoxyguanosine, as well as in points of "puckering" of the pyranosugar. The handedness of the helix differs in that A and B DNA are both right-handed, while the Z form is left-handed. The number of residues per helical turn will differ as a result: B-DNA has 10, while Z-DNA has 12 (6 dimers). The distance between residues therefore will also vary (e.g., 2.5 Å for A, 3.4 Å for B, and 3.7 Å for Z-DNA). Other consequences that arise include differences in the tilt of base pairs (A-DNA 20°, B-DNA 6°) and the actual helical pitch: A = 28 Å, B = 34 Å, Z = 43 Å.

These conformational DNA differences serve to illustrate the intrinsic flexibility of the double helix. Interactions with proteins and other molecules can produce additional conformations significant in certain situations. Examples are bends (seen in chromosomes), doubling over foldbacks, and other tertiary structures. *Supercoiling* is important to the present discussion. It is seen in DNA molecules that are constrained in their topology as a result of being circular and thus covalently closed. This tertiary structure, of course, does not allow the molecule's ends to rotate freely. DNA can form negative (right-handed) or positive (left-handed) supercoils (Fig. 7-2). The negative supercoil contains torsional stress. This creates a tendency to unwind. Positive supercoiling, on the other hand, favors a tightening of the double helix. Supercoiling thus results in a more compact circular duplex structure. This compaction is necessary if one considers that the double-stranded DNA chromosome of an *E. coli* bacterium, for example, is more than 1,000 times longer than the cell (about 1 μm). The DNA molecule is actually contained in the cell's nuclear body, which is even smaller. The way the chromosome is folded within this cell was described by Worcell (1974). He showed supercoiled regions, or domains, of about 20 μm surrounding and attached to an RNA core. Each domain was tightly supercoiled by DNA-gyrase. DNA-gyrase is classified as topoisomerase II, which is able to induce the change of relaxed duplex DNA into a high-energy negative (right-handed) superhelical region (Fig. 7-2). This is accomplished by causing a transient double-strand scission or nicking, an ATP-dependent reaction followed by reclosure. If ATP is not available, then DNA-gyrase can catalyze superhelical formation, where molecular components are held together by repulsive rather than the usual attractive forces.

It was found that nalidixic acid, which causes abnormal increases of single-stranded DNA in *E. coli*, does so by inhibiting chromosome replication through the prevention of enzyme-catalyzed resealing to circular DNA.

DNA-gyrase was identified as the enzyme responsible for nicking the double-stranded DNA, introducing the negative supercoil, and then resealing the nucleic acid. The topoisomerase II of *E. coli* consists of 4 subunits: two α-monomers (subunits A) and two β-monomers (B). Thus a tetrameric A_2B_2 complex exists. The β-subunits are probably responsible for inducing negative supercoiling induction of double-stranded DNA, initially nicked by the action of α subunits, which introduce incisions into each strand of the DNA molecule and subsequently reseal the molecule. It has been further proposed that the

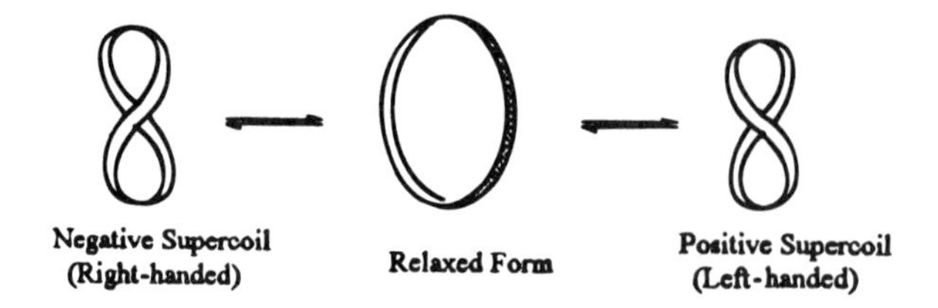

Figure 7-2. Negative and positive supercoil.

4-quinolones also inhibit the resealing of the A-subunit—induced breaks in the DNA molecules. It is therefore now apparent that the A-subunit of DNA-gyrase is the primary site of action of 4-quinolone compounds. Parenthetically, the antibiotic novobiocin[4] that acts on the B-subunit of DNA-gyrase, has been shown to act synergistically with NX against bacteria.

In summary, it can be stated that inhibition of DNA synthesis by 4-quinolones initially involves the inhibition of ATP-dependent DNA supercoiling by binding to subunit A of DNA-gyrase. Secondarily, the drugs also inhibit the relaxation of supercoiled DNA, a reaction not dependent on ATP. Finally, 4-quinolones also block the DNA nicking-closing enzyme that, in the absence of drug interference, is ultimately responsible for DNA elongation.

An interesting and potentially useful idea is the development and evaluation of dual-action (mutual pro-drug?) compounds synthesized by coupling the 3′-hydroxymethyl of several cephalosporins with the 4′-secondary amine function of ciprofloxacin via a carbamate linkage. The cleavage of the β-lactam ring during its covalent binding to the transpeptidase likely cleaves the carbamate as well, in essence making the quinolone a leaving group in the classic sense.

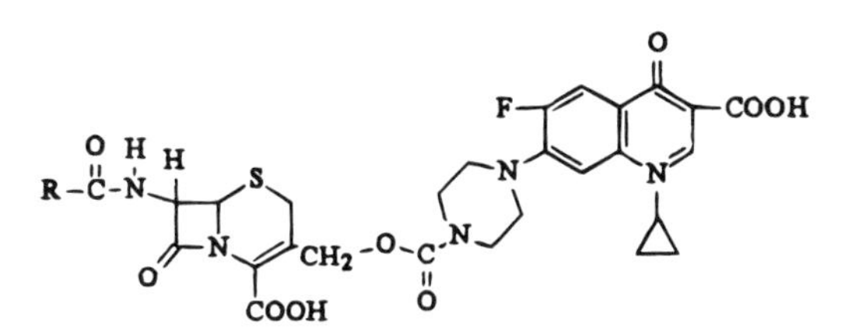

Dual activity cephalosporin-quinolone carbamates

Coupling the quinolone's carboxyl group afforded the ester congener. Such compounds exhibited broad antibacterial activity, as might be expected from the combination of the components. However, the authors suggest a possibility that the intact bifunctional molecule exhibits both cell wall inhibition and the quinolone mechanism of action. Although the evidence was inconclusive, the fact that these compounds showed activity against methicillin-resistant *S. aureus* and *E. colaoe*—the latter resistant to third-generation cephalosporins—lends some credence to this possibility.

7.2. Nonbenzenoid Nitro Compounds

Medicinal chemists have tended to avoid the aromatic nitro group as a structural component of potential drugs because of the well-known ability of compounds such as trinitrotoluene (TNT), used in munitions, to cause a high incidence of methemoglobinemia following skin absorption. The mechanism involved is presumably in vivo reduction to nitroso and phenylhydroxylamine intermediates. However, many nonbenzenoid nitro compounds have been utilized in clinical practice since the introduction of the nitrofuran drugs in 1944.

[4] A 4-hydroxycoumarin antibiotic of relatively narrow Gm+ spectrum.

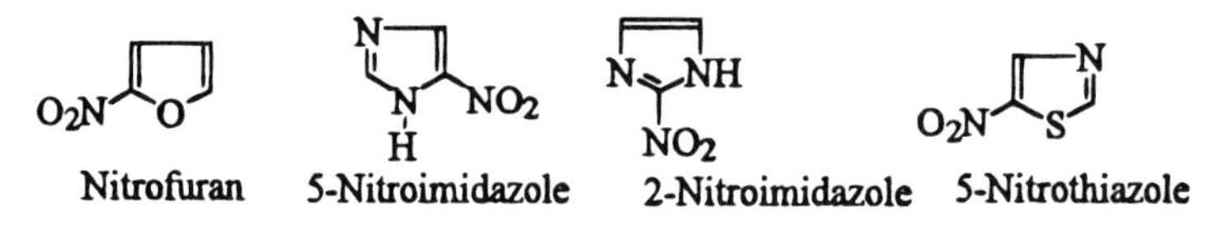

Figure 7-3. Nonbenzenoid nitroheterocycles.

7.2.1. The Nitrofurans

The 5-nitrofuran derivatives were the first nitroheterocycles of therapeutic value (Fig. 7-3). They are all synthesized from 5-nitro-2-furancarboxyaldehyde (5-nitrofurfural) or its diacetate by azomethine (Schiff base) formation with the appropriate amino compound (Eq. 7.1). The important nitrofurans in use today are shown in Figure 7-4. Since the first report of their antibacterial activity about 4,000 nitrofuran compounds have been synthesized and tested. Those that reached commercial use in human and veterinary medicine, including farm animal husbandry, did so because of their broad spectrum of activity including Gm+ and Gm– bacteria (except *P. aeruginosa* and some *Klebsiella* and *Proteus* strains), relatively low toxicity, and infrequent development of resistance.

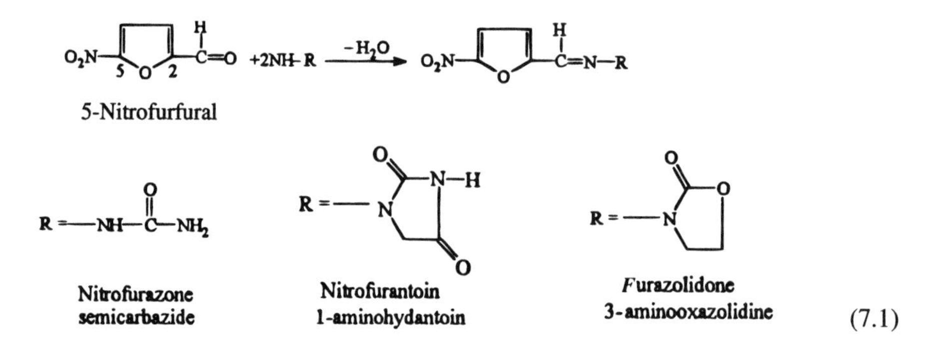

(7.1)

The semicarbazone nitrofurazone (Furacin) is used topically (cream and surgical dressing) to treat and prevent surface infections of Gm+ and Gm– bacteria, including anaerobes. Its antibacterial spectrum, low sensitizing potential, and lack of development of resistance has made this product particularly useful for patients with second- and third-degree burns and skin graft procedures. Because of its additional effectiveness against *Giardia lamblia* and the trichomonal organism, furazolidine has been particularly useful against bacterial and protozoal diarrhea and enteritis.

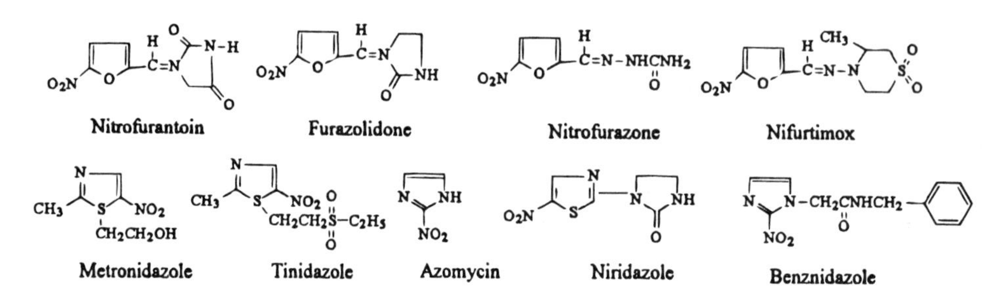

Figure 7-4. Nonbenzenoid antiparasitic nitrocompounds.

The most widely used nitrofuran is nitrofurantoin (Furadantin and Macrodantin), primarily in acute urinary tract infections, since sufficient tissue levels against systemic infections are not achieved. Resistance is rare and not transferable.

It is of interest that the nausea and vomiting experienced by some patients has been minimized by the development of a macrocrystalline form whose rate of absorption on oral administration is decreased.

7.2.1.1. Mechanism of Action

The nitroaromatic and nitro-nonbenzenoid compounds considered in this chapter share a commonality of cytotoxic mechanisms that initially involve one or more reductive reactions on the nitro group, followed by various enzyme inhibitions necessary for the parasitic cell's energy requirements. An understanding of some of the differences between mammalian and microbial cells may explain the favorable selective toxicity these drugs exhibit.

Mitochondria are rod-shaped structures in all eucaryotic cells where oxidative phosphorylation, the citric acid (Krebs) cycle, and fatty acid oxidations happen. Thus the electron-transfer, energy-generation, and storage processes all occur in them. The structural organization of mitochondria consists of a double lipoprotein membrane system, with the outer membrane enclosing an inner membrane that in turn, by a number of invaginations, results in a series of compartments called *cristae*. Within these invaginations are the respiratory assemblies, or oxysomes, containing all the enzymes of the electron transport system in the same sequential order in which the reactions occur: flavoproteins, coenzyme Q, cytochromes b, c_1, c, a, and a_3. Ferrodoxins (iron and sulfur atom–containing proteins involved in redox reactions) are also in the cristae. The Krebs-cycle-associated enzymes, succinic, malic, and α-ketoglutaric dehydrogenase are also located here. Thus the mitochondria have all but one of the tools necessary to carry out the cell's aerobic life functions. The citric acid cycle transforms the acetyl-CoA from carbohydrate and fat metabolism to water and CO_2, yielding energy. The enzymes needed to catalyze the metabolism of the fatty acids and the enzymes involved in the transmission of electrons from various substrates to molecular O_2 are also here, as is the necessary ATP-ase needed to complete the respiratory chain, by storing some of the resulting calories in the form of ATP by the final step of oxidative phosphorylation (Eq. 7.2). Only the Embden–Myerhof scheme of carbohydrate metabolism occurs in the cell's cytoplasm. The inner mitochondrial membrane has limited permeability. The outer membrane is permeable to most smaller molecules.

$$\mathrm{ADP} + \mathrm{Pi} \rightleftharpoons \mathrm{ATP} + \mathrm{H_2O} \qquad (7.2)$$

Bacteria have no mitochondria. In them, the mitochondrial functions described earlier are all carried out within the plasma membrane, where all the needed enzymes are anchored. This relatively poor protection of the enzyme systems (compared with the host's cell) is the likely reason for the selective toxicity of these agents. The situation in protozoa is somewhat different. Some species have a few mitochondria; more often, some contain simpler tubules rather than cristae.

In other protozoa, such as the genus *Trypanosoma*, the organisms have a kinetoplast whose DNA molecule contains the information to produce a mitochondrion once in its next (insect) host in one or more of their stages. Several antiprotozoal agents (e.g., pentamidine) bind irreversibly to the kinetoplast causing it to disintegrate. Some protozoa contain hydrogenosomes instead of mitochondria where pyruvate is converted to acetyl Co-A by ferredoxin-containing enzymes. Metronidazole affects this conversion.

The nitroaromatic compounds such as the nitrofurans are known to be activated by a bacterial nitroreductase system in susceptible microorganisms. Intermediate, highly reactive species such as free radicals produced during the reduction process are likely responsible for damage to DNA strands that lead to bacterial and protozoal cell death. Thus a reduced nitroaryl anion radical could be oxidized by O_2 to produce superoxide anions (Eq. 7.3). Under the influence of superoxide dismutase (SOD) (Chapter 4) hydrogen peroxide can be produced (Eq. 7.4), which, in turn, interacts with additional superoxide anion radical, producing ionizing toxic hydroxyl radicals (Eq. 7.5).

Several events described here may be microbiocidal. The oxygen free radical attacking DNA strands may break them faster than repair mechanisms can restore them. Inhibition of synthesis of enzymes needed to repair the lesions is also a possibility. Organisms devoid of catalase (the enzyme that rapidly decomposes H_2O_2) such as *T. cruzi* would lead to elevated levels and therefore larger concentrations of toxic OH^- radicals.

$$\text{Ar-NO}_2 \underset{2O_2^{\cdot} \;\;\; 2O_2}{\overset{\text{NADPH} \;\;\; \text{NADP}^+}{\rightleftharpoons}} \text{ArNO}_2^{\cdot} \tag{7.3}$$

$$O_2^{\cdot} + O_2^{\cdot} \xrightarrow{\text{SOD}} 2H_2O_2 + O_2 \tag{7.4}$$

$$2H_2O_2 + O_2^{\cdot} \xrightarrow{Fe^{2++}} OH^{\cdot} + OH^{-} + O_2 \tag{7.5}$$

T. cruzi causes South American trypanosomiasis (Chagas' disease). One form of the parasite reproduces *intracellularly*, making its treatment extremely difficult. The experimental drug nifurtimox (Bayer 2502), a 5-nitrofuran derivative (Fig. 7-4), although the current drug of choice, is not satisfactory in spite of high activity against the parasite in vitro. The alternative 2-nitroimidazole benznidazole (Fig. 7-4) has similar shortcomings; namely, at tolerable levels of toxicity, neither drug can prevent the reestablishment of the organism at new tissue sites once released by the drugs from prior sites. Suppressive results, however, are achievable. It was shown that the rate of reductive activation of nifurtimox in the trypanosome is at a higher rate than in the mammalian host cell, which accounts for some of the selective toxicity.

7.2.2. Nitroimidazoles

Only 2- and 5-nitroimidazoles (Fig. 7-3) appear to be of possible clinical interest. An early study with soil streptomyces cultures produced several antibiotic-containing extracts, one of which exhibited activity against several protozoans, specifically trichomonads. The responsible compound was identified as 2-nitroimidazole or azomycin (Fig. 7-4). This compound then became the chemical "lead" for extensive synthetic studies of over 100 compounds. The most effective compound with low toxicity was 1-(β-hydroxyethyl)-

2-methyl-5-nitroimidazole, which was introduced in Europe (1960) and the United States (1963) as metronidazole (Flagyl) (Fig. 7-4). It became the drug of choice against *Trichomonas vaginalis*. It is effective against *Entamoeba histolitica*, thus extending the drug's indication to both enteric and hepatic amebiasis. *Giardia lamblia* is also susceptible. An earlier discovery that metronidazole had significant antibacterial activity against obligate anaerobes such as *Clostridia* and *Bacteroides* species further increased the usefulness of this unique drug considerably.

The antiprotozoal mechanism of action is believed to be common with other nonbenzenoid nitroaromatics. The antibacterial mechanism appears to depend on a reduced oxygen concentration (rather than all growth), thus explaining activity limited to anaerobes. Unlike aerobic bacteria, where oxygen is the only terminal electron acceptor, obligate anaerobes must have anaerobic electron transport systems of low redox potential. Ferredoxinlike proteins are likely candidates. These undergo one-electron oxidation–reduction reactions. Under aerobic conditions of glycolysis, O_2 acts as the electron acceptor of pyruvate oxidation; anaerobic conditions necessitate an alternate pathway whereby pyruvate and phosphate yield acetyl phosphate, carbon dioxide, and protons, two of which form molecular hydrogen on accepting electrons (Eqs. 7.6, 7.7). Metronidazole has been shown to inhibit the production of H_2 in *T. vaginalis* and *Clostridia*, possibly by interfering with the electron-transfer process in ferredoxin. Intracellular reduction of the nitro group and the formation of radical anions on the 1-ethanol chains, $\text{-CH}_2\text{-CH}_2\text{-O}^{-***}$, leads to covalent DNA interactions with fatal consequences to the cell. Tinidazole, a more lipid-soluble analog with less extensive metabolism, produces higher serum levels and longer duration of activity.

$$\underset{\text{Pyruvate}}{CH_3-C(=O)-C(=O)-O^-} + H_3PO_4 \longrightarrow \underset{\text{Acetyl Phosphate}}{CH_3-C(=O)-O-P(=O)(OH)-OH} + CO_2 + 2H^+ \qquad (7.6)$$

$$2H^+ + 2e^- \longrightarrow H_2 \qquad (7.7)$$

The usual suspicion of carcinogenicity and mutagenicity that nitro compounds elicit have been documented with metronidazole in mice and rats. High-dose, long-term (lifetime) studies in these rodents have resulted in pulmonary, mammary, and hepatic tumor.

7.3. Parasitic Diseases

Before considering the drugs currently in use against parasitic infestations of humans, it may be useful to overview the various parasitic diseases that affect humans (and animals). In the sense of a strict definition, *parasites* should include all pathological invasions of mammalian and other species. The present discussion, however, will concern itself with protozoal and helminthic organisms. (Fungal diseases will be considered later in this chapter.) Many of the diseases are endemic in certain parts of the world, with varying degrees of morbidity and mortality. Prevalence is difficult to determine since asymptomatic infections may be of many years duration in some diseases. Table 7-1 summarizes the most prevalent dis-

Table 7-1. Human Parasitic Diseases

I. Protozoal diseases	Causative organism(s)	Geography	Prevalence[a] (millions)
Giardiasis	*Giardia lamblia*	worldwide	unknown
Trichomoniasis	*Trichomonas vaginalis*	worldwide	unknown
Leishmaniasis	*Leishmania brasiliensis*	South and Central America[b]	
	Leishmania donovani	S. America and S. Russia	
	Leishmania tropica	Asia, N. Africa, and S. Europe	
	Leishmania mexicana	Mexico, Central America, and Amazon basin	
Trypanosomiases:			
-South American (Chagas' disease)	*Trypanosomus cruzi*	S. America and Central America	12[c]
-African (sleeping sickness)	*T. brucci gambiense*	Lower Central African cross section[d]	
	T. brucci rhodiense		
Amebiasis			
-Intestinal	*Entamoeba histolytica*	worldwide	unknown
-Tissue dwelling (hepatic)	*Entamoeba histolytica*	worldwide	unknown
-Malaria			
	Plasmodium falciparum	S. E. Asia, tropical	200[c]
		Subtropical Africa	100[e,f]
	P. malariae, falciparum	Asia, subtropical; South America	
	P. vivax		
		Tropical Africa, Western Pacific	
	P. ovale		
Toxoplasmosis	*Toxoplasma gondii*	worldwide[d]	
Pneumocystosis	*Pneumocystis carinii*	Europe, North America	

II. Helminthic diseases	Causative organism(s)	Geography	Prevalence[a] (millions)
Schistosomiasis (blood flukes)	*Schistosoma mansoni*	Africa	400[g]
	S. haematobium	Far and Middle East	
	S. japonium	South and Central America	
Giant intestinal fluke	*Fasciolopsis buski*	Eastern Asia	15[g]
Chinese liver fluke	*Opisthorchis sinensis* or *chlonorchis sinensis*	S. E. Asia	28[g]
Lung fluke	*Paragonimus westermani*	S. E. Asia	5[g]

(Continued)

Table 7-1. *(Continued)*

II. Helminthic diseases	Causative organism(s)	Geography	Prevalence[a] (millions)
	P. africanus and uterbilaterus	W. Africa	
Tapeworm			
-Pork	*Taenia solium*	worldwide	65[g]
-Beef	*T. saginata*	worldwide	65[g]
-Fish	*Diphillobothrium latum*	Scandinavia, Canada	13[g]
-Dwarf	*Hymenolepis nana*	worldwide	29[g]
Intestinal			
-Roundworm	*Ascaris lumbricoides*	worldwide	986[g]
-Whipworm	*Trichuris trichiura*	worldwide	538[g]
-Pin(thread)worm	*Enterobius vermicularis*	worldwide	291[g]
-Hookworm	*Ancylostorna duodenale* or *Necator americanus*	tropics and subtropical areas	716[g]
	Strongyloides stercoralis	worldwide	56[g]
-Strongyloides			
Tissue			
-Filariasis	*Wuchereria bancrofti*	Africa, Asia	296[g]
-Malayan	*Brugia malayia*	Asia	
River blindness (Onchocerciacis)	*Onchocerca volvulus*	S. Africa	39[g]
Calabar swellings	*Loa loa*	Africa	26[g]
Guinea worm	*Dracuncelus medinensis*	Africa, India	79[g]
Trichinosis	*Trichinella spiralis*	worldwide	30[g]

[a] Estimates.
[b] Marinkelle, 1980.
[c] Bloom, 1979.
[d] Beaver et al., 1984.
[e] WHO, 1984.
[f] See text.
[g] Peters and Gillens, 1982.

eases and their known endemic areas. The number of humans presently afflicted is staggering (e.g., malaria, schistosomiasis, roundworm, hookworm, and whipworm). Overall estimates for the world are as high as 3 billion people. That would easily make parasitic diseases the world's leading health problem. Though many of these diseases were thought of as a "third world" problem, this is not so anymore. The ease of air travel and large migrations of people into more temperate climates now presents physicians in the so-called developed nations with these "foreign" diseases. Even though most cannot spread there due to absence of vectors, the patients once present must be knowledgeably treated.[5]

[5] The reader should consult Beaver et al., 1984, for detailed information of clinical aspects of parasitology.

7.4. Chemotherapy of Malaria

Malaria is probably the most-studied and best-understood parasitic disease. Like several of the other protozoal diseases, it has more than one causative species. The organism has a complex life cycle involving several phases with different susceptibilities to therapy in general, or to specific drugs. The genus *Plasmodium* has more than three dozen known species. Of these, only four cause disease in humans (Table 7-1); other species infect lower primates, rodents, birds, and amphibians. *P. falciparum* accounts for 50% of total cases; *P. ovale* about 1%. *P. falciparum* results in the most severe form of the acute stage of malaria and is most often fatal, especially to children, whose resistance is likely to be least developed. The vector of transmission is exclusively by female Anopheles mosquito species whose bite injects the infectious *sporozoite* form from its salivary gland. The resistant spores containing the sporozoites easily survive the vector-to-host transition. Once reaching the liver, the exoerythrocytic cycle begins by growth, segmentation, and finally sporulation. During this preerythrocytic phase of the infection the victim has no symptoms. The mature form leaving the liver—the *merozoites*—invade erythrocytes, in which they transform to *trophozoites*. These mature within the red blood cells (blood schizonts), which ultimately burst, releasing merozoites, at this point producing the characteristic high-fever episode in the patient. These released merozoites reenter erythrocytes, repeating the cycle. This process of *schizogony*, the asexual phase, repeats until suppressed by the host's immunity—or drugs.

In a subset of the infected erythrocytes the trophozoites do not divide; rather, they develop into male and female *gametocytes*. These enter the vector (mosquito) during its subsequent blood meal, where they conjugate in the stomach of the mosquito (sexual phase), then migrate to its salivary gland ready to infect the next host. Figure 7-5 is a schematic representation of the described events.

The high effectiveness of chlorinated insecticides introduced in the post-World War II period made possible control, if not eradication, of the Anopheles mosquitoes in many parts of the world. When combined with the use of effective drugs then available (chloroquine, pyrimethamine), it appeared to some experts that the malaria "problem" was essentially solved (Kikuth, 1954). Beginning in the mid-1960s, however, mosquito strains resistant to DDT–type insecticides, and plasmodia, particularly *P. falciparum*, which is highly resistant to previously curative drugs and even drug combinations, have arisen to the point where malaria has had worldwide resurgence. The World Health Organization (WHO) now estimates 100 million acute clinical cases per year in addition to a large chronic infection reservoir. Worldwide annual mortality is estimated at 2 million, half of them children. Most disturbing is the WHO judgment that 40% of the world's population is at risk (WHO, 1984). Elimination of malaria now looks less hopeful than it did 30 years ago. In fact, chloroquine-resistant *P. falciparum* has now spread to over 40 countries in Asia, Africa, and South America.

Table 7-2 is a summary of drugs used against various protozoal diseases. Antimalarials should be able to accomplish a cure. If that is impossible, suppressive therapy might be achieved. The ideal goal, of course, is prophylaxis. The prospect of a malaria vaccine was a utopian ideal until now. The discovery of monoclonal antibodies represented a major breakthrough in immunologic technology. Research with this and other tools has now brought a malaria vaccine into the realm of possibility, even though a recent (1995) field test in Gambia appears to have failed.

It is conceivable that drugs may exist that can interfere with the plasmodial cycle at any of its various stages. The most effective would be the destruction of injected sporozoites,

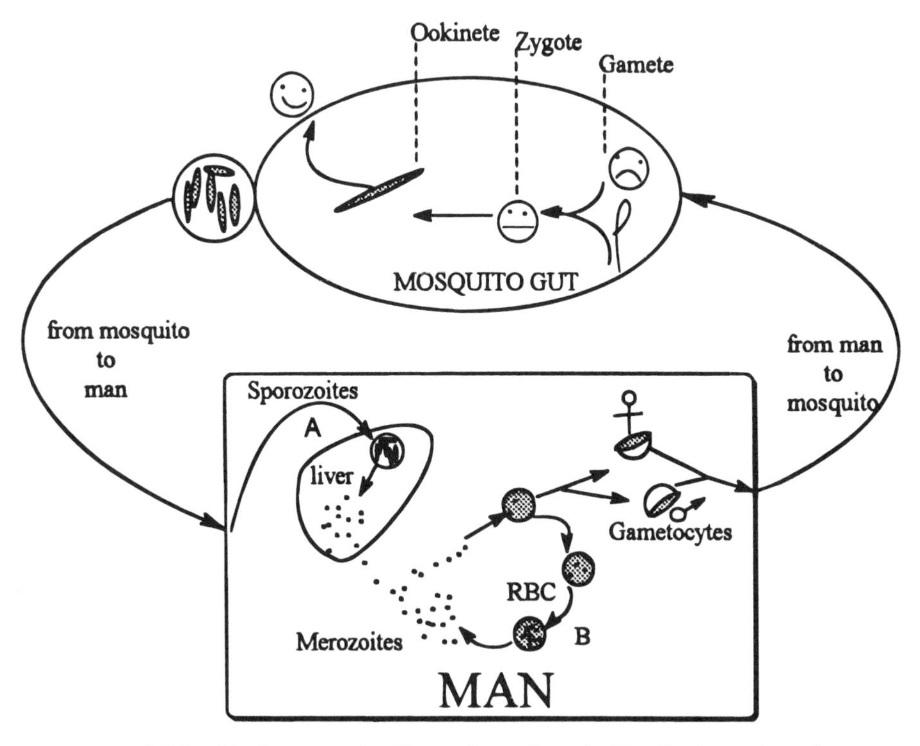

Figure 7-5. Life cycle of malarial protozoans.

Table 7-2. Drugs Used Against Certain Protozoal Diseases

Diseases	Drugs
Amebiasis	
Intestinal	Arsenicals
	Diloxanide (as furoate)
	Furazolidine
	Hydroxyquinolines
	Metronidazole
	Paramomycin
	Tetracycline
	Tinidazole
Tissue (e.g., hepatic)	Chloroquine
	Emetine
	Metronidazole
Giardiasis	Amodiaquine
	Mepacrine
	Metronidazole
	Paramomycin
	Quinacrine
Leishmaniasis	Amphotericin B
	Meglumine antimonate

(*Continued*)

Table 7-2. *(Continued)*

Diseases	Drugs
	Pentamidine
	Stibogluconate sodium
	Tartar emetic
Malaria	Amodiaquine
	Cycloguanil
	Chloroquanide
	Chloroquine
	Halofantrine
	Mepacrine
	Mefloquine
	Pamaquine
	Primaquine
	Proguanil
	Pyrimethamine
	Quinine
	Sulfonamides
	Sulfones
	Tetracyclines
Trichomoniasis	Metronidazole
	Tinidazole
Trypanosomiasis	
Chagas' disease (S. American)	Benznidazole
	Nifurtimox
Sleeping sickness (African)	Diminazine (veterinary)
	Melasoprol
	Pentamidine
	Suramin
	Tryparsamide
Toxoplasmosis	Clindamycin
	Pyrimethamine
	Spiramycin
	Sulfonamides
	Tetracyclines
Pneumocystosis carinii	Cotrimoxazole
	Pentamidine

before they settle in the host's liver. Such sporozotocides have not been found. The next opportunity arises during the primary exoerythrocytic, or liver, phase in which the exoerythrocytic schizonts arise. Drugs able to effect treatment here are mainly 8-aminoquinolines (Fig. 7-6). Even though the plasmodia may not be eradicated, clinical symptoms are suppressed. Merozoites can return following treatment, causing a relapse.

On entering red blood cells the merozoites, which are released from the ruptured parenchymal liver cells, mature to form primary blood schizonts. Several drugs act rapidly as blood schizontocides during this erythrocytic stage of the cycle against all four species. These include quinine, mefloquine, chloroquine, amodiaquin, and quinghaosu. Only *P. falciparum*, however, can be eradicated and cured since this species does not recycle into the liver to produce secondary tissue schizont. Primaquine is probably the only clinically useful drug as a (secondary) tissue schizonticide. The drug is also gametocidal, thus offering an additional point of blockade in the cycle. Table 7-3 summarizes the preceding information.

Table 7-3. Action of Antimalarial[a]—Classification

Drug	Chemical	Pharmacological	Plasmodial species
Chloroquine	4-aminoquinoline	Blood schizontocidal[b]	falciparum, vivax
Mefloquine	4-quinolinomethanol	Blood schizontocidal[b]	falciparum, vivax
Quinine	4-quinolinomethanol	Blood schizontocidal[b]	falciparum, vivax
		gametocidal[c]	vivax, malariae
Primaquine	8-aminoquinoline	Blood schizontocidal[b]	vivax
		Tissue schizontocidal	vivax
		Gametocidal	vivax, falciparum
Pyrimethamide[d,e]	2,4-diaminopyrimidine	Blood schizontocidal	falciparum, vivax
Proguanil[d,e]	Biguanide	Blood schizontocidal	falciparum, vivax
Sulfonamides[f]	Sulfonamides	Blood schizontocidal	falciparum, vivax
Sulfones[f]	Sulfones	Blood schizontocidal	falciparum, vivax
Quinghaosu	Sesquiterpene peroxide	Blood schizontocidal	falciparum ?
Tetracycline	Tetracycline antibiotic	Blood schizontocidal	falciparum ?

[a] Compiled from Dietrich and Kern (1981) and other sources.
[b] Suppressive drugs acting on asexual erythrocytic stage, achieving radical cure in *P. falciparum*, "clinical" cure in other species.
[c] Destroys sexual forms of *P. vivax, malariae, ovale.*
[d] Can be causal prophylactics in preerythrocytic stage or prevent relapse; radical cures possible.
[e] Inhibitors of dihydrofolate reductase.
[f] Inhibitors of dihydropteroate synthetase.

The repetition of the erythrocytic cycle of *P. falciparum* results in the observed clinical manifestations. Either drugs or a sufficient host defense can stop the cycle. *P. vivax* and *ovale*, however, cause relapsing symptoms. The reason is that a fraction of merozoites remains dormant for long periods with these species. Clinical symptoms can reappear—even after years—when ultimately released *P. malariae* infections also can activate after years of dormancy.

One result of these complex differences of plasmodial life cycles is that drugs affecting only asexual blood forms can cure susceptible strains of falciparum but can only suppress the relapsing types of the disease.

Quinine is the main alkaloid obtained from the bark of the cinchona tree (*Cinchona officinalis*, and other hybrids). It is a levorotatory compound (its dextrorotatory diastereoisomer quinidine, obtained from the same source, is used primarily as a cardiac antiarrhythmic, Chapter 10). The cinchona trees cultivated on the island of Java, Indonesia, have the highest total alkaloidal content in its bark, 7–10%. About two-thirds of this total is quinine. The other significant alkaloids are quinidine, the diastereoisomeric pair cincho-

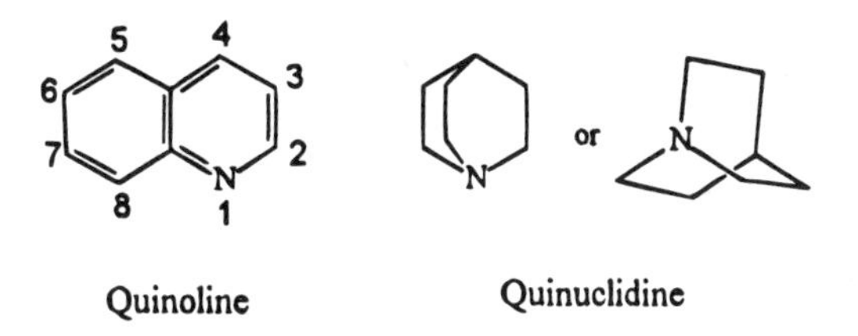

Quinoline Quinuclidine

nine and cinchonidine (6′-desmethoxy), cupreine (6′-OH), dihydrocupreine (the vinyl group at C-3 is reduced to C_2H_5), and an additional dozen alkaloids.

The quinine structure (Fig. 7-6) has two "components": the *quinoline* portion, which is bonded to the 8 position of the *quinuclidine* ring via a methanol bridge at its 4′ position. The stereochemistry of quinine, (–) 8S:9R, differs from quinidine, (+) 8R:9S; the chiral

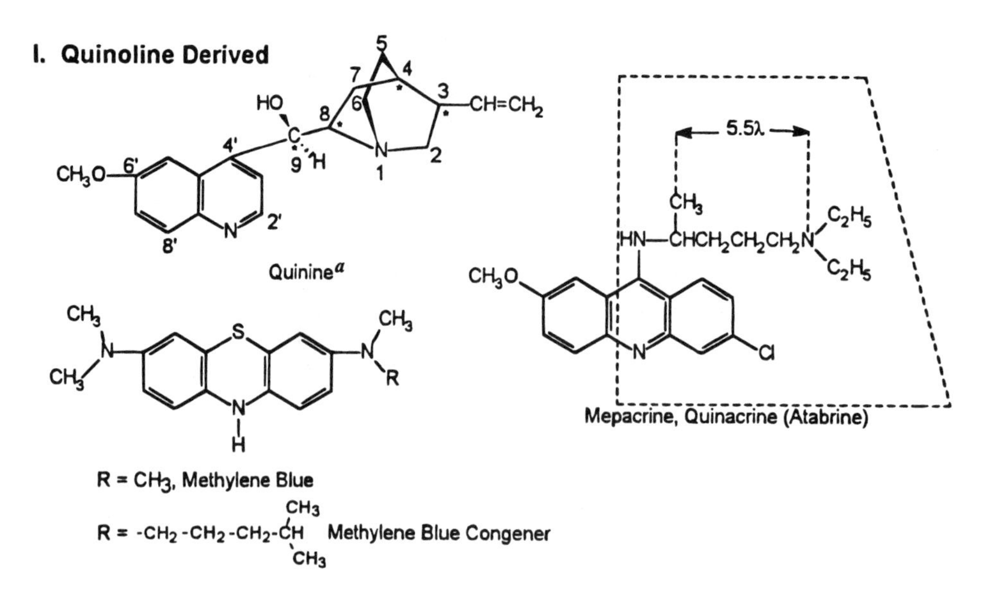

8-Aminoquinolines

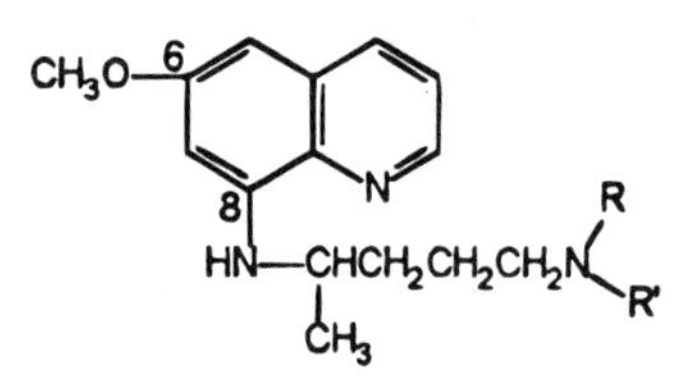

R, R' = C_2H_5 - Pamaquine
R, R' = H - Primaquine

4-Aminoquinolines

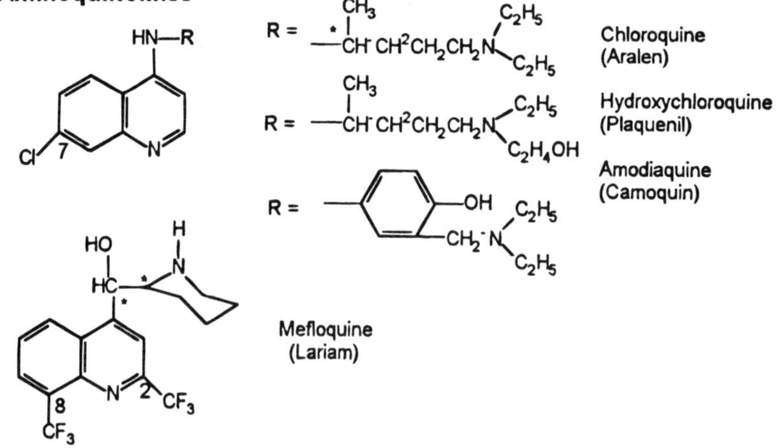

Figure 7-6(A). Antimalarial drugs.

II. Antifolates

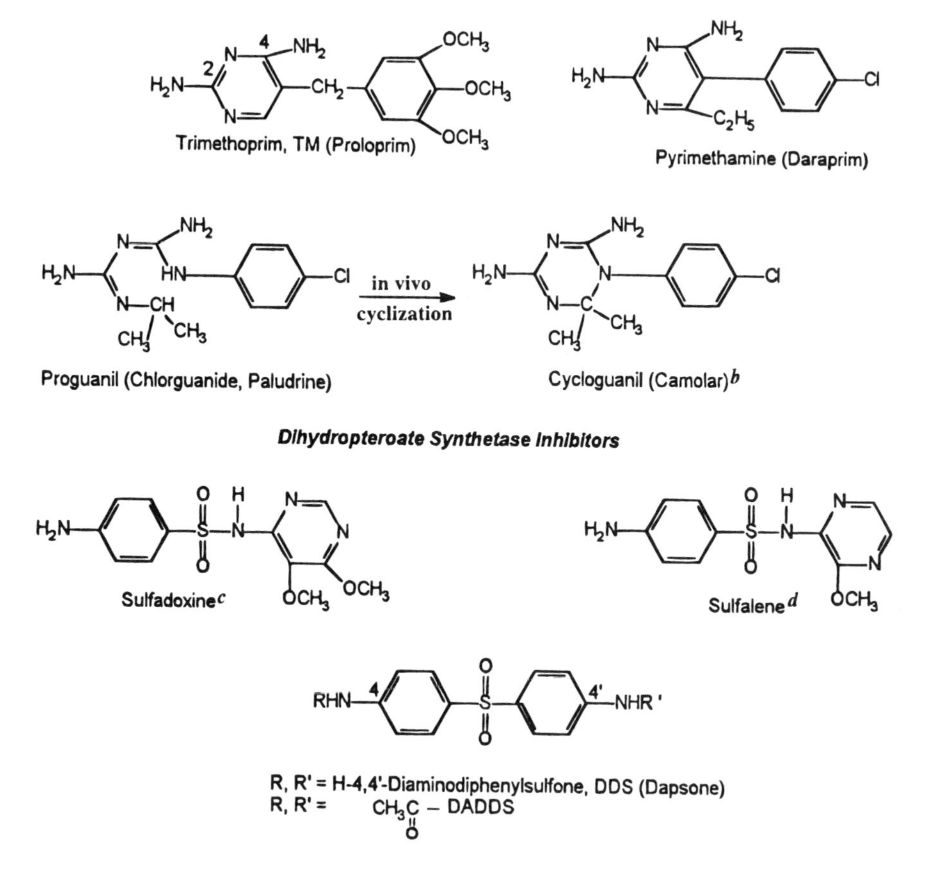

III. Terpene Peroxide

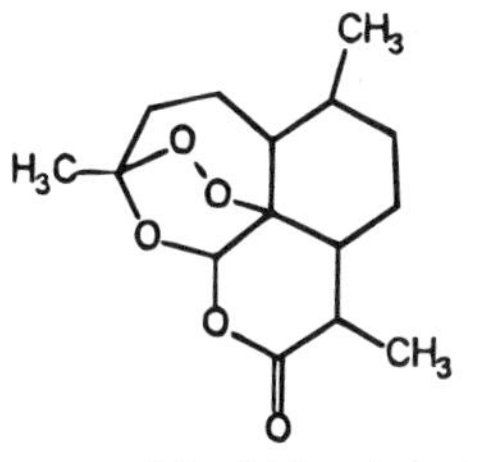

[a]Asterisks designate chiral centers. [b]Available as insoluble pamoate salt for long-acting IM use. [c]Combined with pyrimethamine as Fansider. [d]Used in combination with pyrimethamine (see text).

Figure 7-6(B). Antimalarial drugs.

centers at C-4 and C-3 are the same. Hence, the two drugs are diastereoisomers (not enantiomers). The C-9 epimers (OH only inverted) of both compounds are inactive. The two nitrogens generate dramatically different basicities: the aromatic (quinoline) nitrogen, pKa[6] = 4.2, and the tertiary aliphatic nitrogen (quinuclidine), pKa = 8.8. (The bicyclic quin-

[6] pKas given are for protonated amine.

uclidine might be viewed as a triethylamine in which the three ethyl groups are "tied" behind the back of the nitrogen.) The 9-OH seems essential to activity; oxidation or esterification decreases drug activity dramatically. As in many other biologically active alkaloids, the active cinchona drugs contain the characteristic four-atom -O-C-C-N- grouping, represented by 9-OH, C-9, C-8, and 1-N in quinine. The nitrogen atom can function in a simpler environment than quinuclidine (see mefloquine later).

The Spaniards found the natives using powdered cinchona bark to treat fevers in the seventeenth century in South America and brought it back to Europe. Quinine alkaloid was isolated from the bark in 1820 by Caventou and Pelletieri, and its structure was determined about a century later (1949). By the mid-1920s quinine and related alkaloids were still the only *specific* antimalarials available. The introduction of synthetic pamaquine (Fig. 7-6) offered the first alternative. Quinacrine (mepacrine) followed in 1933. Primaquine and chloroquine (Fig. 7-6) developed during World War II (see later), and became widely used following the war. Both drugs, but particularly chloroquine, were less toxic and more effective than quinine. In fact, by the early 1960s quinine became almost obsolete and might have become a historical footnote if the high and serious worldwide resistance of *P. falciparum* had not developed to the newer drugs. In spite of its shortcomings—rapid hepatic hydroxylation at the quinolinyl 2′ position to an almost inactive metabolite—quinine is used today in combination with other drugs to treat chloroquine-resistant infections. Thus, even though the potency is low and the duration of action relatively short (3× per day dosing), quinine can still be effective.

7.4.1. The 8-Aminoquinolines

In the late nineteenth century, Guttmann and Ehrlich (1891) reported a low level of antimalarial activity by the basic dye methylene blue (Fig. 7-6). This resulted from Ehrlich's early work with dye staining of microorganisms. Prompted by a World War I quinine shortage in Germany, synthesis research utilized this clue to find that dialkylaminoalkyl moieties, replacing one methyl group of the dye's diemthylamino function, intensified the activity. Introducing such groups into the quinolines, particularly aminoquinolines, resulted in more active compounds. After several dialkylaminoalkyl substitutions on 8-aminoquinolines resulted in some active compounds, the 8[(4-diethylamino-1-methybutyl)]amino derivative of 6-methoxyquinoline—pamaquine—was introduced in 1926. Unlike quinine, this drug was able to reduce the relapsing episodes of vivax malaria. Later, structural modifications produced the unsubstituted primary aminoalkyl derivative, primaquine, whose higher activity and considerably lower toxicity (particularly the tendency toward hemolysis) essentially replaced pamaquine as the tissue schizonticide of choice.

By experiments showing that drugs such as pamaquine and primaquine were active only in vivo but not in vitro and that plasma levels and therapeutic values were not necessarily related, it was established that the compounds, as administered, were actually pro-drugs (Fig. 7-7). It was established that O-demethylated-6-hydroxy, and the further hydroxylated 5,6-dihydroxymethabolites 1 and 2 (Fig. 7-7), probably via an equilibrium with quinone structures 3 and 4, constitute the bioactive species. In fact, the metabolites were active in vivo.

After the first success with the "synthetic" quinine primaquine, the search focused on 9-dialkylaminoalkyl acridines. The result was quinacrine (mecaprine, outside the United States). Figure 7-6 illustrates how both methylene blue and quinine can be viewed as structural leads for this drug that achieved extensive use by the Allied forces during the early

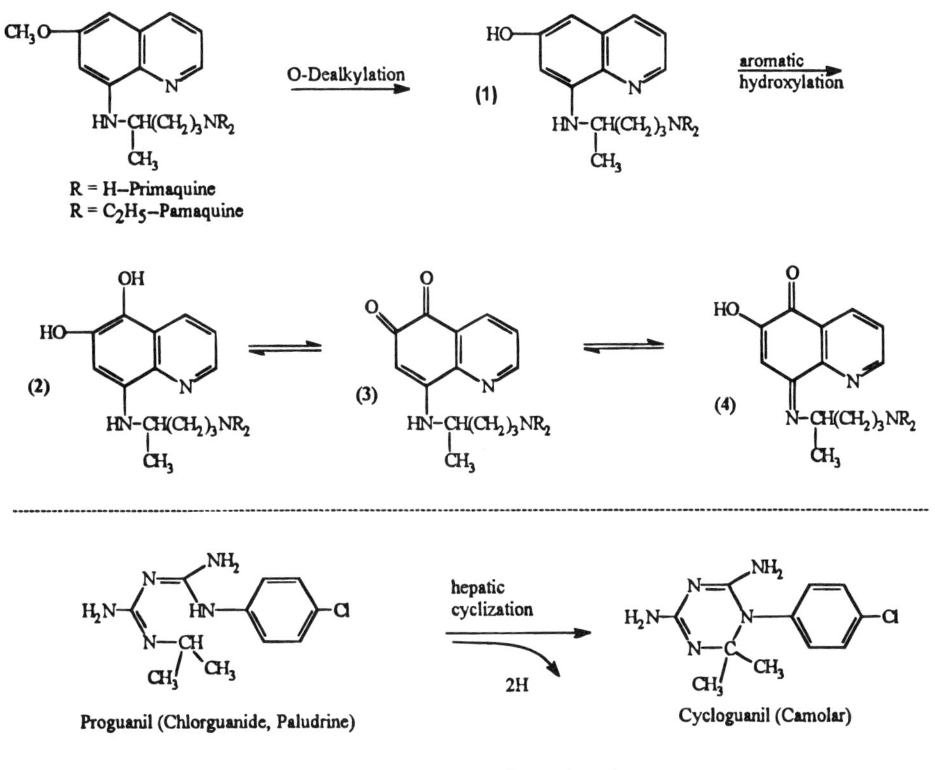

Figure 7-7. Metabolic activation of antimalarial pro-drugs.

years of World War II as a result of the unavailability of quinine. Like quinine, quinacrine was an exoerythrocytic schizontocide that also controlled symptoms in *P. vivax.* Its yellow skin discoloration due to collagen binding in the dermis did not endear the drug to German troops. With the discovery of the "colorless" and otherwise superior chloroquine (see later) later in the war, quinacrine use waned. Its utility today is limited to giardiasis and as an anthelmintic (tapeworms).

7.4.2. The 4-Aminoquinolones

Chloroquine, CQ (Fig. 7-6), was undoubtedly the best antimalarial drug to result from the intense search carried out during World War II for synthetic quinine substitutes by combatant nations on both sides. The Germans apparently synthesized the drug first, as a non-coloring, simplified structural variant of Mepacrine; it was then rediscovered in the United States. CQ, a blood schizonticide, readily achieved radical cures against *falciparum* and clinical cures (i.e., symptoms elimination) in other species such as *vivax.* The drug is also effective as a suppressive in endemic areas or as a prophylactic agent before traveling to and being in such areas. One measure of its effectiveness is that doses for both these indications is as little as 300 mg of the base *once* a week. Treatment of the acute disease requires higher doses. Its potent and rapid action on all species of human malaria and relatively low toxicity made it the drug most responsible for the premature pronouncements that malaria had been eliminated. Unfortunately, the appearance of chloroquine-resistant *P. falciparum* strains in most endemic malaria regions previously described dramatically changed the picture, which continues to deteriorate. In fact, many of these strains have now become multiple drug resistant, including CQ, mefloquine, quinine, and even the structurally unrelated quinghaosu.

On the positive side, it should be recalled that the CQ's gametocidal action in the other three malarial species is still useful in reducing the human reservoir of those organisms. Combination therapy of CQ with sulfonamides and/or dihydrofolate reductase inhibitors can overcome resistance in some cases (see later).

A potentially momentous event was the discovery that the calcium channel blocker verapamil (Chapter 10) seems capable of *reversing P. falciparum* resistance to CQ, even in strains resistant to multiple drugs. It is particularly interesting that the cardiovascular drug reversed *falciparum* resistance at similar concentrations at which it reversed resistance to multiple-drug-resistant cancer cells. This suggests a similar mechanism. For neoplasms this was demonstrated to be an inhibition of the rapid outflow of carcinolytic agents from the cell (the previously resistant tumor cells apparently resensitized them. Whether the molecular mechanism by which verapamil achieves this inhibition in cancer cells, and putatively in plasmodial cells, is similar to the drug's ability to inhibit Ca^{2+} efflux from myocardial cells has not been established. In any case, if the in vitro results seen with resistant cultures can be achieved at the clinical level, then a method to overcome worldwide resistance of *P. falciparum* infections may be feasible.

The other 4-aminoquinolines, hydroxychloroquin, HCQ, and amodiaquine, are of use primarily as alternatives to CQ, offering no advantages in terms of resistance, although HCQ may be somewhat less toxic.

CQ and HCQ have been found useful in the suppressive treatment of autoimmune inflammatory diseases such as rheumatoid arthritis (RA), and systemic lupus erythematosus (SLE) (see Chapter 5). The higher dosages required increase retinopathy as a complicating toxicity.

Mefloquine (Lariam) (Fig. 7-6), may be viewed as a chemically "simplified" congener of quinine wherein the quinuclidine ring is replaced by piperidine. The CF_3 at the 2 position of the quinoline portion will prevent oxidative metabolism there. The resulting half-life is an extraordinary 15 days. It has been effective in some resistant *P. falciparum* infections with as little as a single 0.5–1.5 g dose. An effective treatment has been a short course with quinine followed by a single mefloquine dose. Still, resistance can arise rapidly. In fact, resistant strains of *falciparum* were reported in Thailand even before the drug was used there. Its mechanism of action is still not clear. It does not bind to DNA. In chloroquine-resistant organisms where that drug does not accumulate, mefloquine does, which would give it access to putative receptors and help explain its effectiveness. Somewhat arguing against this is that all four optical isomers exhibited approximately the same level of activity.

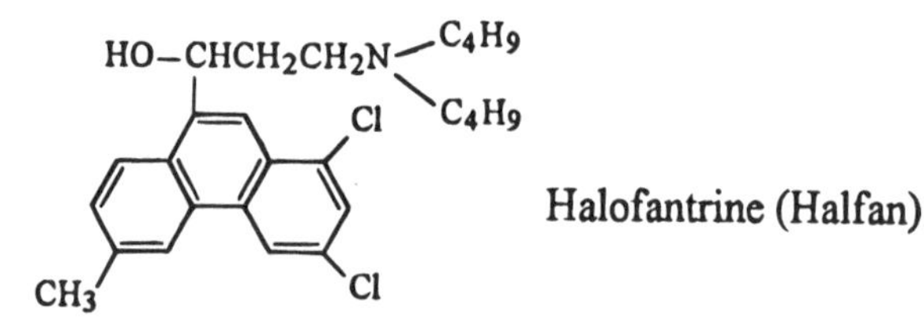

Halofantrine (Halfan)

An interesting entry (1992) is the phenanthrene compound halofantrine (Halofan). Earlier clinical trials exhibited 85% or better effectiveness, including some chloroquine-resistant *P. falciparum* strains. Since the drug does not eliminate exoerythrocytic parasites, the likelihood of relapse in the treatment of *P. vivax* is significant.

7.4.3. Dihydrofolate Reductase Inhibitors

The chemistry and significance of dihydrofolate reductase (DHFR) was discussed in Chapter 4. Analogs of folic acid such as methotrexate (MTX) were found to bind tenaciously (but not covalently) to the enzyme's active site, exhibiting intense but nonselective cytotoxicity and thereby becoming a useful anticancer drug. However, the inability of MTX to enter microbial cells (due to lack of a carrier mechanism that vertebrate cells have) and its considerable human toxicity made it useless as an antimicrobial agent. In a classic series of experiments to discover the molecular component of MTX responsible for the enzyme inhibition, Hitching's group found it to reside in the 2,4-diaminopyrimidine part (Falko et al., 1951). To improve microbial cellular uptake lipoidal groups were added to this component to facilitate this property. Specifically, 2,4-diamino-6-ethyl-5-(*p*-chlorophenyl)pyrimidine resulted in the major antimalarial drug pyrimethamine (PM) (Figure 7-6). The choice of a phenyl ring substituent at C-5 was particularly fortuitous. Over a decade later DHFR binding studies showed the phenyl substituent on the pyrimidine-folate analog Ia was a 12–20 times better inhibitor of the enzyme than the methyl group Ib. It was soon established that this improvement was primarily due to increased hydrophobic (van der Waals) forces. Although highly effective as an antiplasmodial agent, pyrimethamine exhibited inferior antibacterial activity. Further structural modifications that essentially provided the phenyl ring with some "peripheral" hydrophilicity (3,4,5-trimethoxy) resulted in the excellent broad-spectrum antibacterial trimethoprim (TM) (Fig. 7-6).

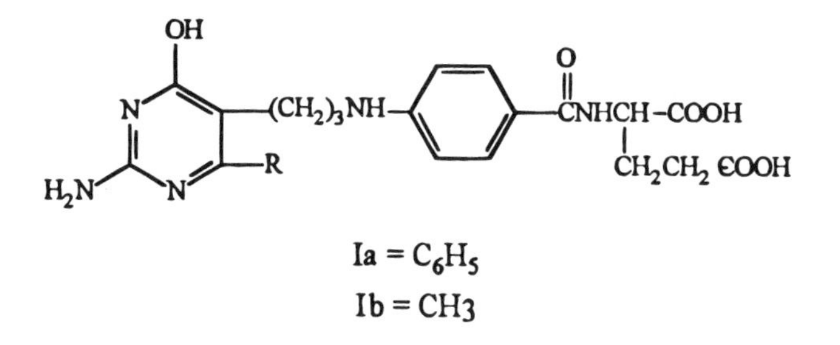

$Ia = C_6H_5$
$Ib = CH_3$

In order to understand the selective toxicity of drugs such as PM and TM, and their significant differential activity between protozoa and bacteria, a brief discussion of multiple forms of enzymes is in order.

Isoenzymes (or isozymes) are multiple forms of an enzyme that differ from each other in such properties as substrate affinity, maximum activity, or regulatory properties. They may be found in different tissues or portions of the same cell. For example, thymidine kinase catalyzing phosphorylation occurs as two isoenzymes—one in the cytoplasm and the other associated with the mitochondria of the same mammalian cell. Lactic dehydrogenase, which catalyzes the reduction of pyruvic acid to L-lactic acid exists in five isozymic forms. These are tetramers formed by the association of two polypeptides of equal size: H (heart) and M (muscle).

More germane to the present discussion are differences between enzymes that carry out the same functions in different organisms. These have been referred to as *analogous* or *homologous* enzymes. The structural differences between these enzymes may be more significant than those encountered within the same organisms. It is precisely the greater significance of these differences—such as varying amino acid composition at or near the active site—that can become the basis for designing superior drugs, once known. Such analogous enzymes have been found to exhibit significant differences in kinetics, substrate, inhibitors, or cofactor specificity, and electrophoretic properties.

The enzyme DHFR has been particularly well studied over the years with regard to differences in substrate affinities as a function of source. The impetus was the development of reversible inhibitors such as PM, TM, cycloguanil, and its pro-drug, proguanil (Fig. 7-6), which have antimalarial and antibacterial activities, and the carcinolytic drug MTX. Table 7-4 summarizes some enzyme inhibition data. It can be seen that inhibitory concentrations of MTX required to produce a 50% inhibition of DHFR from different sources varies over a relatively narrow range, while those of the 2,4-diamino-pyrimidines sometimes do so dramatically. Thus, in the case of PM one sees a 3,600-fold difference in inhibitory ability between the enzyme from a plasmodial species and the human liver. This is sufficient to account for the selectivity of toxicity seen between the malarial parasite and the mammalian cell. The 60,000-fold difference for TM and *E. coli* is even more dramatic, easily explaining the clinical effectiveness of TM for uncomplicated UTIs caused by this bacterium. The data also show that although TM would be active against the plasmodia species, PM is not likely to have antibacterial activity (50,000-fold differential in inhibiting ability).

The plasmodial enzyme must have some significant differences from that of other sources. Certain variations of particular amino acids have been identified and even related to tertiary structural features of the "pockets" into which the pteridine nucleus fits. Stereoscopic representations generated from X-ray diffraction data have been obtained from DHFR co-crystallized with MTX, and the co-enzyme NADPH has helped elucidate "goodness of fit," or its absence, of the inhibitor.

Another concept that should be considered at this point is *sequential blockade* as it relates to chemotherapy. Considering the outline of the folate biosynthesis scheme (Fig. 7–8), the likelihood that the blockade with selective agents of more than one reaction in sequence will increase the therapeutic value of treatment is apparent.

The two steps involved are the biosynthesis of dihydropteroic acid catalyzed by dihydropteroate synthetase and inhibited by sulfonamides and sulfones (Chapter 2), and the reduction of dihydrofolic acid by DHFR, which can be inhibited by MTX, PM, TM, and other DHFR inhibitors. Hitchings proposed such combinations, which ideally should pro-

Table 7-4. Selectivity of Dihydrofolate Reductase Inhibition (nM Concentration for 50% Inhibition)

Enzyme source	Pyrimethamine	Trimethoprim	Cycloguanil	Methotrexate
Plasmodium berghei	0.5	70	3.6	0.7
E. coli	25,000	5	—	1
Liver, human	1,800	300,000	—	90
Liver, rat	700	260,000	—	2
Erythrocyte, mouse	1,000	1,000,000	1,600	—

Data compiled from: Burchall and Hitchings (1965); Ferone et al. (1969); Ferone (1970); Jaffe and McCormack (1967).

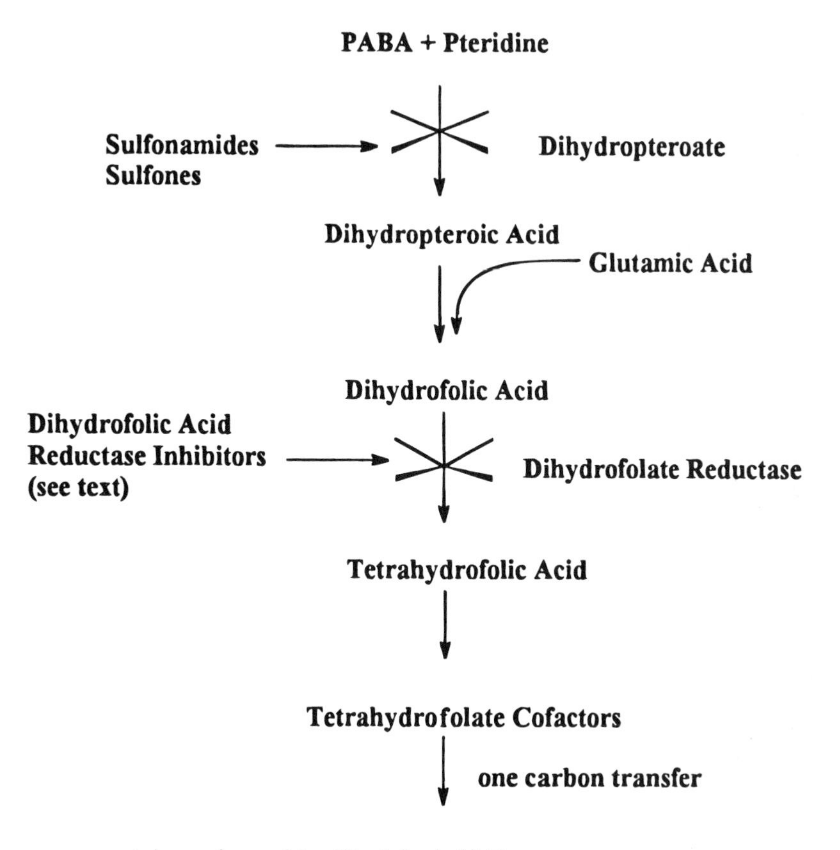

Figure 7-8. Outline of folate scheme. (For the chemical details of this scheme, see Figures 4-13 and 2-3 and accompanying discussion.)

duce two clinical advantages. The first is *synergism*, which is a combined effect greater than the additive effects of the individual components of the drug product. In fact, a combination of TM and sulfadiazine at one-eight of their respective ED_{50s} had an antimalarial effect equivalent to the ED_{50} of either drug singly. Further tests with similar combinations have corroborated this effect. The second potential advantage is to prevent, or at least delay, the emergence of resistance. The theory is that even though a single mutation in a microorganism resulting in resistance to either drug alone can arise readily, the probability of multiple simultaneous mutations is much less likely. This has in fact been found to be the case, even with some multidrug-resistant malarial organisms. Some of the sequential blocking combinations currently in use are listed in Table 7-5. The long half-life of PM (ca. 4 days) requires it be combined with sulfone or sulfonamide of a comparable, or at least overlapping, long excretion rate, so that each drug's peak effectiveness would more or less coincide. The Dapsone-PM combination (Maloprim^R, for example, has been useful in chloroquin-resistant *P. falciparum* infections). The PM-sulfadoxine product (Fansidar^R) contains a 20:1 ratio of the two drugs and is also useful in similar situations. It is now considered to be effective against some strains of chloroquine-resistant *falciparum* yet toxic (*P. vivax* does not respond). However, strains resistant to such combinations are being encountered.

Table 7-5. Sequential Blockade Combinationsa

Dihydropteroate synthetase inhibitor	Dihydropteroate reductase inhibitor	Product name	Primary use
Dapson (28)	Pyrimethamine (83)[b]	Maloprim[R]	Malaria Leprosy
Sulfadoxine (75)	Pyrimethamine (83)	Fansidar[R]	Malaria
Sulfadiazine (7)	Pyrimethazine (83)	—	Malaria
Sulfamethoxazole (66)	Trimethoprin (11)[b]	Co-Trimoxazole Bactrim[R] Septra[R]	Bacterial infections Pneumocystitis[c] Leishmaniasis[d] Isosporiasis[e]

[a] Numbers in parentheses are median elimination $t_{1/2}$ in hours.
[b] See Figure 7-6 for structures.
[c] Caused by *Pneumocysti carinii.*
[d] Caused by species of the protozoal genus *Leishmania.*
[e] Caused by *Isospora belli,* an intracellular coccidial protozoan.

Even though the synergism observed with antifolate–sulfonamide combinations appears real, whether or not the mechanism is truly a sequential blockade is questionable. For example, it was shown that a potent DHFR inhibitor, 2,4-diaminopteroyl aspartate, is not synergistic with sulfamethoxazole. In addition, it was found that DHFR isolated from *E. coli* could be inhibited by sulfonamides, suggesting a multiple simultaneous inhibition of DHFR by both drugs. Curiously, it was also found that the TM potentiates sulfonamides that alone are resistant to the bacterium tested.

It has also been called into question whether TM if used alone is more likely to induce resistant bacterial strains than in the cotrimoxazole product. A decade-long experience of treating UTIs with TM alone in Finland found no emergence of widespread resistance. Opposite results were apparently found in a British study with species of *Enterobacteriaceae.*

Chinese herbal medicine may have made a contribution to the treatment of malaria. The herb Quinghao (*Artemisia annua*), in use over a millennium, has finally yielded its active principle quinghaosu or artemisinine (Fig. 7-6), a sesquiterpene lactone containing a peroxide bridge. Clinical evaluation of it and several derivatives in the late 1970s was reported to have achieved successful treatment in over 2,000 patients, many with chloroquine-resistant *falciparum.* Their activity is blood schizonticidal. The mechanism does not appear to be antifolate or intercalation. The peroxide is essential, which raises a suspicion of free radical damage to the parasite.

7.5. Other Antiprotozoal Drugs

Table 7-2 lists the various drugs that have been useful for a variety of protozoal diseases. It will be noted that several are indicated for more than one condition (e.g., pentamidine against leishmaniasis, trypanosomiasis, and pneumocystosis). Several of these not yet discussed will now be considered.

Protozoa being single-celled are the most primitive members of the animal kingdom. There are probably 45,000 species known. It is fortunate that only some parasitize humans and higher animals. Protozoostatic drugs suppress clinically relevant development stages, but not latent stages. Thus relapses are very characteristic for protozoal diseases (e.g., malaria). Most of the useful drugs available have a narrow safety margin.

Our knowledge to date frequently makes predicting the efficacy of a particular drug impossible. In a few cases, such as the antifolates, the site of action is known, yet even here we do not always understand the difference in effectiveness between protozoan species. In vitro screening is frequently no predictor of clinical results. Several antiprotozoal drugs are effective against nonprotozoal diseases since they were originally developed against other infective organisms and only later discovered to be active against protozoa. Certain antibiotics and antimetabolites are primary examples.

Amebiasis is primarily an intestinal disease caused by *Entamoeba histolytica*, which is a cosmopolitan organism usually transmitted by ingestion of materials contaminated by cysts that descend the GI tract. Trophozoites develop and remain in the colon, caecum, and sigmoidal area. Ulceration of the intestinal wall results, causing symptoms of dysentery. Metastatic lesions sometimes develop in other organs such as the liver (hepatic amebiasis). This is more difficult to treat. Amebicides in use today have all been developed empirically. The nitroimidazoles metronidazole, tinidazole, and the toxic nitro heterocycle niridazole (Fig. 7-4) are effective at all sites.

Emetine (Fig. 7-9) in the form of the crude drug obtained from the roots and rhizomes of Ipecac (*Cephaelis ipecacuanha*) has been in use since the seventeenth century. The alkaloid, as the hydrochloride, has been used parenterally to treat amebic dysentery. It is also effective in hepatic infestation, but not against amebic cysts. Because of its cardiac toxicity and emetogenic properties, it has been superseded by metronidazole and chloroquine, but it is still used as an alternative. The amebicidal mechanism of emetine is protein synthesis inhibition by interference of peptidyl–RNA translocation. Since this action is general to eukaryotic cells, its relative selectivity in the presence of mammalian cells is not well understood. Unrelated uses of Ipecac (presumably due to its alkaloid content) are as an expectorant in cough preparations and an emergency emetic (Syrup of Ipecac).

Halogenated 8-hydroxyquinolines such as 5,7-diiodo-8-hydroxyquinoline (Diiodohydroxyquin, Fig. 7-9) and 5-chloro-7-iodo-8-hydroxyquinoline (Vioform) are useful as oral amebicides. Only the former is available in the United States for this purpose. (Iodochlorhydroxyquin is available in topical formulation for its dubious antifungal action.) The mechanism of action of these quinolines has not been established.

Diloxanide, a dichloroacetylaminophenol, as the highly insoluble furoate ester (Fig. 7-9) is amebicidal and effective in eliminating cysts refractory to other drugs. It is interesting that esters other than of furoic acid are less effective and less well tolerated. Cure rates range from 83 to 95%. The mechanism of action is not known.

Leishmaniasis is a group of diseases caused by several species of *Leishmania* (Table 7-1). They include visceral, cutaneous, and mucocutaneous versions. Because the amastigot or unflagellated form is an intracellular infectant (in reticuloendothelial and macrophage cells), these diseases are difficult to treat satisfactorily. The macrophage penetration reduces the host's immune defenses; the intracellular protection afforded the protozoan makes direct drug contact unlikely. Available drugs are not specific for this affliction. The treatment of choice for the various forms is pentavalent antimony. However, the first antileishmanial drug was the more toxic antimony potassium tartrate (tartar emetic) (Fig. 7-9). Even though this chemical has been known for centuries, its specific antiparasitic effects were not recognized until the beginning of this century. At first the toxic effects of this trivalent antimony compound were attributed to its irritant local effects (e.g., emesis), which could be obviated by the intravenous route. However,

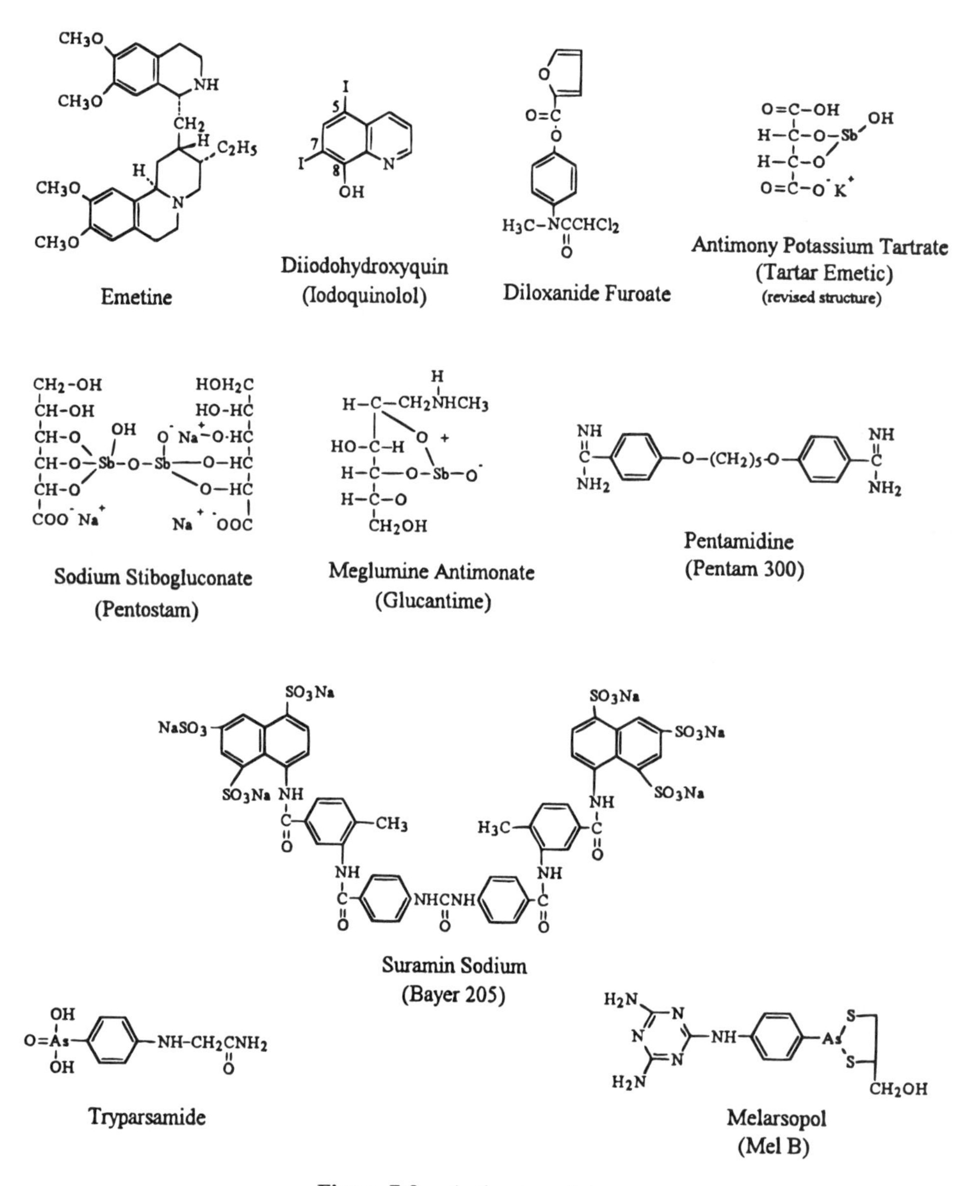

Figure 7-9. Antiprotozoal drugs.

cumulative renal, hepatic, and, particularly, cardiac damage were soon recognized. Pentavalent antimonials, being less toxic, are probably safer and more effective. They may be viewed as derivatives of the antimonide ion $[Sb(OH)_4^-]$. Sodium stibogluconate (Fig. 7-9) can be prepared from Sb_2O_5 and gluconic acid in the presence of solubilizing NaOH solution. The exact structure and composition are not known, but are based on tartar emetic by analogy. Polymeric properties of antimonic acid makes product uniformity difficult. It is a stable product. Meglumine antimonide (Fig. 7-9) is another stable pentavalent antimony product prepared from antimonic acid with N-methyl-glucamine, a glucose derivative.

The pentavalent compounds have high leishmanicidal activity. As they do not share the irritancy of the trivalent compounds, they can be given by an intramuscular route. Even though the mechanism is still unknown, the protozoan begins to be cleared once the drug aggregates within the macrophage cell at a sufficient level. It is still a mystery why the macrophage's lysosomal enzymes have no apparent destructive effect on the engulfed amastigote form. The visceral forms of the disease found in Asia and South America respond to these drugs. The variant seen in East Africa is frequently resistant. Mucocutaneous lesion often requires several courses of treatment.

Alternate drugs that are suitable when the disease is not responsive to antimony include the antifungal antibiotic amphotericin B (Chapter 2) and diaminidines such as pentamidine (see later) (WHO, 1984).

Trypanosomiasis also occurs in several forms, and each is caused by a different species of *Trypanosoma* (Table 7-1). The African Trypanosomiasis is also known as sleeping sickness because the symptoms of the terminal stages of the disease, when the parasite reaches the CNS, includes constant sleeping. The two types of the disease—Rhodesian and Gambian—are now initially treated with pentamidine (Fig. 7-9) and suramin, respectively. The arsenicals melarsopol and tryparsamide (Fig. 7-9) are reserved for the late-stage involvement of the CNS in both cases. The aim is to eliminate the parasite in both types of the disease. Pentamidine and suramin are useful only in the early stages of sleeping sickness, since they do not cross the blood–brain barrier. The usefulness of the organic arsenicals in the "sleeping" stage of the disease is due to their ability to enter the CNS. Prophylactic chemotherapy during the lymphatic or tissue stage is safer and usually more successful when possible. Treatment with arsenic, once the protozoan has entered the CNS, offers a much narrower safety margin to the patient.

Suramin is a colorless dye that evolved from Ehrlich's early work with azo dyes such as Trypan Red that cured mice infected with trypanosomes (Ehrlich and Shiga, 1904). It has therefore been viewed as the first synthetic chemotherapeutic drug (it was inactive in humans). A decade later, suramin, which is, however, trypanocidal in humans, was introduced. It is still in use today for this purpose. Even though the drug has since been shown to inhibit various enzyme systems in parasitic protozoa and worms such as filaria (see avermectin later), its mechanism of action and basis of selective toxicity has still not been elucidated.

The early history of Ehrlich's work with organic arsenicals was mentioned (Chapter 2). Arsenicals were important in the treatment of tropical protozoal diseases as well as syphilis until the advent of the sulfonamides in 1935. Today they are utilized mainly in trypanosomiasis. They are among the least well-tolerated drugs in use today and usually require hospitalization to manage toxic effects.

All organic arsenic compounds fall into one of three categories: pentavalent, trivalent, and arseno (i.e., arsenic to arsenic double bonded). Equation 7.8 relates these three forms. All readily penetrate the CNS and are therefore potentially effective in the late form of the disease.

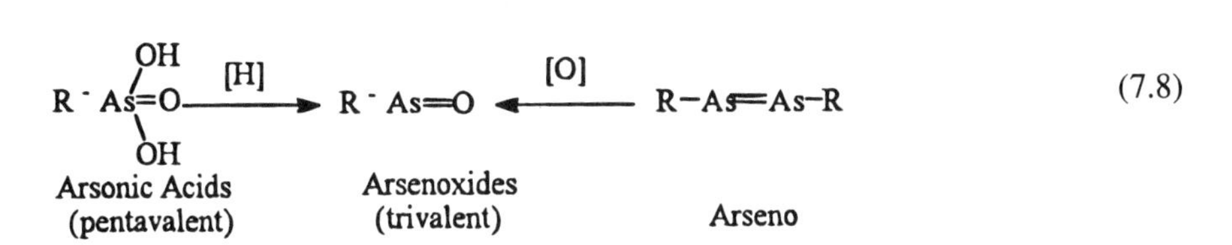

(7.8)

Ehrlich realized that his first drug Atoxyl (phenyl arsonic acid, $R = C_6H_5$, Eq. 7.8) was reduced in vivo to the trivalent arsenoxide, which he believed to be the active form. His originally postulated mechanism that the arsenoxide covalently and irreversibly bonds to sulfhydryl (-SH) groups in the trypanosome essential to the organism is basically correct (Eq. 7.9).

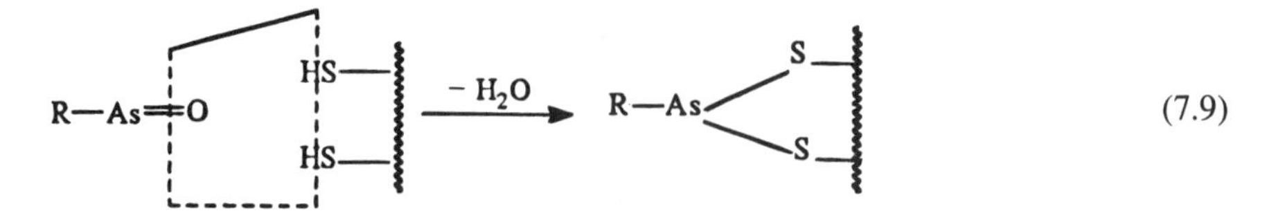

(7.9)

The fact that administration of organic sulfhydryl compounds can both reverse the toxicity of these arsenic drugs and their activity tends to support this simple idea. It seems almost incredible that the organism can develop resistance to such a nonselective protoplasmic poison as arsenic, but it does, apparently by decreased uptake.

Tryparsamide, the somewhat less toxic successor to atoxyl, and melarsopol are both considered pro-drugs today, with the former reduced in vivo to the active trivalent arsenoxide form, and the latter is metabolized back to melarson oxide, from which it is prepared by condensation with 1-hydroxy-2,3-dithiol (BAL, see Chapter 1). It is presumed that this temporary sulfhydryl derivatization serves to reduce the clinical toxicity to the host, thereby increasing the safety of this treatment. It may be interesting to speculate why the triazine-containing arsenical, melarsopol, is more trypanocidal than tryparsamide. It is possible that the symmetrically distributed six nitrogens around the aromatic ring system (N 1,3, and 5 as part of the ring, and NH_2 groups 2,4,6) gives the whole molecule greater affinity to and/or penetrability of the trypanosomal membrane. In any case, melarsopol is the drug of choice for sleeping sickness with CNS involvement. Melarsopol's significant superiority over trypanocidal arsenicals that do not have the melamine (triaminotriazine) ring is its usefulness against tryparsamide-resistant organisms.

Trypanosomiasis, which is caused by *T. cruzi*, and is primarily indigenous to the South American continent, is also known as Chagas' disease. Its symptoms differ from the African variety considerably, with cardiomyopathy being the most prominent feature. The only drugs exhibiting any useful activity against this disease are nifurtimox and benznidazole (Fig. 7-4). Nifurtimox, released in 1976, is considered the drug of choice. Even though these drugs appear curative during the acute stage, long-term administration is necessary to prevent the cleared organisms from reestablishing themselves at new sites. Poor tolerance makes neither drug completely satisfactory.

Pneumocystis carinii is a cosmopolitan opportunistic protozoal (although some view it as a fungal organism) producing lung inflammation (pneumonitis) exclusively in immunocompromised patients. It is always fatal if untreated in its acute stage. Immunodeficiency, until recently encountered primarily in patients who either had it congenitally or when induced by drugs (e.g., carcinolytics) or severe malnutrition, has become a worldwide health problem since the early 1980s. The epidemic now known as acquired immunodeficiency syndrome, or AIDS, is a disease now known to be caused by a retrovirus, the human immunodeficiency virus (HIV). The disease, by primarily affecting T-cells and thereby compromising and ultimately neutralizing the subject's cell-mediated immunity totally, makes them easy targets for opportunistic viral, fungal, and protozoal infections. Approximately 56% of AIDS patients develop a *P. carinii*–caused lung infection, pneumocystosis. Pentamidine, cotrimoxazole, and diflucan are currently used extensively to treat *P. carinii* infections. Cotrimoxazole, which is equieffective with pentamidine, but

Table 7-6. Risk Factors Predisposing to Systemic Mycoses

AIDS
Cancer chemotherapy
Dialysis (renal)
Hyperalimentation
Illicit drug use
Intravascular catheters
Organ transplantation
Prosthetic heart valves
Steroid therapy (high dose)
Urinary catheters

somewhat less toxic, may initially be the drug of choice. Because of the high doses needed, it has resulted in serious adverse reactions in 15% of patients. Pentamidine is the alternate. In the case of AIDS patients, however, the situation appears to be the opposite, with adverse reactions to cotrimoxazole being as high as 65% while to pentamidine being somewhat less at 47%. More recent reports, utilizing pentamidine in a nebulized inhalant dosage form at much lower doses than IV, indicate higher lung tissue levels of the drug than the parenteral route and show effectiveness with less systemic toxicity.

The specific mechanism of action of pentamidine on *P. carinii* is not known. A plausible mechanism involving interference with oxidative phosphorylation in the parasite has been suggested.

7.6. Antifungal Drugs

Fungi used to be the "neglected" microorganisms of medicine. A pharmacology text consisting of 2,200 pages devoted less than 10 of them to mycotic infections and their treatment (Bowman and Rand, 1980). The emphasis was primarily on the nonlife-threatening, commonly encountered topical mycotic infections that were easily treatable. The more serious systemic infections were relatively rare, difficult to treat—due to lack of effective and safe drugs—and frequently incurable and fatal. Beginning with the 1980s the picture changed dramatically for the worse. Systemic mycoses[7] now occur with rapidly increasing morbidity in the bloodstream and various organs. Such infections can account for 5% of all systemic infections in hospitals; the majority are nosocomial in origin. The rate is increasing dramatically. Only a small fraction of this rise can be attributed to better diagnosis. The main reason is a steady increase of patients especially susceptible to such infections: the immunocompromised patient. Table 7-6 lists some conditions and risk factors that predispose hospitalized patients to these hazards. Such infections tend to involve the bloodstream, lungs, the brain, spinal cord, and areas of the CNS in general. The rapidly expanding population of immunocompromised patients is resulting in corresponding increases in meningitis and brain abcesses caused by yeasts and other fungi.

Fungi, like plants and animals, are eukaryotic organisms with organized nuclei contained within a nuclear membrane. By and large, the various cytoplasmic structures and

[7] Literally meaning mushroomlike infections.

biosynthetic pathways are similar to those of mammalian cells. Of course, they do not carry out photosynthesis and therefore, as heterotrophic organisms need preformed sources of energy. There are probably 100,000 species of fungi, of which fewer than 100 are known to be directly causative of human and/or animal diseases. A small subset of these can only survive on protoplasm and are therefore obligate parasites. These constitute the majority of pathogens of mammals and plants.

Fungi occur in two basic morphologic forms—yeasts and molds. Yeasts are less complex and tend to reproduce by asexual budding (blastospores). A sexual process can also occur, resulting in ascospores. The molds consist of multicellular, branching filaments known as *hyphae*, within which individual cells are separated by a crosswall or septum. Some fungi exhibit dimorphism; they may grow as yeasts at one temperature and molds at another.

Fungi can also be classified as lower (Phycomycetes) and higher (Basidiomycetes and Ascomycetes). Molds for which a sexual phase of reproduction has not yet been determined are called Deuteromycetes or *Fungi Imperfecti*.

Fungal cell membranes are bilayered, consisting primarily of protein and lipid material. They also contain the sterol ergosterol (ca. 6%), an important characteristic not found in bacterial membranes. The cell wall is about 85% carbohydrate, with the remainder being lipid and protein; a sterol is woven into this fabric.

The virulence of the fungus determines the pathogenesis of the disease. Infectious fungi may be antiphagocytic, irritant, and inflammatory or have tissue-reactive enzymes. Hyphae may penetrate blood vessels or lymphatics. Yeasts can be transported anywhere in the body by way of the bloodstream or lymphatic system.

There are three ways in which fungi can cause human disease. *Allergies* may result from sensitization to specific fungal antigens due to constant exposure of the respiratory tract by air-carried spores. *Aspergillus* species are commonly encountered this way. Following inhalation, invasive aspergillus in the bronchi and tissues may or may not occur. *Mycotoxicoses* can arise from fungi that secrete substances directly toxic to man. These secondary metabolites include various mushroom toxins and ergot alkaloids resulting from ingestion of *Claviceps purpurea* found on infected rye and other food grains. The third and most common mechanism is *infection*. Because infection involves actual growth of a fungus on or in a host, it should be the most amenable to drug treatment.

Infections range from mild to fatal and may be grouped into two categories. One category is the *superficial mycoses*, infections of keratinized tissues such as skin, nails, and hair. The most prevalent are dermatophytoses (i.e., cutaneous infections). Three genera of *Deuteromycetes* are responsible for dermatophytoses; *Trichophyton* (21 species) ordinarily affect only hair and skin, and *Epidermophyton floccosum* invades only skin and nails (Table 7-7). The yeasts, *Candida albicans* and other species, can infect the skin, and mucous membranes in the mouth (thrush), vagina (vaginitis), and the intestinal tract (intestinal candidiasis).

The second type of fungal infection is the *systemic* and generalized (deep mycoses). There are two types of life-threatening infections encountered in humans. The first is systemic fungal infection by pathogenic fungi in normal hosts. The second is infection suffered by immunocompromised victims (Table 7-6) due to opportunistic fungi that are normally not particularly virulent. These organisms would not ordinarily be infectious.

The systemic fungal pathogens cause diseases that are difficult to diagnose except by X-rays and immunological screening. They are frequently geographically localized and follow specific patterns (e.g., organs, sex, and race). The opportunistic species sometimes exhibit predelictions to particular hosts (e.g., leukemia patients, diabetics). These are

Table 7-7. Common Topical Mycoses

Causative agents	Infection	Principal pathology	Useful treatment[a] topical oral
Pityrosporum orbicultare (Malassezia furfur)	Tinea versicolor	Superficial, asymptomatic, cosmetic sequelae	Ciclopirox Miconazole SeS_2, Haloprogin
Dermatophytes *Epidermophyton floccosum* *Microsporum sp.* *Trychophyton sp.*[a]	Tinea pedis Tinea Corporis Tinea Capitis Tinea unguium Tinea barbae Tinea cruris	Athletes foot, toes and sole; ringworm (skin), scalp and hair; finger and toenails, axilla, inguinal and perianal; "Barber's itch" face and neck	Clotrimazole Miconazole Tolnaftate Econazole Haloprogin Ciclopirox Griseofulvin[d]
Candida albicans (other species)	*Candidiases* Thrus, Vulvovaginitis, cutaneous candidiasis Intestinal	Mucocutaneous infections of mouth, vagina; white patches; skin: vesicles and lesions diarrhea	Miconazole Clotrimazole[e] Ciclopirox Nystatin[b] Candicidin[b] Ketoconazole

[a] Drugs of choice and alternatives are not differentiated; listing is not intended to be complete.
[b] Includes vaginal tablets and creams.
[c] *T. rubrum* most resistant to treatment.
[d] Oral dosage form only.
[e] Also as oral lozenges.

generally organisms of the host's normal microbial flora, or are present in the patient's environment (e.g., hospital room). Table 7-8 summarizes systemic mycoses.

Antifungal Agents. The chemistry, mechanism of action, and clinical utility of the antifungal polyene antibiotics was discussed in Chapter 2.

Griseofulvin (Fig. 7-10) is an antifungal antibiotic produced by a fungus, actually several species of *Penicillium* including *griseofulvum*, from which it was first isolated in 1939. It was initially used to protect plants against fungal infestation and was not found to be orally active against dermatophytes until almost two decades later. The compound is a phenolic benzofuran joined to a cyclohexenone via a spirocarbon atom. Because of its very poor solubility in water, polar solvents such as ethanol and even dimethylsulfoxide (DMSO), as well as lipid solvents such as benzene, resulted in inefficient and erratic absorption from gastric areas. Newer dosage forms are manufactured from ultrafine crystalline forms (e.g., Fulvicin U/F). The resultant increased absorption allows lower doses to reach the required plasma levels. Taking the drug with fat-containing foods (e.g., milk) improves absorption even more.

Clinically, the drug is effective orally (but apparently not topically) against most of dermatophyte species (Table 7-7). The physiologic mechanism by which the drug acts appears to be its concentration from the circulation and incorporation into the precursor cells of newly forming keratin of the skin and nails to which it is then bound. The fresh keratin then becomes resistant to further fungal infiltration.

Even though early work indicated that the drug was having effects on the fungal cell wall, nucleic acid synthesis inhibition, and alterations of cytoplasmic microtubles, later studies all point toward a specific inhibition of mitotic spindle formation.

It is of interest that griseofulvin has been shown to cause liver tumors in male mice. Since therapy with this drug in humans is prolonged—from weeks to months—caution is warranted.

Table 7-8. Common Systemic Mycoses

Causative agents	Infection	Principal pathology	Available treatment[a,b]
Pathogenic			
Blastomyces dermatitis	Blastomycosis	Chronic suppurative lesions; skin, granulomatous and lungs, then disseminated; can be fatal	Amphotericin B Ketoconazole Hydroxystilbamidine
Coccidioides immitis (dimorphic)	Coccidiodomycosis (San Joaquin Fever)	Primary: pulmonary benign; then disseminating to various organs and bone	Amphotericin B
Cryptococcus neoformans (encapsulated yeast)	Cryptococcosis	Meningitis with relapses; and metastasizes to brain other organs; 100% fatal if untreated	Amphotericin B Flucytosine Ketoconazole Miconazole
Histoplasma capsulatum	Histoplasmosis	Pulmonary; initially flulike, then severe symptom necrotic and calcified lesions in lungs, disseminates to other organs; tulminating, can be fatal.	Amphotericin B Ketoconazole
Paracoccidioides brasiliensis	Paracoccidio-idomycosis (South American blastomycosis)	Chronic primary pulmonary disseminates to granulomatous and ulcerative disease in nasal, anal and intestinal mucosa, internal organs; often fatal	Ketoconazole Amphotericin B Miconazole
Sporotrix Schenckii	Sporotrichosis	Chronic, subcutaneous abscess, disseminates to bones, joints, in lymph nodes and internal organs; prognosis grave	Oral and topical potassium iodide[c] Amphotericin B
Opportunistic			
Aspergillus fumigatus (and other A. species)	Aspergillosis	Bronchopneumonia, blood vessel obstruction, lung infarcts, and granuloma	Amphotericin B Ketoconazole
Candida albicans[d]	Candidiasis	Bronchial, urinary tract, heart, bloodstream, and other organ infections	Amphotericin Flucytosine Ketoconazole Miconazole
Mucor **sp.**	Mucormycosis (Phycomycosis)[e]	Usually acute paranasal inflammation and necrosis, may invade brain, then fatal	Amphotericin B Ketoconazole

[a] Drugs of choice and alternatives not differentiated.
[b] Combinations of drugs often indicated.
[c] The cutaneous–lymphatic infections respond.
[d] Normally nonparasitic; becomes so in immunocompromised patients.
[e] Common molds such as *Rhizopus* sp.

There are a number of nonspecific topical drugs, some some of which have been in use for most of this century. Early therapy of superficial mycoses utilized topical fungicides such as salicylic acid, salicylanilide, iodine tincture, and even weak fungistatic compounds such as benzoic acid. Salicylic acid is mainly a keratolytic agent that also removes the softened, loosened infected epithelial layers. In combination with benzoic acid (Whitfield's ointment), it has been a useful product for years.

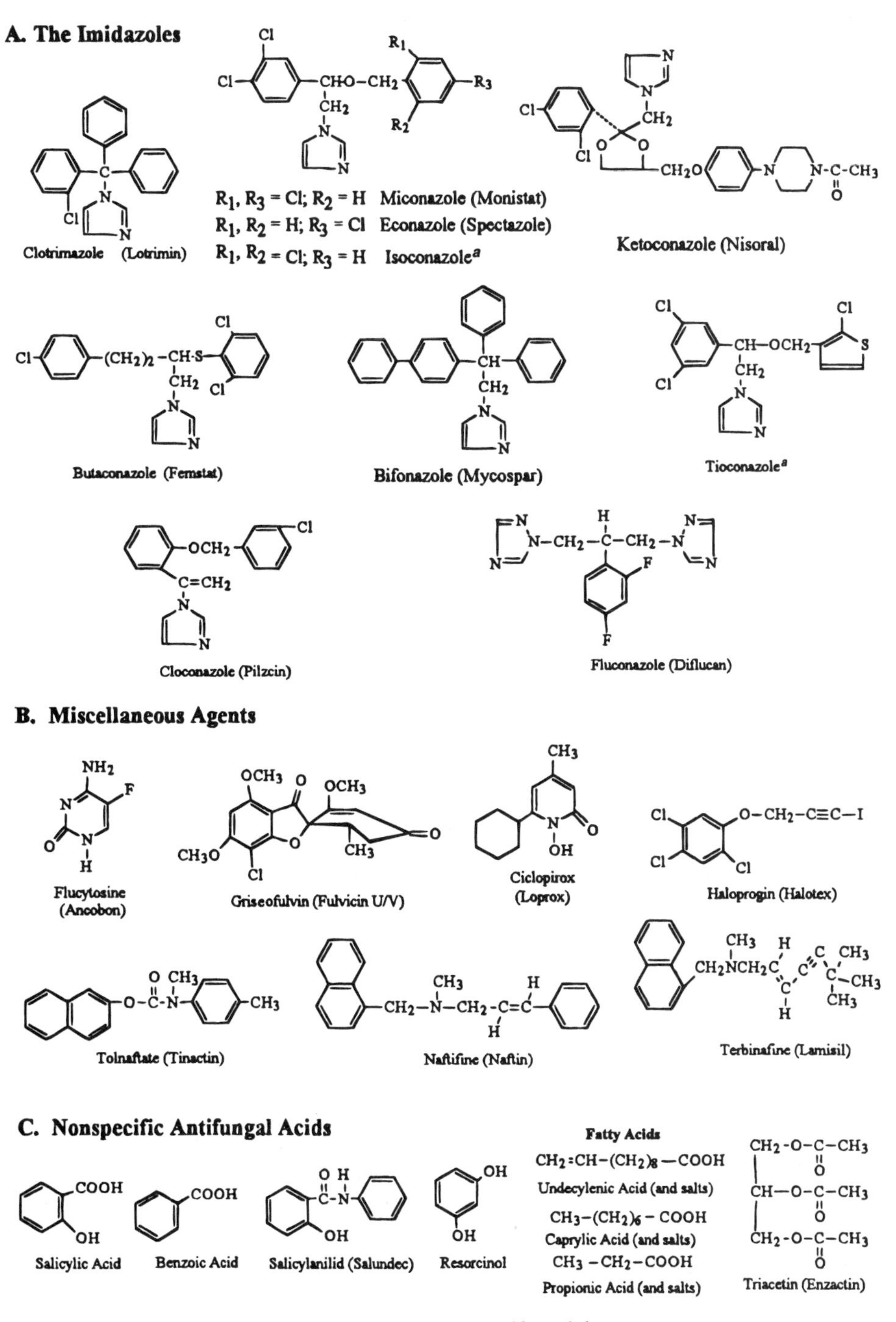

Figure 7-10. Synthetic antifungal drugs.

A number of fatty acids possess antifungal properties limited to dermatophytes (Fig. 7-10). Their mechanism of action is not understood. Acetic acid, as glyceryl triacetate (Triacetin), is used in cream form against tinea pedis. It apparently acts as a pro-drug, slowly releasing the acetic acid on hydrolysis, and is presumably catalyzed by esterases present in the skin and pathogens. Propionic acid and its Na, Ca, and Zn salts are available in varying combinations of creams, ointments, powders, and solutions for superficial tinea infections.[8] Caprylic acid and its salts are similarly available. Undecylenic acid and its Ca, Zn, and Cu salts (Desenex, Cruex) are probably the most popular of the fungistatic fatty acids for the topical treatment of "athlete's foot" (tinea pedis) and ringworm infections. These aliphatic acids are not useful against hair and nail infections. They have no effects on yeast cells.

Tolnaftate (Fig. 7-10) is a naphthiocarbamate of N-methyl-*p*-toluidine that has high activity against dermatophytes such as *Trichophyton* species, including *T. rubrum*, which is resistant to the fatty acids and griseofulvin. The drug is not useful against yeasts such as *Candida* spp.; bacteria do not respond. It is nontoxic and nonsensitizing. Scalp and nail infections are not treatable.

Haloprogin (Fig. 7-10) is a halogenated phenolic ether with iodopropargyl alcohol. It has been considered an analog of the antibiotic lenamycin: H_2NOCC $CCONH_2$. Its in vitro activity includes the dermatophyte fungi (MIC 0.25 μg/ml), yeasts such as *Candida* spp., some Gm+ cocci, and even *Mycobacterium tuberculosis*. In vivo dermatophyte activity is similar to tolnaftate. Haloprogin is effective against superficial *Candida* infections and tinea versicolor. The drug is fungicidal by an unknown mechanism.

Another useful addition to the antifungal armamentarium is the hydroxypyridone ciclopirox. Its broad fungicidal activity includes all of the dermatophytes, yeasts, and *M. furfur* (versicolor). The neutral cream is therefore indicated for the various tinea infections (see Table 7-7) and superficial candidiasis. The mechanism of ciclopirox has not been fully determined, but appears to involve inhibition of uptake from the medium of precursors by the pathogen for the biosynthesis of the biopolymers needed for survival.

A newer addition in this group of antidermatophytic agents is the naphthalene methylamine derivative naftifine. As a topical cream this drug is indicated for tinea cruris and corporis. The latest member is the related allylamine terbinafine, which is indicated as a topical dermatologic agent as well. Its mechanism includes the inhibition of squalene epoxidase, which catalyzes a crucial step in the biosynthesis of fungal sterols. It is, however, the accumulation of squalene in the fungal cells that is responsible for their death. The drug appears to be a broadly effective antidermatophyte agent.

7.6.1. Imidazole Compounds

The imidazole compounds (see Fig. 7-10) today constitute the most versatile and valuable source of antifungal compounds. They appear to transcend the chemotherapeutic boundaries of other antiparasitic drugs with a spectrum of activity that includes the majority of fungi, as well as many bacteria, protozoa, and even helminthic species.

A report in 1944 showed that benzimidazole possessed antibacterial and antifungal activity. The first marketed compound was 1-(*p*-chlorobenzyl)-2-methyl-benzimidazole

---

[8] Sodium and calcium propionate are commonly used as mold inhibitors in baked foods.

(Chlormidiazole) (1958). As a 5% cream it exhibited activity against dermatophytes and some Gm+ bacteria. Thiabendazole, which was introduced subsequently as an anthelmintic, also has antifungal properties that have not been exploited.

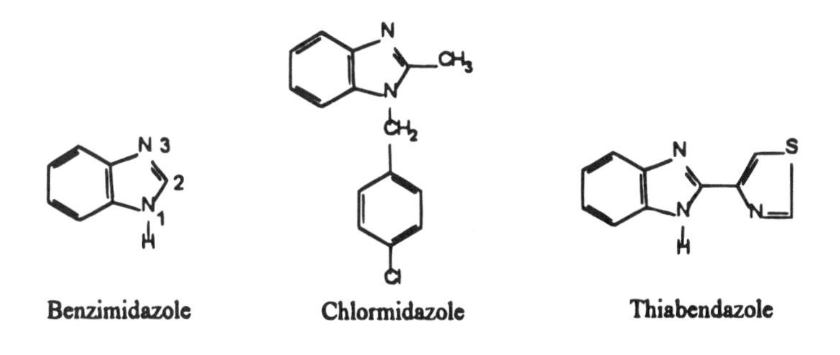

Benzimidazole **Chlormidazole** **Thiabendazole**

Three of the presently available imidazole antifungals were reported the same year, 1969: clotrimazole in Germany and miconazole and econazole in Belgium (Fig. 7-10).

Clotrimazole is a triphenylmethylimidazole, a broad-spectrum drug that is mainly restricted to topical applications (creams, lotions, lozenges) though it is orally active because of rapid metabolism, gastric disturbances, and other untoward effects. Its low MIC against most dermatophytes (0.5–2 μg/ml) and many isolates of *C. albicans* (0.1–0.5 μg/ml) make this drug widely used. Both cutaneous and vaginal Candida infections respond well.

Miconazole, which is also a broad-spectrum antifungal, is effective topically and intravenously against dermatophytes and yeasts, and intravenously against systemic Coccidoides infections. CSF penetration is negligible.

Ketoconazole is one of the more prominent antifungals. It is active orally and topically and has an extremely broad antifungal spectrum. The drug has been approved for the treatment of topical and systemic candidiasis, paracoccidiomycosis, blastomycosis, and histoplasmosis. It has been effective against coccidiomycosis (amphotericin B preferred with meningeal involvement) and nonmeningeal cryptococcosis. Other infections are investigational. Aspergillosis responds poorly. Resistance thus far is uncommon. At the upper dosage range of systemic treatment (800 mg per day) ketoconazole has been shown to inhibit synthesis of adrenal steroids. It is therefore being investigated for use against Cushing's disease and hyperadrenalism and small-cell lung cancer. In addition, the drug suppresses testicular production of testosterone at lower doses. This fact has led to trials of ketoconazole against refractory cases of prostatic carcinoma.

The evolution of the modern imidazole drugs can be traced to the early benzimidazoles that led to phanacylimidazoles (Structure I, Fig. 7-11), followed by substituted benzylamines II that exhibited some in vitro antidermatophyte activity. The replacement of the nitrogen atom in II with oxygen afforded the corresponding benzyl ethers III. This produced broad-spectrum antifungals with excellent in vivo activity as well. Miconazole, econazole, and isoconazole (Fig 7-10) were among the compounds. That same study also reported the synthesis of the ethylene glycol ketals of several phenacylimidazoles (IV, $R = H$) and their in vitro only activity against dermatophytes. This lead was not further pursued for 12 years. It was then expanded by synthetic studies with substituted aryl-1,2-ethanediols producing 79 1,3-dioxalanes of structure IV; the one selected for clinical introductions is ketoconazole (Fig. 7-10).

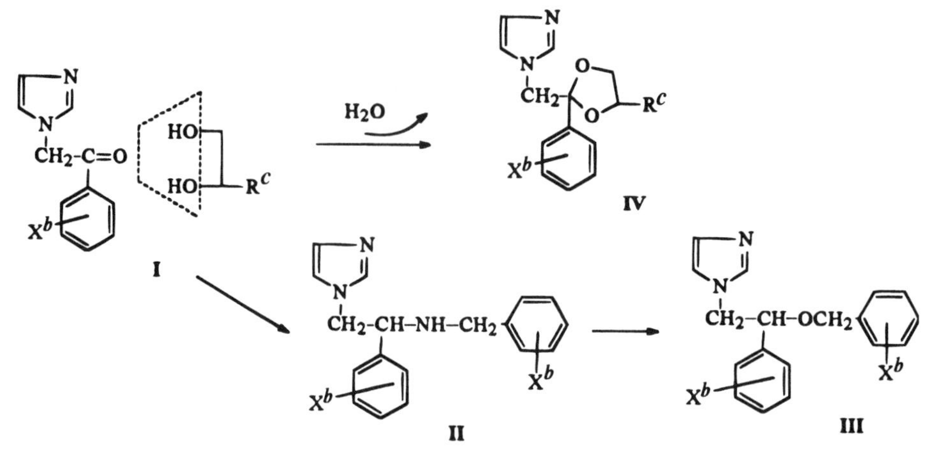

[a]See text for explanation.
[b]X = one or more halogens, usually Cl.
[c]R = $-(CH_2{-}CH_2)-_n$, C_5H_5X, or $-CH_2{-}OC_6H_5X$.

Figure 7-11. Synthetic evolution of antifungal imidazoles.[a]

Extensive SAR studies showed that the imidazole ring is crucial for broad-spectrum, high-potency antimycotic activity in vivo. It is ironic that even replacement with benzimidazole, the original lead compound, exhibited very low effectivness. Exchanging one nitrogen with carbon (i.e., pyrazoles and piperazines) also resulted in less active, or even inactive, compounds.

Additional imidazoles have since been introduced into clinical use (Fig. 7-10). Butaconazole has been marketed primarily in vulvovaginal anticandidal cream. Tioconazole, where the benzylic phenyl ring has been isosterically exchanged with chlorinated thiophene, has been introduced as a topical product (England) against *Candida* and dermatophytes; bifonazole has been marketed in Germany for several years against topical dermatophyte infections including tinea versicolor. Its structure is not only simpler than most active imidazoles, but it is also unique as it is devoid of any aryl halogen atoms (Fig. 7-10). Although also a relatively simple-structured imidazole drug, clocanazole nevertheless seems to have the necessary features for treatment of topical human dermatomycoses. Fluconazole has the unique structural features of triazole rings (instead of imidazole) and predictably effectiveness-enhancing fluorine atoms on the only benzene ring in the molecule. It is currently utilized in AIDS patients who have succumbed to life-threatening systemic fungal infections. Finally, oxiconazole has become available in topical dosage forms.

7.6.1.1. Mechanism of Action of Imidazoles

The previously mentioned ergosterol content of fungal cell membranes differentiates it from membranes where cholesterol is the sterol involved in permeability regulation. The imidazoles have been shown to inhibit the biosynthesis of ergosterol in all susceptible organisms including *C. albicans*. It was established that ketoconazole affected sterol metabolism significantly more in yeast cells than it did in mammalian cells. Furthermore, ergosterol synthesis was 67 times more sensitive to miconazole inhibition than was cholesterol. These differences may well account for the selective toxicity of the imidazoles (and triazoles) observed between fungal and mammalian cells.

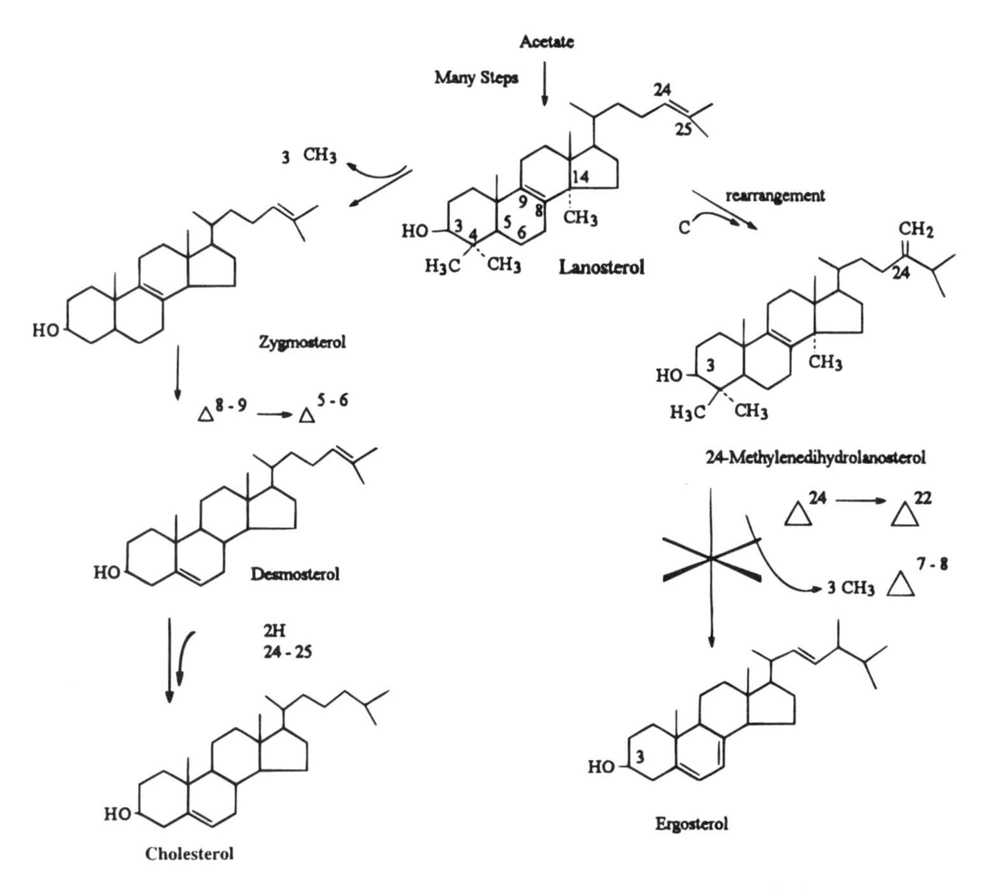

Figure 7-12. Terminal steps of cholesterol and ergosterol synthesis.

Figure 7-12 outlines the terminal steps in the complex sequence of both the biosynthesis of cholesterol and ergosterol. The synthesis, starting with acetate, leads to the common steroid precursor lanosterol.[9] Thus, in the case of cholesterol, the scheme involves steps for the reduction of the Δ^{24} double bond, and the oxidation and removal of the 14α methyl group and of both methyls at C-4. This is followed by isomerization of the Δ^{8} double bond of the resultant zygmosterol to the Δ^{5} position, affording desmesterol, and finally cholesterol.

Ergosterol also arises from lanosterol by a one-carbon addition to C-24 followed by a shift of the Δ^{24} double bond to the *exo* position, forming a Δ^{24} methylene function. There follows oxidative removal of the three methyls at C-14 and C-4 with concomitant shift of the Δ^{24} double bond to Δ^{22} position; isomerization of Δ^{8} position to Δ^{7} and formation of an additional Δ^{5} double bond yielding ergosterol.

Ergosterol synthesis inhibition has been found to coincide with an accumulation of 14 α-methylsterols in yeast cells. The corresponding inhibition in rat liver cells could only be achieved at a sixfold increase in ketoconazole concentration. The enzyme initiating the oxidation of the 14α-CH_3 function is lanosterol-14α-methyldemethylase. It has been

[9] For review of cholesterol biosynthesis, see Zubay, 1983.

established to be a cytochrome P-450–dependent enzyme, and, furthermore, that miconazole and ketoconazole affect this enzyme in yeast microsomes. The corresponding enzyme obtained from rat liver microsomes was much less sensitive. Of course, this relates well with the selective toxicity observed with these drugs. That lanosterol binds to (i.e., is a substrate for) the microsomal cytochrome P-450 system had already been previously demonstrated for *Saccharomyces cerevisiae* (a yeast). It remained to be established that increased levels of methylated sterols and decreased availability of ergosterol had the deleterious effects on fungal membranes observed; namely, increased permeability and loss of cell components such as glucose.

It is now apparent that accumulation of methylated steroids, particularly 14α-methyl, accounts for the detrimental alterations in membrane structure and functions induced by the imidazole drugs. The changes in membrane permeability with the resulting leakage of cellular components is the primary cause of cellular death. It has been suggested the *axial* position of the 14α-methyl group of lanosterol prevents interaction between the flat face of the sterol and the fatty acid chains in the phospholipid membrane.

In addition to inhibition of ergosterol biosynthesis, it is now reasonably well established that the imidazole drugs also have a *direct* action on fungal membranes. This may in part explain the existence of certain Gm+ bacteria susceptible to imidazole drugs. Spectrophotometric evidence indicated interactions between imidazoles and mixtures of saturated and unsaturated fatty acyl groups occurring naturally as part of a fungal membrane as phospholipid complement. The imidazole compounds have a duality of antimicrobial action. One involves interference with the iron atom of fungal cytochrome P-450 enzyme. The result is a blockade of oxidative removal of lanosterol's C-14 methyl group (Fig. 7-12). This mechanism is probably responsible for the fungistatic effect. The fungicidal effect observable at higher drug concentrations is likely the result of direct interactions between these compounds and unsaturated fatty acids in the membranes. Ketoconazole does not appear to share this interaction, thereby helping to explain the absence of fungicidal effects of this particular agent.

Flucytosine (5-Fluorocytosine, Ancobon, Fig. 7-10) was first synthesized as a potential antimetabolite for use in leukemia treatment, but it exhibited no anticancer effects. It was found to have in vivo antifungal activity, however. In vitro activity includes *Cryptococcus neoformans*, many isolates of *C. albicans*, and other *Candida* species and some isolates of *Aspergillus* spp. Fungicidal concentrations range from 0.2 to 12.5 μg/mL. The activity of flucytosine against filamentous fungi is only fungistatic. The drug is not toxic in experimental animals and is readily absorbed. It can therefore be orally administered. Clinical use of flucytosine is limited to susceptible systemic infections and cryptococcal meningitis.

Resistance emerging during therapy can be a major problem, so that even though the drug is less toxic than amphotericin B, the two drugs should be prudently used together in serious mycoses to minimize the likelihood of resistance emergence and to take advantage of the additive, or even synergistic, effects.

The reason for the relative safety of flucytosine, and the lack of carcinolytic effect in humans, is that the enzyme cytosine deaminase does not appear to be present. Conversion to 5-FU occurs to an insignificant degree of about 4% (Eq. 7.10). However, following permease-catalyzed uptake by the fungal cell, the deaminase present there effectively converts the drug to the cytotoxic 5-FU through several steps to 5-fluorouridine triphosphate (see Fig. 4-18). This results in its incorporation into the fungal cell's RNA, leading to the synthesis of miscoded proteins. This faulty RNA (containing 5-FU) ultimately accounts for a major portion of the cell's nucleic acid. In addition, 5-fluorouridine monophosphate can also be reduced to the corresponding 5-fluorodeoxyuridilic acid inhibiting the cell's thymidylate synthetase, thus affecting DNA synthesis as well (see Fig. 4-19).

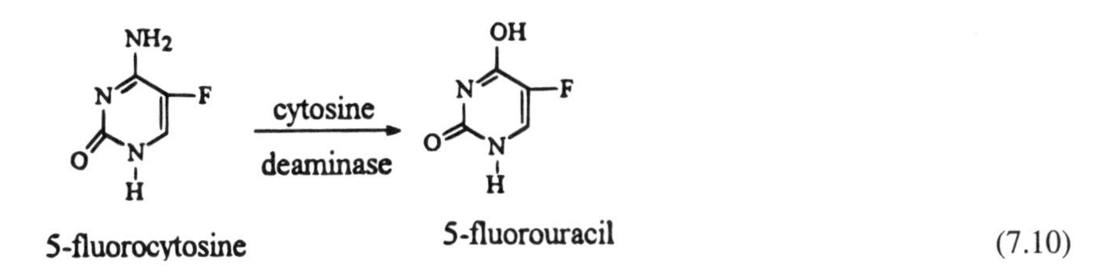

(7.10)

7.7. Anthelmintics

Parasitic helminth (worm) infections are widespread and may be the most common disease in the world. Estimates by the WHO and others are that one type of schistosomiasis, the blood flukes, has 200 million people infected. Ascaris, a large roundworm, accounts for at least 650 million, hookworm 450 million, and the several filarial worm species add another 250 million victims to the total. The situation has improved little in the past decade.

Table 7-9 is a condensed classification and nomenclature of the important parasitic helminths that affect humans. The worms that infect humans are the flatworms and the roundworms. Helminths are the only parasites discussed here that are not microscopic. Quite the opposite, their dimensions are of the order of millimeters and, in the case of certain tapeworms, reach lengths of meters.

There are two types of flatworms; the flukes, whose appearance can best be described as lanceolate or leaflike, and the tapeworms, which are ribbonlike. The latter consist of repeating egg-containing segments, or *proglottids* (numbering up to several thousand) linked in a chain called a *strobilia* and attached to a "head" or *scolex*. They have no digestive system. Their nutrients are absorbed directly from the host's intestines; they are true parasites.

The second helminth type are the roundworms. Their shape is cylindrical. They vary in proportions, size, and structure.

There is no totally broad-spectrum antihelminthic drug. There are groups of drugs that are useful against blood flukes; others against liver flukes or lung-infesting flukes. There are some agents well suited to the treatment of tapeworms. Another category is useful against intestinal roundworms and still another against roundworms infesting tissues or blood. There may be considerable overlaps of activities.

Parasitic helminths infect by transmission of eggs or larvae. Once developed into adult worms, they reproduce by releasing their own eggs or larvae, which then leave the host, where they develop "on the outside," and/or intermediate hosts (e.g., snails). When the larvae become infective and again enter the human host, the life cycle is complete (Fig. 7-13).

Pathology in the host can be produced by the adult worms, the larvae, or the eggs, and the intensity is usually directly related to the worm load. Unlike protozoal disease, clinical symptoms with worm infections are not well defined. Diagnosis usually depends on recovery and identification of the parasite or its eggs. Immunoserological tests are sometimes available.

Drugs. Several formerly popular remedies, while they may be effective in some case, have been abandoned as obsolete due to unacceptable toxicity. The natural lactone san-

Table 7-9. Classification and Nomenclature of Helminths[a]

Class	Common name	Usual dimensions
Phasmidia		
Ascaris lumbricoides	roundworm	20–35 cm
Ancylostoma duodenale	hookworm	10 mm
Enterobius vermicularis (also called Oxyuris vermicularis)	pinworm or seatworm	12–13 mm
Necator americanus	hookworm	5–10 mm
Strongyloides stercoralis	threadworm	2 mm
Trichinella spiralis	trichina (pork roundworm)	2–5 mm
Trichuris trichiura	whipworm	30–50 mm
Loa loa	Filarial worms[c]	
Onchocerca volvulus		
Wuchereria species		
Trematoda		
Fasciola hepatica	liver fluke	30 mm
Fasciolopsis buski	intestinal flukes, giant	75 mm
Heterophyes heterophyes	intestinal flukes, small	2 mm
Metagonimus yokogawai		
Clonorchis sinensis flukes	Chinese liver	25 mm
Paragonimum species	lung flukes	
Schistosoma haematobium		
Schistosoma japonicum	blood flukes	6–26 mm
Schistosoma mansoni		
Cestoda		
Diphyllobothrium latum	fish tapeworm	10 m
Hymenolepsis nana	dwarf tapeworm	2–5 cm
Taenia saginata	beef tapeworm	5–25 m
Taenia soilum	pork tapeworm	3 m

[a] Selected for the purpose of this discussion. Families listed alphabetically.
[b] When sexually differentiated the female is invariably longer than the male.
[c] Commonly encountered in Africa, Asia, and Mediterranean Europe, but not in U.S.

tonin is one example. Other popular vermifuges were the harsh oleoresin from *Aspidium* (male fern) and chenopodium (American wormseed oil), which contained an organic peroxide that exploded on heating. In spite of some new developments, many anthelmintics still in use are anything but "modern" drugs.

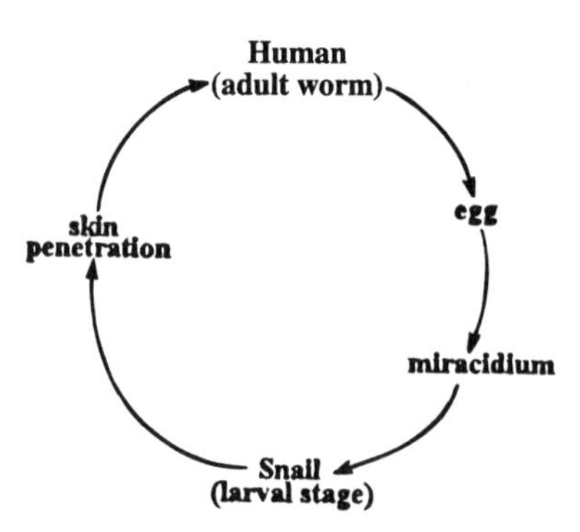

Figure 7-13. The schistosome life cycle.

Whether particular drugs are well tolerated or not, anthelmintic therapy frequently requires adjunctive therapy. Thus allergic reactions to tissue infestation by parasites may require antihistamines, or even steroids. Iron-deficiency anemia (due to blood losses) sometimes has to be counteracted. Gastric effects have to be treated traditionally, as do any secondary bacterial infections. With some agents cathartics may be indicated to expel dead or incapacitated worms.

Certain drugs, although safe and effective, are not available in the United States. They can be obtained by physicians for their patients from the Parasitic Disease Drug Service, Center for Disease Control (CDC), U.S. Public Health Service (USPHS), Atlanta, Georgia.

Anthelmintics as a group share few common features of chemical similarity or mechanism of action. Table 7-10 lists many of the available drugs and the disease or organism against which they have been found useful. (Structures given elsewhere are not repeated.)

Schistosomiasis (Bilharziasis). The schistosome species listed in Table 7-9 parasitize humans widely in Africa, South America, the Far East, and the Middle East. *S. mansoni* is the most widespread species, *S. haematobium* resides mostly in Africa, and *S. japonicum* is found primarily in the Far East. The latter is the most difficult to treat. The intermediate host, the snail, is not prevalent in urban areas of the U.S. mainland; the disease is not endemic here. However, *S. mansoni*, for example, is seen in New York City because of its prevalence in the Caribbean Islands.

Historically, the first treatments were antimonial compounds. The medicinal properties of antimony have been known (but not understood) for thousands of years. Paracelsus prescribed it in the sixteenth century. Antimony potassium tartrate was probably the first specific antischistosomal treatment. Organic antimony compounds are effective. However, the need for lengthy treatment and high toxicity makes their obsolescence a desirable goal, if safer and more effective agents can be developed. Coadministration of mercapto compounds such as penicillamine (dimethylcysteine) (Chapter 5) can reduce toxicity. The chelate with tartar emetic is apparently better tolerated. At the molecular level antimony compounds produce an accumulation of glucose-6-phosphate and fructose-6-phosphate while the level of fructose-1,6-diphosphate decreases in the parasites. This chemical alteration therefore implicates the enzyme phosphofructokinase (PFC) as the target being inhibited. Since schistosomes (and possibly other antimony-sensitive parasites) appear to have their major energy source—glycolysis—inhibited, the parasiticidal activity of these drugs becomes explicable.

The basis of the selective toxicity depends on the enzyme's source. The schistosomal enzyme has been shown to be 80 times more responsive to antimonial inhibition than is PFC from a mammalian source. It is also of interest that the schistosomal enzyme is activated by -SH compounds such as cysteine, penicillamine, glutathione, and even 2-mercaptoethane, and that the inhibitory effect on the enzyme is not lessened by these compounds. This would rule out the likelihood of irreversible inhibition of enzyme SH groups as the underlying mechanism of drug action. Nevertheless, by taking advantage of the attenuated toxicity via its complexation with three molecules like dimercaptosuccinate [$HO_2CCH(SH)CH(SH)CO_2H$], drugs like stibocaptate (Table 7-10) remain as useful, though secondary, antischistosomal drugs.

Niridazole, a nitrothiazole, is another toxic drug that was selected from a drug development program. The drug has some effect on all three species; however, it is primarily useful clinically against *S. hematobium*, especially in children. The antiparasitic effect is directly on the reproductive system: disruption of and abnormal egg productions and

Table 7-10. Anthelmintic Drugs

Drug[a]	Structure[a]	Disease (organism)
Antimony potassium tartrate (tartar emetic)	Fig. 7-9	blood flukes (*schistosoma japonicum*)
Bephenium[c] (Alcopara)	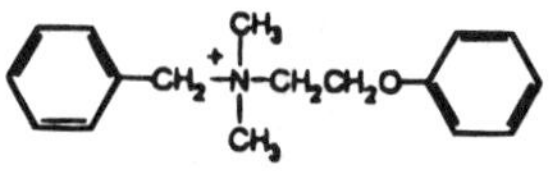	Hookworm (*Ancylostoma Duodenale*)
Bithionol[c,d] (Actamer)	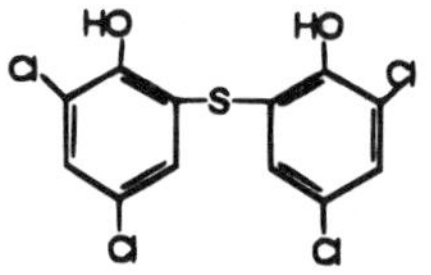	Lung fluke (*Paragonimus westermani*)
Chloroquine (Aralen)	Fig. 7-6	Chinese liver fluke (*Clonorchis sinensis*)
Diethylcarbamazine[d] citrate (Hetrazan)	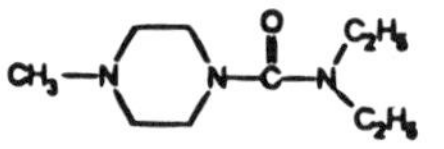	Various filariases—elephantiasis, river blindness (*Wuchereria spp.,Onchocerca volvulus*), and others
Hexylresorcinol (Crystoids)	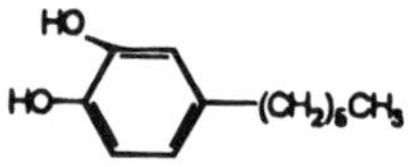	Mixed roundworm diseases (intestinal nematodes)
Hycanthone[a] (Etrenol)	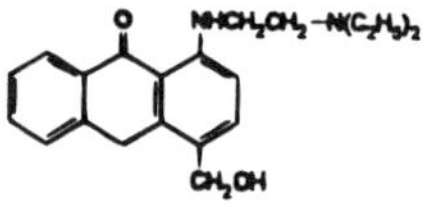	Intestinal and urinary tract blood flukes (*S. mansoni* and *haematobium*)
Ivermectin (Avermectin)	Fig. 7-15	River blindness (*Onchocerca volvulus*)
Levamisole[j] (Ergamisole)	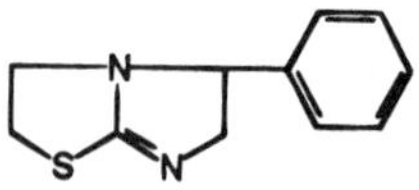	Large roundworm (*Ascaris lumbricoides*)
Mebendazole[e,f] (Vermox)	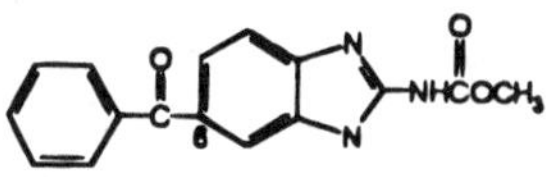	Whipworm (*Trichuris trichiura*) and other spp.
Metrifonate[d] (Bilarcil)	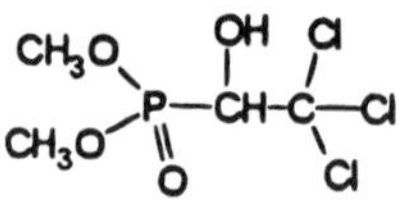	Schistosomiasis (primarily *S. haematobium*)

(*Continued*)

Table 7-10. (*Continued*)

Drug[a]	Structure[a]	Disease (organism)
Niclosamide (Nicloside)	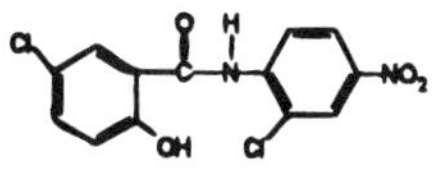	Tapeworms (*T. saginata* and *solium*, *D. latum*, *H. nana*)
Niridazole[c,d] (Ambilhar)	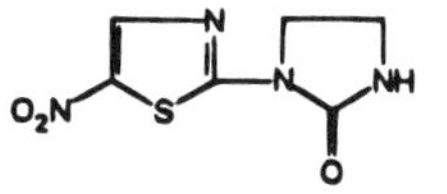	Urinary tract blood fluke (*S. haematobium*)
Oxamniquine (Vansil)	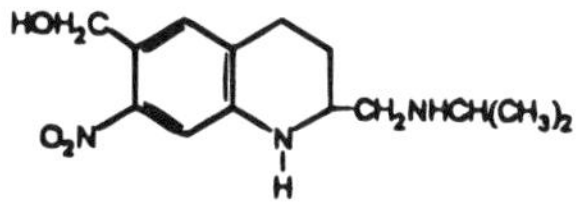	Schistosomiasis (primarily *S. mansoni*)
Paromomycin (Humatin)	Chapter 6	Cestodiasis
Piperazine (citrate) (Antepar)	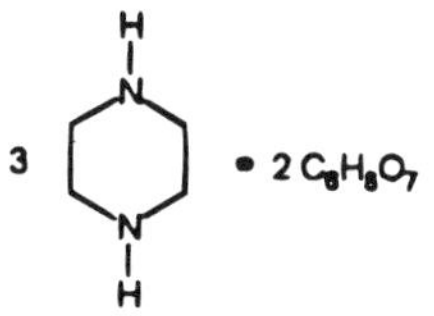	Roundworm (*A. lumbricoides*), pinworms (*E. vermicularis*)
Praziquantrel (Biltricide)	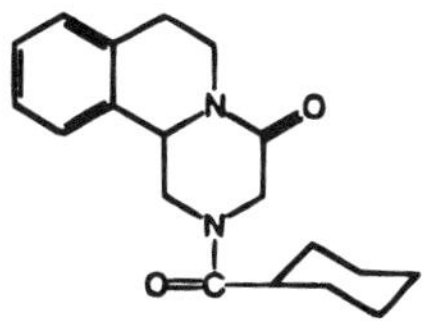	All 3 Schistosoma species[g]
Pyrantel (pamoate)[c] (Antiminth)	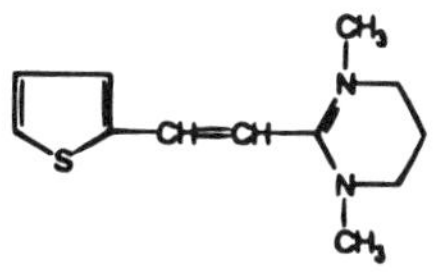	Roundworm (*A. lumbricoides*) and pinworm (*E. vermicularis*)
Pyrivinium pamoate[e] (Povan)	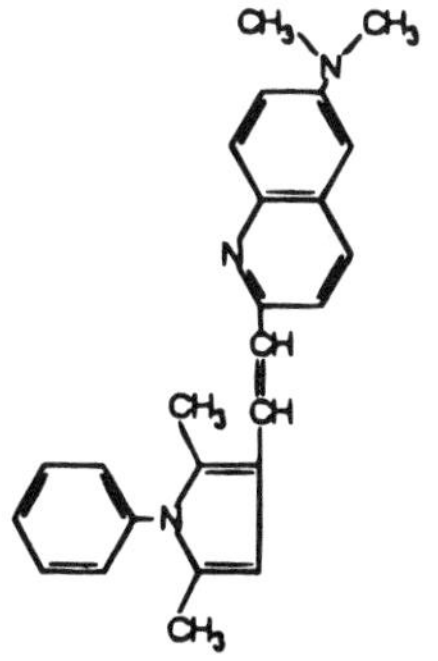	Pinworm (*E. Vermicularis*) Threadworm (*Strongyloides stercoralis*)

(*Continued*)

Table 7-10. *(Continued)*

Drug[a]	Structure[a]	Disease (organism)
Stibocaptate[d] (Astiban)	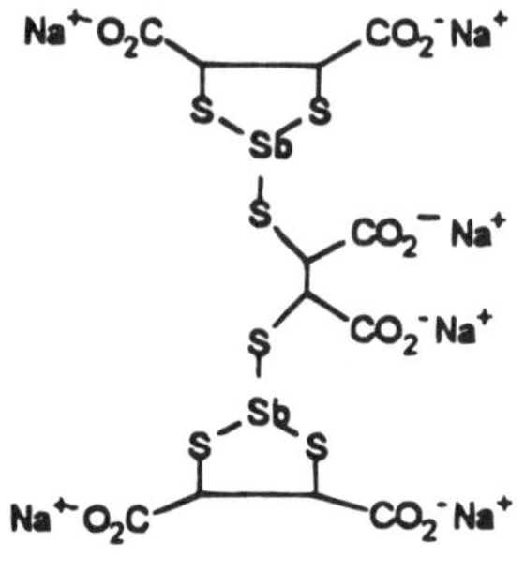	Schistosomiasis, all three species
Suramin[i] (Germanin)	Fig. 7-9	River blindness (*O. olvulus*)
Tetrachlorethylene (Nema)	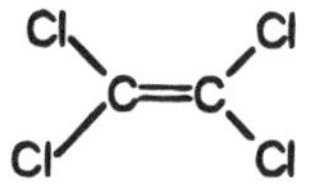	Hookworm (*Necator americanus*), liver and intestinal flukes (*F. hepatica, F. buski*)
Thiabendazole (Mintezol)	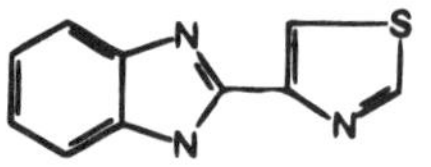	Threadworm (*S. stercoralis*); pinworm (*E. vermicularis*)[h]

[a] Trade name in parentheses.
[b] Number in parentheses indicates where structure can be found.
[c] Very water-insoluble drug, either itself or as a salt such as hydroxynaphthoate or pamoate.
[d] Available in the United States only through CDC, UPHS.
[e] Not available in the United States.
[f] There are several other benzimidazole carbamates under investigation that vary the 6-substituent: *p*-F-C_6H_5CO, flubendazole, cyclopropyl carbonyl, ciclobendazole, and $CH_3CH_2CH_2$-S-, albendazole.
[g] Investigational against *Fasciola* spp. and other flukes.
[h] Investigational in trichinosis (*Trichinella spiralis*).
[i] Not yet approved in the United States.
[j] Now approved as an immunostimulant in cancer chemotherapy.

uptake by the gonads of the male affecting spermatogenesis. There is also a loss of glycogen in the organism. This has been shown to be due to the potentiation of schistosomal, but not mammalian, glycogen phosphorylase by inhibiting phosphorylase phosphatase, the enzyme that catalyzes the inactivation of phosphorylase. Whether this has a contributing effect toward worm death is not clear. The increased wasteful utilization of glucose possibly produces a form of "malnutrition." As with the previously considered nitroheterocycles, the nitro group seems essential, and the drug appears activated by its reduction.

Metrifonate, which was originally introduced as an organic phosphate insecticide, is a pro-drug for dichlorovos,[10] a potent acetylcholinesterase inhibitor (Chapter 8) to which it is nonenzymatically metabolized spontaneously in vitro even at neutral pH (Eq. 7.11). The drug's clinical application is exclusively against *S. hematobium* infections. Although some effectiveness against other schistosomes exists, this is not achieved at safe doses.

[10] Dichlorovos is a commercial ectoparasitizide for veterinary use and against household pet pests.

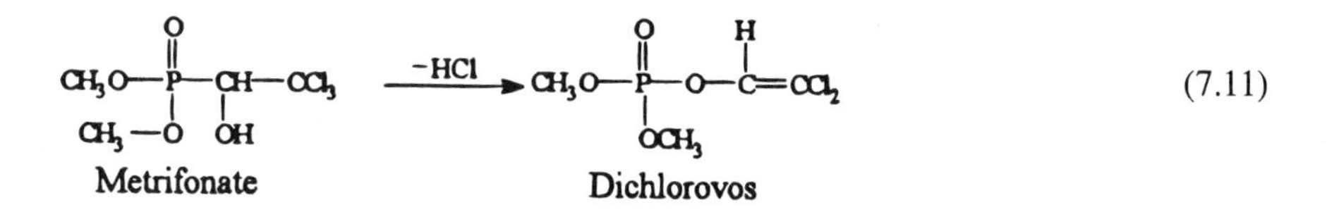

(7.11)

The acetylcholinesterase in schistosomes acts as an inhibitory neurotransmitter so that inhibition of the enzyme results in paralysis. The suckers holding the worm to the wall of the vein let go, and the parasite is swept out (to the lungs) with the blood flow. Shortly after oral administration of the drug, serum cholinesterase levels in the patient are essentially eliminated while erythrocyte levels drop by about three fourths. These take several weeks to be restored.

Hycanthone (Fig. 7-10), although relegated to secondary status as an antischistosomal drug, is of historical significance and pedagogical value. The drug is a thioxanthone tricyclic compound that was developed following the discovery that it was the active metabo-

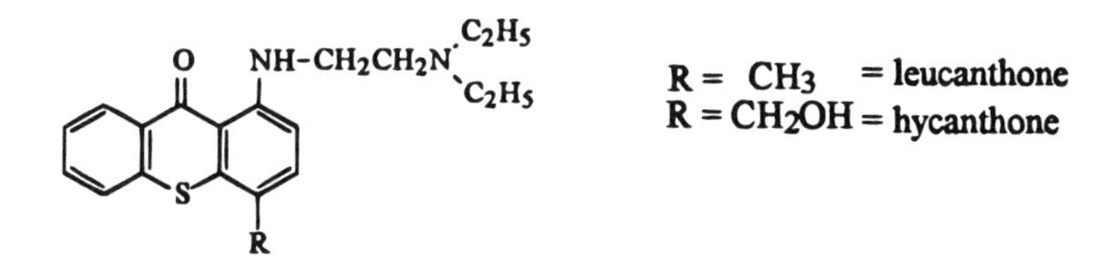

lite of leucanthone, a drug developed in Germany in the 1930s but not published until after World War II. The significance of leucanthone is multiple. First, it was a member of a series of compounds (the miracils) designed by structural variations and modifications of quinacrine. It was the first significant oral antischistosomal drug devoid of heavy metal atom. Studies of the effects of fungi on various drugs showed that an *Aspergillus* mold oxidized the drug to several compounds, one of which—the hydroxymethyl analog, named *hycanthone*—was shown to be both the active metabolite of leucanthone and also vastly superior to it as to effectiveness and toxicological profile. The mechanism of action at the physiological level included a decline in egg production and motility control. At the biochemical level, however, two concepts have been put forward to explain this loss of neuromuscular control by the parasite. The first was that the drug decreased the affinity for and uptake of serotonin, and, second, that hycanthone competed with acetylcholine (ACh) at ACh receptors.

The final lesson to be learned from these two thioxanthones is the significance of considering the three-dimensionality of drug molecules rather than just the classic "blackboard" representation. This is especially important when evaluating two such similar compounds with such "minor" structural differences as CH_3 and CH_2-OH. The concepts of conformational analysis take on a more realistic meaning with the availability of the three-dimensional structural information from X-ray crystallography. Examination of such comparative information for leucanthone and hycanthone is very illuminating. For example, the data indicate two important dissimilarities: The flatness of the tricyclic portion of the two drugs differs in that in the former it is planar, whereas in the latter compound the two benzene rings face each other at a 160-degree angle (i.e., 20 degrees off perfect flatness). The second difference regards the ethylenediamine side chain, which is in a *gauche* relationship to the tricyclic nucleus in leucanthone, but which adopts a *trans* conformation in hycanthone that is almost perpendicular to the phenyl ring to which it is bonded.

Oxamniquine offers a valuable insight to the chemical–pharmacological reasoning that goes into the development of a modern parasiticide. In the early 1960s available drug treatment against schistosomiasis revolved around the very toxic antimonials, the toxic niridazole and hycanthone, which has a 50% incidence of nausea, showed hepatotoxicity, and had reports of mutagenicity and carcinogenicity. A new approach was in order. An extensive drug development project using mirasan (Fig. 7-14) as the jumping-off point was undertaken.

Mirasan exhibited high activity against schistosomes in mice, but not higher species such as the monkey or humans. Cyclization of the 2-diethylaminoethylamine moiety into a piperazine ring (Structure I, Fig. 7-14) doubled the activity—but only in mice. The increased activity can be rationalized by assuming that rigidizing the ethylenediamine side chain produced a better stereochemical fit by a more favorable configuration. If that is correct, then the inactivity of the compound in primates must have other, more complex explanations, including such factors as species differences in pharmacokinetics, absorp-

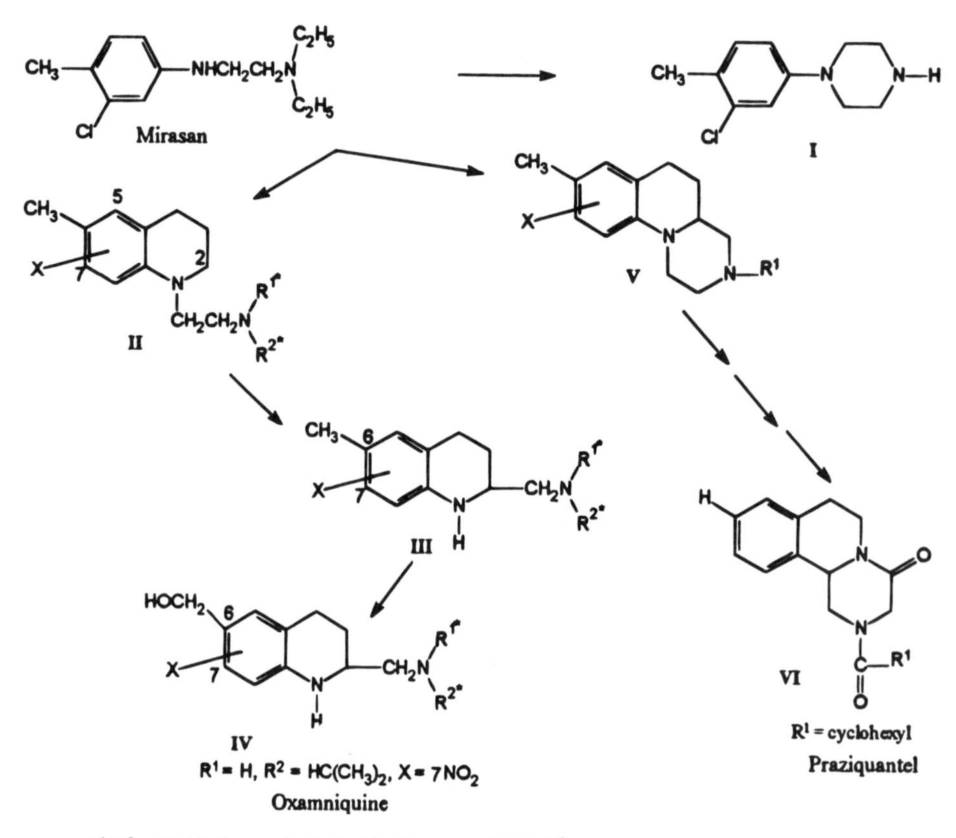

Figure 7-14. The evolution of oxamniquine and praziquantel.

tion, and elimination rates, as well as body distribution patterns. Differences in metabolism, although frequently minor, are sometimes the major factor, as in this case. It was reported that unlike the methyl group (para to the diamine) of leucanthone, which was rapidly oxidized to the hydroxymethyl group of hycanthone, the corresponding metabolic reaction of the CH_3 group in mirasan, simply did not occur in the monkey. However, the synthetically prepared putative hydroxymethyl metabolite was highly active in hamsters that were also unable to metabolize mirasan itself significantly. Metabolic species differences clearly played the dominant role here. A different route to cyclization, where a tetrahydroquinoline structure (Fig. 7-14) was synthesized (II, R^1, $R^2 = C_2H_5$, $X = 5$ or 7 Cl), showed good activity in the mouse and the monkey, but not definitively in humans. Many variations of II (e.g., X = other than Cl, variations of R^1 and R^2) produced active compounds—but not in humans. Nevertheless, the biological data of compounds I, II, and III certainly indicated the significance of side chain stereochemistry. Subsequently, additional cyclization of II to compounds of structure IV produced even more configurational rigidity in the molecule. Compounds of type III, position isomers that maintained the 2-carbon distance between the aniline and the aliphatic amine nitrogen, proved to be the most active compounds if the 6-methyl remained (even the C_2H_5 was inactive) and X was an electron-withdrawing group (NO_2> CN> F> Cl> Br) while R^1 was *isopropyl* or *tert*-butyl; R^2 could be anything from H to C_4 alkyl). Finally, converting the C-CH_3 to the 6-CH_2-OH by microbial oxidation with *Aspergillus sclerotium* affords the drug oxamniquine (IV), which is highly effective against *S. mansoni*, although it has lesser effects against the other species. The primary mechanism is not yet known, but it does not appear to involve acetylcholinesterase.

Praziquantel (PQ) (Table 7-10), a pyrazinoisoquinoline, can be viewed as having evolved from mirasan by optimizing structural features of compound V (Fig. 7-14). For all intents and purposes it is the first really broad-spectrum anthelmintic that is effective against all schistosomes and most other flukes, such as liver, intestinal, and lung (see Table 7-9 for species). PQ is rapidly becoming the drug of choice for many of the preceding infections, as well as for the cestodes (tapeworms).

The mechanism of action has not yet been established. Among the noted effects are an escape of glucose through the worm's covering (tegument). Increased permeability of the plasma membrane is one possibility. Increased permeability to Ca^{2+} has been reported. The drug appears to cause muscular contraction in the parasites, followed by paralysis. The host's immune cells attach to the worm's tegument, vacuolization follows, the parasite enters, and disintegration and death follow.

At the biochemical level energy metabolism might be implicated; however, unlike niclosamide (see later), oxidative phosphorylation is not involved.

PQ is extensively metabolized on passing through the liver to inactive hydroxy derivatives. Its half-life is short, the drug peaking in the serum in less than 2 hours. The high cure rates achieved with a single dose given in one day are thus a good indication of how effective PQ is. The most frequent side effect is drowsiness. Untoward effects are generally mild and transient. Considering the wide effectiveness against flatworms and low toxicity, PQ may be viewed as a major advance in the treatment of parasitic diseases.

Cestodes have generally responded well to niclosamide (NC, Nicloside) (Table 7-10). This salicylanilide is highly curative of all four species of tapeworms that depend on anaerobic metabolism for most of their energy. That means that their mitochondria form ATP by incorporating it from inorganic phosphate. NC inhibits this reaction at low concentrations. The classic anaerobic phosphorylation in mammalian cells would also be uncoupled, but at higher doses. This, and the fact that NC is totally water insoluble and is not systemically absorbed, accounts for the selective toxicity of the drug.

Bithionol (Fig. 7-10) is useful for the treatment of *Fasciola hepatica* (sheep liver fluke) and *Paragonimus westermanii*, the lung fluke. It is highly effective. The drug impairs egg formation, uncouples oxidative phosphorylation, and chelates iron, possibly from crucial enzymes. None of these facts has been related to a molecular mechanism, which remains obscure. In any case, it would appear that PQ will likely replace this *bis*-dichlorophenol as the drug of choice for these two fluke infections.

Hexylresorcinol is another phenolic compound of one-time value against roundworms but is now considered obsolete as an anthelmintic; its usefulness essentially remains as a topical bactericidal antiseptic.

Chloroquine (Fig. 7-6), the antimalarial antiamebic drug, has had some application in the difficult-to-treat Chinese liver fluke, since it will reduce the egg output of the worm. With the high cure rates reported with PQ, the value of CQ will certainly diminish.

A really broad-spectrum drug against nematodes is not yet on the horizon. Bephenium (BP) as the insoluble 2-hydroxynaphthoate salt, has been useful against both species of hookworm (Table 7-9). BP was the best of a series of compounds designed to mimic the ability of acetylcholine (Chapter 8) to cause Ascaris roundworm muscle to contract. When the worms let go of their intestinal attachment, they are fecally excreted. With cure rates as high as 100% with the more sensitive *A. duodenale*, the drug has been popular due to its low cost, although mebendazole is replacing BP especially in mixed infections, including the large roundworm *A. lumbricoides*.

Mebendazole (MB) is somewhat broader in its anthelmintic spectrum. In addition to the two hookworms and *Ascaris*, the drug also eliminates pinworms, whipworms, and possibly threadworms. This spectrum and essentially no toxicity makes it a particularly valuable drug in mixed infections. MB is even effective against the intestinal trichinosis organism, but it may not be in the muscle form.

MB is known to inhibit glucose uptake by the parasite, which then depletes its stored glycogen, leading to ATP failure and death. The drug has also been shown to interact with helminthic tubulin, the structural protein of the microtubles; mammalian tubulin is affected to a much lower extent. This differential effect, combined with the lack of systemic absorption, may account for the selective toxicity. It has not been established whether the tubulin effect or the interference with carbohydrate metabolism is the primary cytotoxic event, or what the relationship is between the two phenomena, if there is any.

Pyrvinium (Table 7-10) is an unsymmetrical red cyanine dye whose amidinium ion system exhibits unusual resonance (Eq. 7.12). Its mechanism of action is not fully understood, nor has it recently been investigated. It was shown to interfere irreversibly with glucose absorption and has been effectively used in the treatment of pinworms and threadworms, which both reside in the low-oxygen environment of the human intestine. Under more aerobic conditions of filarial worms similar dye compounds have been found to inhibit oxygen consumption.

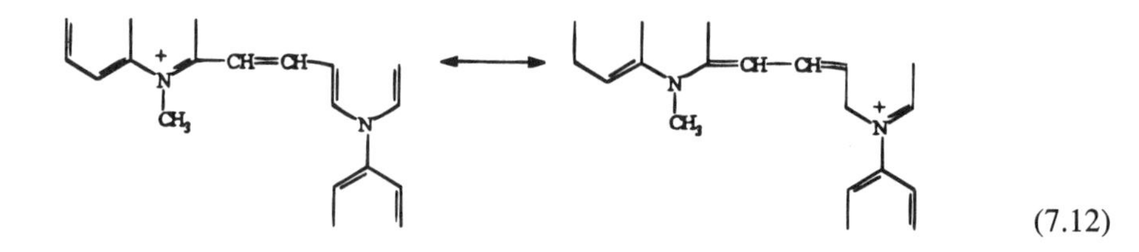

(7.12)

Pyrantel, as pyrvinium, is available as the extremely insoluble and nonabsorbed pamoic acid salt; one oral dose cures pinworm and *Ascaris* infections. It is curious that whipworm and threadworm infections do not respond. The drug acts as a depolarizing neuromuscular blocking agent (Chapter 8) on worms, which are then paralyzed. In *Ascaris* muscle strips it is 100 times more active than acetylcholine.

Thiabendazole (TB) (Table 7-10), another benzimidazole anthelmintic, can be almost considered a broad spectrum drug in the normal roundworm (nematode) area. It can kill larva at the astoundingly dilute level of 10^{-10} M. Even though it is effective against many roundworms, it is used more selectively, especially against strongyloidiasis (threadworm) in humans. The exact mechanism of action has not been established. The drug has been demonstrated to affect strongly anaerobic metabolism by inhibiting *fumarate reductase*, the enzyme that generates ATP as it converts fumarate to succinate (Eq. 7.13). Unlike MB, TB is rapidly absorbed systemically and rapidly converted ($t_{1/2}$ = 1.7 h) to the inactive 5-hydroxymetabolite. Blocking this position and replacing the thiazole ring with a carbamate resulted in MB and several of its variants (see footnote *f*, Table 7-10).

(7.13)

TB has also been found to suppress microtubule assembly selectively in susceptible worms, thus inhibiting acetylcholinesterase. Again, it is not immediately apparent which effect, the one on fumarate reductase or the one on the microtubules, is primarily responsible for the parasiticidal action of the drug. It is possible that it is the combination that is so devastating. As previously mentioned, TB also has broad fungicidal activity. In addition, TB has been found to have clinically significant analgetic–antiinflammatory–antipyretic properties that may be relevant to the useful effects the drug seems to have in trichinosis.

Levamisole (and its racemate tetramisole) leads to worm paralysis by ganglionic stimulation via acetylcholine. It is curious that the drug produces a depolarizing neuromuscular blockade in vertebrates. Levamisole is used in ascariasis outside the United States. The drug also is said to have immunostimulant properties, and it has been approved in the United States (1990) for such indications. As a thymomimetic in mice it tends to restore T-cell immune responses to normal.

Tetrachloroethylene (TCE) is the only halogenated hydrocarbon anthelmintic holdover from the nineteenth century because of its specific efficacy against hookworm. Of course, it has all the hazards attributed to alkyl halide use in humans. Safer drugs should now be utilized. It is interesting that although TCE is still approved for human use in the United States, it is only available as a veterinary product (Nema).

There are two piperazine anthelmintics available (Table 7-10). Piperazine citrate has been highly effective against nematodes, particularly pinworms (*E. vermicularis*) for over 30 years with cure rates usually over 95%. The drug paralyzes the worm, which is then peristaltically expelled, requiring no additional cathartics. The mechanism has been narrowed to two possibilities: an action similar to that of curare on skeletal muscle in mammals [i.e., inhibition of depolarization by acetylcholine at the neuromuscular junction (Chapter 8)], or by nerve stimulation involving inhibitory neurotransmitters such as gamma-amino butyric acid (GABA). Anaerobic glycolysis in *Ascaris* is inhibited affecting motility.

Filariasis results when tissues become infected with adult filarial worms (Table 7-9). The multitude of larvae produced within the tissues, called *microfilaria*, then enter the skin and the bloodstream. Filariases are mosquito-borne diseases that present special difficulties of treatment, because available drugs either kill the adult worm (e.g., the very toxic suramin) or the microfilaria. It was shown that the diethylcarbamoyl derivative of piperazine, diethylcarbamazine (DEC), was selectively effective against filarial nematodes, but only in vivo, possibly by sensitization to phagocytic action. One of the diseases that is the most difficult to treat is onchocerciasis, caused by *Onchocerca volvulus*. The disease always progresses to ocular involvement and is a leading cause of blindness in endemic areas. The only treatment used to consist of Suramin to kill the adult *O. volvulus*, followed by DEC to kill the microfilaria. For mass treatment IV suramin is not practical and has serious nephrotoxicity. DEC can also present the patient with major untoward effects, including encephalitic allergic reactions, which may be minimized with antihistamines and steroids. DEC also increases both live and dead microfilaria in the cornea, producing opacities and optic neuritis, leading to permanent eye damage. Thus treatment of ocular onchocerciasis has been limited to sight-threatening conditions. The reappearance of microfilaria following DEC therapy is also disturbing unless the adult worms are surgically removed.

One of the most promising new microfilaricides is the remarkable group of antiparasitic antibiotics called Avermectins, particularly Avermectin B_1, now generically named Ivermectin (Fig. 7-15).

Avermectins are a group of 16-membered macrocyclic lactones discovered from a previously unknown actinomycete species, now named *Streptomyces avermitilis*. Its antiparasitic activities were discovered during animal screens against *N. dubius* in mice. (They are not active against cestodes or trematodes.) Unlike other macrolides such as erythromycin (Chapter 6), avermectins have no antibacterial or antifungal activity. They do have endoparasitic activity against phytoparasites, as well as free-living nematodes. It is equally significant that avermectin activity against ectoparasites, particularly arthropods such as lice, mites, ticks, and fly larvae, are all species encountered in veterinary practice. The drug, which was first reported in 1978, is widely used in large-animal veterinary medicine worldwide as a broad-spectrum nematode agent with ectoparasiticidal activity as an addi-

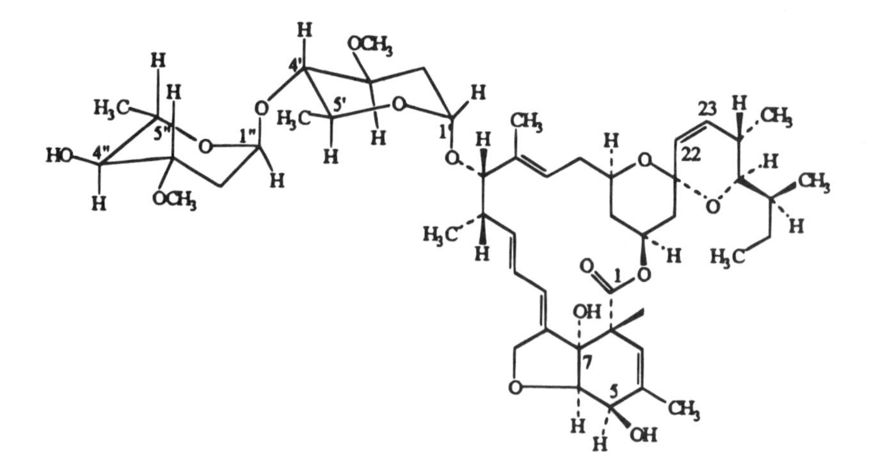

Figure 7-15. Ivermectin.

tional desirable feature. Furthermore, the drug affects nematodes in the larval, juvenile (but not adult) stages in both intestinal and extraintestinal sites. Good effects have been obtained with cattle, horses, sheep, dogs, and even fowl at remarkably low doses (e.g., dog heartworms were eradicated at 1 μg/kg), although doses may range to 0.2 mg/kg.

More recent human clinical research has focused on the blinding ravages of onchocerciasis (river blindness).[11] Comparative studies against DEC (and placebo) have shown ivermectin to clear *O. volvulus* from the skin and eyes with single 10–12 mg doses with fewer (or no) ocular complications than DEC and better tolerance and effectiveness. The only disadvantage of ivermectin was a slower onset of results. A report found ivermectin-treated patients infected the vector flies at a lower rate for at least six months, thus in effect reducing the potential for transmission of the disease and ultimately reducing the need for insecticides.

Mechanism of Action. Avermectin B_{1a}, AVM, paralyzes helminths (*Ascaris*) by affecting interneurons and inhibitory interneurons of the worm. Early work with lobster walking legs showed the drug eliminating inhibitory postsynaptic potentials, followed by a slow decrease in excitatory postsynaptic potentials; resistance of muscle fibers decreased. The authors hypothesized that AVM acts to reduce membrane resistance at the neuromuscular junction by opening the membrane's Cl^- channels regulated by GABA at the inhibitory synapse. It was demonstrated that AVM blocked transmission between interneurons and excitatory motor neurons in the *Ascaris* nerve cords. It is interesting that AVM causes an increase in GABA release from rat brain synaptosomes. Stereospecific binding affinities of the various analogs seem to correlate with anthelmintic activity. Therefore, AVM activity may be either directly GABA-ergic or due to increased tonic GABA release from nerve terminals.

The chemistry of the Avermectins is complex. There are eight major natural AVMs, which are divided into series. The basic structure consists of a 16-membered lactone, glycosidically bound at C_{13} to two L-oleandrosyl sugar moieties. The B series (5-OH) was generally more potent than the A series (5-OCH_3). Types 1 and 2 differ in C-22–23 double bond or C-23-OH. Catalytically, hydrogenating the C-22–23 olefin had little effect; but reducing the C-3–4 olefin caused a large drop in activity. It is particularly intriguing, considering the structure, that the absence of any antimicrobial activity, is due to protein synthesis inhibition such as erythromycin exhibits, or is due to ionophoric activity such as amphotericin B has.

7.8. Antiviral Chemotherapy

The pathogenic virus is the quintessential parasite. It is a totally obligate, *intracellular* parasite that replicates only within the host cell diverting components of the host's nucleic acid synthesizing machinery to build more of itself. While carrying out this rapacious process, the virion is actually protected by the victimized cell's plasma membrane.

Since viral disease constitutes such a large proportion of human illness, from the "trivial" incurable common cold to the so far totally deadly AIDS, a brief overview of the virus and its characteristics is essential to understand the unique problems encountered in viral chemotherapy and the apparently impossible obstacles to develop even specific antiviral

[11] Transmitted by insects living along river courses, hence the name. The disease is found in sub-Saharan Africa, Central and South America, and Arabia.

drugs, much less broad-spectrum compounds. One need only realize that after almost four decades of research there are only a handful of approved drugs for limited indications in the United States. Even these were discovered as a result of screening programs involving thousands of compounds, rather than by design.

The basic goal in developing any chemotherapeutic agent is that it produce its desired effect without compromising host functions by unacceptably severe side effects. In antimicrobial and antiparasitic chemotherapy, this has largely been achieved by the development of drugs with highly selective modes of action. Looking back historically, this was relatively easy because, paradoxically, the target—the bacterial, fungal, or protozoal cell—provided the researcher with many significant biochemical and morphologic differences from the host cell. Once understood, these differences were exploited and selectively inhibited.

The situation is different in the chemotherapy of cancer and viral infections. In the former, we are attempting to arrest the growth of, or to destroy, certain cells of the host. In the latter, the invader becomes a component of the host cell, actually integrating with and coupling to the cell's biosynthetic machinery. Therefore, it should not be surprising that success in both these areas has been minimal and achieved slowly. The reason is the intimate association of biosynthetic processes of the host cell with those of the invading virus. Research in molecular virology, however, has begun to identify certain viral functions that can be delineated from those of the host cell and thus become, potentially at least, amenable to specific chemical interference.

A virus may be considered to be an organism consisting of a nucleic acid core (the *genome*) surrounded by a protein-containing coat. It reproduces exclusively within the living cell. Energy and building materials for the production of the new viral particle are supplied by the infected host cell, as is the synthesis machinery (e.g., ribosomes for the production of virus-specific proteins). Table 7-11 lists some important pathogenic viruses and the human diseases they cause.

The most important characteristic of a virus is the type of nucleic acid it contains. Thus the first division of viruses can be based on whether they are DNA- or RNA-containing.

Table 7-11. The Animal Viruses

Family[a]	Representative genus (species)[a]	Genome[b]	Symmetry[a]	Diseases in human
Papoviridae	Papillomavirus (wart virus)	DNA-SS	I	warts
	Polyomavirus (simian vacuolating virus)	DNA-SS	I	leukoencephalopathy
Adenoviridae	Masadenovirus (many serotypes)	DNA-DS	I	Respiratory, eye infections
Heptoviridae	Herpes virus (Herpes simplex type I)	DNA-DS	I	Skin and eye infections, encephalitis
	(Herpes simplex type II)	DNA-DS	I	Genital infections, skin eruptions, disseminated disease
	(B virus)	DNA-DS	I	Encephalitis
	(Varicella zoster virus)	DNA-DS	I	Varicella, herpes zoster
	(Cytomegalovirus)			Mononucleosis, hepatitis

(Continued)

Table 7-11. *(Continued)*

Family[a]	Representative genus (species)[a]	Genome[b]	Symmetry[a]	Diseases in human
		DNA-DS	I	
	(Epstein–Barr virus)			Infectious mononucleosis, Burkitt's lymphoma (?)
		DNA-DS	I	
Poxviridae	Orthopoxvirus (Variola major)	DNA-DS	I	Smallpox
Picornoviridae	Enterovirus (poliovirus serotypes)	RNA-SS	I	Poliomyelitis
	Coxsackievirus A and B, many serotypes	RNA-SS	I	Aseptic meningitis, respiratory infections
Togoviridae	Alphavirus (equine encephalitis virus)	RNA-SS	I or H	Encephalitis, Eastern, Western, Venezuelan
Bunyaviridae	Rubivirus (rubella virus)	RNA-SS	H	German measles
Orthomyxoviridae	Influenza Virus (Type A, B, C)	RNA-SS	H	Influenza
Parmyxoviridae	Paramyxovirus (Parainfluenza 1)	RNA-SS	H	Respiratory tract infections
				Parotitis, orchitis
	(Mumps)	RNA-SS	H	
				Measles, panencephalitis
	Morbillivirus (measles virus)	RNA-SS	H	
				Respiratory tract infection
	Pneumovirus (respiratory syncytial virus)	RNA-SS	H	
Rhabdoviridae	Lyssavirus (rabies virus)	RNA-SS	H	Rabies
Arenaviridae	Arenavirus (lymphocytic choriomeningitis)	RNA-SS	H	Meningitis
Coronaviridae	Coronavirus (infectious bronchitis)	RNA-SS	H	Lassa fever, hemorrhage
Retroviridiae	Human immunodeficiency virus	RNA-SS	H	AIDS[d]
Reoviridiae	Orbivirus (Colorado tick fever virus)	RNA-DS	I	Encephalitis
Unclassified	Hepatitis B (serotypes)	DNA-DS (?)		Hepatitis B

[a] According to Fenner, 1976.
[b] Genome: SS = single-stranded; DS = double-stranded.
[c] Nucleocapsid symmetry, I = icosahedral (20 sided); H = helical.
[d] Acquired immunodeficiency syndrome.

The nucleic acid in DNA viruses is invariably double-stranded; most RNA viruses have singe-stranded RNA. There are exceptions in both cases.

Not all viruses have a protein coat or envelope. However, a protein structure called the *capsid* is present, and it becomes the only protection for the nucleic acid core in the absence of an envelope. Envelope proteins, when present, are frequently glycosylated. The most important virus-specific proteins are the *transcriptases*. In studying viral replication pathways by utilizing various inhibitors as molecular probes, some processes unique to viruses have been discovered. This may allow the designing of specific replicative inhibitors that might function clinically at acceptable toxicity levels. For example, RNA viruses either have or induce RNA-dependent RNA polymerase (or replicase). The uniqueness of such an enzyme stems from the fact that the host cell's mechanism for producing RNA from its DNA template uses DNA-dependent RNA polymerase (transcriptase). Since the virus in this case cannot utilize its victim's transcriptase, it must use its own replicase. This difference is potentially exploitable in finding a chemotherapeutic agent, one that would specifically inhibit its function.

Another group of RNA viruses called *retro*viruses also contain a unique nucleic acid polymerizing enzyme called *reverse transcriptase* or RNA-dependent DNA polymerase, since the genetic code here is taken from RNA, which is reverse of the host cell's normal mechanism. The uniqueness in this case should also theoretically permit selective control, thereby preventing retrovirus replication.[12]

A DNA-dependent RNA transcriptase is found in the core of poxviridae. Unlike all other DNA viruses, the poxvirus multiplies in cell cytoplasm; with the other DNA viruses, transcription takes place in the nucleus, where the host cell polymerases are utilized. Thus the majority of DNA viruses need no replicative enzymes of their own.

Viral envelopes, when present, are derived from the various host cell membranes. These include plasma, nuclear, vacuolar, and endoplasmic reticular membranes. Thus viral envelopes resemble cellular membranes in regard to structural features (bilayer) and chemical composition (lipo- and glycoproteins). The lipids are the same as those of the host cell; however, the protein components are usually virus-specific.

An understanding of the viral replicative cycle can give us some insight into possible areas of selective chemical interference that leads to successful chemotherapy.

The mode of viral reproduction is uniquely different from the more complex cellular microorganisms. This can be presented in brief outline as follows:

Phase I: Initiation of Infection

1. Adsorption

The first step is necessarily adsorption of the virion onto the cell surface. Physical contact, collision frequency, contact area size, and the ionic conditions of the environment are all factors, but they are not usually sufficient to account for successful adsorption. For many viruses, specific cellular receptor sites have been demonstrated on the plasma membranes. For example, the adsorption of the influenza virus to red blood cells has been shown to be through binding of virion hemagglutinin, in the form of surface spikes on the virion envelope, to cell receptors.

There are no useful drugs to prevent the adsorption of viruses on membranes presently available.

2. Penetration

The second step of the first phase is penetration. With animal viruses, the most common method is pinocytosis (membrane invagination), which leads to gradual engulfment of the virion. Direct penetration can occur. Fusion of the envelope with the plasma membrane has also been observed. Once penetrated, the virion is moved to sites in the cell, where the replicative steps will occur.

Amantadine HCl (Symmetrel) (Fig. 7-16) appears to be an inhibitor of viral penetration or some other early step in the replicative cycle. This has been established with rubella

---

[12] Certain rifamycin derivations can inhibit reverse transcriptase.

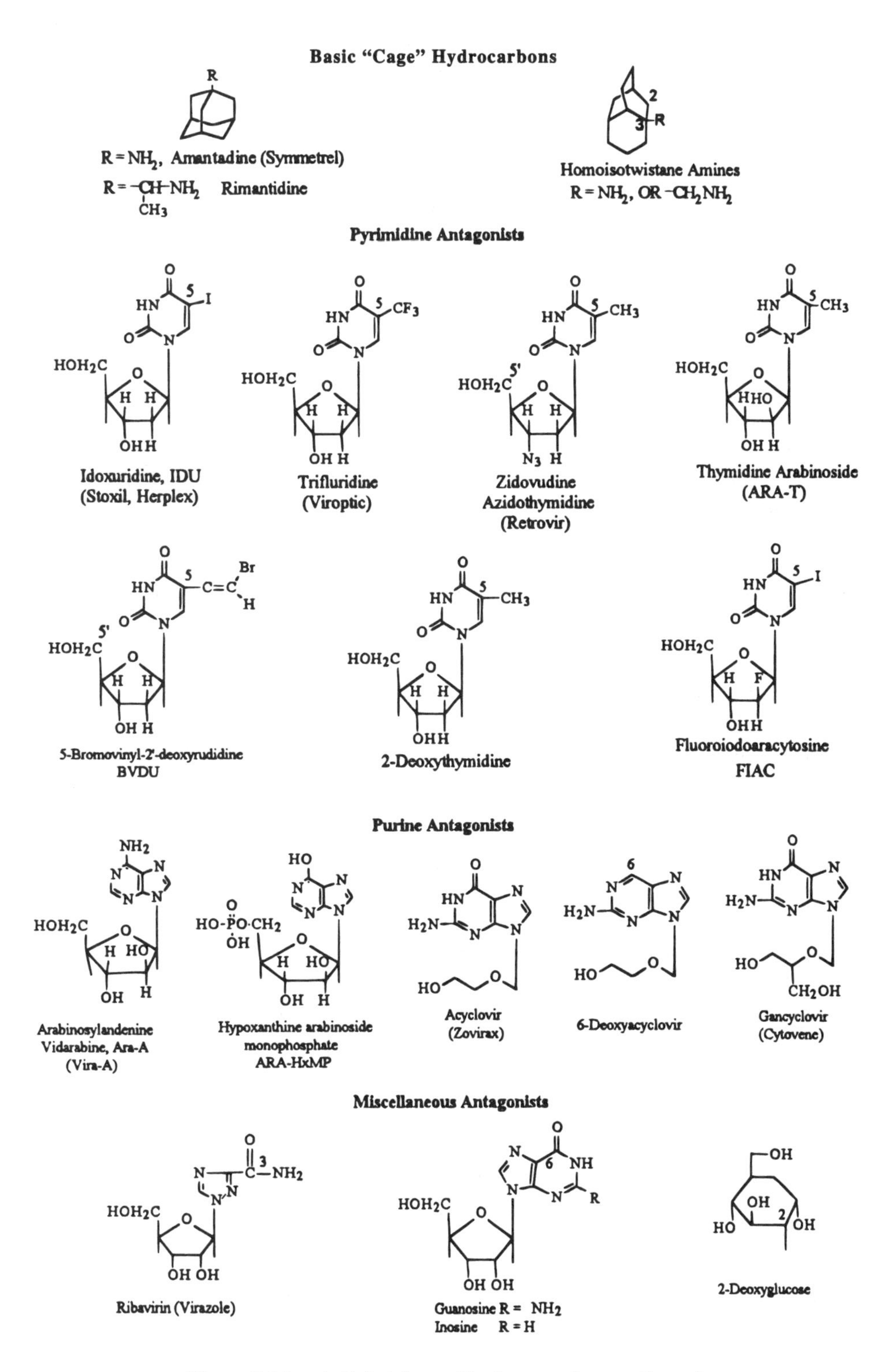

Figure 7-16. Antiviral drugs. (Trade names in parentheses.)

virus, influenza virus A_1 and A_2, and parainfluenza pseudorabies virus. The only officially approved antiviral clinical use of amantadine in the United States is for prophylaxis of influenza A_2. Amantadine is not active against influenza B. Efficacy has been established. Effectiveness of the drug in preventing Hong Kong flu was reported in 1970. A more recent report also demonstrated the prophylactic effectiveness of amantadine against Russian influenza.

An amantadine derivative, rimantidine, may actually be a more active drug, although it has not been as thoroughly evaluated (Fig. 7-16). The amantadine cage-type compounds are not unique. A series of homoisotwistane compounds were synthesized and tested in vitro against Newcastle virus. It was found that the 3-amino and aminomethyl derivatives were 30–50 times more active than amantadine (Fig. 7-16).

3. Uncoating

The third step of the initial phase of viral reproduction involves the shedding of the protective protein envelope and/or capsid. The uncoating can be quite complex, involving a two-stage process in some cases (e.g., poxviridae).

The uncoating process has been shown to be carried out by virus-specific enzymes, synthesized at the direction of viral *m*RNA. The *m*RNA in turn has used a portion of the viral DNA as the template. Experimentally, at least, protein synthesis inhibitors can inhibit the production of these uncoating enzymes. However, due to a lack of significant specificity, the compounds were of no clinical value. The antimalarial drug chloroquine is the only agent of any consequence found to have any inhibiting effect on the uncoating of a virus, namely, Newcastle disease virus.

Phase II: Synthesis Steps

1. RNA or DNA synthesis

The second phase of the viral infective cycle is the synthesis of viral components. The genome, as well as the viral proteins, are synthesized here. This phase is potentially the most promising area for developing selective inhibitors of chemotherapeutic value. As our understanding of viral molecular biology increases, it can be increasingly coupled to the ingenuity of the medicinal chemist. Ideas need to be exploited to develop antiviral agents of low toxicity with the ability selectively to inhibit the synthesis of virus-specific RNAs and DNA by affecting the transcriptase or other important enzymes such as the kinases. Agents capable of acting on polymerase enzymes in some direct manner are possible. Phosphonoformic and phosphonoacetic acids are the result of extensive screening. These acids have antiherpes activity by binding to the pyrophosphate binding site of the viral DNA polymerase.

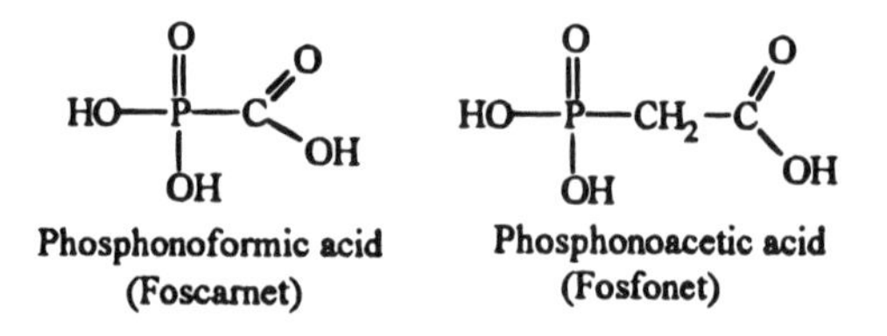

Phosphonoformic acid (Foscarnet) **Phosphonoacetic acid (Fosfonet)**

Another approach is to utilize the antimetabolite concept. The synthetic strategies include two types of chemical modifications of nucleosides. One is to alter the base. In the case of pyrimidines (Fig. 7-16) this usually involves replacement of the 5-CH_3 group of 2-deoxythymidine with a halogen or CF_3 function. The resulting 2-deoxythymidine analog can do one of several things. If the 5-substituent is small (e.g., 5-fluorouracil or its deoxynucleoside, Chapter 4), then the compound inhibits DNA synthesis by effectively competing with deoxyuridine for the active site of thymidylate synthetase. If a bulky substituent such as I (van der Waals radius 2.15 Å) replaces CH_3 (2.00 Å) of deoxythymidine, it becomes an "imposter" molecule. Although originally intended as carcinolytic, Idoxuridine (IDU) became one of the first successful topical treatments for corneal, conjunctival, and eyelid infections by the DNA virus herpes simplex. By overcoming poor solubility with DMSO, the drug is also successfully used against cutaneous herpetic lesions of herpes simplex and zoster. IDU is used primarily against herpetic keratitis of the eye. The drug blocks thymidine utilization by becoming phosphorylated to the ′5-monophosphate.

In the cell IDU is phosphorylated by both cellular and viral thymidine kinases to active triphosphorylated derivatives that inhibit both cellular and viral DNA synthesis, since IDU is incorporated into both viral and cellular deoxynucleic acids. This altered DNA-type material cannot then be replicated with fidelity, nor can the resultant "code" be transcribed properly to *m*RNA. Even though defective virions are also produced, it is the adverse effects on rapidly proliferating host cells that produce the serious adverse effects, following IV administration, on bone marrow, GI tract, leukopenia, thrombocytopenia, and alopecia, all of which are the side effects seen with the antimetabolite carcinolytics (Chapter 4). In topical use these toxicities are not significant so that the drug becomes useful to treat herpes simplex keratitis in ophthalmic infections. Trifluridine with analogous properties systemically, however, is a superior drug topically in treating ocular infections, primarily because it is a more soluble compound, which allows higher concentrations and potency.

Two experimental compounds are of interest. 5-Bromovinyl-2′-deoxyuridine (BVDU) (Fig. 7-16) is more specific, as it is phosphorylated in the 5′ position by herpes simplex virus type I (HSV-1)-induced thymine kinase in cells thus infected. It is effective orally against herpes zoster, but unlike acyclovir (see later) is not active against HSV-2. However, because of development of mutants with high resistance due to lower kinase levels, and because of the appearance of animal tumors on prolonged administration, this drug is not being pursued. Thymidine arabinoside, Ara-T, represents another approach, namely, altering the sugar moiety of nucleosides to produce antimetabolites, rather than the base portion.

Ara-T, with significant activity against HSV-1 and -2, and varicella zoster virus (but little against cytomegalovirus) is selectively phosphorylated by HSV-induced deoxypyrimidine kinase, but not by thymidine kinase from uninfected cells. Thus inhibition of DNA synthesis affects viral replication, but not the growth of normal cells, which explains the observed selective toxicity.

A particularly interesting antiviral drug first marketed in 1987 is zidovudine (azidothymidine, Retrovir, AZT, Fig. 7-16). This compound, synthesized over 25 years ago as a potential carcinolytic that was ineffective, has been found to be one of very few drugs to have any clinically relevant effect on the HIV retrovirus (Table 7-11). The 3′-hydroxy group of thymidine has been replaced by the active azido group (Fig. 7-17). The drug is apparently triphosphorylated (at C-5′) by *cellular* enzymes. The triphosphate compound is a competitive inhibitor of reverse transcriptase and is incorporated into nascent DNA strands that result in termination of DNA chain synthesis. The 3′-azido substituent prevents further 5′- to 3′-phosphodiester linkages from forming since hydrogen bonding with the complementary adenine base is impossible. The difference in susceptibility between mammalian and viral transcriptase is only about 100 times, which clinically translates into potentially serious adverse effects.

Two other agents that block replication of the human immunodeficiency virus causing AIDS should be mentioned. Dideoxyinosine (ddI, DDI, Videx) was approved in late 1991 in

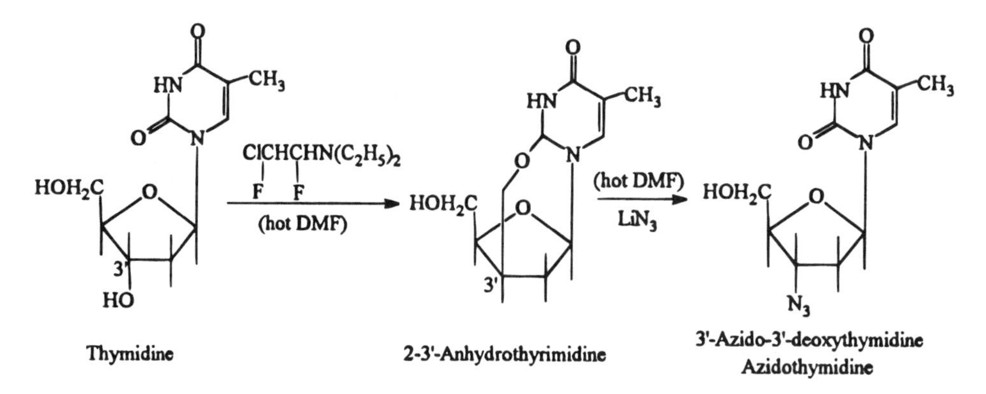

Figure 7-17. Synthesis of Zidovudine (Azidothymidine).

spite of incomplete knowledge of its long-term adverse effects. The dire need for additional therapy against AIDS has accelerated drug approvals for this disease. Dideoxycytidine (ddC, Hivid) is another drug marketed in 1992. Combinations of both agents with AZT are also being tried to reduce toxicity. So are combinations with α-interferon to produce synergism.

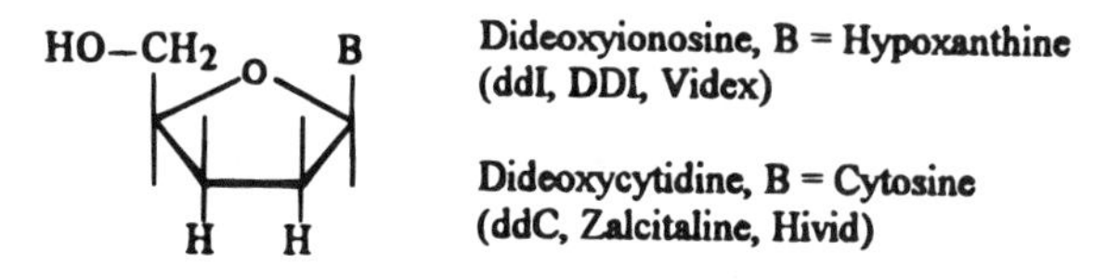

Until recently, much of the research efforts in antiviral therapy has been understandably driven by the almost desperate need to control and eradicate HIV and with it the disease AIDS. The basic premise has been to develop antiretroviral agents, but for the most part the few modestly effective compounds clinically introduced have been nucleoside-based inhibitors of the viral enzyme reverse transcriptase (RT). However, the rapid mutations of the virus, which resulted in drug resistance, combined with significant toxicity of the agents, conspire to negate any reasonable expectation that this approach alone has much more to offer. The possibility that some non-nucleoside RT inhibitors may be superior should be pursued but not depended upon. Other essential viral proteins, known and yet to be discovered, should in principle also be suceptible to inhibition and thus offer alternative modes of action. One candidate in particular came to our attention in the mid-1980s (Ratner et al., 1985; Pearl and Taylor, 1987), HIV-1 protease (proteinase). This enzyme is an aspartyl protease having the catalytic site -Asp-Thr-Gly- known to occur in fungal and mammalian proteases. For example, the idea pursued in the design of renin inhibitors (Chapter 10) could be and were applied to the search for HIV-1 protease inhibition (Greenlee, 1990).

Among the replicative differences between eukaryotic cells (of the viral host) and the pathogenic virions is that the latter need virus-dependent proteinases to produce the mature infectious virus. That is, HIV protease is an essential enzyme utilized by the virus to process proteins in order to reproduce itself. Retroviruses (and RNA viruses) produce the structural and enzymic proteins as large multidomain polyproteins. The specific cleavage of these various domains into the final products is a posttranslational event. These end-products enable the virion to develop into the actual infectious agent (Kransslich and Wimmer, 1988).

In the case of HIV such a protease has now been known and characterized for some time. Well over a dozen of its polyprotein substrates, including their cleavage sites (scissile bonds), have been identified. Thus HIV-1 protease is a virus-specific target against which inhibitors (i.e., drugs) are being designed, synthesized, and experimentally tested. Such peptidomimetic inhibitors of these cleavage reactions would lead to the formation of immature and non-infectious virion particles.

The approaches to drug development include:

1. *Peptide-Derived HIV Protease Inhibitors*. The designing of these compounds is based primarily on mimicking the transition state of peptide bond hydrolysis.

2. *Nonpeptide HIV Protease Inhibitors*. Here, more traditional methods and rapid throughput screening of natural products as well as a large array of synthetic compounds from database libraries are pursued. Some compounds deemed to have potential are derived from structure-based design strategies (e.g., complexes with HIV protease).

A particularly interesting report (Lam et al., 1994) used the computer-aided drug design strategy, based on knowledge of the protein's structural features. Complementarity of potential inhibitors of the protease's active site was designed by computer. Knowing the aformentioned features makes it feasible to design and construct a three-dimensional model of the functional groups needed in a candidate for binding (i.e., the *pharmacophore*). A search of molecular structure databases for "frameworks" having such groups in the desired stereochemical configurations yielded a compound that, with structural refinements, bound tightly to the protease's catalytic site. This was a cyclic urea molecule. Ureides are polar and metabolically stable; the cyclic nature apparently provided the "goodness of fit."

Recently (1996) the FDA (U.S.) has given "fast track" approval to three peptidomimetic HIV-protease inhibitors. These agents were released based on fewer and shorter clinical trials than would be required if the urgency of AIDS were not so acute. As a result, drug interactions and adverse effects are not fully known at this time.

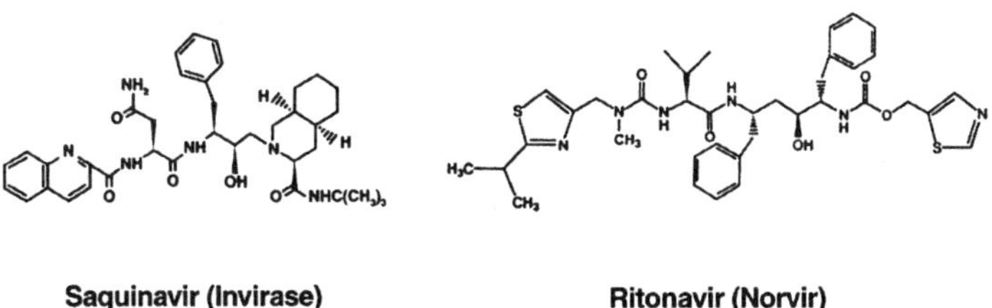

Saquinavir (Invirase) **Ritonavir (Norvir)**

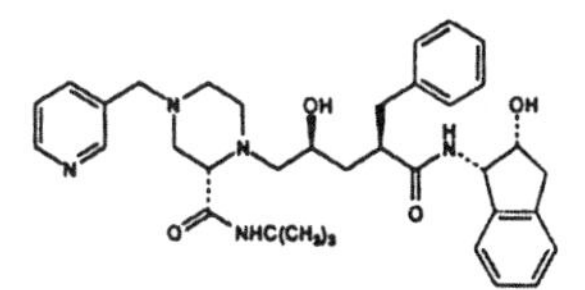

Indinavir (Crixivan)

3. *Combination Therapy*. The concept of utilizing two drugs with different mechanisms of action to postpone the appearance of resistance and/or to produce synergistic effects and decreased toxicities, as seen in cancer chemotherapy, is also a viable clinical approach. One example will be illustrative. A potent, specific peptide-derived HIV protease inhibitor, RU31-8959, with activity against HIV-1 and -2 has been found to be synergistic when tested in certain T4 cells infected with HIV-1 in combination with AZT, ddC, or alpha-Interferon in in vitro experiments and Phase I clinical trials (Muirhead et al., 1992; Johnson et al., 1992; Craig et al., 1993)

The obstacles to achieving rapid breakthrough results in viral chemotherapy are formidable. The main obstacles are complexity at the molecular/biochemical level, poor oral bioavailability (RO 31-8959, 4%), severe toxicities, and rapidly emerging resistance. This has been particularly disappointing with several of the nonnucleoside RT inhibitors. The HIV protease approach, improved RT inhibitors, or a combination of these and with cytokines, will hopefully improve AIDS therapy in this second decade of this worldwide pandemic.

Fluoroidoaracystosine (FIAC, Fig. 7-16) is a potent and selective inhibitor of herpes viruses. Like acyclovir and BVDU, it, too, must be phosphorylated by viral thymidine kinase to act. Its selectivity toward the viral enzyme is several thousand times higher so that it is active against HSV-1 and -2 and herpes zoster virus (HZV) with a good therapeutic index. Clinical studies seem promising.

Several purine antagonists have also been developed that are effective against DNA viruses. Replacing the ribose sugar of adenine with the 2′-epimeric arabinose afforded the 9-β-D-arabinofuranosyl-adenine, Vidarabine (A) (Fig. 7-16). It was first synthesized in 1960 as a potential carcinolytic. It was later found in *S. antibioticus*. Although it works best against various herpetic species clinically in cell cultures its activity against DNA viruses is broad. Nevertheless, it has dramatically reduced mortality of herpes encephalitis from 70% to 28%. The drug's poor solubility (1.8 mg/ml) requires large-volume slow infusions. However, the more soluble monophosphate (Ara-AMP) may be used. As with cytarabine (Chapter 4), Ara-A is phosphorylated to the triphosphate by adenosine and deoxycytidine kinase, yielding Ara-ATP. DNA synthesis is selectively inhibited. Levels of Ara-ATP correlate with viral replication.

Cellular enzymes deaminate Ara-A and its monophosphate to the corresponding hypoxanthine Ara-HxMP, which retains some activity, possibly because of reamination to Ara-AMP.

The development of a guanine nucleoside analog where the glycosidically bonded 2-deoxyribose is replaced by the acyclic 2-hydroxyethoxymethyl group represented the next significant step in viral chemotherapy. The drug's spectrum is limited to the herpes virus genus, but it encompasses all those listed in Table 7-11. Its low toxicity permits safe systemic administration. The drug is available in parenteral, oral, and topical dosage forms. Previously untreatable internal and deep eye infections are now curable.

The high therapeutic index of acyclovir is based on an exquisite selectivity of the first activation step (Fig. 7-18). This phosphorylation of 2′-hydroxyl group to acyclovir monophosphate is catalyzed HSV-specific thymidine kinase at a rate several million times faster than host cell thymidine kinase. The apparent reason is a 200 times higher binding affinity of the drug for the viral enzyme. Further phosphorylation by guanosine monophosphate kinase from the host cell results in the diphosphate. Finally, the triphosphate is obtained by still uncharacterized cellular enzymes. Acyclovir triphosphate then competes with deoxyguanosine triphosphate for the virus-specific DNA polymerase without affecting the host cell's DNA polymerase. Once acyclovir is incorporated in HSV DNA, the effect is chain terminating, since there is no 3′-OH on which to elongate the nucleic acid. Acyclovir has poor solubility in body fluids (1.3 mg/ml) and may crystallize in the urine, causing hematuria during IV therapy. Adequate fluid administration may minimize nephropathy. A better solution may be 6-deoxyacyclovir (Figs. 7-16 and 7-18). 6-Deoxycyclovir is a more soluble compound that is metabolized by xanthine oxidase to acyclovir on *oral* administration at a rate sufficient to match the serum levels attainable by IV acyclovir, and much higher than achievable by oral acyclovir. It is an almost ideal pro-drug.

Another more soluble analog of acyclovir is 9-(1-1,3-dihydroxy-2-propoxymethyl)guanine, gancyclovir (Cytovine) (Fig. 7-16). The mechanism is apparently analogous, yet this

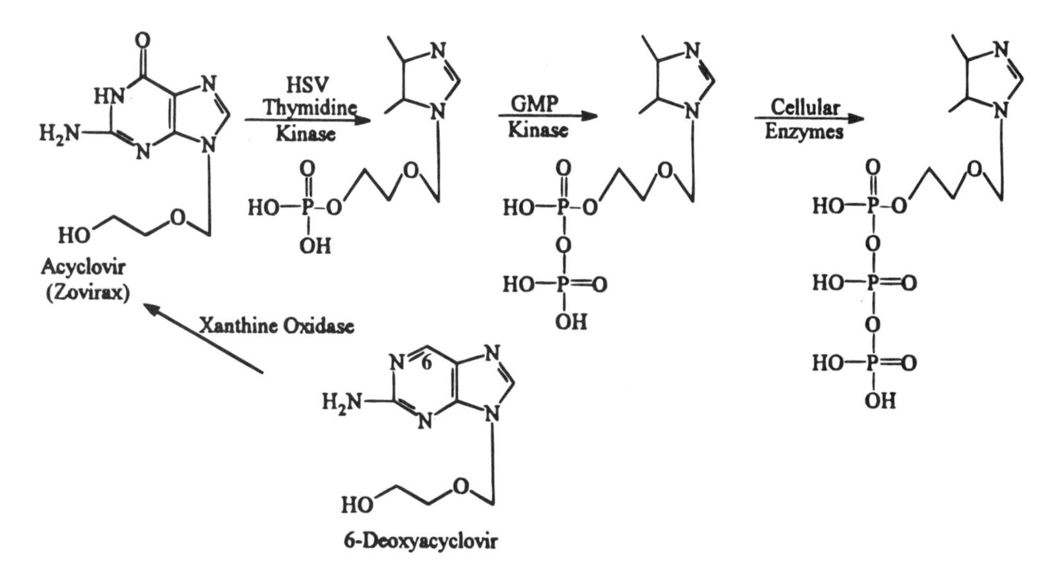

Figure 7-18. Activation of acyclovir and 6-deoxyacyclovir.

drug is more effective than acyclovir against cytomegalovirus and against some HSVs resistant to acyclovir.

Ribavirin represents a drastic remodeling of the purine structure. This compound was first reported in 1972. It is a 1-β-D-ribofuranosyl-1,2,4-triazole-3-carboxamide (Fig. 7-16). A broad spectrum of activity is claimed for this drug against DNA viruses (e.g., HSV, vaccinia) and RNA viruses (e.g., influenza A_1, A_2, and B). Later reports confirmed this. X-ray crystallographic studies have indicated striking similarities between the carboxamide group at position 3 of the drug and the carbonyl oxygen at position 6 and nitrogen at position 1 of guanosine (see Fig. 7-16) and inosine. One important action by ribavirin-5′-phosphate is the inhibition of inosine monophosphate dehydrogenase, essential for DNA synthesis.

In the United States ribavirin has been approved (1987) only for the treatment of respiratory syncytial virus (RSV) in children, while the drug is marketed in over 20 countries for many other viral diseases. Clinical potentials of ribavirin are being evaluated against influenza, measles, AIDS, dengue, and lassa fever.

The fact that ribavirin is a broad-spectrum antiviral exhibiting activity against both DNA and RNA viruses is established; the exact mechanism of action is not. More than one enzyme is obviously involved. It may also vary from one viral type to another.

Table 7-12 lists the various sites in the complex viral replication sequence where selective intervention with drugs is theoretically possible. Inhibition of *m*RNA processing as part of ribavirin's mechanism has been proposed. One other potential area that has only begun to be explored is in the final assembly steps, where the newly synthesized viral nucleic acids and structural proteins come together on newly formed membranes. Inhibition of such viral membranes thus becomes a viable chemotherapeutic approach.

Since viral membranes contain glycolipids and glycoproteins, imposter sugar molecules are good prospects as chemical troublemakers. Both glucosamine and deoxyglucose (2-deoxy-D-glucose) have been tested, with the latter showing the most promise. Studies have shown that inhibition of glycosylation by incorporation of the analog is the mechanism.

The most interesting and successful drugs to date have been those inhibitors of viral

Table 7-12. Potential Sites for Selective Drug Intervention

Adsorption—to cell being infected
Envelopment—of newly synthesized virion
Inhibition—of *m*RNA processing
Penetration—into cell being infected
Release—of completed virion
Transcriptases
1. RNA → RNA
2. DNA → RNA
3. RNA → DNA
Translation
1. Viral Genome replication (RNA or DNA)
a. by polymerase enzyme inhibition
b. by action as antimetabolites
Viral Protein Effects
a. Proteolytic cleavage
b. Glycosidation
Uncoating—of virion following penetration

replication processes whose specificity rests on their ability to be phosphorylated (i.e., activated) primarily by virus-induced kinases.

There appears to be a considerable worldwide effort to develop treatments for viral diseases. The prophylaxis of some diseases by vaccines, where available, remains a valuable method. The acceleration of work with human interferon is gratifying. The concept of using small organic compounds of nuclease-resistant polymeric polynucleotides as interferon inducers has had dubious results in humans, but may still be viable if therapeutic ratios can be improved. Immunoenhancing substances such as tilorone, levamisole, BCG vaccine, and others under investigation have had a limited impact thus far. Neither approach represents chemotherapy in the classic sense and was therefore not discussed here.

A beginning has been made. Since we are dealing with life's basic molecules within the mammalian cell, the potential for mutagenicity and carcinogenicity is high in this area. Nevertheless, progress is visible, and its rate is accelerating.

Finally, a report of potential future clinical significance is of the isolation and identification of a group of cationic, arginine-rich peptides with antimicrobial properties that have been obtained from mammalian phagocytic cells such as pulmonary phagocytes and polymorphonucleated neutrophils. These endogenous antibiotic substances, now named *defensins*, have been identified in human, guinea pig, rabbit, and rat cells. They show considerable homology of structure.

In vitro studies with purified defensins have shown microbiocidal activity at micromolar concentrations against both Gm+ and Gm– bacteria, fungi, and certain enveloped viruses such as herpes simplex. How do they work? Preliminary studies with human and rabbit defensins indicate that they increase the permeability of first the outer and then the inner membranes of *E. coli.* Whether it will be feasible to produce these substances in quantity by gene-splicing techniques in order to evaluate therapeutic efficacy by clinical studies remains to be seen.

References

Bloom, B. R., *Nature* **279**:21, 1979.

Bowman, W. C., Rand, M. J., *Textbook of Pharmacology*, 2d ed., Blackwell Scientific Publications, Oxford, 1980.

Burchall, J. J., Hitchings, G. H., *Mol. Pharmacol.* **1**:126, 1965.

Craig, J. C., Duncan, I. B., Whittaker, L., Roberts, N. A., *Antivir. Chem. Chemother.* **4**:161, 1993.

Dietrich, M., Kern, P., *Antibiot. Chemother.* **30**:224, 1981.

Ehrlich, P., Shiga, K., *Berl. Klin. Woch.* **41**:302, 1904.

Falko, E., Goodwin, L., Hitchings, G. H., et al., *Brit. J. Pharmacol.* **6**:185, 1951.

Fenner, F., *Virology* **71**:371, 1976.

Ferone, R., Burchall, J. J., Hitchings, G. H., *Mol. Pharmacol.* **5**:49, 1969.

Ferone, R., *J. Biol. Chem.* **245**:850, 1970.

Glinski, R. P., Khan, M. S., Kalamas, R. L., Sporn, M., *J. Org. Chem.* **38**:4299, 1973.

Greenlee, W. J., *Med. Res. Rev.* **10**:173, 1990.

Guttman, Ehrlich, P., *Berl. Klin. Woch.* **28**:953, 1891.

Jaffe, J., McCormack, J., *Mol. Pharmacol.* **3**:359, 1967.

Johnson, V. A., Merill, D. P., Chou, T. C., Hirsch, M. S., *J. Inf. Dis.* **166**:1143, 1992.

Kikuth, W., *Dtsch. Klin. Woch.* **79**:1401, 1954.

Kransslich, H. G., Wimmer, E., *Ann. Rev. Biochem.* **57**:701, 1988.

Lam, P. Y. S., Jodhav, P. K., Eyermann, C. N., et al., *Science* **262**:380, 1994.

Marinkelle, C. J., *Bull, WHO* **58**:807, 1980.

Muirhead, G. J., Shaw, T., Williams, P. E. O., et al., *Br. J. Pharmacol.* **34**:170P, 1992.

Pearl, L. H., Taylor, W. R., *Nature* **328**:482, 1987; *ibid* **329**:351, 1987.

Peters, W., Gillens, H. M. *A Colour Atlas of Tropical Medicine and Parasitology*, Wolfe Medical Books, London, 1982.

Ratner, L., Haseltine, R., Patarca, R., et al., *Nature* **313**:277, 1985.

Worcel, A., in *Mechanisms of Regulation of DNA Replication*, Kolberg, A.R., Kohiyama, M., Eds., Plenum Press, New York, 1974, p. 201.

World Health Organization, *Chemotherapy of Malaria*, 2d ed., WH Monograph No. 27, WHO, Geneva, 1981.

World Health Org. Stat. Quart. **37**:130, 1984.

Zubay, C. W., *Biochemistry*, Addison-Wesley Publishing Co., Reading, MA, 1983, pp. 547–54.

Suggested Readings

Archer, S., The Chemotherapy of Schistosomiasis. *Ann. Rev. Pharmacol. Toxicology* **25**:485, 1985.

Beaver, P. C., Jung, R. C., Cupp, E. W., *Clinical Parasitology*, 9th ed., Lea Febiger, Philadelphia, 1984.

James, D. M., Gilles H. M., *Human Antiparasitic Drugs—Pharmacology and Usage*, John Wiley & Sons, New York, 1985.

Mansfield, J. M., Ed., *Advances in Parasitic Diseases*, Marcel Dekker, New York, 1984.

McCalla, D. R., Nitrofurans, in *Antibiotics*, Vol. 1, Hahn, F. E., Ed., Springer Verlag, New York, 1986.

Pratt, W. B., Fekety, R., *The Antimicrobial Drugs*, Oxford University Press, New York, 1986.

Sanders, W. E., Sanders, C., Eds. *Fluoroquinolones in the Treatment of Infectious Diseases* Physicians and Scientists Publishing Co., Inc., Glenview, IL, 1990.

Wang, C. C., Parasite Enzymes as Potential Targets for Antiparasitic Chemotherapy, *J. Med. Chem.* **27**:1, 1984.

CHAPTER

8

Drugs Affecting Cholinergic Mechanisms

8.1. Introduction

This and the next chapter on adrenergic drugs involve chemical agents that derive their therapeutic use from effects on the peripheral autonomic nervous system, primarily by mimicking or blocking, directly or indirectly, the neurotransmitters of that system: acetylcholine, norepinephrine, and dopamine. Drugs affecting these transmitters centrally, rather than peripherally, will also be considered because of chemical or mechanistic similarities. It is therefore necessary to discuss some fundamental aspects of the nerve cell and to have a basic understanding of the function of the nervous systems, particularly the autonomic.

The nerve cell, or *neuron*, is an unusual looking cell (Fig. 8-1). A neuron may be only 0.1 mm in diameter but, may reach lengths of a meter or more. The threadlike extension from the cell body is the *nerve fiber*. These fibers interconnect in various special areas of the body, including the brain, the spinal cord, and the peripheral nerves. The function of the neuron is to transmit signals, or *impulses*. The *dendrites* specialize in receiving excitations, which may be from environmental stimuli (i.e., from peripheral parts of the body, or from another cell). The *axon* conducts the excitation away from the dendrite area (e.g., to the brain).[1] This excitation, or *impulse transmission*, may be on a subconscious level. Transmissions of which the subject is not consciously aware are the type that control various body functions such as temperature, respiration, blood pressure, and peristaltic movements of the gastrointestinal tract. In addition to the traditional functions of a cell membrane, the membrane of the nerve fiber (the axon) has the role of transmitting the electrochemical impulses that are received from the dendrites. The additional protective device called the *myelin sheath* basically serves as an electrical insulator.

[1] Nerve fibers, of course, also carry impulses from the brain to peripheral areas of the body (e.g., to muscles). Also, the neurons are involved in the thought processes within the brain.

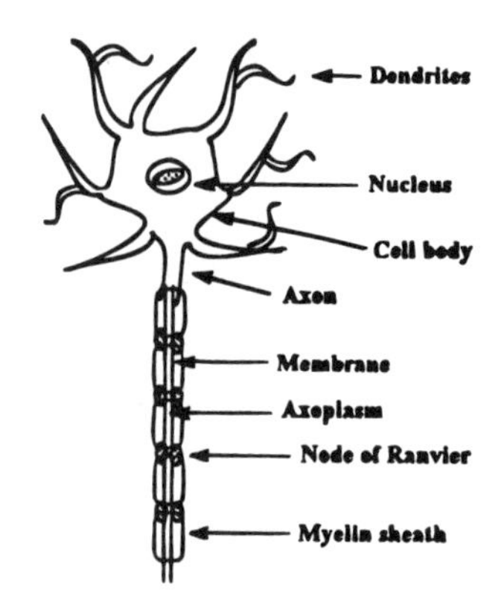

Figure 8-1. The neuron.

The nervous systems acts as if it were a single continuous transmission system, yet it consists of billions of individual units; that is, it really is a discontinuous system. Thus, a junction mechanism of some type must exist. These junctions are called *synapses* (Fig. 8-2). A *synapse* may be defined as the area where the presynaptic cell (neuron) is nearly in contact with the postsynaptic cell. The transfer of an impulse from one neuron across the synapse to the other neuron is termed *synaptic transmission*. Unlike impulse conduction along a neuron that is purely electrical in nature, synaptic transmission across the synaptic cleft is a chemically mediated process. The synaptic space is about 250 Å wide.

Transmission across the cleft is effected by the release of minute quantities of a chemical transmitter (stored in synaptic vesicles) across the presynaptic membrane into the synaptic space from which the transmitter can then interact with the postsynaptic membrane by binding to receptors on it. As impulse is then set up for electrical conduction along the fiber. Impulses generally travel in only one direction—from presynaptic to postsynaptic cells. However, a given nerve fiber *can* conduct impulses in either direction if, for example, it is electrically stimulated in the middle.

The present understanding of the way in which an electrical impulse travels along a nerve fiber is the theory of nerve conduction. The theory states that impulse conduction is governed by the difference in permeability of the nerve cell membrane to Na^+ and K^+.

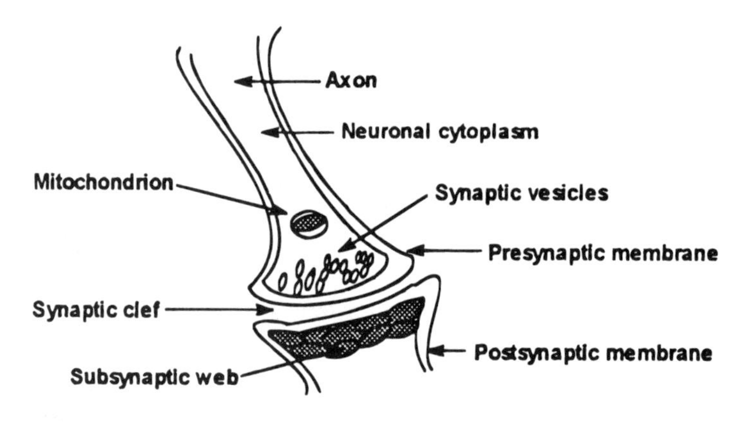

Figure 8-2. Asynapse.

Furthermore, this permeability is regulated by the electrical potential (voltage) across this membrane. The primary requirement for impulse transmission, then, is a permanent potential across the membrane of the axon.

All cellular membranes have the ability to actively transport ions from inside the cell to the outside. The reverse process also occurs. In the nerve we are primarily concerned with the active transport of Na^+ out of the fiber into the surrounding fluid. This phenomenon has been referred to as the "sodium pump." Thus, in the resting state we find a situation in which the outside of the membrane has an Na^+ concentration 10–14 times that found within the neuron. The K^+ concentration, however, is 35 times greater within the nerve cell. These concentrations of Na^+ and K^+ are maintained relatively constant by the sodium pump, which operates in a manner such that Na^+ is pumped to the outside of the membrane, leaving a deficiency of positive charge inside the membrane. This results in an electronegative interior environment with respect to the outside. The *membrane potential* so created has been measured at –85 millivolt (–0.085 V). Figure 8-3A illustrates this "resting" potential.

If we think of nerve fiber as analogous to an electrical cable, we find that, unlike a copper wire, the axon is not well suited for passive transmission along its length. A weak stimulus applied to an isolated nerve dissipates within several millimeters. Some method to amplify the signal in order to move it along itself must therefore exist. This occurs because the excitatory process from the original stimulus, even though effective only over a short distance, in turn stimulates excitation in adjacent portions of the nerve; in effect, the signal is self-boosting, ultimately propagating along the whole length of the fiber.

To understand this process we must consider what happens in the immediate area of the membrane after a stimulus is applied. A temporary increase in the permeability of the membrane to Na^+ occurs, so that Na^+ can easily diffuse into the nerve fiber, which causes a sudden localized positive charge inside and a corresponding negative charge outside (Fig. 8-3B). This situation, of course, is opposite to that prevailing during the membrane's resting state. Such a reversal of electrical charges on a localized area is called *depolarization.* As this depolarized region extends over short distances from its point of origin, it generates a similar increase in permeability to Na^+ in neighboring areas. The cycle of depolarization repeats itself (Fig. 8-3C and D), resulting in increased permeability and thus electric current along the membrane of the nerve fiber. A wave of depolarizations is thus generated. This is the *nerve impulse*, which moves down the fiber in both directions until the ends are reached. There is a threshold voltage called the *action potential* below which excitation and permeability changes will not occur, as well as the *refractory period* of a nerve fiber, which is the period during which the fiber is transmitting an impulse and therefore cannot accept a second stimulus. This period lasts approximately 0.5–2.5 msec. The time is utilized for the process of *repolarization.* When depolarization is complete (i.e., when the

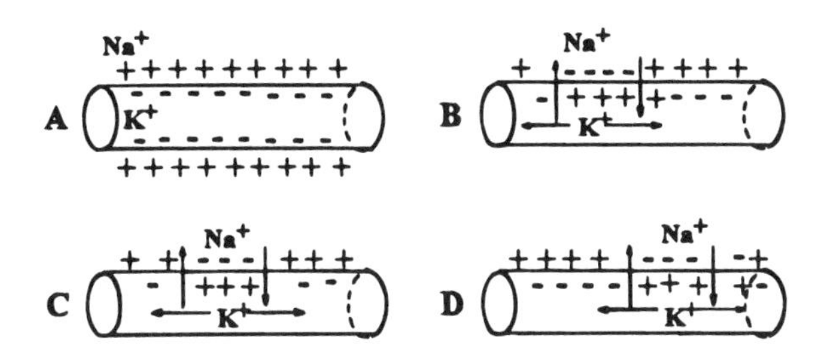

Figure 8-3. Movement of depolarization wave.

impulse has traveled down the whole fiber), the inside becomes fully positively charged. Further diffusion of Na^+ stops; the membrane once again becomes impermeable to Na^+. However, K^+ continues to diffuse easily through the pores, regenerating the normal electronegative conditions inside, and thus repolarizing the membrane.

In schematically illustrating the morphology of a neuron Figure 8–1 shows two features, the myelin sheath and nodes of Ranvier, which have functions that require explanation. Myelin surrounds all large nerve fibers.[2] It is a white lipid material, which in our cable analogy may be thought of as insulating material. It prevents leakage of current and speeds up the transmission of impulses. At frequent intervals (usually 1 mm) the continuity of the myelin is interrupted by a node of Ranvier, an exposed area of membrane where depolarization can occur, unlike underneath the myelin. Thus impulse transmission along a myelinated fiber actually jumps from node to node. This is called *saltatory conduction.* Two advantages result from this process. Impulses travel faster along the myelinated fiber than along an unmyelinated nerve, and depolarization of extensive areas of fiber is prevented, which in turn minimizes leakage of large Na^+ concentrations into the fibers.

As mentioned, the nervous system is a discontinuous system. Therefore, a junction mechanism for relaying an impulse from one nerve to a neighboring fiber exists. Before discussing the mechanism of this synaptic transmission in some detail, let us briefly review some fundamental aspects of the *autonomic nervous system.* This is the segment of our nervous system that is responsible for the control of involuntary body functions such as heart rate, blood pressure, digestive tract motility, glandular secretions, and pupillary size in the eye. It actually consists of two divisions, the *sympathetic* system and the *parasympathetic* system. Most organs are innervated by both systems and are actually often affected in an opposite manner when stimulated by one system or the other (Table 8-1). Both systems have bulblike enlargements at various intervals along the nerves. These are *ganglia* that contain the cell bodies of the neurons. Thus there are preganglionic and postganglionic neural fibers in each system.

A signal is transmitted from the spinal cord, a *central* control system, to peripheral areas through two successive neurons. The first neuron (preganglionic), which originates in the spinal cord, will synapse with the second neuron (postganglionic) in a ganglion. The postganglionic fiber then innervates the organ it controls.[3] Thus in the autonomic system signals must pass over two distinct fibers before stimulation can occur.

The significant difference between the two systems is that their postganglionic fibers secrete different neurotransmitters. Those of the parasympathetic system secrete acetylcholine (ACh) (hence the name *cholinergic*), whereas the postganglionic fibers of the sympathetic system release norepinephrine (NE) (the synonym noradrenaline accounts for the name *adrenergic*). At the ganglia, of course, both systems utilize ACh (Fig. 8-4); therefore, both preganglionic fibers are cholinergic.

The synaptic transmission mentioned previously is the mechanism by which a nerve impulse is sent across the synaptic cleft (Fig. 8-2). Unlike the electrical mechanism in the axon, it is a chemical process. The preganglionic fibers of both systems release ACh at their endings when stimulated; it is the neurotransmitter (neurohormone) at this point. The chemical transmitter replaces "cable" transmission here. ACh, previously synthesized in the nerve

---

[2] Unmyelinated fibers are numerically greater but usually much smaller in size.

[3] In the skeletal system a motor neuron innervates the muscle fiber directly with a single continuous neuron (See Fig. 8-4).

Table 8-1. Effects of Autonomic Stimulation on Body Organs

Organ	Sympathetic	Parasympathetic
Heart muscle	Increased activity	Decreased activity
Coronary vessels	Vasodilation	Vasoconstriction
Systemic blood vessels:		
Abdominal	Constriction	None
Muscle	Dilated (cholinergic)	None
Eye pupil	Dilated	Contracted
Lungs:		
Bronchi	Dilated	Constricted
Blood vessels	Constricted (mild)	None
Gut	Decreased peristalsis and tone	Increased peristalsis and tone
Kidney	Decreased output	None
Liver	Release of glucose[a]	None
Bladder body	Increased	None
Blood glucose	Increased	None
Adrenal glands	Increased cortical secretion	None
Mental activity	Increased	None[b]
Intestinal glands	Vasoconstriction	Increased secretion of digestive enzymes

[a] By glycogen breakdown.
[b] See discussion on memory.

endings (Eq. 8.1) by the catalytic action of the enzyme choline acetylase (choline acetyl transferase, ChAT) on choline and (activated) acetic acid is stored in synaptic vesicles and is then released upon stimulation of an impulse. It then diffuses into the synaptic space.

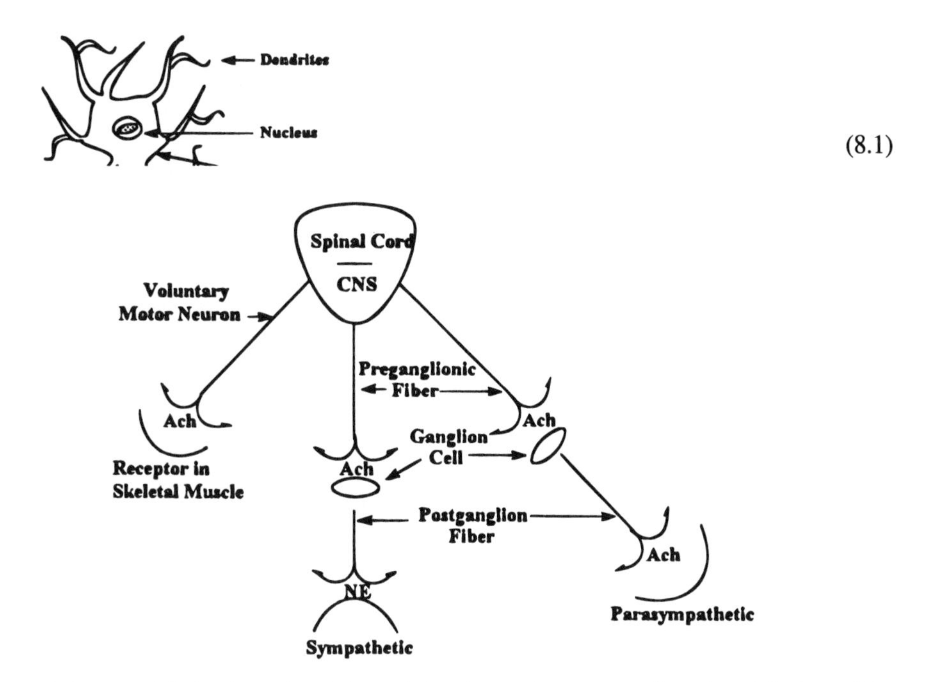

Figure 8-4. Simplified schematic of the peripheral neural system. ACh = Acetylcholine; NE = Norepinephrine.

Some of the ACh molecules successfully diffuse across the cleft and bind to a special area on the surface membrane of the dendrite of the adjacent cells that can be referred to as a *chemoreceptor*.[4] The binding of the ACh with the receptor changes the membrane's permeability, which causes depolarization and finally a new action potential. As this newly generated impulse (which is actually a continuation of the preganglionic impulse) passes down the axon and reaches the terminal, it stimulates the release of another chemical mediator (or the same). It in turn may combine with receptors in the next cell—either another nerve cell or one in the organ or tissue innervated by the fiber. Such chemical transmission at the synapse is believed to be a general phenomenon in mammals. An alternate mechanism of total electrical impulse transmission where pre- and postsynaptic membranes touch is, of course, conceivable and has been found to exist in lower life forms, such as the crayfish.

The transmissions of impulses along the interior of an axon are well protected from outside chemical interference. However, this is not true at the sites of synaptic transmissions or at the junctional sites between a nerve and effector cells. In both instances chemical intervention by drugs can bring about physiologic effects. The transmitter substances (ACh or NE) released by a nerve impulse from the presynaptic nerve endings brings about a response on interaction with postsynaptic or organ receptors. It therefore stands to reason that such *neurohumoral* transmission can be greatly affected by drugs in several ways. First, a drug may bring about changes in the formation of the neurohormone itself; that is, it can inhibit its actual synthesis. Second, the ability to store the chemical transmitters may be altered. Third, the release of mediator can be influenced by changes in their release rates or even total prevention of their release.

Alterations in the type of intensity of interaction between the transmitter and the postsynaptic or junctional receptors can be brought about by exogenously supplied chemicals. There are drugs that only mimic the action of the natural neurotransmitter. Other drugs, however, can actually enhance the receptor binding or, conversely, can prevent combination. Furthermore, there are drugs whose effectiveness is based on a third mechanism. They modify the methods by which the transmitters are removed from the regions of activity. These are compounds that frequently inhibit or enhance (induce) the various enzymes normally involved in the metabolic breakdown and therefore inactivation of the neurotransmitters. We will encounter agents that affect the following important enzymes: acetylcholinesterase (AChE), which catalyzes the hydrolysis of acetylcholine (Eq. 8.2), monoamine oxidase (MAO), which oxidizes norepinephrine and other important biogenic amines (Eq. 8.3), and catechol-O-methyl-transferase (COMT), which methylates the phenolic OH of norepinephrine and other catecholamines[5] in the meta position to the side chain (Eq. 8.3).

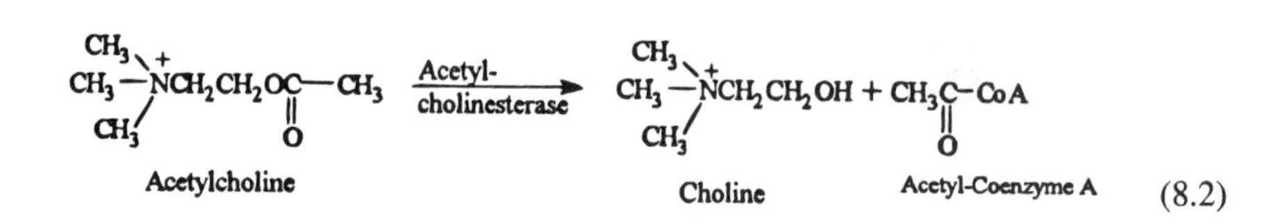

(8.2)

[4] At this point the receptor and drug–receptor interaction discussions in Chapter 1 might be reviewed.

[5] Catechol is 1,2- or *ortho*-dihydroxybenzene.

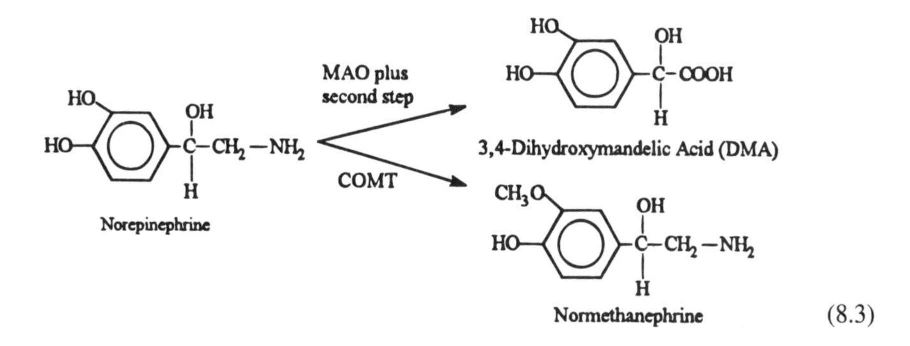

(8.3)

A simplified chart illustrates the peripheral neural systems. It indicates the voluntary motor neurons to the skeletal muscles of the body and the autonomic system that utilizes chemical transmission at five synapses: the two ganglia and the three sites where innervation to muscle or organ occur (Fig. 8-4). It will be noted that ACh is the transmitter at four of these sites, while only one utilizes NE. It must also be appreciated that while ACh is the neurotransmitter at most of the synapses, its effect at each synapse is different. A significant difference is that the response to ACh at the neuromuscular junction (NMJ) to the skeletal muscles, as well as at both the autonomic ganglia is instantaneous, very short, and of a localized nature. The effects obtained by ACh following release at the postganglionic nerve ending of the parasympathetic system tend to have a slower onset of action. The action is also more prolonged and over a more extensive area. All three types of fibers transmit impulses to cells that presumably have cholinergic receptors. However, when ACh is applied to each of the three receptor sites in question, the pharmacologic response is different. Not unexpectedly, therefore, drugs that mimic the effects of ACh (cholinomimetics) will show qualitatively different responses at different cholinergic junctions. A similar situation is encountered with cholinergic blocking agents when experiments are carried out. Effects ranging from total blockade at certain cholinergic receptors to almost total inability to inhibit impulse transmission may be encountered. It is therefore apparent that more than one type of cholinergic receptor exists, and that these differ in topographical and electron distribution characteristics that are not yet fully understood. What is understood is that these differences enable us to develop, initially by serendipitous discovery, and now hopefully by modification and design, drugs to affect these various functions for therapeutic advantages with a high degree of selectivity. This chapter will deal with drugs that affect the action of ACh at each of these junctions, both those having direct or indirect agonist action, and those used in therapy because of their selective antagonist pharmacology.

8.2. Aspects of the Cholinergic System

Acetylcholine is synthesized by the transfer of the acetyl group from its activated acetyl coenzyme A form to the aminoalcohol choline (Eq. 8.1). The enzyme, ChAT, is not actually highly substrate-specific and can acetylate other basic alcohols or transfer acyl groups other than acetyl. The rate of synthesis can be as high as 4,000 μg/g of tissue per hour.

Following synthesis, the ACh is taken up and stored in synaptic vesicles (Fig. 8-5). The ACh is protected from metabolic degradation in storage until it is released into the synaptic cleft after stimulation of the preganglionic axon. The synaptic vesicles ordinar-

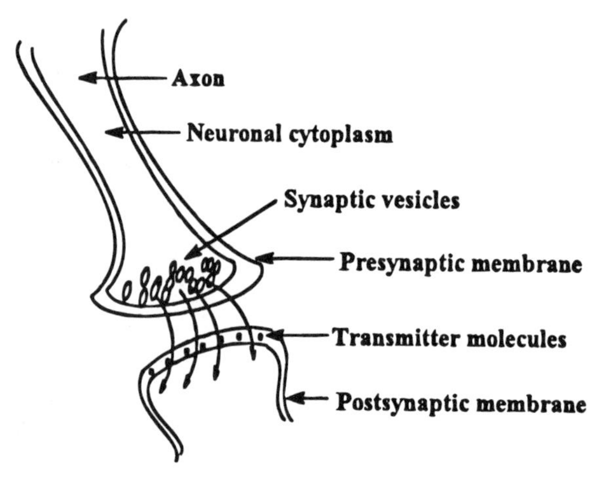

Figure 8-5. Cholinergic neuron system.

ily cannot approach the presynaptic membranes because of the electrostatic repulsion between the positively charged trimethylammonium "head" of ACh and the similarly charged nature of the interior of the membrane. It is also believed that a hydration barrier may be operative. Depolarization of the presynaptic membrane following an impulse removes this electrostatic impediment temporarily, thus allowing the vesicles to approach the membrane's interior and perhaps even to fuse with it. ACh molecules are released into the synaptic cleft and travel across to the postsynaptic membrane where they interact with the cholinergic receptors.

There are few agents that have the ability to interfere with the presynaptic release of ACh. The botulinum toxin is one, probably by binding to a presynaptic site where it blocks Ca^{2+}-stimulated ACh release. Other toxins such as α-bungarotoxin also have this ability. A partial inhibition may exist with morphine and d-tubocurare. The release of ACh from presynaptic neurons can be affected by some drugs including neostigmine, certain Veratrum alkaloids, and scorpion venom.

The advent of newer and more sensitive analytical procedures for determining subnanogram (10^{-9}), even picogram (10^{-12}), levels of substances in physiological fluids and tissues over the past two decades, has revolutionized neurochemical studies. Improved fluorometry, mass spectrometry, radioimmunoassay, enzymatic, and various chromatographic methods are at the forefront. As a result of this more recent research, the highly complex neurochemical web has become clearer. However, these developments also indicate that our chemical–neurological–receptor system is even more baffling than was suspected. Some previously held concepts have been found to be too simplistic. Unravelling all the secrets of the mammalian nerve–brain complex may be a truly Herculean and unending process.[6] A brief overview follows.

Depending on conditions, the turnover rate for ACh in a single cervical ganglion (in the cat) during stimulation can reach 25 ng/min, which represent 8% of the total available at

[6] Greek mythology tells of Hercules fighting Hydra whose heads grew back faster than the sword could lop them off.

rest. There is also evidence that ACh synthesis is simultaneously stimulated. Synthesis and storage require Na^+; release involves Ca^{2+}.

There appear to be three distinct ACh stores: an intracellular type of surplus ACh that is *not* released following stimulation; an ACh depot that is released on stimulation and accounts for about 85% of the neurotransmitter, and a stationary remainder that is also not released.

The rate of ACh synthesis is limited by the availability of choline, which cannot be synthesized in neurons. It ultimately comes from plasma sources, even though 60% may be recaptured from ACh hydrolysis by nerve ending through a reuptake. It is also known that freshly synthesized ACh is released in preference to the stored neurohormone.

ACh is synthesized in the cerebral cortex as well as in all parts of the autonomic system. Intracellular ACh levels regulate the rate of synthesis, thus illustrating a negative feedback mechanism. One concept that may explain the manner of ACh release from storage vesicles is *exocytosis*, an opening in the vesicle membrane to the surrounding medium and subsequent resealing.

ChAT is known to be in the cytoplasm of the nerve endings and not in the vesicles as previously thought (Fig. 8-5). This, of course, necessitates transporting synthesized ACh into the vesicles for storage. Unequivocal proof of such transport is still not at hand; neither is a mechanism for such uptake.

The role of ACh as a neurotransmitter at ganglia, NMJ, and postganglionic parasympathetic nerve endings is well established. Motor neurons of the central nervous system (CNS) are also known to be cholinergic. It is likely that other CNS synapses are too.

It is interesting that peripheral cholinergic nerve endings cannot take up ACh (only choline); however, they can in brain tissue.

Even though some functions of ACh are now well understood, others are not definite. For example, is ACh involved in pain perception or as a sensory transmitter of heat (thermal receptor)? It has been suggested that taste fiber endings are cholinergic. Certain actions of ACh may affect cellular membranes by stimulating the incorporation of phosphate into phospholipids, which are crucial membrane components. Other not fully understood but interesting effects of ACh includes: release of catecholamines from certain cells, a depolarizing effect on smooth muscle (e.g., intestines) that appears not related to nerve fibers, a direct hyperpolarization of certain cardiac muscles, and finally the functions, if any, of ACh presence in such diverse places as the lens of the human eye, the placenta, several bacteria, protozoa, and fungi. These are a complete mystery.

8.2.1. Cholinergic Receptors

The nature of the cholinergic receptor has been the subject of considerable study for over 75 years. Aspects of its shape, conformations, and active sites have, over the years, slowly and enticingly revealed themselves to our prying. The efforts were initially by empirical–intuitive methods, then by SAR analogies, receptor mapping, and other indirect methods. Generous applications of X-ray crystallography and molecular receptor probes followed, and most recently, the actual isolation of the active receptor preparations were achieved.

The puzzling results obtained by early workers who elicited different pharmacological responses by applying external ACh or cholinomimetic drugs to various cholinergic nerves, finally led Dale (1914) to separate these differing effects into two categories: *muscarinic* and *nicotinic*. Muscarine, an alkaloid obtained from the poisonous mushroom

Amanita muscaria, produces the effects predictable from stimulation of postganglionic parasympathetic fibers.[7] Receptors thus affected are therefore referred to as muscarinic. Drugs mimicking the effects of muscarine are cholinergic–muscarinic agents. All effects of muscarinelike drugs are prevented by the alkaloid atropine, an *anticholinergic* drug (cholinergic blocker). Furthermore, neither atropinelike nor muscarinelike drugs show effects at the NMJ and, at best, only slight responses at autonomic ganglia.

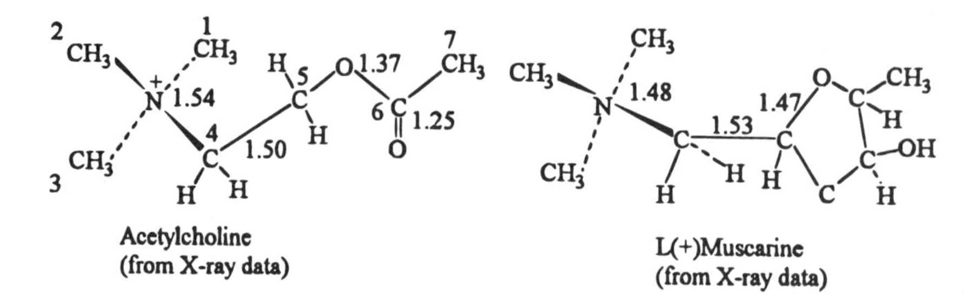

A chemical view of ACh and muscarine may be appropriate at this point. Its structural features were determined by X-ray crystallography of the bromide salt. Among other things the analysis showed that the two oxygens in the ACh molecule were coplanar with carbons 7, 6, and 5 (in the preceding representation), which corresponds with the coplanarity of muscarine's comparably positioned oxygen and carbon atoms to its right and left. It is understood that solid-state configurational characteristics need not remain in similar steric relationships in solution, unless they are restrained by cyclic structures such as those in muscarine, yet it was found that NMR studies of ACh (in D_2O solutions) partially supported the X-ray conclusions. Considering the ACh cation from basic principles of molecular structure, it is axiomatic that in spite of the coplanarity of the atomic groupings on either side of the molecule (i.e., C2-N-C4-C5 or C5-O1-C6-C7, as pointed out earlier), rotation of atoms around the single bonds C4-C5 and C5-O1 can occur, giving the ACh molecule considerable conformational freedom. Since the ester group atoms must be coplanar, a certain restraint on torsional deformation exists. Some restraint to total "freedom" is also undoubtedly induced by the CH_3 groups in the cationic "head" [$(CH_3)_3N^+$]. Combining these partial restraints with the inducement stability that results from ionic and hydrophobic interactions with corresponding binding sites on the receptor surface begins to offer us an understanding for the ACh receptor interactions that occur.

Because of the known molecular dimensions of the rigid, cyclic muscarine [only the L(+) naturally occurring isomer of the eight possible stereoisomers is cholinergically active], it has been a useful tool in studying the cholinergic receptors' dimensional requirements for a substrate. Structural variations established the absolute requirements for the quaternary nitrogen and ring oxygen atom (the sulfur analog is inactive presumably because of decreased hydrogen bonding ability to the receptor).

Table 8-2 illustrates data from early work attempting to evaluate the structural requirements of cholinergic agonist activity directly and thereby indirectly glean some insight into the receptors' geometrical parameters for active compounds. These types of data, when

[7] Postganglionic sympathetic fibers innervating sweat glands are cholinergic; hence, muscarine also causes sweating.

Table 8-2. Relative Potencies of ACh Congeners[a]

	Compound		Relative potency
1.	$(CH_3)_3\overset{+}{N}-CH_2-CH_2-O-C(=O)-CH_3$	(ACh)	1,000
2.	$(CH_3)_3\overset{+}{N}-CH_2-CH_2-CH_2-O-C(=O)-CH_3$		83
3.	$(CH_3)_3\overset{+}{N}-CH_2-CH_2-O-CH_2-CH_3$		15
4.	$(CH_3)_3\overset{+}{N}-CH_2-CH_2-CH_2-CH_2-CH_3$		14(21)[b]
5.	$(CH_3)_3\overset{+}{N}-CH_2-CH_2-C(=O)-CH_2-CH_3$		6.2
6.	$(CH_3)_3\overset{+}{N}-CH_2-C(=O)-CH_2-CH_2-CH_3$		1.6
7.	$(CH_3)_3\overset{+}{N}-CH_3$		0.05
8.	$(CH_3)_3\overset{+}{N}-CH_2-CH_3$		0.07
9.	$(CH_3)_3\overset{+}{N}-CH_2-CH_2-CH_3$		3.0
10.	$(CH_3)_3\overset{+}{N}-CH_2-CH_2-CH_2-CH_3$		21(14)[b]
11.	$(CH_3)_3\overset{+}{N}-CH_2-CH_2-CH_2-CH_2-CH_2-CH_3$		0.05

[a] Modified from Welsh and Taub, 1950, 1951.
[b] The same compound gave differing results at different times.

combined with known features of proteins, basic chemical principles, and intuition q.s.,[8] led to some proposals of hypothetical models of cholinergic receptor—ACh interactions. Even though more sophisticated models appeared in the literature, the more simplistic one represented in Fig. 8-6 is appealing for the present discussion since it conveniently accounts for many of the structural features of ACh.

Table 8-2 clearly shows that ACh molecule with the cationic nitrogen has optimum activity (except for muscarine). Other necessary features for cholinergic activity are the polarized carbonyl oxygen, which is approximately 5.9 Å (see Table 8-3) from the nitrogen, and probably the "ether" oxygen as part of the chain separating the two polar segments

[8] Quantum sufficiatum.

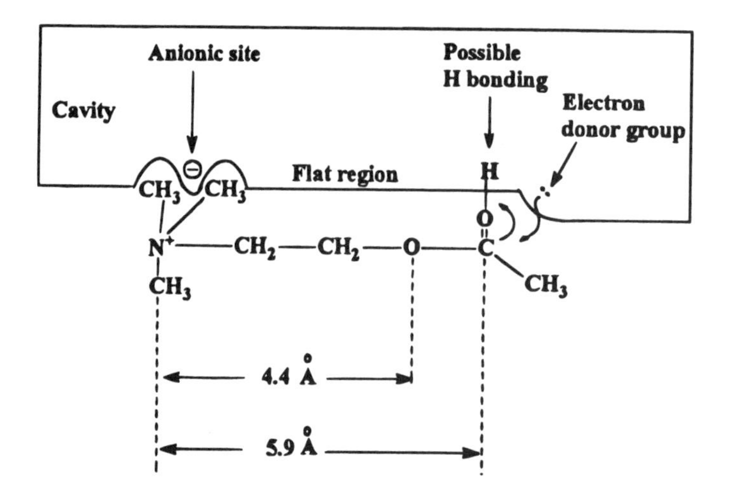

Figure 8-6. Hypothetical acetylcholine receptor complex. (Modified from Goldstein et al., 1974.)

(onium ion and ester carbonyl) of the molecule. Replacement of this oxygen with a CH_2 (compound 2, Table 8-2) decreases activity to 8% that of ACh. Thus maximal bioactivity requires the ester moiety. The carbonyl functions and their position in relation to the charged nitrogen atom is also critical. Deleting it, but retaining the ether oxygen, which affords ethoxycholine (Compound 3), drops the activity to between 1.5% and 10%. It is interesting that replacing the ether oxygen with a CH_2 (Compound 4) elicits no additional change. Compounds 5 and 6, where the carbonyl is displaced closer to the nitrogen, causes some additional diminution of intensity, thus showing that its position in ACh is optimal. It is noteworthy that compound 4 devoid of any oxygen function retains about 2% of the activity of ACh (surprisingly more than compound 5), whereas compounds 7, 8, 9, and 11 are essentially inactive. It will be noted that those compounds, which show respectable activity (remembering that ACh is extremely potent), have a 5-atom chain in common,[9] where carbons, oxygens, or nitrogens are valid. Early workers noted that acyl groups larger than acetyl (e.g., butyrylcholine) were essentially inactive, yet nitrate esters exhibited significant muscarinic activity (Dale, 1914). The British pharmacologist Ing (1949) proposed a so-called rule of five, which held that for significant direct cholinomimetic (and antagonist?) action a five-atom grouping must emanate from the quaternary nitrogen.

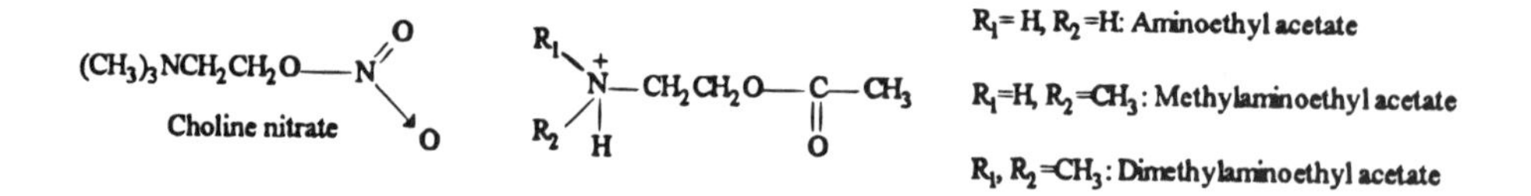

The cationic head characteristics were also explored. Ing (1949) demonstrated that exchanging the methyl groups with either smaller groups (H) or larger groups (ethyl, propyl, etc.) lowers both muscarinic and nicotinic activity considerably, but not nearly as

[9] Not considering H atoms.

Table 8-3. The Atomic Distances of Acetylcholine

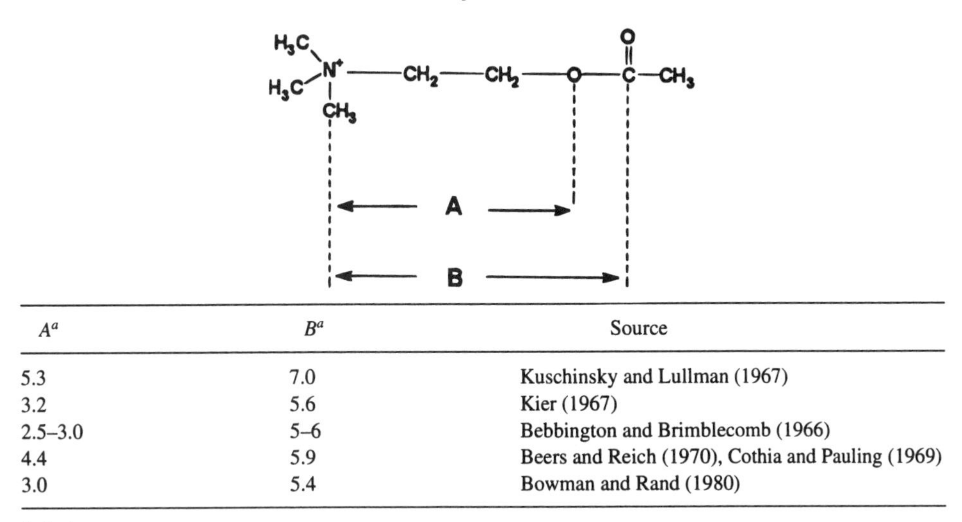

A^a	B^a	Source
5.3	7.0	Kuschinsky and Lullman (1967)
3.2	5.6	Kier (1967)
2.5–3.0	5–6	Bebbington and Brimblecomb (1966)
4.4	5.9	Beers and Reich (1970), Cothia and Pauling (1969)
3.0	5.4	Bowman and Rand (1980)

[a] In Angstroms.

[b] The values were obtained at different times by different methods. The partial double-bond character of the ester may not always have been taken into account. Some were theoretically calculated (Kier).

much as the removal of two methyl groups. The primary amine—aminoethyl acetate—is inactive. Tertiary amines such as dimethylaminoethyl acetate are strong enough bases to be almost totally ionized at physiological pH (7.4), thus retaining the cationic character. The putative cavity into which the trimethylammonium head of ACh "fits" interacts with that site on the receptor by two modalities. An ionic interaction of the cation with an anionic site that is likely to be a carboxylate ion from one of the dicarboxylic amino acids, aspartate or glutamate. The methyl groups are also believed to stabilize this complex additionally by van der Waals–type hydrophobic bonding. Even in ACh only two of the three CH_3 groups would be accommodated in the concave site since the third methyl would necessarily have to point in a direction opposite to the receptor surface. In fact, it is actually somewhat surprising that the dimethyl analog of ACh is so much weaker an agonist. The precipitous drop in activity with larger alkyl groups may be explicable in terms of "poorness of fit" into the cavity (Fig. 8-6).

$$CH_3-C(CH_3)_2-CH_2-CH_2-O-C(=O)-CH_3$$

3,3-Dimethylbutyl acetate

3,3-Dimetylbutyl acetate differs from ACh in only one respect: the cationic nitrogen has been replaced by a quaternary carbon. The dimensions of the "quaternary" head would essentially be as in the quaternary amine since C and N have similar volume and tetrahedral bonding. The difference remaining, then, is the absence of the charge. The fact that cholinergic activity of this compound on guinea pig ileum is only 0.003% that of ACh attests to the importance of the charge to the molecule's interaction with the putative anionic site. However, the fact that the compound exhibits any measurable ACh-like activity indicates that at least an ancillary role is also assigned to hydrophobic interactions. The preceding discussion may now allow us to arrive at some simplistic, but very likely only, tentative assumptions about the ACh–muscarinic interactions. Maximal bioactivity requires an ester linkage (choline itself shows weak agonism). Its separation from the charged nitrogen by an optimum distance is important. The positively charged nitrogen is held in favorable position by electrostatic forces to a presumed anionic function or region. Two of the three methyl groups contribute stability to the substrate–receptor complex via hydrophobic bonding. The so-called flat region probably does not contribute much unless the ether oxygen is present (Fig. 8-6).

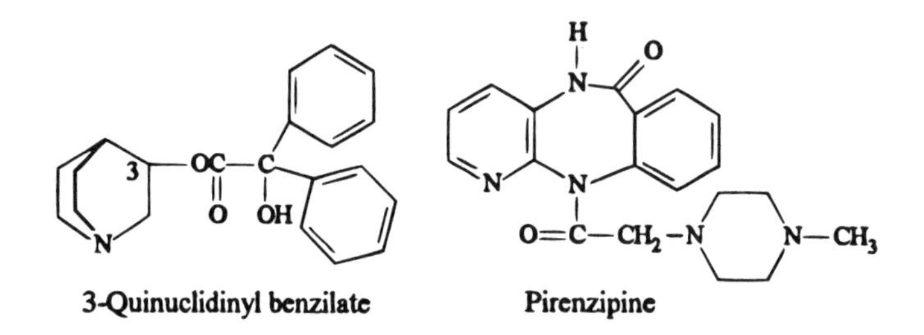

Some newer developments may either support these conclusions, or they may negate some of them. The existence of cholinergic receptors (or any receptors for that matter) that were thought by some to be convenient mental crutches to comprehension not too long ago are now being chemically characterized in greater and greater detail. It is no longer in doubt that they are integral membrane proteins associated with membrane lipids. Solubilized muscarinic (and nicotinic) receptor preparations retaining much of their agonist and/or antagonist specificity are being used to investigate receptor properties using reversibly and irreversibly binding probes labeled with fluorescent or radioactive tags. This enables visualization of location, distribution, and identification of receptor populations. Most useful have been ACh analogs and certain neurotoxins derived from snakes. Much of the work has been with nicotinic receptors, the richest source for which have been the electric fish (see Chapter 1). Muscarinic receptor preparations have now been obtained from the caudate nucleus of bovine brains using the bile acid salt sodium cholate as the extractant and the specific muscarinic receptor binding tritium-labeled quinuclidinyl benzilate (a hallucinogenic anticholinergic) as the isolating tool. A more recent success has been the purification (to homogeneity) of the muscarinic receptor from porcine heart tissue. Its molecular weight is 78,000. Both a high- and low-affinity site have been identified by employing carbamoyl-choline as the agonist probe. Using radioactive muscarinic agonists, the locations of brain muscarinic receptors have been mapped by autoradiography. An even better radiolabeled specific probe for studying muscarinic receptors is ^{3}H-propylbenzylcholine mustard, which is an irreversibly binding antagonist, unlike quinuclidine benzilate. With such affinity labels recognition sites of peripheral muscarinic receptors were identified and quantitatively characterized. Dissociation constants, K_D, with these probes indicated that these receptors were homogeneous. Heterogeneity to such antagonist binding in mouse brain was demonstrated by using a reversible benzilate antagonist as a probe ([^{3}H]4-N-methylpiperidinyl benzilate). Even more distinctions among muscarinic receptors in the brain was established with the 1,5-benzodiazepinone pirenzepine. A definite subdivision of cholinergic receptors into subtypes now referred to as M_1 and M_2 was established.

Pirenzepine appears selective to M_1, which has been demonstrated in the hippocampal area of the brain and sympathetic ganglia. It has not been established whether M_1 and M_2 receptors are interconvertible by ionic charges[10] or represent separate molecular forms.

The ^3H-labeled muscarinic oxotremorine, which is an experimental research compound, has also been shown to bind differentially to what may be subclasses of muscarinic receptor populations.

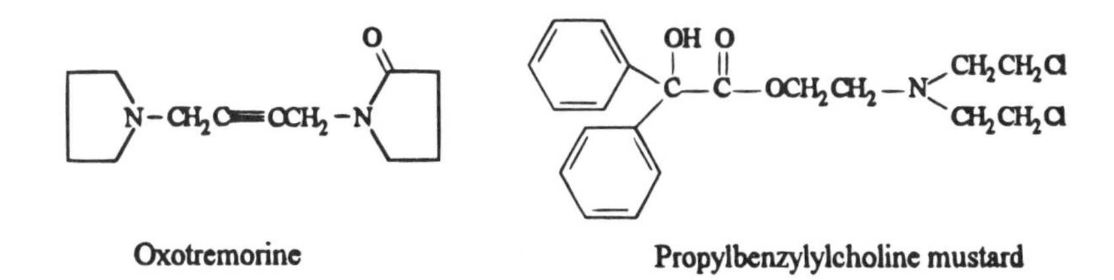

Oxotremorine **Propylbenzylylcholine mustard**

Muscarinic receptors appear to be coupled to Ca^{2+} channels at the interface with smooth muscle and probably with K^+ as well. ACh application to the myocardium slows the heartbeat and reduces Ca^{2+} influx that normally occurs during the action potential. The result is contraction. The process can be inhibited by calcium-channel blocking drugs such as verapamil (Chapter 10). Renewed research interest in the area of muscarinic receptors has expanded over the past decade. New agonist and antagonist probes continue to unravel additional complexities of this system. In addition to the M_1 and M_2 division discussed earlier, new subtypes have recently been proposed. These include M_3 (cardiac) and further subdivision of M_2 into $M_2\alpha$ and $M_2\beta$ with respect to new ligand probes. More discriminatory ligands are likely to emerge that will undoubtedly lead to further modification of the present system of classification. There are currently (1995) human recombinant muscarinic receptor prepararations available from M_1 to M_5.

Nicotinic cholinergic receptors have been intensely studied over the years. Their isolation predates that of their muscarinic counterparts by about a decade. The finding of high concentrations of such receptors in the *electroplax* of the electric eel and ray, with the aid of specific-bonding neurotoxins (α-bungarotoxin, BTX) and solubilizing detergent techniques, led to their isolation (Chapter 2). More recently the purification of the nicotinic ACh receptor from fetal calf muscle was accomplished. Receptor purification, carried out by sodium dodecylsulfate gel electrophoresis, showed a composition of five peptides that appears to be a complex of four homologous peptides. The fifth one has been identified as actin, copurified with the receptor.

The receptor complex appears to be formed of these polypeptide strands into a rosette-shaped aggregate surrounding an ion pore visible at 250,000 magnification. The structure is embedded in the postsynaptic membrane and has been shown to have permease functions by demonstrating that the passage of cations was facilitated by ACh and blocked by the neuromuscular blocking agent, tubocurarine.

The amino acid sequence of the α-subunit of the receptor (of which there are two functioning as the receptor) has been determined. The ACh-binding site appears to involve aspartic acid and histidine at positions 138 and 134, respectively, in this protein containing a total of 437 amino acid residues.

---

[10] See opiate-receptor discussions in Chapter 5.

It requires two (or more) ACh molecules interacting with the α-subunits to elicit an instant, short response by direct effects on the micropore (more later).

This research is particularly significant to myasthenia gravis (MG, human paralytic syndrome), which is now understood to be a disease in which an autoimmune reaction against nicotinic ACh receptor occurs. In MG these receptors at the neuromuscular junction are being destroyed by specific autoantibodies that cross-link them. More detailed knowledge of the construction of these mammalian receptors is vital to understanding this dread disease.

Snake venom extract BTX used to be the primary probe for labeling mammalian nicotinic receptors in the brain. It is a potent and almost irreversible binding agent to nicotinic receptors in the muscle endplate, autonomic ganglia, spinal cord, and certain sites in the brain. Carrying the radioactive ^{125}I label, it has been a useful, but not perfect, tool. Some newer and more selective probes have now also been developed. For example, a cobra α-toxin (*naja taxin*) from the Siamese cobra (*Naja naja Siamensis*), like BTX, also selectively blocks NMJ receptors peripherally as well as the same central sites in the brain. A peripheral antagonist dihydro-erythroidine (DHE) has now also been shown to bind to nicotinic sites in rat brain that seem to be different from BTX sites. The question then arises whether there is a subdivision of nicotinic receptors centrally, analogous to the several types seen peripherally. Peripherally, hexamethonium (HXM) and decamethonium (DCM) (see later) will block ganglionic and neuromuscular nicotinic sites, respectively.

It is possible that use of agonists rather than antagonists as probes will be useful. Racemic, tritium-labeled nicotine itself was shown to bind stereospecifically to rat brain tissue with high affinity, which binding could be inhibited by other nicotinic agonists, but not antagonists.

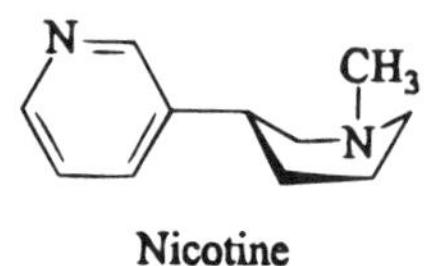

Nicotine

[^{3}H] ACh was used to demonstrate similar binding in the presence of esterase inhibitors and muscarinic blockers to "cover" muscarinic receptors. These studies, utilizing both tritiated ACh and nicotine, further showed that these probes bound the same high-affinity sites in the rat brain, while BTX affected an apparently different subpopulation of central nicotinic receptors. More striking evidence of this dissimilarity of nicotinic receptor distribution in the brain was demonstrated by comparing the autoradiographs of the three radiolabeled ligands on various rat brain sections. It is becoming evident now that the ligands ACh and nicotine have a similar anatomical binding pattern on the same molecular sites of ganglionic-type receptors (i.e., they are analogous to the peripheral receptors blocked by HXM). In addition, it has been shown that in the brain areas that bind densely with the tritium radioligands, yet do not show [^{125}I] BTX binding, the actions of nicotine can be blocked by HXM-type antagonists, but not by BTX. It is equally interesting, but not surprising, that mecamylamine, a ganglion-selective antagonist, has no peripheral effect on nicotinic receptors known to bind readily to BTX on the medullary and hippocampal areas of the brain.

Some facts of potential future clinical significance regarding Alzheimer's disease have begun to emerge. The degeneration of cholinergic afferent neurons to the cortex has been established as a factor in the disease based on autopsies. Recently, it was demonstrated that the cerebral cortex in these patients has a significant decrease in tritiated agonist binding.

The lost sites are most likely presynaptic nicotinic sites. These facts may lead to a more rational therapeutic approach.

8.2.2. Acetylcholinesterase

Before considering drug categories affecting cholinergic neurons and receptors, it is important to speculate on the nature of the enzyme AChE and its interactions with ACh and various drugs.

The function of the enzyme is rapidly and efficiently to catalyze the hydrolysis of ACh (Eq. 8.2). The significance of rapid destruction of ACh is to deactivate the neurotransmitter after it binds to the receptors so as not to accumulate to levels that produce a continuous barrage of impulses by repetitive interactions. After all, the depolarization produced by the ACh–receptor binding must be terminated so that the excitability of the postsynaptic membrane and its permeability can be restored by repolarization. Inhibition of AChE levels would then exist in the vicinity of effector cells. Extreme inhibition, such as may occur following irreversible blockade of the enzyme, would lead to cholinergic intoxication with fatal results.

The enzyme, although available in a highly pure state, has not been crystallized. It is a large molecule (320,000 Daltons) consisting of two types of polypeptide chains, which are each present twice ($\alpha_2\beta_2$). AChE, or *true* acetylcholinesterase, is actually a general ester-hydrolyzing enzyme that is highly effective against ACh. In that sense it is more specific than other esterases such as *pseudo* or *butyryl* cholinesterase (also known as serum esterase). Of course, AChE is present in neurons, the NMJ, erythrocytes, and certain other tissues. The *pseudo*-ChE can be found in the plasma, liver, and other nonneuronal tissue. It, and various other nonspecific cholinesterases, all of whose functions are not known, at least participate in the metabolism of various ester-containing drugs (meperidine, procaine, etc.).

AChE was first described in the late 1930s. Subsequent work with different substrates and inhibitors led to a conceptualization of the enzyme's active center involving two binding sites for ACh: an *ionic site* (most likely the *gamma*-carboxylate anion of a glutamate residue) to which the cationic ACh "head" presumably is attracted and a serine residue, which, in cooperation with the imidazole ring of a histidine, constitutes an *esteratic* site. Here, the serine's hydroxyl group, as the primary nucleophile, initiates hydrolysis (Fig. 8-7). The distance between the two sites has been variously reported to be as little as 2.5–7Å. (The superficial similarity with the cholinergic receptor ends with the realization

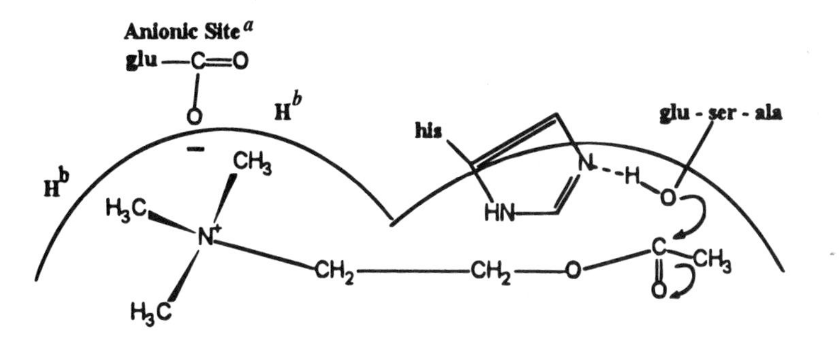

Figure 8-7. Initial acetylcholinesterase–acetylcholine interaction. [a] May be more significant as a lipophilic area than as an ionic site (see text). [b]Hydrophobic van der Waab interactions.

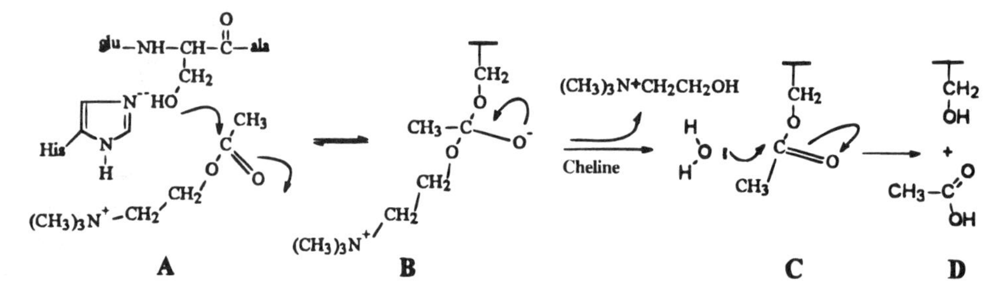

Figure 8-8. Mechanism of acetylcholine hydrolysis (see text for explanation).

that the receptor does not hydrolyze ACh.) The covalent phosphorylation of the serine hydroxyl by organophosphates (see later), which totally inactivates the enzyme, proved the absolute importance of the serine residue to catalytic activity. In fact, careful degradation of the "poisoned" enzyme led to the isolation of the phosphorylated serine residue intact. This also established the identity of the neighboring residues on each side as glutamic acid and alanine. As in other "serine" enzymes such as the proteolytic chymotrypsin, trypsin, thrombin, and others, a histidine residue in a proximal position participates in the catalysis via its imidazole ring (see the following). The fact that 3,3-dimethylbutyl acetate is as readily hydrolyzed by AChE as ACh indicates that the term *ionic site* may be a misnomer. As in the preceding discussion regarding ACh–receptor binding, here, too, the van der Waals forces of the methyl groups around the cationic nitrogen may be more significant than the presumed ionic bonding. It should possibly be viewed primarily as a lipophilic interaction. It is interesting that as was seen in the case of cholinergic receptors, the very weak dimethylbutyl acetate interaction did not lead to significant stimulation.

Figure 8-8 illustrates a simplified mechanism of ACh hydrolysis catalyzed by AChE. The initial reversible enzyme–substrate complexation enables the histidine ring, by acting as a proton acceptor for the hydroxyl hydrogen of serine, to convert it to a very potent nucleophile (as an alkoxide ion) that simultaneously attacks the polarized carbonyl carbon of ACh (Fig. 8-8A). The result is a tetrahedral transitional intermediate (Fig. 8-8B) whose oxyanion's unshared electron pair then re-establishes the carbonyl function. It can do so with facility by either breaking the C-O bond to the serine residue, thus reversing the initial reaction, or by leaving behind an acetylated serine (Fig. 8-8C).[11] Thus in essence the first step is an acetyl transfer reaction. The acetylated serine can now be readily attacked by the weak nucleophile, H_2O, probably activated by participation of another histidine imidazole ring (not shown). Alternatively, the assistance of imidazole may be viewed as another transacetylation from the acetylated serine to the imidazole, which is then easily hydrolyzed to acetate regenerating the imidazole to be used again in classic catalytic fashion. In fact, the latter step is chemically more appealing since it has been shown that imidazole itself can act as a general base catalyst for hydrolysis of labile (activated?) acetates (Fig. 8-9A) or via an N-acetylimidazole intermediate (Fig. 8-9B).

AChE is one of the "fastest" enzymes known. The enzyme is located in the synaptic cleft bound via collagen to the postsynaptic cell. It has been estimated that there is more than enough AChE at the NMJ than is needed to hydrolyze all the ACh molecules used to

---

[11] By considering the events as concerted steps a formal oxyanion need not be invoked.

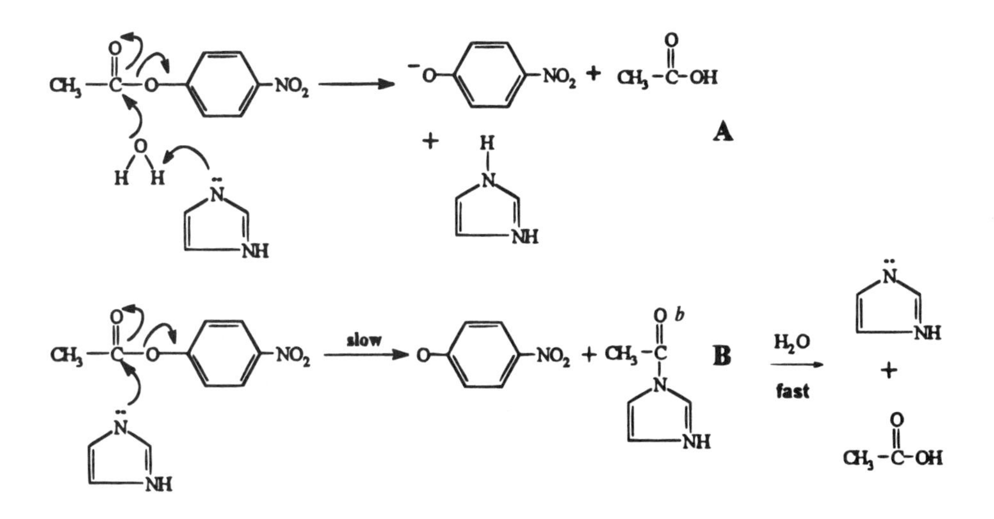

Figure 8-9. The catalytic properties of imidazole. [a] See text and Bruice and Schmir, 1957. [b] N-acetylimidazole has been isolated in the B reaction sequence.

depolarize the end plate. The enzyme's specific activity is striking: the turnover number has been determined at 25,000 sec^{-1}; its turnover time, variously calculated at 30–60 milliseconds, meaning, on average, an ACh molecule is cleaved in that time period. Another way of viewing the activity of the enzyme is to state that it is capable of hydrolyzing 960 millimoles of ACh per milligram of protein per hour (1 millimole = 146 mg).

8.3. Cholinergic Drugs

Cholinergic drugs can be divided into two main categories: (1) the true cholinergics that mimic the action of ACh at the receptor site and presumably interact with it to form a drug–receptor complex, and (2) the indirectly acting anticholinesterase drugs that inhibit the ability of AChE to hydrolyze ACh, thus preserving the endogenous supply and resulting in stimulation of cholinergic receptors. These AChE inhibitors can be further subclassified into reversible and irreversible types.

It can be speculated that the true cholinergic agents exert an action analogous to that of ACh because of considerable structural similarities. A cationic head a proper distance from a degradable polar moiety is the primary structural requirement for ACh-like activity. The remaining parts of the molecule are modifiers and thus may be viewed as of secondary importance. They have pronounced effects, however, on such parameters as degree of activity, duration of action, tissue distribution patterns, and metabolism.

The direct-acting drugs, also termed *parasympathomimetic*, are generally considered as stimulating *muscarinic* peripheral sites. Under clinical conditions these effects are never "pure." It should be remembered that such peripheral sites may also exist in effector cells that were not innervated by parasympathetic neurons but rather in some instances by post ganglionic sympathetic neurons (e.g., sweat glands, certain blood vessels in the skin of the head and neck). In addition ACh itself has *nicotinic* action; to varying degrees, this may carry over to some of the synthetic congeners.

ACh, even though it is a potent cholinergic agonist, is clinically an inferior therapeutic agent. Its shortcomings include poor chemical stability as a result of rapid hydrolysis both in vitro and in vivo. Its actions are brief when it is applied topically. At one time IV doses up to 100 mg were used to abort atrial paroxysmal tachycardia. Overall, its actions are too broad and fleeting. An ophthalmologic use for the drug remains. A few drops of a freshly prepared 1% solution is used to obtain instant and complete miosis for cataract surgery and iridectomy in the eye.

The designing of direct-acting parasympathetic drugs entails circumventing several of these obstacles. Approaches to the solution have included the synthesis of compounds with labile acetyl groups that might be somewhat protected from the catalytic activity of AChE, possibly by not fitting the esteratic site of the enzyme quite as well, or exhibiting steric hindrance to it. The drug methacholine illustrates an early attempt. Methacholine's rate of hydrolysis is slower than that of ACh. Furthermore, it has been shown that this compound, unlike ACh, is only hydrolyzed by AChE, and not by the pseudoesterase enzymes. As a result a more sustained parasympathetic stimulation is observed clinically. Methacholine, with its methyl group introduced in the β-position to the nitrogen atom, is a drug that has predominantly muscarinic cholinomimetic activity. Shifting the CH_3 group to the α-position yields a compound with more prevalent nicotinic action. The β-carbon of methacholine is an asymmetric center generating two stereoisomers. One of these isomers—the + isomer—is more active and is the one with a stereochemical structure resembling that of the most active form of muscarine. The commercial product is racemic. The drug is considered obsolete even though it has been of use to slow the heart rate in supraventricular tachycardia and to dilate spastic blood vessels in peripheral vascular disease (Raynaud's) by subcutaneous injections.

Carbachol is an example of a direct-acting parasympathetic drug in which the stability of ACh is increased by replacement of the acetyl function by the acyl-like group—the carbamyl ester—which has considerable resistance to hydrolysis. Although enzymatic hydrolysis of carbachol occurs, the reaction is so slow that for practical purposes it may be considered insignificant.

Carbachol has all the pharmacological properties of ACh, muscarinic and nicotinic. It produces peripheral vasodilation, slowing of the heart, increased smooth muscle tone, salivation, tearing, and sweating. It also stimulates the ganglia of the autonomic nervous system and skeletal muscles. Because of this lack of specificity and high systemic toxicity, its use today is limited to topical ophthalmology as eyedrops for patients with glaucoma.

Bethanechol incorporates the stabilizing structural features of both methacholine and carbachol. Thus it is like carbachol in that no appreciable cholinesterase hydrolysis takes place, and like methacholine it does not stimulate ganglia or skeletal muscle. This drug appears to show some specificity of action in the gastrointestinal and urinary tracts. Thus, it has clinical applications in postoperative distention of the abdomen and bladder; in the latter case it relieves the urinary retention that results following surgery. The drug is even orally effective.

It will be noted that some drugs in Table 8-4 are tertiary not quaternary amines. Simple tertiary amines would not be expected to have cholinergic activity, even though they would be considerably ionized at physiological pH. If other components of such a molecule contained interactive capability to bind with a receptor it might function as an agonist.

The alkaloid arecoline is illustrative of such a situation. The drug, which is obtained from the areca or betel nut (*Areca catechu*), has been chewed as a euphoric stimulant in southern Asia for centuries. It is used today mainly as a large-animal anthelmintic because of its strong peristaltic and cathartic action. Having a pKa of 7.6, it is 60% protonated at

Table 8-4. Direct-Acting Cholinergic Agonists[a]

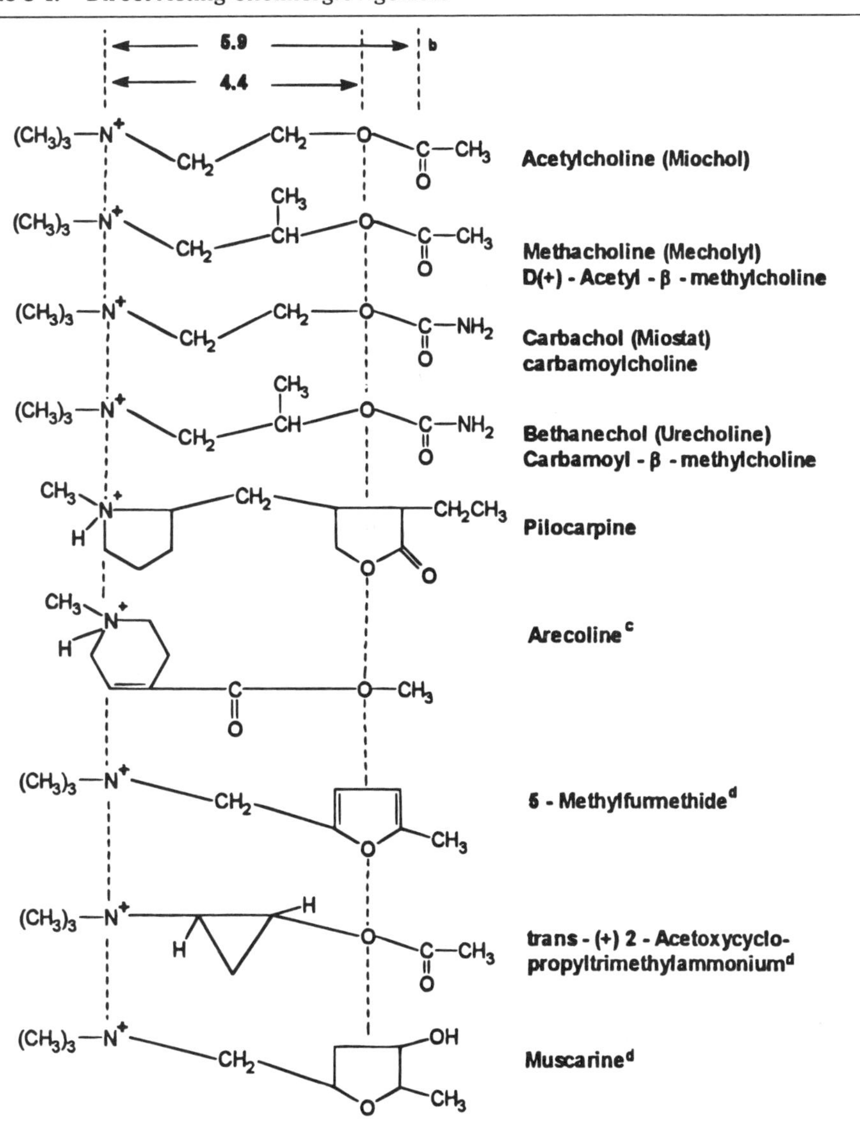

[a] Trade names in parentheses.
[b] Distances in Å vary with source; see text.
[c] Veterinary product.
[d] Not used medically; a research tool.

physiologic pH. In that cationic form it has potent muscarinic properties. This becomes understandable if we view the molecule as a "reversed" ACh-like ester (i.e., a methyl ester of an aminopropionic acid rather than an acetic ester of an aminoethyl alcohol). A glance at Table 8-4 will indicate that the distance between the cationic nitrogen and the ether oxygen of the ester is "just about" right. In interesting experiments, using guinea pig ileum over a 6–9.3 pH range showed the compound to be more muscarinic than ACh at pH 6 (totally protonated), 80% as active at pH 7.4, and only 2% as active as ACh at pH 9.3 (essentially a neutral tertiary amine).

Pilocarpine is an alkaloid obtained from the leaves of several *Pilocarpus* species. It is available as HCl and HNO_3 salt in ophthalmic preparations (drops, ointments and Ocusert^R, which is a slow-release ocular insert device). The drug has been in use over a century, but today is almost exclusively utilized as a miotic in glaucoma. At one time, its strong effect on the sweat glands producing profuse diaphoresis (sweating) was "clinically" valued. Also its ability to produce salivation found application to counteract the oral dryness caused by therapy with ganglionic blockers. Pilocarpine's pKa of 7 allows it to be about 30% cationic at body fluid pH. Its ether oxygen (in the lactone ring) is probably about 4A distant from the nitrogen, therefore meeting structural prerequisites for cholinomimetic activity. The nicotinic component is very weak. An interesting set of stereoisomers of acetoxycyclopropyltrimethylammonium iodides were synthesized by Armstrong and Cannon (1970) as very rigid small ring analogs of ACh to test the conformational parameters of nicotinic and muscarinic activity. This molecule must be viewed as ACh where the two methylenes of ACh are locked into one or the other of two very definite inflexible configurations: the *cis* and *trans*. The *trans* (+) was almost as potent as ACh as a muscarinic agonist, but it showed only 1% of such activity at nicotinic sites. Hydrolysis by AChE was facile. Particularly interesting was that the absolute configuration of the active (+) *trans* isomer was the same as the muscarinically active methacholine.

8.4. Anticholinesterase Agents

Anticholinesterase agents are enzyme inhibitors that preserve endogenous ACh. Therefore, they must be considered indirect cholinergic agents.

Ordinarily the enzyme inhibition that produces reversible effects involves the inhibitor competing with the normal substrate for complexing sites on the enzyme, binding in some fashion, and releasing the free enzyme again. Compounds with only a quaternary nitrogen such as tetraethylammonium and edrophonium inhibit AChE reversibly by binding only to the anionic site of the enzyme, thereby blocking complexation with ACh. Because of the single-site binding inhibition, intensity is relatively low and the duration of the effect is short. In the case of edrophonium, a diagnostic agent for myasthenia gravis, this transient period is useful in differentiating it from other conditions characterized by muscular weakness. Edrophonium (Table 8-5) is also useful as an antidote for curare overdoses or to terminate its effect in the operating room when desired. AChE inhibition is not the operative mechanism in this case. Direct displacement of curariform drugs from the endplate may be involved and may also be an improvement in neuromuscular transmission not involving the nerve–endplate junction.

Edrophonium, which is a short-acting AChE inhibitor, has been found useful against supraventricular tachyarrhythmias. The effectiveness in reducing the number of supraventricular impulses is due to increased ACh levels at the atrioventricular A–V junction of the heart. This in turn reduces A–V conduction.

The classic type of competitive reversible enzyme inhibition can be expressed simply as:

$$\text{Inhibitor} + \text{Enzyme} \underset{B}{\overset{A}{\rightleftharpoons}} \text{Enzyme}-\text{inhibitor complex} \tag{8.4}$$

The complex is based on the relatively weak affinities that do not involve covalent bond formation. Initially, since the inhibitor concentration is relatively high and since the

Table 8-5. Structural Features of Reversible Cholinesterase Inhibitors[a]

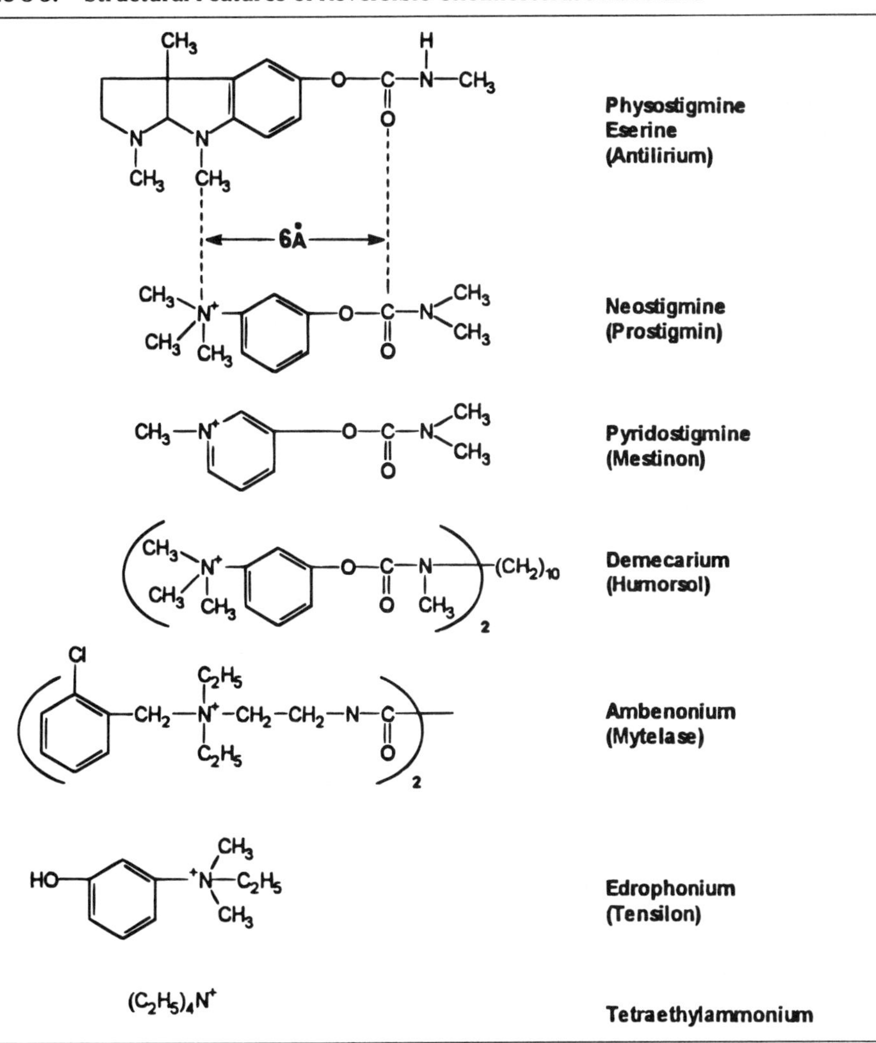

[a] U. S. trade names in parentheses.

inhibitor has good affinity for the binding site, enough sites are temporarily occupied to exclude some ACh molecules and prevent their degradation. The result is cholinergic stimulation. As free inhibitor is removed by metabolism, diffusion, and excretion, more of the complex dissociates until the preexisting normal condition is restored (Eq. 8.4B).

With the carbamate esters, however, the circumstance is quite different. It actually resembles the situation existing with ACh. By using neostigmine as the prototypical drug in this category (Fig. 8-10), it can be shown that binding takes place at both the anionic and esteratic sites at a rate of similar magnitude as ACh (Fig. 8-6, Eq. 8.4). The enzyme inhibition appears to be competitive. The hydrolysis mechanism is analogous to that described for ACh, except it is much slower. That is, the alcoholic (actually phenolic) portion of the molecule is liberated; the serine residue is acylated (carbamoylated) (Fig. 8-10B), and a water molecule nucleophilically attacks (hydrolyzes) the carbamoyl linkage, regenerating

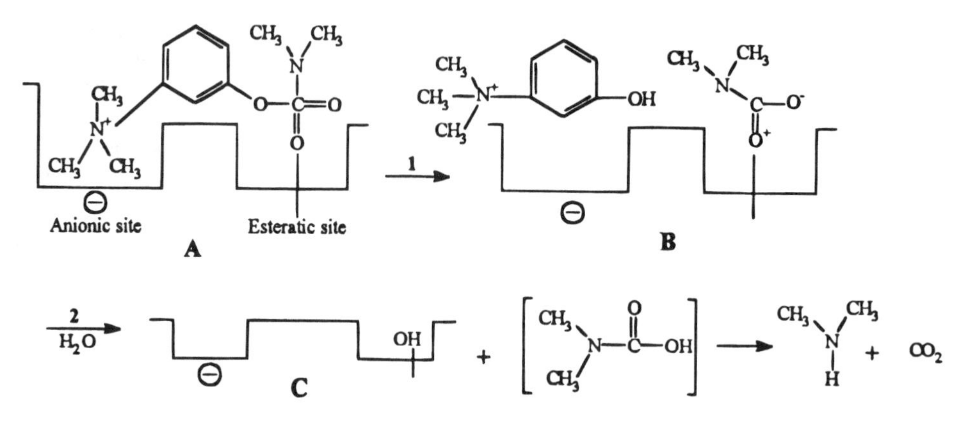

Figure 8-10. Neostigmine–acetylcholinesterase interactions.

the free enzyme (Fig. 8-10C) and liberating diemthylcarbamic acid, which spontaneously decomposes to dimethylamine and CO_2.

The difference between ACh and neostigmine-type drug lies in the much slower (4×10^7 times) rate of hydrolysis of the carbamoylated enzyme (Fig. 8-10B) compared with the acetylated form (Fig. 8-8C). The clinical duration of action of this drug, during which the AChE is presumably unable to process ACh, is 3–4 hours.

Certain differences between physostigmine and neostigmine should be appreciated. Physostigmine (pKa 8.1) is 83% cationic at physiological pH, which is sufficient for binding to the enzyme's anionic sites. However, since it is not a quaternary amine like neostigmine and pyridostigmine, which are 100% ionic at any pH, physostigmine has properties it does not share with the others. Most prominent of these are good oral absorption and crossing of the blood–brain barrier, permitting CNS effects. The quaternary drugs, although they are usable orally, require up to 30 times higher doses by that route than by parenteral administration. Thus oral dosing may increase the likelihood of adverse effects such as abdominal cramps, nausea, and vomiting. On the other hand, central effects are less likely.

The clinical application of reversible cholinesterase inhibitors is mainly limited to stimulating smooth musculature of the intestinal tract and bladder, where tonicity is low or lost (atrophied), and in the treatment of primary glaucoma, where their miotic effects relieve intraocular pressure by promoting drainage of the eye's aqueous humor in both the wide- and narrow-angle versions of the disease. These drugs are presently the only viable chemical methods of alleviating, at least temporarily, the symptoms of MG by way of stimulating (indirectly) cholinergic receptors at NMJs. Terminating the prolonged effects of neuromuscular blockade, such as postsurgically, is another therapeutic use for these drugs. Other special applications, such as in cardiology (edrophonium) and antidoting poisoning with phenothiazine tranquilizers and tricyclic antidepressants (physostigmine), should be mentioned.

The main use of physostigmine has been as a miotic, especially in glaucoma. Demacarium, which may be viewed as two neostigmine molecules joined by a decamethylene chain (Table 8-5) and therefore a long-acting agent, is utilized exclusively in glaucoma at intervals as long as 48 hours.

Neostigmine and pyridostigmine, where the latter has the quaternary nitrogen incorporated into a pyridine ring, are the primary agents in the management of myasthenia. Pyridostigmine with longer duration (up to 6 vs. 2 h) and lesser visceral adverse effects is probably the superior agent. It should be considered that these two drugs (unlike physostigmine) have the potential for direct stimulation of cholinergic receptors at NMJs.

Ambenonium is another *bis*-quaternary compound that is useful against myasthenia. It is not a carbamate; rather it is a much more stable oxamide. It has a prolonged inhibitory action against AChE, with which it binds reversibly without being hydrolyzed by it. It may also act at the skeletal muscle NMJ receptors directly.

Irreversible Cholinesterase Inhibitors. These are a large group of organophosphorus compounds, only a few of which have therapeutic applications. They share the following structure:

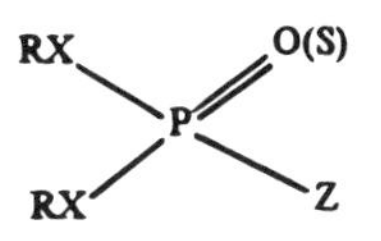

where the phosphorous atom is bonded to an oxygen or sulfur, *X* is usually an oxygen, *R*, *R′* = alkyl or aryl, and *Z* = a strong electron-withdrawing group that is easily displaced, including fluorine (F), cyano (CN), or other good leaving groups such as oxy- or thioalkyl or aryl functions.

These compounds interact only at the esteratic site of AChE (Fig. 8-11A) by acylating (phosphorylating) the serine OH group. This covalently phosphorylated enzyme (Fig. 8-11B) is not normally reactivated by hydrolysis under biological conditions. There is little spontaneous regeneration of the enzyme; it is poisoned and totally inactivated.

The mechanism for this type of enzymatic inactivation was first investigated with diisopropylphosphofluoridate (Table 8-6) using a ^{32}P-tagged molecule. The mechanism is such that the phosphorous atom, due to the polarization of the P–O bond, is highly electron deficient. The activated, strongly nucleophilic serine hydroxyl (as previously explained) attacks the phosphorous atom, forming a covalent bond at the expense of the P–F bond. Hydrogen fluoride is a byproduct. The very facile nature of this reaction is likely aided by the tetrahedral nature of the phosphate, which may be comparable to the geometry of the transition state occurring during acetyl ester hydrolysis. The fact that diisopropylphosphorylserine can be isolated from the "poisoned" enzyme indicates clearly that the postulated reaction is correct.

These types of compounds were initially synthesized in Germany during the early 1930s as insecticides. The Nazis, realizing the potential of these agents as "nerve gas" for chem-

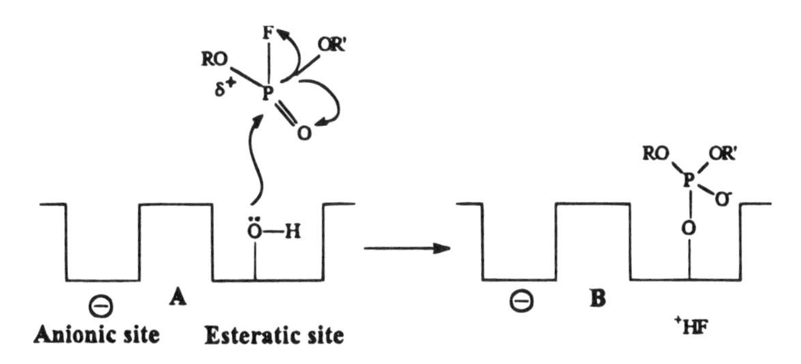

Figure 8-11. Irreversible cholinesterase phosphorylation.

Table 8-6. Representative Bioactive Organophosphates

Structure	Names[a]	Comments, main use
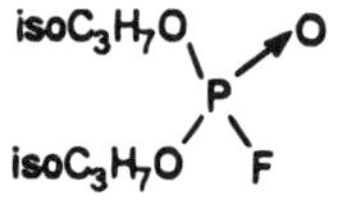 $(isoC_3H_7O)_2P(\rightarrow O)F$	Diisopropyl-phosphofluoridate DFP, isoflurophate (Floropryl)	Miotic in glaucoma Potent, irreversible
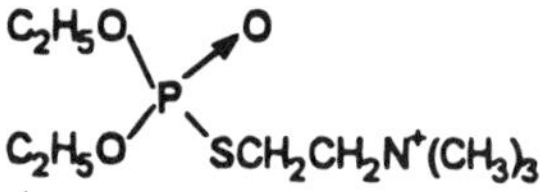 $(C_2H_5O)_2P(\rightarrow O)SCH_2CH_2N^+(CH_3)_3$	Echothiophate (Phospholine)	Very potent miotic in glaucoma; somewhat stable in water
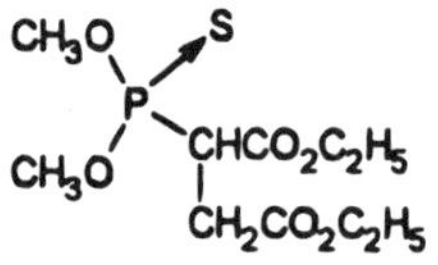 $(CH_3O)_2P(\rightarrow S)CHCO_2C_2H_5$ with $CH_2CO_2C_2H_5$ on the CH	Malathion (Prioderm)	Pediculicide and ovocide agricultural insecticide "Relatively" safe[b]
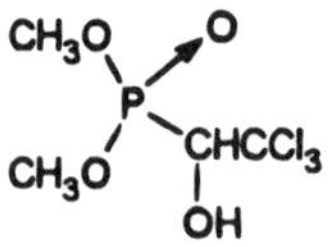 $(CH_3O)_2P(\rightarrow O)CHCCl_3$ with OH on the CH	Metrifonate (Bilarcil)	Anthelmintic-antischistosomal only[c]
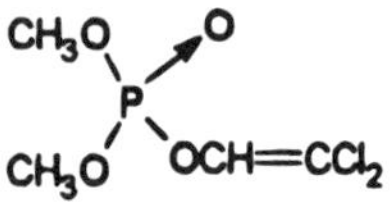 $(CH_3O)_2P(\rightarrow O)OCH{=}CCl_2$	Dichlorvos	Metrifonate metabolite insecticide, pets, bedbugs, farm animal anthelmintic
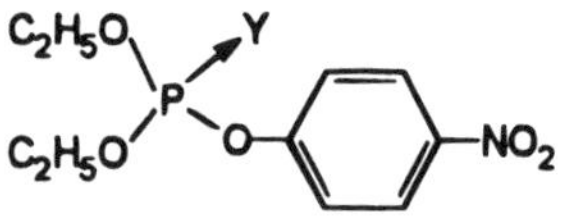 $(C_2H_5O)_2P(\rightarrow Y)O{-}C_6H_4{-}NO_2$	Y = S, Parathion Y = O, Paraoxon, Parathion metabolite	Agricultural insecticide (both compounds)
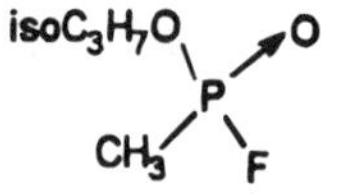 $(isoC_3H_7O)(CH_3)P(\rightarrow O)F$	Sarin Isopropyl methylphos phonofluoridate	Chemical weapon "nerve gas"
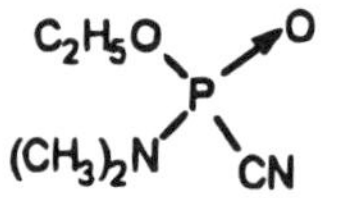 $(C_2H_5O)((CH_3)_2N)P(\rightarrow O)CN$	Tabun Ethyl N-dimethylami docyanidate	Chemical weapon "nerve gas"

[a] Trade name in parentheses. [b] See text. [c] Chapter 7.

ical warfare, developed them for this purpose (Tabun and Sarin, Table 8-6). They were not used in warfare but were used against humans in the gas chambers of concentration camps during World War II. Other organophosphates have since been developed as agricultural insecticides and anthelmintics that exhibit some degree of selectivity of toxicity between the parasite and the host. The selectivity was the result of some compounds being inactive until activated to the neurotoxin by an enzyme unique to the parasite or one more active there than in the vertebrate host. Parathion (Table 8-6), was the first compound to show a margin of safety due to activation of the P = S bond to the P = O function by an insect enzyme. In the case of malathion a dual protection appears to be operative, insect activation as in parathion *and* deactivation by a carboxyesterase catalyzed hydrolysis, which occurs only in the mammalian host. Malathion is therefore used as a home-gardening insecticide and was approved for use against head lice in the United States on prescription. Isoflurophate has been used as an extremely long-acting antiglaucoma drug. Metrifonate, previously discussed as an antischistosomal drug (Chapter 7), is rapidly metabolized to dichlorvos, which is widely used itself as an insecticide on pets and as an anthelmintic in large animals. Only two other compounds are utilized today therapeutically in humans. Isoflurophate, a potent, long-acting irreversible AChE inhibitor still finds occasional use in the management of glaucoma. Because of its rapid decomposition by water, the 0.025% ophthalmic preparation is in an ointment composed of an anhydrous polyethyleneglycol–mineral oil base. Miotic effects may persist for weeks.

Echothiophate is the other organophosphate used in glaucoma when shorter-acting compounds have failed. It should be recognized as a derivative of thiacholine. As in ACh, the quaternary nitrogen undoubtedly aids in the interaction with the enzyme's anionic site—a unique feature. It also makes the drug a water-soluble compound with a usable stability in this solvent of at least a month. The totally ionic character of this drug precludes any significant systemic absorption from the eye.

8.5. Antidotes for AChE Inhibitors

Poisoning by organophosphate may occur by overdose or more likely by accidental ingestion or skin absorption of insecticides. The resultant inhibition is, of course, irreversible, at least under normally encountered biological conditions (i.e., hydrolysis of the serine-phosphate ester by body water with available nucleophiles is extremely slow at best). Early attempts to counteract such poisoning were limited to the use of potent anticholinergic compounds such as atropine. However, this did not represent a true antidote, since such drugs essentially only block ACh interaction at muscarinic receptors. The problem is that we are dealing with an excess ACh situation. Preventing *some* of it from reaching only one type of receptor will at best only relieve the symptoms of muscarinic stimulation (pupillary constriction, nasal discharge, wheezing, nausea, cramps, etc.). Nicotinic symptoms produced by excess ACh accumulating at the NMJ of skeletal muscles are not antagonized. Death, which can result within 5 minutes to several hours following exposure, results from respiratory muscle paralysis. CNS effects complicate the syndrome.

A relatively nontoxic compound of sufficient nucleophilicity to break the phosphate–serine bond and reactivate the enzyme rapidly is required. Drugs that simply "compete" with ACh, even at both muscarinic and nicotinic receptor sites, may partially reduce symptoms temporarily but not necessarily change the fatal outcome. The large excess of ACh accumulating due to lack of unpoisoned AChE precludes totally satisfactory antidoting with such compounds. Thus regenerating the functioning AChE enzyme is really the only satisfactory approach. Hydroxylamine, $_2$HN-OH, can act as such a "reactivator" of the phosphorylated serine hydroxyl in the poisoned enzyme, but is too slow when the victim is seriously poisoned. Time is critical since the longer the victim's phosphorylated enzymes have "aged," the less feasible reactivation becomes (see later). More effective antidotes are oximes or hydroxamic acids derived from hydroxylamine. Such derivatives have a much higher nucleophilicity due to the so-called *alpha*-effect that results from the repulsive forces of two adjacent electronegative atoms—the nitrogen with an unshared electron pair and the oxygen anion bonded to it: $-CH{=}N{-}O^-$.

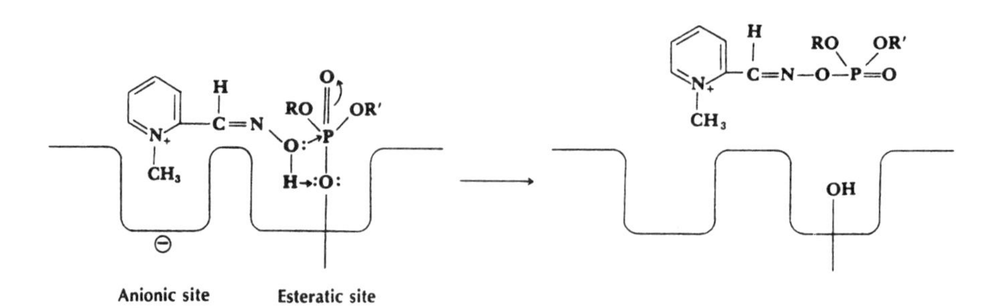

Figure 8-12. Possible mechanism of phosphorylated cholinesterase reactivation.

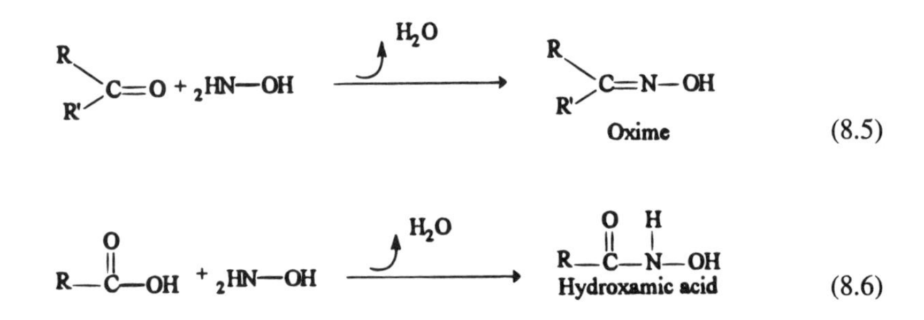

An additional design feature, taking advantage of the neighboring unoccupied *anionic site* of the poisoned enzyme, is to factor a cationic head into the antidote drug molecule. It is to be separated from the "reactivator" portion so as to "fit" and position the nucleophilic end of the drug for maximum effectiveness. One such drug is pralidoxime (Protopam chloride^{R}, 2-PAM), which is easily synthesized from pyridine-2-carboxaldehyde (Eq. 8.7). Figure 8-12 schematically represents such a reactivation of the poisoned enzyme. The mechanism of reactivation may be viewed as a concerted action. The covalent phosphorous–oxygen bond is broken by the oxime oxygen of 2-PAM, which has been weakened by simultaneous hydrogen bonding to the serine residue oxygen atom. The quaternized pyridine nitrogen atom of 2-PAM, by binding to the anionic site of the enzyme, can be presumed to aid in the specificity and effectiveness of the displacement reaction.[12]

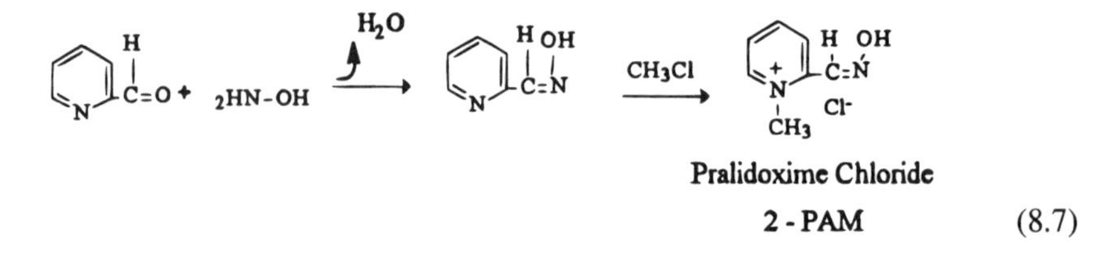

Enzyme regeneration is 10^5 times faster than it is with $_2$HN-OH alone. Under actual poisoning conditions treatment includes 2-PAM as well as heroic doses of atropine (up to 50 mg in 24 hours) to counteract the muscarinic intoxication symptoms. Intravenous barbiturates to stop convulsions may also be utilized.

---

[12] Oximes can exist in two stereochemical forms, *syn* and *antl*, which roughly correspond to *cis* and *trans* in olefins; pralidoxime is in the *anti* configuration.

Other agents useful as cholinesterase reactivators include nicotinhydroxamic acid methoiodide and the *bis*-compound obidoxime (not available in the United States). Toxic overdoses of carbamate AChE inhibitors are not reversible with pralidoxime-type antidotes.[13] In fact, because of some anti-AChE activity of their own, they would actually aggravate the toxicity of physostigmine and neostigmine overdoses.

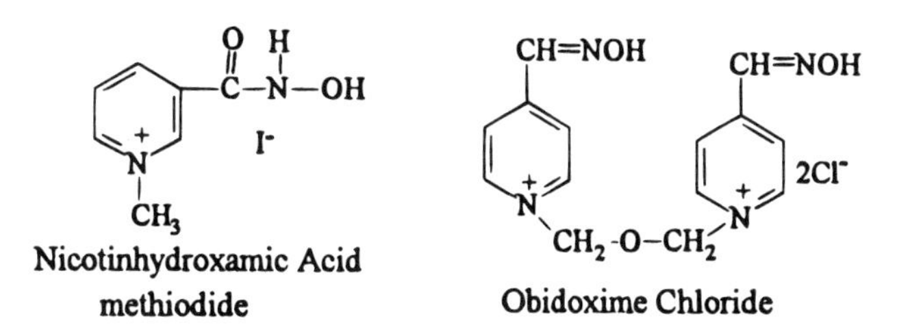

The rate at which AChE reactivators function (or for that matter spontaneous reactivation by in vitro or in vivo hydrolysis) depends very much on the size and degree of branching of the alkyl groups *R*. If the groups are small, such as methyl or ethyl, then active enzyme regeneration may be a matter of hours. When the bulkier and branched isopropyl group is reached (e.g., isofluorophate), no measurable spontaneous in vivo reactivation is likely to occur. Steric hindrance due to "bulk" is the apparently significant factor here.

A somewhat related concept is the decreased ability of reactivators to regenerate the free enzyme from its poisoned serine–organophosphate ester linkage as a function of time, a phenomenon called *aging*. Here too the effect is related to structural variations of the alkyl or alkoxy groups bonded to the phosphorous atom. This effect was noted to be more pronounced, (i.e., arising sooner) with esters of secondary (and tertiary) alcohols. It then became apparent that "aging" actually involved the loss, by hydrolysis, of an alkoxy group from the phosphorylated enzyme, at a rate proportional to its degree of branching. Thus a 2-propoxy group from an isofluorophate-inactivated enzyme would "leave" more readily than would a methoxy from malathion or metrifonate (Table 8-6) (Eq. 8.8). One plausible explanation for the decreased ability of the antidote to reactivate the "aged" poisoned enzyme is that nucleophilic approach to the phosphate ester is now electrostatically repulsed by the negative oxyanion.

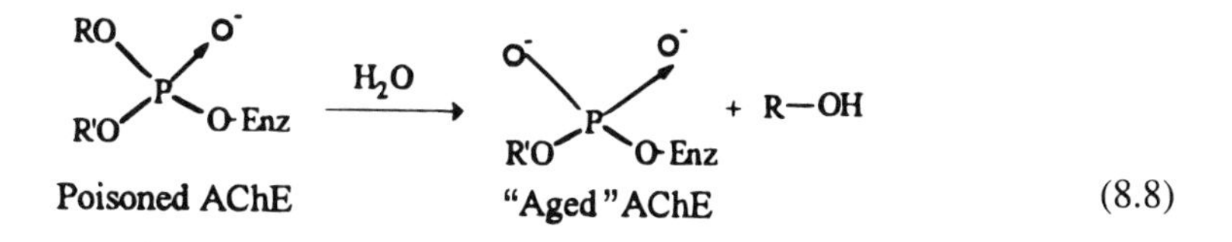

(8.8)

8.6. Memory and Alzheimer's Disease

Declining memory and cognition has long been associated with aging. Recent memory is particularly affected. With increasing public awareness of Alzheimer's disease and other

[13] Several texts erroneously state that they are (Bowman and Rand, 1980).

senile dementias, the tempo of research has greatly accelerated to find causes and discover drugs to treat and, ideally, to reverse the pathology. The tools for such studies have generally improved over the past decade.

The search for an underlying chemical basis has concentrated on central cholinergic involvement. It has been known for some time that muscarinic—cholinergic antagonists capable of crossing the blood–brain barrier (BBB) can impair cognitive functions and produce a temporary state of amnesia. The belladonna alkaloids in the form of extracts of the various Solanaceae plants have been used as mind-altering potions (and poisons) for millennia. Therapeutic doses, or overdoses, of these drugs today [e.g., atropine (AT), Scopolamine (SP)] can produce numerous adverse effects including blurred vision, slurred speech, drowsiness, impaired motor function, confusion, disorientation, memory impairment, and hallucinations—all ascribable to central effects. The significant relationship of cortical cholinergic innervation to memory has been established by showing that drugs capable of stimulating central cholinergic functions can reverse recent memory loss and other performance deficits induced by anticholinergics. Experiments with aged and young monkeys demonstrated more consistent improvements with young animals when physostigmine was used.

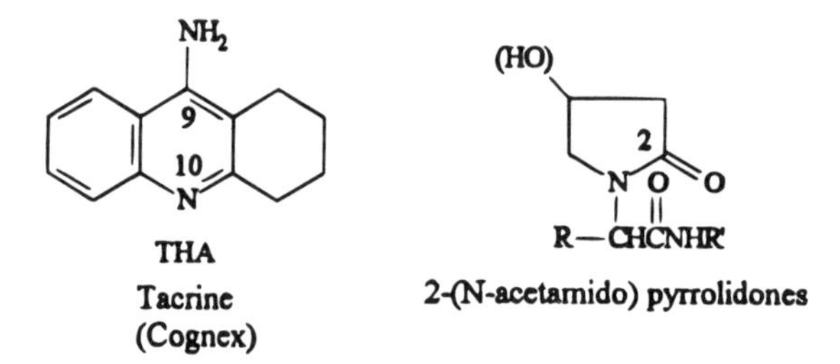

Cholinomimetics that can increase central cholinergic activity have produced *some* improvements in Alzheimer's disease; AChE inhibitors, particularly physostigmine, have been most studied. A study with an experimental potent, centrally acting anticholinesterase compound 1,2,3,4-tetrahydro-9-aminoacridine (THA, Tacrine) involving Alzheimer's patients over a year-long period showed encouraging improvement without some side effects.

Other approaches that are not related to the cholinergic system that are used to treat learning, memory, and cognitive impairment irrespective of disease type are with so-called nootropics (Schindler et al., 1984). These are compounds without either sedative or stimulatory properties that may positively affect blood flow and metabolism in the brain. They tend to share the ability to protect (rats) against chemically induced disruptions of memory deficits and learned avoidance behavior. Among the promising compounds are several 2-(N-acetamido)-pyrrolidones.

Potential modulators of memory storage are also under investigation. A diversity of substances appear to facilitate memory including Substance P, naloxone, amphetamine, and vasopressin.

8.7. Cholinergic Blocking Agents

Drugs capable of blocking, or otherwise antagonizing, the effects of ACh at muscarinic cholinergic receptors and at nicotinic receptors of the NMJ of skeletal muscle will now be considered. Compounds capable of blockading nicotinic receptors at ganglia of the autonomic nervous system will be dealt with in the discussions on antihypertensive drugs (Chapter 10).

8.7.1 Anticholinergics (Antimuscarinics)

Drugs in use today because of their ability to elicit mydriasis[14] (pupillary dilation) and cycloplegia (ciliary paralysis), reduce spasticity of smooth musculature such as the small intestine, inhibit gastric secretions of HCl, and minimize the symptoms of Parkinson's disease and other conditions involving tremors, either still are or have evolved from a group of plants of the potato family—the *Solanaceae*. The species include *Atropa belladonna* (deadly nightshade), *Hyoscyamus niger* (black henbane), and *Datura strammonium.* All have been in use for centuries, if not since antiquity. Until modern times these plants were utilized as galenical pharmaceutical preparations, [i.e., as various alcoholic (tinctures, fluid extracts), hydroalcoholic (elixirs), and even aqueous (infusions) extracts].

These medicinal plants are now understood to owe their pharmacological and toxicological properties to several extremely potent alkaloids (Fig. 8-13). The first to be isolated in 1831 was atropine. Other solanaceous alkaloids are atroscine (which like atropine is racemic), hyoscyamine and scopolamine, and the levorotatory enantiomers of racemic atropine and atroscine, respectively.

The natural alkaloids are esters of the bicyclic amino alcohol tropine (3-tropanol or 3-hydroxytropane, Fig. 8-13) and tropic acid (Table 8-8). The stereochemical basis of these alkaloids should be reviewed at this point. Even though it appears at first glance that 3-tropanol might have a chiral center, on closer inspection, it will be noted that it does not. Empirically, looking at the "flat" representation beginning with C-3 and viewing in both the clockwise (C-2, 1, 7) and counterclockwise (C-4, 5, 6) directions, it will be seen that the chemical environment is the same (i.e., symmetrical). (Conceptually, a plane of symmetry can be established, leading to the same conclusion.) Nevertheless, two stereochemically distinct forms of this amino alcohol can exist. Tropine, where the 3-OH is *axial* and therefore *trans* to the nitrogen bridge (referred to as α), and pseudotropine, where the 3-OH is *equatorial* and therefore *cis* to the nitrogen bridge. It has the β-configuration.[15]

If N-methyl-4-hydroxypiperidine is considered, it will be recognized that the molecule can exist in two conformations: the chair form and the boat form. The boat form, where the H atoms at position 4 and on the CH_3 group are in much closer juxtaposition than they are in the chair form and therefore repel each other, will be the less stable conformer (but still interchangeable at ambient temperatures). If one then views the two-carbon ethane moiety (C6–7) as locking the chair piperidine–ring into the rigid tropine molecule, by bonding into the 1 and 5 positions, then it becomes apparent that two rigid, sterically noninterchangeable molecules can arise. Tropine and pseudotropine do arise. On esterifying the 3-hydroxyl group with tropic acid, as are the solanaceous alkaloids, it is found that the antimuscarinic activity resides in those esters derived from the amino alcohol having the *axial* hydroxyl, namely, tropine. Since tropic acid brings a chiral center to these molecules, it is not surprising that the alkaloids in the plant are one enantiomer (levo), but can racemize readily,

---

[14] Both properties are useful in ophthalmology.

[15] Molecular models would be helpful in satisfying oneself as to the described relationships.

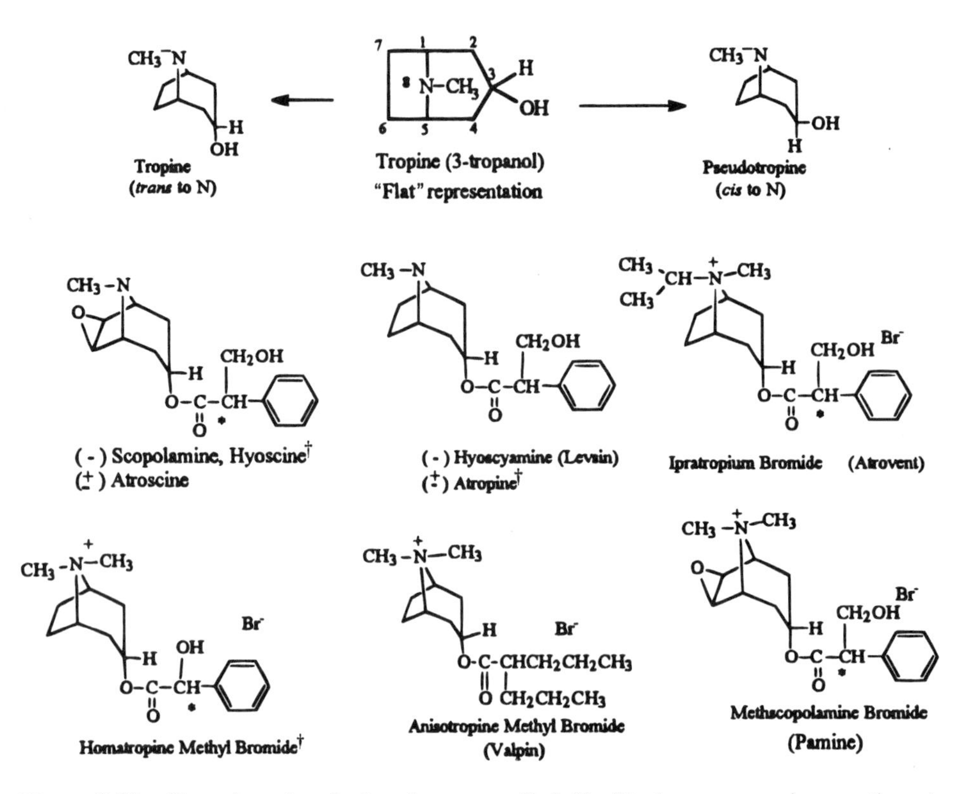

Figure 8-13. Natural semisynthetic solonaceous alkaloids (Trade names are in parentheses). *Asymmetric (chiral) center. †Also used in ophthalmology.

possibly as a result of the conditions of isolation and purification. In fact, the ability to block the muscarinic effects of ACh resides primarily in the (–) isomers. Thus the racemate atropine is actually less active than (–) hyoscyamine. It is curious that stereospecificity of action does not extend to the stimulatory effects on the CNS.

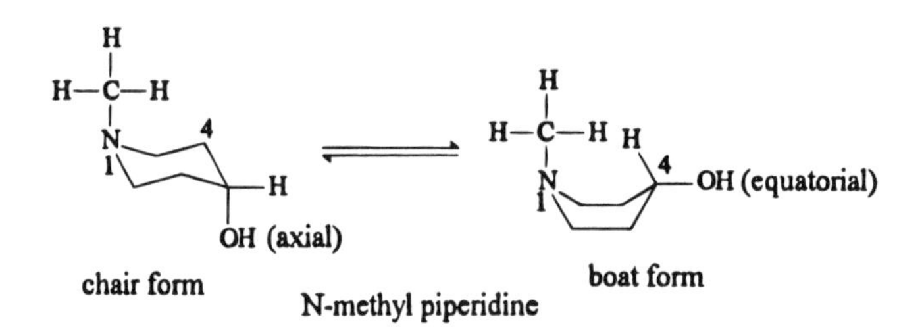

The main pharmacological influences of the naturally occurring alkaloids were already well recognized in the late nineteenth century. This was the triad of antispasmodic, antisecretory, and mydriatic effects. The fact that these drugs antagonized the effects of cholinergic alkaloids like pilocarpine and physostigmine was known but, of course, was not comprehended in chemical terms. That understanding could not begin until the chemical structure of atropine was elucidated. That atropine on hydrolysis yielded a base (tropine)

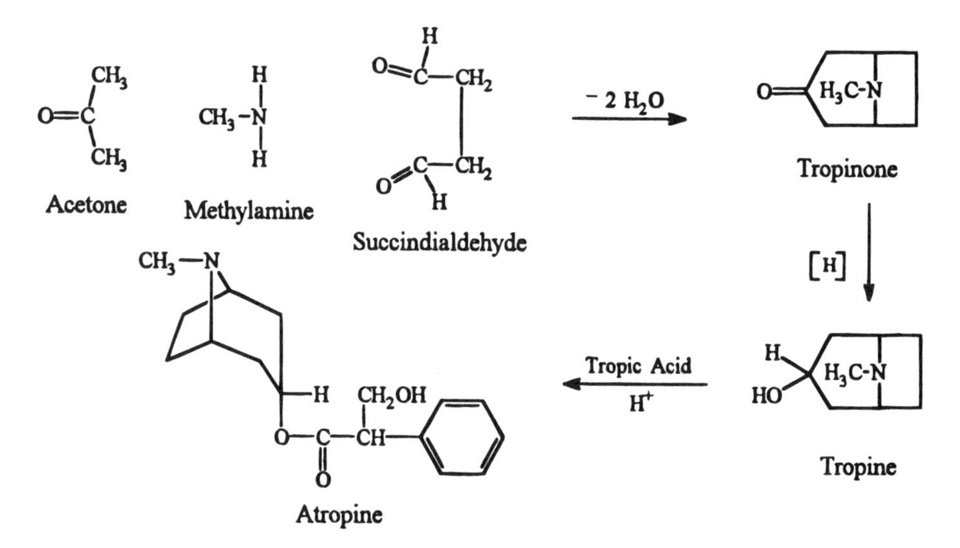

Figure 8-14. Facile synthesis of atropine (from Robinson, 1917).

and racemic tropic acid was determined by early 1868. The structure of tropic acid was later proved by synthesis. The structure of tropine presented difficulties, but was finally established and confirmed by synthesis by Willstatter (1901). It must be appreciated that before the days of UV, IR, NMR, and mass spectrometry total synthesis, when feasible, constituted the only way of absolutely establishing an organic structure. The elegant but simple synthesis of tropinone by Robinson (1917) (Fig. 8-14), yielding tropine on reduction, made preparation of atropinelike compounds by varying the acid moiety easy and inexpensive (e.g., homatropine; Fig. 8-13).

Dale (1914) demonstrated that atropine antagonized all muscarinic actions of ACh. There is little question today that this represents blockade at cholinergic—muscarinic receptors. Using the (–) hyoscyamine as the prototypical model (it being more active than racemic atropine) and considering the oversimplified receptor interactions with ACh previously discussed (Fig. 8-7), it can be argued that the alkaloid occupies, and therefore blocks, ACh from interacting with the very same sites to produce its antagonism. The tertiary nitrogen is 99% protonated at physiological pH ($pK_a = 9.7$); therefore, it is almost totally cationic.[16] Considering the N-methyl group as well, it is convincing then to expect a coulombic interaction with the cation augmented by van der Waals bonding of the methyl group. Analogously to ACh, the ester C = 0 interacts via hydrogen bonding to complementary electron-deficient loci on the receptor, as well as the hydrogen atom of the $-CH_2OH$ group. Additional hydrophobic interactions can be expected in the "flat region" from aliphatic carbons 1 and 2 of the tropine portion of the molecule and from the benzene ring of the tropic acid part at some further point. It might even be suggested that the reason for a weaker effectiveness of the racemic atropine that, of course, is "diluted" 50% with the weakly active (or inactive) (+) hyoscyamine is that the $-CH^2OH$ group of that enantiomer is now 180 degrees in the opposite direction and is unable to interact with the receptor.

[16] It should be noted that unlike a quaternary derivative, this does not prevent atropine from reaching central receptors.

Finally, since no covalent bond formation is involved in this drug–receptor complexation, *competitive* antagonism with ACh, or other muscarinic agonists, may be predicted. The prediction is valid.

Therapeutic applications of atropinelike drugs generally include their use as antispasmodics to relieve spasms of smooth muscles innervated by postganglionic parasympathetic fibers, particularly in the biliary, urinary, and intestinal tract. When combined with the antisecretory effect on gastric juice, the latter has been the basis of peptic ulcer therapy (frequently with the addition of nonabsorbable antacids). However, with the introduction of more specific inhibitors of HCl production such as H_2-histamine receptor antagonists and, more recently, proton pump inhibitors (Chapter 13), anticholinergics have been relegated to secondary importance over the past decade. In fact, at common dosage levels the achievable decrease of acidity is probably not significant. Higher doses, while possibly more effective in this respect, will most likely increase the predictable side effects to unacceptable levels for the patient. Such effects include annoying dryness in the mouth, blurred vision, constipation, and micturition difficulties. Drugs capable of suppressing HCl secretion, but devoid of anticholinergic effects, would be more desirable. The agents mentioned earlier possibly approach that ideal better. They do not, however, reduce hypermotility and spasms, so that true anticholinergics remain valuable where gastric acid is not the only problem.

Bronchodilation is another property of belladonna alkaloids of potential usefulness. Due to relaxation of the smooth musculature of bronchial passageways, this effect has found applications in asthma and other pulmonary obstructive conditions. At one time the smoking of cigarettes made of stramonium leaves was popular. More recently, atropine, quaternized with isopropyl bromide (Ipratropium, Fig. 8-13) has been introduced as an aerosol for inhalation. This renewed interest in antimuscarinics resulted in part from clarifications of the role of the parasympathetic system in bronchial obstructions. The availability of a potent agent such as atropine in a poorly absorbed form (i.e., quaternized) would minimize systemic effects following inhalation. Ipratropium bromide does not cross the BBB. It is longer acting and more bronchoselective than atropine methylbromide and exhibits no CNS effects.

The ability to dry up bronchial secretions and reduce laryngospasms (induced by some general anesthetics) has been the reason for using atropine and scopolamine as presurgical medication. It should be mentioned that scopolamine, which differs from atropine by the β-6, 7-epoxy bridge (Fig. 8-13), while it generally parallels atropine's pharmacological spectrum, does not share its cerebral and medullary stimulation; rather it exhibits CNS depression and amnesia, properties applied to anesthesia in an adjunct capacity.

The mydriatic and, in larger doses, cycloplegic effects of atropine in the eye, which is useful in ophthalmology, has been mentioned. Cardiac effects are transient decreases in rate and tachycardia with larger doses. Atrial arrhythmias are possible. Clinical applications are few and specialized.

The realization that atropine and some of its analogs were able to relieve the rigidity and tremors of Parkinson's disease (paralysis agitans) was empirical. It, and several simplified analogs (see later), were the only drugs available that were partially useful in the treatment of Parkinsonisms and other tremor conditions (e.g., those induced by antipsychotic drugs; see Chapter 12) until about two decades ago. The mechanism of action here is not clear. A central antimuscarinic effect is the most plausible. Since there is a gradual neuronal degeneration in the *substantia nigra* in the brain in Parkinsons disease that results in reduced levels of dopamine, the inhibitory neurotransmitter, the normal balance with ACh, and the excitatory neurotransmitter there is upset. It is this balance that apparently controls motor centers. Thus, when ACh gets the "upper hand," tremors ensue. There are obviously only two biochemical means to attempt to restore the balance. One way is for anticholinergic drugs to cross the BBB and act centrally; however, many patients do not respond satisfactorily. In addition, in the elderly, a majority of patients are particularly sensitive to anticholinergic side effects, further decreasing the usefulness of the drugs. The second way is to increase dopamine levels in the brain. This is discussed in Chapter 9.

Several other lesser applications should be mentioned. The use of atropine as part of the antidoting or organophosphate poisoning was discussed. Toxic doses of reversible AChE inhibitors (physostigmine, neostigmine) can be similarly managed; so can mushroom poisoning by muscarine-containing mushrooms. Finally, nausea induced by motion sickness (but not cancer chemotherapy) can be prevented effectively with scopolamine (transdermal patches).

8.7.2. Structure–Activity Relationships and Atropine Simplification

The multifaceted complexity of the belladonna alkaloids has always presented difficulties to the clinician, not to mention the patient. For example, the lack of selectivity for cholinergic receptors in the gastrointestinal area has made therapy less than satisfactory. The intensity of untoward effects and outright toxicity precluded the use of the higher doses that might have been more effective. The search for more selective anticholinergics has understandably been continuous. Results have been modest. The efforts will now be considered because of the biochemical rationales and intuition that went into their development, not because of their meager success. In fairness, the complexities of the human nervous system, especially of the autonomic and central divisions, must be considered before criticizing these efforts. Much of the nervous system still baffles workers in this field.

Once the structure of atropine (and other alkaloids) was known, exploratory chemical modifications become possible to ascertain the structural parameters within which anticholinergic properties existed and to exploit this knowledge to "improve" atropine.

A major goal of medicinal chemistry in the early decades of the twentieth century was the achievement of increased potency.[17] In the case of these highly potent alkaloids, where single human doses are in fractions of a milligram, this is obviously irrelevant. Thus the search was for increased selectivity of cholinergic blockade that it was hoped would go hand in hand with decreased side effects.

Side effects such as mental confusion and hallucinations with atropine are the result of penetration into the CNS. It was therefore found that quaternizing the nitrogen atom resulted in derivatives that could not readily traverse the lipoidal BBB. Compounds such as atropine methobromide, and more recently ipratropium bromide, are therefore useful as inhalation bronchodilators without central involvement. Applying this concept to ophthalmology was useless since these drugs would not readily penetrate the conjunctiva. In terms of gastrointestinal use as antispasmodics the results were a mixed blessing. Oral absorption of such quaternary salts is predictably erratic. A comparison following parenteral administration will show an absence of central effects that, it must be remembered, would make the quaternary drugs useless in Parkinsonism.

---

[17] Today, of course, it is greater efficacy that we seek.

Quaternization of atropinelike drugs likely prolongs their duration of action. Whether they also represent more active drugs compared with the nonquaternary tertiary amine analogs is debatable. The tradeoff for the elimination of central effects, however, is the appearance of nicotinic receptor blockade, particularly in the autonomic ganglia. Even though it may be argued that this augments antispasmodic effects, the price is a new set of side effects due to ganglionic blockade, particularly in the higher dosage ranges. Postural hypotension and impotence can then occur. In case of toxic overdoses, as in poisonings, even blockade at the NMJ will be seen (curariform action). Respiratory depression and arrest can then ensue.

Replacement of tropic acid in atropine with the homologous mandelic acid (α-hydroxyphenylacetic, Table 8-8) produces homatropine, whose quaternized methobromide is still in use, as is the methobromide of scopolamine.

Medicinal chemists, when viewing a bioactive molecule of a complex structure such as polycyclic natural products, instinctively tend to question the necessity for all that "organic shrubbery" to achieve the desirable pharmacologic properties. The question might be rephrased to: What are the *minimum* structural requirements contained in such a compound at least to qualitatively exhibit the activity sought? The problem here becomes one of simplifying the molecule to the "bare essentials" needed to retain antimuscarinic activity. It might be referred to as the minimum spasmophoric structure that will have a measurable degree of activity when synthesized. In the case of atropine, if the circled portion is excised, this minimum is a phenylacetic acid esterified to a cycloalkyl—or dialkylaminoethyl or propyl alcohol. Such a compound, then, where *R* might be ethyl, becomes the starting point for designing clinically effective antispasmodics, it is hoped, with the desired specificities. The 23 drugs described in Table 8-7 essentially represent the fruit of this labor without having achieved the perfect drug.

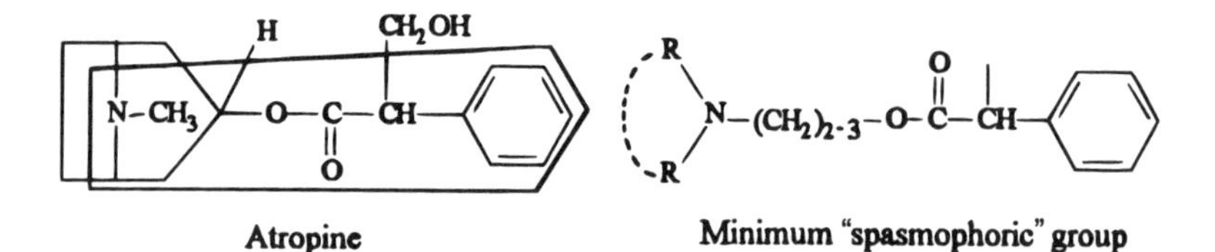

Atropine

Minimum "spasmophoric" group

Parenthetically, the overall similarity of ACh to the minimum structure should be noted. If *R* groups were CH_3 and the benzene ring replaced by an H atom, then this would essentially be ACh. The absence of the third CH_3 does not alter the comparison since, it will be recalled, the third CH_3 of ACh points away from the hydrophobic cavity around the anionic site of the receptor in any case.

The prototypical structure of a phenylacetic acid amino ester shows only minimal anticholinergic activity. Clinically useful activity only becomes apparent with an increase in the bulk of the acid portion as is apparent from the first 12 compounds (Table 8-7), yet the first marketed compound—amprotropine—retained the tropic acid moiety from its "parent." Its modest clinical activity still showed that a complex rigid aminoalcohol such as tropine was not essential. The basicity of the tertiary nitrogen, whether as part of a ring or simply bound to alkyl groups as a diethylamino group, appears to be the significant factor. An examination of amino esters in Table 8-7 shows them about evenly divided between these two types of moieties. Among the early marketed drugs, adiphenine (No. 1) also exhibited significant local

Table 8-7. Simplified Synthetic Atropine Analogs[a]

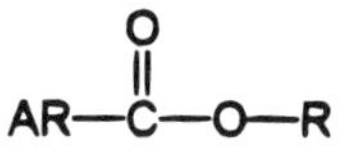

AR—C(=O)—O—R

Tertiary Amino esters:

No.	AR	R	Names	Comments
1.	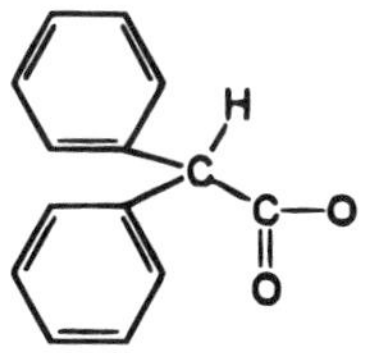 (diphenyl)CH—C(=O)—O	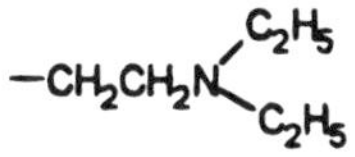 $-CH_2CH_2N(C_2H_5)_2$	Adipehnine[b] (Trasentine)	First such agent (1937), now obsolete
2.	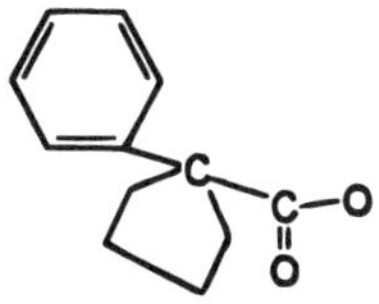 1-phenylcyclopentyl—C(=O)—O	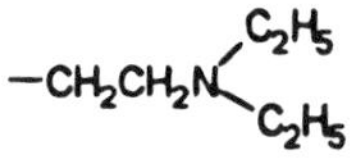 $-CH_2CH_2N(C_2H_5)_2$	Caramiphen[a]	Used only in cough preparations
3.	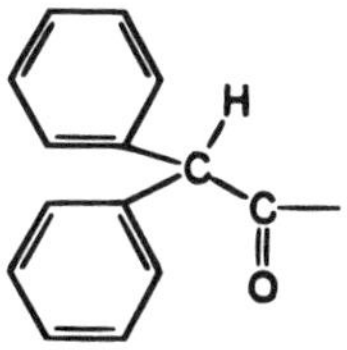 (diphenyl)CH—C(=O)—	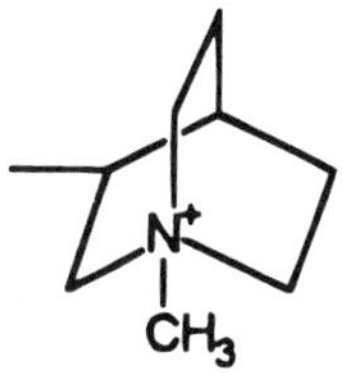 N-methylquinuclidinium (N^+, CH_3)	Clinidiumd (Quarzan)	Quaternized
4.	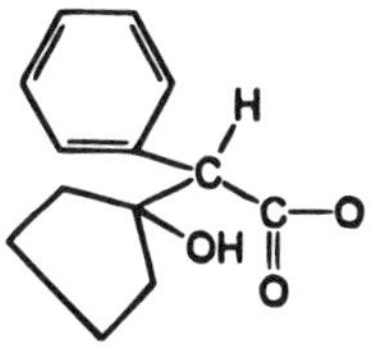 phenyl(1-hydroxycyclopentyl)CH—C(=O)—O (H, OH)	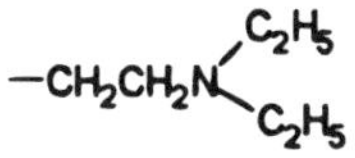 $-CH_2CH_2N(C_2H_5)_2$	Cyclopentolate[b] (Cyclogyl)	Used only as mydriatic and cycloplegic
5.	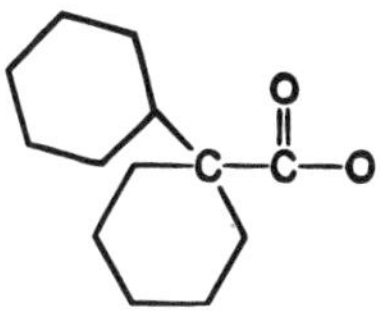 bicyclohexyl—C—C(=O)—O	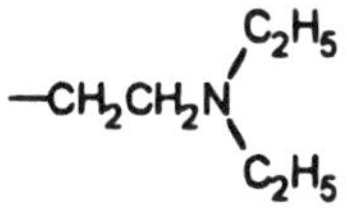 $-CH_2CH_2N(C_2H_5)_2$	Dicyclomine[b] (Bentyl)	Little anticholinergic action
6.	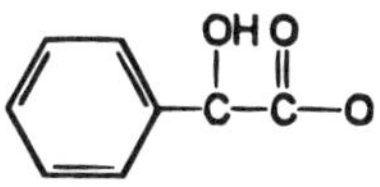 phenyl—CH(OH)—C(=O)—O	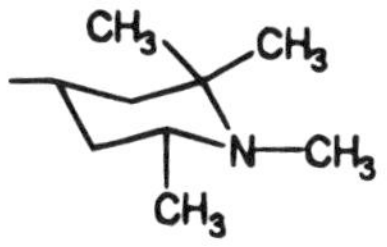 1,2,2,6-tetramethylpiperidin-4-yl (CH_3, CH_3, $N—CH_3$, CH_3)	Eucatropine[b]	Used only as mydriatic; no cycloplegia
7.	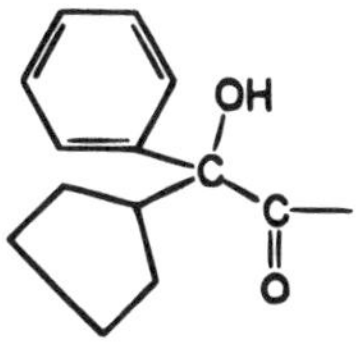 phenyl(cyclopentyl)C(OH)—C(=O)—	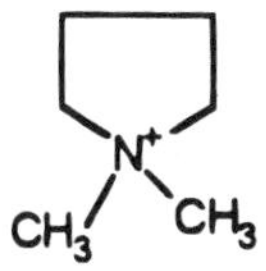 1,1-dimethylpyrrolidinium (N^+, CH_3, CH_3)	Glycopyrrolate[d] (Robinul)	Quaternized

(*Continued*)

Table 8-7. (*Continued*)

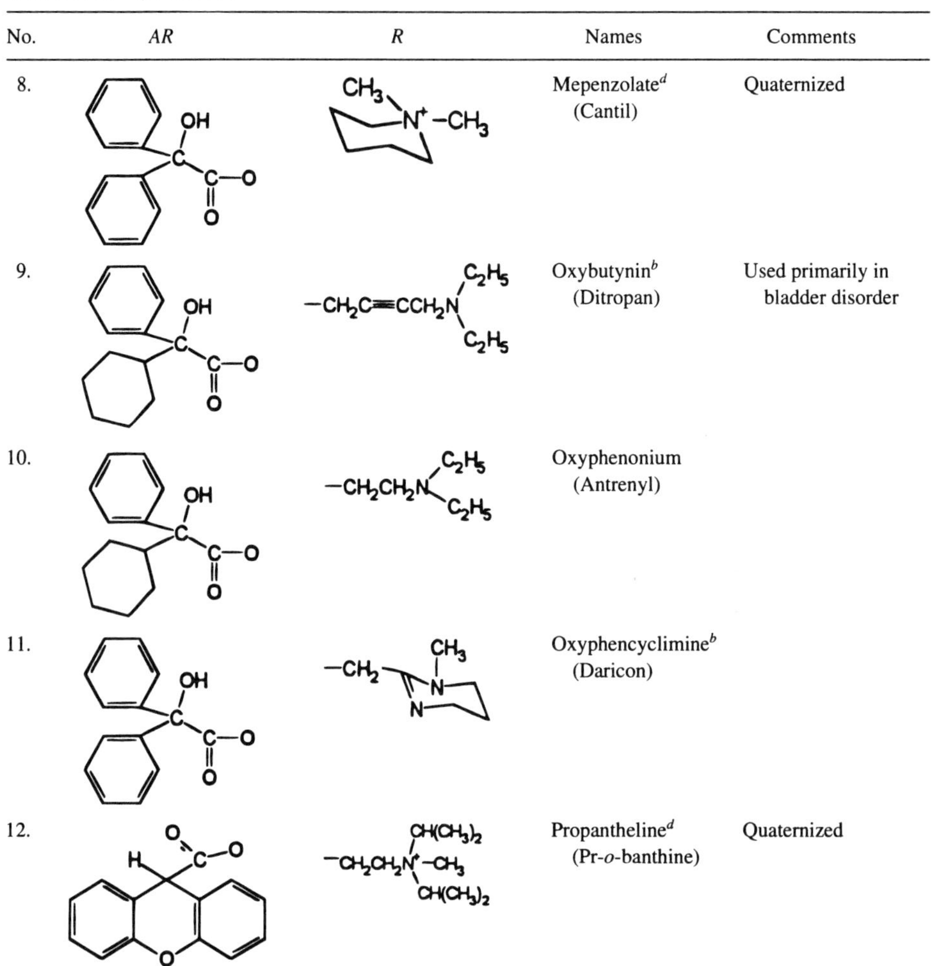

No.	*AR*	*R*	Names	Comments
8.			Mepenzolate[d] (Cantil)	Quaternized
9.			Oxybutynin[b] (Ditropan)	Used primarily in bladder disorder
10.			Oxyphenonium (Antrenyl)	
11.			Oxyphencyclimine[b] (Daricon)	
12.			Propantheline[d] (Pr-*o*-banthine)	Quaternized

B. Tertiary Amino Alcohol Esters: AR-O-R

No.	*AR*	*R*	Names	Comments
13.			Benztropine[a] (Cogentin)	Used in tremor and rigidity conditions

(*Continued*)

Table 8-7. (*Continued*)

No.	*AR*	*R*	Names	Comments
14. and 15.	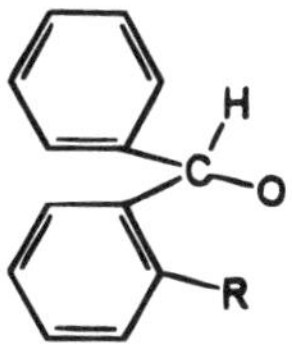 (diphenylmethoxy structure: C_6H_5–C(H)–O, with second ring bearing R)	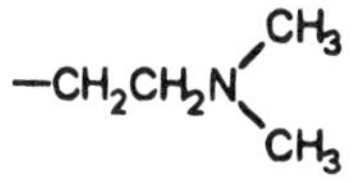 $-CH_2CH_2N(CH_3)_2$	*R* = H, Diphenhydramine[b] (Benadryl)	Also antihistamine and cough supressant
			R = CH_3, Orphenadrine[b,f,g] (Disipal,[b] Norflex[f])	Parkinson's

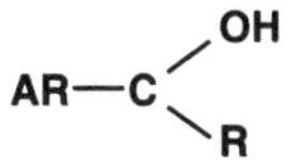

$$AR-C(OH)-R$$

C. Tertiary Amino Alcohols:

No.	*AR*	*R*	Names	Comments
16.	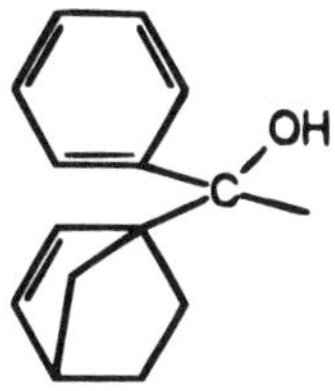 (phenyl and bicycloheptenyl, C–OH)	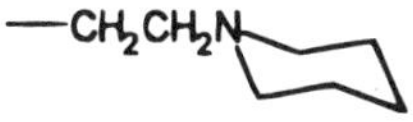 $-CH_2CH_2N$ (piperidine)	Biperiden (Akineton)	Primarily Parkinson's
17.	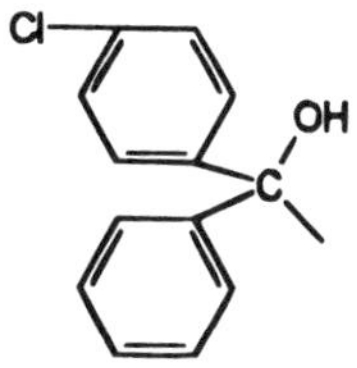 (Cl-phenyl and phenyl, C–OH)	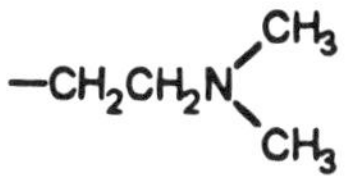 $-CH_2CH_2N(CH_3)_2$	Chlorphendianol (ULO)	Used only in cough preparations
18.	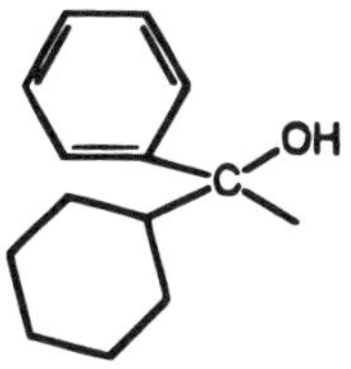 (phenyl and cyclohexyl, C–OH)	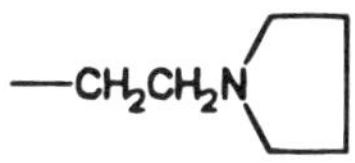 $-CH_2CH_2N$ (pyrrolidine)	Procyclidine[b] (Kemadrin)	Primarily Parkinson's
19.	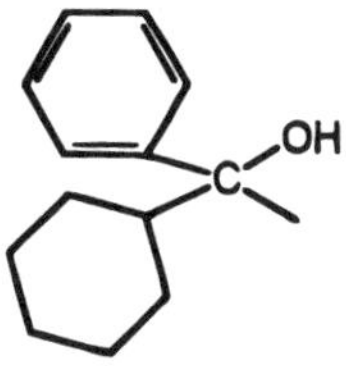 (phenyl and cyclohexyl, C–OH)	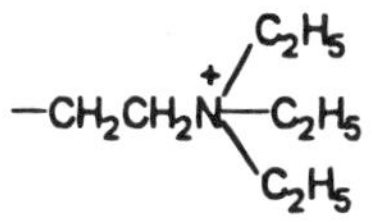 $-CH_2CH_2\overset{+}{N}(C_2H_5)_3$	Trihexethyl[b] (Pathilon)	Quaternized

(*Continued*)

Table 8-7. (*Continued*)

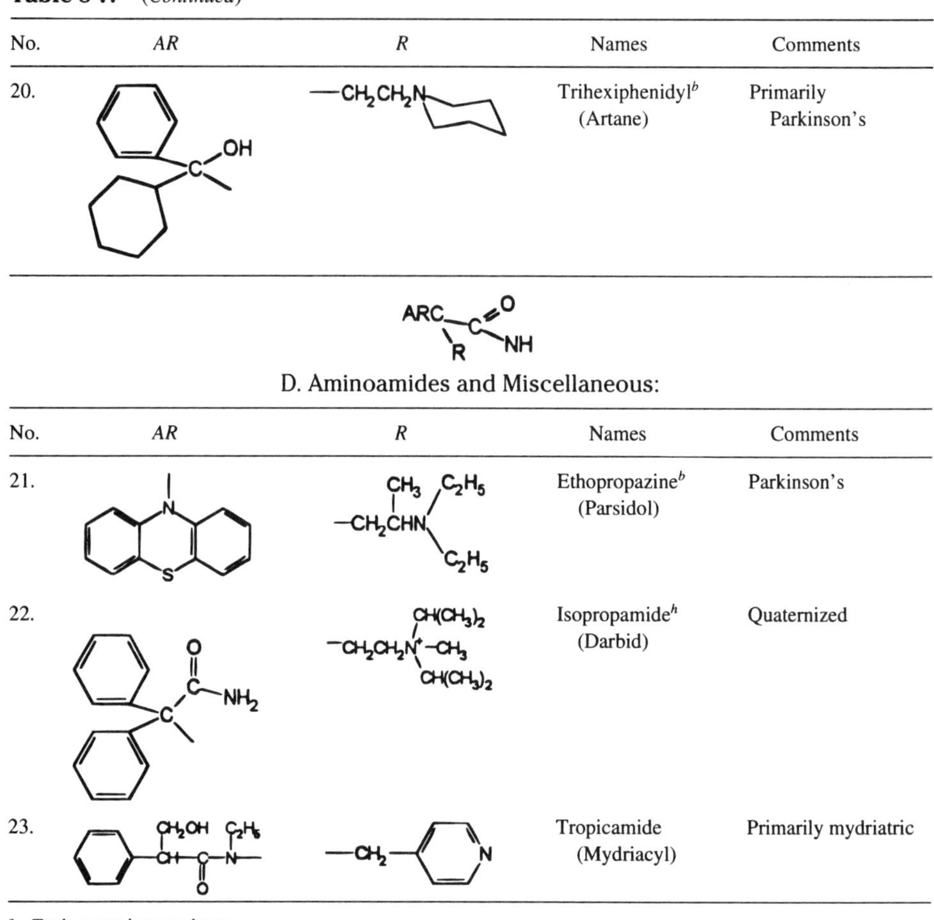

No.	*AR*	*R*	Names	Comments
20.		$-CH_2CH_2N$	Trihexiphenidyl[b] (Artane)	Primarily Parkinson's

D. Aminoamides and Miscellaneous:

No.	*AR*	*R*	Names	Comments
21.			Ethopropazine[b] (Parsidol)	Parkinson's
22.			Isopropamide[h] (Darbid)	Quaternized
23.		$-CH_2-$	Tropicamide (Mydriacyl)	Primarily mydriatric

[a] Trade names in parentheses.
[b] Hydrochloride.
[c] Edsylate.
[d] Bromide.
[e] Mesylate.
[f] Citrate.
[g] Also see Chapter 4.
[h] Iodide.

anesthetic activity. This should not be surprising when the structures of local anesthetics such as procaine and butacaine are examined (Chapter 13). These dialkylaminoalkylesters of aromatic acids were similarly developed by simplification of the cocaine molecule, which is an ester of pseudotropine. The bulkier-type acids such as diphenylacetic, benzilic, and xanthene-9-carboxylic (Table 8-8) appear to enhance antispasmodic action, especially if the nitrogen is quaternized as in propantheline (No. 12), mepenzolate (No. 8), glycopyrrolate (No. 7), and clidinium (No. 3). Less bulky (or simpler) acids such as mandelic seem to favor primarily mydriatic–cycloplegic activity as is seen with eucatropine (No. 6) and homatropine (nonquaternized). On the other hand, cyclopentolate (No. 4), which is a bulky acid derivative, is used exclusively as a mydriatic–cycloplegic. Of course, it does have antispasmodic activity. Total separation of mydriatic and antispasmodic activity is not achieved. In all these drugs there will be varying degrees of overlapping, manufacturers' claims notwithstanding.

Table 8-8. Aromatic Acids Esterified in Anticholinergics

Acid	Names	Examples[a]
CH_2OH / CH—COOH	Tropic	Atropine
H / C—COOH / R	Phenyl acetic[b] *R* = cyclopentyl *R* = cyclohexyl	Glycopyrrolate (7) Oxyphenonium (10)
OH / CH—COOH	Mandelic	Eucatropine (5)
R / C—COOH	*R* = H, Diphenylacetic *R* = OH, Benzylic	Adiphenine (1) Clindinium (3) Oxyphencyclimine (11)
H COOH / O	Xanthene-9-carboxylic	Propantheline (12)

[a] Numbers refer to Table 8-7.
[b] Where α-carbon also carries OH, may be named as a substituted glyoxylic acid, $HOCH_2COOH$.

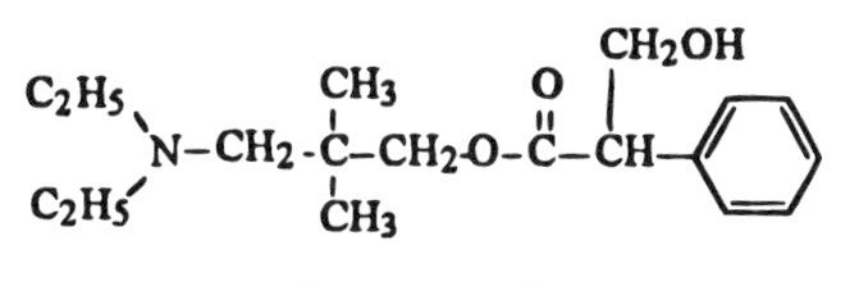

Amprotropine

Another pharmacological (though not clinical) complication should be pointed out. Many of the tertiary amine compounds do not owe all their antispasmodic properties to antimuscarinic (neurotropic) activity. To varying degrees they also possess a direct spasmolytic (musculotropic) action, similar to that observed with papaverine (see subsequent discussion). Dicyclomine (No. 5) is a particularly good example. The compound is not antisecretory and does not produce the expected characteristic anticholinergic (neurotropic) effects in the eye, nor on the sweat and salivary glands. Its antispasmodic actions appear to be mainly musculotropic, more so, in fact, than papaverine. It will be noted that the acid moiety, bicyclohexyl-1-carboxylic acid, does not even have an aromatic component. Anisotropine methyl bromide (Fig. 8-13) has a seven-carbon aliphatic acid esterified to tropine, yet appears to be neurotropic. Oxybutynin (No. 9), however, with a bulky phenylacetic acid function appears to have some selectivity for the bladder musculature; however, it is also mainly musculotropic. Compounds 16–20 show that even esterification is not essential in that the α-carbon of the formerly arylacetic acids can be directly "connected" to the tertiary nitrogen via an alkyl chain of usually two carbons. The polarity of the ester linkage is now simply represented by a tertiary alcohol function. It should be noted that the bulkiness around the α-carbon, so characteristic in the ester-type drugs, is retained in this group of compounds. In a chemical sense these drugs are aminoalcohols. It will be noted that several of the compounds (Nos. 16, 18, 20) are used primarily as anti-Parkinsonism agents. As in tridihexethyl (No. 19), quaternization, of course, reverts this drug to antispasmodic activity only.

Compounds 13–15 are tertiary aminoalcohol ethers, a type of structure typically associated with the early antihistamines such as diphenhydramine. It is known that one of the most prominent and annoying side effects of such drugs (dryness of the mouth) are effects associated with anticholinergic mechanisms. In fact, this property has led to its use in Parkinsonism and particularly for drug-induced extrapyramidal movement disorders. The *o*-methyl analog, orphenadrine, was introduced because of its lower antihistaminic activity and higher anticholinergic action. Because it also reduces spasticity in voluntary muscles it is marketed as a skeletal muscle relaxant as well (under a different trade name).

A particularly interesting study relating muscle relaxant, anticholinergic, and antihistaminic activity of diphenhydramine, orphenadrine, and nefopam, a cyclic analog, is illustrative of the influence of apparently "small" structural changes[18] on pharmacological properties and their coexistence or overlapping within the same molecule (Table 8-9). Nefopam has been in clinical use outside the United States as a nonaddictive analgetic whose side effects include some anticholinergic effects. It is difficult to draw definitive conclusions from such limited data, but one can postulate some general ideas. The authors suggest that cyclization produced a molecular topography of rigidity with a better fit of receptor areas controlling motor functions, hence a 20-fold increase in skeletal muscle relaxation. The reverse—that the extended amino alkyl chain conformation allows the protonated cationic nitrogen to "reach" a putative anionic site on the histamine receptor better than when locked in as in nefopam—may explain the observed 90-fold decrease in antihistaminic activity.

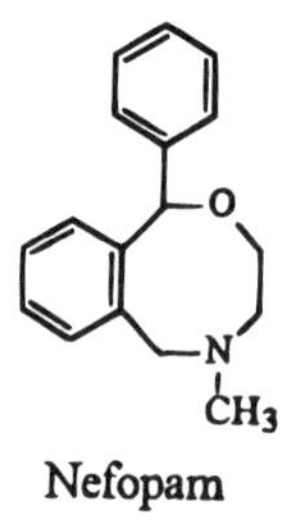

Nefopam

Another dramatic example of significant pharmacological differences between compounds that are "almost" alike is the use of the phenothiazine antihistamine ethopropazine (No. 21, Fig. 8-7) to treat the extrapyramidal Parkinsonlike syndrome *produced* by antipsychotic drugs such as the phenothiazine neuroleptic chlorpromazine (Chapter 12). An examination of the structures of the two types of drugs indicates that the ring nitrogen in the antihistamine is separated from the chain nitrogen by two carbons, whereas the tranquilizers *all* have a three-carbon distance—a "small" difference. The former, however, is an anti-

[18] Are there such changes in bioactive molecules?

Table 8-9. Relative Activity Profile of Three Related Aminoethers[a]

Drug	Muscle relaxant	Anticholinergic	Antihistamine
Diphenhydramine	1	1	1
Orphenadrine	2–3	1.6	0.067
Nefopam	10–30	0.167	0.011

[a] Adapted from Bassett et al., 1969.

histamine with a major anticholinergic component to its actions, while the latter is antipsychotic with a significant capability of inducing tremors.[19]

It is apparent from the foregoing discussion that relatively little success has been achieved in developing anticholinergic drugs capable of meaningful selectivity in the clinically important areas such as smooth muscle spasticity, gastric acid secretion, and cardiac rate. With the work on muscarinic receptor subtype (M_1, M_2) and clarification of differences in antagonist and substrate affinities, it may now be feasible to discover compounds able to take advantage of these dissimilarities. Pirenzepine shows that it will be possible.

The previous reference to the direct effects on smooth musculature causing its relaxation, a musculotropic effect not mediated via parasympathetic neurons (neurotropic), needs a brief elaboration at this point. A large number of chemically unrelated drugs can be shown to have a direct relaxant effect on various smooth muscle types. The ability to do so on coronary, cerebral, and arteriolar blood vessels would theoretically be useful in the treatment of angina, hypertension, peripheral vascular disease, and cerebral ischemias. Similarly such effects on the myocardium may beneficially affect arrhythmias. Spasmolytic effects on intestinal muscle would be useful in the treatment of hypermotile and ulcered gastrointestinal areas. All these properties have at one time been attributed to the opium-derived alkaloid papaverine (Fig. 5-7). The effects can be demonstrated experimentally in vitro. Animal experiments have attributed the ability to inhibit phosphodiesterase in cellular membranes to the drug, thus increasing the levels of cyclic-AMP (see Fig. 5-4). This in turn is believed to lead to the desired relaxant effects. Some dopaminergic properties have also been suggested. Unfortunately, none of these properties has been demonstrated to be of therapeutic value for any of the indications claimed. Because papaverine has been known for over a century, much of our clinical "knowledge" regarding this drug is undoubtedly anecdotal and has been perpetuated over decades.

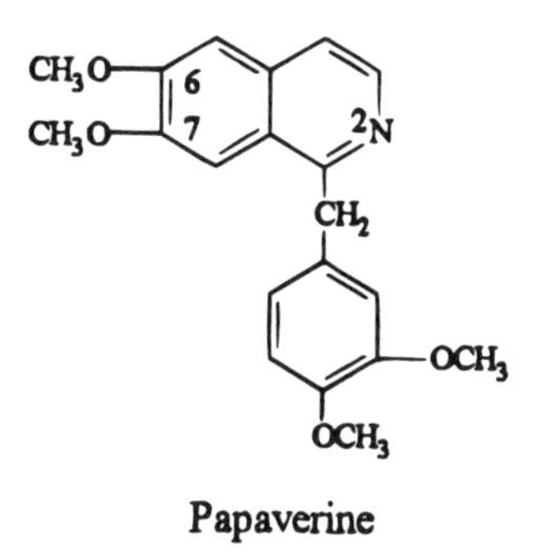

Papaverine

[19] The reader, on completion of this book, may wish to postulate an explanation for this phenomenon.

8.8. Neuromuscular Blocking Agents—Nicotinic Antagonists

These are compounds capable of blocking the neural stimulation of skeletal muscle fibers by the action of ACh, released at the NMJ, on cholinergic–nicotinic receptors. This receptor complex includes ion-conducting channels. By interacting with the receptor, released ACh activates it as a result of electrical events that include the development of postjunctional endplate potentials (EPP). This leads to a conformational change in the receptor's topography, resulting in the opening of the ion-conducting channels in the postsynaptic membrane. The EPPs cumulate to the level of triggering the action potential of the muscle fiber. AChE then terminates the process by hydrolyzing the ACh, preventing reoccupation of the receptor sites. It will be recalled that activation of the receptor required the binding of 2 ACh molecules, one on each α-subunit. It is thus apparent that interference (blockade) of this process will result in muscle relaxation.

There are several clinical applications for neuromuscular blockade. The most important by far is the induction of muscle relaxation during anesthesia for effective surgery. Without such drugs deeper anesthesia, requiring more anesthetic, would be needed to achieve the same degree of muscle relaxation; tracheal intubation would also be impossible because of strong reflex response to tube insertion. It is the decreased need for anesthetics, however, that represents increased surgical safety.

Neuromuscular blockers also find limited utility in convulsive situations such as those precipitated by tetanus infections and to minimize injury to patients undergoing electroconvulsive therapy. Manipulation of fractured or dislocated bones may also be aided by such drugs.

Historically, Bernard (1851, 1856) demonstrated that curare (from arrow poisons used by South American Indians) blocked nerve impulse transmission at the junction of the nerve and skeletal muscle. It was later shown that many alkaloids, such as morphine, atropine, nicotine, and even strychnine and brucine (the latter two normally causing convulsions), will become muscle relaxants when quaternized by methylation. It soon became apparent that a great many quaternary ammonium compounds qualitatively share the ability to produce neuromuscular blockade. In fact, the "onium" ion need not be a nitrogen atom. Thus sulfonium (R_3S^+), phosphonium (R_3P^+), and arsonium (R_3As^+) ions have been shown to be "urariform", although of lesser activity than ammonium ions.

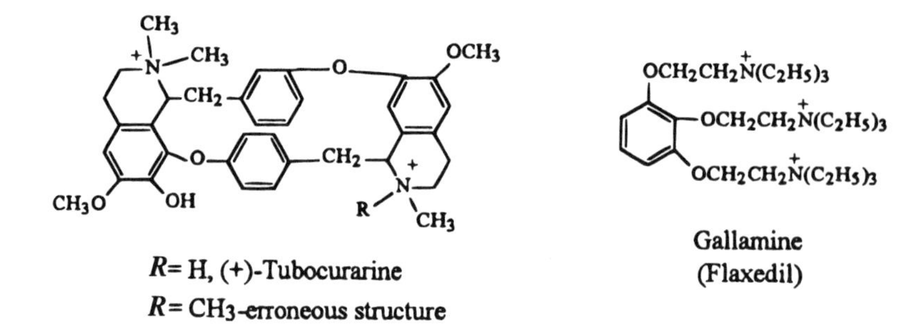

R= H, (+)-Tubocurarine
R= CH_3-erroneous structure

Gallamine
(Flaxedil)

The original name *curare* represented crude, but somewhat purified, extracts from various species of *Strychnos* and *Chondodendron*, which were in use into the 1960s[20] even after the pure alkaloid was already available. This pure alkaloid, now obtained from *C. tomentosum*, was first isolated in crystalline form from a botanically unidentified curare source by King (1935, 1948), who named it *d*-tubocurarine (TB) and proposed the preceeding structure except as a bis-quaternary compound wherein both nitrogens are quaternized. The fact that one of the nitrogens was actually tertiary (but mostly ionized, and therefore also cationic at physiologic pH) was not established until 1970. All early attempts at simplification and otherwise modifying TB to "improve" it were based on the assumption that two quaternized centers separated by an analogously optimum distance (now known to be about 10 Å, but then variously believed anywhere from 12 to 15 Å) was necessary. Still, almost all such attempts were successful, leading to several clinically useful drugs. Parenthetically, quaternizing the tertiary nitrogen will increase potency. Metocurine (dimethylchondrocurarine, Metubine) obtained by treating crude (+) TB with excess CH_3I, which methylates the nitrogen as well as the phenolic hydroxyls, is actually two to four times more potent. It has erroneously been referred to as dimethyltubocurarine.

It was realized that even though the various newly synthesized compounds having two quaternary nitrogens separated by the "proper" distance were clinically effective, they did not necessarily produce this effect by the same mechanism. These compounds are now seen as belonging to two categories: competitive or nondepolarizing and depolarizing. The first type are understood to *compete* with ACh for occupation of the receptor site, thus preventing the neurotransmitter from depolarizing the postsynaptic membrane. These drugs do not possess any intrinsic ACh-like action. (+)TB is the prototype of this group. The other agents share the feature of bulk and having the cationic nitrogen sterically shielded. Thus a *reversible* drug–receptor complex forms involving coulombic attraction of the two cationic areas to anionic centers on the receptors and hydrophobic van der Waals bonding with the fitting hydrocarbon portions of the molecule. Since no depolarization arises from this complexation, there is no initial stimulation. For these reasons any compound capable of depolarizing the postjunctional membrane of the motor endplate should be able to *antagonize* the blockade. ACh, if its concentration could be increased, would competitively reverse this blockade. The reversible AChE inhibitors, such as neostigmine, are therefore used to terminate the effects of the competitive blockers when the surgery is completed, or to reverse respiratory paralysis in cases of overdoses. An anticholinergic such as atropine is usually also given to prevent muscarinic stimulation.

In addition to gallamine, which does not quite fit the geometric parameters outlined, several other synthetic competitive antagonists have been clinically introduced. Pancuronium (PO) is a competitive blocker that represents a separation between two ACh molecules at the desirable distance (11.1 Å), interspaced by a *rigid* steroid framework. (The ACh components are outlined by thickened "bonds" along rings A and D.) The drug's developers attribute the fivefold higher potency of PO vis-a-vis (+)TB to the locked in configuration of the ACh moiety (in ring D). The companion desmethyl vecuronium (VO) is

---

[20] *Chondodendron* Tomentosum Extract, Purified, NND Intocostrin®.

of somewhat shorter duration, but has the identical mechanism. The lack of a quaternizing CH_3 at ring A does not affect this since that nitrogen will be 97% ionized at physiologic pH (pKa = 9).

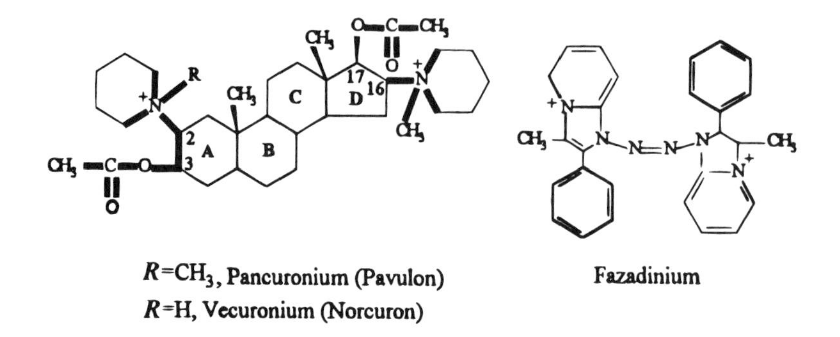

R=CH_3, Pancuronium (Pavulon)
R=H, Vecuronium (Norcuron)

Fazadinium

The ACh component in these two drugs offers an additional advantage over (+)TB, which undergoes little if any metabolism: facile metabolic inactivation by ACh-like hydrolysis. Serum half-life values bear this out: $t_{1/2}$-min-(+)TB = 173, PO = 107, VO = 71. The advantage is especially significant in patients with renal failure. Excretion rates expectedly decrease dramatically with both drugs: (+)TB = 330 min, PO = 257 min. However, with (+)TB, this means a long extension of neuromuscular blockade, whereas with PO two of three metabolites (17-desacetyl and 3,17-desacetyl) retain only 2% of the activity of the parent drug. The remarkable fact is that the metabolites have retained the *bis*-quaternary structure.

Fazadinium, a competitive blocker not available in the United States, represents a different approach to shortened duration of action by facilitating metabolism. Here, hepatic azoreductase cleavage (Chapter 3) will result in an innocuous metabolite and N_2. However, the rate of cleavage in humans is much slower than it is in some animals.

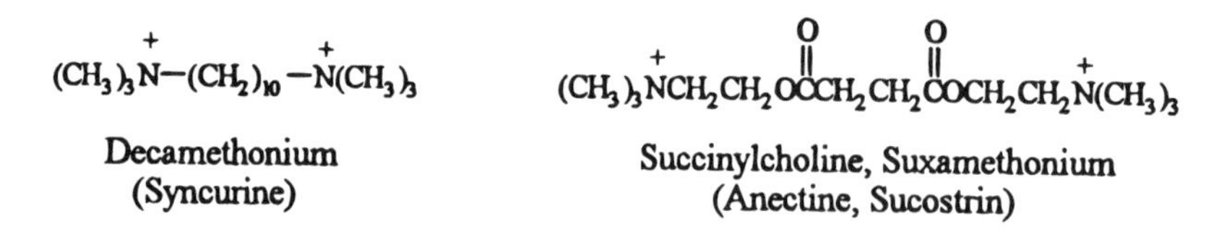

Decamethonium (Syncurine)

Succinylcholine, Suxamethonium (Anectine, Sucostrin)

In addition to the competitive, nondepolarizing compounds, the second category of neuromuscular blocking agents to be considered is the depolarizing drugs. In the early attempts at stripping (+)TB to its minimum a series of *bis*-trimethylammonium polymethylene compounds were synthesized and evaluated. This series of compounds structures varied only by the number of methylenes between the quaternary nitrogens. Optimum curariform activity occurred with the 10 CH_2 group, decamethonium. It was pointed out that decamethonium, unlike (+)TB, could not be antagonized by prostigmine; therefore, a fundamental biochemical difference must exist. This difference has since been established as a depolarization of the postsynaptic membrane in a manner similar to ACh. More recent research has established that such "narrow" compounds interact with the neuromuscular nicotinic receptors similarly to ACh itself; however, instead of being rapidly hydrolyzed by AChE they remain and block the ion channels from further activation by ACh. It should be pointed out that the *initial* effect with such drugs is a brief agonist action producing channel opening, then reverting to antagonist. Even though X-ray crystallography has established that the two nitrogens in decamethonium are 13.7 A apart, this represents the extended molecule in the crystal. In solution, an accordion like looseness is likely. Succinylcholine, which is the other depolarizing drug, shows variable distances in crystalline form (7.8–11.9 Å), depending on the nature of the negative counterion, yet will, in solution, assume flexibility.

The fact that succinylcholine is a choline ester makes it susceptible to cholinesterase hydrolysis, in this case serum or pseudocholinesterase. Rapid hydrolysis first to succinylmonocholine, whose blocking activity is only a little over 1% of that of the parent drug, results in a short duration of action, on the order of 2–5 minutes, depending on dose. For certain procedures this is a very desirable feature. In cases of extreme plasma cholinesterase deficiency due to genetic factors, diminished hepatic synthesis, or even exposure to organophosphates, the blockade can be increased and greatly prolonged. Since AChE inhibitors with antidote competitive blockers are not effective with depolarizing compounds, a hazardous situation can arise.

The fact that termination of full blockade is dependent on renal and biliary excretion mechanisms as well as enzyme levels, any of which can be decreased or damaged because of age, disease, or even genetic abnormalities, represents a therapeutic limitation. Of course, another is the lack of total specificity of this type of drug to the neuromuscular junction. These drugs can and do have an effect on ganglionic nicotinic receptors producing degrees of blockade of both sympathetic and parasympathetic neurons. Except for succinylcholine, these drugs can also blockade the parasympathetic neuron innervating the heart—the vagus nerve—causing increased heart rate and hypotension.

These limitations can be addressed by designing a compound on chemical principles—and empirically acquired knowledge—which will circumvent these difficulties.

The ultimate goal is the ideal neuromuscular blocking drug whose characteristics should include:

1. A *bis*-quaternary structure with a sufficient separation to produce a satisfactory level of blockade. Previous experience has shown this to be in a range of 10–11 Å (about 10–12 atoms).
2. A competitive antagonist would be preferred. Again, prior experience indicates that a bulky, large hydrocarbon environment around the cationic nitrogens effectively denies ACh access to the postsynaptic receptor areas on the motor endplate.
3. No or minimal action at cholinergic receptors other than nicotinic subtypes at the NMJ. Thus quaternization precludes access to centrally located receptors, and the proper distance separating the cationic areas in the drug minimizes activity on ganglionic receptors. The original work with the *bis*-onium compounds had shown that a shorter separation (five to six atoms) of nitrogen-carrying small substituents (i.e., CH_3) favors ganglionic blockade (see Chapter 12). It is not yet understood which structural and electronic factors lead to vagal blockade.
4. Molecular designs that would lead to efficient metabolic degradation and/or excretion patterns totally nondependent on enzyme catalysis.[21] Such a drug would always exhibit

---

[21] A drug Albert (1985) aptly describes as "self-canceling."

short duration and rapid recovery from blockade irrespective of renal or hepatic dysfunction or pharmacogenetic effects.

The drug PO incorporates aspects of the first three items in the preceding shopping list with some success. Acetate hydrolysis, as the metabolic inactivation route, however, fails to meet the requirements of item 4. The introduced atracurium besylate (Tracrium, Fig. 8-15), however, does meet the requirements. The following discussion then will pertain to the chemical logic that went into the design of atracurium.

8.8.1. The Organic Chemistry of Atracurium

The synthesis (Fig. 8-15) of such a complex molecule is surprisingly straightforward. It involves the formation of a diacrylate ester from the corresponding alcohol and the acid chloride of acrylic acid. The acrylate, an α, β-unsaturated ester, then undergoes the classic Michael addition, which is where a nucleophilic species, in this case the secondary amine of tetrahydropapaverine, attacks the activated β-carbon (Eq. 8.9). The net result is the addition of -N-H across the double bond.

Quaternization can be done with several methylating reagents. However, since the product would be used parenterally, a soluble salt was necessary to dissolve the dose in a small volume. The usual methyliodide produced a salt of insufficient solubility. The more hydrophilic benzenesulfonate anion (besylate) proved to be satisfactory.

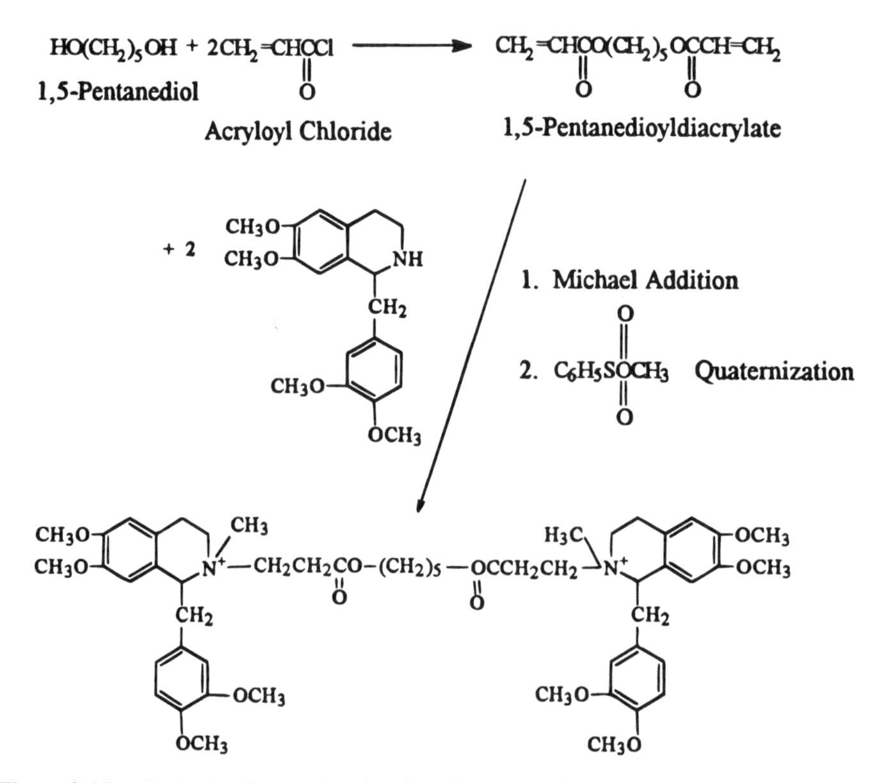

Figure 8-15. Synthesis of atracurium besylate (benzene sulfonate salt). (From Stenlake, 1985.)

The choice of the Hofmann elimination is somewhat more involved. This classic reaction, discovered by Hofmann (1851), shows that quaternized amines on treatment with base (or simply heated in solution if the anion is ^{-}OH) will eliminate the smallest alkyl group (larger than methyl) that is sterically least hindered, as an olefin, leaving a tertiary amine (Eq. 8.10).

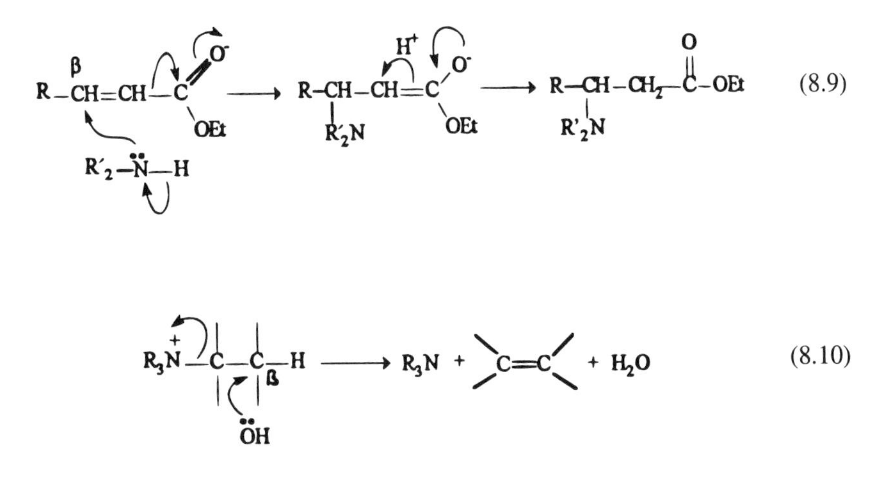

The positive nitrogen causes an electron "drift," which weakens nearby C–H bonds. The nucleophilic OH^- then abstracts the least hindered β-proton, resulting in C–N fission to the olefin and a molecule of water.

This reaction, usually carried with strong base and at elevated temperatures, was an important tool in structure determination of alkaloids in the late nineteenth and early twentieth century before modern instruments were known when repeated (i.e., remethylating to a quaternized salt, then elimination of the second *R* group, and so on).

The ability to build this mechanism into a drug molecule so it would "autometabolize" under physiological conditions (37°C, pH 7.4) presented both a temptation and a problem. The solution was the introduction of an additional electron-withdrawing moiety on the same β carbon that is already subject to nucleophilic attack, as has been explained. This would weaken the β-C–H bond even more, further promoting fission under less rigorous conditions of pH and temperature. The ester group *beta* to the quaternized nitrogen serves that function admirably. Atracarium has a plasma elimination half-life in humans of 21 minutes, which is essentially unchanged in cases of renal failure (18 minutes). Its potency is more than twice that of (+)TB!

Age does not seem to alter the pharmacokinetics. Tests in patients with genetic deficiency of pseudocholinesterase also showed no significant difference in $t_{1/2}$. The half-life of succinylcholine, on the other hand, increased from 2.6 minutes to 4 hours in patients devoid of this enzyme. It appears that the ester group activation of the Hofmann elimination reaction mutually promotes ester hydrolysis as well. The goal of *chemical-* over enzyme-catalyzed metabolism in vivo seems to have been achieved. The possibility that ester hydrolysis may be catalyzed by nonspecific esterases is likely; that plasma cholinesterase is *not* utilized is established.

The design, synthesis, and development of atracurium as outlined illustrates decisively that application of chemical "know-how" can succeed (Fig. 8-16).

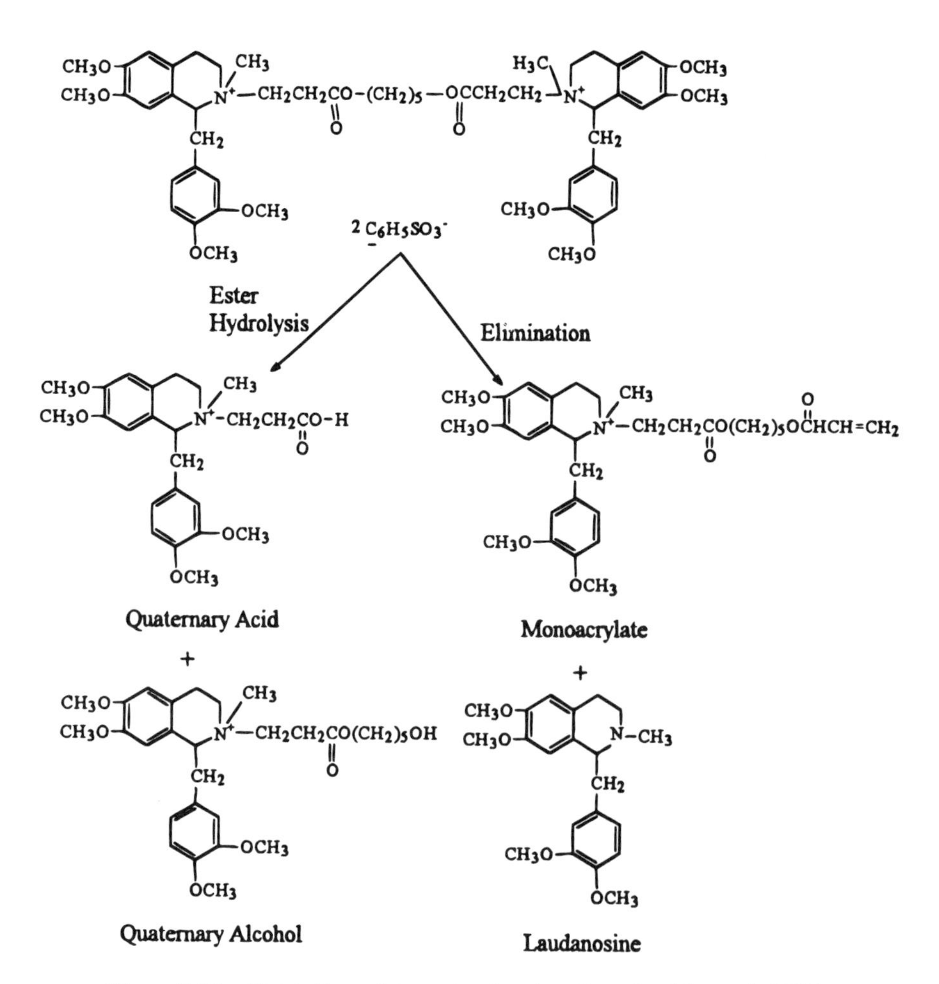

Figure 8-16. Metabolism of atracurium besylate (from Stenlake et al., 1983).

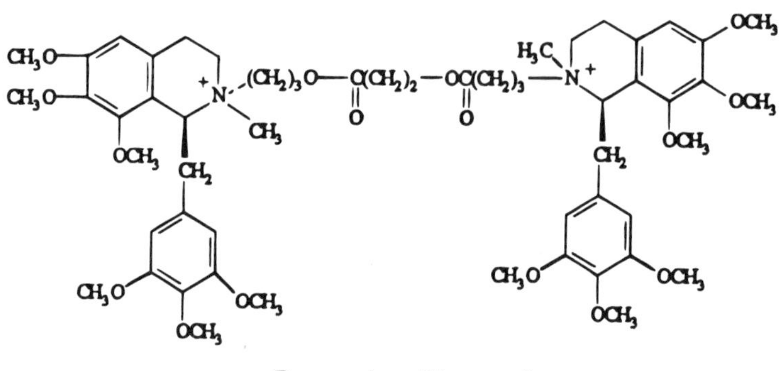

The studies done to determine the structural parameters within which the sought properties of short duration and high potency would exist were later also utilized to develop another homolog—doxacurium, which was introduced into clinical practice (1991). Its main attribute is a long duration of action, which is necessary for especially lengthy surgical procedures.

Both chain length and chain structure were investigated. Extending the polymethylene chain until the two quaternary nitrogens were separated by 14 atom units increased potency but did not significantly alter the duration of action. As discussed earlier, since it is apparent that duration of action is dependent on how facile the Hofmann elimination is, it should be possible to decrease that facility by lengthening the distance from the quaternary locus to the ester carbonyl—in the case of doxacurium it changes from two CH_2 groups to three. This increased separation was shown to decrease the propensity for both Hofmann and the ester hydrolysis reactions. A three- to fourfold increase in recovery time (from onset of blockade) resulted. Parenthetically, by decreasing the polymethylene chain from five CH_2 in atracurium to two, the overall separation between the two cationic nitrogens was not altered; neither was the potency of the drug reduced. In fact, the additional methoxy groups on both the benzyl groups and the aromatic portion of the tetrahydroisoquinoline rings was likely to enhance it.

Doxacurium, the result of this research, is a potent, long-acting noncumulative neuromuscular blocker with good cardiovascular stability. That is, it does not alter the heart rate, blood pressure, or cardiac output. No active metabolites appear in the urine or bile. In fact, the drug is essentially excreted unchanged.

References

Albert, A., *Selective Toxicity*, 7th Ed., Chapman and Hall, New York, 1985, p. 96.

Armstrong, P. D., Cannon, J. G., *J. Med. Chem.* **13**:1037, 1970.

Bassett, J. R., Cairncross, K. D., Hacket, B., Story, M., *Br. J. Pharmacol. Chemother.* **37**:69, 1969.

Bebbington, A., Brimblecomb, R., and Shakeshaft, D. *Brit. J. Pharmacol. Chemother.*, **26**:56, 1966.

Beers, W. H., Reich, E., *Nature* **228**:917, 1970.

Bernard, C., *Royal Soc. Biol.* **2**:195, 1851; *Compt. Rend. Acad. Sci. Paris* **43**:825, 1856.

Bowman, W. C., Rand, M. J., *Textbook of Pharmacology*, 2d ed., Blackwell Scientific Publications, London, 1980.

Bruice, T. C., Schmir, G.L., *J. Am. Chem. Soc.* **79**:1663, 1957.

Cothia, C., Paulin, P., *Nature* **223**:919, 1969.

Dale, H., *J. Pharmacol.* **6**:147, 1914.

Goldstein, A., Aranow, L., Kalman, S. M., *Principles of Drug Action: The Basis of Pharmacology*, John Wiley & Sons, New York, 1974, p. 26.

Ing, H., *Science* **109**:264, 1949.

Kier, L., *Mol. Pharmacol.* **3**:487, 1967.

King, H., *J. Chem. Soc.* 1381, 1935; 265, 1948.

Kushminski, G., Lullman, H., *A Short Textbook of Pharmacaology* (in German), Thieme, Stuttgart, 1967.

Robinson, R., *J. Chem. Soc.* **111**:762, 1917.

Schindler, V., Rush, D. K., Fielding, S., *Drug Dev. Res.* **4**:567, 1984.

Stenlake, J. B., *Pharm. J.* **229**:116, 1982.

Stenlake, J. B., in *Medicinal Chemistry. The Role of Organic Chemistry in Drug Research*, Roberts, S.M., Price, B. J., Eds., Academic Press, New York, 1985, p. 159.

Stenlake, J. B., Waigh, J., Dewar, G. H., et al., *Eur. J. Med. Chem.-Chim. Ther.* **16**:515, 1981.

Stenlake, J. B., Waigh, R. D., Urwin, G. H. *et al.*, *Br. J. Anaesth.* **55**, Suppl. 3, 1983.

Welsh, J. H., Taub, R., *J. Pharmacol. Exp. Ther.* **99**:334, 1950; ibid. **103**:62, 1951.

Willstatter, R., *Berichte* **34**:129, 3163, 1901; *Annalen* **317**:307, 1901.

Suggested Readings

Coyle, J. T., Price, D. L., DeLong, M. L., Alzheimer's Disease: A Disorder of Cortical Cholinergic Innervation. *Science* **219**:1184, 1982.

Dahlbom, R., Stereoselectivity of Cholinergic and Anticholinergic Agents, in *Stereochemistry and Biological Activity of Drugs*. Areines, E. J., Soudijn, W., Timmermans, P.B.M.W.M., Eds., Blackwell Scientific Publications, Oxford, 1983, pp. 127–42.

Fox, S. I., *Human Physiology*, 4th Ed., William C. Brown Publishers, Dubuque, Iowa, 1993.

Lamble, J. W., Ed., *Towards Understanding Receptors*, Elsevier/North Holland, New York, 1981.

Lamble, J. W., Ed., *More About Receptors*, Elsevier Biomedical Press, New York, 1983.

Shepherd, G. M., *Neurobiology*, 2nd Ed., Oxford University Press, New York, 1988.

Taylor, P., Insel, P. A., Molecular Basis of Pharmacological Selectivity, in *Principles of Drug Action. The Basis of Pharmacology*, 3d Ed., Pratt, W. B., Taylor, P., Eds., Churchill Livingston, New York, 1990, pp. 1–102.

CHAPTER

9

Drugs Affecting Adrenergic Mechanisms

9.1. Adrenergic Concepts and Synthesis

The biochemical properties of the sympathetic (adrenergic) system are better understood than some of the ramifications of its physiological functions. As already mentioned, post-ganglionic sympathetic nerve endings elaborate norepinephrine (NE), which is eliminated from storage vesicles within the nerve terminal. Another important biogenic catecholamine[1] is epinephrine (EP, adrenalin), which is produced by N-methylation of NE in the adrenal medulla, a gland on top of the kidneys. Unlike NE, EP is not considered a neurotransmitter. It can, however, interact with the same receptors NE does. EP is believed to be the activator molecule preparing the body for the so-called fight or flight syndrome. The third biogenic amine is dopamine (DA). DA was at one time believed to have as its only function being a chemical precursor for NE and EP (see Fig. 9-1) because of its low levels in peripheral tissues. Its much higher concentrations in central areas of the brain, however, soon led to its evaluation as a major neurotransmitter there. Evidence of dopaminergic fibers and receptors are now well established.

All the steps up to NE outlined in Figure 9-1 occur within the adrenergic neurons and in certain neurons of the CNS. The requisite enzymes to catalyze the various reactions are produced in the cell body of the neuron. It is in the terminal axons, however, that the actual synthetic steps occur. In sympathetic and certain other neurons, the synthesis proceeds to NE; in other central neurons, the dopaminergic neurons, the synthesis stops at DA. Concentrations of DA are particularly high in the caudate nucleus. At this point the steps and enzymes of the biochemical pathways will be considered. Compounds are available to interfere with the various steps, both the synthetic and the degradative. The metabolism reactions will be considered later in this chapter.

[1] Catechol is *ortho*-dihydroxybenzene, hence the name.

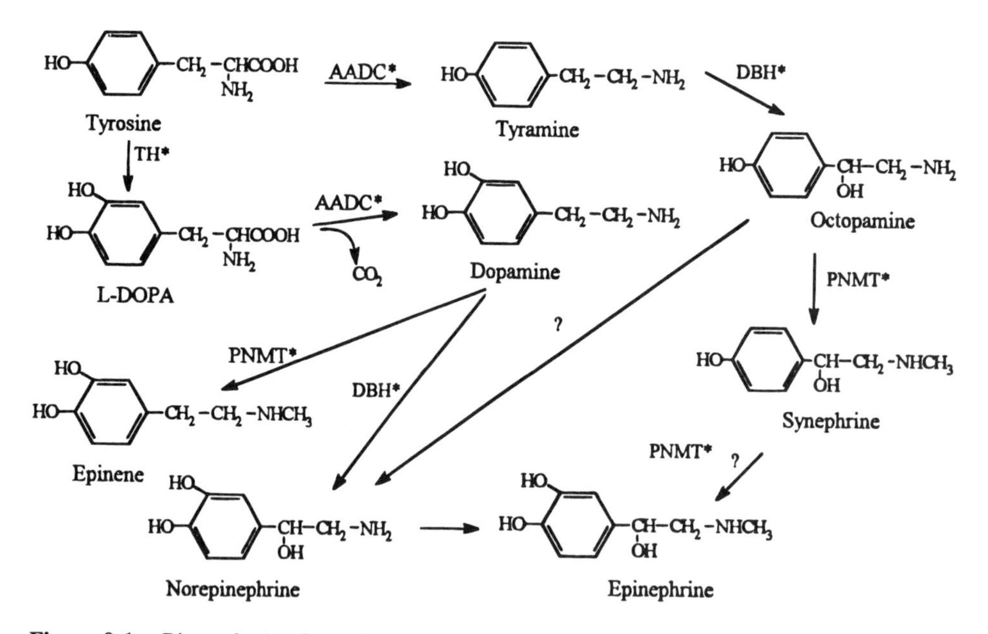

Figure 9-1. Biosynthesis of catecholamines. *Denotes enzyme in transformation: AADC = aromatic *L*-amino acid decarboxylase; COMT = catechol-*o*-methyl transferase: DBH = dopamine-B-hydroxylase; MAO = monoamine oxidase; PNMT = phenylethanolamine-N-methyl transferase; TH = tyrosine hydroxylase.

Many drugs in use known to interfere with a particular synthetic or degradative step of a particular neurotransmitter can have their clinical properties ascribed totally to this blockade or stimulation. None of the enzymes involved is completely specific for the reactions described. Although susceptible to attack by these enzymes, some drugs (e.g., structural analogs of NE) may owe part of their pharmacological properties to other factors. The actual picture, then, is likely to be more complex than will be described; in some cases, it may not be fully understood.

Tyrosine is generally considered the starting point, although in the unlikely event of tyr deficiency, phenylalanine hydroxylase can hydroxylate phe to tyr.[2] Tyrosine hydroxylase (TH) is the rate-limiting enzyme in this pathway. Its addition of the 3-OH yielding L-3, 4-dihydroxyphenylalanine (L-DOPA) requires O_2, tetrahydropteridine, and Fe^{2+} as cofactors. Because this is the rate-limiting step, inhibition of this enzyme (rather than subsequent steps) is the most likely way to reduce NE, DA, and EP levels significantly. Particularly effective are α-methyltyrosine analogs, including esters, especially those containing an iodine atom in the benzene ring. The latter, such as α-methyltyrosine (metyrosine, Demser), are particularly effective. This drug has the ability to reduce EP, NE, and DA dramatically, making it particularly useful in the management of malignant hypertension and, preoperatively, in pheochromocytoma. The latter is a chromaffin cell tumor (usually benign) that produces and spills copious amounts of NE and EP into the circulation. This produces severe hypertension and associated symptoms.

[2] A deficit of that enzyme results in the congenital disease phenylketonuria, a cause of mental deficiency.

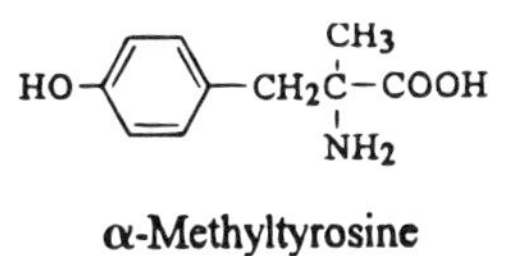

α-Methyltyrosine

The next step is the decarboxylation of L-DOPA to DA. The enzyme catalyzing this is L-aromatic amino acid decarboxylase (AADC).[3] The high activity of the enzyme explains the negligible quantities of L-DOPA found in tissues. The enzyme is not specific to L-DOPA as once thought, but decarboxylates various L-aromatic amino acids, including tyrosine, histidine, 5-hydroxytryptophan, and others. The cofactor for this enzyme is pyridoxal (the aldehyde form of pyridoxine, vitamin B_6). A "suicide substrate" inhibitor for AADC is available as the highly specific *rac-* α-monofluoromethyldopa (MFMD). The enzyme "accepts" the compound as a substrate and decarboxylates the compound into a highly reactive tertiary carbonium ion that then alkylates the enzyme, irreversibly inactivating it. As a chemical probe to study catecholamine steady-state levels and other research, MFMD has been highly useful (Eq. 9.1).

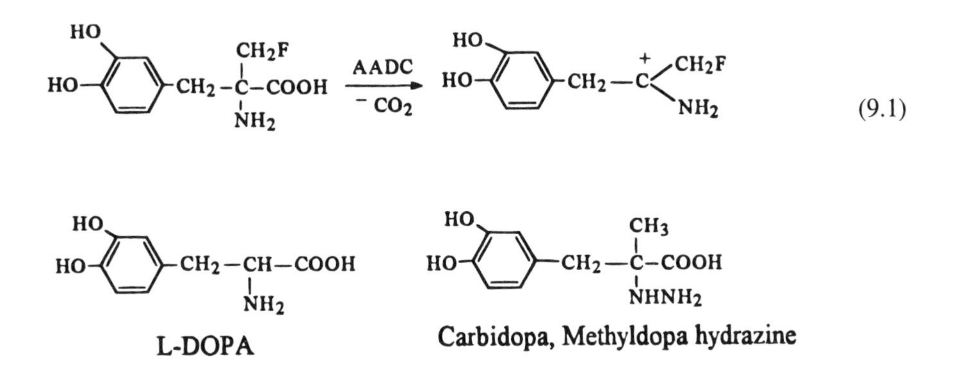

L-DOPA

Carbidopa, Methyldopa hydrazine

Parkinson's disease, and the modest benefits of treating its early stages with centrally acting anticholinergic agents, was briefly discussed in the previous chapter. In simplest terms Parkinsonism can be characterized as entailing a DA deficiency in the brain; a progressive neurological disorder of the extrapyramidal system. The deficiency is concentrated in the caudate nucleus and putamen of the brain. The pathology can be traced to certain large neurons in the *substantia nigra*, whose degeneration is directly related to DA deficiency.

Logic would dictate that increasing brain levels of DA should ameliorate symptoms permanently (a cure?), or at least temporarily. Direct parenteral DA administration is useless since the compound does not penetrate the blood–brain barrier (BBB) (however, see Chapter 10 for cardiovascular applications). It was shown that oral dosing with L-DOPA (levodopa, Dopar) could successfully act as a pro-drug to the extent it entered the brain (on a specific carrier) and was then decarboxylated to DA there. The clinical results in terms of decreased tremors and rigidity were dramatic. However, there were complications

[3] Once called L-Dopa decarboxylase.

conspiring to produce intense side effects, including nausea and vomiting, that were presumably due to chemoreceptor trigger zone stimulation by large amounts of DA produced peripherally. Neurologic, respiratory, and cardiovascular effects are also produced. Many of these adverse effects were the result of the high doses (up to 8 g daily) needed to achieve the desired results. The single most significant reason for this situation is the relatively high peripheral levels of decarboxylase enzyme compared with brain concentrations. Thus 95% of a given oral dose was converted to DA before reaching the brain to be decarboxylated there.

One way to circumvent this problem would be to administer concomitantly an agent able to inhibit the AADC peripherally, but is a compound that will not penetrate the BBB itself. Several such inhibitors have been developed experimentally. The compound in clinical use, α-methyl-α-hydrazino-3,4-dihydroxyphenylpropionic acid (α-methyldopahydrazine), is officially named *carbidopa*. Carbidopa is not generally available for clinical use alone but as a 1:10 and 1:4 combination product Levodopa–Carbidopa (Sinemet). This product alleviates some clinical difficulties associated with L-DOPA. Much lower doses of levodopa (by 75%) now achieve the desired results.

A different approach to Parkinsonism would be the use of inhibitors of monoamine oxidase (MAO), the enzyme that oxidatively deaminates catechol and other monamines, including DA, NE, and EP (see subsequent discussion of metabolism and other uses). Such a drug would tend to preserve brain DA and be effective itself and/or potentiate levodopa. Early attempts at such combinations produced hypertension. Subsequently, with the discovery that two MAO isozymes, A and B, exist (more discussion later), it was realized that the drugs tested were nonselective. It is now understood that MAO B primarily metabolizes DA. Selegiline (Eldepryl) appears to be a specific type B MAO inhibitor, and does extend the duration and increase the efficacy of levodopa. The drug is promising, probably as an adjunct to levodopa or in levodopa refractory patients.

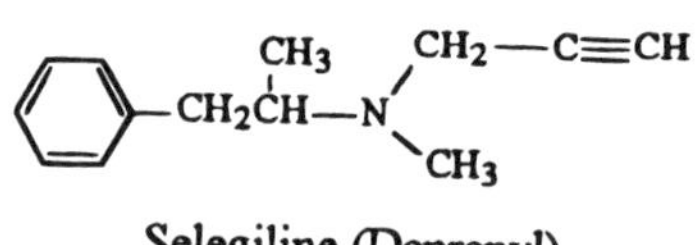

Selegiline (Deprenyl)

An even more satisfactory approach may be the use of compounds that act as dopamine *agonists* themselves and do not rely on endogenous or exogenous supply of DA. Such drugs then would stimulate central DA receptors, now believed to be subtype D_2 (see later).

Apomorphine (APO) (Fig. 9-2), a long-known emetic prepared by drastic (conc. HCl, 140 degrees) rearrangement of the morphine molecule (Chapter 5) and devoid of analgesia, was first reported to have marked anti-Parkinson activity in 1970. On examining the structure of APO, the DA moiety (actually, epinine, the N-methyl derivative, which is equally potent) is clearly discernible (C-11, 12, 7, 6N). The dopaminergic agonist properties of APO are not in doubt. Its clinical utility is greatly hampered by its emetogenic intensity, however. The finding nevertheless gave impetus to the idea of "rigidizing" DA into a ring system in the proper configuration for maximal efficacy. 2-Aminotetralines with properly placed OH groups was one possibility. In a synthetic study of 77 aminotetralines (McDermed et al., 1975), it was shown that compound *a* (Fig. 9-2) represented the optimum structure, being 50 times more potent than APO. Structure *b*, though weaker, was still

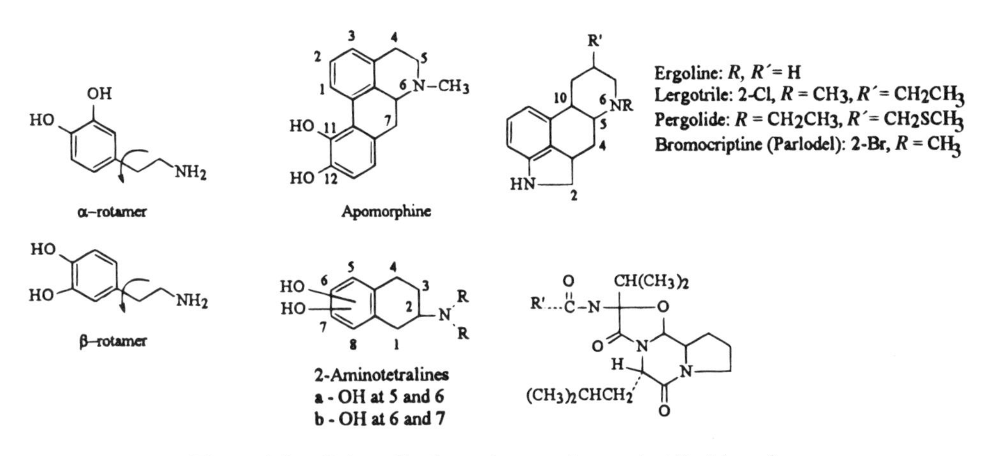

Figure 9-2. Polycyclic dopamine agonists against Parkinsonism.

active. Based on this and additional tetralin compounds it was suggested that the DA moiety in each is locked in a different configuration, corresponding to two conformations of DA called *rotamers* alpha and beta. Furthermore, the α-rotamer is the preferred conformation of DA receptors in the caudate (D_2).

Bromocriptine (Parlodel) (Fig. 9-2) is prepared from α-ergocryptine by bromination with N-bromosuccinimide. Bromination of the ergot alkaloid greatly enhances its dopaminergic effectiveness. It and other semisynthetic ergot alkaloids are considered ergoline derivatives (Fig. 9-2). In addition to its anti-Parkinson activity the drug is also used to treat galactorrhea, female infertility, and acromegaly. All these properties, and its ability to decrease prolactin secretion, are believed to stem from dopamine receptor (D_2) stimulation. Several other derivatives appear promising as anti-Parkinson drugs. They are lergotrile and pergolide (Fig. 9-2). Pergolide, which appears to have D_1 and D_2 receptor agonism (unlike the others), is longer acting and more potent than bromocriptine. Its effectiveness, even when combined with levodopa–carbidopa, diminishes after 6 months.

Even though many investigators consider the ergoline derivatives to be dopaminergic because of the rigid DA component (Fig. 9-2: ring A plus atoms 10, 5, and 6), a proposal suggests that it is the pyrroloethylamine portion (ring B plus atoms 4, 5, and 6,) that correspond to the rigid DA portion of APO and is therefore responsible for the DA agonist activity citing stereochemical reasons. The antiviral drug amantadine (Chapter 7, Fig. 7-17) was serendipitously discovered to relieve Parkinson symptoms and has been used as an adjunct drug since. The mechanism may include augmenting DA release from storage sites, delaying DA reuptake and even some anticholinergic activity. It appears less effective than levodopa. Effectiveness, however, rarely lasts longer than 6 months.

The next step in the catecholamine biosynthesis is side–chain hydroxylation of DA to NE. The enzyme dopamine β-hydroxylase (DBH) catalyzes this reaction. This enzyme, like TH, is a mixed-function oxidase utilizing molecular O_2, in this case to add the OH onto the β-carbon of the phenelthylamine side chain. DBH is a Cu^{2+}–containing enzyme that, with ascorbic acid (Vitamin C) as a cofactor, carries out the necessary electron transfers.

DBH acts on other β-phenethylamines as well. For example, it oxidizes tyramine to octopamine (present at low levels in mammals) and epinine to EP (Fig. 9-1). DBH inhibitors are various phenethylamine isosteres such as hydrazines and oxyamines, which act as ligand imposters. In addition, metal chelating sulfur–containing compounds, by complexing with the Cu^{2+}, remove the enzyme's prosthetic moiety, thus preventing it from acting. Such inhibitions can result in decreased NE and increased DA tissue levels. The effects are fleeting, however, and of no clinical utility.

The last step is the conversion of NE to EP. This reaction is catalyzed by phenylethanolamine-N-methyl transferase (PNMT) and occurs primarily in the brain and heart tissue.

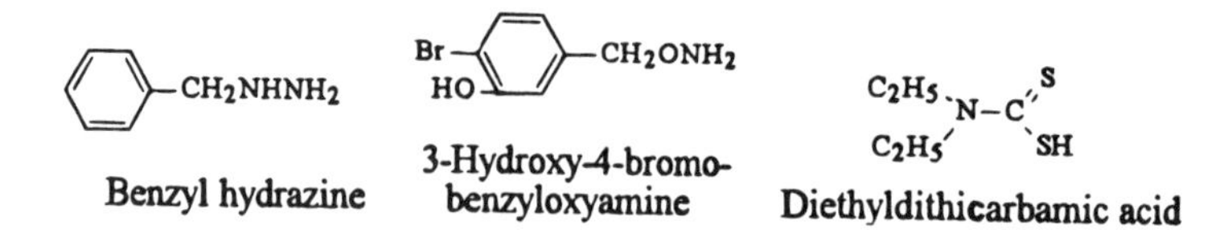

Benzyl hydrazine — 3-Hydroxy-4-bromo-benzyloxyamine — Diethyldithicarbamic acid

As the name implies, PNMT is not specific to NE. It will, however, also convert DA to epinine and octopamine to synephrine (Fig. 9-1). The source of the methyl group is S-adenosylmethionine (Chapter 3) and not, as once believed, methyltetrahydrofolate. The synthesis of PNMT in chromaffin cells of the adrenal glands is stimulated by adrenocortical steroids.

The minor biogenic amines such as epinine, octopamine, synephrine, and tyramine represent a tiny fraction of the total in the tissues synthesizing catecholamines under normal circumstances. However, under certain conditions their levels may increase, such as when MAO is inhibited. Some of the ramifications of this will be considered subsequently.

Once a neurotransmitter molecule is synthesized in the neuronal cells, it does not remain just diffusing throughout the cytosol fraction. The major portion of DA, for example, is concentrated and stored in vesicles or "granules" in the nerve endings, where it most likely is held by ionic interaction (DA would be mostly cationic) with anionic phosphate groups of ATP. Concentration of catecholamines into vesicles is an ATP-driven process linked to the vesicles, membrane potential, or proton pump. DBH in the vesicles converts DA to NE. It should be pointed out that since neither the uptake process nor the substrate requirements of DBH are very specific, entrance of other biogenic amines (e.g., tyramine), or a drug such as amphetamine, can also do so. The former can be converted to octopamine, and both substances, of course, compete for storage space with the endogenous amine DA. Drugs capable of irreversibly inhibiting this vesicular uptake mechanism produce an intense intraneural catecholamine depletion. Reserpine is such a drug.

The soluble contents of the vesicle are released when the axonal action potential reaches the terminal. This opens ionic channels into the terminal, permitting an influx of Ca^{2+}. The elevated Ca^{2+} level then induces a fusion of the storage vesicles with the neuronal membranes, leading to a spillage of NE, ATP, and DBH into the synaptic cleft. The actual egress is believed to be by exocytosis. The release of NE following neuronal uptake of such substances as tyramine and amphetamine is a much simpler process. As mentioned, they displace the endogenous amine from the vesicles into the cytosol, from which they apparently leak out into the extraneural area. Whether by action potential or displacement, the released NE stimulates postsynaptic sympathetic receptors. For this reason such drugs or amines can be referred to as *indirect sympathomimetics* (more on this later).

9.2. Catabolism

Just as the neurotransmitter ACh had to be removed once released across the synaptic cleft and having interacted with its postsynaptic receptor, so must the catecholamine transmitters. Termination of transmitters here is not by a single method; but rather it involves primarily a reuptake mechanism back into the adrenergic axons and into nearby effector cells. As mentioned, this uptake process is not limited to the endogenously or exogenously supplied neurotransmitters DA, NE, and EP, but will occur with other sympathomimetic amines of biogenic or synthetic origin as well. It was demonstrated that two uptake mechanisms exist. One is a process occurring at low NE levels, whereas the other is a threshold mechanism that only becomes saturated at very high NE levels, and is probably extraneural.

The neuronal reuptake process, sometimes called *Uptake* 1 (U_1), is an energy-consuming active transport mechanism requiring ATP, Na^+, and K^+ in low concentrations. In the case of noradrenergic neurons the transport process appears selective for NE, while the carrier in dopaminergic neurons is selective toward DA. This selectivity in reuptake between noradrenergic and dopaminergic neurons allows for selective inhibition of reuptake by different drugs, as will be seen. In addition, by binding to the carrier protein certain other phenethylamine-type drugs (e.g., amphetamines) will be concentrated in these neurons and may compete with the catecholamine reuptake process.

The catecholamine neurotransmitter uptake rate is proportional to low concentrations. Once taken into the axonal cytosol it is again stored in the synaptic vesicles and is "reused." Prior to entering the vesicles, however, the amines are subject to destruction by mitochondrial-associated MAO in the cytoplasm. It is predictable that when vesicle storage is inhibited by drugs such as reserpine, most of the NE or DA will be destroyed. In the presence of MAO-inhibiting drugs such as phenelzine, NE will simply accumulate. In general, the catecholamine reuptake process will inactivate (i.e., remove) 60% or more of NE, depending on the tissue involved.

A second, nonchemical, NE-terminating process is simply diffusion away from the synaptic area. The quantitative significance of this is difficult to gauge since diffusion also brings the NE molecules into the clutches of the extraneural metabolism enzymes catechol-O-methyltransferase (COMT) and MAO. Here the catecholamines are taken up, Uptake 2(U_2), into extraneural tissue and degraded. Unlike the neuronal reuptake process (U_1), this does not represent amine preservation.

COMT is widely distributed in the body and is primarily found on the outer plasma membrane of cells. Even red blood cells have the enzyme. As with PNMT, S-adenosylmethionine is the source of the methyl group. The enzyme's specificity is toward the catechol nucleus; the side chain chemistry seems irrelevant. It is only the *meta*-OH that is methylated. Magnesium is a requirement.

MAO is a more fascinating enzyme. It is also widely distributed throughout the body tissues. However, unlike COMT, it is also associated with neuronal mitochondria. This is significant, because catecholamines in the cytosol of the neuronal cell are not immune from its oxidative ravages unless protected within the synaptosomal vesicles. In addition, unlike COMT, inhibitors of MAO have significant effects on amine levels with physiological consequences. Several such MAO inhibitors have pharmacological applications in the treatment of depression and hypertension, as will be seen.

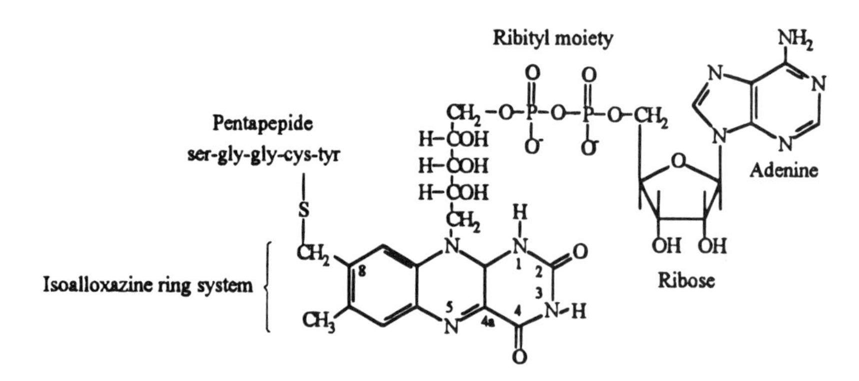

Figure 9-3. Possible structure of monoamine oxidase.

MAO catalyzes the oxidative deamination of catecholamines, 5-hydroxytryptamine (serotonin), and other monoamines, both primary such as NE, and secondary such as EP. It is one of several oxidase-type enzymes whose coenzyme is the flavin-adenine-dinucleotide (FAD) covalently bound as a prosthetic group (Fig. 9-3). The isoalloxazine ring system is viewed as the catalytically functional component of the enzyme. In a narrow view N-5 and C-4a is where the redox reaction takes place (i.e., $+H^+$, $+1e^-$ or $-H^+$, $-1e$), although the whole chromophoric N-5–C-4a–C-4–N-3–C-2–N-1 region undoubtedly participates. Figure 9-3 is a proposed structure of MAO isolated from pig brain (Salach et al., 1976).[4]

The steps of the oxidative deamination are probably still not agreed on, although several ways have been suggested. Equation 9.2 implies the removal of a proton from the nitrogen atom, which in turn reduces FAD. Molecular oxygen oxidizes the product to an intermediate hydroxylamine, which apparently "spontaneously" divests itself of ammonia, thus yielding the aldehyde product. Equation 9.3 is possibly a more defensible mechanism. Here, removal of two protons in sequence from the α-carbon and amino groups, oxidative steps that reduce FAD, leads to the imine formation. On release from inside binding the unstable Schiff base is easily hydrolyzed by H_2O to give ammonia and the aldehyde. It should be understood that the oxygen atom derives from molecular oxygen, which is in turn reduced to peroxide. Thus Eq. 9.4 more accurately represents the overall reactions.

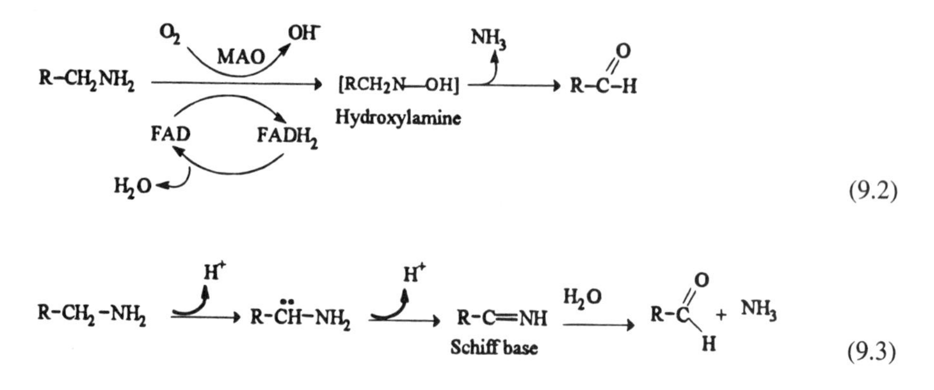

[4] See also Chapter 2 for discussion of irreversible MAO inhibition by acetylenic amines such as pargyline and deprenyl.

$$RCH_2NH_2 \xrightarrow{2H^+} RC{=}NH \xrightarrow[O_2]{H_2O} R\text{-}\overset{O}{\overset{\|}{C}}\text{-}H + NH_3 + H_2O_2 \qquad (9.4)$$

The aldehyde metabolites formed are generally rapidly oxidized to the corresponding carboxylic acids. To some extent, in certain peripheral tissues, but particularly in the CNS, reduction of the aldehyde MOPGA by aldehyde reductase to the glycol MOPEG becomes significant (Fig. 9-4). It is suggested at this point that the reader become familiar with the metabolic schemes shown for EP–NE (Fig. 9-4) and DA (Fig. 9-5), which will be referred to in subsequent discussion. Even though it is not shown in the schemes, the major metabolites found in the urine and CSF may, at least in part, be glucuronide or sulfate conjugates, particularly of the phenolic products.

It should be appreciated that an understanding of biogenic amine metabolism can be important in the management of certain drug therapies and may even aid in diagnosis. For example, CSF levels of HVA (Fig. 9-5), the major metabolite of brain DA, is understandably low in Parkinsonism. Changes resulting from drugs may indicate the degree of therapeutic success. Similarly, the CSF levels of MOPEG (Fig. 9-4) are indicative of NE levels, which can be related to the intensity of depression and the degree of improvement that can be expected with antidepressants before the clinical symptoms improve, which may take weeks.

MAO is now understood to exist as at least two types: MAO-A, with apparent specificity for serotonin and NE, and MAO-B, which exhibits specificity for a majority of phenethyl and benzylamines. An analogous situation exists with respect to MAO

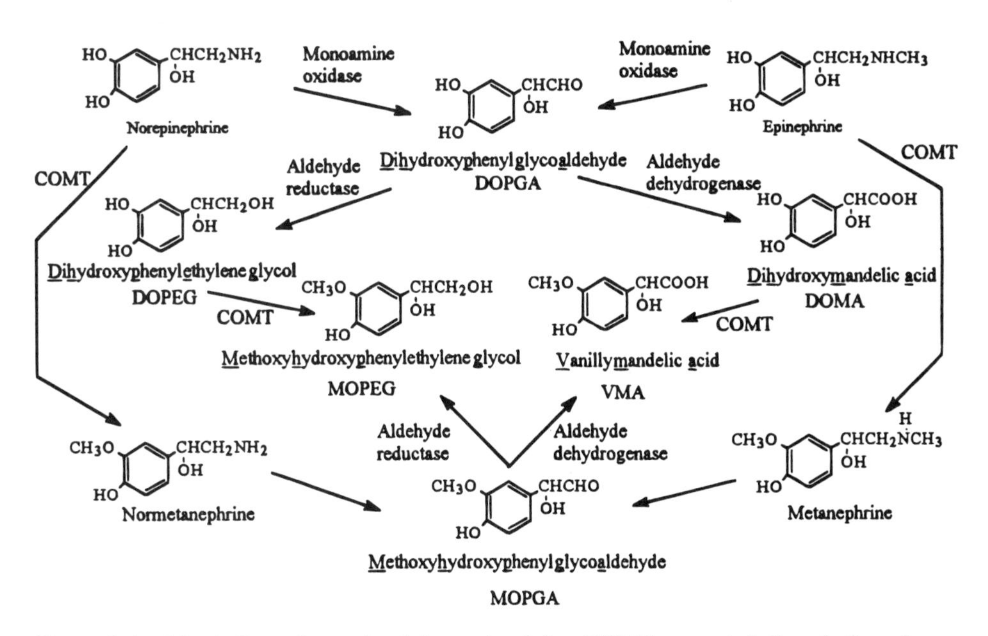

Figure 9-4. Metabolism of norepinephrine–epinephrine (COMT = catechol-O-methyltransferase; acronyms derived from underlined letters, except H = O, i.e., hydroxy = oxy).

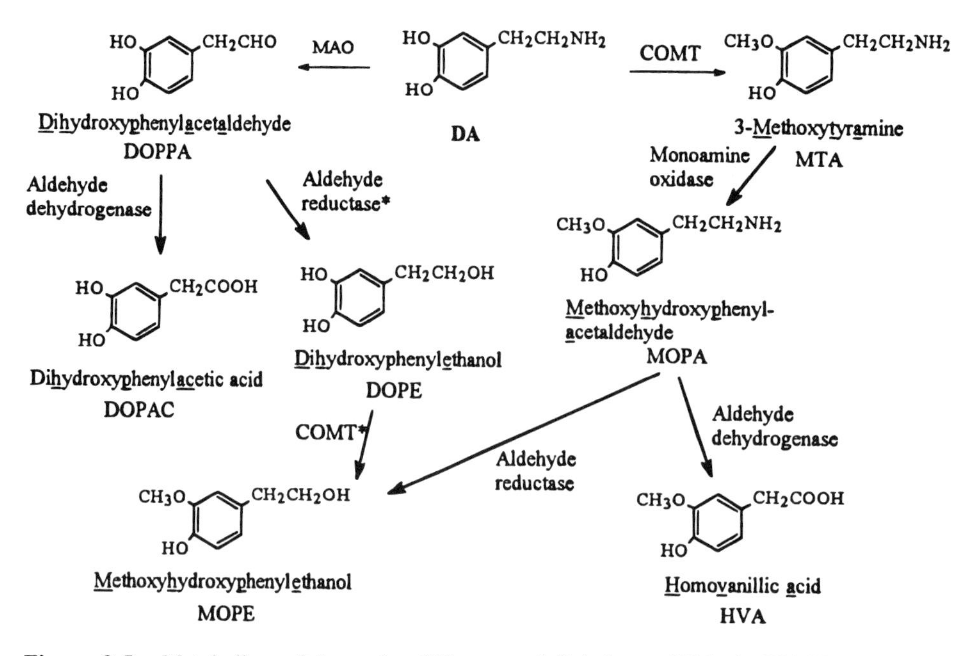

Figure 9-5. Metabolism of dopamine (*Step not definitely established; COMT = catechol-O-methyl transferase; acronyms derived from underlined letters, except H = O, i.e., hydroxy = oxy).

inhibitors. For example, deprenyl and pargyline, both of which are benzylacetelynic compounds, are irreversible inhibitors of MAO-B[5] clorgyline of MAO-A.

In summary, unlike with cholinergic neurons, where termination of action is rapidly accomplished by a single efficient process, hydrolysis of the neurotransmitter by AChE, in the case of catecholamines, a multiple of processes occur simultaneously. A major intraneural reuptake process, a dilution effect by diffusion away from the synaptic cleft, which includes uptake (U_2) into extraneural tissue, oxidative deamination by MAO and *m*-methylation of the catechol moiety by COMT.

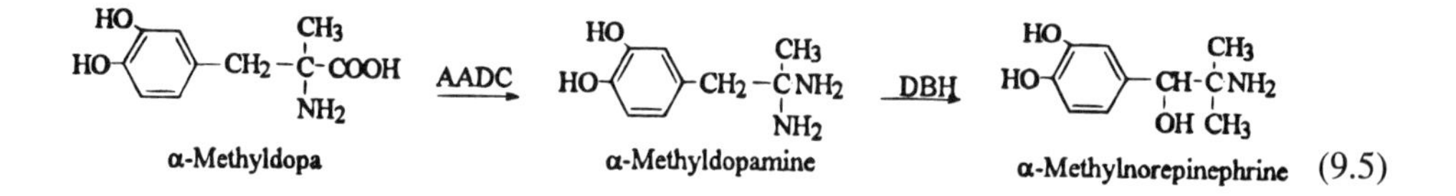

(9.5)

The specificity of these two enzymes is rather broad or, more precisely, relatively non-specific. That is, they are capable of acting on natural substrate analogs. α-Methyldopa (Aldomet) represents a particularly interesting case in point. The compound originated from a search for AADC inhibitors. Its hypotensive properties, however, led to its clinical introduction as an antihypertensive drug. In spite of the fact that its decarboxylase inhibition was not strong, its blood-pressure-lowering ability was nevertheless explained as due to decreased peripheral NE levels. A later theory held the α-methyldopamine resulting from the loss of CO_2 by α-methyldopa within the adrenergic neuron (Eq. 9.5) was stored

[5] Figure 9-3 is a proposed structure.

with NE in the storage vesicles to act as a *false neurotransmitter*. As such, it was suggested that when released into the synaptic cleft after nerve stimulation, its "weaker" agonist properties on postsynaptic receptors would bring about a weaker response, hence decreased blood pressure. However, this hypothesis is also invalid for several reasons. One was the demonstration that α-methyldopamine was not a hypotensive compound. Neither is the next metabolite, α-methyl NE, which is administered peripherally. The latest evidence, and current theory, is that metabolism of α-methyldopa to α-methyl NE also occurs centrally. It is by activating *inhibitory* α-2-adrenergic receptors in the CNS the α-methyl NE reduces sympathetic outflow (more later). Thus interference with peripheral adrenergic mechanisms appears not to be involved.

The preceding negation of the "false neurotransmitter" idea with respect to methyldopa does not invalidate the concept as such. In fact, it is probably valid in explaining the hypotensive activity of certain MAO inhibitors (Chapter 10).

9.3. Catecholaminergic Receptors

The organs innervated by adrenergic nerves (Fig. 9-6) can be stimulated experimentally by EP and NE with the same qualitative results as those obtained by nerve stimulation (see Table 8-1). Thus we speak of these and related drugs as being *sympathomimetics*. It is now generally accepted that the reaction between the adrenergic neurotransmitter NE and adrenoreceptive sites (adrenoceptors) on the effector organs is responsible for these effects.

It became apparent early in the century that some anamolous and some seemingly paradoxical results could be obtained by experimenting with sympathomimetic agents and their antagonists. For example, IV EP causes an increase in systemic blood pressure, increased tone in arterioles, a faster heart rate, and greater force of cardiac contractions. If laboratory animals are pretreated with ergot extracts, blood pressure falls without, however, affecting

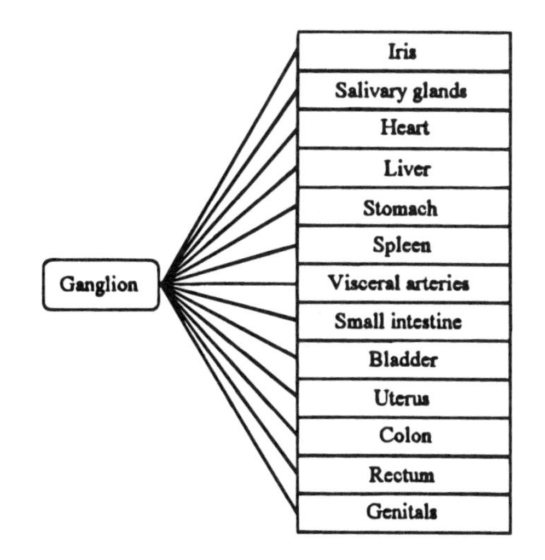

Figure 9-6. Postganglionic adrenergic innervation.

the other two parameters.[6] It can be demonstrated physiologically that EP constricts some blood vessels while dilating others. The vasoconstriction is blocked by ergot; the vasodilation is not. Thus not only is the pressure increase caused by vasoconstriction negated, but, because the vasodilation is the net effect, blood pressure actually falls.

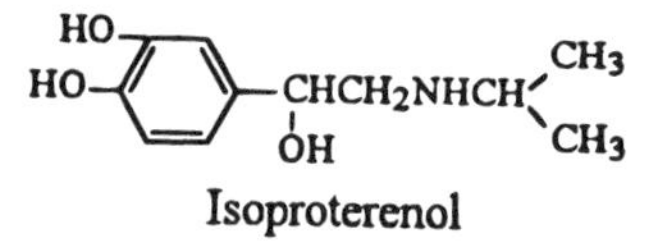

Isoproterenol

To complicate the scheme further, NE has primarily vasoconstricting properties on arterioles. Preadministration of ergot alkaloids only reduces the pressor response of NE. The reaction is not reversed. (The stimulant properties on the heart muscle are the same for both agents.) What makes the situation even more intriguing is the fact that the EP homolog isoproterenol (Isuprel), has no vasoconstricting properties. It causes marked vasodilation while maintaining a strong stimulant effect on the heart; neither effect is counteracted by ergot.

Because of these various facts, along with the study of potency relationship of various sympathomimetic amines, Ahlquist (1948) proposed the existence of two types of adrenoreceptors—designated α and β. His classification was based on the nature of the primary response of the innervated organ. If the response was excitatory, he considered it an α-receptor; if inhibitory, a β-receptor. Two decades later (Lands et al., 1967) a further subclassification became necessary for β-receptors into β_1 and β_2 types. Thus β_1 agonists act on receptors primarily located in cardiac areas, whereas β_2 agonist compounds affect receptors in bronchopulmonary, vascular, and renal areas. Of course, antagonists (blocking agents) can be similarly classified. This expanded classification was based on the assumption that (-)isoproterenol was the prototypical β-agonist with a β_1-to-β_2 ratio of 1. This is by no means certain, however, since other values have been reported.

One of the key observations by Land's group was that increased bulk on the catecholamine nitrogen atom increased β-agonist activity more at some β-sites (now called β_2) than it did at others (β_1). This led to further delineation of the selectivity for bronchial (and other) smooth muscle. Thus the heart now is classified as having primarily β_1 subtype receptors, smooth bronchial muscle (lungs), the uterus, and skeletal muscle chiefly have β_2, and vascular smooth muscle (blood vessels) has both β_2 and β_1 receptors.

A brief historical overview to see "how we got here" may be of interest. Dale (1906) provided the first experimental evidence of two types of sympathetic responses by establishing that one of them, the pressor effect of EP (in the spinal cat) could be abolished with ergot alkaloids (ergotoxine), leaving only the depressor action, which is an apparent reversal. The interpretation then was that pressor activity (i.e., vasoconstriction) is the predominant effect of EP, and the vasodilation (i.e., blood pressure fall) appears when ergot blocks the vasoconstrictor "component." Dale further established that excitatory responses of var-

[6] This is sometimes referred to as *Epinephrine reversal*.

ious organs to EP and sympathetic nerve stimulation were generally blocked by ergot; inhibitory responses were not affected. It was later concluded that the sympathetic neuron—organ juncture "portions" concerned with inhibition and excitation—had different affinities toward EP and the related substances studied. Dale found that the substance secreted from the adrenal medulla, EP (adrenalin),[7] and amine products obtained from putrefaction had the identical effects obtainable by electric stimulation of postganglionic sympathetic fibers.

In the 1930s the idea arose to refer to the circulating sympathetic neurotransmitter as "sympathin" to indicate that it was *not* totally identical to adrenal EP. Because of the apparent dual action of "sympathin" established by Dale earlier, the terminology sympathin E (excitatory) and sympathin I (inhibitory) appeared. It took almost two decades to establish that "sympathin" was EP and NE as a mixture. Von Euler (1956) finally established that NE was *the* neurotransmitter at all adrenergic nerves. Of course, EP has "similar" agonist properties if it reaches these sites from the circulation by diffusion. In a search for improved bronchodilators based on the catecholamine moiety, Konzett (1940) synthesized *iso*propylnoradr*enaline* (isoprenaline, IPR, isoproterenol in the United States). The drug was a more potent bronchodilator than was EP, but it was not a pressor amine. To the contrary, it is a vasodilator, which results in a blood pressure drop and increased heart rate by direct cardiac stimulation and reflex induction of cardiac acceleration. Other effects include increased myocardial contractile force (inotropic), skeletal muscle tremors, and increased lipolysis and glycogenolysis by the liver.[8] Many of these effects only arise on systemic absorption. The drug (like EP) is not absorbed from the gut (but is sublingually) because of high polarity and rapid COMT and sulfate metabolism in the intestinal musculature and liver, if it reaches the portal circulation. IPR is used primarily by aerosol inhalation. Unless used excessively, this avoids much of the cardiostimulatory side effects.

The drug is not significantly affected by MAO, but it is 3-methylated by COMT in the lung and other extraneural tissue. Frequently, repeated administration can lead to "fastness" or a loss of bronchodilating effect, which may in part be due to the accumulating 3-methoxy metabolite, a weak β-blocker theoretically capable of causing bronchoconstriction. In any case, IPR has become the prototype compound against which all the newer bronchodilators are compared.

The stage was then set for Ahlquist's (1948) classic experiments leading to the proposal of α and β adrenoceptors. The research was based on sensitivities to NE and EP and related substances. In essence, he discovered an inverse potency relationship between producing excitation in smooth muscles (peripheral vasoconstriction, uterus, ureter) and intestinal smooth muscle inhibition—Set A—and the production of vascular inhibition (vasodilation) and of the uterus, and cardiac excitation—Set B. The experimentation utilized dogs, cats, and rabbits as well as isolated animal tissues. The amines tested showed a *decreasing* potency in the order listed for Set A effects, and the opposite for Set B effects:

Set A: EP > NE > α-MeNE > α-MeEP > IPR
Set B: IPR > EP > αMeEP > αMeNE > NE

---

[7] Adrenalin is a trade name only in the United States; it is a generic name elsewhere.

[8] As was later determined, by activating the β_2 receptors in that organ.

These very different sets of potencies indicated that two distinct adrenoceptor types are likely to be associated with each set of pharmacological effects. Set A effects could be blocked by adrenergic blockers available in 1948 (dibenamine, ergot alkaloids). These were named as *alpha*-adrenoceptors; Set B effects, where IPR was the most potent agonist, was then said to interact with *beta*-adrenoceptors. In fact, IPR was found to be devoid of any α-agonist effects. Ahlquist's work also established that NE possessed primarily α-adrenoceptor activity, while EP acted on both types of receptors.

There was skepticism of Ahlquist's concept since no compounds capable of blocking β-receptor responses were known. This occurred ten years later with the synthesis of 3,4-dichloro-α[(isopropylamino)-methyl]-benzyl alcohol, which is erroneously referred to as dichloroisoproterenol (DCI). DCI appeared to be a competitive antagonist to IPR. For example, reversal of bronchoconstriction by IPR could be negated by DCI.

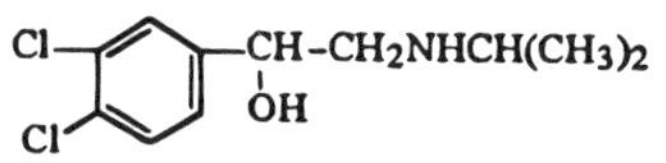

Cl, Cl — $CH-CH_2NHCH(CH_3)_2$, OH

"Dichloroisoproterenol"

The ability to show at least partial β-receptor blockade in a way constituted the missing piece of "evidence" needed for wide acceptability of the α–β receptor concept as reality. Further work ultimately showed the compound was not a "pure" blocker since it had some agonist activity as well. It is referred to today as partial *intrinsic sympathomimetic activity* (ISA). This property eliminated the compound's clinical potential. However, as will be seen, it did ultimately lead to a new group of therapeutically important β-blockers.

One of the tools not available to Ahlquist in the 1940s that has so accelerated receptor research (as discussed in previous chapters) is ligand binding with either agonist or antagonist compounds, or both. These are frequently radioligands, that is, they contain a radionuclide atom whose "cold" counterpart is part of the compound's structure, such as ^{14}C, ^{3}H, ^{32}P, ^{35}S; or they have a radioactive atom bonded to the desired compound (e.g., iodination with ^{131}I or ^{125}I). The availability of very "hot" ligands with high specific radioactivity and high affinity for the receptors under study has been a critical component to success. Of course, newer separation techniques and identification technologies have also been essential to this "revolution." The older techniques were in a sense more physiological than chemical (i.e., studies were carried out on isolated, but intact, organs such as ileum, aorta, or even a whole heart using agonists or antagonists). Results, of course, are more difficult to interpret correctly since measured results such as stimulation or inhibition of contractile tissues, changes in chemical levels, ions, enzymes, hormones, and so on cannot always be definitively attributed to receptor interactions. Other events affect the drug being tested that cannot always be unequivocally assigned because they may not be known. Such events may include absorption, distribution, and metabolism. In the case of antagonists difficulties are even greater since they may exhibit antagonism without necessarily competing with or blocking the effects at the putative receptors at all.

The nature of the β-adrenoceptor (both β_1 and β_2) is now reasonably well understood. Stimulation of the receptor is coupled to the generation of 3′,5′-cyclic AMP (*c*-AMP) by

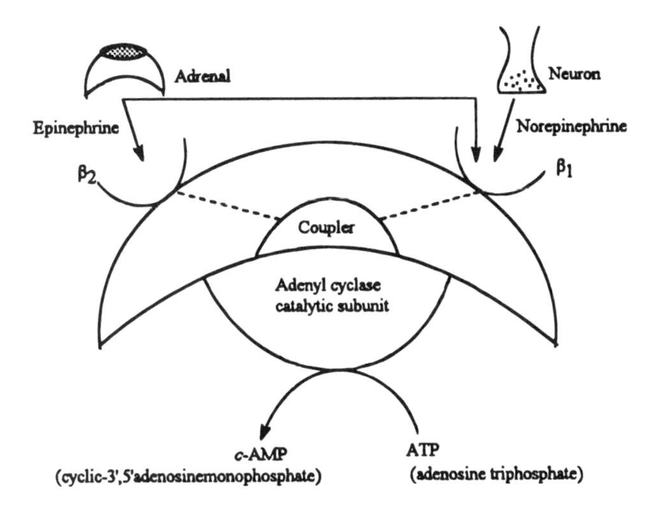

Figure 9-7. Schematic of β-adrenoceptor complex (EP = epinephrine; ATP = adenosine triphosphate; *c*AMP = cyclic -3,5-adenosine monophosphate).

triggering *adenyl cyclase*, the enzyme producing *c*-AMP (Eq. 9.6A); see also Fig. 5-4 and pertinent text in Chapter 5). Figure 9-7 represents a simplified schematic of a β-adrenoceptor–ligand interaction. Sutherland and Rall (1960) first suggested that the β-receptor is a component of a regulatory unit situated on the outside of the cell membrane. The function of the unit is to control the activity of the enzyme *adenyl cyclase*, a secondary messenger carrying out intracellular functions. The coupling subunit (transducer) that allows the agonist—receptor binding event to initiate cyclase activity on ATP is a complex formed by a guanine–nucleotide binding protein (G_m) with guanosinetriphosphate (GTP). G_m-GTP (Gilman, 1984).[9] Thus there is a rapid three-event sequence: (1) agonist binds to receptor complex, (2) coupling subunit transmits effects via GTP through cellular membrane to adenyl cyclase, and (3) the enzyme catalyzes synthesis of *c*-AMP. It must be remembered that β-antagonists (blockers) can also bind in step 1, which presumably prevents steps 2 and 3 from occurring.

It may be useful to elaborate on the coupling process from the NE-binding site on the extracellular surface of the membrane to the catalytic site of the cyclase on the "inside" leading to enzyme inactivation. The mediating third protein (the other two being the receptor protein and the enzyme) is named G since it binds to *guanyl nucleotides* N-glycosidically at C-1 via the ε-guanido group of arginine. A complex of G with GDP (n = 2) is inactive. A reaction whereby the guanyl diphosphate is exchanged for the triphosphate results in GTP-G (n = 3), the protein capable of activating adenyl cyclase to produce *c*-AMP. This exchange reaction is catalyzed by the catecholamine -β-receptor complex, but *not* if the receptor is "empty" or blocked. Equation 9.7 then summarizes the sequence (R = receptor, G = G protein):

---

[9] Work for which Gilman received the Nobel Prize in 1994.

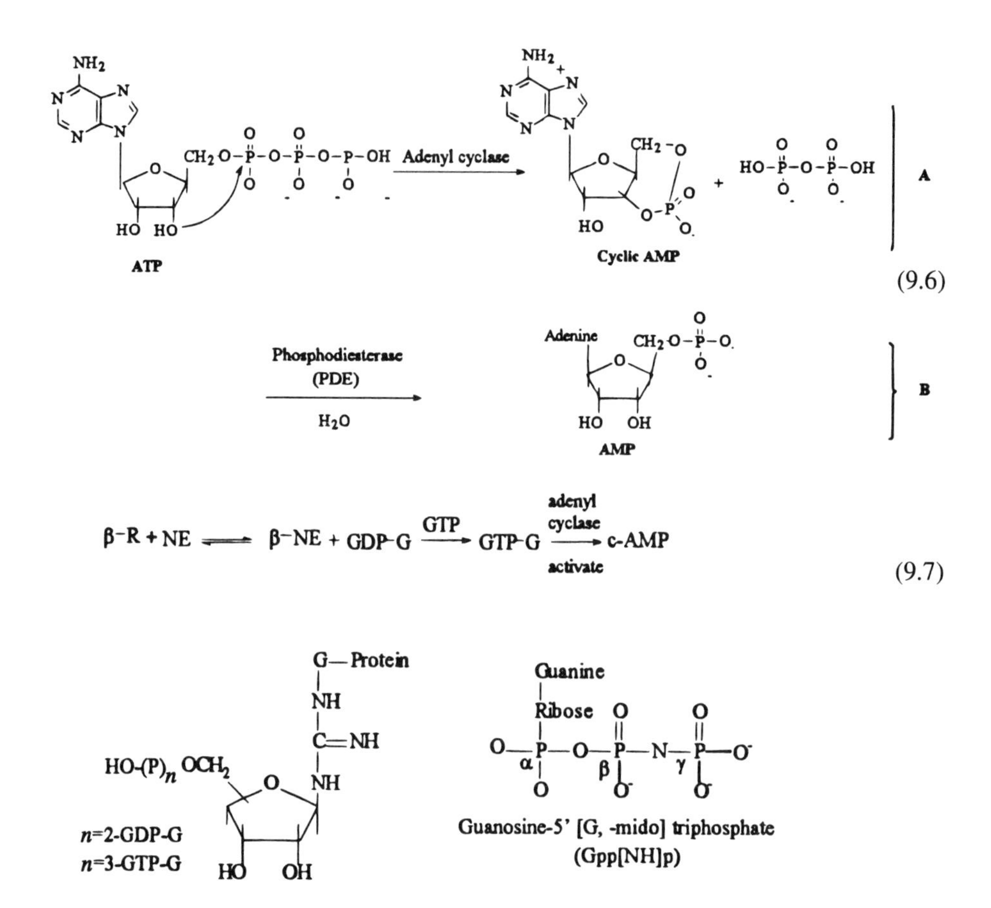

Although much detail at the molecular level is still not fully known, it is clear that the majority of effects seen as a result of catecholamine actions at β-adrenoceptors are mediated via adenyl cyclase activation and the resultant increase of *c*-AMP within cells. Thus the excitatory cardiac effects or increased inotropic response are the result of β_1 stimulation. β_2 stimulation leads to inhibitory effects on smooth muscles.

The correlation between higher *c*-AMP levels and β-adrenergically induced smooth muscle relaxation—and subsequent decreased airway resistance—is well established. The smooth muscle myosin light chain kinase has been shown to be phosphorylated by *c*-AMP-dependent protein kinase, resulting in slower phosphorylation of the P-light chain (of myosin). In addition, it has been proposed that *c*-AMP production may lead to decreased Ca^{2+} levels to interact with phosphorylated contractile proteins such as is found in smooth muscle sarcoplasm. *Beta*-stimulation may decrease the Ca^{2+} concentrations by one of several mechanisms. One possibility is that Ca^{2+} influx into the cells is inhibited by decreased permeability (through the Ca^{2+} channels?). There are other possibilities (e.g., increased efflux) for which there is some evidence from increased Na^+–K^+ pump activity that would extrude Ca^{2+} by an exchange with Na^+. None of the possibilities has been absolutely established. However, the pharmacologic result—bronchodilation—is established and is the underlying basis for the therapeutic use of β-agonists (particularly β_2) in obstructive pulmonary diseases such as asthma, emphysema, and chronic bronchitis.

In in vitro experiments it was shown that the addition of guanylylimidodiphosphate (Gpp[NH]p), a nonhydrolyzable analog of GTP (to G protein), inhibits the ability of the β-adrenoceptor to bind to agonists, presumably by altering receptor conformation. The ability to bind to β-antagonists, however, was not affected.

The β-adrenergic receptor was successfully isolated from frog erythrocytes in solubilized form, purified by ion exchange, affinity chromatography, and gel electrophoresis. It had a molecular weight of 58,000 daltons and was capable of binding stereospecifically to both agonists and antagonists.

Once produced within the cell, *c*-AMP seems to have the activation of protein kinase as its main function. This activation in turn phosphorylates particular postsynaptic proteins in the cell membrane. This results in the alteration of pores facilitating cation passage. As a second messenger it facilitates the action of more than a dozen hormones besides NE/EP, including ACTH, insulin, lipotropin, calcitonin, and vasopressin. To the extent that they are tested experimentally *hormonal* effects could be mimicked by exogenous addition of *c*-AMP.

Beta-agonists, indirectly via *c*-AMP, act on mast cell β_2-receptors inhibiting the release of bronchoconstrictor mediators such as histamine. Slow-reacting substance of anaphylaxis (SRS-A), now known to be leukotriene D_3 and D_4 (Fig. 5-5), and eosinophil chemotactic factor of anaphylaxis (ECF-A) may also have their release inhibited by *c*-AMP. Leukotriene D, which is a mediator of hypersensitivity reactions, may still have an important role in causing asthmatic symptoms. In fact, it is 100 times more effective than histamine in producing vascular permeability.

The enzyme phosphodiesterase (PDE) (Chapter 5), degrades *c*-AMP to inactive AMP (Eq. 9.6B). Inhibitors of this enzyme also have the effect of raising *c*-AMP levels and therefore should act synergistically with β-agonists and other hormones utilizing this secondary messenger. Methylxanthines such as theophylline and caffeine are weak PDE inhibitors. Theophylline is orally effective in relieving bronchoconstriction and is frequently used in the treatment of asthma and emphysema.

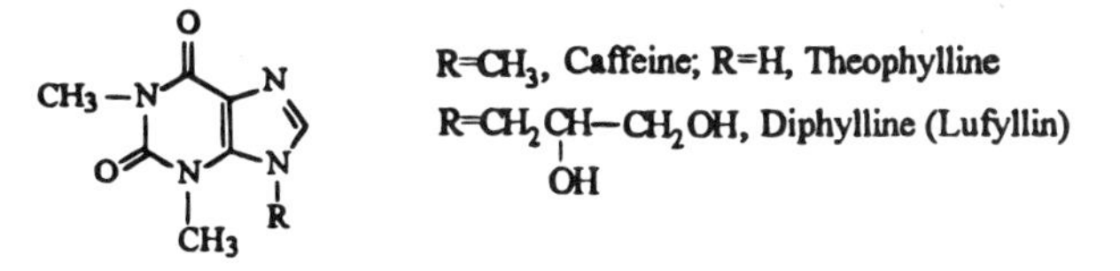

Asthma alone affects millions of people in the United States, at least 4% of which are children. Other pulmonary obstructive diseases such as emphysema and chronic bronchitis add to the total. Bronchial asthma is probably better understood with its wheezing and shortness of breath caused by bronchoconstriction. As the disease progresses viscous secretions add physical obstruction to the clinical problems. The exact underlying pathophysiology is not fully defined by the concept of an imbalance between cholinergic, α-adrenergic, and β-adrenergic control mechanisms of airway diameters, yet they must be a major contributor to the *c*-AMP deficiencies known to exist. As mentioned previously, *c*-AMP mediates smooth muscle tone in bronchial areas and also impedes the release of the other known constrictive mediators: histamine, SRS-A, and ECF-A. All these physiological effects, and therefore symptoms, are reversible with β-agonists and PDE inhibitors.

Prior to 1940 the only effective agents available were EP for acute attacks; ephedrine, offering poor quality, but sustained relief with oral use (see later); and anticholinergics such as stramonium leaf cigarettes (see Chapter 8). The introduction of IPR was helpful to a degree since the drug was not a pressor. However, cardiac effects, now understood to be β_1-mediated, and lack of oral activity, as previously discussed, left the therapeutics of asthma still very deficient. The goal was the development of β_2-stimulant agonists with high bronchial selectivity, long duration of action, and good oral activity as well as effectiveness by inhalation. Table 9-1 lists direct-acting sympathomimetic amines in clinical use today, while Table 9-2 shows sympathomimetics believed to act primarily indirectly (see later). It will be noted that they are all built on the same foundation—the β-phenylethylamine framework. It is not surprising that the successful molecular modifications over the years were modeled after the natural transmitters whose structures have been known for seven decades. The "building block" was an ideal structure for the type of SAR work carried out during the formative years of medicinal chemistry: structural variations on a "lead" compound to determine the effects obtained. In this case, manipulations were possible on four points of the molecules: the phenyl ring, the β-carbon, the α-carbon, and the nitrogen atom.

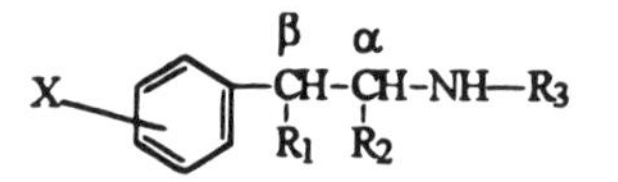

The β-phenethylamine building block

The knowledge that all the major phenethyl biogenic amines were catechols is the likely reason that the early products all contained the 3,4-dihydroxy relationship intact (e.g., IPR, isoetharine, ethylnorepinephrine). Deletion of one (e.g., phenylephrine) or both (e.g., ephedrine) hydroxyls retained adrenergic activity but at a lower intensity. It should be understood that even though the drugs in Table 9-1 may be listed as having only β-activity, that does not signify total selectivity, but only that the β/α ratio is highly in favor of β. Similarly, a β_2 classification means a large value for β_2/β_1. In fact clinically, mild tachycardia can occur, especially at higher dosages.

Once the sacrosanct catechol relationship was violated, several benefits accrued. Thus metaproterenol (orciprenaline, Metaprel), the first drug to do this, not only retained the β-activity that the bulkier N-isopropyl group bestowed earlier to IPR, but also evolved some β_2 selectivity. In addition, not a catecholamine but a resorcinol derivative (1,3-dihydroxybenzene), the drug is not affected by COMT or sulfatase enzymes. The result is a drug that, by inhalation, is equipotent to IPR, has fourfold peak effect (60 vs. 15 minutes for IPR) and a two- to fourfold increase in effective duration. The drug is also used orally (ca. 40% absorption) offering sustained prophylactic effect. These advantages metaproterenol offers all appear to result from a single molecular alteration—moving the 4-OH to the 5 position. Terbutaline retains the resorcinol feature of metaproterenol, but it increases the bulk on the nitrogen to a *tert*-butyl group, thereby retaining the previously mentioned improvements.

Fenetrol is another resorcinol derivative with high β_2 selectivity and long duration of action that is not yet available in the United States (Svedmyr, 1985). The nitrogen bulk in this drug is further increased, and polarity added to the isopropyl group, with the addition of a *p*-hydroxyphenyl group. The aralkyl type of substituent appears to increase β_2 specificity.

Table 9-1. Direct-Acting Sympathomimetic Amines with β-Activity[a]

Drug	*A*	*B*	*C*	R_1	R_2	R_3	Selectivity[b]
1. Dopamine (Intropin)	OH	OH	—	H	H	H	β_1 primarily[n,d]
2. Norephinephrine (Levarterenol)	OH	OH	—	OH	H	H	β_1 (and α)
3. Epinephrine (Adrenaline)	OH	OH	—	OH	H	CH_3	β_1, β_2, α
4. Phenylephrine (Neo-Synephrine)	OH	H	—	OH	H	CH_3	α_1, weak β
5. Ethylnorephinephrine (Bronkephrine)	OH	OH	—	OH	C_2H_5	H	β_2 some α
6. Dobutamine (Dobutrex)[c,d]	OH	OH	—	H	H	$-CHCH_3$, $(CH_2)_2$, phenyl-OH	β_1
7. Isoetharine (Bronkometer)	OH	OH	—	OH	C_2H_5	$CH(CH_3)_2$	β_2, some β_1
8. Colterol [Bitolterol] (Tornalate)[a]	OH	OH	—	OH	H	$C(CH_3)_3$	β_2
9. Isoproterenol (Isuprel)	OH	OH	—	OH	H	$CH(CH_3)_2$	$\beta_2/\beta_1 = 1$
10. Metaproterenol (Alupent, Metaprel)	OH	—	OH	OH	H	$CH(CH_3)_2$	$\beta_2/\beta_1 = 1$[i]
11. Terbutaline (Brethine, Bricanyl)	OH	—	OH	OH	H	$C(CH_3)_3$	β_2
12. Albuterol [Salbutamol] (Proventil)	CH_2OH	OH	—	OH	H	$C(CH_3)_3$	β_2
13. Fenetrol (Berotec),[f,g]	OH	—	OH	OH	H	$HC-CH_3$, CH_2-phenyl-OH	β_2
14. Malbuterol (Broncholin)[f]	CF_3	NH_2	Cl	OH	H	$C(CH_3)_3$	β_2
15. Formterol (Atock)[f]	NHCHO	OH	H	OH	H	CH_2-phenyl-OCH_3	β_2
16. Ritodrine (Yutopar)[j]	H	OH	—	OH	CH_3	$(CH_2)_2$	β_2
17. Pirbuterol (Maxair)[k]	CH_2OH	OH	—	OH	H	$C(CH_3)_3$	β_2

[a] Drugs with primary β-agonist activity (lesser α activity noted).
[b] Qualitative approximation.
[c] Peripherally.
[d] See Chapter 12.
[e] Marketed as the 3,4-di-*p*-toluic acid ester prodrug (Bitolterol) only; see text.
[f] Not marketed in United States.
[g] Highly β_2 selective, long acting (Svedmyr, 1985).
[i] Sources appear contradictory as to degree of β_2 selectivity.
[j] Used to control premature labor and fetal distress; inhibitory uterine effects predominate.
[k] C_5 is an N atom pyridine ring.

Table 9-2. Indirect-Acting Sympathomimetics*

Drug	*A*	R_1	R_2	R_3	R_4	Comments
1. Tyramine	4-OH	H	H	H	H	No drug use
2. Hydroxyamphetamine (Paredrine)	4-OH	H	H	CH_3	H	Decongestant, mydriatric
3. Phenylpropanol-amine[b]	H	OH	H	CH_3		Decongestant, anorexiant
4. The ephedrines	H	OH	H	CH_3	CH_3	See text
5. Amphetamine (Benzedrine, Dexedrine)	H	H	H	CH_3	H	Anorectic, narcolepsy, hyperkinesis[d]; marketed as racemic and dextro forms
6. Methamphetamine[a] (Desoxyn, Methedrine)	H	H	H	CH_3	CH_3	Anorectic
7. Phentermine (Ionamin, Adipex-P, Fastin)	H	H	CH_3	CH_3	H	Anorexiant, available as HCl and resin complex
8. Chlorphentermine (Pre-Sate)	4-Cl	H	CH_3	CH_3	H	Anorexiant
9. Clortermine (Vorans)	2-Cl	H	CH_3	CH_3	H	Anorexiant
10. Mephentermine (Wyamine)	H	H	CH_3	CH_3	CH_3	Decongestant, vasopressor
11. Metaraminol[a]	3-OH	OH	H	CH_3	H	Vasopressor only
12. Benzphetamine (Didrex)	H	H	H	CH_3		Anorexiant
13. Fenfluramine[f] (Pondimin)	$3CF_3$	H	H	CH_3	C_2H_5	Anorexiant
14. Diethyloropion (Tenuate, Tepanil)	H	O (keto)	H	CH_3	$N(C_2H_5)_2$	Anorexiant
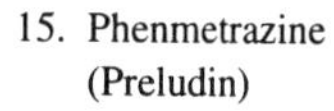 15. Phenmetrazine (Preludin)	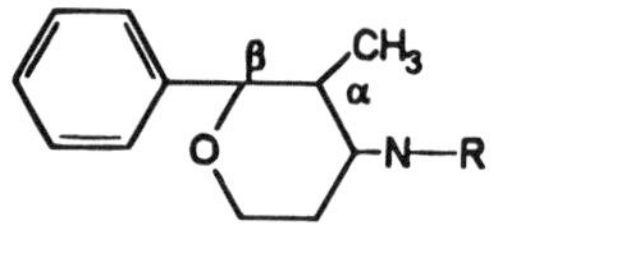 R = H					Anorexiant
16. Phendimetrazine (Plegine, Anorex)	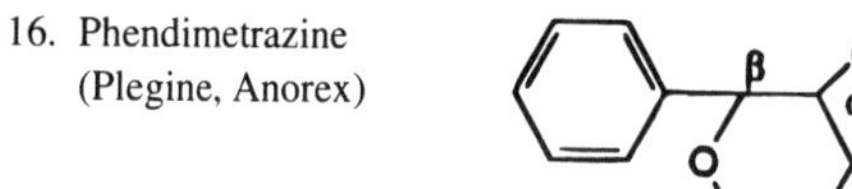 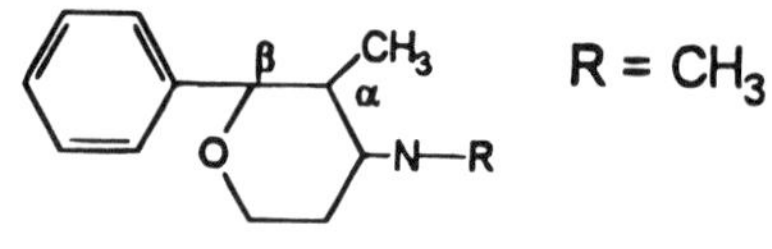 R = CH_3					Anorexiant
17. Methylphenidate (Ritalin)						Narcolepsy hyperkinesis[d]

(Continued)

Table 9-2. *(Continued)*

Aliphatic Sympathomimetics*

$$R\text{—}\underset{\displaystyle NH\text{—}R'}{\underset{|}{CH}}\text{—}CH_3$$

Drug	*R*	*R′*	Comments
18. Tuaminoheptane (Tuamine)	$CH_3CH_2CH_2CH_2CH_2$—	H	All these compounds are used only as nasal decongestants now. They do have vasoconstrictive properties. Believed to have indirect α-type activity.
19. Methylhexaneamine (Forthane)	$CH_3CH_2\text{-}\underset{CH_3}{\underset{\vert}{C}H}CH_2$—	H	
20. Cyclopentamine (Clopane)	cyclopentyl—CH_2—	CH_3	
21. Propylhexedrine (Benzedrex)	cyclohexyl—CH_2—	CH_3	

* Predominantly indirect-acting.
[a] Trade name in parentheses.
[b] Can also be named norephedrine.
[c] Also named desoxyphedrine.
[d] Also called minimal brain dysfunction or attention deficit disorder.
[e] Both indirect and direct acting, but weaker than NE; no CNS stimulation.
[f] Not a CNS stimulant but rather a CNS depressant; causes drowsiness (serotoninergic?); the dextro isomer, dexfenfluramine (Redux), was approved in 1996.

Colterol at first appears to be a step backward. It is again a catechol that differs from IPR by changing the isopropyl to a *tert*-butyl group. In fact N-*tert*-butylnorepinephrine is another name for it. The drug, however, is marketed as a pro-drug where both catechol OHs are esterified with *p*-toluic acid. The di-*p*-toluate ester (bitolterol) which is available for inhalation therapy only, accumulates selectively in the lungs and is hydrolyzed there by esterases slowly enough to the active moiety (colterol) to afford sustained bronchodilation for at least 5 hours (with one-fourth of the patients obtaining 8 hours of relief) (Eq. 9.8). The highly lipoidal *p*-toluoyl groups apparently inhibit the efficiency of the esterases and, of course, protect the OH of the unhydrolyzed fraction of the pro-drug from COMT. The large N-*tert*-butyl group increases the β-selectivity (over the *iso*propyl group) and would also tend to inhibit MAO. It is apparent that good design went into this drug to maximize all presently known factors toward an efficacious agent. Two factors should probably be ascribed to fortuity, however. The rare development of tolerance and the fact that duration of action does not decrease with continuous use.

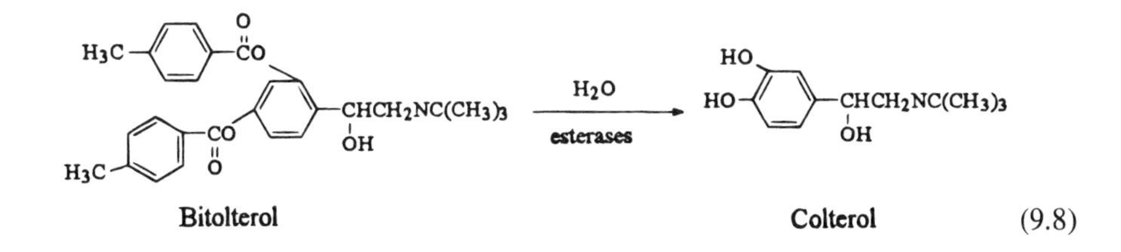

(9.8)

Albuterol (Salbutamol) represents a slightly different approach to noncatechol β-receptor agonists: the substitution of an alcoholic methylol (CH_2OH) group for the *meta*-OH of catechol. This change makes the compound resistant to COMT and unaffected by sulfatase so that it is useful by the oral and inhalation route. Its N-*tert* butyl group assures high β_2-selectivity, as predicted by Lands' work (1967).

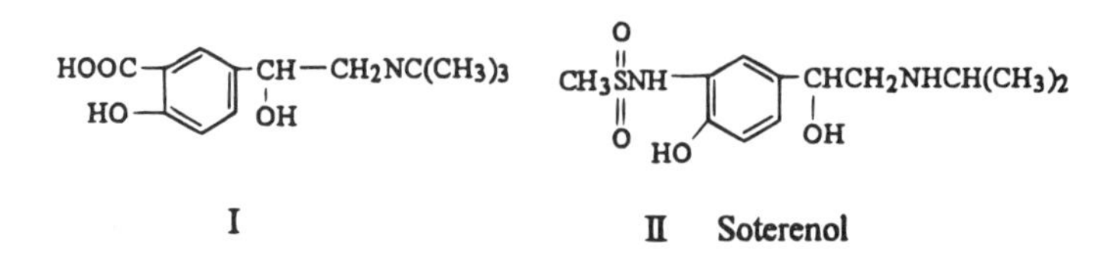

I   II Soterenol

The selection of the CH_2OH at the *meta* position was based on several possibly significant factors relevant to the *meta*-OH in catechol adrenergic amines. These factors include the ability to hydrogen bond, acidity (phenolic), chelation ability (in conjunction with the 4-OH), and others. Since a COOH group might fulfill all three requirements, compounds such as I (including its ester and amide) were prepared and tested. Surprisingly they turned out to possess β-blocker properties (particularly the amide). A failure? Not at all. The target was not missed; overshot, maybe. The implied prediction in the synthesis of these three compounds was that they would affect β-adrenoceptors, which they did. That is a positive result. We simply do not have the sophistication yet to foretell with certainty whether a compound will be an agonist (i.e., bind to a receptor and elicit a response or interact with it and block access to its natural ligand). An effect involving the receptor, however, was correctly foreseen. It is interesting that this lead was later pursued and led to the development of the marketed β-blocker labetalol. The result was the β-mimetic soterenol, II. It is curious that the position isomer of II with the substituents at positions 3 and 4 reversed was inactive. This indicates the significance of the *para*-hydroxyl and, possibly, the unimportance of acidity at position 3. Albuterol, then, resulted with placement of the neutral, but hydrogen bondable methylol group at that position, making the drug a salicyl alcohol derivative. Albuterol is a potent, β_2-selective, long-acting (4 hours) sympathomimetic agonist with extremely low cardiac involvement. It is not affected by COMT or sulfatase and is therefore orally effective as well. Figure 9-8 outlines a synthesis beginning with a Friedel–Crafts acylation of aspirin. It will be noted that enantiomeric resolution is done after the β-keto reduction is accomplished. As with most β_2-adrenoagonists, the levo form is the more active enantiomer, by a factor of 68 in this case.

Formoterol uses a formamido group at position 3 and a *p*-methoxyaralkyl function on the N atom to achieve its orally active β_2-selectivity. The molecule has two asymmetric carbons, generating four enantiomers. Bronchodilating potency varied over a 14:1 range, with the most active being the – *R,R* isomer. The development of malbuterol, a highly selective β_2-mimetic agent, illustrates clearly that lipophilic electron-withdrawing groups can effectively replace phenolic hydroxyl groups without sacrificing any of the desirable β_2-activities.

Finally, the replacement of the catechol nucleus by other ring systems should be addressed. A series of 8-hydroxycarbostyril derivatives were tested, one of which, procaterol, is in clinical use[10] as a potent β_2-mimetic with long duration. Its activity is explic-

[10] Investigational in the United States.

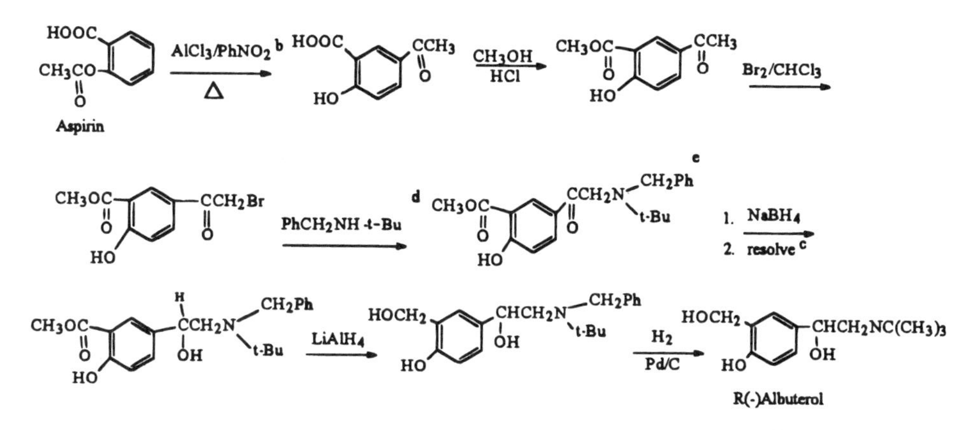

Figure 9-8. A synthesis of albuterol (Salbutamol). [a]Adopted from several sources. [b]Fries rearrangement with (–) *p*-ditoluyltartaric acid. [d]*t*-butylamine afforded poor results. [e]Ph = phenyl.

able in terms of the keto tautomer where two active (weakly acid) hydrogens—phenolic and N-1—approximately correspond to those of the catecholamine hydroxyls. The results are predictably in agreement with other studies. The 3-bromoisoxazole nucleus of broxaterol must be viewed more in terms of an isosteric replacement for benzene; so should the pyridine ring of pirbuterol.

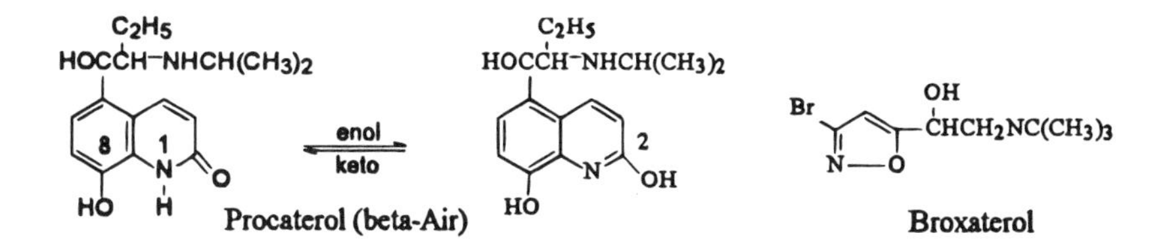

The bronchospasms in asthma are believed mediated by the release of various agents such as the D-leukotrienes (SRS-A) and ECF-A. Histamine, released from mast cells, is also a constrictive mediator; however, its effect alone is too brief to account for a total asthmatic attack. Still, the ability of the chromone drug cromolyn sodium to prevent histamine release into the lungs following allergen exposure (allergic asthma) has been of some help when given prophylactically as an aerosol. The drug is not absorbed orally, has no bronchodilating or antiinflammatory properties, and is not effective in acute asthma attacks. Its mechanism of action is believed to be blockade of Ca^{2+} influx into the mast cells, which, by inhibiting degradation, prevents histamine release and, possibly, that of other mediators as well. Not explained by this, that cromolyn is also beneficial in exercise-induced asthma, is not explained by this theory.

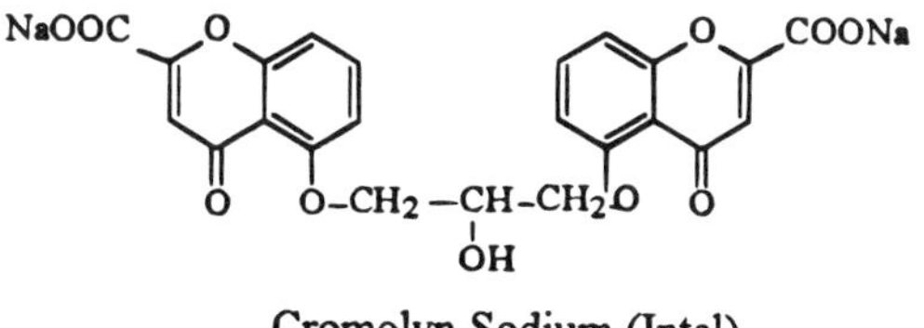

Cromolyn Sodium (Intal)

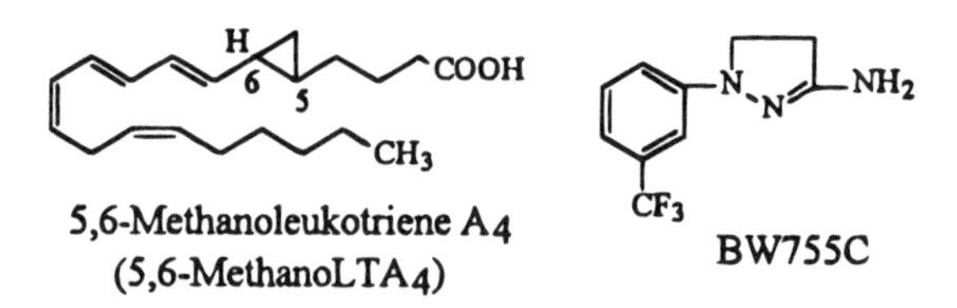

5,6-Methanoleukotriene A4 (5,6-MethanoLTA4)

BW755C

Another biochemically logical approach to asthma would be selective inhibition of leukotrienes. Replacing the 5,6-epoxy oxygen of LTA_4 with a CH_2 bridge affords a potent inhibitor of LTC_4, the next product obtained when the SH group of glutathione attacks the labile 5,6-epoxy bridge nucleophilically to produce the LTC_4 component of SRS-A (Eq. 9.9). The pyrazoline compound BW755C also inhibits thiol ether leukotriene synthesis. Unfortunately, no antileukotriene compounds have thus far proven clinically superior to presently available drugs.

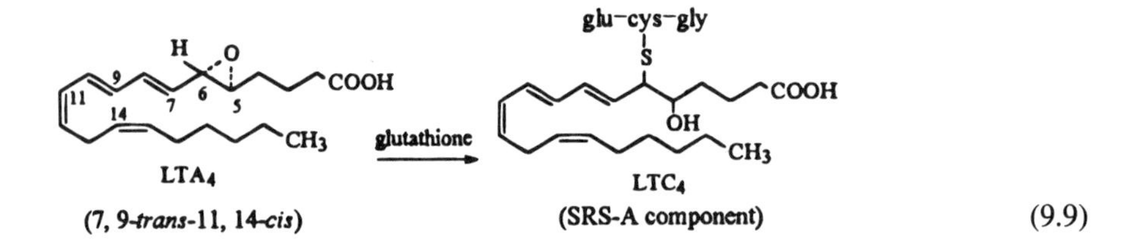

LTA4 (7, 9-*trans*-11, 14-*cis*)

LTC4 (SRS-A component)

(9.9)

The eye, and the ocular disease glaucoma, a malady characterized by increased intraocular pressure (IOP) that can ultimately lead to blindness, is particularly illustrative of the complexity of the autonomic nervous system. The various drugs used to achieve the primary goal in treating glaucoma—to reduce IOP in order to prevent harm to the optic nerve—at first appear to be wrong, or at least contradictory. The miotic effect of indirect cholinergic drugs (physostigmine, echothiophate) was discussed (Chapter 8). By stimulating parasympathetic effector cells, the pupils constrict and the ciliary muscles contract. The resultant reduction in resistance to the outflow of humor decreases IOP.

EP also produces humor outflow by the same mechanical steps. Since pilocarpine and EP are on opposite sides, autonomically speaking, does this represent an actual pharmacological contradiction? Physiological sympathetic stimulation produces the expected opposite effect—pupillary dilation. When applied topically to the eye, however, sympathomimetic amines constrict blood vessels, actually reducing absorption from the conjunctival sac. The fraction of such agents that penetrates the cornea will cause mydriasis inside the eyeball and reduce IOP, which is likely due to local vasoconstriction, leading to a reduced rate of humor production. An alternate or additional mechanism may involve the existence of adrenoceptors in the outflow channels that, when activated by EP, will facilitate outflow. It has not been established whether α or β activation is involved. A study showing that IOP produced by EP is in part inhibited by oral indomethacin, a potent prostaglandin synthesis inhibitor, raises the possibility that the lowering of IOP may also involve some stimulation of PG synthesis.

To overcome several of the pharmacological and pharmaceutical shortcomings of EP as an ophthalmic agent, the pro-drug approach has been successfully applied in the form of the 3,4-dipivaloyl ester, Dipivefrin, DPE (Eq. 9.10).

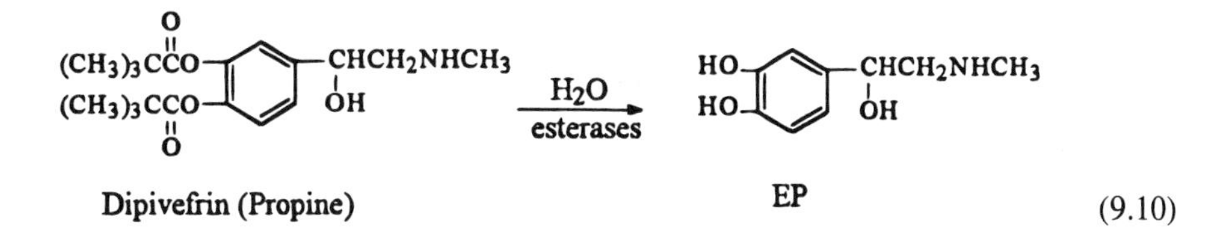

(9.10)

Most of the advantages of this pro-drug over EP itself stem from improved bioavailability. The greatly increased lipophilicity allows much greater penetrability into the eye through the corneal epi- and endothelial layer. The stroma in between requires hydrophilicity for penetration. DPE has that, too, due to the β-OH and cationic nitrogen (the eyedrops contain the HCl salt). This dual solubility permits much greater penetrability into the eye than the very hydrophilic EP hydrochloride. This high bioavailability translates into increased potency such that the 0.1% ophthalmic solution is approximately equivalent to a 2% EP solution. Increased duration of action is also achieved.

As with bitolterol blocking the phenolic OH protects the unhydrolyzed fraction of the drug from rapid enzymatic degradation in vivo and atmospheric oxidation in vitro. One of the pharmaceutical deficiencies of EP solutions is their discoloration once exposed to air forming colored and dark oxidation products (Eq. 9.11).

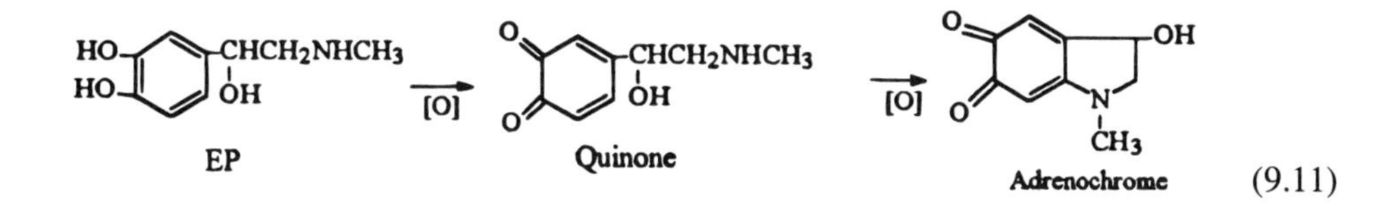

(9.11)

To make the ocular puzzle even more enigmatic β-antagonists (blockers) are now widely used as effective hyoptensives in the management of glaucoma. Since β-receptors are now known to exist in the ciliary body and iris, it would be tempting to accept this as the mechanism. However, since IPR can also lower IOP, and β_1 selective and -nonselective compounds work equally well, the actual mechanism remains elusive.

Two other methods capable of reducing IOP should be mentioned. One is by inhibition of the enzyme carbonic anhydrase. The three drugs acetazolamide, dichlorphenamide, and methazolamide were originally introduced as sulfonamide diuretics, but they are now used mainly as antiglaucoma drugs for short-term treatment preceding surgery. The ocular hypotensive effect is not related to diuresis; rather it is related to carbonic anhydrase blockade in the ciliary epithelium. Effectiveness can be increased by concomitant administration of osmotic agents.

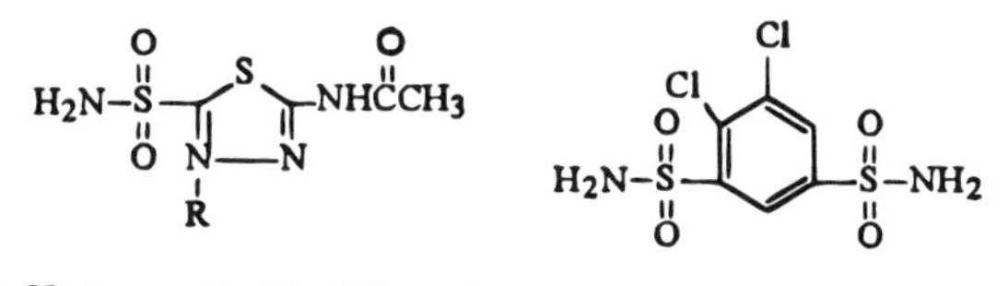

R=H, Acetazolamide (Diamox)
R=CH_3, Methazolamide (neptazane)

Dichlorphenamide
(Daranide)

Osmotic agents are generally hypertonic solutions of sugar alcohols such as mannitol, isosorbide, and glycerin (glycerol). They cause a rapid decrease of fluid from within the eye because the large amounts of compounds administered (c. 1.5 g/kg) increase the osmolarity of the blood. These agents are used mostly for brief treatments prior to ocular surgery and occasionally in some secondary glaucoma patients. An intravenous infusion of a 30% solution of urea is similarly used.

9.4. Indirect Sympathomimetics

Indirect sympathomimetics are drugs capable of actively entering the adrenergic neuron and its vesicles to displace NE from its binding sites. Thus the sympathetic effects observed with such compounds are actually elicited as a result of receptor recognition of NE, and not because of any affinity for these drugs. This is particularly true the more unrelated the compounds are to catecholamines or phenylethylamines in structure. Such agents may retain partial direct activity because they have *some* of the functional moieties needed for receptor interaction (e.g., phenolic OH, or β-ethyl OH). The ephedrines present a particularly interesting situation (see later). Compounds can have a mixed mechanism of action if, for example, they have a β-OH in the right configuration, such as D(–) ephedrine (Table 9-2, No. 4). Metaraminol (No. 11) and phenylpropanolamine (PPA, No. 3) are others. Tyramine (No. 1) and *p*-hydroxyamphetamine (*p*-hydroxy-α-methyltyramine) (No. 2) have the *p*-OH. Only the latter has clinical applications as a nasal decongestant and mydriatic. At one time it was also used for its pressor effect during hypotensive emergencies.

The β-phenethylamine structure is subject to MAO oxidation. The addition of an alkyl (usually CH_3) substituent on the α-carbon confers considerable resistance to this enzyme on the molecule. This both prolongs the indirect peripheral and β adrenergic actions of these compounds, and results in considerable CNS stimulant properties.

These clinical properties have been utilized as psychostimulants (an alerting effect), antidepressants (especially before the development of the real antidepressants discussed in Chapter 2), and appetite suppressants (anorexiants or anorectics), the last mentioned ideally part of weight loss regimens for short periods of two months or so. The compounds in which all these properties are most pronounced are amphetamine (No. 5) and methamphetamine (No. 6). Unfortunately, these are the two most abused and habituating compounds in this category. As the doses tend to increase, stereotypical behavior arises in patients as well as increased locomotor activity. Release of DA from central dopaminergic neurons may also be implicated. At even higher doses psychotic behavior and loss of perception becomes manifest. Because of these bizarre consequences and dependence potential, strong controls have been placed on the use of these and several related drugs.[11] Their complete ban, at least as anorexiants, has been frequently proposed. It may be argued that they should only retain two official indications: the treatment of narcolepsy and attention deficit disorder. Methylphenidate (No. 17) has been shown equally effective for both indications, but it is also considered equally abusable. Its CNS stimulation may be somewhat less, possibly because of the polar group (methyl carboxylate) on the β-carbon. Phenmetrazine (No. 15), in spite of an ether oxygen on the β-carbon, is in the same controlled category as amphetamine.

[11] In the United States parenteral amphetamines have no legal clinical use.

The CNS stimulant properties of α-alkyl-β-phenethylamines just described can be considerably attenuated by the expedient of placing a polar group (usually an OH) on the β-carbon. This increased polarity, of course, lessens penetration onto the CNS. It does tend to increase β-adrenorecptor agonist activity, however. This explains both the antiasthmatic and decreased, but still viable, CNS stimulant properties of ephedrine. Since ephedrine has 2 chiral centers, the α and β carbon, there are really four ephedrines, one stereomeric pair is (–) and (+) ephedrine, the other is named pseudoephedrine (Ψ) (Fig. 9-9). The pressor activity of the enantiomers extends over a 36:1 range (see Table 1-5). To a large degree the ephedrines have an indirect action. The isomer with the most direct sympathomimetic activity appears to be D(–) ephedrine, the major isomer found in the natural source (see later). It is not a coincidence that the configuration of the β-OH is the same as in (–) NE. The clinical effects include orally effective bronchodilation, nasal decongestion, mydriasis, CNS stimulation (less marked than amphetamine), urinary incontinence, and elevation of blood pressure—caused by vasoconstriction *and* cardiac stimulation. It appears that *D*(–) ephedrine is the only isomer with direct α-agonist activity, although the β-receptors are also stimulated. Release of NE (indirect action) probably accounts for a portion of its peripheral pharmacology. The enantiomer *L*(+), which is not in commercial use, most likely acts indirectly. The racemic compound, racephedrine, is available as a nonprescription product, as is ephedrine. Pseudoephedrine, the *L*(+) isomer (Sudafed), which has lesser CNS stimulation, is a popular orally used decongestant.

The *Ephedra* genus whose various species (*major, gerardiana, sinica*, and others) grow in various parts of the world, including Spain, Sicily, Pakistan, Mongolia, Tibet, and China, has been used medicinally in China probably for five millennia to treat colds, fever, and other maladies in the form of herbal teas and "pills." Yamanashi first isolated ephedrine in 1885 in Japan; Nagai (1887) purified and named the alkaloid. A synthesis was accomplished in 1920. It was Chen's work in the early 1920s that attracted worldwide attention to the drug (Chen and Schmidt, 1924). Some species, *E. major*, contain as much as 2.5% alkaloids, most of it ephedrine [i.e., *D*(–)]. Others, such as *E. gerardiana*, have most of it as *L*(+) ephedrine, even though they have a low alkaloidal content (ca. 1%).

The chemistry of ephedrines is summarized in Figure 9-9. It will be noted that the four isomers generated by the two chiral centers (2^2) exist as two nonsuperimposable image pairs. The two ephedrines are enantiomers. Epimerization of the α-carbon produces the other pair—the Ψ-ephedrines. Each Ψ-ephedrine is an enantiomer of the other; however,

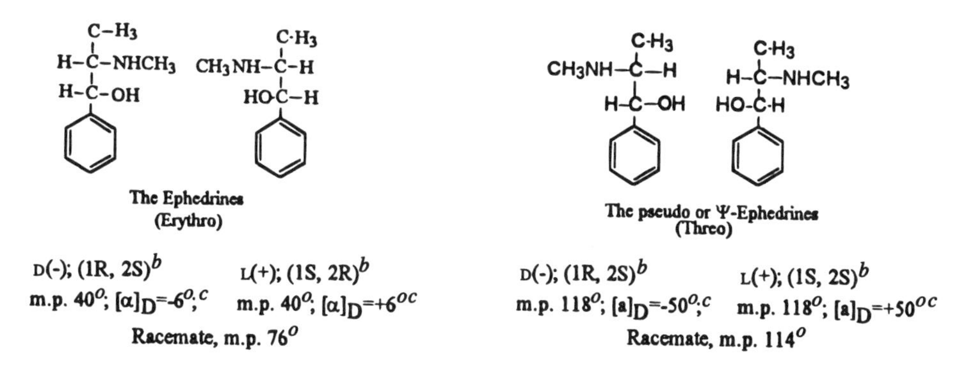

Figure 9-9. Chemistry of the ephedrines. [b]Sequence rule for configuration identification. [c]Specific optical rotation.

the relationship of a given ephedrine to either of the Ψ-ephedrines is that of a *diastereoisomer*. In this case, since the difference involves an inversion of configuration of the centers, they can also be referred to as *epimers* of each other. The difference between an enantiomer and a diastereomer can be seen from some of the physical data in Figure 9-9. Thus the melting points of enantiomers are identical; however, for diastereomers, they are different. Of course, they are really very different molecules. The specific rotations indicate the same fact. It should also be noted that unlike a simple physical mixture of two random compounds where a "mixed melting point" results in lower melting points (i.e., one compound is an impurity to the other), in the case of racemates it can go either up or down. It went up in the case of ephedrine, indicating the probable formation of a *racemic compound*, where an equal number of molecules from each enantiomer interacted to crystallize together as a single solid phase. Physical properties other than m.p., such as density and solubility, will also differ from the (–) and (+) isomer.

Metaraminol's structure (No. 11), which has the *meta*-OH of a catechol, an α-CH_3 of an amphetamine, and the attenuating β-OH of the classic sympathomimetic, should result in some predictable and interesting pharmacology. One would certainly foresee pressor activity. The treatment of hypotensive states constitutes its main therapeutic application today. Its ability to enter neuronal vesicles and displace NE gives indirect activity. It is also able to stimulate α-adrenoceptors directly. Both actions account for the pressor activity.

It is interesting that a hypotensive relapse occurs once the IV administration is discontinued. One explanation may be that the stored metaraminol being released from the NE-depleted neurons, since it is less effective as a pressor amine than NE, now acts as a *false transmitter*, actually having a hypotensive effect. Because of its α-CH_3 it is not significantly affected by MAO (while the NE in the synaptic area is). Metaraminol will reaccumulate, be restored, and be re-released, thus prolonging the hypotensive effect.

Mephentermine (Table 9-2, No. 10) is a phenthylamine with indirect and direct sympathomimetic activity. Its main use is as a pressor amine and decongestant (as an inhaler). Even though its structure greatly resembles methamphetamine (it actually has two α-methyls), its CNS stimulant properties, at usual doses, are surprisingly low. The N-dimethyl homolog phentermine (No. 7) and its *para*- and *ortho*-chloro analogs (Nos. 8 and 9) are used exclusively as anorexiants because of lesser CNS stimulation (and abuse potential) than the amphetamines. They may, however, not be as effective. PPA, as explained, is an orally useful decongestant. It also appears to have some anorexiant properties with modest CNS effects, presumably due to the β-OH polarity. Phenmetrazine (No. 15) and phendimetrazine (No. 16) may be viewed as PPA analogs, wherein the β-OH (and amine nitrogen atom) are rigidized into the morpholine ring. Even though these two agents are very likely more effective anorectic drugs than PPA, they are fraught with abuse potential and CNS effects similar to amphetamines. Unlike in PPA, however, there is no β-OH here; rather, the oxygen should be viewed as an ethoxy-ether, whose lipophilicity should encounter little difficulty in crossing the BBB.

Benzphetamine is a methamphetamine whose tertiary amine also carries a benzyl group. The likelihood of metabolic N-debenzylation yielding methamphetamine would, of course, explain its effects. Diethylpropion (No. 14) is an interesting compound because of its particularly involved metabolic degradation, which is initiated by stepwise N-deethylation and keto reduction to a β-OH. These active metabolites, presumably with the parent compound, explain both anorexiant and CNS effects, as well as the reduced level of the latter when compared with amphetamine. Subsequent oxidations all the way to benzoic, hydroxybenzoic, and mandelic acids all lead to inactivation.

Fenfluramine (No. 13) is unique among amphetamine-type anorexiants. In spite of its amphetamine structure and the presence of the highly lipophilic *m*-CF_3 group, this compound possesses a sedative rather than CNS stimulatory effect. The drug has been shown to cause serotonin depletion. Serotonin (and NE) are involved in sleep regulatory mechanisms. Thus a relationship seems possible.

Mazindol (Sanorex) produces typical amphetaminelike anorexia and CNS stimulation. The drug inhibits NE uptake in a manner identical to the tricyclic antidepressants. Its very potent appetite suppressant ability was found while being evaluated as an antidepressant. Mazindol does not contain the β-phenethylamine foundation within its frame. It is structurally the enol form of an imidazole-substituted benzophenone, probably in equilibrium with its isoindole form (Eq. 9.12) in solution.

(9.12)

Table 9-2 list several aliphatic amines. The sympathomimetic activity of 2-aminoheptane (Tuamine, No. 18) was already known to Barger and Dale. Along with methylhexaneamine (No. 19), 2-aminoheptane has been used exclusively as a nasal decongestant, although both have pressor effects systemically. Apparently they act indirectly. The cycloaliphatic compounds cyclopentamine (No. 20) and propylhexedrine (No. 21) present a more complex situation. They may be viewed structurally as nonaromatic amphetamine analogs. As might be expected, a considerable drop in α and β agonism occurs. In fact, since prophylhexedrine is primarily an indirect agent, displacing NE, its activity resembles an α-stimulant. Its use as a decongestant points this out. Some pressor activity undoubtedly remains. CNS stimulation, which is less than ephedrine, makes the abuse potential low, especially as an inhaler. In fact, propylhexedrine was developed as a replacement for inhalers containing racemic amphetamine, which were widely abused in the 1940s.

Because of the low level of CNS stimulant effects, it is interesting that propylhexedrine has been used as an anorexiant in Europe at doses of up to 100 mg. It is obvious that effects other than simply CNS stimulation are involved in the complex anoretic mechanisms.

Cycloaliphatic amines such as these bring up the question of the importance of aromaticity in adrenergic mechanisms. Although not addressing this question directly, a study of nonaromatic analogs of phenylethanolamine as substrates for the enzyme PNMT found that cyclohexyl and cyclooctane analogs were more effective in binding to the enzyme than were the corresponding aromatic compounds. The conclusion was that the hydrophobic character was a more significant factor than electron density and the postulated charge-transfer complex previously assumed for such enzyme interactions with aryl amines.

Several imidazole compounds with peripheral α-adrenoceptor stimulant properties, but no corresponding β-effects, are listed in Table 9-3. Unlike with other direct and indirect agonists, there is no significant neuronal uptake mechanism involved. These first four drugs are used mainly as decongestants on nasal mucosa as well as on ocular membranes. A degree of α_2 central stimulation likely accounts for some of the sedation encountered

Table 9-3. Imidazole Derivatives Affecting Adrenoceptors

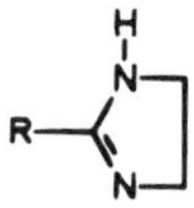

Drug	R	Comments and uses
A. Agonists		
Naphazoline (Privine)	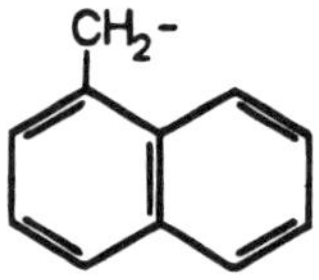 CH_2-	α, peripheral nasal and conjunctival decongestant; central α_2 effects, sedation
Tetrahydrazoline (Tyzine)	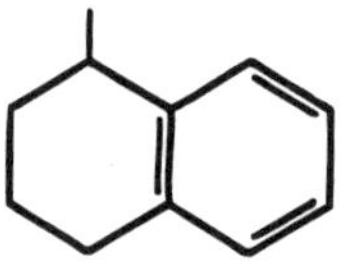	α, peripheral nasal and conjunctival decongestant; central α_2 effects, sedation
X = H, Xylometazoline (Otrivin) X = OH, Oxymetazoline (Afrin)	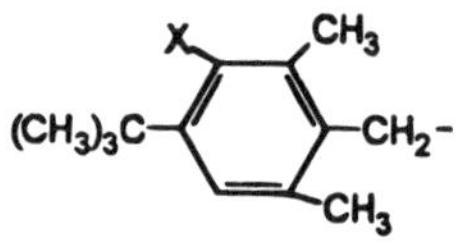 X, CH_3, $(CH_3)_3C$, CH_2-, CH_3	α, peripheral nasal decongestant
Clonidine (Catapress)	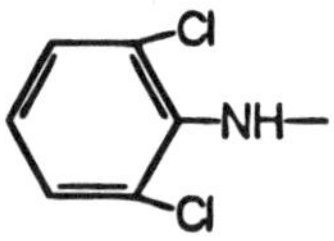 Cl, NH—, Cl	α_2, centrally primarily α_2, peripherally slightly antihypertensive, other uses[b]
B. Antagonists		
Tolazine (Priscoline)	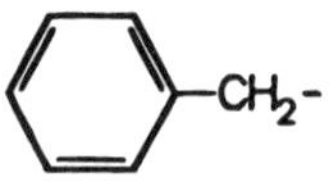 CH_2-	Peripheral α blockade, short plus direct vasodilation[b] pulmonary antihypertensive
Phentolamine (Regitine)	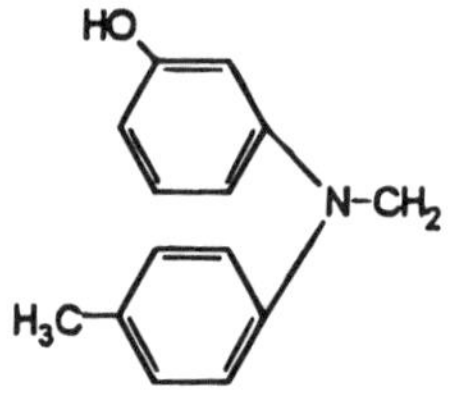 HO, N-CH_2, H_3C	α-blocker peripherally antihypertensive[b]

[a] Trade names in parentheses.

[b] See Chapter 10 on antihypertensives.

with these agents. It will be noted that the peripheral α-agonists (first four compounds, Table 9-3) have the β-phenethylamine framework within the structure.

Due to inhibition of sympathetic outflow by central α_2-receptor stimulation and the weak peripheral tolazoline and phentolamine, clonidine, an antihypertensive, will be addressed more fully in the next chapter.

9.5. The α-Receptors

Comparatively little was known about α-receptors or their structural requirements as recently as the 1970s. A 60-page review on molecular geometry and adrenergic activity dealt with α-receptors in less than one page (Patil et al., 1974). More recent research has increased understanding of this area considerably. Better understanding of receptors and their subtypes allows the development of ever more selectivity in agonists and antagonists, initially for increased depth of receptor research and, subsequently, of clinically useful drugs. In areas where multiple receptors seem to abound, it is multiple interactions of non-selective drugs that may be responsible for the bulk of side effects that arise.

In the brief overview to follow it is not possible to detail all the research.[12] Until the mid-1970s α-adrenoceptors in smooth muscles were viewed as functioning by increasing cell membrane permeability to cations on stimulation. Intestinal muscle cells show increased K^+ entry. Relaxation then follows due to hyperpolarization. Elsewhere it is the influx of Na^+ and Ca^{2+} that increase, the voltage drops, and contraction results from the excitation. More recently two concepts regarding α-adrenoceptors developed more or less simultaneously. One was that two types of α-receptors exist, and are now named α_1 and α_2. The other was the demonstration that in vitro and in vivo NE released by a sympathetic neuron following its stimulation was itself inhibited by this liberated NE by activating *inhibitory* α-adrenoceptors located on the varicosities of this prejunctional neuron. In other words, an inhibitory feedback mechanism was operative. Such autoreceptors were initially classified as α_2. In this arrangement autoreceptors must by their very function be *presynaptic* since they respond to NE molecules released into the synaptic cleft. It is therefore now apparent that synaptic NE levels are under dual control: neuronal stimulation and feedback inhibition. It seemed obvious—and convenient—to refer to the presynaptic inhibitor adrenoceptors as α_2, and those on postsynaptic effector cells (the "classic" α-receptors) as α_1. The assumption was simply that *all* postsynaptic α-adrenoceptors would be subtype α_1. This anatomical classification, however, did not hold up long. It was soon shown not to be true in vascular smooth muscle of rats and cats. α-Adrenoceptors that mediated vasoconstriction as expected were not blocked by the known selective α_1-adrenoceptor drug prazocin (see Chapter 10). The finding was soon verified on human artery tissues. It thus became apparent that not all α_2-receptors were presynaptic. This has since been substantiated by other researchers. Figure 9-10 schematically illustrates the system as it is presently believed to function. The presynaptic fiber releases NE into the synaptic cleft following stimulation. The neurotransmitter produces the expected vasoconstriction and other effects by stimulating postsynaptic α-receptors (α_1 or α_2). The NE can also stimulate the presynaptic α-receptors (α_1 or α_2), inhibiting NE

[12] For such detail see Timmermans, 1987.

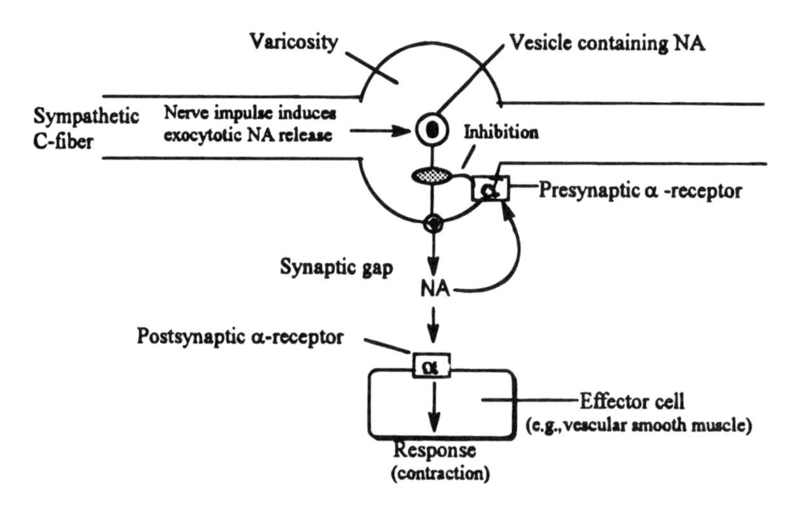

Figure 9-10. Pre- and postsynaptic α adrenoceptors of noradrenergic synapses. (From Timmermen and Van Zwieten, 1972). NA = norepinephrine.

release from subsequent impulses. These presynaptic receptors reside on the membranes of vesicles or varicosities of the presynaptic neuron. The process, then, constitutes a negative feedback system and is not unique to adrenergic neurons. There is now also evidence of similar effects in cholinergic, dopaminergic, β-adrenergic, and other systems. The presynaptic α-adrenoceptors exist at most of the noradrenergic neurons they were expected to be. Activity ratios between pre- and postsynaptic agonists and antagonists have been determined for many tissues. Measurements for presynaptic agonists determine the concentration of drug that will inhibit tritium-labeled NE (^{3}H-NE) overflow; postsynaptic activity measures the concentration of drug needed to induce a smooth muscle constriction (Starke, 1981).

It was determined that methoxamine and phenylephrine are predominantly (but not totally) postsynaptic α-agonists. Methyl-NE, tramazoline, and clonidine show presynaptic preference (α_2). Prazocin has very definite postsynaptic antagonist (α_1) action. The alkaloids yohimbine and rauwolscine exhibit high presynaptic α-adrenoceptor affinity, thereby acting as antagonists. It must be remembered that α-adrenoceptors whose characteristics are those of presynaptic α_2 receptors have been found on nonadrenergic nerve terminals and even at postjunctional sites. In general, though, presynaptic α_2-receptors occur at nonadrenergic neurons where there are postsynaptic α-receptors. Sympathetically innervated organs contain postsynaptic α_1 adrenoceptors. Postsynaptic α_2-receptors have been identified in both vascular smooth muscle as well as in the pancreas (islets), platelets, *and* the CNS, as well as in the eye and kidney (Timmermans and van Zwieten, 1982).

The CNS is now recognized as having a complex control system over the autonomic nervous system by way of both α_1 and α_2 adrenoceptors. Their stimulation or blockade by drugs results in a variety of effects that can have clinical import. Many of these are not yet well understood, and those that are will be addressed.

A limited representative (Table 9-4) lists a few drugs in the various α_1, α_2, and mixed α_1/α_2 categories for both agonists and antagonists. Those whose chemistry was not previously considered (antihypertensives) will be dealt with in the next chapter.

Table 9-4. Representative α_1 and α_2 Agonist and Antagonist Drugs

Agonists		
α_1	α_2	α_1/α_2
Phenylephrine	Clonidine[a]	Guanfacine
Methoxamine	Guanabenz	Noradrenaline
		Adrenaline
		α-Methylnor-adrenaline[a]
		Tramazoline[b]
		Naphazoline
		Oxymetazoline
Antagonists		
α_1	α_2	α_1/α_2
Prazocin	Yohimbine[f]	Phentolamine
Trimazocin	Rauwolscine[g]	Tolazocine
Labetalol[d]		Phenoxybenzamine
Corynthanine[e]		Piperoxan[h]

[a] There are disagreements on exact classification; thus clonidine is variously listed a α_2 or α_1.
[b] A nasal decongestant related to naphazoline.
[c] Active metabolite of methyldopa.
[d] Also β_1-blocker.
[e] Yohimbine epimer.
[f] Yohimbine and related alkaloids in this are powerful adrenergic blockers not used in therapy now.
[g] α-Yohimbine.
[h] Obsolete adrenergic antagonist to circulating EP (Nickerson, 1949).

References

Ahlquist, R. P., *Am. J. Physiol.* **153**:586, 1948.

Chen, K. K., Schmidt, C. F., *J. Pharmacol. Exp. Ther.* **24**:339, 1924.

Dale, H. H., *J. Physiol.* **34**:163, 1906.

Konzett, H., *Arch. Exp. Pathol. Pharmacol.* **197**:27, 1940.

Lands, A. M., *Nature*, **214**:597, 1967.

Lands, A. M., Luduena, F. P., Buzzo, H. J., *Life Sci.* **6**:2241, 1967.

McDermed, J., McKenzie, G. M., Phillips, A. P., *J. Med. Chem.* **18**:362, 1975.

Nagai, M., *Pharm. Ztg.* **32**:700, 1887.

Nickerson, M., *Phamacol. Rev.* **1**:27, 1949.

Patil, P. N., Miller, D. D., Trendelenburg, V., *Pharmacol. Rev.* **26**:323, 1974.

Salach, J. I., Minamura, M., Yasunobu, K. T., et al., in *Flavin and Flavoproteins*, Singer, T. P., Ed., Starke, 1981. Amsterdam, Elsevier, 1976, p. 605.

Starke, K. *Rev. Physiol. Biochem. Pharmacol.* **88**:199-236, 1981.

Sutherland, E. W., Roll, T.W., *Pharmacol. Rev.* **12**:265, 1960.

Svedmyr, N., *Pharmacother.* **5**:109, 1985.

Timmermans, P. B. M. W. M., van Zwieten, P. A., *J. Med. Chem.* **25**:1389, 1982.

von Euler, U. S., *Noradrenaline: Chemistry, Physiology, Pharmacology and Clinical Aspects*, C.C. Thomas, Springfield, IL, 1956.

Suggested Readings

Cooper, J. R., Bloom, F. E., Roth, R. H., *The Biochemical Basis of Neuropharmacology*, 6th ed., Oxford University Press, New York, 1991.

Timmermans, P. B. M. W. M., in *Receptor Pharmacology and Function*, Williams, R. A., Glennon, R. A., Timmermans, P. B. M. W. M., Eds., Marcel Dekker, New York, 1987.

CHAPTER

10

Drugs and the Cardiovascular Diseases

10.1. Introduction

Cardiovascular disease (i.e., disease of the heart and blood vessels) constitutes the largest single cause of death in the industrialized countries. According to the U.S. Department of Health and Human Services, the combined death toll from heart disease and strokes for 1986 exceeded 910,000, which was an improvement over 1970 (943,000) when population growth is factored into the comparison. There were 560,000 heart attacks and 400,000 strokes in 1986.[1] Whatever the cause—changes in lifestyle, diet, or stress—this can only be described as an epidemic. As with cancer, which is a distant second in terms of mortality (465,000 in 1986), cardiovascular disease morbidity increases with age, accounting for about two-thirds of all deaths in persons over 75 years of age.

Even though some of the diseases affect primarily the heart and others the vascular system, they cannot be divorced from each other. This obvious interdependence makes a unified view imperative. One of the major diseases, atherosclerosis (typically accompanied by high blood pressure), affects and ultimately damages the heart, kidneys, and other organs.

The major cardiovascular diseases will be briefly outlined. There will be further elaboration on them when drug groups used to treat them are discussed.

10.2. Cardiovascular Diseases

10.2.1. Hypertension (High Blood Pressure)

Hypertension is termed the silent killer since it is asymptomatic (i.e., without direct characteristic symptoms). Hypertension is a leading cause of death. It is estimated that 20% of

[1] In 1900 heart disease killed 27,000 and strokes 21,000; U.S. population was about half of today's.

the American population has some elevation of blood pressure. Untreated high blood pressure can result in strokes, congestive heart failure, or kidney failure. All are high mortality risk conditions. Approximately half of the people who are hypertensive do not know it, and half of these again are receiving no treatment. Of those who are aware of their condition, half are receiving inadequate or ineffective treatment. This means that only one-eighth of the 23 million Americans believed to have high blood pressure are having their disease adequately controlled.

At this point it may useful to review some of the fundamental aspects of blood pressure, its hemodynamic properties, and its normal control mechanisms. It is, of course, beyond the scope of this discussion to consider the total complexity of physiological factors of the cardiovascular system, except to state that its function is to provide the organism with a homeostatic environment. The ultimate supply of essential nutrients and gases to the cells in exchange for waste materials takes place in the capillary region, or microcirculation.

The arterial blood pressure within the mammalian organism, whether normal or abnormal, is ultimately determined by cardiac output and peripheral resistance to flow. Of the several physical and chemical variables of the cardiovascular system, blood pressure is ordinarily the most constant. This is especially remarkable if one considers the large variations encountered in cardiac output and in heart rate. This feat is accomplished by a negative feedback system. The chief mechanism involves the baroreceptor reflex.

Baroreceptors are stretch receptors in the arterial portion of the cardiovascular system, specifically in the carotid region and in the aortic arch. They monitor (sense) hydrostatic pressure through specialized nerve endings that are sensitive to mechanical deformation of the blood vessel wall. Nerve impulses over these afferent fibers to the central nervous system increase whenever the arterial pressure in the vicinity of the baroreceptors increases. This *inhibits outflow of sympathetic impulses*, reducing vasomotor tone and decreasing the heart rate. Thus vasoconstrictor tone and cardiac output are reduced as a result of the initial pressure increase that caused these series of events to occur. Both factors tend to return the pressure to previous, lower levels.

However, if the initial event is a decrease in pressure, the opposite effects occur. These then are the essential aspects of a negative feedback system. When pathological factors affect the system, hypertension may result.

Hypertension is classified as either essential (primary) or secondary. Essential hypertension comprises approximately 90% of all cases. Its causes are still unknown. The disease is incurable but treatable (i.e., it is controllable, with drugs). The remaining 10% constitute a number of hypertensive diseases with causes that are known. Some can be cured. The following outline is a classification of hypertension.

I. Essential (primary) hypertension

II. Secondary hypertension

  A. Renal hypertension
    1. Renovascular disease
    2. Renal parenchymal disease
    3. Congenital lesions

  B. Neurogenic hypertension
    1. Brain tumors
    2. Cerebrovascular accidents
    3. Psychogenic hypertension

C. Endocrine hypertension
  1. Acromegaly
  2. Adrenal cortical hyperfunction: Cushing's syndrome, primary aldosterism
  3. Pheochromocytoma
  4. Thyrotoxicosis

D. Other
  1. Metabolic and hereditary disorders
  2. Aortic coarctation

10.2.1.1. Essential Hypertension

A clearcut understanding of the exact pathological mechanism has not yet emerged. Cerebral blood flow, cardiac output, kidney and endocrine gland functions all appear to be normal, at least in the early stages of the disease. There has been no dearth of theories over the years regarding the etiology of the disease. A considerable pool of knowledge and understanding about the illness that, although not complete, has improved treatment. Much of it is due to better drugs and their proper utilization. For example, the benefits or even necessity of treating severe and even moderate hypertension was never seriously questioned. The risk factors of nontreatment were known for some time and are now well documented by large well-executed studies involving thousands of persons, primarily men. Cardiovascular complications ascribable to long-term untreated hypertension include congestive heart failure, coronary artery disease, and stroke. Progressive kidney failure can be added to the list. Genetic predisposition has been intensely studied, both as to the disease itself and to the relationship of salt (Na^+) intake to blood pressure. In industrial countries, where diets tend to be high in sodium, essential hypertension is substantially higher than it is in "third world" areas where salt intake tends to be much lower. Research with specially inbred rats, one strain of which developed hypertension when its salt intake is raised, and another strain that is resistant to Na^+ showed certain analogies to humans. These were (1) genetic susceptibility, (2) youth sensitivity (adult rats seemed resistant to high salt intake), and (3) inability to reverse hypertension once initiated. This helps us to understand that high Na^+ intake in childhood may help initiate essential hypertension in adulthood. It may at least help explain why such hypertension, if of long duration, is frequently so resistant to low-salt diets and saluretic diuretics such as the thiazides (see later). It was shown that when these salt-sensitive rats were made hypertensive by salt feeding, reducing salt intake after several weeks did not reduce the blood pressure increase produced earlier. The exact mechanistic relationships between total body Na^+ levels, the sympathetic nervous system and genetic susceptibility is still a mystery.

The relationship of obesity to hypertension was found to be quite direct in the famous Framingham study. Subjects whose weight was 20% above ideal showed a 30% increase in hypertension. It is interesting that weight loss, irrespective of whether Na^+ intake was also reduced or not, showed a significant lowering of blood pressure and a dramatic drop in NE levels (to near that of nonobese subjects). It has been suggested that the increased sympathetic activity, as demonstrated by high norepinephrine levels in obese patients, is, at least in part, related to high caloric consumption. Moderate (90–114 mmHg diastolic pressure) and certainly mild (85–90 mmHg) hypertension could, in principle, be managed by weight reduction diets. Under controlled conditions NE and pressure decreases, illus-

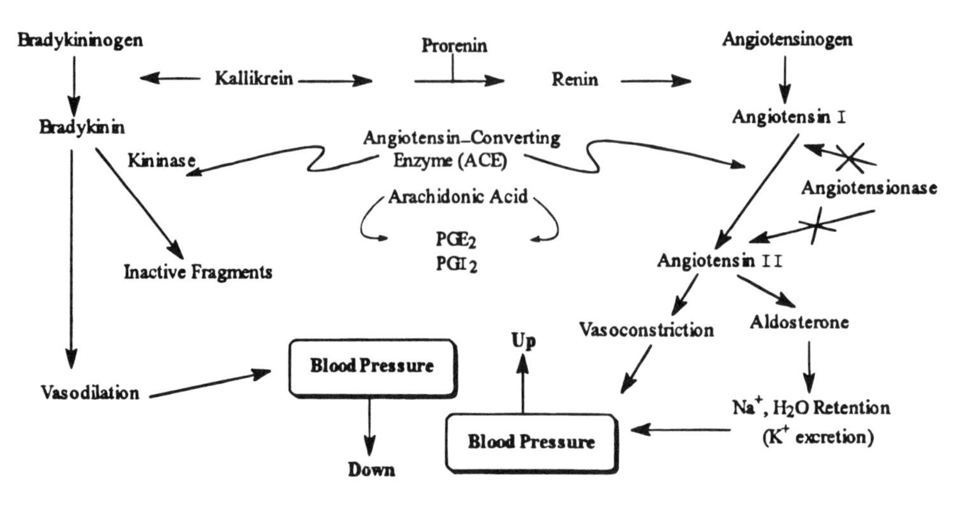

Figure 10-1. The renin–angiotensin system (RAS).

trating a *neurogenic* relationship between hypertension and obesity. It is also clear that the simple volume relationship between Na^+ and blood pressure, once assumed, is not valid. There are now understood to be other factors involved of which salt intake, obesity, stress, and aging may only be some.

No explanation of hypertension will ever be complete without considering the role of the kidney. It has been postulated that the kidney may be the main controller of arterial pressure by means of its function to regulate fluid volume in the body. Higher sympathetic activity (NE levels) sustains blood pressure by inhibiting renal Na^+ excretion. This hypothesis would explain the usefulness of natriuretic (Na^+ excreting) diuretic effects.

10.2.1.2. Secondary Hypertension

Secondary hypertension is hypertension whose causes are understood. The four main categories are renal, neurogenic, cardiovascular, and endocrine.

10.2.1.2.1. Renal Hypertension

Renal hypertension is the most frequently encountered type of secondary hypertension. It is believed to be related to the action of the proteolytic enzyme renin,[2] which is elaborated by the renal cortex following some damage to the kidney (e.g., scarring or renal artery occlusion). Once in the blood renin acts on a specific substrate angiotensinogen (a globulin). Cleavage results in a decapeptide, angiotensin I. This substance has no pressor activity. A converting enzyme splits off two C-terminal amino acids to yield the octapeptide angiotensin II, which is a potent pressor agent (40 times more active than epinephrine). It increases the force of the heartbeat and constricts arterioles. The enzyme angiotensinase normally inactivates angiotensin. These concepts are summarized in Figure 10-1.

[2] Not to be confused with rennin, the milk-curdling enzyme found in the stomach of ruminant animals.

10.2.1.2.2. Neurogenic Hypertension

Cerebral ischemia caused by intracranial hemorrhage or by a brain tumor may result in elevated pressure because of attempts to maintain cerebral flow. A net increase in sympathetic flow results. Such a pressure elevation may even be diagnostic of intracranial bleeding.

10.2.1.2.3. Cardiovascular Hypertension

Cardiovascular hypertension is caused by either increased peripheral vascular resistance or increased cardiac output. The latter may be due to hyperthyroidism. The cause, however, is usually decreased elasticity of blood vessels with age that results in systolic pressure increases. Aortic constriction is occasionally at fault.

10.2.1.2.4. Endocrine Hypertension

Endocrine hypertension actually includes several diseases characterized by elevated blood pressure with causes that can be traced to an endocrine abnormality or actual injury. Two of the best known are pheochromocytoma and Cushing's syndrome. The former is characterized by a tumor that secretes excessive amounts of catecholamines, primarily epinephrine. This in turn increases vasomotor tone, resulting in a considerable pressure increase. Surgical removal of the tumor is curative. The latter is a disease of the adrenal cortex where tumors may also develop. The cortex is the site of corticosteroid production. Thus excessive production of aldosterone will act on the kidneys, causing retention of sodium and water. This increased blood volume is usually manifested as mild, or sometimes moderate, hypertension.

10.2.1.2.5. Arteriosclerosis

The term *arteriosclerosis* is derived from the Latin *arteriola*, meaning small artery, and the Greek *sclerosis*, meaning hardening. Hence it is referred to as hardening of the arteries. There are actually a number of pathological conditions characterized by an accumulation on the inner wall of rough-surfaced deposits consisting of fatty materials, fibrin (a blood clotting factor), cellular debris, calcium, and cholesterol. The arteries harden, thicken inside, and lose the elasticity of their walls. (Atherosclerosis is one type of arteriosclerosis.)

10.2.2. Heart Attack

Even though apparently sudden, a heart attack is the end result of a buildup of causes over years, particularly atherosclerosis in the coronary arteries. The immediate cause is a blood clot (thrombus) that, once having formed, blocks the narrowed lumen of a coronary artery. This denies the affected part of the myocardium blood and therefore oxygen. Technically, this may be referred to as a coronary thrombosis or myocardial infarction (MI). Angina pectoris, which should also be mentioned, is characterized by severe pain and oppression above the heart, radiating to the shoulder and down the left arm.

10.2.3. Stroke

A stroke (sometimes termed a "cardiovascular accident" or CVA) is any occurrence that limits or stops the blood supply to a part of the brain. This can happen by either a cerebrovascular occlusion or a cerebral hemorrhage. An occlusion is the more common event. It is caused by the blockade of a cerebral artery by a thrombus that has formed inside of it.

Such a cerebral thrombosis is believed to occur as a result of atherosclerotic damage in the arterial wall. Narrowing of the arterial lumen slows the blood flow. The projection of rough-surfaced deposits into what is by then a trickle of blood may serve as a nucleus around which the thrombus will form, thereby completing the occlusion. A variation is the "wandering" of a blood clot, or embolus, through the vasculature until finally it becomes wedged in a cerebral artery. This is a cerebral embolism.

A cerebral hemorrhage is the result of rupture of a sclerosed or otherwise diseased blood vessel in the brain. The brain cells in the vicinity of the burst vessel are deprived of blood and therefore nutrients. They will mostly die within several minutes. The symptoms may be mild to severe, including loss of speech and memory, as well as an inability to walk. It is believed that atherosclerosis, combined with hypertension, predisposes a person to such a CVA. Intracranial bleeding can also result from a head injury or from the interaction of certain antidepressant drugs with other drugs or even particular foods.

The classic profile of the stroke victim is one with high blood pressure, evidence of arterial hardening, and high levels of cholesterol and other lipid substances in the blood. Diabetes or gout are frequently present as is a long history of smoking. All of these factors have been found to compound the risk of morbidity and mortality greatly.

10.2.4. Congestive Heart Failure

Congestive heart failure is the reduction of the heart's pumping ability to subnormal levels. This in turn results in fluid accumulation in the lungs and body extremities, which is a condition called *edema*. The edema is accompanied by pulmonary congestion and shortness of breath. The actual cause of congestive heart failure (CHF) is damage to the myocardium. This damage may in turn be due to severe hypertension, atherosclerosis, rheumatic fever, birth defects, or heart attacks.

10.2.5. Rheumatic Heart Disease

Rheumatic heart disease, which is relatively rare in Western countries today, is characterized by damaged (scarred and deformed) heart valves. It is always a result of rheumatic fever, which is an infection caused by group A β-hemolytic streptococci. Children aged 5 to 15 are the usual victims. This is a preventable condition, if the initial attack of rheumatic fever is correctly diagnosed and treated. Treatment and prevention of a relapse is carried out by long-term (years) administration of penicillin G, V, or erythromycin (if allergic) and occasionally sulfonamides.

10.2.6. Congenital Heart Defects

With few exceptions (e.g., German measles during pregnancy), the causes of abnormal prenatal cardiac development are unknown. Overall incidence is probably less than 1%. Treatment is usually surgical.

10.3. Drugs

The following groups of drugs will be considered:

1. Antihypertensive
2. Cardiotonics

3. Antiarrhythmics
4. Anticoagulants, antithrombotics, thrombolytics
5. Antianginals: coronary vasodilators
6. Hypolipemics
7. Drugs controlling and modifying diabetes

10.3.1. Antihypertensives

Before considering the drugs, it may be useful to summarize some relevant neurophysiology in the path of an adrenergic impulse originating in a given brain center. The impulse will pass through the brain stem via a preganglionic fiber to the spinal cord. There, it will synapse via ganglia with various postganglionic neurons by releasing ACh from the preganglionic fiber. Postganglionic conduction proceeds to the various nerve endings in the walls of blood vessels and in the myocardium where NE is released. The neurohormone will interact with and stimulate both α- and β-adrenoceptors.

The primary effect of peripheral α-stimulation is vasoconstriction, whereas vasodilation is that of β-stimulation. In the heart α-receptors are less significant. β-Receptors, however, cause augmentation of both chronotropic (rate) and inotropic (contractile) activity. The excessive force with which the heart ventricle expels blood is pitted against a less stretchable arterial system, caused by arterial constriction. The net result is an increase in blood pressure—both systolic when the heart is contracting, and diastolic when it is dilated and relaxed.

It has been established that hypothalamic, and possibly other areas of the brain, also contain both types of adrenoceptors on neurons. The effect on blood pressure caused by receptor stimulation is opposite to that in the peripheral smooth musculature. That is, α-stimulation of these brain receptors results in a decrease in stimulation due to the inhibitory effect of sympathetic outflow (see Chapter 9).

The overall mechanism is actually considerably more complex since the adrenergic system interrelates with other pressor mechanisms involving, for example, angiotensin and serum sodium levels.

By stimulating the kidney to elaborate renin and thereby angiotensin II, adrenergic impulses in turn stimulate the release of aldosterone from the cortex, resulting in water and Na^+ retention (see Fig. 10-1).

Antihypertensives pharmacologically are the most diverse and interesting group of drugs. They are therapeutically the most promising of the cardiovascular agents in spite of the fact that essential hypertension is an incurable disease with an unknown cause. Most major complications traceable to high blood pressure are currently preventable by antihypertensive drug treatment. What may have once been considered benign essential hypertension (a modest elevation above normal) is not really benign: It does increase the risk of disability and death. It should therefore be treated.

The controversy as to whether antihypertensive therapy really prevents complications and alters the mortality prognosis has definitely been settled in favor of drug treatment. The historic Veterans Administration Cooperative Study on antihypertensive agents showed conclusively that major complications in patients with severe hypertension (diastolic pressure of 115–129 mmHg) were reduced by a factor of 27 (Veterans Administration, 1967, 1970). Even patients with moderate hypertension (105–114 mmHg) were found to have morbidity decreased by 4:1.

The existence of a large number of chemically and mechanistically unrelated drugs illustrates the complexity of the disease and its etiology.

The drugs discussed in this section are primarily, but not exclusively, those that have a depressant effect somewhere in the sympathetic nervous system. The diuretics that have definite antihypertensive properties (as well as beneficial effects in CHF) are dealt with separately in this chapter.

Table 10-1 summarizes some of the basic mechanisms by which blood pressure may be lowered and gives examples of drugs that accomplish this, at least to some degree.

Since the cause of hypertension is unknown, the development of useful antihypertensives has been largely empirical. Developments in research seeking to understand some of the functions of the cyclic nucleotides, such as *c*-AMP, are beginning to be applied in a more rational design of drugs. Some of these efforts will be described.

In general, the ideal antihypertensive drug should:

1. Be active on oral administration;
2. Have a useful duration of action;
3. Be effective in all hypertensive patients;
4. Have minimal side effects;
5. Prevent pathological organ changes;
6. Reverse pathological organ changes that have occurred;
7. Have reasonable cost.

Items 4 and 5 are particularly desirable since therapy, once begun, is likely to be lifelong.

At the risk of oversimplification, it can be stated that chemotherapeutic attempts to lower blood pressure are achieved by one or both of the following pharmacological procedures:

1. Reduction of peripheral resistance by reduction of sympathetic activity, vascular reactivity, or NE-mediated vasoconstriction;
2. Reduction of cardiac output by lowering cardiac tone, reactivity, or the return of fluid volume.

Table 10-1. Antihypertensive Drug Mechanisms

Mechanism	Drug example
Ganglionic blockade	Trimethaphan
Central stimulation of α-receptors	Clonidine
Postganglionic neural depletion of NE	Reserpine
Postganglionic adrenergic blockade	Guanethidine
β-Adrenoceptor blockade	Propranolol
β-α_1 blockade	Labetalol
Catecholamine synthesis blockade	Metyrosine
α-Adrenoceptor blockade, reversible–irreversible	Prazocin Phenoxybenzamine
Vasodilation, direct-acting	Minoxidil
Angiotensin-converting enzyme inhibition	Captopril
Calcium slow-channel blockade	Nifedipine
Monoamine oxidase inhibition	Pargyline
Combined blood volume reduction (initially) and reduced peripheral resistance (long term)	Thiazide Diuretics
Renin inhibition	None yet

10.3.1.1. Inhibition of Peripheral Sympathetic Function

10.3.1.1.1. Ganglionic Blocking Agents

As with infectious diseases, hypertension was in essence not treatable until the late 1930s, and then only with sedative drugs. The first surgical sympathectomies were done in the 1940s; also the salt-free diet was "invented" then. Sodium nitroprusside (see later), which was known since the 1920s, was not recognized for its clinical hypotensive potential for over two decades. About that time veratrum alkaloids began to be used, and the hypotensive actions of dibenamine (see later) were first reported in 1947. That autonomic ganglia were stimulated by the alkaloid nicotine, and later tetramethylammonium [$(CH_3)_4N^+$], was known in the late nineteenth century. It is curious that the homolog tetraethylammonium [$(C_2H_5)_4N^+$, TEA] has the opposite effect. Pharmacological applications, however, were not proposed for another three decades. Corollary work led to the *neuromuscular* blocker decamethonium (Chapter 8). It also led to the finding that two quaternary nitrogens separated by a shorter distance, one equivalent to five or six CH_2 groups, produced primarily ganglionic blockade and a clinically significant drop in blood pressure. Pentolinium (Table 10-2) was soon added. For sustained therapy, however, these *bis*-quaternary drugs were much less satisfactory. Sympathetic blockade, of course, did produce the predicted vasodilation and fall in blood pressure, but oral absorption was poor and erratic, making efficacy unpredictable. Furthermore, since these drugs cannot differentiate between sympathetic and parasympathetic ganglia, the effects produced by blockade of the latter also plagued

Table 10-2. Ganglionic Blockers with Antihypertensive Properties[a]

Drug	Structure
Tetraethylammonium (Etamon)	$(C_2H_5)_4N^+$
Hexamethonium	$(CH_3)_3N^+\text{-}(CH_2)_6\text{–}N^+(CH_3)_3$
Pentolinium (Ansolysen)	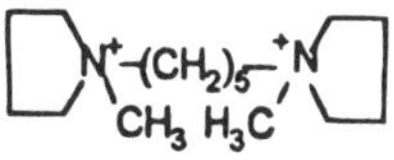
Trimethaphan (Arfonad)	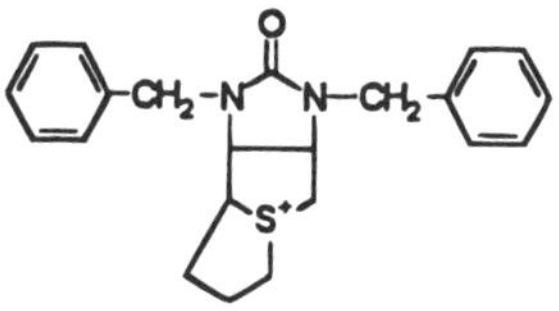
Mecamylamine (Inversine)	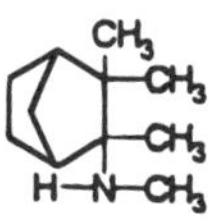
Pempidine	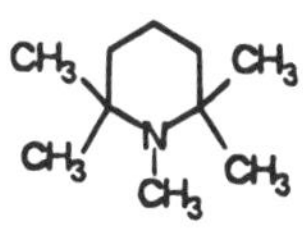

[a] Trade name in parentheses.

patients. Thus tachycardia, mydriasis, constipation, urinary distention, and blurred vision all arose.

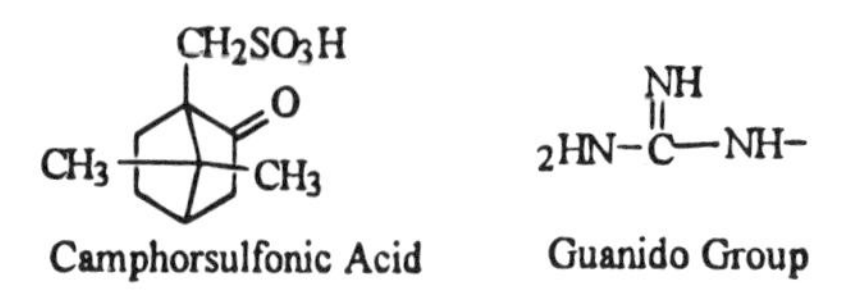

Camphorsulfonic Acid Guanido Group

The development of several noncompetitive drugs that were not quaternary amines solved the oral availability problem, but not the others. Mecamylamine, a secondary, and pempidine, a tertiary, amine were two such compounds, yet the side effects remained. For example, mecamylamine (Inversine) can produce ileal paralysis and had to be administered with cathartic agents. The only justifiable use, in view of the newer drugs, for ganglionic blockers today is for hypertensive emergencies such as malignant hypertension, or for neurosurgical procedures requiring controlled hypotension. Trimethaphan, given intravenously, permits this. Its short duration allows careful titration. As the camphor sulfonate (camsylate) salt it is about the only ganglionic blocker still in use in the United States. Trimethaphan is a sulfonium, rather than an ammonium drug. The other agents are considered obsolete.

10.3.1.1.2. Rauwolfia and Its Alkaloids

The dried roots of the plant *Rauwolfia serpentina*[3] and other species have been used in India for many centuries to treat various maladies. Its real value to Western medicine (tranquilizer and antihypertensive) was not appreciated until the 1950s following the identification of its alkaloidal contents (Table 10-3). Today the alkaloids are used in their pure state or as the product alseroxylon (Rauwiloid), a fat-soluble alkaloidal fraction from *R. serpentina* containing reserpine rescinnamide. Standardized powdered whole root (Raudixin) is also still in use. Combinations of these alkaloids with diuretics may be more effective antihypertensive products.

The antihypertensive mechanism of the rauwiloid alkaloids primarily involves depletion of catecholamines from stores by blocking their reuptake mechanism and thereby storage in neuronal vesicles. The effect is widespread both peripherally and centrally, and includes NE, DA, and 5-HT (serotonin). Depletion of the amines affects blood vessels, the heart, adrenal medulla, and possibly other tissues as well. Since it is the reuptake and not the release of NE that is inhibited in the postganglionic adrenergic neuron, the existing pool must be fully depleted before antihypertensive effects become apparent. The drug also binds to vesicular membranes for days, accounting for the irreversibility of the process.

Although the drug is usually well tolerated in the elderly at low doses, sedation is not uncommon at higher doses. Parkinson-like symptoms can arise, presumably due to lowered central dopamine levels. The most serious potential hazard is psychic depression with suicidal tendencies.

Other sympatholytics (i.e., neuronal blocking compounds that interfere with postganglionic adrenergic endings) include two older drugs, guanethidine and bretylium, as well as debrisoquin and, most recently, guanadrel.

---

[3] Named after the sixteenth century German physician–botanist Rauwolf.

Table 10-3. The Rauwolfia Alkaloids

Alkaloid[a]	*R*	*R′*	Botanical source
Reserpine (Serpasil)	$-C(=O)-C_6H_2(OCH_3)_3$	OCH_3	*R. serpentina, micrantha, tetraphylla*, and *vomitoria*[b]
Rescinnamine (Moderil)	$-C(=O)-CH=CH-C_6H_2(OCH_3)_3$	OCH_3	*R. vomitoria*[b] and other spp.[c]
Deserpidine (Harmonyl)	$-C(=O)-C_6H_2(OCH_3)_3$	H	*R. canescine and recanescine*

[a] Trade name in parentheses.
[b] *R. vomitoria* contains resins that must be separated.
[c] Can also be semisynthesized from the more abundant reserpine by hydrolyzing off the trimethoxybenzoate and reesterifying with 3,4,5-trimethoxycinnamic acid.

Despite its serious clinical shortcomings, guanethidine, which was introduced in 1960, still finds use in situations of severe or refractory hypertension, both essential and due to secondary causes (e.g., renal stenosis). Its main antihypertensive effect is now believed to be neuronal blockade by prevention of NE release from storage sites. Like the rauwolfia alkaloids, there is also NE depletion, but this only becomes significant with very large doses. Because the guanido group imparts such high basicity to this and other guanido-containing drugs (pKa = 12), guanethidine is 99.99% protonated at physiological pH. It does not cross the BBB and, unlike reserpine, exerts no CNS-mediated sedation. Since it does not affect autonomic ganglia, it rapidly superseded the ganglionic blockers clinically and can be viewed as the evolutionary bridge to the newer drugs that followed. The high intensity of its effectiveness and its mechanism, however, are still responsible for its very disconcerting untoward effects. These include a high incidence of orthostatic hypotension and faintness, constipation, and ejaculatory disturbances in men. The more recently introduced

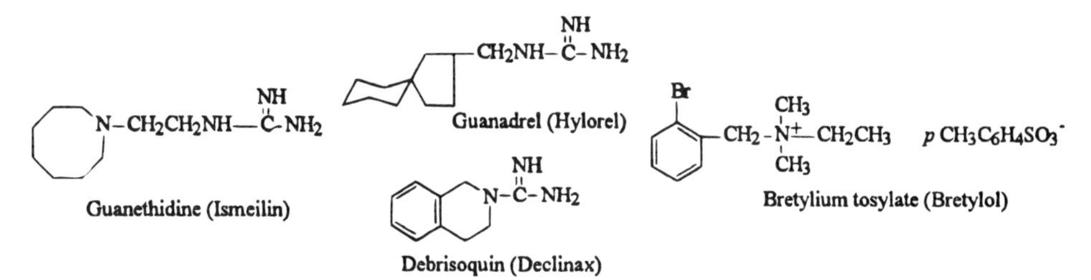

guanadrel shares guanethidine's mechanism, but it may have certain advantages in a shorter duration and more prompt onset of action. Bretylium has adrenergic neuronal blocking activity that antagonizes the response of effector organs to postganglionic nerve action without, however, antagonizing injected catecholamines. They actually reverse the effects of bretylium. The drug is hypotensive. As it is a quaternary amine, however, oral absorption is erratic and tolerance is also rapidly induced. Bretylium has recently enjoyed a renaissance as an antiarrhythmic (see later). The utility of α-methyltyrosine (metyrosine) as an antihypertensive in pheochromocytoma was presented in Chapter 9.

10.3.1.1.3. β-Adrenoceptor Blocking Agents

In Chapter 9 the nature of the β-adrenoceptor was developed. The discussion briefly delved into the concept of β-blockade or, more correctly, partial blockade, of the misnamed dichloroisoproterenol (DCI). Further discourse on the blocking agents was postponed on this clinically important category of drugs, which are generally referred to as β-blockers, to this chapter. The pharmacology of this type of blockade finds clinical application in the therapy of arrhythmias, angina pectoris, and hypertension. Indications outside this area such as glaucoma should also be mentioned.

Before proceeding with the chemical aspects of these compounds, a brief review of cardiac biochemistry and pathophysiology will be helpful in understanding how these uniquely useful drugs evolved by design.

Initial therapeutic utility of the β-blockers was as antianginal drugs. The symptoms that characterize an angina pectoris attack are known to be caused by or related to oxygen imbalance. The pain is believed to result from myocardial anoxia. Agents with the ability to reduce oxygen consumption of the heart muscle should provide relief. The effect is achieved primarily because β-adrenergic blockers decrease the rate and work of the heart, which subsequently reduces oxygen demand. A factor that can increase oxygen consumption through β_1-receptor stimulation is the trigger of fear, anxiety, anger, or other forms of stress, resulting in an anginal syndrome. The β-blockers are useful both in acute situations and prophylactically.

The effects that are beneficial in angina pectoris include a decreased heart rate, decreased myocardial contractility, and, apparently, also an improved blood flow to hypoxic (oxygen-starved) areas of the heart. The latter is probably a mechanical effect. It is interesting that until the advent of β-adrenergic blockers, the only useful drugs used to treat angina were the so-called coronary vasodilators such as organic nitrates (nitroglycerin and others). Ironically it now appears that their salutary action is not due to vasodilation but rather primarily to a decrease in myocardial oxygen consumption.

In Chapters 5 and 8 some consideration was given to *c*-AMP and adenyl cyclase with respect to the adrenergic receptor. They will now be considered in greater detail.

Cyclic 3′,5′-adenosine monophosphate (*c*-AMP) is known to be of key significance as a regulator of cellular metabolism. In mammals it functions as a second messenger for a number of hormones. Abnormalities in *c*-AMP formation, or in its action, seem to be involved in the pathology of various hormonal and nonendocrine diseases.

The original work by Sutherland and co-workers (Robinson et al., 1971)[4] involved glucagon and epinephrine and their relationship to *c*-AMP and phosphorylase A and B

---

[4] E. W. Sutherland received the Nobel Prize in 1971 for this and related work.

enzymes. They were able to demonstrate that the hormones acted by increasing the levels of *c*-AMP. *c*-AMP is formed from ATP in the presence of Mg^{2-} under the catalytic influence of adenyl cyclase (adenylate cyclase) (Fig. 10-2). Its level is additionally regulated by hydrolytic degradation (to the 5′-monophosphate) by phosphodiesterase (PDE), an enzyme more widely distributed than the cyclase. Adenyl cyclase activation can be achieved either by stimulation of β-adrenergic receptors or by blockade of stimuli to α-adrenoceptors (e.g., with phentolamine). The latter apparently removes the normal inhibitory effect of α-adrenoceptors on the cyclase. After the *c*-AMP is formed, in the presence of Ca^{2+}, it gives the physiologic results observed, namely an increased cardiac rate and contractile force as well as a relaxant effect on smooth musculature that lowers blood pressure.

We actually have three ways to achieve increased levels of c-AMP in tissues: (1) activate β-receptors, which results in cyclase stimulation, (2) block α-receptors, or (3) inhibit phosphodiesterase. These then are some, but, as will be seen, not the only, approaches to improve or correct pathophysiological defects in our cardiovascular system.

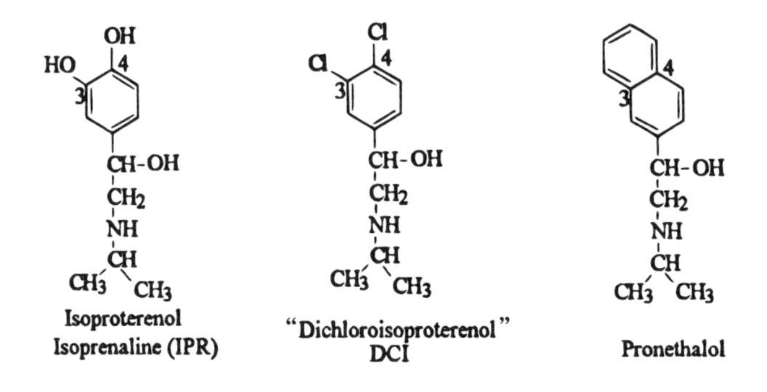

It will be recalled (Chapter 9) that Ahlquist's proposal for the existence of α- and β-adrenoceptors to explain apparent anomalies of sympathomimetic compounds was initially roundly ignored, if not rejected. It was the synthesis of DCI that established that β-recep-

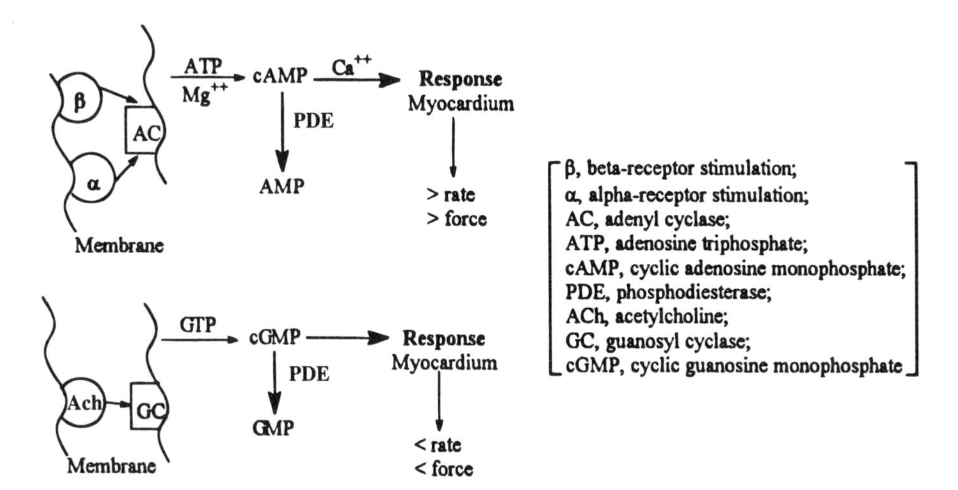

Figure 10-2. Relationship of autonomic receptors and cyclic nucleotides to musculature.

tors, like α-receptors, could also be chemically blocked and therefore exist. Since IPR and DCI differ "only" in that the latter has Cl atoms replacing phenolic hydroxyls, it may be tempting to speculate that increased liposolubility may be the reason for the partial appearance of β-antagonism. The difference should be considerable (just compare phenols's aqueous solubility of 6.7% with chlorobenzene's immiscibility). The inductive and resonance characteristics of these two functions may also give a clue. It should be remembered that while both aromatic hydroxyls and halogens are *ortho–para* directors for electrophilic substitution reactions, halogen deactivates the ring toward further substitution reactions, while the OH groups do the opposite. Total substitution at positions 3 and 4 with additional aromaticity produced pronethalol, which was the first potent β-antagonist.

It is interesting that at the time there was no clear appreciation of the potential clinical applications for β-adrenergic blockade until the proposal outlined later. This was that inhibition of stress-induced sympathetic stimulation of β-adrenoceptors in cardiac tissue would decrease its oxygen demand and effectively treat angina. The clinical introduction of pronethalol for the treatment of angina pectoris proved the drug to be a more potent antagonist of β-receptors than DCI. However, it still retained a degree of agonism referred to as intrinsic sympathomimetic activity (ISA). It soon became apparent that pronethalol had two additional highly desirable pharmacological features: suppression of cardiac arrhythmias and the ability to lower blood pressure in hypertensive persons. The drug was withdrawn soon after its inauguration due to results of long-term feeding studies that indicated lymphoid tissue tumors in mice (but curiously not in rats or dogs).

Ar–CH(OH)–CH_2NH–R   Ar–OCH_2–CH(OH)–CH_2–NH–R

Arylethanolamines   Aryloxypropanolamines

Research for additional improved β-blockers concentrated on two areas. Structural variations on arylethanolamine type that, it will be remembered, was successfully represented in the β-stimulants, and introduction of linking atoms between the aryl portion of the molecule and the ethanolamine side chain. The most successful compounds from this effort were those where the bridging was the oxymethylene -OCH_2- moiety. These are the aryloxypropanolamines, which now represent the overwhelming majority of β-blockers (Table 10-4). The first replacement for pronethalol was propranolol (No. 9), whose potency it exceeds 10 times. Propranolol has become the prototype drug for comparison of subsequently developed compounds in terms of potency as a β-blocker, degree of retained (partial) β-agonism and various other pharmacological parameters considered later.

Partial agonism, or ISA, represents the degree to which the compounds retain the ability to stimulate β-receptors directly while also having the "new" property of blocking (i.e., binding) but not eliciting a direct response. Eleven of the 16 drugs listed in Table 10-4 have this ability to a degree. Evaluation is in rats whose NE levels have been depleted with reserpine and are then challenged by the blocker to determine the absolute increases in their heart rates. Thus propranolol caused no effect while pronethalol caused an increase of 70 beats per minute (bpm). IPR produces a 200 bpm increase. In the case of pindolol (No. 8) sympathomimetic agonism is less than 50% of the maximum response of IPR.

Almost half the β-blockers listed in Table 10-4 also exhibit a membrane-stabilizing action (MSA) on myocardial muscle fibrils. It has also been described as a depressant action on excitable tissue. Propranolol, among clinically available compounds, exhibits this property to the highest degree. The effect can be viewed pharmacologically as a local

Table 10-4. β-Adrenergic Blocking Agents[a]

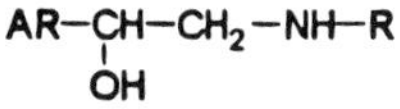

$AR{-}\underset{OH}{CH}{-}CH_2{-}NH{-}R$

Drug	AR	R	β_1-selective	MSA[b]	ISA[a]	Lipo[k]
1. Labetalol (Trandate, Normodyne)	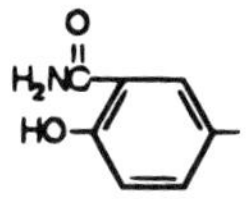 O, H_2NC, HO	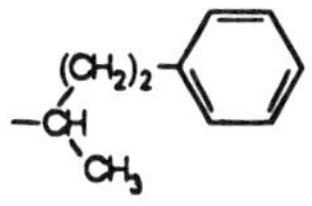 $(CH_2)_2$, –CH, CH_3	No[d]	Yes	No[c]	high
2. Sotalol[f] (Stotalex)	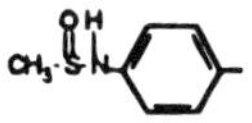 CH_3-S, O, H	-$CH(CH_3)_2$	No	No	No	low
$AR{-}O{-}CH_2{-}\underset{OH}{CH}{-}CH_2{-}NH{-}R$	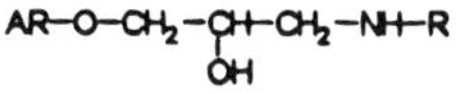					
3. Alprenolol[f] (Aptine)	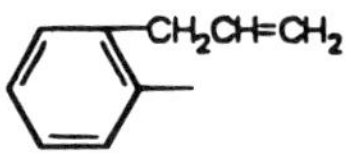 $CH_2CH{=}CH_2$	-$CH(CH_3)_2$	No	Yes	Yes	moderate
4. Carteolol[f] (Cartrol)	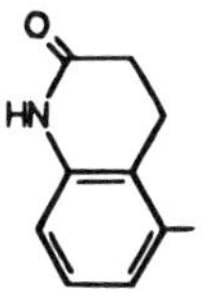 O, HN	-$C(CH_3)_3$	No	No	Yes	—
5. Levobunolol[g] (Betagan)	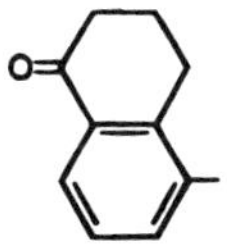 O=	-$C(CH_3)_3$	No	No	No	moderate
6. Nadolol (Corgard)	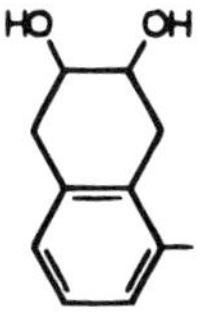 HO, OH	-$CH(CH_3)_2$	No	No	No	low
7. Oxoprenolol[f] (Trasicor)	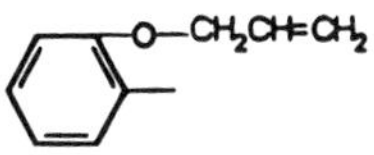 O–$CH_2CH{=}CH_2$	-$CH(CH_3)_2$	No	Yes	Yes	moderate
8. Pindolol (Visken)	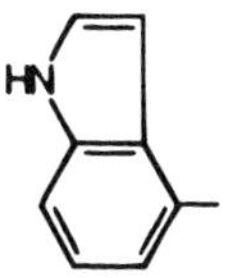 HN	-$CH(CH_3)_2$	No	Yes	Yes[h]	moderate
9. Propranolol (Inderal)	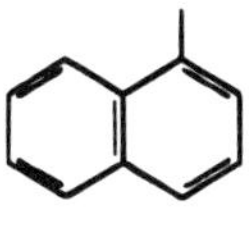	-$CH(CH_3)_2$	No	Yes	No	high

(Continued)

Table 10-4. *(Continued)*

Drug	AR	R	β_1-selective	MSA[b]	ISA[a]	Lipo[k]
10. Timolol (Blocadren, Timoptic)	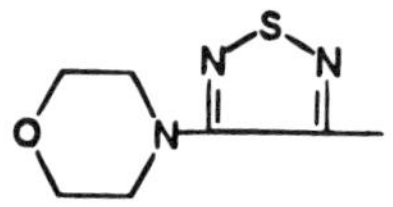	-C(CH$_3$)$_3$	No	No	No	low
11. Acebutolol (Sectral)	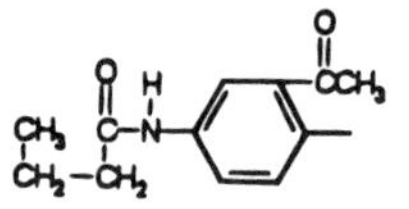	-CH(CH$_3$)$_2$	Yes	Yes	Yes	low
12. Atenolol (Tenormin)	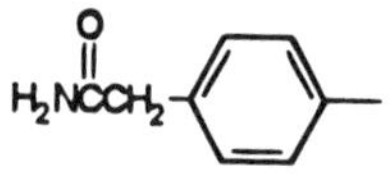	-CH(CH$_3$)$_2$	Yes	No	No	low
13. Betaxolol (Betoptic)	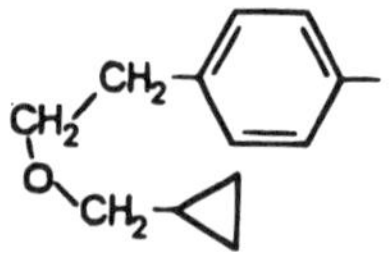	-CH(CH$_3$)$_2$	Yes	Yes	No	low
14. Esmolol[j] (Brevibloc)	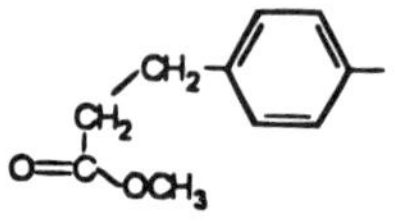	-CH(CH$_3$)$_2$	Yes	No	No	?
15. Metoprolol (Lopressor)	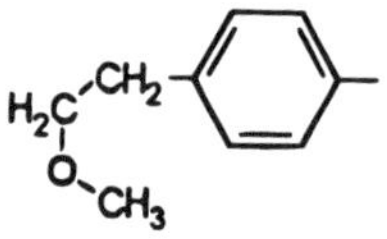	-CH(CH$_3$)$_2$	Yes	Yes	No	moderate
16. Penbutolol (Levatol)	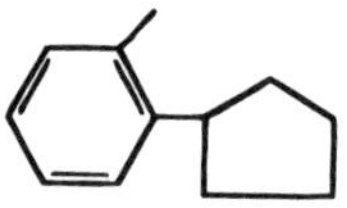	-CH(CH$_3$)$_2$	No	?	Some	High

* Data compiled from numerous sources.
[a] Trade names in parentheses.
[b] Membrane-stabilizing activity (see text).
[c] Intrinsic sympathomimetic activity (partial agonist; see text).
[d] Also α_1-blocker; see Chapter 9.
[e] Only shown in rat uterus, not cardiac and tracheal tissue.
[f] Not yet on U.S. market.
[g] Used in glaucoma only.
[h] Ophthalmic glaucoma formulations.
[j] Self-canceling ester, $t_{1/2} = 9$ min. IV use only (see text).
[k] Lipo = lipophilic.

anesthetic property. In fact, propranolol and pronethalol are experimentally twice as potent as procaine. Procainamide and quinidine, of course, are widely used as cardiac antiarrhythmics (see later). Therefore, when propranolol was also found to have clinically useful antiarrhythmic activity, it was assumed, not surprisingly, that this was the β-blockers' mode of action, too. However, it became apparent that this electrically effected cell membrane stabilization was not demonstrable at the usual clinical doses. In fact, not all β-blockers share this property. In a study of those that do it was shown that MSA was directly related to partition coefficients. In fact, propranolol, the most lipid-soluble drug in the series studied, was listed as an effective local anesthetic. A quick perusal of Table 10-5 will make the relationship of MSA to lipophilicity apparent. Those compounds possessing MSA exhibit it only at 100 times the levels needed to demonstrate β-blockade. This, and the fact that β-blockers devoid of MSA are still clinically antiarrhythmic, brought into question the significance of MSA as an explanation for this cardiac depressant mechanism.

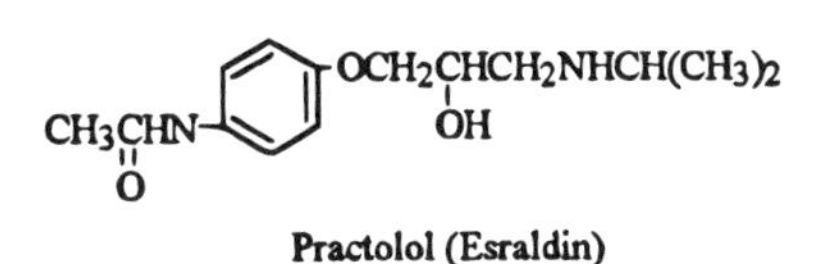

Practolol (Esraldin)

Recognition of relative lipophilicities of these drugs as a significant factor in their varied and complex pharmacology became more important. The suspicion that propranolol's centrally induced side effects and their high incidence may be in part due to its large log *P* value of 3.65 (i.e., a 4500:1 octanol/H_2O partition rate) appeared well founded. It certainly seemed prudent to synthesize some analogs with hydrophilic moieties propitiously placed and see if the side effects would decrease. Selecting *p*-acylamino groups as the hydrophilic moiety, Coleman (1979) synthesized a group of 12 *p*-acylphenoxyethanol- and propanolamines, selecting one for clinical trials (Table 10-5). It was later marketed as practolol. This compound, though less potent than propranolol and exhibiting some ISA, had one property not previously seen with β-blockers: the ability markedly to inhibit IPR-induced tachycardia, while having a minimal or no effect on the IPR depressor (hypotensive) response. In other words, this drug exhibited *cardioselectivity*. For the first time it was shown that inhibition of β-adrenergic response could be restricted to certain sites. It is interesting that the study showed that moving the acylamino group to the *meta* or *ortho* positions on the benzene ring caused loss of this selectivity, but not, of course, β-blockade itself (Table 10-4). An examination of subsequently developed cardioselective β-blockers in Table 10-4 illustrates the significance of *para*-substitution (Nos. 12, 13, 14, 15). After several years practolol was withdrawn due to rare, but serious, ophthalmic and dermatologic toxicities that have resulted in blindness and fatalities. Neither animal studies nor clinical trials have predicted such toxicities, nor have they been explained. None of the other congeners of practolol currently in worldwide use has encountered similar difficulties. Practolol may have been a unique case.

Table 10-5. Cardioselectivity versus Hydrophilicity of β-Blockade[a]

R	log *P*	β_2/β_1
p-$NHCOCH_3$	0.79	133
m-$NHCOCH_3$	0.74	2.9
p-CH_2CONH_2	0.23	40
m-CH_2CONH_2	0.22	0.5
p$CH_2NHCOCH_3$	0.32	17
m-$CH_2NHCOCH_3$	0.34	1.5

[a] Adapted from Coleman et al., 1979.

Table 10-6 represents a summary of empirically arrived at structure–activity relationships. The exception of *para*-aryl substitutions for cardioselective β-blockers (IIB 1 and 4) should be noted but cannot be satisfactorily explained. In fact, not all the *para* substituents are as lipophilic as the amide ones in practolol, acebutolol (No. 11), atenolol (No. 12), or the ester group in esmolol (No. 14). The methoxyethyl group of metoprolol (No. 15) and the large hydrocarbon nature of the cyclopropylmethoxyethyl function of betaxolol (No. 13) would hardly be expected to increase aqueous solubility of these compounds (see Table 10-4). Still, all the β_1-selective compounds are much less lipophilic than propranolol. Such apparent data correlations may have contributed to a belief that a cardioselectivity–hydrophilicity relationship exists due to some putative hydrophilic site on the β_1 receptor. This notion, however, was dispelled by a study comparing hydrophilicity, substituent positions, and cardioselectivity in three sets of 1-(2-propylamino)-3-phenoxy-2-propanols. The data clearly show that for each set of compounds even with identical lipophilicities (log *P*) the β_1-activity resided primarily in the *para* isomer.

Finally, it will be noted that of these five cardioselective agents, only acebutol (No. 11) possesses ISA. The others, which are devoid of this property, all have at least one CH_2 group inserted between the aromatic ring and the hydrophilic moiety, be it amide, ester, or even an ether oxygen. This appears to be a general feature.

Table 10-6. Empirical SARs of β-Blockade

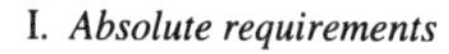

I. *Absolute requirements*

a. aromatic ring[a]
b. β-ethanolamine

II. *Variables*

3 2 4 1 5 6 Y β α –X–CH–CH$_2$–NH–R OH

A. *Side chain*

1. X = direct link or -OCH_2-
   (CH_2CH_2, CH = CH, SCH_2, NCH_2 = < or no activity)
2. R = secondary substitution only; branched optimal
   (all drugs are *i*Pr or *t*-Bu; may include phenyl[b])
3. Alkyl (CH_3) substituents on α, β, or γ (if $X = OCH_2$) all < β-blockade, especially α

B. *Aromatic substitution*

1. Generally $o > m > p$
2. Large p substituents usually < ; o retains some activity[c]
3. Polysubstitution: 2,6-inactive; 3,5-some activity
4. Cardioselectivity, primarily para[d] $> m > o$

C. *Stereochemistry*

1. All β-blockade is in one isomer: ethanolamine-*R*, aryloxypropylamine-*S*;
   absolute configuration is same (active isomers are levo rotatory as well)

[a] Can be benzoheterocyclic such as indole (pindolol, No. 8), or heterocyclic such as thiadiazole (timolol, No. 10).
[b] Ring systems larger than naphthyl (e.g., anthracene) are not useful.
[c] The larger *o*-substituents have <ISA.
[d] Apparent exception to B_1; however, compared to nonselective blockers they are less potent.

Esmolol (No. 14), the first β-blocker pro-drug, has been introduced for IV use when rapid, but not continual, residual blockade is desirable (e.g., acute situations where prolonged cardiac depression must be avoided). Ultrashort duration of activity could solve such a problem.

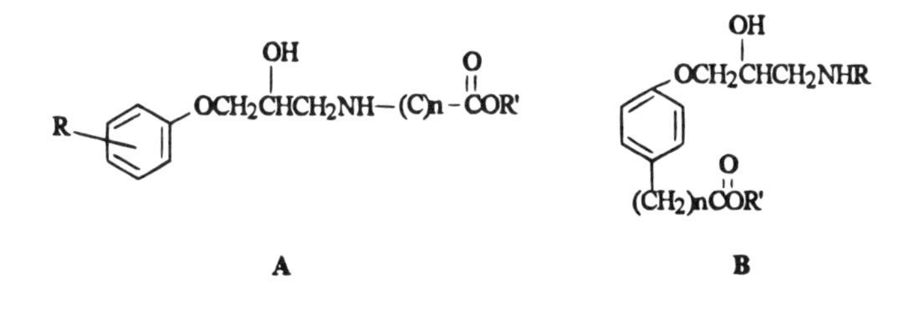

The rationale to develop such a drug involved several considerations. The basic idea was to incorporate a structural component into the frame of known β-blockers that will (1) be susceptible to rapid metabolic degradation in serum or "first phase" hepatic metabolism, (2) by its chemical similarity, in electron distribution (isosterism) or steric dimensions, mimic the pharmacology desired, and (3) following degradation lead to a product that is itself inactive or nearly so and will not be toxic. These requirements, of course, are quite different from those usually sought in a pro-drug. In many cases site selectivity of metabolism is desired, such as stability in gastric juice, but not in serum (e.g., pivampicillin, Chapter 7), or modifying structures (usually by steric hindrance) to actually prolong duration of action (e.g., tocainamide, this chapter). It is a requirement in the majority of situations that the pro-drug be inactive and the metabolite be the pharmacologic species, while here the reverse was the goal.

The logical choice, then, was an ester linkage. First, it is almost always bioisosteric with amides. In addition, the presence of nonspecific esterases in serum and other body tissues should lead to rapid metabolism. In reverse situations where the goal is increased in vivo stability amidic substitution for an ester is frequently successful (e.g., procaine → procainamide).

With traditional β-blockers the β-receptor site interacts with either a protonated cationic nitrogen or an unprotonated nitrogen carrying bulky, lipoidal alkyl groups (e.g., *t*-butyl). The ester pro-drug here on hydrolysis will, of course, bear a very hydrophilic negatively charged carboxylate ion (COO-) that is unlikely to interact with the requisite receptor site and will therefore be essentially without effect.

The initial attempts were the synthesis of compounds represented earlier by structure A, where the ester function was "built" into proximity of the propanolamine. Even though shortened duration was achieved when a short-term infusion (40 min) was carried out, their duration tended to increase considerably after the administration of the infusion itself was prolonged (ca. 3 hours). A particularly novel spontaneous inactivation mechanism independent of enzyme activity was found to occur with compounds of structure A, where *n* was at least three (Eq. 10.1). This *intramolecular* cyclization reaction (which also occurs in vitro at neutral or basic pH), while not producing a carboxylate, does result in a predictably inactive tertiary amide product. In vivo testing of the gamma-lactam derivative did in fact show it to be inactive.

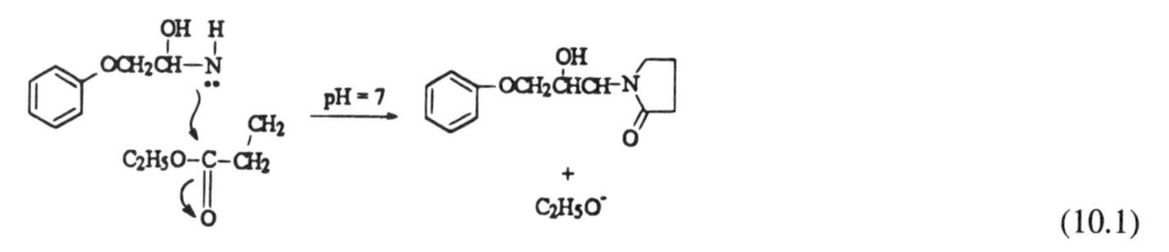

(10.1)

The subsequent approach was to model the synthesis after practolol and similar cardioselective compounds by isosteric replacement of amidic moieties with esters. Esmolol (No. 14) was the compound meeting all the needed requirements. It is interesting that the number of CH_2 between the benzene ring and the ester group did not affect the β-blocking activity of the compounds, but did have a significant effect on the duration of action, measured as the number of minutes needed to achieve 80% recovery from a 50% blockade in dogs. Thus the time with one CH_2 was 20 minutes, $2CH_2$ 12 minutes, and $3CH_2$ 40 minutes. Esmolol is the 12 minute compound.

Research in the area of β-blockers presently seems to increase selectivity and potency toward β_1 activity. Aryloxypropanolamines, where the traditional isopropyl and *tert*-butyl groups on the nitrogen atom give way to arylureas and acetamides (later), offer some promise.

One of the newest clinical additions to this series is penbutolol (No. 16). Its β-blockade is about four times that of propranolol. It is not β_1 selective as the addition of the lipoidal cyclopentane would likely predict. Its main clinical difference is once-a-day dosing and a prolonged onset of action (2 weeks or more).

On the face of it, high cardioselectivity (β_1) in blockers would seem clinically desirable, especially in antiarrhythmic and anginal applications, because β_1 receptors in the myocardium, conducting tissues, and the pacemaker appear to be involved. Furthermore, simultaneous β_2-blockade may lead to bronchoconstriction in asthmatic patients. It should be remembered, however, that the currently available cardioselective drugs are not as cardiospecific and not as potent as one might wish. Therefore, at elevated doses as β_2-antagonism rises, the selectivity will decrease. The solution would be a combination of higher selectivity and potency in the same molecule.

As stated earlier, the complexity of our nervous system is baffling. Research seems to corroborate this statement. Examination of human atrial tissue (from biopsies) by means

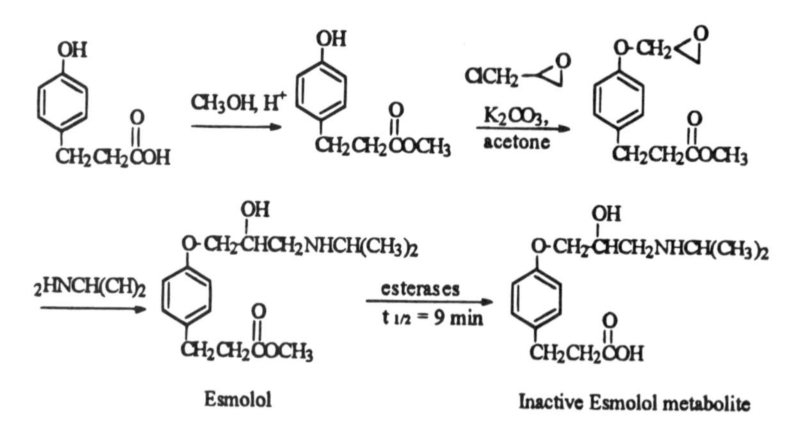

Figure 10-3. Synthesis and metabolism of Esmolol.

of radiolabeled β antagonists has shown it to contain both β_1 and β_2 receptors. This coexistence helps to explain the increased heart rate in asthmatics even if given highly selective β_2 agonists. Much of the effort for highly specific β_2-agonists has been based on the convenient belief that the heart had only β_1-receptors. Furthermore, it was suggested that the facilitation of K^+ influx from extracellular areas into cardiac cells resulting from β_2-stimulated activation of ATPase is not counteracted by cardioselective β_1 blockers, but can be by nonselective drugs such as propranolol. This raises the possibility that cardioselectivity in a β-blocker might not be necessarily desirable. Research by several groups also indicates certain effects of β-blockers on the release of presynaptic NE, possibly involving β_2-receptors. The ramifications of this, if verified, remain unknown.

It has been demonstrated quantitatively that increasing doses of some β-agonists decreases the number of β-receptors; the reverse is true for β-blockers, where β-adrenoceptors actually increase with higher doses. Results were obtained with human lymphocytes and rat heart and lung tissue. These findings may explain the decreased effectiveness in restoring normal breathing in active asthmatics with frequent and continuous use of β-agonists. At the very least such information will require some reassessment of drug design goals in this area and the clinical applications of such compounds.

The information on the actions of β-blockers is extensive, but not total. Some of their many mechanisms are understood, at least at the physiological if not at the molecular level. Others remain unknown. Herewith an outline:

1. β-blockers reduce platelet aggregation induced by catecholamines and ADP. Therefore, transient blocking of coronary arteries already stenosed can be inhibited.
2. The lipolysis induced by catecholamines increases free fatty acid (FFA) levels, which, when entering the myocardium during an infarct, increases the heart's oxygen needs to metabolize these acids when the myocardium can least afford it. β-Blockers reduce this O_2 consumption and workload, apparently by shifting metabolism toward carbohydrates.
3. Our understanding of the antihypertensive effects of β-blockers still has major gaps, probably because many factors leading to hypertension are still unknown. Those that are known include the inhibition of various cardiovascular effects induced by excess catecholamines, and also the release of renin. The significance of that latter will become clearer later in this chapter. It is apparent, however, that there is more to it.

Extensive trials have shown some β-blockers to be effective in preventing persons from reinfarcting after a first MI or heart attack. In particular, a 3-year multicenter study in Norway involving 1,884 men with confirmed MIs showed an overall mortality reduction of over 39%. Other than to say that the effects of increased sympathetic activity are reduced on the heart, no mechanism has been put forward.

Other conditions that may in some way relate to excess sympathetic activity are also being successfully treated with β-blockers. These include akathesia, anxiety, hyperthyroidism (where tremors may be due to overproduction of catecholamines), menopause, migraine, and other types of tremors (it may be recalled that excessive β-agonist dosing in asthmatics can cause tremors).

A novel application of propranolol as a spermicide is under investigation. The mechanism may involve sperm immobilization by way of the drug's MSA. The less effective *R* enantiomer is used to minimize side effects.

Three β-blockers are currently in use in the treatment of glaucoma in the form of eye preparations: timolol (Figure 10-5, No. 10), levobunolol (No. 5), and betaxolol (No. 13).

The actual mechanism by which these drugs reduce intraocular pressure (IOP) is not known. It seems likely that formation of aqueous humor is suppressed in the ciliary body. One explanation is that IOP reduction is achieved by blockade of β-receptors in the ciliary bodies. The theory would predict that such a blockade would prevent endogenous NE from raising *c*-AMP levels in the eye, thereby inhibiting humor production.

β_2-selective antagonists, if specific enough, would be useful as research tools to delineate the role β_2-adrenoceptors play in greater detail even though presently no clinical applications for such compounds are apparent. The first compound to have this property was α-methyl DCI.

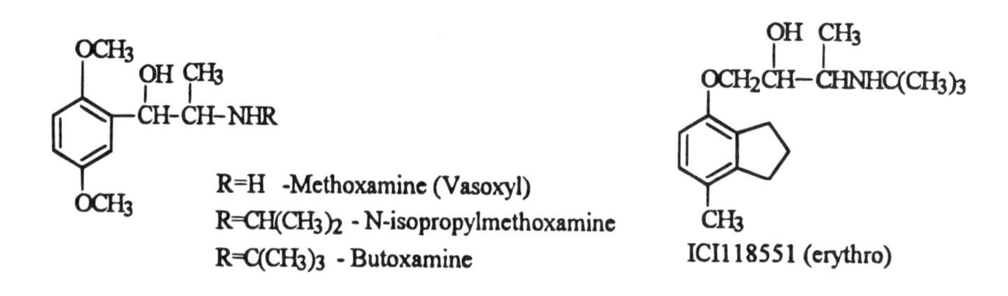

The sympathomimetic amine methoxamine, an α-agonist exhibiting some β_2 blockade, is useful as a pressor and decongestant with little cardiac stimulation. Its direct α-agonism causes vasoconstriction. It is curious that its N-isopropyl homolog, which might be expected to show β-agonism (or antagonism), does not since it is N-dealkylated to methoxamine. Butoxamide, which is not dealkylated, is a β_2-selective antagonist, although it is not very potent. A very selective and potent β_2-antagonist was reported; the compound, ICI118551, exhibits a high β_2/β_1 ratio of 250:1.

10.3.1.1.4. α–β Adrenergic Blockers

There is only one drug presently in clinical use having this unusual "mix" of adrenoceptor pharmacology, labetolol (Table 10-4, No. 7). Structurally, it illustrates the fine line between agonist and antagonist seen in the adrenergic area. Its nonselective β-blockade is considerably less potent than that of phentolamine. The antagonism is competitive in both cases. Some MSA can be demonstrated at high doses. The mechanism of its antihypertensive effect is achieved by vasodilation, effected by decreased peripheral resistance of smooth musculature.

10.3.1.1.5. α-Adrenoceptor Blocking Agents

α-Adrenergic antagonists could be a more descriptive term. Other nomenclature has permeated this area. The term *adrenolytic* predates the recognition of NE as the primary neurotransmitter. *Sympatholytic* became the successor name, but *lytic* implies dissolution or destruction, which is not what happens to adrenoceptors. All of this nomenclature variety, besides adding to the confusion, still leaves compounds that prevent NE release or synthesis (e.g., metyrosine) without a category or name. To add to the obfuscation is the lack of structural commonality, or any resemblance to NE, of drugs capable of antagonizing the physiological effects of α-adrenoceptor stimulation.

α-Adrenoceptors exist in most blood vessels, particularly in cutaneous resistance vessels. Since their stimulation leads to their constriction and therefore blood pressure elevation, it stands to reason that blocking such stimulation would lead to a diminution of blood pressure.

The subtypes α_1 and α_2 have been discussed earlier. Both types are found postsynaptically, where they control smooth muscle contractility (they are, of course, also located presynaptically and centrally).

The two imidazolines, tolazoline and phentolamine, are primarily of historical interest. Tolazoline was first developed in 1939. The structures of these two drugs resemble the α-adrenoceptor agonists listed in Table 9-3. They are competitive antagonists. Tolazoline is rather weak as a vasodilator and its α-blockade is transient. The drug actually has a multifaceted pharmacological profile. For example, it can exhibit sympathomimetic activity or enhancement of response to sympathetic nerve stimulation.

A glance at the structure indicates a β-phenethyl moiety present within the molecule. A more sophisticated rationalization, in view of the agonist activity at low doses, is to suggest that the drug effectively inhibits prejunctional (α_1?) receptors, thereby decreasing available NE. Cholinergic and histaminelike effects can also be discerned. This would explain increased gastric acidity and some of the direct vasodilation. In spite of such an array of fascinating pharmacology tolazoline has been relegated to one clinical application; persistent pulmonary hypertension of the newborn (PPHN).

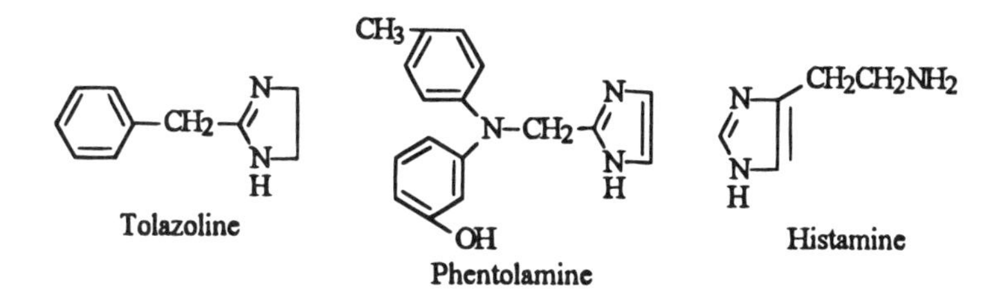

Phentolamine is useful in hypertensive conditions precipitated by excessive catecholamines in circulation. It is therefore still considered in the management of pheochromocytoma (preceding or during surgery on the tumor), accidental overdoses and hypertensive crises from drug or food interactions with MAO inhibitors, or the abrupt withdrawal of clonidine therapy. Both drugs, however, are rapidly becoming obsolete in view of new developments.

Another "survivor" from the first generation of antihypertensives is dibenzyline, which in turn was the successor of the even earlier dibenamine. Dibenamine was included in a patent by Eislieb[5] in 1934; its clinically usable adrenergic properties were not described for more than a decade. Dibenzyline (Fig. 10-4) soon replaced dibenamine because of its superior oral activity. The proposed mechanism of irreversible, and therefore long-acting, α-adrenoceptor alkylation is outlined in Figure 8-4. A spontaneous intramolecular nucleophilic cyclization leads to the aziridinium (ethylene immonium) ion, which interacts with an anionic site on the receptor by coulombic forces, aided by van der Waals interactions of the aromatic ring at a presumed adjacent ancillary site. Ring scission, then, results in covalent alkylation. It is interesting that these mechanisms involving alkylation via ethylene immonium ions have been formulated and were accepted for decades on what is

[5] The developer of meperidine, 5 years later (Chapter 5).

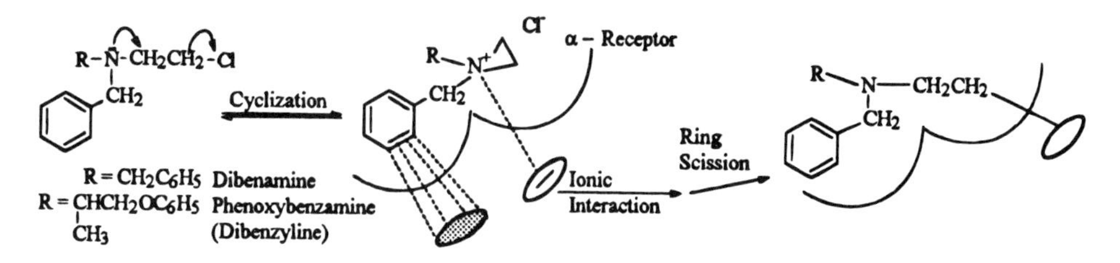

Figure 10-4. Haloalkylamine α-adrenoceptor blockade (see alkylation mechanisms of nitrogen mustards, Chapter 4).

actually indirect evidence. It is therefore gratifying that unequivocal proof was established in 1976 (Heinkel et al.). They synthesized stable perchlorate salts of the aziridinium ions derived from both dibenamine and phenoxybenzamine and showed their α-adrenergic blocking potencies to be essentially identical to that of the parent compounds. Thus the postulated mechanism is correct.

Today, phenoxybenzamine is not used as a general antihypertensive, but rather, it finds application in prolonged pheochromocytoma therapy. If tachycardia occurs, a β-blocker can be added to prevent it. In addition, it can be a useful drug in urinary retention cases due to paralytic bladder and other outlet obstructions (also see urecholine for the cholinergic approach to such problems, Chapter 8). Dibenzylene may also be of value in voiding disturbances due to prostatic obstruction.

10.3.1.1.6. The Ergot Drugs

This is a group of alkaloids, and semisynthetic derivatives, that have literally come down to us from the Middle Ages, when the oxytocic[6] action of ergot was known to sixteenth century midwives. Ergot is a fungus, *Claviceps purpurea*, that infects rye plants forming *sclerotia*, which are purple club-shaped compact mycelia that replaces the grains of rye during the resting or mature stage. The dried sclerotium, 1–5 cm long, contains numerous substances that include various indole alkaloids. The first U.S.P. (1820) recognized this crude drug. Several of these alkaloids in pure crystalline forms are still (170 years later) official in the current pharmacopoeia. Some of the alkaloids have α-adrenoceptor antagonist activity. They will be briefly considered. The unraveling of the complex chemistry began early in this century with the isolation of ergotoxine in 1906.[7]

The fundamental chemistry is outlined in Figure 10-5. All the alkaloids can be considered as derivatives of (+)lysergic acid. The epimeric (+)isolysergic acid (epimerized on C-8) and its corresponding alkaloids produced during extraction are essentially inert.

The important natural alkaloids obtained from ergot are usually 'subdivided' into two categories based on aqueous solubility. A water-soluble fraction consists of ergonovine. It was not discovered until the 1930s. Two semisynthetic drugs, methylergonovine and methysergide, have since been added to this group. Ergonovine is a strong oxytocic agent; methylergonovine is even more potent. Their main use is to stop postpartum and postabortional bleeding. They are rapid acting (<1 min, IV), causing uterine contractions without

---

[6] Stimulating uterine contractions and accelerating childbirth.

[7] Later found to be an equal tripartite mixture of ergocristine, ergocryptine, and ergocornine (Fig. 10-5).

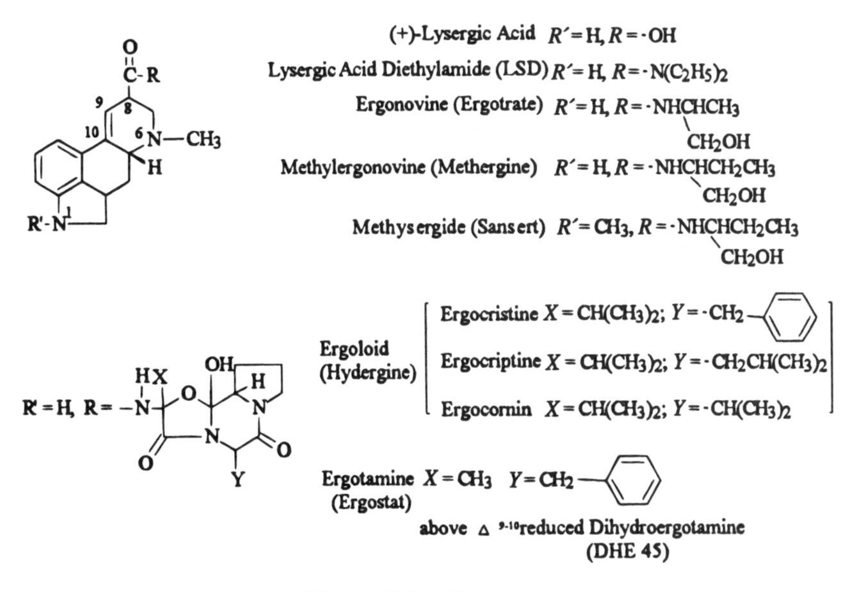

Figure 10-5. The ergots.

affecting nonuterine smooth muscle, including blood vessels. They have weak α-antagonist activity. Methysergide seems to have little oxytocic activity. It is, however, a potent peripheral serotonin antagonist that has been found to be an effective prophylactic for migraine headaches (it will not abort an acute attack). The antiserotonin activity may be relevant to its unknown mechanism of action. The chemically related DA agonist bromocryptine (Fig. 9-2) should be mentioned. The diethylamide, LSD, which is a potent hallucinogen producing bizarre effects, will not be considered here, since it has no legitimate clinical applications.

Ergotamine and ergotoxine (now known to be a group of three alkaloids) constitute the water-insoluble fraction of ergot. Ergotamine has a high affinity for α-adrenoceptors, but it produces a lower response than does NE (i.e., it can be viewed as a partial agonist).

α-Blockade sets in after initial pressor activity. Its principal use is in the treatment of migraine. The semisynthetic dihydro derivative (obtained by hydrogenation of the C-9–C-10 double bond) is said to have lower side effects when used to abort migraine and cluster headaches. Combination products of ergotamine with caffeine (Cafergot) are claimed to have increased effectiveness due to increased absorption.

Finally, a mixture of dihydroderivatives of the ergotamine group [i.e., dihydroergotoxine (ergoloid)] is on the market to alleviate cerebrovascular insufficiency in geriatric patients. A rationale for the alleged benefits of this product has yet to be developed.

Among the more recent α-adrenergic blocking agents introduced for the treatment of high blood pressure was prazocin.

The drug was originally "designed" on the basis of biochemical pharmacology of the early 1970s. Its synthesis evolved from attempts to incorporate structural features of cyclic nucleotides, such as the 4-aminopyrimidine moiety, from *c*-GMP and the *o*-dimethoxybenzene component of papaverine. Portions of theophylline were also considered. The prazocin molecule that emerged did what was expected of it. The drug inhibited PDE much better than did theophylline. Animal studies demonstrated a direct vascular relaxation resulting in vasodilation. Some interference with peripheral sympathetic function was noted, but it was downplayed. Most important, the drug clinically lowered blood pressure effectively. Later, based on considerable additional work with α-adrenoreceptors, prazocin has been reclassified as a selective α-blocking agent, which accounts for the decreased peripheral resistance. Unlike the older α-blockers such as phentolamine, prazocin is highly selective. In fact, following sympathetic denervation, prazocin has no vasodilating effect.

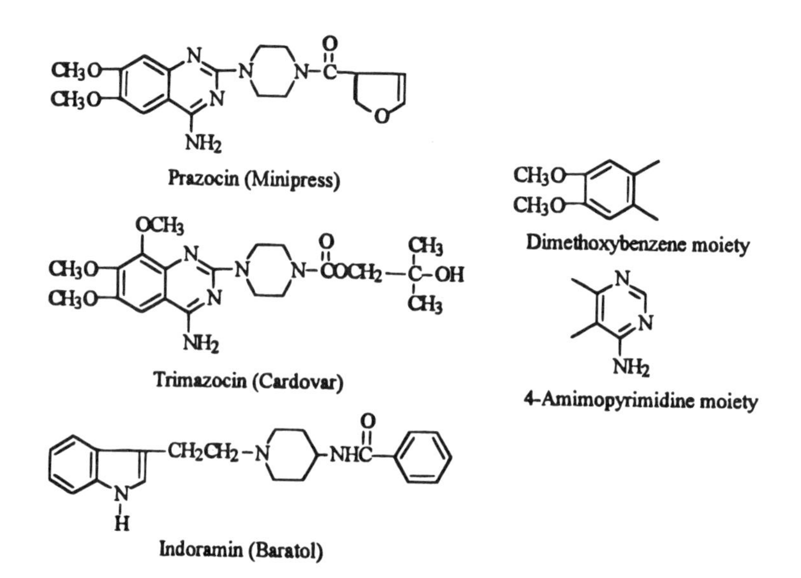

Trimazocin shares the polymethoxyquinazoline and piperazinecarbonyl portions of the prazocin molecule, except that the *tert*-butanol is linked via a carbamate ester. It is not surprising that there are some nuances in its mechanism that still involve inhibition of α_1-adrenoceptors. In addition, there is a direct action on the vascular musculature leading to dilation.

Indoramine is a highly selective α_1-antagonist. Its pharmacology has since been shown to include MSA as well as antiserotonin and antihistaminic properties. The drug's sedative CNS side effect had postponed its consideration for clinical use until its high α_1 selectivity was recognized.

10.3.1.2. Central Intervention of Cardiovascular Output

This can be affected by some old drugs whose mechanisms had been previously "misdiagnosed" (e.g., methyldopa; see Chapter 9 for details). The obsolete veratrum alkaloids and some new agents are being considered. In the past 15 years clonidine has become the prototype of the centrally acting antihypertensive. Its chemistry and development will therefore be discussed in some detail.

Table 9-3 lists a group of imidazole derivatives that affect adrenoceptors. Some are agonists, which are now used primarily as nasal decongestants; two were antagonists of mostly historical antihypertensive interest. All have a -CH_2-function bridging it to an aromatic system (I). In effect, this is a cyclic amidine. However, one compound (clonidine II) has a nitrogen atom binding the two rings (II). Either of the two tautomeric forms has a guanido

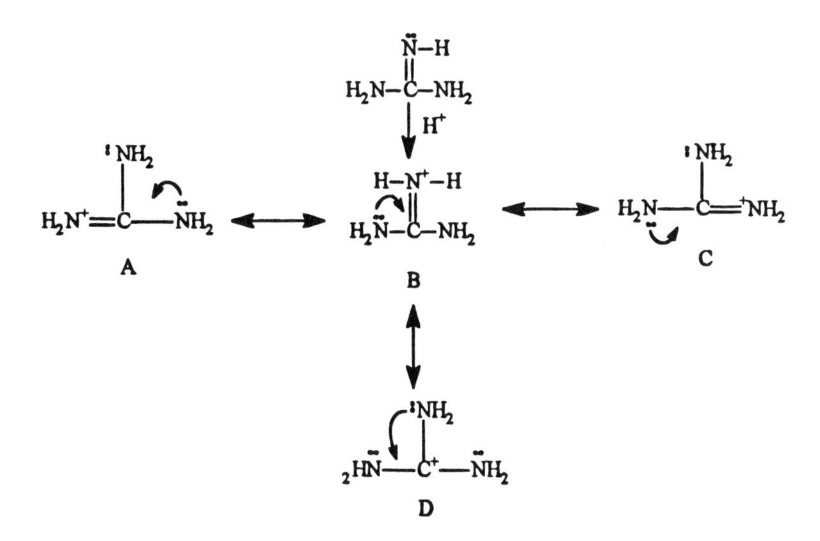

Figure 10-6. Resonance stabilization of guanidine.

function built into the molecule. Spectroscopic equilibria studies indicate that the *imino* tautomer predominates.

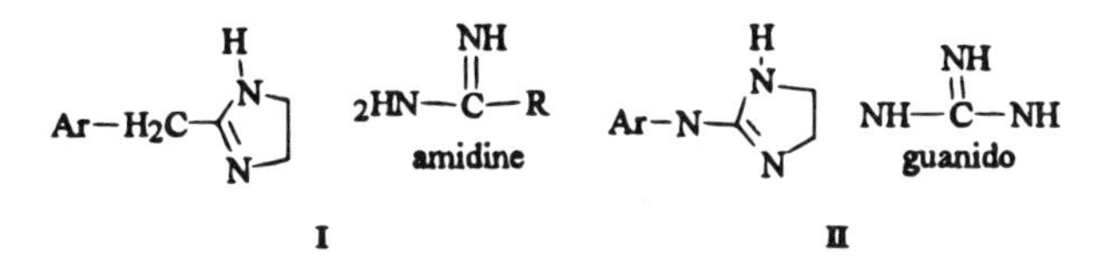

The guanido group imparts strong basicity to a molecule[8] because guanidine is the strongest organic nitrogen base known. Its pKa is 13.6. The imino (sp^2) nitrogen protonates to yield a highly stable conjugate acid. The stability of this protonated acid is due to the equivalence of all three NH_2 groups through electron delocalization. The resonance forms that can be written (Fig. 10-6) to illustrate this point are only valid as long as the nitrogen atoms and the carbon are coplanar. Moving any one, or all, of these atoms out of plane will destabilize resonance and weaken the basicity.[9]

The two lipid-solubilizing chlorine atoms at C-2 and C-6 ensure clonidine's entry into the CNS (no substitutents or polar OHs are inactive). Judging from scale molecular models, it is highly unlikely that the two C1 atoms would permit coplanarity of the benzene and imidazole rings without the C1 atom "grinding" into the N–H bond as indicated. Stability for clonidine, then, would increase the more out-of-plane the two rings can be, ideally 90° (i.e., perpendicular). Such a torsional rotation can only occur around the bond between the imino nitrogen and the benzene ring, as the double bond to the imidazole ring would not permit such a move. The actual angle of rotation, at least in the crystal (of the hydrochloride), has been reported as 75° (Byer et al., 1976).

---

[8] For example, guanethidine, pKa = 11.9.

[9] Also refer to discussion on β-lactam ring stability, Chapter 6.

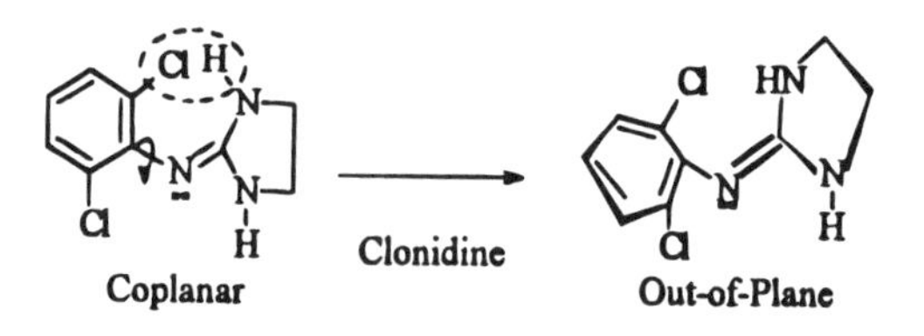

In the case of the protonated (salt) form of clonidine, using the analogy with guanidine (Fig. 10-6), in order to retain optimum molecular stability, the proton must be found on the bridge nitrogen. That way, as the following composite resonance structure shows, the electrons of the double bond can be delocalized so that the positive charge is distributed over all three nitrogens.

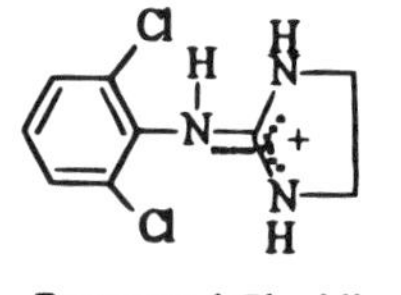

Certain corresponding features between clonidine and NE have also been established, as both are substances that interact with α-adrenoceptors. These are schematically represented in Figure 10-7, where NE, the prototype adrenergic neurotransmitter, and clonidine may be visualized as both interacting analogously with the receptor. The distances between the center of the aromatic ring and the protonated nitrogen of NE or, the delocalized positively charged area center of clonidine (as discussed earlier) on average 5.1 Å. The angles of the bonds leading to these positive centers rise 1.2–1.4 Å above the plane of the aromatic portion of the molecules.

The original basis for the synthesis of clonidine was to improve on the imidazoline decongestants available in the early 1960s (Table 9-3). The substitution of the CH_2 span to the aromatic ring by a nitrogen bridge had already been accomplished without any pharmacological surprises. However, the same structures with *ortho* substitutents on the aromatic ring had not been achieved probably because standard synthetic procedures were thwarted due to steric hindrance. Thus a new synthesis, outlined in Figure 10-8, was

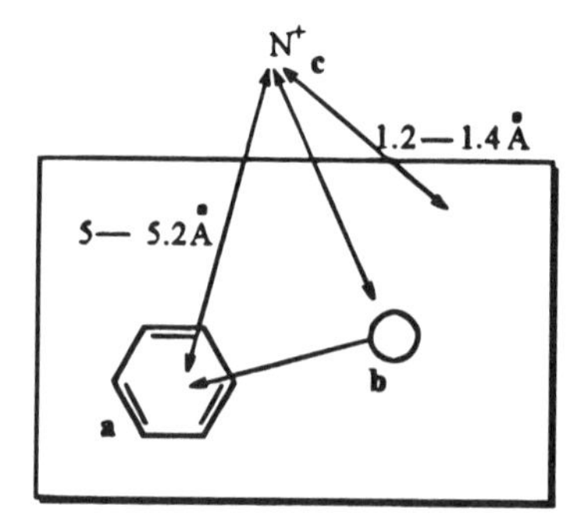

Figure 10-7. α-Adrenoceptor interactive site overlaps for clonidine or norepinephrine.

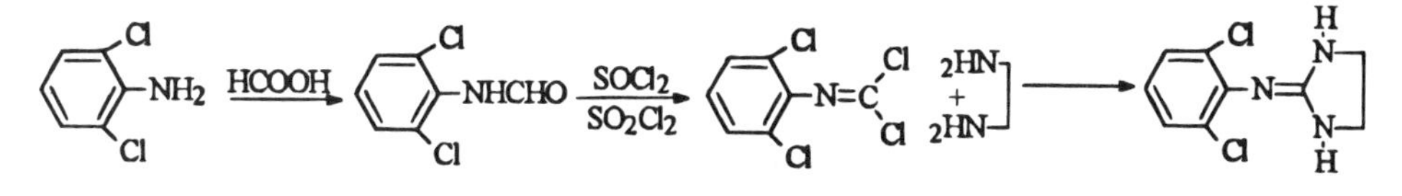

Figure 10-8. Synthesis of clonidine.

devised. N-Formyl-2,6-dichloroaniline, obtained by formylation of the aniline with formic acid, is chlorinated to N-(dichloromethylene) 2,6-dichloroaniline with a mixture of sulfonyl and sulfuryl chlorides. Condensation with ethylene-diamine affords the desired product.

Clonidine has some decongestant activity. The important discovery—and surprise—was its very significant antihypertensive properties. It turned out to be the first antihypertensive with a demonstrable central site of activity.

Placement of the two halogens in the *ortho* positions was particularly fortuitous for two reasons. It increased the compound's lipophilicity sufficiently to cross the BBB and enter central sites. Second, by forcing the two rings out of coplanarity, it apparently satisfied steric (and other) requirements for a high "goodness of fit" into the α-adrenoceptor. Engagement of the central receptors at medullary sites results in a decreased sympathetic outflow, lowering peripheral sympathetic tone and increased vagal reflex action. Since the immediate initial response following IV administration of clonidine is transient pressor activity, it can be explained that the initial flush of drug into the circulation leads to interaction with peripheral receptors (raising blood pressure) before the drug crosses the BBB to central sites. There are no good reasons to believe that structural engagement features differ much in central and peripheral α-adrenoceptors. That clonidine exerts both vasoconstrictive (peripheral) and hypotensive (central) effects has been known for some time. In fact, each effect can be ascertained separately. The peripheral α-adrenergic activity of clonidine and a number of congeners (including most of those listed in Table 10-7) were determined by IV administration to pithed rats. The centrally mediated hypotensive effects were determined on the intact rat either intravenously or intracerebrally.

A hypothetical α-adrenoceptor model may be envisioned where stimulation by clonidinelike protonated compounds involves a binary interaction. One is a coulombic attraction of one of the positively charged imidazole nitrogens to the electron-dense site on the receptor, and the other is a charge-transfer-type reaction between the π-electron clouds of the benzene rings and an electron-deficient area on the receptor. The latter would only occur to a significant extent if the geometric fit-pattern between drug and receptor meet some apparently stringent requirements.

An examination of Table 10-7 illustrates some of this stringency. In fact, of 19 compounds only 3, the 2-Br, 6-Cl, and the 2,6-dibromo congener, show high central hypotensive activity (where $X = N$).

There are obviously several parameters being juggled here. As already mentioned, adequate liposolubility to reach central receptors is one factor, and 2,6-disubstitution is another. Consider No. 1, which is the compound where no substituents are on the benzene ring. Here, liposolubility and activity are both extremely poor. Compounds with low or negative log P values (1, 9, 11, 13, 14) could be expected to fare poorly, yet this is not the only factor since compounds 7 and 10 have high and adequate lipophilicity, respectively. Compound 6, which has a high log P' value, shows meager hypotensive effects in the rat. Disubstitution at 2 and 6 is imperative, but so is the bulk of the substituent. Comparing

Table 10-7. SARs of Clonidine and Congeners

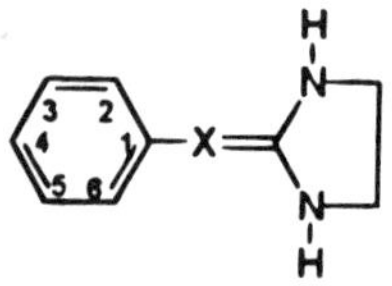

Compound	X	Conformation	log P^{a}	$ED_{30}{}^{b}$	$ED_{20}{}^{c}$	pKa^{e}
1. H	N	Coplanar	−1.92	25,000	>3.00	9.30
2. 2,6-Cl_2*	N	Nonplanar	0.83^{d}	2.7	0.01	7.11
3. 2-Br, 6Cl	N	Nonplanar	0.87	2.9	—	
4. 2,6-Br_2	N	Nonplanar	1.15	5.5	0.045	
5. 2-Cl, 6-F	N	Nonplanar	0.52	12.5	—	
6. 2,4,6-Cl_3	N	Nonplanar	1.47	21	0.09	
7. 2,4,6-Br_3	N	Nonplanar	2.24	580	—	
8. 2-Cl	N	Coplanar	−0.67	240	1.00	
9. 2,4,-Cl_2	N	Coplanar	0.29	61	0.17	
10. 2,5,-Cl_2	N	Coplanar	0.71	150	0.25	
11. 2,6-F_2	N	Coplanar	−0.16	600	>3.00	
12. 2,6-(CF_3)	N	Nonplanar	—	—	0.06	
13. 2,4,6-$(CH_3)_3$	N	Nonplanar	−1.28	280	0.03	
14. 2-CH_3	N	Coplanar	−1.82	2,000	1.20	9.83
15. 2-C_2H_5	N	Nonplanar	—	—	0.40	
16. 2,6-Cl_2	S	Nonplanar	—	50^{e}		
17. 2,6-Cl_2	CH_2	Nonplanar	—	72^{e}		
18. 2,6-Cl_2	O	Nonplanar	—	5750^{e}		
19. 2,6-Cl2,4NH2**	N	Nonplanar	—			9.20^{f}

* Clonidine.
** Apraclonidine (Iopidine).
[a] Apparent partition coefficient octanol/0.1M phosphate buffer, pH 7.4.
[b] μg/kg following IV administration in normotensive rats (Timmermans and van Zwieten, 1977; Rouot et al., 1976).
[c] mg/kg to produce a 20 mmHg drop in blood pressure in normotensive dogs (Hoefke, 1976).
[d] Also reported as 0.62 and 0.48 (Jen et al., 1975).
[e] Timmermans et al., 1980.
[f] Dunn, D., personal communication, March, 1989.

compounds 6 and 7 makes this apparent. Compound 11 has 2 and 6 substituents, but the F atom has a much smaller diameter (1.28 Å) than does Cl (1.98 Å), and it has a much lesser effect on increasing lipophilicity (compare log P' value with No. 2, 3, or 4). Thus compound 11 would have lesser ability to penetrate into central medullary areas and, probably more importantly, the small F atoms would not impede the benzene and imidazole rings from assuming an almost coplanar relationship to each other, permitting the resonance-stabilizing effect that decreases the hypotensive activity precipitously.

More recently it was shown that introduction of a third substituent in the *meta* position offers a way of increasing hypotensive potency of clonidine. Thus the 2,3,6-Cl_3 compound was three times more active in both the rat and the cat. With a log P value of 1.91 it is likely that this increase is due mainly to improved penetrability into central areas. The question remaining, then, is why moving the *meta* Cl to the *para* position (No. 6, Table 10-7), a compound with even more lipophilicity than clonidine, is eight to nine times *less* potent. It is possible that the *para* position is less tolerant of large substituents, adversely effecting the "goodness of fit," again illustrating the limitations of this particular receptor.

A glance at compounds 16, 17, and 18 quickly indicates that exchanging the bridging N atom for O, S, and CH_2, even though not changing the distance between the two rings significantly, still caused a precipitous drop in activity. The oxygen analog is essentially inactive. Similar results are obtained if the imidazole is expanded by additional carbons.

Clonidine has also been found to exhibit several other pharmacologic properties that, in some cases, do not seem to be related to its α-adrenoceptor mechanism. Effectiveness in glaucoma is one that is. It is experimental, but may be a property shared by other α_2-agonists.

Sudden withdrawal of clonidine from patients on antihypertensive therapy produce symptoms clinically similar to opiate withdrawal. These include headache, nervousness, tachycardia, stomach pains, and sweating. It may be this apparent similarity that gave someone the idea to treat heroin withdrawal symptoms with clonidine. It seems to work with some addicts and is now experimental for this application.

Opiate withdrawal symptoms can be reversed without inducing a euphoria of its own (like methadone). The tie-in may be related to findings that, at least in rats, long-term administration of morphine increases the brain's α-adrenoceptor population. If opiate withdrawal symptoms can be somehow related to a portion of the excess receptors being "unoccupied," then clonidine's activation of them may, in part, explain its effectiveness.

Clonidine also appears to have significant analgesic properties to which tolerance does not develop. Antinociceptive effects have been demonstrated in many of laboratory tests used to demonstrate analgesia. The effect is centrally mediated. For example, the hot plate assay that demonstrated analgetic properties for clonidine orally failed to do so for other imidazole α-mimetics such as oxymetazoline and tetrahydrazoline, which do not cross the BBB unless administered intracerebrally.

It should be recognized that the synthesis and introduction of clonidine in the early 1960s was done without the benefit of prior knowledge of the receptor technologies, their subdivisions, and present detailed understanding of the autonomic system discussed in this and several of the previous chapters. In fact, much of it was developed since then. In addition to having been a valuable drug over the years, clonidine was also an important research tool that helped unravel much of what is known about hypotensive mechanisms.

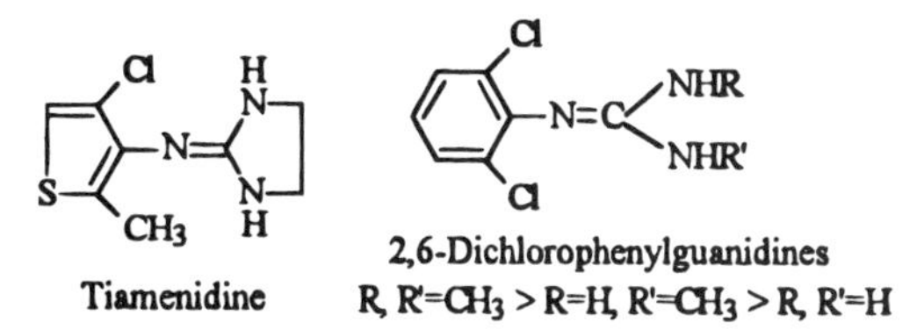

Tiamenidine

2,6-Dichlorophenylguanidines
R, R'=CH_3 > R=H, R'=CH_3 > R, R'=H

Isosteric replacement of the benzene ring by heterocyclic isosteres such as pyrroles, isoxazoles, and thiophenes resulted in compounds with antihypertensive activity. Tiamenidine was found to have a clonidinelike pharmacological spectrum.

Noncyclic guanidines were also investigated. Synthesis of the preceding three 2, 6-dichlorophenylguanidines showed them to have very modest antihypertensive properties (rat, dog), with dimethyl substitution having the most effect.

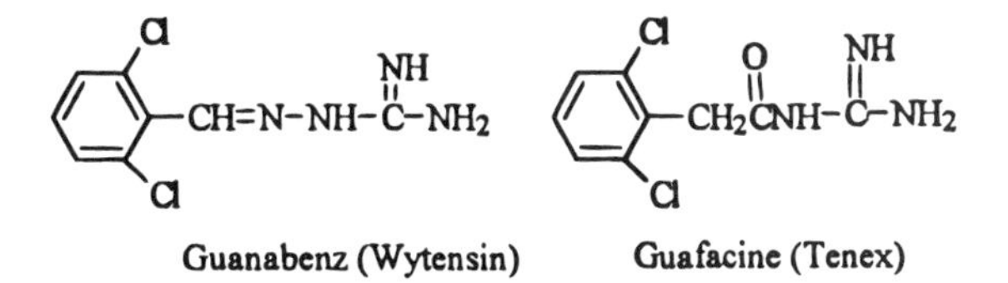

Guanabenz (Wytensin)

Guafacine (Tenex)

More extensive variations of the guanidine function resulted in the potent antihypertensive guanabenz, whose activity is primarily due to inhibition of sympathetic outflow from the brain by activation of central α_2-adrenoceptors. There is some peripheral blockade. Its clinical pharmacology parallels that of clonidine. The half-lives are comparable (ca. 8 hours). In guafacine, a methylenecarbonyl (-CH_2CO-) function separating the guanido group produces more separation than seemed permissible with imidazole compounds such as clonidine. However, as with guanabenz, this compound also stimulates central α_2-adrenoceptors resulting in reduced peripheral resistance and blood pressure. Side effects are again qualitatively similar to the other centrally acting guanidine drugs, including the hazard of rebound hypertension on sudden withdrawal. Animal tests indicate a lesser degree of sedation that may be true with human subjects as well.

In Chapter 8 possible relationships of central cholinergic neuronal functions, cognitive decline as a function of aging, and the pathological condition of Alzheimer's disease were mentioned. The utility of cholinergic drugs to arrest, or at least slow down, the decline was briefly discussed. The results with this type of ACh indirect "replacement" therapy have been mostly disappointing with aged nonhuman primates as well as Alzheimes patients. Direct-acting agonists may be interpreted to have very limited beneficial effects. More recently, there has been some initial evidence that α_2-adrenergic agonists, particularly clonidine, may show some drug-mediated improvement in cognitive deficiencies in aged nonhuman primates. The study indicates a relationship existing between NE levels in certain cortical regions and α_2-adrenoceptor population densities in the areas involved in working memory. It has been hypothesized that the memory functions through actions at postsynaptic α_2-receptors. It must be emphasized that the results are preliminary, but they do seem to suggest possible treatment of age-related memory disorders with adrenergic drugs.

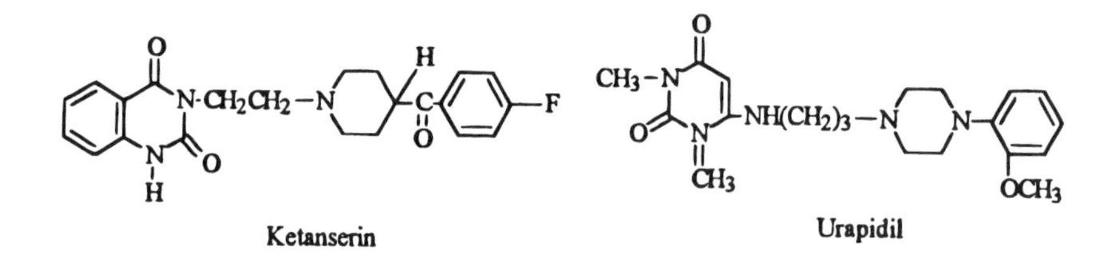

Urapidil, like labetolol, dabbles in different receptors simultaneously. It blocks postsynaptic α_1-adrenoceptors (like prazocin) and seems to have central clonidinelike α-agonist quality as well. Of course, this represents a desirable combination of effects, both additively, if not synergistically, lowering blood pressure.

Ketanserin is another investigational hypotensive quinazoline compound. Its blood pressure, lowering ability may differ significantly from the other compounds even though it has been shown to possess postsynaptic α_1 antagonism. The difference lies in the different biochemical features of this drug, which may make it impossible to assign it a single clearcut mechanism. Ketanserin has been shown to inhibit the peripheral effects of the neurotransmitter serotonin at S_2 ($5HT_2$) receptor sites. This would reverse serotonin-produced vasoconstriction and platelet aggregation (also bronchoconstriction). The role that each of these factors plays in the clinical reduction of blood pressure is not clear.

10.3.1.3. Direct-Acting Arteriolar Dilators

This group of antihypertensive agents is represented by only a few compounds that have nothing structurally in common.

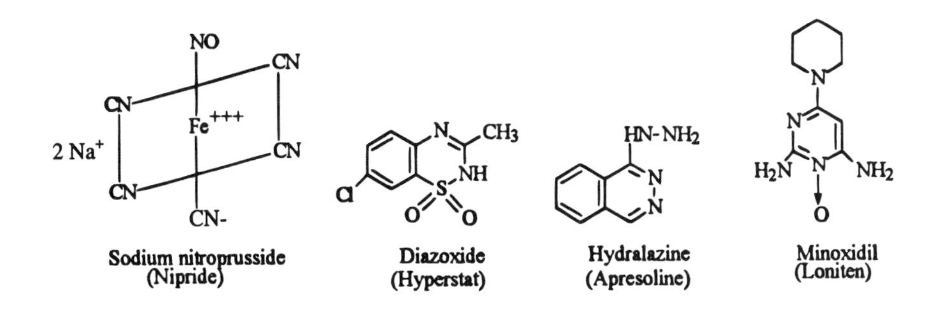

The antihypertensive properties of nitroprusside have been known since the late 1920s but they were not clinically used until the 1950s.[10] It can be easily synthesized from sodium ferrocyanide and strong nitric acid by heating.

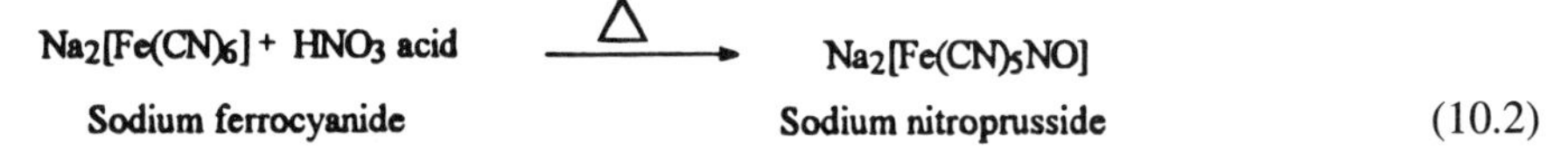

(10.2)

At first the drug was used orally as an antihypertensive, but it was phased out with the advent of the ganglionic blockers and reserpines. Sodium nitroprusside was subsequently resurrected as a parenteral emergency drug. It is now a most powerful vasodilator, given by IV infusion to obtain almost immediate reduction in pressure in hypertensive crises. By regulating the rate of administration minute by minute, blood pressure control becomes possible.

The physiological mechanism is by direct vascular dilation, both arterial and venous. The mechanism is not at all clear at the molecular level. The hypotension has been attributed to the thiocyanate ion, SCN^-, which is produced metabolically by rhodanese in the liver. The sulfur atom is supplied by thiosulfate:

$$CN^- + S_2O_3: \xrightarrow{\text{Rhodanase}} SCN^- \tag{10.3}$$

This reaction normally represents hepatic detoxification of cyanide ions produced from nitroprusside in erythrocytes. The SCN^- is 99% less toxic than CN^-. NaSCN was also still official in the NF X (1960) as an antihypertensive drug. A nitroprusside mechanism on vascular smooth muscle by interference with the role of Ca^{2+} in the muscle contraction process has also been suggested.

Hydralazine (Apresoline) is a phthalazine derivative that has been in use for more than three decades, primarily in combination with other types of antihypertensives. One innovation was to combine it with β-blocking propranolol, which effectively prevents reflex cardiac stimulation and allows lower dosages. Adding a diuretic to the regimen prevents the usual sodium and water retention. Its molecular mechanism of action is not fully understood. Its toxic effects at higher doses can resemble a syndrome of symptoms simulating

---

[10] It was marketed in the United States in 1974, a 45-year "drug lag."

rheumatoid systemic lupus erythematosus. The use of lower doses, by combination with the agents mentioned above, minimize the hazards.

Diazoxide is a parenteral, rapid, and direct-acting vasodilating antihypertensive used in hypertensive emergencies. An IV injection can drop blood pressure by as much as 80 mmHg in 5 minutes. Unlike sodium nitroprusside, however, venous dilation is not part of its mechanism. Chemically it is a benzothiadiazide without the sulfamoyl function at the 7 position (see diuretics). In fact, diazoxide is not a diuretic. Chronic use of diazoxide reflexly increases renin release, which actually counteracts the antihypertensive effect of the drug by expanding the volume of circulating fluid.

The diamino pyrimidine derivative minoxidil became available in the United States in 1980. It is probably the most potent oral vasodilator in use, making it a very important drug in severe hypertension, especially when refractory to other agents. It is best used in combination with a β-blocker (to prevent reflex tachycardia) and a diuretic (to prevent fluid accumulation).

The mechanism of minoxidil appears to be the same as hydralazine. It is longer acting ($t_{1/2}$ 4 vs. 2 h), thus affording effectiveness up to 24 hours from a single dose. The N-oxide function is integral to the drug's activity. The reduced form is inactive in humans (but not in cats or dogs) since humans cannot oxidize these compounds to their active $N^+ \rightarrow O^-$ form.

One side effect should be mentioned: hypertrichosis. Within 2 months of initiating therapy increased hair growth begins on the face and the palms of the hands. It affects three-fourths of patients and may have psychological ramifications in women and children. The effects may persist for several months after the drug is withdrawn. Attempts to utilize this side effect to restore hair growth in cases of baldness with topical preparations of minoxidil (Rogaine) have succeeded.

Slow Calcium Channel Blockers (SCCBs), which are also referred to as calcium entry blockers and even, erroneously, as calcium antagonists, are a number of marketed drugs, several of which have no chemical relationship to the others. They will be discussed in some detail with the anti-anginal drugs (see later). At this point, suffice it to state that their antihypertensive effect is due to their ability to modulate the influx of Ca^{2+} across the cell membrane of the arterial smooth muscle. This will reduce Ca^{2+}-promoted activation of the cells' contractile proteins, leading to vasodilation.

However, another dimension may have been added with reports that in animals SCCBs also interact with *peripheral* α_2-adrenoceptors involving Ca^{2+}. Thus the normal vasoconstrictive effects of agonist ligand interactions with the *peripheral* α_2-adrenoceptors may be obtained. It should be remembered that endogenous NE is a nonselective α-agonist. Therefore, when SCCBs are also circulating (presumably at therapeutically effective levels) α_2-stimulation by NE is blocked; α_1-stimulation is not. Thus vascular tone and peripheral resistance are reduced—and blood pressure with it.

10.3.1.4. Angiotensin-Converting Enzyme (ACE) Inhibitors

The earlier discussion on renal hypertension briefly dealt with the renin–angiotensin system (RAS) (Fig. 10-1). Before considering the several drugs on the market, as well as some others under active consideration, that do or may interfere in some manner with either ACE or renin, it will be useful to survey briefly some present understanding of what the RAS is and what its functions might be. It should be pointed out at the outset that much about this complex system is still not known.

The RAS should probably be viewed as a cascade of proteolytic reactions under integrated controls. The cascade produces peptides with pressor- and aldosterone-producing effects.[11] The RAS controls blood pressure, blood volume, and electrolytic balance.

The beginning of the cascade is considered to be angiotensinogen, an α_2-globulin, one of an abundant group of glycoproteins in plasma produced in the liver. It is the only substrate for renin, a plasma enzyme synthesized by the juxtaglomerular cells of the kidney from which it is released. This rate-limiting enzyme cleaves the protein between the tenth and eleventh amino acid from the amine terminus (arrow) to produce an inactive decapeptide, angiotensin I. The more recent discovery of other local areas of renin synthesis such as the brain has led to suggestions of separate RAS in the CNS (Murakami et al., 1984)

1 2 3 4 5 6 7 8 9 10 | 11 12 13 14
2HNAsp-Arg-Val-Tyr-Ile-His-Pro-Phe-His-Leu-Val-Ile-Tyr-Ser-Protein

Angiotensinogen

1 2 3 4 5 6 7 | 8 9 10
2HN-Asp-Arg-Val-Tyr-Ile-His-Pro-Phe-His-Leu-OH
Angiotensin I

1 2 3 4 5 6 7 8
2HN-Asp-Arg-Val-Tyr-Ile-His-Pro-Phe-OH
Angiotensin II

The prohormone Angiotensin I is then cleaved at the Phe^8–His^9 peptide bond by the relatively nonspecific ACE (also known as kininase II or dipeptidyl carboxypeptidase). ACE is a membrane-bound glycoprotein, found to be in the epithelial cells of pulmonary capillaries, the urogenital tract, the spleen, and the intestine. Incidentally, by removal of the terminal two amino acids from the carboxyl end of des-Asp-Angiotensin I, ACE can produce seven-membered peptide angiotensin III. The kininase responsible for the breakdown of bradykinin is now known to be identical to ACE.

The substrate specificity of ACE is minimal, requiring at least a tripeptide with a COOH terminal and pretty much any L-amino acid, other than proline as an intermediate. Thus ACE is incapable of hydrolyzing angiotensin II further.

A glance at Figure 10-1 makes it apparent that angiotensin II can raise blood pressure in several ways. One is by directly acting on receptors in smooth muscle cells of arterioles causing vasoconstriction. Second is by stimulating aldosterone production and release, which promotes tubular reabsorption of Na^+ by acting on the distal tubules of the kidney (Fig. 10-13). The increased Na^+ levels, to maintain osmolarity, will retain water, thereby increasing blood volume and pressure. Finally, ACE, which is the enzyme degrading the vasodilating bradykinin, reduces its hypotensive efficacy, thus helping to raise blood pressure as well.

The sites of action of angiotensin II are varied. Some, directly or indirectly, result in increased blood pressure by vasoconstriction (e.g., arteriolar smooth muscle in general

[11] Actually, angiotensin III leads to aldosterone, which is not indicated in Figure 10-1.

and efferent arterioles in the kidneys). The zona glomerulosa of the adrenal glands are stimulated to synthesize aldosterone by ANG II; the medulla to release catecholamines. Interaction with the area postrema in the brain probably also results in blood pressure elevation.

ANG II is known to be implicated in many forms of hypertension other than renal. For example, the antihypertensive effect of β-blockers is now understood also to involve a reduction of renin release. The octapeptide salarasin antagonizes its effect competitively at target cells and, after a brief initial pressor effect (because of residual agonist activity), will lower blood pressure.

Sar[12]–Arg–Val–Tyr–Val–His–Pro–Ala–OH
Salarasin

Considering the left side of Figure 10-1 another subsystem—the kallikrein–kinin system–should be noted. Kallikrein, which is an an enzyme that converts bradykininogen to the vasodilating bradykinin when activated, also activates prorenin to renin.[13] Bradykinin also secondarily aids vasodilation by stimulating PGE_2, a dilatory prostaglandin. Thus the two systems, RAS and what may be termed a kallikrein–kinin–prostaglandin system, are interconnected by two enzymes, ACE and kallikrein. A constant tuning between the two systems would then be needed for normal vascular tone and electrolyte (Na^+) balance. Additional systems may also be involved in this homeostasis.

ANG II is also known to induce release of catecholamines from the adrenal medulla and to suppress the reflex inhibition of the heart rate, with the latter probably the inhibiting vagal (cholinergic) neurons to the heart. This has a specific clinical effect that is useful in the use of ACE inhibitors as antihypertensives (see later) namely, the absence of reflex tachycardia seen with other types of blood pressure-lowering drugs such as α-blockers, diazoxide, hydralazine, and ganglionic blockers.

ANG II has been shown to have a central pressor action via receptors in the anterior hypothalamus and other areas of the brain in some animals—but not the rat. It is curious that these areas are not protected by the BBB and can therefore be affected by ANG from the brain RAS, establishing its role in blood pressure regulation, as well as by the peripherally circulating enzyme. Thus the hormone ANG II, originally thought of as only important to blood pressure relating to renal factors, has now become another of Hercules' hydras, an enzyme interfacing with other systems and areas (e.g., the brain). The following scenario, then, may represent a summary of events responsible for homeostatic blood pressure maintenance. Released renin produces ANG I by specific proteolysis of angiotensinogen. The inactive prohormone is then rapidly hydrolyzed by the highly active hormone ANG II, which is catalyzed by the converting enzyme ACE. The hormone then acts on various receptors located in vascular smooth muscle causing constriction, the CNS, *increasing* sympathetic outflow,[14] and from the adrenal cortex, resulting in arteriolar constriction. In addition, ANG II reduces vagal tone and stimulates aldosterone secretion. The combined effects of all these actions is increased peripheral resistance to blood flow, increased heart rate and output, and increased water (and Na^+) reabsorption. The "bottom line" then is to

---

[12] Sar=Sarcosine or N-methylglycine.

[13] Which also is a blood clotting factor that activates factor XI.

[14] These are not central α-adrenoceptors where, it will be recalled, stimulation caused *decreased* sympathetic outflow.

restore or maintain blood pressure that may be below normal due to low Na^+ levels or decreased blood volume. Both are factors that precipitate renin release, thus initiating the described sequence of events. This stimulation of renin release is effected primarily by decreases of kidney perfusion (i.e., lowered pressure) possibly triggering baroreceptors. Situations leading to this point can result from hemorrhage or dehydration; so will stenosis of the renal artery. Such physical obstruction to blood flow by clamping the renal artery in a laboratory animal has been used as a research tool to produce "renal hypertension." It should be mentioned that the sympathetic system, and other factors, may also regulate renin secretion.

The development of ACE inhibitors began with the discovery that venom of a pit viper, *Bothoropa jararaca*, potentiated the activity of bradykinin, a vasodilator, and of ACE, a vasoconstrictor. This venom was soon found to consist of at least nine active peptides. The next year Ondotti et al. (1971) isolated, determined the structure of, and developed syntheses for several of these peptides. The most potent, the nonapeptide teprotide, soon established experimentally that ACE inhibition can control raised blood pressure. Teprotide was a nonapeptide, with the unusual *pyro*-glutamic acid at the N-terminal. It was the terminal proline that inhibited ACE from hydrolyzing the peptide. The lack of oral effectiveness of such a peptide as a drug (gut digestion and no meaningful absorption) precludes the use of such compounds clinically. Thus it became necessary to "design out" the proteinlike quality and yet maintain both the binding and inhibiting qualities that teprotide has for the active site of the enzyme. Synthesizing peptoids wherein nonprotein-type amino acids in the putative binding region should be a useful approach. Detailed SAR studies (with snake venom components) and their modifications were undertaken.

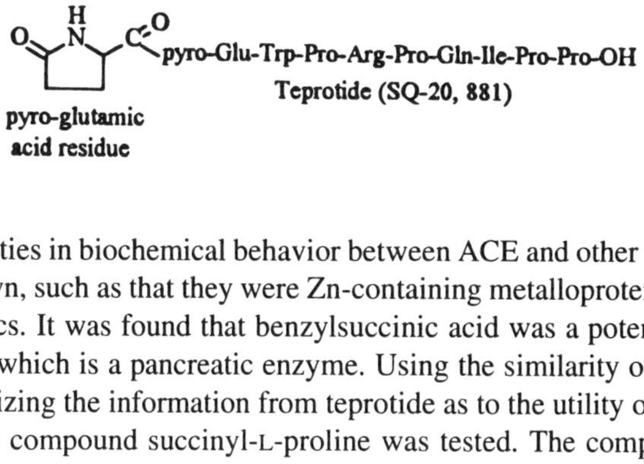

pyro-Glu-Trp-Pro-Arg-Pro-Gln-Ile-Pro-Pro-OH

Teprotide (SQ-20, 881)

pyro-glutamic acid residue

Certain similarities in biochemical behavior between ACE and other carboxypeptidases were already known, such as that they were Zn-containing metalloproteins and had certain other characteristics. It was found that benzylsuccinic acid was a potent inhibitor of carboxypeptidase A, which is a pancreatic enzyme. Using the similarity of the two enzymes as a basis, and utilizing the information from teprotide as to the utility of the terminal proline, the prototype compound succinyl-L-proline was tested. The compound was a weak but specific inhibitor of ACE, confirming the innovativeness of this approach. The remaining research then was to build on this lead and "improve the design." A key component in this work was recognizing the similarities in hydrolytic mechanisms between ACE and carboxypeptidase A and using this as a starting point.

Figure 10-9 schematically represents the assumed subsites of the reactive site of ACE to which various potential inhibiting substrates were "fitted." Large numbers of dipeptides without strong zinc ligands were included, among which val-trp was the most competitive inhibitor. This work clearly illustrated that the C-terminal dipeptide portions and the larger peptides bind to the same part of the active site of the enzyme. The next group was dipeptide analogs with strong zinc ligands. One was a mercaptoacyl amino acid, which ultimately became captopril, 1-[(2S)-3-mercaptopropionyl]-L-proline. It was the most potent

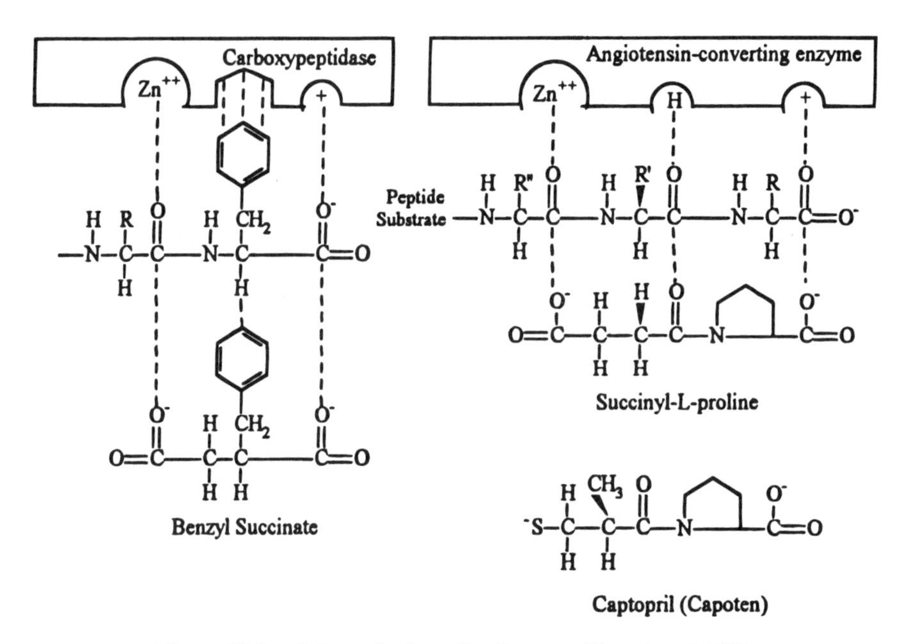

Figure 10-9. Schematic sites of carboxypeptidase A and ACE.

member of this class with an I_{50} of only 0.005 μM using rabbit lung ACE. Table 10-9 illustrates through SAR studies how specific the requirements for binding to a particular subsite can be, and this will be briefly discussed.

This group of compounds has functional groups that are assumed to have some binding capability to the Zn atom, the ability to hydrogen bond, and to interact with a cationic area—all subsites of the reactive site of ACE, as illustrated in Figure 10-9. Captopril (No. 1), which is the most potent in this category, is thus used as the prototype. Epimerizing the methyl group from the R to the S configuration decreases potency 100 times, indicating that the methyl participates in site interaction of captopril. Removing it completely (No. 5), however, causes only a 10-fold drop in activity (but may also affect lipid solubility). A polar acid function on the pyrrolidine ring capable of ionizing to an anion at physiological pH is critical to ionic binding to the cationic site on the enzyme. Removal of captopril's COOH function (No. 9) leads to a precipitous 48,000-fold drop in activity. Compound 6, which is 360,000 times weaker than captopril, illustrates two points: the modest drop of the methyl group multiplied by the inability of the inverted COO^- (of D-proline) to "reach" the cationic site. It is interesting that the dehydro analog (No. 7), by making the ring planar, restores that capability greatly. Compounds with ionizing acid function other than carboxyl, such as phosphonic acid (No. 8), hydroxamic acid (No. 3), and the tetrazole ring (No. 4) result in only modestly diminished activity. Isosteric replacement of the 4-carbon in the pyrolidine ring with a sulfur atom (the 4-thia analog of captopril, YS-980, Table 10-8) shows identical activity to captopril. It is a more lipid soluble compound than is captopril.

Increasing clinical experience with captopril has led to the recognition of certain side effects. Some, such as neutropenia, although rare (0.04–0.06%), occur with captopril as well as other ACE inhibitors, such as enalapril (Table 10-8). Others such as skin rash appear to be somewhat higher with captopril (4–7% vs. 1.5%). Disglusia (taste distur-

Table 10-8. Representative ACE Inhibitors[a]

Name	Structure	Comments
Teprotide (SQ 20,881)	Glu-Trp-Pro-Arg-Pro-Gln-Ile-Pro-Pro	Venom from viper snake *Bothrops jaracara*
Captopril (Capoten)	CH_3; $HS-CH_2-C-C-N$; H O; COOH	Marketed in the United States
YS980	CH_3; $HS-CH_2-C-C-N$; H O; S; COOH	Experimental; more lipid soluble than captopril
WY-44,221	CH_3; $HS-CH_2-C-C-N$; H O; COOH	Experimental; more potent than captopril
Rentiapril (SA446)	HOOC; H; $HS-CH_2-C-C-N$; H O; S; OH	Experimental drug in Japan (Yamauchi et al., 1987)
Enalapril (Vasotec)	CH_3; $-CH_2CH_2CHHNCHC-N$; C; O; O OC_2H_5; COOH	Marketed in the United States Pro-drug for enalaprilic acid (eater hydrolyzed to COOH)
Ramipril (Altace)	H; H; CH_3; $-CH_2CH_2CHHNCHC-N$; C; O; O OC_2H_5; COOH	Pro-drug, better oral availability
Benzapril (Lotensin)	$-CH_2CH_2CHHN-$; C; O; O; N; OC_2H_5 CH_2COOH	Orally effective product. More potent than enalapril. Once-a-day dosage

(*Continued*)

Table 10-8. *(Continued)*

Name	Structure	Comments
Lisinopril (Zestril)	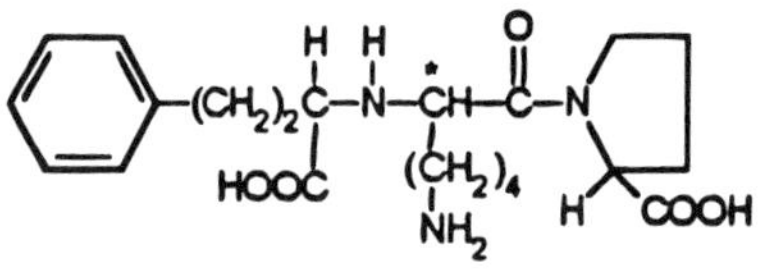 H H * O (CH₂)₂C–N–CH–C–N HOOC (CH₂)₄ NH₂ H COOH	Long-acting; $t_{1/2}$ = 13 h; once-a-day dosage; not a pro-drug
Fosinopril (Monopril)	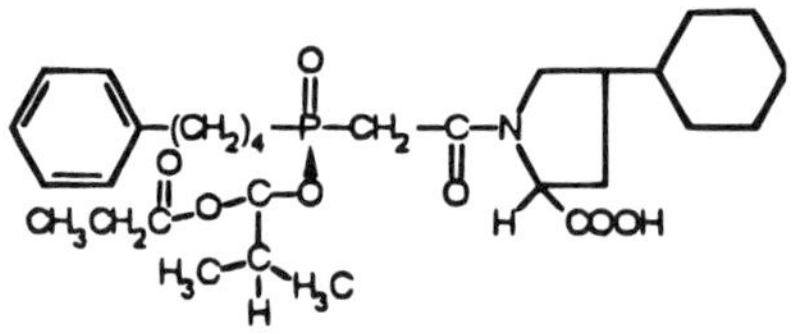 O (CH₂)₄–P–CH₂–C–N O O CH₃CH₂C–O–C–O H₃C–C–H₃C H O H COOH	Long-acting; once-a-day dosing; Useful in patients with renal insufficiency. Pro-drug
Quinapril (Accupril)	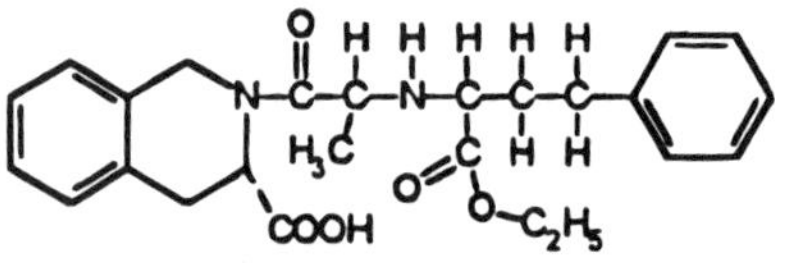 O H H H H H N–C–C–N–C–C–C H₃C C H H O= O–C₂H₅ COOH	Pro-drug; 80% ACE inhibition with once daily dosing of 20 mg

[a] Trade name or corporate designation in parentheses.
* Chiral centers.

Table 10-9. SAR of Captopril and its Analogs[a]

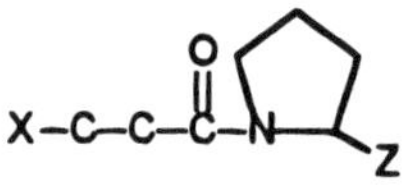

X–C–C–C(=O)–N ... Z

X	*R*	*Z*	I_{50}[b]
1. HS	CH_3(S)	COOH(L)	0.005[c]
2. HS	CH_3(R)	COOH(L)	0.50
3. HS	CH_3(R)	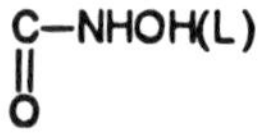 C(=O)–NHOH(L)	0.022[d]
4. HS	CH_3(R)	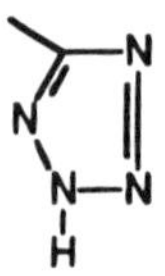 N, N, N–N, H (tetrazole ring)	0.26
5. HS	H	COOH(L)	0.20
6. HS	H	COOH(D)	1800
7. HS	H	COOH($\Delta^{2,3}$)	0.65
8. HS	H	PO_3H_2	1.7
9. HS	CH_3(R)	—	240
10. HOOC	H	COOH(L)	135
11. HOOC	CH_3(R)	COOH(L)	12
12. Teprotide	(SQ 20,881)		1

[a] Data from Ondetti et al., 1977; Petrillo and Ondetti, 1982.
[b] *In vitro* concentration μM inhibiting rabbit lung ACE.
[c] Captopril.
[d] R-2-methylsuccinyl-L-proline.

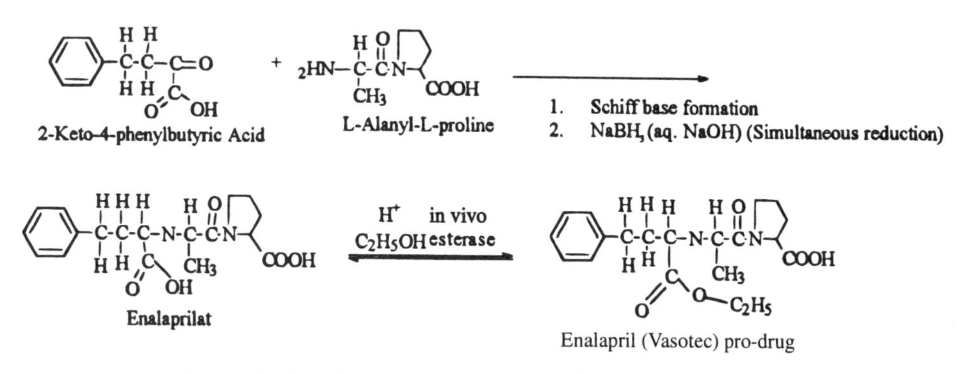

Figure 10-10. Synthesis of enalaprilat and its pro-drug enalapril.

bances, e.g., metallic), particularly at higher doses, is similarly high with captopril. The taste loss and rashes resembling those of penicillamine have implicated the SH group as the culprit. Even though lowering the dose will frequently reduce this problem, it has led to the development of nonsulfhydryl ACE inhibitors, of which enalapril has become the prototype.[15] The SH group is replaced by substituted α-amino-—phenylbutyric acids, thus resulting in carboxyalkyl dipeptides, such as enalaprilic acid, which is a dicarboxylic amino acid. The carboxylate anion here becomes the Zn ligand, which gives the compound an I_{50} of 1.2 μM.

Because of the zwitterionic highly polar nature, very poor oral bioavailability made the drug clinically useless by that route. Thus the ester, enalapril, is used as the pro-drug, as are the other nonsulfhydryl ACE inhibitors in Figure 10-8. Figure 10-10 gives the synthesis of enalaprilat and its pro-drug enalapril. Enalapril has a slower onset of action than does captopril, possibly because the drug must first be metabolically cleaved and, even though it is better than the enalaprilat, it still has inadequate oral absorption. This drug also has a longer duration of action. It should be noted that part of the impetus to development of nonsulfhydryl ACE inhibitors was to avoid captopril-type side effects, yet some of the same reactions have been reported.

The bicyclic drug ramipril (Table 10-8) was stereoselectively synthesized from L-alanine to produce S configurations at three of the five chiral centers. In vitro assays showed enalaprilat and the diacid of ramipril (i.e., the parent compound) to be equally potent in rats and the dog. Still, orally ramipril was 10 times more effective than enalapril. This indicates a much better bioavailability for the pro-drug. Studies with stroke-prone spontaneously hypertensive rats showed high, long-acting activity. Inhibition of ACE activity in the kidney and mesenteric artery persisted for 1 week. Clinical studies in healthy human volunteers corroborated long duration and high potency. Expansion of the proline ring into a seven-membered lactam ring and fusing it with benzene into a benzolactam (or benzazepine) produced the ACE inhibitor (as a pro-drug) benzapril (Lotensin), which was marketed in 1991 (Table 10-8).

The introduction (1991) of fosinopril (Monopril), which is an ester pro-drug of fosinoprilate, illustrates several features leading to some clinical advantages. The drug has a phosphinate group claimed to be capable of specific binding to the active site of ACE.

[15] It was the only one marketed in the United States until 1988.

Excretion is evenly divided between renal and hepatic routes. This results in patients with renal insufficiency clearing the drug at a similar rate since the rate of hepatobiliary elimination is increased to compensate. Thus no dosage adjustments are needed.

Renin is the first enzyme in the RAS. Chemical interference in this system can lower blood pressure. That has been amply demonstrated experimentally and clinically with the ACE inhibitors over the past decade. Since ACE is not a highly specific enzyme—it has several natural substrates—its inhibition with drugs is likely to have a greater potential for side effects. Inhibiting renin would avoid that. It has only one specific substrate—angiotensinogen—which is a large-molecular-weight α_2-globulin. Intervention at this earlier point in the cascade then may offer improvements, or at least alternatives, to ACE inhibition.

Renin is classified as an aspartic proteinase (i.e., a proteolytic enzyme structurally and mechanistically in the same class as pepsin and similar enzymes). The *active site* of renin, and related aspartic proteinases, is in a cleft that can bind to and accommodate substrate peptides of 8 to 14 amino acids. The *scissile bond* (i.e., hydrolyzable) in the substrate that human renin can cleave is leu-val (it may be different in other mammals). It will be recalled that angiotensinogen, renin's normal substrate, is hydrolyzed at the scissile leu^{10}-val^{11} peptide bond (from the amine terminal). The 14-amino acid portion, beginning with Asp, must be the substrate "portion" of this large zymogen that "fits" into the cleft to bind to the enzyme's active site. Much of the previous work revolved around designing analogs of the renin polypeptide substrate (usually 8 to 14 amino acid residues) called RIPs (renin inhibitory peptides) in which the scissile peptide bond is modified to noncleavable dipeptide analogs. The hydroxyethylene linkage can then be viewed as a *transition-state* analog. Several such RIPs have achieved in vitro inhibition in the nonamolar (10^{-9}) range. An interesting naturally occurring, competitive RIP called *pepstatin* has been known since 1971. Even though it was potent against pepsin, it was a weak inhibitor against human renin. Several synthetic analogs, however, were in the nanomolar range. It was then suggested that the central statine in this inhibitory peptide might function as an analog of a putative leu-val *transition state* believed to bind normally to the active site of the protease (in this case renin) during its catalysis of scission. However, since Sta does not offer the enzyme a scissile bond, but only the binding capacity to inactivate its active site temporarily, molecules containing Sta in the proper molecular position become competitive protease inhibitors. Thus statine could be a suitable replacement for the ileu-val hydroxyethylene isostere, in spite of the fact that this gamma-amino acid lacks valine's alkyl side chain. Figure 10-11 summarizes the essential chemistry.

Pepstatin-A *Iva -Val-Val-Ala-Sta-Ala-Sta *Iva = Isovaleric acid

Statine (Sta) (3S, 4S)

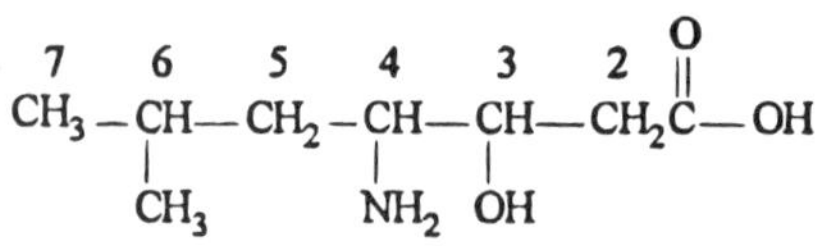

$$\overset{7}{CH_3}-\underset{\underset{CH_3}{|}}{\overset{6}{CH}}-\overset{5}{CH_2}-\underset{\underset{NH_2}{|}}{\overset{4}{CH}}-\underset{\underset{OH}{|}}{\overset{3}{CH}}-\overset{2}{CH_2}\overset{\overset{O}{||}}{C}-OH$$

These ideas, then, became the strategy for the design for numerous peptoids resembling known renin substrates. The scissile dipeptide was replaced with nonscissile statine and analogous amino acids such as 4-amino-5-cyclohexyl-3-hydoxypentanoic acid (ACHPA) and fluorostatinone to act as presumed transition-state analogs and to become potent renin inhibitors.

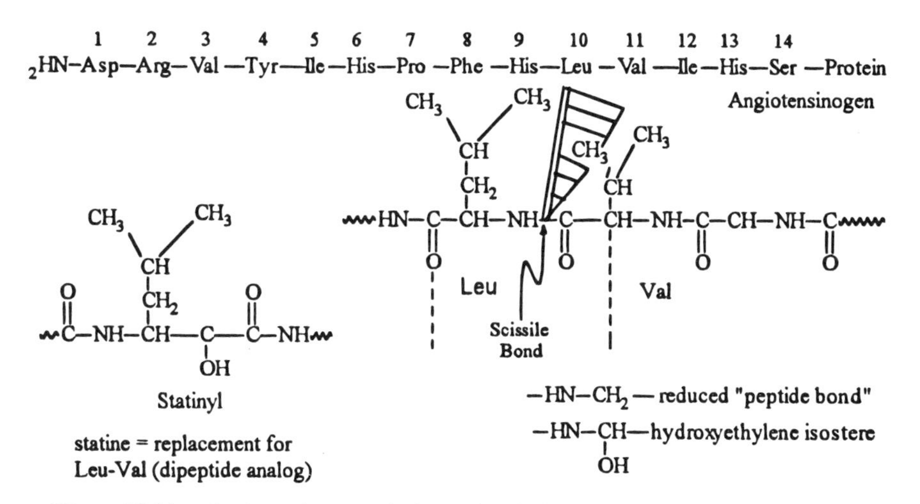

Figure 10-11. Angiotensinogen scission and renin inhibition (see text for discussion).

Many transition-state (or cleavage-site) mimics have been synthesized and tested, some with high inhibitory potency against human renin in vitro. Several also exhibited antihypertensive properties in animals.

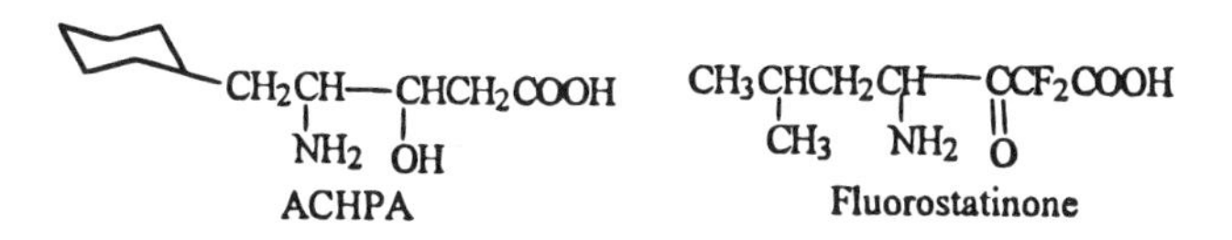

The clinical problem is not a shortage of potent human renin inhibitory substances. Some even show in vivo effectiveness. The problems are intrinsic to any potential drug that is primarily peptidyl in nature: poor oral absorption, rapid hydrolysis by proteolytic enzymes, and fast biliary excretion of undegraded molecules. The latter two factors lead to short duration of action. Thus drug design in this area is turning to reducing the peptide nature of these compounds by synthesizing smaller molecules and embellishing a nonscissile angiotensinogen-binding fragment with nonpeptide organic moieties, which can be tailored toward desirable physicochemical properties and potencies.

Plattner's group at Abbott Laboratories (Bolis et al., 1988) has recently "miniaturized" human renin antagonists down to one peptide bond, yet they have maintained high binding ability to the appropriate subsite of the enzyme based on understanding of the normal substrate's functioning at that site. Figure 10-12 illustrates how this scaled-down drug was "conceived." Structure A represents the fragment of the large angiotensinogen prohormone (specifically, Nos. 8, 9, 10, 11, 12, and 13) understood to bind into the cleft, or to fit into the pocket in which the active hydrolytic site, and presumed binding subsites of the renin enzyme, reside. The phe-his dipeptide unit was retained (see A, Fig. 10-12) as important to the binding process. As discussed, the leu-val scissile bond of angiotensinogen was replaced by the nonhydrolyzable hydroxyethylene isosteric moiety (see B), which had been shown to be superior to the reduced "peptide bond" as an inhibitor (see Fig. 10-11). The terminal amine function is protected by a BOC (*tert*-butoxycarbonyl) group at C. The cyclohexylmethyl at D replaces the renin-binding capacity of the corresponding isobutyl

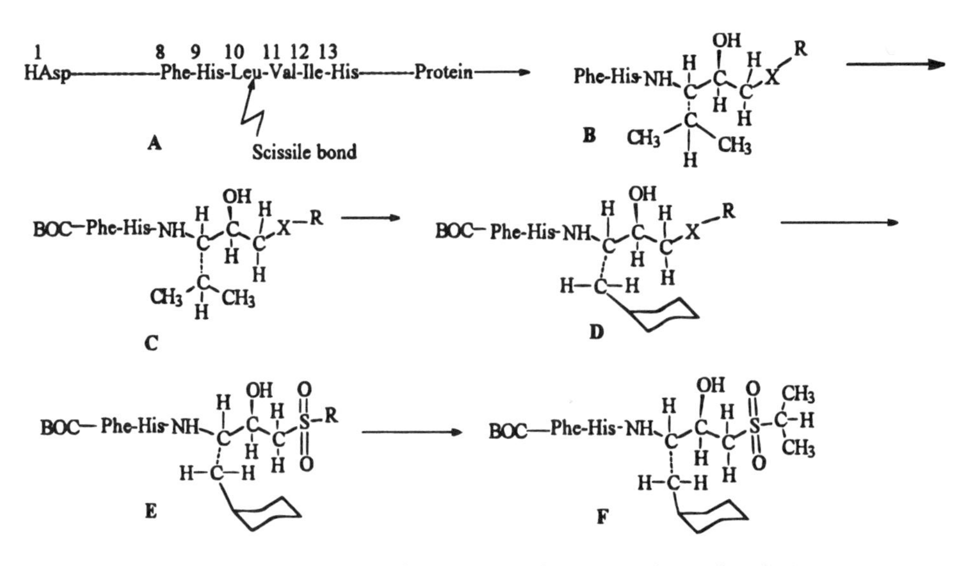

Figure 10-12. Design of renin antagonist (see text for explanation).

group of Leu in the natural substrate with a 20-fold increase in inhibitory potency. The SO_2 group, E, represents a bioisosteric replacement of the peptide bond of the Val-Ile region. The most remarkable aspects of this study were probably the successful replacement of the bulk of the angiotensinogen polypeptide with a small hydrophobic *iso*-propyl group, F. This compound, which was tested in salt-depleted cynomolgus monkeys, intravenously showed high potency and may be orally active as well.

The rationale applied in miniaturizing renin inhibitors to achieve clinically useful compounds by eliminating their protein nature and increasing inhibitory potency has, of course, wider ramifications. Judicious use of bioisosteric replacements and the relevancy of the transition-state analogy concept can also be applied with similar success to polypeptide neurotransmitters.

A new agent has been put into clinical play (1995) in the renin–angiotensin cascade game. Losartan (Cozaar) does not counteract the effects of renin; nor does it affect ACE. Rather, it blocks the product of the ACE-catalyzed reaction—Angiotensin II—from binding to AT_1 receptors found in various tissues. These AT_1 receptors, which exist in vascular tissue and the adrenal glands, trigger the vasoconstriction that normally helps to maintain blood pressure when bound to Ang II, and as a pathological event become components of the hypertension equation. The drug appears to be a reversible competitive inhibitor, without partial agonist activity. Its primary active metabolite acts similarly, but it is noncompetitive at AI_1. It is considerably more potent.

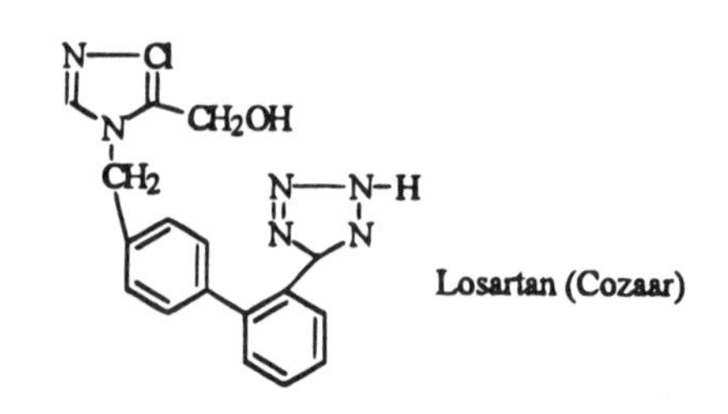

Table 10-10. Major Clinical Uses of Diuretics[a]

Diabetes insipidus
Edema
  of the brain
  cardiopulmonary, acute
  due to cirrhosis with ascites
  due to chronic congestive heart failure[b]
  idiopathic
  due to pregnancy[c]
Glaucoma
  open angle (long-term treatment)
  acute angle (preoperative)
Hypokalemic alkalosis[d]
Hypertension
Mountain sickness, acute
Nephrotic syndrome
Renal failure
  chronic
  acute, oliguric
Urolithiasis, calcium idiopathic

[a] Conditions listed alphabetically, not in order of clinical significance.
[b] Subsequent to digitalis therapy initiation.
[c] Now often considered unnecessary by many clinicians.
[d] Treated only with K^+-sparing diuretics (see text).

Losartan is mechanistically the most specific interrupter of the renin–angiotensin reaction sequence presently in use. Its clinical advantage resides in its once-a-day dosing (improved patient compliance) and its very gradual onset of action. The latter almost eliminates the hypotensive "overshoot" which is a common side effect during early treatment with most other agents.

The tetrazole ring on the biphenyl system is used here as a carboxyl bioisostere, with the sole hydrogen atom being neutralized to a potassium salt, the marketed form of the drug.

Losartan and analogs that are now in development resulted from a search for nonpeptide AT receptor antagonists. Use of peptide SAR information, pharmacophore models for AT, were proposed and subjected to molecular modeling overlay strategy to find potential synthetic candidates exhibiting the desired activity. Once found, molecular modifications optimized the pharmacology.

10.3.1.5. Diuretics

The simplest definition of a *diuretic* is that it is an agent that increases the rate of urine formation. This means that water is involved, as is the excretion of ions (i.e., salts). Terms like *natriuretic* (sodium excreting) or *saluretic* (salt excreting) are encountered. The following discussion will not attempt a complete consideration of diuretics and their varied clinical applications (see Table 10-10). Interest at this point is primarily turned to agents that have been useful in the management of cardiovascular disease, namely, congestive heart failure and hypertension. The kidney plays a major role in regulating the volume and chemical composition in the body's internal environment. The nephron is the basic structural unit. Each kidney contains approximately 1 million such units (Fig. 10-13).[16]

[16] The reader may wish to consult recent physiology texts for additional background.

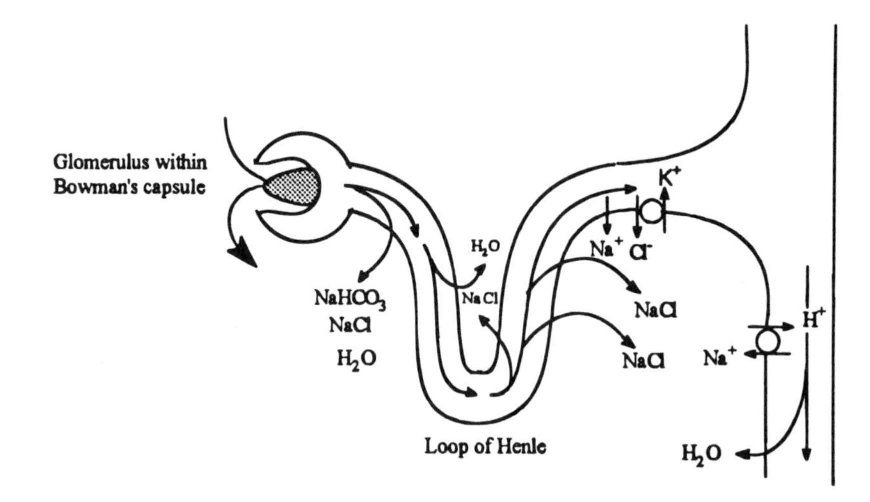

Figure 10-13. Schematic nephron.

The kidneys in the normal healthy young adult handle a blood flow of approximately 1,200 ml/minute. This allows for a high rate of glomerular filtration. Approximately one fifth of the flow is filtered through the glomeruli into the spaces of Bowman's capsule. The major portion of this filtrate is reabsorbed by the renal tubules (proximal and distal). Unwanted waste or excess materials are left behind to be excreted, together with an osmotically required amount of water. Urine is the result. Thus in addition to glomerular filtration, tubular reabsorption and tubular secretion are the factors involved in urine secretion. The factors governing movements of substances across glomerular (and other) membranes have been discussed elsewhere.

Compounds with high diuretic potency resulted from the observation in 1919 that an antisyphilitic organic mercury compound, merbaphen, also had strong diuretic properties. It was demonstrated the next year that the drug increased the excretion rate of Na^+ and Cl^- (i.e., it was saluretic).

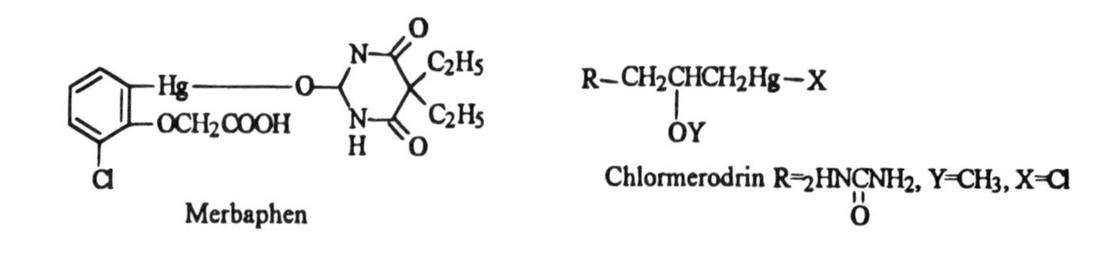

Merbaphan, which is an organomercury derivative of barbital (Chapter 11) and a chlorophenoxyacetic acid, was too toxic for general use, but it became the initial lead for the development of a group of highly potent, and somewhat safer, organomercurials, most of which were derivatives of 1-mercury-2-propanol, where X is a variety of substituents from Cl to theophylline, $Y = H$ or CH_3, and R is a variety of moieties including the urea of chlormerodrin. These drugs were the mainstay in the treatment of edema associated with CHF from 1920 to well into the 1950s, when they were rapidly replaced by the orally effective and much safer (no kidney damage or fatal ventricular fibrillations) benzothiadiazine

(Table 10-11). Chlormerodrin survives as a radiopharmaceutical containing the ^{197}Hg isotope ($t_{1/2} = 64$ h) used for the visualization of renal parenchyma. The chlorophenoxyacetic acid portion was later modified into another type of diuretic (see later). It is also of interest that the diuretic properties of mercurous chloride, Hg_2Cl_2 (calomel), were known to Paraselsus in the sixteenth century. The exact mechanism of action of the organomercurials has remained unsettled. There is general agreement, however, that a breakdown to ionic mercury at the acidic urinary pH may be involved. The bonding of a Hg atom to an organic residue decreased the toxicity of the inorganic compounds to "acceptable" levels. In addition, it can be reasoned that as an organic ligand the probability of cellular penetration to the sulfhydryl enzymes in the proximal tubules is greatly improved so that renal enzymes involved with tubular reabsorption processes can be inactivated, thus producing diuresis.

Table 10-11 lists several groups of diuretics that have become the clinical successors to the earlier organomercurials. Even though they differ in their exact site and mechanism of action in the nephron, they all share one particular chemical feature: They are aromatic sulfonamides. There is a historical reason for this. Soon after the introduction of the sulfonamides as antibacterial agents in the mid-1930s (Chapter 2), changes in the electrolyte balance of patients were noted as was systemic acidosis accompanied by an alkalinization of the urine due to increased rate of HCO_3^- excretion. Proposals by several workers and others established that inhibition of the enzyme carbonic anhydrase (CA) accounted for the electrolytic imbalances produced. Since the antibacterial sulfonamides were relatively weak inhibitors, a successful search for more potent CA inhibition ensued. This effort resulted in several 1,3,4-thiadiazole-2-sulfonamides whose in vitro inhibitory potency exceeded sulfanilamide by several hundred times (but were devoid of antibacterial activity). The 5-acetylamino derivative, acetazolamide (Table 10-11), became the first successful drug introduced into clinical use.

$$CO_2 + H_2O \xrightleftharpoons{\text{Carbonic anhydrase}} H_2CO_3 \rightleftharpoons H^+ + HCO_3^- \qquad (10.4)$$

Equation 10.4 illustrates reactions that occur whenever CO_2 and water molecules interface. The enzyme CA simply increases the rate at which equilibrium is reached. CA is found in the luminal membrane of the epithelial cells that line the proximal tubules. Within these cells the reaction is primarily in the direction of carbonic acid dehydration (i.e., $H_2CO_3 \rightarrow CO_2 + H_2O$). The CO_2 then readily penetrates into the proximal cell, where it rehydrates with H_2O. In this case CA catalyzes the reaction more toward hydration (i.e., $CO_2 + H_2O \rightarrow H_2CO$).[17] The H^+ produced by the spontaneous ionization of the carbonic acid $H_2CO_3 \rightarrow HCO_3^- + H^+$ is secreted through the luminal membrane into the tubular lumen *in exchange* for Na^+, which enters the tubular cell. The HCO_3^- ions produced in the tubular cell, together with the exchanged Na^+, then enter the peritubular blood supply. The H^+ ion, now in the tubular lumen, combines with the HCO_3^- there, $H^+ + HCO_3 \rightarrow H_2CO_3$, which under the influence of CA then again dehydrates to CO_2, which then reenters the tubular cell. The next result is reabsorption of most of the bicarbonate. These concepts are schematically represented in Figure 10-14.

---

[17] It will be recalled that enzymes' "direction" of equilibrium can vary with pH.

Table 10-11. Structures, Type, Site, and Mechanism of Sulfonamide Diuretics[a]

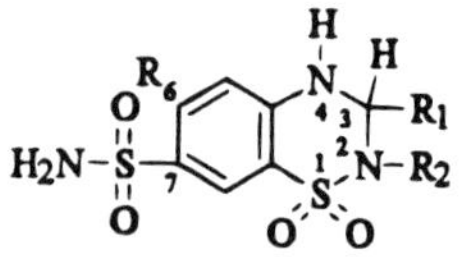

A. Benzothiadiazines (thiazides)

R	*R*$_1$	*R*$_2$	Name	Site
Cl	H	H	Chlorthiazide (Diuril)[b]	Inhibition of Na^+ (and Cl^-) reabsorption in early distal tubule and ascending limb of loop of Henle
Cl	CH_2SCH_2-C_6H_5	H	Benzthiazide (ExNa)[b]	
Cl	H	H	Hydrochlorothiazide (Hydrodiuril)	
Cl	$CHCl_2$	H	Trichlormethiazide (Naqua)	
Cl	CH_2Cl	CH_3	Methylclothiazide (Enduron)	
Cl	$CH_2SCH_2CF_3$	CH_3	Polythiazide (Renese)	
Cl		H	Cyclothiazide (Anhydron)	
CF_3	H	H	Hydroflumethiazide (Saluron)	
CF_3	CH_2-C_6H_5	H	Bendroflumethiazide (Naturetin)	

Thiazide isosteres $O=S=O$ changed to $C=O$

R	*R*$_1$	*R*$_2$	Name	Site
Cl	CH_2CH_3	H	Quinethazone (Hydromox)	Inhibition of Na^+ (and Cl^-) reabsorption in early distal tubule and ascending limb of loop of Henle
Cl	CH_3	$_pC_6H_5$-CH_3	Metolazone (Zaroxolyn)	
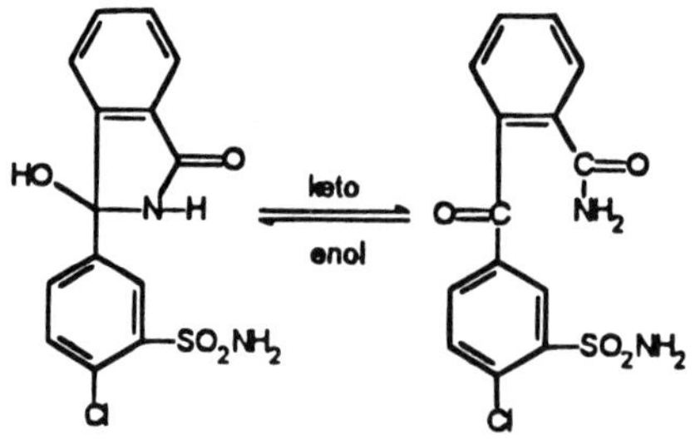			Chlorthalidone (Hygroton) enol (lactam) predominant	

keto ⇌ enol

(Continued)

Table 10-11. *(Continued)*

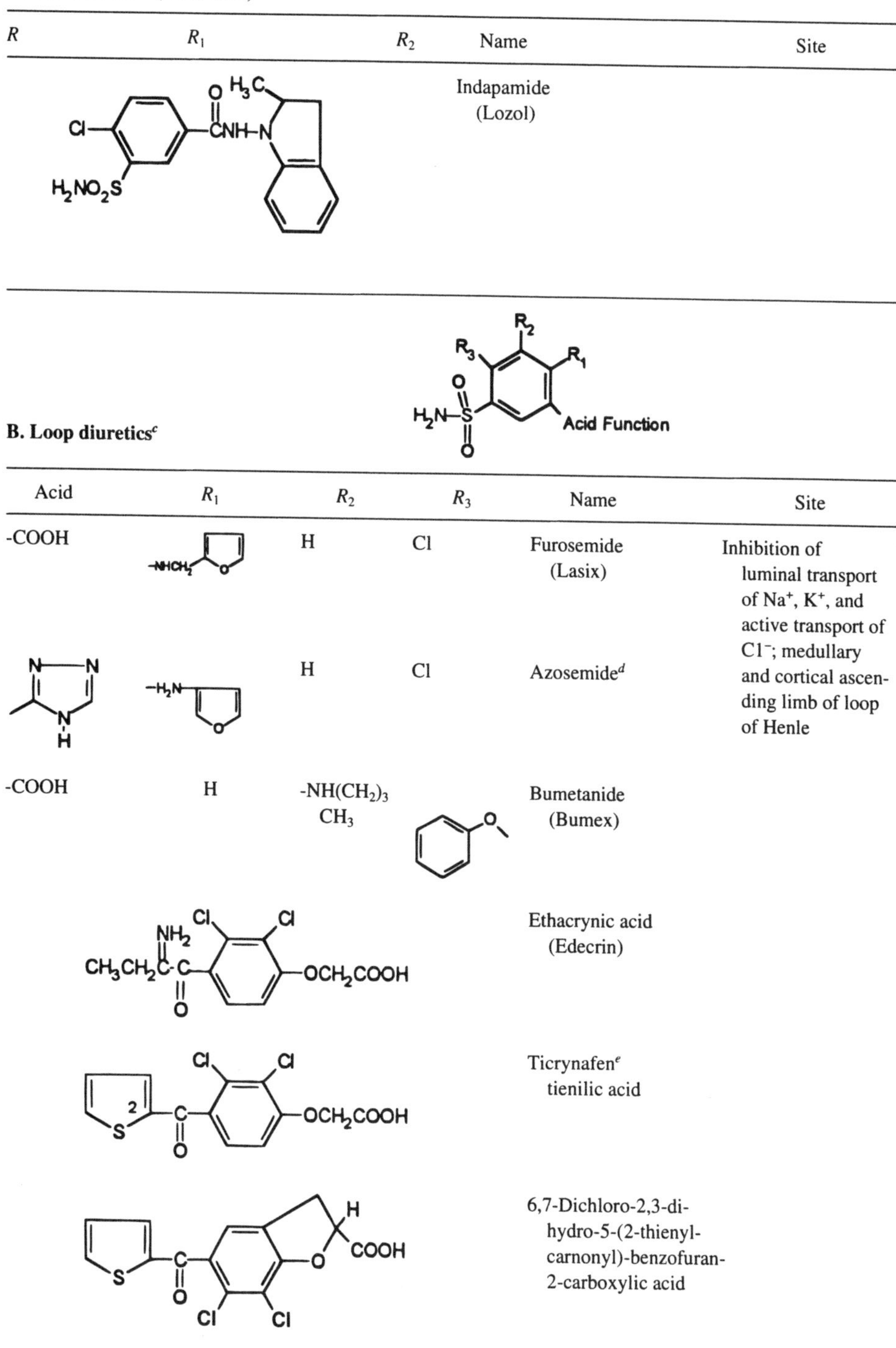

R	R_1	R_2	Name	Site
			Indapamide (Lozol)	

B. Loop diuretics[c]

Acid	R_1	R_2	R_3	Name	Site
-COOH		H	Cl	Furosemide (Lasix)	Inhibition of luminal transport of Na^+, K^+, and active transport of Cl^-; medullary and cortical ascending limb of loop of Henle
		H	Cl	Azosemide[d]	
-COOH	H	$-NH(CH_2)_3CH_3$		Bumetanide (Bumex)	
				Ethacrynic acid (Edecrin)	
				Ticrynafen[e] tienilic acid	
				6,7-Dichloro-2,3-dihydro-5-(2-thienylcarnonyl)-benzofuran-2-carboxylic acid	

(Continued)

Table 10-11. *(Continued)*

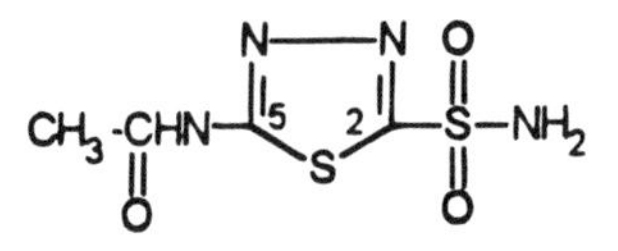

C. Carbonic anhydrase inhibitors

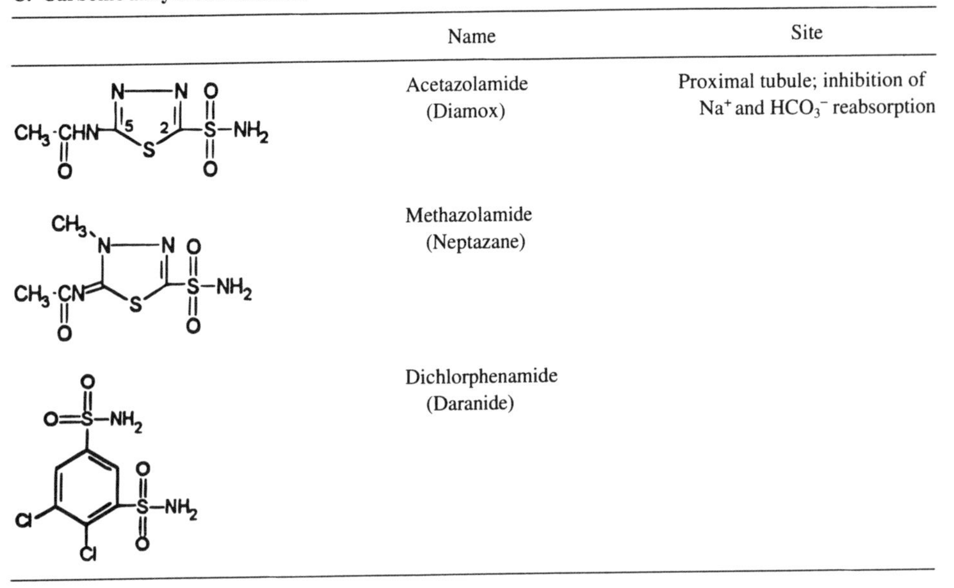

	Name	Site
	Acetazolamide (Diamox)	Proximal tubule; inhibition of Na^+ and HCO_3^- reabsorption
	Methazolamide (Neptazane)	
	Dichlorphenamide (Daranide)	

[a] Trade names in parentheses.
[b] Double bond between 3 and 4.
[c] Also referred to as high-ceiling diuretics because of higher peak diuresis.
[d] Not on U.S. market; investigational.
[e] Withdrawn from market due to liver toxicity in 1980.

The proximal tubule is where up to 75% of the glomerular filtrate is reabsorbed. Reabsorption of Na^+ occurs simultaneously with isoosmotic amounts of H_2O.

A hydrogen-bonding mechanism that acts competitively explains the action of certain sulfonamide carbonic anhydrase inhibitors that have diuretic and antiglaucoma properties. Carbonic acid is thought to be the normal substrate that fits into a cavity of and complexes

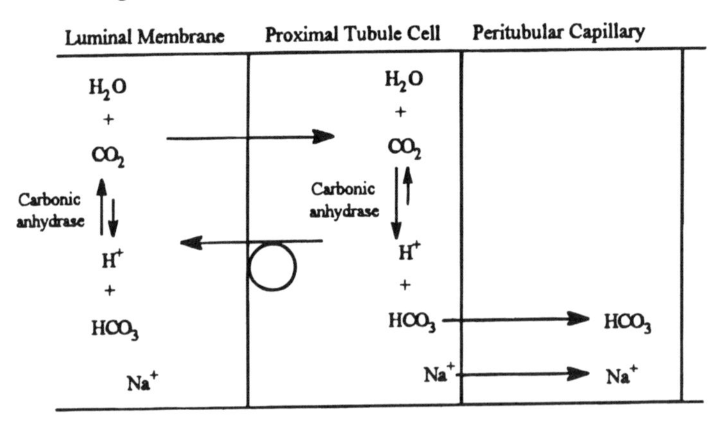

Figure 10-14. Bicarbonate reabsorption in proximal tubule.

with the enzyme CA (Fig. 10-15A). This complex is strongly stabilized by *four* hydrogen bonds. The sulfonamide agents that have a geometric structure permitting an equally good fit into the cavity of the enzyme also effectively bond, presumably to the same four areas by hydrogen bonds (Fig. 10-15B). Thus these sulfonamide agents competitively prevent the carbonic acid from binding at this site and consequently inhibit the action of the enzyme, which leads to an acid-base imbalance (Eq. 10.4) and ultimately diuresis. The increased systemic acidity that results also manifests itself in anticonvulsant properties for reasons not yet completely understood. Sulfonamides of the type in which even one of the two hydrogens has been replaced (i.e., bond R–SO_2NHR') and in which the possibilities for hydrogen bonding have been reduced from four to three are inactive compounds. (Compounds of the type $RSO_2NR'R''$ are, of course, also totally inactive.) A reduction of hydrogen-bonding potential of only 25% is sufficient to eliminate the pharmacological action, which definitely exemplifies the significance of this phenomenon.

CA inhibitors then, were historically the first orally effective nonmetallic diuretics. Because their primary effect of blocking the NA^+–H^+ exchange is in the proximal portion of the nephron, these drugs inhibit the reabsorption of $NaHCO_3$ and H_2O. This explains both the diuresis they produce as well as the resultant alkalinity of the urine. A significant proportion of the Na^+ not reabsorbed in the proximal tubule, however, is reabsorbed further down in the loop of Henle region. Therefore, CA inhibitors are not particularly potent diuretics. Furthermore, because of little Cl^- excretion, a mild hyperchloremic acidosis occurs, leading to an acid–base imbalance in which the diuretic does not function well. Thus a tolerance or lack of systemic effectiveness develops. As a result CA inhibitors are not used as primary therapy; rather, they are relegated mostly to glaucoma treatment where their benefits are not dependent on diuresis. It is CA inhibition in the eye's ciliary epithelium that can lead to considerable IOP reduction when added to the systemic acidosis produced. The application of all three drugs listed in Table 10-11 can be used in the treatment of long-term open-angle glaucoma even in patients refractory to timolol and EP. As previously mentioned, acetazolamide has sometimes been helpful in epilepsy management, although tolerance may rapidly develop.

The era of benzothiadiazines began with the synthesis (1957) and discovery of the valuable diuretic properties of chlorthiazide (CTZ). Today, the thiazides and several related isosteres and analogs (Fig. 10-16) constitute a major drug group in the treatments of CHF and hypertension. Many such cyclic sulfamoyl compounds have been synthesized and tested since then; about a dozen are currently in use and are listed in Table 10-11.

As a group they are the most widely used oral diuretics, both in the treatment of edema, such as in CHF, and hypertension. Their pharmacology is qualitatively the same as CTZ; namely, they inhibit the membrane in the cortical section of the ascending limb of Henle's loop from reabsorbing Na^+. On prolonged use, particularly at higher doses, K^+ losses can become significant and present a potentially hazardous arrhythmia situation to the

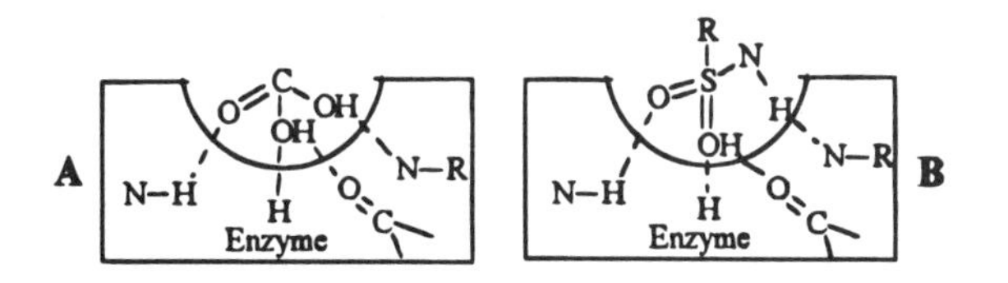

Figure 10-15. Interactions at hypothetical reactive sites of carbonic anhydrase.

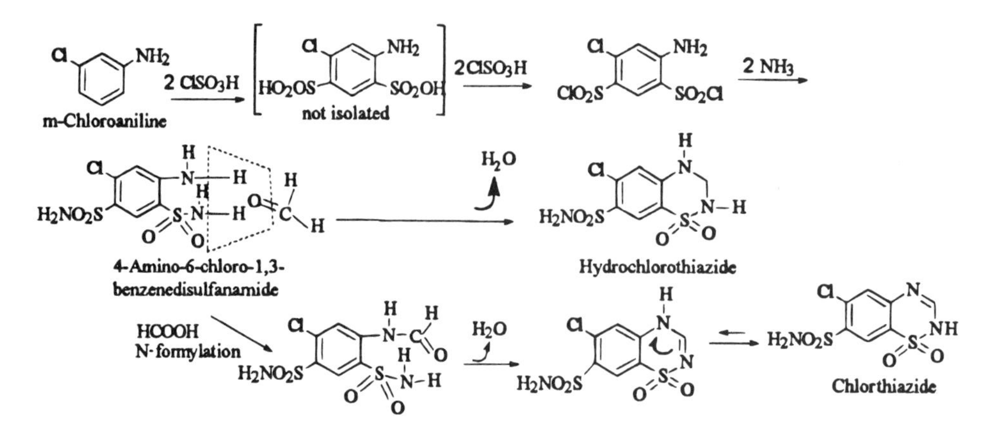

Figure 10-16. Synthesis of chlorthiazide and hydrochlorothiazide.

myocardium. In many patients liquid potassium supplements to replenish the losses are necessary. Another untoward effect presenting clinical problems is decreased excretion of uric acid due to competition for limited secretion capacity. This may precipitate attacks of gouty arthritis. Increased blood sugar and cholesterol levels are also often seen.

SAR studies have empirically shown that the unsubstituted sulfamoyl group at position 7 is an absolute requirement for diuretic activity (note its absence in diazoxide); so is a lipoidal electron-withdrawing group at position 6. This is invariably a Cl or CF_3 function. The role of the SO_2 group at position 1, although found in the majority of the drugs here, can be isosterically replaced with a >C = O. Such compounds tend to be drugs of lesser potency (e.g., quinethazone and metalazone, Table 10-11). Unsaturation, such as with the 3–4 double bond, seems to offer lower potency as is well exemplified by CTZ (daily doses of 500–2000 mg) versus hydrochlorothiazide (HCTZ) (25–50 mg), yet this can be overcome by highly lipid solubilizing substituents at C-3. Thus, even though benzthiazide has a 3–4 double bond, the 3-benzylthiomethyl group more than counteracts it. Nevertheless, all the other thiazides are saturated at this position. Addition of lipid soluble moieties at C-3 of such compounds increases potency even further so that cyclothiazide and bendroflumethiazide can be effective in daily doses as low as 1–2 mg. The dosage range of clinically used compounds is therefore from 1 to 2,000 mg. The only other, but more important, clinical variable within which a choice is present is duration of action. There is CTZ with a low of 6 hours to several compounds in the 24+ hour range (e.g., methyclothiazide).

Chlorthalidone, which is not strictly speaking a thiazide, has the longest duration of action (60 hours), permitting alternate-day dosing. This can be useful with patients having medication compliance difficulties.

The thiazides exhibit some CA inhibition, yet it is not fundamental to their mechanism of action, which is still unclear at the molecular level. In fact, HCTZ, which is 10 times more potent than CTZ, has an inhibitory activity toward CA, which is actually 10 times less ($I_{50} = 2 \times 10^{-5}$ vs. 2×10^{6}).

Indapamide, although differing structurally from thiazides, is not so pharmacologically. It can be viewed chemically as consisting of a polar sulfamoylchlorobenzamide and a highly lipoidal methylindolyl moiety. It is a potent drug that may owe only part of its antihypertensive effectiveness to natriuresis since it also has vasodilator properties. Its full potential in blood pressure reduction can take several weeks to develop.

A third group of sulfamoyl derivatives are the high-ceiling loop diuretics. These compounds have a much higher peak diuretic activity than the previously discussed groups. Their remarkable activity, with up to 40% of the NaCl load filtered out into the urine, occurs mostly in the ascending limb of Henle's loop. This is typical. Potassium loss and increased uric acid levels may indicate some similarity of mechanism with the thiazides. As a group they are rapid; acting (within 30 minutes intravenously) and in large doses can produce copious urinary flow.

There are currently three such drugs in the United States (Table 10-11). Furosemide, which is the prototype, is an anthranilic (2-aminobenzoic) acid derivative. It retains the traditional sulfamoyl group and the Cl atom. Bumetanide, a 3-aminobenzoic acid, also retains the sulfamido moiety and is even more potent (1 mg = 40 mg furosemide). The phenoxy group (in lieu of Cl?) may account for this in part through increased lipophilicity. A more recent, still investigational drug, is azosemide. It is of interest because of its successful replacement of the COOH with the isosteric tetrazolyl group.[18]

Ethacrynic acid (Table 10-11) is a powerful loop diuretic whose molecular mechanism of action is not fully clear. However, it has marked pharmacodynamic similarities to the mercurial diuretics such as merbaphen (see above) and mersalyl, both of which are also phenoxyacetic acid derivatives, as well as in vitro and in vivo comparability in its reaction with SH groups: It competes with them for the same receptors. It is not surprising that an analogy, if not identity, of mechanism of action at the cellular level has been proposed. Equation 10.5, which illustrates a Michael-type addition, might represent a possible enzyme inactivation.

An SAR comparison of nine acryloylphenoxyacetic acids appears to demonstrate that an activated double bond susceptible to a nucleophilic attack is imperative to effective diuresis. However, reduction of the double bond, making 1,4-addition of an SH group impossible, reduced but did not eliminate saluretic activity. In fact, the same workers reported potent diuresis in a series of 5-indanyloxyacetic acids, which were incapable of reacting with sulfhydryl compounds. It is therefore difficult to assess the significance of sulfhydryl binding in the mechanism of diuretic action of ethacrynic-type drugs, except possibly to assign it a secondary status. The exact molecular mechanism therefore remains to be established.

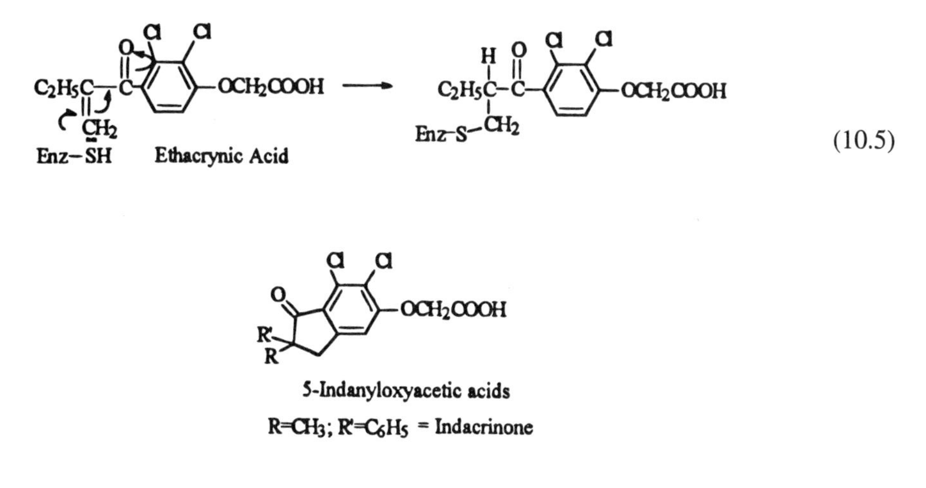

[18] See Chapter 2 for a comparison of COOH and tetrazolyl groups.

Tycrinafen (tienelic acid), which bears a structural resemblance to ethacrynic acid, was briefly on the market (1979–1980). Its diuretic profile resembled that of the thiazides, except that it was also an effective uricosuric agent, thus eliminating the elevated uric acid levels that plague the thiazides. Unfortunately, the appearance of serious hepatotoxicity caused the drug to be withdrawn.

It is hoped that indacrinone (earlier) will represent a safer uricosuric diuretic. It has a uricosuric feature in both enantiomers; the diuretic potency appears to reside mostly in the levorotatory isomer. An experimental approach whereby a 90%(+)/10% (–) isomer mixture is utilized to achieve effective diuresis without increased urate levels has been reported.

Tycrinafen's cyclic homolog of benzofuran-2-carboxylic acid (Fig. 10-11), is even more interesting. Here, the phenoxyacetic acid portion of the molecule is "rigidized" via a CH_2 bridge to the ortho position of the benzene ring. A complete stereoisomeric separation between diuretic and uricosuric activity has been demonstrated in several animal species. Such a compound may give us some insight into the steric requirements of diuretic/uricosuric activities.

10.3.1.5.1. Potassium-Sparing Diuretics

There are three diuretics available that impede the outflow of K^+ rather than promote it as do the thiazide and loop diuretics. Even though hypokalemia with the potent saluretics can usually be prevented with oral potassium supplements, they can present problems of palatability and are sometimes not reliable in maintaining desirable K^+ levels. It may therefore be preferable with certain patients to achieve diuresis without potassium depletion (Fig. 10-17).

Two such direct-acting drugs are the pteridine derivative triamterene and the pyrazine carbonyl-guanidine compound amiloride. These two drugs are unique among diuretics in not being acidic; in fact, they are basic (amiloride's pKa = 8.7). Their potency, however, is at best moderate. For this reason, and the fact that if used alone both drugs can actually produce hyperkalemia, they are most frequently used together with a lower dose of a thiazide. Combinations with HCTZ have been particularly popular.

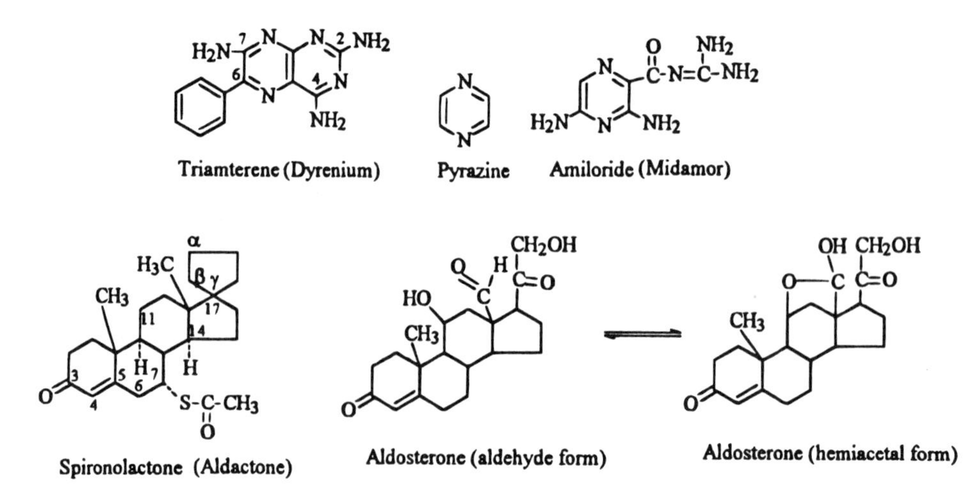

Figure 10-17. Potassium-sparing diuretics (trade names in parentheses).

Both drugs appear to have the same mechanism, acting on the distal tubule to block reabsorption of Na^+ and secretion of K^+. The mechanism of the third drug, spironolactone, is better understood since it acts as a true competitive inhibitor of aldosterone (Fig. 10-17), whose function has been known for some time. Aldosterone (probably physiologically present in the hemiacetal form) is a mineralocorticoid produced by the adrenal medulla. Mineralocorticoids have a direct effect on increasing the rate of Na^+ reabsorption by the epithelium of the renal tubule. Since aldosterone is 3000 times more effective in causing Na^+ retention than is cortisone, it is therefore the important antidiuretic hormone. When large quantities of aldosterone are secreted, almost all the Na^+ in the glomerular filtrate may be reabsorbed. This increased reabsorption of Na^+ from the tubules causes water retention and a decreased reabsorption of K^+, which flows through the tubules into the urine.

Spironolactone specifically competes with aldosterone at its receptor sites, which trigger the synthesis of the enzyme(s) that catalyze Na^+ transport when stimulated. Thus it *reverses* these electrolyte changes by blocking the renal tubular action of the hormone. By inhibiting Na^+ reabsorption the drug produces diuresis and decreases K^+ excretion. The drug is ineffective in clinical situations known to have high circulating aldosterone levels (e.g., cirrhosis with ascites). In general the drug is not potent and has a slow onset of action. The time lag (up to several days) is presumably the time needed for the renal enzyme levels to decrease. Spironolactone is also frequently used in conjunction with HCTZ.

The *gamma*-butyrolactone that shares the C-17 carbon with the cyclopentano ring of the steroid portion of the drug (spiro carbon) apparently is sufficiently similar to the hormone's C-17 "superstructure" to permit enough receptor binding to act as an antagonist. The 7-thioacetate is not critical since a 6–7 double bond without any substituent (canrenone) is also active.

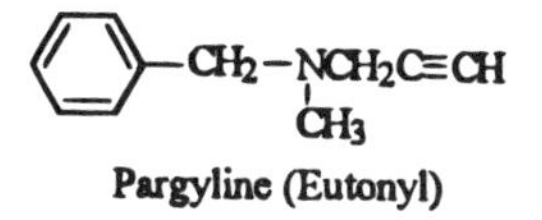

Pargyline (Eutonyl)

MAO inhibitors that have been, and still are, occasionally used to treat depression (Chapter 12) have long been known to have a hypotensive side effect. One such drug, the nonhydrazine MAO inhibitor pargyline, has at times been utilized as an antihypertensive, possibly in depressed patients, more to avoid the depressant side effects encountered with other antihypertensives (e.g., reserpine, methyldopa). How this apparent paradoxical effect might arise has not been definitely established. One hypothesis states that tyramine formed by decarboxylation of tyrosine in the gut would normally be destroyed by intestinal MAO. During its inhibition, however, tyramine would enter the circulation (Fig. 9-1). There, entering adrenergic neurons it would, on release, act as a *false neurotransmitter*. It should be understood that because of the potential for serious adverse effects characteristic of MAO inhibitors, the drug is now rarely used to lower blood pressure.

Prostaglandins as antihypertensives have so far not materialized as first hoped. PGs of the E and A series are potent vasodilators, as is PGI_2 (see below). They decrease vascular resistance and systemic blood pressure. PGEs work directly on smooth muscle to produce this effect. PGA_1 and PGA_2 decrease pressure in essential hypertension to normal levels, increase cardiac output, and lower peripheral resistance. Renal circulation also improves.

Prostyacyclin (PGI_2), which is a potent vasodilator, also dilates coronaries and can reduce blood pressure dramatically. Clinical applications, other than for emergencies, must await chemically stable analogs.

10.3.1.5.2. Special and Minor Diuretics

There are several compounds with diuretic effects that have special applications. These are usually sugars or sugar alcohols—the osmotic diuretics. They simply inhibit Na^+ and H_2O reabsorption in the proximal tubules, the Loop of Henle, and the urine-collecting duct of the nephron. These agents find applications to reduce IOP and the volume of vitreous fluid in the eye preceding ocular surgery. They are also utilized in reducing intracranial pressure during neurosurgical procedures, where edema of the brain is a complication. Simply by using large quantities (e.g., 150 g for an adult) of the water-soluble, nonmetabolizable compounds to produce a hypertonic condition (i.e., high osmolarity), water is withdrawn from tissues such as the eyeball, reducing pressure in it. The osmotic compounds include glycerin (glycerol), mannitol, isosorbide, and urea. Such agents also have potential use in acute poisoning with substances that are readily reabsorbed by tubules (e.g., salicylates, inorganic bromides). Osmotic agents have also been used to prevent necrosis following severe injury.

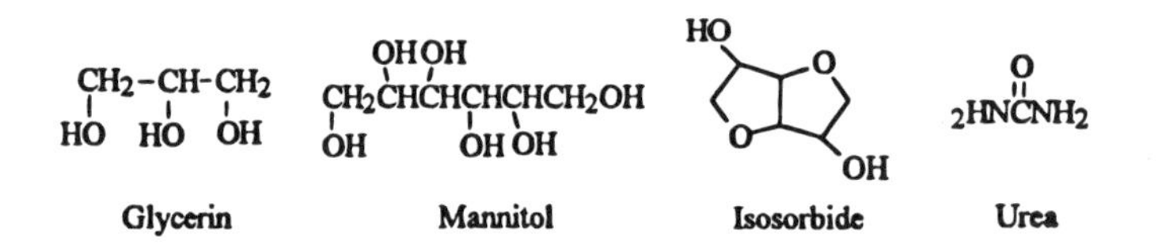

Other compounds with relatively mild diuretic properties are the xanthines, theobromine and theophylline; the former usually as a soluble complex with sodium benzoate, the latter with ethylene diamine (aminophylline). Theophyline's present use is primarily as an antiasthmatic (see Chapter 9).

Finally, certain inorganic salts when used in large doses were once used for short-term (several days) diuresis and acidification. These included NH_4NO_3, NaH_2PO_4, and NH_4Cl. Once absorbed, ammonium chloride has its NH_4^+ converted to urea in the liver. Its excretion produces some osmotic diuresis. The Cl^- displaces the HCO_3^-, which is excreted in the lungs via H_2CO_3 and finally CO_2. In essence the Cl^- decreases its buffering capacity, resulting in acidosis by lowering blood bicarbonate. Oral doses of NH_4Cl are 8–12 g/day. Because these quantities cause gastric irritation, the tablets should be enteric coated.

10.3.1.6. Cardiotonic Agents

Cardiotonic agents are a group of drugs utilized for their ability to increase the contractile force of the myocardium. In addition, they have, to varying degrees, the ability to affect the excitability of the heart tissue, automaticity, velocity of impulse conduction, and refractory period. The main therapeutic indications for such agents are CHF, atrial fibrillation and flutter, and atrial tachycardia, especially if paroxysmal. The drugs in this category are generally the digitalis glycosides.

Digitalis (foxglove) is a plant in which the leaves are the source of the active ingredients. Several species of the genus *Digitalis* are utilized, particularly *D. purpurea* and *lanata.* The plant was known botanically since the sixteenth century, but it was not until the

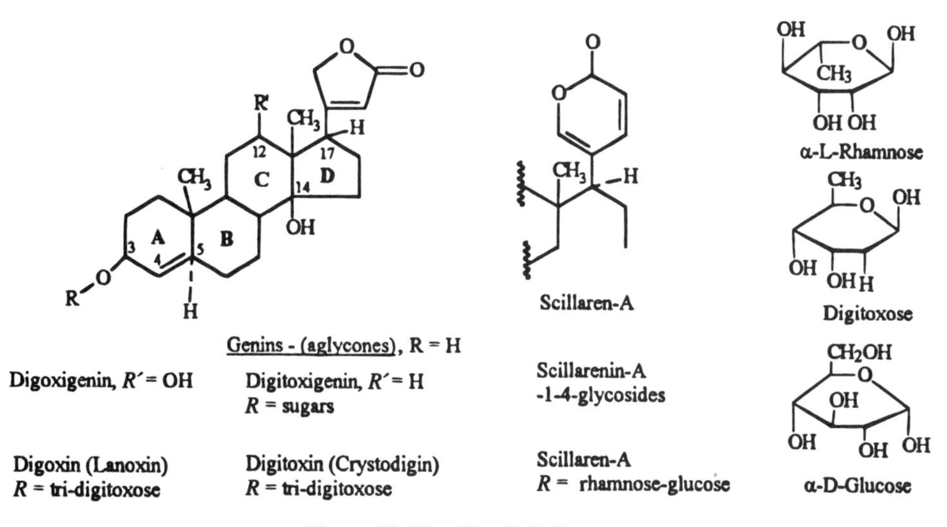

Figure 10-18. The digitalis group.

late eighteenth century that its real medicinal value was recognized in the treatment of CHF (actually, the accompanying edema, or dropsy) by W. Withering in England. After a decade of study and use he wrote a treatise in 1785 on the use of foxglove for dropsy and other diseases. For the next 150 years digitalis was used in the form of decoctions, infusions, tinctures, fluid extracts, and even as the powdered dried leaf.

Today, the active constituents have all been isolated and structurally identified as glycosides of a steroid nucleus with two unusual features: a hydroxyl function at C-14 and an unsaturated five-membered lactone called an α-β-unsaturated butenolide at C-17 (Fig. 10-18). Both features appear essential to cardiotonic activity. Enzymatic or chemical cleavage of the glycosidic bond at C-3 removes the various sugars attached, yielding the aglycone (e.g., digitoxin → digitoxigenin). The sugars such as digitoxose and rhamnose, which are deoxyhexoses, have no biological activity, but they do increase and modify the action of the aglycones. The number of alcoholic functions they bring to the total molecule predictably affects solubility in body fluids, transport ability across membranes, binding tenacity to cardiac (and other) tissues, and ultimately the onset and duration of action. Thus, digoxin, carrying the additional OH at C-12, is slightly more water soluble than digitoxin. One result is a somewhat better oral absorption for digitoxin, where lipid solubility is slightly enhanced.

Active cardiac glycosides have also been obtained from species of *Strophanthus* and squill (a sea onion). The former is a source of glycoside ouabain, which has an intense but short action; the latter provides scillaren A and other glycosides that are not widely used anymore. The glycosides most frequently used today are digoxin (Lanoxin) and digitoxin (Crystodigin, Purodigin).

The geometry at the fusion points of the various rings of the steroidal aglycone are significant. For example, the A/B ring fusion of the active glycosides is *cis* (i.e., the CH_3 at C-10[19] and the H at C-5 are β, or above the general plane of the steroid ring system).[20] The

[19] The CH_3 carbon itself is No. 19; it is bonded to C-10.

[20] In most other naturally occurring steroids such as sex hormones, the configuration is *trans*.

hydrogen atoms on the two carbons representing the B/C ring juncture (C-8 and C-9) are *trans* to each other, with the C-9 H being α, or below the ring system. This is characteristic of all bioactive steroids. The C/D ring juncture has *cis* stereochemistry, where the CH_3 group at C-13 (itself the No. 18 carbon) and the C-14 OH are both β. This C/D configuration is again unique to the cardiac glycosides. These geometric requirements are "strict." Thus removal of the C-14 OH or inversion to the α-configuration eliminates or greatly reduces the activity. Epimerizing the configuration of the C-17 cardenolide from β to α abolishes activity.

To better understand certain aspects of the mechanism of digitalis drugs, it would be useful to outline briefly their cardiovascular properties. The increased force of myocardial contraction produced by these glycosides is by far their most dramatic pharmacodynamic property. This positive inotropic (increased contractile force) action translates into increased cardiac output and effects on cardiac size and blood volume through diuresis (i.e., the relief of the edema that accompanies CHF). The rate of tension development is apparently affected, not the length of time during which contraction is maintained by the muscle fiber. Digitalis exerts its effect even in the presence of β-blockers or reserpine.

For many years it was thought that factors other than the increase in contractile power were responsible for the beneficial effects of these agents (Withering believed foxglove had a renal effect). One explanation was that since an enlarged heart was inefficient, digitalis' benefits were derived from its ability to reduce cardiac size by some "tonic" effect. The situation is actually the reverse—cardiac inefficiency causes the enlargement as a compensatory effect. The ultimate benefit of digitalis is the increased ability of the myocardium to do the work (that is, pump) at any given filling pressure.

The heart rate is decreased by digitalis in the patient with CHF, but not significantly in the normal individual. The effect is only partially mediated through the vagus nerve to the heart. Improved work capacity is not dependent on a decreased rate; a decreased rate should really be viewed as a secondary therapeutic effect. It is of interest that in spite of centuries of use of digitalis against heart failure, it was not until recently that its long-term beneficial effects were demonstrated.

Digitalis can be shown to cause some increase in the electrical excitability of both atrial and ventricular fibers in low doses. In excessive doses, this effect may increase to dangerous levels, thus constituting one of the toxic effects of these drugs. Digitalis compounds are very toxic and have a very small therapeutic index. Clinical dosages are frequently within 60% of toxic levels. Digoxin dosages were revised downward in the middle 1970s. Age, renal failure, electrolyte imbalance (particularly K^+), and cardiac ischemia all tend to increase greatly the risk of digitalis toxicity.

The rate of electrical impulse conduction through different cardiac tissues is variably affected. Thus low doses increase the velocity in the atrium and ventricle but toxic doses decrease velocity, sometimes even producing a block. Digitalis affects the refractory period by shortening it in the atrium, but markedly prolonging it for atrioventricular (AV) transmission.

Automaticity is the ability of cardiac tissue fibers to depolarize spontaneously. This enables them to contract again without external nerve stimulation. It is not surprising that digitalis affects this ability profoundly. When digitalis causes an increase in the rate of spontaneous depolarization during diastole of the ventricle, automaticity is increased at the pacemaker site. This may result in premature beats that increase as higher doses reduce excitability.

When the heart develops an arrhythmia because of defective electrical transmission from the atrium to the ventricle (AV transmission), an atrial fibrillation will occur. The cause is likely to be failure of the impulses to reach the AV node. Changes in the refractory period may aggravate the condition even further. By a very complex series of events, including prolongation of the refractory period, digitalis can restore the rhythm to almost normal. Coronary circulation does not appear to be significantly affected by digitalis in either normal patients or patients with congestive heart failure; nor is oxygen consumption by cardiac tissue increased.

Heart failure results in edema either because of increased hydrostatic pressure in capillaries or because of a compensatory renal mechanism. As the effective output of the heart decreases, renal flow, and therefore glomerular filtration, also decrease. Reabsorption of sodium and water is more complete, resulting in increased tissue retention of both substances. Digitalis has a diuretic effect to the extent that it relieves the underlying factors in the heart.

Although the various mechanisms by which digitalis glycosides exert these actions have been intensely studied for more than four decades, a total understanding is still not forthcoming. Certain aspects, however, have been established.

Cardiac glycosides are specific inhibitors of the pumping mechanism that transports Na^+ and K^+ across cell membranes against the electrochemical gradient (Chapter 8). The pivotal role in the active transport of these two ions across the membrane of the cardiac cell has been attributed to a membrane-bound enzyme, adenosine triphosphatase (ATPase, or more properly Na^+–K^+-ATPase), which is in a steric orientation that apparently allows for interaction with the digitalis steroid nucleus and the lactone at C-17. This interaction affects the enzyme's activation by causing variations in the concentrations of Na^+ and K^+. It is agreed that this inhibition occurs when digitalis is administered in toxic doses. It has not been well established that the same mechanism is also operative in producing increased cardiac contractility. Of course, it can be postulated that suppression of the pump would result in higher intracellular Na^+ (and decreased K^+) levels, resulting in a lower action potential. Impulse-sensitive cells (neuronal or myocardial) exchange these ions with a net loss (three Na^+ for two K^+) that obviously establishes the electrical potential between the internal and external membrane surfaces.

Extensive SAR studies, as well as a wealth of clinical experiences, have shown digitalis glycosides to be very potent, very selective, and very specific—facts that suggest that a receptor interaction is involved in producing the physiological results observed. Interactions of Na^+–K^+-ATPase with its various substrates are complex. Thus binding affinities with ATP, cofactor Mg^{2-}, Na^+, K^+, and a digitalis glycoside are all important to the overall effect. The assumption can safely be made that binding will bring about conformational changes. It is now accepted that a digitalis receptor is one or more of the conformations of Na^+–K^+-ATPase that occur during the ion pump's operation, possibly during a state in which the drug helps to stabilize one of the intermediate states of the enzyme (for example, during phosphorylation).

Evidence suggests that the entire glycoside molecule participates in the proposed drug–receptor interaction. The steric relationship of the lactone ring (β) to the steroid nucleus is absolute. The double bond is also critical since saturation results in an almost total loss of activity. The required stereochemical positioning of rings C and D in relation to each other (*cis*) and of A and B (also *cis*), and the configuration of the OH at C-14 have all been established. Figure 10-19 represents a highly simplified version of a proposed interaction of the C-17 lactone side chain with such a digitalis receptor.

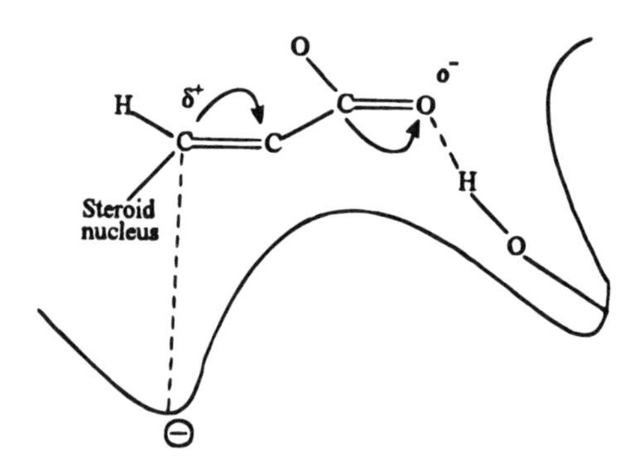

Figure 10-19. Hypothetical digitalis–receptor interaction.

If we "view" the butenolide as being perpendicular to this page with the double bond at eye level lying within an imaginary cavity formed by the enzyme's surface, a two-point binding can be visualized: (1) the polarized carbonyl group, with its electron-rich oxygen, hydrogen bonds to a hydroxyl group on the enzyme's surface (a serine residue?) and (2) the carbon atom bonded to the steroid nucleus is attracted to an anionic (or at least electron-dense) site at a secondary location. The electron deficiency (δ^+) on that carbon can be easily justified by the resonance concept, which requires an overall shift of electrons toward the oxygen in a conjugated system as exists in this case.

The cyclic C-17 α-β-unsaturated butenolide (lactone) could be replaced with a small group of what can be considered noncyclic bioisosteres. The requirements seem to be at least an α-β-unsaturated system that also permits a partial positive charge on the β-carbon, as long as the system is not extended (as in D) or bulky (as in C, when $R = C_2H_5$).

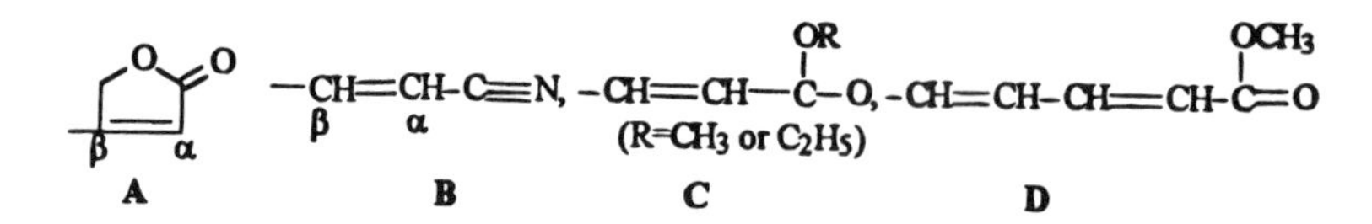

The centrality of this mechanism has by now been established. Na^+–K^+-ATPase is the glycoside receptor. This receptor—two catalytic enzymes plus glycoprotein and phospholipid—maintains the Na^+ and K^+ gradient.

It is tempting to rationalize this Na^+–K^+-ATPase digitalis–receptor complex as the biochemical explanation for drug-induced myocardial contractility. Much of the experimental evidence looks promising. However, some of it may be circumstantial. One disturbing fact is that substances exist that can inhibit the enzyme, yet do not affect myocardial contractility. In the final analysis we do not yet have a fully satisfactory explanation for the actions of digitalis glycosides. The fact remains that after two centuries this botanical drug is still absolutely essential for the treatment of the cardiac patient.

Other inotropic drugs are necessary not to replace digitalis as much as to supplement in cases of severe heart failure refractory to the glycosides, and in situations where their efficacy is insufficient and the therapeutic index is too low. The clinical goal is to improve the symptoms of cardiac failure by improving left ventricular function or by reducing resistance to the heart's output against the peripheral circulation—or even better—to achieve both.

An approach to improved inotropic drugs was to re-examine β-adrenergic agonists. The problem with most β-agonists was a duality of action: inotropic activity on the myocardium, which is desirable in CHF, and chronotropic activity (i.e., increased rate or tachycardia and the high associated risk of arrhythmias). In addition, of course, any significant pressor activity would also be detrimental. Isoproterenol is not a pressor; it has inotropic activity and because of strong peripheral β_2 agonism would, by its skeletal muscle effects there, lower resistance (reduce diastolic pressure) to the point of diminishing myocardial perfusion. A synthetic SAR study was therefore undertaken systematically to modify IPR to reduce its chronotropic, arrhythmogenic, and vascular effects while retaining, primarily and selectively, increased cardiac contractility. The compound sought was one with maximum cardiac β_1 agonist activity and minimal peripheral β_2 and α agonism. The result was dobutamine (Dobutrex, Tables 9-1 and 10-12). The drug acts mostly on cardiac β_1-adrenoceptors, much less so on peripheral β_2- and α-receptors, and not at all on renal and mesenteric dopamine receptors (the compound is a dopamine derivative, having no β-OH; see structure). Its chronotropic effect is only 25% that of IPR. Dobutamine is thus a beginning, having been useful for short-term therapy to improve cardiac output in severe, chronic cardiac failure. It should be pointed out that DA itself, within a narrow dosage range, has also been useful, especially in cases of cardiac failure accompanied by

Table 10-12. Dobutamine and Experimental Analogs as Inotropes[a]

R_1, R_2–C$_6$H$_3$–CH(R$_3$)–CH$_2$–NH–R

R_1	R_2	R_3	R	Name
OH	OH	H	–H	Dopamine (Intropine)
OH	OH	H	–CH(CH$_3$)(CH$_2$)$_2$–C$_6$H$_4$–OH	Dobutamine (Dobutrex)
OH	H	OH	–(CH$_2$)$_2$–C$_6$H$_3$(OCH$_3$)$_2$	Denopamine[b] (TA-064)
OC(=O)CH(CH$_3$)$_2$	OC(=O)CH(CH$_3$)$_2$	H	–CH$_3$	Ibopamine[c]
OH	H	OH	–(CH$_2$)$_2$–C$_6$H$_3$(OCH$_3$)$_2$	Butopamine[d]
OH	OH	H	theophyllin-7-yl (1,3-dimethylxanthine, N–CH$_3$, N–CH$_3$)	D-4975e

[a] Tradename or code number, if experimental, in parentheses.
[b] Ikeo et al., 1986; orally effective in animal studies.
[c] Benassi et al., 1987; in clinical trials, a pro-drug of N-methyldopamine.
[d] Nelson and Leier, 1981; orally effective, since it is not a catecholamine.
[e] McCaig and Parratt, 1979; a covalent "combination" of dopamine and theophylline. Five times more potent than dobutamine, but raises blood pressure.

hypotension where its α-agonism actually increases peripheral resistance; desirable under those conditions. Table 10-2 illustrates a structural comparison of clinically available and several experimental drugs that may offer better and prolonged therapeutic efficacy. However, neither these nor others not included in Table 10-12 currently being tested appear to be a breakthrough solution.

There are some interesting drugs in clinical use and in stages of development that can only be characterized as having mixed activity (e.g., inotropic with vasodilatory) or one of undefined mechanisms. Two are the bipyridines, amrinone and milrinone. Various studies seem to offer different answers as to how these compounds produce their desirable effects. The mechanisms that are not involved should be pointed out. Amrinone is not a β-adrenergic agonist as dobutamine; it is not an inhibitor of Na^+–K^+-ATPase as digitalis. α-Adrenoceptor or cholinoreceptor stimulation is not involved; neither are effects on autonomic ganglia. The evidence strongly points to increased *c*-AMP levels as being responsible for both the direct positive inotropy and the vasodilation that seen. Studies since then indicate that PDE inhibition is the main mechanism for milrinone. Unfortunately, patient survival is not high. Enoximone and a related compound, although more potent than dobutamine, could not sustain the initial improvements; mortality was high.

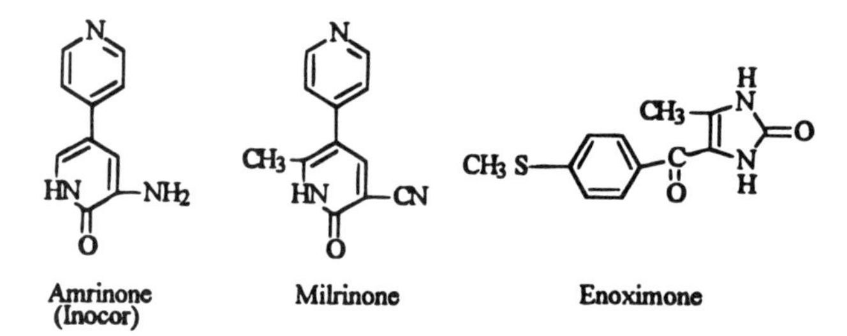

Because of the high prevalence of CHF, estimated at 4 million and increasing, and the poor prognosis for long-term survival, a considerable effort to develop new drugs and re-evaluate other cardiovascular agents such as ACE inhibitors, dopaminergic drugs, and calcium channel blockers is going on. It is somewhat ironic that a published Veterans Administration cooperative study was the first controlled study statistically to demonstrate a significant (25%) mortality reduction over 2 years using two drugs that have been around for decades: hydralazine and isosorbide dinitrate.

A significant study using extracts of rat cardiac atrial tissue produced copious excretion of urine, Na^+ and Cl^- at levels 30 times above normal on being injected into rats. The discovery of a saluretic peptide hormone thus established the heart as an endocrine gland with strong effects on water and electrolyte balance (Na^+ and Cl^-). Now named Atrial Natriuretic Factor (ANF), it may be useful in treating CHF patients in whom levels have been shown to be increased, yet are associated with decreased cardiac output. The normal effect of ANF infusion is to increase water, Na^+, and K^+ excretion and inhibit it for aldosterone and cortisone. Most of these effects do not occur in CHF patients, which may be the reason for the edema accompanying CHF. Why that is, however, remains to be seen. When the reasons become known, it may be possible to find a way around the CHF patient's inability to respond to ANF. That should solve many problems.

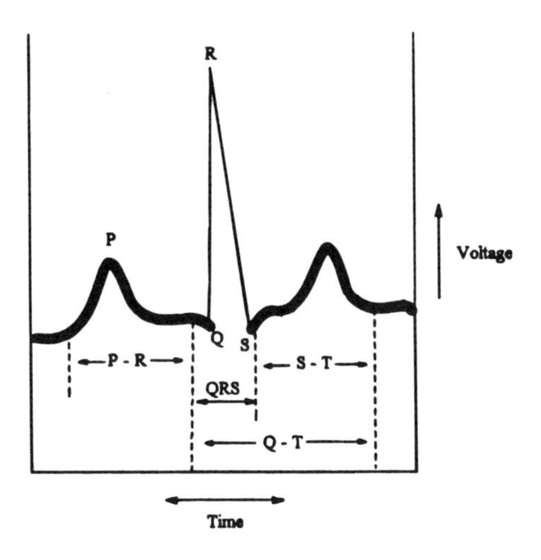

Figure 10-20. Electrocardiogram tracing.

10.3.1.7. Antiarrhythmic drugs

Antiarrhythmic drugs are those capable of reverting any irregular cardiac rhythm or rate to normal. An arrhythmic situation is one in which either initiation or propagation of a heartbeat stimulus is abnormal.

Before discussing the various arrhythmias and the drugs used to treat them, it is advisable briefly to review a normal sinus rhythm and some of the electrical activities of the heart.

The normal orderly sequence of events in cardiac contraction is initiated by a primary pacemaker, the sinoatrial (SA) node,[21] which is located near the surface at the junction of the right atrium and the superior vena cava. One of its properties is automaticity. The normal firing frequency is 60–100 impulses/minute. The established rhythm is conducted to the atrioventricular (AV) node. This node serves to slow the beat somewhat so that atrial contraction can occur before the ventricle is stimulated. The AV node is in the septum (dividing wall) between the atria. The impulse is conducted from the AV node to a common bundle of fibers (Bundle of His) that cross the right atrium to the left ventricle. From there a division of fibers directs impulses along the septum dividing the ventricles, down to the lateral walls and the apex of the heart. The branching of the common bundle leads into the Purkinje fibers that innervate the heart musculature of the ventricles.

The electrical activity generated by the depolarization and repolarization of myocardial tissue, especially the nerve fiber cells, can be detected by electrodes and graphed as a function of intensity (millivolts) versus time (seconds); this is illustrated in a tracing—the electrocardiogram (ECG or EKG) (Fig. 10-20). When properly interpreted, an ECG represents the cardiac cycle, normal or abnormal. The P wave represents the electrical activity passing over the atrial surface; the ventricles produce waves Q, R, and S (the QRS complex); and the repolarization of ventricular muscle fibers results in the T wave. Irregular distances between the various wave peaks (e.g., a shortened ST interval, can then be related to particular rhythmic abnormalities and be invaluable to the physician in the diagnosis and in the choice of the proper drug.

---

[21] A node is a "knot" or small rounded structure, which in this instance consists of special nerve fibers.

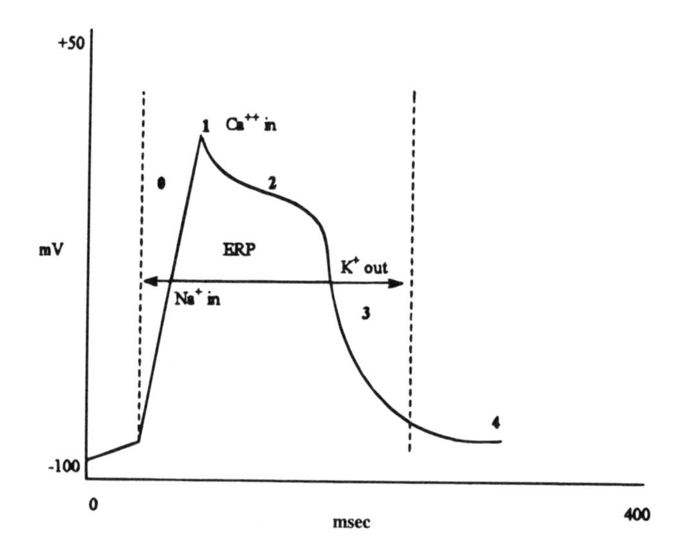

Figure 10-21. Schematic of cardiac electrical activity. (0–4 represents phases. ERP = effective refractory period. See text for discussion.)

Any disturbance to the orderly conduction of the electrical impulses, or their initial formation becomes the basis of an arrhythmia.

The electrophysiology of the heart is concerned with the *transmembrane resting potential.* Thus the potential inside a Purkinje fiber cell at rest, with respect to the outside, is about 90 mV. This potential difference is maintained by an active transport system (pump) that keeps a higher extracellular Na^+ concentration than the intracellular K^+ level. On excitation, this very rapidly reverses to a positive voltage, spiking at maybe +30 mV. There is a rapid simultaneous movement of Na^+ into the cell, as if a *gate* had suddenly opened a channel.[22] Recovery from excitation results in restoration of the resting potential in phases. Beginning with phase 4, the resting potential is followed by the rapid depolarization and reversal. Phase O is then continued by a series of three repolarization phases; 1, 2, and 3. Figure 10-21 is an idealized representation of the preceding. Thus, depending on the areas measured (with intracellular microelectrode recorders), the separations of phases are not always distinct; the voltages vary among the major cell types of the heart. In addition to the Purkinje fibers, these are ventricular tissue, the AV node, atrial cells, and the SA node.

The rapid influx of Na^+ through the so-called gates or channels during phase O results in the cell rapidly depolarizing, which in turn "closes" the gate behind them to further influx (but now favors a release of K^+, phase 1). A slow influx of Ca^{2+} is also triggered to balance the K^+ loss and to maintain a relative voltage plateau (see phase 2 in Fig. 10-21). As Ca^{2+} entry slows, the membrane potential drops rather rapidly to predepolarization levels (phase 4). The net gain of Ca^{2+} is needed for coupling to the actinomyosin mechanism (for muscle contraction). In heart muscle electrical activity is coupled to mechanical activity by Ca^{2+} as the trigger.

A *refractory period* (ERP, Fig. 10-21) of several hundred milliseconds follows during which no further stimulus can propagate an impulse. Automaticity, or impulse initiation, is

[22] In fact, this is frequently referred to as a *gating mechanism.*

a property of cardiac fibers that enables them to generate action potentials spontaneously and therefore impulses.

From what has thus far been gleaned from the complexity of cardiac functions, it should not be surprising that there is a large number of arrhythmias. The major ones are as follows:

1. *Sinus bradycardia*: A sinus rhythm of less than 60 beats/minute caused by defective impulse generation because of decreased SA node automaticity.
2. *Sinus tachycardia*: Increased SA automaticity results in a sinus rhythm in excess of 100 beats per minute. Sinus tachycardia is usually not pathological and is caused by simple anxiety or stimulants (caffeine, amphetamines).
3. *Atrial fibrillation*: A common arrhythmia with a maximum of 650 beats/minute. It exhibits irregular nonsynchronous contractions of the atria. Impulses do not originate at the SA node; rather, they originate at the atrium, where rapid impulses bombard the AV node. Most of these impulses will not reach the ventricle, resulting in an irregular beat there also. Thus there is actually an AV block.
4. *Atrial flutter*: Atrial flutter also originates in the atrium (not the SA node), generating more than 200 beats/minute. The pattern, however, is regular. Normal sinus rhythm can usually be restored with electrical shock treatment.
5. *Paroxysmal atrial tachycardia* occurs with a sudden onset of 140–220 beats/minute, which may cease abruptly. The focus is ectopic—it can be anywhere in the atrium.
6. *Junctional tachycardia* originates in, or close to, the AV node. Beats are 160–250/minute, which may be paroxysmal.
7. *Atrioventricular block*. There are many types of AV block, such as a first-degree block in which atrial impulses are delayed in the AV conduction to the ventricle. This is shown on the ECG as a prolongation of the PR interval. Digitalis toxicity or excessive K^+ levels can be causative.
8. *Ventricular premature contractions*: Ventricular premature contractions are ectopically focused from anywhere in the ventricles and are not initiated in the SA node. It is a very common arrhythmia, frequently associated with myocardial infarction. It can be the precursor to potentially fatal ventricular tachycardia and fibrillation.
9. *Ventricular tachycardia*: An arrhythmia with a rate of 180–250 beats/minute. It occurs in patients with myocardial infarction and also in cases of digitalis overdose. It results in severe CHF and is life threatening–the mortality is 60–100%.
10. *Ventricular flutter fibrillation*: Ventricular flutter fibrillations with ventricular contractions of 180–250 beats/minute cause sudden death. There are no coordinated contractions, resulting in loss of blood pressure and pulse. Unconsciousness and respiratory arrest follow. The usual cause is coronary arterial disease (i.e., myocardial infarction).

The Vaughan-Williams (1984) classification of antiarrhythmics categorizes these agents into four major groups[23]; Class I can be further subdivided into three subgroups based on the effect on the action potential duration (Harrison, 1985). Class I drugs may also be viewed as sodium channel blockers, particularly those compounds with local anesthetic properties. These classifications are based on predominant properties with several drugs that have some properties fitting another class also (Table 10-13).

Procainamide (Table 10-14, No. 1), the amide isostere of the first synthetic local anesthetic procaine, was first found useful in the treatment of arrhythmias in the early 1950s. It followed earlier observations that procaine could raise the threshold at which EPI–chloro-

[23] Other classifications have five and even six categories.

Table 10-13. Classification of Antiarrhythmic Drugs[a]

Class	Action	Drugs
I	Inhibit depolarizing effect by blocking Na^+ influx through channels; local anesthetic effect	
IA	Increase refractory period[b]; impede conduction; moderate effect on Phase O[c]	Diisopyramide, procainamide, quinidine
IB	Decrease refractory period; effect on conduction rate and phase minimal	Lidocaine, mexiletine, phenytoin, tocainide
IC	Repolarization effect minimal; conduction strongly slowed	Encainide, flecainide, lorcainide, propafenone
II	Antiadrenergic	β-blockers (e.g., propranolol, bretylium)[d]
III	Increase refractory period	Acecainide, amiodarone, bretylium,[d] sotalol[e]
IV	Slow calcium channel blockers	Diltiazem, verapamil

[a] Based on Vaughn-Williams, 1984, and Harrison, 1985.
[b] Prolong repolarization.
[c] See Figure 10-21.
[d] There is disagreement in literature on class.
[e] See Figure 10-4.

form-induced ventricular fibrillations in animal experiments occur. Procainamide became only the second drug clinically available for the treatment of cardiac arrhythmias since quinidine (the dextrorotatory isomer of quinine) was introduced for that purpose early in this century. Both orally effective drugs are today classified as Is. They have local anesthetic properties, depress automaticity, slow impulse conduction, and increase the refractory period of the atrium. Even though the first seven compounds in Table 10-4 seem closely structurally related (variants of procainamide), there are some surprising differences both mechanistically and clinically. The major metabolite of procainamide, N-acetylprocainamide (NAPA, No. 2), is classified as a selective class III drug. In addition, the lupuslike symptoms frequently encountered with the parent drug are apparently not induced with the metabolite.

Lidocaine (No. 3) is a particularly interesting compound. It was first synthesized in Sweden (1937) as a "reverse" amide of procaine with two *o*-methyls added for steric purposes. By substitution of an amide for the ester linkage and the additional protection of the ortho "blockers," the goal of achieving an active and stable local anesthetic was achieved. For example, neither refluxing lidocaine with 50% H_2SO_4 or ethanolic KOH for hours resulted in significant hydrolysis. Its potential as an antiarrhythmic was inexplicably not put to clinical use until the 1960s. In spite of its desirable properties, it could not be given orally, and its plasma $t_{1/2}$ was short (<2 h). N-Deethylation by the liver is so efficient that most of the drug is cleared from the blood on the first pass (i.e., almost as fast as the blood flow through it). This requires large and continuous IV dosing and caution, since lowered blood flow, such as in CHF, can be down 50%. In such cases lidocaine doses should be cut in half to maintain needed plasma levels. Tocainide (No. 4) solves several clinical problems. Without altering the pharmacology (both drugs are Class IB) it is an orally active drug with almost 100% bioavailability. Because the drug does not have any–alkyl groups for hepatic enzymes to remove, first pass metabolism is negligible. Metabolism is about 60%; 40% of the drug is excreted unchanged. A comparison of the metabolic steps that the two drugs undergo illustrates some significant differences (Fig. 10-22). In the case of lidocaine, microsomal N-dealkylation is a much more facile reaction than is amide hydrolysis.

Table 10-14. Classification of Antiarrhythmic Drugs[a]

Structure	Name[b]	V[c]	SV[c]	$t_{1/2}$ (h)	Class
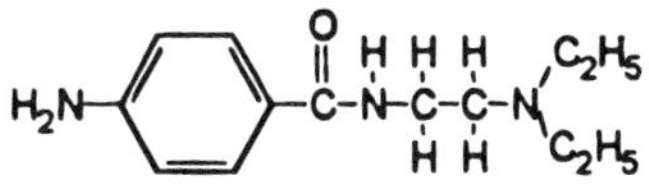	Procainamide (Pronestyl)	yes	yes	3–4	I-A
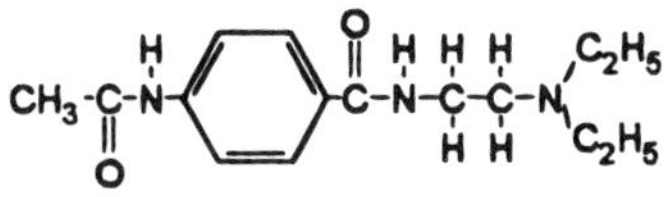	Acecainide NAPA (Investigational)	yes	yes	6–42	III
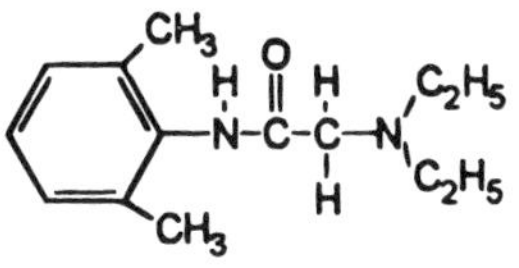	Lidocaine (Xylocaine)	yes	no	1–2	IB
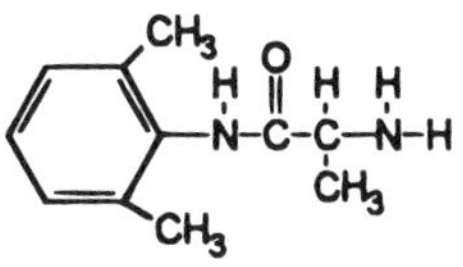	Tocainide (Tonocard)	yes	no[d]	11–15	IB
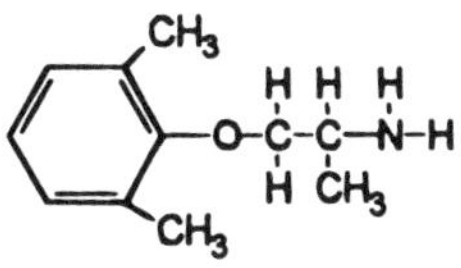	Mexiletine (Mexitil)	yes	no[d]	6–13	IB
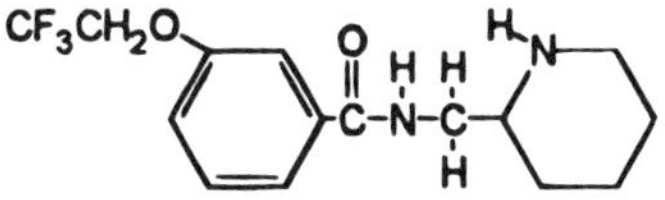	Flecainide (Tambocor)	yes	no	12–27	IC
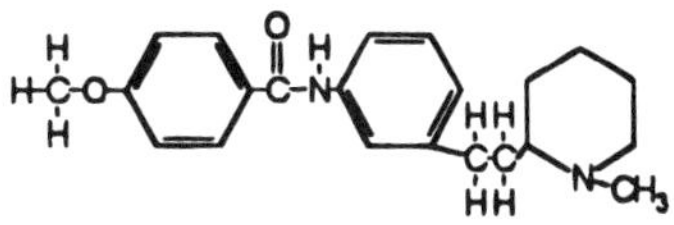	Encainide (Enkaid)	yes	some	4–11	IC
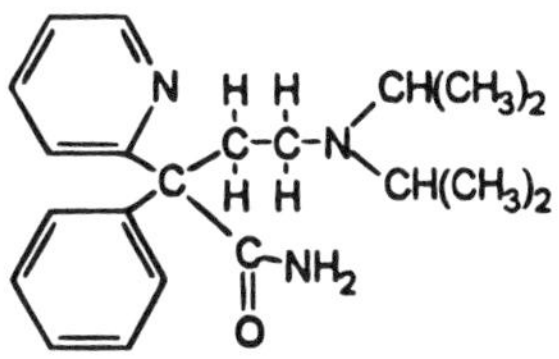	Disopyramide (Norpace)	yes	yes	6–8	IA
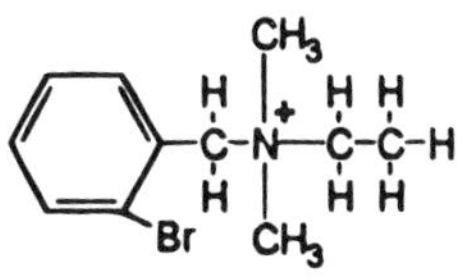	Bretylium (Bretylol)	yes	no	4–8	III

(*Continued*)

Table 10-14. *(Continued)*

Structure	Name[b]	V[c]	SV[c]	$t_{1/2}$ (h)	Class
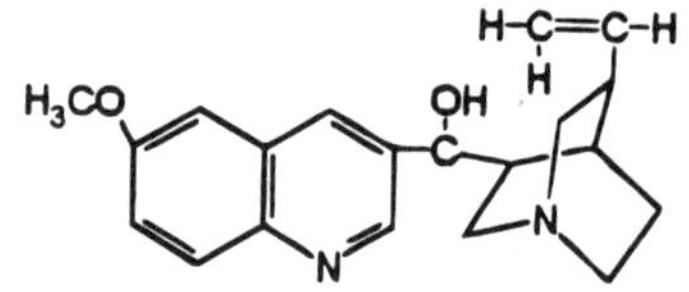	Quinidine (+) isomer	yes	yes	6	IA
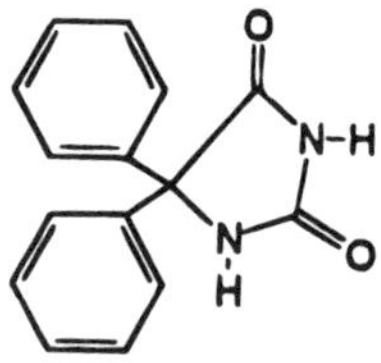	Phenytoin (Dilantin)	some	no[d]	24	IB
See Table 10-4	Propranolol (Inderal)	yes	yes	8	II
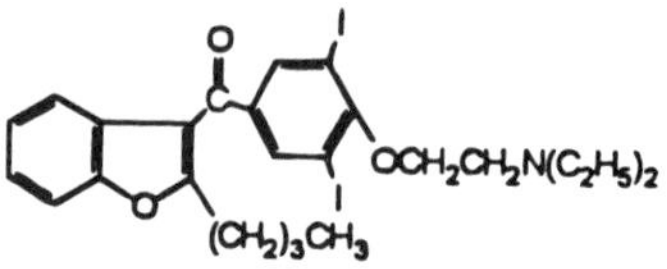	Amiodarone (Cordarone)	yes	yes	14–100 days	II
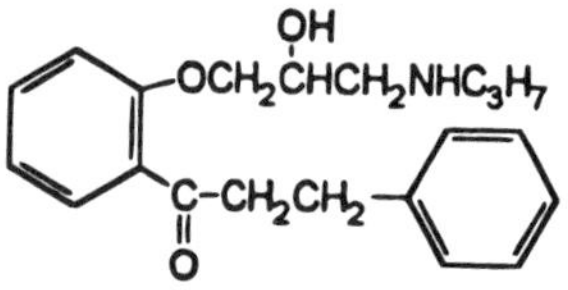	Propafenone (Rhythmol)	yes	yes	7+	IC
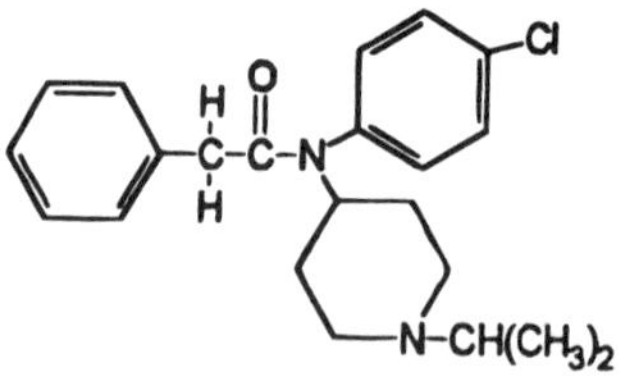	Lorcainide (Investigational)	yes	yes	8	IC
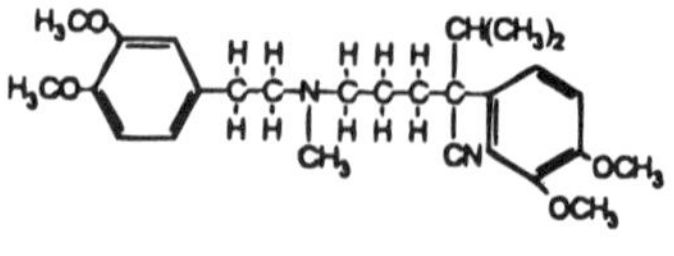	Verapamil (Isoptin, Calan)	—	yes	3–7	IV
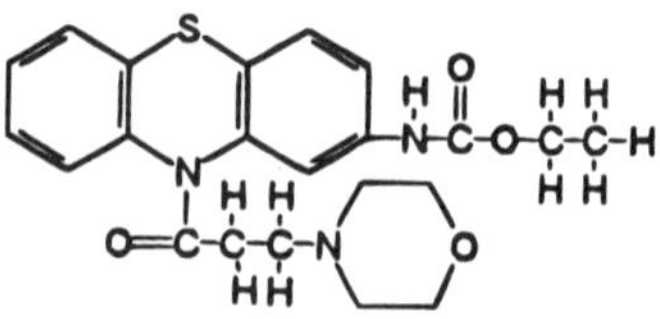	Morizicine (Ethmozine)	yes	some	1.5–3	I-A

[a] β-blockers are listed in Figure 10-4.
[b] Trade names in parentheses.
[c] Effectiveness in ventricular and supraventricular arrhythmias.
[d] Possibly in digitalis-induced atrial arrhythmias. V = ventricular, SV = supraventricular

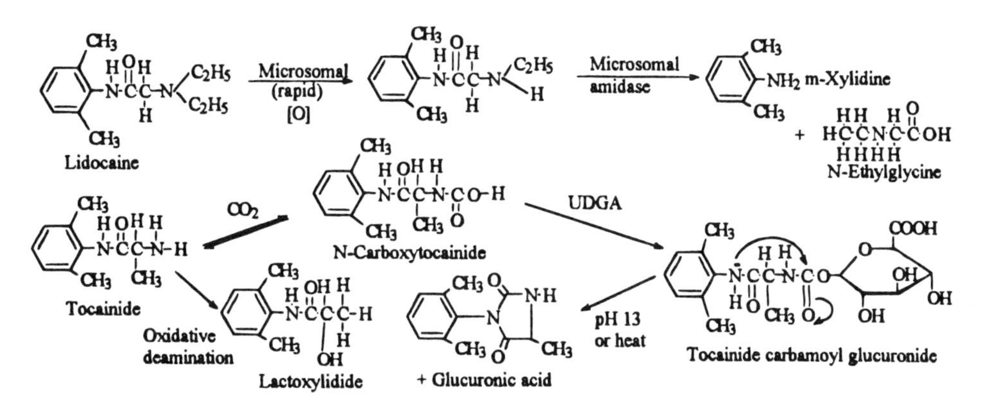

Figure 10-22. Comparative metabolic pathways of lidocaine and tocainide.

The N-deethylated metabolite, although still bioactive, is then rapidly cleaved by microsomal amidases to inactive *m*-xylidine and N-ethylglycine. This readily explains the short half-life of the drug.

Tocainide, on the other hand, presents a different picture. Its oral bioavailability is primarily due to slow hepatic metabolism on the first and subsequent passes. The large percentage of the drug excreted without enzymic attack attests to its resistance. An examination of its structure can be helpful. Oxidative deamination of the primary amine function that does occur and yields lactoxylidide can, however, be expected to be sluggish due to the α-CH_3 group, which, as was seen with amphetamines, tends to inhibit MAO to a degree. As was demonstrated, little amide hydrolysis occurs. The microsomal amidases have both the stubborn amide linkage to contend with as well as a double dose of steric hindrance: the 2,6-dimethyl groups exerting a protective block and the α-methyl offering additional restraint to the enzyme's effectiveness. As a result a novel metabolic alternate pathway has been established; the glucuronidation of a theoretically very unstable carbamic acid intermediate, N-carboxytocainide, which is presumably formed by reversible reaction with CO_2. Simply hydrolyzing tocainide glucuronide would, of course, yield tocainide (presumably via the decomposition of the putative N-carboxytocainide); however, strong base readily catalyzes an intramolecular glucuronic acid (the leaving group) and results in cyclization to 4-methyl-1(2,6-dimethylphenyl)-hydantoin. This reaction confirmed the existence of a carbamoyl glucuronide rather than an N-glucuronide (see Fig. 10-22).

The third member of this class IB trio, mexiletine (No. 5), which was marketed in 1986, was first found to have anticonvulsant properties. Examination of its electrophysiological behavior led to its characterization as a strong ventricular antiarrhythmic. As the other drugs in this subclass, mexiletine blocks Na^+ influx and reduces the maximal impulse velocity of phase O of the action potential.

As with tocainide, first-pass metabolism is not a problem. Since there is not an amide or ester to hydrolyze, major metabolites are hydroxyl derivatives (of CH_3 groups) by mixed-function oxidases. They are inactive.

Two additions to the IC class are flecainide (No. 6) and encainide (No. 7). They block sodium channels of the Purkinje fibers. Although experience with these compounds is still relatively short, they appear to be effective in patients who were (or have become) refractory to other agents. That is, after all, one of the purposes of developing newer agents.

Quinidine is the dextro isomer of quinine and is obtained together with quinine from the cinchona bark. Although it is a depressant of skeletal as well as cardiac muscle, quinidine is more effective than procainamide is, but it appears to be more toxic. Its cardiac actions include the ability to increase the electrical potential necessary for excitation, which may explain why the drug can abolish the generation of ectopic impulses. Quinidine has demonstrated the ability to increase the effective refractory period in the atrium, ventricular muscle, and Purkinje fibers without increasing the duration of the action potential. Thus even after the resting potential of the myocardial cell membrane is restored, the tissue remains refractory in the presence of quinidine. This ability may constitute the most significant action of this and similar drugs. Additional properties of these drugs are the ability to decrease the velocity of impulse conduction and to reduce spontaneous frequency at the SA node (i.e., pacemaker activity). Premature beats are also reduced, which decreases the cell membrane's responsiveness to incomplete repolarization.

Quinidine has anticholinergic properties that can result in the blockade of impulses from the vagus nerve to the heart. This would prevent the slowing of the heart rate that is ordinarily produced by normal vagal stimulation. The resulting increased heart rate counters the direct depressant effect on the pacemaker. This can understandably complicate therapy from the clinical standpoint. The effectiveness of quinidine and procainamide treatment can be traced on ECG by the increase in the Q-T interval, which can be attributed to the increased duration of systole, as well as decreased intraventricular conduction velocity.

Quinidine is a toxic, dangerous, but useful drug. In digitalized patients it may cause unpredictable changes in cardiac rhythm. Depressed cardiac contractility is also a potentially toxic effect.

Diisopyramide (No. 8), a nonquaternized pyridine analog of the antimuscarinic isopropamide (Fig. 8-17), has antiarrhythmic properties almost identical to those of quinidine. Like most drugs in class I, it shows negative inotropic properties. It is not considered a first-line antiarrhythmic and must be used with caution in patients with CHF.

Phenytoin (diphenylhydantoin, Dilantin) has been in use as an antiepileptic agent since 1938. Its ability to abolish ventricular tachycardia was noted much later. It is currently used in the treatment of several clinical arrhythmic disorders but especially in disorders induced by toxic dosage levels of digitalis, in ventricular tachycardia, and in atrial tachycardia with block. It is probably not effective in other types of atrial arrhythmias.

The use of propranolol (Inderal) as an antiarrhythmic is useful in the treatment of atrial tachyarrhythmias. Its mechanism is somewhat controversial. Since propranolol is a β-adrenergic blocking agent, it seems logical to study this feature. The drug can be shown to block the cardiac effects of catecholamines. The effective refractory period of the AV node has been shown to increase. However, some quinidinelike properties have also been observed. The suggestion has been made that the antiarrhythmic action is caused by interference with the cellular transport of Ca^{2+}.

The major component of the antiarrhythmic mechanism of a β-blocker undoubtedly is the termination or prevention of those tachyarrhythmias caused by excess sympathetic tone, which may in turn be the result of increased circulating catecholamines. The control of ventricular flutter that is not effectively controlled with digitalis therapy alone can also be achieved by the addition of β-blockers since they prolong the AV refractory period.

Bretylium (No. 9), a quaternary bromobenzylic amine, had been in use three decades ago as an oral antihypertensive because of its ability to inhibit NE release. It is now considered obsolete for that purpose, but it has re-emerged over the last 10 years as an effective, parenteral class III antiarrhythmic when its direct actions on the myocardium were recognized and evaluated. It is currently used to prevent and to treat ventricular fibrillations and tachycardia that do not respond to standard drugs. It is very rapid-acting (IV). Bretylium is probably most useful during emergencies.

The release of a class III drug, amiodarone (No. 13), may represent another small advance in cardiac therapy. The drug has been in use as an antiarrhythmic and antianginal overseas for several years. The smooth muscle relaxant khellin probably served as the initial model for amplifying that property since both are benzofuran-type compounds. The iodinated phenoxy moiety is reminiscent of triiodothyronine (T_3). Since T_3 is known to cause an increased rate and a more forceful heartbeat, the variance of its structure in amiodarone might result in inhibition of such an effect. Even though the drug is classified as III, its complex and still incompletely understood electrophysiology includes more than simply increasing the heart's refractory period. For example, its effective blockade of sodium channels is limited to those in an inactivated state (quinidine also blocks activated channels). The prolongation of the membrane action potential may also involve K^+ channel blockade. Some Ca^{2+} channel blockade and noncompetitive β-adrenoceptor inhibition have also been suggested.

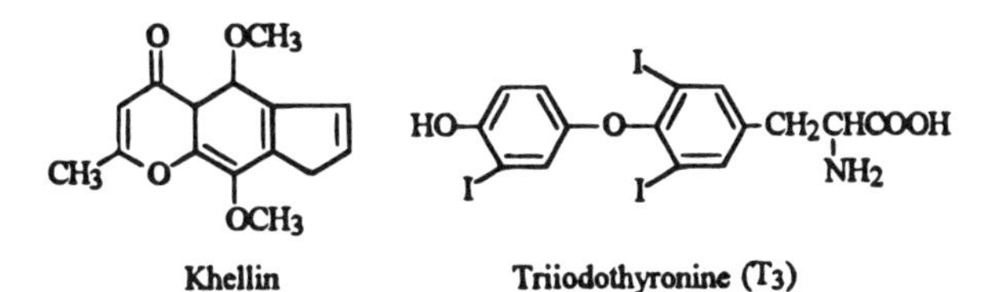

Khellin Triiodothyronine (T3)

Two new IC compounds should be mentioned. Propafenone (Rhythmol) (No. 14), introduced in the United States in 1989, has the structural components to be a β-blocker. It is a weak one. It also possesses Ca^{2+} blocking properties and slows conduction in the atria and ventricles, the node between them, and the Purkinje–His areas. This earns the drug IC status. Over 80% of first-pass metabolism following oral absorption gives it poor bioavailability. Nevertheless, it is used orally as well as IV for ventricular arrhythmias.

Lorcainide (No. 15) is another local anesthetic-type antiarrhythmic. Its pharmacokinetics are illustrative of the interpatient variabilities encountered. N-dealkylation to norlorcainide is extensive on the first pass. The parent drug's plasma clearance $t_{1/2}$ is about 8 hours; with the metabolite, which is active, it is 27 hours. Individual $t_{1/2}$ for the drug varied from 2.5 to over 15 hours. A single, small oral dose is only 4% bioavailable. However, with larger and repeated doses this can increase to over 20% (metabolic saturation).

Finally, morizicine (No. 17) represents a particularly interesting new entry (1990) in the continuing battle against arrhythmia-caused mortality. It represents the first Russian contribution in this area.[24] The compound was synthesized in the early 1960s, but only approved by the U.S. Food and Drug Administration in the 1980s. At first glance its structure appears to be a minor variant of the phenothiazine antipsychotics of that time (Chapter 12), but the drug exhibits no central (or peripheral) dopaminergic properties. Its first-pass metabolism is extensive; at least 26 metabolites have been identified, with each representing less than 1% of the administered dose. Morizicine's most startling attribute

[24] Its use in the United States essentially has been discontinued.

probably came to light in the Cardiac Arrhythmia Suppression Trial (CAST, 1989), in which the drug was compared with encainide and flecainide in its ability to suppress arrhythmias following MIs. While the latter two drugs were actually found to increase mortality rates two to three times (and were dropped from the study), morizicine was superior to the placebo.

10.3.1.8. Antianginal Drugs: The Coronary Vasodilators

Angina pectoris, or ischemic heart disease, is the name given to the symptomatic oppressive pain resulting from myocardial ischemia. In simplest terms it results when the oxygen demand of myocardial tissues exceeds the circulatory supply. Once a local anemia due to an obstruction (i.e., ischemia) exists, biochemical changes are inevitable. Metabolic products will accumulate in the area, contractility declines, and NE release occurs from sympathetic neurons. The end result is pain. This series of events can arise by one of two physical mechanisms. One mechanism, which produces a localized oxygen starvation, is stenosis i.e., a constriction or narrowing of a coronary vessel or valve, which may include atherosclerosis. This produces *stable angina*. Episodes can be precipitated by emotional stress or exercise, but they usually cease rapidly with rest or nitroglycerin (NTG) (see later). It can probably be stated that a narrowing of a coronary of 50% or more would be necessary to initiate the process.

The second mechanism responsible for angina is a series of coronary artery spasms, without necessarily involving any occlusion. This is referred to as *unstable angina*. Painful episodes can arise with little effort or even when the patient is at rest. There are several types, but the best known is *Prinzmetal's* or *variants angina*. These patients have a risk of MIs. Other conditions that may result in reduced coronary flow or greater myocardial oxygen consumption (but need not involve coronary disease) can be aortic valve disease, mitral valve stenosis, and cardiac myopathy. Other factors are known to aggravate or even precipitate angina pectoris. These include anemia, CHF, drugs (noncardiac), hyperlipidemia, hypertension, hyperthyroidism, obesity, smoking, and tachyarrhythmias. It has been recognized more recently that coronary vasospasms, possibly induced by autonomic factors, may also be contributory to stable angina. Increased O_2 demands arising from tachycardia may also produce unstable angina episodes. All this, of course, complicates therapy. Nevertheless, the goals remain the same. The short-term objective is to abort the anginal attack. The long-term aim must still be the prevention of ischemic episodes.

Increasing the O_2 supply to the myocardium cannot be meaningfully achieved by raising the O_2 content of the blood nor by increasing the fraction the heart muscle can extract from it (which is about three-fourths). That leaves only one modality: amplifying the coronary blood flow.

To a degree, it is possible to reduce myocardial O_2 demand and thereby *delay* the onset of ischemic pain with β-blockers. It will be recalled from earlier discussions that these drugs can decrease the heart rate, decrease myocardial contractility, and reduce blood pressure (especially during exercise), all beneficial effects for the stable (exertional) angina patient, especially if hypertension is also present.

It would theoretically be expected that cardioselective β_1 blockers would be preferable agents. However, at the clinical level, other than in patients with pulmonary obstructive diseases, the favorable difference of β_1 over nonselective compounds may be more apparent than real.

Table 10-15. Organic Nitrites and Nitrates

Drug	Trade name	Structure
Ethyl nitrite	"Sweet Spirits of Nitre"	$CH_3-CH_2-O-N=O$
Amyl nitrite	Vaporole	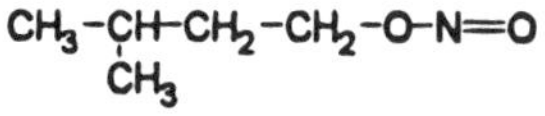 $CH_3-CH-CH_2-CH_2-O-N=O$ CH_3
Nitroglycerin, glyceryl trinitrate	Many	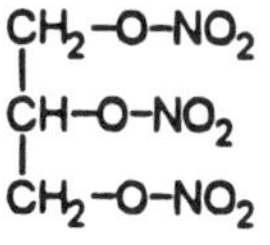 CH_2-O-NO_2 $CH-O-NO_2$ CH_2-O-NO_2
Erythrityl tetranitrate	Cardilate	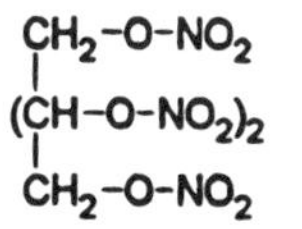 CH_2-O-NO_2 $(CH-O-NO_2)_2$ CH_2-O-NO_2
Isosorbide dinitrate	Isordil	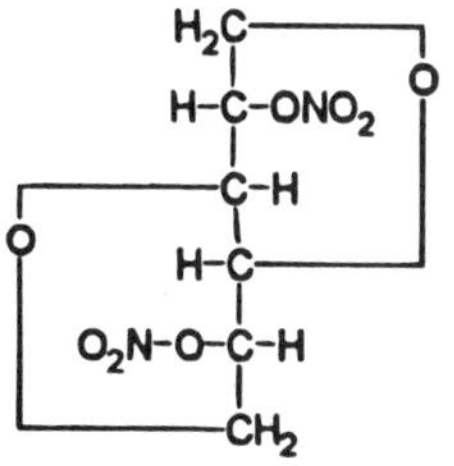 H_2C $H-C-ONO_2$ O $C-H$ O $H-C$ $O_2N-O-C-H$ CH_2
Pentaerythritol tetranitrate (PETN)	Peritrate	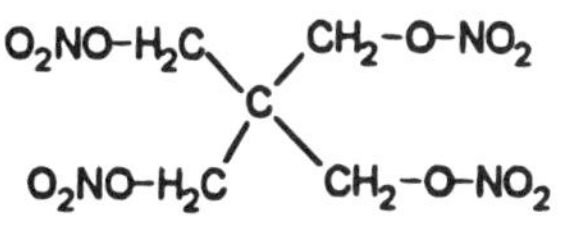 O_2NO-H_2C CH_2-O-NO_2 C O_2NO-H_2C CH_2-O-NO_2
Isosorbide mononitra	Imdur,Ismo	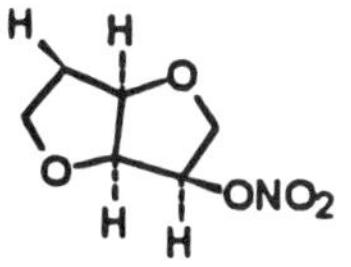 H H O O ONO_2 H H

10.3.8.1. The Nitrates/Nitrites

Table 10-15 lists the several organic nitrites and nitrates available. Nitrites, *R*-O-N=O, are esters of nitrous acid, HNO_2, whereas nitrates, *R*-O-NO_2, are esters of nitric acid, HNO_3. Ethyl nitrite was used as a popular nonprescription diaphoretic for decades (as a 4% alcoholic "spirit"), but it is no longer sold in the United States. Sodium and other inorganic nitrites, although used in the nineteenth century, have only a fleeting effect. (Today, $NaNO_2$ has as its only medical use being a component of the cyanide antidote kit.) Amyl nitrite, in spite of its high boiling point of 96°C, is very volatile, even at room temperature. It is packaged in glass pearls that can be crushed and inhaled in an emergency. Its duration of action is only a few minutes, but the onset is within seconds. Glyceryl trinitrate, which is often erroneously referred to as nitroglycerin,[25] is a dense (Sp. G = 1.6) sweet-smelling

[25] Nitroglycerin, though an incorrect name (since it is an ester), has been ingrained by usage of over a century and will remain so.

oil that has been crystallized (m.p. 13°C), but is usually obtained as a liquid. Its synthesis from glycerol and nitrating acid (a mixture of HNO_3 and H_2SO_4) is carried by cooling since the reaction is exothermic (Eq. 10.6). The drug is not very effective orally as a single dose, probably due to ester hydrolysis in the intestinal mucosa, and further reductive degradation by hepatic glutathione-organic nitrate reductase (first pass). Nevertheless, so-called sustained-release tablets and capsules are available. Their effectiveness, if any, as long-term prophylactic products may be due to an accumulation of the small amount of drug not affected after each dose by the first-pass phenomenon. On continuous absorption it may reach clinically significant blood levels. In essence this would "saturate" hepatic degredative capacity for nitrates and result in reduced anginal attacks and increased exercise tolerance.

$$\begin{array}{l} CH_2-OH \\ | \\ CH-OH \\ | \\ CH_2-OH \end{array} + HNO_3 \xrightarrow[10°C]{H_2SO_4} \begin{array}{l} CH_2-ONO_2 \\ | \\ CH-ONO_2 \\ | \\ CH_2-ONO_2 \end{array} + 3\,H_2O \qquad (10.6)$$

The most utilized dosage form of NTG for the majority of this century has been the sublingual (SL) tablet. It is used to abort an imminent or actually occurring attack. Buccal absorption is rapid, offering almost instantaneous relief of sufficient duration (<30 min) for the emergency.

Nevertheless, because of the valuable properties NTG is now known to have in angina pectoris and CHF, continuous blood levels of the drug are highly desirable. Therefore, different and innovative dosage forms are being developed. A transmucosal tablet formulation that is placed under the lip (not SL) on slow dissolution has been shown to enhance exercise tolerance for hours. Such tablets thus have both rapid onset and long duration. A 2% NTG ointment has been available for some time. NTG is efficiently absorbed through the skin. After about an hour applying large doses (e.g., ointment equivalent to 30 mg) will provide effectiveness of 4–8 hours. The introduction of NTG skin patches (Nitro-Dur, Transderm Nitro) was more recent. The patches contain the drug in a form resulting in continuous release onto (and through) the skin, purportedly for 24 hours for a total dose of 5 to 10 mg. However, variability of actual absorption and therefore patient response can be considerable.

The most recent introduction (1991) of NTG is as an intravenous infusion dosage form. It is being used in patients with CHF, unstable angina especially when accompanied by acute MIs, preoperative hypertension, and postoperatively in cardiac surgery. In this form NTG acts as fast as nitroprusside and, of course, has a short life so that effects can be controlled almost by the minute. The fact that no CN^- or SCN^- is produced may be an advantage. Finally, an aerosol NTG (nitrolingual) product in lieu of SL tablets for very rapid onset is now available.

The therapeutic effects of NTG (and all nitrates) are predicated on its ability to relax vascular smooth muscle, both arterial and venous. The complete mechanism of action is still evolving. It has now been established that these, and probably other N-O-type compounds (e.g., nitroso) can activate guanyl cyclase. This enzyme catalyzes the cyclization of GTP to *c*-GMP, which, in turn leads to smooth muscle relaxation. The enzyme is not activated by the intact organic nitrate (or nitrite) considering its rapid metabolism. Thus a preceding reaction of the nitrate with two reduced SH groups, probably of an enzyme embedded in vascular tissue (nitrate receptor?), oxidizes them to a disulfide form while itself being reductively cleaved to inorganic nitrite. This is certainly a type of reaction encountered with various dehydrogenase enzymes in which their sulfhydryl groups are oxidized to disulfide linkages. An interesting but further complicating proposition was put forth that the beneficial action of NTG, at least in part, may be due to analgesia, which is produced when the final NTG metabolite nitric oxide (NO), a substance known to modulate nociception by direct effect on pain fibers, is produced. To substantiate this idea further, these workers were able to produce analgesia in rats by administering L-arginine, a substance known to increase NO levels. A clinically encountered tolerance to nitrates also occurs at the vascular smooth muscle level. The reason may well be a depletion of free SH groups since the effect is reversible experimentally by the potent disulfide-reducing agent dithiothreitol.

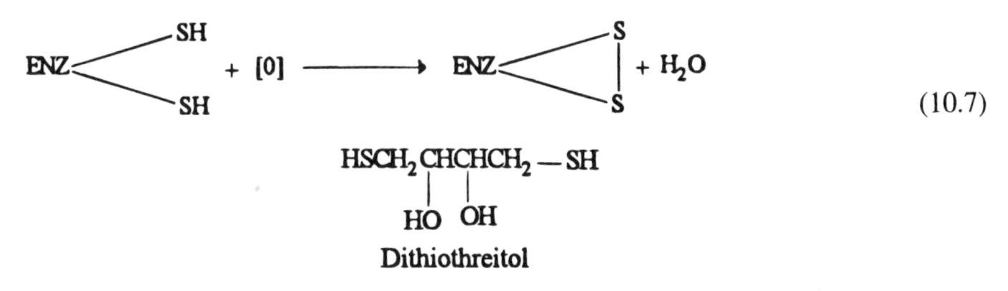

Whether the enzyme action is a result of some type of nitroso thiol formation, the conversion of the NO_2 to a more reactive nitric oxide free radical (NO°) or other species is not yet established. Figure 10-23 summarizes these possible events. It should be understood that all the pieces in the puzzle are not yet in place.

The 4-carbon erythritol tetranitrate (ETN), its 5-carbon analog pentaerythritol tetranitrate (PETN), and isosorbide dinitrate differ from NTG primarily by slower onset and particularly longer duration of action. The molecular mechanisms are the same. (PETN, ETN, and NTG are all used as industrial explosives; NTG is used in the form of dynamite.)

The newest addition to the nitrate group is the major active metabolite of isosorbide dinitrate, namely, isosorbide mononitrate. Much of clinical activity of dinitrate can be attributed to this metabolite.

The nitrates/nitrites have multiple physiological effects. NTG, which has been in use for well over a century and is probably the most studied, is considered the prototype. It is an arterial and venous dilator. However, venous dilation is more pronounced at low nitrate levels, resulting in a pooling of blood in the veins. The ensuing decrease in venous return reduces ventricular volume, preload, which in turn reduces left ventricular end diastolic

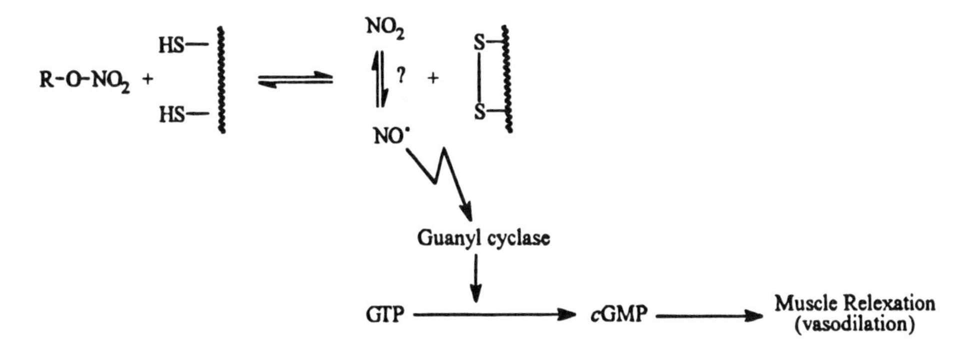

Figure 10-23. Possible nitrate mechanism of action.

pressure—after load. Such decreases in preload or afterload reduce wall stress and O_2 utilization by the heart muscle. The fact that NTG dilates both large and small arteries, which results in lowered systemic vascular resistance and blood pressure, is also to be considered. Both factors, then—reduced preload and afterload—reduce total O_2 demand by the heart muscle and more than offset the reflex tachycardia induced by the nitrate-caused hypotension. It is curious that nitrate effects on the coronary circulation itself—once thought to be the only reason for the antianginal effects of NTG—is now the least well understood.

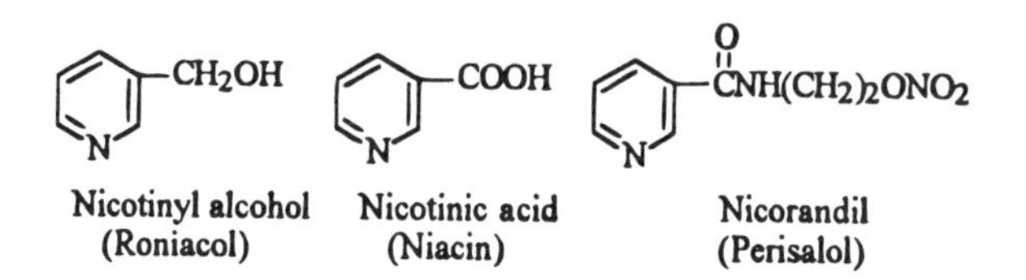

Nicotinic acid and its reduced form, nicotinyl alcohol, have been used for years in attempts to manage peripheral vascular disease. However, nicotinic acid and the alcohol (which is metabolized to the acid) have weak vasodilating activity. At tolerated doses they probably exhibit some activity on dermal blood vessels. The nitrate ester of N-(β-hydroxyethyl)nicotinamide, nicorandil, was developed in Japan as an antianginal agent. The drug has coronary and peripheral vasodilating properties as well as spasmolytic effects. It thus acts as a classical nitrate. Whether the molecular mechanisms are the same has not been established.

10.3.8.2. Calcium Channel Blockers

This is a group of structurally diverse compounds, the first two members of which—prenylamine and verapamil (Table 10-16)—have been in clinical use since the early 1960s as antianginal drugs. However, at that time little of the physiological significance of Ca^{2+} was understood, nor was the relationship of these two drugs to the ion fully appreciated. For example, verapamil's introduction (in Germany) as an antianginal drug was based solely on its vasodilator properties; its supraventricular antiarrhythmic properties were not recognized for a decade. One reason these drugs had such a long "drug lag" before U.S. approval (e.g., verapamil, 1962–1981) was that most of the clinical studies were done outside the country and many of these were not controlled and randomized clinical evaluations. Therefore, they could not be evaluated by the U.S. Food and Drug Administration.

In addition to calcium channel blockers (CCB), the drugs considered in this section are also referred to as calcium entry blockers or calcium antagonists. Clinical applications of this group of drugs presently include angina pectoris, hypertension, and supraventricular and ventricular arrhythmias of various types. Some of the mechanisms of these drugs are not fully clear. Much is still being learned of the bioregulatory functions of Ca^{2+}. It should be recognized that the calcium channels being considered are those activated by membrane polarization (i.e., gated). Channels linked to membrane receptors and controlled by agonist–receptor-type interactions, although believed to play a role in both smooth muscle and secretory cells, have been largely uncharacterized and are not well understood.

External stimuli are believed to trigger Ca^{2+} to be "pumped" in, which can then bind to calcium-binding proteins within the cell, activating them. These in turn can then interact with various target proteins, likely enzymes, which subsequently set off the various cellular

Table 10-16. Calcium Channel Blockers[a,b]

Structure	Name	Comment
A. Dihydropyridines		
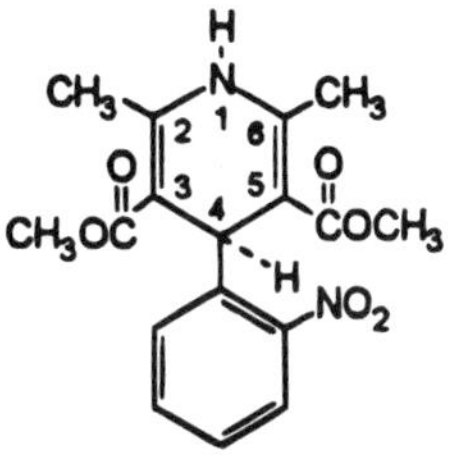 H; N; CH_3; CH_3; 1; 2; 3; 4; 5; 6; O; O; CH_3OC; $COCH_3$; H; NO_2	Nifedipine (Procardia)	Introduced in United States in 1982
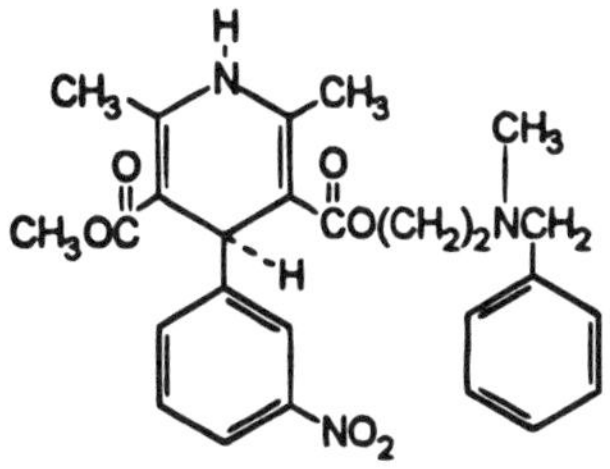 H; N; CH_3; CH_3; O; O; CH_3OC; $CO(CH_2)_2NCH_2$; CH_3; H; NO_2	Nicardipine (Cardene)	Also inhibits PDE[c]
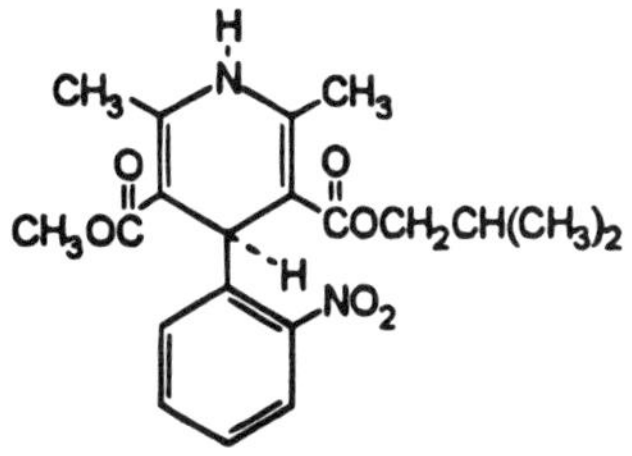 H; N; CH_3; CH_3; O; O; CH_3OC; $COCH_2CH(CH_3)_2$; H; NO_2	Nisoldipine	Investigational; may also inhibit K^+ -induced contraction
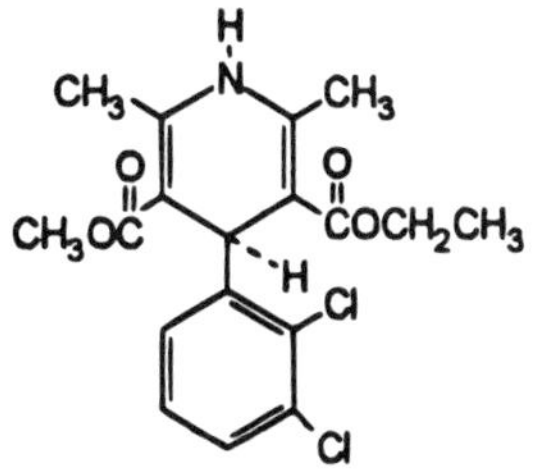 H; N; CH_3; CH_3; O; O; CH_3OC; $COCH_2CH_3$; H; Cl; Cl	Felodipine (Plendil)	Marketed in 1991 only as extended release formulations for once-a-day dosing
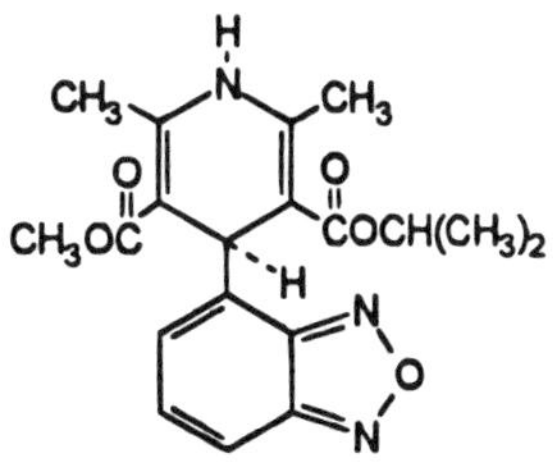 H; N; CH_3; CH_3; O; O; CH_3OC; $COCH(CH_3)_2$; H; N; O; N	Isradipine (Dyna Circ)	Exhibits some diuretic activity

(Continued)

Table 10-16. *(Continued)*

Structure	Name	Comment
B. Arylakylamines		
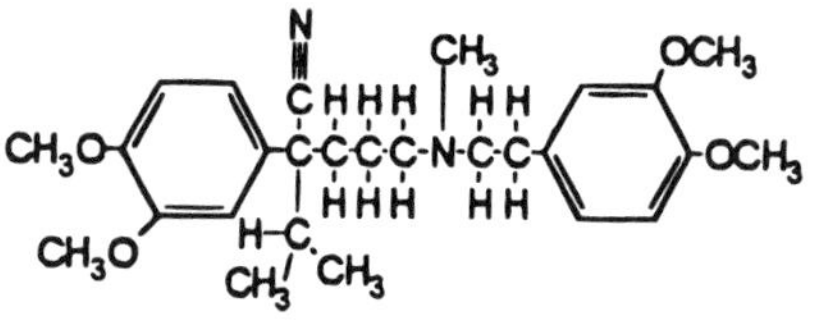	Verapamil (Isoptin, Calan)	Introduced in 1962; in the United States, 1981
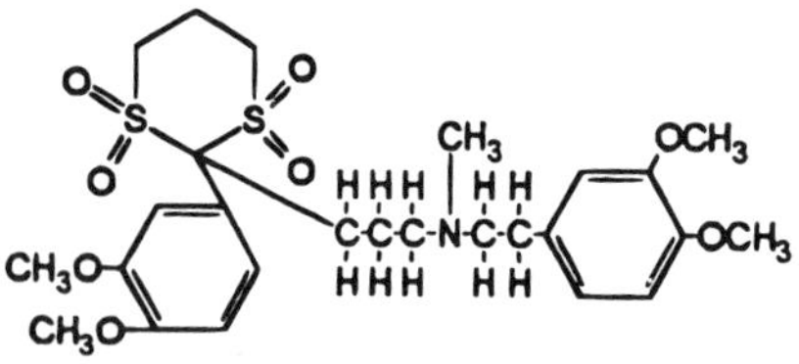	Tiapamil	Investigational
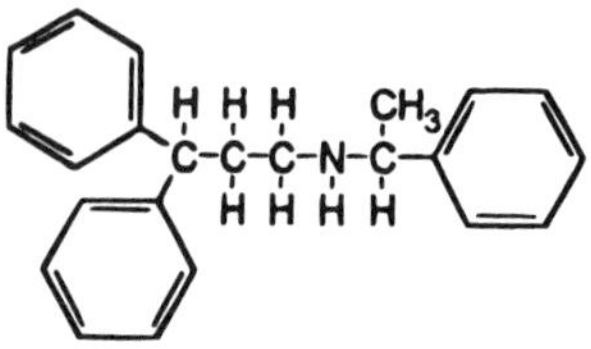	Prenylamine	First calcium channel blocker, 1960; never introduced in the United States
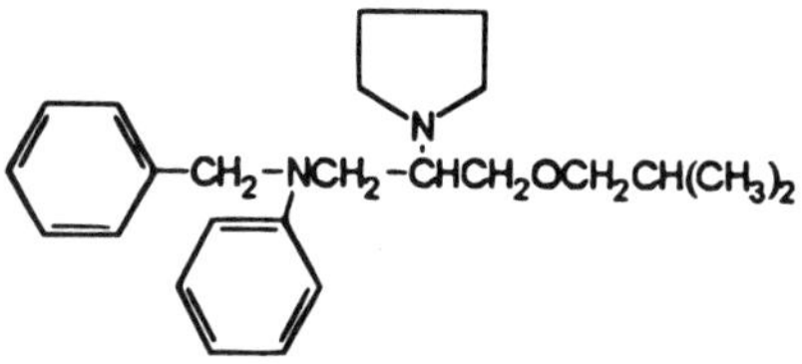	Bepridil	Investigational; also inhibits fast Na+ channels

Structure	Name	Comment
C. Piperazine derivatives		
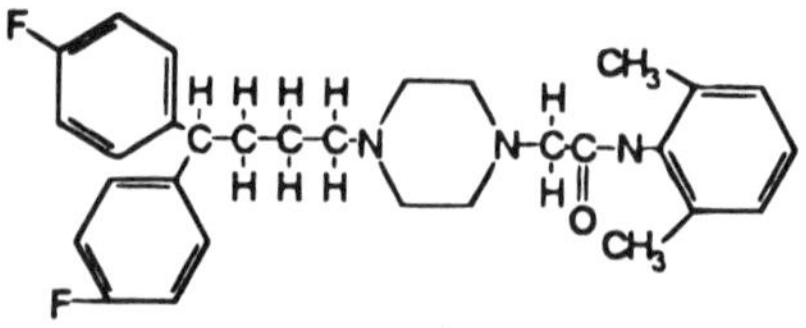	Lidoflazine (Angex)	Inhibits Na^+ as well; potentiates adenosine
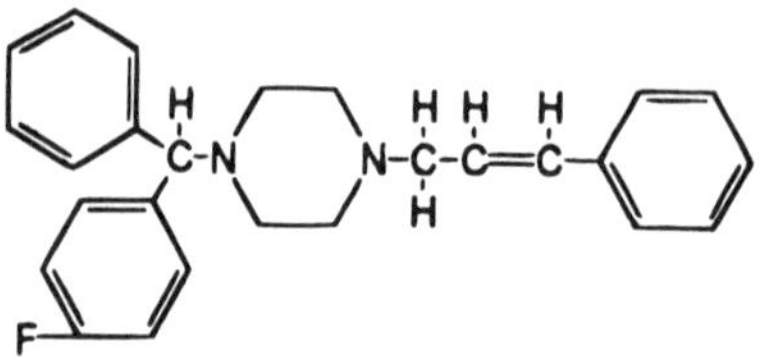	Flunarizine	Investigational

(Continued)

Table 10-16. *(Continued)*

Structure	Name	Comment
D. Benzothiazepines		
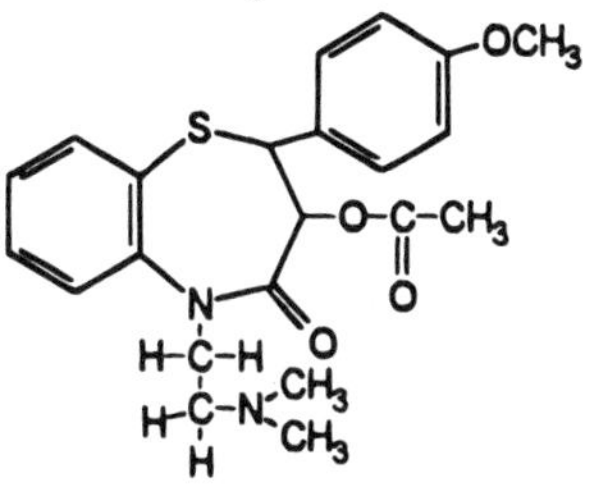	Diltiazem (Cardizem)	Available in the United States since 1983

[a] Also known as slow channel inhibitors and calcium channel inhibitors.
[b] Trade names in parentheses.
[c] Phosphodiesterase.

events. The Ca^{2+}, in these circumstances, can be said to act as a second messenger for muscle contraction (both smooth and striated) and possibly other processes such as neurotransmitter release. The myocardial slow inward calcium channels are energized by ATP, and can be activated either directly by depolarization, *c*-AMP–mediated phosphorylation, or inhibition of dephosphorylation catalyzed by phosphatase. The term *slow* is relative to the very rapid inward movement in Na^+ channels; operative voltages differ also. These Ca^{2+} channels can be blocked by the drugs considered here, as well as by CN^- and divalent cations larger than Ca^{2+} such as Mn^{2+}, Co^{2+}, and Ni^{2+}. It is evident that the CCBs do not all share a common site of action in spite of the fact that they all inhibit the influx of extracellular Ca^{2+}. Their specificities and potencies expectedly differ.

A comparison of the structures in Table 10-16 illustrates their heterogeneity. From a *chemical* standpoint they can be divided into four main groups: the dihydropyridines (e.g., nifedipine), the arylalkylamines, the piperazines, and the benzothiadiazines. There are others.

The arylalkylamine, verapamil, and the drug diltiazem have selectivity on Ca^{2+} channels in cardiac muscle. The diphenylpiperazine subgroup seems to be less selective, being capable of inhibiting Ca^{2+} and Na^+ channels in the myocardium. Verapamil and diltiazem seem active on Ca^{2+} channels in both cardiac and vascular muscle. The dihydropyridines and diphenylpiperazine derivatives show preference for vascular smooth musculature.

It has been demonstrated in numerous studies that the effects of CCBs are stereospecific and are receptor mediated within the channels. High-affinity binding for heart muscle tissue channels in guinea pigs using a (^3H)-nitrendipine (a nifedipine analog) has been established. Similar studies were done with brain tissue and smooth muscle. Stereospecificity, a probable indication of receptor involvement, has been shown for various dihydropyridines (DHPs) and several other agents. With some exceptions, it was the levorotory S form that was active (e.g., verapamil). An even more convincing piece of evidence for the existence of receptors, at least for DHP-type of CCBs, is the development of an antagonist to them.[26] By the chemical expedient of replacing the methylester at C-3 of the dihydropyridine ring of nifedipine (Table 10-16) with a NO_2 group (compound Bay-K-8644), competitive antagonism to nifedipine was demonstrated. This compound was found to interact with the same binding sites as CCBs, to exhibit positive inotropic and vasoconstrictor effects, and

[26] Such a compound can be viewed as calcium channel agonist.

to enhance the flow of Ca^{2+} through its channels. All these effects, of course, are diametrically opposite to those produced by the typical CCB.

It has not yet been determined what this putative CCB receptor is, nor even what the full characterization of its properties are. The widespread calcium-dependent protein calmodulin has been proposed as a candidate because of its role in smooth muscle contraction and other factors. However, the evidence for this proposal so far is not established.

Qualitative SAR studies of nifedipine determined the components of the DHPs essential for activity. The DHP ring is an absolute requirement; oxidation to a pyridine destroys the activity, which indicates the secondary nitrogen >NH in the heterocyclic ring is a requisite to CCB activity. At physiologic pH the nitrogen is not protonated to a salt. A bulky substituent in the 4-position (a benzene ring) is also essential. At least one ester at the 3-position is necessary. Whether a second ester at C-5 is needed is not definite, although it would certainly add to lipophilicity. The 2,6-dimethyls appear, by their position and steric bulk, to assure noncoplanarity of the two rings: DHP and phenyl.

X-ray crystallographic studies of a series of nifedipine analogs that differed by type and position of phenyl ring substituents showed the DHP ring in each case to be in a boat conformation. The *degree* of puckering (or distortion), however, varied with the position and size of the aryl ring substituents, which, in turn, influenced intensity of activity. Thus placing a small F on each of the phenyl ring's five carbons caused little distortion, while adding a bulky -$N(CH_3)_2$ group on the 4-carbon that also carries the phenyl ring created the most distortion. In this limited study the pentafluoro compound was most active, whereas the 4-$N(CH_3)_2$ was the least active.

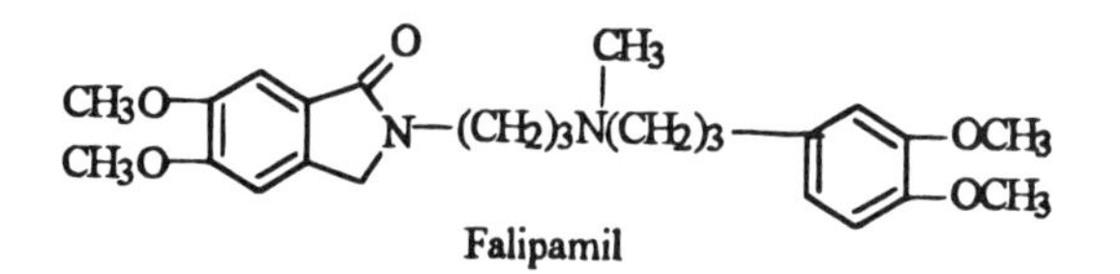

Falipamil

Some of the structural parameters within which verapamil-type drugs function have been determined. Both aromatic rings are essential, although they can be heteroaromatic as they are in falipamil. The tertiary amine function is also a requirement. The isopropyl group and substituents on the aromatic rings are not essential, although the positions on the ring influence potency. For example, *meta* substituents offer the best potency, whereas *ortho* groups decrease it. The suspicion here is strong that bulky moieties at these points may sterically interfere with optimum drug–receptor interactions. Because the drug's S (–) isomer is the more potent, the substituents on atoms adjacent to the chiral center influence the effectiveness of the compound more. For example, varying substituents and positions on the benzene ring to the left of that center affect potency significantly, while modifications on the right ring, six atoms away, do not. Unlike nifedipine, where the DHP nitrogen remains unchanged, at pH 7.4 verapamil (pKa 8.73) is over 95% protonated and cationic in body fluids. This, and reasons not yet recognized, is probably responsible for some puzzling differences between the three marketed compounds. Thus clinically only verapamil and diltiazem affect AV nodal conduction and nifedipine does not. Nifedipine has the most potent vasodilating properties of the three, but it does not depress the myocardium significantly. The vasodilating properties of diltiazem are more pronounced on the coronary arteries than they are on peripheral vessels; the reverse is the case with verapamil and nifedipine. One would intuitively not expect a group of compounds with such dissimilar structures and physicochemical properties (pKa, log P', etc.) to behave identically clinically even if their mechanism at the molecular level might be the same.

Regarding the examples of investigational drugs listed in Table 10-16, tiapamil, unlike verapamil, acts preferentially on coronaries, yet clinically the drug's profile is qualitatively similar to verapamil. The two highly polar SO_2 groups, however, do significantly increase hydrophilicity (three times verapamil), yet, because of the insulation of the electronic effects from the nitrogen atom by four saturated carbons, the pKa (8.5) is minimally decreased. The compound is 92% ionized at pH 7.4.

Another interesting DHP agent is amlopidine (Norvasc). The "traditional" NO_2 group on the *o*-phenyl position is here replaced by a Cl atom, thus avoiding the photolability of similar drugs. The use of the basic aliphatic amine function is probably unique. It seems to improve the bioavailability of the drug, and it produces a long-acting drug with a much slower rate of achieving a maximum effect (3 hour vs. 0.5 h for nifedipine). This is likely due to a threefold lower receptor association rate. The utility of this fact is avoidance of acute hypotensive effects such as reflex tachycardia.

Bepridil does not seem to have much structural relationship to verapamil. In addition to slow Ca^{2+} channels, it also inhibits fast Na^+ channels and exerts useful antiarrhythmic properties as well as antianginal effects.

In a relatively short time CCBs have become a factor of major significance in the management of hypertension and arrhythmias. They may already have revolutionized the treatment of angina pectoris.

10.3.1.9. Miscellaneous Drugs

Dipyridamole (Persantine) evolved from a synthetic study of pyrimido (5,4-d) pyrimidines as a coronary vasodilator over three decades ago. Some early studies indicated that the increased coronary flow alleviated ischemic O_2 demand and was therefore useful prophylactically in angina. With the advent of well-controlled as well as long-term studies in more recent years, it became apparent that the drug cannot be convincingly shown to have the efficacy it was believed by some to have—this in spite of the fact that it does dilate coronary resistance vessels, but not the conductance vessels the nitrates dilate.

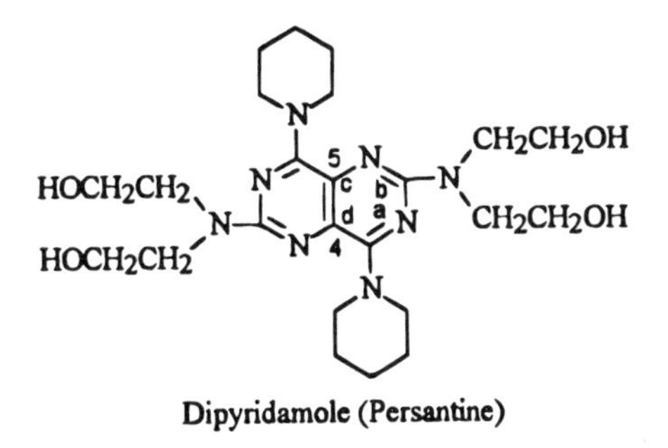

Dipyridamole (Persantine)

Dipyridamole, however, also has some ability to inhibit platelet aggregation, thus being potentially useful in preventing thromboemboli (see Chapter 11). Even here, however, the clinical evidence is unimpressive, unless aspirin is added to the regimen.

References

Benassi, M. G., Modena, G., Mattioli, G., *Arzneim-Forsch.* **36**:390, 1986.

Bolis, G., Fung, A. K. L., Greer, J., et al., *J. Med. Chem.* **31**:532, 1988.

Byer, G., Mostad, A., Romming, C., *Acta Chem. Scand.* **30**:843, 1976.

CAST, *N. Engl. J. Med.* **321**:406, 1989.

Coleman, A. J., Paterson, D. S., Sommerville, A. R., *Biochem. Pharmacol.* **28**:10, 1979.

Harrison, D. C., *Am. J. Cardiol.* **56**:185, 1985.

Heinkel, J. G., Portoghese, P. S., Miller, J. W., Lewis, P., *J. Med. Chem.* **19**:6, 1976.

Hoefke, W., in *Antihypertensive Agents*, Engelhardt, E. L., Ed., ACS Symposium Series, Vol. 27, Gould, R. F., Ed., American Chemical Society, Washington DC, 1976, p. 27.

Ikeo, T., Nagao, T., Murata, S., *Arzneim-Forsch.* **36**:1063, 1986.

Jen, T., Van Hoeven, H., Groves, W., et al., *J. Med. Chem.* **18**:90, 1975.

Keenan, R. M., Weinstock, J., Finkelstein, J. A., et al., *J. Med. Chem.* **35**:3858, 1992.

McCaig, D., Parratt, J. R., *Br. J. Pharmacol.* **67**:239, 1979.

Murakami, E., Eggena, P., Barrett, J. D., Sambhi, M. P., *Life Sci.* **34**:385, 1984.

Ondetti, M. A., et al., *Biochem.* **10**:4033, 1971.

Ondetti, M. A., Rubin, B., Cushman, D. W., *Science* **196**:441, 1977.

Robinson, G. A., Butcher, R., Sutherland, E. W., *Cyclic AMP*, Academic Press, New York, 1971.

Rouot, B., Leclerc, G., Wermuth, C. G., *J. Med. Chem.* **19**:1049, 1976.

Stewart, T. A., Weare, J. A., Erdos, F. G., *Peptides* **2**:145, 1981.

Timmermans, P. B. M. W. M., Hoefke, W., Stahle, H., van Zwieten, P. A., *Progr. Pharmacol.* **3**:1, 1980.

Timmermans, P. B. M. W. M., van Zwieten, P. A., *J. Med. Chem.* 1636, 1977.

Vaughn-Williams, E. M., *J. Clin. Pharmacol.* **24**:129, 1984.

Veterans Administration Cooperative Study Group on Antihypertensive Agents, *J. Amer. Med. Assoc.* **202**:1028, 1967; ibid. **213**:1143, 1970.

Yamaguchi, H., Nishimura, K., Nakata, K., et al., *Arzneim-Forsch.* **37**:157, 1987.

Suggested Readings

Brater, D. C., The Pharmacological Role of the Kidney, *Drugs* **19**:31, 1980.

Bristol, J. A., Evans, D. B. Agents for the Treatment of Heart Failure, *Med. Res. Rev.* **3**:259–87, 1983.

Carboeuf, E., Ionic basis of electrical activity in cardiac tissues, *Am. J. Physiol.* **234**:101, 1978.

Janis, R. A., Triggle, D. J., New developments in Ca^{2+} channel antagonist, *J. Med. Chem.* **26**:775–85, 1983.

Ondotti, M. A., Cushman, D. W., Enzymes of the Renin-Angiotensin System and their Inhibitors. *Ann. Rev. Biochem.* **51**:283–308, 1982.

Pettinger, W. A., Minoxidil in the Treatment of Severe Hypertension. *N. Engl. J. Med.* **303**:922, 1980.

Petrillo, E. W., Ondotti, M. A., Angiotensin-Converting Enzyme Inhibitors: Medicinal Chemistry and Biochemical Actions, *Medicinal Res. Rev.* **2**:1–41, 1982.

Rahwan, R. G., Mechanism of Action of Membrane Calcium Channel Blockers and Intracellular Calcium Antagonists, *Medicinal Res. Rev.* **3**:21–42, 1983.

Stanaszek, W. F., Kellerman, D., Brogden, R. N., Romankiewicz, J. A., Prazocin Update; A Review of its Pharmacological Properties and Therapeutic Use, *Drugs* **25**:339–84, 1983.

Timmermans, P. B. M. W. M., Hoefke, W., Stahle, H., van Zwieten, P. A., Structure activity relationships in clonidine-like imidazolines and related compounds, *Progr. Pharmacol.* **3**:1–104, 1980.

CHAPTER

11

Drugs and the Cardiovascular Diseases II

This chapter, which is essentially a continuation of the preceding discussion, will deal with the chemotherapeutic approaches to such cardiovascular conditions as result in thrombus formation, especially by involving thrombocytes or platelets, and the control of blood lipids, particularly certain types of cholesterol-containing lipoproteins. Since diabetes ultimately involves cardiovascular pathology, some of the chemicals used to control and ameliorate the so-far inexorable diabetic processes will also be considered. Sickle cell anemia, although still not having any effective approved treatment, is being actively researched. Some progress exists.

11.1. Clotting Prevention and Lysis

From any viewpoint blood is chemically the most complex tissue in the body. In addition to the multitude of cells and platelets, it contains inorganic ions (electrolytes), various plasma proteins, hormones, lipids, vitamins, a large variety of enzymes, nucleic acid breakdown products, a large number and unknown types of environmentally ingested chemicals (drugs, food additives, pollutants) at varying stages of metabolic conversion, gases, and, of course, water. The list of specific entities found in human blood continues to grow. Among these is a group of more than a dozen chemical factors, which will cause the blood to coagulate, or clot when properly triggered. Of course, this is the mechanism that prevents us from bleeding to death in the event of external or internal injuries.[1]

An endogenous substance that apparently functions to counteract the coagulating tendency of blood is the mucopolysaccharide heparin. Heparin is a polymeric organic acid. The many oxygen–and nitrogen–sulfate linkages, which have strong acidity resulting in

[1] A hereditary disease, hemophilia, is characterized by the absence of one of these factors (factor VIII), which causes the failure of blood to clot. A minor cut, for example, can be fatal.

their being ionized at physiological pH, are believed to react with certain clotting factors, thereby inhibiting the complex clotting sequence. It is used as a rapid, but short-acting, anticoagulant, especially in emergency situations.

A pathological situation arises when thrombi and emboli develop, either as causative factors in such conditions as strokes or in other thromboembolic disorders following a cardiovascular ischemia (e.g., a myocardial infarction, MI). These, then, continue to threaten the patient's life.

Two therapeutic approaches should therefore be possible. Clots could be prevented from forming initially (or additional clots following the primary incident) by the use of anticoagulant compounds or other modalities. Thrombin, once formed, could be destroyed with thrombolytic agents. Heparin, with modest antithrombotic properties, has been used prophylactically and therapeutically. A third approach is in reality an extension of the prophylactic concept. It is the use of drugs capable of preventing platelet aggregation, which is believed to be the primary step in the pathogenesis of arterial thrombosis.

To further understand the drugs in this area, it would be useful to have an overall view of the clotting process as it is presently understood (Fig. 11-1).

The mechanism by which human blood clots is another example of the "hydra" phenomenon. Once thought to be relatively simple and well understood, discoveries of various exogenous activation factors have revealed that the clotting (and lysis) process is very complex. The following discussion, though sufficiently detailed to be a foundation for the drug products to be considered, will understandably be condensed.

Coagulation of blood is initiated by blood contact with damaged tissue such as a ruptured or cut blood vessel. The purpose is to initiate a train of complex events—the clotting mechanism—the body's system for minimizing blood loss. Considering the final phase of

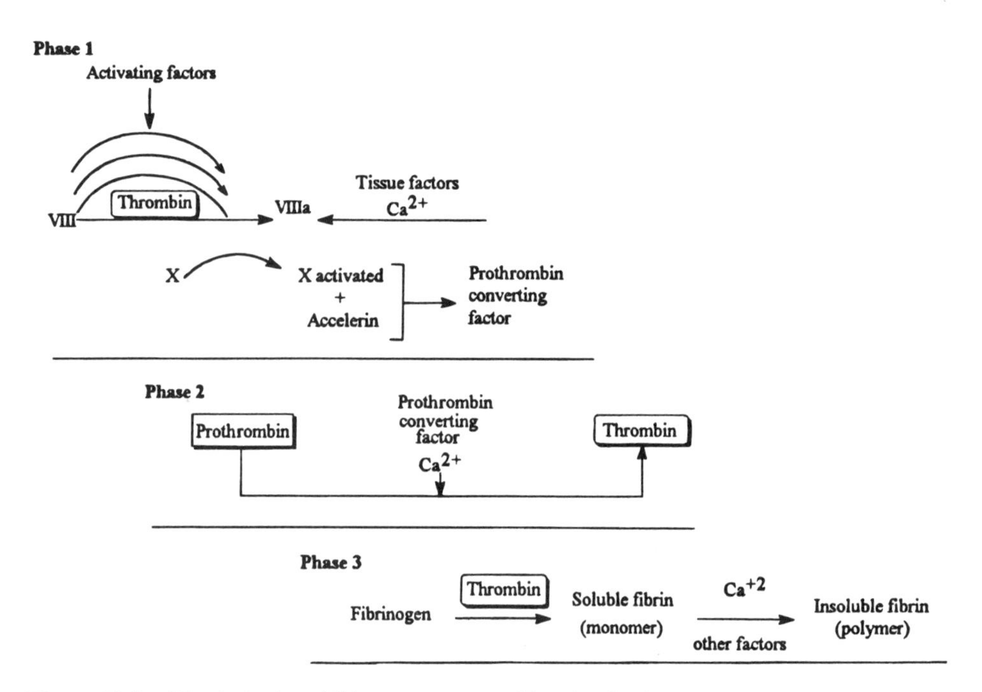

Figure 11-1. Blood clotting. (This represents an abbreviated scheme. Roman numerals indicate clotting factors as listed in Table 11-1).

the blood-clotting process, it involves the thrombin-catalyzed removal of two small peptides from fibrinogen, a soluble, circulating, very large (350,000 daltons) special protein. This results in a still soluble fibrin monomer that is subsequently polymerized to an insoluble fibrous substance called *fibrin*. The irregular mesh formed entraps red cells, platelets, and plasma, forming the clot. The trigger for this hemostatic process following injury is probably the exposure of the blood to the collagen underlying the endothelium layer. The initial plug follows the "sticking" of platelets to the exposed collagen and their release of serotonin, EP, ADP, and other platelet factors. Serotonin and EP cause vasoconstriction; ADP strongly induces additional platelets to adhere. As the thrombocytes continue to aggregate, their membranes dissolve, forming a sticky jelled mass capable of arresting the bleeding (i.e., hemostasis). The fibrin forming in phase 3 (Fig. 11-1) forces this loose plug, binding it into a *clot*. As it matures the clot retracts, squeezing out most of the plasma devoid of fibrinogen (serum). Coagulation initiated by blood contact with damaged tissue is sometimes referred to as the *extrinsic clotting system*.

In Chapter 5 it was pointed out that release of arachidonic acid from membrane-bound "storage" initiated the AA cascade, resulting in the various prostaglandins (PGs). The previously mentioned alterations of platelet membranes that activate phospholipase enzymes A_1 and A_2, which in turn liberate AA from its membrane phospholipid bondage, are of particular significance in this discussion. The resultant PGs, thromboxane A_2 (TXA_2) and prostacyclin (PGI_2), have powerful and opposite effects on the vasculature (see Fig. 5-3). TXA_2, which is probably the most potent (but short-lived) vasoconstrictor known, also induces platelets to aggregate dramatically. PGI_2, on the other hand, is formed from PGG_2 and PGH_2 in the walls of vessels, and is an extremely potent vasodilator and the most potent inhibitor of platelet aggregation known. Thus the continuous tendency to aggregate platelets and thereby to coagulation and clot formation is continuously counteracted by PGI_2. Hence, a balance between these two PGs to regulate platelet functions is imperative. Any imbalance in favor of TXA_2 by mechanisms such as atherosclerosis would prevent PGI_2 in the vessel walls from reaching and neutralizing coagulating nuclei forming in the vessel lumen. This can precipitate thrombus formation and lead to an embolism. In diseased blood vessels prone to such coagulatory events, thrombus formations can be venous, which might lead to pulmonary emboli. Thrombi containing a high proportion of platelets can also occur in arteries. If the vessels are coronaries and lead to occlusion, the result is an MI. An occlusion in the brain and the resultant clinical sequelae is termed a *stroke*.

There are also mechanisms for thrombolysis that occur normally (see later). Thrombolytic drugs will also be considered.

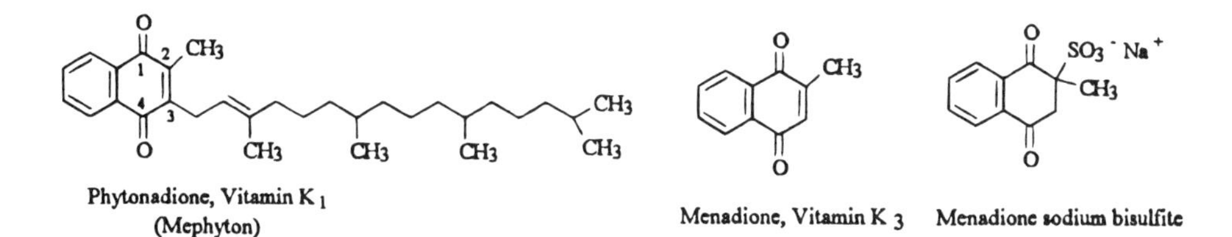

Phytonadione, Vitamin K 1 (Mephyton) Menadione, Vitamin K 3 Menadione sodium bisulfite

The fibrinogen-to-fibrin monomer reaction is catalyzed by the proteolytic enzyme thrombin. Thrombin is in turn produced by the prothrombin-converting factors V, X, and Ca^{2+} from circulating prothrombin (Phase 2). *Prothrombin* is a globular protein made in the liver (as is fibrinogen). Vitamin K is an essential factor in its synthesis. Prothrombin-converting factor in the presence of Ca^{2+} cleaves the molecule; one fragment is the active enzyme thrombin. Table 11-1 lists the clotting factors.

The first phase of the multiphased sequence has the production of the prothrombin-converting factor as its ultimate purpose. The synthesis of the activated form of factor X begins with the activation of factor XII (the contact factor) to XI, then IX followed by VIII. This represents the *intrinsic pathway*. An *extrinsic pathway* to activated factor X also exists. It depends on the release of thromboplastin (factor III) into the blood from the damaged blood vessel wall (or other tissue). This tissue thromboplastin, in the presence of Ca^{2+} and activated factor VII, is also activated to factor X. The two pathways are essential. They operate at very different rates.

Vitamin K_1 is an oil-soluble vitamin occurring in most green-leaved plants, including edible vegetables of the human diet. It is a participant in the liver's synthesis of prothrombin, specifically the carboxylation of the γ-carbon of a glutamic acid residue in the protein that results in that carbon then carrying two COOH groups (i.e., a γ-carboxyglutamate). The vitamin, in its naphthahydroquinone form, in which the 1- and 4-keto functions are reduced to OH groups, is the active cofactor for the carboxylating enzyme (Eq. 11.1). γ-Carboxy residues on prothrombin are essential for binding to Ca^{2+} necessary for its activation to thrombin in Phase 2 (Fig. 11-1).

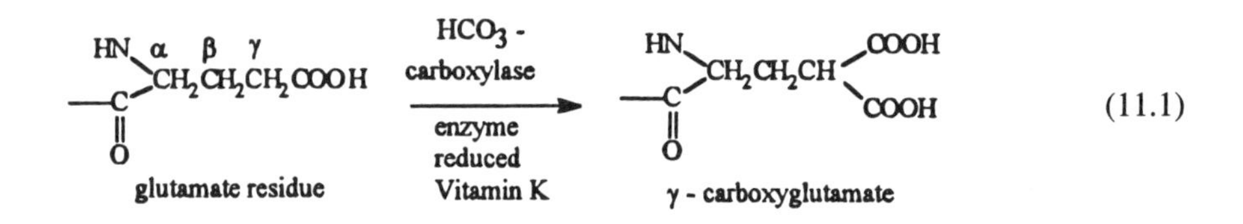

(11.1)

The vitamin K structure shown earlier is that of K_1, primarily found in green plants. Vitamin K_2 contains from three to six isoprene units (depending on the source), each of which has a double bond. Isoprene, 2-methyl-1,4-butadiene, is found in polymerlike repetitive "isoprenoid" units in natural rubber, carotenoids, steroids, and smaller compounds

Table 11-1. Factors in Blood Clotting

International designation	Common nomenclature
I[a]	Fibronogen
II	Prothrombin[b]
III	Thromboplastin (tissue)
IV	Calcium ion
V	Prothrombin accelerator (and others)
VI	Obsolete
VII	Proconvertin
VIII	Antihemophilic factor (AHF)[c]
IX	Christmas factor, autoprothrombin II
X	Stuart factor
XI	Plasma thromboplastin antecedent (PTA)
XII	Hageman factor (contact factor)
XIII	Fibrin-stabilizing factor
Platelet factor	

[a] Sometimes fibrin monomer and polymer are referred to as factors I′ and I″, respectively.
[b] When activated becomes thrombin or Factor IIa.
[c] Also named antihemophylic globulin, AHG.

called *terpenes*. Vitamin K_2 is usually found in animal and bacterial sources. Synthesis by intestinal bacteria is one of its "nutritional" sources.

For coagulant activity the isoprenoid side chains are not mandatory. Thus menadione, which is a naphthaquinone carrying only a methyl group at the 3-position, is considered a synthetic K vitamin (K_3). It has the same actions as the natural products. It, and two derivatives—the water-soluble sodium bisulfite addition compound and the disodium salt of the phosphate ester of reduced menadione—are all used clinically as hemostatics, and to antidote overdoses of anticoagulants.

The first orally active anticoagulant, dicoumarol, which is a molecule consisting of two 4-hydroxycoumarin moieties bonded at their 3-position via a CH_2 bridge (3,3′-methylenebis-4-hydroxycoumarin), was isolated from decomposed clover. It was discovered and identified to be a cause of hemorrhagic death of cattle ingesting this improperly stored feed in the early 1920s (sweet clover disease). It was two decades before applications to human therapeutics occurred. This followed demonstration that oral use of this compound increased clotting time and decreased the incidence of postsurgical intravascular thrombus formation.

Because of the striking structural similarity of the 1,4-naphthaquinone and 4-hydroxycoumarin ring systems, the likely mechanism of action of dicoumarol and the subsequent synthetic derivatives introduced (Table 11-2) was viewed from the antimetabolite perspective (i.e., as vitamin K antagonists). They can be shown physiologically to interfere with the vitamin's action, and, in turn, its effects (e.g., hemorrhages from overdoses) can be reversed by vitamin K_1.[2] At the molecular level it is now known that the mechanism involves inhibition of reduction of the 2,3-epoxide of vitamin K that was produced by oxidation during its involvement in the γ-carboxylation of glutamate residues on prothrombin. Thus it is the recycling of the vitamin to its reduced active cofactor status that results ultimately in decreased thrombin levels since less γ-carboxylation diminishes Ca^{2+} bonding and thereby prothrombin activation. Of course, clinically coagulation is inhibited.

Since the epoxidized form of vitamin K is inactive, and the coumarin drugs therefore hinder the carboxylation reaction (Eq. 11.1) indirectly, it is theoretically incorrect to refer to them as vitamin K antagonists. On the other hand, it can be assumed that it is the structural congruity between the coumarin drugs and the K vitamins that enable them to inhibit the reducing enzyme diaphorase.

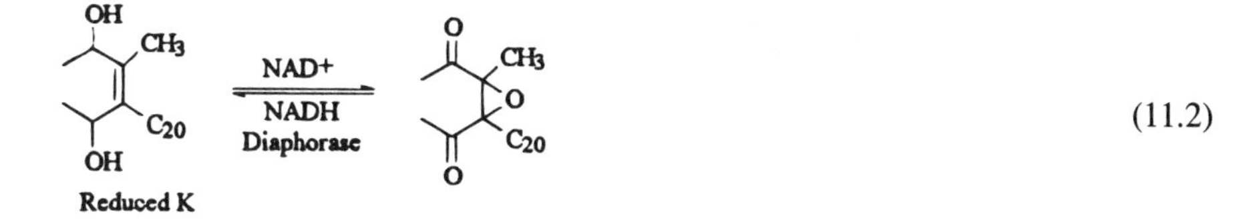

(11.2)

The structure–activity requirements (SAR) for the coumarin-type anticoagulants are simple: The 4-hydroxycoumarin ring system is minimally needed for activity. Alkyl substituents at the 3-position will enhance activity over an H atom. Dicoumarol, the prototype, is of relatively low potency, with a slow onset of up to 5 days for peak activity and hypoprothrombinemia. The anticoagulant effect may persist for more than 1 week after stopping the drug. Even though overdoses can be antidoted with IV vitamin K_1, clinical adjustment of anticoagulation, particularly downward, is difficult. Phenprocoumin offers mainly lower

---

[2] The menadiones are not nearly as effective for this particular application.

Table 11-2. Oral Anticoagulants

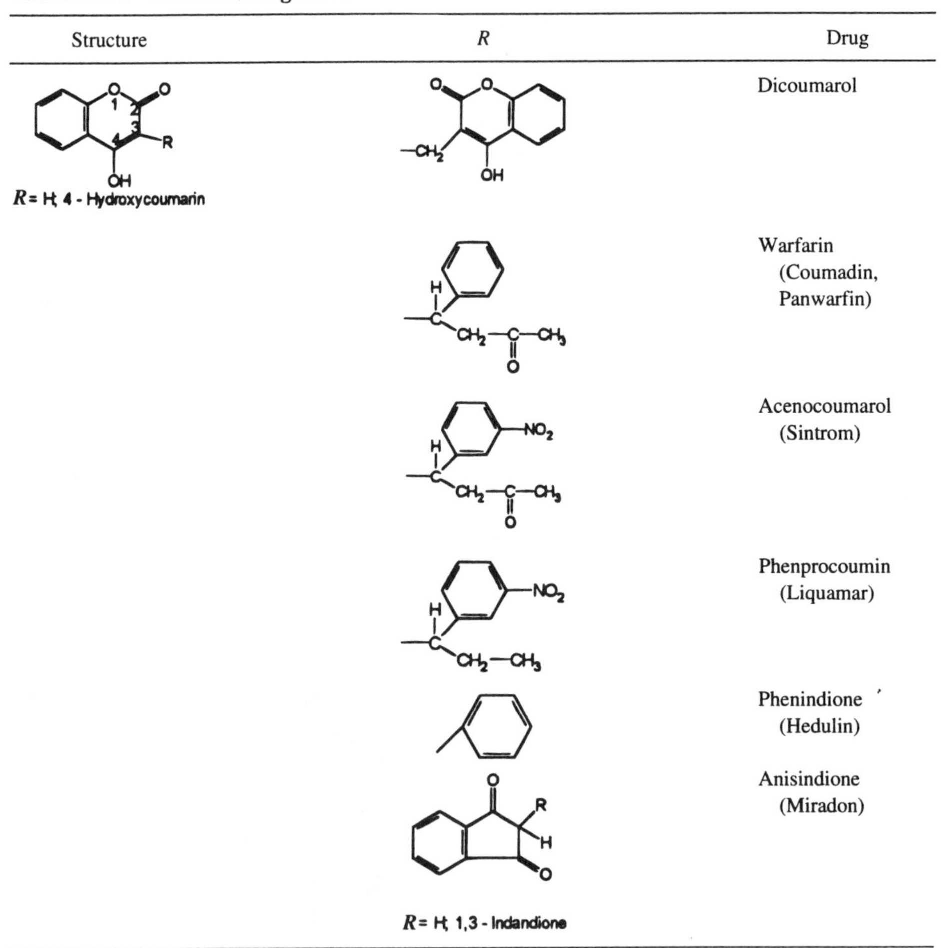

dosages, a dubious advantage. Warfarin has become the most widely used of the coumarin drugs. It is the most potent, with many patients being maintained on only 5 mg/day.

Warfarin was initially introduced as a rodenticide because it was thought too dangerous for human use. It is still used in pest control. The drug, as sweetened pellets, causes rats to die from internal bleeding. In the early 1960s rats resistant to warfarin were noted in London. They were nicknamed "super rats." A mutation had arisen. Several years later this phenomenon appeared in humans (identical twin brothers). It has been shown to be an inherited autosomal dominant trait. Persons with this trait require a 20-fold increase in the drug to achieve anticoagulation—easily fatal to normal patients. Explanations for this unusual phenomenon have been proposed. One is that a tissue protein regulating the synthesis of one or more of the clotting factors in Table 11-1 became genetically altered. Another is a mutation in the enzyme diaphorase that makes it less susceptible to coumarin-drug inhibition.

The 1,3-indanedione ring system was developed into a few orally active anticoagulants on the basis that it has an even closer analogy to the 1,4-naphthaquinone system than does

4-hydroxycoumarin. Its SAR requirement appears to be an enolizable (acidic) hydrogen at position 2, with the other substituent being one of lipid-soluble character, a benzene ring. Phenindione, which is no longer available because of rare but serious untoward effects, has been replaced by anisindione. The drug offers no advantages in potency or peak effects except in patients who may not tolerate the coumarin drugs. The mechanism of action of the indanediones and coumarins is identical.

Important indications for oral anticoagulants are long-term prophylaxis of deep vein thrombosis (DVT) and pulmonary emboli (PE). Theory would predict greater efficacy for these compounds against fibrin-produced thrombi of veins than arterial thrombi where aggregated platelets are a major component of the clot. This is because the vitamin K antagonists, by inhibiting thrombin production, diminish fibrin yield. They have no particular effect on platelet functions. Other indications include control of atrial fibrillations in patients with diseased mitral heart valves and those that have artificial plastic and metal valves installed (lifelong therapy). Using anticoagulants to treat cerebrovascular insufficiencies poses great hazards of hemorrhage in the brain, with the possible exception of patients having transient ischemic attacks (TIAs). The drugs are sometimes used where strong proof of efficacy at acceptable risk is lacking. An example of such a controversial application might be peripheral vascular disease. There are also definite contraindications for anticoagulant use. These include cerebral thrombotic infarctions, severe hypertension, pregnancy, and patients with potential bleeding sites such as acute pericarditis, gastrointestinal ulcers, blood dyscrasias, severe renal or hepatic disease, open wounds, and recently performed eye and spinal surgery.

Heparin is only effective parenterally and has a very rapid onset of action, which makes it a frequent choice in initiating therapy (later followed by oral agents). Its mechanism of action involves enzyme inhibition, not cofactor (vitamin K) antagonism. It prevents prothrombin activation (Phase 2, Fig. 11-1) by interaction with factors XII, IX, X, XI, and others, importantly blocking catalysis by thrombin. Since all these reactions require the plasma cofactor antithrombin III (an α-globulin) and possibly others that complex with thrombin irreversibly, a viable mechanism for heparin can be formulated. Heparin, unlike the oral anticoagulants, exhibits some clinically useful ability in preventing platelet aggregation by inhibiting this partially thrombin-induced effect. The drug reduces platelet adhesiveness as well.

11.2. Antithrombotics

Antithrombotic drugs are a diverse group of compounds capable of suppressing platelet functions. This properly makes them potentially useful in the treatment of *arterial* thrombotic disease. The anticoagulants, as mentioned earlier, are more applicable to the control of venous thromboemboli.

As was seen, hemostasis is a mechanism by which bleeding from vascular damage is stopped. The various events bringing this about include vasoconstriction, platelet adhesion to the blood vessel wall, and aggregation to each other. The initial platelet plug formed is somewhat tenuous. Its consolidation by spontaneous aggregation with fibrin monomer leads to a fragile clot. Thrombus formation (thrombosis) is an exaggerated hemostatic process—and it is a pathological event. Arterial thrombosis can be seen as initiated by platelet aggregation. Subsequent cross-linkage of the fibrin monomer by covalent transamidation, specifically between glutamine side chains of one fibrin molecule and a lysine residue of another, further stabilizes (matures) the thrombus (Eq. 11.3).

$$\text{Fibrin-}CH_2CH_2\overset{O}{\overset{\|}{C}}NH_2 + H_3\overset{+}{N}CH_2CH_2CH_2CH_2\text{Fibrin} \xrightarrow{\text{transamidase}} \text{Fibrin-}CH_2CH_2\overset{O}{\overset{\|}{C}}\text{-}NHCH_2CH_2CH_2CH_2\text{Fibrin} + NH_4^+ \quad (11.3)$$

There are now two therapeutic avenues possible. One is to inhibit the clot or thrombus from forming (i.e., antithrombotic therapy). Drugs capable of doing this will be now considered. The second approach, which becomes necessary if the first and preferable method is not undertaken (or was unsuccessful), is to lyse the clots once formed with thrombolytic agents. In the case of coronary occlusions surgical intervention (bypass) or angioplasty ("balloon") may physically intervene to remove the thrombus.

Screening for antithrombotic drugs in laboratory animals tests the effect they have on bleeding time. Thus, timing the cessation of bleeding following a skin incision or a tail cut or puncture, though a useful test, does not really measure antithrombotic effects directly; actually it determines hemostatic efficiency, which is assumed to be related to thrombogenesis. For example, aspirin prolongs bleeding time and is known to have antiaggregation properties. However, hydroxychloroquin, which is known to have antithrombotic activity, does not prolong bleeding time. Another type of screening would be to induce thrombus formation by surgically damaging blood vessels, or by electrical stimulation.

The increase in current threshold to produce thrombi that a test drug elicits becomes the measure of its antithrombotic efficiency. Preliminary in vitro tests with a platelet aggregometer that measures light transmission through platelet-rich plasma can also be done.

The clinical interest in effective and safe platelet-active drugs grows as our understanding of the complexity of platelet functions expands. To varying degrees platelets are now believed to be involved in acute occurrences such as angina (unstable), MIs, sudden cardiac death. TIAs, and stroke. Implication in the pathogenesis of atherosclerosis is also strong. The preceding, when added to hypertension, easily accounts for the majority of cardiovascular mortalities in this and other nations.

11.3. Cyclooxygenase Inhibitors

It will be recalled (Chapter 5, Fig. 5-3) that NSAIDs are the most important drugs inhibiting the enzyme cyclooxygenase (PG-synthetase), thereby hindering the generation of the endoperoxides PGG_2 and PGH_2. Deficiency of these two endoperoxides in turn depletes thromboxane TXA_2 and PGI_2. TXA_2 formed in platelets is an extremely potent platelet-aggregating compound on release and can form thrombi. Thrombus formation begins when platelets are exposed to collagen, following arterial injury. If TXA_2 production in platelets is blocked by a drug, it can appear that such a drug inhibits collagen-induced platelet aggregation. Of course, NSAI compounds also simultaneously inhibit vessel wall production of PGI_2, which activates platelet adenylate cyclase as one of its properties, thereby raising *c*-AMP levels. This inhibits adhesion and aggregation of platelets. Since NSAIDs in vivo have an antiaggregating effect, it can be concluded that it is the predominant effect.

The details of cyclooxygenase inhibition by NSAIDs were developed in Chapter 5. It should now be reiterated that only aspirin inhibits the enzyme in platelets irreversibly, even at low doses, thus preventing the production and release of TXA_2 from them. Aspirin's inhibition is irreversible, and it persists for the life of the platelet[3] (ca. 12 days). Other NSAIDs, even potent ones such as indomethacin and phenylbutazone, only inhibit the enzyme reversibly, which produces a transient effect. This is also true of salicylic acid, the weakly active metabolite of aspirin, following removal of its acetyl function.

The concurrent inhibition of PGI_2 by aspirin, which is transient since epithelial cells continue to produce fresh cyclooxygenase, has produced curious, even apparently contradictory, results in early clinical (and some animal) studies. The significance of dosage was not appreciated in the early attempts to assess aspirin's potential as a prophylactic for TIAs and MIs in the post-MI period, as well as against arterial thrombosis. The ideal dosage size and schedules for aspirin remain to be determined. Thus high daily doses (over 1–2 g) were indistinguishable from the placebo in some studies. A puzzling animal study showed that high aspirin doses in rabbits were actually thrombogenic, yet a long-term followup study of rheumatoid arthritis patients on high-dose aspirin showed no increase of thromboembolic incidents.

One factor that is not always taken into account may be the individual variations in circulating platelet recovery following aspirin that is achieved by new platelets entering the bloodstream. Because endothelial cells that produce prostacyclin recover rapidly by synthesizing new enzyme while platelets do not, a single dose, given once daily, has been suggested. The size of that dose may well be in the 100–600 mg range; 325 mg or less has been appearing in the clinical literature. Some clinicians now recomend alternate-day medication or daily doses well below 100 mg.

Although devoid of analgesic–antiinflammatory properties, the 4-phenylsulfinyl derivative of pyrazone retains potent uricosuric effectiveness and is used in the treatment of gout. The drug also inhibits cyclooxygenase, reversibly. It apparently does not affect PGI_2 synthesis in endothelial cells. Due to the drug and its reduced sulfide metabolite, these effects last only as long as blood levels persist; bleeding time is not prolonged. In spite of several studies for MI and TIA prophylaxis, the drug has not been approved in the United States as an antithrombotic.

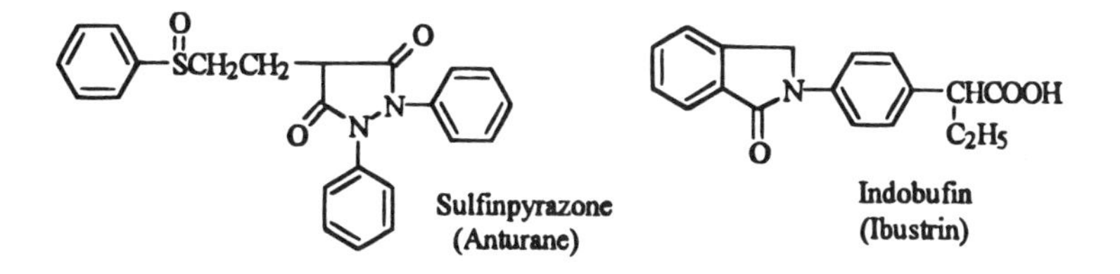

Sulfinpyrazone (Anturane)

Indobufin (Ibustrin)

Dipyridamole was briefly mentioned in Chapter 10. In addition to vasodilator effects, the drug inhibits phosphodiesterase, thereby increasing intraplatelet *c*-AMP. Although these events do lead to decreased platelet aggregation, the drug's effectiveness clinically seems poor unless synergistically combined with either aspirin or warfarin.

An isoindolinylphenylacetic acid, indobufen, marketed overseas, has potent antiaggregating activity through inhibition of collagen-induced thrombus formation.

There are many other types of compounds with the ability to inhibit platelet aggregation, at least experimentally. Since serotonin induces platelet aggregation, it is not surpris-

---

[3] Being devoid of ribosomes, platelets cannot synthesize additional cyclooxygenase.

ing that serotonin antagonists inhibit this effect. Cyproheptadine (which also has H_1 antihistaminic properties) has this effect. It may also inhibit platelet clumping induced by ADP, collagen, and thrombin. Adenosine, ATP, and AMP are specific inhibitors of ADP-induced aggregation at the membrane site. α-Adrenergic blockers should inhibit epinephrine-induced aggregation. At low levels, heparin blocks thrombin-induced platelet clumping.

Substances that can raise *c*-AMP levels, either by stimulating adenyl cyclase (e.g., PGE_1) or by inhibiting phosphodiesterase (e.g., theophylline) may have varying abilities to reduce platelet aggregation. Verapamil has been shown to inhibit ATP-induced platelet aggregation.

Thromboxane synthetase inhibition is an appealing approach to antithrombotic therapy, since it would be more specific and not "shut off" the whole AA cascade as do the cyclooxygenase inhibitors such as aspirin. Such a drug should result in an accumulation of PGG_2 and PGH_2 that could then be diverted to increased prostacyclin production, itself an antithrombotic, thus offering a double-barreled effect. This actually occurs.

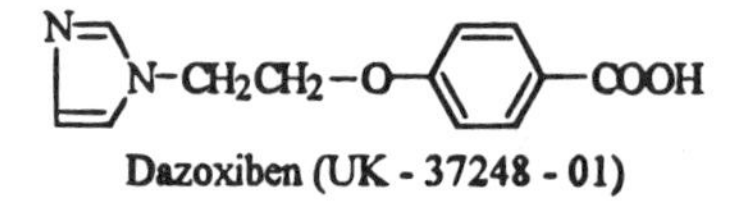

Dazoxiben (UK - 37248 - 01)

Imidazole has a selective but low-level inhibiting effect on TXA_2 synthetase. Structural modifications increased the inhibitory potency dramatically. Dazoxiben [4-(2-imidazolylethoxy)benzoic acid], was a promising derivative. Doses of 100–200 mg produced a 90% enzyme inhibition after 1 hour. Transient, small increases in bleeding time occurred that related to TXA_2 levels. A subsequent study indicated a high human platelet TXA_2 synthetase inhibition ($IC_{50} = 3 \times 10^{-9}$ M), while PGI_2 synthetase was 100,000 times less (IC_{50} > 10^{-4} M), a dramatic difference.

A more recent clinical evaluation found that even this potent inhibition still permitted AA-induced aggregation, which could be prevented by coadministration of small aspirin doses. This apparently prevented cyclooxygenase product accumulation capable of platelet aggregation in spite of TXA_2 absence. Dazoxiben may become the forerunner of this new type of antithrombotic.

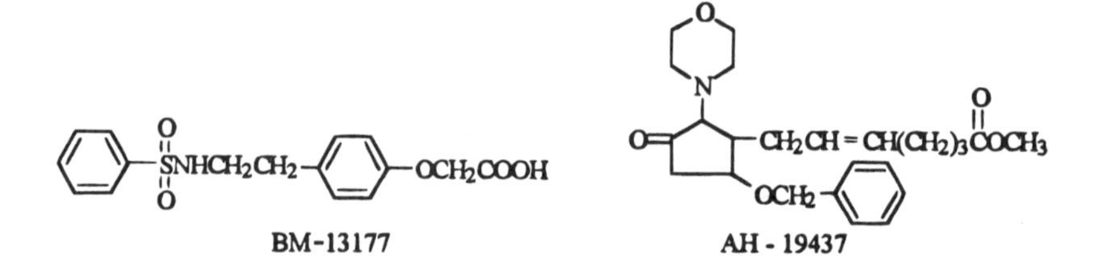

Thromboxane A_2 antagonists whose antiaggregant properties of platelets depends on blocking TXA_2 receptors are also being discovered, so far without any structural relationships. Two of many such compounds are cited as examples. The sulfonamide phenoxyacetic acid BM-13177 has antiplatelet effects not mediated by AA release, cyclooxygenase inhibition, or blockade of TXA_2 synthesis. That left TXA_2 receptor antagonism as the only viable mechanism. A very different nonacidic but lipophilic AH-19437 has also been shown to have TXA_2 antagonist activity. One difficulty is that these compounds are competitive antagonists. Since it is at points of arterial plaque or otherwise damaged sites that TXA_2 is copiously produced by adhering and aggregating platelets, the activity of the antagonist drugs may be overwhelmed. Research on thromboxane synthetase inhibitors (TXSIs) is accelerating.

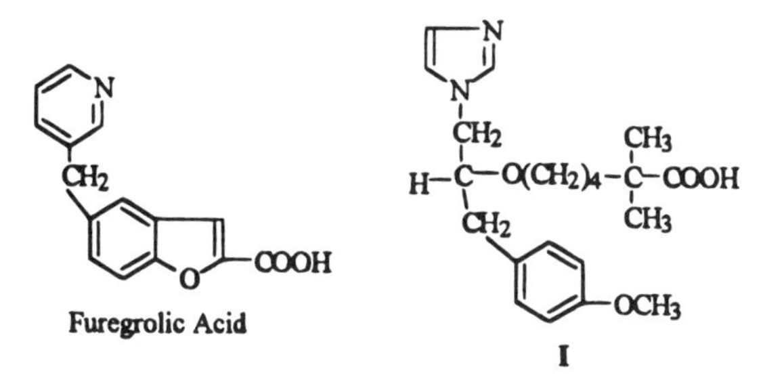

The dazoxiben analog furegrolic acid was reported to have clinical potential, especially when tested in animal models together with CCBs or TXA_2 antagonists in induced angina and coronary thrombosis. More recently, a series of complex imidazolylalkanoic acids that can also be viewed as having "evolved" from dazoxiben have been reported as potential TXSIs. The most potent, the preceding hexanoic acid (I), has an enzyme-inhibitor complex dissociation, $K_i = 9.6 \times 10^{-8}$; the S enantiomer is 12 times as effective as the R isomer. The drug is thought to inhibit the enzyme at the 4 PGH_2 binding sites in the active center postulated to contain an iron-heme site to which the endoperoxide bridge of PGH_2 would normally bind. Since imidazole (and pyridine) nitrogens are known to coordinate with heme Fe atoms also, and both PGH_2 and the drugs considered here have a carboxylate anion 10 Å distant that can be assumed to bind to a secondary ionic site in proper conformation, a viable explanation for TXSI activity is at hand (Fig. 11-2).

Interestingly, the ethyl ester of the preceding imidazole drug is 500 times less potent when tested in intact human platelets in vitro. This phenomenon, not seen with esters in

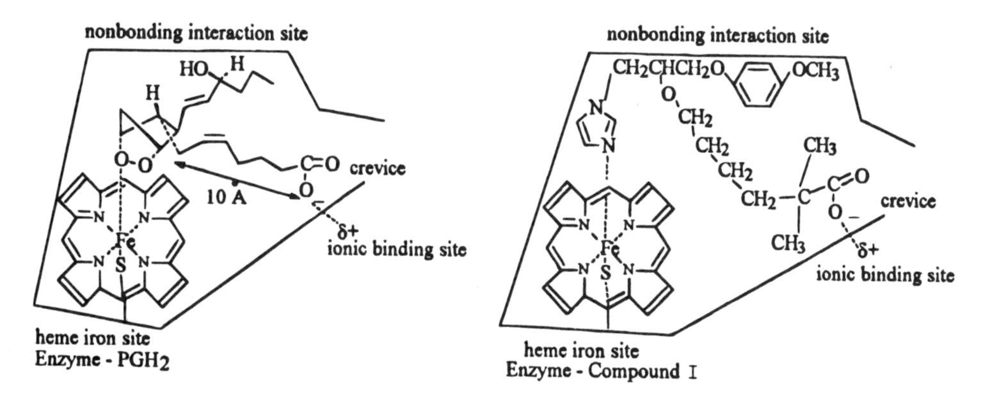

Figure 11-2. Concept of TXA_2 synthesis inhibitor and PGH_2 interactions (adapted from Kato et al., 1985).

this series that do not carry geminal methyls on the carbon α to the COOH group, illustrates two points. One is the effectiveness of steric hindrance in inhibiting platelet esterase, which easily hydrolyzes the sterically unhindered esters. Another is that it established the carboxylate as a necessary and active pharmacophore, presumably for the second ionic bonding interaction (Fig. 11-2).

Pentanoic or heptanoic acid analogs were both inferior. The optimum alkyl spacer between the imidazole and carboxyl pharmacophores in this limited series is the equivalent of six methylenes. Selective inhibition of the TXA_2 synthase was established by demonstrating a 16-fold increase of PGE_2, which is presumed to be due to an expanded pool of PGH_2. Increased levels of 6-keto PGF_1, the inactive metabolite of PGI_2, was also shown. This, too, is explained in terms of higher PGH_2 levels that arise by not being diverted toward TXA_2 isomerization. It also demonstrates that the drug does not inhibit PGI_2 synthetase.

11.4. Prostacyclins

Prostacyclin (PGI_2) was found to be the strongest inhibitor of platelet aggregation and a smooth muscle relaxant even before its exact structure was known. In fact, it was briefly named PGX. Much has been learned about this fascinating molecule, which is now classified as a hormone, since then. It is continuously produced both in the walls of blood vessels as well as in kidneys and lung tissue. Tests of this property have demonstrated that infusion of PGI_2 produces a potent antithrombotic effect following electrically stimulated thrombus formation in rabbits. The compound, which is now commercially available, is named *epoprostenol* (Cyclo-Prostin). Intravenous solutions of the sodium salt (from a synthetic freeze-dried powder) has a very short half-life (ca. 3 min.); therefore, it must be continuously infused. Nevertheless, the drug is being successfully used as an antithrombotic (in lieu of heparin) during dialysis in renal disease and other extracorporeal circuits (heart–lung machine). The drug has been effectively used in advanced arteriosclerosis obliterans of the lower extremities. Pain ceased after three days of drug infusion; ischemic ulcers healed. In peripheral vascular disease (Raynaud's), intermittent infusion of epoprostenol produced long-term symptomatic improvement.

The in vivo metabolic instability of the 15-OH group against dehydrogenase-catalyzed oxidation to the inactive 15-keto derivative must be added to the deficiency of rapid hydrolysis in solution both in vitro and in vivo. This led to the design and development of prostacyclin analogs to overcome these shortcomings. Table 11-3 lists eight selected examples that have been comparatively evaluated against PGI_2. None is yet marketed in the United States, but they were chosen to illustrate drug design ideas used in their conception.

In carbacyclin (Table 11-3, No. 2) the 6,9 oxygen bridge of PGI_2 is replaced with a carbon atom. A comparative study found it to be more stable, but only one-fifteenth as active. Its overall spectrum of activity is similar to PGI_2. The addition of a 9β-CH_3 (ciprostene, No. 3) offered increased hydrolytic stability as well, but the activity was only 4% that of prostacycline's. Neither of these two molecular changes protects the 15-OH from oxidations. Of course, that would require "protective" substituents on, or near, C-15. Converting the last 5 carbons into a cyclopentane ring (Compound No. 4) achieves sufficient stability and oxidative resistance to add oral activity. The presumed reason is a degree of steric effects around C-16. The drug's antiplatelet activity is only one-tenth that of PGI_2 in vitro but it is considerably more potent than carbacyclin. Thiaprostacycline, where a sulfur atom

Table 11-3. Prostacyclin and Designed Analogs

Structure	Name[a]	Uses[b]	Comments
1. 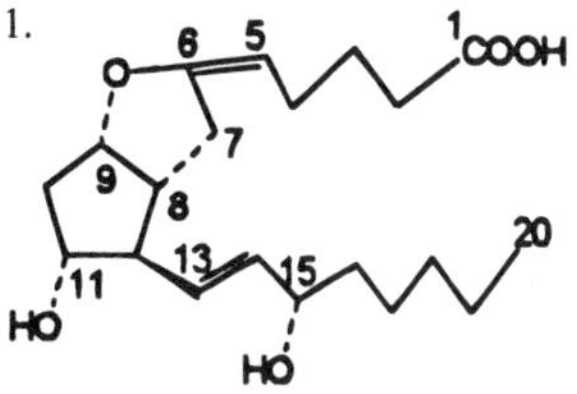	Prostacyclin Na[c] PGI_2 Na Epoprostenol Na (Cyclo-Prostin)	Cardiopulmonary bypass surgery; perfusion for kidney patients; peripheral vascular disease; preinfarction angina stroke	Short $t_{1/2}$ in plasma and in vitro; 2–3 min at 37°C
2. 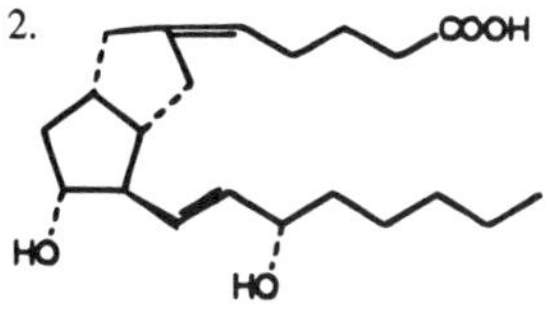	Carbacyclin[d] 6-carba-PGI_2	Inhibits platelet adhesion; increased platelet *c*AMP; being evaluated for clinical applications mentioned for PGI_2 but less activity	ca. 3 times longer $t_{1/2}$ than PGI_2
3. 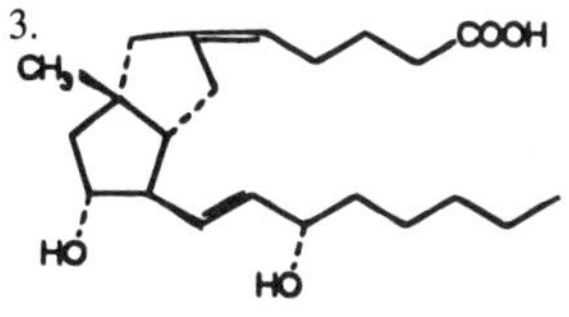	β-Methylcarbacyclin[d] (Ciprostene)	See carbacyclin	Chemically stable
4. 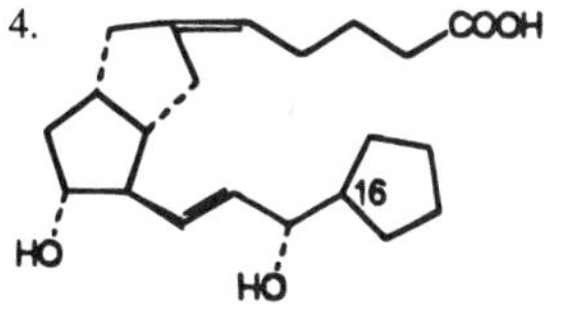	Ono-41483[d]	See carbacyclin; more stable than PGI_2; antiplatelet activity in vitro 10% of PGI_2, but orally active due to stability	Chemically stable
5. 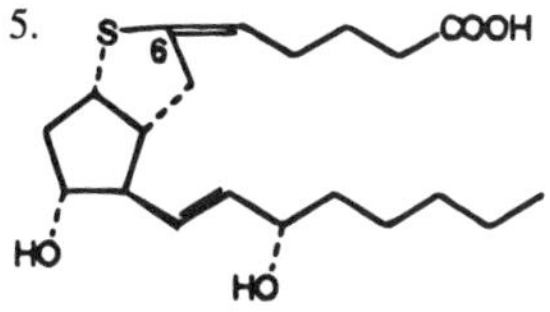	6,9-Thiaprostacyclin[d] PGI_2-S	Platelet antiaggregant potency 50% of PGI_2	Causes coronary vasoconstriction, not dilation; more stable than PGI_2
6. 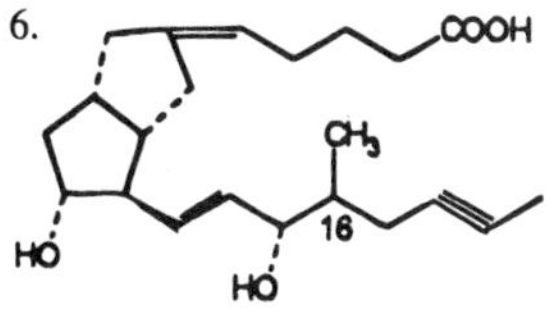	16-Methyl-18,-18,19,19-tetradehydrocarbacyclin (Ciloprost)	See carbacyclin; activity close to PGI_2	More stable than PGI_2
7. 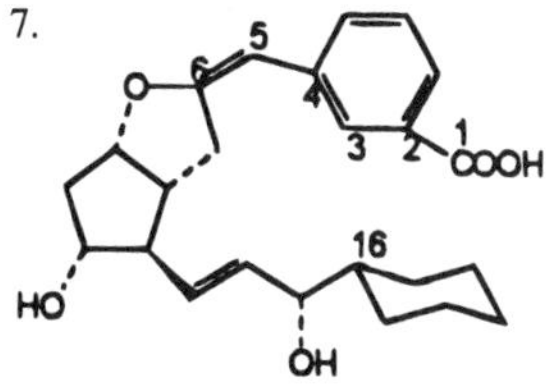	CG-4203[d]	Antiplatelet activity 1/5 of PGI_2; COOH isomers *ortho* and *para* not effective	

(Continued)

Table 11-3. *(Continued)*

Structure	Name[a]	Uses[b]	Comments
8. 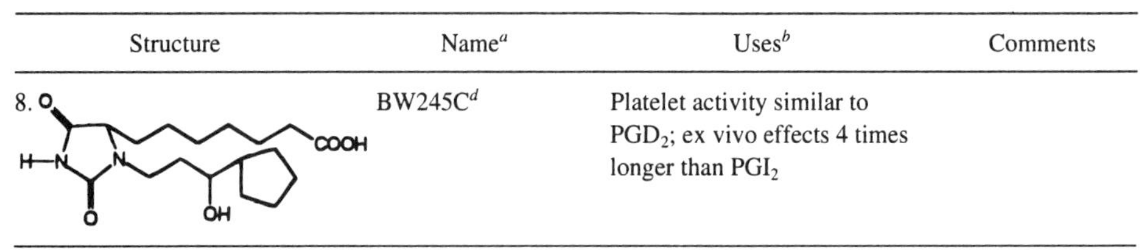	BW245C[d]	Platelet activity similar to PGD_2; ex vivo effects 4 times longer than PGI_2	

[a] Trade name or proposed name in parentheses.
[b] Actual or potential.
[c] Marketed.
[d] Experimental.

replaces the oxygen at C-6, produces a much more stable prostacyclin that retains about half of the latter's activity. It is curious that it constricts, rather than dilates, coronary arteries, which may negate clinical usefulness.

Ciloprost (No. 6) is a multiple-point modification of prostacyclin. Replacement of the oxygen with carbon (as in No. 2) stabilizes the rapid hydrolysis; the 18–19 triple bond's effect is not clear. However, the overall result is a hydrolysis- and oxidation-stable compound whose antiaggregating properties are close to PGI_2. It shows clinical promise.

Compound 7 (CG-4203) shows some interesting structural innovations. It simulates the C_1–C_4 prostacyclin portion with a carboxyphenylene moiety. Since carbons 2, 3, and 4 of PGI_2 have been replaced with those of the phenyl ring, degradation in vivo by β-oxidation (to which aliphatic carboxylic acids are prone) cannot occur. This increases the molecule's stability. The C-16 carbon, which is a component of a chair-form cyclohexane ring, helps to protect the 15-OH from 15-hydroxyprostaglandin dehydrogenase (as would other substituents on or near C-15; e.g., ciloprost, or a CH_3 sharing C-15 with the OH).

The 10 Å separation between the COOH and the oxa ring, or the analogous area of PGH_2 discussed previously with regard to Table 11-3, is very crucial here also. In vitro tests of AA-induced platelet aggregation (IC_{50} values)[4] showed CG4203 to be 27,000 times more potent than either the *ortho*- or *para*-carboxyl position isomers. Of course, these would constitute a one-carbon shortened or lengthened distance from the COOH to the oxa ring, respectively. An ionic receptor bonding to an electron-poor site on the receptor presumably would not "reach" well. The drug has been successfully used experimentally in dialysis patients.

Compound 8 (BW-245c) is an imidazole analog with potent antiplatelet activity. The drug binds selectively to PGD_2, whose cardiovascular effects it mimics. Its activity may be mediated via PGD_2 receptors in or on platelets.

The pyrazolinone derivative Bay g6575 may be the first example of a drug stimulating PGI_2 release in vivo from wall vessels. The compound does *not* affect coagulation, fibrinolysis, or platelet aggregation in vitro, but it appears to be antiaggregant in vivo in rabbits and humans. Its antithrombotic effect appears to be accompanied by some thrombolytic properties as well. Its potency is much higher than is that of aspirin.

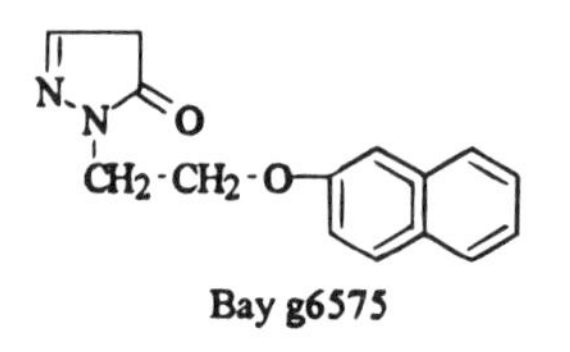

Bay g6575

[4] Concentration in μmoles to achieve 50% inhibition.

Bay g6575 is potent and has extremely low toxicity—the LD_{50} in rats could not be determined, but it was estimated >10,000 mg/kg p.o. It lacks analgesic, antiinflammatory, autonomic, or cardiovascular effects other than antithrombosis.

11.5. Thrombolytics

Thrombolytics are agents capable of lysing thrombi in the cardiovascular system. As was seen earlier, clot formation is governed by the coagulation scheme. Understanding of that scheme has led to agents able, to a degree, to inhibit clot formation prophylactically. Anticoagulants and platelet aggregation inhibitors considered in this chapter are prime examples. Lysis of clots is regulated by another complex series of interrelated biochemical reactions referred to as *the fibrinolytic system* (Fig. 11-3). The reason that exaggerated systemic thrombotic events, or hemorrhage, is not a continuous occurrence is due to a fine balancing act between the two strategems. Any imbalance resulting in one or the other gaining dominance can result in a pathologic situation such as hemophilia or an MI.

The key enzyme in humans is plasmin, which breaks the fibrin component of a clot (see Fig. 11-1 and related discussion) down into fibrin degradation products (FDPs) by proteolytic action. This action is not fibrin specific. When in circulation, plasmin will also hydrolyze its precursor fibrinogen as well as clotting factors such as V, VIII, and XII. There are circulating plasmin inhibitors such as α_2-antiplasmin and α_2-macroglobulin to prevent or minimize this reaction (i.e., to balance the system). These plasmin inhibitors complex with plasmin in the circulation, very rapidly inactivating it. The $t_{1/2}$ of free plasmin in plasma is probably less than 0.1 seconds. The presence of these natural inhibitors in plasma also prevents the use of plasmin *directly*. Therefore, the physiologic process of plasminogen activation to plasmin is utilized clinically by accelerating it via the expedient of increasing the level of enzymes capable of catalyzing the reaction.

Activation of plasminogen to plasmin produces several events. The primary event is the dissolution breakup of any fibrin-containing clots encountered, including those found obstructing blood vessels. However, clots formed to prevent bleeding caused by injuries, such as cuts (surgical), puncture wounds, or pathological sites such as bleeding ulcers, would also be degraded. This constitutes an undesirable interference with the integrity of the vascular system. Excessive circulating plasmin is continuously destroyed by the plasmin inhibitors to control or to minimize such effects. As mentioned, circulating plasmin continuously breaks down fibrinogen and other clotting factors. Another factor to be considered is that the plasmin-produced FDPs prevent the production of more fibrin by inhibiting the conversion of fibrinogen to fibrin by thrombin (Fig. 11-1). This will reduce available circulating fibrinogen, thereby producing an anticoagulant effect that FDPs will continue to have even after plasminogen activators cease to function. In addition, anticoagulation will continue as the available pool of circulating plasmin inhibitors becomes

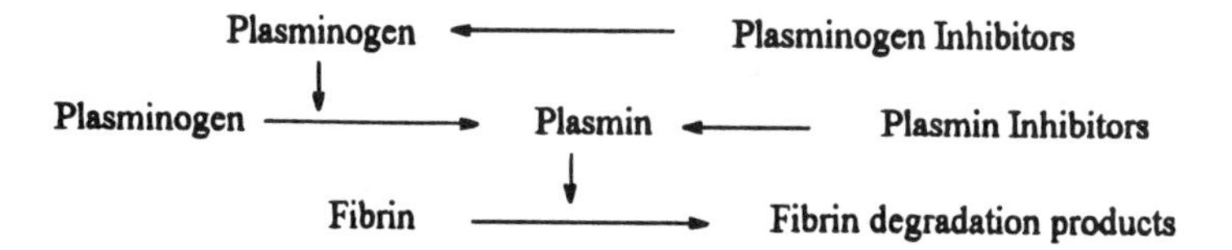

Figure 11-3. The fibrinolytic system. (See also Fig. 11-1.)

exhausted. Such a rise in the systemic lytic state during thrombolytic therapy increases the risk of complications due to bleeding. Though several highly effective plasminogen activator enzymes are already available for clinical use (see later and Table 11-4), research continues to develop new or modified agents.

Human plasminogen is a single-chain glycoprotein containing 790 amino acid residues. There are at least two forms: glu-plasminogen and lys-plasminogen. The latter form is converted to plasmin by cleavage of the Arg560–Val561 bond. Glu-plasminogen is also convertible to plasmin in a much slower, possibly two-step process.

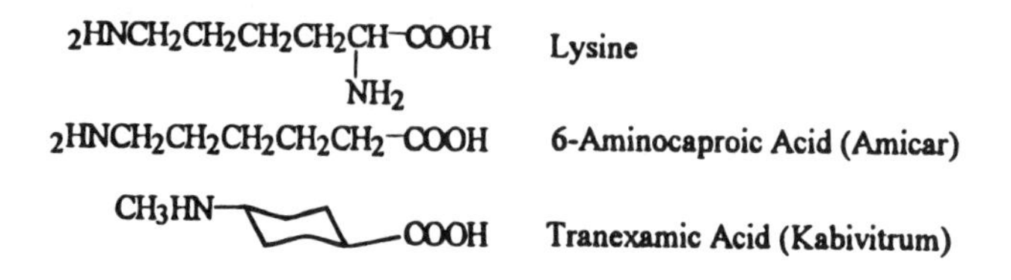

Plasminogen contains lysine-binding sites. These are structural components on the enzyme that interact specifically with lysine and certain other synthetic amino acids whose amino and carboxyl functions are about equally separated. The synthetic lysine analogs aminocaproic and tranexamic acid [*trans*-(4-aminomethyl)cyclohexane carboxylic acid] are inhibitors of fibrinolysis. By interacting with the lysine binding sites, these acids act as competitive inhibitors of plasminogen activators and, less effectively, plasmin. Because the effect of inhibiting plasminogen activation is decreased plasmin, fibrinogen levels increase, permitting the use of these specific antidotes for plasminogen activator overdoses. They are also useful as hemostatic adjuncts in hemophiliacs, in postsurgical gastrointestinal bleeding, following prostate removal, and after other surgery. Hemorrhaging due to platelet deficiency and coagulation defects are not controlled.

Table 11-4. Thrombolytic Agents[a]

Product[b]	Advantages[c]	Disadvantages[d]
Streptokinase (Streptase)	Low cost, effective, decade of clinical experience	Antigenic, allergic reaction, not very clot selective
Urokinase (Abbokinase)	Nonantigenic,[e] effective, some clot selectivity	High cost
Acylated (or anisoylated) plas minogen streptokinase activator complex APSAC (Eminase)	Effective, prolonged action, modest cost	Antigenic, allergic reactions
Recombinant tissue plasminogen activator, rt-PA, t-PA, alteplase (Activase)	Effective, nonantigenic; highly clot selective	Short $t_{1/2}$ (4–6 min)[f]; may require long infusion and heparin coadministration, very costly
Recombinant Pro-urokinase	Effective, highly clot selective, nonantigenic	Short $t_{1/2}$; heparin coadministration; very costly

[a] Commercial or experimental [b] Trade names in parentheses. [c] Rapid perfusion (1 h) is feature of all five drugs. [d] Increased bleeding risks, particularly with invasive procedures, is common to all thrombolytics; so are hemostatic defects in clotting and platelet activity. [e] Since obtained from human urine or human kidney tissue cultures. [f] In humans; in rabbits only 2–3 minutes.

11.6. Plasminogen Activators

Plasminogen activators (PAs) are serine proteases with high plasminogen specificity. Hydrolysis of the proenzyme yields an active two-chain serine protease—plasmin. Plasminogen can also be physiologically activated by the *intrinsic* system (see earlier), specifically by prekallikrein, factor XII, and a high-molecular-weight kininogen found in the plasma. More relevant to clinical applications, however, are the enzymes of the *extrinsic* system: the serine proteases to be considered. These enzymes, which are present in various tissues including the vascular endothelium, are released from the cells holding them in response to thrombus development. They bind to *fibrin* following their release. Plasminogen will then bind to this fibrin–activator complex, where, following the proper peptide bond cleavage, it becomes plasmin. It should be noted that while fibrin is complexed with plasmin (through its lysine-binding sites), it is protected from the inactivator molecules, which require the now-occupied sites for their adsorption. This, then, permits the fibrin-bound plasmin to lyse the fibrin-containing clot.

The clinical applications of PAs are in life-threatening situations resulting from clot formations. It is now known that reconstituting coronary blood flow and reperfusing the infarcted area of the heart as soon after an MI as possible is essential to prevent or, at least, to minimize myocardial damage to the cells involved. It has been demonstrated that initially reversible damage following a myocardial ischemia becomes rapidly irreversible if the flow of blood is not restored. The acute damage caused by total coronary occlusion will become necrotic tissue within 3–6 hours. It is possible to salvage not-yet-necrotized heart tissue only if reperfusion can be accomplished as early as possible within this time frame.

It should be appreciated that PA restoration of an artery's patency (time vessel remains open) is an acute emergency procedure having no effect on the underlying pathology. In addition, disintegration of a clot may still not restore full blood flow due to stenosis, ultimately risking a reocclusion. The effectiveness of PAs dissipate rapidly once the drug is discontinued (e.g., 1–2 hours). As a practical matter, to lower the risk of reinfarction, anticoagulants are instituted. This may be a week or more of heparin followed by several months of a hydroxycoumarin agent or an antithrombotic drug.

There are additional indications for PAs. These may include pulmonary embolism, deep vein thrombosis, acute thromboembolism of peripheral arteries, thromboses precipitated by prosthetic heart valves, and other occlusions. Some of the applications are still investigational.

The two main routes for PA administration are intravenous and intracoronary. The former is simpler; however, the latter may minimize circulatory effects and offer faster results.

Streptokinase (SK), in spite of its name (-ase suffix), is a nonenzyme protein (mol. wt. 47,000). It is a catabolic product of group C β-hemolytic streptococci. Unlike the real enzymes to be considered, it activates the fibrinolytic system indirectly. More recent research has shown that necessary steps require a prior formation of a 1:1 complex with plasminogen. This alters its conformation, thereby exposing an active site in the modified proenzyme. It is this modified complex that becomes the actual PA. SK is still a widely used thrombolytic agent, primarily because it is the least costly, is easy to obtain and had been in use for years before other PAs became available.

Shortcomings of SK can include antigenic and, rarely, anaphylactoid reactions. This is because it is a foreign protein. For example, fever can occur in one-third of patients. Resistance to the effects of SK can occur. The usual reason is a high titer of antibodies to this protein developed by the patient either by a previous streptococcal infection or through previous therapy with this drug. Another reason can be that continuous infusion of the drug has dramatically lowered plasminogen levels for subsequent doses to activate. Cerebral bleeding is a possible, but rare (<0.3%), occurrence.

Anisoylated plasminogen streptokinase activator complex (APSAC)(Table 11-4) is an anisoyl (*p*-methoxybenzoyl) derivative of the active (lysine) site of the plasminogen component of this complex. Acylation inactivates the enzyme but does not decrease the affinity of the complex for fibrin. The resultant slow deacylation ($t_{1/2}$ = 40 min) should achieve relative selectivity for the fibrin in the clot over circulating plasma fibrinogen. Hydrolysis of anisoyl amide following binding onto the thrombus (i.e., activation) leads to fibrinolysis. APSAC has been reported to produce a 60–80% reperfusion rate following IV administration.

Urokinase plasminogen activator (U-PA) activates plasminogen directly. It is a serine-type protease that is isolated from human urine or that can be obtained from cultured human embryonic kidney tissue. The main molecular form, S_2, consists of two polypeptide chains (20,000 and 34,000 daltons) connected by a single disulfide bridge. A lower-molecular-weight (31,600) form, S_1, which represents a proteolytic degradation product, has also been identified U-PA, which is a human protein, is nonantigenic and, unlike SK, does not cause allergic reactions. This makes it a useful alternative in SK-allergic or -resistant patients.

Since the human urokinase gene has now been cloned and expressed in *E. coli*, the pro-urokinase enzyme has been produced by recombinant DNA technology. Here, too, the proenzyme binds to fibrin, where, on hydrolytic action, it may exhibit good clot selectivity. The drug is investigational.

Recombinant pro-urokinase, a single-chain urokinase (also named kidney-plasminogen activator), is another, still investigational, thrombolytic drug produced by recombinant DNA procedures. The single-chain enzyme has also recently been identified in urine and other tissues. It is a glycoprotein containing 411 amino acid residues, 24 of them cysteine. Prourokinase is apparently reformed to the two-unit urokinase (see earlier) by plasmin present around the clot, which then activates plasminogen to generate plasmin at the thrombus. Like urokinase, this version does not produce circulating antibodies or allergic reactions, nor is it antigenic. Animal studies indicate considerable clot specificity comparative to urokinase.

Recombinant tissue plasminogen activator (rt-PA) is possibly the most exciting product in this area. The native protease enzyme consists of 527 amino acid residues in a single chain. As with single-chain urokinase limited plasmin activation converts the enzyme to a two-chain activator molecule joined via a single disulfide linkage. When cleaved, the two chains afford a heavy (31,000 daltons) chain derived from the amino-terminal of original molecule, and a carboxyl-terminal-derived light chain (28,000 daltons). The catalytic site resides in the light chain component of t-PA. It contains His at position 322, Asp at 371, and Ser at 478. The amino acids around the active site area are very homologous to corresponding areas of other serine proteases such as trypsin, thrombin, plasmin, and urokinase. Both the one- and two-chain version of t-PA have been shown to have almost equal clot-lysing and plasminogen-activating effectiveness.

The suspicion that the t-PA found in blood arises from the endothelial cells of blood vessels seems to have been verified in endothelial cell culture experiments.

t-PA is a serine protease that can bind to fibrin while fibrin clots are forming. In the absence of fibrin the enzyme shows poor activity. Fibrin, however, enhances the plasminogen activation dramatically. The suggestion that fibrin-bound t-PA increases affinity for plasminogen may explain this. The functional areas involved in binding to fibrin and catalytic action of t-PA have been located, but not fully identified.

Recombinant t-PA was first clinically evaluated by IV and intracoronary infusion in 1984, resulting in reperfusion of occluded arteries in 83% of patients with little systemic lysis. Plasma $t_{1/2}$ was 8–9 minutes. This is insufficient time to produce complete thrombus lysis following a single large IV dose (a bolus). Thus the drug is infused over varying periods (45–60 minutes). More extensive trials have been undertaken by the U.S. National Institutes of Health since then with the Thrombolysis in Myocardial Infarction trials (TIMI, 1985). A randomized comparative trial of rt-PA versus SK found a higher patency rate for rt-PA (70% to 55%) as well as less systemic fibrinolysis. The comparative results of TIMI trials were so striking (66% reperfusion with rt-PA vs. 36% SK) that a preliminary report was published before the trials were completed. More recent clinical experiences, however, seem to indicate a lesser advantage over SK.

11.7. Hypolipidemic–Hypocholesterolemic Drugs

The fact that even young men can develop atherosclerosis was reported following autopsies of German soldiers during World War I. However, it was a U.S. Armed Forces Institute of Pathology report toward the end of the Korean War (Enos et al., 1953) showing that young men (average age 22) had evidence of coronary heart disease (CHD) that made the medical profession take note of the epidemic that was CHD. The problem is still extensive in spite of considerable, but slow, progress since the 1950s. By the early 1980s there were 5.4 million Americans with symptomatic diagnosed CHD. The number of undiagnosed cases is, of course, unknown. In any case, the results were 1.5 million heart attacks and over 550,000 deaths. A review of the medical literature since the late 1950s shows the name *cholesterol*, and particularly *hypercholesterolemia* (HPC), to be associated with CHD with increasing frequency. By the early 1970s HPC was implicated as the primary CHD risk factor. Hypertension and smoking are now also recognized as important risk factors.

More recently the term *hyperlipoproteinemia* has become useful. Saturated fatty acids, usually esterified to glycerol as triacylglycerides (i.e., fats), can be related to cardiovascular disease since they increase plasma cholesterol levels. This cholesterol is now known to be primarily associated with a low-density lipoprotein (LDL) fraction. Lipoproteins are complex particles consisting of proteins, triacylglycerol (fat), phospholipids, cholesterol, and cholesterol esters. LDL has density of 1.00–1.06 and contains at least 45% cholesterol. It is the cholesterol found in the LDL fraction that is the harbinger of human arterial plaque manifested as atheroma. An inverse relationship has been established between cholesterol in high-density lipoprotein (HDL) ($d = 1.20$, 18% cholesterol) and CHD. HDL transports cholesterol from accumulation in arterial walls to the liver for biodegradation.

In the popular literature HDL is frequently referred to as "good cholesterol" and LDL as "bad cholesterol." Blood chemistry reports often state total cholesterol levels, even though LDL and HDL values (or their ratios) would probably be more relevant to a patient's clinical evaluation. Total serum cholesterol levels, normally reported in milligrams/deciliter (mg%) do not indicate the fraction that is free, esterified, or bound up in the various lipoprotein fractions.

Developments in genetic research, when combined with population studies, have made it clear that low levels of HDL actually increase the risk of cardiovascular disease. At the same time, a high level of cholesterol in the form of LDL is the primary etiological factor

Table 11-5. Lipoprotein Abnormalities

	Type	Lipoprotein Abnormalities[a]	Other Abnormalities
I	Familial lipoprotein lipase deficiency	Increased chylomicrons; decreased HDL levels	Absence of certain apoproteins
IIa	Familial hypercholesterol emia (FH)	Greatly increased LDL levels	Defects in LDL receptors
IIb	Familial combined hyperlip idemia	Increased LDL and VLDL levels	Receptor defects, increased cholesterol levels
III	Broad beta disease	Increased β-VLDL levels	Imperfect clearance of lipoprotein remnants
IV	Familial hypertriglyc eridemia	Increased VLDL levels	Increased VLDL secretion
V	Familial hypertriglyc eridemia	Increased chylomicrons and VLDL	Defects in remnant clearance and increased VLDL secretion

[a] HDL = high-density lipoproteins; LDL = low-density lipoproteins; VLDL = very-low-density lipoproteins.

of atherosclerotic diseases, particularly CHD. Of course, there are other risk factors as well (e.g., hypertension and smoking). The medical consensus is now focused on the necessity for reductions in total fat, saturated fats, and cholesterol in the blood. There are still some disagreements as to the numeric goals. Cholesterol levels as high as 240 mg/dL, considered the upper "normal" only several years ago, are now viewed as too high. Values of 200 mg or less are now believed more desirable.

The genetic research of Goldstein and Brown[5] (1983) that led to the discovery of cell surface LDL receptors and their mutations elucidated the biochemical basis of familial hypercholesterolemia (FH) types IIa and IIb (Table 11-5) and greatly increased the understanding of the controlling mechanisms of plasma cholesterol levels.

Lipoprotein metabolism will be outlined. Cholesterol is transported in plasma as a component of the lipid–protein complexes synthesized in the gut and liver. Dietary triglycerides (fats) are digested in the gut to monoglycerides and fatty acids that are resynthesized into triglycerides and integrated into chylomicrons following intestinal absorption. Once they have entered the circulation, specific lipase enzymes hydrolyze the chylomicrons into glycerol, free fatty acids (FFA), and remnant particles. The liver synthesizes a very-low-density lipoprotein (VLDL), which has a high triglyceride content and about 15% cholesterol. The fats of VLDL also undergo lipolysis, the remnants reentering the circulation. Their destiny is to either be removed by the liver or converted to LDL—the major cholesterol-containing lipoprotein. LDL also has one of two fates. About three-fourths is removed by the recently discovered LDL-specific cell surface receptors, primarily in the liver, although a fraction may be cleared by extrahepatic tissues. The remaining quarter is thought to be eliminated by nonreceptor methods.

The HDLs, which obtain cholesterol primarily from nonhepatic tissues, then transfer it to VLDLs and LDLs, from which it is hepatically eliminated. The degree to which the various lipoproteins are responsible for the buildup of arterial athera varies, but the major potential to produce atherosclerotic plaque is by LDL. These small cholesterol-rich (45%) particles readily penetrate the arterial wall and have the highest propensity to form the thick cholesterol and cholesterol ester deposits on the inner arterial surfaces. This precipi-

[5] Nobel laureates, 1985.

tation and sequestration is facilitated by apo-β proteins present in LDL, but not in HDL. Our concern with hypercholesterolemia (hyperlipidemia), then, should be properly focused on LDL levels.

11.7.1 Hypercholesterolemia

Severe HPC is most often encountered as FH (Type IIa and IIb, Table 11-5). This is a hereditary disease characterized by an LDL receptor absence, or defect, specifically caused by an abnormality or mutation of the gene coding for the receptors (Goldstein and Brown, 1983). The majority of affected individuals are *heterozygous* (i.e., they have inherited only half the number of properly functioning LDL receptors). The others may be either absent or defective in function and thus unable to bind and transfer the LDL complex into the cell. The result is severely elevated LDL levels, aggravated by low clearance rates. Clinical symptoms can occur by the late thirties. With only half the normal receptor population cholesterol levels are at least double (300–500 mg/dL). Treatment should theoretically include methods to stimulate LDL receptor synthesis by chemically or physically interdicting enterohepatic reabsorption of bile acids. This will increase diversion of cholesterol to the synthesis of new bile acids. This has been partially achieved with the polymeric resins (Table 11-6) that sequester bile acids in the gut. The resultant decrease in hepatic cholesterol levels has been found to stimulate LDL receptor synthesis. Preventing absorption of dietary cholesterol from the gut can be achieved with oral neomycin (Chapter 6) to a degree.[6] Finally, blockade of cholesterol biosynthesis should also reduce its serum levels and lead to increased LDL receptor production. In the late 1950s, the diphenylcarbinyl derivative triparanol (MER/29) was available. It specifically inhibited the conversion of desmosterol to cholesterol (Figs. 11-4 and 7-13).[7] Lovastatin (see later) also blocks cholesterol synthesis, but at an earlier stage (Fig. 11-4). It appears more promising.

An even more serious form of FH is the *homozygous* type. Here, no functional LDL receptors are produced. Plasma cholesterol levels can reach over 1,000 mg/dL. Clinical disease arises in the early teens, frequently followed by fatal MIs by age 20 (Goldstein and Brown, 1983). Since cholesterol-lowering drugs to date owe their effectiveness to stimulation of LDL receptors, they have not been very useful here. More drastic but hazardous methods of treatment are being attempted. These may include plasmapheresis of LDL with LDL antibodies and liver transplant that restored LDL receptors to about half normal levels, thereby permitting follow-up drug treatment with lovastatin to maintain normal cholesterol levels.

FH represents only about 2% of HPCs. All others, although likely to involve some genetic factors, have a major dietary component. It is now understood that any increased liver cholesterol levels will decrease LDL receptor synthesis by a biofeedback mechanism. Therefore, any reason for elevated cholesterol levels such as increased absorption or decreased bile acid synthesis can lead to primary HPC not of the familial type. In addition, reduced receptor activity may complicate the picture, either because there are fewer of them or they function less efficiently (e.g., lowered LDL affinity). Drug therapy is potentially useful in such patients.

There seems to be a consensus that serum cholesterol levels above 200–240 mg/dL should be lowered. Mild elevations should be managed by diet. Even modest reductions of

---

[6] See also charcoal later.

[7] The drug was withdrawn due to serious toxicities.

10% reduced coronary risk by 15%. At levels of 300 mg/dL or more, where the cholesterol consists primarily of LDL, diet alone is likely to be inadequate; drugs become necessary.

There are at least 10 pro-drugs and analogs of clofibric acid (*p*-chlorophenoxyisobutyric acid) on the world market; two are available in the United States. The ethyl ester

Table 11-6. Lipid-Lowering Drugs

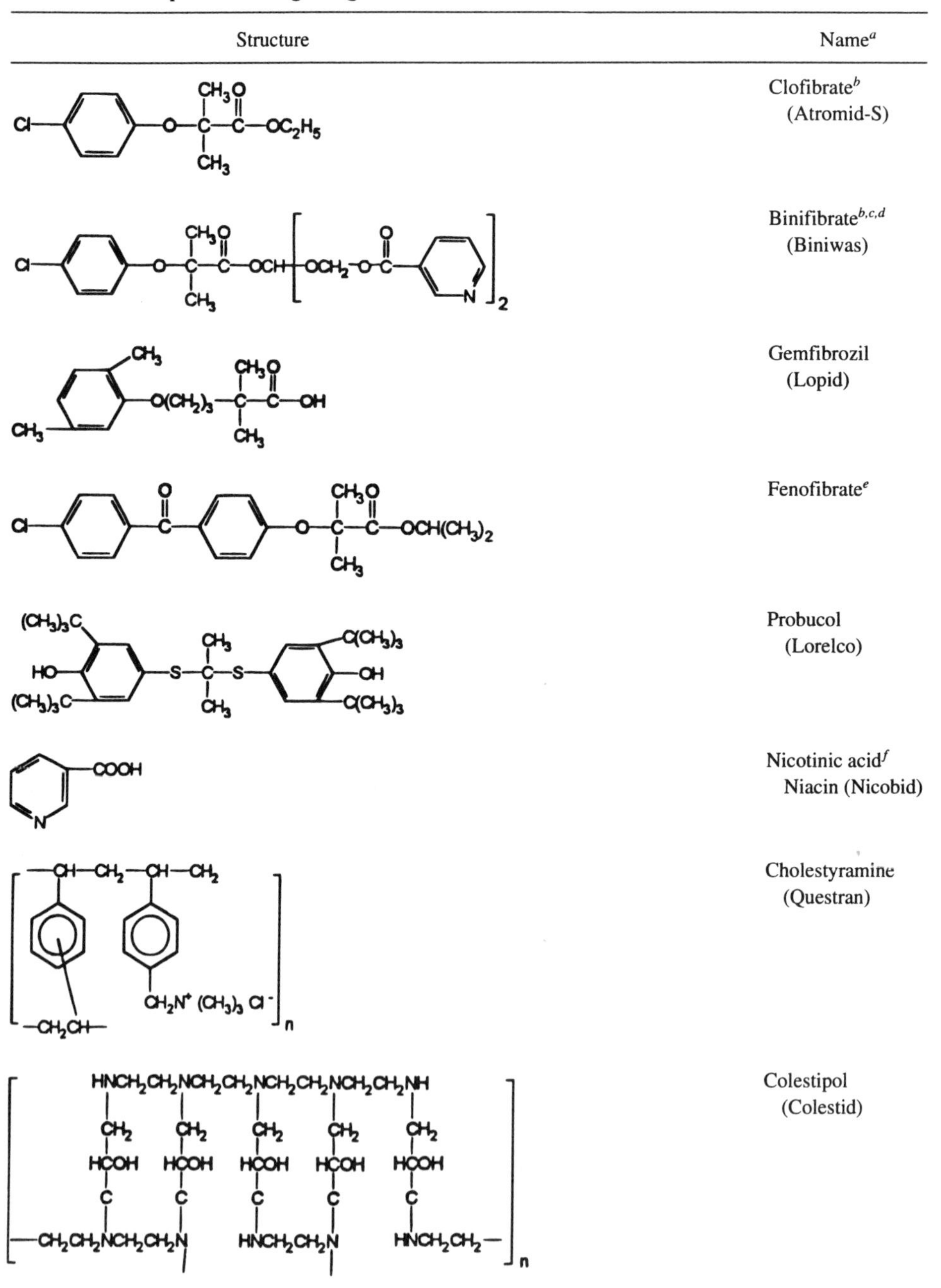

Structure	Name[a]
	Clofibrate[b] (Atromid-S)
	Binifibrate[b,c,d] (Biniwas)
	Gemfibrozil (Lopid)
	Fenofibrate[e]
	Probucol (Lorelco)
	Nicotinic acid[f] Niacin (Nicobid)
	Cholestyramine (Questran)
	Colestipol (Colestid)

(Continued)

Table 11-6. (*Continued*)

Structure	Name[a]
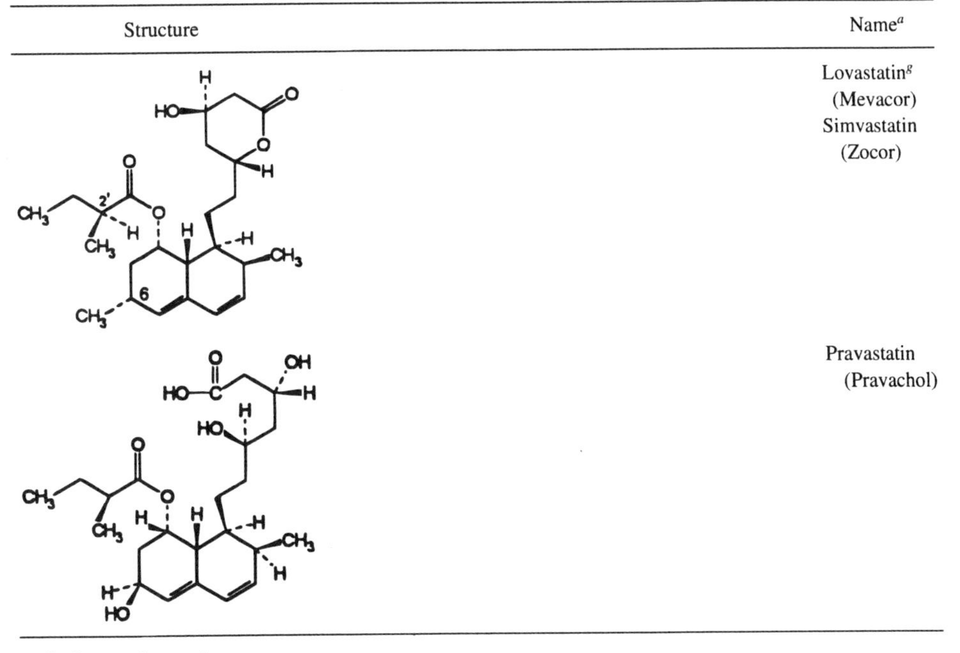	Lovastatin[g] (Mevacor) Simvastatin (Zocor) Pravastatin (Pravachol)

[a] Trade names in parentheses.
[b] May be considered a pro-drug of clofibric acid.
[c] Also a mutual pro-drug with nicotinic acid.
[d] Marketed overseas.
[e] Investigational.
[f] Nicotinyl alcohol (Roniacol) is oxidized to the acid, and therefore can be viewed as pro-drug.
[g] Originally named mevinolin.

(Table 11-6) clofibrate has been in use for more than 20 years. Following absorption the drug is rapidly hydrolyzed to the parent drug in serum and liver. There is still no general agreement as to how clofibric acid lowers blood lipids. Research has implicated triglyceride synthesis, thyroid displacement from albumin binding, and blockade of cholesterol synthesis. Even the point of blockade has been in contention. Nevertheless, by consistently lowering triglycerides and thereby VLDL levels, clofibrate is useful in hyperlipoproteinemias where elevations of these serum components are characteristic: Type IV, II, and familial dysbetalipoproteinemia (Table 11-5).

Gemfibrozil (Table 11-6) is a clofibric acid analog. Unlike clofibrate it raises HDL levels while triglyceride levels are reduced. The drug is used in diet-refractory hypertriglyceridemia. An interesting mutual pro-drug in which clofibric acid is esterified with a nicotinic acid ester (binifibrate, Table 11-6) has been introduced in Spain. Of course, such a compound takes advantage of the lipid-lowering properties of clofibric and nicotinic acid.

Nicotinic acid (niacin) was first shown to lower serum cholesterol levels over 40 years ago (1955). This activity is not related to its vitamin functions (nicotinamide does not share its action). At daily doses of 3 g niacin's hypocholesterolemic effects are about equivalent to clofibrate. In addition to lowering triglycerides, it can also raise HDL levels. Unfortunately, at these high doses nicotinic acid can produce intense flushes and itching, resulting in poor patient compliance. Circumventing the problem with potential pro-drugs such as nicotinyl alcohol, fructose tetranicotinate, and others were not very successful.

Probucol is a highly lipid-soluble sulfur containing *bis*-phenol. It is capable of lowering LDL-cholesterol up to 20% without affecting triglyceride levels. The drug has the dual action of increasing bile acid secretion and increasing the degradation of LDL-apoβ. Probucol, however, also lowers HDL as part of the overall cholesterol reduction.

A more physicochemical mechanism for lowering serum cholesterol is by the sequestering of bile acids in the gut ionically, causing their fecal excretion rather than allowing reabsorption. The quaternary ammonium anion-exchange resin cholestyramine (Table 11-6), which is a copolymer styrene and divinylbenzene, does this effectively; so does colestipol, a copolymeric resin without aromatic components. The resins, though hydrophilic, are not water soluble, are not absorbed, and are not degraded.

The interruption of enterohepatic recirculation of bile acids by the resins effectively lowers plasma cholesterol levels since cholesterol must now be diverted to de novo synthesis of bile acids. In addition, intestinal absorption of dietary cholesterol, normally facilitated by bile acids, is also reduced due to their excretion. Two significant compensatory mechanisms are called into action; increased activity of hydroxymethylglutaryl coenzyme A reductase (HMG CoA reductase), which is the rate-controlling enzyme in the hepatic synthesis of cholesterol (see Fig. 11-4 and discussion to follow), *and* an increase in the number of LDL receptors. The latter mechanism offered the first meaningful treatment of heterozygous FH. Homozygous FH patients lacking LDL receptors, of course, do not respond.

Moderate LDL reductions may be achieved with oral administration of up to 2 g of neomycin. It forms insoluble complexes with intestinal bile acids, preventing their recirculation as occurs with the resins. This action is unrelated to the drug's antibiotic properties. Combinations with nicotinic acid have been effective.

11.7.1. Hydroxymethylglutaryl-CoA Reductase Inhibition

Enzymes, as targets of drug design, continue to tantalize medicinal chemists. Several have been described in this book: dihydropteroate synthetase (sulfonamides), dihydrofolic acid

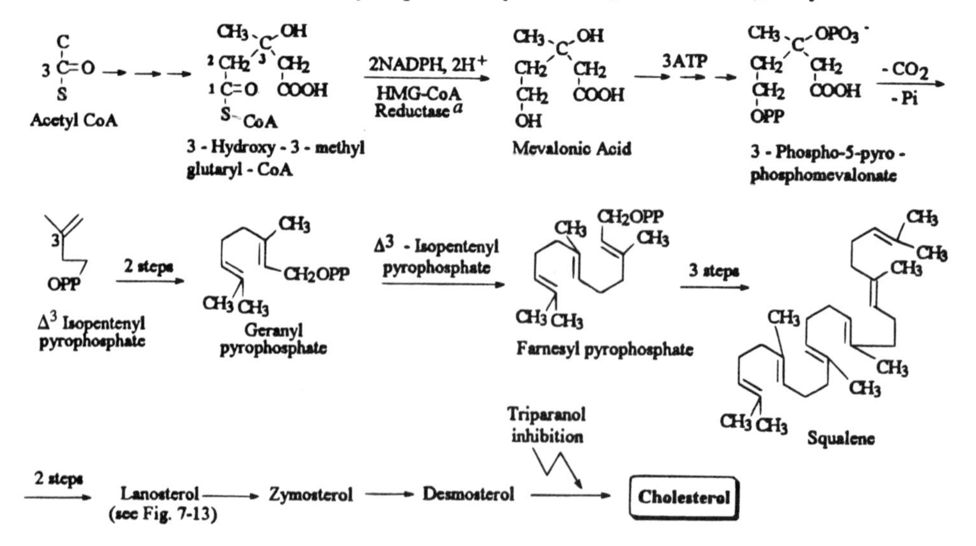

Figure 11-4. Abbreviated cholesterolbiosynthesis. [a]Hydroxymethylglutary-Coenzyme-A reductase catalyzes the rate-limiting step in overall synthesis of cholesterol and is inhibited by it. This key intermediate is also utilized in biosynthesis of steroid horrmones, bile acids, and vitamins A, D. E, and K.

reductase (trimethoprin), transpeptidase (β-lactam antibiotics), thymidilate synthetase (fluorouracil), and others. In humans, 50% of body cholesterol is obtained by de novo synthesis. The biosynthetic sequence has been known in some detail for three decades. The development of inhibitors of the various enzymes involved therefore offered attractive therapeutic potentials for lowering serum cholesterol. Figures 11-4 and 7-13 outline abbreviated schemes of cholesterol biosynthesis.

The triparanol experience was an unfortunate but sobering event. The drug, which was developed in the late 1950s, effectively lowered cholesterol levels by inhibiting the final step: reduction of the 24–25 double bond of desmosterol to yield cholesterol.

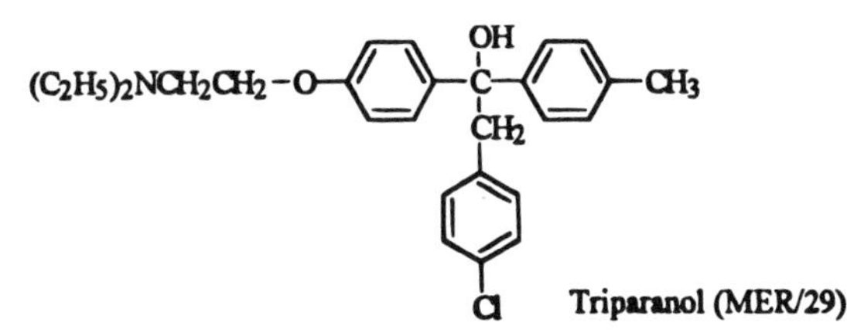

However, opacities in lenses and dermatological side effects led to withdrawal from the market in 1962. Suspicion for development of the serious toxicities pointed to desmosterol, which, with no metabolic alternatives, simply accumulated.

This tragic occurrence illustrates the pitfalls of interfering with important multistep biochemical pathways. In such a complex system, where intermediates are diverted to various critical products (cholesterol itself is a starting point for the synthesis of other important biosteroids), it is difficult to foresee all the ramifications of interference in the mammalian organism (Fig. 11-5).

As previously mentioned, the rate-limiting enzyme in the biosynthesis of cholesterol is HMG-CoA reductase. It catalyzes the reduction of 3-hydroxy-3-methyl-glutaryl-coenzyme A to mevalonic acid (Fig. 11-4). The enzyme is also the point in the sequence where the end product cholesterol acts as its own feedback inhibitor. 7-Oxocholesterol, a non-bioactive steroid, retards the cell growth that cholesterol promotes and strongly inhibits HMG-Co-A reductase activity. The compound thus acts as a false feedback inhibitor. This inhibition is reversible by the addition of either cholesterol or mevalonic acid.

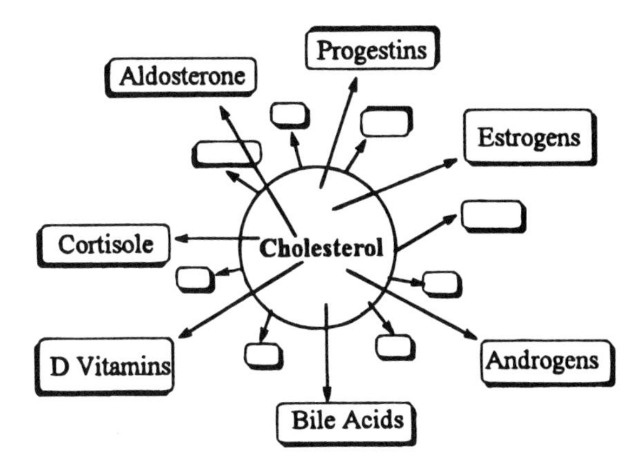

Figure 11-5. Pivotal role of cholesterol.

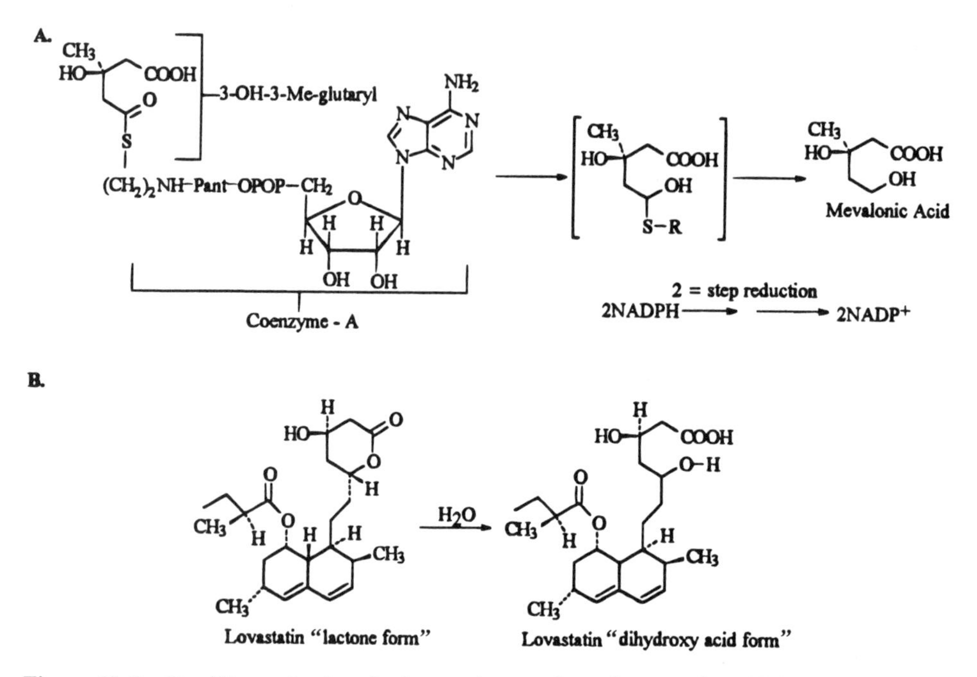

Figure 11-6. Possible mechanism for lovastatin-type drug. See text for additional explanation. Pant = pantothenic acid.

The regulatory function of this enzyme at such an early phase in the cholesterol synthesis, and the findings that a defect in this regulation exists in FH patients, rekindled interest in this enzyme as a target for cholesterol inhibition after years of "hands off" because of the triparanol experience.

In screening for steroid biosynthesis inhibitors, Japanese workers reported the isolation of a potent inhibitor of HMG-CoA reductase. ML-236B, from *Penicillium citrinum* (compactin, 6-demethyllovastatin, Fig. 11-6). With interest thus revived, workers at Merck and Noko University reported the isolation and characterization of lovastatin[8] from the fungus *Aspergillus terreus*. This highly potent and selective inhibitor of HMG-CoA reductase has been recently marketed in the United States (1987). The 2′-methyl homolog simvastatin (Zocor) was launched in 1988 (U.S., 1992). Cholesterol reductions from 18% (with a 2.5 mg dose) to over 40% (with an 80 mg dose) have been achieved with lovastatin in heterozygous FH patients. Triglyceride levels fell as well. Combining the drug with cholestyramine increased LDL catabolism even more by inducing hepatic apo-β receptors and selectively inhibiting direct synthesis of LDL apoprotein B in miniature pigs. The effect may be synergistic.

Figure 11-6 represents a simplified "analogy" mechanism of action for a lovastatin-type drug. 3-Hydroxy-3-methylglutaryl-CoA is reduced to mevalonic acid in two steps, catalyzed by HMG-CoA reductase. It may be suggested that a hypothetical, partially reduced intermediate (in brackets) is reversibly displaced from an assumed enzyme-bound complex by a dihydroxy acid form of lovastatin (derived by hydrolysis of lactone portion of the

[8] Lovastatin was originally named *mevinolin* in the United States; the Japanese named it *monacolin*.

drug). This deprives the cholesterol synthesis scheme of mevalonate. The dihydroxy acid portion of the drug and the upper portion of the "intermediate" are certainly structurally very similar, thus serving as a recognition site for the enzyme. By stretching credulity somewhat, the Co-A chemical appendage (pantothenate–adenylate) might be viewed as analogous to hexahydronaphthalene region of the drug. It is not possible at this time to take this tenuous discussion much further since little is known of the enzyme's chemistry or the other aspects of the reduction. For example, such details of the reduction as the steps involved in the enzyme's acceptance of the 3-hydroxy-3-methylglutarate and subsequent release of the reduced form are not established. Still, these frail mechanistic threads may be helpful as a starting point for rational design improvements of the natural compounds found to be active. The traditional approaches have yielded some interesting compounds. Among these are hydroxyl analogs at the 6 position achieved by microbial oxidation. Functional parameters can also be delineated and potencies related to a putative transition-state analog by synthesis of the various stereoisomers. Simplified structures have also shown activity. The dichlorophenylpyranose, I, and other lactone analogs of the natural products have been reported, II.

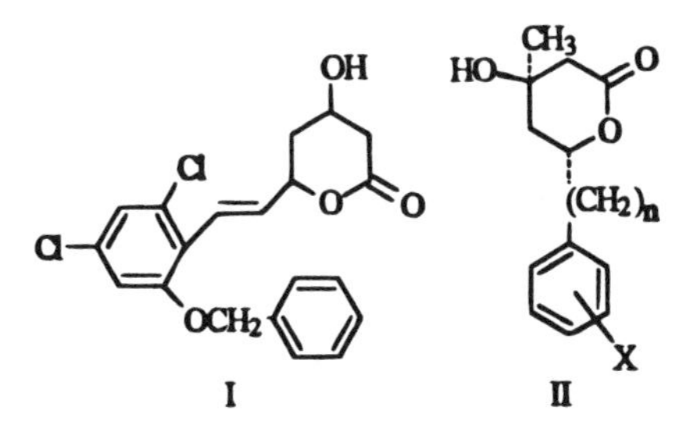

Another analog marketed in the United States (1991) is the hydroxy acid provastatin (Pravachol) as the sodium salt. It differs from simvastatin in having an OH group replace the methyl function at the 6 position of the hexahydronaphthalene ring. However, clinical efficacy is comparable. More important is a report that provastatin is more selective in inhibiting HMG-CoA reductase in the hepatocytes of the liver than in other tissues. The higher ratio of provastatin distribution between liver and intestinal tissues, as well as between kidney and testicular tissues, may in part explain this potential advantage when compared with the earlier agents.

Two other totally synthetic variants are of interest. Fluvastatin (Lescol) (marketed in 1993) has been shown to lower LDL cholesterol significantly via competitive inhibition of HMG-CoA reductase. Here an indole ring replaces the decalin structure of lovastatin. With the additional phenyl ring it easily supersedes the hydrophobic requirement for the reductase inhibition. The *para*-fluorine atom undoubtedly enhances potency. Another experimental entry in this effort is RG12561.

Other potentially major developments in the cholesterol "area" are several closely related fungal products named *zaragozic acids* that are effective in inhibiting squalene synthase. This enzyme controls the first step committing the synthesis to cholesterol (Fig. 11-4). This fact also give rise to the alternate name *squalestatins*. The total synthesis was accomplished by two groups (1994), possibly enabling modification in the structure to improve further on the already powerful cholesterol-lowering properties of the natural compounds.

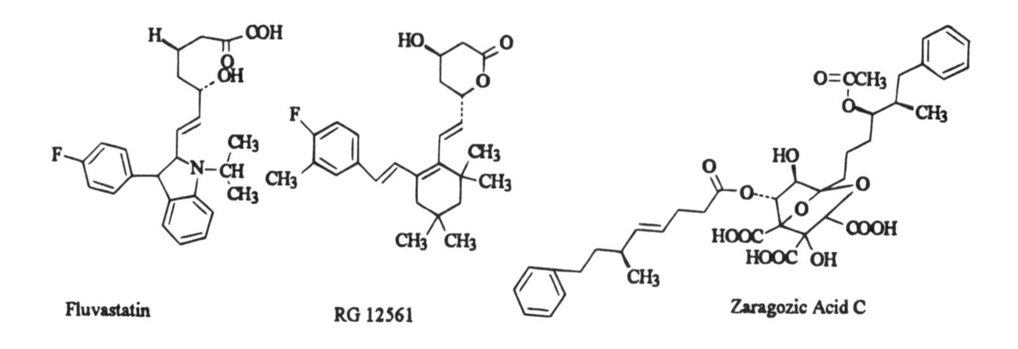

Some novel cholesterol-lowering substances are of interest. Recently, a study in Finland has demonstrated clinically significant (25–41%) reductions of cholesterol, particularly LDL, in patients given 8 g of activated charcoal three times a day for 4 weeks.

Another product that took decades to develop and might be termed a "medical food" is based on the discovery that fatty acids arising from their mono, di, tri, and tetra esters were readily absorbed from the gut of rats. However, as the ester group number increased, the

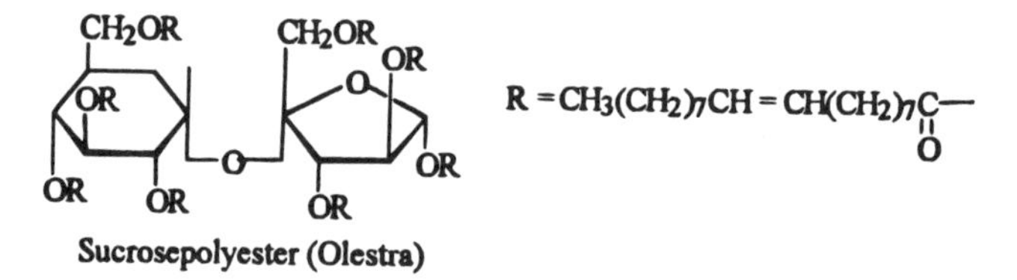

Sucrosepolyester (Olestra)

absorbability decreased. Using the monounsaturated oleic acid, it was found that while glyceryl trioleate or sucrose monooleate resulted in complete oleic acid absorption, and xylitol pentaoleate gave partial absorption, sucrose octaoleate produced none on oral ingestion. The reason was shown to be the inability of the several lipase enzymes to hydrolyze the presumably highly "bulky" polyesters. Sucrosepolyester SPE (Olestra) has been evaluated as a dietary fat substitute in obese and hypercholesteremic patients. The ester has the oral "feel" of shortening, and since it is totally nonabsorbed, it is noncaloric. In addition, it has been found also to inhibit cholesterol absorption. A clinical study with 10 obese hypercholesterolemic patients (including several diabetics) using SPE to reduce caloric intake by replacing 75% of the dietary fat brought cholesterol levels down 20% (and LDL 26%). Olestra was approved for certain food products in the United States in 1996.

11.8. Drugs and Diabetes

Diabetes mellitus is a chronic disease whose characteristics include defects in the metabolism (or utilization) of insulin, carbohydrate, fat, and proteins. The structure and functions of blood vessels are also adversely affected as the disease progresses.

To consider diabetes "only" as a pancreatic endocrine disease is an oversimplification. The pancreatic cells known as islets of Langerhans found throughout the gland produce at least four hormones. The largest proportion of islet cells are the beta cells that produce insulin (actually proinsulin and C-peptide; see later). The *alpha* cells, representing about one-fifth of the islet cells, produce proglucagon and glucagon, which is a hyperglycemic factor. Its function is to mobilize glycogen stores. The *delta* cells produce the hormone somatostatin, which is sometimes referred to as the universal secretory cell inhibitor. It is also produced in and secreted by the hypothalamus. Finally, a small number of islet cells produce pancreatic polypeptide, which is a substance involved in digestion but is not well understood.

The current classification of diabetes was introduced by the National Diabetes Data Group (1979), a division of the National Institutes of Health. Type I, or insulin-dependent diabetes mellitus (IDDM) (formerly named juvenile-onset or ketosis-prone diabetes) and Type II or noninsulin-dependent diabetes mellitus (NIDDM) (formerly known as maturity-onset or adult-onset diabetes). Type II is encountered as a nonobese and obese type, with the latter predominating.

IDDM, which represents less than 10% of cases, is the more severe disease. Type I diabetes, occurring most commonly in juveniles, is characterized by an almost total absence of circulating insulin. Glucagon levels are high. The *beta* cells do not produce insulin and cannot be successfully stimulated to produce any. It is suspected that this apparent destruction of the *beta* cells' capacity to produce the hormone has resulted from a toxic insult arising from a prior viral infection or some environmental exposure. The only way to reverse the catabolic state, reduce excessive blood glucose and glucagon, as well as prevent ketosis, is with exogenously administered insulin. Prior to the discovery of insulin by Banting and Best (1992) in the summer of 1921 at the University of Toronto, there was no meaningful way to treat Type I diabetes. The disease therefore had a 100% mortality. The relationship of the pancreas to diabetes was shown by Minowski (1889), who produced the disease in dogs by removing the pancreas. The original experiments demonstrated that pancreatic extracts (bovine) could lower blood glucose in dogs in whom the pancreas had been removed. In fact, such animals could be kept alive for months with daily injections. Following human testing during 1922, the interest of the Eli Lilly Company in the insulin research of Banting and Best led to the first commercial insulin product the following year (a highly impure preparation of amorphous hormone). Following the crystallization of insulin (Abel, 1926), improved "regular" insulin products were made from crystals of constantly improved purity over the years. Stability of the solutions was also greatly increased. Since the duration of action of "regular" insulin solutions is relatively short (5–7 hours) ways to develop preparations with more sustained effect were sought. The result was a precipitated compound of insulin and protamine,[9] to which Zn^{2+} has been added. The resulting protamine Zn insulin (PZI), today offers the long-acting (24–36 hours) dosage form. Another product of very long duration is now also available. This is a highly insoluble Zn-insulin prepared by adding a 10-fold excess of Zn^{2+}. By admixture with amorphous forms, intermediate-acting forms have also been formulated (ultralente and lente).

In addition to the traditional single injections of insulin that Type I diabetics self-administer at prescribed intervals, there are now also available continuous subcutaneous insulin infusion pumps that deliver the drug (by peristaltic pump or motor-driven syringe) on a continuous basis at a fixed or variable rate. Since they cannot sense blood glucose levels (as does the pancreas) and the patient must monitor them, this is referred to as the "open-

[9] A mixture of six similar very basic proteins (rich in arginine) isolated from trout sperm.

loop" system. More recently, automated "closed-loop" infusion systems have been developed for special acute emergencies such as during ketoacidosis episodes or surgery. The blood glucose–controlled devices depend on a continuous blood supply (by aspiration) to an external sensor. Another novel idea that is not yet widely used is an aerosol nasal insulin product wherein the drug is formulated with a detergent to produce rapid absorption through the nasal mucosa.

11.8.1. Insulin Chemistry

At the end of a decade of prodigious work Sanger and co-workers determined the complete structure of bovine insulin (1955) and that of other animals. The insulin molecule consists of two polypeptides, the A chain containing 21 amino acid residues and the B chain with 30 residues. Two interchain disulfide bridges hold the two chains together; a single intrachain linkage exists on the A chain (Fig. 11-7). The insulins isolated and sequenced from several species have been found to be extraordinarily similar. For example, the A chain in humans, pig, dog, and sperm whale are identical; the B chain of the cow, pig, and goat are identical. In fact, human and porcine insulin differ only in that the B-chain's carboxy terminal alanine in the pig is threonine in humans. Thus, by exchanging these two amino acids, human insulin is being produced from porcine insulin.

Three groups of workers independently reported the total chemical synthesis of insulin during the 1960s. The methods used were to synthesize both chains separately and then to couple them. Since coupling was predictably random, the yields were low. The preparations took years. It is ironic that, utilizing Merrifield's development of solid-phase peptide synthesis with present automated peptide synthesis machines, insulin could be synthesized today in less than 200 hours.

As was seen with other bioactive proteins, insulin is synthesized in an inactive pro-form (i.e., proinsulin). It is a single polypeptide chain of 78–86 residues (depending on the species) (Fig. 11-7). It is stored in this form in granules within the β-cells. Activation is

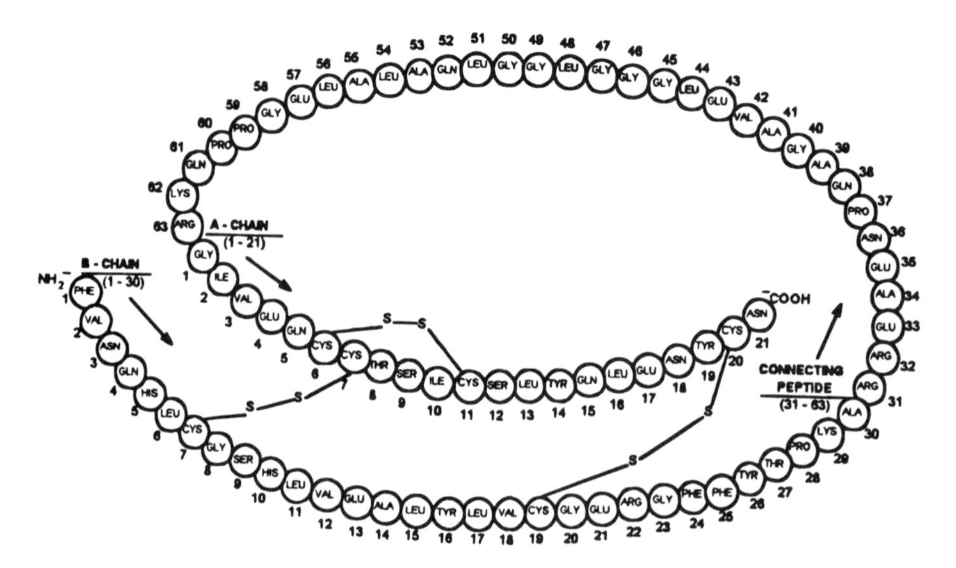

Figure 11-7. Porcine proinsulin.

through excision of connecting (or middle) peptide at Arg 31 and Arg 63 (Fig. 11-7). The two terminal portions of proinsulin then become insulin's A and B chains. At least two enzymes have now been implicated in the activation process, plasminogen activator and a protease activated by thiol function.

The metabolic actions are complex. It will be stated here only that they include metabolism of fats, proteins, and, particularly relevant to diabetes, carbohydrates, specifically D-glucose. The effects insulin produces include increased membrane transport of glucose, amino acids, and K^+. The activities of many intracellular enzymes are also affected, yet even polymer-coupled insulin molecules that are clearly unable to penetrate into the cell themselves still control these intracellular functions. The explanation with the most evidence is that all these insulin–mediated effects interact with specific insulin receptors on the surface of the cellular membranes. The receptor has been characterized to a considerable degree. It consists of two α and two β subunits connected by disulfide bonds. The search for the identity of putative second messengers has not yet established a proven candidate.

Resistance to insulin is encountered in some patients, with needed doses being very high (up to 1,000 units daily). Acute resistance can accompany infections or inflammatory conditions. Resistance has been attributed to high levels of circulating insulin-binding antibodies (e.g., IgG), and a tissue insensitivity that frequently accompanies obesity known to synchronize with increasing serum levels of insulin. It should be understood that the affinity and/or number of insulin receptors can be affected by many "regulators" besides insulin levels. These include sulfonylureas (see later) pH, *c*-AMP, exercise, meals, diet, antibodies, and even other hormones.

The bulk of insulin in the United States is bovine. The somewhat less antigenic porcine hormone—and mixtures of the two—have also been in use for decades. The desirability of using human insulin therapeutically is more than its lack of antigenicity. It is also that the supply from meat animals may not keep pace with the doubling in the number of type I diabetics predicted for the year 2000 (base period mid-1980s). Commercial quantities of pure human insulin by traditional chemical synthesis would be hopeless. Enzymatic conversion of porcine to human insulin by substituting the β-30 ala with thr has been done for some time (Novolin), but it is no solution to projected shortages. However, recombinant DNA techniques are now producing commercial quantities of human insulin. *E. coli* is the workhorse organism into whose plasmid the genes for the A or B chain are introduced and cloned. The A and B chains, separately produced have the cysteine's SH groups oxidatively "bridged" with sulfite.[10]

Type II diabetes (NIDDM) is a milder form of diabetes that occurs primarily in adults and accounts for the vast majority of diabetics (ca. 90%). The β cells, though still functioning, respond inadequately to glucose stimulation. The insulin levels are low, but they are usually adequate to prevent the dreaded ketoacidosis syndrome. There is also tissue resistance to insulin.

There are several potentially practical ways to coax partially functioning islets of Langerhans to function more effectively. One is to stimulate the β-cells to release stored insulin from their granules. (Some β-cell function exists in NIDDM patients.) Inhibition of insulin metabolism is another possibility. Other ways to potentiate the existing, but inadequate, levels of insulin in Type II diabetics might be developed. If achievable, the suppression of glucagon release (or synthesis) would constitute another approach. Increasing the number of insulin receptors is, theoretically, also a viable approach to treat NIDDM.

---

[10] See Hopwood, D.A., 1981.

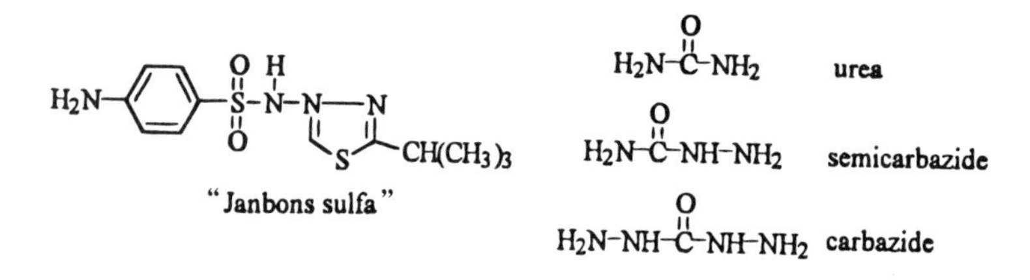

The ability of antibacterial sulfonamides to lower blood sugar levels was first noted with a thiadiazole analog of sulfanilamide, "Janbons Sulfa" (1942), while treating typhoid fever. Several mortalities were diagnosed as caused by severe hypoglycemia. Studies more than a decade later with a series of thiadiazoles indicated that the $ArSO_2NHC(=N\text{-})S$ represented a "minimum" glucose-lowering moiety. This led to the synthesis of arylsulfonylureas, and the introduction of 1-*n*-butyl-3-sulfonylurea (Carbutamide, BZ55, Table 11-7) in Germany, as the first orally active hypoglycemic drug. In spite of the strong "family resemblance" to sulfanilamide, carbutamide no longer exhibits antibacterial properties. The appearance of liver toxicity and blood dyscrasias soon precluded further clinical use of the drug. Even though it is useful as a lead compound, it led to intense SAR studies that designed out antibacterial action emphasizing hypoglycemic activity. The first clinical success of this effort was accomplished by the replacement of the *p*-NH_2 of BZ-55 with a CH_3 group to produce tolbutamide. Because of the facile hepatic oxidation of this $CH_3 \rightarrow COOH$, which yields an inactive benzoic acid, tolbutamide is a relatively short-acting drug requiring frequent administration at high (up to 3,000 mg/day) doses. Although it also possesses an oxidizable CH_3 group, tolazamide is metabolized more slowly, affording a somewhat longer duration of action and considerably higher potency. Tolazamide is technically a sulfonylsemicarbazide rather than urea.

Acetohexamide presents a more interesting metabolic profile. The carbonyl oxygen of the acetyl group (which replaces the CH_3 group of the compounds mentioned earlier) is rapidly reduced to the secondary alcohol (the 1-hydroxy-hexamide), which is actually two to three times more hypoglycemic than the parent compound. It is also three times longer acting. The clinical outcome of the biphasic pharmacokinetics is a more potent drug that is effective with a once-a-day dosing. Two other points should be stated. The active metabolite produced is the L(–) enantiomer. There is also additional metabolism at the 4′-position of the cyclohexane ring and both the parent compound and the active metabolite. The final product excreted is:

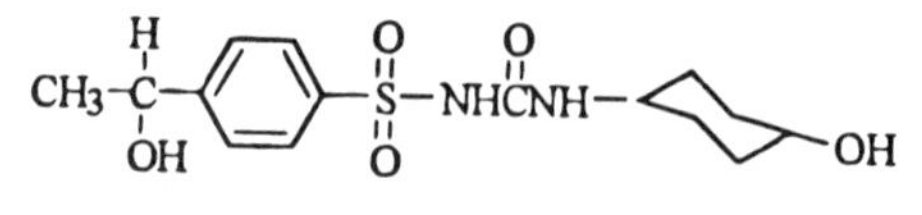

L(–)-4′-*trans*-hydroxyhydroxyhexamide

Chlorpropamide further illustrates how an understanding of drug metabolism can be applied to design into the molecule desirable pharmacokinetic properties. A lower homolog of tolbutamide, the chlorine atom now in the paraphenyl position previously occupied by CH_3, is essentially totally resistant to oxidation or displacement. The result is a long-acting, more potent drug that offers blood glucose-lowering effectiveness with a single daily dose. Excretion of unmetabolized drug (c. 60%) is slow. Hepatic hydroxylation of the middle carbon of the n-propyl group has been shown to occur.

Table 11-7. Oral Hypoglycemic Agents

Structure	Name[a]	$(t_{1/2})^b$ (h)	Comments
$H_2N-C_6H_4-SO_2-NHCONHC_4H_{9n}$	Carbutamide BZ-55	—	Never marketed in the United States
$H_3C-C_6H_4-SO_2-NHCONHC_4H_{9n}$	Tolbutamide (Orinase)	6–12 (4–8)	Low potency 500–3,000 mg daily dose; avg. 2,000 mg
$H_3C-C_6H_4-SO_2-NHCONH-N$(azepane)	Tolazamide (Tolinase)	10–14 (5–8)	Average dose 250 mg
$H_3CC(=O)-C_6H_4-SO_2-NHCONH$-cyclohexyl	Acetohexamide (Dymelor)	12–24 (1.5)[c]	Average dose 1,000 mg L-(–)-hydroxy-metabolite
$Cl-C_6H_4-SO_2-NHCONHC_3H_{7n}$	Chlorpropamide (Diabinese)	35–70 (30–35)	Excreted unchanged up to 10 days
Cl, OCH_3-C_6H_3-C(=O)-$NH(CH_2)_2-C_6H_4-SO_2-NH-C(=O)-NH$-cyclohexyl	Glyburide[d] (DiaBeta, Micronase)	10–24 (4–12)	Very potent; dose range 1.25–20 mg
H_3C-pyrazinyl-C(=O)-$NH(CH_2)_2-C_6H_4-SO_2-NH-C(=O)-NH$-cyclohexyl	Glipizide (Glucotrol)	10–24 (2–4)	Dose range 2.5–40 mg
$C_6H_5-CH_2$ $H_2NHC(=NH)-NH-C(=NH)-NH_2$	Phenformin[e] (DBI)	4–5	Not metabolized
$(CH_3)_2N-C(=NH)-NH-C(=NH)-NH_2$	Metformin (Glucophage)	6–7	Not metabolized

[a] Trade name in parentheses.
[b] Plasma half-life.
[c] Active metabolite ca. 5 hours.
[d] Overseas name glibenamide.
[e] Available in United States only on investigational basis (see text).

The newer "second generation" compounds, glyburide and glypizide, have a greatly increased potency and sometime efficacy (i.e., they may produce clinical responses in patients who have failed to exhibit one with earlier agents).

It is curious that the mechanism of action of the sulfonylureas is still not fully understood. It is established that the ability to secrete some endogenous insulin must exist for the sulfonylureas to work (they do not function in pancreatectomized animals). Sulfonylurea stimulation of preformed insulin release had been established early. However, this is now seen as occurring only during the initial period of therapy (ca. 2 weeks). The exact mechanism, however, is not known. Inhibition of β-cell *c*-AMP is possible. It was later found that after 4–5 weeks of therapy the pancreatic insulin content actually decreases and the hormone's secretion rate in response to food stimulation (e.g., breakfast) also diminishes. Insulin tissue levels even diminish. The drugs, nevertheless, remain effective over prolonged periods (years). It has been demonstrated that sulfonylureas can reduce the biosynthesis of proinsulin. It is therefore apparent that the mechanism(s) of action of these drugs must involve mainly extrapancreatic effects.

It has been shown that prolonged treatment with glipizide in NIDDM patients significantly potentiated the action of circulating insulin and that this effect is most likely produced by an increased insulin receptor population. Reduction of serum glucagon levels with sulfonylurea administration suggests that its secretion is suppressed, thus contributing to the observed hypoglycemic effect. The result, then, is a physiologic outline of a mechanism. However, an exact mechanism at the molecular level is still elusive.

Other than the sulfonylureas, only guanidines have the ability to lower blood sugar values. This led to several clinically useful biguanides such as the phenethylbiguanide phenformin (Table 11-7). The drug does not stimulate insulin secretion. Its mechanism of action has not been elucidated. Though effective, it was withdrawn (1978) from the U.S. market because of an occurrence of lactic acid acidosis in some patients that resulted in high mortality. A dimethyl analog, metformin (Glucophage), has been introduced (1995) with a somewhat better safety profile (Table 11-7).

11.8.2. Complications of Diabetes

The discovery of insulin and its development as a therapeutic agent for diabetes allowed Type I diabetics to survive for many years for the first time in history. The later introduction of the sulfonylureas greatly increased survival rates of Type II diabetic patients. Quality of life also improved. It was naively hoped that control of blood sugar thus achieved, though not curative of the disease, would at least "solve" the problem; however, it did not. It was, ironically, the prolonged and increased survival rates that permitted the development of long-term complications in these patients. The causes for these complications are not known. They seem to occur in tissues in which glucose transport is not dependent on insulin. Furthermore, the enzyme aldose reductase (AR) appears to be intertwined in some of the pathological processes that occur.

Long-term diabetic complications are serious and disabling. Among these are blindness (12% of total in the United States), cataracts, gangrene, renal disease, neurological complications, and cardiovascular complications—the reason diabetes and its treatment, which of necessity also involves treating these conditions, is considered in this chapter. The following, then, are pathologies to which diabetics are predisposed: coronary artery disease, cerebrovascular disease, and peripheral vascular disease. In the case of microvascular (small vessel) disease an association with diabetes is definitely established; its role in atherosclerotic (macrovascular) disease is not as strongly established. Thus diabetics surviving an acute MI have double the risk of subsequent morbidity. The incidence of stroke fatality is also double in diabetics. An increased predisposition to peripheral vascular disease has also been reported.

The ocular complications result in pathologies to the retina (diabetic retinopathy). The lens damage by osmotically induced swelling causes myopia and cataracts; it has been shown to be related to polyol metabolism. Over half of diabetic patients present abnormalities of the corneal epithelium.

Various peripheral neuropathies are caused by diabetes. Even though nerve damage may be discernible in only about 10% of patients when their diabetes is first diagnosed, the percentage rises to 50% once the disease has progressed 25 years.

The IDDM-induced complication leading to the highest mortality rate is kidney disease. It is likely that 50% of Type I diabetics die of renal failure within 20 years of developing diabetes. Statistics on Type II diabetics are not available, but they are likely less severe and less frequent. Type II patients are more likely to die from an MI.

Insulin treatment prolongs the life of the diabetic. However, the development of diabetes-associated complications are not prevented.

Until the early 1970s there was no understanding of the chemical underpinnings on which the relationship of the preceding pathologies and diabetes rested. They have since been found to be based, at least in part, on abnormalities of glucose metabolism in the diabetic patients. This is referred to as the polyol pathway (Fig. 11-8).

The bulk of intracellular glucose is used to produce energy, by way of glucose-6-phosphate into the Krebs cycle (Fig. 11-8). The phosphorylation, catalyzed by hexokinase, is the crucial step. An alternate route exists—the polyol pathway—whereby glucose can be converted to D-fructose, which after phosphorylation proceeds through glycolysis. As shown in Figure 11-8, the D-glucose to D-fructose transformation passes through the sugar-alcohol sorbitol. The reduction of D-glucose is catalyzed by AR. NADPH is the hydride source. Sorbitol dehydrogenase subsequently utilizes NAD^+ as the hydrogen acceptor to oxidize sorbitol to fructose.

The enzyme AR and the polyol pathway it initiates was first reported to occur in seminal vesicles. It was soon discovered in the lens. It is now known to exist in many tissues. This and other studies of the eye led to the postulation of the *osmotic hypothesis* of cataract formation. This polyol concept will be briefly outlined.

AR and hexokinase coexist in the eye (and other tissues) and, in principle, compete for the same substrate, D-glucose. Since hexokinase has a much greater affinity for glucose, it will be rapidly converted to the 6-phosphate; at physiologically available intracellular levels (such as in the nondiabetic), little or no sorbitol is produced. However, at the excessive sugar levels of the diabetic the ability of the hexokinase present to phos-

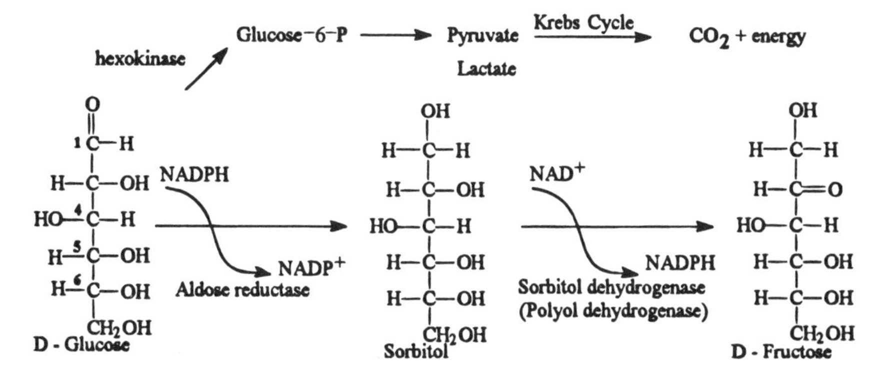

Figure 11-8. The polyol pathway. D-Galactose (4-OH is inverted) is similarly reduced to galacticol, but it is not further oxidized. Also known as the citric acid cycle. For an excellent chemical and historical review of this most important of metabolic cycles, see Lehninger et al., 1993.

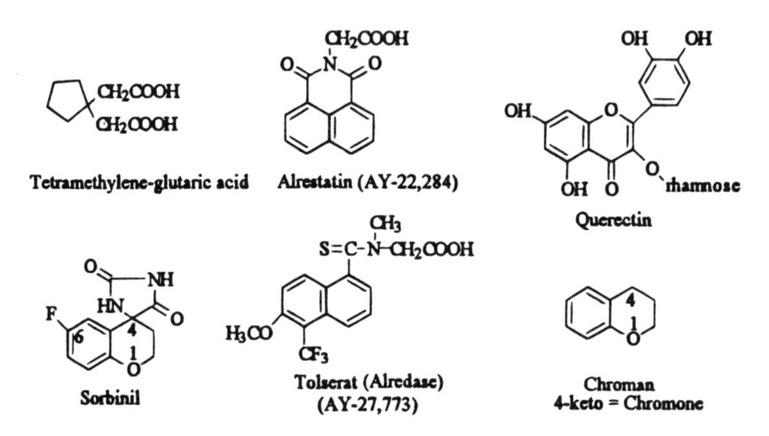

Figure 11-9. Aldose-reductase inhibitors.

phorylate this excess glucose is exceeded. The excess is then metabolized by AR to sorbitol. In fact, sorbitol levels may accumulate faster than polyol dehydrogenase can oxidize it to fructose.

The accumulating intracellular sorbitol in the lens or neuron, for example, does not readily diffuse through the cellular membrane because of its high polarity. Intracellular polyol accumulation produces a hyperosmotic condition, thus causing an influx of fluid into the cells. The hypothesis then holds that this produces changes in membrane permeability and the subsequent pathologies. The glucose-to-sorbitol reaction will also deplete NADPH levels, thereby affecting redox potentials of the cells involved. This may also increase the pathology observed.

It is becoming apparent that inhibiting reductions to polyol with suitable AR inhibitors should be potentially useful in at least delaying cataract formation by reducing the osmotic swelling that has been shown to swell fibers that rupture and ultimately lead to opacity and cataracts. The pain of peripheral diabetic neuropathy should also benefit from polyol reduction with such drugs. Animal studies have shown that osmotic swelling damages Schwann cells of nerve fibers and can result in their demyelination. The fact that AR only becomes significant in the nonphysiologic states (i.e., at elevated blood glucose levels) makes inhibition of this enzyme a particularly intriguing biochemical approach.

Although not able to penetrate membranes, tetramethyleneglutaric acid (Fig. 11-9) could, in in vitro experiments, prevent the structural changes occurring in sugar cataract formation, thus helping to establish the osmotic hypothesis by pointing to the involvement of AR. The benzoisoquinoline compound alrestatin was the first water-soluble compound able to delay appearance of cataracts in rats on a 30% galactose diet by 80%.

Flavonol glycosides (flavinoids) such as rutin, hesperidin, and querectin, which contain the chromone ring system, were found to have AR-inhibiting activity. Querectin was the most potent.

Combining the chroman ring with hydantoin afforded a series of potent spirohydantoins.[11] Sorbinil (Fig. 11-9) is now in advanced phases of clinical trials. One study has already demonstrated the drug's ability to relieve symptoms of chronic diabetes neuropathy.

[11] In spiro compounds two-ring systems share a carbon atom.

An even more promising drug is the naphthalenethiocarboxylic acid amide of N-methylglycine, tolserat (Alderase). This drug is also in clinical trials. A preliminary report of a 24–week study claimed significant improvements in patients with diabetic peripheral neuropathy.

Even though many chemically diverse compounds have demonstrated activity, studies indicate certain structural commonalities. The interaction of these drugs with AR seems to be at a common stereospecific site in the enzyme. The molecular requirements appear to be a planar structure with two hydrophobic (aromatic) areas and a carbonyl moiety susceptible to (reversible) nucleophilic attack. An inhibitor site proposes a charge-transfer complex formation between the carbonyl and nucleophile residue.

The use of AR inhibitors to ameliorate the tragic consequences of long-term diabetes look promising from a biochemical and early clinical viewpoint. However, the pathological processes involved are complex and probably only partially understood at this time. Results of short trials to assess the value of drugs to block events that take a decade or more to develop have to be viewed with great caution.

11.9. Sickle Cell Disease or Anemia

This is an inherited genetic disease. It was first described by a Chicago physician (Herrick, 1910) who believed it was caused by syphilis. It affects about 60,000 Americans, the majority of whom are black. In India, Iran, and around the Mediterranean, it affects nonblacks as well. Nearly two decades after Herrick's report susceptible erythrocytes were demonstrated to change to sickle cells when deoxygenated. This reversed on oxygenation. It took another two decades before Pauling et al. (1949) demonstrated by electrophoresis that the hemoglobin of sickled cells (HbS) was different from normal hemoglobin (HbA).

HbA: $_2$HN-Val-His-Leu-Thr-Pro-**Glu**-Glu-Lys · · ·
HbS: $_2$HN-Val-His-Leu-Thr-Pro-**Val**-Glu-Lys · · ·

Pauling also proposed, correctly, that it was HbS molecules that formed rigid structures by interacting with each other, resulting in red cells having distorted membranes. By not readily passing through arterioles, these deformed erythrocytes impair circulation and result in the characteristic symptoms of sickle cell anemia: susceptibility to infections, pain episodes due to occlusion crises, organ degeneration, and anemia because sickling cells have a shorter life span.

It was proposed that sickling develops from an autosomal recessive gene (S gene) carried by persons having the trait. Such individuals having the S gene and the normal (A) gene are heterozygous (AS); of course, homozygous persons with both genes S (i.e., SS), are those having the disease.

Following Pauling's pioneering work, a peptide "fingerprinting" technique was developed that precisely established the location of the mutation as the sixth position from the N-terminus of the β-hemoglobin chains. There, the hydrophilic amino acid residue Glu is substituted by the lipophilic valine. This results from a point mutation in the globin gene. Substitution of thymine (T) by adenine (A) converts Glu (Codon GAA or GAG) to Val (GUA or GUG). This single amino acid substitution in both β-chains of Hb produces the pathology of the disease. It was subsequently shown that even though both HbA and HbS have identical crystalline structures *and* solubilities when oxygenated, the solubility of

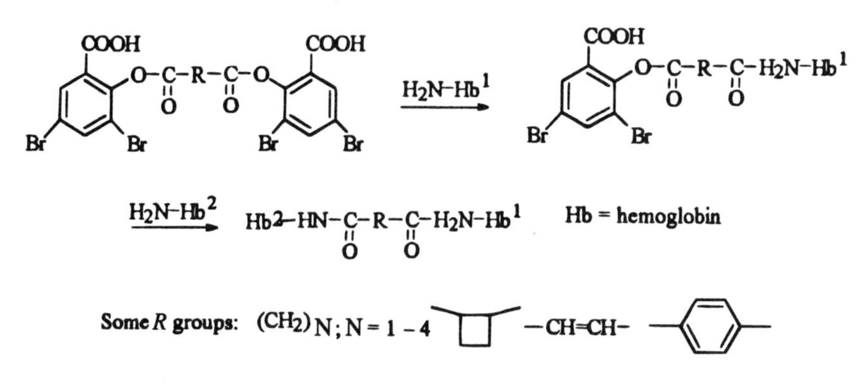

Figure 11-10. Cross-acylation of hemoglobin by diaspirins.

HbS when deoxygenated decreases dramatically. This is now believed to be due to polymerization of deoxyHbS tetramer into high-molecular-weight fibers.

Despite the fact that most of the molecular details of the disease are probably known and many proposed remedies exist, there are to date no clinically useful antisickling drugs available for treatment. Compounds that seemed promising in in vitro models failed in clinical trials.

Agents under investigation fall into two categories: (1) those able to prevent the polymerization of deoxyHbS, and (2) those capable of modifying erythrocyte membrane structures without affecting Hb.[12] The vast majority of compounds being proposed and tested are in the first category. The attempts to chemically modify Hb can be undertaken by noncovalent and covalent means. The aim is to reverse the association of HbS molecules into the insoluble fiber arrangement that, ideally, should eliminate the symptoms of the disease. Many of the early compounds that were claimed to have some clinical efficacy based on their purported ability to interfere with hydrophobic bonding (between HbS molecules) could not be verified. Subsequently, in well-controlled trials, compounds such as urea and sodium cyanate ($Na^+N = C = 0$) have been abandoned.

Agents that could be shown to have antisickling properties (in vitro) by covalently modifying HbS and leading to possible reversal of the aggregating process are of potentially greater interest. Here, interaction between HbS molecules would be directly blocked since the sites at the contact interfaces would be covalently occupied by derivatization. Thus compounds capable of acylating, carbamoylating, and forming Schiff bases have shown in vitro activity. Mechlorethamine (HN_2), the carcinolytic alkylating agent, strongly inhibits sickling, but toxicity would obviously preclude oral or parenteral use.

A particularly novel approach was to take advantage of the ability of aspirin to acetylate the ε-NH_2 of lysyl residues of proteins (Chapter 5). Amplifying the acylating potency by placing electron-withdrawing halogens (Br) in the 3 and 5 positions and adding a cross-linking potential between Hb chains (ideally both β) led to a series of "double-headed" or diaspirin-type compounds that were able to penetrate the erythrocyte membranes and were shown to cross-link between Lys 82 of β_1 and β_2 hemoglobin chains (Fig. 11-10). A most recent (1995) but modest development was the approval of hydroxyurea (Hydrea, Chapter 4) as a useful palliative for sickle cell disease.

---

[12] A third method, gene modification or replacement, is beyond present capabilities.

Hemorrcheologic agents are compounds that are able to affect blood flow by lowering its viscosity and therefore be potentially useful in treating intermittent claudication, a condition of inadequate blood flow to the muscles of the lower limbs, that essentially results in an ischemic situation. Although once believed to have been caused by only arteriosclerotic obstruction, rheological factors affecting blood flow are also significant. Poiseuille's law relates flow to viscosity by considering vessel radius, length, volume, and pressure. These factors can be applied to blood flow by approximation. One other factor affecting blood viscosity is erythrocyte flexibility.

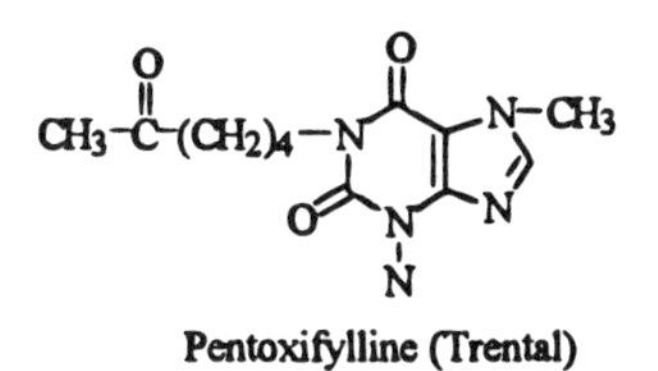

Pentoxifylline (Trental)

The drug pentoxifylline, a 2-hexanoyl analog of such xanthines as theophylline and caffeine, has the ability to lower blood viscosity by increasing the flexibility of red blood cell membranes in patients. This is most probably due to PDE inhibition, which increases *c*-AMP levels. Unlike the vasodilators considered earlier, pentoxifylline is effective in claudication.

11.10. Thyroid Functions and Drugs Affecting Them

The thyroid gland, a 20 g tissue structure on the upper segment of the trachea, controls essential physiological function by its ability to make, store, and secrete two thyroid hormones. These functions include regulation of growth and general development, primarily by controlling DNA transcription and therefore protein synthesis. This in turn results in various enzymatic activities. Deficiencies in thyroid hormones have catastrophic effects in early life, the extremes of which can be manifested in cretinism by dwarfism and severe mental retardation. Another action is referred to as the calorigenic effect. This is evidenced by the increase in basal metabolic rates in organs such as the kidneys, heart, and liver, as well as in skeletal muscles. Valid mechanisms by which these effects can be satisfactorily explained have been elusive. Thermal regulation should be included in this area, as shown by the fact that thyroid secretions can be stimulated by a decrease in temperature.

Cardiovascular effects can be dramatically derived from thyroid hyperactivity and must be given therapeutic consideration when dealing with that system. For example, heart rates accelerate, and the beats become more forceful, thereby increasing cardiac output. Even though these symptoms strongly resemble CNS stimulation, increased catecholamine levels cannot be demonstrated. This might be explained in part by direct hormonal effects on the myocardium, making it very sensitive to normal catecholamine levels. An increase in β-adrenergic receptors has been suggested. The fact that drugs, such as propranolol are effective in reversing the thyrotoxic symptoms lends credence to this notion. However, some studies did not verify this. Nevertheless, the inotropic effects of thyroid agents is not in question.

Another demonstrable effect of thyroid hormones is their stimulation of the biosynthesis of bile acids from hepatic cholesterol. Hypercholesterolemia is known to accompany conditions of low thyroid levels. More evidence (Salter et al., 1988) that these hormones will increase LDL binding to its receptors in hepatocytes would seem to give these agents some metabolic functions as well. Other involvements of thyroid hormones in lipolysis enhanced by catecholamines are known. Furthermore, effects on carbohydrate metabolism have also been established.

11.10.1. The Hormones

Figure 11-11 depicts the several structures relevant to this discussion. The thyroid hormones and precursors are all iodinated tyrosine derivatives, and their iodine content is undoubtedly a unique occurrence in mammals.

After ingestion and absorption, iodine, either organically bound or as iodide ion, I, is concentrated in the thyroid gland by a process called *iodide trapping*. In addition, other monovalent anions with similar spherical or tetrahedral shape and sizes will also accumulate in the gland. Ions such as astadite, At, bromide, Br, and perchlorate, ClO_4^-, act as competitive inhibitors of iodide ions and can be shown to have antithyroid activity. The linear thiocyanate, SCN^-, is a similarly competitive agent. Once in the gland the iodide is oxidized to iodine by H_2O_2, and is catalyzed by the membrane-bound enzyme thyroid peroxidase. The reactive iodinating species, which is not free iodine, is likely an iodosulfenyl cysteine protein that will then iodinate the tyrosine residue in mammalian thyroglobulin in

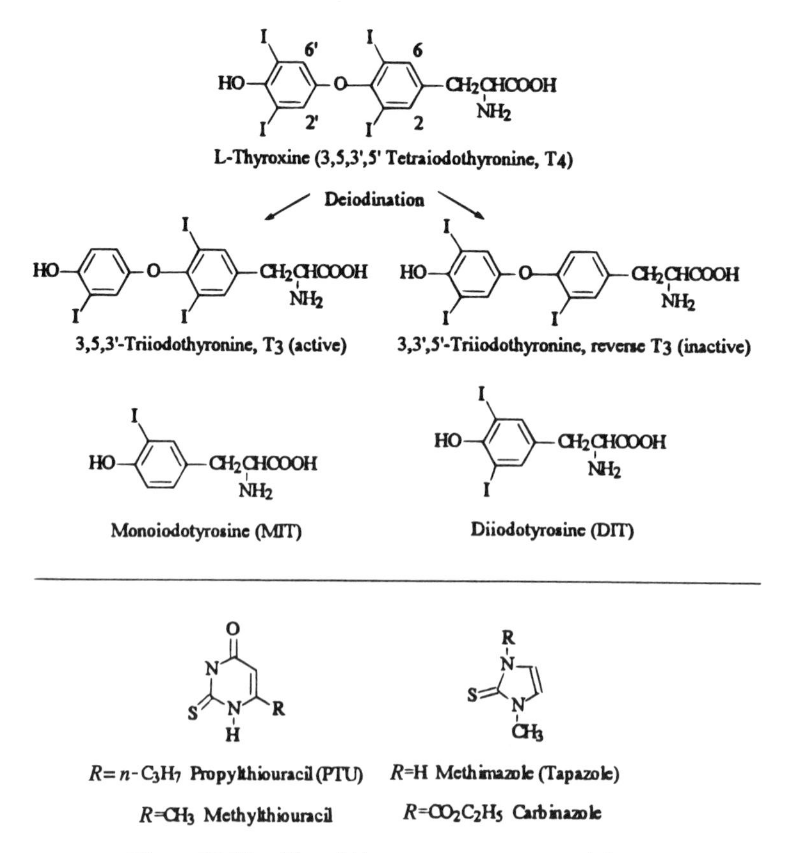

Figure 11-11. Thyroid hormones: precursor and drugs.

the predicted position next to the phenolic OH group. This then yields both the protein-bound 3- and 3,5-iodotyrosil residues (MIT and DIT, Fig. 11-11).

The next step, leading to the actual hormones L-thyroxine (T_4) and the 3,5,3-triiodothyrosine (T_3), is carried out by coupling the iodotyrosine residues to their biphenyl ether structures, which are catalyzed by the same thyroid peroxidase. There is a probability that this is an oxidative step where the peroxide forms highly reactive free radicals that will form the stable covalent ether bond when favorably juxtaposed within the tertiary structure of the thyroglobulin. Additional T_3 can then also be produced by deiodination of the 5′position of the T_4 molecule. It should be pointed out that T_3 is several times more potent than thyroxin. An alternate deiodination at the 5 position also occurs, yielding the inactive 3,3′,5′-triiodotyrosine or *reverse*-T^3. The thyroglobulin serves as a reaction vessel for hormone synthesis, and can also act as a storage site for the hormones and iodide from which diffusion occurs only sparingly.

11.10.2. Pathology

Thyroid-gland-related pathologies are those traceable to either hypo- or hyperfunction of the gland. Thyroid hypofunction (i.e., hypothyroidism) initially results in a release of excess thyroid stimulating hormone (TSH). The resulting hypertrophy leads to gland enlargement (nontoxic goiter) as a compensatory mechanism. When the condition increases in severity, demonstrable atrophy and glandular breakdown becomes apparent. As mentioned, a congenitally defective gland leads to irreversible cretinism.

In hyperthyroidism excessive T_4/T_3 secretion leads to severe and prominent symptoms. Graves' disease, which is a toxic goiter, is expressed by thyrotoxicosis and ophthalmic pathologies such as eyeball protrusion, which is called *exophthalmos*. Treatment of pathologies and symptoms of hypothyroidism is conceptually simple, namely, hormonal replacement. There are several products available. The oldest and least costly is Thyroid, U.S.P., which is a desiccated extract of the thyroid glands from mainly bovine and porcine sources whose iodine content (mainly in thyroglobulin) must be 0.17–0.23%. The T_4/T_3 ratio is not designated but is assumed to be approximately that of the gland. It is available in tablet form in a wide range of strengths. A mixture of synthetic L-thyroxine (T_4) and levothyronine (T_3) is also available as sodium salts, Liotrix (Euthroid, Thyrolar). Thyroglobulin itself (Proloid) is obtained by fractionation processing of porcine glands. It has identical therapeutic properties, but possibly with fewer hypersensitive reactions. The pure L-thyroxine as the sodium salt (Synthroid), which is either obtained from glandular tissue or by total synthesis, is also in widespread use. Finally, liothyronine (T_3) (Cytomel) is also on the market. Even though this product is more potent and has a faster onset of action, it does have a shorter half-life and a higher cost. Laboratory blood monitoring difficulties are also associated with it.

Reducing thyroid activity and its hormonal effects is the goal of treating hyperthyroidism. Several approaches are available theoretically: hormonal synthesis inhibitors, blockade of iodide transport mechanisms, high iodide levels to suppress the gland, and partial or total glandular destruction with radioactive^{131}I. This isotope, whose beta radiation of 0.6 MeV is absorbed within 2 mm of its point of origin, can produce localized tissue destruction. Surgical ablation is another option.

The use of ions other than iodide has not been widely employed. For example, perchlorate (ClO_4^-) tends to have serious toxicities, such as aplastic anemia. The thiocyanate ion (SCN^-) has a dual mechanistic appeal. It blocks iodide concentration processes and

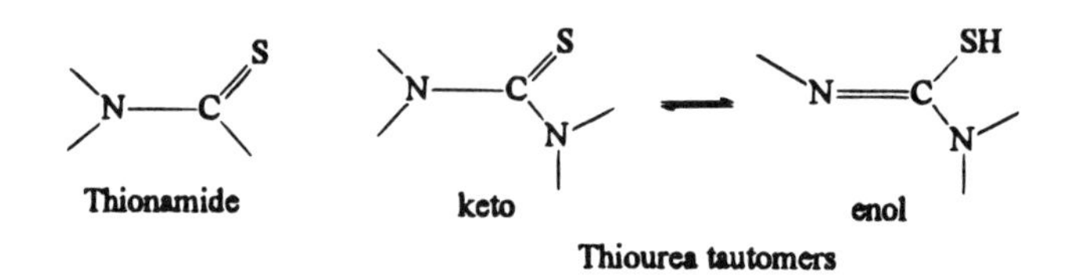

Figure 11-12. Thiourea tautomers.

also inhibits iodination of tyrosine residues in thyroglobulin by inhibiting thyroid peroxidase enzymes, yet here, too, toxicities override its usefulness. This leaves primarily several agents that attenuate the overactive thyroid gland by nondestructive means. These are essentially a few cyclic thionamides (thiocarbamides). The most effective compounds contain an additional nitrogen atom bonded to the carbon atom carrying the sulfur and are thus thioureas (Fig. 11-12). Their dual mechanisms of inhibiting the peroxidases that catalyze the iodination of the tyrosine residues of thyroglobin as well as the subsequent ether-forming coupling was established over 20 years ago (Taurog, 1976). Prior to World War II thiourea itself was briefly used as a goitrogenic agent; 2-mercaptouracil and thiobarbital soon followed. Many compounds with goitrogenic properties were found and evaluated, yet only two are currently the mainstay: the 6-propyl derivative of 2-thiouracil, propylthiourea (PTU), and methimazole, which is a 2-thioimidazole (Fig. 11-11). The latter is the more potent and has a longer half-life. However, since both drugs are efficiently concentrated in the thyroid gland, this difference has little effect on the duration of action. Adverse effects with both drugs approach 10%, the most common being a macopopular rash. More seriously, but also more rarely, lupus-like symptoms may arise. The incidence of life-threatening agranulocytosis occurs at a rate of about 0.5%, particularly in older patients. Carbimazole is a methimazole pro-drug, which on ester hydrolysis loses CO_2.

References

Abel, J, *Proc. Natl. Acad. Sci. USA* **12**:132, 1926.

Banting, G. F., Best, C. H., *J. Lab. Clin. Med.* **7**:251, 1992.

Enos, W. F., Holmes, R. H., Beyer, J., *J. Am. Med. Assoc.* **152**:1090, 1953.

Goldstein, J. L., Brown, M. S., in *The Metabolic Basis of Inherited Disease*, Stanbury, J.B., Ed., McGraw Hill, New York, 1983, pp. 672–712.

Herrick, J. B., *Arch. Int. Med.* **6**:517, 1910.

Hopwood, D. A., *Sci. Amer.* **245**:67, 1981.

Kato, K., Ohkawa, S., Terao, S., et al., *J. Med. Chem.* **28**:287, 1985.

Lehninger, A. L., Nelson, D. L., Cox, M. M., *Principles of Biochemistry*, 2nd Ed., Worth Publishers, Inc., New York, 1993, p. 446.

Minkowski, O. *Diabetes*, **38**:1, 1889.

National Diabetes Data Group. *Diabetes* **28**:1039, 1979.

Pauling, L., Itano, H. S., Singer, S. J., Wells, I., *Science* **11**:543, 1949.

Salter, A. M., Fisher, S. C., Brindley, D. N., *Atherosclerosis* **71**:77, 1988.

Taurog, A., *Endocrinology* **98**:1031, 1976.

TIMI Study Group. *N. Engl. J. Med.* **312**:932, 1985.

Suggested Readings

Bliss, M., *The Discovery of Insulin*, University of Chicago Press, Chicago, 1983.

Edelstein, S. J., Sickle Cell Anemia, *Ann. Rep. Med. Chem.* **20**:247, 1985.

Jacobs, S., Cuatrecasas, P., Insulin Receptors, *Ann. Rev. Pharmacol. Toxicol.* **23**:461, 1983.

Jorgensen, E. C., Thyromimmetic and Antithyroid Drugs, in *Burger's Medicinal Chemistry*, 4th ed., Part III, Wolff, M. E., Ed., John Wiley and Sons, New York, pp. 103–145, 1981.

Kador, P. F., Kinoshita, P. F., Sharpless, N. E., Aldose Reductase Inhibitors: A Potential New Class of Agents for the Pharmacological Control of Certain Diabetic Complications, *J. Med. Chem.* **28**:841–849, 1985.

Klotz, I. M., Haney, D. N., King, L. C., Rational Approaches to Chemotherapy: Antisickling Agents, *Science* **213**:724–731, 1981.

Tissue Plasminogen Activators in Thrombolytic Therapy, Sobel, B. E., Collen, D., Grossbard, E. B., Eds., Marcel Dekker, New York, 1987.

Verstraete, M., Collen, D., Thrombolytic Therapy in the Eighties. *Blood* **67**:1529–1541, 1986.

CHAPTER
12
Psychoactive Drugs—Chemotherapy of the Mind

It is difficult to exaggerate the complexity of the human brain's chemistry and functions, many of which are only beginning to be understood. Much is still undiscovered.

Before embarking on a discussion of mental diseases and mechanistic aspects of the drugs used in their treatment, it is important to define certain terms frequently encountered.

Psychosis. A mental disorder of sufficient magnitude to result in personality breakdown and loss of contact with reality. Delusions and hallucinations are characteristic. Institutionalization is invariably necessary. There are several types of psychoses.

Depressive. Characterized by extreme depression and melancholia.

Involutional. Usually occurring during involutional period of a person's physical decline (age 45–65).

Manic-depressive. Characterized by alternating periods of severe depression and excessive well-being; sometimes even alternating with periods of apparent normalcy.

Senile. "Caused" by old age.

Toxic. Caused by a toxic agent.

Traumatic. Resulting from head injuries.

Neurosis. A mental disorder that is not caused by demonstrable disease of any part in the central nervous system. Life adjustments are impaired, but contact with reality is not lost. Treatment does not require hospitalization but it is complicated in that patients usually have mixed psychoneuroses of two or more of the following types: fatigue, anxiety, phobic, hysterical, obsessive-compulsive, hypochondriasis, and others.

Psychotropic agent. A chemical that affects psychic functions and behavior. This would include drugs used to treat mental disorders, as well as drugs that may cause them (e.g., hallucinogens).

Schizophrenia. A group of severe mental disorders involving disturbances of thinking, mood, and behavior, including delusions and hallucinations. Four main types are usually considered.

Simple. Characterized by emotional dullness, loss of ambition, and withdrawal tendency.

Paranoid. Characterized by delusions of persecution.

Catatonic. Marked by excitement or stupor, yet lucidity; exhibits so-called split personality.

Hebephrenic. Marked by speech anomalies, silly childish behavior, delusions, and hallucinations.

Tranquilizer. A drug capable of reducing mental tension without interfering with normal function.

Ataractic. A drug producing ataraxia, which is a state of mental calm and tranquility.

Neuroleptic. A drug producing symptoms similar to those of depressant diseases of the central nervous system (CNS). Frequently used to refer to an antipsychotic agent.

Thymoleptic. A synonym for an antidepressant drug.

It is difficult to find a widely acceptable system of classification of psychotropic agents. One classification that is useful is the following:

I. General Sedatives
  A. Sedatives
  B. Hypnotic
  C. Narcotics
II. Tranquilizers
  A. Minor
    1. Carbamates
    2. Benzodiazepines
    3. Diphenylmethanes
  B. Major
    1. Phenothiazines
    2. Reserpines
    3. Butyrophenones
    4. Miscellaneous agents
III. Antidepressants
  A. Tricyclics
  B. MAO inhibitors
  C. Others
IV. Psychotostimulant
  A. Psychotoanaleptics (amphetamines)
  B. Psychotodyleptics (hallucinogens)

Some of this classification is not "pure" in that agents in some subcategories have multiple pharmacological properties that are frequently of clinical usefulness and not necessarily related to psychotropic effects. For example, some of the minor tranquilizers include members that have excellent skeletal muscle relaxant properties; several are used as antiepileptics (e.g., diazepam). The diphenylmethane derivative hydroxyzine is a particularly good example of multiple pharmacology. In addition to its anxiolytic (antianxiety) property, it also exhibits the following clinically significant assets: antihistaminic, adrenolytic, antiemetic, antispasmodic, hypothermic, and sedative effects.

12.1. Historical Overview

Humanity's first "mood-altering" substances were herbal sedatives and alcohol. Plants, which are now known to contain opium and belladonna alkaloids, were discovered millennia ago but entered "mainstream" medicine during the Renaissance and were widespread by the eighteenth century. The nineteenth century saw the introduction of inorganic bromides as antiepileptics in 1853 and as sedatives in 1864. The chemical "chloral," which was discovered by Liebig in 1832, entered medical practice after 1869 as chloral hydrate (Eq. 12.1). Paraldehyde, which is a trimer of acetaldehyde, came into use about a decade later. The long-acting barbiturate barbital was developed in 1903, followed by phenobarbital in 1912. The middle 1930s saw the introduction of short-acting barbiturates such as seco- and pentobarbital. Over the years probably hundreds of barbiturates with varying durations of action were developed, and many were introduced. Many of these, in less than hypnotic doses, became popular as daytime sedatives to treat anxieties and neuroses (the term *tranquilizer* was not yet in use). Sulfonal and related sulfones developed in the late nineteenth century were powerful hypnotics whose toxicities caused their discontinuation some years later.

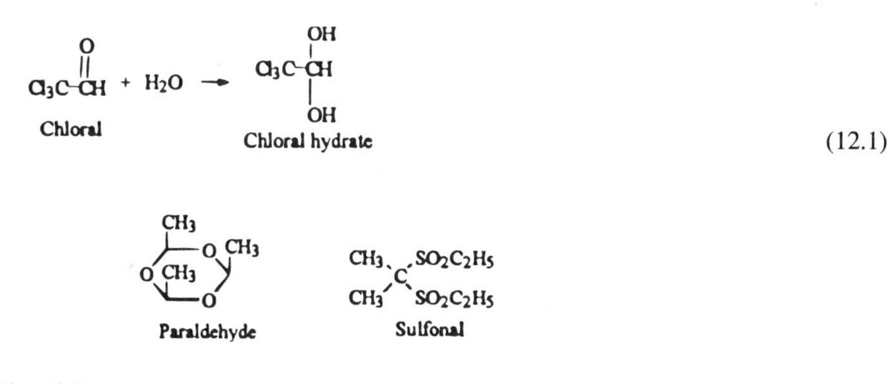

The 1950s were particularly prolific. Psychopharmacology may be said to have had its beginning with the synthesis and later establishment of chlorpromazine's antipsychotic properties. Several barbituratelike compounds such as glutethimide and methyprylon were marketed. These two compounds did not offer any real therapeutic advantages. The carbamates, beginning with meprobamate (1954), were the first antianxiety compounds (anxiolytics) whose pharmacology included more than just CNS depression. These drugs are now viewed as so-called minor tranquilizers. What may be viewed in retrospect as the second-generation sedative anxiolytics began with the benzodiazepine (BZD) chlordiazepoxide in 1960.

The treatment of mental disease before 1950, such as it was, consisted mainly of the administration of CNS depressants—the barbiturates or the bromides. The drugs did little more than depress the agitated psychotic into a condition of stupor, which made it difficult for the patient to function; it did nothing to alter the prognosis of the disease. Psychoanalysis was of some value in the treatment of neuroses.

However, even with this type of disorder, the annual increase in the number of patients made such treatment of limited general value. Of course, in the case of institutionalized psychotic patients it was not even a viable method. The number of patients in mental hospitals in the United States (and other Western countries) was increasing at an alarming rate in the post–World War II years. Figure 12-1 illustrates the dramatic annual increases in the number of such patients prior to the introduction and widespread use of effective psychopharmacological agents—and the equally impressive decline of their numbers following their introduction.

The difficulty in evaluating potential new analgesic drugs with animals in simulated or real pain situations was mentioned in Chapter 5. One can imagine how much more difficult

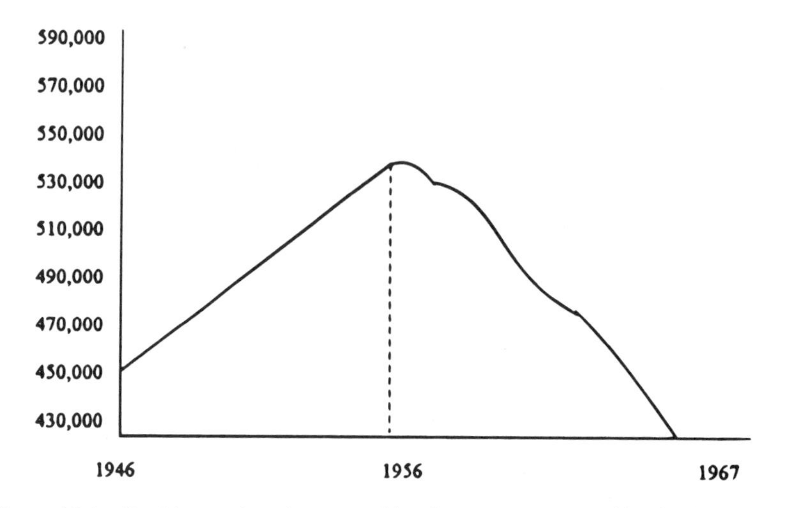

Figure 12-1. Resident patients in state and local government mental institutions in the United States.

the area of psychopharmacology would be. The establishment of animal models in which results would be readily interpreted and agreed upon is still a major problem. Is it possible for the experimental and clinical pharmacologist to come to terms? How much analogy or identity is there in the little that is known in the area of brain chemistry? How do we evaluate the effects of drugs on the emotional and mental states of humans when we experiment on animals that have no comparable circumstances? Do experimental animal studies with psychotropic drugs have any validity or significance for humans?

Some workers in the field argue that experimental studies in animals have no real significance in humans; however, they are in the minority. After all, the functional circuits of the human brain are quite similar to those of the animals frequently used in psychopharmacological studies (rat, cat, monkey). In addition, certain emotional factors controlling behavioral patterns are common to many species. Efforts are constantly being made to devise animal models of changed animal behavioral patterns, which may be analogous to human psychopathological alterations.

Two types of behavior are generally studied: general and specialized. *General behavior* may include evaluation of motor activity and muscle strength; *specialized behavior* may point out muscle relaxant properties in those psychotropic drugs that have them. A complete profile, including behavioral, autonomic, and neurological factors, will frequently be done on animal models. Table 12-1 lists an abbreviated cross-section of parameters that might be evaluated.

Various tests have been devised to screen drugs for potential psychopharmacological activity. Several of interest will be briefly discussed. Apomorphine (Fig. 9-12) is a powerful emetic. This emesis is strongly antagonized by the phenothiazine tranquilizers. Hence, experimental compounds able to block this effect may also have potential neuroleptic properties in humans. Since one important qualitative difference between neuroleptics and the classical CNS depressants (e.g., barbiturates) is the ability of the former to reduce spontaneous motor activity without significant sedation, various methods to test an animal's alertness are used; among these are the so-called alertness cages for small rodents. The ability to negate the stimulant effects of amphetamines has also been interpreted as a neuroleptic property.

Table 12-1. Profile of Animal Parameters Useful in Psychotropic Drug Evaluation

Behavior	Autonomic system	Neurological effects
Stupor and arousal	Size of eye pupils	Tremors
Space orientation	Urination	Muscle twitching
Pain response	Salivation	Gait
Touch response	Piloerection	Grasping power
Spontaneous activity		Body tone
Restlessness		Corneal reflex
Irritability		Righting reflex
Aggressiveness		

Since certain psychotic patients exhibit manic and aggressive behavior that can be controlled by drugs, the ability to subdue aggressive behavior in animals (ideally without sedation) with experimental compounds can be considered an antipsychotic property. One of the ironies encountered is the difficulty in finding animals that are intrinsically aggressive.[1] Thus methods to make animals aggressive (even if only temporarily) had to be devised. One method is the induction of viscousness in cats by inflicting lesions in the limbic system of their brains. Rodents can be made aggressive by long periods of isolation. Neuroleptic drugs reduce or eliminate such induced aggressive behavior. The correlation of this effect in animals with clinically antipsychotic effectiveness appears to be valid.

The conditioned response induced by the classic Pavlovian techniques can also be used to screen antipsychotic drugs (Pavlov, 1927). Thus an animal conditioned to avoid an unpleasant stimulus (e.g., electric shock) following a particular sound (e.g., a bell) can have this conditioned avoidance blocked by neuroleptic drugs.

Another evaluation method is the reinforced behavioral response test devised by Skinner (1955). In this test, a combination of positive reinforcements (rewards such as food) and negative reinforcements (punishment such as electric shock) is given depending on the animal's choice of lever pressed. Response rates vary with the reinforcements applied. Psychopharmacological drugs alter these responses greatly. Interpretation of results is difficult. Mazes of differing complexities, patterns, and size are also used as test parameters for psychotropic drugs.

Pavlovian conditioning can be used to induce experimental neurosis. For example, a dog can be conditioned to salivate when shown a particular geometric shape, such as a circle, and to crease salivation when shown a square object. The animal is thus actually taught to discriminate between the two shapes. The animal is then shown a number of squares that become more and more rounded until the animal reaches a point at which it cannot decide whether it sees a circle or a square. The behavior of the dog at this point—rapid heart beat, ticlike movements, pupil dilation, changed respiration—is strongly analogous to a neurotic response in humans. Drugs that reduce or eliminate these symptoms frequently display antineurotic activity in humans. Sleep deprivation can also produce behavioral changes in animals that seem to correspond to psychotic behavior in humans.

Results obtained from humans directly would, of course, be ideal. Applying some of the described methods to humans, however, is greatly limited by ethical considerations.

[1] A characteristic humans apparently share with few other species.

Sedatives can be defined as agents that exert a soothing or tranquilizing effect. In the context of therapeutics they induce mild depression and calm anxiety, even excitation. Ideally, they should not cause drowsiness or impair performance. A hypnotic is a drug that induces sleep at a minimum. Hypnotics also cause a partial or complete loss of consciousness and can therefore produce insensibility to pain, probably by inhibiting afferent impulses and/or the cortical centers in the brain. These can be sedatives used medically at higher than sedative doses, to induce sleep in treating insomnia. For that purpose the names *somnifacients* or *soporifics* have also been applied. Anesthesia can be produced at even higher doses.

Until the 1930s sleep was believed to be a simple, passive process. Studies with the electroencephalogram (EEG) first characterized sleep as existing in separate and distinct time patterns. Periodic appearances of rapid eye movements (REM) were subsequently identified during sleep. These appeared to coincide with dreaming. Sleep is now viewed as an active and complex process occurring in stages and cycles. Briefly, there are two kinds of sleep. One is an orthodox sleep, which is a nonrapid eye movement state (NREM). The other is a paradoxical sleep that does involve REM, and is sometimes called *active* or *fast sleep*.

NREM sleep progresses through several stages. Initial drowsiness produces relatively high-frequency (8–12 Hz) alpha waves. This is followed by Stage I theta waves (3–7 Hz) lasting less than several minutes. External stimuli diminish, thinking drifts away from reality, and dreaming is possible. The second stage produces high-frequency (12–14 Hz) "spindles" and slower complex patterns. The sleep now is less shallow, but it is not as deep as in the next stage, called delta sleep, where the EEG shows slow waves of only 0.5–2 cycles per second (cps). This stage is usually reached in half an hour or so and may persist for an hour to be followed by a stage of REM sleep, with its characteristic EEG tracings. The cycles then repeat themselves beginning again with Stage 2 sleep. Thus alternating NREM stages cycle with the REM stages throughout the sleep period.

12.2. Neurotransmitters

In Chapter 8, peripheral aspects of the cholinergic system were considered. The basics of the neuron (Fig. 8-1), specifically the cholinergic neuron (Fig. 8-5), the synapse (Fig. 8-2), and depolarization (Fig. 8-3), were also presented. The significance and chemistry of acetylcholine (ACh) was discussed. Chapter 9 continued with the remainder of the peripheral autonomic nerve plan (i.e., the sympathetic, or adrenergic, system). The biosynthesis of the catecholamines DA, NE, and EP were discussed and illustrated (Fig. 9-1), and the metabolism schemes for NE and EP (Fig. 9-3) and DA (Fig. 9-4) were outlined. Much of the early (before 1960) neurochemistry elucidated, and neurotransmitters identified, were central to the functioning of the *peripheral* nervous system.

The discovery of the newer psychotropic drugs beginning in the early 1950s, and the recognition that their mechanisms were not explicable by the then-known brain chemistry, gave part of the impetus for the neurochemical progress that ensued over the next 20 years. Newly developed techniques provided the tools for studying in greater detail the synthetic sequences of neurotransmitters and their identification in central neuronal processes. Even the establishment of new transmitters, which were not previously known to function as such, was accomplished (e.g., DA, serotonin, and certain amino acids). Much new information on how neurotransmitters are taken up into synaptic vesicles or granules and subsequently released was developed during that period.

Table 12-2. Summary of Sites of Drug Action at Synapse

1. Interference with neurotransmitter synthesis
   a. stimulate synthesizing enzymes
   b. blocking synthesizing enzymes
2. Precursors
   a. to produce increased neurotransmitter levels
   b. to produce less effective "false" transmitter
   c. to produce more effective "false" transmitter
3. Effects on neurotransmitter uptake and vesicle storage
   a. blockade—resulting in nerve ending depletion; can be a drug mechanism
   b. augment—very unlikely
4. Stimulation of neurotransmitter release from vesicles
5. Catabolic enzymes
   a. block in cytoplasm—prevents transmitter breakdown
   b. block in synpatic cleft—prolongs transmitter action
6. Direct action on receptors
   a. activation (i.e., agonists)
   b. blockade (i.e., antagonists)
   c. activation and blockade—mixed agonist/antagonists
7. Direct action on synaptic membranes
   a. usually changes permeability—affects ionophores

The function of Ca^{2+} in the complex activation of the storage vesicles, which resulted in their fusion to the neuronal membrane and the release of their contents into the synapse following membrane depolarization, was also elucidated.

Table 12-2 is an outline summary of the various points in the synaptic complex that drugs can affect. A psychotropic drug will rarely express a single effect. Our interest here is the principal influence, with the understanding that the others may well be responsible for the observed side effects and toxicities.

One of the factors that greatly increases the difficulty of unraveling the complex mechanisms of psychopharmacological agents is the relatively recent recognition of *autoreceptors*.[2] Unlike postsynaptic receptors, autoreceptors are affected by transmitter substances released into the synaptic cleft by the same neuron that produced and released it. They function to inhibit further release (or synthesis) of themselves.

In the case of drug effects, they may be agonist or antagonist, depending on whether they affect pre- or postsynaptic receptors more. In turn, that will likely depend on drug concentration (i.e., dose). For example, a low dose of a given drug might block the presynaptic receptors of a dopaminergic neuron that is preventing DA from producing its normal feedback inhibition of DA synthesis. This allows intrasynaptic levels of DA to rise and interact increasingly with its postsynaptic receptor. The net result of our hypothetical drug would then be a dopaminergic effect. On the other hand, a higher concentration (dose) of the same compound would block the postsynaptic receptor, thereby producing an antidopaminergic effect. This concept has explained some previously puzzling "contradictory" reports of drugs at different doses producing opposite results.

Another regulatory mechanism involving postsynaptic receptors is the phenomenon where subnormal stimulation—produced by denervation or simply decreased levels of sub-

---

[2] Also see discussion on clonidine, Chapter 10.

strate agonists—results in increased receptor synthesis and/or increased receptor sensitivity. This phenomenon can be observed with β-blockers (Chapter 10) and LDL receptors (Chapter 11).

The opposite is decreased receptor availability as a result of excessive stimulation (e.g., β-agonists such as isoproterenol).

Autoreceptors have been identified for NE, DA, serotonin, and gamma-aminobutyric acid (GABA)–containing neurons. In the case of NE autoreceptors have been well characterized peripherally and centrally (in the brain) as being mainly—but not exclusively—α_2 types (see Chapter 10).

12.2.1. Specific Neurotransmitters

Acetylcholine (ACh) and cholinergic receptors were considered in some detail in Chapter 8. Table 12-3 illustrates the distribution of ACh in the (rat) brain. Cells in various parts of the brain are sensitive to ACh applications. In the cat the microapplication of ACh in the cerebral cortex causes a response in a significant fraction of neurons.

The synthesis, storage, release, and breakdown of ACh were discussed (Chapter 8). It should be allowed that cholinoreceptors may be post- or presynaptic, with the latter probably near transmitter-releasing sites, thereby permitting feedback inhibition. Various areas of the brain have cholinoreceptors; they may be excitatory or inhibitory. The cortex, hippocampus, caudate nucleus, thalamus, hypothalamus, and possibly the spinal column have muscarinic cholinoreceptors.

In Chapter 8 certain relationships between memory loss (e.g., Alzheimer's disease) and central cholinergic mechanisms were addressed. Cholinomimetics acting either directly or indirectly are capable of crossing the BBB to exert some beneficial effects. The cholinesterase inhibitor physostigmine is the only such marketed agent that can enter the brain, thereby increasing central cholinergic activity. This is the reason that physostigmine has been so useful to treat the adverse effects of excessive central (as well as peripheral) blockade by anticholinergic muscarinic drugs such as atropine and its synthetic tertiary amine analogs. In fact, the drug is considered an antidote for poisoning with such drugs by pharmacologically reversing central muscarinic blockade. Many of the antipsychotics such as the phenothiazines and tricyclic antidepressants that will be discussed later have a strong central (as well as peripheral) anticholinergic component in their pharmacological profile,

Table 12-3. Distribution of Biogenic Amines in the Rat Brain[a]

Brain area	Serotonin[b]	Norepinephrine[b]	Dopamine[b]	Acetylcholine[b]
Whole brain	0.37	0.28	0.49	[c]
Hemispheres	0.25	0.17	0.50	
Hypothalamus	0.65	0.73	0.52	
Thalamus	0.36	0.35	0.22	
Medulla oblongata	0.55	0.32	0.09	12.3
Cerebellum	N[d]	0.10	N*d*	2.3
Olfactory bulbs	0.21	0.26	0.19	16.4
Mesencephalon	—	—	—	16.7
Diencephalon	—	—	—	22.3

[a] Modified from Valzelli, L., et al. (1968)
[b] μg per gm of tissue (standard deviations omitted).
[c] White matter, 7.9 μg/mg gray matter, 3.3 μg/gm.
[d] Nonmeasureable.

as do many H_1 antihistamines (Chapter 13). Physostigmine can therefore be used successfully to reverse the central symptoms of overdose.

Catecholamines. The peripheral aspects of NE and EP as neurotransmitters were considered in Chapter 9. NE exists at significant levels in the hypothalamus and in lesser amounts in the medulla oblongata, midbrain areas, and the pons. EP, the adrenal hormone, may possibly have some central transmitter functions. DA, however, has major involvement in central mechanisms and will be considered further here.

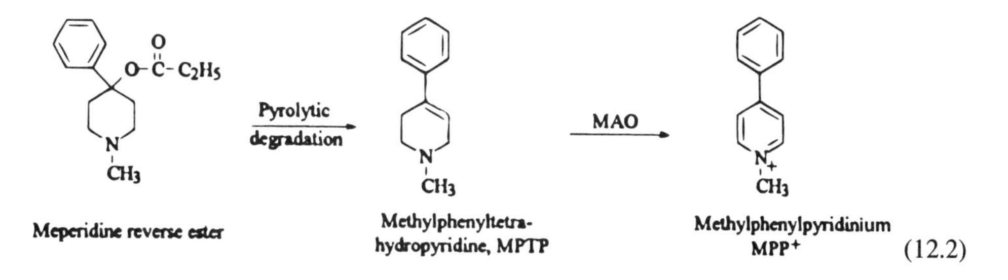

(12.2)

The relationship of Parkinson's disease to DA deficiency that results from a slow degeneration of dopaminergic neurons in the *substantia nigra* area of the brain was brought out in Chapter 9. The biochemical rationale of its treatment was also discussed. One of the difficulties hampering research in the area was lack of a suitable animal model. A byproduct contaminating an illicit product batch of "reverse" meperidine that was apparently produced by excessive temperatures (Eq. 12.2) led to the identification of the potent neurotoxin N-methyl-4-phenyltetrahydropyridine (MPTP) as the agent inducing and irreversible Parkinsonism in several young drug abusers. The toxin produced similar symptoms in monkeys and other animals, thus enabling them to become models for studying the disease and its treatment. Further research showed MPTP to be a "protoxin"; that is, it first had to be extraneuronally oxidized to N-methyl-4-phenylpyridinium (MPP^+) by type B glial cell monoamine oxidase (MAO) for which it has a high affinity. Premedication with MAO inhibitors blocked MPTP from becoming neurotoxic.

The highest brain concentrations of DA have been shown to be in the caudate nucleus. About two-thirds of the neurons in this structure respond to DA; most are depressed by the transmitter, and a minority are excited. The caudate nucleus is a component of the extrapyramidal system that controls skeletal muscle. Of course, this fact relates to Parkinson's disease. Drugs that reduce dopaminergic functions by blocking DA receptors or in some other way interfere with DA may produce Parkinson-like symptoms as side effects, irrespective of what other activities they possess. This is what occurs with many antipsychotic drugs such as the phenothiazines.

The elucidation of dopaminergic pathways in the brain—and catecholamine pathways generally—became possible only after the development of immuno- and cytofluorescent techniques in the 1960s. These enabled the identification and mapping of pathways and receptor locations of each neurohormone separately. DA receptors and neurons have now been identified in the vomiting center (area postrema), which explains why dopaminergic blockade results in antiemetic effects (e.g., metoclopramide, phenothiazines).

Even though initial receptor-binding studies led to the belief that four distinct DA receptors existed, it is currently accepted that there are only two receptors for DA: D-1 and D-2; however, each may occur in high- and low-affinity states.

The present view is that D-2, as a high-affinity species, represents the presynaptic autoreceptor in the CNS. The low-affinity D-2 receptor, then, is postsynaptic. D2 receptors are not coupled to *c*-AMP as a secondary messenger. Their mass is estimated as 136,700 daltons. The structural variety of D-2 *antagonists* varies considerably and includes many clinically important groups of antipsychotic drugs: the phenothiazine "tranquilizers" and several of their bioisosteres (the butyrophenones), a dibenzodiazepine (clozapine), the indole derivative molindone, and a benzamide (sulpiride), all to be discussed later. The ergot alkaloids represent D-2 agonists.

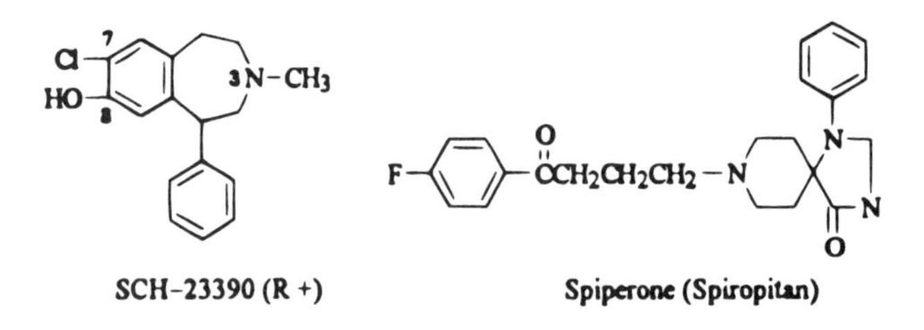

SCH-23390 (R +) **Spiperone (Spiropitan)**

D-1 receptors are distinct. In fact, only one class of specific D-1 antagonists is currently known: a group of 1-phenyl-3-methyl benzazepines, particularly SCH-23390. It was initially noted because of its potent inhibition of DA-sensitive adenyl cyclase at 2,000 times lower concentrations than necessary to displace tritium-labeled [^{3}H]-spiperone, which is a butyrophenone that is the most potent neuroleptic and D-2 agonist found thus far. The selectivity of a D-1 receptor antagonist ligand SCH-23390 was demonstrated by using the tritium-labeled compound in [^{3}H]-spiperone binding studies. The "cold" compound had a 2,500 times greater ability to displace its [^{3}H]-labeled counterpart than did [^{3}H]-spiperone. Conversely, spiperone's affinity for D-2 receptors was 70,000 times higher than for D-1. Both binding and inhibition studies of DA-sensitive adenylcyclase reinforce the notion of two distinct DA receptors. The molecular weight of D-1 receptors is also significantly different at 79,500. Any early doubt regarding the existence of D-1 receptors was dispelled with the demonstration of some behavioral responses attributable to SCH-23390, particularly blockade of stereotypy[3] induced by apomorphine- and amphetamine-type drugs. Even greater interest was generated by the finding that SCH-23390 and possibly other D-1 antagonists are also potent blockers of the conditioned avoidance response (CAR) in rats and monkeys. The CAR has long been considered a predictor of clinically useful antipsychotic properties in humans. Such psychopharmacology was previously believed the sole domain of D-2 antagonists. It is, of course, a characteristic of all neuroleptics in clinical use; however, it is now apparent that in some at least, this property also extends to D-1 receptors. Thus D-1/D-2 mixed antagonisms exists.

The most troublesome untoward effect with the antipsychotic drugs in use are the extrapyramidal effects (EPE). These are mainly Parkinson-like motor effects: tremors, abnormal repetitive movements, loss of muscle movement (akinesia), tardive dyskinesia (TD), rhythmic stereotypical movements, and drooling.

Other side effects characteristic of most neuroleptic–antipsychotic drugs include autonomic effects such as postural hypotension and impaired sexual function. Increased plasma prolactin represents an endocrine effect.

---

[3] Persistent, meaningless movements characteristic of catatonic stupor.

DA agonists (particularly D-2) are also of therapeutic and theoretical interest. They could add therapeutically to the few available drugs to treat Parkinson's disease, acromegaly, and hyperprolactinemia (e.g., bromocriptine, pergolide, Chapter 9, Fig. 9-2). Such compounds theoretically may aid in the characterizations of DA–receptor geometry and topography that have not yet been satisfactorily established.

A large number of DA agonists have been synthesized and their DA-ergic activity has been evaluated regarding comparative potency, types of receptor affinity (DA1, DA2, autoreceptor), and SARs, including stereochemical factors. The types of structures involved in these studies are depicted in Figure 12-2. Note that incorporating the phenethyl moiety into the framework of ring systems produces semirigid analogs of DA (e.g., the aminotetralins, ATN). The benzoquinolines also offer semirigid position isomerism: octahydrobenzo[g]quinolines and octahydrobenzo[f]quinolines each in turn also exhibits *cis–trans* isomerism, depending on ring fusion geometry. It will be recalled (Chapter 9) that apomorphine (APO) has been shown to mimic DA activity in the CNS and has since been reviewed as a prototype for potential SAR development. The ATNs are particularly interesting. Their synthesis as potential DA agonists in a systematic manner received impetus by a suggestion that the DA pharmacophore within the APO structure is the dihydroxy ATN framework. Both 2-amino-6,7-dihydroxytetralin (Fig. 12-2, $R,R' = H$) and the 5,6-dihydroxy isomer were potent DA-agonists. Disubstitution on the N (unlike with DA itself) produced increased potency. Of particular clinical antipsychotic potential is that the N,N-dimethylamino-6,7-dihydroxytetralin (Fig. 12-2, $R,R' = CH_3$, $X = 7\text{-OH}$) was shown to be a presynaptic DA receptor agonist. Its ability therefore to control (suppress) DA release as doses are increased may eliminate the extrapyramidal side effects so characteristic of the currently used postsynaptic DA antagonist drugs. If further studies corroborate this information, a new generation of antipsychotic drugs may emerge.

Serotonin, or 5-hydroxytryptamine (5-HT), is another monoamine whose important central effects have only been recognized recently. It had previously been known as a vasoconstrictor in the plasma. Once identified chemically, it was found to be widely distributed in the body. After determination of significant brain levels in the hypothalamus, medulla, midbrain, and other areas (Table 12-3), and after establishing its biosynthetic paths, serotonin became recognized as a neurotransmitter. 5-HT is presently less well understood than are the catecholamines. Figure 12-3 outlines its biosynthesis from the essential amino acid tryptophan. Try enters the brain by active transport (as is L-dopa) and is hydroxylated there by tryptophan hydroxylase, which is an enzyme similar to if not identical to, tyrosine hydroxylase.

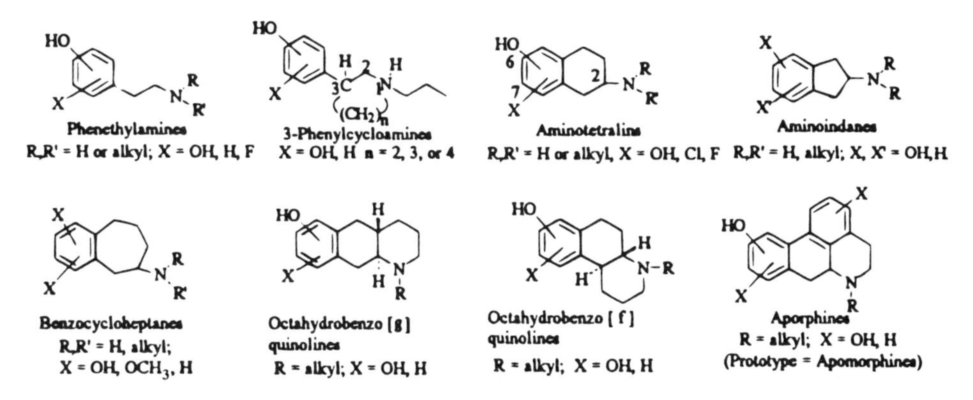

Figure 12-2. Experimental analog classes. (Adapted from Katorinopoulos and Schuster, 1987.)

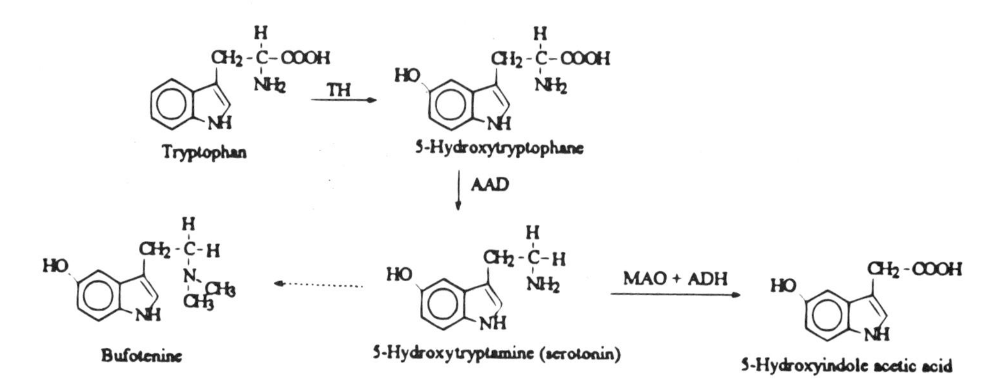

Figure 12-3. Synthesis and metabolism of serotonin. TH = tryptophane hydroxylase; AAD = L-aromatic amino acid decarboxylase; MAO = monoamine oxidase; ADH = aldehyde dehydrogenase. (Broken arrow indicates possible aberrant pathway.)

Decarboxylation by L-aromatic amino acid decarboxylase intraneurally (as does L-dopa) affords serotonin. Even though the same reaction occurs peripherally, this neurotransmitter, like DA and NE, does not cross the BBB. In a manner analogous to the catecholamines, 5-HT is taken up and stored in presynaptic granules. Again, analogously to NE, reuptake into the neuron is the major route of inactivation in the synapse areas. In addition, 5-HT is also a suitable substrate for MAO-A, which catalyzes its oxidation to 5-hydroxyindoleacetaldehyde for further oxidation by alehyde dehydrogenase to 5-hydroxyindoleacetic acid (5-HIAA). Of minor consequence is a possible reduction to the alcohol and sulfate conjugation of the 5-OH group.

The possibility that an aberrant pathway that utilizes N-methyltransferase might convert the serotonin to 5-hydroxy-N,N-dimethyltryptamine (bufotenine) in schizophrenic patients has fascinated researchers. Bufotenine, which is an alkaloid, was first isolated from the skin glands of toads (of the genus *Bufo*) as well as of several plant sources. N,N,-Dimethyltryptamine is a reported psychotomimetic and hallucinogen. Since it was found in the CAR test, the possibility that this 5-HT metabolite, if formed in schizophrenics, might be, at least in part, the cause of the disease could not be overlooked. Reports that IV infusion of bufotenine was hallucinogenic in humans, however, have not been corroborated by others, even with high doses.

Serotonin-containing neurons were visualized and mapped by utilizing the histochemical techniques previously mentioned. Serotoninergic receptors have been identified on both neurons containing and devoid of 5-HT. Their existence in multiple forms in the CNS is almost certain. Binding studies indicate two or more receptors. Two distinct postsynaptic receptors are likely: 5-HT_1 and 5-HT_2 (5-HT_1 is inhibitory and possibly linked to adenyl cyclase). A presynaptic receptor controlling serotonin release has also been proposed. As it is inhibitory, it has been considered as a 5-HT_1 subtype.

More recently, 5-HT_3 receptors have become an established entity. Actions attributable to stimulation and inhibition of this receptor subtype, both centrally and peripherally, have been described, and some potential antagonists are in clinical trials. Therapeutic benefits are expected in the CNS and in gastrointestinal disorders.

Other examples of 5-HT_3 agonists besides HT are 2-methyl-HT, bufotenine, and its methyl quaternized derivative. More antagonists of clinical interest are under investigation.

Cocaine, which has been known as a serotonin antagonist since 1953, appears to have 5-HT_3 antagonism, albeit of low specificity. 5-HT_3 receptors in the brain are of particular interest. Direct evidence of their existence was established with the tritium-labeled potent indole antagonist. This study identified a single high-affinity site. Such a high-density locus was the area postrema, containing the chemoreceptor trigger zone (CTZ). Although not the only area in the brain that is well endowed with 5-HT_3 receptors (e.g., also the hind brain, substantia gelatinosa, and dorsal spinal cord), its importance in mediating anticancer drug- and radiation-induced emesis seemed to offer the best promise of a new type of highly effective drug against this dread condition. Even though other therapeutic applications for 5-HT_3 antagonist seem viable (e.g., inflammatory pain), antiemesis research has already borne fruit with the introduction (1991) of ondansetron (Zofran), which is a drug effective against cisplatin- and phosphoramide-induced emesis.

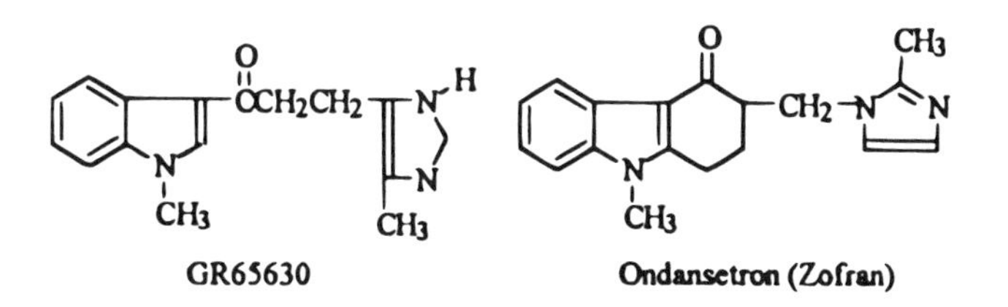

LSD (Fig. 10-9), as the D-isomer, psilocin, and dimethyltryptamine are all potent hallucinogens containing the indole nucleus. They may therefore be viewed as serotonin analogs. They are 5-HT antagonists.

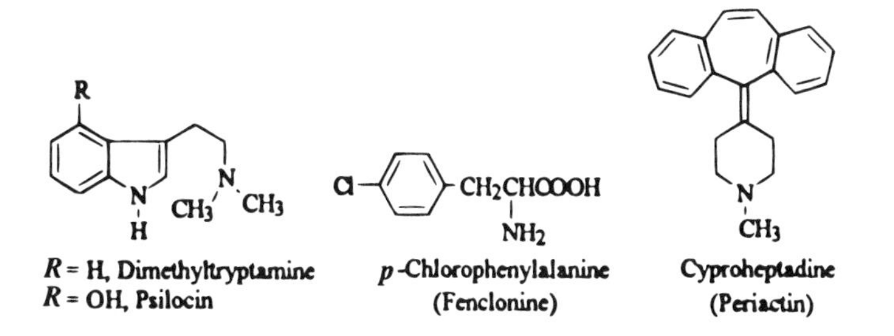

As might be expected, other semisynthetic ergot derivatives, such as methysergide (Fig. 10-9) and 2-bromo-LSD, act similarly. Other unrelated structures that have antitryptaminergic (antiserotoninergic) actions while simultaneously exhibiting other unrelated actions are cyproheptadine and ketanserin.

Cyproheptadine is also a potent H_1 antihistamine (Chapter 13). In addition, it exhibits antimuscarinic and sedative effects. Kitanserin is an inhibitor of peripheral 5-HT_2 receptors, and it exhibits postsynaptic α_1 antagonism as well. It is unlikely to have psychotropic activity because of its peripheral nature.

Serotonin itself has no clinical applications. However, its several physiological effects suggest that serotoninergic drugs might modify these for therapeutic advantage. Among the established or putative roles of 5HT in the brain is anterior pituitary control. Sleep control is under investigation. The neurotransmitter has been shown to initiate NREM sleep; it also curbs the appetite. The latter two effects go a long way in helping to explain the apparently paradoxical sedation of the amphetamine-structured anorectic fenfluramine (Chapter 9), an effective inhibitor of 5HT reuptake. By depleting serotonin from the central neurons involved, these drugs will raise synaptic levels, essentially increasing those two physiologic actions of serotonin. By acting on serotoninergic neurons, possibly as an agonist, or by another unknown mechanism, the psychotomimetic actions of dimethyltryptamine (and other indole compounds) may be partially responsible for its behavioral effects.

It should be mentioned that *p*-chlorophenylalanine (fenclonine) is an irreversible inhibitor of tryptophan hydroxylase, the first and rate-determining step in serotonin synthesis. Even though the efficiency of this decrease in 5HT may be as high as 90%, sedation does not result. Reserpine alkaloids, which also deplete 5HT and NE, are sedative. That leaves catecholamine depletion as relevant to sedation. It is interesting that *p*-chlorophenylalanine is used to treat carcinoid syndrome, which is a "nonmalignant" intestinal mucosal tumor pouring prodigious amounts of 5HT into the circulation.

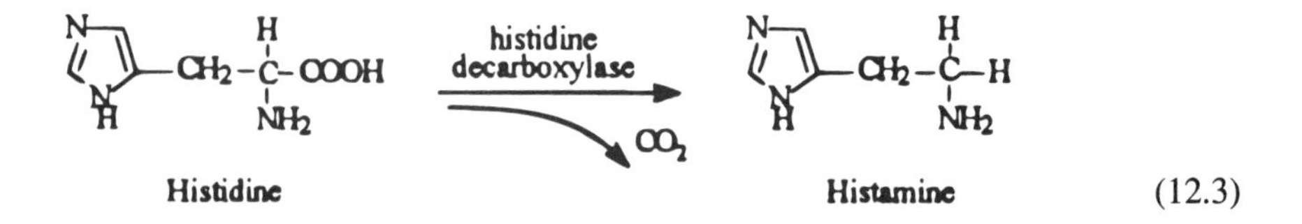

(12.3)

Histamine, as a biologically important amine, has been known since the beginning of this century. It arises in vivo by decarboxylation of the amino acid histidine (Eq. 12.3). Its involvement in allergic reactions and gastric acidity—both peripheral phenomena—will be discussed (Chapter 13). The extent of its role in the brain is still largely unclear. It is believed to be implicated in the central regulation of water intake, emesis, arousal, thermoregulation, and probably some pituitary functions. The question of central histaminergic neurons in the CNS, their putative functions, and what, if any, neurotransmitter functions histamine may have has not been determined with any certainty. Information is sketchy at best. For example, H_1 antagonists have been known to have considerable CNS depressant action (sedation) for decades. It is interesting that H_1 antagonists do not seem to have a similar effect on wakefulness during nighttime sleep. It should be remembered that these drugs also have blocking effects on 5HT, α-adrenergic, and cholinergic receptors. Therefore, simplistic interpretations of their effects may easily be incorrect.

In addition to the monoamines NE, DA, 5HT, and possibly histamine that function as neurotransmitters in the CNS, a number of amino acids (and peptides) with likely neurotransmitter credentials exist. GABA among these is the best documented. Others may be considered as putative or even simply as transmitter candidates at this point. Amino acids such as glycine (gly), serine (ser), glutamate (glu), aspartate (asp), taurine, substance P (SP), and others may be included here.

Gamma-aminobutyric acid. This is probably the most important, or at least the most studied, central amino acid neurotransmitter. It has been established as inhibitory with ubiquitous distribution in most areas of the brain (e.g., in the substantia nigra and globus pallidus levels are as high as 1 mg/g). The enzymes that synthesize and degrade it are likewise distributed. Like other transmitters, GABA is concentrated in specific synaptosome populations. Finally, autoradiography with radiolabeled GABA and immunocytochemical studies have shown distinct GABA-containing neurons in the brain and spinal cord. The compound has the "qualifications" of a neurotransmitter.

Figure 12-4 outlines GABA's metabolic interrelationships. Its biosynthesis, by decarboxylation of L-glutamic acid, occurs only in the CNS (and retina) since the enzyme, glutamic acid decarboxylase (GAD), only occurs there. Once formed, it cannot cross the BBB. Thus, it remains centrally. GABA-transaminase inactivates GABA to succinic semialdehyde, which is then rapidly oxidized to succinate. The NH_2 group released from GABA serves to transaminate the α-ketoglutarate to glu, which in turn again decarboxylates to GABA, thereby establishing a "shunt" into the Krebs cycle. Since α-ketoglutarate then essentially becomes the precursor of GABA, and its availability is "guaranteed" through its tie-in to brain glucose metabolism (glycolysis), the process assures a continuous supply of neurotransmitter.

GABA-ergic receptors must be assumed, and much evidence for them exists. Microiontophoretic application of GABA and the release of endogenous transmitter (by nerve stimulation) produced postsynaptic hyperpolarization in the cerebral cortex and other brain areas probably caused by increased Cl^- conductance of the postsynaptic membrane. The convulsant alkaloid bicuculline (Fig. 12-5), which can be viewed as containing a GABA isosteric component within its structure, *selectively* blocks the GABA effect. This suggests a competitive reaction at a GABA receptor.

Continued studies indicate that the synaptic GABA receptor is a multifaceted structure with several interrelated sites. Research utilizing labeled agonists such as [^{3}H]-GABA and [^{3}H]-muscimol has pinpointed recognition sites; so has the apparent antagonist [^{3}H]-bicuculline. Other antagonists may bind at nonrecognition sites such as associated chloride channels (see Fig. 12-6). It appears that low- and high-affinity GABA recognition sites

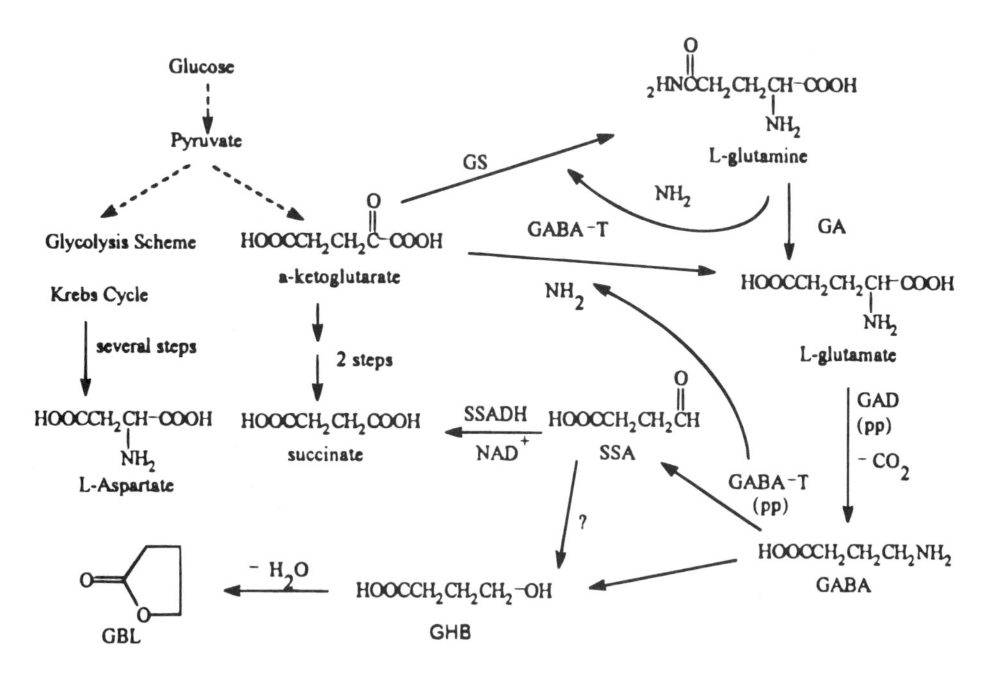

Figure 12-4. Gamma-aminobutyric acid metabolic interactions. GA = glutaminase; GABA = γ-aminobutyric acid; GABA-T = GABA: α-oxaloglutarate transaminase; GAD = glutamic acid decarboxylase; GS = glutamic synthetase; NAD^+ = nicotinamide adenine dinucleotide; PP = pyridoxal phosphate (vitamin B_6); SSA = succinic semialdehyde; SSADH = succinic semialdehyde dehydrogenase; GHB = γ-hydroxybutyric acid; GBL = γ-butyrolactone.

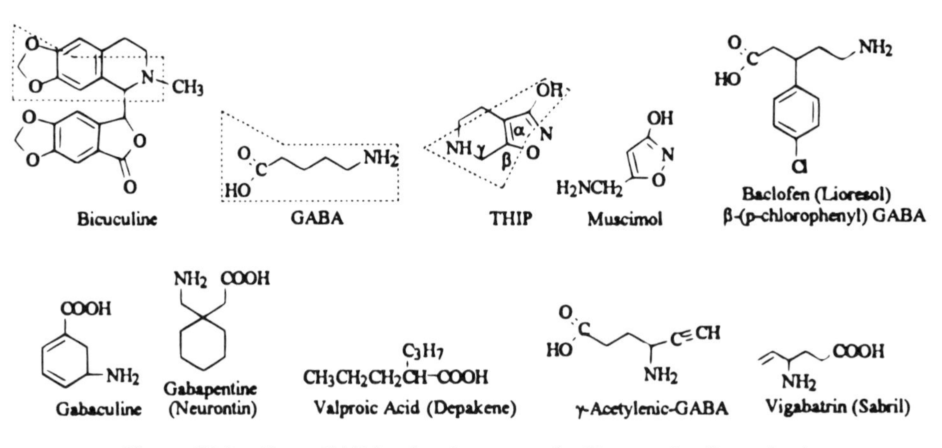

Figure 12-5. Some GABA-related compounds. (See text for discussion.)

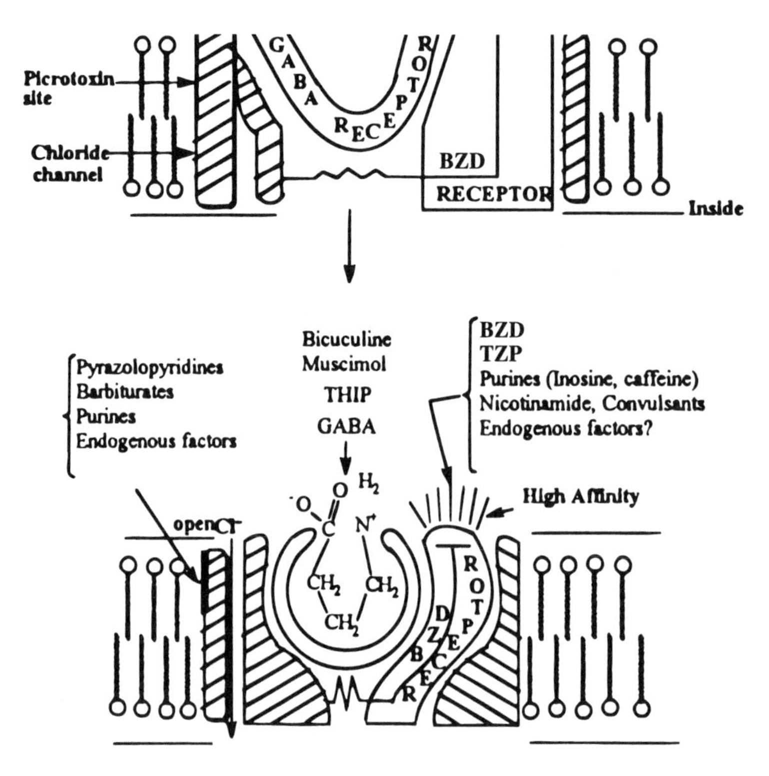

Figure 12-6. Schematic model of benzodiazepine-GABA-receptor chloride ionophore complex. BZD = benzodiazepine; TZP = triazolopyridazine; THIP and GABA; see Figure 12-5. (From Skolnick and Paul, 1981.) See text for discussion.

exist. The latter are "masked" by a modulator substance that is removable with detergent treatment (Triton-X at 0°). This "unmasking" may have led to the finding that the BZD anxiolytics (see later) may remove this modulator, thus activating this site and enhancing BZD binding to it—the BZD receptor. The wide GABA receptor distribution in the mammalian CNS makes it almost certain that subtypes exist. It is not important whether the activated BZD binding sites should be considered as such ($GABA_2$). The recognition that a group of sites in proximity interact in concert (affect each other) while having different recognition patterns is significant.

The current understanding is: Barbiturates (see later) in low doses and BZDs increase GABA-ergic transmission throughout the CNS, thereby potentiating both pre- and postsynaptic inhibition mediated by GABA. The sites are very likely pivotal in mediating the antianxietal properties of BZDs (Skolnick and Paul, 1981). It should be understood that compounds acting on the GABA receptor component ($GABA_1$?) such as GABA itself, muscimol, THIP, and bicuculline do not inhibit the BZD recognition site. This establishes that different recognition sites, or subtypes, exist. Actually, GABA, GABA-mimetics, barbiturates, and even ethanol increase the affinity of the BZD-binding sites, possibly by a neighboring perturbation effect (Skolnick and Paul, 1981). Is this also an unmasking?

Earlier investigations to relate BZD actions to known or putative neurotransmitters in the brain have shown that NE, DA, 5HT, ACh, glu, gly, and others were all ineffective in their ability to affect BZD (diazepam) binding. The studies, however, did point to GABA as the probable mediator of BZD effects. The chloride ion channels (ionophores) can open and close, thereby participating in influencing the topographic features of the receptor complex in response to either endogenous ligands or drugs. The receptor–chloride–ionophore complex (Fig. 12-6) evolved as a concept.

Because of the growing recognition of GABA's importance, the search for GABA–receptor agonists, antagonists, uptake inhibitors, and inactivators of the GABA-T enzyme began. The last mentioned would potentiate GABA transmissions. Such compounds may have therapeutic potential in the treatment of conditions characterized by demonstrated or suspected GABA function deficiencies. Diseases that might benefit from such drugs are Parkinsonism, epilepsy, and schizophrenia.

GABA-T utilizes pyridoxal as the cofactor in the transamination reaction (Fig. 12-7A). Pyridoxal 5-phosphate (Vitamin B_6, the cofactor) forms a Schiff base with GABA's NH_2 group. The adjacent C–H bond has its proton abstracted by the enzyme; reprotonation results in the tautomeric Schiff base, which on hydrolysis affords succinic semialdehyde and pyridoxamine. The pyridoxamine then forms a Schiff base with the carbonyl of α-ketoglutarate, reversing the steps, whereby hydrolysis of the tautomeric base yields L-glutamic acid, and the pyridoxamine, which has given up its NH_2 function, reverts to the cofactor aldehyde form to repeat the cycle.

Isolation of the simple hexadiene ring analog of GABA, *gabaculine* from *Streptomyces toyocaensis*, provided the most potent GABA-T inhibitor yet. It also inhibits the pyridoxal-dependent enzyme GAD, but it is 99.9% less effective since GABA levels in mice rise dramatically in its presence. Gabaculine is a suicide enzyme or K_{cat} inhibitor (see Chapter 2). It will be recalled that such inhibitors have latent reactive functions that the target enzymes activate on binding, thereby becoming irreversibly inhibited. Such inhibitors tend to be very specific since they must be acceptable as substrates for a particular enzyme in their *inactive form*. In the case of gabaculine it can be seen (Fig. 12-7B) that the expected Schiff base forms and also equilibrates via a 1,3-proton shift to produce the nonhydrolyzable

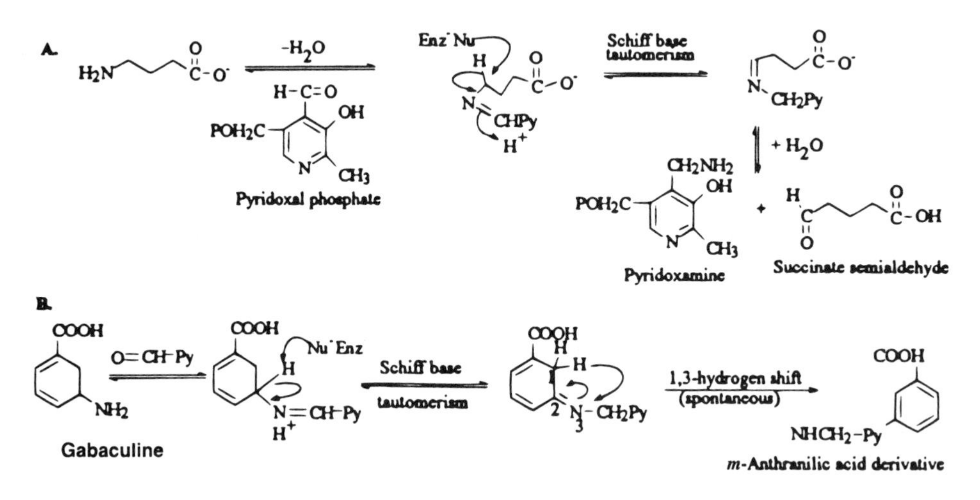

Figure 12-7. GABA-T mechanism and its "suicide" inhibition.

anthranilic acid derivative shown (Rando, 1977). γ-Acetylenic-GABA may be considered as a noncyclic analog (Fig. 12-5) that may also act this way.

Baclofen, β-(*p*-chlorophenyl)-GABA (Lioresal) was designed, by the addition of the lipid solubilizing benzene ring, to cross the BBB better. Its GABA-mimetic action appears indirect, most likely by being able to stimulate GABA release stereoselectively [(+)isomer]. The drug has clinical applications as an oral muscle relaxant in cerebral palsy and other muscle spasticities. Valproic acid (Fig. 12-5) as the sodium salt affects GABA metabolically by inhibiting succinic semialdehyde dehydrogenase (Fig. 12-4). Since the aldehyde backs up, GABA-T activity decreases, thus raising brain GABA levels. The result is seizure inhibition. The drug is used as an antiepileptic.

Muscimol is a psychotomimetic found in *Amanita muscaria*, which is a poison mushroom. Like THIP, it is a selective GABA agonist. Both compounds share the 3-isoxazolyl moiety, which may be viewed as a masked carboxyl function that is "recognized" by GABA receptors, but not other GABA-interacting sites.

Glycine (aminoacetic acid) is the simplest of the amino acids. Because of the many metabolic pathways in which it is found, its central inhibitory neurotransmitter properties went unrecognized until its spinal cord distribution was examined. The convulsant alkaloid strychnine is believed to block postsynaptic glycine receptors. Presynaptic release impairment may also be part of its action. While the peripheral chemical reactions of glycine are reasonably well known, its conversion to pyruvate via serine—the biochemistry of gly in central neuronal fibers—is still not fully understood. High-affinity uptake systems for gly (and several other transmitter amino acids) were soon demonstrated in CNS tissue. Serine, of course, acts as a precursor to gly. Whether it also has inhibitory neurotransmitter properties is in doubt.

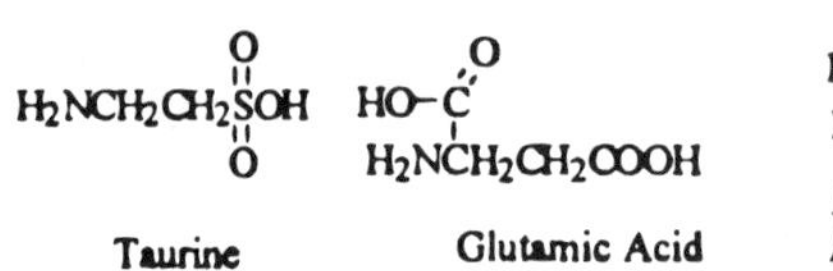

$RHNCH(COOH)CH_2COOH$

R = H, Aspartic Acid
R = CH_3, N-methyl-D-aspartic acid

Taurine (2-aminoethanesulfonic acid) may have certain central inhibitory properties that qualify it for neurotransmitter status. At present only strychnine appears as a viable antagonist (as it is for gly), making it difficult to differentiate the central effects of these two amino acids.

Glutamic and *aspartic* acids, which are viewed as excitatory neurotransmitter amino acids, are present in the CNS in relatively large amounts. Kainic acid, which is viewed as a rigid and therefore conformationally restricted glutamate analog, has been proposed as an agonist of it, if not as a direct receptor activator, then by sensitizing these receptors to glu by some allosteric interaction.

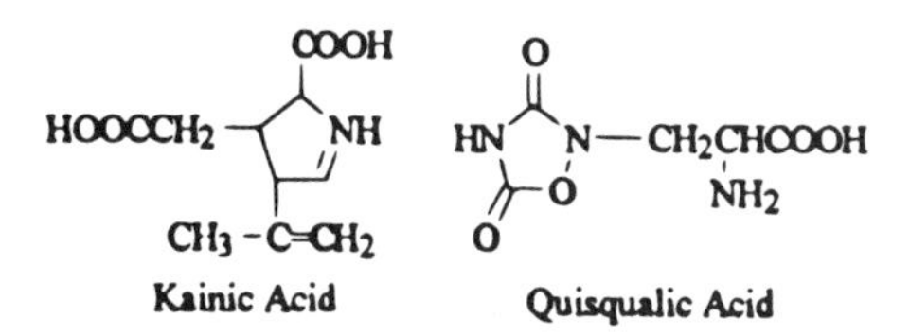

Kainic Acid   Quisqualic Acid

To complicate the picture, kainic acid is a neuronal toxin. In immature animals, in the presence of the compound, either glu or asp will cause nerve cell degeneration. Direct intracranial injection of kainic acid will destroy neuronal cells around the site of injection. Quisqualic acid and several other cyclic analogs also appear to exhibit glutamate agonism. It is possible that glu and asp antagonists may be neuroprotective in some situations. Be that as it may, by producing selective brain lesions kainic and similar acids have been useful neurobiological research tools in exploring the physiology of excitatory CNS transmission.

A third excitatory amino acid (EAA) receptor that is now receiving attention is the N-methyl-D-aspartate (NMDA) receptor. This receptor may be involved in epileptic discharges and hypoxia damage. There are also selective agonists (NMDA and cyclopentane glutamate) and a selective antagonist (D-APV_5) available for sophisticated studies to demonstrate the extent brain processes are affected by NMDA activation of its receptors.

EAA postsynaptic receptors have been divided into three types. One subtype is NMDA activated, one is activated by kainic acid, and the third by quisqualic acid. However, this separation has already been challenged by newer experiments that suggest additional subdivisions based on apparent pharmacological differences at different brain sites.

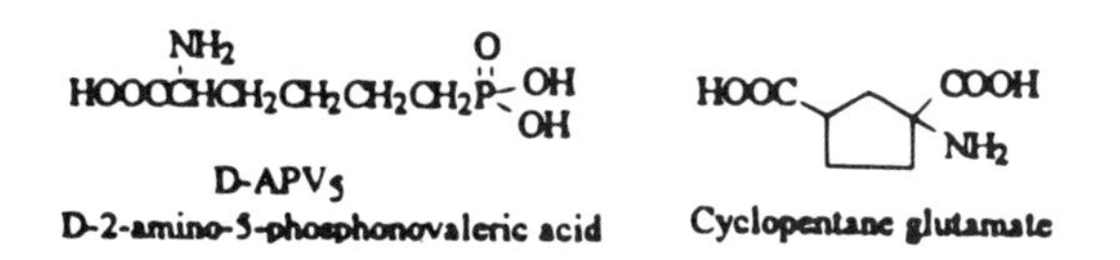

D-APV_5
D-2-amino-5-phosphonovaleric acid   Cyclopentane glutamate

Substance P (SP), which is the undecapeptide involved in pain mediation (Chapter 5), even though it is not fully characterized due to lack of specific antagonists, is considered an excitatory neurotransmitter. It occurs in all brain areas and the spinal cord. Neuroactive peptides are frequently found to coexist in amino acid or monoamine secreting neurons (e.g., SP in cholinergic and serotoninergic fibers, and somatostin with GABA).

Peripheral neurotransmitters almost constitute a "mature" area of understanding. The situation is centrally very much in the "quagmire" phase. Thus, while NE, DA, ACh, 5HT, and GABA are definitively established transmitters, as are gly, glu, asp, and probably taurine and

SP, a number of neurotransmitters that are still putative include other neuronal peptides (enkephalins, endorphins) and peptides that act as hormones elsewhere (e.g., angiotensin I and II, prolactin, somatostatin) as well as other amino acids. The number easily exceeds 50 substances, with no end in sight. Other small molecules that may be called the "also suggested" category include such candidates as ATP, *c*-AMP, GTP, *c*-GMP, estrogen, corticosterone, and even several prostaglandins. They will not be considered in this brief overview.

Even though it has been definitely established that certain drugs known to affect behavior do alter neurotransmitter levels (for example, MAO inhibitors), others that are equally effective experimentally and even clinically, do not appear to bring about measurable fluctuations of levels in the brain. This does not necessarily mean that no such changes are effected. The difference might be too small for current methods to detect; also, they may be extremely localized. A given drug may not cause direct changes on normal levels, yet it may be very significant when the dynamics of a given transmitter's synthesis and degradation are considered. For example, it is entirely possible to maintain a steady-state concentration of a given biogenic amine whose synthesis or degradation, or both, have multiple pathways and rates. It is feasible for a drug to affect the rate of synthesis of a neurotransmitter, which in turn will stimulate or depress its metabolism or vary the rate of synthesis by feedback mechanisms. This will result in unmeasurable changes in concentrations even though dynamic (turnover) effects may be drastically affected. Simple concentration values are unlikely to be informative in such situations. Figure 12-8 schematically illustrates these relationships.

12.3. CNS Depressants

In this section drugs used because of their CNS depressant properties will considered. The volatile general anesthetics (gases and low-boiling liquids), fixed general anesthetics (e.g., certain barbiturates), and sedative hypnotics (as defined earlier) will be included. In subse-

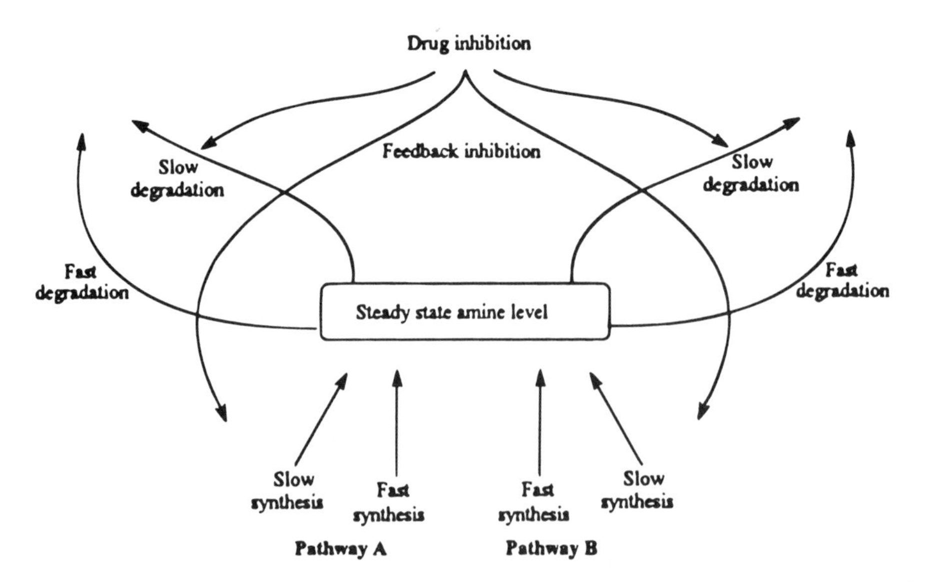

Figure 12-8. Schematic relationships between synthesis and metabolic degradation of a biogenic amine.

Table 12-4. Representative Volatile General Anesthetics

Structure or formula	Name	Comments
1. N_2O	Nitrous oxide "laughing gas"	Shallow anesthesia; some use in dentistry; given with 20% O_2; analgetic
2. $_2HC{=}CH_2$	Ethylene	Historical interest; need 80% for surgical anesthesia; obsolete; explosive and flammable
3. $CH{=}CH$	Acetylene	Obsolete; similar to ethylene
4. Δ	Cyclopropane	Potent anesthetic; used as 20% mixture for surgery; explosive
5. CH_3CH_2-O-CH_2CH_3	Diethyl ether	Potent at 6–8% concentration; good safety margin; high postoperative nausea and vomiting; flammable; forms explosive peroxides
6. $CH_2{=}CH$-O-$CH{=}CH_2$	Vinyl ether (Vinethene)	Short induction, rapid recovery used at 4%; polymerized, hepatotoxic
7. CF_3CH_2-O-$CH{=}CH_2$	Fluoroxene	Flammable; fast induction at 4–6%; useful muscle relaxation
8. Cl_2CHCF_2-O-CH_3	Methoxyfluorane (Penflurane)	Nonflammable or explosive; skeletal muscle relaxant; no bronchial irritation; slow induction and recovery
9. F_2CH-O-CF_2CHFCl	Enflurane (Ethrane)	Nonflammable; 2–5% concentration; little or no cardiac effect
10. $CHCl_3$	Chloroform	Potent anesthetic; cardio and hepatotoxic; obsolete for human use; some veterinary applications
11. $CF_3CHClBr$	Halothane (Fluothane)	Nonflammable; nonirritant, used at 1–2% concentration; little nausea; may depress circulation and respiration
12. $Cl_2C{=}CHCl$	Trichloroethylene	Obsolete except for trigeminal neuralgia pain relief
13. Xe	Xenon	Excellent anesthetic, but not readily available; very costly, not really used clinically

quent sections of this chapter dealing with anxiolytics and the major tranquilizers, CNS depression should be considered primarily as an undesirable side effect that, with the possible exception of buspirone, has not been successfully "designed out" of such compounds.

12.3.1. General Anesthetics (Volatile)

One of the earliest applications of organic chemistry to therapeutics was the synthesis of compounds that have narcotic (sleep-inducing) and anesthetic properties free of the dangers of morphine-containing preparations that had been in use for these purposes prior to modern chemistry.

In general, substances depressant to our CNS tend to be structurally nonspecific, relatively inert organic compounds.[4] Their CNS activity appears to depend primarily on physical properties that are not related in any obvious way to their chemical geometry. An examination of Table 12-4 makes this apparent.

Early hypotheses at the turn of the century (Meyer, 1901) recognized a relationship of the then-known inert substances that had depressant action to lipid solubility, and that this

[4] Several inert inorganic gases have anesthetic properties.

effect was more pronounced in cells with a high lipid content. The fact that this could be quantified by determining oil–H_2O partition coefficients was also appreciated; however, the fact that the linear relationship did not extend to compounds whose extreme lipid solubility precluded any aqueous solubility was not. It can be assumed that the likely reason is that such extreme lipid solubility in a compound would preclude its partitioning itself out of the first site it enters and not significantly affect the nerve fibers.

Ferguson (1939) proposed that the bioactivity of various chemically unrelated inert compounds would be the same if they reached the same relative saturation point. This means it would be informative to determine the ratio of the anesthetic vapor pressure in the gas (mixture) being administered to that of saturation vapor pressure of that substance; in effect, measure the thermodynamic activity of the gas. A later refinement measured alveolar concentrations of volatile anesthetics at equipotent levels. It demonstrated that the oil–gas partition ratio corresponded better to anesthetic activity than did other factors.

Consideration of the bilayer membrane structures of mammalian cells led to a suggestion that anesthetics function by displacing the phospholipid layers, as long as the molecular volume of such compounds (gases) is larger than that of O_2 and H_2O (vapor). This would presumably disrupt certain membrane structures and thereby interfere with impulse transmission.

A novel theory envisaged that the water, not lipid molecules, should be used to explain anesthesia. This notion stated that those volatile compounds that depress the CNS do so by forming hydrates or clathrates (inclusion compounds) that then ensnare the ions normally involved in the impulse conduction and therefore impeded depolarization. The difficulty with this idea is that even though several hydrocarbons (e.g., cyclopropane) and inert gases (e.g., Xe) do form actual hydrates (but not at body temperature and 1 atmosphere), other known anesthetics such as methoxyfluorane, ether, and halothane could not be induced to form them in vitro. Considering the variety of chemical compounds able to produce anesthesia—from inert gases (Xe), inorganic substances (N_2O), simple organic compounds ($CHCl_3$), and more complex ones (enflurane)—and the lack of any apparent SAR, it is unlikely that one of the many theories propounded this century will be proven correct. The best generality that may be acceptable is that volatile anesthetics do affect the structural lipoidal components of the biological membranes.

Many of the early anesthetics such as ethylene, chloroform, and cyclopropane are considered obsolete in modern medicine. The reasons vary. Flammability and explosive potential apply to ethyl ether and cyclopropane. The inability of some to produce the skeletal muscle relaxation necessary for surgery, except at very deep levels of anesthesia, doomed them with the introduction of the muscle relaxants in the 1940s, beginning with curare and the synthetic analogs that followed (Chapter 8). This permitted the use of shallower levels of anesthesia and increased safety. Chloroform is quite hepato- and cardiotoxic, yet it was still widely used well into the twentieth century. Since ether was such an effective and toxicologically safe drug and had been in use for so long (since the mid-1840s), the idea of retaining the ether structure, but decreasing its flammability by substituting halogen atoms for hydrogens, was valid. Compounds 7, 8, and 9 (Table 12-4) are prime examples of this approach. Fluoroxene (No. 7), which is an ethyl vinyl ether in which only 3 H atoms have been replaced with the lightest halogen fluorine, is still very volatile (b. 43°) and very flammable. It also needs a suitable stabilizer to prevent polymerization via the vinyl moiety. The drug does have positive clinical attributes (see comments, Table 12-4). Enflurane (No. 9) contains 5 F atoms plus a heavier Cl atom. It is nonflammable and has a higher boiling point (57°). Methoxyflurane is the least volatile of the group (b. 105°) but the most potent anesthetic.

Halothane (No. 11) was introduced in 1956 and was soon among the most-used anesthetics; it became a prototype. It is quite volatile (b. 50°) and nonflammable, which are both desirable attributes. It has a good safety record over several decades. This is somewhat surprising considering halothane's similarity of physical and anesthetic properties to chloroform. Even though no SARs can really be carried out, halothane's low toxicity must be attributed to the fluorine atoms. In addition, the various fluorinated refrigerants (Freons) developed earlier were also found to be nontoxic.[5]

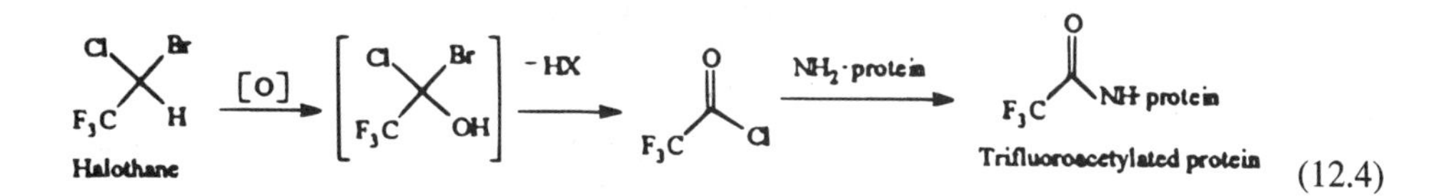

(12.4)

Since halogenated hydrocarbons are viewed as relatively inert, it is not surprising that up to 80% of the inspired (and absorbed) halothane that is exhaled is unchanged. Most of the remainder is also excreted unchanged. However, a small fraction is reductively dehalogenated. As expected, it is the weaker C–Cl and C–Br bonds that yield their halides as ions. The C–F bond appears undaunted. Thus trifluoroacetic acid (from the acid chloride) is identified in the urine (Eq. 12.4). However, in the case of enflurane and its analog isoflurane, trace quantities of F^- have been also identified.

The discovery that even inert gases such as N_2 and Ar can produce a narcotizing effect under pressure led to the evaluation of the heavier and more lipid-soluble inert gases, particularly Xe. Clinical trials demonstrated that Xe (with 20% O_2) had many of the desirable features sought in an anesthetic: rapid induction, muscle relaxation, and rapid recovery. This clearly showed that inert gases, if sufficiently fat soluble, achieve CNS depression by purely physical means.

Because lipophilicity is critical and SARs are not, relatively simple explanations of anesthetic activity are still appealing. Thus considering the known fact that such compounds do enter cellular membranes means that they are very likely to become intimately associated with lipoidal components of the membrane matrix that, in addition to the genuine lipids (e.g., myelin membranes in the brain are ca. 80% lipid), will include the larger alkyl and aryl portions of the amino acids of the protein components of the membranes. This may in turn result in a partial closure of the ion channels, thereby inhibiting polarization–depolarization of the membrane.

The theory that volatile anesthetics may act by specific binding to lipoprotein components of nerve tissue membranes or of the membrane itself has some experimental evidence. The use of ^{19}F nuclear magnetic resonance spectroscopy with halothane indicated that saturable anesthetic sites for halothane exist in living rats at 2.5% inspired gas. The authors consider this to support the idea that volatile anesthetics do act specifically even stereospecifically (Moody *et al.*, 1994).

Fixed (Intravenous) Anesthetics. The addition of nonvolatile (fixed) anesthetics to the anesthetists armamentarium would increase flexibility to the patient's management in

[5] Would CHF_3—fluoroform—be a safe gas anesthetic?

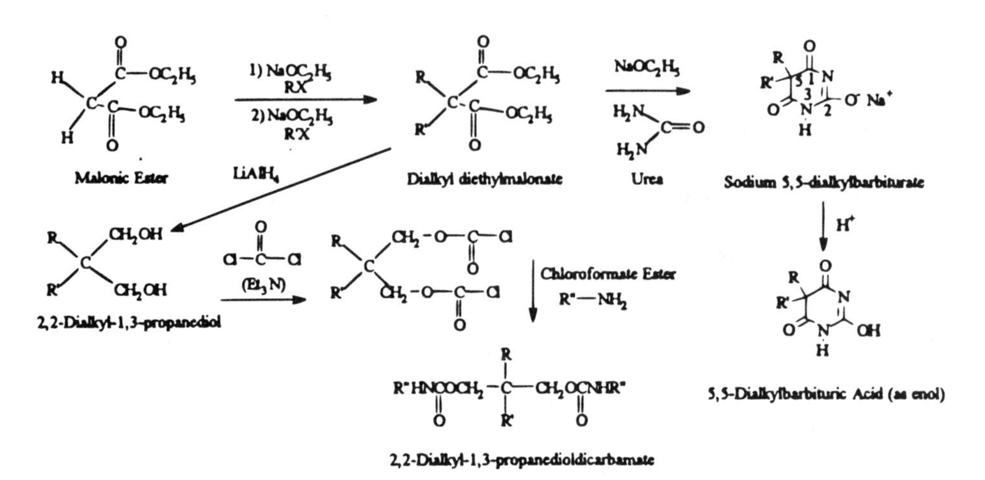

Figure 12-9. Synthesis of barbiturates and meprobamate-type drugs. (The use of thiourea, ($_2$HN)-C-5, would produce 2-thiobarbiturates; see text.)

surgery. This became possible with the introduction of short-acting barbiturates in the mid-1930s. Barbiturates, or barbituric acid derivatives, are ureides (triketopyrimidines) that are synthesized by condensing urea, or thiourea, with appropriately substituted malonic esters (Fig. 12-9). They undergo enol–keto tautomerism and thereby exhibit acidic properties (Eq. 12.5). Because the barbituric acid can also be viewed as a cyclic amide, the lactam terminology (analogous to lactone for cyclic esters) can be applied. The corresponding tautomer is therefore called a *lactim*.

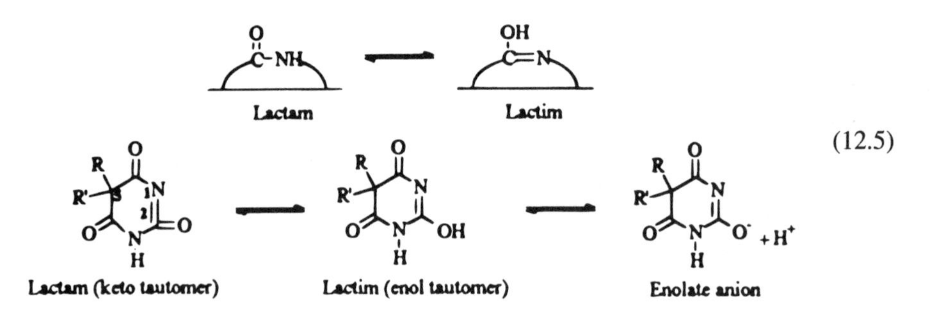

(12.5)

In the solid (crystalline) form the barbituric acid exists in the triketo or lactam form. In aqueous solution tautomerism to the enolic lactim occurs; the enolic hydroxyl (at C-2) is acidic acid and is ionized according to the particular drug's pKa (e.g., pentobarbital = 8.0, phenobarbital = 7.5). Titrating such a solution with a stoichiometric equivalent of base such as NaOH will convert the lactam quantitatively to the barbiturate's sodium salt, which can be isolated. Many barbiturates are commercially produced both in the lactam and in the sodium enolate salt form. Of course, the salts are water soluble and thus are used to formulate injectable dosage forms including those for IV anesthetic use.

Barbiturates have now been in use since the beginning of the century (barbital, 1903) and were used worldwide as sedative hypnotics. Prior to the advent of "real tranquilizers" they were used in subhypnotic doses as daytime sedatives. With the advent of antianxiety (anxiolytics) agents, meprobamate, and later chlordiazepoxide in the 1952–1960 period, barbiturates decreased in popularity, even though they were less costly. Their comparatively higher toxicity, abuse and addiction liability, and their inferior quality of sleep induction all contributed to their slow demise as sedatives and hypnotics over the years. Those with specialized properties—such as anticonvulsants, Nos. 5 and 8 (Table 12-5) and IV anesthetics, Nos. 6, 10, and 11—continue to be widely used for those indications. The latter three drugs, used intravenously (as solutions of their enolic sodium salts), are employed for anesthesia because of their extremely high lipid solubility. For example, partitioning between heptane and acidified water (pH 1, to simulate gastric juice) shows pen-

Table 12-5. Representative Barbiturates in Clinical Use

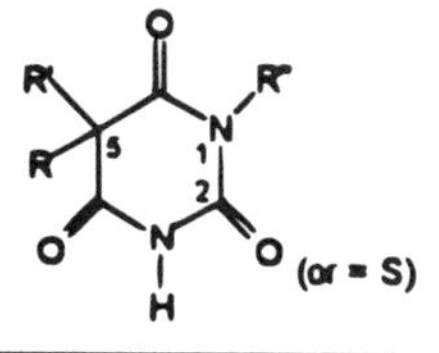

Barbiturate[a]	*R*	*R′*	*R″*	Duration (h)	$t_{1/2}$ (h)
1. Barbital (Veronal)[b]	C_2H_5	C_2H_5	—	8–12	—
2. Amobarbital (Amytal)	C_2H_5	$CH_2CH_2CH(CH_3)_2$	H	2–8	8–42
3. Aprobarbitol (Alurate)	$CH_2{=}CHCH_2$	$HC(CH_3)_2$	H	2–8	14–34
4. Butabarbital (Butisol)	C_2H_5	$HC(CH_3)CH_2CH_3$	H	2–4	34–42
5. Mephobarbita (Mebaral)	C_2H_5	$-C_6H_5$	CH_3	1–4	11–67
6. Methohexital (Brevital)[c]	$CH_2{=}CH\text{-}CH_2$	$HC(CH_3)C{\equiv}CCH_2CH_2CH_3$	CH_3	—	4
7. Pentobarbital (Nembutal)	C_2H_5	$HC(CH_3)CH_2CH_2CH_3$	H	2–4	15–48
8. Phenobarbital (Luminal)	C_2H_5	$-C_6H_5$	H	4–12	80–120
9. Secobarbital (Seconal)	$CH_2{=}CHCH_2$	$HC(CH_3)CH_2CH_2CH_3$	H	1–4	15–40
10. Thiamylal (Surital)[a,d]	$CH_2{=}CH\text{-}CH_2$	$HC(CH_3)CH_2CH_2CH_3$	—	—	—
11. Thiopental (Pentothal)[c,d]	C_2H_5	$HC(CH_3)CH_2CH_2CH_3$	H	—	11

[a] Trade names in parentheses.
[b] Of historical interest.
[c] Ultrashort-acting, used as intravenous anesthetic only.
[d] O at C-2 replaced by S atom.

tothal, No. 11, to be 33 times more lipid soluble than secobarbital, No. 9, and 3,300 times more so than barbital. In the case of the two thiobarbiturates, this high fat solubility results from the 2-sulfur substituent. Methohexital, No. 6, is devoid of S, and has a higher hydrocarbon side chain content than does pentothal (9 vs. 7). The CH_3 on N^1 also eliminates hydrogen bonding with and enolization of the H atom bonded to that nitrogen in the other barbiturates. In any case, there is a rapid distribution into adipose tissue and the CNS following IV administration. Anesthesia begins within 30 seconds or so. However, the duration of action of a single dose is also relatively short. One suggestion is that the drugs will also redistribute rapidly to peripheral fatty tissues. Even though hepatic metabolism is facile, due to branched and/or unsaturation of the C-5 substituents, it is not fast enough to account for the short duration of action (10–30 minutes). Long-hour infusion is not recommended since the elimination half-lives would predict a cumulative effect with exacerbation of side effects such as respiratory depression. Thus the ultrashort barbiturates are utilized primarily for short procedures (e.g., oral surgery) or rapid induction of anesthesia followed by a volatile agent such as halothane with N_2O.

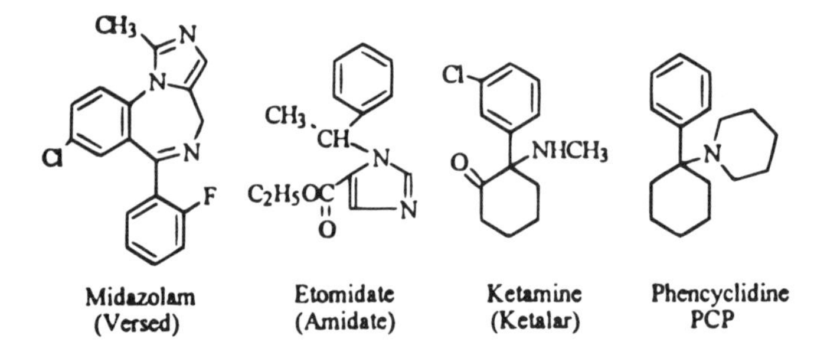

Other fixed anesthetics given intravenously such as the BZDs diazepam (see later) and midazolam are also in use. Midazolam, with its imidazole ring fused onto the "normal" benzodiazepine system, adds basicity (pKa 6.2) and thereby stability in solution to its salts. Its rapid onset of action is attributed to high lipophilicity at physiologic pH. Midazolam induces anesthesia within 2 to 2.5 minutes and even faster if a preanesthetic narcotic is administered. Duration of action following a single IV administration is 4.5 minutes. The drug is 99.97% metabolized hepatically; the major metabolite is the hydroxylation of the methyl group to 1-hydroxymethylmidazolam. This inactive metabolite is excreted as the glucuronide conjugate.

Another nonbarbiturate anesthetic is the imidazole carboxylate etomidate. Even though it produces no analgesia, it is used with fentanyl (see later) to produce very rapid anesthesia induction. The drug's appeal is its minimal effect on respiration and the cardiovascular system. Unlike many of the volatile compounds, it produces no annoying histamine release. It does cause intermittent muscle twitching.

Ketamine is a potent analgesic–anesthetic that is also effective intramuscularly. One particular property, production of cardiovascular stimulation, is of special advantage in elderly patients and those in shock (e.g., from burns). However, its propensity to precipitate hallucinations, delirium, disorientation, and other perceptual illusions postoperatively in about 12% of patients has led to its infrequent use in the United States. Ketamine's close structural analogy to the notorious and dangerous hallucinogen, phencyclidine (PCP, "angel dust"), should be noted. This drug, which was first also introduced as an anesthetic, produces such psychotic-type bizarre behavior that its use is limited to certain veterinary procedures.

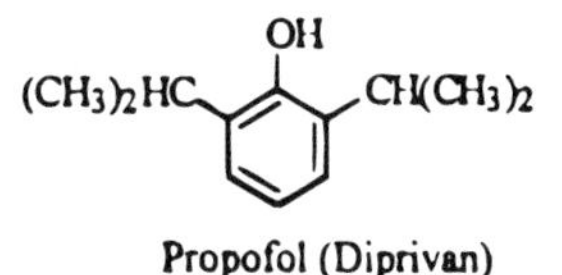

Propofol (Diprivan)

Finally, the latest nonbarbiturate fixed anesthetic to be introduced (1989) was propofol; 2,6-diisopropylphenol. It has a rapid intravenous induction (ca. 40 sec) as well as a quick recovery period of about 8 minutes once the drug is stopped. Both are features of obvious advantage. The drug can be combined with other anesthetics.

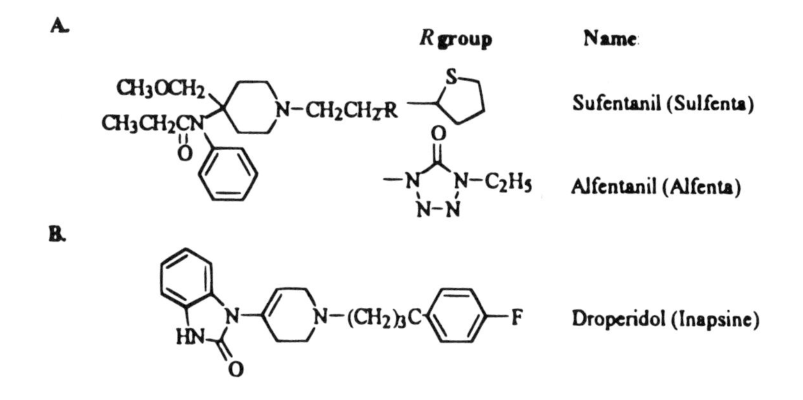

It has been possible to produce a "balanced" type of anesthesia wherein low-potency N_2O is combined with an opioid analgesic, usually the meperidine analog fentanyl (Fig. 5-10), or the more recent analogs sufentanil and alfentanil. When a butyrophenone-type neuroleptic (see later), particularly droperidol, is added, a state of *neuroleptoanesthesia* is produced where, without loss of consciousness, analgesia sufficient for surgery is achieved during which the patient still responds to commands.[6]

12.3.2. Hypnotics and Antianxiety Drugs

12.3.2.1. Ureides

Acyclic ureides, particularly those containing halogen atoms, were introduced as sedative hypnotics in the latter half of the nineteenth century following the recognition that halogenated hydrocarbons (e.g., $CHCl_3$) had anesthetic properties. Several retained popularity through the 1950s as short-acting alternatives to the more commonly used barbiturates, in the belief that they were "safer" and less habituating.

---

[6] A fixed ratio of fentanyl (0.05 mg/ml) with droperidol (2.5 mg/ml) is marketed as Innovar.

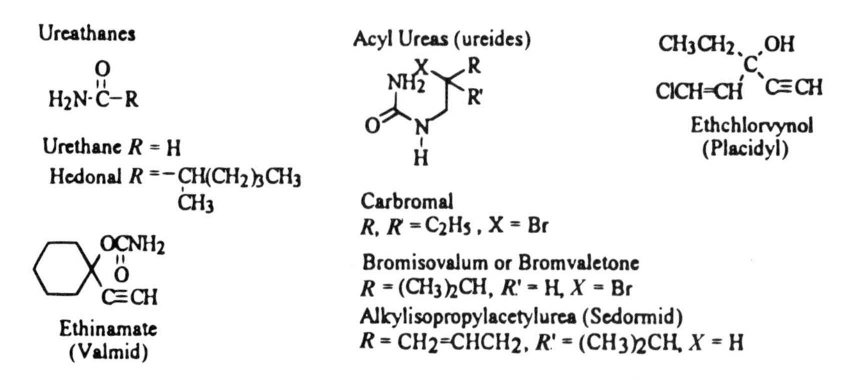

Figure 12-10. Early and simple-structured sedative-hyponotics. [These drugs are either obsolete (hedonal) or rarely used in the United States.]

Schmiedberg (1886) first reported that urethane (ethyl carbamate) produces rapid and deep narcosis in dogs. (The drug was subsequently used as a veterinary anesthetic.) More alkyl esters were soon introduced as hypnotics or anesthetics. For example, hedonal, which is more active than urethane, was introduced by Dreser[7] (1899) as a fixed anesthetic for intravenous use. Figure 12-10, illustrates some early compounds, such as urethane and hedonal, which are only of historical interest, along with several others that, while still available, are rarely used in the United States.

The barbiturates, some of which have special uses as mentioned, are listed in Table 12-5. Their synthesis (Fig. 12-9) was previously explained. Barbiturates have been traditionally classified by duration of action: ultrashort, short, intermediate, and long-acting. The ultrashort, as general anesthetics, were discussed. The long-acting, such as phenobarbital and mephobarbital, now used primarily against epilepsy, illustrate the overlap of hours of duration of action that makes this classification of little clinical utility. In those still used as hypnotics ("sleeping pills"), such as secobarbital (short) and amobarbital (intermediate), it should be understood that the values shown are those that might be expected from a single, occasional dose. The serum half-lives ($t_{1/2}$), with their long times and extreme ranges (individual response variations), will predictably alter these values dramatically on repeated dosing (e.g., over a week) due to cumulation of effects. All these differences are at least partially explained by the ease of, or resistance to, metabolism of the substituents at C-5. Figure 12-11 summarizes some of the empirically determined SARs based on thousands of (animal) tested compounds.

Attempts to circumvent the undesirable properties of barbiturates, particularly physical dependence and addiction, have led to the synthesis of "nonbarbiturate" analogs as sedative-hypnotics during the 1950s. The two piperidinediones, glutethimide and methyprylon, and the quinazolone, methaqualone, are particularly noteworthy. The first two have a particular structural resemblance to barbiturates. Unfortunately, pharmacologically there is no meaningful distinction. In fact, methaqualone has now been banned in several countries, including the United States (1984), because the abuse potential has been found to be higher than even the short-acting barbiturates.

---

[7] The same physician who discovered aspirin that year (Chapter 5).

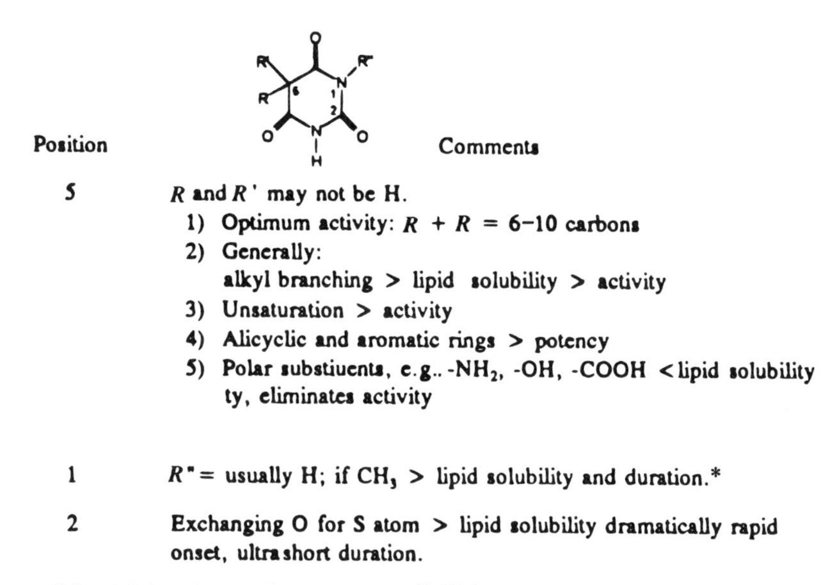

Position	Comments
5	R and R' may not be H. 1) Optimum activity: $R + R$ = 6–10 carbons 2) Generally: alkyl branching > lipid solubility > activity 3) Unsaturation > activity 4) Alicyclic and aromatic rings > potency 5) Polar substiuents, e.g., $-NH_2$, -OH, -COOH <lipid solubility ty, eliminates activity
1	R'' = usually H; if CH_3 > lipid solubility and duration.*
2	Exchanging O for S atom > lipid solubility dramatically rapid onset, ultrashort duration.

*However, N-demethylation to longer-acting parent compound is likely.

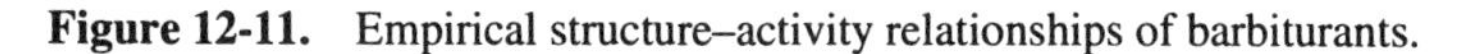

Figure 12-11. Empirical structure–activity relationships of barbiturants.

The mechanisms of action of barbiturates, and presumably of the related drugs, definitely involve the benzodiazepine–GABA–receptor–chloride ionophore complex (Fig. 12-6 and related discussion thereof in this chapter). The reticular activating system (RAS) (see Fig. 12-14) of the mesencephalon is very sensitive to these drugs, as are other areas at doses less than anesthetic. Barbiturates appear to facilitate synaptic inhibitions of GABA by prolonging rather than enhancing them. There is disagreement, however, as to whether these compounds also depress the action of excitatory neurotransmitters and thereby affect nonsynaptic membranous functions as well, or have no influence on monoaminergic or glycine-mediated processes. Nevertheless, it is apparent that barbiturate effects are not limited to GABA-ergic inhibition.

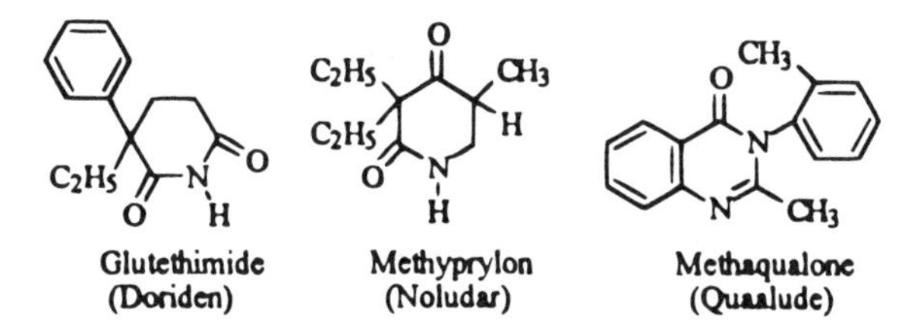

Barbiturates have presented some hydrolytic problems in regard to formulation of liquid dosage forms (elixirs, parenterals). The difficulty is ^{-}OH-catalyzed degradation of the ureide rings. Acid solvolysis does not cause similar difficulties. It can be shown that cleavage can occur at the 1,2 and/or 1,6 positions. Figure 12-12, which uses phenobarbital as the representative, illustrates the degradative routes. The 1,6-cleavage is the preeminent pathway of ionized barbiturates (e.g., aqueous solutions of sodium salts). Such solutions would have to be stabilized. Elixir phenobarbital, a hydroalcoholic-flavored solution of the triketo form, has

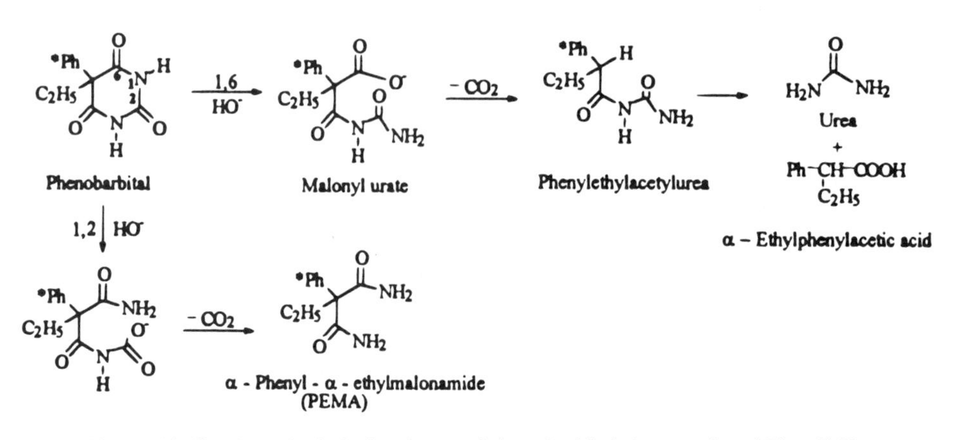

Figure 12-12. Base hydrolysis scheme of phenobarbital ring opening. *Ph = C_6H_5.

a pH of 6 and would be stable as formulated.[8] Various neutral organic solvents such as ethanol, glycerol, and sorbitol as aqueous cosolvents have a considerable stabilizing effect on barbiturate solutions. The accepted explanation is that the decreased dielectric constants of such solvents inhibits reactions between ions of like charge (i.e., the barbiturate anion and OH^-).

12.3.2.2. Stereochemistry

Some interesting facets of stereochemistry on barbiturate effects have come to light. Depressant activity generally resides more in the S(–) form, while the R(+) isomers have a predominantly excitatory effect. In one example the R(+) as well as the RS (racemic) forms actually were convulsant.

12.3.2.3. Polyols and Their Carbamates

Certain glycerol and propanediols, particularly as carbamates, with both central depressant and skeletal muscle relaxant properties, have been in use since the 1940s. Their employment soon expanded to sedative applications as alternatives to barbiturates. The term *central relaxants* was applied to them. Several compounds in this category additionally exhibit anticonvulsant properties. They act by decreasing neuronal impulses at the interneurons of the polysynaptic reflex arc at the spinal cord level. Interneurons are components of the CNS that act as connecting links between receiving (sensory) and effector (motor) elements of the reflex arc (Fig. 12-13). Thus a stimulus at the receptor generates an impulse that travels down the sensory neuron, across the interneuron, down the motor neuron, and stimulates the effector muscle to contract. Overstimulation causes a spastic condition. Thus blockade or depression results in clinically useful alleviation of such conditions as strains, sprains, and direct trauma involving the neck, back, and joints.

Since depression of neuronal components of the CNS is involved, these drugs also produce sedation, especially in large doses. However, in doses well below these levels, these drugs can alter emotional responses. It is currently believed that this effect is mediated

[8] The 15–20% alcohol has a stabilizing effect.

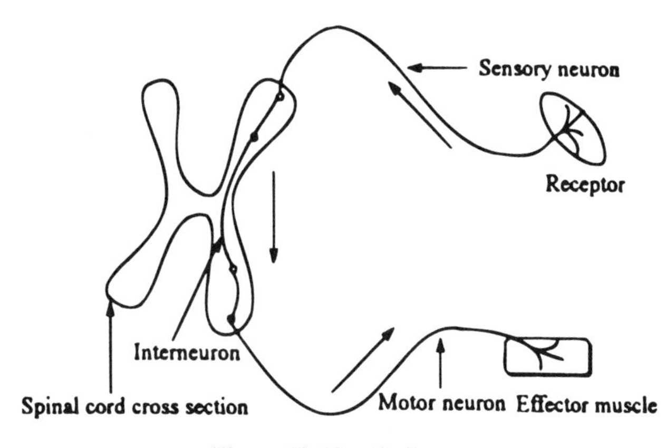

Figure 12-13. Reflex arc.

through the limbic system of the brain (including the amygdala and hippocampus), where these drugs probably exert a depressant effect (Fig. 12-14). The cortex and hypothalamus are not affected; neither are peripheral nerve ganglia. The CNS depressant qualities of these drugs (both of the phenoxypropanediols and meprobamate types) are not complicated by antihistaminic and anticholinergic side effects, as in the case with certain other neuroleptics.

The phenoxypropanediols (Table 12-6) were introduced first as centrally acting muscle relaxants. Although it is no longer in use, the phenylglyceryl ether 3-phenoxy-1,2-propanediol (Antodyne) was briefly used early in this century as an analgetic and to treat muscle spasms. It is predictable now (but not then) that its action, due to rapid conjugation and excretion, was too brief. More than three decades later, a more systematic synthetic study of glyceryl ethers led to the first therapeutically useful compound 3-(*o*-toloxy)-1,2-propanediol, mephenesin. Its rapid hepatic oxidation to the inactive 1-carboxy metabolite, soon led to the production of mephenesin carbamate (Table 12-6). The guaiacol analog, methocarbamol, followed as a somewhat more effective agent. Its noncarbamate analog, glyceryl guaiacolate or guaifenesin, is widely used in cough syrups as an expectorant. Chlorphenesin carbamate is

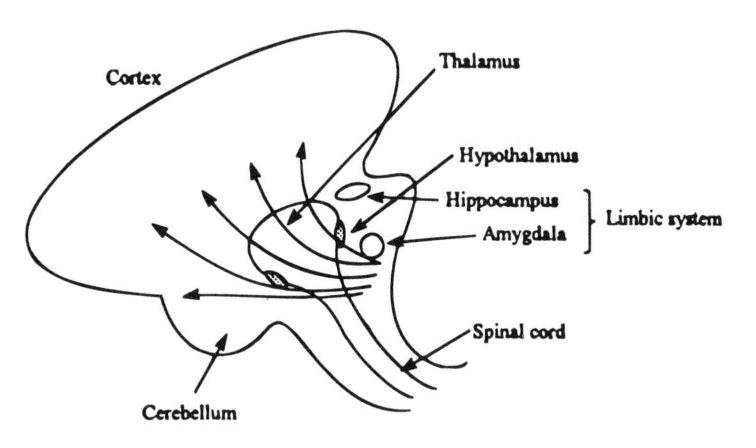

Figure 12-14. The reticular activation system. As indicated by the curved arrows, the RAS emanates upward from the reticular areas of the mesencephalon.

Table 12-6. Phenoxypropanediol-Type Muscle Relaxants

Structure	Drug	Comments
O-CH₂-CH-CH₂-OH; OH	Antodyne	First used in 1910 as an analgetic–antipyretic
O-CH₂-CH-CH₂-O-C-NH₂; OH; O; CH₃	Mephenesin carbamate (Tolseram)	Basically a muscle relaxant with some tranquilizing properties
O-CH₂-CH-CH₂-O-C-NH₂; OH; O; OCH₃	Methocarbamol (Robaxin)	Longer acting and more effective than mephenesin, but more toxic; exhibits a significant sedative effect
Cl; O-CH₂-CH-CH₂-O-C-NH₂; OH; O	Chlorphenesin (Maolate)	Used only as a skeletal muscle relaxant
OCH₂; NH; O; O; OCH₃	Mephenoxalone (Trepidone)	Essentially a cyclized analog of methocarbamol
CH₃; OCH₂; NH; O; O; CH₃	Metaxalone (Skelaxin)	Exhibits analgesia in rodents, but, not in humans

also useful in treating the pain of muscle spasms. The two oxazolidones, mephenoxalone and metaxalone (Table 12-6), may be viewed as a cyclic carbamate analog.

$$\overset{3}{ZO\text{-}CH_2}-\overset{R}{\underset{R'}{\overset{2|}{C}}}-\overset{1}{CH_2\text{-}OZ}$$

R, R' = alkyl; R" = H or alkyl

2,2-alkyl-substituted-1,3-propanediols (Z = H) or carbamates (Z = C(=O)-NH*R*′)

Extensive synthetic SAR work that began with mephenesin established that an aromatic portion to these drugs was not essential to muscle relaxant activity; aliphatic propanediols (and their carbamates) were actually more effective. The introduction of meprobamate (Table 12-7) soon followed. One of several syntheses of this type of compound is outlined in Figure 12-10. Note that the initial steps parallel those of the barbiturates. Meprobamate is today considered as a useful antianxiety agent, though it is inferior to the BZDs. In a general sense it can be considered as a "minor" tranquilizer. Its muscle relaxant properties, though weaker than mephenesin, are of much longer duration. Carisoprodol, however, is superior on both counts.[9]

Phenaglycodol, which is now obsolete, is included in Table 12-7 to illustrate that glycols with "unprotected" OH groups qualitatively have the same properties, albeit of lower

[9] Carisoprodol is also marketed in combination with aspirin and codeine.

Table 12-7. Meprobamate and Related Propanediol Derivatives

Structure	Drug	Comments
$H_2N{-}C(=O){-}O{-}CH_2{-}C(CH_3)(C_3H_7){-}CH_2{-}O{-}C(=O){-}NH_2$	Meprobamate (Miltown, Equanil)	Exhibits antianxiety muscle relaxant and sedative properties
$H_2N{-}C(=O){-}O{-}CH_2{-}C(CH_3)(C_3H_7){-}CH_2{-}O{-}C(=O){-}NHCH(CH_3)_2$	Carisoprodol (Soma, Rela)	Mainly a muscle relaxant
$H_2N{-}C(=O){-}O{-}CH_2{-}C(CH_3)(CH_3{-}CH{-}CH_2{-}CH_3){-}CH_2{-}O{-}C(=O){-}NH_2$	Mebutamate (Capla)	Has some hypotensive activity
$H_2N{-}C(=O){-}O{-}CH_2{-}C(CH_3)(C_3H_7){-}CH_2{-}O{-}C(=O){-}NH{-}C_4H_9$	Tybamate (Tybatran, Solacen)	Same spectrum of activity as meprobamate
$Cl{-}C_6H_4{-}C(OH)(CH_3){-}C(OH)(CH_3){-}CH_3$	Phenaglycodol (Ultran)	A mild sedative with muscle relaxant properties; short-acting

intensity and shorter duration. It should be pointed out that the polyol compounds are generally not effective in spasticity due to dyskinesia.

There are several chemically unrelated skeletal muscle relaxants. Chlorzoxazone is an example of a drug that evolved from a discovery that certain benzimidazoles had mephenesinlike activity. Chlorzaxazone, the benzoxazole isostere, appears to have the same spectrum of activity and mechanism(s) as the polyol carbamates.

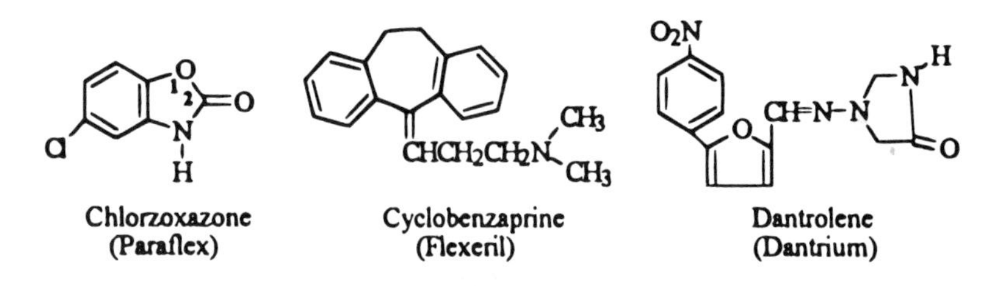

Chlorzoxazone (Paraflex) Cyclobenzaprine (Flexeril) Dantrolene (Dantrium)

Cyclobenzaprine is structurally closely related to the tricyclic antidepressants (see later) and therefore shares many of their adverse reactions. These include drowsiness (CNS effects), dry mouth (anticholinergic effects), and dizziness. The potential for serious interactions with MAO inhibitors also exists. Nevertheless, the drug's centrally acting skeletal muscle relaxation is useful for short-term treatment of local muscle spasms.

Dantrolene, which is chemically related to the nitrofuran antibacterials (Chapter 7), was introduced as a myorelaxant with a unique direct action on skeletal muscle that reduces contractility by effecting Ca^{2+} release from the sarcoplasmic reticulum of certain muscle fibers. The CNS is not involved. The drug can control spasticity caused by spinal cord injuries, multiple sclerosis, cerebral palsy, and even strokes, but it is not indicated for arthritis or local acute spasms.

Malignant hyperthermia (or hyperpyrexia) is a rare, genetically determined syndrome consisting of sudden high body temperatures, severe acidosis, and general muscular rigidity, which are symptoms precipitated during inhalation anesthesia (halothane) by the addition of neuromuscular blocking agents (Chapter 8). In susceptible individuals these conditions may lead to an unpredictable sudden large-scale release of Ca^{2+} from the myoplasm, believed to be due to impaired binding of the cation. By its ability to act directly on the reticulum, dantrolene inhibits Ca^{2+} release, thereby inhibiting the excitation–contraction trigger of this life-threatening emergency. Untreated, body temperatures may exceed 42°C (107.5°F), at which the mortality rate can be high.

Other drugs for different spasticities include baclofen (this chapter, earlier), which is useful in multiple sclerosis and spinal cord trauma but not rheumatoid conditions; the centrally acting orphenadrine (Chapter 8, Fig. 8-4) for Parkinson-like symptoms; and the BZDs, particularly diazepam (considered later).

12.3.3. The Benzodiazepines

The early 1950s began the modern era of the treatment of anxieties and neuroses. We moved from the use of simple daytime sedation with drugs that were essentially hypnotics, albeit at much lower subhypnotic doses, to compounds that became known as minor tranquilizers—the various polyol carbamates. In addition, the development of "major" tranquilizers that, for the first time in history, had more than just a custodial impact on schizophrenic–psychotic patients occurred simultaneously. These initial anxiolytics, with meprobamate now considered the prototype, were the first drugs whose pharmacologic profile included more than just CNS depression.

Because of the shortcomings of what may be referred to as the "first-generation anxiolytics,"[10] and the greatly expanded market for psychotropic drugs, many pharmaceutical concerns began to actively research the field. Some took the traditional approach to drug development, namely, modifications of a lead compound to optimize and determine the structural parameters of the desired activity. This led to several of the meprobamate homologs and congeners illustrated in Tables 12-6 and 12-7. It is interesting that some of these drugs became clinically more successful as muscle relaxants than as anxiolytics (e.g., methocarbamol).

The other approach, taken by Hoffman–LaRoche, was to attempt to discover a new, chemically unique lead structural type. Considering the sparsity of knowledge and understanding of the chemical processes of the brain, much less their relationship to behavior and mental disease, such an undertaking might have been akin to a search for the Fountain of Youth. In practical terms, such discoveries had been made based on random screening with the expectation of an "accidental" hit or, less colloquially, a serendipitous discovery. Tests were developed to determine the ability of compounds to

---

[10] More aptly sedative–anxiolytics or tranquillosedatives.

prevent or to minimize convulsions induced by electroshock, strychnine, or pentylenetetrazol and characteristics of muscle paralysis and relaxation. The screens essentially looked for compounds with a meprobamate like profile. The expectation was that structurally different compounds that have muscle-relaxant activity would also exhibit anxiolytic properties.

The medicinal chemist behind this adventurous undertaking was Leo H. Sternbach, who decided, based on chemical "intuition," to pursue a ring system he had worked on two decades earlier in a search for dyes. The ring, which is a benzoheptoxdiazine (Fig. 12-15), was first "synthesized" in 1891 from *o*-aminoacetophone ($R = CH_3$ and using $ClCH_2COCl$ in the second step). This ring system had several appealing features. By using aminobenzophenones (R = phenyl) and treating the resultant amine with chloroacetyl chloride (as in Fig. 12-15), the purported benzoheptoxdiazine, by carrying a chloromethyl group ($-CH_2Cl$), could then be reacted with various amines, producing basic amino derivatives that, from empirical experience with other drug types, would be more likely to have biological activity. In addition, since it should be possible to produce compounds with numerous substituents on either benzene ring without too much synthetic difficulty, a large number of compounds would be available for screening. Furthermore, the heptoxdiazine structure had been "correctly" established.

The heptoxdiazine structure was in fact incorrect and was soon determined to be a quinazoline-3-oxide. The products expected following substitution reactions with secondary amines were then in fact 2-N,N-di or cycloalkylamine derivatives of such quinazolines (Fig. 12-15). Unfortunately, pharmacological testing of these compounds showed them to be without value. The project was abandoned. One product, which was the only one produced by a primary amine (CH_3NH_2) rather than a secondary amine and was therefore believed to have been a methylamino derivative of the quinazoline-3-oxide, was not screened. That compound, which was "found" on the bench during a cleanup two years later, was subsequently tested and, amazingly, gave positive results in a battery of six screens for anticonvulsant, muscle relaxant, antiagression (taming), and other properties. This "quinazoline" derivative appeared to be superior to phenobarbital, meprobamate, and chlorpromazine.

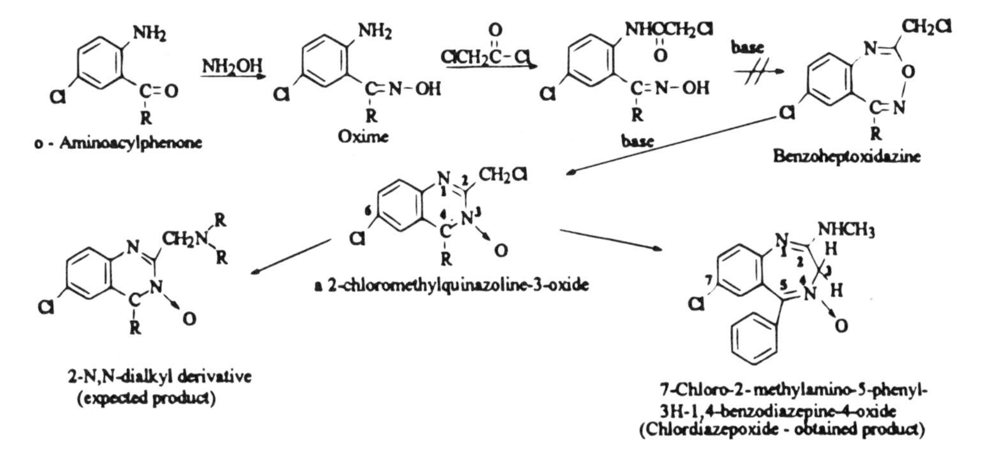

Figure 12-15. Discovery of benzodiazepines—the synthesis of chlordiazepoxide.

The lack of similarity of UV and IR spectra to the 2-dialkylamino or 2-chloromethyl derivatives indicated chemically that this compound was not a quinazoline. Further determinations, including degradation studies, found the compound to be a then-unique 1,4-benzodiazepine (BZD) derivative. A ring-expanding rearrangement had taken place, apparently by an initial attack of the primary amine on the electron-deficient 2-position rather than the nucleophilic displacement of halogen on the CH_2Cl group. The resultant insertion of the CH_2 (with loss of HCl) produced the new benzodiazepine structure. This product, chlordiazepoxide, was marketed in 1960, thus becoming the first member of a new type of anxiolytic drug. Over the next several decades two dozen or more additional BZDs have been marketed (Table 12-8).[11] To a considerable extent BZDs have replaced barbiturates and related compounds as anxiolytics and hypnotics for a number of reasons, including effectiveness, less potential for dependence and abuse, as well as interaction with other drugs, and a greater safety margin in case of overdose. Some of the newer compounds that have faster elimination rates have been found to be effective at lower doses, offering even greater therapeutic advantages, including safety margins. Diazepam, which was introduced in 1963, soon became the world's most prescribed drug.

This does not mean, of course, that BZDs represent the ultimate in anxiolytic–hypnotic agents. In spite of their usefulness and apparently greater safety, they are not single-effect drugs. If one were to use them as hypnotics, then one must still consider their muscle-relaxant properties and interactions with alcohol and other CNS-depressant compounds.

When treatment of anxiety conditions is undertaken, the patient using BZDs is additionally burdened with varying degrees of undesirable daytime sedation, potential safety hazards when operating machinery and driving a vehicle, and the ever-present possibility of dependency and abuse on prolonged therapy.

With an understanding of the mechanisms and metabolic reactions of these drugs, newer drug design concepts were applied in later synthetic phases. It is not surprising that compounds were developed in which certain pharmacologic properties became elevated or subdued (but never eliminated). Flurazepam (No. 7, Table 12-8), nitrazepam (No. 13), and triazolam (No. 22) became widely used as hypnotics; clonazepam (No. 3) is used almost exclusively as an anticonvulsant, while diazepam (No. 6), clorazepate (No. 4), lorazepate (No. 19), prazepam (No. 16), and aprazolam (No. 19) are marketed primarily as anxiolytics.

Some agents (e.g., diazepam) also have wide use as skeletal muscle relaxants. Even though such may be the officially approved indications for various members of the BZD group, several clinically significant effects should be expected. The preponderance of one effect or the other may vary from one drug to another. Thus skeletal muscle relaxant, sedative–hypnotic, antianxiety, and anticonvulsant properties are all present.

It is important for the clinician to remember that there are no neuropharmacologic differences between "hypnotic" and anxiolytic BZDs. All are antianxiety agents at doses below those that induce sleep. The FDA-approved indication in the United States may be based, in part, on pharmacokinetic properties, but even more so on the direction of clinical research during the development of a particular drug. This multifaceted pharmacologic spectrum is to varying degrees shared by some of the "first-generation" hypnotics. Thus, in testing for anticonvulsant effects by the classic methods of blocking elec-

[11] For a personal account of the BZD discovery, see Sternbach (1979).

troshock-, strychnine-, or pentylenetetrazol-induced seizures, activity can be demonstrated by all the BZDs as well as by phenobarbital, pentobarbital, and glutethimide. Experiments demonstrated muscle-relaxant properties of various hypnotics, diazepam, flurazepam, and phenobarbital—in laboratory animals. Actually all hypnotic compounds have this effect with potency variations over a 1–1000 range, depending on the testing procedure.

Table 12-8. Bensodiazepines[a]

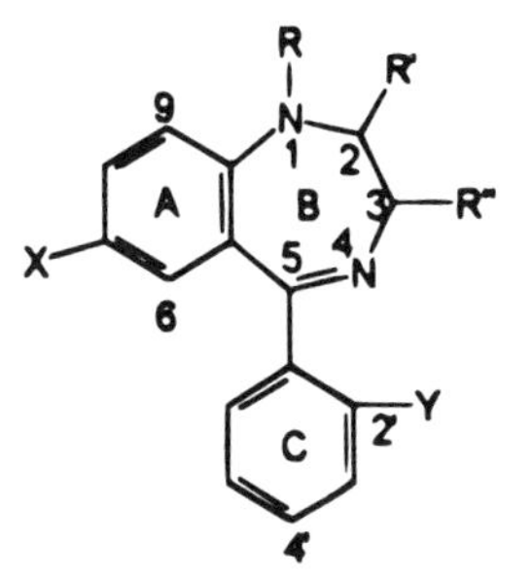

Drug	*R*	*R′*	*R″*	*X*	*Y*	*M*[b]	Comments
1. Bromazepam* (Leoxtani)	H	=O	H	Br	—	+	C-5 is α-pyridyl
2. Chlordiazepoxide (Librium)	C	-$NHCH_3$	H	Cl	H	+	$N^4 \rightarrow O$
3. Clonazepam (Klonopin)	H	=O	H	NO_2	Cl	–	Use only as antiepileptic
4. Clorazepate (Tranxene)	H	=O	-COO^-	Cl	H	+	pro-drug -CO_2
5. Demoxepam*	H	=O	H	Cl	H	+	$N^4 \rightarrow O$
6. Diazepam (Valium)	CH_3	=O	H	Cl	H	+	
7. Flurazepam (Dalmane)	$(CH_2)_2$ $N(C_2H_5)_2$	=O	H	Cl	F	+	Used only as hypnotic
8. Fludiazepam*	CH_3	=O	H	Cl	F	+	
9. Halazepam (Paxipam)	CH_2CF_3	=O	H	Cl	H+		
10. Lorazepam (Ativan)	H	=O	OH	Cl	H	–	
11. Medazepam* (Nobrium)	CH_3	=O	OH	Cl	H	+	
12. Nimetazepam*	CH_3	=O	H	NO_2	H	+	
13. Nitrazepam* (Mogadon)	H	=O	H	NO_2	H	+	Hypnotic use
14. Oxazepam (Serax)	H	=O	OH	Cl	H	–	
15. Pinazepam*	$CH_2C{=}CH$	=O	H	Cl	H	+	
16. Prazepam (Centrax)	$CH_2\triangleleft$	=O	H	Cl	H	+	Pro-drug
17. Quazepam (Doral)	CH_2CF_3	=S	H	Cl	F	+	
18. Temazepam (Restoril)	CH_3	=O	OH	Cl	H	+	

(*Continued*)

Table 12-8. *(Continued)*

Triazolo and Related Benzodiazepines+

Drug	*R*	*Z*	*R′*	*X*	*Y*	*Y*	Comments
19. Alprazolam (Xanax)	CH_3	N	H	Cl	H	H	
20. Estrazolam* (Enrodin)	H	N	H	Cl	H	H	
21. Midazolam (Versed)	CH_3	C	H	Cl	F	F	Intravenous anesthetic only
22. Triazolam (Halcion)	CH_3	N	H	Cl	Cl	Cl	Marketed as hypnotic
23. Flumazenil (Mazicon)	H	C	H	F	—	—	
24. Estazolam (Prosom)	H	N	H	Cl	H		

* Not clinically available in the United States.

[a] Trade names in parentheses.

[b] Metabolites + = active; – = inactive, short lived or weak.

[c] Double bond N^1 to C_2.

The intensive synthetic and screening efforts that followed the discovery of chlordiazepoxide resulted in many thousands of 1,4-benzo and heterodiazepinone compounds being evaluated (over 4000 at Hoffman-LaRoche alone).

Substituent variations on rings A, B, and C enabled the development of SAR dos and don'ts that empirically determined the important relationships of substituents and their positions to activity; positions 1, 7, 2′, and 4′ were particularly significant. Figure 12-16 summarizes the salient facts.

More recent innovations involve the bridging of the N^1 nitrogen and 2-keto of the azepine ring to form a third fused ring such as in the triazolo BZDs triazolam (No. 22) and alprazolam (No. 19) and the imidazole ring in midazolam (No. 21). The results are generally a significant increase in potency compared with the corresponding 1-methyl compounds. It also shows that a 2-keto function is not essential.

The synthesis of 1,5-benzodiazepinediones was undertaken to determine the parameters of structural modifications that will retain and possibly improve the anxiolytic profile. Two such drugs have been developed outside the United States: clobazepam and triflubazepam. Replacement of benzo portion of BZDs with heteroaromatics was also accomplished. Ripazepam represents a pyrazole congener. It was equipotent to diazepam yet showed a striking absence of sedation in animal tests. In clinical tests, however, sedation was present, even though it was less.

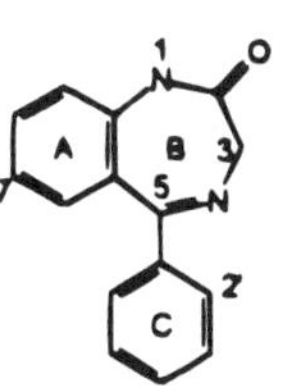

Ring A Substitution:

1. Optimum position #7
2. Only e^- -withdrawing groups > activity; Cl, Br, CF_3, NO_2
3. Any substitution on positions 6, 8, 9 < activity

Ring B Substitution:

1. N^1 > activity by N-alkyl, haloalkyl, alkynyl and small cycle or aminoalkyl groups[a]
2. Position 3: hydroxylation < duration and potency due to rapid conjugation; still clinically useful[b]

Ring C Substitution:

1. e^- withdrawer at 2' > activity (e.g., Cl, F)[c]
2. Any substitution at 4' < activity dramatically
3. C-ring can be heteroaromatic,[d] even cycloalkenyl[e]

Figure 12-16. Summary of empirical benzodiazepine SARs. [a]The highly branched *t*-butyl derivative is inactive and not metabolically removed (Gilman and Sternbach, 1971). [b]e.g., oxazepam and lorazepam. [c]Combining proper substituents at C-7, 2′ and N^1 produces good additive properties. [d]2-pyridyl (Bromazepam). [e]Cyclohexenyl (Tetrazepam).

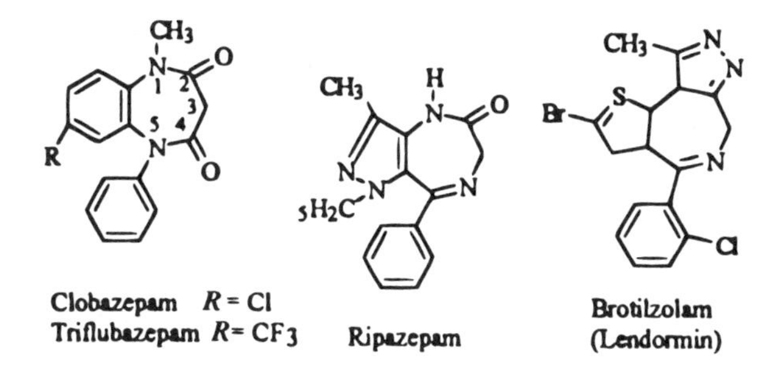

Brotilzolam represents a twofold modification: the fused benzo ring is replaced by the bioisosteric thiazole (the Br substituent apparently in the No. 7 equivalent position), and a triazolo fusion also is carried out. The drug, which is marketed in Germany and Portugal, is about equipotent to triazolam.

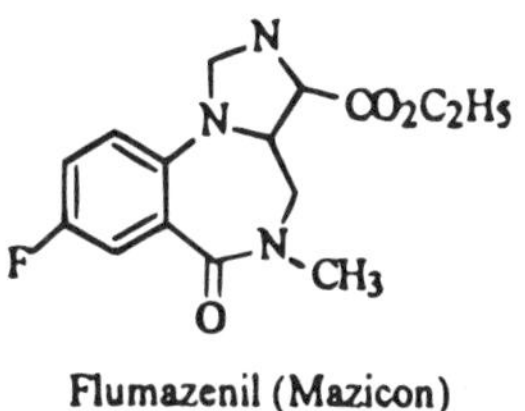

Flumazenil (Mazicon)
RO – 151788

Triazolo-type 1,4-benzodiazepines lacking the 5-phenyl group began to be researched in the mid-1980s. Some exhibited enhanced anticonvulsant and anticonflict activity in animals, while having a weaker muscle relaxant action and atoxic properties. Monkeys treated for two weeks with several such compounds did not show withdrawal symptoms on challenge with a then-experimental BZD antagonist RO-151788. This drug had been shown to be effective in reversing the effects of BZDs that were used as general anesthetics in humans; yet it was also successfully demonstrated to have anticonvulsant efficacy in grand mal epilepsy without concomitant drowsiness or muscle relaxation in other patients. This would tend to indicate at least a partial agonist activity. Nevertheless, RO-151788 was introduced as a clinical BZD antagonist in 1987 (1991 in the United States) as flumazenil. Indications include complete or partial reversal of BZD sedation in general anesthesia or its induction and in cases of overdose. It is curious that a strong warning of seizures in subjects on long-term BZD therapy or tricyclic antidepressant overdose is stated. Has the vast research and clinical development efforts with BZD-type compounds since the initial discovery of chlordiazepoxide over three decades ago been a success or failure? The answer is probably *neither*. Potency, safety, and clinical management of patients compared with the pre-BZD era has certainly improved. However, a major goal of drug development since chlordiazepoxide and diazepam must certainly have been a narrowing of the pharmacological spectrum to, ideally, only anxiolytic effects. This separation from muscle relaxant, anticonvulsant, and particularly sedative (side) effects has not been achieved by this approach; neither has the development of tolerance and dependence been eliminated.

12.3.3.1. Metabolism

It is by now quite apparent that the available BZDs qualitatively have almost identical neuropharmacologic effects. This makes it very unlikely that any one particular BZD will be clinically totally superior to any other. Therefore, is there a clinical justification for marketing so many similar compounds? This question was asked editorially over the years when far fewer BZDs were available than now. If the statements preceding the question are correct, is the question valid? Even so, the second statement is not necessarily sound. Efficacy clinically involves more than the basic mechanism(s) by which a drug functions. Questions such as the type of metabolic reactions a drug undergoes, how rapidly they occur, are the metabolites active or inactive, rates of drug absorption, and excretion of the drug and its major metabolites, must all be answered. After that, all the answers to these questions must be factored into an evaluation of a drug's actual clinical efficacy.[12] In other words, the metabolic profile and pharmacokinetics must be taken into account before a final judgment can be made.

To a large degree, pharmacokinetics of a drug depend on its metabolic profile and the chemical properties of both the drug and its metabolites. Thus long-acting drugs may be so because they are metabolized with difficulty and/or slowly excreted (i.e., long serum $t_{1/2}$). This is the case with several of the barbiturates such as barbital and phenobarbital. On the other hand, they may be readily metabolized, but to long-acting active metabolites that will accumulate with continuing therapy. In cases of impaired or diminished liver functions, such as in the elderly, decreased metabolism without compensating dosage reductions will

[12] Of course, difference in toxicity and allergic reactions should also be considered.

necessarily result in higher sedation levels masquerading as hangovers and increased toxicity (poor efficacy?). In other patients longer duration may be desirable (better efficacy?). The differences in kinetics and metabolism among the various BZDs will dictate dosing schedules and drug accumulation in long-term therapy.

The BZDs can be broadly characterized into two types. One group includes the long-acting compounds. These are essentially those that are metabolized to active metabolites whose serum half-lives are long. Figure 12-17 shows that seven BZDs are metabolized by N-dealkylation to *desmethyldiazepam* (DMD). The $t_{1/2}$ of this major metabolite in humans is 30–120 hours; however, in some cases, it reaches 200 hours (ca. 8 days). Thus on repeated dosing DMD will accumulate, allowing for anxiolytic therapy on a once-a-day basis. DMD is then slowly oxidized to the 3-hydroxide metabolite, oxazepam. Oxazepam, when clinically used, is a member of the short-acting BZD category since it will be efficiently conjugated to the inactive, rapidly excreted 3-hydroxyglucuronide. The same direct rapid conjugation to a 3-O-glucuronide applies to lorazepam (Fig. 12-17; Table 12-9) and tamazepam, which accounts for their short activity.

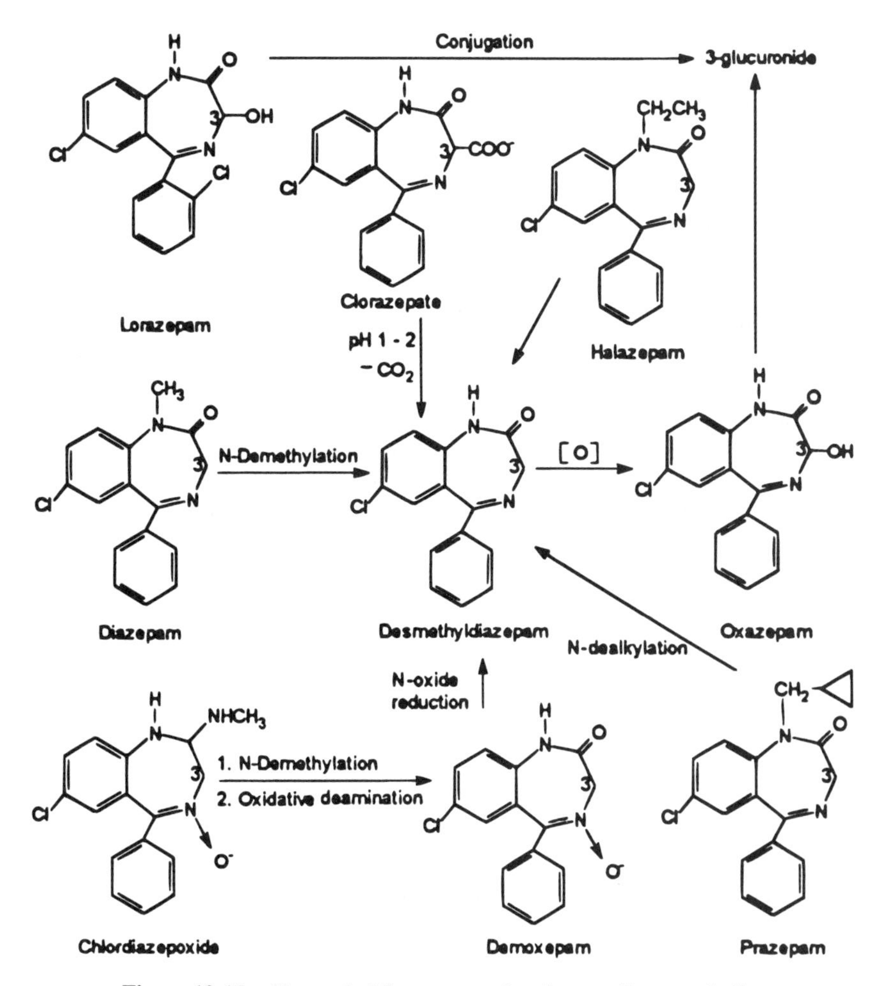

Figure 12-17. Desmethyldiazepam—a key intermediate metabolite.

Table 12-9. Benzodiazepine Metabolic Comparisons

Drug	$t_{1/2}$ (h)	Active metabolite	Metabolite $t_{1/2}$ (h)	Comments
Long acting				
Chlordiazepoxide	5–30	Desmethyldiazepam	30–120[b]	
Clorazepate	Pro-drug[c]	Desmethyldiazepam	30–120[b]	In stomach, decarboxylation
Diazepam	20–70	Desmethyldiazepam	30–120[b]	
Flurazepam	1–4	Desalkylflurazepam	50–100	
Halazepam	Pro-drug[c]	Desmethyldiazepam	30–120[b]	
Prazepam	Pro-drug[c]	Desmethyldiazepam	30–120[b]	Slow oral absorption
Estazolam	13–35	4′-Hydroxyestazolam	Long	
Short acting				
Alprazolam	12–15	α-Hydroxyprazolam[d]	Short	Rapid oral absorption
Lorazepam	10–20	There is none	—	
Oxazepam	3–10	There is none	—	Slow oral absorption
Temazepam	5–10	Oxazepam (minor)	3–10	Slow oral absorption
Triazolam	3–5	α-Hydroxytriazolam	Short	Rapid oral absorption

[a] Desmethylchlordiazepoxide is prior major metabolite, $t_{1/2}$ also 5–30 hours.
[b] May reach 200 hours (see text).
[c] See text for significance.
[d] Hydroxylation of triazolomethyl group affords an active but short-lived metabolite.

An interesting situation arises when the two long-acting anxiolytics, prazepam (No. 16) and clorazepate (No. 4), are considered. Neither drug reaches the systemic circulation in meaningful quantities; they are essentially pro-drugs that are converted to DMD with great facility.

It has been found that prazepam is nearly totally N-dealkylated to DMD in its "first pass" through the liver following oral absorption into the enterohepatic circulation. An earlier belief that 3-hydroxyprazepam might also make a contribution to the drug's activity was not substantiated since that compound has been identified only as its glucuronide in the blood. Clorazepate, which is also a pro-drug of DMD, probably never reaches the circulation, being rapidly decarboxylated in solution by the low pH of gastric juice. DMD then rapidly enters the systemic circulation. In spite of this apparent similarity of metabolic profile, the rates of this metabolic conversion differ between these two drugs. Clorazepate's loss of CO_2 is rapid; the resultant DMD is then rapidly absorbed. Prazepam must first be absorbed to reach the liver, where it loses the N^1-cyclopropylmethyl group relatively slowly. DMD from this drug therefore reaches the general circulation more slowly than it might from a sustained-release formulation. However, both drugs can be dosed on a once-a-day basis. One difficulty that can be encountered with long-acting BZDs used for hypnotic purposes (e.g., flurazepam, No. 7) is that the persistent active metabolites (in this case desalkylflurazepam) (Table 12-9) may result in residual daytime sedation.

Lorazepam (No. 10) and oxazepam (No. 14) represent short-acting BZDs ($t_{1/2}$ = 15 and 10 hours, respectively). As 3-hydroxy derivatives they are directly derivatized to inactive glucuronides. Since there are no active metabolites, cumulative effects on continuing dosing are not likely. Like diazepam, IV lorazepam produces anticonvulsant and amnesic effects. However, these effects persist longer because drug distribution throughout the body is less rapid and extensive. These properties may be predictable since a comparison of structures with diazepam shows lorazepam to be more polar: the absence of lipid-solubilizing N^1-CH_3 and presence of water-solubilizing 3-OH. Lorazepam, therefore, is also used as a preanesthetic drug.

12.3.3.2. Mechanisms of Action

Interest in understanding the biochemical profile of such a wide variety of compounds that are able to induce sleep as discussed in this chapter has not led to a unifying concept. However, during the past decade research has implicated various neurotransmitters. Serotoninergic neurons appear involved in the sleep–wake cycle. The turnover rate of serotonin is decreased by BZDs and barbiturates. Barbiturates, in low doses, and BZDs have been shown to increase GABA-ergic transmission throughout the CNS, thereby potentiating both pre- and postsynaptic inhibition, which seem to be mediated by GABA, the inhibitory neurotransmitter.

A significant breakthrough in the late 1970s however, was the discovery of high-affinity saturable and stereochemically specific receptors for BZDs in the mammalian CNS. It is now believed that these sites are pivotal in mediating the antianxiety properties of the BZDs (Skolnick and Paul, 1981). Although early reports indicated that only BZDs exhibiting demonstrable pharmacologic activity were capable of displacing radiolabeled diazepam from high-affinity binding sites, several unrelated structures were found that also did so. One such compound is zopiclone, which was marketed in Britain (1989) as a sedative–hypnotic. Its pharmacologic profile parallels that of the short-acting BZDs ($t_{1/2}$ = 4.5 h).

Compounds known to act on GABA receptors do not inhibit the BZD receptor. GABA receptors are not identical recognition sites. Complicating the picture is that GABA, GABA-mimetics, barbiturates, and even ethanol have been shown to increase the affinity of BZD receptors, possibly by a neighboring perturbation effect.

Earlier investigations to relate BZD actions to one or more known or putative neurotransmitters in the brain have found that NE, DA, 5-HT, ACh, Glu, Gly, and others were all inactive in their ability to affect diazepam binding to BZD receptors. However, these studies did point to one substance as a probable mediator of BZD effects: GABA. It appears that BZDs act to facilitate GABA-ergic inhibitory transmissions in the CNS. In other words, BZD and GABA receptors, although not identical, are physiologically and biochemically linked.

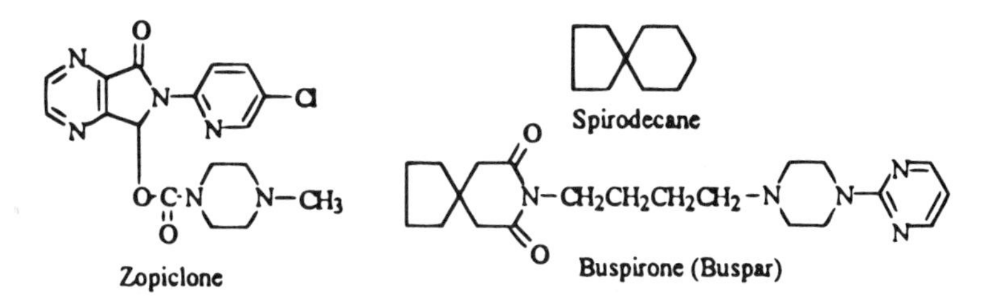

There appears to be an additional factor in the BZD–GABA receptor picture, chloride ion channels that can close and open in the neuronal membrane. These channels influence the topographic features of the receptors in response to either endogenous ligands or drugs. The model that has evolved is the BZD–GABA–receptor–chloride ionophore complex (Fig. 12-6) that has been previously considered. It can be helpful to consider possible coupling between BZD receptors, GABA receptors, and chloride channels, which are postulated to be in the double-layered lipoidal structure of neuronal plasma membranes. Thus one can conceptualize that a drug (or endogenous ligand) binding to the GABA receptor, or chloride ionophore, can allosterically alter the BZD receptor's features, resulting in altered affinity for the drug (e.g., diazepam) or the endogenous ligand (Skolnick and Paul, 1981). This would be analogous to an allosteric site on an enzyme binding a modulator molecule and thus having a regulatory function.

Anxiolytic properties of ethanol, barbiturates, and certain experimental compounds and their effect on the BZD–GABA–receptor–chloride ionophore complex may indicate a commonality of mechanism, yet phenobarbital does not enhance the binding affinity of diazepam to its receptor. Another apparent inconsistency is the strong increase in diazepam binding caused by a benzyl barbiturate that is a convulsant (Skolnick and Paul, 1981). Flumanezil (vide infra) is a selective BZD antagonist; it binds to the receptor, but it does not trigger the typical effects when it does so. The important differential in structure of this imidazolo BZD is the replacement of the 5-phenyl moiety with a keto function and an N^4-CH_3. The compound will displace ^{3}H-diazepam (and other labeled BZDs) from synaptosomal brain fractions. It will antagonize sedative and anticonvulsant effects of diazepam, but not of phenobarbital in mice. This drug has been a research tool for studying BZD-binding characteristics before its clinical utility had been developed.

12.4. Buspirone—Is Anxioselectivity Possible?

The question arises whether anxiolytic activity must always be accompanied by concomitant skeletal muscle relaxant and anticonvulsant activity as well as strong sedation. Can an anxioselective drug exist that will not interact significantly and additively with CNS depressant compounds, particularly alcohol? More significantly, both from a medical and sociologic viewpoint, will it be possible to treat anxiety and stress without the added complication of potent sedative effects, dependency, and abuse?

Evaluation of a series of cyclic imides as potential psychotropic agents yielded three candidates with tranquilizing action and very low sedative effects. The compound chosen for detailed pharmacologic evaluation, MJ9022-1, buspirone (BS), was marketed in 1986 as the first member of a new class of azaspirodecanediones. Early screening led to the erroneous belief that the compound might have antipsychotic properties. However, when further testing revealed that the compound had a taming effect without muscular incoordination (ataxia) in aggressive rhesus monkeys, its potential for anxiolytic activity was postulated. Clinical studies soon established the efficacy of BS to be equal to that of diazepam and clorazepate. A comparison study of the effects of ethanol on BS and lorazepam showed that BS did not interact with alcohol. BS does not share diazepam's anticonvulsant activity nor its sedative or muscle relaxant properties. Both animal studies and human trials with "recreational" drug abusers showed BS not to be addictive or to reinforce illicit drug use. It appears that BS does not mimic any of the effects of BZDs except the antianxiety activity. If continued clinical experience bears this out, then the question remaining is: By what molecular mechanism(s) is the narrow psychotropic mechanism achieved?

The mechanism, at this writing, is far from clear. It is known that BS does not bind to BZD or GABA receptors.

BS has been shown to interact with dopamine receptors in vitro. Pharmacologically both DA agonists and antagonist properties have been demonstrated. There is also evidence that serotonin receptors (type 2) are involved as well as interaction with NE and ACh systems. Thus any overall mechanism presently proposed will likely be speculative. However, some evidence appears to zero in toward serotonin receptor antagonism, particularly the $5HT_{1A}$ subtype, which likely mediated inhibitory effects in raphe neurons and pyramidal cells in the hippocampus. Affinity for BZD binding sites is very low. This fact goes a long way to explain the lack of sedation, ataxia, muscle relaxation, and the anticonvulsant properties of BS. The fact that BS, unlike diazepam, produces little or no loss of memory may be a considerable advantage in long-term anxiolytic therapy.

The metabolism of buspirone affords the inactive 5-hydroxybuspirone (OH on the 5-position of the pyrimidine ring). It is interesting that if the position is blocked, prolonged antipsychotic effects do occur. This may explain buspirone's lack of this effect. The 1-pyrimidinyl piperazine metabolite, which results from N^4-dealkylation of BS, retains some activity and is being evaluated.

12.5. Antiepileptic Drugs

Epilepsy is a category of CNS disorders characterized by recurrent transient attacks of disturbed brain function that result in motor (convulsive), sensory (seizure), and psychic episodes. Convulsions are not a consistent finding. Consciousness may be lost or altered depending on the type of disease (see Table 12-10). Seizures may be clonic, alternating muscle contraction and relaxation, or tonic (i.e., muscular tension).

The incidence of epilepsy in the United States has been estimated at 3–10 per 1000 population. The major types of epilepsy are briefly outlined in Table 12-10. It should be pointed out that each category listed can be further subdivided, based on types of seizure and detailed EEG tracings.

Several of the clinically encountered generalized seizures correlate well with those chemically inducible in animals. The drug giving good reproducible effects, particularly in

Table 12-10. Abbreviated Classification of Epileptic Seizures[a]

I. Partial seizures (focal)	Comments
A. Simple partial seizures	Elementary symptoms—consciousness not impaired; for example, Jacksonian epilepsy, convulsion of single limb or muscle group; older children and adults
B. Complex partial seizures	Loss of consciousness general; behavior confused and bizarre motor activity as well; characteristic EEG; adults and older children
C. A and B evolving into generalized seizures	
II. Generalized seizures (with or without convulsion)	Comments
A. Absence seizures (*petit mal*)	Onset preteen years; abrupt short loss of consciousness (<30 sec); mild clonic motions (e.g., blinking, jerking)
B. Tonic–clonic seizures (*grand mal*)	Major convulsions; maximal tonic spasm (<30 sec) then clonic phase (<1 min); seizures last 2–5 minutes, unconsciousness variable; occurrence at any age
C. Clonic and myoclonic seizures	Repetitive clonic jerks (early childhood) or sudden generalized or focal contractions (myoclonic) of face, limb, or trunk
D. Atonic seizures	Loss of postural tone

[a] Adapted and greatly simplified from Commission on classification (1981).

precipitating absence seizures, is pentylenetetrazol (PTZ, Metrazol). Similarly, partial seizures in humans can be mimicked in animals by the use of maximal electric shock (MES) test. The ability of experimental compounds to inhibit seizures produced by these two methods can then be used to screen for potential anticonvulsant drugs. Furthermore, these tests can also be predictive against the type of epilepsy the test substance is likely to affect. For example, a compound able totally to protect a rat from seizures produced by a dose of metrazol known to precipitate them in 95% or more of the animals is likely to possess anti–petit mal efficacy. On the other hand, a compound capable of preventing convulsions in the rat produced by MES (e.g., 150 mA for 0.25 second) might be a candidate for a drug against grand mal or psychomotor seizures. Other test variations may use different chemical inducers such as strychnine or bemegride (a convulsant barbituratelike compound) or different electrical methodology (e.g., AC vs. DC currents or different pulse rates).

Chemotherapy of epilepsy may be said to have begun with the use of inorganic bromides in the 1850s. However, the utilization of just-introduced phenobarbital seven decades later is the beginning of meaningful treatment of epilepsy. No further significant progress was made until the studies of Merritt and Putman (1937, 1938), who were looking for drugs able to suppress electrically induced animal convulsions, discovered 5-substituted hydantoins able to do this. The best and least sedative (a most desirable feature considering then-available therapy) was 5,5-diphenylhydantoin (phenytoin, Fig. 12-18). Over the next two decades—building on the "common denominator" structure (Fig. 12-18)—a dozen or so chemically related agents were marketed. Viewed from this vantage point all these compounds share a ureido-type structure as outlined. The oxazolidones and succinimides fit this concept if one is willing to allow for bioisosteric replacement of one of the nitrogens with O and C, respectively. (Phenacemide may be viewed as an acyclic hydantoin.)

The cause of epilepsy is still unknown. The general statement that excessive neuronal firings in the cortex or temporal lobe is a manifestation is not very helpful in reasoning out SARs of drugs able to suppress them. Nevertheless, some empirical attempts can be made.

The hydantoins, phenobarbital, mephobarbital, and pyrimidone, are effective in generalized tonic–clonic seizures (grand mal) and partial seizures that are secondarily generalized. They are in the class effective in preventing maximal electroshock-generated convulsion. Carbamazepine and its analog oxcarbazepine has the same efficacy pattern. It is therefore tempting to ascribe this analogy of action to the presence of two, or at least one, aromatic ring. However, the succinimides, and the less-used oxazolidines, whose clinical indications are mostly in the treatment of absence seizures, show remarkable effectiveness in blocking PTZ-induced convulsions in animals, yet two of the three succinimides also have a phenyl group as hydrophobic sites. The best that can presently be generalized is that they contain two hydrophobic regions, one of which may be an alkyl group. A heteroatom system also exists, usually cyclic, that atom invariably being a nitrogen. Two electron-donor moieties similarly oriented at an optimal separation need to exist. A carbonyl (or equivalent) is essential. The opposite activities, convulsant and anticonvulsant, of the barbiturate R and S isomers, respectively, which were described earlier, illustrates that molecular conformation, not only binding, is significant since it can be reasonably assumed that the intrinsic binding capacities would be the same for either enantiomer if steric factors such as those imposed by a receptor's topography were not a factor.

Still, we find empirically that superior activity against partial seizures requires at least one phenyl ring for its hydrophobic region. Diphenylhydantoin would seem to indicate

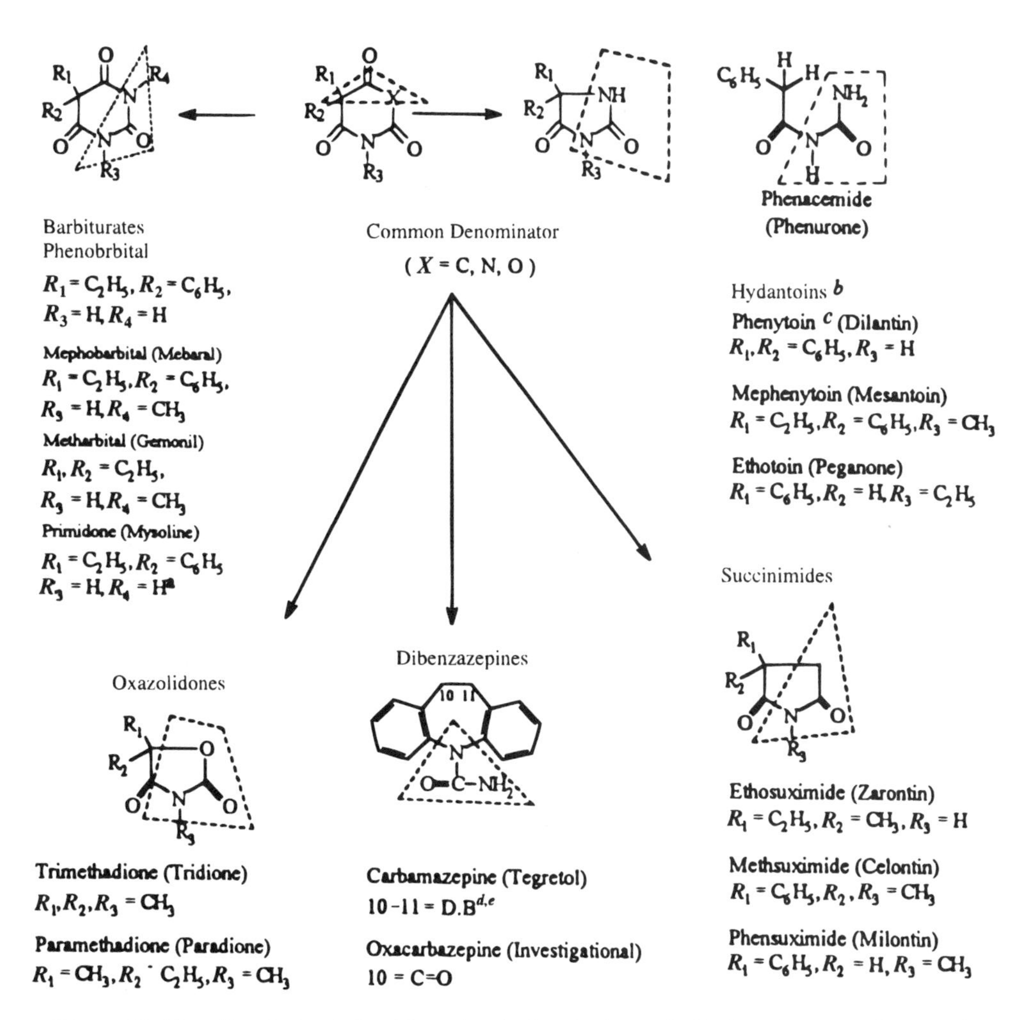

Figure 12-18. Chemical relationships of antiepileptic drugs. [a] The keto between nitrogens is reduced to CH_2. [b] Imidazole-2,4,-diones. [c] Formerly diphenylhydantoin. [d] Double bond. [e] The 10–11 epoxide is an active metabolite.

this. Therefore, 5,5-diphenyl-barbituric acid should be superior to phenylethylbarbituric acid (phenobarbital). The compound does inhibit experimental seizures in animals well. However, the additional benzene ring decreases aqueous solubility so much that it precludes clinical usefulness in humans. A complicating factor in understanding a drug such as phenytoin is its multiplicity of actions. It is of interest that replacing a 5-phenyl with an alkyl group produces sedation.

Phenytoin's five-decades long existence as a clinical entity makes it one of the most studied anticonvulsants. In addition, it has antiarrhythmic activity (Chapter 10). At the physiological level the drug has been shown to influence ion conductance, neuronal membrane potentials, and the levels of most of the neurotransmitters discussed, including GABA. The extent to which any of these effects affect epilepsy is still not clear. Nevertheless, more recent research seems to "point the finger" toward a potentiation of GABA-mediated inhibition, at least as part of the answer. Valproic acid (Fig. 12-5) and the two BZDs indicated for epilepsy, diazepam and clonazepam, have also been shown to enhance this inhibition. The fact that certain chemoconvulsants such as bicuculline (Fig. 12-5), PTZ, and picrotoxin decrease GABA activity supports this hypothesis. In fact, GABA has become a strong candidate for the neurotransmitter whose dysfunction is most involved as the precipitant of spontaneous electrical discharges leading to convulsions and seizures. There are several ways such a dysfunction could be manifested. One is that inhibition of GABA transaminase is a mechanism. Another supposition, that GABA receptors are involved, has led to the possibility of a new generation of anticonvulsants. Since GABA itself does not cross the BBB, other receptor-specific agonists are being sought. The rigid GABA-analog agonist THIP (Fig. 12-5) has undergone clinical trials. Even more promising GABA agonists are progabide and *p*-nitrophenyl esters of nipecotic and isonipecotic (piperidine-4- and piperidine-3-carboxylic) acids. Progabide significantly decreased seizures in epileptic patients deemed refractory to standard therapy. Piperidine-4-carboxylic (isonipecotic) acid can, of course, be conceived as a "rigidized" GABA molecule. As GABA, it and the 3-COOH isomer (nipecotic acid) are too polar to cross the BBB. Converting them to lipid-solubilizing esters with good "leaving groups" such as *p*-nitrophenolate (Chap. 8; Fig. 8-10) enabled these compounds to traverse into the CNS and there be readily hydrolyzed to the GABA-mimetic (i.e., they were essentially pro-drugs). Both compounds exhibit protection in mice against PTZ convulsions, but less so than phenobarbital. It seems that some parts of the puzzle are slowly beginning to fall into place.

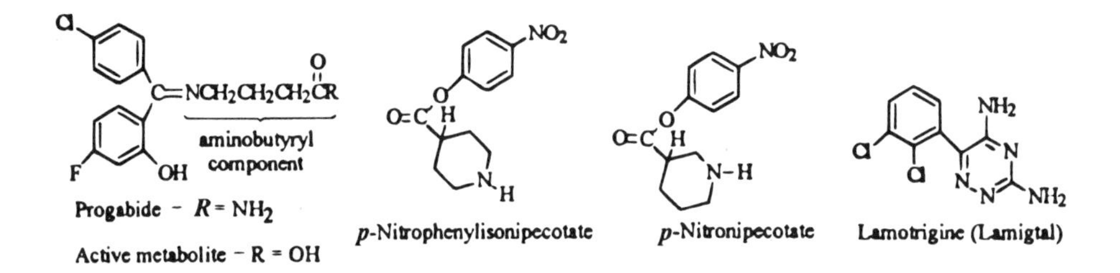

Lamotrigine (Lamigtal) was originally synthesized as a dihydrofolate reductase inhibitor. The compound was much better as an epileptic than it was in its original mission. Its talent is apparently in stabilizing presynaptic neuronal membranes by blocking their sodium channels as well as inhibiting excitory neurotransmitters. These features make it useful in the treatment of partial onset epilepsy.

Gabapentin (Neurontin, Fig. 12-5) was designed as a GABA-mimetic able to cross the BBB to alleviate or prevent seizures by compensating for a perceived GABA deficiency. It has been introduced (1993) as an effective agent. Curiously, it has no effect on the GABA system, but is effective nevertheless with a useful half-life of 5–7 hours. The drug is excreted in the urine without metabolism.

A different approach led to the development of vigabatrin (Sabril). Its ability selectively and irreversibly to inhibit GABA-transaminase from metabolically inactive GABA (as a suicide inhibitor) makes it an effective agent (Fig. 12-5), particularly in complex partial seizures.

Drugs with particularly low water–lipid partition can predictably result in poor and erratic absorption from oral dosage forms. Additionally, such compounds are also more likely to have their absorption adversely affected by formulation difficulties. Thus the pharmaceutical additives used in the dosage form manufacture (e.g., fillers, disintegrating agents, preservatives, etc.) as well as the particle size and crystalline type of the drug itself, can all affect the bioavailability of the active ingredients in a pharmaceutical product. This can lead to variability of drugs in the same dosage form made by different manufacturers, or even different production lots from the same source. Two such products, both containing the same drug in the same quantities (pharmaceutically equivalent), may not be *therapeutically equivalent.* This problem has occurred with few drugs, yet several important ones have presented potentially serious problems in clinical situations. Phenytoin is such a drug. Others have been nitrofurantoin, digoxin, and griseofulvin.

12.6. Neurochemistry of Mental Disease

Can abnormalities in the "steady state" of biogenic amines (NE, DA, 5-HT) and other neurotransmitters of the CNS be causative factors in mental diseases such as schizophrenia, endogenous depression, and manic behavior? Only 50 years ago arguments in favor of such a notion would have elicited, at best, benign smiles of amusement from an audience of psychiatrists.

It is interesting that some experiments in the 1880s had suggested a possibility of chemical factors affecting the pathology of psychoses. If the state of organic and analytical chemistry is considered, it is understandable that some of these early "results" were not reproducible. However, things changed dramatically by the 1950s, mainly due to the discovery of drugs that were able to affect significantly and positively—even if palliatively—mental illness.

Research over the past several decades has resulted in interesting theories involving these amines and their receptors. It should be pointed out, however, that these concepts are still theories, not established explanations of the disease state.

Following the empirically based introduction of psychotropic drugs in the early 1950s, it was observed that some tended to normalize mood swings. It was further noted that brain amine levels of patients suffering from affective pathophysiology (e.g., endogenous depression) were also normalized.

Furthermore, the rauwolfia alkaloids (reserpine and others), which also began to be used at that time in psychiatry (and as antihypertensives; see Chapter 10), were found to be potent depletors of brain NE and 5-HT in animals. These and additional findings over the next decade led to the (mono)amine hypothesis of mental disorders. The DA hypothesis of schizophrenia is currently the one with the most supportive evidence.

The neuroleptic phenothiazines (available then) to treat schizophrenic symptoms were shown to decrease central DA transmission. The DA synthesis inhibitor α-methyltyrosine can potentiate these drugs. On the other hand, drugs known to enhance central DA activity can worsen symptoms and even produce psychotic-type signs in normal individuals. MAO inhibitors, by retarding biogenic amine breakdown, are good examples. The antipsychotic phenothiazines (see Table 12-11), as well as apparently chemically diverse drugs with clinically effective antipsychotic activity, all have the ability to affect DA negatively. This includes the butyrophenones and diphenylbutylamines (see Table 12-12), dibenzoxazepines (see Table 12-13), and dihydroindolone compounds (see later). It is now reasonably well established that all these drugs are capable of blocking DA receptors in the brain.

There are three major central DA pathways: the *nigrostriatal,* which is affiliated with motor effects produced by antipsychotic drugs; the *tuberoinfundibular* tract, which is asso-

Table 12-11. The Neuroleptic Phenothiazines and Thioxanthenes

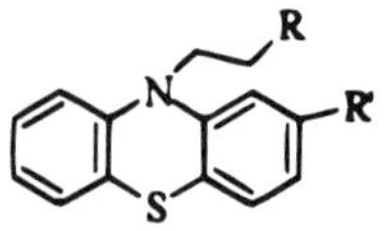

Phenothiazines

Drug[a]	*R*	*R′*	Other effects[b]
A. Dialkylaminopropyl group			
1. Chlorpromazine (Thorazine)	$-CH_2N(CH_3)_2$	Cl	Antiemetic, sedation
2. Promazine (Sparine)	$-CH_2N(CH_3)_2$	H	Antiemetic
3. Triflupromazine (Vesprin)	$-CH_2N(CH_3)_2$	CF_3	Antiemetic, extrapyramidal effects
B. Alkylpiperidines			
4. Mesoridazine (Serentil)	CH_3-N (piperidine)	$S(=O)CH_3$	Sedation
5. Thioridazine (Mellaril)	 CH_3-N (piperidine)	SCH_3	Sedation
6. Piperacetazine (Quide)	 CH_3-N (piperidine), CH_2CH_2OH	$C(=O)CH_3$	
C. Alkylpiperazines			
7. Acetaphenazine (Tindal)	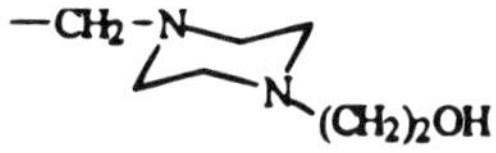 $-CH_2-N$ (piperazine) $N-(CH_2)_2OH$	$C(=O)CH_3$	
8. Fluphenazine (Prolixin)[c]	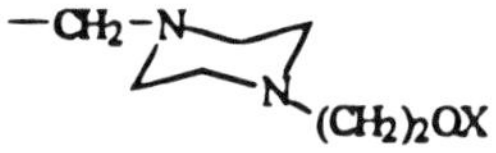 $-CH_2-N$ (piperazine) $N-(CH_2)_2OX$	CF_3	Extrapyramidal; antiemetic
[Fluphenazine *X* = H, Enanthate *X* = $C(=O)(CH_2)_8CH_3$ Decanoate *X* = $C(=O)(CH_2)_5CH_3$]			
9. Perphenazine (Trilafon)	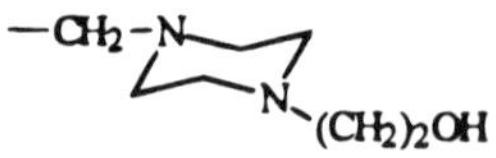 $-CH_2-N$ (piperazine) $N-(CH_2)_2OH$	Cl	Extrapyramidal; antiemetic
10. Prochlorperazine (Compazine)	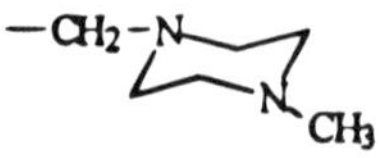 $-CH_2-N$ (piperazine) $N-CH_3$	Cl	Antiemetic; extrapyramidal
11. Triethylperazine (Torecan)	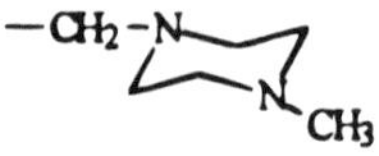 $-CH_2-N$ (piperazine) $N-CH_3$	SCH_2CH_3	Used only as antiemetic
12. Trifluperazine (Stelazine)	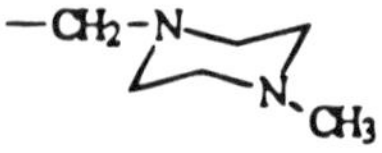 $-CH_2-N$ (piperazine) $N-CH_3$	CF_3	Extrapyramidal

(*Continued*)

Table 12-11. (*Continued*)

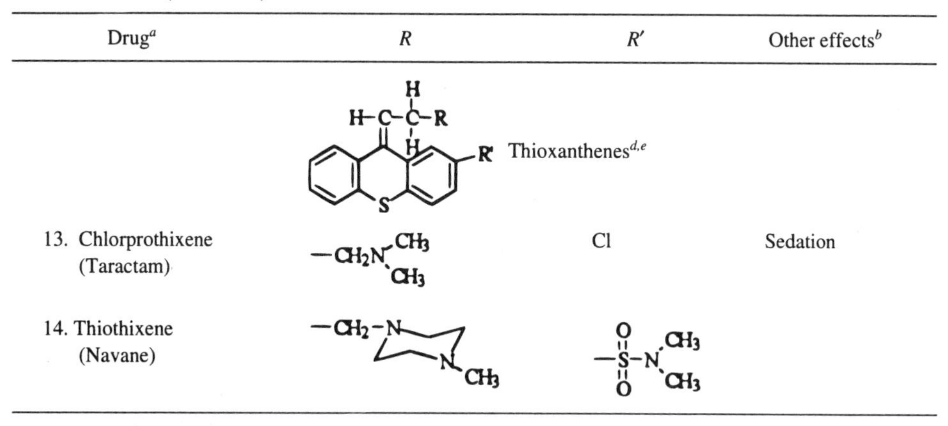

Drug[a]	*R*	*R′*	Other effects[b]
Thioxanthenes[d,e]			
13. Chlorprothixene (Taractam)	$-CH_2N(CH_3)_2$	Cl	Sedation
14. Thiothixene (Navane)	$-CH_2-N$ (piperazine) $N-CH_3$	$-SO_2-N(CH_3)_2$	

[a] Trade name in parentheses.
[b] May be either of clinical value or considered as adverse effects.
[c] These alkanoic acid esters in edible (sesame) oil solution give 1–3 weeks of continuous effect following a single IM injection.
[d] The ring N is isosterically replaced with sp^3 carbon
[e] The *cis* (Z) isomer shown is much more potent than *trans* (E) isomer.

ciated with the endocrine effects of neuroleptics; and the *mesolimbic* pathway, which is the most likely to relate to the symptoms of schizophrenia. Of the three central DA receptor subtypes D_1, D_2, D_3, the D_2 receptor, which has been shown to bind selectively to [^{3}H]haloperidol and [^{3}H]spiroperidol (Table 12-12), is believed to be most relevant to antipsychotic drug action. The relation of clinical potencies to D_2 receptor affinity correlates exceedingly well.

Other evidence is mounting that supports the DA hypothesis of schizophrenia in addition to the parallelism of potency and receptor affinity. For example, amphetamines are known to exacerbate schizophrenia in patients and even to produce temporarily schizoid states in normal individuals. The very potent cholinergic alkaloid arecoline (Chapter 8) can produce short periods of lucidity in highly psychotic patients. Because cholinergic and adrenergic effects frequently oppose each other in many situations, even such a brief respite of symptoms is additional support for the biogenic amine mechanism. The evidence suggests an aberration in DA activity in psychotic diseases. An alternative possibility is either hypersensitive or excess D_2 receptors.[13]

Binding studies with tritiated neuroleptics have established the existence of two types of central DA receptors: D_1 and D_2; their numbers and ratios vary in brain areas. Most interesting, however, is that both are altered in drug-naive schizophrenics. Emphasis has concentrated on the D_2 subtype that seems implicated in the observable clinical responses. The role of the postsynaptic D_1 receptor is not clear. It is known, however, that even though neuroleptic drugs are D_1/D_2 antagonists, in vitro D_2 effects are achieved at 10^3 lower concentrations. D_2 receptors are also located in central areas outside the BBB. One is the CTZ in the medulla. Presynaptic stimulation, which leads to DA inhibition there, may be the reason that many of the phenothiazine and butyrophenone neuroleptic drugs also excel as antiemetics (see comments, Table 12-11). It also explains why several DA_2

[13] Nomenclature identifies DA_1/DA_2 as peripheral, renal, and cardiovascular receptors and D_1/D_2 as central dopaminergic sites.

Table 12-12. The Evolution of Neuroleptic Butyrophenones

Name	Structure	Morphoid activity [a]	CPZ-like activity [a]
1. Meperidine	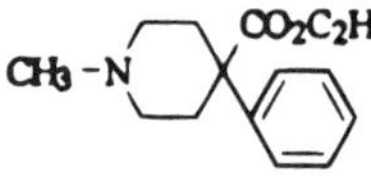 CH_3-N, $CO_2C_2H_5$	+	0
2. Propiophenone Analog of meperidine (R-951)	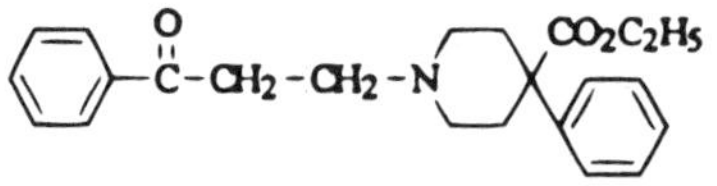 O, C-CH_2-CH_2-N, $CO_2C_2H_5$	+++	0
3. Butyrophenone Analog of normeperidine (R-1187)	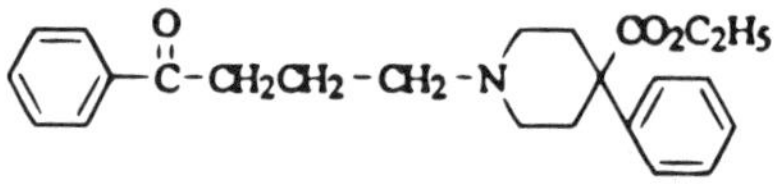 O, C-CH_2CH_2-CH_2-N, $CO_2C_2H_5$	++	+
4. R-1472	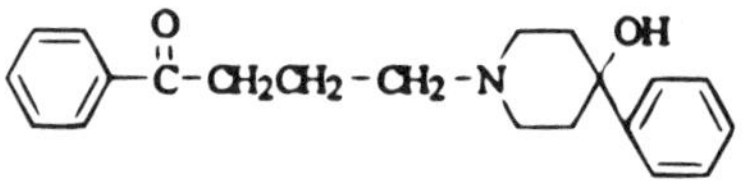 O, C-CH_2CH_2-CH_2-N, OH	0	++
5. Peridol (R-1589)	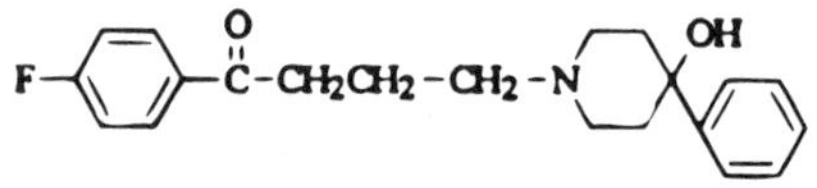 F, O, C-CH_2CH_2-CH_2-N, OH	0	+++
6. Haloperidol (R-1625)	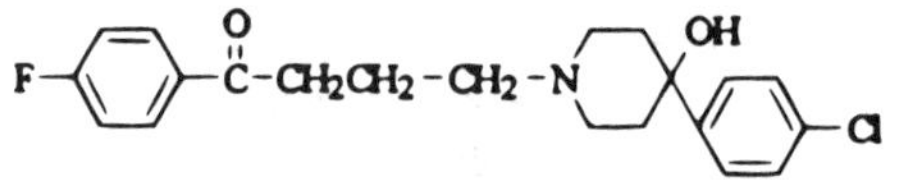 F, O, C-CH_2CH_2-CH_2-N, OH, Cl	0	++++
7. Fluspiperone	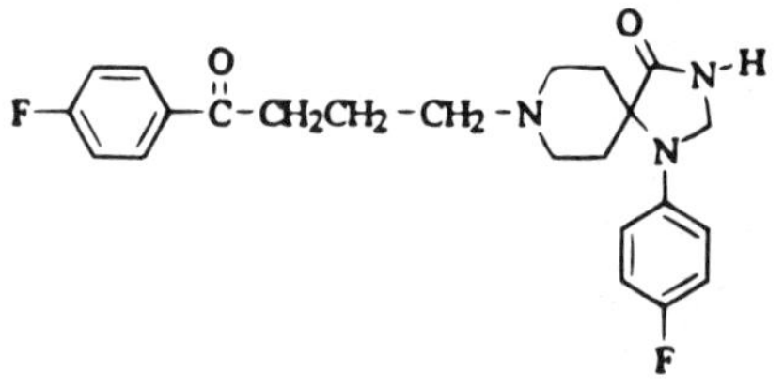 F, O, C-CH_2CH_2-CH_2-N, O, N-H, N, F	0	6000X

[a] Morphine = +++; CPZ = ++.

antagonists that do not cross the BBB well are antiemetics devoid of significant antipsychotic effects. Metoclopramide and the investigational drug domperidone are prime examples. The latter's structural similarity to droperidol and spiperone should be noted (see Table 12-17).

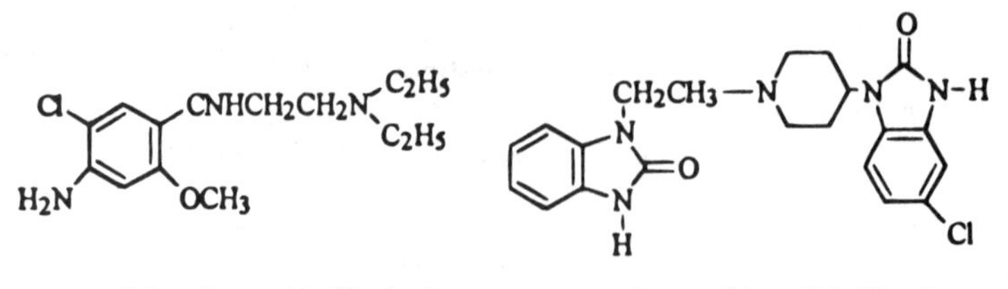

Metoclopramide (Reglan)

Domperidone (Motilium)

Table 12-13. Butyrophenones and Related Antipsychotics

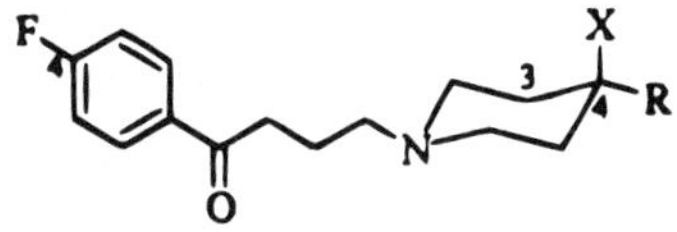

Drug[a]	X	X/R	R	Comments
1. Bromperidol[b]	OH		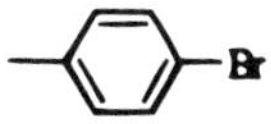 —C6H4—Br	
2. Clofuberol[b]	OH		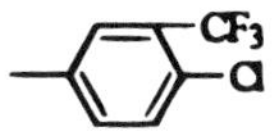 —C6H3(CF3)(Cl)	
3. Droperidol[c] (Inapsine)		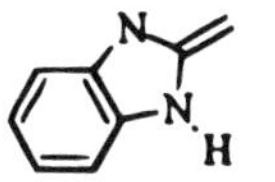 (benzimidazolone, N–H)		Used for neuroleptoanesthesia
4. Haloperidol (Haldol)	OH		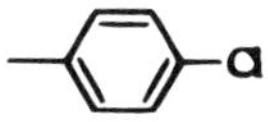 —C6H4—Cl	
5. Spiperone[b,d]		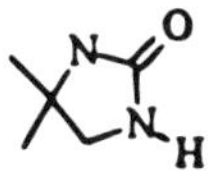 (spiro imidazolidinone, N–H, O)		3000 times the potency of CPZ
6. Trifluperidol[f,b]	OH		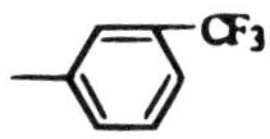 —C6H4—CF3	

Diphenylbutylamines

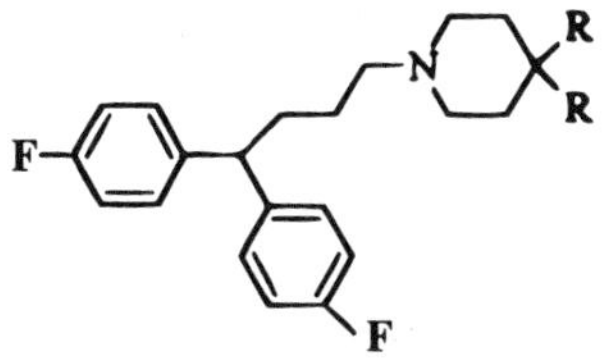

Drug[a]	R group	Comments
7. Fluspirilene[b]	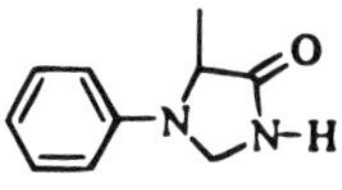 (spiro phenylimidazolidinone, N–H, O)	
8. Penfluridol (Semac)	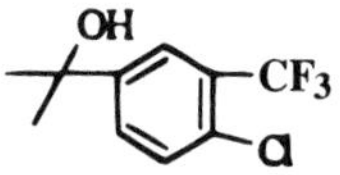 OH; —C6H3(CF3)(Cl)	
9. Pimozide (Orap)	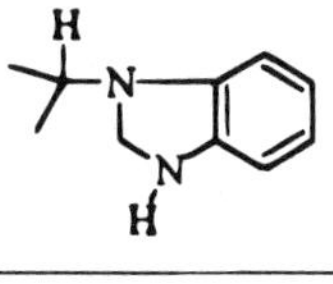 H; (benzimidazolinone, N, N–H)	Only against Tourette syndrome[e]

[a] Trade names in parentheses.
[b] Not available for clinical use in the United States.
[c] Piperidine ring has 3–4 DB (i.e., tetrahydropyridine).
[d] See earlier discussion in this chapter.
[e] In patients refractory to other drugs.

The other central area outside the BBB that is rich in D_2 receptors is the anterior pituitary gland. Since stimulation of these receptors will result in decreased prolactin secretion, DA_2 inhibitors will produce galactorrhea as a side effect because of increased prolactin levels.

It is a reasonable supposition that D_2 receptors at these various central locations may differ. This would explain why, for example, not all phenothiazines are equally good antiemetics.

The paradoxical effects of certain DA agonists exhibiting an antipsychotic effect at very low doses is explicable when one considers the ideas presented earlier regarding presynaptic (auto)receptors. In such cases low agonist levels do actually inhibit DA release by stimulating these receptors. At normal or higher doses the "expected" psychotic side effects are noted. DA blockade in basal ganglia produces the extrapyramidal effects also known as the *Parkinson syndrome*. Along these lines, at elevated doses sufficient levels of metoclopramide can reach the ganglia to develop these adverse effects as well. Other side effects, such as orthostatic hypotension, that result from α-adrenergic blockade, are also encountered in the clinical use of antipsychotic drugs.

The question of excess D_2 receptors in schizophrenic patients is in a controversial state of flux. That the numbers increase is not in contention. An earlier study showed that phenothiazines and butyrophenones can lead to an increase of this receptor population with long-term use. Using positron emission tomography (PET) scans (see later) on drug-naive patients also finds increased numbers of D_2 receptors. More definitive was a study using PET scanning to quantify receptor density and affinity in the brains of living human subjects (rather than postmortems as some earlier studies). The radioligand used was (3N-^{11}C methyl)spiperone, which is a high-potency D_2 antagonist antipsychotic. The study used three groups of subjects: normal volunteers, patients never treated with neuroleptics, and patients treated with these drugs. The results showed that D_2 DA receptor density in the caudate nucleus and putamen were higher in both groups of patients than they were in the normal subjects. The results indicate that schizophrenia, rather than prior neuroleptic drug exposure, is responsible for the D_2 receptor increase seen in schizophrenics.

PET scanning is where images of the distribution and areas of concentration (in this case, the brain) are made by γ radiation as they are with the better-known CAT scanning technique. However, the γ particles result from positron, β^+, decay from positron-emitting nuclides, colliding with and annihilating a beta particle, β^- (Eq. 12.6). This is γ annihilation. The resulting energy emanates as a pair of 0.511 million electron volt (MeV) photons in exactly opposite directions (Eq. 12.6). The photons are detected and amplified.

$$\beta^+ + \beta^- \longrightarrow \gamma\ (0.511\ \text{MEV}) \tag{12.6}$$

The other end of the psychiatric spectrum (Fig. 12-19) is depression. A catecholamine hypothesis that evolved here during the mid-1960s essentially stated that most of depression is associated with a relative or absolute catecholamine deficiency, especially NE at functionally important adrenergic receptor sites in the brain (Schildkraut, 1965). It is presumed that the opposite situation, which is excess catecholamines, may produce mania. Even though overly simplistic, the hypothesis served as a useful initiation into the developing complexity that followed. The NE deficiency, of course, can arise in any of several ways: (1) decreased synthesis, (2) impairment of receptor binding, (3) storage impairment, (4) increased intracellular release, (5) increased oxidative metabolism rate, and (6) decreased receptor sensitivity. It is possible that different depression subtypes relate to dif-

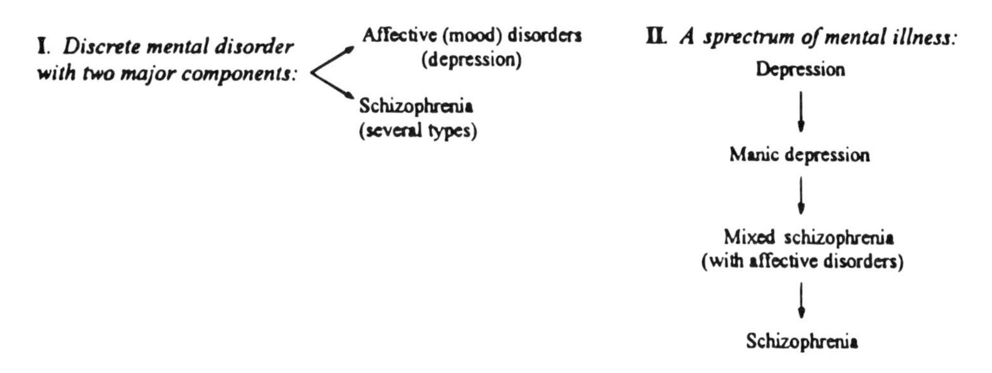

Figure 12-19. Two concepts of psychosis.

ferent mechanisms of deficiency. Examination of the metabolic reactions of NE (and EP) (Fig. 9-3) led to the recognition that oxidation by MAO initially to the aldehyde dihydroxyphenylglycolaldehyde, is then primarily reduced to the alcohol 3-methyoxy-4-hydroxyphenylethylene glycol (MHPG or MOPEG) in the CNS, rather than oxidized further to the corresponding acid, as it is peripherally. MHPG crosses the BBB so that urinary levels, as conjugates, can be measured. In bipolar depression—where patients exhibit alternation between manic and depressive phases—urinary MHPG levels rose in the former and decreased during the latter phase. Patients with unipolar symptoms showed inconsistent MHPG excretion patterns. There also seemed to be some puzzling relationships of metabolic excretion patterns, depending on seemingly minor structural variations in the antidepressant drugs used in treatment. The initial hypothesis simply could not account for apparent anomalies and had to be modified to include other biogenic amines, particularly the indoleamine serotonin, 5HT. (This amine's metabolism is outlined in Figure 12-3.) The situation clarified somewhat, as will be seen.

Evaluation of central amines, by monitoring the plasma values, urinary concentrations, or even cerebrospinal fluid levels, is full of pitfalls. Correlation with turnover and types of depression can be poor. Postmortems, of course, present other problems such as time elapsed since death and whether it was a violent suicide or following a lengthy coma. Assay procedures, depending on techniques used, can still technically give disparate results. Many data seem to lack consistency in the "biochemistry of depression." Even though emphasis has been on NE and 5HT, there are other substances either not yet well understood or even identified. Octopamine (Chapter 9) can accumulate following the use of certain antidepressant drugs and likely affects neuronal functions. What effect might it have on synapses? Where does ACh fit in? Diet? Many studies do not necessarily support the amine hypothesis. Why can biochemical "abnormalities" remain even though the clinical situation improves in some patients?

12.7. Antipsychotic Drugs—The Neuroleptics

The tricyclic compound phenothiazine, which has been known since the late nineteenth century, is a toxic compound of little therapeutic interest except as a large-animal anthelmintic at one time. An N-10 derivative, diethazine, was synthesized by French workers in the late 1940s as a possible antihistamine. The added anticholinergic properties it had brought it briefly into use against Parkinsonism. It was soon superseded by better agents (Chapter 8). In a search for antimalarials there was synthesized the dimethylamino homolog, promethazine, a superior antihistamine with sedative and antiemetic properties as well. Additional systematic modifications by researchers at Rhone-Poulenc in France led to the synthesis of a 2-chloro-substituted phenothiazine, where the 10-N was separated from the side-chain N atom by three, rather than the two, carbon atoms of the previous compounds. It was named chlorpromazine (CPZ). The drug was found to have some sedative properties, the ability to potentiate other CNS depressants, and, historically more important, a type of "artificial hibernation"—a conscious state exhibiting an indifference to one's surroundings. Some described it as a pharmacological lobotomy. Its psychotropic properties were first tested in manic, then schizophrenic, patients. CPZ was introduced clinically in 1952.[14] The era of psychopharmacology had begun.

Even though several dozen antipsychotic phenothiazines and several more isosteres had been introduced worldwide over the next 10–15 years (but few if any afterwards), CPZ, the first drug, is still widely used and is considered the prototype. These drugs have settled into four subcategories based on apparent relationships of the chemical nature of the side chain on the N^{10} to some nuances of observed clinical effects and side effects. Among the several features almost all these compounds share are electronegative substituents at the 2-position.[15] These are usually Cl, CF_3, SCH_3, or $SO_2N(CH_3)_2$. An absolute requirement is the separation of the ring (actually the N atom in phenothiazines, the sp^3 carbon in isosteres) from the side chain N atom by three saturated carbons without branching on the middle carbon. A 2-carbon chain invariably restores antihistaminic activity with total loss of neuroleptic effect. A 3-carbon separation with a middle carbon substituent (e.g., trimerazine, Chapter 13) produces a primarily antihistaminic drug with good antipruritic activity, but no meaningful antipsychotic properties.

As a broad generalization it can be stated that the alkylaminopropyl group (compounds 1–3, Table 12-11) possess useful antipsychotic properties, marked sedation, and relatively modest antihistaminic and anticholinergic activities. Promazine is the least potent of the trio. The CF_3 group of triflupromazine imparts the highest potency here; all three are effective antiemetics. None is antinauseant in cases of motion sickness, however. Hypotension is encountered, as is a high tendency toward extrapyramidal effects.

The alkylpiperidine group (Nos. 4–6) tends to be less effective and exhibits little antiemetic influence. However, they appear to have less extrapyramidal effect and may be considered to have the least toxicity of the four groups. Thioridazine has been reported to have the possibly unique property of inhibiting premature or nocturnal ejaculation in males without impairment of potency or orgasm. The active metabolite mesoridazine has not been similarly evaluated.

The third group, the alkylpiperazines, is the most antipsychotic and has the most effective antiemetics (Nos. 7–12). Thiethylperazine (No. 11) is marketed exclusively as an antiemetic. As a group these drugs tend to have a high incidence of extrapyramidal side effects and may also be the most toxic group. However, it must be considered that because of their considerably higher potency, the clinical significance of this may not be great.

---

[14] In the United States in 1954.

[15] Except promazine, which has none.

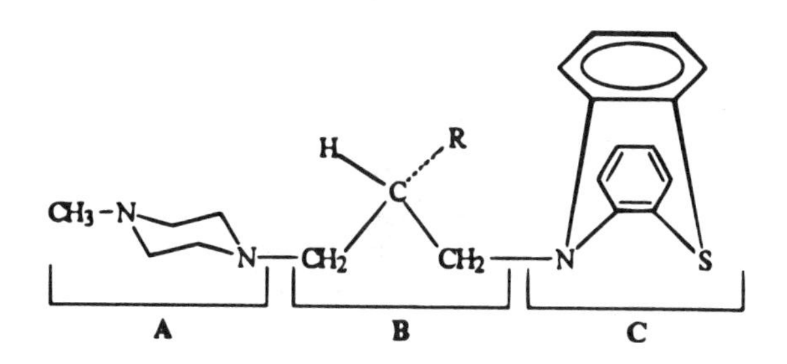

Figure 12-20. Relationship of piperazine-type drug to neuroleptic receptor. Phenothiazine in "boat"-shaped projection.

For example, fluphenazine (No. 8) can be useful in daily doses as low as 2 mg. The derivatization of this compound as long-chain fatty acids, such as the decanoate and enanthate, affords pro-drugs whose oleaginous solutions, on intramuscular injection of as little as 12.5 mg, can produce sustained clinical effects for as long as four weeks. Such a dosage form offers a distinct advantage for maintenance therapy for noncompliant patients.

The mammalian metabolism of the phenothiazines has been studied in greatest detail in the prototype chlorpromazine. A partial profile is outlined in Figure 3-5. It will be noted that all the predictable metabolic reactions occur: aromatic hydroxylation, followed by glucuronide and sulfate conjugation, N-dealkylation, followed by possible MAO attack to yield N^{10}-aldehydic side chains, and sulfoxidation.

When it is considered that all these chemical processes occur simultaneously at different rates and, furthermore, that the 2-Cl (or any 2-substituent) eliminates the plane of symmetry of the phenothiazine ring system, the number of possible metabolites and conjugates is very large. The initial proposal of 168 metabolites has since been expanded to over 190. Most, if not all, have been actually identified. There has been little evidence over the years for the existence or in vivo accumulation of active metabolites.

Thioridazine (No. 5) undergoes an additional sulfoxidation at the 2-thiomethyl group. However, in this case the metabolite produced is active and marketed as mesoridazine (No. 4).

The piperazine group of drugs offers additional metabolic opportunities such as N-oxidations at the N^1 and N^4 positions of the piperazine ring as well as N-dealkylations at these points. A partial piperazine ring degradation also occurs.

The synthesis and evaluation of a large number of phenothiazines in the two decades following the introduction of CPZ has led to the recognition of SARs that outlined the parameters within which clinically effective antipsychotic activity resides. These will be briefly considered because they can give the student some insight into how medicinal chemists and pharmacologists sometimes reach their empirical conclusions.[16]

Figure 12-20 schematically represents a putatively preferred conformation for the piperazine type of phenothiazine antipsychotic. As numerous analogs were tested, it became apparent that the B site of these molecules had the strictest specificity requirements for optimum activity. The activity screened to establish an index of comparison to CPZ (the chlorpromazine index, CI) was the ability of a compound to block the conditioned escape response of rats. It is expressed as a decimal fraction or multiple of CPZ. The absolute

---

[16] For additional detail, see Gordon (1967).

requirement of a three-carbon separation between the N atoms was mentioned. A comparison of CPZ, CI = 1.0, with prochlorperazine showed a CI = 3.0 (Fig. 12-21). This established the superiority of the piperazine ring in the side chain. The crucial atom is the middle carbon. A methyl substituent at this position reduced activity so that only good antihistaminic properties predominated (e.g., see trimeprazine, Chapter 13). Larger alkyl substituents eliminated meaningful activities. The CIs of compounds A_2, A_3, and A_4 (Fig. 12-21) further illustrate the intolerance of this position to modifications. It seems likely that more than substituent size at this position is at stake here; a certain degree of freedom for the 2-carbon to "rotate" was suspected as well. The prediction that compounds C, D, and E (Fig. 12-21) would have drastically lower activity than would CPZ due to the immobilization of this carbon in a ring structure was borne out. The predictably much higher activity of compound F was experimentally established as well.

The assumption of the model in Figure 12-20 is a three-point receptor attachment. Site C, where the phenothiazine ring interacts with the receptor topography, seems to require an electron-withdrawing substituent at the 2-position. The nature of *X* was already stated. Substituents at positions 1, 3, and 4, reduce neuroleptic effectiveness dramatically. Even a parallel substituent on the other aromatic ring (i.e., at 2 and 8 is detrimental). The electron-withdrawing potency does not seem relevant. The fact that all three rings of phenothiazine are not coplanar must be appreciated. If one considers the nitrogen to sulfur axis as a "hinge" on which the two benzene rings "flutter," it can be understood that any 2-substituent cannot reach and therefore not interact with site C directly. Figure 12-22 depicts some of the steric requirements of CPZ-type drugs discussed here.

The specificity of activity contributed by the part of the drug in a position to interact with site A depends on its transverse rather than longitudinal characteristics. In the case of the dialkylamino-containing compounds the unfettered rotating ability of alkyl groups can sweep in wide arcs. *R* groups larger and bulkier than CH_3 invariably diminish activity considerably. This has been ascribed to steric interference with proper A site interfacing. The effective width of N^4-substituted piperazine affords an even better accommodation (prochlorperazine, CI = 3) than CPZ. Even the pyrrolidine derivative maintains good activity (CI = 0.7). This has led to the speculation that the A site may have a narrow slotlike geometric requirement. By increasing that nitrogen's basicity (compared with the pyrrolidino),

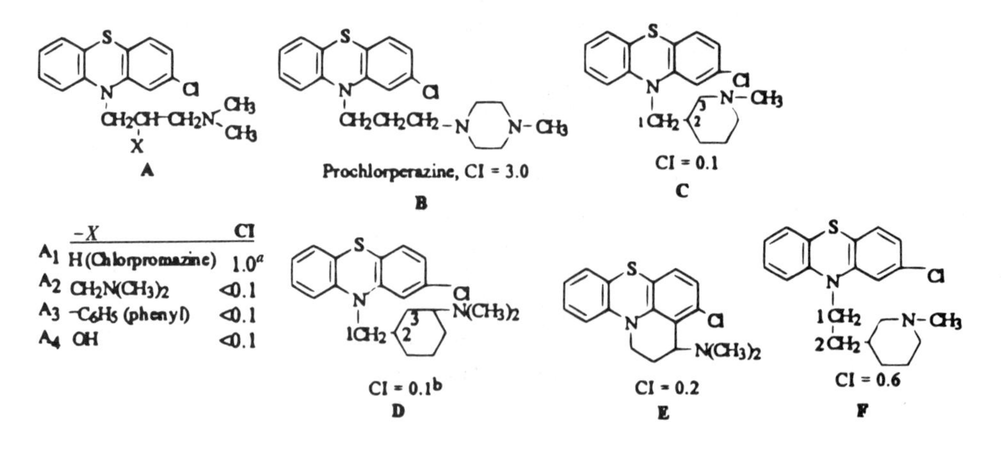

Figure 12-21. Potency SARs of some phenothiazines. [a]CI = chlorpromazine index where CPZ = 1.0. [b]*cis* or *trans*.

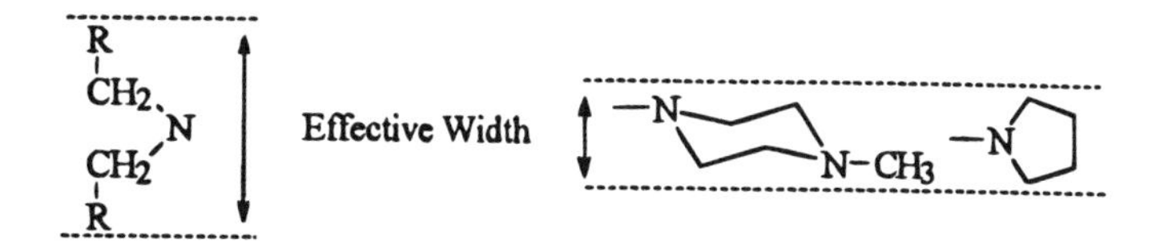

Figure 12-22. Steric comparison of dialkylamino and cycloalkylamino groups.

the N-substituent of the piperazines may account for the enhanced activity. The fact that a *p*-aminophenethyl group at N^4 increases the activity even more was taken to mean that the A site is quite long. Finally, the lowering of the activity by the addition of a CH_3 group on the 2 or 3 positions of the piperazine ring may constitute further evidence to the "effective width" rationalization.

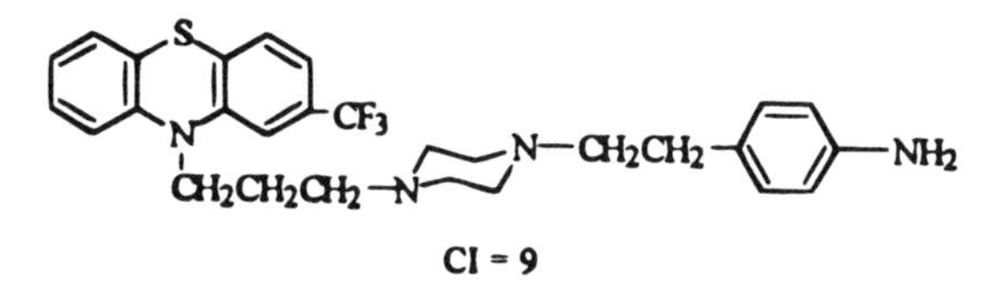

Table 12-11 lists two thioxanthenes. Chlorprothixene represents the bioisostere of CPZ where the nitrogen of the phenothiazine ring is replaced with an sp^3 carbon. Thiothixene (No. 14) carries a dimethylsulfonamido moiety at the 2-position of the thioxanthene ring. Both drugs are primarily active in the *cis* or Z geometric configuration (i.e., the side chain is folded toward the ring carrying the 2-substituent), as X-ray crystallography has shown it is in the solid state for CPZ. The *trans* or E forms of these two drugs, which cannot assume the same conformation, are pharmacologically much weaker. It is interesting that the parallelism of the pharmacology of these two thioxanthene drugs with the corresponding phenothiazines extends to similar dosages and effectiveness, as well as to the side effect characteristic of the dialkylaminopropyl and piperazino side chains, respectively.

12.8. The Butyrophenones—Serendipity and Drug Development

This important group of potent neuroleptics (and the diphenylbutylamines that evolved somewhat later) are classic examples of how an apparently serendipitously discovered lead in the late 1950s was successfully exploited to evolve into a major category of psychotropic drugs. Paul Janssen[17] undertook a systematic study of the 4-phenylpiperidine molecule, particularly to evaluate the limits of N-substitution of normeperidine. It will be recalled (Chapter 5) that replacement of the methyl of meperidine with a *p*-aminophenylethyl moiety (anileridine) and a phenylaminopropyl group (piminodine) afforded both active analgetics and more potent ones as well. There are other examples that are not marketed in the United States.

---

[17] Founder of Janssen Pharmaceutica in Belgium.

Table 12-12 outlines the sequence of molecular modifications that on first attempt produced a normeperidine propiophenone analog where analgetic potency exceeded that of morphine 100 times. Lengthening the chain by one CH_2 produced a butyrophenone whose analgesia was still much higher than morphine, but, surprisingly, that also exhibited a low-level CPZ-like pharmacology in animal screens. Even more unexpected was that replacement of the 4-carboethoxy (ester) group with a simpler OH led to the disappearance of opioid properties while simultaneously acquiring neuroleptic activity that was indistinguishable from CPZ both in potency and quality. It did not take long to determine the proper positions (para) for halogen substituents in both aromatic rings to increase the effectiveness of R-1472 dramatically. First was peridol, with its *p*-F atom on the ketonic phenyl ring, followed by haloperidol, which carried halogens on both benzenes. Further modifications of the 4-piperidine position resulted in spirotriazolodecanes carrying a keto function where the 4-OH was on the earlier compounds. Here potencies exceeded CPZs by several thousand times. Table 12-13 gives some additional examples of butyrophenones in various clinical applications, including psychoses and as antiemetics, alone and in conjunction with anesthetics.

SARs of butyrophenones have some unusual features that sometimes do not seem to conform to many of our empirical "rules." The 4-F atom on the benzoyl portion of the molecule seems mandatory for optimum activity. When combined with 1-phenyltriazospiro group as in the spiperones, it achieves extreme potencies. Absence of substitution on the benzoyl ring or substituents other than F, especially at the 2 or 3 positions, diminishes activity considerably. Isosteric substitutions, such as thiophene for benzene, are also not useful. The following series illustrates the diminution of activity. It is particularly surprising how much inferior the Cl atom is to F.

$$F-C_6H_4- > C_6H_5- > C_4H_3S- > CH_3-C_6H_4- > Cl-C_6H_4- > CH_3O-C_6H_4-$$

Attempts to modify the keto function, even by isosteric S for O replacements that would not alter the distance between the aromatic ring and the N atom, were futile, leading to drastic cuts in potency. The only exception is methine, -CH-, when it is carrying a *p*-F-phenyl moiety. Even though such a substituent does diminish potency, it results in neuroleptic compounds with considerably increased duration of action. With such a 4,4′-*bis*-fluoro-substituted system the decreased activity is still adequate to take advantage of the new pharmacokinetics obtained. These drugs, then, represent the diphenylbutylamines, three of which are listed in Table 12-13 (Nos. 7, 8, and 9). They may be viewed as a type of subgroup of the butyrophenones.

Combining the two *p*-fluorophenyl groups with the N-phenylspirotriazolodecane into the still investigational diphenylbutylamine fluspirilene (No. 7) resulted in a potent, very long-acting antipsychotic injectable whose prolonged effectiveness is due to uniformly slow release from the intramuscular injection site. It is a microcrystalline drug in aqueous suspension. Unlike the fatty acid esters such as fluphenazine and haloperidol decanoate, where drug release is by erratic ester hydrolysis, fluspirilene gives consistent plasma levels for 1 week from a single 2–8 mg dose.

Penfluridol, which is a diphenylbutylamine analog of clofuberol (Table 12-13, No. 8), increases the flexibility of treatment of psychiatric outpatients even further since it is orally effective for at least 7 days following a single 20–160 mg dose. Compliance may be even more improved since injections are avoided.

Pimozide is available as an antipsychotic limited to use as an alternate to haloperidol in the treatment of Tourette's syndrome (TS). It is indicated for patients with severe symptoms who cannot tolerate or do not respond to haloperidol. The adverse reaction profile is similar to that of other neuroleptics; however, serious ventricular arrhythmias can occur; sudden unexpected deaths have been reported with high doses.

TS is a psychotic obsessive–compulsive disorder of lifelong duration. Its salient characteristics are recurring motor or vocal tics. The latter may be sounds such as barking and obscene language in between normal conversation (hence the name "cursing disease"). The vocal tics are repetitive and explosive. They involve facial movements and abnormal head and limb activities. This compulsive behavior is repetitive and ritualistic. Because about 80% of TS patients respond to haloperidol, and sometimes other antidopaminergic drugs, the "dopamine hypothesis" has been invoked here. However, there have been reports where long-term neuroleptic therapy may have actually induced TS in schizophrenic patients.

Even though many drugs are used to treat TS, only three may be clinically useful: haloperidol, penfluridol, and pimozide. Clonidine, which is the imidazole antihypertensive (Chapter 10), may be useful in about 70% of patients.

The purported mechanism of antipsychotic action of the butyrophenones and diphenylbutylamines is believed to be as that of the phenothiazines, namely, at least in part, antagonism at postsynaptic DA receptors. It is not surprising that the spectrum of side effects is also similar, with extrapyramidal dystonic reactions at the top of the list.

An aspect of neuroleptic therapy that has been puzzling is the delay of three or more weeks before subjective improvement is noted, even though it can be shown that the drugs block D_2 receptors at the beginning of therapy. It has been known for some time that schizophrenics have much higher levels of homovanillic acid (HVA), which is the major DA metabolite, than do healthy controls. The fact that HVA concentrations parallel alterations in DA turnover in the brain has also been understood. It was demonstrated that fluphenazine steadily decreased HVA levels after several weeks and can ultimately reach normal values. Moreover, in patients that were evaluated, improvement correlates well with the HVA levels. This delayed effectiveness, then, may indicate that decreased DA turnover is the important factor rather than the immediate DA receptor blockade. It should be realized that in humans antipsychotics may have increased HVA levels in the CSF over the first few weeks, apparently because of DA blockade, which likely increases DA synthesis as a compensatory mechanism in the presynaptic neuron. After that initial 3-week period, however, DA turnover decreases to normal levels while the drug is continuing.

An addition (1994) to the antipsychotic armory is risperidone (Risperdal). Although still a member of the second generation, it may well be a member of its elite because of its alleged ties to both DA, D_2 and serotonin type 2 receptors. Histamine and α receptors may also have connections to this agent. The extent of affinities of the latter three, however, are not known. The 9-hydroxy metabolite of risperidone is equally effective.

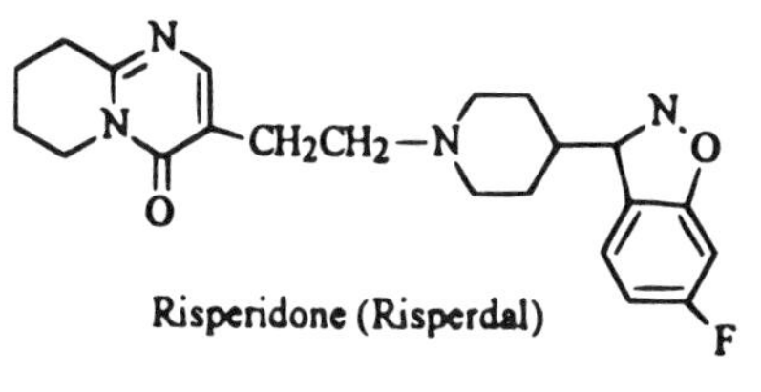

N N O CH_2CH_2–N N O F

Risperidone (Risperdal)

Table 12-14. Psychotropic Dibenzazepines[a]

Drug[a]	X	R'	R''	R	Comments
Amoxapine (Asendin)	O	H	Cl	H	Antidepressant[b]; a dibenzoxazepine
Loxapine (Loxitane)	O	H	Cl	CH_3	Antipsychotic
Clozapine (Clozaril)	NH	Cl	H	CH_3	Neuroleptic; a dibenzodiazepine[d]
Metiapine[c]	S	H	Cl	CH_3	Antipsychotic; a dibenzothiazepine
Fluperlapine[c]	CH_2	H	F	CH_3	Antipsychotic[d]

[a] Trade names in parentheses.
[b] See following discussion.
[c] Not available in the United States.
[d] Agranulocytosis may occur.

Table 12-14 shows several dibenzazepines. Of the two dibenzoxazepines (where $X = O$) loxapine, marketed in 1975, is antipsychotic. The topographical similarity of this tricyclic antipsychotic to the others may be appreciated when viewed as presented in Table 12-15. Its pharmacology, including side effects, mimics those of the phenothiazines as well as

Table 12-15. Topographical Comparison of Major Neuroleptics

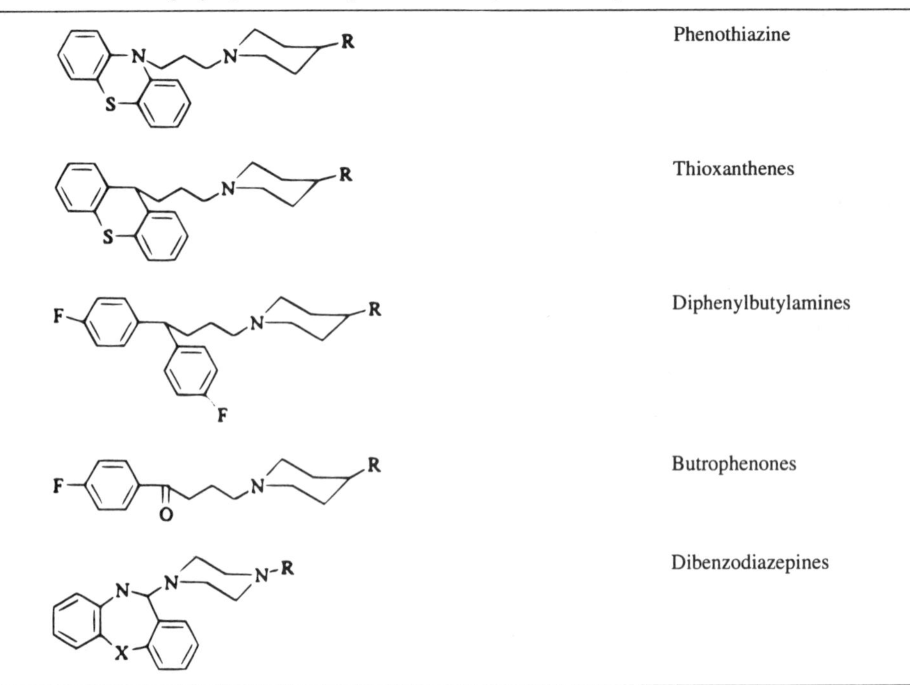

thioxanthenes. Its 8-hydroxy metabolite retains antipsychotic activity until it is also converted by N-demethylation to 8-hydroxyamoxapine, which, as amoxapine itself, possesses antidepressant activity (see later).

The dihydroindolone molindone, though structurally different from the phenothiazine and butyrophenones, shares most of their antipsychotic properties and side effects. Its clinical effects resemble those of the piperazine phenothiazines. The drug's specific advantages, if any, are not readily apparent.

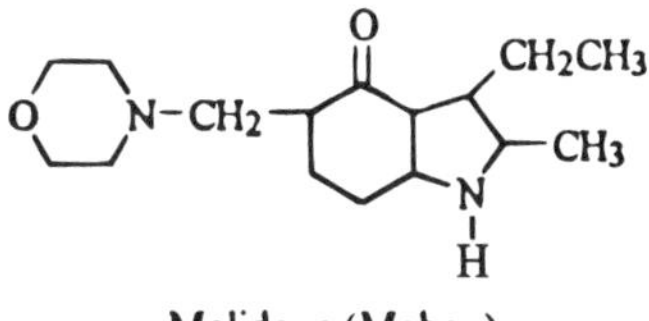

Molidone (Moban)

Tardive dyskinesia, once rarely encountered (e.g., Huntington's disease), has become commonplace since the introduction of antipsychotic drugs. Such drug-induced dyskinesia (rhythmic involuntary movements) can be attributed to supersensitive postsynaptic DA receptors. (Such symptoms, of course, can also result from decreased cholinergic activity.) It is doubtful whether any of the antipsychotic drugs considered are significantly lower (better) in producing these symptoms. Table 12-16 lists many nonneuroleptic effects of the phenothiazines and their analogs, many of them quite undesirable.

The dibenzodiazepine clozapine (Table 12-14, $X = NH$), which is an antischizophrenic, has been approved on a limited basis (1989). Its atypical properties are that it does not bind to DA receptors and, as a result, does not appear to cause extrapyramidal symptoms. It may even reverse tardive dyskinesia. Unfortunately, reports of a high incidence of agranulocytosis severely restricts its use to patients who are refractory to standard antipsychotic medications. Fluperlapine (Table 12-14) is currently being investigated as an alternative

Table 12-16. Non-Neuroleptic Effects of Phenothiazines

A. Effects of potential clinical utility
Anticonvulsant
Antiemetic
Antihistamine
Hypotension
Hypothermia
Local anesthesia
Potentiation of analgetics
Potentiation of general anesthetics
Sedation
B. Effects that are adverse reactions
Adrenolytic
Agranulocytosis (rarely)
Anticholinergic (usually weak)
Cholestatic jaundice (rarely)
Extrapyramidal effects
Tardive dyskinesia
Weight gain

antipsychotic to clozapine. It appears to share clozapine's nonclassic profile. It is hoped that the profile will not include blood dyscrasias.

It is obvious when considering the pre-1950s era that tremendous progress has been made in the treatment of psychotic diseases. It is equally apparent that the problem is nowhere near solved. Every theory of the causes of these mental diseases is flawed. Our understanding of the mechanisms by which even the effective drugs function is incomplete at best.

At the clinical level various obstacles to improved therapy also exist. One in particular is the widespread myth (or misconception) that the antipsychotic drugs are interchangeable. Since patient responses to different antipsychotics can actually differ substantially, the belief of some clinicians that they do not must invariably result in suboptimal treatment for many patients. In fact, it has been shown that the overlap area of responding patients to CPZ, fluphenazine, and trifluoperazine is very small, while perphenazine shows little or no overlapping response at all. In one comparison it was found that 89% of patients improved with thioridazine while only 59% did so when treated with thiothixene. Thiothixene also produced a higher rate of worsening effects (15% vs. 4%).

The antipsychotic agents represent a useful demonstration of the nonequivalence of potency and efficacy. As was stated previously, the antipsychotic drugs affect both DA and, to different degrees, other neurotransmitters (e.g., NE, ACh). It is assumed that lower-potency neuroleptics are those with lower D_2 receptor affinity. It is interesting that they possess a relatively higher affinity for cholinergic receptors, which they also effectively block. The situation is dramatically opposite with high-potency antipsychotics. Rather than view this as an undesirable side effect, some clinicians see it as a benefit since a "balanced" decrease of both central postsynaptic DA and central ACh activities means that these drugs have a "built-in" mechanism against the development of Parkinsonism, hence, greater efficacy. Examples of such low-potency antipsychotics with high anticholinergic activity are piperidine phenothiazines, meso- and thioridazine, and also probably CPZ. High-potency compounds such as the piperazine trifluoperazine and fluphenazine, and particularly haloperidol, have little ability to block ACh, thus producing a higher incidence of dystonic problems. This frequently requires coadministration of anticholinergics such as benztropine, procyclidine, and other anti-Parkinson drugs (Chapter 8). The drugs must be carefully titrated to avoid excessive anticholinergic symptoms. There exists a paradoxical situation when very large doses of high-potency antipsychotics are given. Since the D_2 receptors are likely to be fully occupied (blocked), "excess" molecules will begin to occupy cholinergic receptors even though the drugs may have a low affinity for them. Thus no additional anticholinergic may be needed. It should be remembered that Parkinsonism is only one of the extrapyramidal side effects that neuroleptic drugs can induce. The others, such as acute dystonias, akathisias, and tardive dyskinesia, are not usually ameliorated by the anti-Parkinson drugs. In fact, akathisias, which is a condition of constantly being in motion, is dose related and can be minimized by lowering the doses or switching to a lower-potency agent.

12.9. Antidepressants

Depression, or melancholy, as it was once called, is a sense of lowering mood. It has become a mental health problem of major proportions in modern society. It has been estimated that one-fifth of our population is likely to experience at least one major depres-

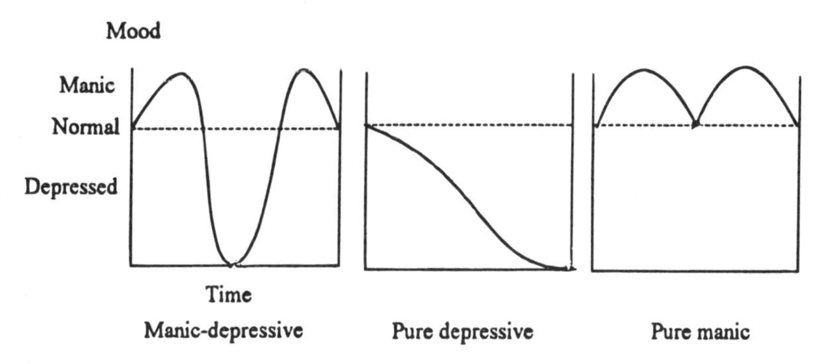

Figure 12-23. Types of manic depressions.

sive episode in his or her lifetime. For many there will be recurrent episodes. *Organic depression* is identifiable by pathological sources, such as degenerative brain processes following trauma, tumor development, or infections. In the majority of cases we are dealing with *endogenous* depression, which is usually characterized by phases of depression alternating with periods of apparent normalcy and manic manifestations, hence a manic–depressive psychosis (Fig. 12-23). In classical endogenous depression, patients show consistent sadness, feelings of hopelessness, helplessness, guilt, sleep disturbances, loss of appetite and weight, decreased libido, and personality changes. All these symptoms are amenable to drug treatment. Feelings of self-pity, anger, resentment, and unhappiness frequently are not.

Before the advent of effective psychochemotherapy in the 1950s, "treatment" consisted of stimulants such as caffeine and later amphetamines to ameliorate the depressive phases, and barbiturates, chloral hydrate, and bromides to allay agitation, anxiety, and insomnia. At best, such attempts at therapy may have offered transient relief to some patients. Suffering usually decreased little. Stimulant drugs more than likely increased agitation. Electroconvulsive therapy (ECT), which was in use by 1940, was the first modality that actually reversed symptoms of depression. Similar short-lived convulsion could also be chemically induced—pentelenetetrazole and insulin. Convulsive therapy, a totally empirical concept, frequently gave favorable results and has had a resurgence of use (now with coadministration of neuromuscular blocking agents to minimize injury during the convulsions). ECT is still an empirical approach in refractory cases. Depression as well as schizophrenia are still dealt with on the basis of the biogenic amine hypothesis. Thus endogenous depression can, at least in part, be explained in terms of low levels of such as amines and possibly other modulators of transmission at central adrenergic receptors in the brain. The reason for lowered concentrations of such amines may be their increased rate of enzymatic destruction and/or their inability to reach these receptors because of excessively efficient neuronal uptake and vesicular storage. Drugs to affect both of these processes exist.

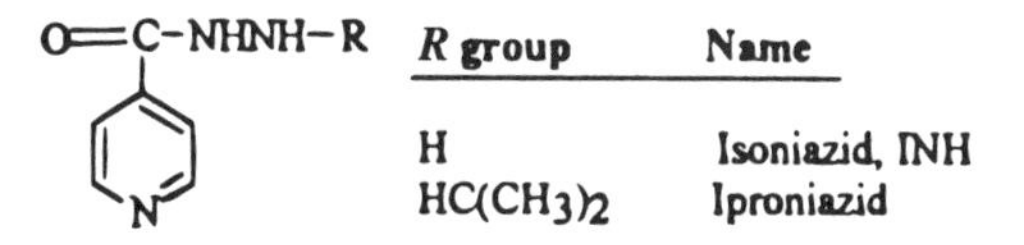

R group	Name
H	Isoniazid, INH
$HC(CH_3)_2$	Iproniazid

Isonicotinic acid hydrazide (isoniazid) and its N-isopropyl analog, iproniazid, were used as antitubercular drugs in the early 1950s. Independent reports showed that iproniazid (but not isoniazid) exhibited CNS stimulant properties in TB patients and had the ability to inhibit the enzyme MAO. Reports showed that the expected depressant effect of reserpine in animals could be prevented (reversed?) by pretreatment with iproniazid and resulted in CNS stimulation. In fact, NE's and 5-HT's neuronal depletion produced by reserpine was actually blocked by this pretreatment. This work, then, led to the proposal that the clinical antidepressant efficacy of iproniazid (and related MAO inhibitors synthesized in the interim) resulted from the increased levels of NE and 5-HT, which were allowed to rise because of MAO inhibition. Many other hydrazine and hydrazide variants were soon introduced into clinical use.

The basic concept regarding MAO inhibitors is straightforward. These compounds prevent MAO-catalyzed deamination of biogenic amines (5-HT, NE, DA) following their reuptake into the nerve terminal from synaptic cleft. As a result, higher concentrations of the neurotransmitters will be stored in the vesicles and become available for release from the presynaptic terminals on demand. Of the two MAO isozymes, type A deaminates NE and 5-HT preferentially; MAO-B seems to exhibit a preference for phenethylamine. Many of the previously marketed MAO inhibitors have been withdrawn because serious hepatotoxicities were demonstrated. Only three have survived. The hydrazine phenelzine, the hydrazide isocarboxazide, and the amphetaminelike structured tranylcypromine (Table 12-17). The problem with these drugs is certainly not lack of effectiveness, but

Table 12-17. The First-Generation Antidepressants[a]

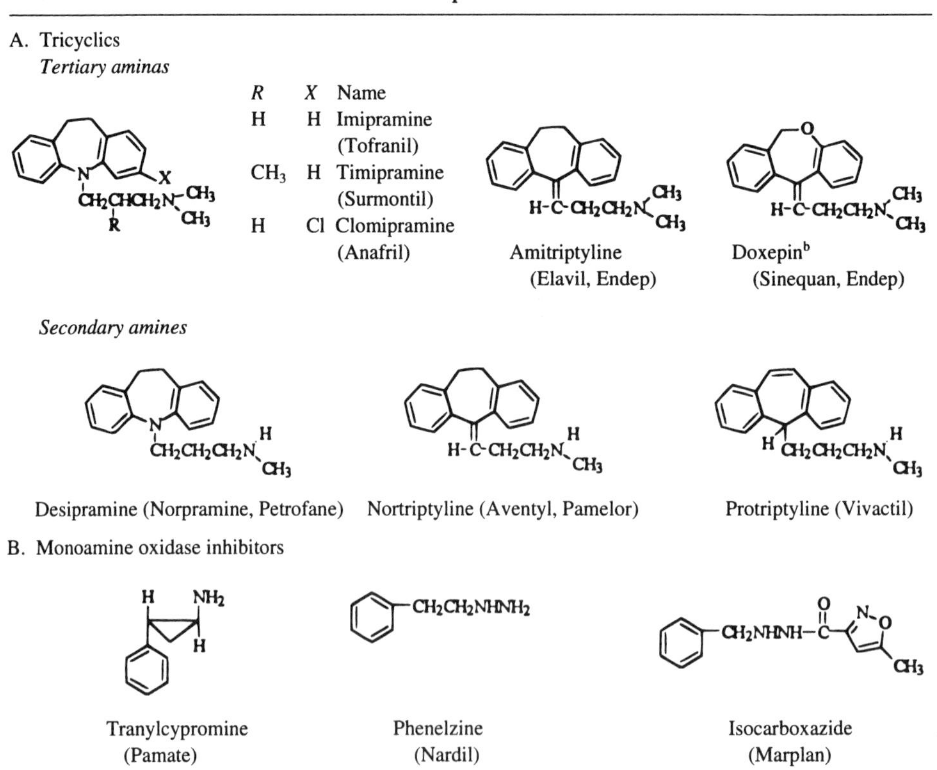

[a] Trade name in parentheses.

lack of selectivity. MAO is a ubiquitous enzyme, not limited to central neuronal tissue. In addition, enzyme inhibition is irreversible, and it leads to very long-term residual effects. This combination of characteristics presents a potentially dangerous situation in terms of serious drug interactions as well as food interactions.

Drug-induced inhibition of microsomal and extrahepatic enzyme systems, such as MAO, will be predictably responsible for clinically significant interactions. Because of decreased metabolic activity, more unmetabolized drug will be in circulation, resulting in prolonged and sometimes greatly increased pharmacological activity.

One of the most significant effects of MAO inhibition is the depression of the metabolism of sympathetic amines. Subsequent release of the resulting increased neuronal stores of these agents may cause an exaggerated adrenergic response such as sudden and extreme blood pressure elevations, which have caused fatalities. The term *hypertensive crisis* has been applied to such a response.

The pressor effect of tyramine presents a uniquely interesting situation. Tyramine is an indirectly acting sympathomimetic amine capable of producing elevated blood pressure in humans. Many foods, including various yellow cheeses (particularly sharp cheddar), beef and chicken liver, pickled herring, and Chianti wines, contain this substance in appreciable concentrations. It is produced in the foods by bacteria elaborating decarboxylase enzymes, which produce tyramine by utilizing tyrosine and phenylalanine as substrates. The intestinal wall ordinarily contains sufficient MAO to inactivate the tyramine in these foods when ingested. However, patients on MAO inhibitors have severely depressed MAO concentrations, permitting the tyramine to be absorbed intact. The adverse reactions encountered vary from skin eruptions and headaches to increased blood pressure and even intracranial hemorrhage.

Interactions of MAOIs with other sympathomimetic drugs are well known. The increased neuronal stores of NE represent a "booby-trap" hazard that can easily be triggered by indirectly acting adrenergic agents such as the amphetamines found in many antiobesity preparations. Drugs such as ephedrine, phenylpropanolamine, and phenylephrine are potentially more dangerous because of their ready availability to the public in over-the-counter medications such as "cold medicines" and nose drops.

The tricyclic antidepressants (TCAs) (see later) represent another group of drugs whose interactions with MAOIs have been widely reported. Clinical significance is usually beyond doubt and well documented. Body temperatures of as high as 107°F have been reported, as have fatalities. The severe reactions are frequently atropinelike in that they produce convulsions, tremors, and delirium. Even though some clinicians think that combinations of MAOIs and tricyclic antidepressants can be used effectively, it is generally recommended that a 2-week interval be allowed after cessation of one type of agent before beginning therapy with the other.

Most antihistamines possess anticholinergic properties that by themselves are of minor intensity but may increase considerably if given together with MAOIs. Interactions of MAOIs with antihypertensives (e.g., reserpine-type drugs) and oral antidiabetic drugs, as well as insulin and L-dopa, have all been encountered. It can be seen why MAO inhibitors are not viewed today as initial therapy except possibly in atypical depression. Certain panic and phobic reactions respond well. An important indication would be in case of therapeutic failures with the tricyclic or other heterocyclic antidepressants.

The TCAs arrived on the scene in the same period as the MAOIs. Imipramine (Table 12-17), which is a dibenzazepine, was synthesized as a logical isosteric extension of the phenothiazines. The antidepressant properties were subsequently discovered and the drug was clinically introduced in 1957. Furthermore, isosterically replacing the N with an sp^3

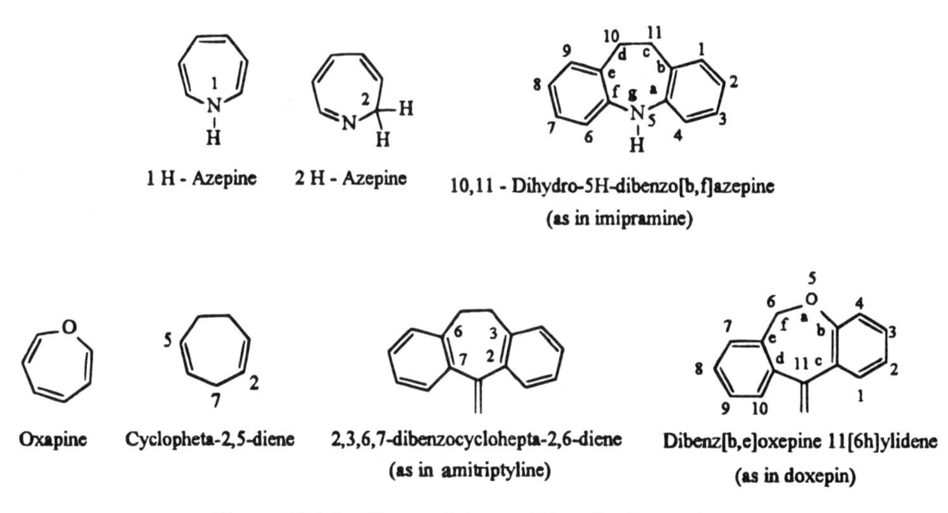

Figure 12-24. Nomenclatures of tricyclic ring systems.

carbon led to amitriptyline. Finally, the isosteric replacement of a carbon with an oxygen led to the introduction of doxepin (1969) as the third logical modification. (The *cis* isomer is the more active one.) Figure 12-24 compares the ring systems and nomenclature of these tricyclic systems. Figure 12-25 outlines the synthesis of this drug.

It will be noted (Table 12-17) that these four first-generation TCAs are tertiary amines. Their clinical activity was initially described as neuroleptic (mood elevating). They also exhibited a degree of anxiolytic effects and reduced agitation in animals and humans. The introduction of clomipramine (2-chloroimipramine) (1990) offered the psychiatrist the first drug effectively to treat obsessive–compulsive disorders of the type serious enough to interfere with social or occupational functions.

When the N-demethylated metabolites (i.e., secondary amines, Table 12-17) were also found to have clinically valuable antidepressant activity, they were in turn also marketed (desipramine, nortriptyline, and protriptyline). However, it was soon realized that the antidepressant effects of these drugs were accompanied by stimulation rather than anxiolysis.

Depression initially was presumed to be the result of a deficiency of NE and/or 5-HT in the CNS, and the increased levels of the neurotransmitters were shown to result from inhibition of their reuptake in the central neurons by the drugs being considered.

With the recognition of two biochemical types of depression each having different amine and metabolism products relating to two sets of symptoms (Table 12-18), came the

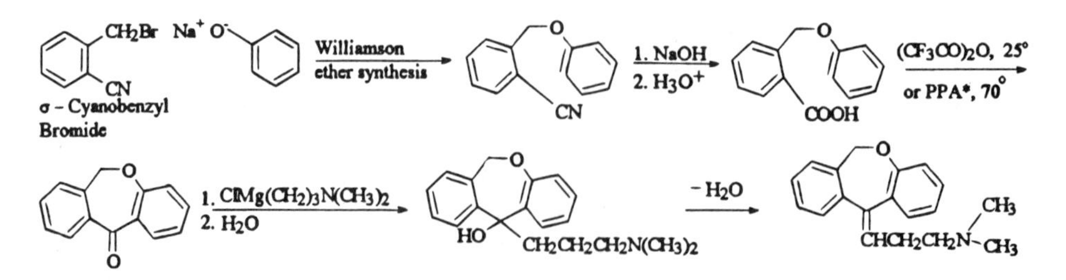

Figure 12-25. Synthesis of Doxepin. *Polyphosphoric acid.

Table 12-18. Relationships of Biogenic Amines to Neuroleptic Activity[a]

Drug	Type of amine	Amine uptake inhibition 5-HT	NE	DA	Biochemical abnormality
Amitriptyline	tertiary	++++	+	0	Low serotonin levels in CNS,
Doxepin	tertiary	++	+	0	CSF levels of 5HIAA[b]
Imipramine	tertiary	+++	++	0	
Desipramine	secondary	0	++++	0	Decreased levels of NE;
Maprotiline	secondary	0	+++	0	decreased urine levels
Nortriptyline	secondary	+	++	0	of MHPG[c]
Protriptyline	secondary	?	+++	0	
D-Amphetamine	—	0	+++	++++	

[a] Compiled from numerous sources.
[b] 5-Hydroxy indole acetic acid.
[c] Methoxyhydroxyphenylglycol.

recognition that each set responded differently to the drugs in use. In essence, the highly anxious, sleep-disturbed patients whose central serotonin levels were low responded best to treatment with tertiary amine drugs. Those patients whose levels of NE and (urinary) levels of MHPG were low responded better to those drugs with a secondary amine side chain. The two subtypes were at times referred to as the "norepinephrine depression" and the "serotonin depression."

Thus there appear to be significant mechanistic differences: the hydrazines inhibit MAO, the tertiary amine tricyclics seem to inhibit the serotonin amine pump, whereas the secondary amine ones seem better in switching off the NE reuptake mechanism. Table 12-18, however, shows that there is some overlap. Nevertheless, there is a thread of commonality—the net increase of amine neurotransmitters in the synaptic area—yet all these drugs require several weeks of treatment before objective results are noted. The biogenic amine hypothesis does not satisfactorily explain this. It is even more difficult to explain the antidepressant action of some of the second-generation drugs, such as mianserin (Fig. 12-26), that seem to have no significant effect on amine reuptake mechanism of either

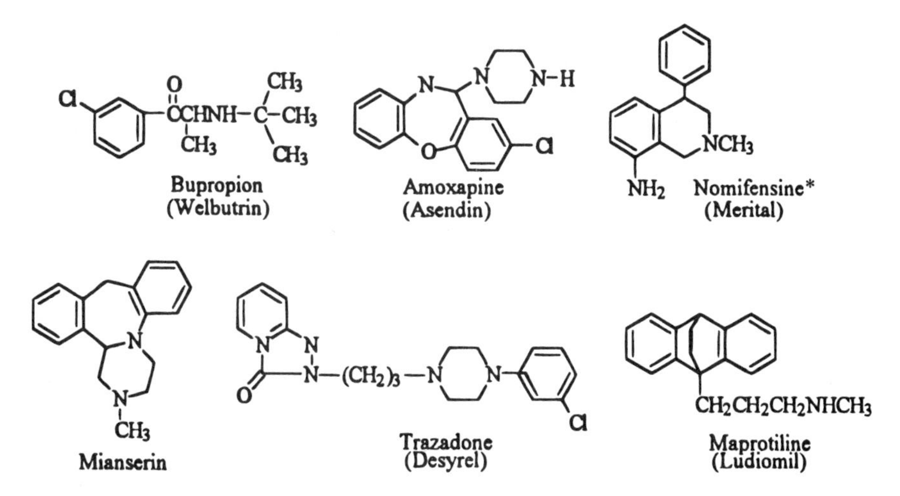

Figure 12-26. Second-generation antidepressants. Trade names in parentheses. *Withdrawn in 1987.

monoamine, while cocaine and amphetamine—which do—are not particularly useful in treating depression.

It may be reasoned that if the acute effect of TCAs and MAO inhibitors is to raise the amine levels and yet no symptom relief is discernible for the first 2 weeks of treatment that the amine levels per se are not really significant; *or* that these compounds are achieving their results by a different technique—possibly a chronic effect, not necessarily involving only adrenergic neurons.

Effects on other receptors are known (e.g., antimuscarinic and antihistaminic), but these appear to elicit mostly side effects. The hypothesis that arises—and may replace the biogenic amine concept partially—is that it may be the chronic exposure of central β_2-adrenoreceptors to these elevated biogenic amines (NE, 5-HT, and maybe DA also) that is the basis of the action. The most likely effect now seems to be decreased sensitivity of these receptors. By measuring *c*-AMP levels, studies using long-term drug administration indicate that central β-receptors become less responsive postsynaptically and also that central 5-HT and α_1 receptors may become enhanced.

With its 2-carbon bridge across the middle, maprotiline (Fig. 12-26) is technically a tetracyclic compound, yet it behaves as a secondary amine TCA (i.e., a relatively selective NE reuptake). The drug's second-generation standing is purely chronological. It has been in the U.S. market since 1981. Amoxapine, which was marketed the same year, is, of course, the N^4-demethylated antipsychotic loxapine (Fig. 12-26). It has been suggested the cumulation of the 8-OH metabolite may be responsible for the inhibition of NE uptake and account for the antidepressant effect. Nomifensine, which is a tetrahydroisoquinoline with potential as an inhibitor of NE and DA, but not 5-HT, was withdrawn in 1987 for toxicity reasons.

Trazodone may be viewed as a triazolopyridine (left portion of the molecule) or as a phenylpiperazine. It is not to be easily categorized. It is inactive in most of the usual animal screens used. Its pharmacology is complex and not well understood. It shows weak serotonin uptake inhibiting ability and may also increase release of NE by blocking α_2-adrenoceptors. The major metabolite is *m*-chlorophenylpiperazine. Toxicities are low and some patients respond well; others obtain little benefit.

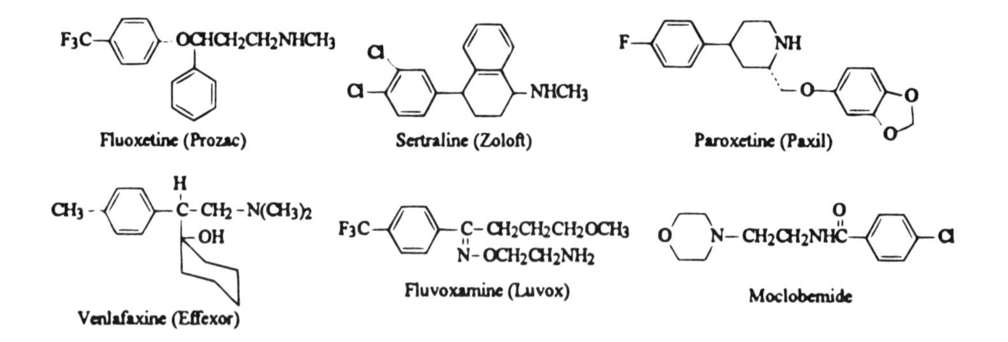

Six other atypical antidepressant drugs must be considered. Fluoxetine, which is the first of the selective serotonin reuptake inhibitors (SSRIs) was introduced (1986) as an antidepressant following extensive neurochemical research on the role of 5-HT in affective disorders as well as its relationship to NE and DA. The efficacy of this new addition to psychiatry is similar to the tricyclics; however, patient tolerance appears to be better due to a different set of side effects. Thus, the TCAs, which typically produce sedation, considerable anticholinergic activity, and certain cardiovascular problems, can now be "traded" for occasional nausea, dizziness, and nervousness, a much more patient-acceptable profile. Both enatiomers are active.

Sertraline, which was introduced in 1992, has a similar mechanism, yet it may offer some advantage over fluoxetine by exhibiting little CNS action (i.e., it has less sedation and anxiety and is shorter acting). Both drugs show promise in the treatment of obesity and obsessive–compulsive behavior.

Three additional agents in this group were introduced in 1993–1994; paroxetine, venlafloxine, and fluvoxamin. They differ from the earlier agents in duration of action, dosages, and metabolism rather than in mechanism or clinical indications.

The modest improvements achieved in selectivity with respect to serotonin reuptake inhibition may also have been achieved with an isoenzyme system. Moclobemide, which was introduced in Sweden, reversibly inhibits monoamine oxidase A (RIMA). It is likely that this eliminates the severe hypertensive drug and food interactions that so severely limit the usefulness of the very effective earlier MAO inhibitors, since tyramine is now metabolized. An additional benefit of such agents may be a lack of cholinergic and cardiovascular effects.

Bupropion (Fig. 12-26) is a phenylethylamine resembling the anorexiant diethylpropion (Chapter 9). Since the *tert*-butyl group is not easily removed metabolically, the compound shows little or no pressor activity. That group and the keto function afford the compound lipid solubility, as does the *meta*-Cl atom. It also probably protects it from facile oxidation of the phenyl ring. The compound is an atypical antidepressant, exhibiting some selectivity in DA reuptake (in rats). Some direct dopaminergic action on DA receptors seems also to exist. It shows no MAO inhibition. Dose-related CNS stimulation (in animals) has been reported. Because seizures have occurred at 450 mg daily doses, the drug is recommended to treat depression that is not responsive to other traditional antidepressants. It is not a first-choice drug.

Lithium. The discovery that Li^+ can control manic behavior in manic–depressive patients was reported before the appearance of CPZ. It is now widely used as Li_2CO_3. No totally acceptable mechanism for its action exists. Postulations involve actions that would likely adjust overactive catecholaminergic activity, which is the accepted occurrence in mania.

Several studies showing that Li^+ has some antidepressant effects are known. Some weak biphasic alterations of NE and 5-HT turnover in the brain were established. Figure 12-27 is a simplified schematic model of a central dopaminergic neuron showing possible sites of drug action as discussed in this chapter and in Chapter 11. Similarly, Figure 12-28 depicts a noradrenergic neuron and its potential sites of drug action.

12.10. Stereochemical Aspects of Psychotropic Drugs

Stereochemical and other geometric factors are significant in determining the type of psychotropic activity obtained. It appears that planar ring systems yield optimal neuroleptic activity. The tricyclic 6-6-6 system (i.e., all three rings contain six atoms) of the phenothiazines seems to prove this. However, compounds with total rigid planarity, as in a 6-5-6 system exemplified by carbazole and fluorene derivatives, are invariably inactive. It might be postulated that neuroleptic receptors are planar but have rather strict size requirements for binding.

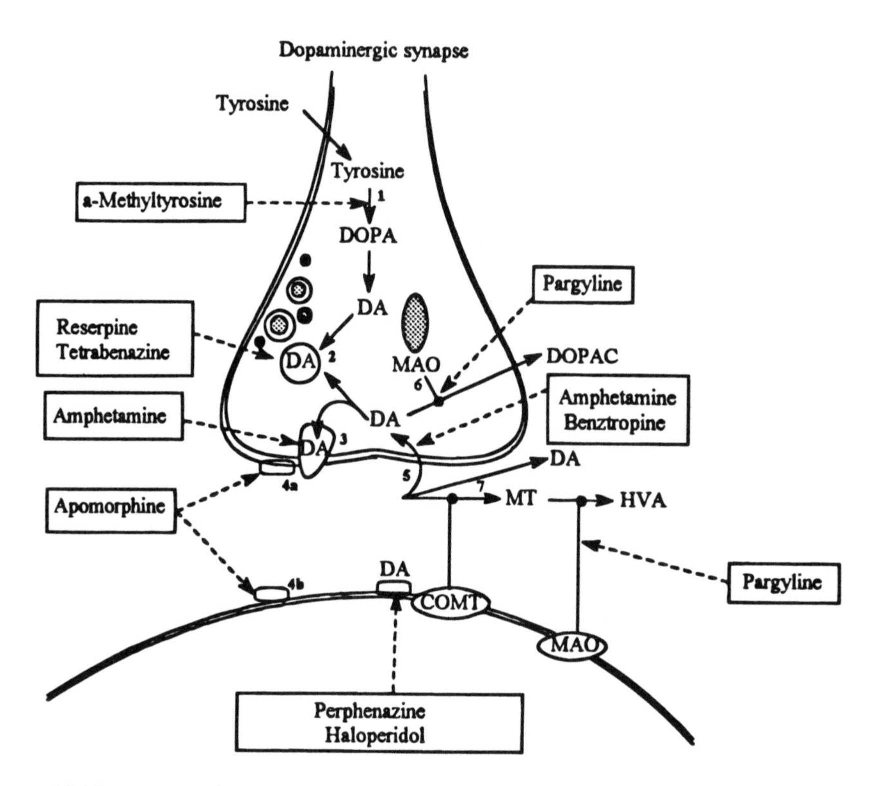

Figure 12-27. Model of central dopaminergic neuron (adapted from Cooper et al., 1991, with permission).

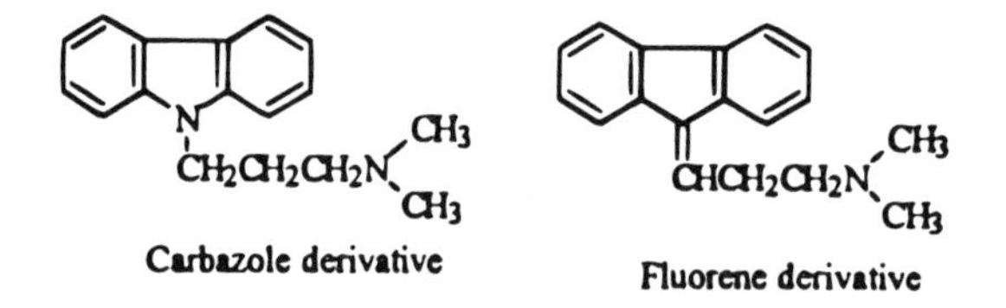

Ring systems with relatively few deviations from planarity, by angling of the atoms in the middle ring, begin to show some thymoleptic effects while still maintaining neuroleptic activity. Amitriptyline and doxepin are likely examples. More pronounced twisting of the ring system usually results in exclusive antidepressant action, such as with imipramine. A nonplanar angled thymoleptic receptor concept might be involved. Another supporting example may be the two acridine derivatives as shown here:

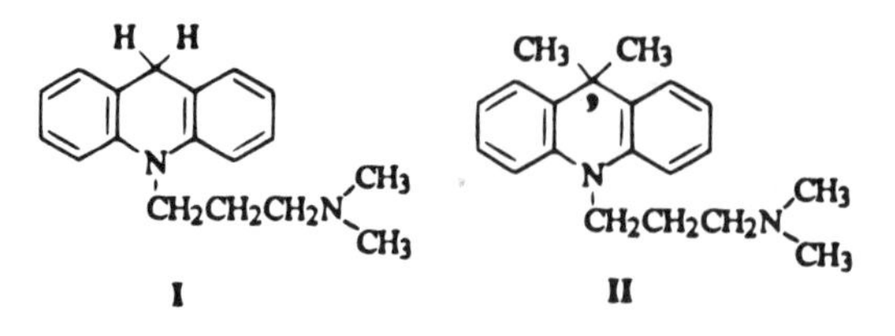

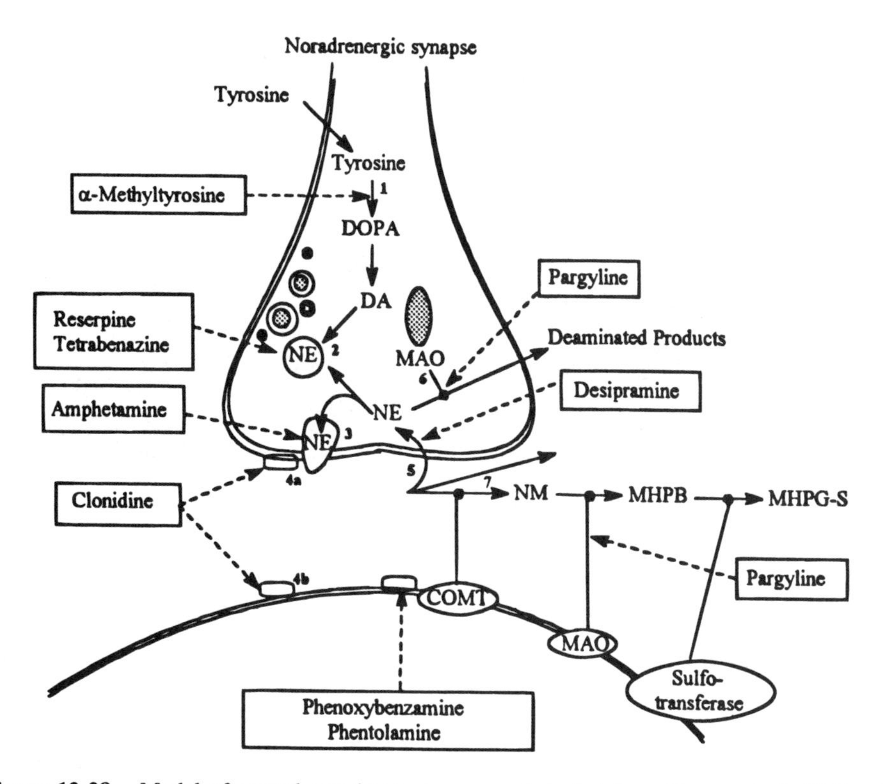

Figure 12-28. Model of central noradrenergic neuron (adapted from Cooper et al., 1991, with permission).

Compound I displays good antipsychotic activity. Compound II, however, shows antidepressant activity analogous to that of imipramine. The two additional methyl groups at position 9 result in a twisting of the three-ring system to a sufficient degree that the requirements of the postulated nonplanar angled antidepressant receptor site are accommodated. Two situations arise in compound II that may explain the clinical observations. First, it can be assumed that the planar acridine ring simply cannot accommodate two bulky methyl groups at the 9 position and maintain flatness. Thus a twisting must occur that results in loss of coplanarity of the three rings. Second, it can be argued that the bulk of the methyl groups prevents a good fit to the planar neuroleptic receptors.

If we look at the tricyclic system at a fundamental level, its framework can be seen as angled or flexed, to varying degrees. *Flexure* represents the degree of nonplanarity between the A_1 and A_2 rings (Fig. 12-29). *Annulation* is the angle at which these two rings are attached to the middle ring. *Torsion* is the degree to which the molecule is twisted out of symmetry. When the angle of flexure is small (approximately 30–35 degrees), the system is relatively flat and clinically neuroleptic. Further twisting out of plane, to a flexure angle of 50–60 degrees, results in thymoleptic properties. Angles of annulation and torsion do not appear to have that much bearing on therapeutic effects.

A third generation of antidepressants will be needed to advance into an era of higher selectivity to yield even safer and especially more effective compounds. The roadblock may well be still insufficient neurochemical knowledge. Even though the second-generation drugs discussed are safer than were the early compounds, their impact therapeutically

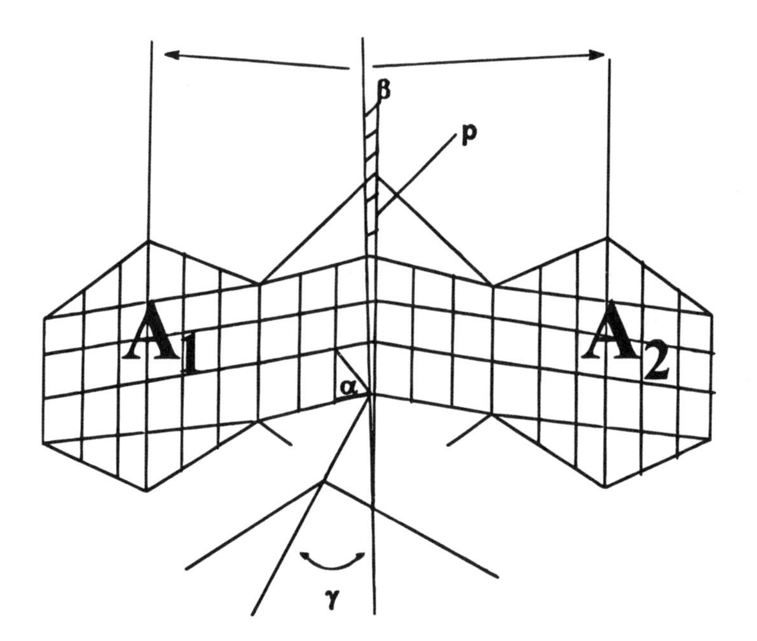

Figure 12-29. Steric parameters of tricyclic psychotropic drugs. (Modified from Wilhelm and Kuhn, 1970.)

has been disappointing. The reason is that their effectiveness, in most cases, is not superior to the tricyclics that are still the mainstay in spite of slow onset of action, side effects, and a significant fraction of refractory patients. Research is active in areas such as α-2 adrenoceptors, several HT receptors, GABA receptors, lithium mimetics, certain anticonvulsant calcium antagonists, neuropeptides (e.g., endorphins, GH), and even corticosteroids.

References

Commission on Classification and Terminology of the International League Against Epilepsy, *Epilepsy* **22**:489, 1981.

Cooper, S. J., Bloom, F. E., Roth, R. H., *The Biochemical Basis of Neuropharmacology*, 6th ed. Oxford University Press, New York, 1991, p. 367.

Ferguson, J., *Proc. Royal Soc. B* **127**:387, 1939.

Gilman, N. W., Sternbach, L. H., *J. Heterocycl. Chem.* **8**:297, 1971.

Gordon, M., *Psychopharmacological Agents*, Volume II of Vol. 4. Academic Press, New York, 1967.

Katerinopoulos, H. E., Schuster, D. I., *Drugs Future* **12**:223, 1987.

Meyer, H., *Arch. Exper. Path. Pharmakol.* **42**:109, 1901.

Merritt, H., Putman, T., *Science* **85**:525, 1937; *Arch. Neurol. Psychiatry* **39**:1003, 1938.

Moody, E. J., Harris, B. D., Skolnick, P., *Trends Pharmacol. Sci.*, **15**:387, 1994.

Pavlov, J. B., *Conditioned Reflexes*, Oxford University Press, New York, 1927.

Rando, R. R., *Biochemistry* **16**:4604, 1977.

Schildkraut, J. J., *Amer. J. Psychiatry* **122**:509, 1965.

Schmiedeberg, K., *Arch. Exp. Path. Pharm.* **20**:206, 1886.

Skinner, B. F., *Trans. NY Acad. Sci.* **17**:547, 1955.

Skolnick, P., Paul, S. M., *Ann. Rev. Med. Chem.* **16**:21, 1981.

Sternbach, L. H., *J. Med. Chem.* **22**:1, 1979.

Valzelli, L., Grattini, S., *J. Neurochem.* **15**:259, 1968.

Wilhelm, M., Kihn, R., *Pharmacosych.-Neuropsychopharmacol.* **3**:317, 1970.

Wong, D. E., Wagner, H. N., Tune, L. E., et al. (a total of 17 authors) *Science* **234**:1558, 1987.

Suggested Readings

Cooper, S. J., Bloom, F. E., Roth, R. H., *The Biochemical Basis of Neuropharmacology*, 6th ed., Oxford University Press, New York, 1991.

Davis, J. B., *The Myth of Drug Addiction*, Harwood Academic Press, Philadelphia, 1992.

Goodwin, K. G., *Manic Depressive Illness*. Oxford University Press, New York, 1990.

Kent, S., Bluthe, R., Kelley, K. W., Dantzer, R., Sickness behavior as a new target for drug design, *Topics in Pharmacol. Sci.*, **13**:24–28, 1992.

CHAPTER

13

Histamine Antagonists and Local Anesthetics

13.1. Histamine Antagonists

By the beginning of the twentieth century the imidazole ring was known to occur in the amino acid histidine, the alkaloid pilocarpine (Chapter 8), and histamine, all of interest to biologically "oriented" chemists at the time. Its synthesis was accomplished in 1907. The vascular effects of this biogenic amine were not fully recognized for another decade; neither was its wide occurrence and distribution in animal tissues. In 1920 it was demonstrated that histamine stimulated secretion of gastric acid. However, it was Best et al. (1927) who identified histamine in high concentrations in the lungs. They observed that it causes vasoconstriction, respiratory distress, and shocklike syndrome following IV administration in animals. Later, histamine's role in the pathogenesis of the anaphylactic reaction was established. An association of histamine with asthma—IV histamine in asthmatic patients decreased vital capacity and caused a parallel release of histamine with bronchoconstriction—briefly led to the false hope that antihistamines might be useful in treating chronic asthma. There is the possibility, of course, that the limitations of dosage imposed by the adverse effects of the earlier ("classic") antihistamines simply did not supply enough drug effectively to antagonize the copious amounts of histamine released during asthma attacks. Now, with newer more potent and less toxic compounds on the horizon, the hypothesis may again be open to verification.

A combination of circumstances came together in the 1930s. The various deleterious effects of excess histamine were being established, and in vitro and in vivo methods were being developed to test chemical effects on contraction of gastrointestinal, bronchial, and other smooth muscle tissues. The stage was thus set for screening compounds for both antihistamine activity and for anticholinergic and antiserotoninergic properties.

$$\text{L-Histidine } (\text{imidazole}-CH_2CH(NH_2)COOH) \xrightarrow[\text{Decarboxylase}]{\text{Histidine}} \text{Histamine } (\text{imidazole}-CH_2CH_2NH_2) + CO_2 \qquad (13.1)$$

Another incentive for the search for antihistamines was the belief at that time that histamine was the mediator of inflammation and shock. Thus a search for antihistamines, it was reasoned, would afford antiinflammatory drugs. It will be recalled that later when serotonin was believed to have been the main mediator of inflammation, indole derivatives were synthesized as potential antagonists for analogous reasons, yet the indoleacetic acid indomethacin, although a potent antiinflammatory, had a very different mechanism of action (Chapter 5).

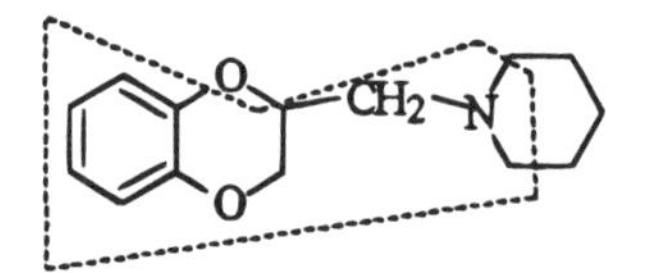

929F: 2-(N-piperidinomethyl) 1,4-benzodioxane

The basic piperidinomethylbenzodioxide, 929F, may be considered as the first compound shown to have antihistaminic properties in vivo, since it protected animals from the bronchial spasms induced by exposing them to histamine vapors (Fourneau and Bovet, 1933). The anticholinergic effects of this compound can be rationalized by viewing it as a basic choline ether (outlined). This property, by the way, became an annoying side effect of almost all the antihistamines introduced over the next several decades.

Drugs developed during the 1940s (e.g., pyrilamine and diphenhydramine) were marketed for the treatment of various allergic conditions such as urticaria and hay fever, for which they are still in use. Many additional drugs joined these two pioneering agents. Ultimately, five chemical classes of antihistamines evolved: ethanolamine derivatives, ethylenediamines, alkylamines, piperazines, and phenothiazine and other polycyclics. Representative examples of each class are illustrated in Table 13-1.

$$(AR)_2X\text{-}C_{2\text{-}3}N(R)_2$$

Optimal antihistamine structure
***AR*=heterocycle or benzyl, *X*=N, CH, C-O, *R*=CH_3 or small ring**

Almost all the "classic" antihistamines can be summarized into a general structure of a diaryl system with a 2- (or 3-) carbon aminoalkyl side chain.

The concept of isosteric replacement within organic compounds to improve or otherwise modify their biologic effects that were introduced in the early 1930s (see Chapter 1) was first successfully applied to antihistamines and to the antibacterial sulfonamides (Chapter 2). It proved to be an extremely useful early tool in what may retrospectively be called rudimentary drug design. For example, by exchanging the oxygen atom in the ethanolamine aryls, ether type of antihistamines (e.g., diphenhydramine; Table 13-1, No. 1) with one of nitrogen, one obtains an equivalent, or possibly superior, histamine antagonist of the ethylene-diamine type, as well as achieves a degree of differentiation of antihistaminic activity from the considerable anticholinergic effects of the former agent.

Table 13-1. Representative Antihistamines—H_1 Antagonists

$$\begin{matrix} AR_1 \\ AR_2 \end{matrix} > XCH_2CH_2N < \begin{matrix} R_1 \\ R_2 \end{matrix}$$

Name[a]	AR_1	AR_2	X	R_1	R_2
A. Ethanolamines					
1. Diphenhydramine (Benadryl) Dimenhydrinate[b] (Dramamine)	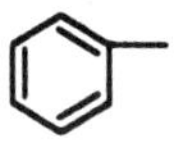	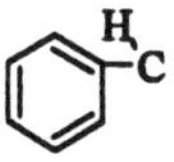 H C	O	CH_3	CH_3
2. Doxylamine (Decapryn, Unisom)	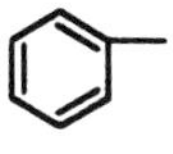	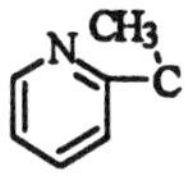 N CH_3 C	O	CH_3	CH_3
3. Carbinoxamine (Clistin)	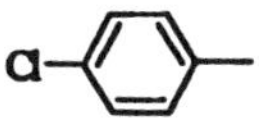 Cl–	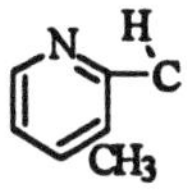 N H C CH_3	O	CH_3	CH_3
4. Clemastine (Tavist)	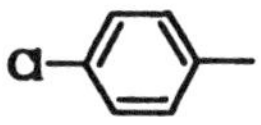 Cl–	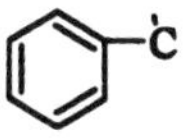 C	O	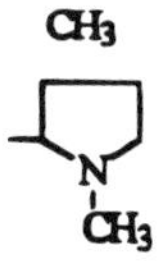 CH_3 N CH_3	
B. Ethylenediamines					
5. Pyrilamine[c]	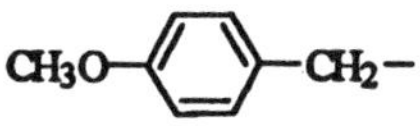 CH_3O– –CH_2–	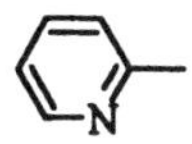 N	N	CH_3	CH_3
6. Tripelennamine (Pyribenzamine)	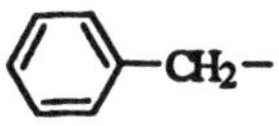 –CH_2–	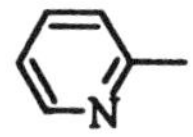 N	N	CH_3	CH_3
7. Methapyriline (Histadyl)	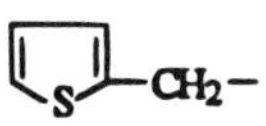 S –CH_2–	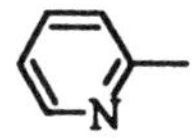 N	N	CH_3	CH_3
C. Alkylamines					
8. Chlorpheniramine (Chlortrimeton) Dexchlorpheniramine[d] (Polaramine)	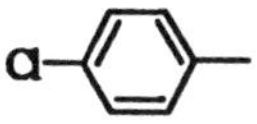 Cl–	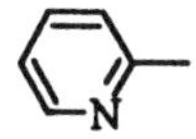 N	CH	CH_3	CH_3
9. Bromphenyramine (Dimetane) Dexbrompheniramine[d] (Disomer)	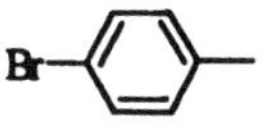 Br–	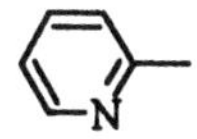 N	CH	CH_3	CH_3

D. Alkenylamine

10. Triprolidine (Actidil)

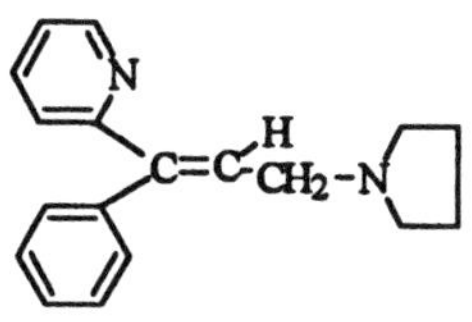

N
C=C H CH_2-N

(*Continued*)

Table 13-1 *(Continued)*

E. Benzhydryl piperazines

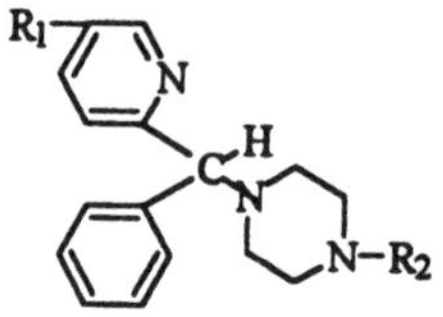

	R_1	R_2
11. Hydroxyzine HCl (Atarax) Hydroxyzine Pamoate[e] (Vistaril)	Cl	$—CH_2CH_2OCH_2CH_2OH$
12. Cyclizine (Marezine)	H	$—CH_3$
13. Chlorcylizine (Mantadil)	Cl	$—CH_3$
14. Meclizine (Bonine)	Cl	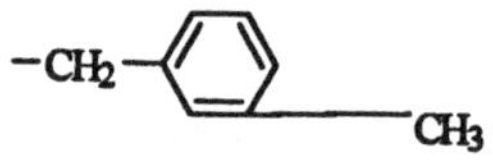 $-CH_2-C_6H_4-CH_3$
15. Buclizine (Softram)	Cl	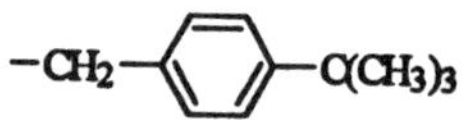 $-CH_2-C_6H_4-C(CH_3)_3$

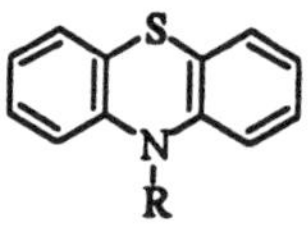

F. Phenothiazines and other ring systems

Name	R	Ring System
16. Promethazine (Phenergan)	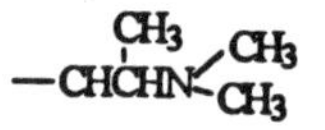 $—CH(CH_3)CH_2N(CH_3)_2$	Phenothiazine
17. Trimeprazine (Temaril)	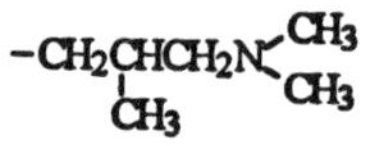 $-CH_2CH(CH_3)CH_2N(CH_3)_2$	Phenothiazine
18. Methyldiazine (Tacaryl)	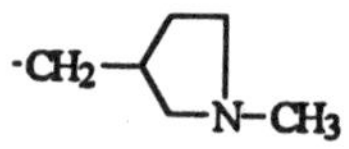 $-CH_2$-(N-methylpyrrolidinyl) $N-CH_3$	Phenothiazine
19. Phenindamine (Theruhistin)	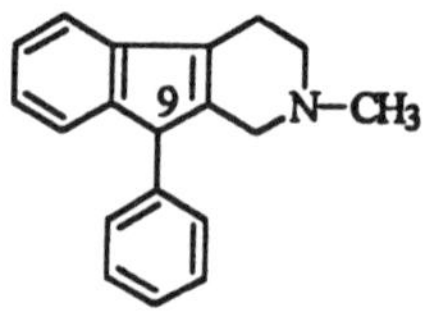 (structure; N-CH3, position 9)	Indenopyridine
20. Azatadine[f] (Optimine)	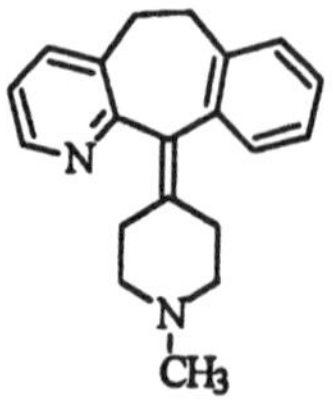 (structure; N, N-CH3)	Benzocyclohepta-pyridine

[a] Trade name in parentheses.
[b] The 8-chlorotheophylline salt of diphenhydramine.
[c] Numerous brands available and in mixtures with other cold medicines.
[d] Dextroenantiomer.
[e] Insoluble salt for sustained release IM injections.
[f] Azo-isostere of cyproheptadine (Chapter 12).

Bioisosteric replacements of benzene ring by thiophene, furan, and pyridine have afforded many clinically useful compounds: doxylamine (No. 2), tripelennamine (No. 6), methapyrilene (No. 7), and other examples given in Table 13-1. The antispasmodic adiphenine (Fig. 8-17) and the antihistaminic diphenhydramine may be viewed as having an isosteric relationship. The etiology of diphenhydramine's anticholinergic activity is thus readily explicable if one considers the ester and ether moiety as isosterically related. In fact, many of the "first-generation" H_1 antihistamines antagonize ACh. Such parasympatholytic activity may be judged as undesirable side effects since it includes dry mouth, voiding difficulties, and even impotence. However, the ability to reduce nasal secretions can be considered a clinical attribute, when included in hay fever and "cold" medications.

Sedation (due to CNS depression), whose intensity can vary among the various compounds, has been a bane accompanying these drugs. It is most marked among the ethanolamine ethers (e.g., diphenhydramine) and phenothiazines (e.g., promethazine). In fact, diphenhydramine and doxylamine are marketed as nonprescription "sleeping pills." Clinical studies sometimes give curious results. In one such study evaluating several antihistamines 30% of the subjects dropped out (because of side effects such as sedation). The irony was that 10% of the patients were not receiving the drug; rather, they received a placebo instead.

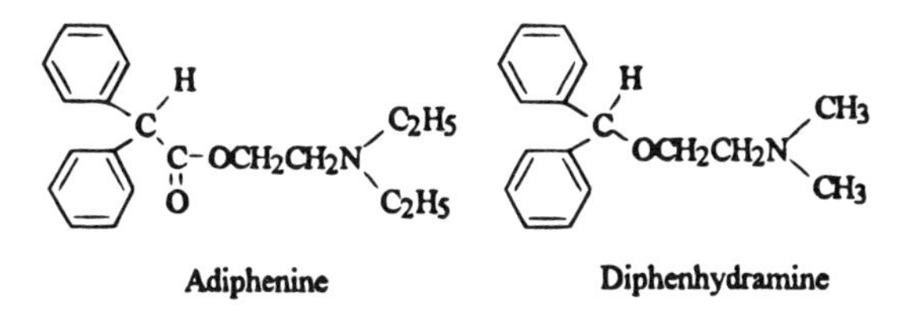

In patients in whom the sedative syndrome is significant, hazards in potentially dangerous circumstances may arise. These would include the operation of machinery and vehicles. Work efficiency may decrease under more ordinary conditions. The potentiation of

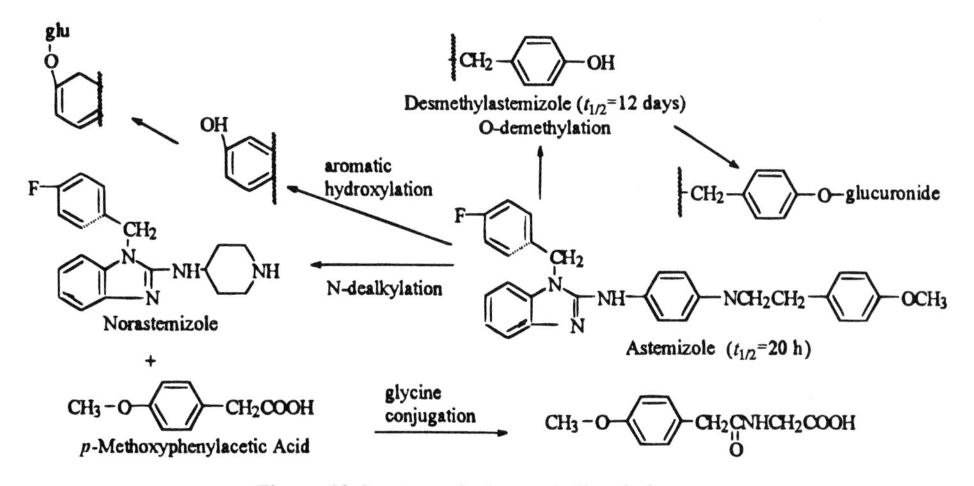

Figure 13-1. Astemizole metabolism in humans.

CNS depressants such as barbiturates, benzodiazepines, and alcohol should also be taken into consideration.

The precise extent of sedation is difficult to determine. Different members of the five classes of the first-generation antihistamines exhibit this property to different degrees at therapeutic dose levels. The alkylamine group (e.g., chlorpheniramine, No. 8, and particularly its dextro stereomer dexchlorpheniramine) may exhibit less sedation. However, clinical studies do not always support this contention, primarily because of placebo effects. Individual patient variation can be considerable. The placebo effect seen in double-blind crossover clinical studies has been reported as high as 68% in one study. It is usually in the 15–30% range.

With the exception previously noted, anticholinergic activity is considered a prominent adverse effect with these compounds. This fact can lead to potentially significant drug interactions with antispasmodic agents and tricyclic antidepressants. Most of the aforementioned adverse effects can probably be ascribed to the relatively high lipid solubility of these compounds, which allows them to penetrate the CNS and cross the BBB, where they can readily bind to H_1 receptors in the brain, in addition to peripheral receptors. In fact, it has been proposed that blockade of such central histamine receptors contributes to their sedative effect. To partially explain the unusually large placebo effects, others have suggested the possibility that patients on placebos in the various studies were "preconditioned" to anticipate sedation from prior experiences with antihistaminic drugs.

From a clinical standpoint in using these H_1 antagonists in the treatment of allergic rhinitis, urticaria in the early part of the pollen season (when the count is still low), these drugs seem highly effective with little or no adverse effects. As the pollen count rises and the period of exposure becomes protracted, symptoms are not as well controlled; for instance, the rhinorrhea, nasal itching, nasal obstruction, and sinus congestion increase. The earlier low doses of antihistamines satisfactorily performed a balancing act between antagonizing the effects of moderate levels of released histamine and the unwanted drowsiness (and mucosal dryness). As the level of released histamine increases (due to greater allergic challenge), those low doses become inadequate. Simply increasing these doses is not satisfactory since the adverse anticholinergic and sedative effects may reach unacceptable levels; the balance of effects achieved before is lost. A partially successful solution is the coadministration of sympathomimetic decongestant drugs such as the ephedrines and phenylpropanolamine (Chapter 9), which can relieve nasal congestion and somewhat counteract drowsiness by their CNS stimulant side effects.

At the pharmacological level, however, the situation is not that simple. Other factors are invariably involved as well. One thing to consider is the probability that blockade of histamine at its receptors is not an adequate response by itself. Even though histamine may be the initial mediator on the scene, other allergic mediators are now also known to be released. The leukotrienes are particularly important in causing allergic responses such as bronchospasms and excessive mucous secretion as they appear to be hundreds of times more potent in this respect than histamine. There is a likelihood that the effectiveness of antihistamines may not be limited to H_1 histamine receptors alone. An interesting study demonstrated that when applied intranasally, the tricyclic antihistamine azatadine (No. 20, Table 13-1) inhibited the copious release of inflammatory mediators normally triggered in sensitive subjects when exposed to ragweed.

Histamine was previously mentioned as a putative neurotransmitter (Chapter 12). It may have a role in sedation and wakefulness. In fact, drowsiness had been implicated previously with a possible inhibition of histamine-N-methyltransferase by antihistamines and subsequently with the blockade of central histaminergic receptors as well. However, it must be understood that antihistamines are not purely H_1 receptor antagonists, since they exhibit blocking effects on cholinergic (M) and α adrenoceptors as well. Finally, the high incidence of placebo sedation remains a puzzle.

Irrespective of what the mechanism of sedation actually is, the drug must be able to traverse the BBB in order to elicit this effect. Most of the H_1 antagonists are sufficiently lipid soluble to do this. This suggests, therefore, what the "ideal" antihistamine can be: a compound with high H_1-receptor affinity and low lipid solubility at physiological pH. The former would assure a poor affinity for cholinergic and adrenergic receptors (minimizing side effects from those interactions) and the latter would assure primarily interaction with peripheral rather than central H_1 receptors, thereby greatly minimizing or even eliminating the therapeutically limiting sedative side effect.

Among the first drugs to meet some of these criteria was terfanidine, which is a piperidine butanol derivative (Table 13-2). It can be considered a second-generation H_1 antagonist. Its structure is a considerable departure from previous H_1 antihistamines. The drug was shown to be a selective H_1-histamine receptor antagonist by in vitro and in vivo studies. Comparative receptor binding studies clearly demonstrated the effects of chlorpheniramine (No. 8, Fig. 13-1) on central receptors in the brain, whereas terfanidine has no comparable effect in guinea pigs. This and the fact that terfanidine shows no interactions with α- or β-adrenergic receptors clearly indicates lack of CNS penetration. Anticholinergic activity (in rats) was also weak. Activity against H_2 receptors (see later) was also absent at H_1-effective drug levels.

The primary metabolite of terfanidine (*R*=OH) was recently (1996) marketed as fexofenadine (Allegra). It may not exhibit the serious drug interactions with ketoconazole and erythromycins.

The second of the nonsedating antihistamine is the benzimidazole derivative astemizole (Table 13-2). This is an extremely long-acting drug ($t_{1/2}$ = 104 hs). The skin reaction to histamine, for example, is modified for several days following a single oral 10 mg dose. A single

Table 13-2 Second-Generation H_1 Antihistamines[a]

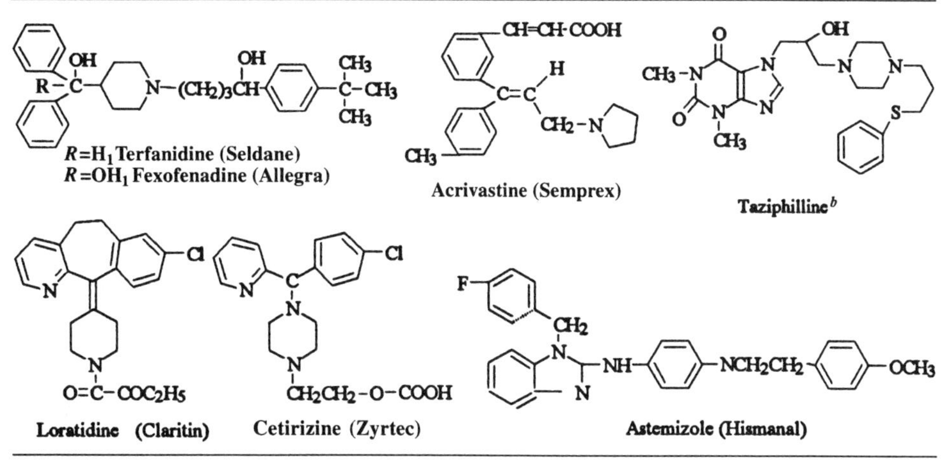

[a] Trade name in parentheses.

40 mg dose extended this effect to almost 1 month. Other aspects of the drug's pharmacokinetics are also unusual. The drug is rapidly and efficiently absorbed, with peak plasma levels occurring in about 1 hour. Yet symptom relief is not apparent for 4–5 days. Duration of action is extremely long. The drug, therefore, would be particularly useful for prophylaxis rather than for treatment of acute symptoms. In fact, one study found it to be somewhat less sedating than was the placebo. These unusual pharmacokinetic properties (especially long duration) are the result of a complex metabolic profile (Fig. 13-1). Astemizole itself has a serum $t_{1/2}$ of 20 hours. It is then metabolized by three major pathways. O-Demethylation produces the most important metabolite, desmethylastemizole, with a particularly long $t_{1/2}$ of about 12 days. Its pharmacological profile is similar to that of the parent drug. N-Dealkylation affords norastemizole that is somewhat, but briefly, active. Hydroxylation of the benzimidazole results in 6-hydroxyastemizole, also a weak, short-acting metabolite.

The toxicity of astemizole also appears to be remarkably low. An acute toxicity study in rats found that at 25,000 times the antihistaminic dose (2,560 mg/kg) the drug produced some transient sedation but no lethality up to 2 weeks after administration—an extreme safety margin. This raises the possibility that antihistamines may have intrinsic antiasthmatic activity, but at such doses that the toxicity of the first-generation drugs precluded such a use at safe doses.

There are additional nonsedating H_1-selective antihistamines at various stages of clinical investigation. Acrivastine is particularly interesting from a drug design standpoint. Rather than serendipitously searching for new molecular entities, this drug does not represent "new chemistry." Rather, the deceptively simple (and obvious) approach of increasing the hydrophilicity of an old compound, triprolidine (No. 10, Table 13-1), by the proper positioning of an acrylic acid moiety produced acrivastine (Table 13-2). As predicted, this reduced lipid solubility dramatically but retained effective H_1 antagonism—peripherally. It was predictable that the $t_{1/2}$ was shortened to 1.7 hours (from 4.6 hours for triprolidine) and the onset of action was short as well: 1–2 hours.

Loratadine, where the lipophilic N-methyl group of azatidine (No. 20, Fig. 13-1) has been exchanged for the much more hydrophilic carbamate, is largely devoid of CNS activity and has been shown to be a nonsedating antihistamine. As azatidine, this drug is also isosterically related to cyproheptadine (Chapter 12), which is also a serotonin antagonist, in addition to being clinically useful as an antihistamine.

Ceterizine should be interpreted as a traditional benzhydryl piperazine H_1 antagonist (see chlorcyclizine, No. 13, Table 13-1), where the N-methyl group has been interchanged for an ethoxycarboxylic acid.

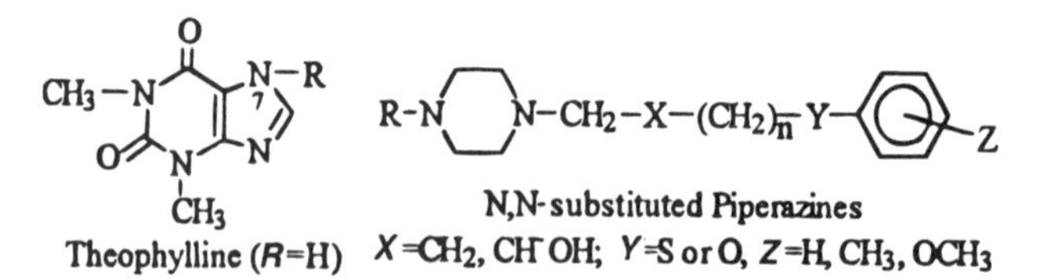

Theophylline (*R*=H)

N,N-substituted Piperazines
X=CH_2, CH·OH; *Y*=S or O, *Z*=H, CH_3, OCH_3

Taziphilline (Table 13-2) represents a novel approach to the design of second-generation H_1-antagonists, hopefully with additional desired properties such as antiasthmatic effects. Theophylline (1,3-dimethylxanthine) has long been known, and used, for its bronchodilator activity and efficacy in asthma. Several N^7-derivatives are also useful [e.g., Dyphilline, Lufylline, R = CH(OH)CH_2-OH]. After finding that N,N′-disubstituted piperazines (see earlier) had potent antihistaminic activity that linked these compounds to the N^7-position of theophylline, hope of achieving both effective antihistaminic and bronchodilating activity arose. Taziphilline represents the compound being clinically useful. It shows a 10-fold potency over terfanidine and astemizole and appears to have no significant sedative effects. Bronchospasms induced in guinea pigs by histamine (but not ACh) were inhibited better than with theophylline itself (95% vs. 56%), which indicates the selectivity of this inhibition.

13.2. Inhibition of Mediator Release

The discussion so far may have given the impression that allergic, particularly asthmatic, attacks (responses) are solely mediated by a sudden release of copious quantities of histamine that are stimulated to do so by some pathological event such as an autoimmune reaction, pollen exposure, or an antigen–antibody reaction. While it is true that histamine is released by mast cells (and also their circulating counterparts, the basophils), causing swelling of nasal passages (by affecting blood vessels), itching (by exciting nerve endings), and precipitating rhinorrhea (by causing mucous glands to release their contents), these tend to be short-lived events. Long-acting inflammatory mediators are also known to exist. One, which was long believed to cause the characteristic bronchospasms of an asthma attack—named slow-reacting substance anaphylaxis (SRS-A), was suspected already in the 1920s, but it was not referred to in the literature until 1930. Its nonidentity with histamine was shown by the slowness with which its spasmogenic effects on smooth musculature developed, and later, by the fact that early antihistamines did not ameliorate the effects. SRS-A's chemical structure remained elusive. Nevertheless, investigations into its physiology with various purified extract preparations continued. It was apparent that the substance was very potent, was obtainable only in minuscule amounts, and was extremely labile. It also became apparent that the likelihood that SRS-A elicited asthmatic symptoms was strong.

Immunologic research slowly unraveled some of the mystery. It was soon found that the immunoglobulin E (IgE) was the basis of SRS-A production. A heat-stable variant of IgG (IgG_2a) was also discovered to stimulate SRS-A production, but IgE was superior in causing histamine release. IgG_2a has been shown to motivate SRS-A release. Finally, it was recognized that mast cells were needed for histamine release following an IgE response, while the cytotropic IgG_2a stimulates SRS-A production by the leucocytes called *polymorphonuclear neutrophils* (PMNs). Finally, a differential was established by demonstrating that cromolyn sodium inhibited histamine release, while diethylcarbamazine (the antifilarial drug, Chapter 7) selectively inhibited SRS-A production. Thus the two substances could now be pharmacologically separated.

The isolation and chemical structure of SRS-A was finally achieved by the early 1980s: They are the 6-glutathione and 6-cysteinyl derivatives of leukotriene C and D, LTC_4, and LTD_4 (see Fig. 5-5 and relevant discussion in Chapter 5).

Even though various antileukotriene drugs have been synthesized, none has reached clinical acceptability. Inhibition of histamine release by an apparent mast cell stabilizing mechanism (in lung tissue, but probably not elsewhere) is achieved with the carboxychromone derivative cromolyn sodium. The mechanism is believed to involve inhibition of histamine release from pulmonary mast cells by blocking Ca^{2+} movement through membrane channels.

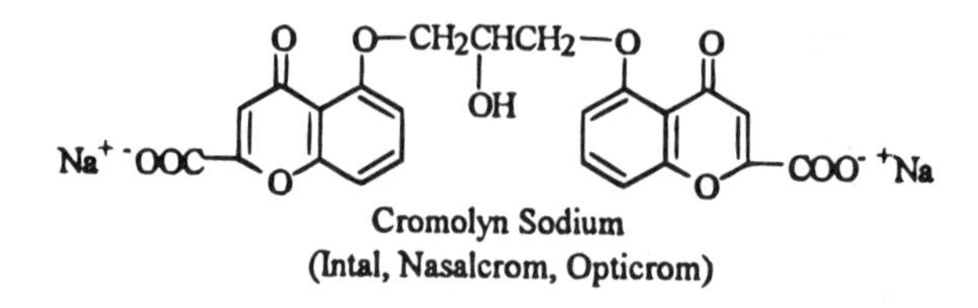

Cromolyn Sodium
(Intal, Nasalcrom, Opticrom)

Cromolyn sodium (Intal) is insufflated into the lungs by a special device ("Spinhaler") as a micronized powder. The drug is strictly prophylactic; it will not abort an asthmatic attack in progress. Oral administration is useless as the compound is not absorbed. The product is also available as a prophylactic nasal spray (Nasalcrom) for seasonal (but not chronic) rhinitis. It can also prevent severe nasal congestion. An eyedrop dosage form (Opticrom) is also available for atopic diseases of the eye. Cromolyn is about equieffective with several of the H_1 antihistamines.

The synthesis of a series of triazolobenzopyranones carrying piperazinoalkoxy substituents on the phenyl ring demonstrated the feasibility of integrating the mast-cell-stabilizing properties of the chromone (benzopyranone) moiety and H_1 antihistaminic properties into one molecule. The most effective compound of a small series was BR-28390. Its antihistaminic potency (guinea pig ileum) was the same as mepyramine; histamine release inhibition effectiveness (tested by rat passive peritoneal anaphylaxis) was somewhat less than cromolyn.

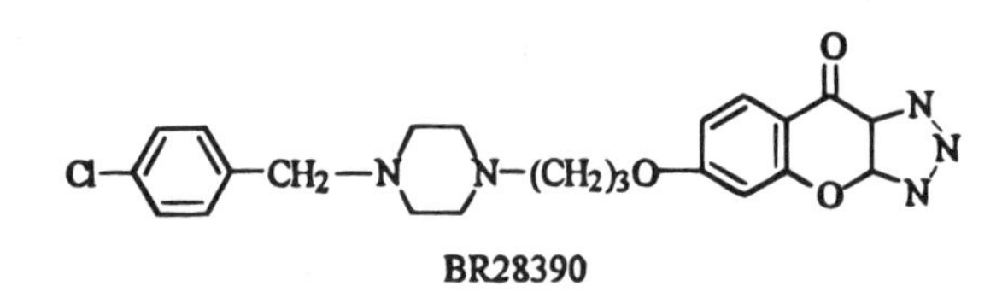

BR28390

13.3. Peptic Ulcer Disease

Peptic ulcer disease (PUD) is a disease in which ulceration occurs in the lower esophagus, stomach (along its lesser curvature), the duodenum, or jejunum. These are all areas concerned with digestion, hence the name *peptic*. The most prominent symptom is gnawing pain that is relieved by food and alkali, but worsened by alcohol and condiments. Painless periods of remission can occur even without treatment.

The proximate cause of PUD is relative or absolute gastric acid hypersecretion. The etiology of the disease is still unknown. Many pathophysiologic abnormalities have been proposed as explanations for the mechanism(s) by which this excess acidity causes gastric and duodenal ulceration. Our understanding of what the cytoprotective controls are that prevent this necrotic process in healthy individuals is simultaneously increasing.

Some of the "basics" have been known for some time. Since the function of gastric secretions is to initiate protein digestion, the wall of the stomach itself must be protected from this destructive process. To a large extent this was believed to be accomplished by abundant secretion of mucous by mucous-producing cells as well as the gastric glands, which also secrete the "gastric juice." It is not likely that all the cytoprotective mechanisms have been identified.

The main digestive substances secreted are the proteolytic enzyme pepsin (as the proenzyme pepsinogen) and hydrochloric acid to produce the low pH at which the enzyme functions.[1] The critical nature of HCl to digestion and ulceration has been acknowledged, but not understood, for decades. In fact, the old dictum clinicians went by—and still do—is: "no acid, no ulcer." Since the molecular basis of what are only beginning to be recognized as complex interacting mechanisms was not well understood, treatment had as its goal the decrease of acidity. That meant initially the use of simple chemical acid neutralizer (i.e., antacids). The earliest—sodium bicarbonate, followed by calcium carbonate—though very efficient (Eq. 13.2) and rapid, are not recommended for long-term use. The former may result in systemic alkalosis and fluid retention, whereas the latter may ultimately lead to hypercalcemia—milk alkali syndrome—since significant amounts of soluble Ca^{2+}, which are produced by the neutralization reaction, will be absorbed. In addition, acid rebound is a problem, particularly with $NaHCO_3$. Calcium tends to be constipating and is therefore frequently combined with magnesium. Aluminum and magnesium hydroxides represent somewhat more "sophisticated" antacids (Eq. 13.3).

$$NaHCO_3 + HCl \longrightarrow NaCl + H_2O + CO_2$$

$$CaCO_3 + 2\,HCl \longrightarrow CaCl_2 + H_2O + CO_2 \tag{13.2}$$

$$Al(OH)_3 + 3\,HCl \longrightarrow AlCl_3 + 3\,H_2O$$

$$Mg(OH)_2 + 2\,HCl \longrightarrow MgCl_2 + 2\,H_2O \tag{13.3}$$

Aluminum-containing antacids are slow, sustained, nonsystemic, and have a buffering capacity at acidic pH values (3–5). The soluble Al^{3+} ion produced is astringent[2] and tends to promote constipation. The Al^{3+} ion will bind to drugs forming unabsorbable complexes or chelates (e.g., tetracyclines). It will also prevent phosphate absorption by forming an extremely insoluble $AlPO_4$ ($K_{Sp} > 10^{-15}$), which has been used itself as a nonabsorbable antacid. It is interesting that decreased phosphate absorption, which leads to decreased serum levels, is useful to patients with renal failure and certain bone diseases.

Magnesium hydroxide (Milk of Magnesia), which is a rapid and efficient acid neutralizer, will exert some laxative effects, due to the soluble Mg^{2+} ion. When combined with aluminum compounds, this will lessen the constipating effect of the latter.

A hydrate complex of aluminum hydroxide with octasulfated sucrose, sucrasulfate, has no useful acid-neutralizing capacity, yet it does have healing effects on gastric and duodenal ulcers. Its action is strictly local and involves the formation of a sticky, viscous gel that appears to protect the gastric mucosa by acting as a physical barrier to acids and pepsin.

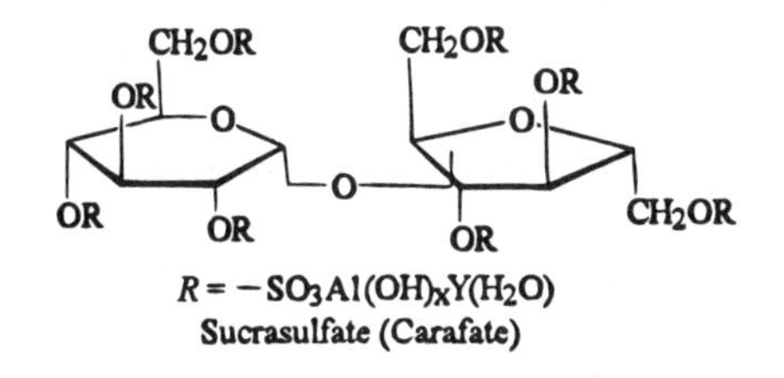

$R = -SO_3Al(OH)_xY(H_2O)$

Sucrasulfate (Carafate)

[1] Pepsin is inactive above pH 4.

[2] The reason aluminum salts are used in antiperspirant deodorant preparations.

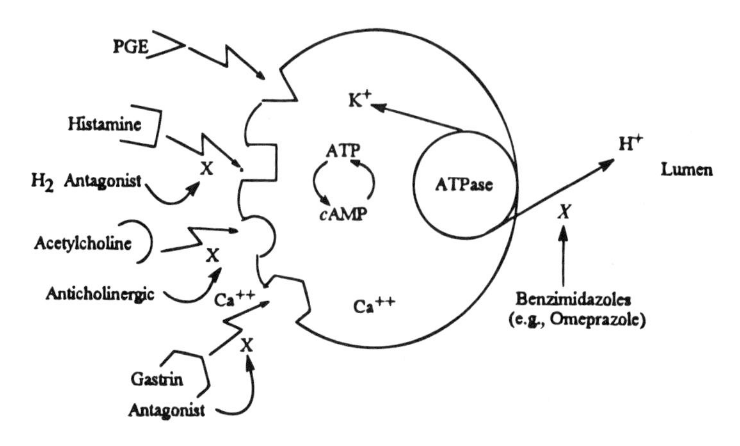

Figure 13-2. The simplified parietal cell.

Both bile acids and pepsin are adsorbed and their damaging activity is reduced. More important, the gel has been shown to adhere with particular affinity to necrotic tissue, such as the cratered ulcers from which it is difficult to dislodge by washing procedures. The ability of sucrasulfate to complex with proteins such as fibrinogen and albumin may be part of the explanation.

The use of antacids and demulcents, although usually alleviating acute symptoms, and of some prophylactic value against their recurrence, is not a satisfactory method of treatment. The use of muscarinic anticholinergic atropinelike compounds (some of which were later shown to be nonantisecretory direct-acting spasmolytics) were of some help and were considered in Chapter 8. The usually tolerated doses used were, in retrospect, insufficiently antisecretory (to HCl). Unfortunately, effective doses would have accentuated the anticholinergic side effects to unacceptable levels in most patients. The development of highly selective M_1 antagonists such as pirenzepine holds out some promise, however.

Successful PUD therapy requires an understanding of all the factors regulating gastric secretions. This will lead in the direction of new therapeutic concepts and drugs. Until recently, three mediators of gastric secretion and their recognized receptors on the parietal cell[3] had been considered: (1) ACh: muscarinic receptor:anticholinergic drugs; (2) histamine: H_2 receptor: H_2 antagonists (see later); (3) gastrin: gastrin receptor:antagonist. Figure 13-2 represents a simplified schematic of a parietal cell. The inhibition of gastric acid secretion, then, should be possible by discovering, or designing, effective antagonists against ACh, gastrin, and histamine. As seen in earlier discussions, since research has often resulted in the discovery of multiple receptor subtypes for a given agonist or mediator, specificity will have to be obtained here, too, to produce high efficacy and low levels of adverse effects resulting from interaction with irrelevant receptor types.

Anticholinergics such as atropine are now believed to act on both M_1 and M_2 cholinergic receptors, accounting for both its antisecretory (M_1) and other anticholinergic properties such as ocular, bladder, and salivary side effects (M_2). As mentioned earlier, the benzodiazepine compound pirenzepine is highly M_1 specific. At doses that inhibit HCl secretion these side effects are not a significant problem. Although not necessarily more effective than cimetidine (see later), it is effective in otherwise treatment-resistant peptic

[3] Large cells on the margin of the peptic glands.

ulcers and Zollinger–Ellison (ZE) syndrome. Several experimental analogs may prove to be more potent and M_1 specific.

Gastrin is a 17-amino-acid residue polypeptide hormone that stimulates parietal cells in the pyloric region of the stomach to secrete HCl. In turn, stimulation of hormone production results from the presence of food. Discovery of substances effective in blocking gastrin receptors on parietal cells should inhibit acid production. Proglumide has some antigastrin properties. It is interesting that in isolated parietal cell experiments it was found that gastrin alone produced only small quantities of HCl, and that furthermore this modest effect is not affected by anticholinergics or H_2-histamine antagonists.

Cholinergic stimulants such as ACh activate M_1 receptors and increase the influx of Ca^{2+} into the parietal cell—an effect antimuscarinics, of course, would block. It must be remembered that a background level of histamine from most cells, as well as ACh from cholinergic neurons, is continuously being replenished, thus stimulating the parietal cells via their respective receptors. This likely potentiates the effect of gastrin. In addition, interplay between histamine and gastrin and/or cholinergic agonism may further modulate the response of parietal cells. This interplay, or modulating effect, has been demonstrated in in vitro parietal cell preparations in that cholinergic and H_2-histamine blockade separately remove the potentiating effects of their respective neuromediators. This suggests indirectly that under in vivo conditions, where such stimulation by ACh and histamine is constant, blockade with antagonists of either receptor will remove the normal sensitization enhancement of parietal cells. The result is a decrease, or elimination of the cycle in which parietal cells exist—a resting state and secreting state.

The antihistamines considered earlier were unable to block all the effects of histamine. Specifically, they are incapable of counteracting the ability of histamine strongly to stimulate gastric acid secretion, inhibit rat uterine contraction, or increase the heart rate. An answer to this apparent puzzle was not forthcoming for almost two decades until Ash and Schild (1966) postulated the existence of multiple receptors in a classic presentation. The structures of the classic (early) antihistamines seem to derive their antagonism to histamine through their bulk, probably by steric blockade of that histamine receptor's "nooks and crannies," obviously unrelated to any structural analogies to the agonist. This receptor type was named histamine-1 or H_1. Thus the antihistamines that are able to counteract the allergic effects attributable to histamine are now referred to as H_1 antagonists. The other receptor type whose stimulation results in gastric HCl was named H_2.

Black et al. (1972) subsequently reasoned that H_2 antagonists should be based on that receptor's recognition of histamine's imidazole ring by altering its structure in such a way as to modify its acidity by affecting its proton tautomerism. Initial studies began with 4-methylhistamine.

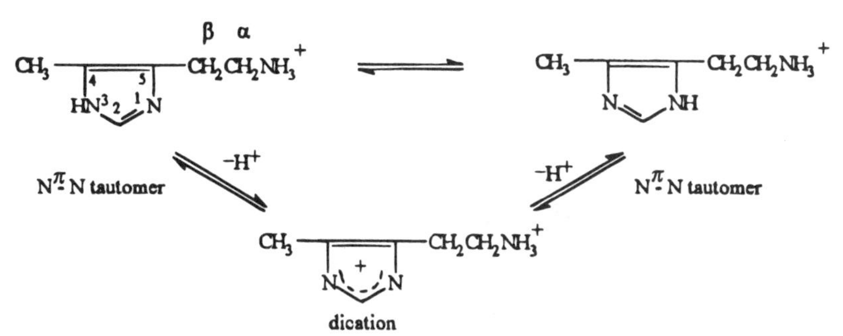

Ion and Tautomer Species of 4-Methylhistamine

Methylation of histamine, depending on the position, produced decreases of H_1-agonist action by as much as 28% (side chain N) to as little as 0.2% (4-methyl), a 1000-fold differential. 4-Methylhistamine was selected to investigate imidazole ring proton shifts (tautomerism)—deemed an important parameter—because it exhibited a 200:1 preference for the then-still, putative H_2 receptor (Black et al., 1972). Other analogs differed as follows: 2-methyl (17%), 3-methyl (0.4%), and α- and β-methyl (0.4% and 0.8%, respectively). The assumption that these differences represented differences in drug—receptor interactions was strong.

Black et al. (1972) undertook a systematic search for histamine antagonists able to inhibit gastric secretion (i.e., H_2 antagonists). Previous workers had also modeled their attempts based on the histamine structure, but they did not stray far from close analogs. They were unsuccessful. The Black group expanded its search further away, but they still retained the essence of the imidazole structure. The underlying hypothesis remained receptor recognition, yet found antagonists that would bind more tenaciously to the H_2 receptor than histamine. By the early 1970s there were conceptual analogies that could be drawn from SAR relationships of agonists to antagonists from other types of receptors, enzyme substrates and their inhibitors, and even antimetabolites.

Remaining "true" to the histamine molecule as a starting point, synthetic attempts included those affecting degrees of ionization and tautomerization, lipophilicity, and geometric rigidity. By analogy to β-adrenergic agonist—blockers (Chapter 10), even fusing aromatic rings to the imidazole was investigated.

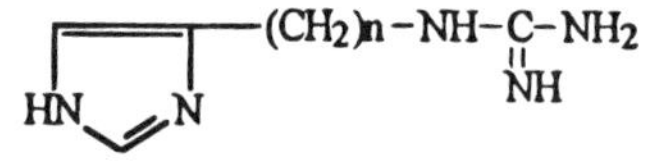

$(CH_2)_n-NH-C(=NH)-NH_2$ (on imidazole ring, HN N)

n=2, N-guanylhistamine
n=3,3-(imidazol-4-yl)propylguanidine

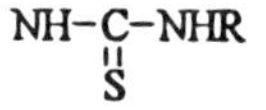

$NH-C(=S)-NHR$

R= H, thiourea
R= CH_3, methylthiourea

A valuable lead compound was found in N^{α}-guanylhistamine. This compound showed weak, but definite, H_2 antagonism; however, it was also a partial agonist. Unlike histamine, pKa 5.9, which would be only 3% ionized at physiological pH, the guanidine group on the side chain (pKa 13.6), would be totally ionized as a side chain cation[4] in the same environment. Unlike the $-NH_3^+$ of histamine, the guanidium cation is a planar species with its electrons delocalized over several atoms. Hydrogen bonding effects are therefore likely to be different. The fact that guanylhistamine showed some antagonism was attributed to the greater distance between the imidazole ring and the terminal nitrogen in the side chain than is the case with histamine. If the slight antagonism exhibited by guanylhistamine is a result of this modest separation, then increasing the distance by insertion of an additional CH_2 should afford better antagonist action. The homolog 3-(imidazolyl-4-yl) propylguanidine (earlier, $n = 3$) was eight times more potent. However, it still retained agonist as well.

At this point an additional assumption was made: The cationic nature of guanidine side chain allows receptor recognition in conjunction with the imidazole ring, and it accounts for the partial agonism seen. Thus, it was reasoned, replacing the very basic guanido group with a neutral, but still polar, group might eliminate the agonist component of activity. The

[4] The strong basicity of guanidine is explained in Chapter 1.

thiourea moiety was chosen (pKa = −1.2). Exchanging the imino, = NH, group with thione, = S, produced an uncharged but still hydrogen-bondable compound devoid of partial agonist activity. However, its H_2 antagonism was extremely weak. This problem was solved by extending the side chain by one additional methylene. The resultant drug was burimamide (Table 13-3). It proved to be a highly specific competitive H_2 antagonist with a 100-fold increase of activity over guanylhistamine. It exhibited no partial H_2 agonism and no H_1 activity at all. Two shortcomings, however, inhibited its clinical potential. It was not significantly absorbed by the oral route, and its effect was weak even on IV administration. This was remedied by two chemical changes. One involved the isosteric replacement of the β-CH_2- of burimamide with -S- which, although lengthening the chain slightly (0.3 Å), sur-

Table 13-3. Evolution of H_2 Antagonists from 4-Methylhistamine*

Basic group	Heterocycle side chain polar group	Name[a]
1.	HN, N ring $-CH_2 \cdot CH_2 \cdot CH_2 \cdot CH_2 \cdot NHCNHCH_3$ (=S)	Burimamide
2.	CH_3–HN, N ring $-CH_2 \cdot S \cdot CH_2 \cdot CH_2 \cdot NHCNHCH_3$ (=S)	Metiamide[b]
3.	CH_3–HN, N ring $-CH_2 \cdot S \cdot CH_2 \cdot CH_2 \cdot NHCNHCH_3$ (=NC≡N)	Cimetidine (Tagamet)
4.	CH_3–HN, N ring $-CH_2 \cdot S \cdot CH_2 \cdot CH_2 \cdot NHCNHCH_2C≡CH$ (=NC≡N)	Etinidine[c]
5.	$(CH_3)_2NCH_2$–O ring $-CH_2 \cdot S \cdot CH_2 \cdot CH_2 \cdot NHCNHCH_3$ (=CH · NO_2)	Ranitidine (Zantac)
6.	$(H_2N)_2C{=}N$–S, N ring $-CH_2 \cdot S \cdot CH_2 \cdot CH_2 \cdot NHCNHCH_3$ (=NC≡N)	Tiotidine[c]
7.	$(H_2N)_2C{=}N$–S, N ring $-CH_2 \cdot S \cdot CH_2 \cdot CH_2 \cdot C \cdot NH_2$ (=NSO_2NH_2)	Famotidine (Pepcid)
8.	N · CH_2–benzene ring–$O \cdot CH_2 \cdot CH_2 \cdot CH_2 \cdot NH$–($CH_3$–N–N, N ring)–$NH_2$	Lamtidine[c]
9.	$(CH_3)_2NCH_2$–S, N ring $-CH_2 \cdot S \cdot CH_2 \cdot CH_2 \cdot NHCNHCH_3$ (=CH · NO_2)	Nizatidine (Axid)

* See text for discussion.
[a] Trade name in parentheses.
[b] Withdrawn.
[c] Investigational.

prisingly did not have a significant effect on lipid solubility. It had electronic effects on the ring (lowering the pKa by unity). The addition of a CH_3 to the 4(5) position had the opposite electronic effect (raising the pKa back to approximately 7.2). This molecular fine tuning resulted in metiamide (Table 13-3)—an orally active drug that with its somewhat lower pKa (7.25 vs. 6.8) led to a 50% lower degree of ionization of physiologic pH (50 vs. 24%). The drug, although highly effective in PUD, was withdrawn shortly after its introduction due to cases of neutropenia (a blood dyscrasia). This hematological toxicity was traced to the thiourea moiety, which had been known to cause blood toxicities in other unrelated drugs. A nonionizing isosteric function was needed. The cyanoguanidine group (Table 13-4) fit the bill initially. This represents the polar guanidine moiety whose basicity is greatly reduced by a strong electron-withdrawing group on the imino N atom—the cyano group. Thus the first widely introduced (1977) H_2 antagonist, cimetidine (Table 13-3), was developed. Over the next several years it became one of the most prescribed drugs in history. The nitro group, where the imino nitrogen atom is replaced by an sp^2 carbon, also seems to function well in the doubly isosterically replaced drug ranitidine (furan for imidazole).

Cimetidine revolutionized the treatment of PUD. It effectively inhibits histamine-stimulated gastric acid secretion in humans and animals. It heals the majority of duodenal ulcers, reduces frequency of reulceration, and is very useful in the treatment of ZE syndrome, wherein pancreatic tumors secrete excess gastrin that in turn stimulates copious quantities of gastric HCl secretion.

Etinidine, a cimetidine analog where a propargyl (methyl ethynyl) group bonds to the terminal guanido nitrogen (Table 13-3), essentially doubles the potency of the compound, allowing a single 300–600 mg night dose to reduce intragastric acidity by 80%. An 800 mg dose of cimetidine achieves similar acid suppression.

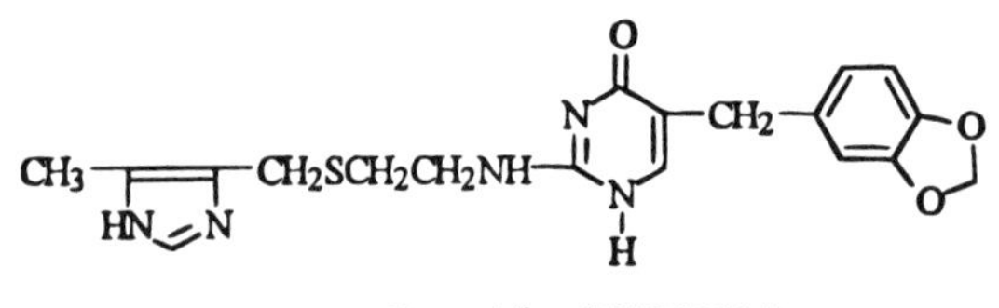

Oxmetidine (SKF 92994)

The cyanoguanidine group found in cimetidine is not essential for H_2 antagonist activity. If viewed as an isosteric replacement for the hemopoetically toxic thiourea group of metiamide, then other modifications should also be effective. The left side of Table 13-4 illustrates a number of neutral, but polar, alternates that have been successfully used. Oxmetidine, which is a potent selective H_2 antagonist that incorporates a 5-substituted isocytosine group, illustrates the latitude and complexity of digression from the urea permitted. The 3-pyridinyl analog (SKF 92564, Table 13-4) is a particularly interesting example since it also exhibits some H_1 antagonism.

Research into H_2 antagonists revolved around the imidazole ring until the late 1970s because of the belief that it may be needed for good receptor recognition. In addition, an inordinate amount of basic and applied research seemed to indicate that a "right balance" of ionic tautomeric species of that ring might be a requirement for high-level H_2 antagonism. All the hypotheses on which the work was based were called into question with the development of two new major drugs in 1979: ranitidine and tiotidine (Table 13-3). In ranitidine (Fig. 13-3) the bioisosteric furan ring replaces the imidazole; the dimethy-

Table 13-4. Bioisosteres for Thiourea and Imidazole Moieties

Name	Structure	Example
Thiourea	—NH-C(=S)—NHR	Metiamide
Cyanoguanidine	—NH-C(=N—C≡N)—NHR	$R = CH_3$ Cimetidine
Nitrodiaminoethene	—NH-C(=CH—NO_2)—NHR	$R = CH_3$ Ranitidine
N-Aminosulfonylimidamide	—C(=N—SO_2NH_2)—NHR	Famotidine
5-Amino-3-methyl-1,3,4-triazole	CH_3—N—N, N, NH_2	Lamtidine
3,4-Diaminothiadiazideoxide	O, S, N 5 1 2 N, 4 3, —HN, NH_2	BL6341[a]; a Famotidine analog
3,4-Diaminocyclobutenyldione	O, O, —HN, NH_2	BMY-25368[b]
Imidazole	HN, H	Cimetidine
Furan	O	Ranitidine
Thiazole	S, N	Famotidine, Tiotidine
3-Pyridinyl	N	SKF 92564
m-Phenylene		Lamtidine

[a] Anonymous, 1984.
[b] Buyniski et al., 1983.

laminomethyl serves to supply the basicity the imidazole ring brought to cimetidine. Ranitidine is actually a stronger base (p*Ka* 8.44). Tiotidine has a thiazole ring replacing the imidazole, while retaining the cyanoguanidine function. Basicity is supplied by the guanido group. Clinical trials stopped following reports of stomach lesions in chronically

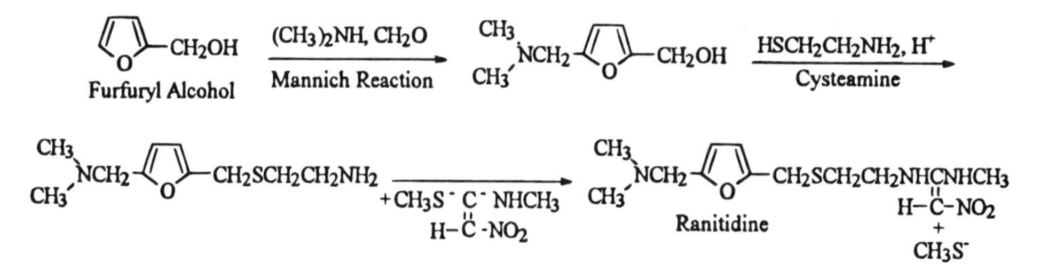

Figure 13-3. Synthesis of ranitidine.

treated rats. However, because of its high potency (i.e., receptor affinity), it may still find use as a radioligand for receptor-binding studies.

Famotidine, which is also a 2-guanidothiazole isostere H_2 antagonist, was introduced in 1986. It lacks the chain NH function of earlier drugs (Table 13-3). The N-aminosulfonylimidamide (Table 13-4) represents the polar moiety on the side chain. Famotidine's high potency found in animals (dogs) was carried over to humans, so that a single daily dose of 40 mg (20 mg as maintenance) is recommended. Another addition is the drug nizatidine, which is a thiazole analog of ranitidine. Zaltidine is a famotidine analog where the side chain and polar groups both seem incorporated into a 2-methylimidazolyl ring. The drug inhibited acid secretion in normal and stimulated subjects as well as PUD patients.

The investigational drug lamtidine (Table 13-3), which is a potent, long-acting H_2 antagonist, in a way represents the ultimate isosteric replacement—the benzene ring. Basicity is supplied by a piperidinomethyl group in the meta position. A diamino-1,3,4-triazole ring represents the polar "urea equivalent."

13.4. Proton Pump Inhibitors (H^+–K^+-ATPase)

Figure 13-2 implicates in simplistic form the various receptor activations of the parietal cell with H^+–K^+-ATPase—the proton pump—and thereby the release of H^+ into the gastric lumen. Cyclic-AMP, and its production from ATP, which is catalyzed by adenyl cyclase, is the significant mediator system in this process. Omeprazole (Fig. 13-4) emerged from a series of benzimidazoles as a compound able effectively and specifically to inhibit the "pump's" production of gastric acid. Cellular components of the gastric digestive system other than parietal cells seem not to be affected. Even its ability to inhibit phosphodi-

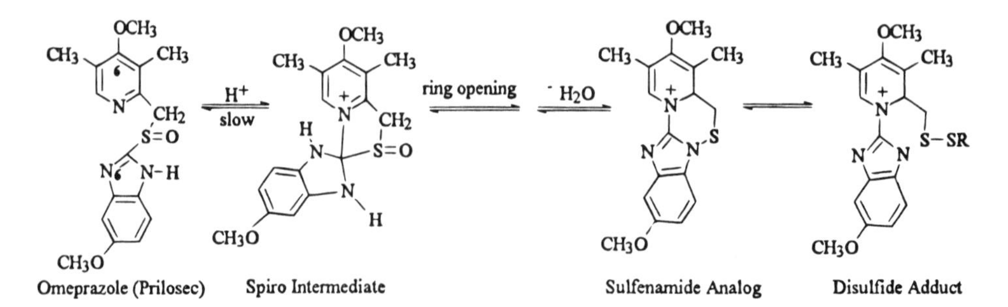

Figure 13-4. Omeprazole and its probable activation (see text for explanation).

esterase (thereby preserving *c*-AMP) is only 0.3% that of H^+–K^+-ATPase inhibition. Clinical studies seem to indicate that the drug, which is effective in ZE patients, is not responsive to H_2 antagonists. Animal metabolic studies show omeprazole undergoing the predictable metabolic conversions (e.g., 6-hydroxylation followed by glucuronidation, O-dealkylation of the methoxy groups, and aliphatic hydroxylation of the methyl functions on the pyridine ring). A potentially disturbing report that high-dose, long-term toxicity studies in rats that developed carcinoid tumors,[5] had initially limited the availability of omeprazole to a compassionate use program (e.g., ZE patients resistant to H_2 antagonists). However, the drug is now marketed in the United States (since 1989).

An interesting proposal of omeprazole's mechanism of action was made. They demonstrated in vitro that under acid conditions as high as 0.5M, HCl, omeprazole cyclized reversibly to a spiro-dihydroimidazole intermediate (Fig. 13-4), which opens to sulfenic acid (not isolated). Cyclization, by the loss of a H_2O molecule, leads to a cyclic sulfenamide that was isolated and identified. Treatment of the sulfenamide with mercaptoethanol ($HSCH_2CH_2OH$) opened the ring to produce the predicted disulfide adduct shown. Since these conditions simulate the gastric environment (at 37°C the reactions were complete in 5–15 minutes) and the enzyme was known to have an essential -SH group, it has been proposed that the sulfenamide produced from omeprazole is the chemical species that forms a covalent drug–enzyme complex with H^+–K^+-ATPase in the acid compartment of the parietal cell, thereby blocking H^+ release.

It is also of interest that a 6-methyl group in the pyridine ring of omeprazole produces an acid-stable (and therefore inactive) analog. Molecular models indicate that such 6-substituents sterically interfere with the imidazole ring. The interpretation is that this will sterically prevent the formation of the spiro intermediate (Fig. 13-4), thus supporting the proposed mechanism for omeprazole.

Analogs of omeprazole where substituents on either the pyridine or benzene portion of the benzimidazole rings were varied produced several potent and long-lasting inhibitors of clinical potential.

13.5. Prostaglandins

The fact that PG-synthetase is present in the gut wall, which, therefore, also contains types E and F PGs, has been known for a long time. The gastric mucosa in particular secretes E series PGs into gastric juice. PGEs (and possibly PGAs) can inhibit histamine-stimulated HCl secretion without diminution of mucosal flow. The ability to maintain the mucosal integrity and thus protect the gastric wall, which is a property now named *cytoprotection*, has been ascribed to PGEs. In fact, such PG-mediated cell protection may even extend to the duodenum, colon, bladder, and possibly other organs as well. This does not mean that it is the sole mechanism of gastric protection. Blood flow, mucus production, and HCO_3^- secretion all undoubtedly participate. When the mucosal barrier fails, the ever-present aggressive factors can lead to ulcer development. These factors include acid, pepsin, and bile salts. But ulcerogenic substances can be causative as well. Ethanol is one. However, the NSAIDs such as aspirin, indomethacin, and others (Chapter 5), are also known to be ulcerogenic. They, of course, are also effective inhibitors of PG synthesis. The increasing

[5] Serotonin-producing tumors arising in intestinal tract.

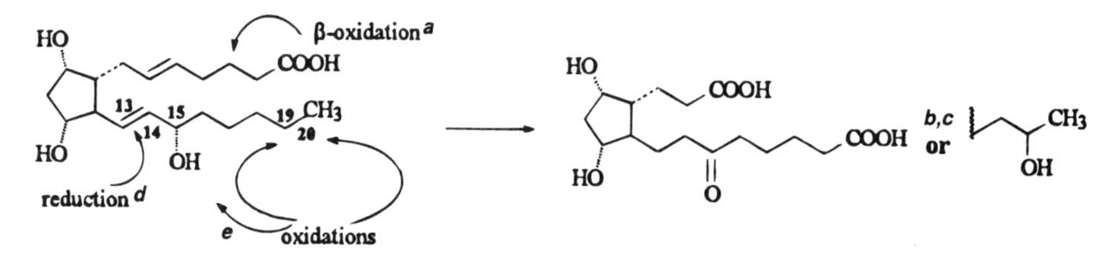

aβ-oxidation is usually sequential, removing C1–C2, then C3–C4. b Either ω-oxidation producing a C_{20} COOH or ω-1 resulting in 19-OH product can occur. c Major metabolite of $PFG_2\alpha$ above will be: 2,3,4,5-tetranor-13,14-dihydro-15-oxo-20-carboxylic acid, for example. d By Δ^{13} reductase. e By 15-hydroxyprostaglandin dehydrogenase.

Figure 13-5. Metabolism of type E and F prostaglandins.

evidence that endogenous PGs are crucial to maintaining gastroduodenal integrity has led to the suggestion that PUD might be a prostaglandin deficiency disease. Along these lines, this also partially explains the ulcerogenic character of all presently available NSAIDS. The fact that the stomachs of gastric ulcer patients contained significantly lower levels of PGEs than did those of normal subjects is an important piece of supportive evidence for the central role of PGs in PUD. Finally, with the recognition of a new receptor in the cell membrane of parietal cells from canine, porcine, and rabbit sources,[6] an explanation for the antisecretory effects of E-type PGs is now at hand (Fig. 13-2).

Natural PGs, it will be recalled, are short-lived compounds in vivo because of rapid metabolism. This means brief duration of action for activities of potential clinical usefulness. E- and F-type PGs undergo three major modes of degradation (Fig. 13-5). The most rapid inactivation is a dehydrogenation (oxidation) of the C-15 OH to a 15-keto analog (e.g., a first pass through lung tissue results in 90% inactivation). Simultaneously, at varying rates, the classic β-oxidation of fatty acids beginning at the COOH group also occurs. This leads to the sequence of C_1–C_2 and C_3–C_4 oxidation of the side chain as acetate units. The result is primarily the tetranor product. Finally, the terminal (ω) position methyl group is oxidized to a carboxyl group; the ω-1 position, C-19, can also be oxidized to a C-19 hydroxyl derivative. The 13–14 double bond is usually reduced following 15-keto formation. The final product is then a 2,3,4,5-tetranor-13,14-dihydro-15-oxo-20-carboxylic acid.

It is therefore apparent that oral administration of PGE_1 or PGE_2 is not effective. IV administration offers some duration of action.

Synthetic modifications of PGs were aimed primarily at protecting the crucial 15-OH group (present in all natural bioactive PGs) from rapid oxidation by 15-hydroxyprostaglandin dehydrogenase. One way is to place a methyl on the carbon with it. In addition, placing two methyls on the adjacent C_{16} has been of some success. Thus 16,16-dimethyl PGE_2 (Table 13-5) and 15(R)-15-methyl PGE_2, arbaprostil, showed early effectiveness as gastric acid inhibitors orally. The acid inhibition was of gastric secretion of basal or nocturnal origin or induced by gastrin, histamine, or food.

In addition to inhibiting HCl secretion directly by agonist action on the newly discovered PGE receptors (Fig. 13-2), E-type PGs are also cytoprotective to the mucosa. This is independent of the antisecretory activity and may involve increased secretion of alkaline substances (HCO_3^-?) and mucus, which both act synergistically. The gel-like mucus adheres to mucosal cells, giving mechanical protection. It also inhibits the back diffusion of H^+, which makes acid neutralization by HCO_3^- easier. Mast cell membrane stabilization by PGs, and possibly other substances, may also occur.

[6] By binding techniques with ^{3}H-labeled PGE analogs.

Table 13-5. Antiulcer Prostaglandins*

Drug[a]	Structure	Comments
Launched[b]		
Prostaglandin E_1, PGE_1 as clathrate (Prostandin)	see Table 5-3	Short effectiveness due to instability; main use cardiovascular (Chapter 5)
Prostaglandin E_2, PGE_2, Dinoprostone (Prostin E^2) also as clathratesee	Table 5-3	Short effectiveness; main use abortion and labor induction
Arbaprostil (15R)15-methyl-PGE_2	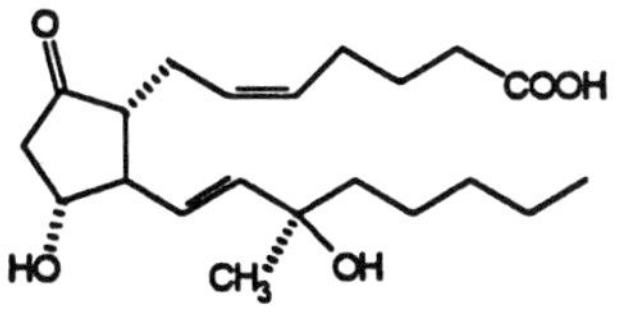 O; COOH; HO; CH_3; OH	
Misoprostol (Cytotec) (+)(16RS)-15-deoxy-16-hydroxy-16-methyl-PGE_1, methyl ester	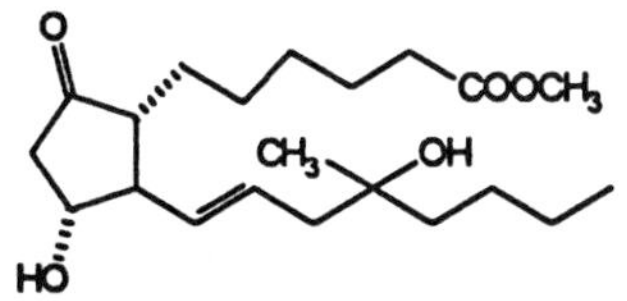 O; $COOCH_3$; CH_3; OH; HO	
Enprostil (Gardin) (+)-4,5-Dide hydro-16-phenoxy-α-tetranor-PGE_2 methyl ester	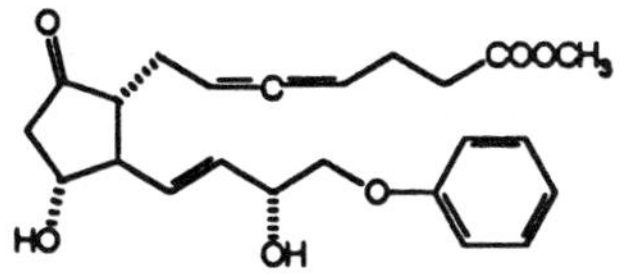 O; C; $COOCH_3$; O; HO; OH	
Under development		
16,16-Dimethyl-PGE_2	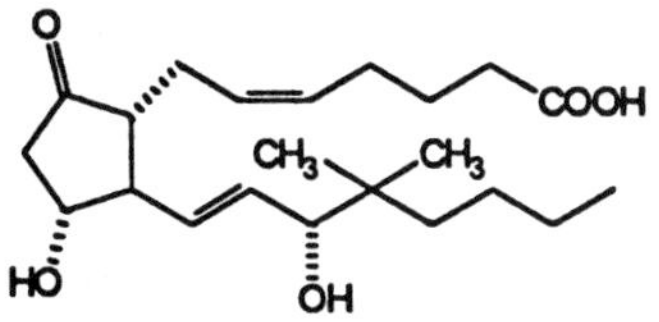 O; COOH; CH_3; CH_3; HO; OH	By Upjohn as potential ulcer prophylactic; causes diarrhea
11-Deoxy-11a,16,16-trimethyl-PGE_2	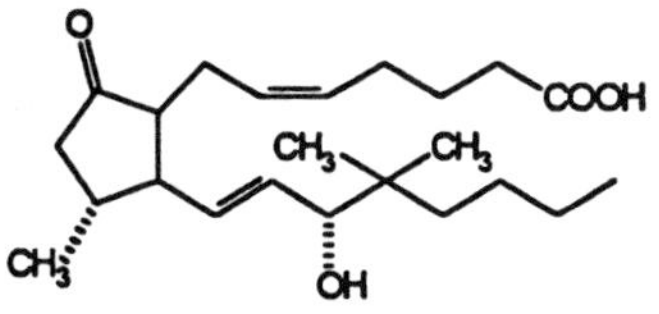 O; COOH; CH_3; CH_3; CH_3; OH	By Roche, as above
(+)-11a,16a,b-Dihydroxy-1,9-dioxo-1-(hydroxymethyl)-16-methyl-13-*trans*-prostene	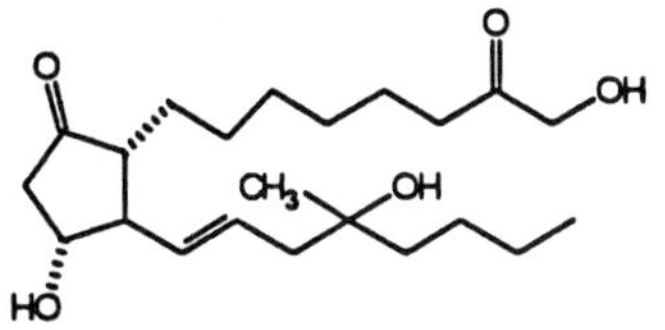 O; O; OH; CH_3; OH; HO	By Lederle, to treat gastrointestinal ulcers and other hypersecretory states
Nocloprost 9-chloro,16,16-dimethyl-PGE_2	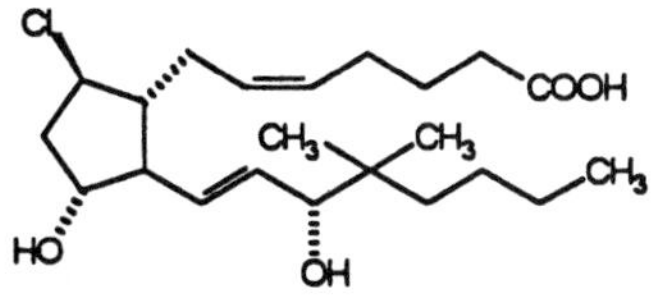 Cl; COOH; CH_3; CH_3; CH_3; HO; OH	No diarrhea produced at doses up to 250 μg/kg

(Continued)

Table 13-5. (*Continued*)

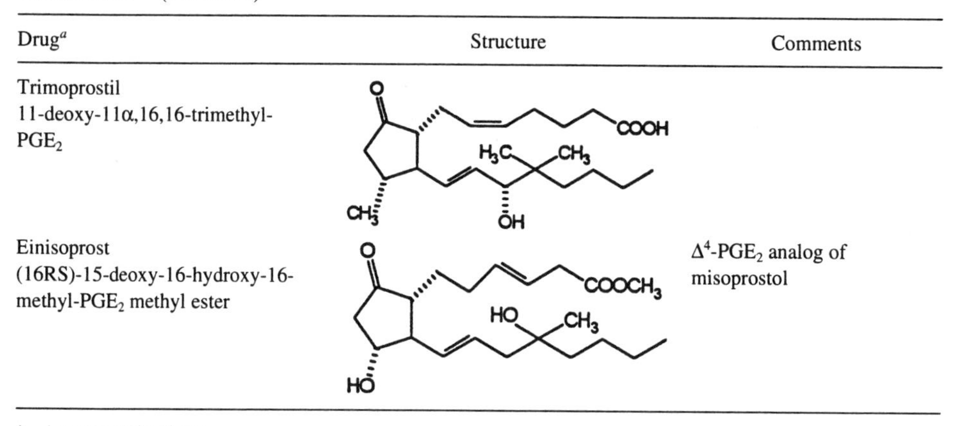

Drug[a]	Structure	Comments
Trimoprostil 11-deoxy-11α,16,16-trimethyl-PGE_2		
Einisoprost (16RS)-15-deoxy-16-hydroxy-16-methyl-PGE_2 methyl ester		Δ^4-PGE_2 analog of misoprostol

* A representative listing.
[a] Trade name in parentheses.
[b] Either in the United States or elsewhere.

As useful as natural PGs may be in principle, they offer almost insurmountable clinical obstacles. As already noted, these obstacles are rapid metabolic degradation, lack of specificity, and other effects. Specifically, uterine effects, which could precipitate abortion in pregnant women, and diarrhea can present problems. Thus by molecular modifications it is hoped to obviate or minimize as many of these shortcomings as possible.

Misoprostol (Table 13-5) is a 15-deoxy-16-hydroxy-16-methyl analog of PGE_1 that is in clinical use in over 40 countries and represents an advance in synthetic PGs for use in the treatment of PUD. Its development arose from systematic synthetic studies to find orally active, effective PGE analogs. It was found that shifting the 15-OH of PGE to C_{16} decreased many effects, but not gastric antisecretory activity (Fig. 13-6). In fact, the methyl ester of the 16-OH PGE_1 position isomer was about equipotent to PGE_1. It was, however, almost inactive orally, presumably because, as PGE_1, it was labile to 15 (or 16?) OH oxidation. It is curious, that placing two CH_3 groups on either C-15 or C-17, decreased antisecretion action dramatically. Placing a methyl on C-16 produced a highly orally effective drug that has both increased potency and duration. The oxidation of the α-chain (carrying the COOH) by β-oxidation (Fig. 13-5) was not affected. Since the natural PGE_2 with a 5–6 double bond was prone to β-oxidation, a shift to the "unnatural" 4–5 position of both

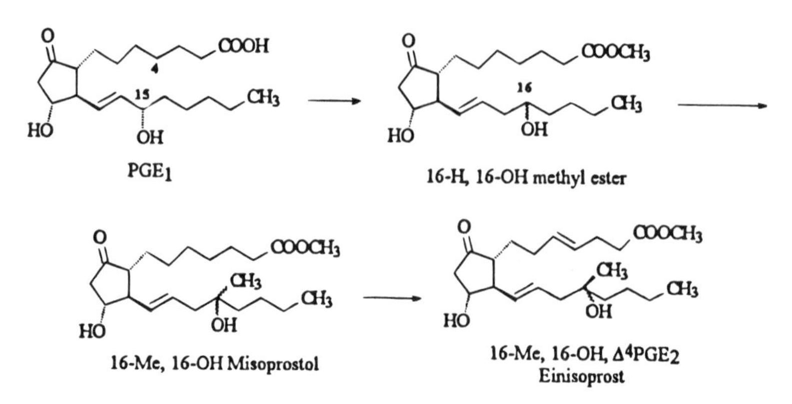

Figure 13-6. Clinical improvements of PGEs.

acetylenic and olefinic bonds was carried out. Only the Z or *cis* isomer showed increased (three times) antisecretory (and decreased diarrheagenic) activity. In fact, the E, or *trans*, was 30 times weaker. This drug is now einisoprost (Table 13-5). The acetylenic compound had no antisecretory effects.

Misoprostol, which is now widely marketed at 100 or 200 μg four times a day, was equieffective to 300 mg cimetidine four times daily in both gastric and duodenal ulcers. It was also effective in patients unresponsive to cimetidine for 10 weeks. A mild diarrhea was self-limiting. Increased uterine toxicity, however, was reported in pregnant women. In some countries a warning against use in pregnancy has been added.

Enprostil (Table 13-5), which is marketed in some countries, has an allenic bond system in the α-chain. It thereby retains the characteristics of PGE_2 (Δ^5) and the β-oxidation protection of the IΔ^4 bond. The phenoxy group at C-16 protects the 15-OH from oxidation while replacing the last 4 carbons of a traditional PG. Nocloprost is an investigational 16-dimethyl PGE_2 where the C-9 oxygen function has been replaced by Cl. It is curious that the corresponding PGE_1 analog (no 5–6 double bond) was not useful.

13.6. Local Anesthetics

Local anesthetics are used to produce a loss of sensation, without a loss of consciousness, at localized areas of the body. When they are applied topically as creams, lotions, and aerosol sprays, they can be useful in alleviating pain of sunburns, severe itching, and other discomforts of the skin. This represents the relatively trivial uses of these drugs. More significant applications are by the parenteral route. Here, solutions of soluble salts such as hydrochlorides are used to alleviate, or actually block, the pain associated with certain surgical procedures, such as dental work, drainage of pus-containing sites, and other incisions. Serious surgical procedures can also be carried out under local anesthesia when a general anesthetic cannot be used or the patient needs to be conscious. The drug is then given by various regional techniques. For example, one of the conventional methods is simply to infiltrate a solution of the anesthetic in proximity to the area of surgery extravascularly. By IV regional anesthesia (Bierblock), it is possible to suppress sensation in an entire area (e.g., a limb).

Peripheral nerve block anesthesia involves injecting the drug near the nerves close to the area to be anesthetized. Epidural anesthesia results when the local anesthetic is injected into the epidural space between a lumbar and sacral vertebra. Several drugs can safely produce useful levels of anesthesia for obstetrics as well as postoperative pain. Another method of utilizing local anesthetics is to inject solutions into subarachnoid spaces (e.g., the spaces between certain vertebrae).

A more recent innovation is the addition of local anesthetics to lowered doses of morphine by epidural infusion for intractable cancer pain involving nerve roots and bone. This has been reported to be useful in terminal patients where the tumor already involves the spine, long bones, or is growing into the nerve root.

The prototype, and historically first, substance utilized to prevent or treat local pain was a South American folk medicine: an infusion made from the leaves of the coca bush (cocaine plant), *Erythoxylum coca* and several other species. Coca is native to the uplands of South America, particularly in what is today Colombia, Ecuador, Peru, and Bolivia. Large-scale cultivation and production is now mainly in Colombia, Peru, and Bolivia at elevations up to 6,000 feet. The alkaloidal content of coca leaves is 0.7–2.5%, the main alkaloids being derivatives of ecgonine: cocaine (Fig. 13-7), cinnamoylcocaine, and α- and β-truxilline (truxillic acid derivatives). Total hydrolysis of cocaine produces ecgonine, benzoic acid, and methanol; cinnamoylcocaine yields ecgonine, cinnamic acid, and methanol, similar treatment of α and β-truxilline also results in ecgonine and methanol, but it also affords the two isomeric truxillic and truxinic acids that are cinnamic acid dimers (see later).

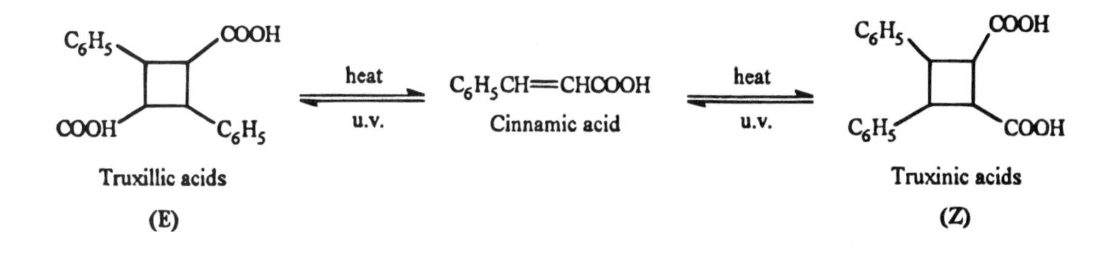

Large-scale production of cocaine (both licit and illicit) essentially involves extraction of the total bases, which are then totally hydrolyzed so that all three ester alkaloids, cocaine, cinnamoyl cocaine, and the truxillines, are converted to ecgonine. The ecgonine is then methylated, followed by esterification with benzoic acid to cocaine. All of the ecgonine in the leaves is converted to the desired product in this way, which is a more efficient process than attempting to extract and separate out only the actual cocaine content from the plant.

The coca leaf was brought to Europe by Spanish physicians in the late sixteenth century. The isolation of the alkaloid occurred three centuries later by Nieman (1860). Even though early native and later European uses of coca leaves included everything from sedative to the treatment of asthma and gastric ailments, modern applications of the alkaloid cocaine began with the discovery of Koller (1884) of the local anesthetic properties of the drug, which were on topical application in the eye. This use soon became widespread. By the end of the nineteenth century, peripheral nerve block and spinal anesthesia came into use. Simultaneously, however, acute and chronic effects of cocaine, including the addictive properties, were recognized.

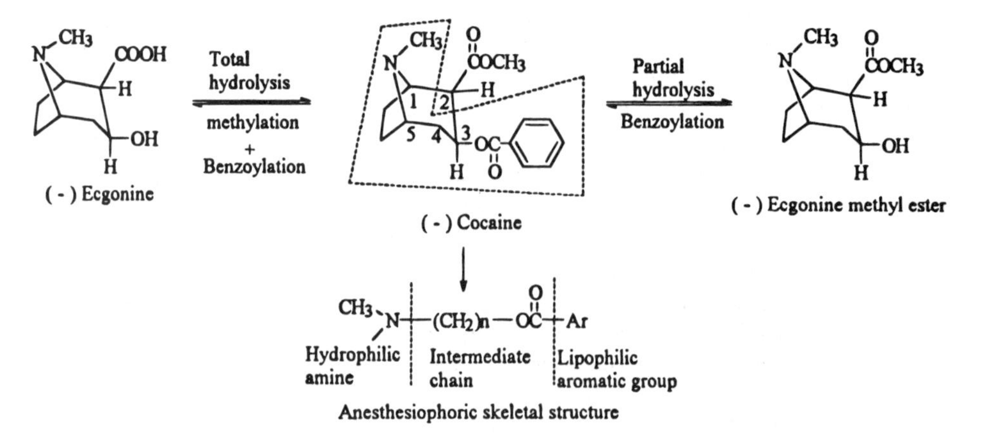

Figure 13-7. Chemical aspects of cocaine.

Today, the drug is known to possess a multiplicity of central and peripheral effects. The hazards include strong CNS stimulation and blockade of DA and NE uptake. Cocaine thus potentiates the action of direct-acting sympathomimetics. After giving cocaine the $t_{1/2}$ of injected NE is actually increased. The drug's psychotostimulant properties strongly resemble those of amphetamine and methylphenidate. Additional hazards of cocaine result from cardiovascular effects. Even though small doses may systemically actually slow the heart, it is the larger doses seen with cocaine abuse that accelerate the heart rate. There is also an initial increase in blood pressure due to vasoconstriction and other factors. Finally, large IV doses have caused fatal cardiac failure. A direct toxicity on the myocardium has been established.[7]

Since the multiple pharmacology of cocaine was recognized before its complete structure was known (the benzoate and methyl ester nature had been confirmed earlier), derivatives with acids other than benzoic and alcohols other than methanol were synthesized. These are only of historical interest today. Once the structure (but not stereochemistry) of cocaine was determined (Willstatter and Muller, 1898), attempts to modify and especially to simplify the structure became possible. The "anesthesiophoric" skeletal structure, which is the minimal structure having local anesthetic properties, was soon determined (see Fig. 13-7). This was an aminoalkylester of benzoic acid, with the alkyl chain being two to three carbons long. Among the earliest successful and widely accepted compounds was 2-diethylaminoethyl-*p*-aminobenzoate (procaine, Table 13-6, No. 2), which was synthesized by Einhorn and Uhfelder (1909).[8] To this day procaine remains the prototype with which subsequent "caines" are compared. Several other compounds of that period were more potent (even than cocaine); however, procaine exhibited considerably lower toxicity and irritation, good tolerance, and superior stability in solution, which are properties not attributable to cocaine.

It should be mentioned that several more drastic simplifications of cocaine appeared on the scene even earlier. The only survivor is ethyl *p*-aminobenzoate (benzocaine, Table 13-6, No. 5). The others of interest were methyl *p*-amino-*m*-hydroxybenzoate (orthoform), which, like benzocaine, was nontoxic, highly insoluble, and therefore not suitable for parenteral administration. Orthoform has no activity on intact skin, but it was useful as a powder on painful wounds. It was superseded commercially by the position isomer *p*-hydroxy-*m*-aminobenzoate methyl ester for reasons of cost since large-scale production of orthoform (as the pure, correct isomer) presented difficulties at the time. It should be pointed out that water-soluble hydrochlorides of the aminobenzoates can be prepared; however, their solutions are much too acidic to inject.

Many more aminobenzoate amino esters were synthesized, and some were developed as local anesthetics. The only ones still of importance are tetracaine, No. 3, and chloroprocaine, No. 1. An examination of the data on the three aminoalkyl *p*-aminobenzoates (Nos. 1, 2, and 3, Table 13-6) illustrates some classic. SARs within that series. For example, the *o*-chloro atom of chloroprocaine would predictably enhance the ester groups instability as exemplified by a four-fold faster rate of hydrolysis—and therefore inactivation (15-fold when compared with tetracaine). The partition coefficients indicate a 570 times greater

[7] See Van Dyke and Byck, 1982, and Anonymous, 1984, for reviews of toxicology and abuse hazards of cocaine.

[8] Procaine was clinically introduced earlier (1905), but its synthesis apparently was not published until 4 years later.

Table 13-6. Local Anesthetics: Structure and Properties[a]

Name	Structure	Use	Part[b] coeff.	Prot. Bndg%	pKa	Duration (min)[c]	LD_{50} [d] (μg/kg)	Hydrol.[e]
A. Esters								
1. Chlorprocaine (Nesacaine)	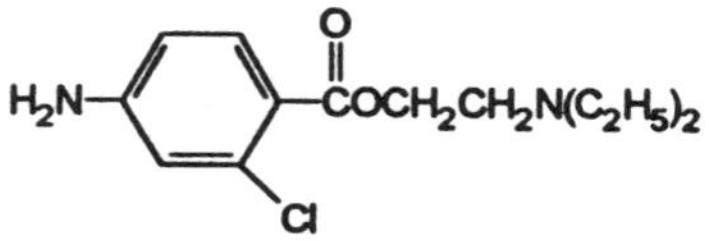 H_2N–, O, $COCH_2CH_2N(C_2H_5)_2$, Cl	Nerve block Epidural Infiltration	0.14	—	8.7	15	1396	4.7
2. Procaine (Novacaine)	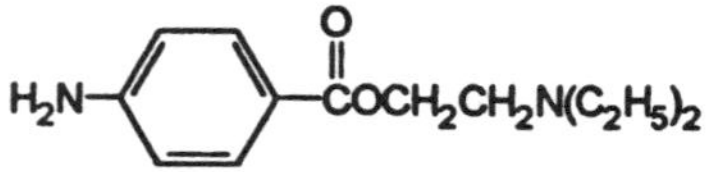 H_2N–, O, $COCH_2CH_2N(C_2H_5)_2$	Infiltration nerve block Epidural subarachnoid	0.6	5.8	8.9	50	615	1.1
3. Tetracaine (Pontocaine	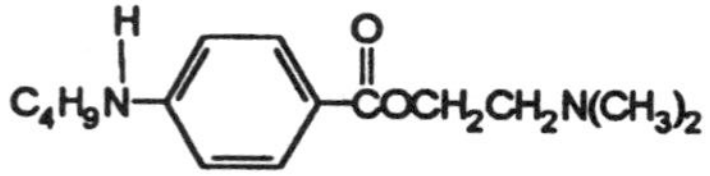 H, C_4H_9N–, O, $COCH_2CH_2N(CH_3)_2$	Subarachnoid (spinal)	80	75	8.4	175	48	0.3
4. Butamben picrate (Butesin)	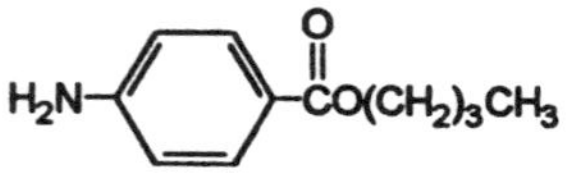 H_2N–, O, $CO(CH_2)_3CH_3$	Topical only	—	—	—	—	—	—
5. Benzocaine (Americane	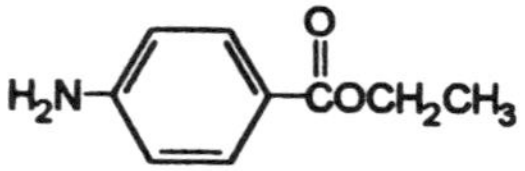 H_2N–, O, $COCH_2CH_3$	Topical only	—	—	—	—	—	—
B. Amides								
6. Lidocaine (Xylocaine)	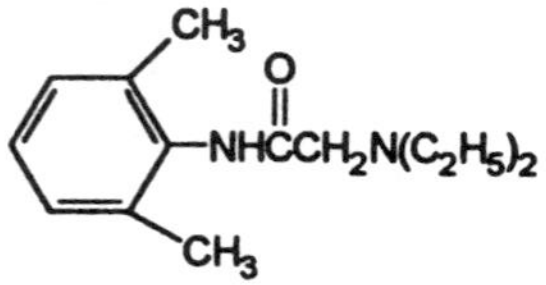 CH_3, O, $NHCCH_2N(C_2H_5)_2$, CH_3	Infiltrationl Nerve block Epidural Topical	2.9	64	7.9	100+	—	—

Table 13-6. *(Continued)*

Name	Structure	Use	Part[b] coeff.	Prot. Bndg%	pKa	Duration (min)[c]	LD_{50}[d] (μg/kg)	Hydrol.[e]
7. Etidocaine (Duranest)	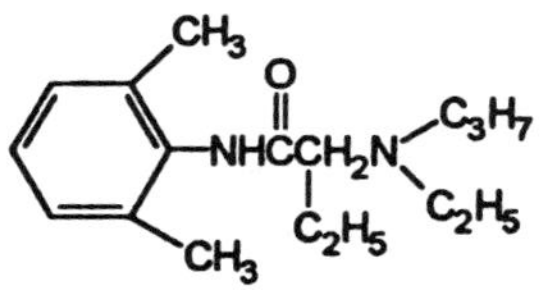	Infiltration Nerve block Epidural	141	94	7.7	200+	—	—
8. Bupivacaine (Sensorcaine)	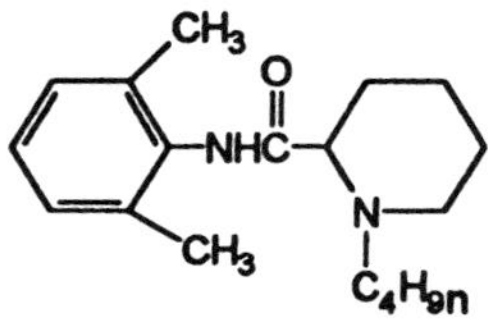	Infiltration Nerve block Spinal Epidural	28	95	8.1	175	—	—
9. Mepivacaine (Carbocaine)	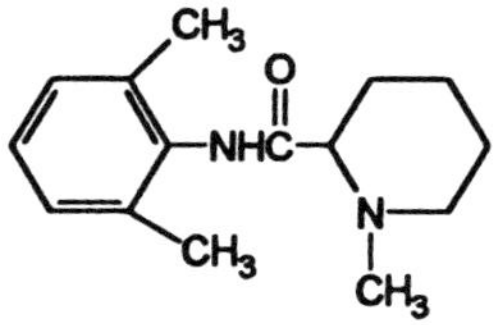	Infiltration Nerve block Epidural	0.8	77	7.6	60[f]	—	—
10. Prilocaine (Citanest)	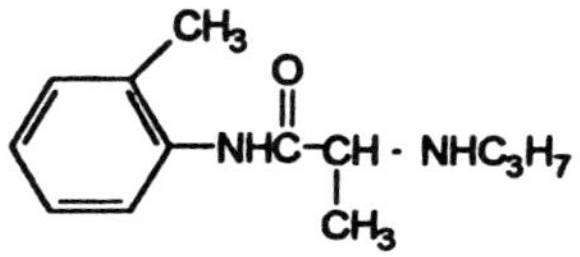	Infiltration Nerve block Epidural	0.9	50	7.9	60–70	—	—

(Continued)

Table 13-6. *(Continued)*

Name	Structure	Use	Part[b] coeff.	Prot. Bndg%	pKa	Duration (min)[c]	LD_{50}[d] (μg/kg)	Hydrol.[e]
11. Dibucaine (Nupercaine)	nC_4H_9O–(naphthalene)–$\overset{O}{\overset{\Vert}{C}}NHCH_2CH_2N(C_2H_5)$	Subarachnoid (spinal only) Topical	—	—	8.3	180–240 (360)[g]	Most toxic	—
C. Miscellaneous								
12. Pramoxine (Tronothane)	nC_4H_9O–(benzene)–$O(CH_2)_3$–N(morpholine)	Topical only	—	—	—	—	—	—
13. Dyclonine (Dyclone)	nC_4H_9O–(benzene)–OCH_2CH_2–N(piperidine)	Topical only	—	—	—	—	—	—

This is not a complete listing. Data presented obtained from many sources.
[a] Trade names in parentheses.
[b] Oleyl alcohol/buffer pH 7.2 or *n*-heptane/buffer pH 7.4.
[c] Approximately.
[d] Mice. Foldes et al., 1965.
[e] μmol/ml/h; Foldes et al., 1965.
[f] From nerve block.
[g] With addition of epinephrine.

lipid solubility for tetracaine than chloroprocaine (133 times compared with procaine). This is presumably a result of the *n*-butyl group on the aromatic nitrogen. With a much slower hydrolysis rate and a high proportion of the drug distributed into the lipid biophase, a much longer duration of action (potency?) can be safely predicted for tetracaine than for chloroprocaine. The actual duration is 12 times (175 vs. 15 minutes). Similarly, the short duration and rapid hydrolysis of chloroprocaine should result in a much lower toxicity. The data (LD_{50s}) show this drug to be about 28 times less toxic than tetracaine. In fact, chloroprocaine is the least toxic local anesthetic in use.

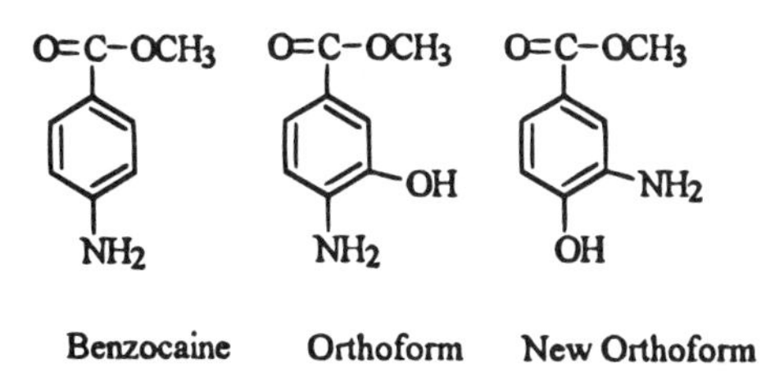

A drawback with the ester-type local anesthetics is the occurrence of allergic reactions, which manifest as dermatitis and systemic effects.

The synthesis and local anesthetic properties of at least one acylanilide compound, nirvanine, was reported even before procaine (1900). However, the amide isostere concept was not seriously pursued for over three decades.

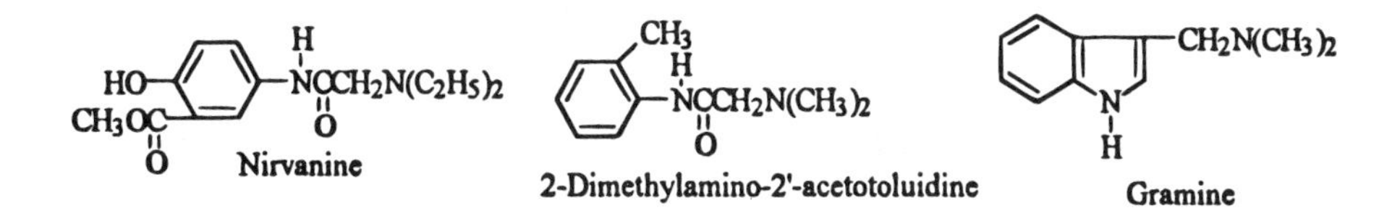

In unrelated structural studies of the hallucinogenic harmala alkaloid gramine an intermediate in the synthesis of the indole ring and 2-dimethylamino-2′-acetotoluidine was found to have typical local anesthetic activity (von Euler and Erdtman, 1935). This led to the development in 1937 of lidocaine, whose two *o*-methyl groups provided remarkable hydrolytic stability for the amide bond against hydrolysis by either hot 50% H_2SO_4 or ethanolic 20% KOH.[9] The advantages of greater stability, relatively low toxicity, and extremely low sensitizing potential exemplified by lidocaine (Table 13-6, No. 6) made it the prototype for the amide-type anesthetics. In fact, no new ester types have been introduced into clinical use since then (except chloroprocaine). Figure 13-8 gives a synthesis for lidocaine and outlines the major metabolic reactions it undergoes. Mepivicaine (No. 9) and bupivicaine (No. 8), where the nitrogen is cyclized into a piperidine ring, were introduced later. Here, again, the increased size of the N-alkyl group (CH_3 vs. C_4H_{9n}) illustrates how significantly that difference increases liposolubility (partition coefficient 0.8 vs. 28, Table 13-6) and the resulting tripling of the duration of action that results from that

[9] See lidocaine discussion in antiarrhythmic section, Chapter 10.

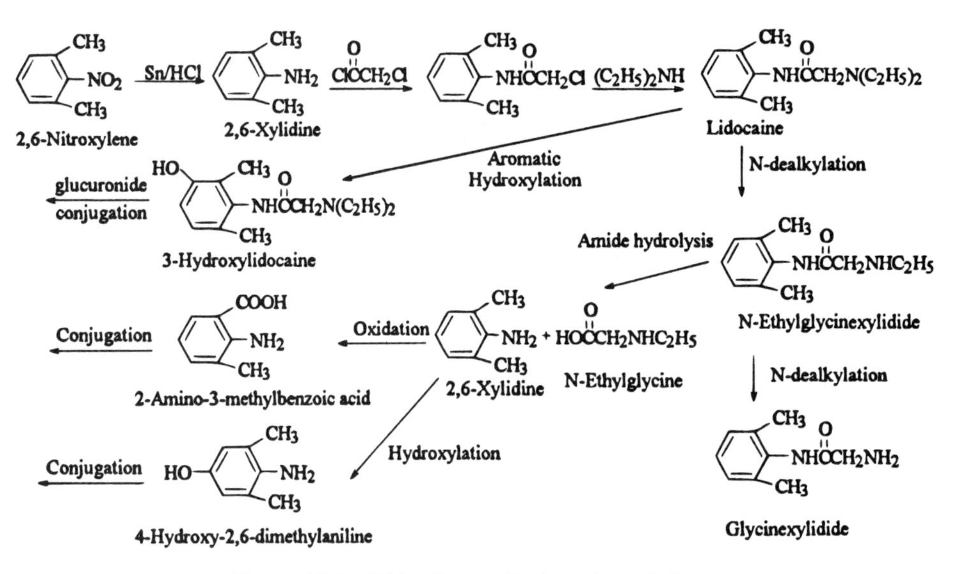

Figure 13-8. Lidocaine synthesis and metabolism.

molecular change. It is interesting that unlike lidocaine, mepivicaine lacks typical anesthetic properties.

Prilocaine (No. 10) represents an interesting situation in that the presence of only one *o*-methyl group has two consequences. A predictably shorter duration of action because of more facile amide hydrolysis, as compared with lidocaine, and a significant likelihood of methemoglobinemia at higher doses being produced by *o*-toluidine that results from this amide hydrolysis. The much lower levels of 2,6-xylidine produced from lidocaine (Fig. 13-8) do not produce this hematological toxicity. However, because of more rapid metabolism and the resultant shorter duration of action, overall toxicity of prilocaine is about 40% less than lidocaine.

Etidocaine (No. 7) is the most recently introduced local anesthetic (1972). The molecular alterations responsible for the 50-fold increase in lipid solubility (and near doubling of protein binding) is the addition of an ethyl group to the α-carbon and, to a lesser extent, the net increase of CH_2 on the nitrogen atom of lidocaine. The result is a doubling of duration of action, a fourfold potency increase, and a quadrupling of toxicity in mice; all (qualitatively) predictable effects.

A clinical comparison of ester and amide types of local anesthetics shows that the differences ensue from different sites and labilities of an ester and an amide-type linkage to metabolism. The benzoate esters are primarily and rapidly hydrolyzed by serum pseudocholinesterase. The resultant *p*-amino benzoic acid has been proposed as the cause of allergic manifestation that may occur in a small percentage of patients. Unhydrolyzed drug will accumulate in individuals having a genetically induced deficiency of this enzyme, producing toxic and particularly long effects. Even in normal subjects, if ester-type neuromuscular blocking agents such as succinylcholine (Chapter 8) are used simultaneously, the two drugs will compete for a limited supply of cholinesterase, thus prolonging the effects of both drugs—a hazard.

Amide-type local anesthetics, being hydrolyzed more slowly in the liver by microsomal enzymes, are not dependent upon cholinesterase and have therefore become the preferred

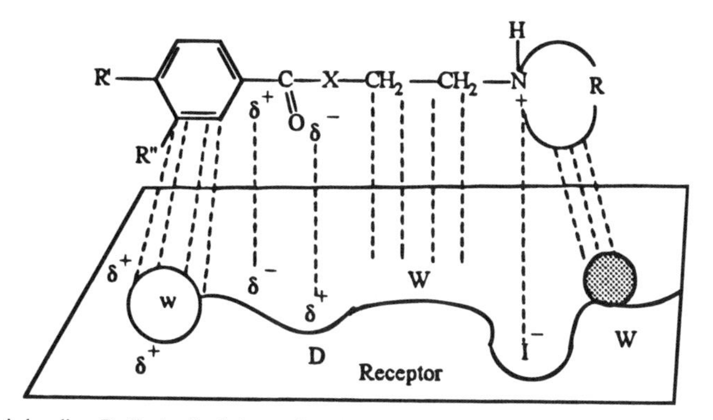

X = O or NH; I, ionic bonding; D, dipole–dipole interactions; W, van der Waals forces.

Figure 13-9. Idealized local anesthetic–receptor interactions. (Modified from Buchi et al., 1966; Buchi and Perlia, 1960.)

agents in many situations. However, in cases of hepatic disease, the reverse may be true. Here, two alternatives exist: use of an ester anesthetic or reduce the dose of the amide drug drastically.

Dibucaine (No. 11) is probably the most toxic anesthetic presently in use, as it is 15 times more toxic than procaine. It is also about 15 times more potent; thus it is usable at low doses. Parenteral use is limited to the spinal route; effects are long lasting (see Table 13-6). This property makes it particularly appealing for topical use (skin) and mucocutaneous areas (e.g., rectal) in the form of creams and ointments.

Two closely related basic esters, pramoxine (No. 12) and dyclonine[10] (No. 13), find a variety of nonparenteral applications. These include topical uses for pain caused by abrasions and burns, and skin and rectal pruritis from any cause (e.g., hemorrhoids, dermatoses). Solutions are too irritant for ophthalmic applications, but they can be used in gargles and lozenges for sore throats (Sucrets).

13.7. Mechanism of Action

The mechanism of action of local anesthetics can be understood primarily on the basis of nerve fiber electrophysiology and physicochemical features of the drugs. In brief, they raise the electric threshold for stimulation, which in turn slows down the rate at which an impulse travels down the nerve fiber. The process by which nerve conduction occurs (Fig. 8-3, and discussion thereof) is essentially impeded. The stabilization of the neuronal membrane does not allow for the depolarization when the action potential reaches the area of blockade.

At the molecular level interaction between the drug and a local anesthetic receptor has been proposed (Fig. 13-9). The interaction involves several of the classic forms considered in Chapter 2; namely ionic, dipole, and van der Waals.

Ion permeability may be affected via drug interaction with such receptors along the sodium channels leading into the neuronal membrane.

[10] Not to be confused with the spasmolytic dicyclomine (Chapter 8).

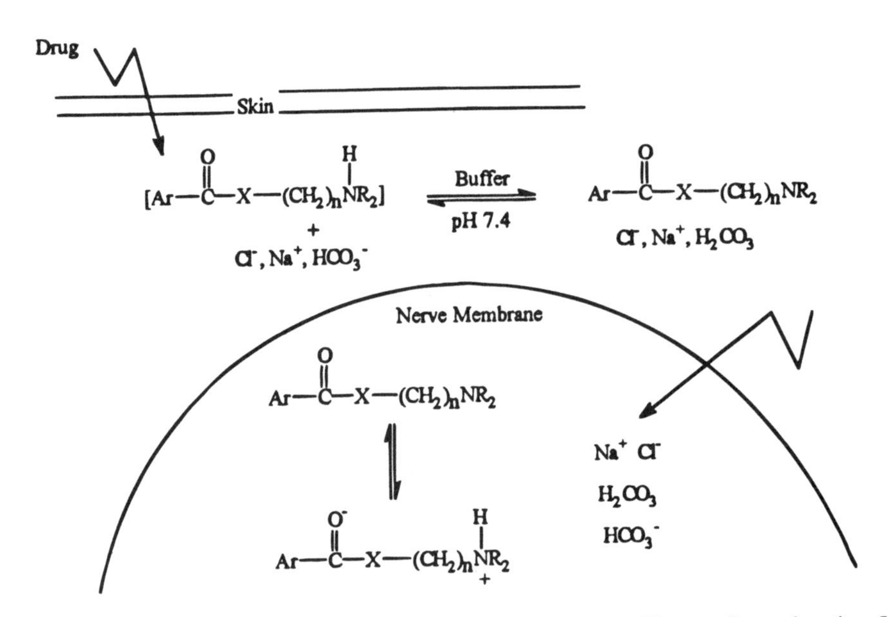

Figure 13-10. The local anesthetic in the physiological environment. (See text for explanation; $X =$ O or N.)

The majority of anesthetics have pKas such that when aqueous solutions of their salts (e.g., hydrochlorides) are injected into the pH 7.4 buffer of the extracellular fluid, they will equilibrate so that the neutral form will readily penetrate through the neuronal membrane (e.g., lidocaine would be 24% nonionized). Once inside, this form would again equilibrate to afford a 75% majority of ionic species, which presumably can then interact with the putative receptor. These concepts are represented in Figure 13-10. It is interesting that since early experiments (1927) appeared to show that these drugs were more active when administered in a basic solution, it was assumed the neutralized form (nonionic) was the active anesthetic species. Later work demonstrated this to be wrong, since nerve preparations that were blocked by an anesthetic that was shown to be bound to the nerve could be easily reactivated by bathing it in an alkaline buffer of about pH 9.5.

References

Anonymous, *Drugs Future* **9**:64, 1984.

Ash, A. S. F., Schild, H. O., *J. Pharmacol.* **27**:427, 1966.

Best, C. H., Dale, H. H., Dudley, H. W., Thorpe, W. V., *J. Physiol.* **62**:397, 1927.

Black, J. W., Duncan, W. A. M., Durant, G. J., et al., *Nature* **236**:385, 1972.

Buchi, J., Perlia, X., *Arzneim-Forsch* **10**:1, 1960.

Buchi, J., Muller, K., Perlia, X., Preiswerk, M. A., *Arzneim-Forsch.* **16**:1263, 1966.

Buyniski, L. et al., SARs among Newer Histamine H_2 Anatagonists. Second Camerino Symposium on *Recent Advances in Receptor Antagonist Chemistry* (Camerino), 1983.

Einhorn, A., Uhfelder, E., *Ann. Chem.* **371**:131, 1909.

Foldes, F. F., Davidson, G. M., Duncalf, D., Kuwabara, S., *Clin. Pharmacol. Ther.* **6**:328, 1965.

Fourneau, E., Bovet, D., *Arch. Exp. Pathol. Pharmakol.* **160**:53, 1933.

Koller, C., *Lancet ii*:990, 1884.

von Euler, H., Erdtman, H., *Ann. Chem.* **520**:1, 1935.

Wilstatter, R., Muller, W., *Chem. Ber.* **31**:2655, 1898.

Suggested Readings

Anonymous. Cocaine out of Control. *Emerg. Med.*, Sept. 1984, pp. 65–87.

Ashton, A. M., Stone, S. N., The Human Cost of Chronic Cocaine Use. *Medical Aspects of Human Sexuality*, Nov. 1984, pp. 122–130.

Van Dyke, C., Byck, R., Cocaine. *Sci. Amer.* **246**:128, 1982.

CHAPTER

14

Steroids

14.1. Introduction

Steroids are polycyclic hydrocarbons that can be viewed as fully saturated (i.e., perhydro) analogs of cyclopentanophenanthrene. The ring designations and numbering of the carbon skeleton are indicated earlier. In the majority of steroids the junctions between ring C/D and B/C are *trans* (i.e., the H atoms at 8 and 9), and the H at 14 and the CH_3 at 13 (itself numbered 18) are one above and one below the general plane of the ring system as shown later. Substituents above the ring systems are termed β (heavy line) and the ones below, α (dotted line). The juncture between rings A/B can be either *cis* or *trans*. By convention, the angular methyl groups (numbered 18 and 19) are written as being above the ring system (i.e., in the β configuration).

The traditional depiction of the steroid system as having all four rings, A,B,C, and D, coplanar or flat is, of course, convenient, but misleading. A more representative description is given next as two forms. One represents the 5 α-steroids where the A/B ring juncture is *trans* so that the 5-substituent, in this case a H, is α and therefore *trans* to the CH_3 bonded to C-10 (but is itself 19), while the other is the 5-β series, where the A/B geometry is *cis* and the 5-H therefore has the β-configuration.

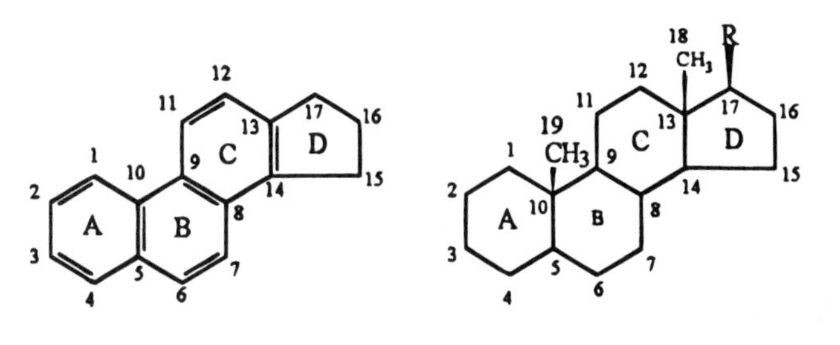

Cyclopentanophenanthrene The Streoid Hydrocarbon

In the naturally occurring steroids where the B/C juncture is invariably *trans*, both the B and C rings must assume the chair conformation. The nature of the *R* group at position 17 determines the "root" name (Fig. 14-1). In these planar representations groups that are *cis* to a substituent on C-10 (i.e., above the ring system) have a heavy line (β), whereas those that are *trans* to it (α) are shown by a dotted line. It should be noted that a double bond involving a carbon participating in a ring juncture, such as A/B, B/C, and so on, eliminates the spatial relationships at those points. The examples of cholesterol and the female sex hormone estradiol will illustrate the nomenclature of steroids shown with their absolute configurations.

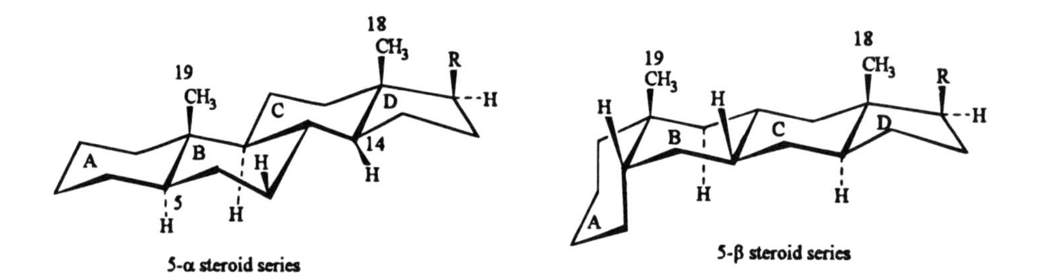

The A ring of the female sex hormone is aromatic; thus a C-10 substituent is not possible. However, for purposes of IUPAC nomenclature, since the ring system is in the estrane group (Fig. 14-1), the A ring is named as having three conjugated double bonds (a triene). The third double bond is between carbons 5 and 10 (not 5 and 6) and so must be indicated by the number 10 in parentheses.

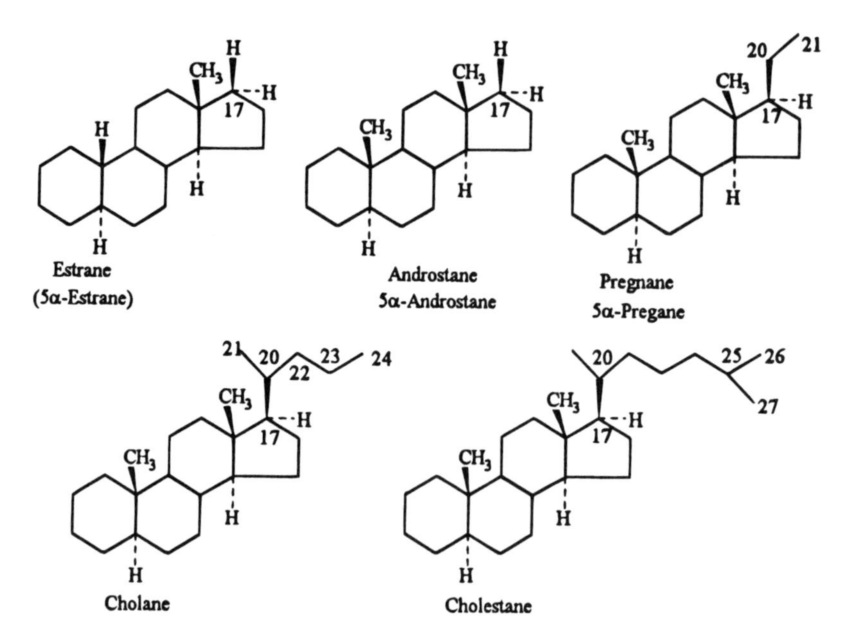

Figure 14-1. The main steroid nuclei.

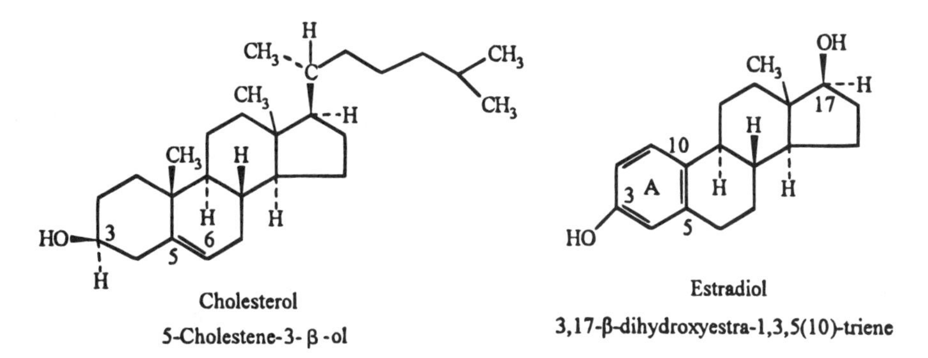

Cholesterol
5-Cholestene-3-β-ol

Estradiol
3,17-β-dihydroxyestra-1,3,5(10)-triene

Our interest in steroids is paramount from the biological viewpoint. Oxygenated derivatives are found in plants and animals that exhibit a fascinating array of effects in humans. These effects can be mimicked by using them, or synthetic variants, to influence physiologic events for therapeutic purposes.

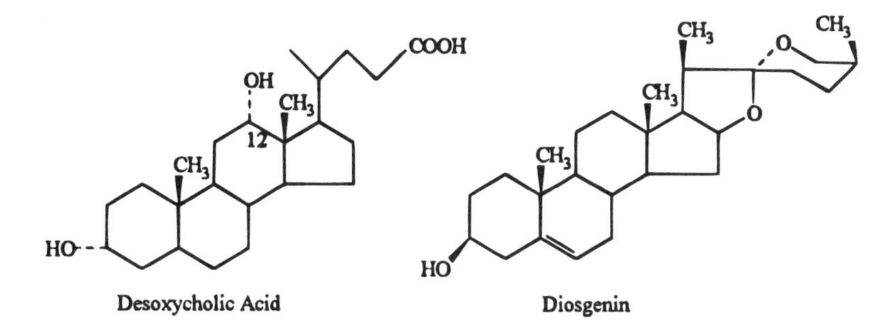

Desoxycholic Acid

Diosgenin

Alcoholic derivatives (3-OH) are found in plants (e.g., sitosterols, Chapter 11), mammals (e.g., cholesterol), and microorganisms. The cardiac glycosides (digitalis and related groups) were considered as cardiotonics (Chapter 10). The D vitamins contain the steroid ring system. Other steroids that are involved in mammalian physiology include the bile acids, the corticosteroid hormones produced by the adrenal glands, the male sex hormones produced by the testes, and the female hormones elaborated by the ovaries: estrogens and progesterone. All these hormones, and the many synthetic modifications developed, have medical applications. The pivotal role of cholesterol as the single biochemical precursor of the steroids involved in mammalian physiology cannot be overemphasized; the complexity of the biosynthetic interrelationships is critical to an understanding of that physiology—and pharmacology. The role of cholesterol and its metabolism in atherosclerotic disease has been considered (Chapter 11). An abbreviated sysnthesis of cholesterol including the last several steps are outlined in Chapters 11 and 7, respectively (Figs. 11-7 and 7-13).

14.2. The Steroid Hormones of the Adrenal Cortex

The adrenal cortex produces at least 50 adrenocortical hormones.[1] Cholesterol may be considered as the "starting point." Some cholesterol is synthesized in the cortex as well as the

---

[1] The adrenal medulla produces the sympathetic hormones (e.g., epinephrine).

liver from acetate. Exogenous cholesterol also enters the system from dietary sources and probably represents the major source for the biosynthesis of the various steroids. The majority of reactions are oxidations catalyzed by mixed function oxidases (vide infra).

Adrenal cortical hormones are controlled by the hypothalamus, which secretes corticotropin-releasing hormone (CRH) in response to stress situations. CRH, then, causes the anterior pituitary gland to release corticotropin into the circulation. Adrenal cortical surface receptors are then stimulated by corticotropin to produce and release the corticoid steroids.

In addition to the several adrenal cortical steroids whose biological effects relate to cortical function, some with androgenic (male sex hormonal) activity are also produced.

Corticosteroids are traditionally classified into three groups. The *glucocorticoids* (GCs) are those adrenal hormones involved in the control of glucose. Thus promotion of gluconeogenesis from amino acids is one of the functions of this type of steroid. Blood glucose is controlled by monitoring liver glycogen deposits and by opposing some of the actions of insulin. Diminishing peripheral glucose usage is another GC control mechanism; so is ketogenesis and utilizing fatty acids. From a therapeutic standpoint the considerable antiinflammatory and antiallergenic properties of GCs can be dramatic (vide infra). The most significant member of this group is cortisol (hydrocortisone). Corticosteroids such as cortisone and the semisynthetic prednisone (Table 14-1) are inactive until the 11-keto function is hepatically reduced to the 11-hydroxy group. Thus the predominant effect of GC steroids is on carbohydrate and protein metabolism.

The condition of excessive GC production is known as *Cushing's disease*, which is characterized by a typical "moon face" appearance. This results from loss of muscle mass and changes in body fat distribution.

Those steroids whose main function is the control of electrolyte balances are known as *mineralocorticoids* (MC). These compounds cause Na^+ retention and promote renal excretion of K^+. The purpose of MCs, then, is the maintenance of salt and water balance. The major mineralocorticoid hormone is aldosterone (Fig. 10-24, and relevant discussion). It should be pointed out that aldosterone has somewhat weak glucocorticoid effects and cortisol exhibits slight mineralocorticoid properties.

Corticosterone, which is the biosynthetic precursor to aldosterone (Fig. 14-2), is an intermediate that has weak properties of both previously mentioned groups. Note that all three have several features in common: the hydroxymethylketone group at C-17, the 11-OH function, and an α,β-unsaturated ketone system from C-3 through C-5.

The antiinflammatory property appears to reside in, or is parallel with, glucocorticoid activity. When synthetic analogs of cortisone are clinically used systemically, side effects such as Na^+ retention and edema can be related to the degree of mineralocorticoid activity that these compounds possess. Therefore, the aim of structural modifications of such synthetic steroids is to maximize the glucocorticoid component while minimizing the mineralocorticoid activity. The success in achieving this double goal simultaneously followed the realization that increasing the unsaturation of ring A by the addition of a second double bond (Δ^1) accomplished both. Thus, while the oral use of cortisone and cortisol can be accompanied by considerable water retention, the additional Δ^1 double bond (prednisone, Table 14-1) reduces this effect dramatically. The same is true with 6-α-methylprednisolone and dexamethasone. In the case of triamcinolone, edema is not demonstrable. The fact that these more unsaturated compounds are also much more potent, that is, equieffective at much lower doses (cortisone 20–70 mg, triamcinolone 4–12 mg daily), is undoubtedly a contributory factor as well. Increased potency of hydrocortisone was first achieved

Table 14-1. The Antiinflammatory Steroids

Structure	Name	Availability
21 CH_2OH, C=O, CH_3, O, OH, CH_3, 11, 17, 3, 4, 5, O 21, 17-dihydroxypregns-4 en-3, 11, 20-trione	Cortisone	Tablets
O, 21 CH_2OCCH_3 (C=O), CH_3, C=O, O, OH, 11, 17	Cortisone acetate	Tablets, suspension (sterile), ophthalmic ointment
21 CH_2OH, C=O, H, CH_3, HO, OH, 11, 17	Hydrocortisone (Hydrocortone)	Cream, ointment, lotion, enema, tablets, ophthalmic solution
O, 21 CH_2OCCH_3, C=O, H, CH_3, HO, OH, 11, 17	Hydrocortisone acetate (Cortef Acetate)	
21 CH_2OH, C=O, CH_3, O, OH, CH_3, 11, 17, 6, O 21, 17-dihydroxypregns-1,4-diene-3, 11, 20-trione	Prednisone	Tablets, syrup
21 CH_2OH, C=O, H, CH_3, HO, OH, 11, 17	Prednisolone and 21-Acetate	Tablets, ointment, cream, suspension
21 CH_2-O-P(ONa)$_2$ (P=O), C=O, H, CH_3, HO, OH, 11, 17	Prednisolone sodium phosphate (Hydeltrasol)	Injectable, ophthalmic, and otic ointment, suspension, and solution
21 CH_2-OCC(CH_3)$_3$ (C=O), C=O, H, CH_3, HO, OH, 11, 17	Prednisolone terbutate (Hydeltra TBA)	Injectable suspension (every 14–21 days duration)
21 CH_2OH, C=O, CH_3, HO, OH, CH_3, 17, 6, O, CH_3 11, 17, 21(11) trihydroxypregna-1, 4-diene-6-α-methyl-3, 20-dione	Methylprednisolone (Medrol)	Tablets

(*Continued*)

Table 14-1. *(Continued)*

Structure	Name	Availability
21 CH_2OCCH_3 (C=O, O, OH, 17)	Methylprednisolone acetate (Medrol Acetate)	Cream, ophthalmic ointment, enema, ointment, injectable suspension
21 $CH_2OC(CH_2)_2COO^- Na^+$ (C=O, OH, 17)	Methylprednisolone sodium succinate (Solu-Medrol)	Injectable solution
21 CH_2OH (C=O, CH_3, HO, OH, OH, 17, 16, CH_3, F, O) 9-Fluoro-11, 16, 17, 21-tetrahydroxy-(11,16)-pregna-1, 4-diene-3, 20 dione	Triamcinolone (Aristocort, Kenacort)	Tablets, ointment, cream
21 CH_2OH (C=O, O, O, CH_3, CH_3, 17, 16)	Triamcinolone acetonide (Aristoderm)	Cream, ointment, lotion, areosol, injectable
21 CH_2O-CCH_3 (C=O, O, OH, 17, 16, O-CCH_3, O)	Triamcinolone acetate	Injectable
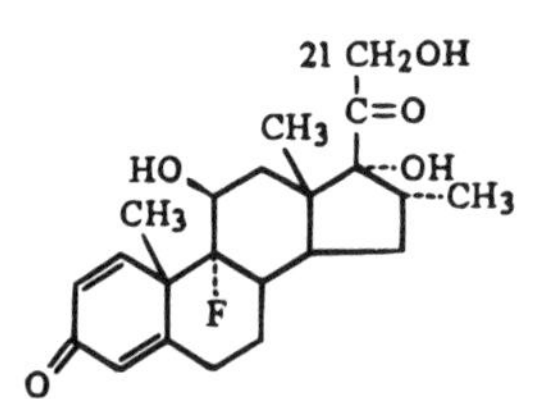 9-Fluoro-11 β,17,20-trihydroxy-16-α methylpregna-1,4-diene-3,20-dione	Dexamethasone (Decadron)	Tablets, elixir, cream, gel aerosol, ophthalmic susp., enema
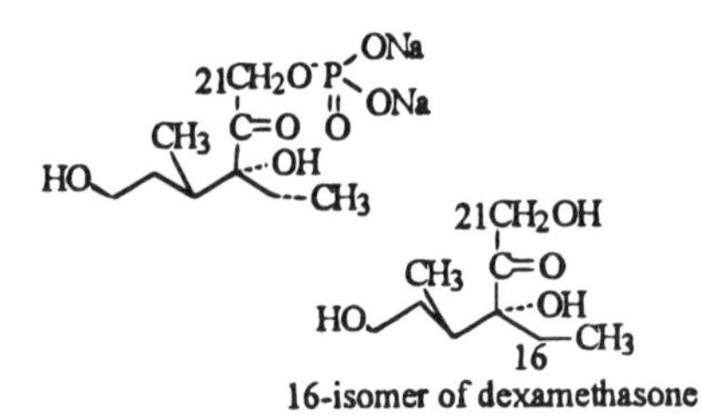	Dexamethasone sodium phosphate (Decadron phosphate)	Cream, opthalmic cream, ointment, and suspension, and injectable
16-isomer of dexamethasone	Betamethasone (Celestone)	Tablets, suspension

(Continued)

Table 14-1. *(Continued)*

Structure	Name	Availability
21 CH_2O $\overset{\parallel}{C}CH_3$ CH_3 C=O O HO ···OH 16 CH_3	Betamethasone acetate	Injectable suspension
21 CH_2O P ONa ONa C=O O CH_3 ···OH HO CH_3	Betamethasone sodium phosphate	Injectable suspension
21 CH_2OH C=O CH_3 HO ···$OC(CH_2)_3CH_3$ 17 O	Betamethasone valerate (Valisone)	Aerosol, ointment, cream, and lotion
21 CH_2 $OCCH_2CH_3$ CH_3 C=O O HO ···OCC_2CH_3 O	Betamethasone dipro-prionate (Diprosone)	Same as valerate
21 CH_2OH C=O CH_3 ···O ···O CH_3 CH_3 HO CH_3 F O F 6, 9-Difluoro-11β, 16α, 17, 21-tetrahydroxypregna-1, 4-diene-3, 20-dione-16, 17-cyclic acetonide	Fluocinolone acetonide (Synalar)	Cream, ointment, and topical solutions
21 CH_2O CCH_3 CH_3 C=O O HO ···O ···O CH_3 CH_3	Fluocinonide (Lidex)	Cream, ointment, and gel
O 21 CH_2-$OC(CH_3)_3$ C=O CH_3 ···OH ···CH_3 HO CH_3 F O F 6, 9, Difluoro-11, 17, 21-trihydrox-16-methyl-pregna-1, 4-diene-3, 20-dione-21-privalate	Flumethasone pivalate (Locacorten)	Cream

(Continued)

Table 14-1. *(Continued)*

Structure	Name	Availability
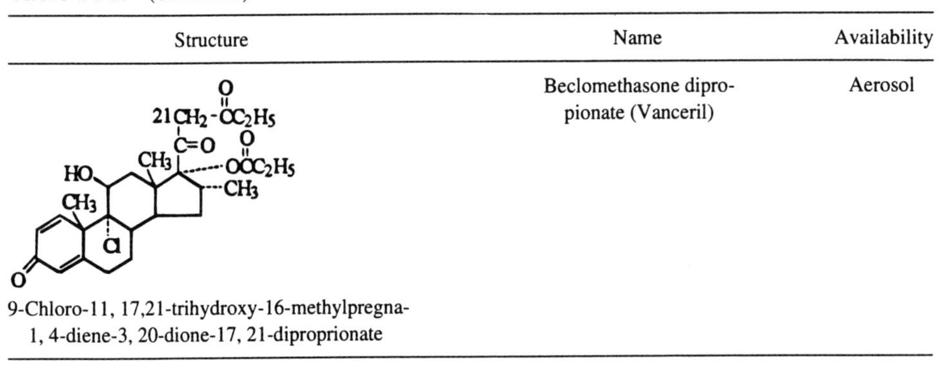 9-Chloro-11, 17,21-trihydroxy-16-methylpregna-1, 4-diene-3, 20-dione-17, 21-diproprionate	Beclomethasone dipropionate (Vanceril)	Aerosol

by the addition of a 9 α-F atom; it increased cortisol's potency tenfold. When further substituted by F (or Cl) atoms at positions 6, 9, and 16, and in some cases a CH_3 at C-6, the prednisolone molecule produced steroids having potency increases as high as 1000-fold (e.g., 6, 9β-diF, 16β-Cl). While it reduces MC effect modestly—ca. 20% when comparing

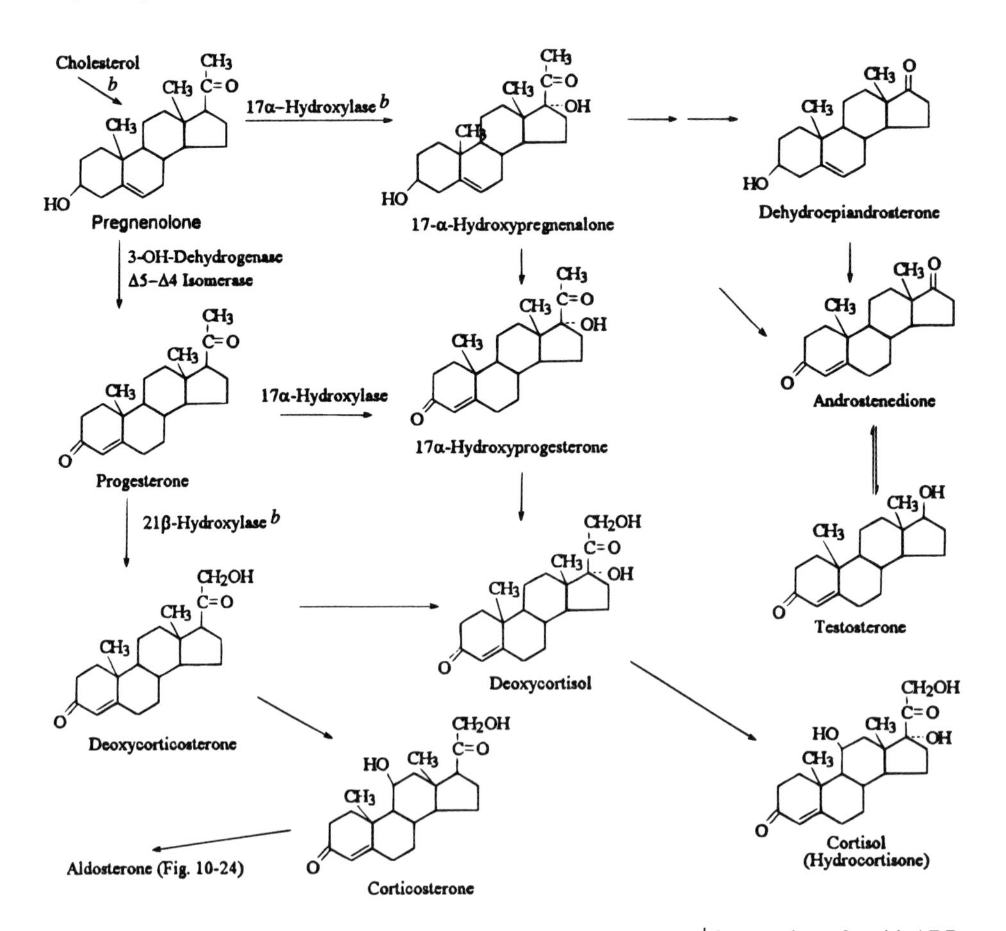

Figure 14-2. Corticosterol biosynthesis major pathways. [a]Abbreviated. [b]Oxygen from O_2 with ADP.

cortisol with prednisolone—the additional unsaturation of Ring A while increases the desirable antiinflammatory activity by 400%.

14.2.1. Cortisone

Addison recognized in the 1850s that the adrenal cortex has life-maintaining functions and characterized the disease (now carrying his name) resulting from a malfunctioning cortex. It was not until the 1930s that cortical extracts were used in patients. During that decade, cortical hormones began to be isolated as crystalline compounds and some of their structures established. The major hindrance to clinical research was the unavailability of the individual compounds in meaningful quantities. One ton of bovine adrenal glands (obtained from about 10,000 cattle) yielded less than 1,000 mg of GC steroids (frequently less than 100 mg). Most of these precious quantities were utilized for structure determinations. The advent of World War II led to intensive efforts to develop synthetic routes to cortisone from readily available compounds.

One such synthesis began with desoxycholic acid (from ox bile), which was laboriously converted to cortisone in 32 steps, many of which were required to transpose the oxygen function from the 12 to the 11 position, to introduce the 17 α-hydroxyl and to construct the α,β-keto arrangement in ring A. Over a 3-year period (1946–1949) some 1,200 pounds of the bile acid laboriously yielded less than 1 kg of cortisone. Workers at the Mayo Clinic (Hensch et al., 1949) reported on their dramatic success in controlling rheumatoid arthritis pain with cortisone. This gave the impetus to find a better way to provide cortisone than by the 32-step synthesis from desoxycholic acid or the laborious extraction from ox adrenals.

The solution came, in part, by the application of biotransformations. These can be defined as biological processes that modify organic compounds via simple chemical reactions (oxidations, reductions) by means of enzymes contained in microbial, plant, or even animal cells. The aim is usually a one-step reaction to a recoverable product in a sequence of steps in which the majority of conversions are chemical steps (i.e., synthesis). In fermentation processes the whole sequence of reactions is carried out by microorganisms, be it a carbohydrate breakdown to alcohol (or other solvents), the production of antibiotics, or even enzymes.

In practice biotransformations are carried out by microorganisms. They offer efficient transformation of substrates both as to yields, and often, more importantly, stereospecificity. Today, the selection of enzymes available through microbial use is vast.

Some early attempts to effect changes by perfusion of animal organs or cell cultures were useful but essentially complex laboratory-scale procedures.

Following Pasteur's work in the 1860s, which elucidated some roles of microbes in acetification (wine turning to vinegar), interest in microbial transformations slowly developed. By the end of the century several processes were in use. Even resolution of some racemates to obtain (at least one) optically pure enantiomer were viable procedures (e.g., lactic and mandelic acids). An observation that yeast (in fermentations) is capable of reducing 17-ketosteroids to 17β-hydroxysteroids in the late 1930s ultimately led to a search for other microbially mediated chemical transformations.

In 1952 workers at the Upjohn Company discovered an efficient one-step 11α-hydroxylation by the use of the mold *Rhizopus arrhizuz*. This reaction became crucial in the commercial production of various adrenocortical hormones and thereby synthetic analogs as well. Large-scale synthesis of other steroidal drugs have also become possible by the discovery of other organisms capable of highly (stereo) specific transformations. Table 14-2

Table 14-2. Microbiological Steroid Transformations

Reactions	Typical organism	
I. Oxidations		
a. hydroxylation	*Rhizopus arrhizius* or *Nigricans*	11 α
Almost all positions either α or β[a]	*Cunninghamella blakesleena*	11 β
	Streptomyces argenteolus	16 α
b. Dehydrogenation	*Streptomyces lavendulae*	Δ^1
(double bond formation)	and *S. fradiae, Fusarium solani*	
at $\Delta^{1,4,7,9,14,16}$	and *F. Caucasicum*	
c. Epoxidations		
d. Alcohols → ketone at		
3β-OH, 11β-OH, 17β-OH		
II. Reductions		
a. Double bond $\Delta^{1,4,5}$		
b. Acid → aldehyde		
c. Ketone → alcohol		
III. Hydrolysis		
a. Ester → alcohol		
b. Epoxide → diol		
IV. Double bond isomerization		
e.g., $\Delta^5 \rightarrow \Delta^4$		
V. Resolution of racemates		
VI. Other reactions[b]		

[a] Using different organisms.
[b] Such as aminations, esterifications, halogenations, and several rearrangement reactions.

illustrates just some examples of the types of reactions that are being carried out microbially on the steroidal system. Yields are frequently in the 80–90% range.

The availability of the saponin glycoside diosgenin from the Mexican yam (*Dioscorea composita*), which has the necessary 3-hydroxy function, enabled production of progesterone (Fig. 14-3) in relatively few steps (5). When combined with the 90% efficient 11α-hydroxylation by *R. arrhizus*, thereby affording the 11-α hydroxyprogesterone when followed by seven (now five) chemical steps, this produces cortisone[2] at the early 1950s price of $3.50/g, while the previous totally synthetic 32-step method cost over $100/g (equivalent to about $1,000 today). Figure 14-3 outlines these steps. Today several thousand tons of diosgenin are used as a substrate for steroid hormone manufacture.

It will be noted that stigmasterol also represents a commercial route to progesterone. This abundant plant sterol that is readily available from soybean with its C_{22}–C_{23} double bond, easily converts to progesterone in four steps and has therefore become a preferred substrate for various steroidal drugs. Cholesterol itself, which is the natural substrate for all mammalian steroid hormones, and is readily available from various sources (e.g., wool grease), is now also convertible to desirable steroid intermediates utilizing biotransformation chemistry. The same is true of β-sitosterol (Chapter 11), which is another plant sterol available from soybean.

The discovery of mutant Mycobacterium species that convert β-sitosterol (or cholesterol and stigmasterol) to either androstenedione or androstadienedione in one step made these sterols ideal substrates for conversion to estrone and spironolactone (Fig. 14-4).

[2] Four steps to hydrocortisone.

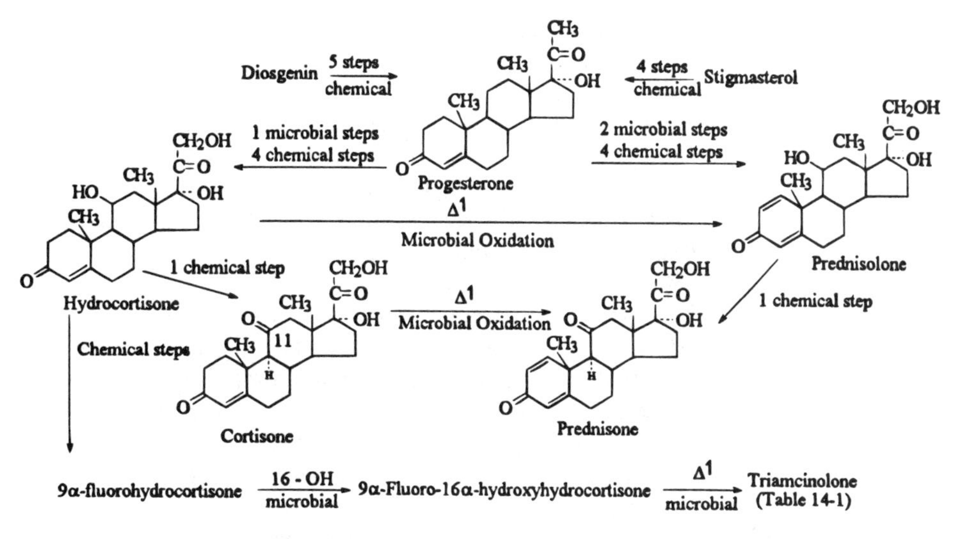

Figure 14-3. Routes to some steroidal drugs.

14.2.2. Glucocorticoid Steroids

Clinical use of corticosteroids, or more correctly those with mostly glucocorticoid activity, encompass many disease states. The obvious one of endocrine origin is Addison's disease. Since diseases such as tuberculosis, which is destructive to the adrenal glands, are not as prevalent as before the antibiotic era, the morbidity of Addison's disease has declined dramatically. The cases most often encountered are idiopathic adrenocortical atrophy, frequently of unknown etiology. In any case, since the symptoms are due to glucocorticoid as well as mineralocorticoid deficiency, patients can be kept alive with mineralocorticoid replacement therapy. However, since any stress factor can cause a collapse, and food deficiency can produce fatal hypoglycemia, such crisis situations require rapid parenteral replacement glucocorticoids such as cortisone or hydrocortisone in the various forms listed in Table 14-1. Salt and fluids must also be given.

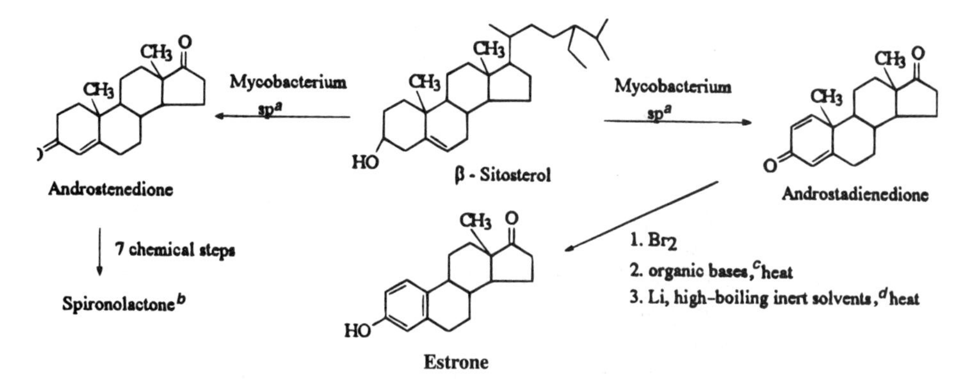

Figure 14-4. Plant sterol transformations. [a]Differing mutant species. [b]Chapter 10. [c]For example, collidines or quinoline. [d]For example, biphenyl.

Other conditions that may cause such primary adrenocortical insufficiency can be autoimmune diseases and cortex-destructive diseases other than tuberculosis, such as histoplasmosis and metastatic carcinoma.

Secondary and tertiary adrenocortical insufficiencies can arise by inadequate ACTH production, especially following surgery around the pituitary gland or actual pituitary disease. Deficiency can also occur following cessation of long-term glucocorticoid therapy. In addition, inherited enzyme deficiencies that lead to corticosteroid synthesis inhibitions occur (e.g., 21-hydroxylase and 3-β-hydroxylase deficiencies). All of these conditions require replacement therapy.

By far the larger utility of adrenal corticosteroids is in nonendocrine diseases. The rationale for their effectiveness is not always apparent. Thus antiinflammatory steroids find widespread use in autoimmune diseases such as rheumatoid arthritis and lupus erythematosus. Various allergic diseases including asthma, hay fever, even drug reactions, and many dermatological conditions such as atopic dermatitis and eczemas, poison ivy, and urticaria (itching), irrespective of etiology, also respond. Topical products of various potencies and efficacies are used.

Hematologic diseases are also treated with GCs. These can include idiopathic and acquired hemolytic anemia and thrombocytopenic leukemia. Gout, bursitis, pulmonary emphysema, and even adrenal hyperplasia may respond to steroid therapy. The reasons for efficacy, when it exists, are not always clear.

Among drugs useful in the treatment of leukemias and lymphomas corticosteroids are in the forefront (see also Chapter 4). The synthetic analogs prednisone as well as 6-α methylprednisolone and dexamethasone (Table 14-1) are superior to the natural GC steroid because of higher potency and decreased MC activity. The latter particularly minimizes electrolyte imbalances and retention of water. In acute lymphoblastic leukemia these GCs are considered major drugs in the combination drug protocols. Gratifying effectiveness (remissions) can also be achieved in chronic lymphocytic leukemia, Hodgkin's disease, malignant lymphomas, and multiple myeloma. Because of the ability of GCs to suppress adrenal pituitary function, metastatic breast cancer also responds to these drugs. The hypercalcemia that frequently accompanies breast cancer can be alleviated by GC therapy. Finally, the edema that results from brain tumors responds as well. Both effects are only palliative, but they are very helpful.

Additional indications for GCs include certain types of shock, hepatic diseases (e.g., hepatic necrosis), renal diseases (e.g., idiopathic nephrotic syndrome), and certain respiratory disorders (e.g., pulmonary sarcoidosis). It should be understood that in many of the conditions GCs are palliative at best and that in some instances their use may be controversial.

In summary, adrenal corticoid therapy is either replacement therapy (in cases of adrenal insufficiency, direct or indirect) or takes advantage of these drugs' antiinflammatory or immunosuppressant properties. In order to take advantage of these two properties, the doses must be in large excess of the body's daily endogenous output of hydrocortisone, which is about 20 mg (or its equivalent in the more potent synthetic analogs). Short-term therapy (less than 1 month), even in excessive doses, is not likely to produce serious adverse effects; prolonged use will. (Some of these are listed in Table 14-3.) One effect of high-dose prolonged use is suppression of the hypothalamic–pituitary–adrenal axis (HPAA). This suppression, by negative feedback, results in adrenal atrophy and pituitary function inhibition. Stopping the drug therapy does not lead to instant recovery of these functions. Recovery is actually slow and prolonged. In stressful situations following drug discontinuance the patient will not secrete normal hormone levels and can be in mortal

Table 14-3. Potential Adverse Effects of Glucocorticoid Steroid Therapy

1. Gastrointestinal effects due to
   a. decreased mucosal protection
   b. increased secretion of HCl
2. Edema due to
   a. sodium retention and potassium loss; little mineralocorticoid activity
   b. hypokalemia
3. Osteoporesis due to
   a. inhibition of osteoblastic formation
   b. inhibition of Ca^{2+} reabsorption
4. Gluconeogenesis
   a. insulin antagonism
   b. elevated serum triglycerides
5. Growth inhibition
   a. decreased dosage in children
6. CNS effects
   a. euphoria
   b. increased mental alertness
   c. irritability
   d. psychotic behavior
   e. manic–depressive (extreme cases)
7. Myopathy
8. Ocular effects
9. Miscellaneous
   a. "Moon face"
   b. acne
   c. hirsutism
   d. purpura (purple skin and mucous membrane patches)

danger from what is essentially an iatrogenically induced Cushing's syndrome. The patient, thus having become dependent on exogenous GCs, must be slowly withdrawn from the therapeutic steroid drug. Since steroids also suppress the immune system, the patient is susceptible to viral and other infections during both the therapy and withdrawal period. Both HPAA and immune suppression hazards can be somewhat minimized, but not eliminated, by such expedients as single daily or even alternate day dosing.

14.2.3. Antiinflammatory GCs

Because of the highly desirable antiinflammatory properties of cortisone and hydrocortisone, the search for more potent analogs began immediately after their introduction. Table 14-4 lists some of the empirically derived SARs that affect glucocorticoid, and therefore antiinflammatory, activity. It is interesting that an oxygen function at C-11 is not an absolute requirement for antiinflammatory action if other activating features exist in the molecule. Thus, as shown in compound A, activity is retained in the absence of an 11-oxygen atom, because of the $\Delta^{1,2}$ double bond and 6 α-CH_3 group, coupled with the 16 α-OH's reduction of the mineralocorticoid activity. When a C-11 hydroxy is present, however, its stereochemistry strongly affects activity. A dramatic example is that of the hydrocortisone epimer, in which the 11-OH function has the α-configuration. It is inactive, at least as far as its glycogen deposition ability is concerned.

In the treatment of localized inflammatory conditions, high-potency compounds are desirable, as is the localization of their effects to the topical areas of application. Thus the

Table 14-4. Some Empirical SARs of Glucocorticoid Steroids[a]

Substituent	Activity[b]	Example
11-O	glucocorticoid	Hydrocortisone
11-deoxy	mineralocorticoid[c]	11-Deoxycorticosterone
11-keto	inactive[d]	Cortisone, Prednisone
$\Delta^{1,2}$	>>> glucocorticoid > duration	Prednisone
9α-F	>> glucocorticoid > mineralocorticoid	9-Fluorohydrocortisone
16α-OH or αCH_3	< mineralcorticoid glucocorticoid unchanged	Triamcinolone Dexamethasone
6αF or αCH_3	> glucocorticoid, but less than at C-9	Methylprednisolone Fluprednisolone
21-O ester	>> topical effectiveness	Prednisolone acetate Flumethasone-21-trimethyl acetate
17-O-ester	>> topical effectiveness	Betamethasone valerate
17-21-O-diester	>> topical effectiveness	Betamethasone diproprionate
16α, 17α ketals[e]	>> topical effectiveness	Triamcinolone acetonide
17α-21-acetonide	>> topical effectiveness	9α, 11β-dichloro-16-β-methyl 17α, 21-dihydroxypregna--1,4-diene-3,20-dione-17,20-cyclic acetonide[f]

[a] Numbered positions refer to pregnane steroid.
[b] Primarily.
[c] See text.
[d] Topically in vivo hepatic reduction to 11-OH activates compounds.
[e] $R = CH_3$, acetonide or dioxolane.
[f] Not in clinical use.

goal with topical corticosteroids is to use drugs that exhibit little or no percutaneous absorption, thereby avoiding the adverse effects of systemic steroids.

Topical applications of hydrocortisone and triamcinolone can be shown to lead to systemic absorption. Large amounts of the more potent analogs produce systemic effects.

The skin consists of six layers. The stratum corneum (upper layer) with its keratin surface must, of course, be penetrated by a steroid if it is to be effective. Since this is a relatively loosely packed layer, its penetration by the drug is easily achieved provided the molecule has sufficient lipid solubility to permeate the lower portion of the stratum, which does constitute a barrier, as well. The lowest layer, called the *dermis*, is also a lipoidal partition. Once the steroid enters through this obstruction, it is in the systemic circulation. The problem in drug design, therefore, becomes one of facilitating steroidal entry into the epidermis, while at the same time minimizing rapid migration through the intermediate layers and the drug's penetration through the final dermal obstacle. A balancing act is needed.

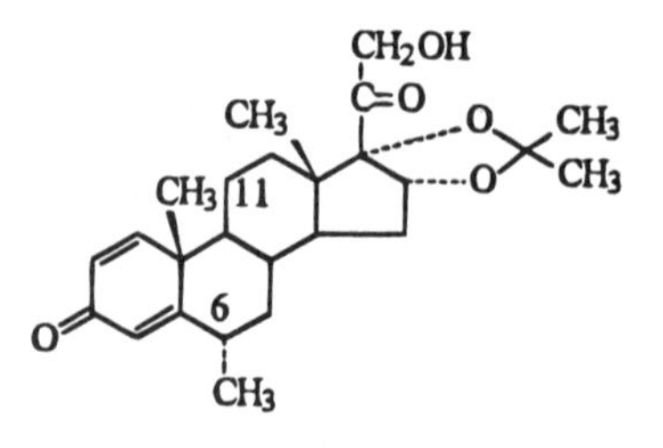

A

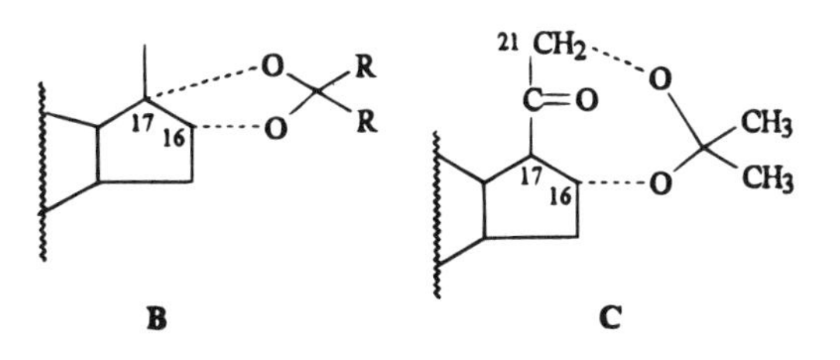

The lipoidal characteristic of the GC steroids needed to facilitate epidermal entry can be achieved by various derivatizations of OH functions at C-16, C-17, and C-21, specifically esters and ketals (structures B and C). Derivatization of even one hydroxyl can be useful. Table 14-4 gives examples of the type and positions of derivatives that impart increased lipoidal characteristics to steroids to enhance their local antiinflammatory activity.

Dioxalones, which are formed from diols that are usually *vicinal* (but can be farther apart) with ketones, represent cyclic ketals. If the ketone is acetone, then they are called *acetonides*. The reaction is carried out under anhydrous conditions with acid catalysis (Eq. 14.1). Ketals are quite stable to alkali, but they are readily hydrolyzed back to the parent diol and ketone by dilute acid. Even though a steroidal acetonide such as triamcinolone acetonide is probably more stable toward acid than a low-molecular ketal, it is likely that the more lipid-soluble acetonide (which is quite soluble in $CHCl_3$) will be slowly hydrolyzed back to the diol (which is not soluble in chloroform) in the slightly acidic environment of the skin (pH 6). The freed, but more polar triamcinolone, its movement through the skin impeded by hydrogen bonding to water, is now less likely to cross the dermis barrier. What one observes, then, is triamcinolone acetonide as a compound with prolonged topical activity and a 5–10 times higher antiinflammatory potency than the parent drug. In fact, triamcinolone itself is rather ineffective. Both compounds, however, exhibit equal potency systemically.

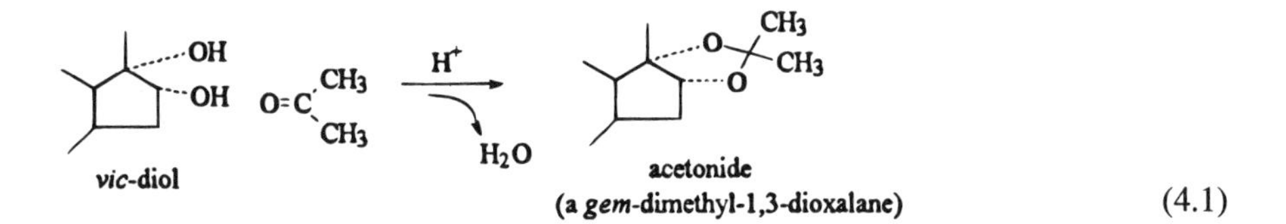

(4.1)

It should be mentioned that product formulation can also influence percutaneous absorption. An ointment vehicle with a high proportion of oils and fats is more likely to ensure skin penetration. A hydrophilic cream base is favorable to the absorption of more hydrophilic compounds. Skin hydration, which is achievable by the incorporation of humectants (e.g., glycerol) or by occluding dressings, can also be helpful. Particle size of substances that are not readily solubilized in the vehicle used must also be considered in developing the dermatological product.

In the case of ester derivatives the degree of esterification (i.e., mono or diesters) and the "size" of the esterifying acid helps to determine the extent of lipophilicity of the product. Thus betamethasone 17-valerate (Table 14-1) would be expected to be more soluble in $CHCl_3$ (1:2) than the 21-acetate (1:16), which in turn is more lipid soluble than the alcohol (1:325). Topical antiinflammatory potency parallels such values.

Branching of the esterifying acid, such as in flumethasone pivalate (trimethyl acetate) may, because of steric hindrance, slow down the rate of hydrolysis, thereby increasing effectiveness. Fluocinolone acetonide (Table 14-1) incorporates all these desirable features. Its topical potency is exemplified by its 0.01% concentration in commercial products (compared with hydrocortisone, 1%).

The replacement of systemic corticosteroid therapy and its attendant side effects in asthmatic and allergy patients with locally acting corticosteroids by way of aerosol inhalation devices is a gratifying success story. Even though dexamethasone phosphate was initially used for brief periods, the development of systemic effects such as adrenal suppression offered no advantage over oral therapy. Beclomethasone diproprionate showed itself to be hundreds of times more potent, and it produced no systemic symptoms even even on prolonged use with 400 μg daily doses. Some systemic symptoms developed at three to four times higher doses.

Intraarticular injections of insoluble corticosteroid depot suspensions directly into painful inflamed joints can give dramatic relief to arthritic patients that can last 3–4 weeks. This can be viewed as localized therapy. Microcrystalline suspensions of 6-methylprednisone acetate, betamethasone diproprionate, and dexamethasone 21-pivalate are among the effective agents.

The treatment of ocular inflammations due to allergy, following eye surgery or other causes, also respond well to topical (ocular) therapy, thereby avoiding much of the systemic toxicity. It is interesting that the lower-potency hydrocortisone (acetate) suspension is less likely to precipitate glaucoma symptoms than the more potent dexa- and betamethasone, which tend to raise intraocular pressure.

14.3. Mechanism of Action

Adrenal cortical hormones readily pass through the cell membranes of target tissue into the cytosol, where they interact with and bind to specific intracellular receptor proteins. It appears that there are high-affinity cytoplasmic receptors for the various physiologic steroids.

The receptor mechanisms discussed in regard to other drug categories thus far involved natural ligands and antagonists that act by initially binding to membrane-bound receptors. Polypeptide and catecholamine hormones, or neurotransmitters, did not need to enter cells. They control metabolic events via the secondary messengers such as *c*-AMP, which is synthesized by adenyl cyclase and hydrolyzed back by phosphodiesterase.

In the case of insulin the effectors and second messenger are not known. It has been established that hormone deficiency effects can exist even in the presence of above-normal levels of hormone. The receptors appear to be resistant to the hormone. One likely reason for this is decreased receptor availability as well as lowered sensitivity by them. Several diseases are believed to involve the development of antibodies to membrane receptors, so they do not function normally toward their ligands. Among such autoimmune diseases are myasthenia gravis, multiple sclerosis, and possibly schizophrenia.

The first demonstration of intracellular receptors was the finding of target tissues in the uterus (Jensen and Jacobson, 1960). A generalized steroid–receptor hypothesis has evolved since then in which the putative intracellular cytoplasmic protein is activated. The specific steroid enters the cell as a free compound, but arrives at its target in a mostly protein-bound form. Following activation the complete steroid–receptor complex is *translocated* to the

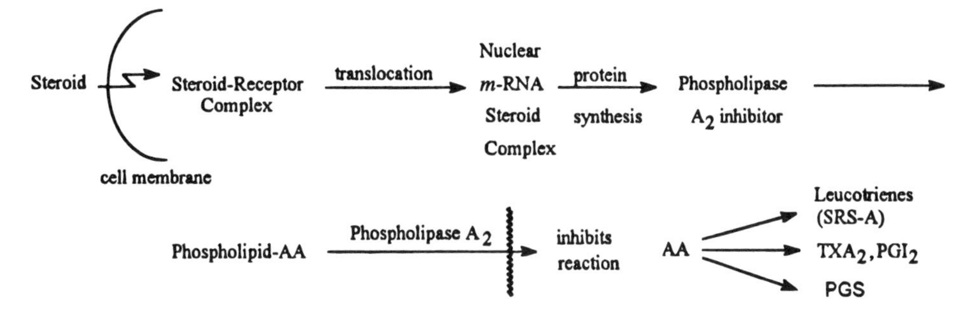

Figure 14-5. Mechanism of action of antiinflammatory steroids. AA = arachidonic acid.

nucleus of the cell, where it binds to its chromatin material and initiates *transcription*, leading to protein synthesis.[3] The most studied steroid receptor is that of estrogen, and this will be discussed later.

Receptor sites for glucocorticoids have been determined in the liver, hepatoma cells, breast carcinoma, lymphoid cells, and in the thymus.

It has been known for some time that antiinflammatory steroids, like NSAIDs, also inhibit prostaglandin synthesis. Cyclooxygenase inhibition, however, is not involved. Rather, they somehow prevent the release of arachidonic acid from its phospholipid ester linkage, a process normally efficiently catalyzed by phospholipase A_2. The decreased level of AA gives the cyclooxygenase-AA cascade less substrate to convert to PGs. It has been shown that GC steroids inhibit the release of TXA_2 by agents such as histamine and serotonin, but not AA, and do so at a potency related to their antiinflammatory activity. The mechanism, therefore, involves the inhibition of the enzyme phospholipase A_2. It was subsequently established that steroids accomplish this feat by inducing the cells to synthesize a phospholipase A_2 inhibitor. The picture that has emerged is summarized in Figure 14-5.

14.4. The Sex Hormones

Sex hormones can be classified into three categories: the female sex hormones, or *estrogens*, the hormones concerned with pregnancy, or *progestins*, and the male sex hormones, or *androgens*.

Estrogens. The biosynthetic relationship of the ovarian female sex hormones to androstenedione (their precursor), to cholesterol, and to each other is outlined in Figure 14-6. Estrone [3-hydroxy-1,3,5(10)-estratriene-17-one] was the first sex hormone to be isolated (Doisy et al., 1930; Butenand, 1930). Estradiol, which has been established as the true female sex hormone, is about 10 times more potent than estrone. It was not isolated for five years and took the extraction of 4 tons of sow ovaries to produce a little more than 10 mg of the hormone. Estrone was initially isolated from human pregnancy urine.

Estrogens, particularly estradiol secreted from the ovaries, promotes the development of secondary female characteristics during onset of puberty. They also stimulate mammary gland development during pregnancy. They have additional functions such as inducing heat, or estrus, in animals. It is curious that estrone has been isolated from several plant species including palm kernels.

---

[3] See Chapter 6 for protein synthesis overview.

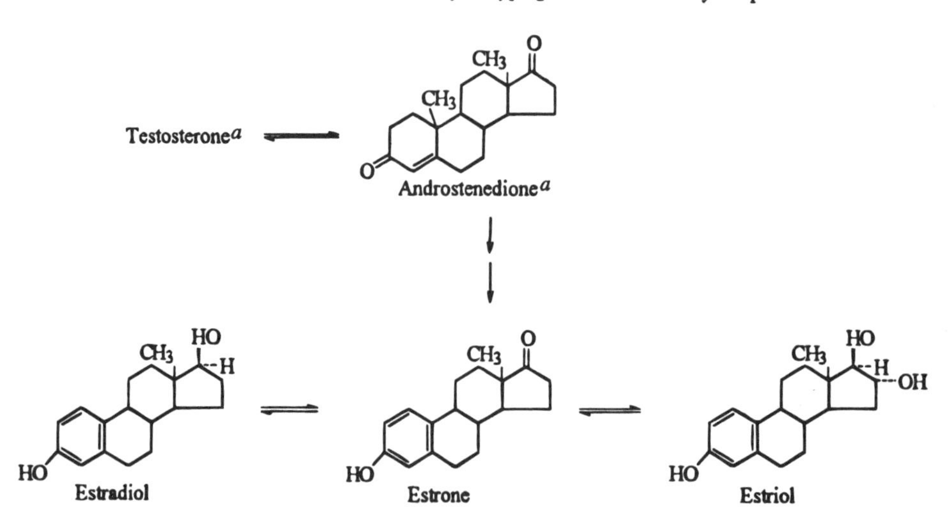

Figure 14-6. The natural estrogens—biochemical pathways. [a]See Figure 14-2 for sequence.

The steroidal estrogens are unique among the steroids considered in having an aromatic A ring. They are 19-nor steroids. Removal of this angular methyl group is preceded by its oxidation to an aldehyde or alcohol.

In the nongravid female primary estrogen production is in the ovaries with additional amounts from the stratum granulosum, corpus luteum, and graafian follicles. During pregnancy major estrogen synthesis shifts to the placenta and may increase 100 times over normal near parturition time. The adrenals and male testes also produce small quantities. Estradiol and estrone, which are the main products from ovaries, coexist in 1:2 or even 1:4 equilibrium, a dynamic state of enzyme catalyzed interconversion.

The varied clinical uses of estrogens include menstrual disorders, alleviation of menopause, associated symptoms (replacement therapy), osteoporosis, postmenopausal breast cancer and prostatic cancer and hypertrophy, lactation suppression, postcoital contraceptives, oral contraceptives (combined with progestins), estrogen deficiencies unrelated to the menopause, ovarian failure, and hypogenitalism.

Table 14-5 gives the natural estrogens and several of their derivatives and salts, synthetic steroidal analogs, and synthetic nonsteroidal agents, the latter depicted in a manner to indicate some structural similarities. It should be mentioned that disturbing reports in the 1970s indicated that prolonged postmenopausal estrogen use may be associated with endometrial carcinoma.

Compounds with *antiestrogenic* activity can include any substance that counteracts or even modifies estrogen activity. In the present context, however, it will be limited to drugs that bind to cytoplasmic estrogen receptors (ERs). The translocation of the ER–drug complex to the cell nucleus follows. Thus endogenous estrogens are precluded from their normal ER interactions since less "open" receptors are now available. Clomiphene is such a drug. Actually, it has weak estrogenic activity. It has apparent structural similarity to chlortrianisene (Table 14-5). Nevertheless, it binds to ERs thereby inhibiting potent estrogens. The normal feedback inhibitory control of estrogen synthesis is now interrupted. The result is increased production of gonadotropins that result in stimulation of the ovaries to ovulate.

Table 14-5. Estrogenic Drugs in Clinical Use[a]

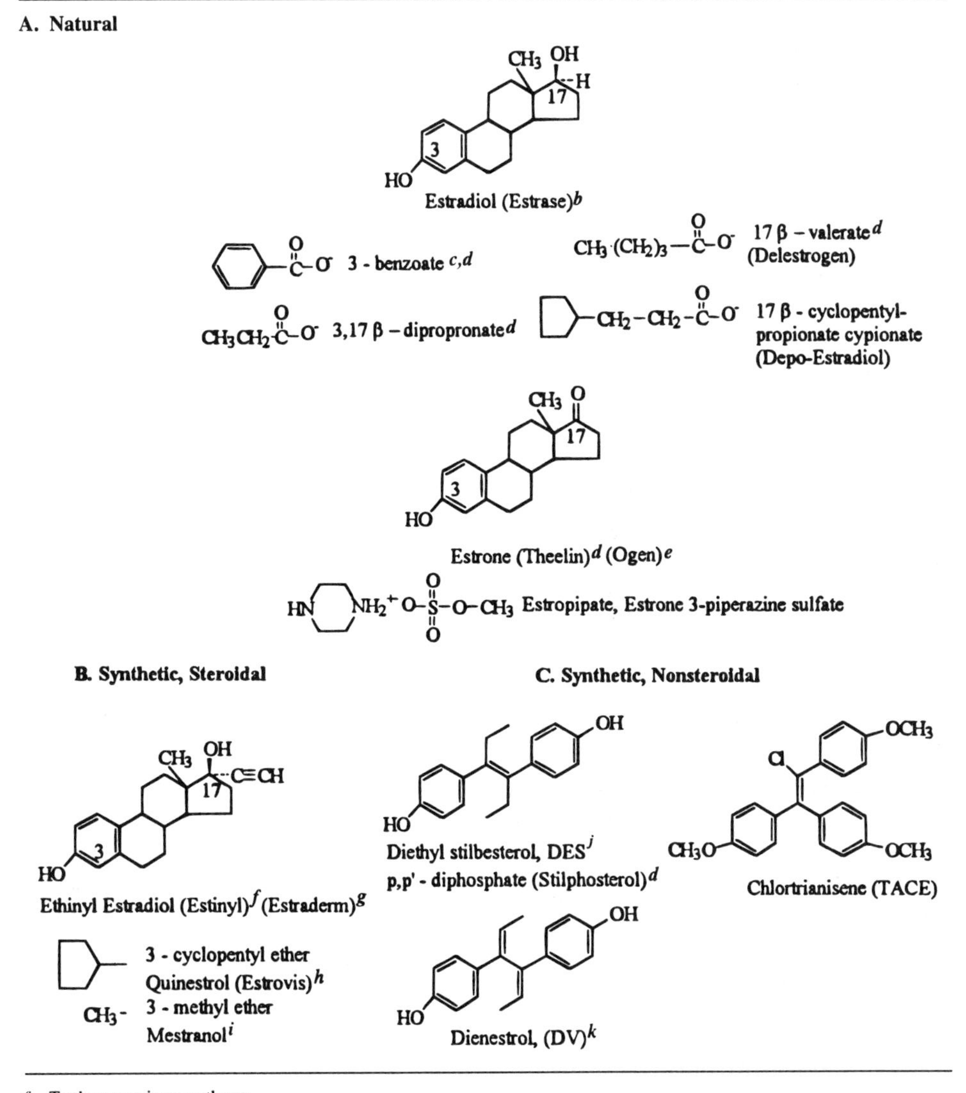

[a] Trade names in parentheses.
[b] May be commercially synthesized.
[c] Tablets of micronized drug.
[d] IM suspension.
[e] Also sodium sulfate ester.
[f] Tablets alone and in oral contraceptives.
[g] Transdermal products.
[h] Pro-drug.
[i] Oral contraceptives.
[j] Tablets and vaginal cream.
[k] Vaginal cream only.

This effect has become the basis of treating infertility in women. In fact, overstimulation of the ovaries occurs in about 8 % of cases that result in multiple births.[4]

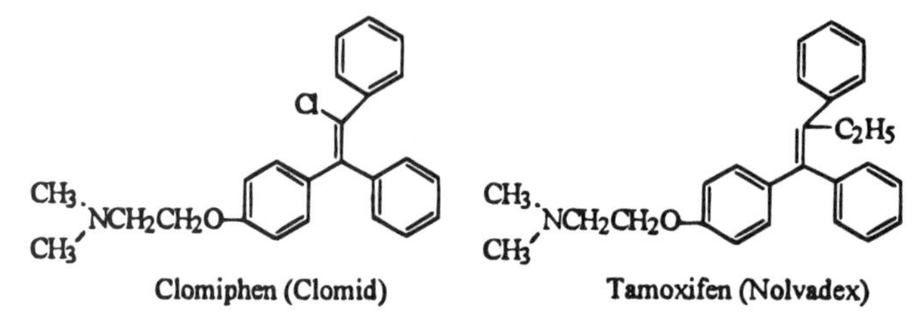

Clomiphen (Clomid) **Tamoxifen (Nolvadex)**

Estrogen receptors have now been established in the uterus, anterior pituitary, hypothalamus, fallopian tubes, vagina, and—of significance to cancer chemotherapy—in breast and breast carcinoma as well as in endometrial carcinoma. In the case of dependent breast cancer tissue, where estrogen enters and binds to the ERs, translocation to the nucleus promotes tumor growth. Oophorectomy (ovary removal) has previously been sometimes successfully used to effect tumor regression. Antiestrogen drugs such as tamoxifen (Chapter 4) have now replaced such surgery. The synthesis of progesterone receptors (PgRs) in breast tissue is also estrogen controlled. Only breast tumors that have ERs (ER^+) are likely to respond to antiestrogen therapy, which explains failures of treatment. The development of detection tests and qualitative assay procedures for ER^+ tissue made predictability of successful therapy possible. In general ER^+ tumors are slow growing, nonaggressive, and tend to be in older women. Tamoxifen is likely to be useful here. ER lesions are mainly aggressive, rapidly growing, and tend to recur more readily. These types of tumors, which appear mostly in younger women, do not respond well to antiestrogen treatments.

14.5. Progesterone, Progestins, and Their Receptors

The steroid hormone progesterone (Table 14-6) is secreted by the ovary, particularly the corpus luteum. The placenta and adrenal glands also produce some progesterone. It changes the uterine endometrium in the second half of the menstrual cycle prior to implantation of the blastocyst.[5] The hormone's responsibility also includes development of the maternal placenta following implantation, and development of mammary glands. In addition to acting on target genital tissue and endocrine glands, cytosol receptor proteins have also been identified in the uterus and breast carcinoma. Furthermore, general systemic effects have also been recognized.

Under physiological conditions estrogens accompany progesterone and may well be synergistic with its actions. This has been referred to as *estrogen priming* of a target organ. During pregnancy progesterone levels are much higher than is normally the case. Since the hormone suppresses the contractility of uterine musculature, this may explain, in part at least, the hormone's importance in maintaining pregnancy, including the likelihood of suppressing an immunological rejection of the developing fetus.

Progesterone is not orally available (except in excessive quantities) because of rapid, hepatic "first-pass" metabolism. The parenteral oil-injectable dose-forms are used even though they are really suspensions of microcrystalline progesterone, the drug being only

---

[4] Sextuplet and septuplet births have been reported.

[5] An early stage in the development of the mammalian embryo.

Table 14-6. Progesterone and Related Progestins[a]

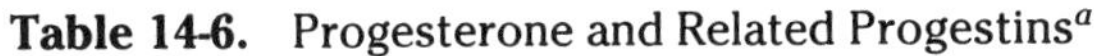

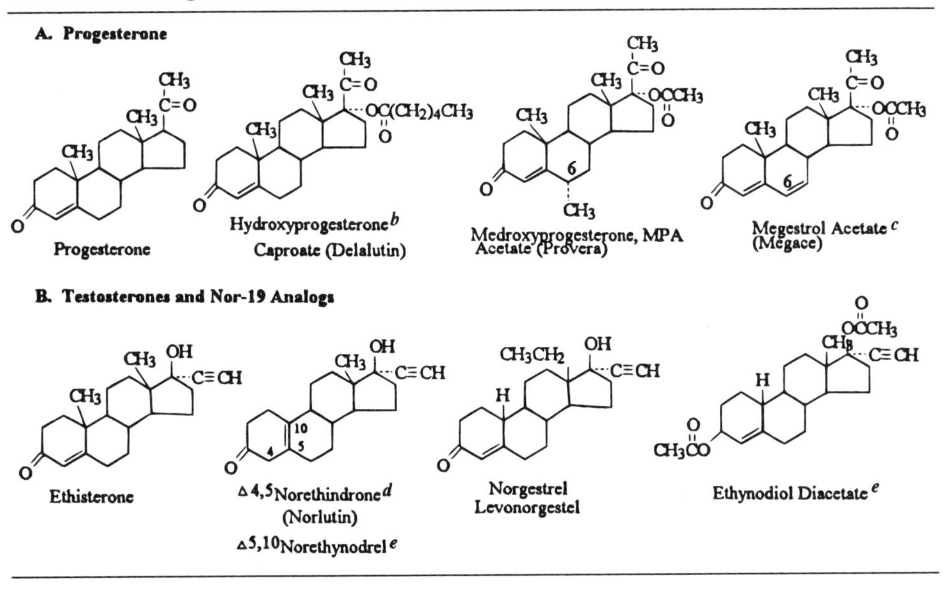

[a] Trade names in parentheses.
[b] Devoid of estrogen activity.
[c] Used primarily against breast cancer.
[d] Also available as acetate (Norlutate).
[e] Available only in combination with estrogen as oral contraceptive (see Table 14-7).

slightly soluble in oil. In addition, the drug is well absorbed from the buccal cavity and the vaginal mucosa, thereby avoiding the initial pass through the liver.

Progesterone has traditionally been indicated to treat habitual threatened abortion. However, its erratic absorption and rapid metabolism ($t_{1/2}$ = 5 min) does not allow effective blood levels to be achieved. Thus its usefulness for this application is questionable.

An innovative progesterone-containing device (Progestasert) has been made available as a long-lasting intrauterine contraceptive. It contains 38 mg of the drug in a silicone oil solution, which on proper insertion releases 63 μg of progesterone per 24 hours for periods up to 1 year. The contraceptive effects of the device itself (an intrauterine device, IUD) is believed to be enhanced by the hormone's local effect on the endometrium.

The term *progestin* has evolved to refer to synthetic analogs that have progestational effects. They have been developed to provide useful oral, as well as longer, duration of action. A generic term to denote all progestational compounds is *progestogens*.

Another important and more reliable use of progestogens is in the diagnosis and treatment of menstrual disorders. Additional therapeutic applications include endometriosis and endometrial carcinoma. A major agricultural use is in animal husbandry to synchronize the estrus in breeding cattle and sheep.

Table 14-6 depicts three derivatives of progesterone. Hydroxyprogesterone caproate is a long-acting caproic acid ester of 17α-hydroxyprogesterone. Administered parenterally, as a suspension, the effects of a single dose intramuscularly (375 mg) may persist for more than 2 weeks. In uterine carcinoma doses as high as 1 g have been used. Unlike other progestogens, this drug does not prevent ovulation. The acetate of the α-methyl analog of progesterone, medroxyprogesterone acetate (MPA) is orally active and can be used as a long-lasting IM suspension against endometrial carcinoma and orally to treat dysmenorrhea or premenstrual tension. Neither this drug nor the more potent megestrol acetate (which has an additional double bond at C-6 to C-7) possesses estrogenic activity, thus necessitating an estrogen priming dose.

Progestational activity can tolerate considerable molecular tinkering to "improve" the pharmacology. Even the change of the β-acetyl at C-17 to the α-configuration does not alter progestational activity. Enlarging the C-18-CH_3 group to an ethyl group (as in norgestrel) surprisingly increases this activity.

Several structural modifications of progesterone have resulted in desirable oral activity. One is the addition of a 17α-hydroxy function (usually esterified); a Δ^6dehydro feature is another. Megestrol acetate incorporates both successfully.

It is curious that the male androgen testosterone, by the addition of a 17 α-ethynyl group (-C≡CH), produced one of the earliest leads to such progestational compounds with the introduction of ethisterone (Table 14-6) in 1938. This drug, 17α-ethynyl-testosterone, retains some androgenic activity, but, surprisingly, exhibited oral progestational properties. Ethisterone metabolism even results in compounds with estrogenic effects.

The next logical approach was to investigate the 19-nor-testosterone structure (devoid of the CH_3 at C-10). Norethynodrel—$\Delta^{5(10)}$-17-ethynylnortesterone—was introduced by Pincus in 1958 as an orally active antifertility agent and became the starting point for the oral contraceptives to follow (*vide infra*). The 13-ethyl substituent (norgestrel) enhanced this activity.

19-Nortesterone itself, which lacks the β-acetyl group of progesterone, has a low binding affinity for the PrR. This affinity is greatly enhanced by the hydrophobic 17α-ethynyl side chain in norethisterone. It also protects the 17β-OH from rapid metabolic oxidation, a feature utilized in many other orally active progestogens (Table 14-6). It is now established that the norethisterone–PgR complex, like the progesterone–PR complex, produces progestational effects. It has been proposed that specific binding interactions between receptor molecule and the functional group of the steroid (3-keto-17-oxygen) of about 3 kcal each, when added to hydrophobic interactions from the steroidal skeleton itself, can account for the high binding energies seen (about 12 kcal total).

14.6. Oral Contraceptives (OC)

Oral contraceptives ("the pill") are usually combinations of an estrogen and progestins in very low doses taken on a daily basis (21 days of menstrual cycle) by women to prevent ovulation. Either type of drug alone is able to prevent conception by inhibiting ovulation. The primary physiological mechanisms are inhibition of follicle-stimulating hormone (FSH) by the estrogen and prevention of the release of luteinizing hormone (LH) by the progestin. In one view, these products "do persistently what nature does intermittently" (Albert, 1985). In the combination of oral contraceptive products available commercially two estrogens are in use: ethynyl estradiol or the 3-methyl ether, mestranol. The choice of progestin is somewhat larger (see Table 14-7). Products containing only a progestin are now also available.

14.7. Androgens

The interstitial or Leydig cells of the testis produce and elaborate the male hormones or androgens, with the principal one being *testosterone* (Table 14-8); lesser quantities are also

Table 14-7. Representative Oral Contraceptive Formulations

Estrogen (μg)	Progestin (μg)	Trade names
	Formulations with less than 50 μg of estrogen	
Ethinyl estradiol 35	Norethindrone 50	Brevicon, Modicon
Ethinyl estradiol 35	Norethindrone 40	Ovcon-35
Ethinyl estradiol 35	Norethindrone 100	Ortho Novum 1/35
Ethinyl estradiol 30	Norethindrone Ac 150	Loestrin 1.5/30
Ethinyl estradiol 20	Norethindrone Ac 100	Loestrin 1/20
Ethinyl estradiol 35	Ethylnol diAc 100	Demulen 1/35
Ethinyl estradiol 30	Norgestrel 30	Lo Ovral
Ethinyl estradiol 30	Levonorgestrel 15	Nordette
	Formulations with 50 μg of estrogen	
Ethinyl estradoil 50	Norethindrone 100	Ovcon-50
Ethinyl estradiol 50	Norethindrone Ac 100	Norlestrin 1/50
Ethinyl estradiol 50	Norgestrel 50	Ovral
Ethinyl estradiol 50	Ethynol DiAc 100	Demulen
Mestranol 50	Norethindrone 100	Norinyl 1 + 50
	Formulations with greater than 50 μg of estrogen	
Mestranol 100	Norethynodrel 250	Enovid E
Mestranol 100	Norethyndrone 200	Norinyl 2
Mestranol 80	Norethyndrone 100	Norinyl 1 + 80
Mestranol 100	Ethynodiol DiAc 100	Ovulen
	Formulations with progestin only (minipill)	
None	Norethindrone 35	Micronor
None	Norgestrel 7.5	Ovrette

produced in the adrenal cortex and ovaries (about 10% that of men). The release of the hormone brings about the maturation of the genital organs and initiates the development of the secondary sexual characteristics such as body hair, voice deepening, and the male pattern of muscular development. The latter effect, which is accomplished by the hormone's influence on nitrogen retention and protein anabolism, has led to the abuse of synthetic androgens with high anabolic activity by athletes to increase muscle mass.

An important use of androgens is in hypogonadism, which is a prepubertal deficiency of androgen production. Unfortunately, this deficiency is usually not recognized until delays in pubertal changes are noted. Another use is to affect positively the body's nitrogen balance and muscle development. Even though the results can be dramatic in hypergonadal men, whether therapeutic doses will exceed muscle development that normal testicular secretions can produce is not established. Attempts, by synthetic modifications, to develop compounds in which a total separation of androgenic and anabolic properties is achieved have not been successful, although variations in the relative ratios seems possible. Thus all these compounds in use, or that have been tested as anabolic agents, retain considerable androgenic properties. This does not mean, however, that they cannot be therapeutically useful for their anabolic activity for short-term catabolic states such as those following

Table 14-8. The Androgenic Steroids[a]

Structure	*R* group	Name
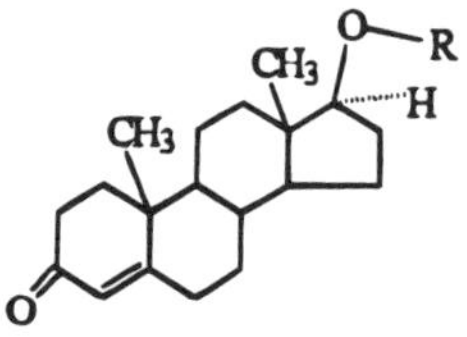	$\overset{\displaystyle O}{\|\|}CCH_3$	Testosterone acetate (various mfg)
	$\overset{\displaystyle O}{\|\|}CCH_2CH_3$	Testosterone propionate (Oreton propionate)
	$\overset{\displaystyle O}{\|\|}CCH_2CH_2$–cyclopentyl 	Testosterone cypionate (Various mfg)
17β-Hydroxyandrost-4-ene-3-one	$\overset{\displaystyle O}{\|\|}$C(CH$_2$)$_4CH_3$	Testosterone enanthate (Delatestryl)

Structure	*X*	*R*	Others	Name
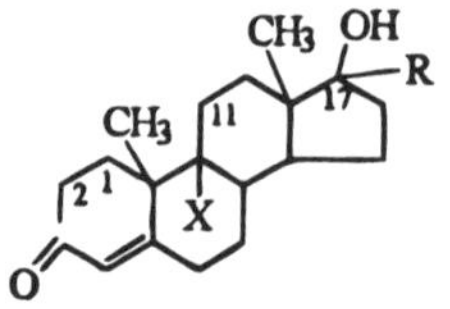	H	CH$_3$		Methyltestosterone (Methyl Oreton)
	H	H	2 = α-CH$_3$ 17-propionate	Dromostanolone propionate (Drolban)
	F	CH$_3$	11=OH	Fluoxymesterone (Halotestin)
	H	CH$_3$	Δ[1] = double bond	Methandrostenolone (Dianabol)
	H	CH$_3$	2 = CHOH	Oxymetholone (Anadrol)

Structure	*R* group	Name
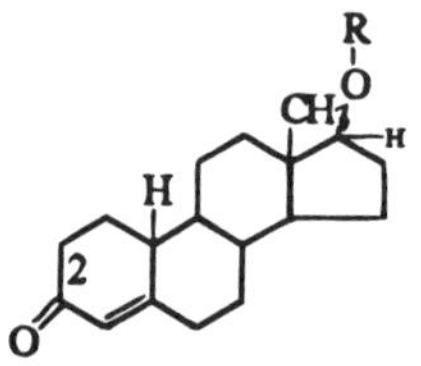	H	Nandrolone
	$\overset{\displaystyle O}{\|\|}$C(CH$_2$)$_8CH_3$	Nandrolone decanoate (Deca Durabolin)
	$\overset{\displaystyle O}{\|\|}CCH_2CH_2C_6H_5$	Nandrolone phenpropionate (Durabolin)

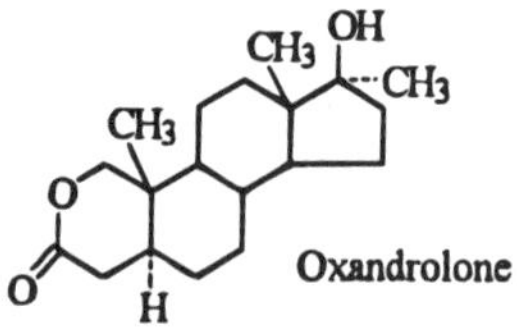

Oxandrolone (Anavar)

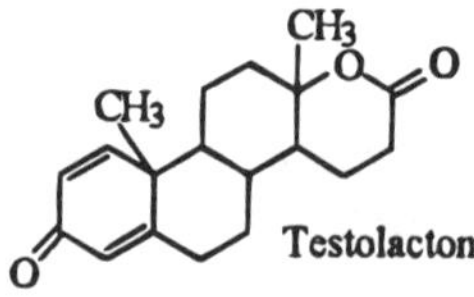

Testolactone (Teslac)

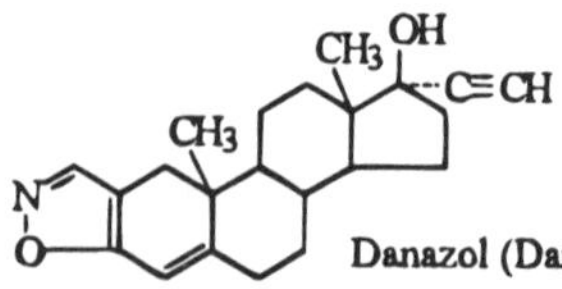

Danazol (Danocrine)

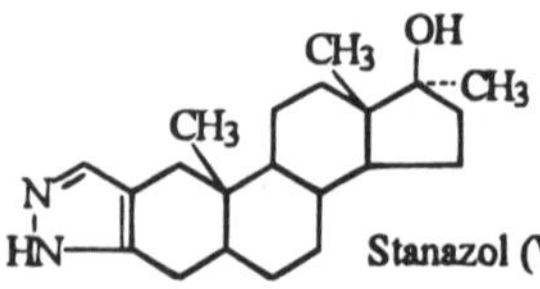

Stanazol (Winstrol)

[a] Trade names in parentheses.

injuries or surgery. Attempts to use androgens as a weight-gaining strategy in elderly debilitated and undernourished patients have not been consistently successful. Other applications have been to stimulate erythropoiesis in cases of anemias associated with renal or bone marrow failure. Hereditary angioneurotic edema seems to be treatable, especially with agents that have weak androgenic activity (e.g., danazol).

Androgens have been used to treat osteoporosis in postmenopausal women. It is unlikely that they are superior to estrogen therapy, which, of course, does not cause masculinization. The most useful indication in women would probably be prolonged androgen therapy of advanced metastatic breast cancer. It is tempting to assume that the reason for the activity might be an antiestrogenic effect, especially in ER^+ patients, yet, overall, androgens here probably are not superior to conventional therapy.

Testosterone itself, although it is absorbed on oral administration, is ineffective because of poor bioavailability (i.e., rapid "first-pass" hepatic metabolism). Even the $t_{1/2}$ of injected testosterone from an oil solution is only about 4–10 minutes. The several esters listed in Table 14-8—acetate, propionate, cypionate, and enanthate—on intramuscular injection of oil solutions, allows for infrequent doses (e.g., every 2 weeks) because of relatively slow absorption and subsequent hydrolysis of these pro-drugs to the parent compound—testosterone. In a practical sense, then, these esters are more active.

The synthetic compounds (Table 14-8) are to a certain degree successful in attempts to increase the $t_{1/2}$ and to decrease side effects such as masculinization (virilism). Even though "pure" anabolic steroids (i.e., devoid of androgenic properties) have not been developed, some of the synthetics at moderate doses may show less masculinization in women and children, and can therefore offer some advantage.

Since the animal testing methodology for determination of the anabolic–androgenic ratio leaves much to be desired and human results are so subjective, conflicting data are often encountered. Thus methandrostenolone has been variably listed as having a 1:1 ratio or being a "strong" anabolic and a "weak" androgen. In the case of testolactone, however, the fact that it is not a true steroid may be the reason it is said to be devoid of androgenic activity. Its use is mainly as a palliative in inoperable breast cancer.

Two drugs of interest from a chemical standpoint are danazol and stanazol (Table 14-8). Both contain an additional (heterocyclic) ring fused to ring A of the steroid skeleton, danazol (an isoxazole) and stanazol (a pyrazole ring). Danazol, which is a weak androgen, but without estrogenic and progestational activity, is useful in suppressing gonadal function, breast pain associated with fibrocytic breast conditions, and may relieve lupus symptoms in some women. Stanazol is claimed to have a good anabolic to androgenic ratio and is probably used primarily to promote nitrogen utilization, yet virilization in children and women does occur. Figure 14-7 outlines syntheses for both drugs.

The outline for the mechanism of action of androgens is analogous to those of the other steroids. They enter the cells of target tissues (e.g., prostate), form a complex with their specific protein, and bind, diffuse, or otherwise translocate to the nuclear chromatin. There, following a "freeing" of a repressed codon from its lock-on on a DNA segment, the necessary *m*-RNA will produce the requisite protein using the ribosomal machinery. It appears that although testosterone binds to this receptor protein in some target tissue to produce the characteristic activity, in most instances it must first be reduced (activated) by 5α-reductase to *dihydrotestosterone*, which then binds to its specific acidic androgen-dependent protein in the prostate gland. It should be pointed out that many of the "finer points" of the complex steroidal mechanisms are still not completely known.

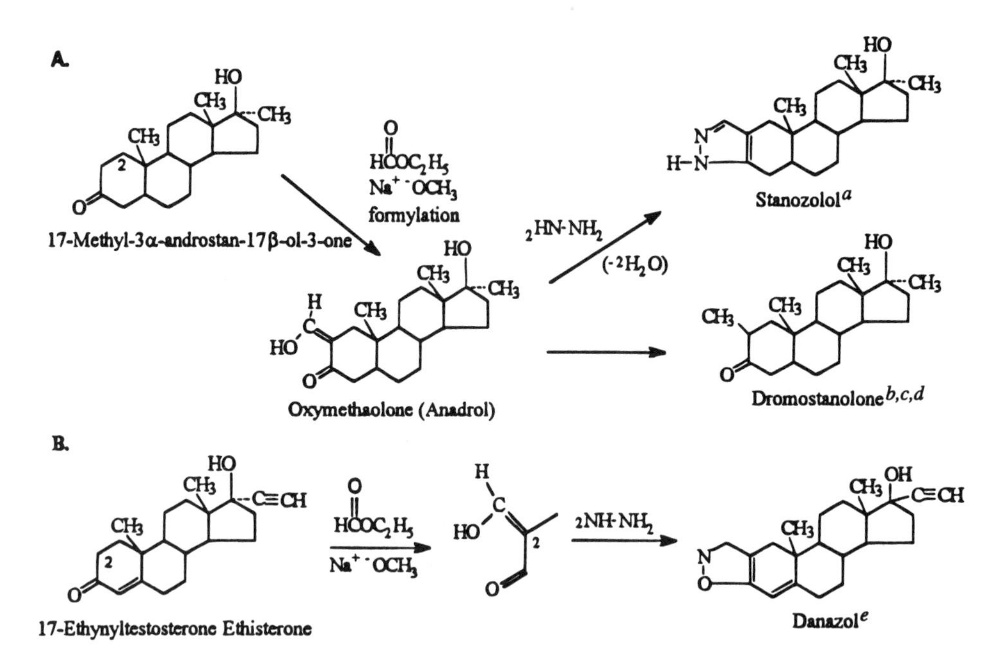

Figure 14-7. Syntheses of certain androgenic drugs. [a]Clinton et al., 1959. [b]Ringold et al., 1959. [c]Reduction affords a-Me, which is isomerized to b-Me. [d]Product is esterified with propionic anhydride. [e]Manson et al., 1963.

14.8. Antiandrogens

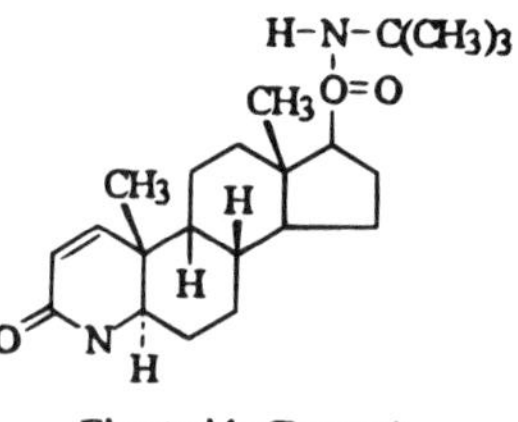

Finasteride (Proscar)

Observations that hermaphroditic children who ultimately turned male were genetically deficient in 5α-reductase led to the idea that competitive inhibitors of this enzyme would greatly decrease levels of dihydrotestosterone. This might in turn reduce the oversize gland in the common affliction of elderly men called benign prostatic hyperplasia (BPH), and avoid the usual therapy—surgery. The synthetic 4-azasteroid finasteride (Proscar) was introduced (1992) to treat BPH in a defined patient population.

References

Albert, A., *Selective Toxicity*, 7th Ed., Chapman and Hall, New York, 1985, p. 358.

Butenandt, A., *Z. Physiol. Chem.* **191**:127, 1930.

Clinton, R. O., Manson, A. J., Stovner, F. W., et al., *J. Am. Chem. Soc.* **81**:1513, 1959.

Doisy, E. A., Veler, C. D., Thayer, S., *J. Biol. Chem.* **86**:499, 1930.

Hensch, P. S., Kendall, F., Slocum, C. H., Follex, H. F., *Proc. Staff Mtg. Mayo Clin.* **24**:181, 1949.

Jensen, E. V., Jacobson, H. J., in *Biochemical Activities of Steroids in Relation to Cancer*, Pincus, G., Vollmer, E. P., Eds., Academic Press, New York, 1960, p. 161.

Manson, A. J., Stonner, F. W., Neumann, H. C., et al., *J. Med. Chem.* **61**:1, 1963.

Ringold, H. J., Batres, E., Halpern, O., Necoechea, E., *J. Am. Chem. Soc.* **81**:427, 1959.

Suggested Readings

Gaspard, U. J., Metabolic Effects of of Oral Contraceptives, *Am. J. Obstet. Gynecol.* **157**:1029–41, 1987.

McGuire, W. L., Steroid Hormone Receptors and Disease: Breast Cancer. *Proc. Soc. Exp. Biol. Med.* **162**:22–5, 1979.

Neumann, F., Topert, M., Pharmacology of Antiandrogens. *J. Steroid Biochem.* **25**:885–95, 1986.

Polca, J., Cherfas, J., *Science* **245**:1319, 1989 (RU 486).

Ulmann, A., Teutsch, G., Philibert, D., RU-486, *Sci. Amer.* **262**:42, 1990.

CHAPTER

15

New Developments and New Problems

15.1. Introduction

The following discussion will be a brief kaleidoscopic overview of new ideas, concepts, and problems arising in drug development and utilization. Some may not relate specifically to any one particular drug area or disease state, but may have wider applications. Others may be of direct relevance to medicinal chemistry as it evolves a broader interface with what is often referred to as biotechnology. The initial achievements of the biotechnology revolution, specifically in the area of natural proteins as drugs, were accomplished by gene splicing techniques. One need only point to such products as human insulin, tissue plasminogen activator, human growth hormone, erythropoietin, several interleukins, and interferons. The fact that these substances were already known and many of their functions understood does not diminish the impressive achievement of their commercial production in sufficient quantity and purity for clinical use. As drugs, of course, they should be considered as new agents and their number expanded. However, it will likely remain a limited list. There will undoubtedly follow some analogs of these natural agents. SARs are already being established. Nevertheless, as large proteins they are burdened with critical clinical shortcoming: lack of oral bioavailability. Large polypeptides generally cannot cross biomembranes. Thus intracellular targets will not be accessible. However, protective measures such as liposomal encapsulation may be able to circumvent this handicap (e.g., insulin). In addition, natural peptides tend to be metabolized rapidly by proteases. Research, therefore, has been shifting toward active small molecule nonpeptide peptidomimetics. Such small molecule agents are simpler to develop and may be clinically superior. They are likely to be more stable, easier to administer, and certainly cheaper to produce. Some companies are deviating from protein-based research toward these new approaches. The addition of new chemical and biological tools to develop and utilize processes such as structure-based computer-aided drug design (based on a knowledge of protein target structures) and combinatorial chemistry provides small quantities of a prodigious number of compounds for evaluation. This opens new vistas of possibilities. The evaluations by rapid screening techniques such as microbinding assays makes these new approaches feasible. Known and newly characterized receptor sites are becoming targets for such small molecule drugs. Those that show promise can then be further developed by more traditional methods.

15.2. Gene Therapy

Gene therapy—the very name seems so futuristic, almost like science fiction. Not long ago one might have put it in the "pipe dream" category; not any more. It is not likely to explode on the scene with cures, or even with amelioration of hundreds of genetic defects and the pathologies resulting from them, yet between 1990 and mid-1995 a total of 87 gene therapy trials were approved by the Recombinant DNA Advisory Committee to the National Institutes of Health. Of the approximately 600 patients involved in experimental trials, about half were accepted in the first months of 1995 (Kolata, 1995).

Conceptually gene therapy involves the repair or correction of the machinery, as it were, that is at the very foundation of genetically induced diseases. Genetic diseases arise because a defective gene does not "turn on" a specific protein. In genetically induced cancers, however, the opposite may occur. That is, a gene that normally suppresses abnormal cell production is defective and fails to do so. This allows uncontrolled cell proliferation in essence, a malignancy.

As a concept gene therapy must certainly be considered simple and elegant. In a practical sense, however, we are still groping around the periphery of the technology that will be needed to carry it out. For example, cystic fibrosis has been known to be caused by an inborn error of metabolism for some time. More recently the identification and even experimental "repair" of those defects has been reported (Rommens et al., 1989; Rich et al., 1990). There are presently at least a half-dozen gene therapy trials underway for this dreaded disease. A partial listing of other disease candidates include: familial hypercholesterolemia (defective LDL receptors), adenosine deaminase (ADA) deficiency in lymphocytes, beta-thalassemia, an erythrocyte beta-globin gene defect, chronic granulomatous disease caused by a defective gene in monocytes and neutrophils, Gaucher's disease involving deficiency in glucocerebrosidase enzymes, and several other rare lysosomal storage diseases. There is no dearth of subjects to attack. In fact, as soon as a new faulty gene is discovered and identified, there is the desire to "fix it," even considering gene therapy was made feasible by dramatically successful research 20–25 years ago. First was the development of techniques to isolate individual genes (segments of DNA) from previously identified chromosomal "neighborhoods." Second, techniques to insert these genes into the genetic material of various microorganisms such as yeasts and bacteria were evolved that then reproduced these genes along with their own genetic material. This so-called gene splicing technique gave us the ability to produce virtually unlimited sources and quantities for research and—as a spinoff—for therapy.

Thus the potential to utilize normal genes to supersede mutated ones, or to replace them if missing, became a reality. How this utilization is to be done efficiently and safely are the problems currently being addressed. Success in developing and producing "correct" genes in and of itself is only part of the equation. If one is to identify and restore a mutant gene to its normal function, technical means to accomplish this must be found. Earlier experiments that resulted in specific sequence alteration in target genes in mammalian tissue cultures and even in transgenic animals were demonstrated (Thomas and Capecchi, 1987; Mansur et al., 1988). A more feasible procedure, as far as human practicality is concerned, might be viewed as augmentation. Here, normal gene segments are introduced into the host cells to restore normal gene expression. This is essentially a random process, yet it works (Friedman, 1989). Finding the optimum procedures to introduce a therapeutic gene poses challenges. One way that has evolved requires packaging it with a virus that then acts as a biological transporter to ferry its replacement cargo to a designated site. Such carriers are called *gene transfer vectors*. Expectations for such vectors are many: efficient transmission, stabilization of the DNA being transported into the host genome, production of the desired expression and control in the targeted tissue, and safety during the transfer process as well as during its subsequent tenure in the host.

A potential safety factor to be monitored with retroviral vectors in human trials is the development of mutations following intracellular insertion of the vector material (i.e., insertional mutagenesis.) Another factor to be wary of is the appearance of replication-competent retroviruses from the same source (Muenchen et al., 1990). Other factors of concern would be vector-induced carcinogenesis by oncogene activation or other effects. One way to circumvent these potential difficulties is to utilize replication-incompetent vectors.

Mechanical approaches to affect genetic material insertion include parenteral injections, utilizations of catheters, and even aerosol atomization. Physicochemical preparation processes are precipitation and/or complexation of genetic material with polymers or polycations and encapsulation in liposomes or other microsphering materials. Such transfection procedures during earlier investigations demonstrated that integration of multiple gene copies into genomes was stable and viable.

Other vectors for gene integration include mutant adenoviruses that have deletions in specific regions that make them replication incompetent. These are presently utilized in some cystic fibrosis therapy trials.

Nonviral vectors are also under investigation. Among these are receptor-mediated gene transfers. By its very nature these would be highly tissue specific. Here particular glycoproteins would target specific organs such as the liver (Wu, 1991). Liposomal encapsulation was already mentioned. In vivo injection of DNA can afford a direct transfection into cells. This nonvector method holds particular promise for muscle cells (Wolff et al., 1990).

Finally, ex vivo gene therapy should be mentioned. In this procedure cells are taken from a patient, genetically altered, and returned. This approach is being tested with certain cancers. It involves the removal of lymphocytes, bone marrow material, or even tumor cells. Cell growth can then be carried out in the laboratory while gene transfection occurs during this cell expansion. This treatment is highly patient specific and may have advantages. Gene therapy is still on the runway. It will ultimately take off. How rapid or steep the ascent will be it is difficult to predict this early. There will be pitfalls and obstacles. For example, cells transfected with new genetic material have been destroyed by the immune system that apparently "recognized" them as foreign. The quantity and efficiency of gene transfection is still low. This will result in clinical trials that will fail, but that may succeed in the future. There will be some successes with several diseases caused by single gene mutations. Maladies involving multiple genetic defects are probably out of reach for the foreseeable future.

Even this brief discussion of gene therapy would be incomplete without stating that we are at the brink of a power with which to alter life's blueprint. Social and ethical questions emanating from this fact must not be ignored.

15.3. Drug Resistance

It is probably a relatively recent recognition that one of the great factors in shaping human history has been infectious disease. Almost whole populations may have disappeared from the map. Bubonic plague killed one-third of Europe's population in the fourteenth century. The first documented pandemic was in the early sixth century. It began in northeast Africa (Egypt and Ethiopia) and spread through Europe. It lasted 60 years and is estimated to have killed 100 million people. Smaller epidemics continued up to the end of the nineteenth century.

As recently as the late eighteenth century tuberculosis (TB) killed one out five people in Europe. Throughout history the staphylococcus bacteria altered trivial wounds into fatal infections. If one adds cholera, typhus, virulent pneumonia, malaria, and trypanosomal diseases as a partial list, it becomes apparent why the average human life expectancy did not increase significantly from Roman to modern times.

In this preantimicrobial era large city hospital patients as recently as 1930 consisted primarily of those with bacteremia, endocarditis, syphilis, rheumatic fever, and tuberculosis, diseases for which no meaningful treatment existed. Tragically, large segments of these patients were children and adolescents—most of whom would die. Today the great preponderance of patients in the hospitals of the industrial countries suffer from cancer, heart diseases, and diabetes-related hypertension and its consequences.

With the advent of the first generally effective antimicrobial sulfonamides in 1934–1935, followed by aminoglycosides and penicillin G in the 1940s, humanity slipped into a period in which it was charmed into a fantasy that the inferno of infectious disease was essentially extinguished. The illusion was fed by some wonderful facts. Between the application of the insecticide DDT, which appeared to eliminate the anopheles mosquito, and the use of the drug chloroquine, which cured already infected patients, malaria receded dramatically. Malaria was eradicated by the early 1950s (see Kikuth, 1954) in the southern United States (the Mississippi Valley) and the island of Madagascar. Also at that time staphylococcus bacteria were almost totally susceptible to Penicillin G. Penicillin-resistant pneumococci constituted less than 0.2% of those encountered as recently as the early 1980s. How myopic of the U.S. Surgeon General in 1969 to have stated to Congress that the time "to close the book on infectious diseases was at hand."

"The good old days" are rapidly vanishing. Established and once curable diseases are back with a vengeance. Staphylococci have been over 90% penicillin resistant since the early 1980s.

Malarial morbidity worldwide is now increasing rampantly due to chloroquine-resistant *Plasmodium falciparum*—with annual mortality rates approaching 2 million people. Streptococcal pneumonia is now over 40% penicillin-resistant. Trypanosomiasis (African sleeping sickness) in Zaire alone is expected to kill 200,000 people in 1995. We are increasingly encountering microorganisms exhibiting multiple drug resistance even when used in combinations. The penicillin-resistant pneumonia organism is now also frequently unresponsive to erythromycin, cefotaxime, and trimethoprim-sulfamethoxazole combinations. None of the previously mentioned facts should come as a surprise to the medical—scientific community. The rate at which it is occurring and the changing pattern of virulence were not predictable. Predictable or not, humanity is now faced with rapid and dramatic increases in antimicrobial resistance. In addition, although this pattern has been appearing primarily in institutions (nosocomial infections), it is now crossing into community settings as well. Exacerbating the situation greatly is the advent of a new disease:

AIDS. HIV-infected persons, to the degree that their immune system is compromised or suppressed, are at high risk of succumbing to various bacterial, fungal, protozoal, and viral diseases, which cannot be controlled due to resistance and virulence. In essence, then, AIDS patients die from drug-resistant infections. TB strains resistant to most available drugs and having a high contagion potential to the community and hospital population is now spreading rapidly. In fact, we are witnessing the previously unimaginable scenario where an epidemic disease—AIDS—is actually fostering the reemergence of another extremely contagious disease—TB.

From a global perspective infectious diseases have never really been eradicated and, as throughout history, remain the largest cause of death. Among these diseases TB remains the leading killer, with approximately 3 million deaths annually (Kochi, 1991). TB mortality in the developing world in the 15–59 age group accounts for nearly 19% of all deaths (Murray et al., 1992). The frightening increase of drug-resistant strains may sooner or later lead to an out-of-control public health disaster. The decline of TB that began in the United States in the early 1880s ceased and began to reverse in 1985 when over 26,000 cases were reported (Morbidity, 1991). One out of three cases in New York City in 1991 was resistant to one or more antitubercular drugs (Frieden et al., 1992).

Many of the molecular mechanisms by which microorganisms affect resistance to drugs have been understood to some degree for some time. Others have now come to light as more sophisticated molecular biological and chemical methods were developed.

Chemotherapy against microorganisms constitutes a selective pressure to which they respond by one or more of the following methods. One involves altering the ability of drugs to reach their intracellular targets. This is usually achieved by a decrease in membrane permeability as a result of gene mutations leading to a decrease (or even elimination) of membrane porins (channels). Another way to decrease intracellular drug accumulation is inner membrane protein changes that lead to an accelerated drug ejection process or efflux. These mechanisms have been identified as having a role in resistance to tetracyclines, fluoroquinolones, and certain antimalarial agents.

The best-understood resistance mechanism is probably drug modification by microbial enzymes, which are frequently induced by the drug itself. Thus β-lactamases of increasing effectiveness are being identified (Jacoby and Medeiros, 1991). Overproduction of various β-lactamases can occur by means of gene amplification (Sanders and Sanders, 1992). In addition, genetically altered penicillin binding proteins (PBPs) have also been shown to be the cause of resistance even to third-generation cephalosporins (Klugman, 1990). Other enzymatic alterations that lead to drug inactivation are conjugative reactions. Chloramphenicol acetyl transferase esterifies the 3-OH group: aminoglycoside modifications are effected by N-acteyltransferase, phosphoryl transferase, and adenylation enzymes (see Chapters 6 and 7 for chemical details). Macrolide antibiotics can be inactivated by such derivatization reactions at the 2′-OH group as glycosilation and phosphorylatyion (Cundliffe, 1992). Lactone hydrolysis is also likely.

A third approach by which microorganisms can develop resistance is by modifying the drug target itself. Such modifications can result in decreased affinity for the drug or even the "opening" of an alternative pathway by which a target continues to function in the presence of its former inhibitor (Chopra, 1992). In this category can be found point mutations in the A subunit of DNA gyrase (the fluoroquinolone target) such as a Leu or Phe replacement of Ser at position 84. The result is ciprofloxacin resistance (Sreedharans et al., 1991). Other examples include dimethylation of adenine by methyltransferases in positions of ribosomal RNA subparticles such as 23S. Such a modification can be assumed to effect a change in ribosomal affinity for various antibacterials whose mechanisms of action depend on inhibition of protein synthesis. It is likely that this would represent a change of ribosomal conformation at the drug binding sites, and not a direct binding interference (Vannuffel et al., 1992).

The preceding is a brief and very incomplete survey of the explosive drug resistance problem. It should be mentioned that resistance to anticancer drugs is also serious and, to some degree, a parallel problem (e.g., increased drug efflux). Intense research in these areas is needed and is going on. Major breakthroughs can be hoped for, but not depended on. After all, the "easy stuff" has been tried. In the short run we may have to settle for a holding action. There has been some progress in vaccines: meningococcal against meningitis and an antiviral against hepatitis A (Havrix,® 1995). However, two serious setbacks more than offset these positive results. The likelihood of an AIDS vaccine seems more and more remote. Also the promising malaria vaccine, SPf66, has been found to offer little significant protection, after field tests in children in Gambia.

15.4. Antisense Drugs

"Antisense drugs" is a catchy name that may make more sense with a technical term such as sequence-defined oligonucleotides (ONs). Perhaps a brief explanation of what it is we are attempting to achieve with these curiously named substances will perhaps be more useful. The first term gives a hint. It contains the word *drug*. Thus we are seeking a new, different, and more specific approach to therapeutics. Most drugs today have a downside in terms of adverse effects. That is, they are primarily small molecules that interplay with endogenous macromolecules by mostly non-sequence-specific interactions. Therefore, compounds that at least ideally interact with our biopolymers in a sequence-specific manner can at least give hope of a meaningful reduction of side effects.

In principle it should be possible to effect changes in cellular gene expression by "treating" the cells with ON analogs. The idea is that such a down regulation of gene expression, if sufficiently specific, can achieve a decrease—or elimination—of disease-causing proteins. In essence, the modified ON becomes a therapeutic agent that treats a disease at its point of inception.

In the case of antisense ONs, binding to specific sites of that protein *m*-RNA would inhibit effective translation of the genes code to the ribosome. The result is a nonfunctional protein. Such a process is based on the likelihood that the single-stranded *m*-RNA is capable of strong (i.e., high-affinity) binding at the specific nucleotide sequence to complementary, or antisense, sequences by hydrogen bonding. Other scenarios are also possible, and even likely. One of particular interest is where sequence specific (complementary) binding to the *m*-RNA leads to activation of the enzyme RNAse H, which catalyzes hydrolysis of the RNA component of the RNA/DNA complex. This results in the loss of the message and thereby cessation of protein synthesis (Ramanafhan et al., 1993). Other mechanisms of protein synthesis inhibition resulting from *m*-RNA–antisense compounds are also likely.

Antisense agents targeted to specific sequences of double-stranded DNA (technically antigene drugs) could also be envisioned. Ultimate availability of both types of agents could be potentially utilized to bind to DNA and RNA viruses. This would hinder the invader from (1) utilizing the host cell's protein synthesizing mechanics or (2) to express foreign genes. It is not apparent that such drugs would have the potential to become ideal antiviral agents, if they can be synthesized and developed.

Clinical trials are still few. Several of the earliest reported to date are only from the early 1990s. To illustrate, a treatment trial on genital warts induced by papilloma virus was reported (Anonymous A, 1993). An AIDS infection trial (Anonymous B, 1993) and an acute myelogenous leukemia test (Anonymous C, 1993) are also in the literature.

Before going on it may be worth exploring the chemistry. It might also be of interest to point out that naturally occurring phosphodiester ONs were not found to be useful at desirable concentrations as drug candidates. The reasons are that they failed two of the most significant characteristics sought. One is poor binding affinity to target sites; the other is very short half-lives due to rapid hydrolytic degradation by nuclease enzymes. Therefore, attention turned to traditional medicinal chemistry techniques as discussed throughout this book. The goals are to increase binding affinity and specificity, and hydrolytic stability, and to improve cell membrane penetratability.

Alterations of the normal oligonucleotide with the 3′ to 5′ phosphodiester backbone connecting the nucleic acid bases to ribose or 2′*deoxy*ribose can be carried out in several ways. One is alteration of the backbone, another is base modification, and a third is the introduction of changes in the sugar component. Some examples of chemical modifications that have been carried out are illustrated in Figure 15-1. Increased stability toward nuclease-catalyzed hydrolysis (compared with the parent phosphodiester in natural DNA) is sought; increased thermal stability as determined by comparative melting points (T_m) and improved binding affinities concomitantly with better site specificity.

One of the simpler backbone alterations with which experience has been gained is the phosphorothioate, where one phosphate oxygen is exchanged for a sulfur atom (Fig. 15-1). It offered resistance to nuclease hydrolysis, but it showed diminished binding affinity. Peptide nucleic acid analogs (PNAs) (Fig. 15-1) have exhibited some interesting properties. The apparently bind-to-target sequences of ONs is either an orientation in which the RNAs amino terminal aligns itself with the 5′-end of the DNA (parallel), or with the 3′-end (antiparallel). The latter form, when hybridized with a mixed purine/pyrimidine nucleic acid sequence, formed the more stable duplex, as demonstrated by an increased T_m (Egholm et al., 1993).

A fascinating study by Bruice et al. (1995) took a somewhat different approach. Instead of a chiral negatively charged backbone (i.e., phosphate) or the uncharged PNA scaffolding, they synthesized a pentameric, positively charged and nonhydrolyzable backbone built of guanidinium linkages. This compound is thymidyl*deoxy*ribonucleic guanidine (DNG) (Fig. 15-2). It resisted hydrolysis, formed specific complementary RNA and DNA complexes in the form of double and triple helical structures (e.g., DNG_2-RNA) that were exceedingly thermally stable (T_m > 93°). The DNG pentamer predictably associated with

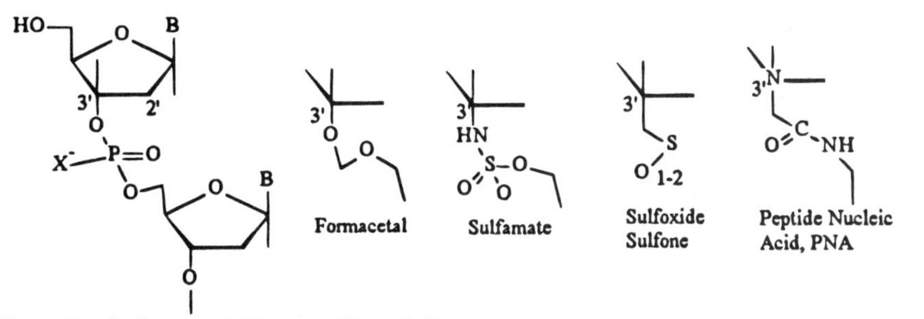

Figure 15-1. Oligonucleotide backbone modifications.

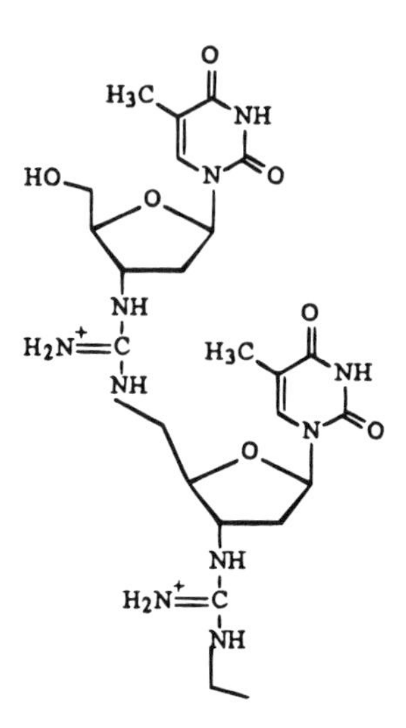

Figure 15-2. Thymidyl-deoxyribonucleic guanidine, DNG.

adenine on either DNA or RNA. In addition, this strong binding occurs at low ionic strength. Other DNA analogs do so only at high ionic strengths. Larger polymers consisting of more than five bases, and RNA analogs that utilize other bases, are planned. Membrane permeability is still a question.

A second point in the molecular modification effort is the sugar component of the oligomer. The 2′ position has been one target. For example, 2′-hydroxyalkyl ethers of varying lengths and degrees of unsaturation have been prepared (Lesnik et al., 1993). A fluorine atom at position 2′ also exhibited an increase in thermal stability.

A third alternative locus for molecular alteration of ONs are the bases. For example 5-substituents on both thymidine and cytosine bases included halogens and a propyne group in both "normal" and phosphorothioate analogs, with or without a 2′-O-allyl function. Results were mixed (Froehlen et al., 1993; Wagner et al., 1993).

This is a new area of research that is barely out of the concept phase. Much work is still being done in pushing aside the webs that block our vision. Higher sequence specificity, better membrane permeability, good tissue distribution, and—the ultimate goal—therapeutic activity with acceptable or low levels of toxicity and other untoward effects will appear from behind these webs.

15.5. Cytokines

Cytokines are protein hormones that differ from other types of known hormones in that they are usually produced by more than one type of cell. They exhibit broad regulatory effects and have a plethora of functions. They have been subdivided more or less on that basis. Thus *interleukins* are secreted proteins that mediate local interactions between leukocytes (white blood cells) but do not bind to antigens (the subgroup produced by lymphocytes only are called *lymphokynes*). *Interferons* have been defined as being produced in response to invasion by viruses and protozoal intracellular parasites. At one time (1960s) it was believed that interferon was a single substance that had viral inhibitory effects. By the 1980s four major categories were known: alpha, beta, gamma, and omega, which have broad biological effects. Subtypes of each have been identified since then.

The advent of gene-splicing technology and other developments and improvements in molecular biology techniques have allowed for commercial production of these and other cytokines discussed here, in quantities adequate for clinical evaluation, as well as for those deemed successful for therapeutical applications as (see Table 15-1). Agents now available have significant antiviral activity. Antineoplastic utility has been demonstrated more

Table 15.1. Cytokine Products[a,b]

Name[c,d]	Uses[f]
Filgastrin, G-CSF (Neupogen), USA[d], 1991*	Myelosuppressant chemotherapy, in bone marrow transplantation, to reduce length of drug-induced neutropenia (aplastic anemia, hairy cell leukemia, AIDS)
Sargamostin, GM-CSF (Leukine, Prokine), USA[d], 1991*	Cancer patients following autologous bone marrow transplants (To > WBC counts in AIDS; to < neutropenia secondary to chemotherapy)
Interferon-alpha 2a[g] (Roferon A), USA[d], 1994*	Chronic myelogenous leukemia, advanced colorectal cancer (with 5-FU), renal-cell cancer, AIDS-related Kaposi's sarcoma
Interferon-alpha 2b (Intron-A), USA[d]	AIDS-related Kaposi's sarcoma, chronic leukemia, hepatitis B
Interferon-alfa NL (Welferon)	AIDS-related Kaposi's sarcoma, papilloma virus
Interferon-beta-1bg (Betaseron) USA*D*, 1993*	To treat relapsing remitting multiple sclerosis; mechanism unknown (potential in melanoma, renal-cell carcinoma, genital warts, hepatitis)
Interferon Gamma-1α (Maruho), Japan[d], 1992*	To treat cutaneous T-cell lymphoma, possibly also ovarian cancer
Interferon Gamma-1b (Actimmune, USA[d], 1991*	To reduce severity and frequency of infections associated with chronic granulomatous disease
Interferon Gamma (Polyferon), USA[d], 1989[a]	Rheumatoid arthritis, immunotherapy adjunct, hepatitis B
Interleukin-1 Receptor Antagonist (Antril) USA[d]	Juvenile rheumatoid arthritis, graft vs. host disease from transplantation
Interleukin-2, Aldesleuk (Proleukin), USA[d], 1989*	Metastatic renal cancer, CNS and brain tumors (colorectal cancer, non-Hodgkin's lymphomas)
Interleukin-3, IL-3 USA[d]	Autologous bone marrow transplants, promote red cell production in congenital red cell aplasia
Erythropoietin, Epoetin-Alfa (Epogen, Procrit), USA[d], 1987	Anemia in chronic renal failure and in AZT-treated HIV patients
Growth hormone, Somatren (Protropin, Somatropen), USA[d], 1987*	Long-term treatment of children with growth failure due to lack of the endogenous hormone
Tumor necrosing factor, TNF-alpha, USA[d]	Antiviral, inhibits HIV replication; in phase I/II clinical trials.
Epidermal growth factor (Gentel), Italy,[d] 1987*	As ophthalmic product to stimulate healing of corneal epithileum

[a] As of mid-1995.
[b] Commercial production by DNA recombinant methods.
[c] Trade name in parentheses.
[d] Country of origin.
[e] Year of introduction (not necessarily USA).
[f] Not USA-approved use, (i.e., off-label).
[g] Orphan drug status.

recently well. Some promising developments involve synergistic activity with anticancer drugs such as 5-fluorouracil.

Another family of cytokines—the hematopoietic growth factors known as colony stimulating factors (CSF)—consists of: granulocyte macrocyte colony-stimulating factor (GM-CSF), the granulocyte colony-stimulating factor (G-CSF), and the macrophage colony-stimulating factor (M-CSF). The first two glycoproteins with molecular weights of 23 and 25 K-daltons, respectively, are being commercially produced and clinically used (Table 15-1). The body's "workhorse" phagocytes, the polymorphonuclear neutrophils (PMNs), and the macrophages arise from a common committed progenitor cell called the granulocyte-macrophage (GM). The PMNs and other granulocytes (basophils and eosinophils) leave the bloodstream after a few hours into extravascular sites and survive only briefly (days). The macrophages, however, survive for months.

The CFSs, which are prime factors in these events, are in turn produced by a variety of cells. These include fibroblasts, lymphocytes, endothelial cells, and also macrophages. Parenthetically, interleukin-3, which is produced by T-lymphocytes and even epidermal cells, accelerate this process by acting synergistically with CSF cytokines on pluripotent stem cells. Many progenitor cells, even on the terminally differentiated factors that are obtainable from their cloned genes, are now being utilized as drugs to fight infections and regenerate hematopoietic tissue (Table 15-1). Fused proteins (e.g., GM-CSF/IL-3) are being experimentally evaluated (Curtis et al., 1991).

The CSFs, as well as the glycoprotein erythropoietin, act at picomolar (10–12) concentrations by binding to their respective cell-surface receptors.

Cytokine receptors deserve to be mentioned. They are not only found on cells of the immune system. This may indicate that cytokines have functions outside the immune and hematopoietic systems. A detailed discussion is beyond the scope of this brief dissertation. A few examples should be of interest from a medicinal chemistry viewpoint, however. Interleukin-1 (IL-1) has the ability to mediate inflammatory occurrences among its activities. Among these is the ability to stimulate the synthesis and secretion of certain prostaglandins and other inflammatory factors (e.g., tumor necrosing factor, TNF). It can also stimulate neutrophil adhesion and be an attractant to them (Dinarello, 1989). The recent discoveries and experimental developments of several soluble cytokine receptor preparations (IL-1, IL-4, and TNF) allowed the determination as to whether these receptor proteins could reduce the negative consequences of these cytokines. Some preliminary positive results were obtained (Fanslow et al., 1990). TNF has inflammatory mediation and the production of cachexia (wasting) in cancer patients among its varied bioactivities. A soluble TNF receptor has been expressed in *E. coli* (Klausner, 1987). Its affinity for TNF offers the possibility of developing a treatment for certain inflammatory conditions as well as toxic shock syndromes.

Another approach that ultimately may produce some important therapeutic results is to modulate cytokine functions. One encounters a labyrinthine maze when one views the known actions and effects of the 10 or so interleukins. First, a given cytokine has several biological functions. Furthermore, frequently any given function is mediated by two or more cytokines. Many functions provide pathological results. However, once their molecular mechanisms and biochemical pathways are understood, then decreasing the cytokine's levels or blocking its effects can produce new therapeutic modalities. Of course, their beneficial attributes can be directly exploited for therapeutic purposes.

An examination of just one cytokine, IL-1, can be illustrative. Its activities include regulation of lymphocyte development and activation, a variety of proinflammatory effects such as prostaglandin production, certain proteases, and synovial and other cell types (Martin and Resch, 1988). Thus, it is an important factor in inflammatory and immune responses and is believed to be a factor in such pathologies as rheumatoid arthritis and inflammatory bowel disease, hemodynamic shock, and even lethal sepsis (Dinarello, 1991).

Various small molecules that can modulate interleukin synthesis and release exist. The antiprotozoal drug pentamidine (Chapter 7) blocks the posttranslational modification of pro-IL-Alpha to the full cytokine; some arylidine acetic acids (classic ANSAIDs) also appear to prevent IL-1 release from certain leukocytes in addition to PG synthesis inhibition via cyclo-oxygenase. It is interesting that synthesis or release inhibitors are also found among the nonclassic antirheumatics such as chloroquine and the gold compounds auranofin and aurothimolate (Chapter 5). Compounds with dual cyclo-oxygenase/5-lipo-oxygenase inhibiting properties also demonstrated activity. Hypercholesterolemic drugs that also have antioxidant properties such as probucol can also inhibit IL-1 production (Ku et al., 1988). Enders et al., (1989) have even demonstrated reduction of both IL-1 alpha and beta levels following fish oil administration (omega-3 acids).

IL-1 was found to offer radiation protection (Neta et al., 1987) and protection against lethal microbial infection by utilizing its beneficial attributes.

Considering all the ILs, it is apparent that some potentially very interesting drugs can be prospected for here.

Finally, erythropoietin should be mentioned. It stimulates the production of the most common and numerous cell in the blood—the erythrocytes. A death of these cells, or a shortage of oxygen, stimulates the kidney cells to synthesize and secrete this stimulating cytokine. The progenitor, which is known as the erythrocyte colony-forming cell or CFC-E, produces 60 red blood cells. Erythropoietin has now become therapeutically invaluable to kidney dialysis and AIDS patients. It is sad that it has also become an underground drug of abuse in athletics, especially in sports such as long-distance running and bicycle racing, where stamina can apparently be increased by the higher oxygen-carrying capacity produced by heavy dosing prior to competition.

15.6. Computers as Drug Design Aids

The traditional method of drug development, at least in this century, has been to develop leads by first using, and then by isolating and identifying, the active chemical constituents from natural products, some of which may have been medicinally in use since antiquity. With the advent of modern organic chemistry some of these purified compounds were used directly (e.g., morphine, cocaine, atropine, quinine), and, once their chemical structures were ascertained, they became leads for hoped-for chemical modifications to achieve improved efficacy, less toxicity, or, at least, higher potency (e.g., dihydromorphinone, homatropine, acetylsalicylic acid).

Another approach would be to make educated guesses as to the minimum pharmacophoric structure, synthesize this simplified structure, and, if some activity resulted, prepare a number of analogs and isosteres to optimize overall clinical properties (e.g., from cocaine to the various "caine" local anesthetics in use today; see Chapter 13). By sheer mass of grunt work, sometimes intuition, and occasionally, some good luck, this led to what may be nostalgically referred to as the golden age of drug discovery (ca. 1940–1965). It was also, of course, the "me too" era of copycat medicinal chemistry, either from natural products, which had no patents, or from competitors who did but who sometimes also had loopholes.

When this era came to an end, most of the "easy stuff" had been done. Then, what became known as rational drug design made its debut on the drug scene. Theories of drug action mechanisms began to apply ideas and facts from tangential disciplines such as physical organic, and biochemistry. Their tools changed "armchair" SARs to quantitative SARs. Physics lent medicinal chemists some of its tools, which had magnificent applications by allowing us to "see" our small molecules. Among these tools were spectroscopies with first names such as IR, NMR, ESR, and MS. As those began to improve in terms of sensitivity, resolution, and speed chemists must have begun to feel somewhat like astronomers peering further and further into the molecular and atomic universe. A quantum leap for medicinal chemistry was the introduction of X-ray crystallography. At first a cumbersome tool whose results were mathematically burdensome to evaluate, it had helped to resolve some pesky structural problems as far back as the early 1940s. Routine use, however, was then out of the question; even occasional application was a luxury available to few practitioners. As the name implies, one shortcoming was that substances had to be highly pure and crystallizable, but this was greatly overshadowed by its ability to let us observe certain biological molecules, especially proteins, with some accuracy. One of the first we were allowed to peek at was the unveiled molecular shape of the sperm whale's myoglobin (Kendrew et al., 1958). As X-ray methods improved their resolution, and early computers became faster at number crunching, results were obtained at an accelerated pace. Our eyes were rewarded more frequently by observing other proteins such as horse hemoglobin (1961) and egg-white lysozome (1965). Our voyeurism became a lust to observe more and more specimens; the gratification was the knowledge and understanding this afforded us. Now in the 1990s it is apparent what an achievement that was.

As techniques and apparatus improved, allowing for higher accuracy in identifying atomic positions through more reliable electron density calculations, the concept of designing drugs by "receptor-fit" models became more of a reality. High-speed computers have enabled these achievements in two ways. One is their ability to carry out the fast and accurate calculations required. The other is their ability, with sophisticated software, both to display three-dimensional structures in various representations (ball and stick, space-filling models, etc.), and to allow us to manipulate them by rotation and various other maneuvers so as to produce on-screen changes at will. These expanded abilities allow the visualization of stereochemical factors in a dynamic manner almost impossible with the molecular models of old.

Quantifying the various molecular variables such as the geometry, electron densities, and other energy factors raises the quality of information even higher. These types of feats are now called *molecular graphics*. One can essentially literally perform on screen what may be viewed as acrobatics with three-dimensional geometric forms. Imagine being able to manipulate by turning, twisting, and rotating two three-dimensional geometric forms—one small and less complex, the other large and much more intricate—while attempting docking maneuvers visually until one succeeds, even if it requires shape modifications. Such modifications may require altering protrusions in shape and size or modifying concave areas in depth and other characteristics. Of course, this could be a useful activity in learning solid geometry. However, if the larger object were to represent an enzyme and the concavity its reactive site, or a receptor protein with its binding site, then one might call this *molecular modeling*.

Some of the early attempts that were successful may be illustrative. Trimethoprin (TM) has been known as a useful inhibitor of dihydrofolate reductase (DHFR) since the mid-1960s. The enzyme's three-dimensional structure had also been fully characterized by X-ray crystallography. In addition, complexes of DHFR with methotrexate as well as TM and NADHP as a ternary complex had also been described from X-ray data. The stage was thus set to apply this knowledge to design newer (and better) TM analogs based on structural information of the enzyme's binding requirements. By successfully synthesizing several analogs superior to TM in binding to the reactive site (one was 55 times so), molecular modeling was clearly established as an instrument for designing drugs (Kuyper et al., 1982, 1985).

Since molecules are three-dimensional entities whose properties (biological and physical) also depend on electron distribution, it might be stated that computer-assisted drug design is actually a four-dimensional exercise. Its largest contribution thus far has really been to decrease the need for empirical experimentation. This should help speed up new drug development and reduce costs—no mean achievement. The potential to design a lead compound, or ideally to design a clinically useful drug de novo, based on knowledge of the receptor molecule's structural requirements seems at first glance not on the horizon. As the quality and information increases, however, that day may arrive sooner. The information databanks on physical aspects such as bond lengths and electronic parameters are rapidly becoming "richer," making the modeler's job more efficient, if not easier.

Computer programs are now available that can rapidly transform the well-known two-dimensional structure into its three-dimensional counterpart, including its various conformers and those of minimum energies. Since two-dimensional databases are numerous and worldwide, this huge amount of data convertible to three-dimensional can then be searched for similarity of features such as surfaces and shapes, electron densities, and even pharmacophore similarities. This technique may also be helpful in finding and evaluating protein cavities. To illustrate one of the strategies in three-dimensional similarity searches, three-dimensional databases can match a three-dimensional structure of interest to a large number of ligands based on this similarity. In essence this represents a search for leads (Willett, 1992).

This brief and incomplete overview of the utility of computers in applying massive data and speed to help us answer what will more likely work as a useful drug can only increase our "luck" ratio in the search for new and better therapeutic agents. The utilization of the newest computer marvels and the programs that go with them will occupy a continuously larger portion of the stage of drug discovery. Computers will not be our panacea. A good lead compound is still the best way to go. Computers will help us to decide which analogs not to synthesize, and which, in a given situation, should be made and evaluated.

De novo drug design is still for the most part out of reach, as long as we retain large patches of ignorance on the goings-on at the molecular level. Activities at many of the receptor interiors and surfaces have not been unraveled. Many diseases, such as cancers, viral infections, and mental afflictions are really still a mystery to us. Meanwhile, we need to retain some of the empiricism as stopgap insurance—slow and inefficient as it may be.

References

Anonymous A, *Cancer Weekly*, July 13, 1993.

Anonymous B, *Gen. Eng. News*, Oct. 15, 1993, p. 7.

Anonymous C, *Cancer Researcher Weekly*, June 7, 1993, p. 7.

Bruice, T. C. Dempcy, R. O., Browne, K. A., *J. Am. Chem. Soc.* **117**:6140, 1995.

Chopra, I., *J. Antimicrob. Chemother.* **30**:737, 1992.

Cundliffe, E., *Antimicrob. Agents Chemother.* **36**:348, 1992.

Curtis, B. M., Williams, D. E., Broxmyer, H. E., et al., *Proc. Natl. Acad. Sci.*, USA. **88**:5809, 1991.

Czuprynski, C. J., Brown, J. F., *Microb. Pathogen.* **3**:377, 1987.

Dinarello, C. A., *Adv. Immun.* **44**:153, 1989.

Dinarello, C. A., *Blood* **77**:1627, 1991.

Egholm, M., Buchard, O., Christensen, L., et al., *Nature* **365**:566, 1993.

Enders, S., Ghorbani, R., Kelley, Ve., et al., *N. Engl. J. Med.* **320**:265, 1989.

Fanslow, W. C., Sims, J. E., Sassenfeld, H., et al., *Science* **248**:739, 1990.

Friedman, T., *Science* **244**:1275, 1989.

Frieden, T. R., et al., Abstract, 41st Annual Epidemic Intelligence Service Conference, C.D.C., Atlanta, Ga, April 1992.

Froehler, B. C., Jones, R. J., Cao, X., Terhorst, T.J., *Tetrahedron Lett.* **34**:1003, 1993.

Huvrix, 1995

Jacoby, G. A., Madeiros, A. A., *Antimicrob. Agents Chemother.* **35**:1697, 1991.

Kendrew, J. C., Bodo, G., Dintzis, H. M., et al., *Nature* (London) **181**:662, 1958.

Kikuth, W., *Dtsch. Klin. Woch.* **79**:1401, 1954.

Klausner, A., *Biotech.* **5**:335, 1987.

Klugman, K. P., *Clin. Microbiol. Rev.*, **3**:171, 1990.

Kochi, 1911

Kolata, G., *The New York Times*, July 25, 1995, p. C3.

Ku, G., Dohert, N., Woolos, J. A., Jackson, R. L., *Am. J. Cardiol.* **62**:77B, 1988.

Kuyper, L. F., Roth, B., Bakanari, D. P., et al., *J. Med. Chem.*, **25**:1122, 1982; *ibid.* **28**:303, 1985.

Lesnik, E. A., Guinosso, A. M., Kawasaki, A. M., et al., *Biochemistry* **32**:7832, 1993.

Mansur, S. L., Thomas, K. R., Capecchi, M. R., *Nature* **336**:348, 1988.

Martin, M., Resch, K., *Trends Pharmacol. Sci.* **9**:171, 1988.

Morbidity Mortality Weekly Reports, **39**:944, 1991.

Muenchen, D. D., Freeman, S. M., Cornetta, K., et al., *Virol.* **176**:262, 1990.

Murray, C. J.L., Styblo, K., Rouillon, A., in *Disease Control Priorities in Developing Countries*, Jamison, D. T., Mosley, D. H., Eds., Oxford University Press, New York p. 50. 1992.

Neta, R., Douches, S., Oppenheim, J. J., *J. Immun.* **136**:2483, 1987.

Ramanathan, M., MacGregor, R. D., Hunt, C. A., *Antisense Res. Dev.* **3**:3, 1993.

Rich, D. P., Anderson, M. P., Gregory, R. J., et al., *Nature* **336**:347, 358, 1990.

Rommens, J. F., Iannuzi, M. C., Kerem, B. et al., *Science* **245**:1066, 1073, 1989.

Sanders, C. C., Sanders, W. E., *J. Clin. Inf. Dis.* **15**:824, 1992.

Sreedharans, S., Peterson, L. R., Fisher, L. M., *Antimicrob. Agents Chemother.* **35**:2151, 1991.

Thomas, K. R., Capecchi, M. R., *Cell* **51**:503, 1987.

Vannuffel, P., DiGgiambattista, M., Morgan, E. A., Cocito, C. *J. Biol. Chem.* **267**:8377, 1992.

Wagner, R. W., Matteuci, M. D., Lewis, J. G., *Science* **260**:1510, 1993.

Willett, P., *J. Chemomem.* **6**:289, 1992.

Wolff, J. A., Malone, R. W., Williams, P., et al., *Science*, **247**:1465, 1990.

Wu, G. Y., *J. Biol. Chem.* **226**:14388, 1991.

Suggested Readings

Antisense Research and Applications, Crooke, S. T., Lebleu, B., Eds., CRC Press, Boca Raton, 1993.

Baum, R., Borman, S., Combinatorial Chemistry, *Chem. Eng. News*, Feb. 12, 1996, pp 28–54.

Bloom, B. R. and Murray, C. L., Tuberculosis: Commentary on a Reemergent Killer. *Science* **257**:1055–62, 1992.

Evolutionary Medicine: Rethinking the Origins of Disease, Lappe, M., Sierra Club Books, San Francisco, 1994, 255 pp.

Humblet, C., Dunbar, J. B., Database Searching and Docking Strategies, in *Ann. Rev. Med. Chem.* **28**:283, 1993.

Medicinal Chemistry for the 21st Century, Wermuth, C. G., Koga, N., Konig, H., Metcalf, B. W., Blackwell Scientific Publications, Oxford, 1992.

Index